T5-BSH-494

Natural Computing Series

Springer
Berlin
Heidelberg
New York
Hong Kong
London
Milan
Paris
Tokyo

Ashish Ghosh • Shigeyoshi Tsutsui (Eds.)

Advances in Evolutionary Computing

Theory and Applications

With 368 Figures and 111 Tables

Editors

Dr. Ashish Ghosh

Indian Statistical Institute
Machine Intelligence Unit
203 B.T. Road
Kolkata 700 035
India
ash@isical.ac.in

Professor Dr. Shigeyoshi Tsutsui
Hannan University
Department of
Management Information
5-4-33 Amamhigigashi
Matsubara, Osaka 580-8502
Japan
tsutsui@hannan-u.ac.jp

Series Editors

G. Rozenberg (Managing Editor)
Th. Bäck, A.E. Eiben, J.N. Kok, H.P. Spaink

Leiden Center for Natural Computing
Leiden University
Niels Bohrweg 1
2333 CA Leiden
The Netherlands
rozenber@liacs.nl

Cataloging-in-Publication Data applied for

A catalog record for this book is available from the Library of Congress.

Bibliographic information published by Die Deutsche Bibliothek
Die Deutsche Bibliothek lists this publication in the Deutsche Nationalbibliografie; detailed bibliographic data is available in the Internet at http://dnb.ddb.de

ACM Computing Classification (1998): F.1, I.2.8, F.2, J.3, G.2

ISSN 1619-7127
ISBN 3-540-43330-9 Springer-Verlag Berlin Heidelberg New York

Springer-Verlag Berlin Heidelberg New York,
a member of BertelsmannSpringer Science+Business Media GmbH

http://www.springer.de

Cover Design: KünkelLopka, Heidelberg
Typesetting: Camera ready by the editors
Printed on acid-free paper SPIN 10869074 45/3142XT - 5 4 3 2 1 0

To our wives
Susmita and Kumiko

Preface

The term evolutionary computing refers to the study of the foundations and applications of certain heuristic techniques based on the principles of natural evolution; thus the aim of designing evolutionary algorithms (EAs) is to mimic some of the processes taking place in natural evolution. These algorithms are classified into three main categories, depending more on historical development than on major functional techniques. In fact, their biological basis is essentially the same. Hence

$$\mathrm{EC} = \mathrm{GA} \cup \mathrm{GP} \cup \mathrm{ES} \cup \mathrm{EP}$$

$$\begin{aligned}
&\mathrm{EC} = \text{Evolutionary Computing}\\
&\mathrm{GA} = \text{Genetic Algorithms}, \mathrm{GP} = \text{Genetic Programming}\\
&\mathrm{ES} = \text{Evolution Strategies}, \mathrm{EP} = \text{Evolutionary Programming}
\end{aligned}$$

Although the details of biological evolution are not completely understood (even nowadays), there is some strong experimental evidence to support the following points:

- Evolution is a process operating on chromosomes rather than on organisms.
- Natural selection is the mechanism that selects organisms which are well-adapted to the environment to reproduce more often than those which are not.
- The evolutionary process takes place during the reproduction stage that includes mutation (which causes the chromosomes of offspring to be different from those of the parents) and recombination (which combines the chromosomes of the parents to produce the offspring).

Based upon these features, the previously mentioned three models of evolutionary computing were independently (and almost simultaneously) developed.

An evolutionary algorithm (EA) is an iterative and stochastic process that operates on a set of individuals (called a population). Each individual represents a potential solution to the problem being solved. Initially, the population is randomly generated. Every individual in the population is assigned, by means of a fitness function, a measure of its goodness with respect to the problem under consideration, which guides the search. The whole process is sketched as:

$Generate[P(0)]$
$t \leftarrow 0$
$WHILENOT \;\; Termination - Criterion$
DO
$Evaluate \;\; [P(t)]$
$P'(t) \leftarrow Select \; [P(t)]$
$P''(t) \leftarrow \;\; Apply - Reproduction - Operatorson \; [P'(t)]$
$P(t+1) \leftarrow Replaceby \; [P(t), P''(t)]$
$t \leftarrow t + 1$
END
$RETURN \;\; Best - Solution$
Skeleton of an Evolutionary Algorithm

It can be seen that the algorithm comprises three major stages: selection, reproduction, and replacement. During the selection stage, a temporary population is created in which fitter individuals have a higher number of instances than less fit ones (natural selection). The reproductive operators are applied to these individuals in this population yielding a new population. Finally, individuals of the original population are substituted by the newly created individuals. This replacement usually tries to keep the best individuals, deleting the worst ones. The whole process is repeated until a termination criterion is achieved. It should be noted that EAs are heuristics, and thus they do not ensure an optimal solution.

Major differences between GA, ES, and EP come from the operators they use and in general from the way they implement the three stages: selection, reproduction, and replacement. EP is closely related to ES. Unlike GA, no crossover operator is used in ES/EP. Moreover, more emphasis is placed on behavioral changes than on the modification of the genetic material. For this reason, the genotype in ES/EP is usually very different (e.g., real numbers for ES, and a finite automaton for EP), and mutation operators are prepared to deal with such representations.

Any EA is composed of a set of common elements, in spite of its differences from other EAs, as follows:

- A population of trial solutions/strings. Typically strings are composed of binary, float, or some complex structure (e.g., a tree) genes.
- A fitness function to be optimized for evaluating strings.
- Some selection/replacement mechanism in order to simulate the survival of the fittest individuals for future generations.
- Nature-inspired operators (like recombination and mutation) for changing a string into a new string.

A large number of researchers, all over the world, have been engaged in developing EC methodologies for designing intelligent decision-making systems for a variety of real-world problems. However, research articles on such topics are sparse.

The present book provides a collection of 40 articles, divided into two parts, containing new material on the theoretical aspects of EC, and demonstrating the usefulness/success of EC for various kinds of large real-world problems. Each chapter represents an article in its own right. Part I contains 23 articles dealing with various theoretical aspects of EC, while Part II contains 17 articles demonstrating the success of EC methodologies. These articles are written by leading experts from many countries.

Part I starts with the article of Vassilev et al. who studied structures of fitness landscapes to provide a suitable mathematical framework for investigating the evolvability of complex systems. Xin Yao et al. (in a related article in Chap. 2) demonstrated the role of search step size in approximating the landscape by using different hybrid EAs. Chapter 3 is intended to provide an introductory review of the existing work done on visualizing EAs, and to identify some of the key issues for future research.

New schemes of EAs or designing their operators are described in Chaps. 4 and 5. A parameter-free GA (called PfGA) is proposed by Sawai et al. in Chap. 4 based on the disparity theory of evolution which exploits different mutation rates and variable population sizes. In Chap. 5 Droste and Wiesmann suggest guidelines for the design of genetic operators and the representation of phenotypic space to solve specific types of problem. The applicability of this concept is shown by a systematic design of a GP system for finding Boolean functions. This system is the first GP system that has reportedly found the 12-parity function.

Eiben demonstrates the utility of multiparent reproduction with successful results in Chap. 6. The traditional debate of mutation and crossover is also considered in the light of multiparent reproduction.

In Chapter 7 Michalewicz and Schmidt propose a test case generator which is capable of creating various test problems with different characteristics, including the dimensionality of the problem, number of local optima, number of active constraints at the optimum, topology of the feasible search space, etc. Such a test case generator is useful for analyzing and comparing different constraint-handling techniques.

Handling real-coded parameters in GAs is an important research topic nowadays. In Chap. 8 Ono et al. propose a new crossover operator named the unimodal normal distribution crossover for real-coded GAs which works efficiently for optimization problems with epistasis among parameters.

Branke and Schmeck provide a good survey of the literature on EC in Chap. 9 for dynamic optimization problems, and offer a classification of the same set of problems. They also suggest a new technique for this task using a multipopulation structure.

In the next chapter Deb proposes a few classical techniques to identify a preferred or compromise solution by introducing a biased sharing technique to find a biased distribution of Pareto-optimal solutions in multiobjective

GAs. The results are encouraging for more complex multiobjective optimization problems.

The utility of gene expression in scalable genetic search is studied by Kargupta in Chap. 11.

Knjazew and Goldberg present an ordering messy GA in Chap. 12 that is able to solve difficult permutation problems efficiently according to the experimental results.

Global optimal solutions are not always acceptable, if they are sensitive to perturbations in the environment. In Chap. 13 Tsutsui and Ghosh suggest ways of detecting robust solutions thereby extending the utility of GAs.

In Chap. 14, EC is used by Spears and Gordon to evolve finite-state machines having an optimal number of states for better performance in resource allocation.

Chapters 15 and 16 try to link EC with statistical inferencing. In Chap. 15 Zhang tries to view EC as a Bayesian inference that iteratively updates the posterior distribution of a population from the prior knowledge and observation of new individuals to find an individual with the maximum posterior probability. Chapter 16 by Aizawa is an attempt to combine experimental design and EC into a single search strategy using a specific type of recombination function called a deterministic crossover operator.

The next three chapters deal with the theoretical understanding of biology and its simulation. Maley's article in Chap. 17 aims to use EAs to extend our theoretical understanding of biology, and to reunite theoretical biology with experimental biology. Kumar and Bentley present a brief survey on using embryology and genetics in developmental biology in Chap. 18. An application of two embryological techniques is also shown in the evolution of certain predefined letters. In Chap. 19 Ray describes an evolutionary approach to synthetic biology which inoculates the process of natural evolution in an artificial medium, and finds the natural form of the living organisms in the artificial medium. He also suggests a possible means of harnessing the evolutionary process for the production of complex computer software.

Chapters 20–22 deal with other heuristic algorithms closely related to EAs, simulating some other natural phenomena. The main goal of Chap. 20 by Glover et al. is to demonstrate the development of scatter search procedures by illustrating how they may be applied to a class of nonlinear optimization problems of bounded variables. They conclude the chapter by highlighting the key ideas and research issues that offer the promise of yielding future advances. In Chap. 21 the application of several ant colony optimization techniques is demonstrated by Carbonaro and Maniezzo on a number of hard optimization problems with specific attention to a new algorithm called ANTS. In recent years, considerable interest and enthusiasm have been generated by the prospect of widespread use of intelligent agent-based systems. A coevolutionary optimization approach for evolving agent groups for multiagent systems and an adaptive system approach are suggested by Sen et al. in Chap. 22.

Schmidhuber studies in Chap. 23 an embedded active learner that can limit its prediction to arbitrary computable aspects of spatiotemporal events using probabilistic algorithms.

In Chap. 24, the first article of Part II, Ku et al. demonstrate a method to combine local search and evolutionary search techniques for neural network learning to reduce the computational time.

Chapters 25–27 deal with designing analog circuits using EC techniques. Analog circuits are evolved using variable-length genetic algorithms in Chap. 25 by Iba et al. Koza suggests a technique in Chap. 26 for applying genetic programming techniques for the automatic synthesis of topology and sizing for analog electrical circuits, the synthesis of placement and routing for circuits, and the synthesis of both the topology and tuning of controllers. In Chap. 27 Cohoon et al. use EC for a physical design problem where the input to the physical design set is a logical representation of the system under design, and the output of this step is the layout of a physical package that optimally or nearly optimally realizes the logical presentation. They also discuss important requirements for evolutionary-based approaches for even greater acceptance within the VLSI community.

The next two chapters (28 and 29) discuss issues related to designing communication channels. In Chap. 28 Zimmermann et al. report results on the application of EAs constrained to multiobjective, large, real-world antenna placement problems for mobile radio networks. EAs are used by Back et al. in Chap. 29 to find a routing table that increases the performance of a communication network by reducing the probability of end-to-end blocking, and which is applied to a non-hierarchical network.

Scheduling by EAs is discussed in Chaps. 30 and 31. Ross et al. present a survey with critical analysis on the application of EC in timetable scheduling problems in Chap. 30. They claim that a wide-ranging investigation is needed on this problem. Chapter 31 by Dorndorf et al. describes techniques to use GAs as metaheuristics to guide an optimal design schedule decomposition sequence for solving the minimum makespan problem for job shop scheduling, resulting in shorter makespans than for other local search algorithms. A scheme for bus driver scheduling along various routes is suggested by Yoshihara in Chap. 32.

Chapters 33 and 34 demonstrate the usefulness of EAs for data-mining problems. Alex Freitas presents an excellent survey of different EAs for data mining problems in Chap. 33. He also discusses whether the tasks of data mining and knowledge discovery in data bases will influence the design of EAs. Interactive EC is used by Teraso and Irada in Chap. 34 to get effective features from the data, and inductive learning is used to acquire simple decision rules from the subset of data for data-mining problems using clinical data.

Chapter 35 by Bhanu and Fonder discusses an approach to image segmentation which is guided by GAs and learns the appropriate subset and spatial combination of a collection of discriminating functions associated with image features. In this context they also suggest techniques for physics-based segmentation evaluation, novel crossover operator and fitness functions, and a system prototype, and they demonstrate experimental results on real synthetic aperture radar imagery of varying complexity.

Cao and Dasgupta propose an immunogenetics approach to recognize spectra for chemical analysis in Chap. 36. Their experimental results show the effectiveness of the approach in finding products responsible for a composite spectrum in which there are multiple, physically mixed products.

Steffen Kremer describe techniques which applied GAs to two-dimensional protein folding in Chap. 37. The results and limitations of the applicability of GAs to the problem of three-dimensional fold prediction are also presented.

Hasegawa and Fukuda suggest a method to control the regrasping motion of a four-fingered robot hand using EP in Chap. 38.

In Chap. 39 Lanzi and Riolo review recent advances and trends in learning classifier systems (LCS). These include credit assigned to rules, alternative LCS architectures like rule syntax and semantics, and the increase in the number and range of LCS.

David Fogel describes a hybrid technique in Chap. 40 for exploiting the advantages of neural networks and EC for designing a program which plays checkers at an expert level.

Kolkata, September 2002 *Ashish Ghosh*
Osaka, September 2002 *Shigeyoshi Tsutsui*

Contents

Part I

Part II

Part I
Theory

Smoothness, Ruggedness and Neutrality of Fitness Landscapes: from Theory to Application

Vesselin K. Vassilev[1], Terence C. Fogarty[1], and Julian F. Miller[2]

[1] School of Computing
Napier University
Edinburgh, EH14 1DJ, UK
E-mail:{v.vassilev, t.fogarty}@dcs.napier.ac.uk
[2] School of Computer Science
University of Birmingham
Birmingham, B15 2TT, UK
E-mail: j.miller@cs.bham.ac.uk

Summary. The theory of fitness landscapes has been developed to provide a suitable mathematical framework for studying the evolvability of a variety of complex systems. In evolutionary computation the notion of evolvability refers to the efficiency of evolutionary search. It has been shown that the structure of a fitness landscape affects the ability of evolutionary algorithms to search. Three characteristics specify the structure of landscapes. These are the landscape smoothness, ruggedness and neutrality. The interplay of these characteristics plays a vital role in evolutionary search. This has motivated the appearance of a variety of techniques for studying the structure of fitness landscapes. An important feature of these techniques is that they characterize the landscapes by their smoothness and ruggedness, ignoring the existence of neutrality. Perhaps, the reason for this is that the role of neutrality in evolutionary search is still poorly understood.

In this chapter some recent results on the spectral properties of the algebraic structures of fitness landscapes are summarized to provide a basis for studying the landscape structure. This approach is further employed to introduce an information analysis that characterizes the structure of fitness landscapes in terms of their smoothness, ruggedness and neutrality. The findings are finally applied in a study of the fitness landscapes generated by evolving digital circuits using an idealized model of a field-programmable gate array. The landscapes of this engineering problem are quite different from many recently studied landscapes that tend to be defined over simplified combinatorial and optimization problems. The difference originates from the genotype representation that is a configuration defined over two completely different alphabets. This makes the study of the corresponding landscapes much more involved. It is shown that the circuit evolution landscapes are products of subspaces with different characteristics. They are landscapes with vast neutrality and sharply differentiated plateau.

1 Introduction

The notion of a fitness landscape being a collection of genotypes arranged in an abstract metric space was introduced by Wright [1]. A fitness landscape has been defined as a space in which each point is a genotype assigned with a value, and two genotypes are adjacent if one can result from the other after a single mutation. Wright's original intent was to visualize biological evolution as a population flow on a surface in which the altitude level of a point qualifies how well the corresponding organism is adapted to its environment. The fact that evolutionary adaptation can be considered as a walk on a landscape implies the importance of developing a suitable mathematical framework for studying the features of landscapes referred to as landscape theory. In evolutionary computation the theory of landscapes has its own history. Related work can be traced back to studies on the genotype epistasis [2,3], schemata and deception [4–9], landscape modality [10,11], landscape ruggedness [12–15]. Various authors have considered that the structure of landscapes affects the ability of evolutionary algorithms to search [12,16–18]. Further, Jones [19] proposed a model of landscapes general enough to represent a variety of mutation and recombination landscapes. A fitness landscape has been defined on a graph for which each vertex consists of one or more configurations, and two vertices are connected by an edge if the configuration(s) from one of these vertices are obtained by applying the move operator to the configuration(s) from the other. Consequently the number of configurations in a vertex is specified by the nature of the move operator. For instance, the vertices of mutation spaces are single configurations while those of "two parent" recombination spaces are all possible pairs of configurations. A major problem associated with the model above is the difficulty in studying the similarities of mutation and recombination spaces [20]. Work on the mutation recombination isomorphism approach can also be found in [21].

A model of landscapes was given in the algebraic (graph) theory of landscapes outlined in [22,20]. In this model the landscape underlying structure is a hypergraph in which each vertex is a configuration, and each hyperedge is the set of all configurations that can be obtained by applying the move operator to a configuration. To construct a recombination space the model employs P-structures [23,24] that are mappings of pairs of configurations to the hyperedges of the hypergraph associated with this space. The fitness landscape results from the combination of the following objects:

1 A set of configurations that are often referred to as *genotypes.*
2 A cost function that evaluates the configurations, known in evolutionary computation as a *fitness function.*
3 A topological structure that allows relations within the set of configurations.

The set of configurations consists of the encoded elements of the "search" space. The relationship between the configurations is defined by a move operator that induces a topological structure on the set of all configurations called a configuration space. This space together with the cost function specifies a fitness landscape.

In this chapter some spectral properties of the algebraic structures of landscapes are summarized that define a basis for studying the landscape structure [22,20]. Analyzes of the ruggedness of landscapes are described following those given in [25,26]. The structure of a landscape can be completely characterized by its smoothness, ruggedness and neutrality. The interplay of these characteristics plays an important role in evolutionary search. To enable a study of a landscape in terms of the interplay of smoothness, ruggedness and neutrality an information analysis is further proposed [27,28]. The idea of the analysis is that a landscape can be presented as an ensemble of objects each of which consists of a configuration and its nearest neighbors. The analyzes are finally applied to study the landscapes associated with the evolution of digital circuits, particularly two-bit multipliers, on an idealized model of a field-programmable gate array. For the purpose of this study, a model for studying these landscapes is introduced, and their algebraic characteristics are investigated. The latter includes the calculation of the eigenvalues of the P-structure graph Laplacian matrix for uniform crossover on many-valued configurations.

Evolving a fully functional two-bit multiplier in a configuration of logic cells is a simple case of digital circuit evolution, recently studied in the nascent field of evolvable hardware [29–33]. Digital circuit evolution refers to the design of electronic circuitry in which configurations of logic gates for some pre-specified computational task can emerge in a population of gate arrays using artificial evolution. The two-bit multiplier is a simple electronic circuit, small enough to be feasible for evolutionary design, and practically useful as a fundamental building block used in the synthesis of many digital systems. Examples of evolved two-bit multipliers and other combinatorial circuits can be found in [34] where it was suggested that the design of digital circuits on a gate array in general can be implemented in terms of the functionality and routing of the array [35,36]. These terms refer to the logic functions utilized in the array's cells, and the connections between the array inputs, cells and outputs, respectively, that are then encoded in a genotype. The motivation behind this study is to understand the feasibility of evolving digital circuits – what are the impediments and the limiting factors?

The fitness landscapes of this engineering problem are quite different from many recently studied landscapes [12,13,15,37]. The difference stems from the genotype representation that is a string defined on two completely different many-valued alphabets that represent the functionality and connectivity of the array of cells. It gives rise to a complicated *mixture* of epistatic characteristics of the genotype. This makes the study of the corresponding landscapes much more convoluted. The results show that the two-bit multiplier landscapes originate from products of three configuration spaces that define landscape subspaces with different characteristics. The two-bit multiplier landscapes are vastly neutral with sharply differentiated plateaus. The results are further employed in a study of the relationship between landscape structure and the search to identify principles of evolving this digital circuit.

The chapter begins with a discussion of results from the algebraic theory of landscapes (section 2) and closes with a discussion and suggestions for future work (section 5). Section 3 introduces an analysis for studying the interplay of landscape smoothness, ruggedness, and neutrality. The theoretical findings so far are applied to study the structure of two-bit multiplier landscapes in section 4 where a discussion of the relationship between landscape structure and search is proposed.

2 Algebraic Theory of Fitness Landscapes

A fitness landscape results from the combination of three elements: a set of configurations $\mathcal{V}$, a cost function f, and a move operator ϕ that induces a topology on the set of configurations. The configurations are structures defined over a finite set of symbols called an alphabet. The cost function assigns to each configuration a real value taken from an interval $\mathcal{I} \subset \mathbb{R}$ as follows:

$$f\colon \mathcal{V} \to \mathcal{I}. \tag{1}$$

The operator ϕ defines a relationship between the configurations from $\mathcal{V}$ in the following way :

$$\phi\colon \underbrace{\mathcal{V} \times \mathcal{V} \times \ldots \times \mathcal{V}}_{n-times} \to \underbrace{\mathcal{V} \times \mathcal{V} \times \ldots \times \mathcal{V}}_{m-times}. \tag{2}$$

Thus the space resulted by the three mathematical objects is the fitness landscape

$$\mathcal{L} = (\mathcal{G}_f, f, \phi), \tag{3}$$

where $\mathcal{G}_f$ is the landscape underlying the *hyper*graph whose vertices are the elements from $\mathcal{V}$ labelled with values given by f, and whose edges are specified by the move operator ϕ.

The move operator specifies the topology of the landscape underlying hypergraph. Landscapes defined on a variety of configuration spaces such as Hamming graphs, Johnson graphs, odd graphs, Cayley graphs and many others were studied in [14,15,22]. In this section mutation and recombination fitness landscapes with configuration spaces that are defined on a Hamming graph are described. A graph, $\mathcal{G}_l^n$, is called *Hamming* when its vertices are all sequences of length n that consist of symbols taken from a finite alphabet with size l, and its edges connect configurations that differ from each other in one position [38].

2.1 Mutation Landscapes

Consider a one-point mutation operator. In this case two configurations are neighbors if they differ from each other in one locus. Furthermore each configuration can result from any of its neighbors by mutation. Therefore the configuration space is a Hamming hypercube, $(\mathcal{V}, \mathcal{E})$, with l^n vertices. The landscape underlying graph is undirected, regular and distance transitive. The adjacency matrix $\mathbf{A}$ is defined as

$$\mathbf{A}_{ij} = \begin{cases} 1, & \text{if } \{v_i, v_j\} \in \mathcal{E} \\ 0, & \text{otherwise,} \end{cases} \tag{4}$$

where $v_i, v_j \in \mathcal{V}$. Since the underlying graph is regular, all vertices have the same degree D. The vertex degree matrix $\mathbf{D}$ is equal to $D\mathbf{I}$ where $\mathbf{I}$ is the identity matrix and the degree D is equal to $(l-1)n$. Hence the graph Laplacian matrix is $-\boldsymbol{\Delta} = D\mathbf{I} - \mathbf{A}$.

The eigenvalues of the graph Laplacian matrix can be derived from the algebraic properties of the collapsed adjacency matrix of the configuration space. The collapsed adjacency matrix $\hat{\mathbf{A}}$ of this space has $n+1$ distinct eigenvalues given by

$$\lambda_k^{\hat{\mathbf{A}}} = (l-1)n - lk \tag{5}$$

for $0 \leq k \leq n$. The corresponding left eigenvectors of matrix $\hat{\mathbf{A}}$ are orthogonal and they are given by

$$\vartheta_k(d) = \frac{1}{(l-1)^k \binom{n}{k}} \mathrm{K}_k^{l,n}(d), \tag{6}$$

where $\mathrm{K}_k^{l,n}(d)$ are the Krawtchouk polynomials

$$\mathrm{K}_k^{l,n}(d) = \sum_{i=0}^{k} (-1)^i \binom{d}{i} \binom{n-d}{k-i} (l-1)^{k-i}. \tag{7}$$

In equations 6 and 7 the parameter d is the distance $d(v_k, v_r)$ defined for an arbitrarily chosen reference vertex v_r. Therefore the eigenvalues of the graph Laplacian matrix are

$$\lambda_k = lk, \tag{8}$$

since the eigenvalues of $\hat{\mathbf{A}}$ are eigenvalues of the adjacency matrix.

The eigenvalues of the graph Laplacian matrix of the one-point mutation landscapes can also be obtained as a sum of the eigenvalues of the graph Laplacian of n complete graphs on l vertices [39]. Note that the hypercube, $\mathcal{G}_l^n$, is a Cartesian sum of n copies of the complete graph on l vertices, $\mathcal{K}_l$. Since the eigenvalues of the graph Laplacian of $\mathcal{K}_l$ are 0 and l, the Laplace spectrum of $\mathcal{G}_l^n$ consists of the numbers lk for $k = 0, 1, ..., n$. The multiplicity of lk in the spectrum of the hypercube is equal to $(l-1)^k \binom{n}{k}$.

The algebraic properties of landscapes associated with other mutation operators can be specified in a similar way. For instance, the configuration space for the uniform mutation and underlying Hamming graph is the complete hypergraph with vertex degree $l^n - 1$.

2.2 Recombination Landscapes

The crossover operator is defined as a map $\chi\colon \mathcal{V} \times \mathcal{V} \to \mathcal{V} \times \mathcal{V}$ defined in the following way: for each four configurations x, y, u and v, $\chi(x, y) = (u, v)$ iff for each locus i either $u_i = x_i$ and $v_i = y_i$, or $u_i = y_i$ and $v_i = x_i$. The operator generates an “offspring” of two configurations for each pair of configurations called “parents”.

A *recombination* operator ϱ is the family of all crossover operators that act on $\mathcal{V} \times \mathcal{V}$. This is an operator that maps every pair of configurations of $\mathcal{V}$ to a subset of $\mathcal{V}$ called a power set of $\mathcal{V}$ and denoted by $P(\mathcal{V})$. The recombination operator obeys the following conditions [20]:

i) $\varrho(u, v) = \varrho(v, u)$
ii) $\varrho(u, u) = \{u\}$
iii) $\{u, v\} \subseteq \varrho(u, v)$
iv) $|\varrho(u, v)| \leq |\varrho(u, w)|$ for each $v \in \varrho(u, w)$.

Note that this definition of recombination can be easily generalized to multiparent recombination [40], and even to the more general form of a move operator given in equation 2.

The configuration space induced by a recombination operator, ϱ, is the pair $(\mathcal{V}, \varrho)$ [20]. This construction is called a *P-structure*, and it is a hypergraph whose vertices are the elements of $\mathcal{V}$ and whose edges are the recombination sets generated by ϱ [23,24].

A P-structure can be identified by its incidence matrix $\mathbf{H}$ defined as

$$\mathbf{H}_{i(jk)} = \begin{cases} 1, & \text{if } v_i \in \varrho(v_j, v_k) \\ 0, & \text{otherwise.} \end{cases} \tag{9}$$

The matrix has l^n rows and l^{2n} columns. The vertex degree matrix $\mathbf{D}$ and the hyperedge degree matrix $\hat{\mathbf{D}}$ are defined as

$$\mathbf{D}_{ij} = \delta_{ij} \sum_{\forall k,p\colon\, v_k, v_p \in \mathcal{V}} \mathbf{H}_{i(kp)} \quad \text{and} \quad \hat{\mathbf{D}}_{(ij)(kp)} = \delta_{(ij)(kp)} |\varrho(v_i, v_j)|, \tag{10}$$

respectively. The number $|\varrho(v_i, v_j)|$ quantifies the different possible offsprings of a pair of parents and it is given by

$$|\varrho(v_i, v_j)| = \sum_{\forall k\colon\, v_k \in \mathcal{V}} \mathbf{H}_{k(ij)}. \tag{11}$$

An important matrix associated with a P-structure is the generalized form of the adjacency matrix of the hypergraph. This is a $|\mathcal{V}| \times |\mathcal{V}|$ matrix, $\mathbf{S}$, with elements

$$\mathbf{S}_{ij} = \frac{1}{|\mathcal{V}|} \sum_{\forall k:\, v_k \in \mathcal{V}} \mathbf{H}_{i(jk)} |\varrho(v_j, v_k)|^{-1}, \tag{12}$$

where in this section $|\mathcal{V}|$ is equal to l^n. This directly leads to the definition of the Laplacian matrix

$$-\boldsymbol{\Delta} = \mathbf{I} - \mathbf{S}. \tag{13}$$

The eigenvalues of the graph Laplacian matrices of one-point and uniform recombination P-structures are currently known only for binary strings (configurations for which $l = 2$). These eigenvalues are

$$\lambda_k = 1 - \frac{n-k}{2(n-1)} \quad \text{and} \quad \lambda_k = 1 - \frac{1}{2^k} \tag{14}$$

for one-point and uniform recombination operators [23,26], respectively (see also [24]). Further results on the algebraic structure of recombination spaces are given in [41].

2.3 Correlation Functions of Landscapes

A fitness landscape can be characterized by its autocorrelation function [13,25]. The correlations of a landscape indicate how similar the configurations are in terms of their fitness values. The higher the correlations are, the less different are the fitness values. The autocorrelation function of a landscape is perhaps the most simple and convenient technique for studying the landscape structure. To calculate the autocorrelation function of a fitness landscape, Weinberger [13] proposed to measure the correlations of a *time series* obtained by a simple random walk on the landscape. Thus the autocorrelation function of a time series $\{f_t\}_{t=0}^{T}$ is

$$\rho(s) = \frac{E[f_t f_{t+s}] - E[f_t]E[f_{t+s}]}{V[f_t]} \tag{15}$$

where $E[f_t]$ and $V[f_t]$ are the expectation and the variance of the time series, respectively. Note, however, that this is an "instance" autocorrelation function that is defined on a single time series and it is strongly dependent on the statistical properties of the landscape.

A landscape can be characterized by its expected autocorrelation function defined as follows :

$$r(s) = \frac{\langle f(v_t) f(v_{t+s}) \rangle_{v_0,t} - \langle f(v_t) \rangle_{v_0,t} \langle f(v_{t+s}) \rangle_{v_0,t}}{\sqrt{(\langle f(v_t)^2 \rangle_{v_0,t} - \langle f(v_t) \rangle_{v_0,t}^2)(\langle f(v_{t+s})^2 \rangle_{v_0,t} - \langle f(v_{t+s}) \rangle_{v_0,t}^2)}}, \tag{16}$$

where the expectations are taken over all random walks and all initial conditions v_0. In [25] this function is referred to as a *random walk correlation function.* It is reasonable to enquire how reliable to the random walk correlation function is as a measure of landscapes. This was answered in [25] for landscapes defined on regular graphs. Consider a fitness landscape with a configuration space $\mathcal{G}_f$ defined on the set of configurations $\mathcal{V}$. The mean fitness and the fitness variance of the landscape are given by

$$\mu_f = \frac{1}{|\mathcal{V}|} \sum_{v \in \mathcal{V}} f(v) \tag{17}$$

and

$$\sigma_f^2 = \frac{1}{|\mathcal{V}|} \sum_{v \in \mathcal{V}} (f(v) - \mu_f)^2, \tag{18}$$

respectively. It is important to note that the quantities μ_f and σ_f^2 are defined over all configurations and hence they have nothing to do with any statistical model of landscapes. A simple random walk on graph $\mathcal{G}_f$ can be considered as a Markov process with a transition matrix $\mathbf{T} = \mathbf{D}^{-1}\mathbf{A}$ where $\mathbf{A}$ and $\mathbf{D}$ are the adjacency and the vertex degree matrices, respectively. Thus for a times series obtained by a simple random walk $\{v_t\}_{t=0}^{T}$ on $\mathcal{G}_f$, the probability $Pr[v_t = w]$ is given by the wth coordinate of vector $\{\mathbf{T}^t p_{v_0}\}$ where the probability $p_{v_0,w}$ is the initial condition $p_{v_0}(w)$ [42]. Then for all t and initial conditions v_0 the mean $\langle f(v_t)\rangle_{v_0,t}$ can be defined in the following way:

$$\begin{aligned}
\langle f(v_t)\rangle_{v_0,t} &= \lim_{T\to\infty} \frac{1}{T+1} \sum_{t=0}^{T} \frac{1}{|\mathcal{V}|} \sum_{v_0 \in \mathcal{V}} \sum_{w \in \mathcal{V}} f(w)\{\mathbf{T}^t p_{v_0}\}_w \\
&= \lim_{T\to\infty} \frac{1}{T+1} \sum_{t=0}^{T} \frac{1}{|\mathcal{V}|} \sum_{w \in \mathcal{V}} f(w)\{\mathbf{T}^t (\sum_{v_0 \in \mathcal{V}} p_{v_0})\}_w \\
&= \lim_{T\to\infty} \frac{1}{T+1} \sum_{t=0}^{T} \frac{1}{|\mathcal{V}|} \sum_{w \in \mathcal{V}} f(w)\{\mathbf{T}^t \mathbf{1}\}_w \\
&= \lim_{T\to\infty} \frac{1}{T+1} \sum_{t=0}^{T} \frac{1}{|\mathcal{V}|} \sum_{w \in \mathcal{V}} f(w) = \lim_{T\to\infty} \frac{1}{T+1} \mu_f .
\end{aligned} \tag{19}$$

Note that $\mathbf{T1}$ is equal to $\mathbf{1}$ only for regular graphs. Following [25] one can represent $\langle f(v_t)^2\rangle_{v_0,t}$ and $\langle f(v_t) f(v_{t+s})\rangle_{v_0,t}$ in forms similar to this in equation 19, and thus obtain the autocorrelation function

$$r(s) = \frac{\frac{1}{|\mathcal{V}|}\langle f, \mathbf{T} f\rangle - \mu_f^2}{\sigma_f^2} \tag{20}$$

avoiding any assumption of a random walk.

The question of whether the autocorrelation function on a single random walk is an appropriate measure of the structure of a landscape was addressed in [13]. It was suggested that the autocorrelation function as given in equation 15 is an appropriate tool for studying the structure of statistically isotropic landscapes. A fitness landscape is statistically isotropic if the fitness values assigned to the landscape configurations form a stationary random process of the assumed joint distribution of fitness values [43].

2.4 Fourier Transforms of Landscapes

Consider a fitness landscape with a configuration space $\mathcal{G}_f$. The Fourier transform of the landscape on $\mathcal{G}_f$ is given by

$$f(v_k) = \sum_{i=0}^{|V|-1} a_i \varphi_i(v_k), \tag{21}$$

where $\{\varphi_i\}$ is a complete and orthogonal system of eigenfunctions that are eigenvectors of the graph Laplacian matrix $-\mathbf{\Delta}$, and $\{a_i\}$ are the Fourier coefficients [44,22]. The elements $\{v_k\}$ are configurations from $\mathcal{V}$ and $\varphi_i(v_k)$ is the kth component of the eigenvector φ_i.

A landscape is *flat* when its configuration space is defined on a constant cost function, i.e. $f(v) \equiv const$ for any $v \in \mathcal{V}$. For such landscapes the cost function is an eigenvector of $-\mathbf{\Delta}$ with eigenvalue $\lambda_0 = 0$.

A landscape is called *elementary* when the cost function over the underlying graph has the form

$$f(v) = c + \varphi(v), \tag{22}$$

where c is a constant, and φ is an eigenvector of $-\mathbf{\Delta}$ with eigenvalue $\lambda > 0$. Consequently the graph derivative of φ is given by $(-\mathbf{\Delta})\varphi = \lambda\varphi$. The random walk correlation function of a non-flat and elementary landscape defined on a regular graph can be written in the form [25]

$$r(s) = \sum_{i \neq 0} A_i (1 - \frac{\lambda_i}{D})^s, \tag{23}$$

where A_i are the amplitudes normalized in the following way :

$$A_i = \frac{|a_i|^2}{\sum_{j \neq 0} |a_j|^2}. \tag{24}$$

This follows from equations 20 and 21, and from the fact that $\mathbf{T}\varphi_i = 1 - \frac{\lambda_i}{D}$. Equation 23 implies that $r(s)$ is an exponential function iff all non-zero amplitudes A_i belong to a single eigenvalue, λ_k, of the graph Laplacian matrix $\mathbf{\Delta}$ [25].

The random walk correlation function of an elementary landscape can provide the nearest neighbor correlation of the landscape. The nearest neighbor correlation is given by $\hat{r} = r(1)$. Note that this can also be written as $\hat{r} = 1 - \frac{\lambda_k}{D}$ where λ_k is a non-zero eigenvalue of the graph Laplacian matrix.

2.5 Smoothness, Ruggedness and Neutrality of Landscapes

The structure of a fitness landscape is completely determined by three landscape characteristics. These are the landscape smoothness, ruggedness and neutrality. In [12,11,25,45] the landscape smoothness and ruggedness relate to the fitness differences between neighbouring points, while the landscape neutrality is determined by the flat fitness landscape areas.

The smoothness, ruggedness and neutrality of a landscape result from the properties of its local optima [11]. A vertex $v \in \mathcal{V}$ is a *local maximum* if its fitness value is higher than the fitness values of all neighbours of v, e.g. $f(v) \geq f(w)$ for all $w \in \mathcal{N}(v)$. Local minima are defined analogously [19]. The number and the fitness distribution of local optima specify the landscape ruggedness [12,11]. It is said that a landscape is rugged if the number of local optima is high. The fitness distribution specifies another property of the landscape ruggedness that relates to the variety of landscape forms such as ridges, valleys, cliffs, peaks and others. For instance, a landscape may have many clusters of very high optima surrounded by significantly lower peaks.

When the number of optima is low, the landscape is either smooth or neutral. A landscape is smooth if its optima are characterized by large basins of attraction. According to Jones [19], a *basin of attraction* of a vertex, v_n, is the set of vertices

$$\begin{aligned}\mathcal{B}(v_n) = \{v_0 \in \mathcal{V} \mid \exists v_1, ..., v_n \in \mathcal{V} \text{ with } v_{i+1} \in \mathcal{N}(v_i) \text{ and } f(v_{i+1}) > f(v_i)\\ (\text{or } f(v_{i+1}) < f(v_i) \text{ if minimizing}) \quad \text{for each } i, 0 \leq i < n\}. \quad (25)\end{aligned}$$

The number of vertices in a basin of attraction of an optimum can be considered as the *size* of the basin. The local optima with smallest basins of attraction are called *isolated.* Hence, the larger the basins of attraction are, the smoother is the landscape. Similarly to the landscape ruggedness, the smoothness also depends on the fitness distribution. Note that the fitness distribution specifies the steepness of the basins of attraction.

The other landscape characteristic associated with landscapes that are characterized by a low number of local optima is the neutrality. The neutral areas of a landscape originate from its plateaus and ridges. A *plateau* is a set $\mathcal{P}$ of two or more vertices such that for each two configurations $v_0, v_n \in \mathcal{P}$ a subset configurations $v_1, ..., v_{n-1}$ exists for which $f(v_i) = f(v_{i+1})$ and $v_{i+1} \in \mathcal{N}(v_i)$ for all i, $0 \leq i < n$ [19]. Similarly, two configurations v and w are *neutral* if $f(v) = f(w)$ [45]. The number of *neutral neighbors* of a configuration v is given by

$$\nu_v(f) = \sum_{w \in \mathcal{N}(v)} X_{v,w}(f), \tag{26}$$

where $X_{v,w}(f)$ are random variables that take the value 1 if $f(v) = f(w)$ and 0 otherwise [46].

3 An Analysis of Landscapes

There are various techniques for studying the smoothness, ruggedness and neutrality of fitness landscapes. The structure of a fitness landscape can be characterized by studying the properties of walks that are defined to capture the statistics of certain landscape characteristics [11]. For instance, to study the landscape smoothness and ruggedness Kauffman and Levin [47] employed an algorithm called *adaptive walk*. It was shown that the adaptive walks of rugged landscapes are short. However, the technique of an adaptive walk does not convey information about the neutrality of landscapes. This landscape feature can be studied by *neutral walks* as was suggested in [45]. The larger the neutral network of a landscape is, the longer are the neutral walks.

The techniques of adaptive and neutral walks can often be computationally expensive. For instance, to add a configuration to a neutral walk that is a list of configurations, the new configuration needs to be a neutral neighbour that results in an increase in the distance from the starting configuration. This may be a long process when considering that this procedure needs to be repeated for all neighbours of the current configuration. Furthermore, in engineering the landscapes may be huge and often defined on many-valued alphabets. They can be vastly neutral with many and different types of local optima.

The structure of a fitness landscape can be studied by measuring the autocorrelation function of a time series obtained by a random walk on the examined landscape (section 2). The autocorrelation function as a measure of landscapes was studied by Weinberger [13] and Stadler [22] who suggested that the autocorrelations can be used to estimate the amplitude spectra derived from the Fourier transforms of the landscapes [26]. Another landscape analysis, based on the Box and Jenkins [48] approach, was proposed by Hordijk [49]. The basic idea, initially suggested by Weinberger [13], was to explore the landscape structure by studying the corresponding autoregressive moving average (ARMA) model which in fact characterises the landscape ruggedness much more precisely. An important feature of these techniques is

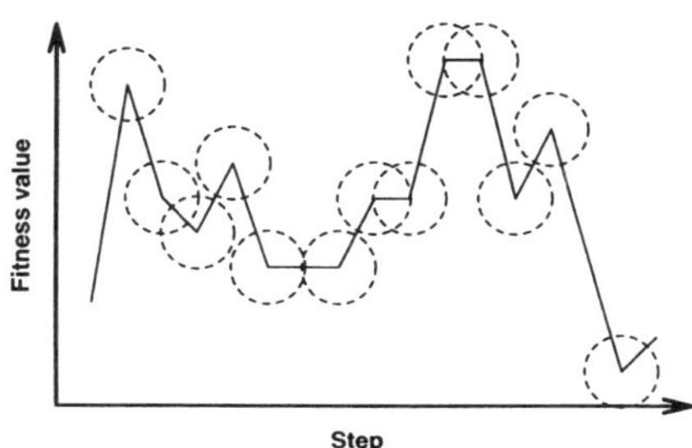

Fig. 1. A sequence of fitness values as an ensemble of objects

that they study the structure of the landscapes in terms of their ruggedness. In this section an information analysis is proposed to investigate the interplay of smoothness, ruggedness and neutrality of the fitness landscapes. Related work on the analysis can be found in [27,28]. The underlying idea is to consider a fitness landscape as an ensemble of objects as is illustrated in Figure 1. Each object consists of a configuration and its nearest neighbors. The figure shows how a landscape path obtained by a random walk on the landscape can be represented as an ensemble of objects. To study how the ruggedness and smoothness of a landscape are related to the landscape neutrality, two subsets of the original ensemble are specified as demonstrated in Figure 2. For each subset an *information function* is constructed that reveals how the

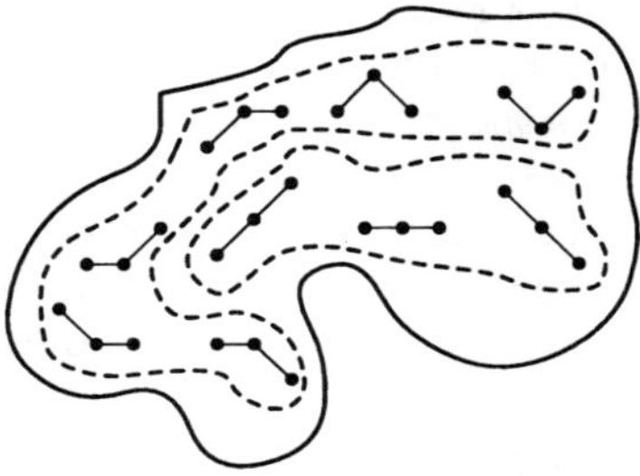

Fig. 2. A fitness landscape as an ensemble of objects in two categories

estimate of the entropy of this subensemble changes when the neutrality of the landscape is increased.

3.1 Ensemble on a Simple Random Walk

Consider time series $\{f_t\}_{t=0}^{n}$ of fitness values from the interval $\mathcal{I}$ that are obtained by a *simple random walk* on a landscape. The time series convey information about the structure of the landscape. The aim is to extract this information, by representing the time series as an ensemble of objects. The ensemble can be defined as a string, $S(\varepsilon) = s_1 s_2 s_3 ... s_n$, of symbols $s_i \in \{\bar{\mathbf{1}}, \mathbf{0}, \mathbf{1}\}$, and they are obtained by the function

$$\begin{aligned} \Psi_{f_t}(i, \varepsilon) &= \bar{\mathbf{1}}, \text{ if } f_i - f_{i-1} < -\varepsilon \\ &= \mathbf{0}, \text{ if } |f_i - f_{i-1}| \leq \varepsilon \\ &= \mathbf{1}, \text{ if } f_i - f_{i-1} > \varepsilon \end{aligned} \tag{27}$$

in the following way :

$$s_i = \Psi_{f_t}(i, \varepsilon) \tag{28}$$

for any fixed ε. The parameter ε is a real number from the interval $[0, l_{\mathcal{I}}]$, where $l_{\mathcal{I}}$ is the length of the interval $\mathcal{I}$. Note that the parameter ε determines the accuracy of calculation of the string $S(\varepsilon)$. If $\varepsilon = 0$, the function Ψ_{f_t} will be very sensitive to the differences between the fitness values and $S(\varepsilon)$ will be determined as precisely as possible. When the parameter ε is $l_{\mathcal{I}}$, $S(\varepsilon)$ will be a string of **0**s.

The string $S(\varepsilon)$ contains information about the structure of the landscape. Note that the function Ψ_{f_t} associates each edge of the walk with an element from the set $\{\bar{\mathbf{1}}, \mathbf{0}, \mathbf{1}\}$. Consequently, each object of the path is represented by a string, $s_i s_{i+1}$, that is a sub-block of length w of $S(\varepsilon)$. According to section 2 the string $S(\varepsilon)$ is a sequence of non-zero elements (a sample) of the *weighted incidence* matrix of the landscape underlying graph. This is a matrix with elements

$$\mathbf{H}_{ij} = \begin{cases} a, & \text{if edge } e_i \text{is incident to vertex} \\ 0, & \text{if otherwise} \end{cases} \tag{29}$$

where $a \in \{\bar{\mathbf{1}}, \mathbf{0}, \mathbf{1}\}$ is determined in the same manner as the elements of $S(\varepsilon)$. The incidence matrix of a landscape is related to the landscape's graph *Laplacian* matrix whose eigenvectors are the orthogonal basis of eigenfunctions of the Fourier transform of the landscape (see section 2).

3.2 Information Characteristics

Two entropic measures of the ensemble of the sub-blocks of length w of $S(\varepsilon)$ can be devised. These are

$$H(\varepsilon) = -\sum_{p \neq q} P_{[pq]} \log_6 P_{[pq]}, \tag{30}$$

and

$$h(\varepsilon) = -\sum_{p = q} P_{[pq]} \log_3 P_{[pq]}, \tag{31}$$

referred to as the first and second entropic measures (FEM and SEM), respectively. The FEM is an estimate of the ruggedness of a landscape with respect to the landscape neutrality, while the SEM relates to the interplay of smoothness and neutrality of the landscape. The probabilities $P_{[pq]}$ are frequencies of the possible blocks pq of elements from set $\{\bar{\mathbf{1}}, \mathbf{0}, \mathbf{1}\}$. They are defined as

$$P_{[pq]} = \frac{n_{[pq]}}{n} \tag{32}$$

where $n_{[pq]}$ is the number of sub-blocks pq in the string $S(\varepsilon)$. The logarithms from equations 30 and 31 are taken with bases 6 and 3, respectively, since these are the numbers of blocks of length 2 composed respectively of two different and two equal symbols taken from $\{\bar{\mathbf{1}}, \mathbf{0}, \mathbf{1}\}$.

The FEM and the SEM characterize the time series $\{f_t\}_{t=0}^n$ with a certain accuracy that can be varied by the parameter ε. This in turn defines the entropic measures as functions of the accuracy. One can think of the parameter ε as a magnifying glass through which the landscape can be observed. For small values of ε, the function Ψ_{f_t} from equation 27 will be very sensitive to the difference between the fitness values, i.e. the glass will make each element of the landscape visible. If ε is zero then the accuracy of the estimations, obtained by $H(\varepsilon)$ and $h(\varepsilon)$, is high. In contrast, for $\varepsilon = l_{\mathcal{I}}$, the FEM and the SEM of $S(\varepsilon)$ are 0, i.e. for such ε the landscape path will be determined as relatively flat.

3.3 Information Analysis

The analysis starts by performing a random walk on the landscape, using the evolutionary operator by which the neighbours of each configuration are specified. For each step, the fitness value of the current current is recorded. Thus, for a certain number of steps a time series will be obtained. Then the information functions $H(\varepsilon)$ and $h(\varepsilon)$ are calculated.

To quantify the entropic measure of a landscape, represented as an ensemble of objects, does not necessarily imply measuring the ruggedness of the landscape. At this point, the information analysis differs significantly from many other statistical approaches. For instance, consider the autocorrelation function, which reveals how the correlation is "spread" over the landscape. The lower the correlations are, the more rugged is the landscape. The information analysis is different. It gives us a notion of what the relation is between the landscape smoothness, ruggedness and neutrality. For a better understanding of the analysis some interesting properties of functions $H(\varepsilon)$ and $h(\varepsilon)$ are given below.

Let $\varepsilon^\star$ be the lowest value of ε for which $S(\varepsilon)$ is a string of 0s. Firstly, the FEM and the SEM of a time series generated by a *regular* walk on the landscape are positive constants for each $\varepsilon \in [0, \varepsilon^\star)$. The term *regularity* is defined as follows: a time series, $\{f_t\}_{t=0}^n$, is generated by a *regular* walk (it has nothing to do with the notion of regular graphs) on a landscape when the time series obeys

$$f_{t+1} = f_t \pm \kappa c, \tag{33}$$

where c is a constant and κ is a variable which can be 0, 1 or -1. If equation 33 is not fulfilled, the landscape path is generated by an *irregular* walk.

In practice regular walks on a landscape are a rare occurrence. Consider equation 33 in the form $f_{t+1} = f_t + \kappa \sum_i c_i$ where c_i are different constants. The degree of regularity of a landscape is the number of different κc_i; that is to say, the number of all possible differences of fitness values. For instance,

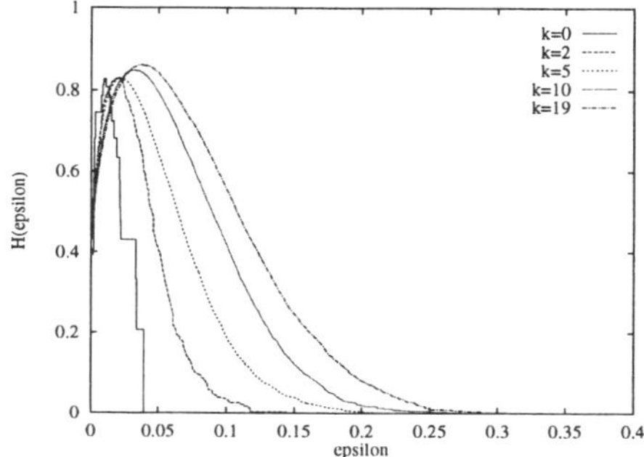

Fig. 3. The information function $H(\varepsilon)$ of Nk landscapes with random neighborhoods for $N = 20$ and different values of k

the degree of regularity of Nk landscapes generated by one-point mutation is low, and it decreases as k increases from 0 to $N - 1$ since the number of possible fitness values increases with a higher than linear rate (Figure 3).

Secondly, there exists a sequence of fitness values, and parameters ε_1 and ε_2, such that $H(\varepsilon_1) < H(\varepsilon_2)$, where $0 \leq \varepsilon_1 < \varepsilon_2 \leq l_{\mathcal{I}}$. The landscapes associated with this class are characterized by relatively low neutrality. For such landscapes, $H(\varepsilon)$ is an increasing function for small values of ε, since the landscape as an ensemble consists mainly of two types of objects. An example of such landscapes is given in Figure 3, which shows the functions $H(\varepsilon)$ of Nk landscapes for different values of k.

Thirdly, if $H(\varepsilon) = \log_6 2$ and $h(\varepsilon) = 0$ for $\varepsilon = 0$, then the explored time series is maximally multimodal or an increasing/decreasing step function.

Lastly, the neutrality prevails over the ruggedness in a time series, if $H(\varepsilon)$ is a decreasing function.

4 Circuit Evolution and the Structure of Landscapes

The section studies the structure of two-bit multiplier landscapes that originate from a simple case of digital circuit evolution. Related work can be found in [50–53].

4.1 Evolving Digital Circuits

In [54,34] digital circuits are evolved using an evolutionary algorithm. A population of digital electronic circuits that are instances of a particular program is maintained, the genotypes of which are initially generated at random. The fitness value of each genotype is evaluated, by calculating the percentage of total correct outputs of the encoded electronic circuit in response to appropriate inputs. A new population is generated by applying uniform crossover with probability 0.5 followed by one-point mutation to genotypes selected from the parent population. The selection is tournament with size 2 in which the winner of the tournament is chosen with a certain probability. For convenience, the fitness values are scaled in the interval [0, 1].

To encode a digital electronic circuit into a genotype, a genotype/phenotype mapping is defined. This is a rectangular array of cells each of which is an atomic two-input logic gate or a multiplexer. Thus the genotype is a linear string of integers and it consists of two different types of genes that are responsible for the functionality and the routing of the evolved array. It is important to note that the representation is restricted to allow only feed-forward circuits since only combinational design is considered. The genotype is characterized by four parameters of the array of cells: the *number of allowed logic functions*, the *number of rows*, the *number of columns* and *levels-back*. The first parameter defines the functionality of logic cells, while the latter three parameters determine the routing of the array. Note that the number of inputs and outputs of the array are specified by the evolved function. An example of an $n \times m$ array with n_I inputs and n_O outputs is given in Figure 4. The array is a composition of cells each of which can be any allowed two-input

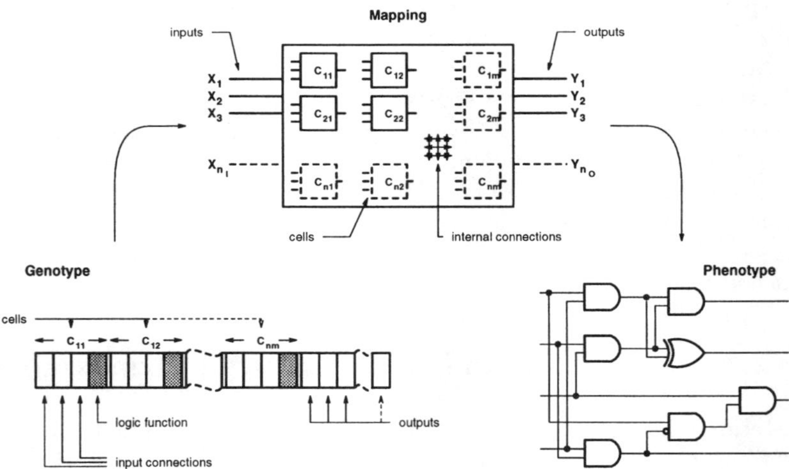

Fig. 4. The phenotype that is a digital circuit is encoded within a genotype by an array of logic cells called a *genotype-phenotype mapping*. The mapping is a $n \times m$ geometry of logic cells with n_I inputs and n_O outputs.

logic gate or alternatively a 2-1 multiplexer (two inputs plus a single control input). The allowed cell functions are listed in Table 1. The first 16 letters in the list represent all boolean functions of two arguments (see [55]). The remainder of the list are the six possible basic *universal-logic-gate 2* functions as given in [56] where functions 16 to 19 are binary multiplexors defined by

Letter	Function	Letter	Function
0	0	11	$a \oplus \overline{b}$
1	1	12	$a + b$
2	a	13	$a + \overline{b}$
3	b	14	$\overline{a} + b$
4	$\overline{a}$	15	$\overline{a} + \overline{b}$
5	$\overline{b}$	16 – 19	$f_L(a, b, c)$
6	$a \cdot b$	20	$f_M(a, b, c)$
7	$a \cdot \overline{b}$	21	$f_H(a, b, c)$
8	$\overline{a} \cdot b$	22	$f_I(a, b, c)$
9	$\overline{a} \cdot \overline{b}$	23	$f_J(a, b, c)$
10	$a \oplus b$	24	$f_K(a, b, c)$

Table 1. Allowed cell functions.

$f_L(a, b, c) = a \cdot \overline{c} + b \cdot c$ with various inputs inverted. The symbols for logic operators used in the table are "$-$" NOT, "$\cdot$" AND, "$+$" inclusive-OR, and "$\oplus$" exclusive-OR (XOR). The corresponding symbols used to represent logic gates on the circuit diagrams below in the chapter are shown in Figure 5. The inversion that may appear in any of the inputs or the outputs of these devices is denoted by little circles. It is important to note that multiplexors can be considered as atomic gates both formally and from an implementational point of view. The gate is atomic in that it is a universal logic module so it can be used to represent any logic function. This advantage is cleverly used in some modern field-programmable gate arrays with a multiplexer based architecture [34].

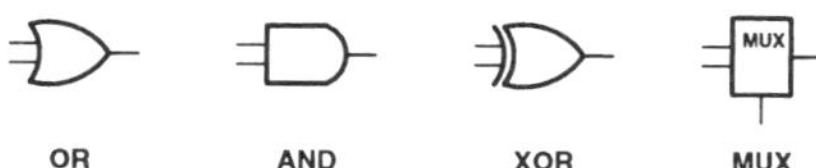

Fig. 5. Binary circuit symbols

The internal connectivity of the array is defined by the connections between the array cells. The inputs of each cell are only allowed to be inputs of the array or outputs of cells with lower column numbers. The array connectivity is also dependent on the levels-back parameter. This parameter defines the neighborhood of each cell as follows: assume that the gate array inputs are outputs of cells with column number 0 and the levels-back parameter is L (L has nothing to do with configuration f_L from Table 1), then the inputs of cells $\{c_{ij}\}_{\forall i,j}$ are chosen from all possible outputs of "preceding" cells with column numbers j' such that $j'' \leq j' < j$ where $j'' = j - L$ if $j > L$, otherwise $j'' = 0$.

The gate array output connectivity is defined in a similar way. The output connections of the array are allowed to be outputs of cells or array inputs. Again, this is dependent on the neighborhood defined by the levels-back parameter. Consider, for instance, the gate array in Figure 4. If the levels-back is $m + 1$ then the array outputs can be any of the outputs of cells $\{c_{ij}\}_{\forall i,j}$ and the inputs of the array. On the contrary, if the levels-back is 1 then only cells from column m can be connected to the outputs of the array.

The construction of the genotype is straightforward. Consider the notation given above. All inputs and cells of the array are associated with indexes in the following way. Each array input X_k is labelled with $k - 1$ for $1 \leq k \leq n_I$. Alternatively, each cell c_{ij} is labelled with an integer given by $n_I + n(j - 1) + i - 1$ for $1 \leq i \leq n$ and $1 \leq j \leq m$. For instance, if the gate array has 5 inputs the first cell has the index 5 since the last array input is labelled with 4. The genotype consists of two parts. The first part consists of groups of four integers where each group encodes one cell of the gate array. The first three values of each group give the indexes of the cells to which the inputs of the encoded cell are connected. If the cell represents a two-input logic function, then the third connection is redundant. This type of redundancy is referred to as *input redundancy*. The last integer of the group represents the logic function of the cell. Cells may also not have their outputs connected in the operating circuit. This is another form of redundancy called *cell redundancy*. The redundancy in the genotype related to the function of the array may also be *functional redundancy*. This is the case in which the number of cells of a digital circuit is higher than the optimal number needed to implement this circuit. The second part of the genotype is a string of integers that represent the indexes of the cells connected to the outputs of the array. The genotype structure is shown in Figure 4.

An example of a genotype of an evolved two-bit multiplier on a 4×4 array with fitness value 1.00 (the highest possible fitness) is given in Table 2. To obtain this genotype, the first 21 logic functions as they are listed in Table 1 were allowed. The levels-back parameter in this experiment was set to 2. The *phenotype* of the represented two-bit multiplier is depicted in Figure 6. The figure shows the evolved circuit with respect to the genotype/phenotype mapping. Each cell in the array is given as a box where the number in the top-right corner is the index of the cell. The dashed lines represent the connections to redundant cells.

The two-bit multiplier, given in Figure 6, does not represent an optimal solution of this optimisation problem. Examples of much better solutions of combinatorial circuits such as “adders”, “multipliers” and “parity functions” can be found in [54,34]. Two better combinatorial circuits of the two and three-bit multiplier; are also represented further in this chapter.

Cell	Allele(Gene)			
4	$0_{(0)}$	$0_{(1)}$	$2_{(2)}$	$15_{(3)}$
5	$0_{(4)}$	$1_{(5)}$	$0_{(6)}$	$4_{(7)}$
6	$0_{(8)}$	$2_{(9)}$	$0_{(10)}$	$19_{(11)}$
7	$1_{(12)}$	$2_{(13)}$	$1_{(14)}$	$19_{(15)}$
8	$2_{(16)}$	$3_{(17)}$	$0_{(18)}$	$4_{(19)}$
9	$0_{(20)}$	$7_{(21)}$	$2_{(22)}$	$6_{(23)}$
10	$3_{(24)}$	$4_{(25)}$	$0_{(26)}$	$14_{(27)}$
11	$3_{(28)}$	$1_{(29)}$	$3_{(30)}$	$6_{(31)}$
12	$4_{(32)}$	$7_{(33)}$	$4_{(34)}$	$11_{(35)}$
13	$11_{(36)}$	$11_{(37)}$	$6_{(38)}$	$16_{(39)}$
14	$9_{(40)}$	$7_{(41)}$	$10_{(42)}$	$17_{(43)}$
15	$5_{(44)}$	$6_{(45)}$	$6_{(46)}$	$16_{(47)}$
16	$15_{(48)}$	$13_{(49)}$	$8_{(50)}$	$9_{(51)}$
17	$10_{(52)}$	$14_{(53)}$	$10_{(54)}$	$6_{(55)}$
18	$14_{(56)}$	$12_{(57)}$	$9_{(58)}$	$2_{(59)}$
19	$15_{(60)}$	$13_{(61)}$	$15_{(62)}$	$8_{(63)}$
Outputs	$19_{(64)}$	$16_{(65)}$	$14_{(66)}$	$13_{(67)}$

Table 2. The genotype of a two-bit multiplier evolved on 4 × 4 array of cells. The genotype has fitness value 1.00. The column "Cell" represents the index of each cell with genetic information listed on the right side at column "Allele(Gene)" where the number of each gene is given in parentheses.

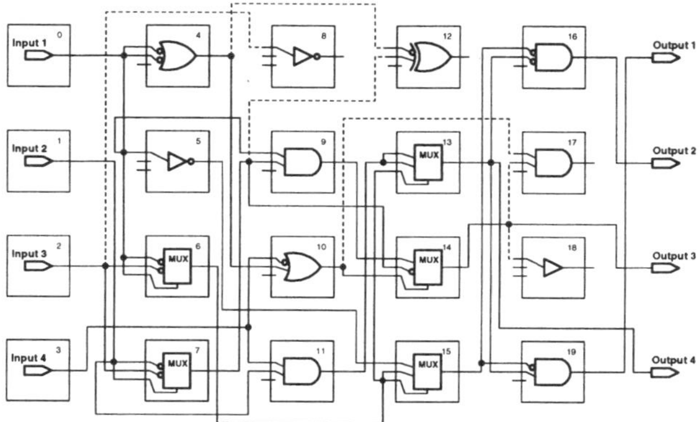

Fig. 6. A two-bit multiplier evolved on a 4 × 4 array of cells. The levels-back parameter is 2 and the allowed functional gates are the first 21 logic functions listed in Table 1.

4.2 The Circuit Evolution Landscapes

The genotype can be considered as a composition of three different parts that are responsible for, first, the functionality of the cells, second, the array internal connectivity, and third, the array output connections. For convenience, each part is termed a *chromosome*. The reason for splitting the genotype into chromosomes is the difference of the purposes of these genotype parts. Thus, each genotype becomes a composition of three chromosomes with different length; that are defined over two completely different alphabets. The "functionality" chromosomes are strings over alphabet α with length the number of cells. The "internal connectivity" and the "output connectivity" chromosomes are defined over alphabet β, and they are strings with length the number of gates and the number of array outputs, respectively. The alphabet α is a set of integers that represents the allowed logic functions listed in Table 1. Therefore, the alphabet size, l_α, is the number of logic functions used in the circuit design. The alphabet β is given in a different way. It is related to the number of neighbors of the cells and the array outputs. Again, the alphabet is a set of integers, but they are reference numbers of the elements of a neighborhood. Hence, the size of β is $l_\beta = \begin{cases} nL, & \text{if } L \leq m \\ nm + n_I, & \text{otherwise} \end{cases}$, where n is the number of rows, m is the number of columns, n_I is the number of inputs of the gate array, and L is the levels-back parameter. For instance, the alphabets α and β of the example given by Table 2 and Figure 6 have sizes $l_\alpha = 21$ and $l_\beta = 8$, respectively.

The genotype structure as given above determines a convoluted *mixture* of epistatic characteristics of the genotype. In a study of the corresponding fitness landscapes, this is an almost insurmountable impediment if the genotype is considered as an inseparable chain. The circuit evolution landscapes are highly *non-isotropic*. This was revealed by measured autocorrelations of random walks. A vast number of random walks were performed each of which had length $100,000$. The autocorrelation functions of these walks differed from each other over a significantly wide range. To overcome this problem, the genotype space is split into three subspaces as was done in [57].

Since each genotype consists of three chromosomes it is assumed that the configuration space of the circuit evolution landscapes is the generalised Hamming graph. Hence, the configuration space is a Cartesian product of three hypercubes defined on the alphabets α and β. The hypercubes are $\mathcal{G}^{nm}_{l_\alpha}$, $\mathcal{G}^{3nm}_{l_\beta}$ and $\mathcal{G}^{no}_{l_\beta}$ and they represent the configuration spaces of the corresponding chromosomes responsible for functionality, internal connectivity and output connectivity, respectively. Consider an evolutionary operator ϕ. For each hypercube a family of landscapes that represents the genotype space with respect to the hypercube can be defined. The landscapes are

$$\begin{aligned} &(1)\ \{\mathcal{G}_{f_i}, f_i, \phi\}_{i=0}^{l_\beta^{3nm+n_O}-1} \\ &(2)\ \{\mathcal{G}_{g_i}, g_i, \phi\}_{i=0}^{l_\alpha^{nm} l_\beta^{n_O}-1} \\ &(3)\ \{\mathcal{G}_{h_i}, h_i, \phi\}_{i=0}^{l_\alpha^{nm} l_\beta^{3nm}-1}. \end{aligned} \tag{34}$$

The hypergraphs $\mathcal{G}_{f_i}$, $\mathcal{G}_{g_i}$ and $\mathcal{G}_{h_i}$ are defined with vertices from $\mathcal{G}^{nm}_{l_\alpha}$, $\mathcal{G}^{3nm}_{l_\beta}$, and $\mathcal{G}^{no}_{l_\beta}$, respectively, assigned with fitness values, and have edges specified by the corresponding evolutionary operator (see also [22,20]). The fitness values are provided by fitness functions $\{f_i\}_{\forall i}$, $\{g_i\}_{\forall i}$ and $\{h_i\}_{\forall i}$ that are defined as follows:

$$\begin{array}{lll} (1)\ \forall i(\mathbf{c}_i \in \mathcal{G}_{l_\beta}^{3nm} \times \mathcal{G}_{l_\beta}^{no}), \forall \mathbf{x} \in \mathcal{G}_{l_\alpha}^{nm} & : & f_i(\mathbf{x}) = F(\mathbf{x} \circ \mathbf{c}_i) \\ (2)\ \forall i(\mathbf{c}_i \in \mathcal{G}_{l_\alpha}^{nm} \times \mathcal{G}_{l_\beta}^{no}), \forall \mathbf{x} \in \mathcal{G}_{l_\beta}^{3nm} & : & g_i(\mathbf{x}) = F(\mathbf{x} \circ \mathbf{c}_i) \\ (3)\ \forall i(\mathbf{c}_i \in \mathcal{G}_{l_\alpha}^{nm} \times \mathcal{G}_{l_\beta}^{3nm}), \forall \mathbf{x} \in \mathcal{G}_{l_\beta}^{no} & : & h_i(\mathbf{x}) = F(\mathbf{x} \circ \mathbf{c}_i). \end{array} \quad (35)$$

The function F is taken over the whole genotype space (the operator "$\circ$" is defined to merge two strings in a special way so that the genotype structure as given in Figure 4 is maintained) and evaluates the percentage of correct output bits of the represented digital circuit. Note that for each family of landscapes there is a group of fitness functions, and each fitness function estimates only a part of the genotype. Hence, its index is uniquely defined by the constant string $\mathbf{c}_i$ that is the remainder of the genotype.

4.3 The Landscape Ruggedness

In this section a study of the ruggedness of the circuit evolution fitness landscapes is proposed. The fitness landscapes originate from the evolutionary design of a two-bit multiplier evolved on a 4×4 array of cells. The evolutionary operators are one-point mutation and uniform crossover with probability 0.5. The levels-back is 2 and the allowed logic functions are the first 21 functions listed in Table 1.

Autocorrelation Functions The structure of two-bit multiplier landscapes is investigated by measuring the autocorrelations of time series obtained by random walks on the landscapes as was suggested in [13,22]. To gather statistics of the circuit evolution landscapes with respect to the outlined landscape families 1,000 random walks per family were performed, each of them with length 100,000. The initial configurations were chosen randomly, with fitness values uniformly distributed in the interval $[0, 1]$. Thus, for each family of landscapes, random walks were performed on 1,000 instances. The random walk on the mutation landscape is implemented as follows: start from a randomly chosen landscape point, generate all neighbours of the current point by mutation and evaluate their fitness values, choose randomly one neighbour and record its fitness, generate all neighbours of the new point, which becomes "current", and so on [13]. The random walk on the recombination landscape is different. In this case, the random walk is implemented by the mapping ϱ defined in section 2 (see also [23] (p. 248)). The algorithm is given by Wagner and Stadler [24] and can be described as follows: start with an arbitrary pair of parents, generate a set of offsprings by ϱ (applying the recombination operator) and evaluate their fitness values, from those choose randomly one (record its fitness) and mate it with a randomly chosen genotype, etc., until the termination conditions are satisfied. The walks were performed with respect to the investigated genotype partitions.

The autocorrelation functions of walks on one-point mutation and uniform crossover landscapes are depicted in Figures 7 and 8, respectively. They are obtained by applying the formula from equation 15 to each time series. Additionally, standard deviations are given since the autocorrelations are averaged over 1,000 random walks per landscape family. The results are depicted with scaled distance, d/n, as was suggested in [57]. The experimental data indicate that the autocorrelation functions and their deviations converge towards the depicted plots in approximately 20 walks after which insignificant changes were observed.

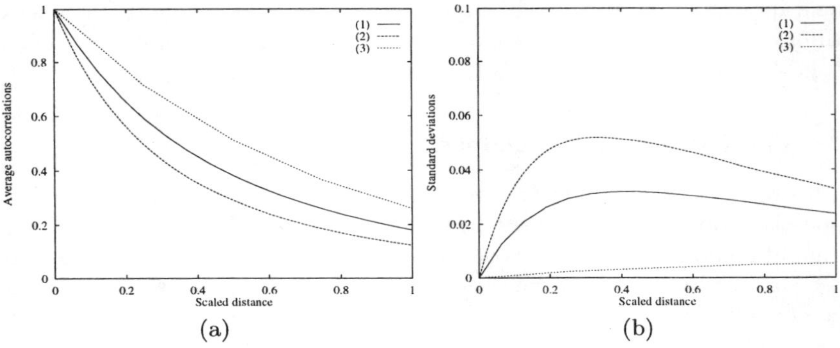

Fig. 7. Mutation: (a) autocorrelations and (b) standard deviations with scaled distance of one-point mutation landscapes with respect to (1) functionality, (2) internal connectivity and (3) output connectivity

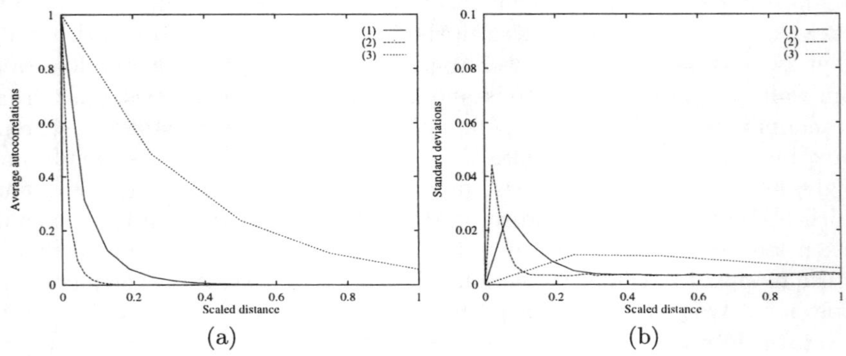

Fig. 8. Recombination: (a) autocorrelations and (b) standard deviations with scaled distance of uniform crossover landscapes (probability 0.5) with respect to (1) functionality, (2) internal connectivity and (3) output connectivity

The figures reveal that the correlations of uniform crossover landscapes are low, and therefore the landscapes might be difficult for evolutionary search since they are rugged. This agrees with Miller *et al.* [54] who suggested that the crossover operator might be dropped in order to improve the search. The plots of the autocorrelations also show that the two-bit multiplier landscapes are compositions of three sub-landscapes with different ruggedness. It can be seen that the output connectivity landscapes are significantly smoother than the functionality and internal connectivity landscapes; this is valid for both mutation and recombination. On the contrary, the internal connectivity landscapes are the most rugged. The scaled standard deviations reveal how the chromosomes within each genotype are related to each other. The figures reveal that the smoother the landscape is, the weaker the dependence on the structure of the other landscapes. Consequently, the chromosomes within the genotype relate to each other differently.

Amplitude Spectra The autocorrelation functions of two-bit multiplier landscapes are studied in greater detail by measuring the amplitude spectra obtained from the Fourier transform of the landscapes. The amplitude spectra analysis was outlined in [22] and studied by Hordijk and Stadler [26] who suggested that the amplitudes of a landscape convey information about the landscape ruggedness. The idea was adopted by Slavov and Nikolaev [58] who showed that better evolutionary search might be attained by investigating sub-populations on sub-landscapes that are smoother than the original one.

Section 2.4 showed how a landscape can be represented as a superposition of elementary sub-landscapes (see also [22]). Each sub-landscape contributes to the structure of the original landscape by its component ψ_k, a normalized eigenvector of the Fourier transform, where the component is amplified by a certain amplitude A_k. The amplitudes, $\{A_k\}_{\forall k}$, uniquely specify the structure of the fitness landscape. Each A_k measures the influence of the interaction of order k on the fitness function. Hence, the landscapes are more rugged when the amplitudes are high for higher values of k. For instance, Stadler and Happel [59] showed that only the first $k+1$ amplitudes contribute to the spectrum of one-point mutation Nk landscapes (the adjacent neighborhood model). The amplitudes of a given landscape can be obtained by solving the system from equation 23 where λ_i are the eigenvalues, D is the vertex degree of the landscape underlying graph, and $r(s)$ is the autocorrelation function. Note that the parameter D from equation 23 becomes 1 for recombination spaces due to normalization in equation 12. The autocorrelation function of a landscape can be estimated by measuring the autocorrelations of time series obtained via random walk [13]. The autocorrelations of one-point mutation and uniform crossover landscapes associated with the evolution of digital circuits are already known from the study above (section 4.3). The eigenvalues of the one-point mutation and uniform crossover landscapes are given by equations 8 and 14. Note, however, that the eigenvalues in equation 14 are calculated for recombination landscapes defined on binary strings. The eigenvalues of the P-structure graph Laplacian matrix of a uniform crossover landscape defined on Hamming graph $\mathcal{G}_l^n$ are

$$\lambda_k = 1 - \left(\frac{l+2}{2l}\right)^n \left(\frac{(l-1)(l+1)-1}{(l-1)(l+2)}\right)^k \tag{36}$$

where n is the length of the genotypes that are defined on an alphabet of l letters. This is calculated in appendix A.

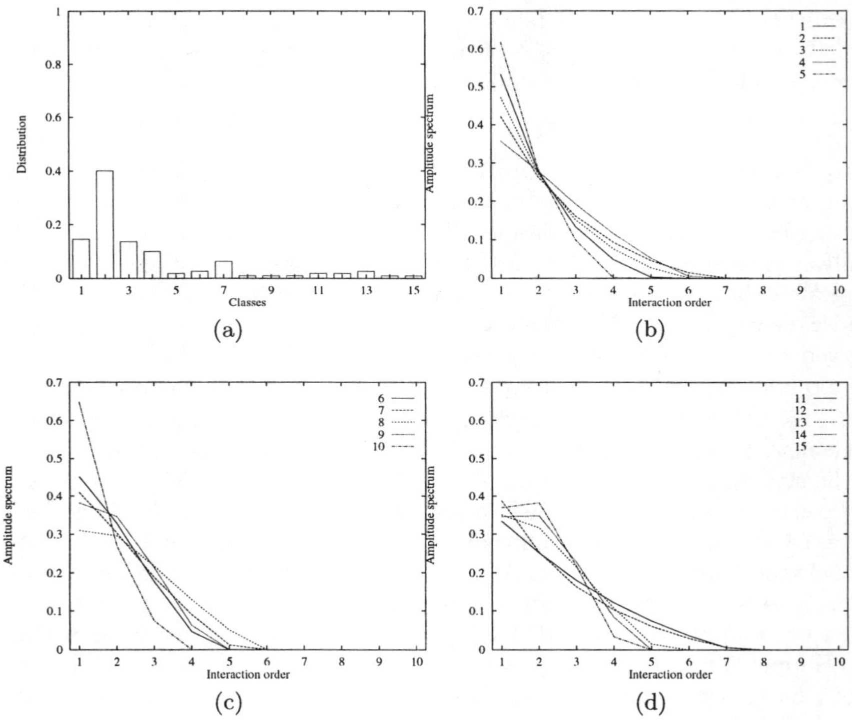

Fig. 9. Functionality landscapes (mutation): (a) distribution by classes, and (b-d) amplitude spectrum

In [26] it was suggested that the straightforward solution of the system from equation 23 does not yield satisfactory accuracy of the obtained results. It was suggested that the steepest descent technique [60] (pp. 614-620) that iteratively minimises the sum of squared errors in the system might be useful to attain accurate approximation. The stop criterion of the steepest descent algorithm is related to a certain tolerance which in this chapter is 10^{-5}. The

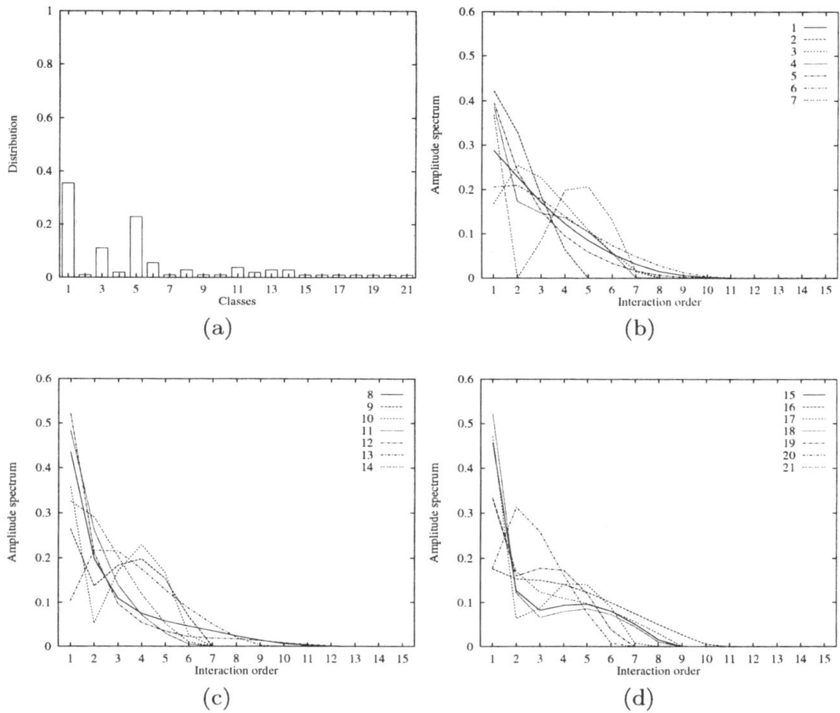

Fig. 10. Internal connectivity landscapes (mutation): (a) distribution by classes, and (b-d) amplitude spectrum

calculated amplitudes, obtained by using the steepest descent method for different starting points, were not noticeably influenced by the starting point. Thus, for each landscape family the amplitudes were found to depend solely on the remainder of the genotype. This was expected, since relations between the functionality, internal connectivity and output connectivity of the array exist as was shown in section 4.3. In order to explore this relationship a classification algorithm was employed by which groups of similar amplitude spectra were identified. The method is based on the *crystal algorithm* given by Soderland *et al.* [61]. The algorithm starts with a class containing a single vector that is an amplitude spectrum. The class expands when the Euclidean distance between a non-classified vector and the *class representative* vector is less than a certain tolerance. The class representative is also a vector with components averaged over all instances of the class. When a vector is classified, a procedure for shrinking the class is started that checks whether the distance between the new representative vector and the elements of the class is less than the tolerance, mentioned above. In the experiments discussed be-

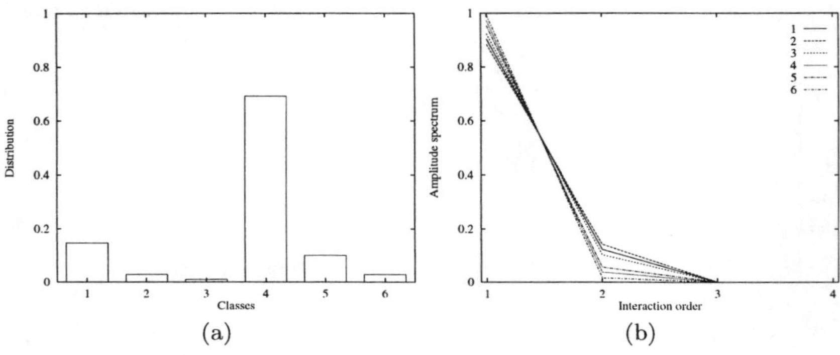

Fig. 11. Output connectivity landscapes (mutation): (a) distribution by classes, and (b) amplitude spectrum

low, the tolerance was set to 10^{-3}. The results for the functionality, internal connectivity and output connectivity landscapes for one-point mutation are given in Figures 9, 10, and 11, while these for uniform crossover are given in Figures 13, 14, and 15 below respectively. The figures represent the distribution of the amplitude spectra by classes (Figures 9a,11a and 13a–15a) and the estimated amplitude spectra versus the interaction order (Figures 9b-d, 10b-d, 11b and 13b–15b). The spectra of each class are represented by the class representative vector that is the *average* over the elements of the class. Standard deviations were also calculated and were found to be in the range from 0 to 10^{-3}.

The amplitude spectra of the mutation landscapes reveal that only the first 7, 12 and 2 amplitudes may respectively contribute to the ruggedness of functionality, internal connectivity and output connectivity landscapes. This explains the difference between the autocorrelation functions of these landscapes studied in section 4.3. It is shown that the internal connectivity landscapes are much more rugged than the functionality and output connectivity ones, since more amplitudes are represented in the corresponding spectra. In [51] it was suggested that the number of non-zero amplitudes of internal connectivity landscapes is strongly dependent on the levels-back parameter. The structure of internal connectivity landscapes is also related to the input redundancy of the genotypes. This is illustrated by Figure 12.

The figure depicts a part of the genotype listed in Table 2. The solid boxes represent genes from the internal connectivity chromosomes while the dashed boxes are genes from other chromosomes. Each box has a label in its top-right corner that is the index of the corresponding gene given in parentheses in Table 2. The filled boxes represent the redundant inputs that are genes 18, 22 and 26. The solid arrows represent the interactions of order 1(top) and 2(bottom) while the dashed arrows are the possible interactions of order 1 and 2 that are less likely to appear in the chromosome. The likelihood of an

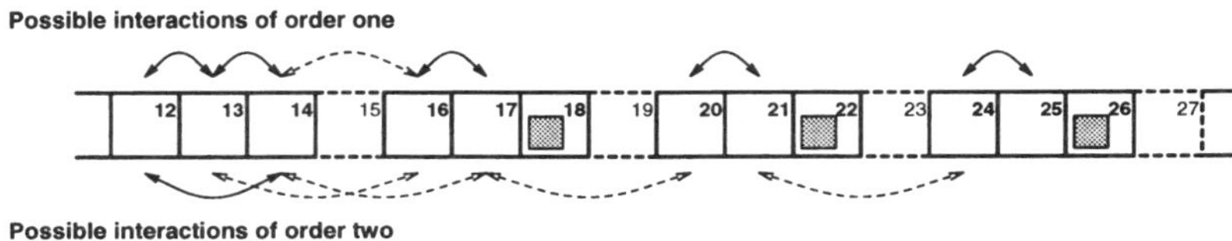

Fig. 12. The possible interactions of order 1 and 2 in the internal connectivity chromosome

interaction appearing is defined by the probability that a cell is connected to its nearest neighbors in the genotype. This probability is 0 for cells that are not taken from the top or the bottom of the corresponding column of the array. The figure reveals that spectra with very low second amplitudes can appear, for instance the spectra from classes 7, 10 and 17 in Figure 10, because for certain functionality the input redundancy can be very high. Perhaps, the input redundancy is also the reason why the spectra 3, 13 and 20 of input connectivity landscapes have lower first amplitudes. The amplitude spectra of the output connectivity landscapes generated by mutation suggest that the landscapes are smooth, and therefore easy to search.

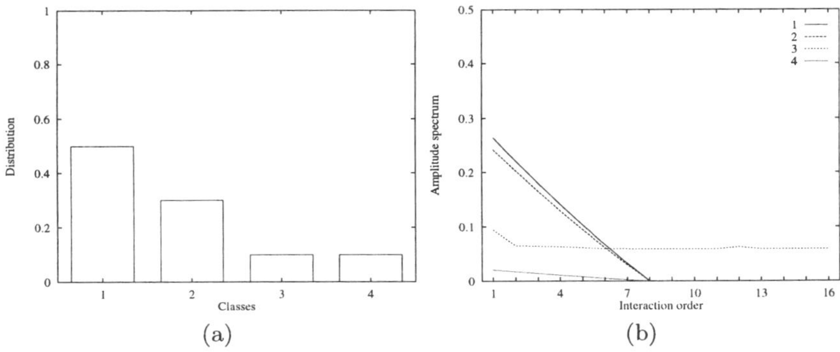

Fig. 13. Functionality landscapes (recombination): (a) distribution by classes, and (b) amplitude spectrum

The amplitude spectra of the recombination landscapes show that the functionality and input connectivity landscapes are rugged since spectra with non-zero amplitudes for all values of k were found (Figures 13b and 14b). This agrees with the results for the corresponding autocorrelation functions obtained in section 4.3. The functionality landscapes are, however, smoother than the internal connectivity landscapes. Note that most of the spectra, shown in Figure 13b, consist of the first 6 and 7 amplitudes while the spectra

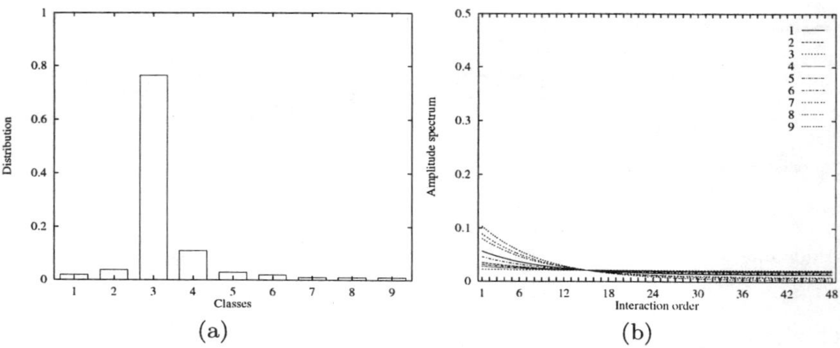

Fig. 14. Internal connectivity landscapes (recombination): (a) distribution by classes, and (b) amplitude spectrum

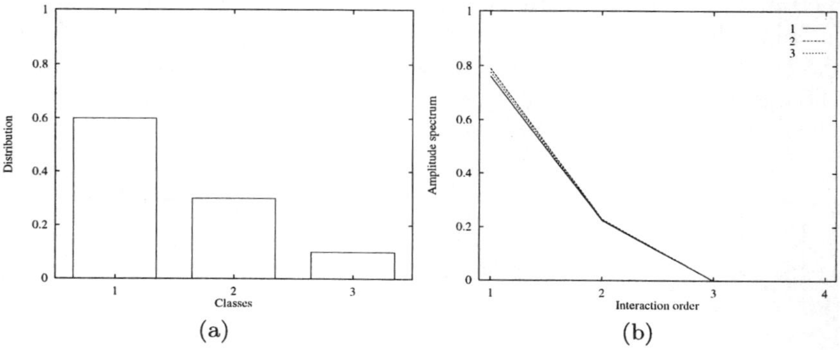

Fig. 15. Output connectivity landscapes (recombination): (a) distribution by classes, and (b) amplitude spectrum

of the internal connectivity spaces are represented by all the 48 amplitudes. Again the spectra of the output connectivity landscapes generated by recombination are only represented by the first two amplitudes. Consequently, these landscapes are smooth.

The relationship between the functionality, internal connectivity and output connectivity of the array is revealed by the ratio of the number of identified classes (Figures 9a–11a and 13a–15a). This is 15: 21: 6 and 4: 9: 3 for mutation and recombination, respectively. The ratio did not appear to change significantly when the classification was performed with a lower tolerance. The findings suggest that the structure of the internal connectivity landscapes strongly depends on the functionality and output connectivity chromosomes, since the number of the identified classes is higher. The relationship between the three types of chromosomes is also revealed by the amplitude spectra of the corresponding spaces. For instance, the amplitude spectra of the internal connectivity landscapes differ from each other in a significant range (Figures 10 and 14). This also holds for the functionality landscapes associated with the recombination operator (Figure 13). The results indicate that there exists a relationship between the structure of the outlined spaces and the interaction between the chromosomes within the genotypes. For instance, the structure of output connectivity landscapes, characterized in Figures 11 and 15, are only slightly dependent on the remainder of the genotype. The results suggest that the more rugged a landscape is, the stronger the dependence on the structure of the other landscapes.

4.4 The Interplay of Smoothness, Ruggedness and Neutrality

The information analysis of landscapes proposed in section 3 is employed to study the interplay of smoothness, ruggedness and neutrality of two-multiplier landscapes. For the purposes of this study, the scenario used to evolve two-bit multipliers is that described in section 4.3. However, the analysis is applied to time series obtained by random walks on six instances of the functionality, internal connectivity and output connectivity landscapes associated with the evolution of the two-bit multiplier that is represented in Table 2 and Figure 6. Hence, six random walks are performed with respect to the studied subspaces. The subspaces originate from the operators one-point mutation and uniform crossover with probability 0.5. The length of each random walk is $100,000$ steps. The starting point is the genotype listed in Table 2. For each time series the information functions $H(\varepsilon)$ and $h(\varepsilon)$ are calculated.

The information functions $H(\varepsilon)$ and $h(\varepsilon)$ of (1) functionality, (2) internal connectivity, and (3) output connectivity landscapes resulting from one-point mutation are represented in Figures 16a and 16b, respectively. The three subspaces are vastly neutral, since the information characteristics $H(0)$ and $h(0)$ of each subspace are significantly higher than $\log_6 2$ and 0, respectively. The plots also imply landscapes with significantly different profiles. It can be seen that the information functions $H(\varepsilon)$ and $h(\varepsilon)$ of the output connectivity landscape increase as ε increases from 0 to 0.0624 while the information functions of the functionality and internal connectivity landscapes decrease as ε increases. Therefore, the neutrality of functionality and internal connectivity subspaces prevails over the landscape ruggedness and smoothness. This is not valid for the structure of the output connectivity landscape, according to the properties of the information characteristics described in section 3.3.

The information functions $H(\varepsilon)$ and $h(\varepsilon)$ of (1) functionality, (2) internal connectivity and (3) output connectivity landscapes resulting from uniform crossover are represented in Figures 17a and 17b, respectively. Again, the information characteristics $H(0)$ and $h(0)$ of the studied subspaces are significantly higher than $\log_6 2$ and 0, which implies the existence of neutrality. The information functions reveal that these subspaces have similar profiles.

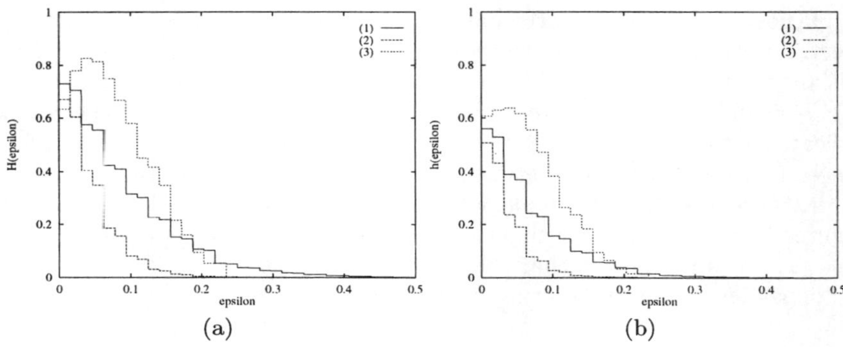

Fig. 16. Mutation: the information functions (a) $H(\varepsilon)$ and (b) $h(\varepsilon)$ of one-point mutation landscapes with respect to (1) functionality, (2) internal connectivity and (3) output connectivity

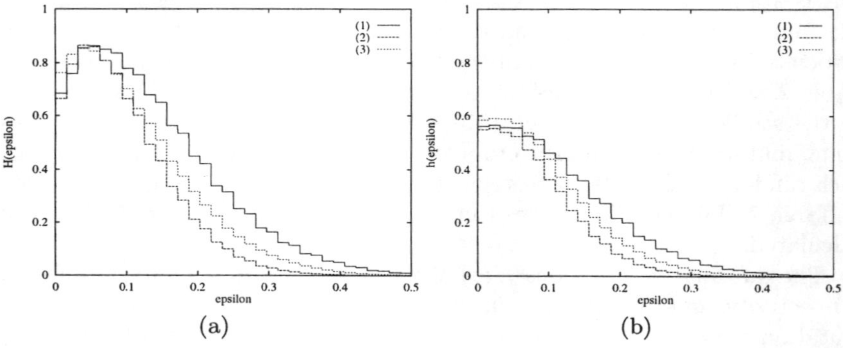

Fig. 17. Recombination: the information functions (a) $H(\varepsilon)$ and (b) $h(\varepsilon)$ of uniform crossover landscapes with respect to (1) functionality, (2) internal connectivity and (3) output connectivity

It is seen that the smoothness and ruggedness of these landscapes prevail over the landscape neutrality, since $H(\varepsilon)$ and $h(\varepsilon)$ increase for small values of ε.

The plots in Figures 16 and 17 suggest a certain similarity between the studied mutation and recombination spaces. The figures reveal that the degree of regularity in these two-bit multiplier landscapes is high. This together with the landscape neutrality implies sharply differentiated plateaun. Perhaps this is one reason of why an evolutionary strategy based algorithm might be much more effective to searching these landscapes as suggested in [62,34].

4.5 Landscape Structure and Search

The results represented so far showed that the two-bit multiplier landscapes are defined on subspaces with different landscape characteristics. The uniform crossover landscapes are rugged, and hence difficult for evolutionary search. In general, it was shown that the two-bit multiplier landscapes are vastly neutral with sharply differentiated plateau. The findings suggest that uniform crossover is not the right recombination operator, and therefore more efficient search might be obtained if mutation only is employed. Furthermore, the mutation operator must not treat the different chromosomes within a genotype equally, since the chromosomes define subspaces with different landscape characteristics. In [34] a simple form of $(1 + \lambda)$ evolutionary strategy [63,64] was used, where λ was chosen to be about 4. The efficiency of this approach was indicated in [62]. This brings the important question: what kind of evolutionary techniques should be used when evolutionary search needs to be performed on vastly neutral landscapes? Unfortunately, an accurate answer to this question cannot be given in this approach however, one can suspect that a small population based search might possibly overcome large flat regions.

In [54] an operator with a certain probability to perform mutation to a genotype was used. The three chromosomes can be mutated with different probabilities since they contribute differently to the structure of the landscapes. Let us decide that the probabilities to mutate genes of functionality, internal connectivity and output connectivity chromosomes as μ_1, μ_2, and μ_3, respectively. To determine the mutation probabilities one can study how they are related to each other. It is possible to determine this by considering the relationship between the functionality, internal connectivity and output connectivity chromosomes, studied in section 4.3. It was shown that the structure of internal connectivity landscapes is strongly dependent on the structure of the functionality and output connectivity landscapes. Consequently, to attain high efficiency of the evolutionary search, the probabilities μ_1 and μ_3 should be lower than μ_2, even though the smoothness of the functionality and output connectivity landscapes tolerates higher values for μ_1 and μ_3. Another reason for this is the vast neutrality of the internal connectivity landscapes that was revealed in section 4.4. Thus the following two hypotheses can be considered:

H_1: $\mu_1 \leq \mu_2$
H_2: $\mu_3 \leq \mu_2$.

Further, if the mutation probability μ_2 is low enough to attain high and sufficiently stable correlations when a genotype is mutated, one can expect that such correlations may occur in the functionality and output connectivity landscapes, since $\mu_1 \leq \mu_2$ and $\mu_3 \leq \mu_2$. The relatively high autocorrelations and low standard deviations shown in Figure 7, and the amplitude spectra analysis given in Figure 11 of the output connectivity landscapes, suggest that the probability μ_3 can be more freely tuned when compared to μ_1. Since the functionality landscapes are more rugged than the output connectivity landscapes (Figures 7, 9 and 11), the inequality $\mu_1 \leq \mu_3$ is more likely to lead to efficient search. This is referred to as hypothesis

H_3: $\mu_1 \leq \mu_3$.

Thus according to hypotheses H_1, H_2, and H_3, the probabilities for mutation can be considered in the following order :

$$\mu_1 \leq \mu_3 \leq \mu_2. \tag{37}$$

To test this assumption, 10 evolutionary runs per triple of mutation probabilities were performed. Each probability was varied from 0.0 to 0.04 with step 10^{-3}; 5,144 solutions that are genotypes with fitness value 1.0 were obtained. The solutions were classified with respect the order of the mutation probabilities in the following way:

i) $\mu_1 \leq \mu_2 \leq \mu_3$,
ii) $\mu_1 \leq \mu_3 \leq \mu_2$,
iii) $\mu_2 \leq \mu_1 \leq \mu_3$,
iv) $\mu_2 \leq \mu_3 \leq \mu_1$,
v) $\mu_3 \leq \mu_1 \leq \mu_2$,
vi) $\mu_3 \leq \mu_2 \leq \mu_1$.

The distribution of the solutions by classes, and the mutation probabilities averaged over all evolutionary runs for which the evolutionary algorithm attained genotypes with fitness values 1.0, are given in Figure 18 and Table 3, respectively. The figure reveals that class 2 dominates over the other classes.

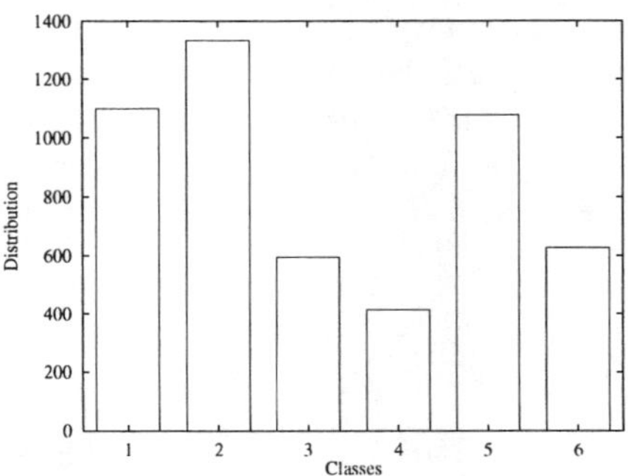

Fig. 18. Distribution of the attained genotypes with fitness values 1.0 by classes

Consequently, the order of mutation probabilities suggested by class 2 define,

Class	μ_1	μ_2	μ_3
1	0.011 ± 0.007	0.023 ± 0.007	0.034 ± 0.006
2	0.011 ± 0.007	0.031 ± 0.006	0.020 ± 0.008
3	0.021 ± 0.008	0.012 ± 0.007	0.032 ± 0.006
4	0.029 ± 0.006	0.012 ± 0.008	0.020 ± 0.008
5	0.019 ± 0.008	0.031 ± 0.006	0.009 ± 0.007
6	0.029 ± 0.006	0.021 ± 0.008	0.011 ± 0.007

Table 3. Mutation probabilities for which genotypes with fitness 1.0 were attained

evolutionary search that is more likely to attain genotype with fitness values equal to 1.0. This agrees with the suggestion stated in equation 37. The hypotheses can also be tested one by one, considering the results shown in the figure. For instance, the number of genotypes with fitness 1.0 from class 1 is high, which is also valid for the genotypes from class 5. On the other hand, classes 3, 4 and 6 consist of less elements, whereas class 4 is the smallest. It appears that the number of the considered hypotheses relates to the success of evolving genotypes with fitness values equal to 1.0. This is illustrated in diagram in Figure 19. The diagram shows how the stated hypotheses relate

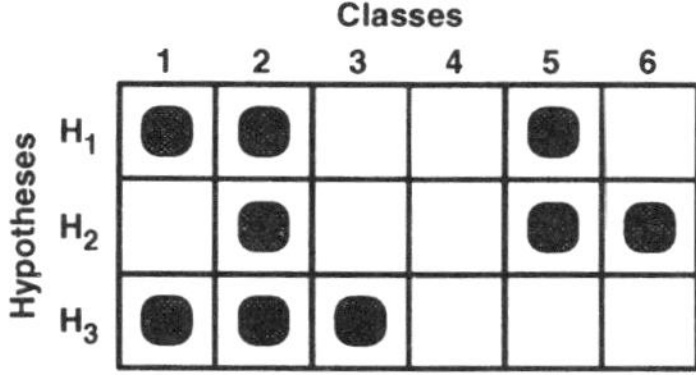

Fig. 19. A diagram of the supported hypotheses by classes $1 - 6$

to the distribution of the attained solutions represented in Figure 18.

5 Discussion

In this chapter methods for studying the structure of fitness landscapes have been discussed and further applied to an engineering optimization problem, particularly the design of a two-bit multiplier on an array of cells. A landscape was defined over a configuration space with topological features determined by the fitness function and the operator used to move on the landscape. Although the model differs from the one proposed by Jones [19] in the connotation of a landscape point, it agrees with the concept "one operator, one landscape", which is to say that each landscape can be associated with an evolutionary operator that defines the neighborhood relationship in the set of all configurations. Thus it became possible to characterize a landscape by a certain correlation function and its amplitude spectrum obtained by decomposing the landscape to elementary sub-landscapes using a Fourier transform. The two landscape analysis are powerful tools for studying the landscape ruggedness. They were applied in a study of the ruggedness of two-bit multiplier landscapes, where it was shown that these landscapes can be considered as products of three configuration spaces with different landscape characteristics. Thus it was possible to explore the relationship between different features of the evolved digital circuit, such as functionality and routing. For the purpose of this study, the eigenvalues of the uniform crossover Laplacian matrix for many-valued configurations were calculated (see appendix A). The analysis also showed that the recombination landscapes generated by uniform crossover might not be suitable for digital circuit evolution. Note that the study of the recombination spaces was implemented on time series obtained by random walks, and therefore it assumes that all configurations can appear in the population with the same probability.

Another technique discussed in this chapter is the information analysis of landscapes, defined to investigate the interplay of the landscape smoothness, ruggedness and neutrality. Again the analysis was applied to time series obtained by random walks that start from a particular configuration. It showed that the two-bit multiplier landscapes have vast neutrality and sharply differentiated plateaun. The information analysis is a potentially powerful technique. Note that it can convey information about the relationship between the landscape characteristics. Another advantage of the analysis is that it can also be used to investigate the *near neutrality* of landscapes. Two configurations v and w are *nearly neutral* if $|f(v) - f(w)| < \varepsilon$ for any small positive value of ε. The role of nearly neutral landscapes in evolution was investigated by [65,66] who suggested that the role of neutrality for adaptive evolution that is to provide a “path” for crossing landscape regions with poor fitness [67–70] is very similar to that of *near* neutrality in the case of large populations (for critical comments on the neutral theory see [71]). Perhaps this should be considered as another indication to believe that a small population might be much more successful for evolutionary search on vastly neutral landscapes. This implies the importance of studying the neutrality and the near neutrality of a landscape and their relation to the landscape smoothness and ruggedness.

An important question that can be raised in this chapter is : what is the role of the theory of landscapes in evolutionary computation. One can argue that evolutionary algorithms can be designed to search without any information about the search spaces associated with the optimization problem that needs to be solved. The answer might be positive when considering a simple optimization problem. Often, however, the optimization problem is difficult enough for it to be worthwhile to take time to understand why an evolutionary technique fails. In the case of evolutionary design of digital circuits, the

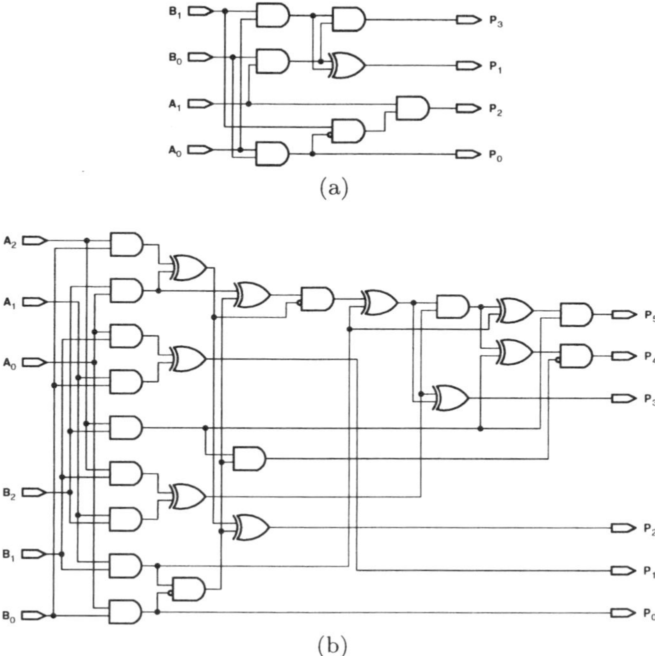

Fig. 20. Two efficient digital circuits evolved by evolutionary techniques: (a) two and (b) three-bit multipliers that consist of 7 and 24 two-input logic gates, respectively

study of landscapes helped and the evolutionary search was improved significantly. Thus bigger and more efficient circuits could be evolved in a shorter time. In [34,53] it was reported that three-bit multipliers were evolved that were 20% more efficient than the conventional circuit design. Two examples of efficient two and three-bit multipliers in terms of gate usage are given in Figure 20.

The chapter showed that the theory of landscapes can be a powerful tool in evolutionary computation. There is unfortunately still a long way to go in this field. Many questions are difficult to answer satisfactorily. Recently, it was shown that many optimization problems might be too difficult for evolutionary search [72]. How can one identify such problems? How can the optimization problems be classified so that one can more easily choose an evolutionary technique that could perform successful search? A theory that could treat these aspects and so help in the application of evolutionary computation is still under development.

A Eigenvalues of The Uniform Crossover Laplacian Matrix

This appendix gives the eigenvalues of the Laplacian matrix of the uniform crossover landscape underlying graph. The eigenvalues are calculated in the same manner as that carried out in [23]. Consider the notation used in section 2.2. For convenience, the distance between two configurations v_i and v_j is denoted by d_{ij}.

Lemma 1. Let $\mathcal{G}_l^n = (\mathcal{V}, \mathcal{E})$ and $d_{ij} = d$ where $v_i, v_j \in \mathcal{V}$. Then,

$$\sum_{\forall k:\ d_{jk}=d'} \mathbf{H}_{i(jk)} = \binom{n-d}{d'-d}(l-1)^{d'-d}. \tag{38}$$

Proof: Since d_{ij} is equal to d, there are at least d letters in v_i from v_k. Hence, the remainder of configuration v_i is obtained by letters from v_j and v_k. This can be re-written in the following way: consider the notation $v_i = v_{i,1}v_{i,2}...v_{i,n}$ where $v_i \in \mathcal{V}$, then the configuration v_i consists of elements from the set $\{v_{i,s_1}, v_{i,s_2}, ..., v_{i,s_d}\} \cup \{v_{i,t_1}, v_{i,t_2}, ..., v_{i,t_{n-d}}\}$ where v_{i,s_m} is equal to v_{k,s_m} for $m = 1, .., d$, and v_{i,t_m} is either v_{j,t_m} or v_{k,t_m} for $m = 1, .., n-d$. Consequently, for given indexes $s_1, s_2, ..., s_d$, there are exactly l^{n-d} possible strings v_k for which $v_i \in \varrho(v_j, v_k)$.

Since $d' \geq d$, it follows that there are two subsets of indexes for the second part of v_k which consists of $n-d$ elements. These are the subset of $d'-d$ elements for which $v_{j,t} \neq v_{k,t}$, and, of course, the remainder that is a set of $n-d'$ elements. The number of genotypes v_k is equal to the number of configurations of length $d'-d$ defined on the set of indexes $\{t_1, t_2, ..., t_{n-d}\}$. This is given by $\binom{n-d}{d'-d}$. Since each configuration accepts $(l-1)^{d'-d}$ strings, the number of different possible offsprings of a pair of configurations is exactly the number given in equation 38. $\diamondsuit$

Lemma 2. Consider a uniform crossover landscape defined on the hypercube $\mathcal{G}_l^n = (\mathcal{V}, \mathcal{E})$. Then,

$$\mathbf{S}_{ij} = \frac{1}{(2l)^n}(1+l)^{n-d_{ij}}. \tag{39}$$

Proof: The generalized adjacency matrix from equation 12 can be re-written as follows :

$$\begin{aligned}
\mathbf{S}_{ij} &= \frac{1}{l^n}\sum_{d'=0}^{n}\Bigg(\sum_{\forall k:\ d_{jk}=d'}\frac{\mathbf{H}_{i(jk)}}{|\varrho(v_j,v_k)|}\Bigg)\\
&= \frac{1}{l^n}\sum_{d'=0}^{n}\Bigg(\sum_{\forall k:\ d_{jk}=d'}\frac{\mathbf{H}_{i(jk)}}{2^{d'}}\Bigg)\\
&= \frac{1}{l^n}\sum_{d'=d}^{n}2^{-d'}\binom{n-d}{d'-d}(l-1)^{d'-d}\\
&= \frac{1}{l^n}\sum_{m=0}^{n-d}2^{-m}2^{-d}(l-1)^m\binom{n-d}{m}\\
&= \frac{1}{l^n2^d}\sum_{m=0}^{n-d}\left(\frac{l-1}{2}\right)^m\binom{n-d}{m}\\
&= \frac{1}{l^n2^d}\left(1+\frac{l-1}{2}\right)^{n-d}
\end{aligned} \tag{40}$$

which leads directly to equation 39. In the calculation above m is a substitution of $d'-d$. ◇

Lemma 3. Let $\mathbf{S}'$ be a $l^n \times l^n$ matrix with elements $\mathbf{S'}_{ij} = q^{d_{ij}}$ where $q = \frac{1}{l+1}$. Then $\mathbf{S}'$ is positive definite and has $n+1$ distinct eigenvalues

$$\lambda_j^{\mathbf{S}'} = \left(1-\frac{1}{(l+1)(l-1)}\right)^j\left(1+\frac{1}{l+1}\right)^{n-j}. \tag{41}$$

Proof: Let $\mathbf{A}$ be the adjacency matrix of the hypercube $\mathcal{G}_l^n$. The distance matrices $\mathbf{A}^{(d)}$ with elements

$$\mathbf{A}_{ij}^{(d)} = \begin{cases} 1, & \text{if } d_{ij} = d \\ 0, & \text{otherwise} \end{cases} \tag{42}$$

can be written as polynomials of order d of $\mathbf{A}$, since the graph is distance transitive [22,73]. The matrix $\mathbf{S}'$ is a linear combination of $\mathbf{A}^{(d)}$. Hence, $\mathbf{S}'$ has the same eigenvectors as $\mathbf{A}$, and the eigenvalues are polynomials of the eigenvalues of $\mathbf{A}$. The eigenvalues of $\mathbf{A}$ are $\lambda_k^{\mathbf{A}} = (l-1)n - lk$, since they are eigenvalues of the collapsed adjacency matrix $\hat{\mathbf{A}}$. The latter is a matrix with elements defined as follows :

$$\hat{\mathbf{A}}_{d'd} = \sum_{\forall i:\ d_{ij}=d'}\mathbf{A}_{ij}, \tag{43}$$

for all v_j so that $d_{jr} = d$ where v_r is an arbitrarily chosen *reference* vertex. The left eigenvectors of the collapsed adjacency matrix are

$$\vartheta_k(d) = \frac{1}{(l-1)^k\binom{n}{k}}\binom{n}{d}K_k^{(l,n)}(d), \tag{44}$$

where $K_k^{(l,n)}(d)$ are the Krawtchouk polynomials (see equations 6 and 7). Since $\vartheta_k(0) = 1$ the eigenvalues of $\mathbf{S}'$ are

$$\lambda_j^{\mathbf{S}'} = \sum_{d=0}^{n} \vartheta_j(d) q^d \tag{45}$$

and they can be re-written as

$$\begin{aligned} \lambda_j^{\mathbf{S}'} &= \frac{1}{\binom{n}{j}} \sum_{d=0}^{n} \frac{1}{(l-1)^j} q^d \binom{n}{d} \sum_{i=0}^{j} (-1)^i (l-1)^{j-i} \binom{d}{i} \binom{n-d}{j-i} \\ &= \frac{1}{\binom{n}{j}} \sum_{d=0}^{n} \sum_{i=0}^{j} q^d \frac{(-1)^i}{(l-1)^i} \binom{n}{d} \binom{d}{i} \binom{n-d}{j-i}. \end{aligned} \tag{46}$$

Since $\binom{n}{d}\binom{d}{i}\binom{n-d}{j-i} = \binom{n}{j}\binom{j}{i}\binom{n-j}{d-i}$ the eigenvalues become

$$\begin{aligned} \lambda_j^{\mathbf{S}'} &= \sum_{i=0}^{j} \left(\frac{-1}{l-1}\right)^i \binom{j}{i} \sum_{d=0}^{n} \binom{n-j}{d-i} q^d \\ &= \sum_{i=0}^{j} \left(\frac{-1}{l-1}\right)^i \binom{j}{i} q^i \sum_{k=0}^{n-j} \binom{n-j}{k} q^k \\ &= \left(\sum_{i=0}^{j} (-1)^i \left(\frac{q}{l-1}\right)^i \binom{j}{i} \right) \left(\sum_{k=0}^{n-j} \binom{n-j}{k} q^k \right) \\ &= \left(1 - \frac{q}{l-1}\right)^j (1+q)^{n-j}, \end{aligned} \tag{47}$$

which is the expression from equation 41. ◇

Theorem 1. Consider a uniform crossover landscape defined on the hypercube $\mathcal{G}_l^n$. The eigenvalues of the corresponding P-structure graph Laplacian matrix are

$$\lambda_k = 1 - \left(\frac{l+2}{2l}\right)^n \left(\frac{(l-1)(l+1)-1}{(l-1)(l+2)}\right)^k \tag{48}$$

Proof: It follows from lemmas 1, 2 and 3. ◇

References

1. Wright, S. (1932) The roles of mutation, inbreeding, crossbreeding and selection in evolution. In Jones, D.F., ed.: Proceedings of the 6th International Conference on Genetics. **1**, 356–366
2. Davidor, Y. (1990) Epistasis variance: Suitability of a representation to genetic algorithms. Complex Systems. **4**, 369–383

3. Davidor, Y. (1991) Epistasis variance: A viewpoint on ga-hardness. In Rawlins, G.J.E., ed.: Foundations of Genetic Algorithms. Morgan Kaufmann, San Mateo, CA, 23–35
4. Goldberg, D. (1987) Simple genetic algorithms and the minimal deceptive problem. In Davis, L., ed.: Genetic Algorithms and Simulated Annealing. Pitman, London, 74–88
5. Goldberg, D. (1989) Genetic algorithms and Walsh functions: Part I, a gentle introduction. Complex Systems. **3**, 129–152
6. Goldberg, D. (1989) Genetic algorithms and Walsh functions: Part II, deception and its analysis. Complex Systems. **3**, 153–171
7. Whitley, D.L. (1991) Fundamental principles of deception in genetic search. In Rawlins, G.J.E., ed.: Foundations of Genetic Algorithms. Morgan Kaufmann, San Mateo, CA, 221–241
8. Whitley, D.L. (1992) Deception, dominance and implicit parallelism in genetic search. Annals of Mathematics and Artificial Intelligence. **5**, 49–78
9. Altenberg, L. (1995) The schema theorem and price's theorem. In Whitley, L.D., Vose, M.D., eds.: Foundations of Genetic Algorithms, Volume 3. Morgan Kaufmann, San Francisco, CA, 23–49
10. Horn, J., Goldberg, D. (1995) Genetic algorithm difficulty and the modality of fitness landscapes. In Whitley, L.D., Vose, M.D., eds.: Foundations of Genetic Algorithms, **volume 3**. Morgan Kaufmann, San Francisco, CA, 243–269
11. Palmer, R. (1991) Optimization on rugged landscapes. In Perelson, A., Kauffman, S., eds.: Molecular Evolution on Rugged Landscapes. Volume IX of SFI Studies in the Sciences of Complexity. Addison-Wesley, Reading, MA, 3–25
12. Kauffman, S. (1989) Adaptation on rugged fitness landscapes. In Stein, D., ed.: Lectures in the Sciences of Complexity. SFI Studies in the Sciences of Complexity. Addison-Wesley, Reading, MA, 527–618
13. Weinberger, E.D. (1990) Correlated and uncorrelated fitness landscapes and how to tell the difference. Biological Cybernetics. **63**, 325–336
14. Stadler, P.F., Happel, R. (1992) Correlation structure of the landscape of the graph-bipartitioning problem. J. Phys. A: Math. Gen. **25**, 3103–3110
15. Stadler, P.F., Schnabl, W. (1992) The landscape of the traveling salesman problem. Physical Letters A. **161**, 337–344
16. Manderick, B., de Weger, M., Spiessens, P. (1991) The genetic algorithm and the structure of the fitness landscape. In Belew, R.K., Booker, L.B., eds.: Proceedings of the 4th International Conference on Genetic Algorithms, San Mateo, CA, Morgan Kaufmann, 143–150
17. Mitchell, M., Forrest, S., Holland, J. (1991) The royal road for genetic algorithms: Fitness landscapes and ga performance. In Varela, J., Bourgine, P., eds.: Proceedings of the 1st European Conference on Artificial Life, Cambridge, MA, MIT Press, 245–254
18. Mathias, K., Whitley, D. (1992) Genetic operators, the fitness landscape and the traveling salesman problem. In Männer, R., Manderick, B., eds.: Parallel Problem Solving from Nature II, North-Holland, Elsevier Science Publishers, 219–228
19. Jones, T. (1995) Evolutionary Algorithms, Fitness Landscapes and Search. PhD thesis, University of New Mexico, Albuquergue, NM
20. Gitchoff, P., Wagner, G.P. (1996) Recombination induced hypergraphs: A new approach to mutation-recombination isomorphism. Complexity. **2**, 37–43

21. Culberson, J.C. (1995) Mutation-crossover isomorphism and the construction of discriminating functions. Evolutionary Computation. **2**, 279–311
22. Stadler, P.F. (1995) Towards theory of landscapes. In Lopéz-Peña, R., Capovilla, R., García-Pelayo, R., Waelbroeck, H., Zertuche, F., eds.: Complex Systems and Binary Networks. Springer-Verlag, Berlin, 77–163
23. Stadler, P.F., Wagner, G.P. (1997) Algebraic theory of recombination spaces. Evolutionary Computation. **5**, 241–275
24. Wagner, G.P., Stadler, P.F. (1998) Complex adaptations and the structure of recombination spaces. In Nehaniv, C., Ito, M., eds.: Algebraic Engineering. World Scientific, Singapore, 96–115
25. Stadler, P.F. (1996) Landscapes and their correlation functions. J. Math. Chem. **20**, 1–45
26. Hordijk, W., Stadler, P.F. (1998) Amplitude spectra of fitness landscapes. Adv Complex Systems. **1**, 39–66
27. Vassilev, V.K. (1997) An information measure of landscapes. In Bäck, T., ed.: Proceedings of the 7th International Conference on Genetic Algorithms, San Francisco, CA, Morgan Kaufmann, 49–56
28. Vassilev, V.K., Fogarty, T.C., Miller, J.F. (2000) Information characteristics and the structure of landscapes. Evolutionary Computation **8** In press.
29. Higuchi, T., Niwa, T., Tanaka, T., Iba, H., de Garis, H., Furuya, T. (1992) Evolving hardware with genetic learning. In: Proceedings of Simulation of Adaptive Behaviour, Cambridge, MA, MIT Press, 417–424
30. Higuchi, T., Iwata, M., eds. (1996) Proceedings of the 1st International Conference on Evolvable Systems: From Biology to Hardware. Volume 1259 of Lecture Notes in Computer Science., Berlin, Springer-Verlag
31. Tomassini, M., Sanchez, E., eds. (1996) Towards Evolvable Hardware: The Evolutionary Engineering Approach. Volume 1062 of Lecture Notes in Computer Science, Springer-Verlag, Berlin
32. Mange, D., Tomassini, M., eds. (1998) Bio-Inspired Computing Machines: Towards Novel Computational Architectures Presses Polytechniques et Universitaires Romandes
33. Thompson, A. (1998) Hardware Evolution: Automatic Design of Electronic Circuits in Reconfigurable Hardware by Artificial Evolution. Springer-Verlag, London
34. Miller, J.F., Job, D., Vassilev, V.K. (2000) Principles in the evolutionary design of digital circuits — part i. Journal of Genetic Programming and Evolvable Machines **1** In press.
35. Miller, J.F., Thomson, P. (1998) Aspects of digital evolution: Geometry and learning. In Sipper, M., Mange, D., Pérez-Uribe, A., eds.: Proceedings of the 2nd International Conference on Evolvable Systems: From Biology to Hardware. Volume 1478 of Lecture Notes in Computer Science, Springer-Verlag, Heidelberg, 25–35
36. Miller, J.F., Thomson, P. (1998) Aspects of digital evolution: Evolvability and architecture. In Eiben, A.E., Bäck, T., Schoenauer, M., Schwefel, H.P., eds.: Parallel Problem Solving from Nature V. Volume 1498 of Lecture Notes in Computer Science. Springer, Berlin, 927–936
37. Hordijk, W. (1997) Correlation analysis of the synchronising-ca landscape. Physica D. **107**, 255–264
38. Hamming, R.W. (1980) Coding and Information Theory. Prentice-Hall, Inc., Englewood Cliffs, NJ

39. Mohar, B. (1997) Some applications of laplace eigenvalues of graphs. In Hahn, G., Sabidussi, G., eds.: Graph Symmetry: Algebraic Methods and Applications. Volume 497 of NATO ASI Series C. Kluwer, Dordrecht
40. Eiben, A.E., Bäck, T. (1997) Empirical investigation of multiparent recombination operators in evolution strategies. Evolutionary Computation. **5**, 347–365
41. Stadler, P.F., Seitz, R., Wagner, G.P. (1999) Evolvability of complex characters: Dependent fourier decomposition of fitness landscapes over recombination spaces. Bull. Math. Biol. Santa Fe Institute Report 99-01-001.
42. Spitzer, F. (1976) Principles of Random Walks. Springer-Verlag, New York, NY
43. Priestley, M.B. (1981) Spectral Analysis and Time Series. Academic Press Inc., London, UK
44. Weinberger, E.D. (1991) Fourier and taylor series on fitness landscapes. Biological Cybernetics. **65**, 321–330
45. Reidys, C.M., Stadler, P.F.(1998) Neutrality in fitness landscapes. Technical Report 98-10-089, Santa Fe Institute Submitted to *Appl. Math. & Comput.*
46. Stadler, P.F. (1999) Spectral landscape theory. In Crutchfield, J.P., Schuster, P., eds.: Evolutionary Dynamics — Exploring the Interplay of Selection, Neutrality, Accident and Function. Oxford University Press, New York, NY (1999) To appear.
47. Kauffman, S., Levin, S. (1987) Towards a general theory of adaptive walks on rugged landscapes. J. Theor. Biol. **128**, 11–45
48. Box, G., Jenkins, G. (1970) Time Series Analysis, Forecasting and Control. Holden Day
49. Hordijk, W. (1996) A measure of landscapes. Evolutionary Computation. **4**, 335–360
50. Vassilev, V.K., Miller, J.F., Fogarty, T.C. (1999) Digital circuit evolution and fitness landscapes. In: Proceedings of the Congress on Evolutionary Computation Volume 2., IEEE Press, Piscataway, NY, 1299–1306
51. Vassilev, V.K., Miller, J.F., Fogarty, T.C. (1999) On the nature of two-bit multiplier landscapes. In Stoica, A., Keymeulen, D., Lohn, J., eds.: Proceedings of the 1st NASA/DoD Workshop on Evolvable Hardware, Los Alamitos, CA, IEEE Computer Society, 36–45
52. Vassilev, V.K., Miller, J.F., Fogarty, T.C. (1999) Digital circuit evolution: The ruggedness and neutrality of two-bit multiplier landscapes. In Harvey, D.M., ed.: Evolutionary Hardware Systems, IEE Press, London, 6/1–6/4
53. Miller, J.F., Job, D., Vassilev, V.K. (2000) Principles in the evolutionary design of digital circuits — part II. J. Genetic Programming and Evolvable Machines **1** In press.
54. Miller, J.F., Thomson, P., Fogarty, T. (1997) Designing electronic circuits using evolutionary algorithms. arithmetic circuits: A case study. In Quagliarella, D., Periaux, J., Poloni, C., Winter, G., eds.: Genetic Algorithms and Evolution Strategies in Engineering and Computer Science. Wiley, Chechester, UK, 105–131
55. Andrews, P.B. (1990) An Introduction to Mathematical Logic and Type Theory: To Truth Through Proof. Academic Press, Orlando, Florida (1986)
56. Chen, X., Hurst, S.L. (1982) A comparison of universal-logic-module realizations and their application in the synthesis of combinatorial and sequential logic networks. IEEE Transactions on Computers. **C-31**, 140–147

57. Stadler, P.F., Grünter, W. (1993) Anisotropy in fitness landscapes. J Theor Bio. **165**, 373–388
58. Slavov, V., Nikolaev, N. (1999) Genetic algorithms, fitness sublandscapes and subpopulations. In Reeves, C., Banzhaf, W., eds.: Foundations of Genetic Algorithms, Volume 5. Morgan Kaufmann, San Francisco, CA, 199–218
59. Stadler, P.F., Happel, R. (1999) Random field models for fitness landscapes. J. Math. Biol. **38**, 435–478
60. Burden, R.L., Faires, J.D. (1997) Numerical Analysis. Brooks/Cole Publishing Company, Pacific Grove, CA sixth edition.
61. Soderland, S., Fisher, D., Aseltine, J., Lehnert, W. (1995) Crystal: Inducing a conceptual dictionary. In: Proceedings of the 14th International Joint Conference on Artificial Intelligence, Morgan Kaufamnn, San Francisco, CA
62. Miller, J.F. (1999) An empirical study of the efficiency of learning boolean functions using a cartesian genetic programming approach. In Banzhaf, W., Daida, J., Eiben, A.E., Garzon, M.H., Honavar, V., Jakiela, M., Smith, R.E., eds.: Proceedings of the 1st Genetic and Evolutionary Computation Conference. Volume 2., Morgan Kaufmann, San Francisco, CA 1135–1142
63. Schwefel, H.P.(1981) Numerical Optimization of Computer Models. John Wiley & Sons, Chichester, UK
64. Bäck, T., Hoffmeister, F., Schwefel, H.P. (1991) A survey of evolutionary strategies. In Belew, R., Booker, L., eds.: Proceedings of the 4th International Conference on Genetic Algorithms, Morgan Kaufmann, San Francisco, CA 2–9
65. Ohta, T. (1992) The nearly neutral theory of molecular evolution. Annual Review of Ecology and Systematics. **23**, 263–286
66. Ohta, T. (1996) The current significance and standing of neutral and nearly neutral theories. BioEssays. **18**, 673–684
67. Huynen, M.A., Stadler, P.F., Fontana, W. (1996) Smoothness within ruggedness: The role of neutrality in adaptation. Proceedings of the National Academy of Science U.S.A. **93**, 397–401
68. Huynen, M.A. (1996) Exploring phenotype space through neutral evolution. Journal of Molecular Evolution. **43**, 165–169
69. Banzhaf, W. (1994) Genotype-phenotype-mapping and neutral variation — a case study in genetic programming. In Davidor, Y., Schwefel, H.P., Männer, R., eds.: Parallel Problem Solving from Nature III, Berlin, Springer-Verlag, 322–332
70. Harvey, I., Thompson, A. (1996) Through the labyrinth evolution finds a way: A silicon ridge. In Higuchi, T., Iwata, M., Liu, W., eds.: Proceedings of the 1st International Conference on Evolvable Systems. Volume 1259 of Lecture Notes in Computer Science. Springer-Verlag, Berlin, 406–422
71. Gillespie, J.H. (1987) Molecular evolution and the neutral allele theory. In Harvey, P.H., Partridge, L., eds.: Oxford Surveys in Evolutionary Biology. Volume 4. Oxford University Press, New York, 11–25
72. Wolpert, D.H., Macready, W.G. (1997) No free lunch theorems for optimization. IEEE Transactions on Evolutionary Computation. **1**, 67–82
73. Biggs, N.J. (1995) Algebraic Graph Theory. Cambridge University Press, Cambridge, UK, second edition

Fast Evolutionary Algorithms

Xin Yao[1], Yong Liu[2], Ko-Hsin Liang[3], and Guangming Lin[3]

[1] School of Computer Science
The University of Birmingham
Edgbaston, Birmingham B15 2TT, UK
Email: x.yao@cs.bham.ac.uk, URL: http://www.cs.bham.ac.uk/~xin

[2] Evolvable Systems Laboratory
Computer Science Division, mbox 1501
Electrotechnical Laboratory
1-1-4 Umezono, Tsukuba, Ibaraki 305-8568, Japan
Email: yliu@etl.go.jp

[3] School of Computer Science
University College, The University of New South Wales
Australian Defence Force Academy, Canberra, ACT, Australia 2600
Email: liangk@cs.adfa.edu.au

Summary. This chapter discusses a number of recent results in evolutionary optimization. In particular, we show that the search step size of a variation operator plays a vital role in its efficient search of a landscape. We have derived the optimal search step size of mutation operators in evolutionary optimization. Based on this theoretical analysis, we have developed several new evolutionary algorithms which outperform existing evolutionary algorithms significantly on many benchmark functions.

Most of the existing work in evolutionary optimization concentrates on different variation (i.e., search) operators, such as crossover and mutation. However, there may be a better way to solve a complex problem by transforming it into a simpler one first and then solving it. The key issue here is how to approximate the problem without changing the nature of the problem (i.e., the optima we wish to find). This chapter will present the latest results on landscape approximation and hybrid evolutionary algorithms.

1 Introduction

Although evolutionary algorithms (EAs) are often introduced from the point of view of *survival of the fittest* and from the analogy to natural evolution, they can also be understood through the framework of *generate-and-test* [1]. The advantage of introducing EAs as a type of generate-and-test search algorithm is that the relationships between EAs and other search algorithms, such as simulated annealing (SA), tabu search (TS), hill-climbing, etc., can be made clearer and thus easier to explore and understand. Under the framework of generate-and-test, different search algorithms investigated in artificial intelligence, operations research, computer science, and evolutionary computation can be unified. Such cross-disciplinary studies are expected to generate more insights into the search problem in general.

A general framework of generate-and-test search can be shown by Figure 1.

1 Generate the initial solution at random and denote it as the current solution;
2 **Generate** the next solution from the current one by *perturbation*.
3 **Test** whether the newly generated solution is *acceptable*:
(a) Accept it as the current solution if yes.
(b) Keep the current solution unchanged otherwise.
4 Goto Step 2 if the current solution is not satisfactory, stop otherwise.

Fig. 1. A general framework of generate-and-test

It is quite clear that various hill-climbing algorithms can be described by Figure 1 with different strategies for perturbation. They all require the new solution to be no worse than the current one to be acceptable. SA does not have such a requirement. It regards a worse solution to be acceptable with certain probability. The difference among classical SA [2], fast SA [3], very fast SA [4], and a new SA [5] lies mainly in their perturbations, i.e., methods of generating the next solution.

EAs can be regarded as a population-based version of generate-and-test search. They use search operators like crossover and mutation to perturb the current solutions, and use selection to decide whether a solution is acceptable. From this point of view, it is clear that we do not have to limit ourselves to crossover and mutation. In theory, we can use any search operators as long as they perform well on the given representation of the problem they are dealing with. This is also true for selection. In practice, a good way to tailor the generate-and-test search to the problem we are interested in is to incorporate problem-specific heuristic knowledge into search operators and selection schemes [6].

Evolutionary programming (EP) has been applied with success to many numerical and combinatorial optimization problems [7–9]. Optimization by EP can be summarized in two major steps:

1 Mutate the solutions in the current population, and
2 Select the next generation from the mutated and the current solutions.

These two steps can be regarded as a population-based version of the classical generate-and-test method [1], where mutation is used to *generate* new solutions (offspring) and selection is used to *test* which of the newly generated solutions should survive to the next generation.

One disadvantage of EP in solving some of the multimodal optimization problems is its slow convergence to a good near-optimum (e.g., f_8 to f_{13} studied in this chapter). The generate-and-test formulation of EP indicates that mutation is a key search operator which generates new solutions from the current ones. A new mutation operator based on Cauchy random numbers is proposed and tested on a suite of 23 functions in this chapter. The new EP with Cauchy mutation significantly outperforms the classical EP (CEP), which uses Gaussian mutation, on a number of multimodal functions with many local minima while being comparable to CEP for unimodal and multimodal functions with only a few local minima. The new EP is denoted as "fast EP" (FEP) in this chapter [10].

In order to explain why Cauchy mutation performs better than Gaussian mutation for most benchmark problems used here, theoretical analysis has been carried out to show the importance of the neighborhood size and search step size in EP. It is shown that Cauchy mutation performs better because of its higher probability of making longer jumps. Although the idea behind FEP appears to be simple and straightforward (the larger the search step size, the faster the algorithm gets to the global optimum), no theoretical result has been provided to answer the question why this is the case and how fast it might be. In addition, a large step size may not be beneficial at all if the current search point is already very close to the global optimum. This chapter shows the relationship between the distance to the global optimum and the search step size, and the relationship between the search step size and the probability of finding a near (global) optimum. Based on such analyses, an improved FEP has been proposed and tested empirically [10].

Finding the global optimum on a large, multimodal, complex, and discontinuous (or nondifferentiable) landscape is usually very hard, even using the evolutionary approach. However, some of these complex landscapes can be approximated and smoothed without changing the nature of the problem, i.e., without modifying the global optimum and its location. The approximated and smoothed landscape is often much easier to search than the original one. In this chapter, we describe a family of algorithms using landscape approximation and hybrid evolutionary and local search. We also list several algorithm design principles. Based on our description of the algorithm framework, two specific algorithms are presented and compared with existing algorithms.

The rest of this chapter is organized as follows. Section 2 describes the global minimization problem considered in this chapter and the CEP used to solve it. The CEP algorithm given follows suggestions from Fogel [8,11] and Bäck and Schwefel [12]. Section 3 describes the FEP and its implementation. Section 4 gives the 23 functions used in our studies. Section 5 presents the experimental results and discussions on FEP and CEP. Section 6 investigates FEP with different scale parameters for its Cauchy mutation. Section 7 analyzes FEP and CEP and explains the performance difference between FEP and CEP in depth. Based on such analyses, an improved FEP (IFEP) is proposed and tested in Section 8. Section 9 describes a framework of those evolutionary algorithms that use approximation and local search. An example algorithm is introduced. Section 10 presents another algorithm which improves some of the drawbacks of the first algorithm. The empirical results and discussions are presented in Section 11. Finally, Section 12 concludes the chapter with some remarks and future research directions.

2 Function Optimization by Classical Evolutionary Programming

A global minimization problem can be formalized as a pair (S, f), where $S \subseteq R^n$ is a bounded set on R^n and $f : S \mapsto R$ is an n-dimensional real-valued function. The problem is to find a point $\mathbf{x}_{min} \in S$ such that $f(\mathbf{x}_{min})$ is a global minimum on S. More specifically, it is required to find an $\mathbf{x}_{min} \in S$ such that

$$\forall \mathbf{x} \in S : f(\mathbf{x}_{min}) \leq f(\mathbf{x}),$$

where f does not need to be continuous but it must be bounded. This chapter only considers unconstrained function optimization.

Fogel [8,13] and Bäck and Schwefel [12] have indicated that CEP with self-adaptive mutation usually performs better than CEP without self-adaptive mutation for the functions they tested. Hence the CEP with self-adaptive mutation will be investigated in this chapter. According to the description by Bäck and Schwefel [12], the CEP is implemented as follows in this study:[1]

1 Generate the initial population of μ individuals, and set $k = 1$. Each individual is taken as a pair of real-valued vectors, $(\mathbf{x}_i, \eta_i)$, $\forall i \in \{1, \dots, \mu\}$, where $\mathbf{x}_i$'s are objective variables and η_i's are standard deviations for Gaussian mutations (also known as strategy parameters in self-adaptive evolutionary algorithms).

2 Evaluate the fitness score for each individual $(\mathbf{x}_i, \eta_i)$, $\forall i \in \{1, \dots, \mu\}$, of the population based on the objective function, $f(\mathbf{x}_i)$.

3 Each parent $(\mathbf{x}_i, \eta_i)$, $i = 1, \dots, \mu$, creates a single offspring $(\mathbf{x}_i{}', \eta_i{}')$ by: for $j = 1, \dots, n$,

$$x_i{}'(j) = x_i(j) + \eta_i(j) N_j(0,1), \tag{1}$$

$$\eta_i{}'(j) = \eta_i(j) \exp(\tau' N(0,1) + \tau N_j(0,1)), \tag{2}$$

where $x_i(j)$, $x_i{}'(j)$, $\eta_i(j)$, and $\eta_i{}'(j)$ denote the j-th component of the vectors $\mathbf{x}_i$, $\mathbf{x}_i{}'$, η_i, and $\eta_i{}'$, respectively. $N(0,1)$ denotes a normally distributed one-dimensional random number with mean 0 and standard deviation 1. $N_j(0,1)$ indicates that the random number is generated anew for each value of j. The factors τ and τ' are commonly set to $\left(\sqrt{2\sqrt{n}}\right)^{-1}$ and $\left(\sqrt{2n}\right)^{-1}$ [12,11].

[1] Swapping the order of Eq.1 and Eq.2 improved CEP's performance on some test functions, but worsened for others.

4 Calculate the fitness of each offspring $(\mathbf{x}_i', \eta_i')$, $\forall i \in \{1, \ldots, \mu\}$.
5 Conduct pairwise comparison over the union of parents $(\mathbf{x}_i, \eta_i)$ and offspring $(\mathbf{x}_i', \eta_i')$, $\forall i \in \{1, \ldots, \mu\}$. For each individual, q opponents are chosen uniformly at random from all the parents and offspring. For each comparison, if the individual's fitness is no smaller than the opponent's, it receives a "win."
6 Select the μ individuals out of $(\mathbf{x}_i, \eta_i)$ and $(\mathbf{x}_i', \eta_i')$, $\forall i \in \{1, \ldots, \mu\}$, that have the most wins to be parents of the next generation.
7 Stop if the halting criterion is satisfied; otherwise, $k = k + 1$ and go to Step 3.

3 Fast Evolutionary Programming

The one-dimensional Cauchy density function centered at the origin is defined by

$$f_t(x) = \frac{1}{\pi} \frac{t}{t^2 + x^2}, \qquad -\infty < x < \infty, \tag{3}$$

where $t > 0$ is a scale parameter [15] (p.51). The corresponding distribution function is

$$F_t(x) = \frac{1}{2} + \frac{1}{\pi} \arctan\left(\frac{x}{t}\right).$$

The shape of $f_t(x)$ resembles that of the Gaussian density function but approaches the axis so slowly that an expectation does not exist. As a result, the variance of the Cauchy distribution is infinite. Figure 2 shows the difference between Cauchy and Gaussian functions by plotting them on the same scale.

The FEP studied in this chapter is exactly the same as the CEP described in Section 2 except for Eq.(1) which is replaced by the following [16]:

$$x_i'(j) = x_i(j) + \eta_i(j)\delta_j, \tag{4}$$

where δ_j is a Cauchy random variable with the scale parameter $t = 1$, and is generated anew for each value of j. It is worth indicating that we leave Eq.(2) unchanged in FEP in order to keep our modification of CEP to a minimum. It is also easy to investigate the impact of the Cauchy mutation on EP when other parameters are kept the same.

It is clear from Figure 2 that Cauchy mutation is more likely to generate an offspring further away from its parent than Gaussian mutation due to its long flat tails. It is expected to have a higher probability of escaping from a local optimum or moving away from a plateau, especially when the "basin of attraction" of the local optimum or the plateau is large relative to the mean step size. On the other hand, the smaller hill around the center in

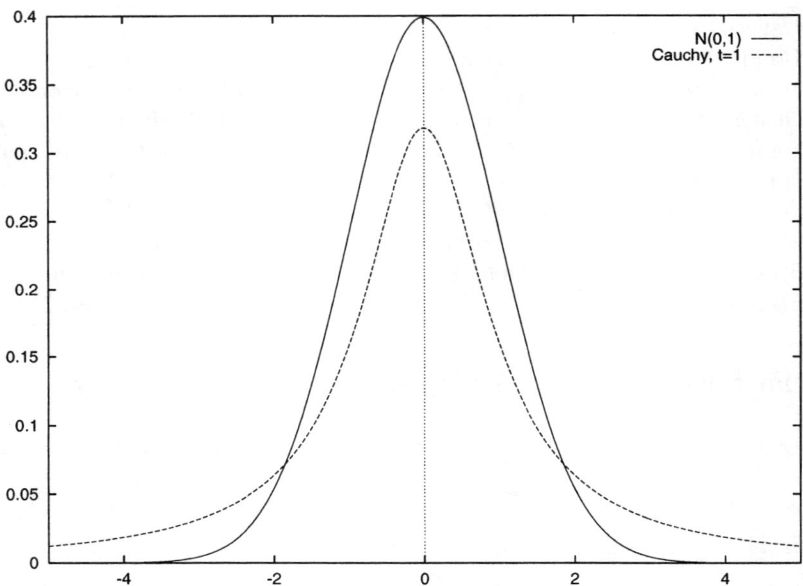

Fig. 2. Comparison between Cauchy and Gaussian density functions

Figure 2 indicates that Cauchy mutation spends less time in exploiting the local neighborhood and thus has a weaker fine-tuning ability than Gaussian mutation in small to mid-range regions. Our empirical results support the above intuition.

4 Benchmark Functions

Twenty-three benchmark functions [7,12,17,18] were used in our experimental studies. This number is larger than that offered in many other empirical study papers. However, this is necessary since the aim here is not to show that FEP is better or worse than CEP, but to find out when FEP is better (or worse) than CEP and why. Wolpert and Macready [19,20] have shown that under certain assumptions no single search algorithm is best on average for all problems. If the number of test problems is small, it would be very difficult to make a generalized conclusion. Using too small a test set also has the potential risk that the algorithm is biased (optimized) towards the chosen problems, while such bias might not be useful for other problems of interest.

The 23 benchmark functions are given in Table 1. More detailed description of each function is given in the Appendix. Functions f_1 to f_{13} are high-dimensional problems. Functions f_1 to f_5 are unimodal. Function f_6 is the step function, which has one minimum and is discontinuous. Function f_7 is a noisy quartic function, where $random[0, 1)$ is a uniformly distributed random variable in $[0, 1)$. Functions f_8 to f_{13} are multimodal functions where the number of local minima increases exponentially with the problem dimension [17,18]. They appear to be the most difficult class of problems for many optimization algorithms (including EP). Functions f_{14} to f_{23} are low-dimensional functions which have only a few local minima [17]. For unimodal functions, the convergence rates of FEP and CEP are more interesting than the final results of optimization as there are other methods which are specifically designed to optimize unimodal functions. For multimodal functions, the final results are much more important since they reflect an algorithm's ability of escaping from poor local optima and locating a good near-global optimum.

Table 1. The 23 benchmark functions used in our experimental study, where n is the dimension of the function, f_{min} is the minimum value of the function, and $S \subseteq R^n$. A detailed description of all functions is given in the Appendix.

Test function	n	S	f_{min}
$f_1(x) = \sum_{i=1}^{n} x_i^2$	30	$[-100, 100]^n$	0
$f_2(x) = \sum_{i=1}^{n} \lvert x_i \rvert + \prod_{i=1}^{n} \lvert x_i \rvert$	30	$[-10, 10]^n$	0
$f_3(x) = \sum_{i=1}^{n} (\sum_{j=1}^{i} x_j)^2$	30	$[-100, 100]^n$	0
$f_4(x) = \max_i \{\lvert x_i \rvert, 1 \le i \le n\}$	30	$[-100, 100]^n$	0
$f_5(x) = \sum_{i=1}^{n-1} [100(x_{i+1} - x_i^2)^2 + (x_i - 1)^2]$	30	$[-30, 30]^n$	0
$f_6(x) = \sum_{i=1}^{n} (\lfloor x_i + 0.5 \rfloor)^2$	30	$[-100, 100]^n$	0
$f_7(x) = \sum_{i=1}^{n} i x_i^4 + \text{random}[0, 1)$	30	$[-1.28, 1.28]^n$	0
$f_8(x) = \sum_{i=1}^{n} -x_i \sin(\sqrt{\lvert x_i \rvert})$	30	$[-500, 500]^n$	−12569.5
$f_9(x) = \sum_{i=1}^{n} [x_i^2 - 10 \cos(2\pi x_i) + 10]$	30	$[-5.12, 5.12]^n$	0
$f_{10}(x) = -20 \exp\left(-0.2\sqrt{\frac{1}{n} \sum_{i=1}^{n} x_i^2}\right) - \exp\left(\frac{1}{n} \sum_{i=1}^{n} \cos 2\pi x_i\right) + 20 + e$	30	$[-32, 32]^n$	0
$f_{11}(x) = \frac{1}{4000} \sum_{i=1}^{n} x_i^2 - \prod_{i=1}^{n} \cos\left(\frac{x_i}{\sqrt{i}}\right) + 1$	30	$[-600, 600]^n$	0
$f_{12}(x) = \frac{\pi}{n} \left\{10 \sin^2(\pi y_i) + \sum_{i=1}^{n-1} (y_i - 1)^2 [1 + 10 \sin^2(\pi y_{i+1})] + (y_n - 1)^2\right\} + \sum_{i=1}^{n} u(x_i, 10, 100, 4)$, $y_i = 1 + \frac{1}{4}(x_i + 1)$, $u(x_i, a, k, m) = \begin{cases} k(x_i - a)^m, & x_i > a, \\ 0, & -a \le x_i \le a, \\ k(-x_i - a)^m, & x_i < -a. \end{cases}$	30	$[-50, 50]^n$	0
$f_{13}(x) = 0.1 \left\{\sin^2(3\pi x_1) + \sum_{i=1}^{n-1} (x_i - 1)^2 [1 + \sin^2(3\pi x_{i+1})] + (x_n - 1)^2 [1 + \sin^2(2\pi x_n)]\right\} + \sum_{i=1}^{n} u(x_i, 5, 100, 4)$	30	$[-50, 50]^n$	0
$f_{14}(x) = \left[\frac{1}{500} + \sum_{j=1}^{25} \frac{1}{j + \sum_{i=1}^{2} (x_i - a_{ij})^6}\right]^{-1}$	2	$[-65.536, 65.536]^n$	1
$f_{15}(x) = \sum_{i=1}^{11} \left[a_i - \frac{x_1(b_i^2 + b_i x_2)}{b_i^2 + b_i x_3 + x_4}\right]^2$	4	$[-5, 5]^n$	0.0003075
$f_{16}(x) = 4x_1^2 - 2.1x_1^4 + \frac{1}{3}x_1^6 + x_1 x_2 - 4x_2^2 + 4x_2^4$	2	$[-5, 5]^n$	−1.0316285
$f_{17}(x) = \left(x_2 - \frac{5.1}{4\pi^2} x_1^2 + \frac{5}{\pi} x_1 - 6\right)^2 + 10\left(1 - \frac{1}{8\pi}\right) \cos x_1 + 10$	2	$[-5, 10] \times [0, 15]$	0.398
$f_{18}(x) = [1 + (x_1 + x_2 + 1)^2 (19 - 14x_1 + 3x_1^2 - 14x_2 + 6x_1 x_2 + 3x_2^2)] \times [30 + (2x_1 - 3x_2)^2 (18 - 32x_1 + 12x_1^2 + 48x_2 - 36x_1 x_2 + 27x_2^2)]$	2	$[-2, 2]^n$	3
$f_{19}(x) = -\sum_{i=1}^{4} c_i \exp\left[-\sum_{j=1}^{4} a_{ij}(x_j - p_{ij})^2\right]$	4	$[0, 1]^n$	−3.86
$f_{20}(x) = -\sum_{i=1}^{4} c_i \exp\left[-\sum_{j=1}^{6} a_{ij}(x_j - p_{ij})^2\right]$	6	$[0, 1]^n$	−3.32
$f_{21}(x) = -\sum_{i=1}^{5} [(x - a_i)(x - a_i)^T + c_i]^{-1}$	4	$[0, 10]^n$	−10
$f_{22}(x) = -\sum_{i=1}^{7} [(x - a_i)(x - a_i)^T + c_i]^{-1}$	4	$[0, 10]^n$	−10
$f_{23}(x) = -\sum_{i=1}^{10} [(x - a_i)(x - a_i)^T + c_i]^{-1}$	4	$[0, 10]^n$	−10

5 Experimental Studies

5.1 Experimental Setup

In all experiments, the same self-adaptive method (i.e., Eq.(2)), the same population size $\mu = 100$, the same tournament size $q = 10$ for selection, the same initial $\eta = 3.0$, and the same initial population were used for both CEP and FEP. These parameters follow the suggestions from Bäck and Schwefel [12] and Fogel [7]. The initial population was generated uniformly at random in the range as specified in Table 1.

5.2 Unimodal Functions

The first set of experiments was aimed to compare the convergence rate of CEP and FEP for functions f_1 to f_7. The average results of 50 independent runs are summarized in Table 2. Figures 3 and 4 show the progress of the mean best solutions and the mean of average values of population found by CEP and FEP over 50 runs for f_1 to f_7. It is apparent that FEP performs better than CEP in terms of convergence rate although CEP's final results were better than FEP's for functions f_1 and f_2. Function f_1 is the simple sphere model studied by many researchers. CEP's and FEP's behaviors on f_1 are quite illuminating. In the beginning, FEP dispays a faster convergence rate than CEP due to its better global search ability. It quickly approaches the neighborhood of the global optimum and reaches approximately 0.001 in around 1200 generations, while CEP can only reach approximately 0.1. After that, FEP's convergence rate reduces substantially, while CEP maintains a nearly constant convergence rate throughout the evolution. Finally, CEP overtakes FEP at around the 1450th generation. As indicated in Section 3 and in Figure 2, FEP is weaker than CEP in fine-tuning. Such weakness slows down its convergence considerably in the neighborhood of the global optimum. CEP is capable of maintaining its nearly constant convergence rate because its search is much more localized than FEP. The different behavior of CEP and FEP on f_1 suggests that CEP is better at fine-grained search while FEP is better at coarse-grained search.

The largest difference in performance between CEP and FEP occurs with function f_6, the step function, which is characterized by plateaus and discontinuity. CEP performs poorly on the step function because it mainly searches in a relatively small local neighborhood. All the points within the local neighborhood will have the same fitness value except for a few boundaries between plateaus. Hence it is very difficult for CEP to move from one plateau to a lower one. On the other hand, FEP has a much higher probability of generating long jumps than CEP. Such long jumps enable FEP to move from one plateau to a lower one with relative ease. The rapid convergence of FEP shown in Figure 4 supports our explanations.

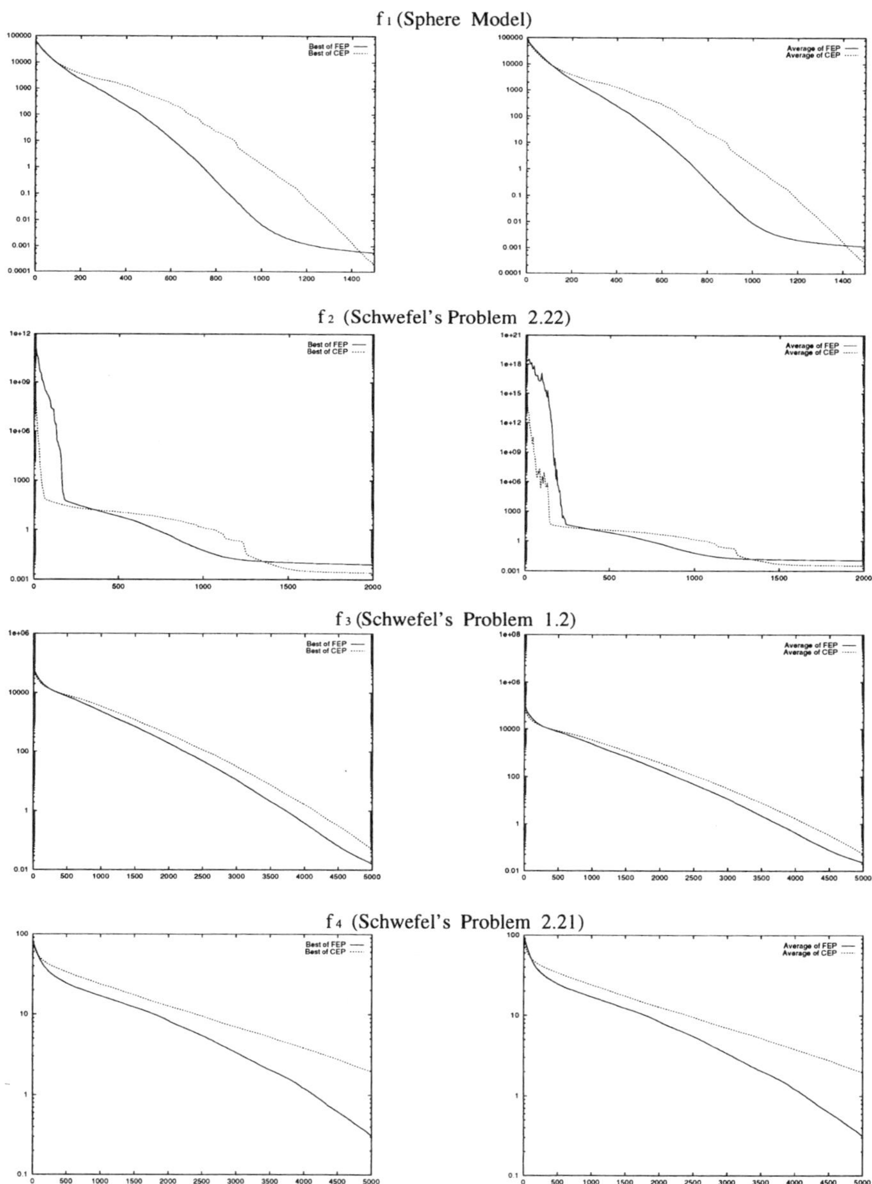

Fig. 3. Comparison between CEP and FEP on f_1–f_4. The vertical axis is the function value and the horizontal axis is the number of generations. The solid lines indicate the results of FEP. The dashed lines indicate the results of CEP. The left figures show the best results. The right figures show the average results. Both were averaged over 50 runs.

Table 2. Comparison between CEP and FEP on f_1–f_7. All results have been averaged over 50 runs, where "Mean Best" indicates the mean best function values found in the last generation, and "Std Dev" stands for the standard deviation.

Function	Number of	FEP		CEP		FEP−CEP
	Generations	Mean Best	Std Dev	Mean Best	Std Dev	t-test
f_1	1500	5.7×10^{-4}	1.3×10^{-4}	2.2×10^{-4}	5.9×10^{-4}	$4.06^{\dagger}$
f_2	2000	8.1×10^{-3}	7.7×10^{-4}	2.6×10^{-3}	1.7×10^{-4}	$49.83^{\dagger}$
f_3	5000	1.6×10^{-2}	1.4×10^{-2}	5.0×10^{-2}	6.6×10^{-2}	$-3.79^{\dagger}$
f_4	5000	0.3	0.5	2.0	1.2	$-8.25^{\dagger}$
f_5	20000	5.06	5.87	6.17	13.61	−0.52
f_6	1500	0	0	577.76	1125.76	$-3.67^{\dagger}$
f_7	3000	7.6×10^{-3}	2.6×10^{-3}	1.8×10^{-2}	6.4×10^{-3}	$-10.72^{\dagger}$

†The value of t with 49 degrees of freedom is significant at $\alpha = 0.05$ by a two-tailed test.

5.3 Multimodal Functions

Multimodal Functions with Many Local Minima Multimodal functions having many local minima are often regarded as being difficult to optimize. f_8–f_{13} are such functions where the number of local minima increases exponentially as the dimension of the function increases. Figure 5 shows the two-dimensional version of f_8.

The dimensions of f_8–f_{13} were all set to 30 in our experiments. Table 3 summarizes the final results of CEP and FEP. It is obvious that FEP performs significantly better than CEP consistently for these functions. CEP appeared to become trapped in a poor local optimum and unable to escape from it due to its smaller probability of making long jumps. According to the figures we plotted in order to observe the evolutionary process, CEP fell into a poor local optimum quite early in a run while FEP was able to improve its solution steadily for a long time. FEP appeared to converge at least at a linear rate with respect to the number of generations. An exponential convergence rate was observed for some problems.

Multimodal Functions with Only a Few Local Minima In order to evaluate FEP more fully, additional multimodal benchmark functions were also included in our experiments, i.e., f_{14}–f_{23}, where the number of local minima for each function and the dimension of the function are small. Table 4 summarizes the results averaged over 50 runs.

Interestingly, quite different results have been observed for functions f_{14} to f_{23}. For 6 (i.e., f_{15}–f_{20}) out of 10 functions, no statistically significant difference was found between FEP and CEP. In fact, FEP performed exactly the same as CEP for f_{16} and f_{17}. For the four functions where there was statistically significant difference between FEP and CEP, FEP performed

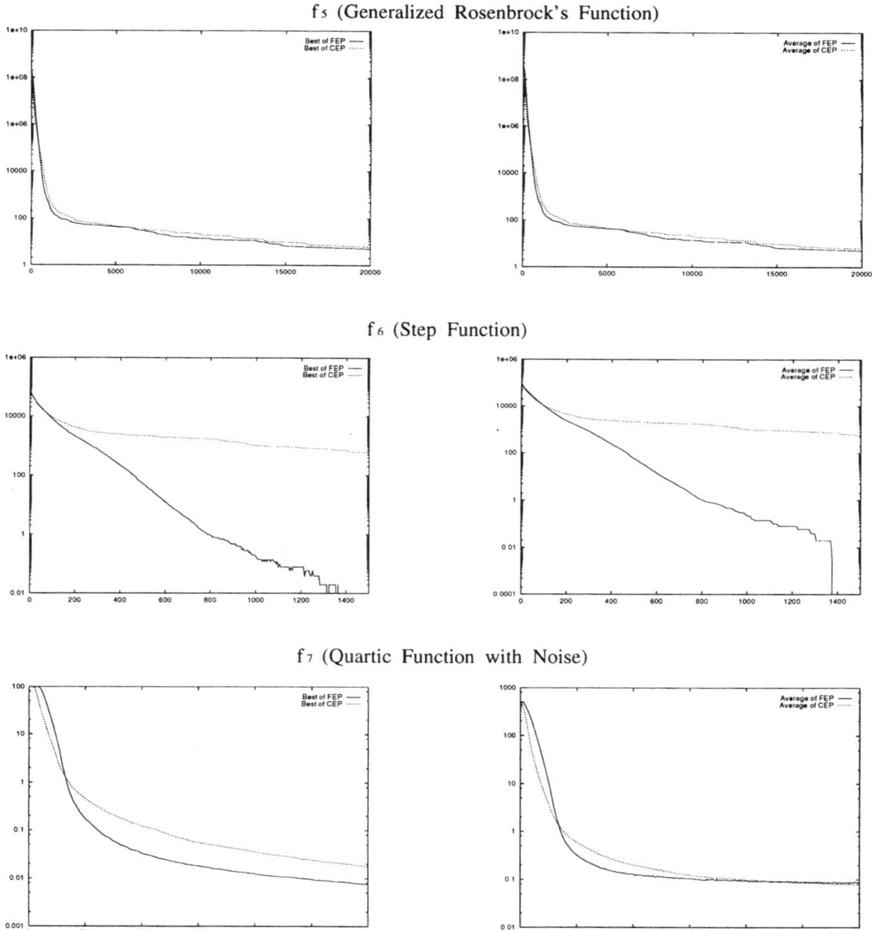

Fig. 4. Comparison between CEP and FEP on f_5–f_7. The vertical axis is the function value and the horizontal axis is the number of generations. The solid lines indicate the results of FEP. The dashed lines indicate the results of CEP. The left figures show the best results. The right figures show the average results. Both were averaged over 50 runs.

better for f_{14}, but was outperformed by CEP for f_{21}–f_{23}. The consistent superiority of FEP over CEP for functions f_8–f_{13} was not observed here.

The major difference between functions f_8–f_{13} and f_{14}–f_{23} is that functions f_{14}–f_{23} appear to be simpler than f_8–f_{13} due to their low dimensionalities and a smaller number of local minima. In order to find out whether or not the dimensionality of functions plays a significant role in deciding FEP's and CEP's behavior, another set of experiments on the low-dimensional ($n = 5$)

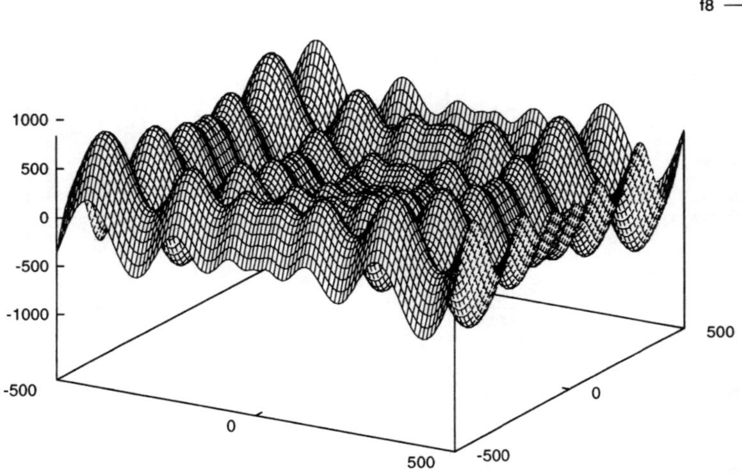

Fig. 5. The two-dimensional version of f_8

Table 3. Comparison between CEP and FEP on f_8–f_{13}. The results are averaged over 50 runs, where "Mean Best" indicates the mean best function values found in the last generation, and "Std Dev" stands for the standard deviation.

Function	Number of	FEP		CEP		FEP−CEP
	Generations	Mean Best	Std Dev	Mean Best	Std Dev	t-test
f_8	9000	−12554.5	52.6	−7917.1	634.5	$-51.39^{\dagger}$
f_9	5000	4.6×10^{-2}	1.2×10^{-2}	89.0	23.1	$-27.25^{\dagger}$
f_{10}	1500	1.8×10^{-2}	2.1×10^{-3}	9.2	2.8	$-23.33^{\dagger}$
f_{11}	2000	1.6×10^{-2}	2.2×10^{-2}	8.6×10^{-2}	0.12	$-4.28^{\dagger}$
f_{12}	1500	9.2×10^{-6}	3.6×10^{-6}	1.76	2.4	$-5.29^{\dagger}$
f_{13}	1500	1.6×10^{-4}	7.3×10^{-5}	1.4	3.7	$-2.76^{\dagger}$

†The value of t with 49 degrees of freedom is significant at $\alpha = 0.05$ by a two-tailed test.

version of f_8–f_{13} was carried out. The results averaged over 50 runs are given in Table 5.

Very similar results to the previous ones on functions f_8–f_{13} were obtained despite the large difference in the dimensionality of functions. FEP still outperforms CEP significantly even when the dimensionality of functions f_8–f_{13} is low ($n = 5$). It is clear that dimensionality is not the key factor which determines the behavior of FEP and CEP. It is the shape of the function

Table 4. Comparison between CEP and FEP on f_{14}–f_{23}. The results are averaged over 50 runs, where "Mean Best" indicates the mean best function values found in the last generation, and "Std Dev" stands for the standard deviation.

Function	Number of	FEP		CEP		FEP–CEP
	Generations	Mean Best	Std Dev	Mean Best	Std Dev	t-test
f_{14}	100	1.22	0.56	1.66	1.19	$-2.21^\dagger$
f_{15}	4000	5.0×10^{-4}	3.2×10^{-4}	4.7×10^{-4}	3.0×10^{-4}	0.49
f_{16}	100	-1.03	4.9×10^{-7}	-1.03	4.9×10^{-7}	0.0
f_{17}	100	0.398	1.5×10^{-7}	0.398	1.5×10^{-7}	0.0
f_{18}	100	3.02	0.11	3.0	0	1.0
f_{19}	100	-3.86	1.4×10^{-5}	-3.86	1.4×10^{-2}	-1.0
f_{20}	200	-3.27	5.9×10^{-2}	-3.28	5.8×10^{-2}	0.45
f_{21}	100	-5.52	1.59	-6.86	2.67	$3.56^\dagger$
f_{22}	100	-5.52	2.12	-8.27	2.95	$5.44^\dagger$
f_{23}	100	-6.57	3.14	-9.10	2.92	$4.24^\dagger$

†The value of t with 49 degrees of freedom is significant at $\alpha = 0.05$ by a two-tailed test.

Table 5. Comparison between CEP and FEP on f_8–f_{13} with $n = 5$. The results are averaged over 50 runs, where "Mean Best" indicates the mean best function values found in the last generation, and "Std Dev" stands for the standard deviation.

Function	Number of	FEP		CEP		FEP–CEP
	Generations	Mean Best	Std Dev	Mean Best	Std Dev	t-test
f_8	500	-2061.74	58.79	-1762.45	176.21	$-11.17^\dagger$
f_9	400	0.14	0.40	4.08	3.08	$-8.89^\dagger$
f_{10}	400	8.6×10^{-4}	1.8×10^{-4}	8.1×10^{-2}	0.34	-1.67
f_{11}	1500	5.3×10^{-2}	4.2×10^{-2}	0.14	0.12	$-4.64^\dagger$
f_{12}	200	1.5×10^{-7}	1.2×10^{-7}	2.5×10^{-2}	0.12	-1.43
f_{13}	200	3.5×10^{-7}	1.8×10^{-7}	3.8×10^{-3}	1.4×10^{-2}	-1.89

†The value of t with 49 degrees of freedom is significant at $\alpha = 0.05$ by a two-tailed test.

and/or the number of local minima that have a major impact on FEP's and CEP's performance.

6 Fast Evolutionary Programming with Different Parameters

FEP investigated in the previous section used $t = 1$ in its Cauchy mutation. This value was used for its simplicity. In order to examine the impact of different t values on the performance of FEP in detail, a set of experiments have been carried out on FEP using different t values for the Cauchy mutation. Seven benchmark functions from the three different groups in Table 1 were used in these experiments. The setup of these experiments is exactly the same as before. In particular, self-adaptation (i.e., the evolution of η's) is used in all experiments. Table 6 shows the average results over 50 independent runs of FEP for different parameters.

Table 6. The mean best solutions found by FEP using different scale parameter t in the Cauchy mutation for functions $f_1(1500)$, $f_2(2000)$, $f_{10}(1500)$, $f_{11}(2000)$, $f_{21}(100)$, $f_{22}(100)$, and $f_{23}(100)$. The values in "()" indicate the number of generations used in FEP. All results have been averaged over 50 runs.

Function	$t = 0.0156$	$t = 0.0313$	$t = 0.0625$	$t = 0.1250$	$t = 0.2500$
f_1	1.0435	0.0599	0.0038	1.5×10^{-4}	6.5×10^{-5}
f_2	3.8×10^{-4}	3.1×10^{-4}	5.9×10^{-4}	0.0011	0.0021
f_{10}	1.5627	0.2858	0.0061	0.0030	0.0050
f_{11}	1.0121	0.2237	0.1093	0.0740	0.0368
f_{21}	−6.9236	−7.7261	−8.0487	−8.6473	−8.0932
f_{22}	−7.9211	−8.3719	−9.1735	−9.8401	−9.1587
f_{23}	−7.8588	−8.6935	−9.4663	−9.2627	−9.8107
Function	$t = 0.5000$	$t = 0.7500$	$t = 1.0000$	$t = 1.2500$	$t = 1.5000$
f_1	1.8×10^{-4}	3.5×10^{-4}	5.7×10^{-4}	8.2×10^{-4}	0.0012
f_2	0.0041	0.0060	0.0081	0.0101	0.0120
f_{10}	0.0091	0.0136	0.0183	0.0227	9.1987
f_{11}	0.0274	0.0233	0.0161	0.0202	0.0121
f_{21}	−6.6272	−5.2845	−5.5189	−5.0095	−5.0578
f_{22}	−7.6829	−6.9698	−5.5194	−6.1831	−5.6476
f_{23}	−8.5037	−7.8622	−6.5713	−6.1300	−6.5364

These results show that $t = 1$ was not the optimal value for the seven benchmark problems. The optimal t is problem-dependent. As analyzed later in Section 7.1, the optimal t depends on the distance between the current search point and the global optimum. Since the global optimum is usually unknown for real-world problems, it is extremely difficult to find the optimal t for a given problem. A good approach to deal with this issue is to use self-adaptation so that t can gradually evolve towards its near optimum although its initial value might not be optimal.

Another approach to be explored is to mix Cauchy mutation with different t values in a population so that the whole population can search both globally and locally. The percentage of each type of Cauchy mutation will be self-adaptive, rather than fixed. Hence the population may emphasize either global or local search depending on different stages in the evolutionary process.

7 Analysis of Fast and Classical Evolutionary Programming

It has been pointed out in Section 3 that Cauchy mutation has a higher probability of making long jumps than Gaussian mutation due to its long flat tails shown in Figure 2. In fact, the likelihood of a Cauchy mutation generating a larger jump than a Gaussian mutation can be estimated by a simple heuristic argument.

It is well known that if N_1 and N_2 are independent and identically distributed (i.i.d.) normal (Gaussian) random variates with density function

$$f_{Gaussian}(x) = \frac{1}{\sqrt{2\pi}} e^{-\frac{x^2}{2}},$$

then N_1/N_2 is Cauchy distributed with density function [21](p.451)

$$f_{Cauchy}(x) = \frac{1}{\pi(1+x^2)}.$$

Given that Cauchy and Gaussian mutations follow the aforementioned distributions, Cauchy mutation will generate a larger jump than Gaussian mutation whenever $|N_1/N_2| > |N_1|$ (i.e., $|N_2| < 1.0$). Since the probability of a Gaussian random variate smaller than 1.0 is

$$\int_{-1}^{1} \frac{1}{\sqrt{2\pi}} e^{-\frac{x^2}{2}} dx = 0.68,$$

Cauchy mutation is expected to generate a longer jump than Gaussian mutation with probability 0.68.

The expected length of Gaussian and Cauchy jumps can be calculated as follows:

$$E_{Gaussian}(x) = 2\int_0^{+\infty} x \frac{1}{\sqrt{2\pi}} e^{-\frac{x^2}{2}} dx = \frac{2}{\sqrt{2\pi}} = 0.80,$$

$$E_{Cauchy}(x) = 2\int_0^{+\infty} x \frac{1}{\pi(1+x^2)} dx = +\infty \quad \text{(i.e., does not exist).}$$

It is obvious that Gaussian mutation is much more localized than Cauchy mutation.

Similar results can be obtained for a Gaussian distribution with expectation $m > 0$ and variance $\sigma^2 > 1$ and a Cauchy distribution with scale parameters $t > 1$. The question now is why larger jumps would be beneficial. Is this always true?

7.1 When Larger Jumps Are Beneficial

Sections 5.2 and 5.3 have explained qualitatively why larger jumps are good at dealing with plateaus and many local optima. This section shows analytically and empirically that this is true only when the global optimum is sufficiently far away from the current search point, i.e., when the distance between the current point and the global optimum is larger than the "step size" of the mutation.

Take the Gaussian mutation in CEP as an example, which uses the following distribution with expectation 0 (which implies the current search point is located at 0) and variance σ^2:

$$f_{G(0,\sigma^2)}(x) = \frac{1}{\sigma\sqrt{2\pi}} e^{-\frac{x^2}{2\sigma^2}}, \qquad -\infty < x < +\infty.$$

The probability of generating a point in the neighborhood of the global optimum x^* is given by

$$P_{G(0,\sigma^2)}(|x - x^*| \le \epsilon) = \int_{x^*-\epsilon}^{x^*+\epsilon} f_{G(0,\sigma^2)}(x)dx, \tag{5}$$

where $\epsilon > 0$ is the neighborhood size and σ is often regarded as the step size of the Gaussian mutation. Figure 6 illustrates the situation.

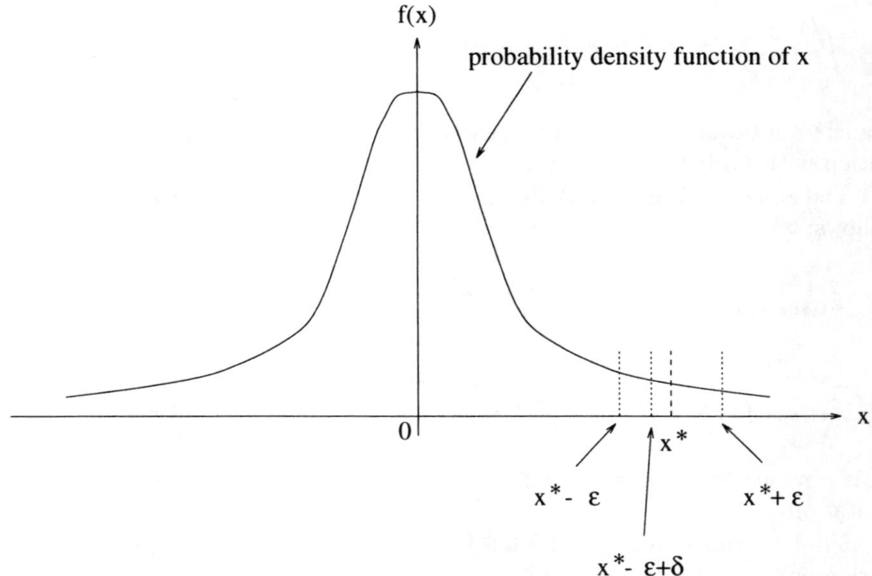

Fig. 6. Evolutionary search as neighborhood search, where x^* is the global optimum and $\epsilon > 0$ is the neighborhood size. δ is a small positive number ($0 < \delta < 2\epsilon$).

The derivative $\frac{\partial}{\partial \sigma} P_{G(0,\sigma^2)}(|x - x^*| \leq \epsilon)$ can be used to evaluate the impact of σ on $P_{G(0,\sigma^2)}(|x - x^*| \leq \epsilon)$. According to the mean value theorem for definite integrals [22](p.322), there exists a number δ $(0 < \delta < 2\epsilon)$ such that

$$\int_{x^*-\epsilon}^{x^*+\epsilon} f_{G(0,\sigma^2)}(x)dx = 2\epsilon f_{G(0,\sigma^2)}(x^* - \epsilon + \delta).$$

Hence,

$$\begin{aligned}
&\frac{\partial}{\partial \sigma} P_{G(0,\sigma^2)}(|x - x^*| \leq \epsilon) \\
&= \frac{\partial}{\partial \sigma} \int_{x^*-\epsilon}^{x^*+\epsilon} f_{G(0,\sigma^2)}(x)dx \\
&= \frac{\partial}{\partial \sigma} \left(2\epsilon f_{G(0,\sigma^2)}(x^* - \epsilon + \delta)\right) \\
&= 2\epsilon \frac{\partial}{\partial \sigma} \left(\frac{1}{\sigma\sqrt{2\pi}} e^{-\frac{(x^*-\epsilon+\delta)^2}{2\sigma^2}} \right) \\
&= 2\epsilon \left(\frac{(x^* - \epsilon + \delta)^2}{\sigma^4\sqrt{2\pi}} e^{-\frac{(x^*-\epsilon+\delta)^2}{2\sigma^2}} - \frac{1}{\sigma^2\sqrt{2\pi}} e^{-\frac{(x^*-\epsilon+\delta)^2}{2\sigma^2}} \right) \\
&= \frac{2\epsilon}{\sigma^2\sqrt{2\pi}} e^{-\frac{(x^*-\epsilon+\delta)^2}{2\sigma^2}} \left(\frac{(x^* - \epsilon + \delta)^2}{\sigma^2} - 1 \right).
\end{aligned}$$

It is apparent from the above equation that

$$\frac{\partial}{\partial \sigma} P_{G(0,\sigma^2)}(|x - x^*| \leq \epsilon) > 0 \text{ if } \sigma < |x^* - \epsilon + \delta|, \tag{6}$$

$$\frac{\partial}{\partial \sigma} P_{G(0,\sigma^2)}(|x - x^*| \leq \epsilon) < 0 \text{ if } \sigma > |x^* - \epsilon + \delta|. \tag{7}$$

That is, the larger σ is, the larger $P_{G(0,\sigma^2)}(|x - x^*| \leq \epsilon)$ will be, if $\sigma < |x^* - \epsilon + \delta|$. However, if $\sigma > |x^* - \epsilon + \delta|$, the larger σ is, the smaller $P_{G(0,\sigma^2)}(|x - x^*| \leq \epsilon)$ will be.

Similar analysis can be carried out for Cauchy mutation in FEP. Denote the Cauchy distribution defined by Eq.(3) as $f_{C(t)}(x)$. Then we have

$$\begin{aligned}
&\frac{\partial}{\partial t} P_{C(t)}(|x - x^*| \leq \epsilon) \\
&= \frac{\partial}{\partial t} \int_{x^*-\epsilon}^{x^*+\epsilon} f_{C(t)}(x)dx \\
&= \frac{\partial}{\partial t} \left(2\epsilon f_{C(t)}(x^* - \epsilon + \delta)\right) \\
&= \frac{2\epsilon}{\pi} \frac{\partial}{\partial t} \left(\frac{t}{t^2 + (x^* - \epsilon + \delta)^2} \right) \\
&= \frac{2\epsilon}{\pi} \left(\frac{1}{t^2 + (x^* - \epsilon + \delta)^2} - \frac{2t^2}{(t^2 + (x^* - \epsilon + \delta)^2)^2} \right) \\
&= \frac{2\epsilon}{\pi} \frac{(x^* - \epsilon + \delta)^2 - t^2}{(t^2 + (x^* - \epsilon + \delta)^2)^2},
\end{aligned}$$

where δ ($0 < \delta < 2\epsilon$) may not be the same as that in Eqs.(6) and (7). It is obvious that

$$\frac{\partial}{\partial t} P_{C(t)}(|x - x^*| \le \epsilon) > 0 \text{ if } t < |x^* - \epsilon + \delta|, \tag{8}$$

$$\frac{\partial}{\partial t} P_{C(t)}(|x - x^*| \le \epsilon) < 0 \text{ if } t > |x^* - \epsilon + \delta|. \tag{9}$$

That is, the larger t is, the larger $P_{C(t)}(|x - x^*| \le \epsilon)$ will be, if $t < |x^* - \epsilon + \delta|$. However, if $t > |x^* - \epsilon + \delta|$, the larger t is, the smaller $P_{C(t)}(|x - x^*| \le \epsilon)$ will be.

Since σ and t could be regarded as search step sizes for Gaussian and Cauchy mutations, the above analysis show that a large step size is beneficial (i.e., increases the probability of finding a near-optimal solution) only when the distance between the neighborhood of x^* and the current search point (at 0) is larger than the step size, or else a large step size may be detrimental to finding a near-optimal solution. The above analysis also show the rates of probability increase/decrease by deriving the explicit expressions for $\frac{\partial}{\partial \sigma} P_{G(0,\sigma^2)}(|x - x^*| \le \epsilon)$ and $\frac{\partial}{\partial t} P_{C(t)}(|x - x^*| \le \epsilon)$.

The analytical results explain why FEP achieved better results than CEP for most of the benchmark problems we tested, because the initial population was generated uniformly at random in a relatively large space and was far away from the global optimum on average. Cauchy mutation is more likely to generate larger jumps than Gaussian mutation and thus better in such cases. However, FEP would be less effective than CEP near the small neighborhood of the global optimum because Gaussian mutation's step size is smaller (smaller is better in this case). The experimental results on functions f_1 and f_2 illustrate such behavior clearly.

The analytical results also explain why FEP with a smaller t value for its Cauchy mutation would perform better whenever CEP outperforms FEP with $t = 1$. If CEP outperforms FEP with $t = 1$ for a problem, it implies that this FEP's search step size may be too large. In this case, using a Cauchy mutation with a smaller t is very likely to improve FEP's performance since it will have a smaller search step size. The experimental results presented in Table 6 match our theoretical prediction quite well.

7.2 Empirical Evidence

To validate the above analysis empirically, additional experiments were carried out. Function f_{21} (i.e., Shekel-5) was used here since it appears to pose some difficulties to FEP. First we made the search points closer to the global optimum by generating the initial population uniformly at random in the range of $2.5 \le x_i \le 5.5$ rather than $0 \le x_i \le 10$, and repeated our previous experiments. (The global optimum of f_{21} is at $x_i^* = 4$.) Such minor variation to the experiment is expected to improve the performance of both CEP and FEP since the initial search points are closer to the global optimum. Note that both Gaussian and Cauchy distributions have higher probabilities in generating points around 0 than those in generating points far away from 0.

The final experimental results averaged over 50 runs are given in Table 7. Figure 7 shows the results of CEP and FEP. It is quite clear that the performance of CEP improved much more than that of FEP since the smaller average distance between search points and the global optimum favors a small step size. The mean best value of CEP improved significantly from −6.86 to −7.90, while that of FEP improved only from −5.52 to −5.62.

Table 7. Comparison of CEP's and FEP's final results on f_{21} when the initial population is generated uniformly at random in the range of $0 \leq x_i \leq 10$ and $2.5 \leq x_i \leq 5.5$. The results were averaged over 50 runs, where "Mean Best" indicates the mean best function values found in the last generation, and "Std Dev" stands for the standard deviation. The number of generations for each run was 100.

Initial Range	FEP		CEP		FEP−CEP
	Mean Best	Std Dev	Mean Best	Std Dev	t-test
$2.5 \leq x_i \leq 5.5$	−5.62	1.71	−7.90	2.85	$4.58^{\dagger}$
$0 \leq x_i \leq 10$	−5.57	1.54	−6.86	2.94	$2.94^{\dagger}$
t-test‡	−0.16		$-1.80^{\dagger}$		

†The value of t with 49 degrees of freedom is significant at $\alpha = 0.05$ by a two-tailed test.
‡FEP(CEP)$_{small}$−FEP(CEP)$_{normal}$.

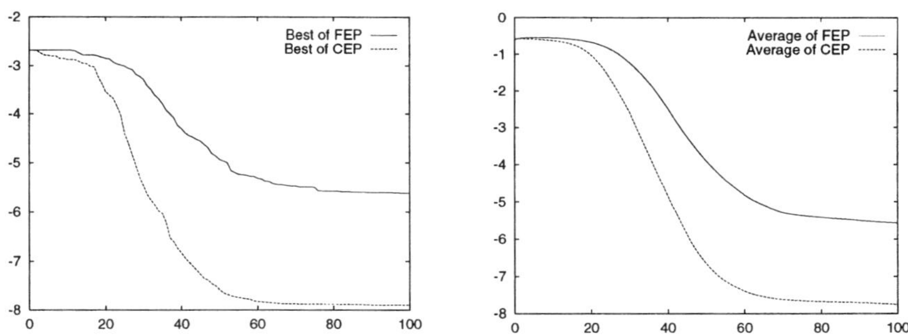

Fig. 7. Comparison between CEP and FEP on f_{21} when the initial population is generated uniformly at random in the range of $2.5 \leq x_i \leq 5.5$. The solid lines indicate the results of FEP. The dashed lines indicate the results of CEP. The left figure shows the best result. The right figure shows the average result. Both were averaged over 50 runs. The horizontal axis indicates the number of generations. The vertical axis indicates the function value.

Then three more sets of experiments were conducted where the search space was expanded 10 times, 100 times, and 1000 times, i.e., the initial population was generated uniformly at random in the range of $0 \le x_i \le 100$, $0 \le x_i \le 1000$, and $0 \le x_i \le 10000$, and a_i's were multiplied by 10, 100 and 1000, respectively, making the average distance to the global optimum increasingly large. The enlarged search space is expected to make the problem more difficult and thus make CEP and FEP less efficient. The results of the same experiment averaged over 50 runs are shown in Table 8 and Figures 8, 9, and 10. It is interesting to note that the performance of FEP was less affected by the larger search space than CEP. When the search space was increased to $0 \le x_i \le 100$ and $0 \le x_i \le 1000$, the superiority of CEP over FEP on f_{21} disappeared. There was no statistically significant difference between CEP and FEP. When the search space was increased further to $0 \le x_i \le 10000$, FEP even outperformed CEP significantly. It is worth pointing out that a population size of 100 and the maximum number of generations of 100 are very small numbers for such a huge search space. The population might not have converged by the end of generation 100. However, this does not affect our conclusion. The experiments still show that Cauchy mutation performs much better than Gaussian mutation when the current search points are far away from the global optimum.

Even if a_i's were not multiplied by 10, 100, and 1000, similar results can still be obtained as long as the initial population was generated uniformly at random in the range of $0 \le x_i \le 100$, $0 \le x_i \le 1000$, and $0 \le x_i \le 10000$. Table 9 shows the results when a_i's were unchanged. The figures of this set of experiments are omitted to save some space. It is quite clear that a similar trend can be observed as the initial ranges increase.

Table 8. Comparison of CEP's and FEP's final results on f_{21} when the initial population is generated uniformly at random in the range of $0 \le x_i \le 10$, $0 \le x_i \le 100$, $0 \le x_i \le 1000$, and $0 \le x_i \le 10000$, and a_i's were multiplied by 10, 100, and 1000. The results were averaged over 50 runs, where "Mean Best" indicates the mean best function values found in the last generation, and "Std Dev" stands for the standard deviation. The number of generations for each run was 100.

Initial Range	FEP		CEP		FEP−CEP
	Mean Best	Std Dev	Mean Best	Std Dev	t-test
$0 \le x_i \le 10000$	−3.97	2.28	−2.60	2.43	$-4.02^{\dagger}$
$0 \le x_i \le 1000$	−5.00	2.96	−5.33	2.76	1.05
$0 \le x_i \le 100$	−5.80	3.21	−5.59	2.97	−0.40
$0 \le x_i \le 10$	−5.57	1.54	−6.86	2.94	$2.94^{\dagger}$
$F(C)EP_{10000}-F(C)EP_{1000}$	$2.73^{\dagger}$		$6.57^{\dagger}$		
$F(C)EP_{1000}-F(C)EP_{100}$	1.63		0.71		
$F(C)EP_{100}-F(C)EP_{10}$	−0.48		$2.10^{\dagger}$		

†The value of t with 49 degrees of freedom is significant at $\alpha = 0.05$ by a two-tailed test.

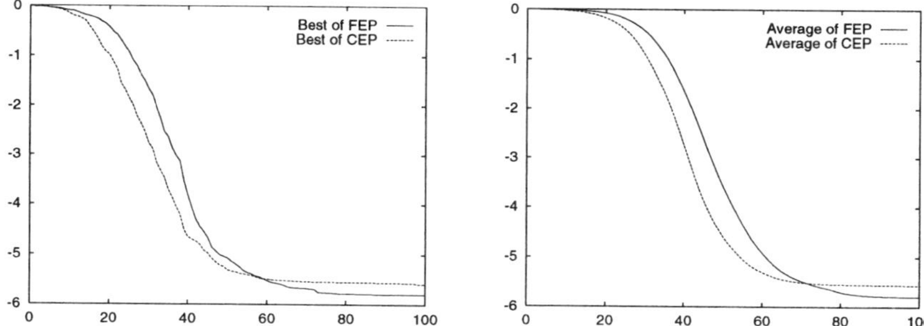

Fig. 8. Comparison between CEP and FEP on f_{21} when the initial population is generated uniformly at random in the range of $0 \leq x_i \leq 100$ and a_i's were multiplied by 10. The solid lines indicate the results of FEP. The dashed lines indicate the results of CEP. The left figure shows the best result. The right figure shows the average result. Both were averaged over 50 runs. The horizontal axis indicates the number of generations. The vertical axis indicates the function value.

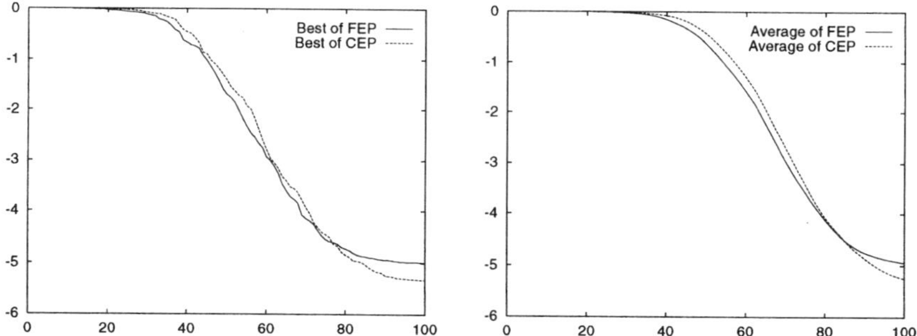

Fig. 9. Comparison between CEP and FEP on f_{21} when the initial population is generated uniformly at random in the range of $0 \leq x_i \leq 1000$ and a_i's were multiplied by 100. The solid lines indicate the results of FEP. The dashed lines indicate the results of CEP. The left figure shows the best result. The right figure shows the average result. Both were averaged over 50 runs. The horizontal axis indicates the number of generations. The vertical axis indicates the function value.

It is worth reiterating that the only difference between the experiments in this section and the previous experiment in Section 5.3 is the range used to generate initial random populations. The empirical results match quite well with our analysis on the relationship between the step size and the distance to the global optimum. The results also indicate that FEP is less sensitive to initial conditions than CEP and thus more robust. In practice, the global optimum is usually unknown. There is little knowledge one can use to constrain the search space to a sufficiently small region. In such cases, FEP would be a better choice than CEP.

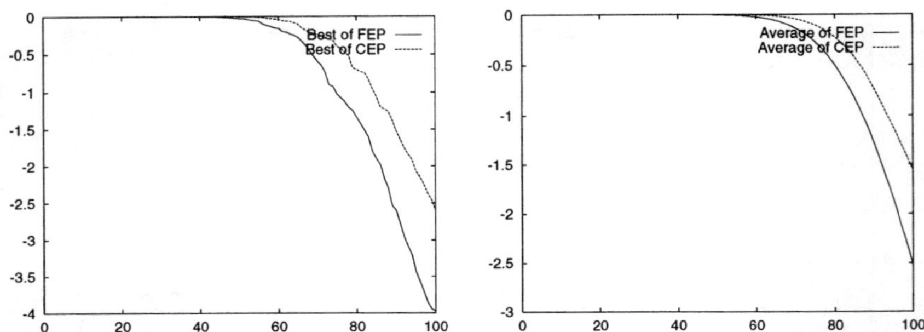

Fig. 10. Comparison between CEP and FEP on f_{21} when the initial population is generated uniformly at random in the range of $0 \leq x_i \leq 10000$ and a_i's were multiplied by 1000. The solid lines indicate the results of FEP. The dashed lines indicate the results of CEP. The left figure shows the best result. The right figure shows the average result. Both were averaged over 50 runs. The horizontal axis indicates the number of generations. The vertical axis indicates the function value.

Table 9. Comparison of CEP's and FEP's final results on f_{21} when the initial population is generated uniformly at random in the range of $0 \leq x_i \leq 10$, $0 \leq x_i \leq 100$, $0 \leq x_i \leq 100$, $0 \leq x_i \leq 1000$, and $0 \leq x_i \leq 10000$. a_i's were unchanged. The results were averaged over 50 runs, where "Mean Best" indicates the mean best function values found in the last generation, and "Std Dev" stands for the standard deviation. The number of generations for each run was 100.

Initial Range	FEP		CEP		FEP−CEP
	Mean Best	Std Dev	Mean Best	Std Dev	t-test
$0 \leq x_i \leq 10000$	−4.12	1.45	−1.40	1.43	$-10.44^{\dagger}$
$0 \leq x_i \leq 1000$	−5.10	0.81	−5.15	1.66	0.19
$0 \leq x_i \leq 100$	−5.16	1.13	−5.61	2.31	1.21
$0 \leq x_i \leq 10$	−5.57	1.54	−6.86	2.94	$2.94^{\dagger}$
F(C)EP$_{10000}$−F(C)EP$_{1000}$	$4.72^{\dagger}$		$12.46^{\dagger}$		
F(C)EP$_{1000}$−F(C)EP$_{100}$	0.33		1.33		
F(C)EP$_{100}$−F(C)EP$_{10}$	1.48		$2.24^{\dagger}$		

†The value of t with 49 degrees of freedom is significant at $\alpha = 0.05$ by a two-tailed test.

7.3 The Importance of Neighborhood Size

It is well known that finding an exact global optimum for a multimodal function is hard without prior knowledge about the function. It might take an infinite amount of time to find the global optimum for a global search algorithm such as CEP or FEP. In practice, one often has to sacrifice discovering the global optimum in exchange for efficiency. A key issue that arises here is how much sacrifice one has to make in order to get a near-optimum in a reasonable amount of time. In other words, what is the relationship between the optimality of the solution and the time used to find the solution? This issue can be approached from the point of view of neighborhood size, i.e., ϵ in Eq.(5), since a smaller neighborhood size usually implies better optimality. It will be very useful if the impact of neighborhood size ϵ on the probability of *generating* a near-optimum in that neighborhood can be worked out. (The probability of finding a near-optimum would be the same as that of generating it when the elitism is used.) Although not an exact answer to the issue, the following analysis does provide some insights into such impact.

Similar to the analysis in Section 7.1, the following is true according to the mean value theorem for definite integrals [22](p.322): for $0 < \delta < 2\epsilon$,

$$\begin{aligned}
&\frac{\partial}{\partial \epsilon} P_{G(0,\sigma^2)}(|x - x^*| \leq \epsilon) \\
&= \frac{\partial}{\partial \epsilon} \int_{x^*-\epsilon}^{x^*+\epsilon} f_{G(0,\sigma^2)}(x)dx \\
&= \frac{\partial}{\partial \epsilon} \left(2\epsilon f_{G(0,\sigma^2)}(x^* - \epsilon + \delta)\right) \\
&= \frac{\partial}{\partial \epsilon} \left(2\epsilon \frac{1}{\sigma\sqrt{2\pi}} e^{-\frac{(x^*-\epsilon+\delta)^2}{2\sigma^2}}\right) \\
&= \frac{2}{\sigma\sqrt{2\pi}} \frac{\partial}{\partial \epsilon} \left(\epsilon e^{-\frac{(x^*-\epsilon+\delta)^2}{2\sigma^2}}\right) \\
&= \frac{2}{\sigma\sqrt{2\pi}} e^{-\frac{(x^*-\epsilon+\delta)^2}{2\sigma^2}} \left(1 + \frac{\epsilon}{\sigma^2}(x^* - \epsilon + \delta)\left(1 - \frac{\partial \delta}{\partial \epsilon}\right)\right).
\end{aligned}$$

For the above equation, there exists a sufficiently small number $\epsilon_1 > 0$ such that for any $\epsilon \leq \epsilon_1$,

$$\left|\frac{\epsilon}{\sigma^2}(x^* - \epsilon + \delta)\left(1 - \frac{\partial \delta}{\partial \epsilon}\right)\right| < 1.$$

That is, for $0 < \epsilon \leq \epsilon_1$,

$$\frac{\partial}{\partial \epsilon} P_{G(0,\sigma^2)}(|x - x^*| \leq \epsilon) > 0,$$

which implies that the probability of generating a near-optimum increases as the neighborhood size increases in the vicinity of the optimum. The rate of such probability growth (i.e., $\frac{\partial}{\partial \epsilon} P_{G(0,\sigma^2)}(|x - x^*| \leq \epsilon)$) is governed by the term

$$e^{-\frac{(x^*-\epsilon+\delta)^2}{2\sigma^2}} = e^{-\frac{(x^*-(\epsilon-\delta))^2}{2\sigma^2}}.$$

That is, $\frac{\partial}{\partial \epsilon} P_{G(0,\sigma^2)}(|x - x^*| \leq \epsilon)$ grows exponentially faster as $\epsilon - \delta$ increases.

Similar analysis can be carried out for Cauchy mutation using its density function, i.e., Eq.(3). Let the density function be $C(1)$ when $t = 1$. For

$0 < \delta < 2\epsilon$,

$$
\begin{aligned}
& \frac{\partial}{\partial\epsilon} P_{C(1)}(|x - x^*| \leq \epsilon) \\
&= \frac{\partial}{\partial\epsilon} \int_{x^*-\epsilon}^{x^*+\epsilon} f_{C(1)}(x)dx \\
&= \frac{\partial}{\partial\epsilon} \int_{x^*-\epsilon}^{x^*+\epsilon} \frac{1}{\pi(1+x^2)}dx \\
&= \frac{\partial}{\partial\epsilon} \left(\frac{1}{\pi}(\arctan(x^* + \epsilon) - \arctan(x^* - \epsilon)) \right) \\
&= \frac{1}{\pi} \left(\frac{1}{1+(x^*+\epsilon)^2} + \frac{1}{1+(x^*-\epsilon)^2} \right) \\
&> 0.
\end{aligned}
$$

Hence the probability of generating a near-optimum in the neighborhood always increases as the neighborhood size increases. While this conclusion is quite straightforward, it is interesting to note that the rate of increase in the probability differs significantly between Gaussian and Cauchy mutation since $\frac{\partial}{\partial\epsilon} P_{C(1)}(|x - x^*| \leq \epsilon) \gg \frac{\partial}{\partial\epsilon} P_{G(0,1)}(|x - x^*| \leq \epsilon)$.

8 An Improved Fast Evolutionary Programming

The previous analyses show the benefits of FEP and CEP in different situations. Generally, Cauchy mutation performs better when the current search point is far away from the global minimum, while Gaussian mutation is better at finding a local optimum in a good region. It would be ideal if Cauchy mutation is used when search points are far away from the global optimum and Gaussian mutation is adopted when search points are in the neighborhood of the global optimum. Unfortunately, the global optimum is usually unknown in practice, making the ideal switch from Cauchy to Gaussian mutation very difficult. Self-adaptive Gaussian mutation [12,7,13] is an excellent technique to partially address the problem. That is, the evolutionary algorithm itself will learn when to "switch" from one step size to another. However, there is room for further improvement to self-adaptive algorithms like CEP or even FEP.

This chapter proposes an improved FEP (IFEP) based on *mixing* (rather than switching) different mutation operators. The idea is to mix different search biases of Cauchy and Gaussian mutations. The importance of search biases has been pointed out by some earlier studies [23](p.375–376). The implementation of IFEP is very simple. It differs from FEP and CEP only in Step 3 of the algorithm described in Section 2. Instead of using Eq.(1) (for CEP) *or* Eq.(4) (for FEP) alone, IFEP generates two offspring from each parent, one by Cauchy mutation and the other by Gaussian. The better one is then chosen as the offspring. The rest of the algorithm is exactly the same as FEP and CEP. Chellapilla [24] has recently presented some more results on comparing different mutation operators in EP.

8.1 Experimental Studies

In order to carry out a fair comparison among IFEP, FEP, and CEP, the population size of IFEP was reduced to half of that of FEP or CEP in all the following experiments, since each individual in IFEP generates two offspring. However, reducing IFEP's population size by half actually disadvantages IFEP slightly because it does not double the time for any operators (such as selection) other than mutations. Nevertheless, such comparison offers a good and simple compromise.

IFEP was tested in the same experimental setup as before. For the sake of clarity and brevity, only some representative functions (out of 23) from each group were tested. Functions f_1 and f_2 are typical unimodal functions. Functions f_{10} and f_{11} are multimodal functions with many local minima. Functions f_{21}–f_{23} are multimodal functions with only a few local minima, and are particularly challenging to FEP. Table 10 summarizes the final results of IFEP in comparison with FEP and CEP. Figures 11 and 12 show the results of IFEP, FEP, and CEP.

Table 10. Comparison among IFEP, FEP, and CEP on functions $f_1, f_2, f_{10}, f_{11}, f_{21}, f_{22}, f_{23}$. All results have been averaged over 50 runs, where "Mean Best" indicates the mean best function values found in the last generation.

F	# of Gen's	IFEP Mean Best	FEP Mean Best	CEP Mean Best	IFEP−FEP t-test	IFEP−CEP t-test
f_1	1500	4.16×10^{-5}	5.72×10^{-4}	1.91×10^{-4}	$-28.06^{\dagger}$	$-2.39^{\dagger}$
f_2	2000	2.44×10^{-2}	7.60×10^{-2}	2.29×10^{-2}	$-51.61^{\dagger}$	$3.47^{\dagger}$
f_{10}	1500	4.83×10^{-3}	1.76×10^{-2}	8.79	$-48.54^{\dagger}$	$-21.26^{\dagger}$
f_{11}	2000	4.54×10^{-2}	2.49×10^{-2}	8.13×10^{-2}	$2.16^{\dagger}$	$-2.19^{\dagger}$
f_{21}	100	−6.46	−5.50	−6.43	$-2.19^{\dagger}$	$-5.46^{\dagger}$
f_{22}	100	−7.10	−5.73	−7.62	$-2.25^{\dagger}$	0.84
f_{23}	100	−7.80	−6.41	−8.86	$-2.14^{\dagger}$	$1.73^{\dagger}$

†The value of t with 49 degrees of freedom is significant at $\alpha = 0.05$ by a two-tailed test.

8.2 Discussions

It is very clear from Table 10 that IFEP has improved FEP's performance significantly for all test functions except for f_{11}. Even in the case of f_{11}, IFEP is better than FEP for 25 out of 50 runs. In other words, IFEP's

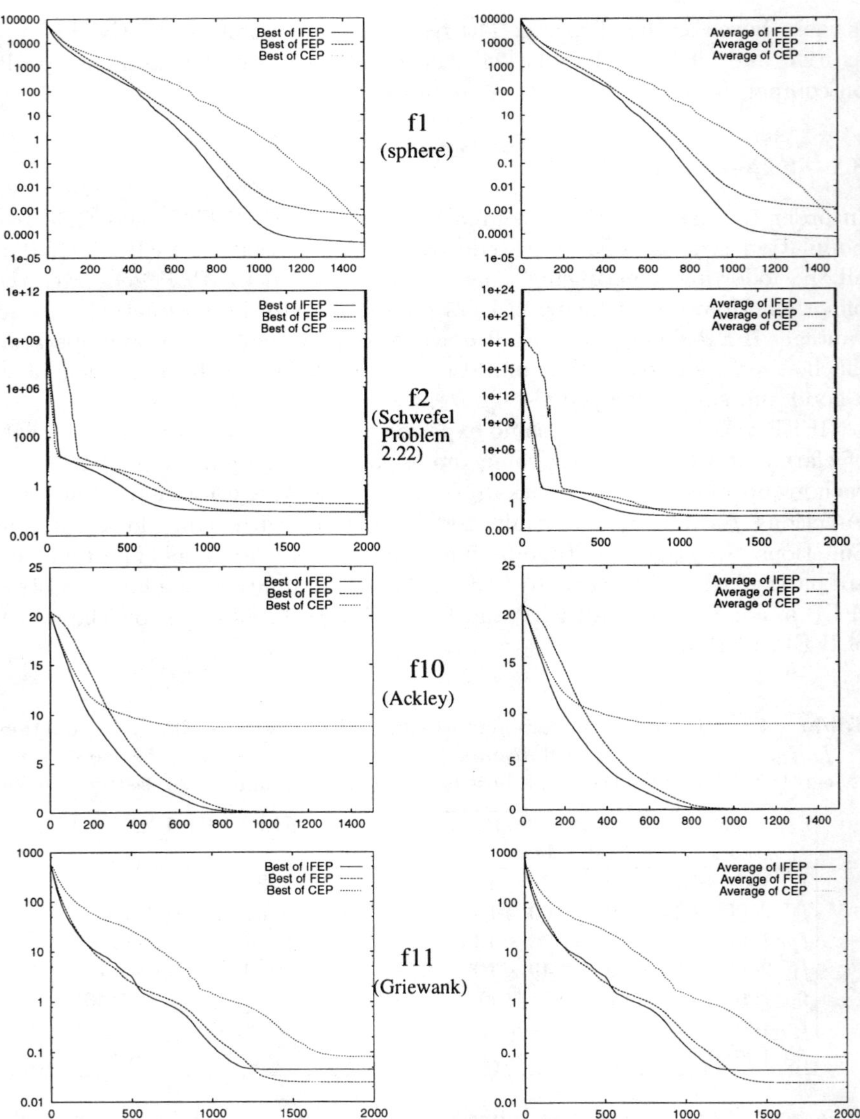

Fig. 11. Comparison among IFEP, FEP, and CEP on functions f_1, f_2, f_{10}, f_{11}. The vertical axis is the function value and the horizontal axis is the number of generations. The solid lines indicate the results of IFEP. The dashed lines indicate the results of FEP. The dashed lines indicate the results of CEP. The left figures show the best results. The right figures show the average results. All were averaged over 50 runs.

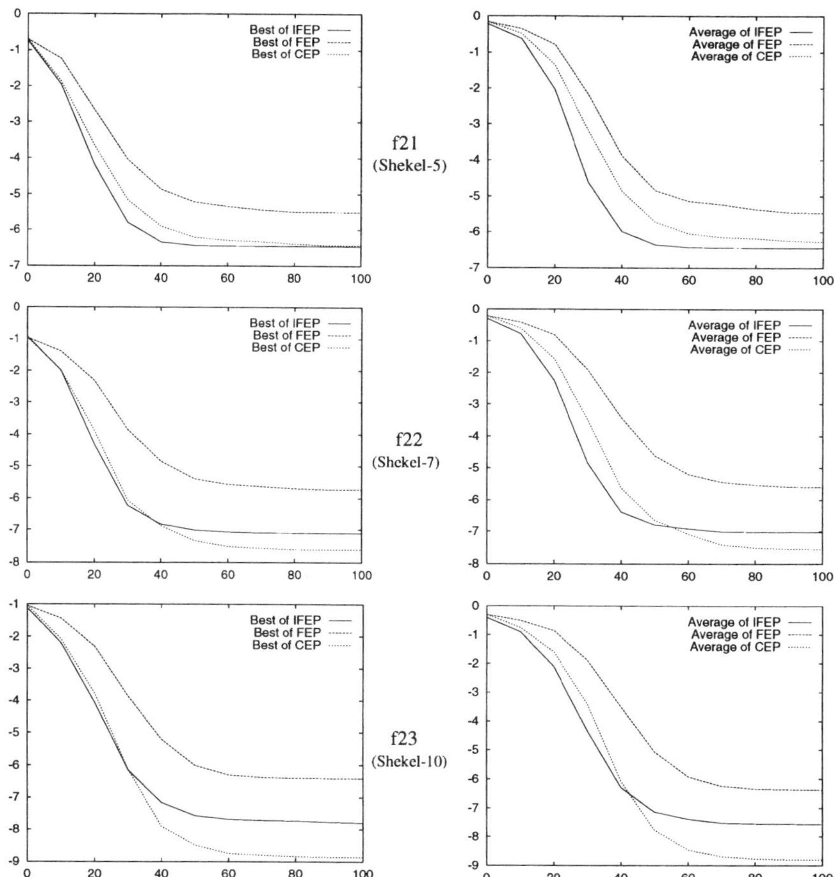

Fig. 12. Comparison among IFEP, FEP, and CEP on functions f_{21}–f_{23}. The vertical axis is the function value and the horizontal axis is the number of generations. The solid lines indicate the results of IFEP. The dashed lines indicate the results of FEP. The dashed lines indicate the results of CEP. The left figures show the best results. The right figures show the average results. All were averaged over 50 runs.

performance is still rather close to FEP's and certainly better than CEP's (35 out of 50 runs) on f_{11}. These results show that IFEP continues to perform at least as well as FEP on multimodal functions with many minima, and also performs very well on unimodal functions and multimodal functions with only a few local minima with which FEP has difficulty handling. IFEP achieved performance similar to CEP's on these functions.

For the two unimodal functions where FEP is outperformed by CEP significantly, IFEP performs better than CEP on f_1, while worse than CEP on f_2. A closer look at the actual average solutions reveals that IFEP found much better solution than CEP on f_1 (roughly an order of magnitude smaller) but only performed slightly worse than CEP on f_2.

For the three Shekel functions f_{21}–f_{23}, the difference between IFEP and CEP is much smaller than that between FEP and CEP. IFEP has improved FEP's performance significantly on all three functions. It performs better than CEP on f_{21}, the same on f_{22}, and worse on f_{23}.

It is very encouraging that IFEP is capable of performing as well as or better than the better one of FEP and CEP for most test functions. This is achieved through a minimal change to the existing FEP and CEP. No prior knowledge or any other complicated operators were used. There is no additional parameter used either. The superiority of IFEP also demonstrates the importance of mixing different search biases (e.g., "step sizes") in a robust search algorithm.

The population size of IFEP used in the above experiments was only half that of FEP and CEP. It is not unreasonable to expect even better results from IFEP if it uses the same population size as FEP's and CEP's. For (μ, λ) or $(\mu + \lambda)$ evolutionary algorithms where $\mu < \lambda$, it would be quite natural to use both Cauchy and Gaussian mutations since a parent needs to generate more than one offspring anyway.

It has been mentioned several times in this chapter that Cauchy mutation performs better than Gaussian mutation because of its higher probability of making large jumps (i.e., having a larger expected search step size). However, according to our theoretical analysis, large search step sizes would be detrimental to search when the current search points are very close to the global optimum. Figures 13, 14, and 15 show the number of successful Cauchy mutations in a population in different generations. It is obvious that Cauchy mutation played a major role in the population in the early stages of evolution since the distance between the current search points and the global optimum was relatively large on average in the early stages. Hence Cauchy mutation performed better. However, as the evolution progressed, the distance became smaller and smaller. Large search step sizes produced by Cauchy mutation tended to produce worse offspring than those produced by Gaussian mutation. The decreasing number of successful Cauchy mutations in those figures illustrates this behavior.

9 Evolutionary Algorithms with Approximation and Local Search

9.1 Local Search in Evolutionary Algorithms

The local search techniques have been applied in variant global optimization methods. The simplest way to benefit from a local search procedure is known as Multistart. This method applies local search from each randomly generated point. However, local searches are the most time-consuming parts in the

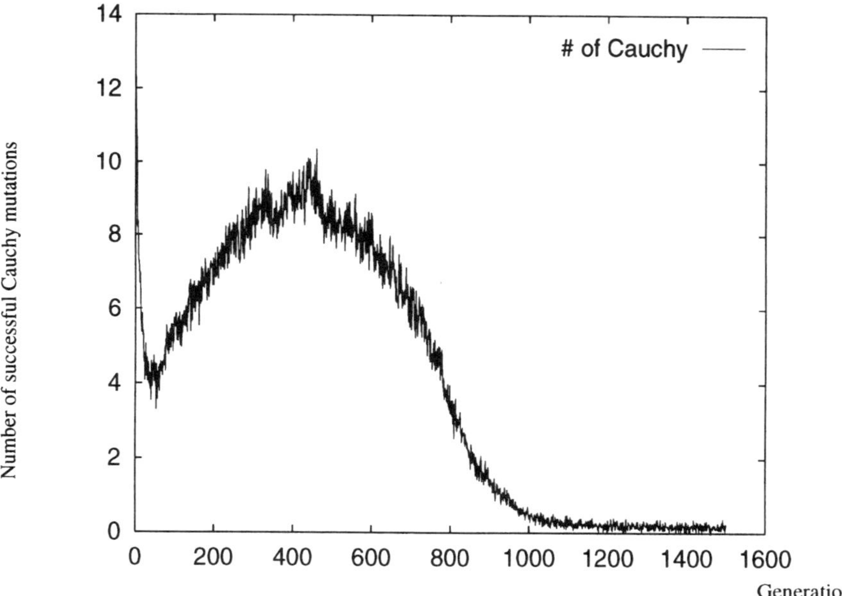

Fig. 13. Number of successful Cauchy mutations in a population when IFEP is applied to function f_1. The vertical axis indicates the number of successful Cauchy mutations in a population and the horizontal axis indicates the number of generations. The results have been averaged over 50 runs.

algorithm. Obviously, the inefficiency is caused by using many local searches to find the same minimum several times. To improve the inefficiency, clustering methods [25] and the Multi Level Single Linkage [26] are designed. They use variant clustering techniques to link points to different groups. Within each group local search is only conducted on one point and the found local optimum is assumed to be the representative of that group. These methods provide impressive results to the benchmark test function in Dixon and Szegö [27]. However, when solving a problem with a large number of minima, these methods may not be suitable [26,28].

In recent years, evolutionary algorithms have become increasingly robust and easy to use as a global optimization method. Applying local search in the GA has substantially improved their performance for some multimodal problems [29,30]. In terms of evolution, the local search can be thought of as a consequence of individual learning during the individual's lifetime. Combining learning and evolution has the effect of changing the fitness landscape. The fitness of an individual will be changed after applying a learning process (i.e., local search). However, the individual's genetic codes (parameters) may or may not be altered depending on the way that learning and evolution interact. They are known as *Lamarckian evolution* and the *Baldwin effect* [31,32],

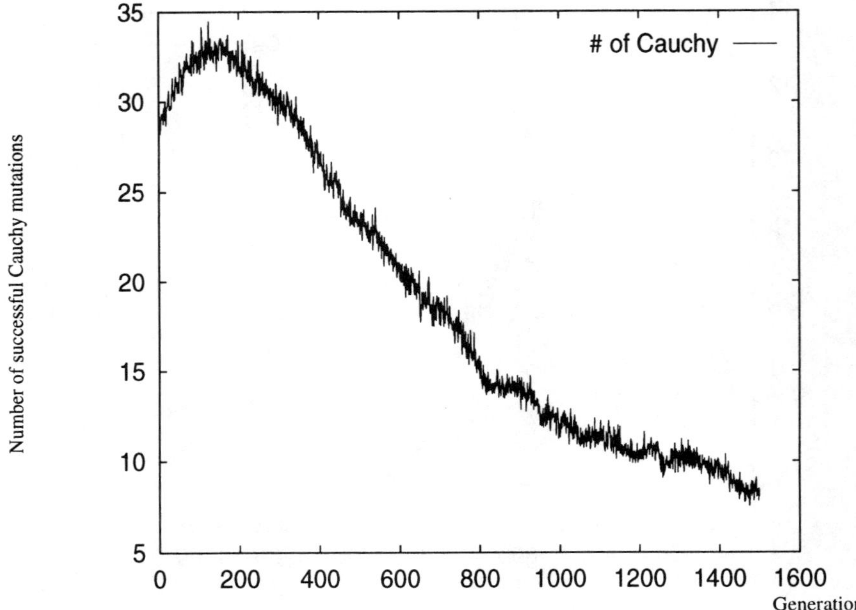

Fig. 14. Number of successful Cauchy mutations in a population when IFEP is applied to function f_{10}. The vertical axis indicates the number of successful Cauchy mutations in a population and the horizontal axis indicates the number of generations. The results have been averaged over 50 runs.

respectively. In Hart and Belew [30], the performance of the Baldwin effect and Lamarckian evolution are compared. It should be noticed that the usage rates of local search to the GA have different impacts on the evolutionary process. To design an efficient search algorithm, the application of the local search should be carefully considered.

9.2 The Idea of Approximation

Although some functions can be very rugged and have many local optima, we are only interested in the global one. It would make the optimization task easier if we could smooth out unwanted local optima and retain only the global one. One idea is to approximate such complex functions crudely by a simple smooth function, such as a quadratic function. Then the minimum of the quadratic function can be used as a starting point (an offspring) for further search.

To illustrate the idea, Figure 16 shows an example that uses quadratic approximation and local search on the one-dimensional case of Ackley's function. In this example, P_1, P_2, P_3 are three local optima, P_4 is the minimum

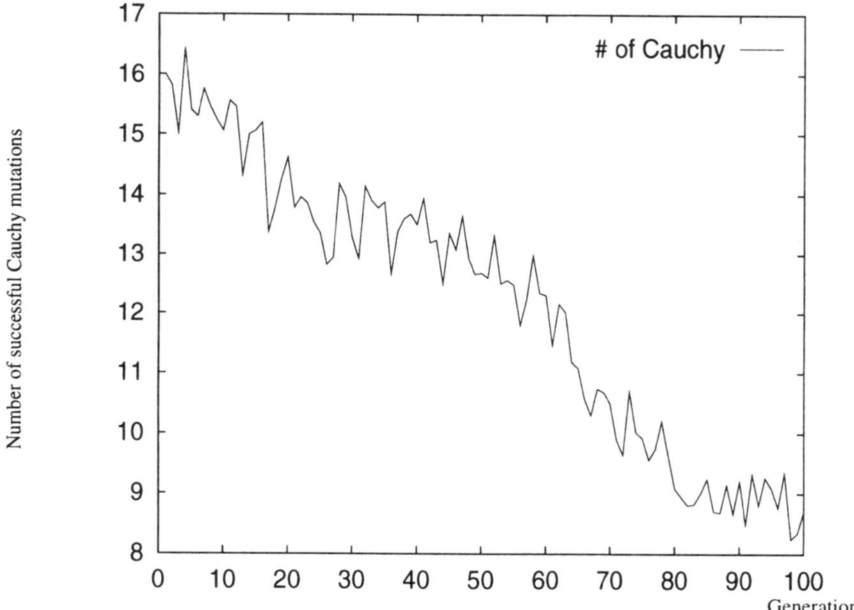

Fig. 15. Number of successful Cauchy mutations in a population when IFEP is applied to function f_{21}. The vertical axis indicates the number of successful Cauchy mutations in a population and the horizontal axis indicates the number of generations. The results have been averaged over 50 runs.

of the quadratic approximation and is mapped to P_4' in the original function. The global optimum is quite easy to locate from P_4'.

9.3 The Basic Framework

Based on local search and approximation, we can derive a family of hybrid evolutionary algorithms that use different approximation and local methods. The basic framework for all such algorithms can be summarized as follows:

Algorithm A0: The Framework

Step 1: Generate μ individuals and conduct local search.
Step 2: Apply λ individuals to provide ρ predicted minimum points using an approximation technique.
Step 3: Conduct local search for ρ individuals.
Step 4: Select next generation from the union of μ, λ, and ρ individuals.
Step 5: Go to Step 2 if termination criteria are not met.

There are a number of practical considerations we have to make in order to design a particular algorithm, e.g., as follows.

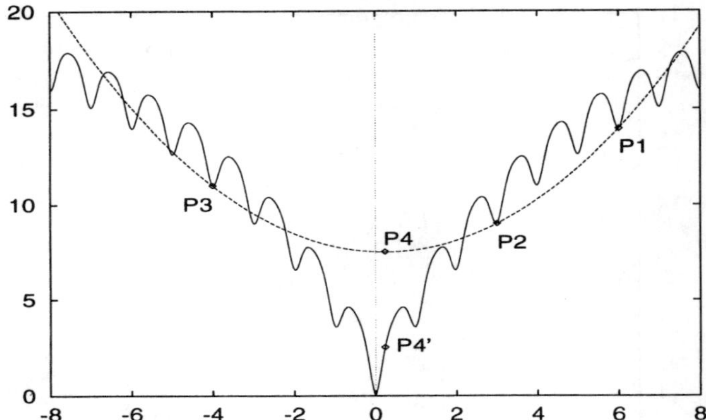

Fig. 16. The one-dimensionae landscape approximation with local search result on Ackley's function

1 **Select an efficient local search technique.** We can categorize local search into gradient or non-gradient-based methods. Many real-world problems have difficulties computing any associated derivatives (gradient or Hessian). For non-gradient-based local search methods, direct search methods [33,34], the linear approximation method [35], and local evolutionary search [36] are all very good choices.

2 **Use a simple approximation method.** A basic principle of applying approximation techniques is that the approximated function can be easily calculated to get the approximated minimum location. A quadratic polynomial with least square approximation is a good example and has been applied in some optimization methods [37–40]. The DACE model based on Bayesian statistics can give a more precise approximation about the landscape; however, other auxiliary global optimization algorithms are needed to obtain the approximated minimum from the approximation function. Different approximation methods also demand different λ individuals. The larger λ is, the more precise approximation can be obtained. But the computation time will be longer.

3 **Add other evolutionary operators.** The operator implied in Algorithm A0 can be regarded as a specialized recombination operator which has λ parents and ρ offspring. However, other evolutionary operators, such as mutations, can also be used in addition to such a recombination.

9.4 LALS: a Hybrid Algorithm Based on Approximation and Local Search

According to Algorithm A0 given in the previous section, we have developed the following hybrid algorithm based on landscape approximation and local search (LALS) [40].

1 Initialize μ individuals at random, each individual performs a local search.
2 REPEAT
  (a) Generate three points P_1, P_2, P_3 by global discrete recombination.
  (b) Perform a quadratic approximation using P_1, P_2, P_3 to produce a point P_4.
  (c) Conduct a local search from P_4 and update P_4 with the search result.
  (d) Place P_1, P_2, P_3, P_4 in the population and do a $(\mu + 4)$ truncation selection.
3 UNTIL termination criteria are met.

The local search method used in this chapter is the Local Evolutionary Search with Random Memorizing (LESRM) from Voigt and Lange [36]. LESRM generates promising new search directions from pre-stored solutions to the current point. A generated good solution is stored in a sequential memory which is randomly accessed later.

For global discrete recombination, each new individual's objective variable is decided at random and copied from one of the parents:

$$x_{i,j} = x_{\chi,j}, \forall j \in \{1, \dots, n\},$$

where $x_{i,j}$ denotes the j-th component of the vectors x_i, $\forall i \in \{1, 2, 3\}$, n is the dimensionality. χ denotes a uniformly distributed random integer in $\{1, \dots, \mu\}$, and μ is the population size.

We approximate the position of P_4 using the quadratic interpolation method as follows:

$$x_{4,j} = \frac{1}{2} \cdot \frac{(x_{2,j}^2 - x_{3,j}^2)f(x_1) + (x_{3,j}^2 - x_{1,j}^2)f(x_2) + (x_{1,j}^2 - x_{2,j}^2)f(x_3)}{(x_{2,j} - x_{3,j})f(x_1) + (x_{3,j} - x_{1,j})f(x_2) + (x_{1,j} - x_{2,j})f(x_3)}.$$

The algorithm terminates when the maximum number of function evaluations used is reached or the differences of the fitness values of P_1, P_2, P_3 are less than 1e-6.

9.5 Experimental Results on LALS

The test functions used in our experiments include all multimodal functions given in Table 1 (i.e., functions f_8–f_{23}) plus two Schaffer's functions [41] as listed in Table 11. Each experiment was run independently for 50 times. The population size was 30. The maximum number of function evaluations was 500000. To evaluate the performance of LALS, we also list the results of IFEP [42] in the result tables.

Table 11. The 18 test functions used in our experimental studies, where n is the dimension of the function, f_{min} is the minimum value of the function, and $S \subseteq R^n$.

Test function	n	S	f_{min}
$f_{24}(x) = 0.5 + \frac{\sin^2 \sqrt{x_1^2+x_2^2} - 0.5}{[1.0+0.001(x_1^2+x_2^2)]^2}$	2	$[-100, 100]^n$	0
$f_{25}(x) = (x_1^2 + x_2^2)^{0.25}[\sin^2(50(x_1^2 + x_2^2)^{0.1}) + 1.0]$	2	$[-100, 100]^n$	0

Results on f_8–f_{13} Table 12 lists experimental results for IFEP and LALS. LALS has better solution quality and global hits on all functions except f_{13}. However, LALS used more function evaluations on f_{10} and f_{12}. For f_{13}, we could achieve 50/50 global hit and 614252 average number of function evaluations if we set the maximal number of function evaluations to 700000.

Table 12. The results of IFEP and LALS on f_8–f_{13}. "Eval." is the number of function evaluations.

F	IFEP					LALS			
	Eval.	Mean	Std Dev	Found at	Global hit	Eval.	Mean	Std Dev	Global hit
8	900000	-10640.18	431.21	7575	0/50	199244	-11834.65	298.78	1/50
9	500000	2.89	1.98	5000	1/50	306280	2.67	1.58	5/50
10	150000	6.33e-4	3.86e-4	1500	50/50	370686	2.08e-12	2.11e-12	50/50
11	200000	1.27e-1	1.90e-1	1765	3/50	173592	1.87e-10	1.31e-9	50/50
12	150000	2.49e-2	5.70e-2	1500	40/50	476245	5.84e-9	2.16e-8	50/50
13	300000	4.42e-8	2.65e-8	1500	50/50	504212	1.19e-2	3.01e-2	33/50

Although good results for f_8 and f_9 were found using LALS, the global hits are not as high as we had hoped. We ran the experiments again with population sizes 50 and 60. The results are shown in Table 13. The global hits improved significantly in both cases at the cost of increased computation time although the improvement for f_8 is less than that for f_9. The reason was given elsewhere [40].

Table 13. The results of LALS on f_8–f_9 with population size 50 and 60

F	Pop. size	Eval.	Mean	Std Dev	Global hit
8	50	327434	-12296.68	119.27	1/50
8	60	391634	-12358.64	166.33	12/50
9	50	508021	2.98e-1	4.97e-1	36/50
9	60	610163	2.19e-1	4.12e-1	39/50

9.6 Results on f_{14}–f_{25}

The final results of IFEP and LALS on functions f_{14}–f_{25} are summarized in Table 14. It is clear that LALS has better global reliability than IFEP. LALS, however, seems to require more computation time than IFEP on some of the test functions.

Table 14. The results of IFEP and LALS on f_{14}–f_{25}. "Eval." is the number of function evaluations.

F	IFEP					LALS			
	Gen.	Eval.	Std Dev	Found at	Global hit	Eval.	Mean	Std Dev	Global hit
14	10000	1.56	0.96	10000	33/50	3052	1.04	0.20	48/50
15	400000	4.17e-4	2.98e-4	235500	44/50	111748	3.0749e-4	2.45e-10	50/50
16	10000	-1.031628	2.70e-8	3500	50/50	2817	-1.031628	2.70e-8	50/50
17	10000	0.3979	8.26e-9	4500	50/50	5496	0.3979	8.26e-9	50/50
18	10000	3	0	4500	50/50	4676	3	0	50/50
19	10000	-3.86	0	6000	50/50	6852	-3.86	0	50/50
20	20000	-3.26	5.90e-2	17500	24/50	17504	-3.32	2.35e-7	50/50
21	10000	-7.02	2.74	10000	21/50	13790	-10.1532	0	50/50
22	10000	-8.42	2.96	9500	34/50	13354	-10.4029	2.16e-7	50/50
23	10000	-9.09	2.87	7500	39/50	14312	-10.5364	4.58e-7	50/50
24	10000	8.86e-3	2.67e-3	8500	4/50	10754	3.89e-4	1.90e-3	48/50
25	10000	2.47e-3	1.27e-3	10000	50/50	15614	1.07e-4	7.51e-4	50/50

f_{24} is an intersting case because of the large difference between IFEP's and LALS's performance. Figure 17 displays the one-dimensional landscape of Schaffer's function f_{24}. For traditional step size control EAs (such as IFEP), it is difficult for individuals to escape from local minima in later evolutionary stages. The way LALS is designed enables it to fly over high barriers even in the later stage of evolution. For functions f_{14} and f_{24}, LALS can easily achieve the global hit of 50/50 when its population size is increased to 40.

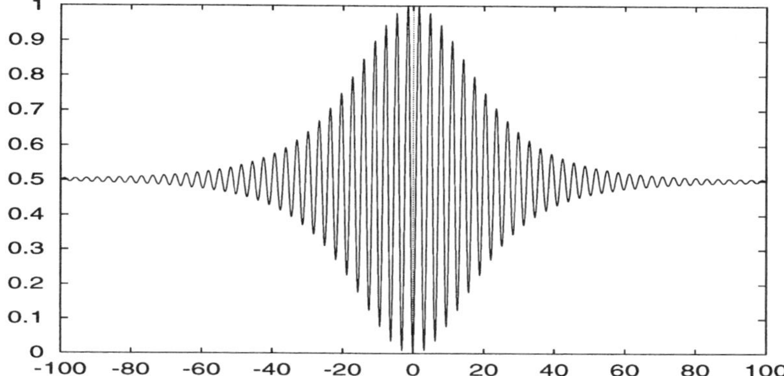

Fig. 17. The one-dimensional landscape of Schaffer's function(f_{24}).

10 EANA: An Evolutionary Algorithm with N-dimensional Approximation

The algorithm described in the previous section, i.e., LALS, approximates an N-dimensional function using N one-dimensional quadratic functions. It is unclear whether this is an appropriate approximation method. It would be interesting to investigate whether an N-dimensional approximation would work better in our hybrid algorithm.

The local search algorithm used in LALS is quite time consuming. In fact, LALS spends more time in local search than on evolutionary search for quite a few benchmark problems. It is desirable to cut down the local search time in our hybrid algorithm. One simple method is to carry out only one or two downhill moves and then start approximation, rather than to search for local optima.

The following algorithm, i.e., EANA (Evolutionary Algorithm with N-dimensional Approximation), incorporates the idea of N-dimensional approximation and limited local search in our hybrid algorithm.

1 Generate μ initial individuals at random.
2 Go downhill for one step only for every individual.
3 Obtain λ individuals using global discrete recombination.
4 Use the n-dimensional quadratic approximation technique to produce one approximated minimum point from λ individuals.
5 Perform local search (using LESRM [36]) from the approximated individual if the approximated minimum is less than any one of its parents, otherwise perform local search with a probability of 0.5.
6 Conduct $(\mu + \lambda + 1)$ selection to produce the next generation.
7 Go to Step 2 if termination criteria are not met.

The next section describes how EANA is implemented in our algorithm.

10.1 Landscape Approximation — the Polynomial Model

A quadratic polynomial model has the form

$$y(\mathbf{x}) = c_0 + \sum_{i=1}^{n} c_i x_i + \sum_{i=1}^{n} c_{i+n} x_i^2 + \sum_{i=1}^{n(n-1)/2} c_{i+2n} x_j x_k, \tag{10}$$

where n is the number of variables, c_i is the i-th polynomial coefficient, and $1 \le j < k \le n$. There are $t = (n+1)(n+2)/2$ unknown coefficients. Thus, we need at least t sample points to obtain a unique solution.

This polynomial model with t sample points expressed in matrix notation is $\mathbf{y} = \mathbf{X}\mathbf{c}$, where $\mathbf{y}$ is the vector formed by t fitness values of the sample points, $\mathbf{y} = [y^1, y^2, \ldots, y^t]^{\mathbf{T}}$, and $\mathbf{c}$ is the vector of unknown coefficients, $\mathbf{c} = [c_0, c_1, \ldots, c_{t-1}]^{\mathbf{T}}$, and $\mathbf{X}$ is the matrix expressed as

$$\mathbf{X} = \begin{bmatrix} 1 \; x_1^{(1)} \ldots x_n^{(1)} \; (x_1^{(1)})^2 \ldots (x_n^{(1)})^2 \; x_1 x_2^{(1)} \ldots x_{n-1} x_n^{(1)} \\ \vdots \; \vdots \; \ddots \; \vdots \quad \vdots \quad \ddots \vdots \quad \vdots \quad \ddots \quad \vdots \\ 1 \; x_1^{(t)} \ldots x_n^{(t)} \; (x_1^{(t)})^2 \ldots (x_n^{(t)})^2 \; x_1 x_2^{(t)} \ldots x_{n-1} x_n^{(t)} \end{bmatrix}$$

The unique least squares solution of $\mathbf{c}$ is

$$\mathbf{c} = (\mathbf{X}^{\mathbf{T}}\mathbf{X})^{-1}\mathbf{X}^{\mathbf{T}}\mathbf{y},$$

if $\mathbf{X}^{\mathbf{T}}\mathbf{X}$ is nonsingular.

Using the quadratic polynomial model of Eq. 10 to obtain an approximated point will need $(n+1)(n+2)/2$ points. This is very time consuming for high-dimensional problems. For a 30-dimensional problem, for example, we would need 496 individuals to accomplish this job. An alternative and simplified approach is to discard the third summation in Eq. 10. Therefore, only $2n+1$ individuals are needed to perform an approximation.[2]

To understand the impact of the simplified version of quadratic approximation, we ran a preliminary experiment to compare the difference on computation time and performance between the original and the simplified quadratic approximation. The experiments are tested on the generalized Rastrigin function with various dimensions using EANA. Figure 18 shows the result. On the left it shows the computation time used to accomplish 20 runs in the experiment. The time used on the 30-dimensional problem is 57 hours from the original approximation, and 10 minutes for the simplified one. The right part of Figure 18 shows the performance of the algorithms. It is clear that the performances are very close to each other although the algorithm using the original approximation was slightly better. In this chapter, only the simplified quadratic approximation will be used.

11 Experimental Results on Hybrid Algorithms

We use all the multimodal functions tested by LALS [40] and two more functions from [43] in our experimental studies. Table 15 lists the two additional functions. To compare the performance of EANA with others, we also list the results of ES and IFEP [42] in the tables in the next two sections. The ES uses n self-adaptive strategy parameters and no correlated mutations. Recombination is discrete on object variables and global intermediate on strategy parameters. The selection uses a $(15, 100)$ mechanism. In IFEP each parent generates two offspring, one with Gaussian and the other with Cauchy mutation. The parameter setup follows the suggestions from [42]. Each problem

[2] The matrix computation is implemented with the public domain package LinAlg available from http:// www.lh.com/~oleg/ftp/packages.html

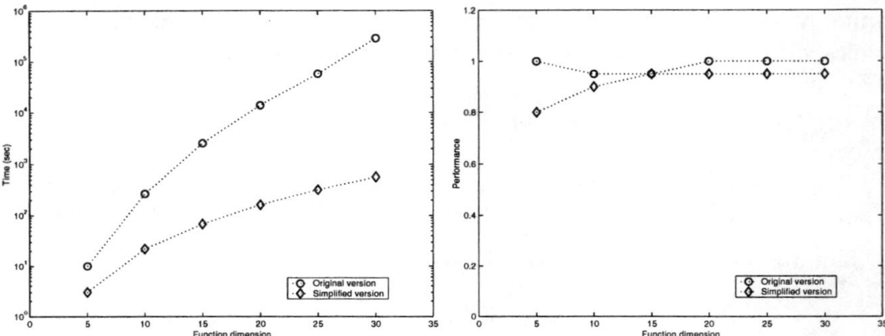

Fig. 18. The difference in computation time (left) and performance (right) between the original and simplified quadratic approximation on solving the generalized Rastrigin function.

for each algorithm is run 50 times. The average number of function evaluations and the best values are recorded. The number of the global minimum found is noted as "global hit."

To compare the three algorithms, we regard the global hits as the primary objective achieved by the algorithm and the number of function evaluations as the second objective.

Table 15. The two additional test functions used in our experimental studies, where n is the dimension of the function, f_{min} is the minimum value of the function, and N_{min} indicates the number of minima. $S \subseteq R^n$.

Test function	n	N_{min}	S	f_{min}
$f_{26}(x) = \sum_{i=1}^{n}(A_i - B_i)^2$	30	2^n	$[-\pi, \pi]^n$	0
$A_i = \sum_{j=1}^{n}(a_{ij} \sin \alpha_j + b_{ij} \cos \alpha_j)$				
$B_i = \sum_{j=1}^{n}(a_{ij} \sin x_j + b_{ij} \cos x_j)$				
$f_{27}(x) = -\sum_{i=1}^{30} c_i[\exp(-\frac{1}{\pi}A) \cdot \cos(\pi A)]$	30	$> 10^n$	$[0, 10]^n$	unknown
$A = \sum_{j=1}^{n}(x_j - a_{ij})^2$				

11.1 Results on f_8–f_{13}, f_{26}, f_{27}

Functions f_8–f_{13}, f_{26}, f_{27} are relatively high-dimensional multimodal functions with many local minima. The number of local minima increases exponentially as the dimension increases. Table 16 summarizes the experimental results for ES, IFEP, and EANA. The numbers in bold indicate the best performance among the three.

For functions f_8, f_9, f_{11}, and f_{27}, EANA outperforms ES and IFEP in terms of effectiveness and/or efficiency. ES is better on functions f_{10} and f_{12},

Table 16. The results of ES, IFEP, and EANA on f_8–f_{13}, f_{26}, f_{27}. "Eval." is the number of evaluations. "G. hit" is the global hits.

F	ES			IFEP			EANA		
	Eval.	Mean	G. hit	Eval.	Mean	G. hit	Eval.	Mean	G. hit
8	241500	-10091.26	0/50	757500	-10640.18	0/50	**102470**	**-12191.21**	**0/50**
9	500000	344.44	0/50	500000	2.89	1/50	**83483**	**1.99e-2**	**49/50**
10	**100000**	**2.66e-10**	**50/50**	150000	6.33e-4	50/50	181578	3.30e-5	50/50
11	100000	0	50/50	176500	1.27e-1	3/50	**9372**	**2.05e-10**	**50/50**
12	**50000**	**1.06e-9**	**50/50**	476245	2.49e-2	40/50	185318	8.14e-11	50/50
13	50000	4.40e-4	48/50	**150000**	**4.42e-8**	**50/50**	349059	1.88e-9	50/50
26	200000	5223.03	0/50	200000	23438.64	0/50	1139316	222.81	0/50
27	200000	-9.20e-3	N/A	200000	-8.55e-2	N/A	**49839**	**-0.24325**	N/A

where it provides better efficiency. IFEP outperforms the others on function f_{13} with a better efficiency than EANA and better effectiveness than ES. Function f_9 is an interesting case which will be discussed in more detail later.

f_{27} is the generalized Langerman function.[3] We do not know the fitness value of its global minimum; hence the global hits are not available.

11.2 Results on f_{14}–f_{25}

The final results of ES, IFEP, and EANA on functions f_{14}–f_{25} are summarized in Table 17. For functions f_{15}–f_{24}, EANA performed better than ES and IFEP in terms of the global hits and/or the number of function evaluations. Function f_{14} is the only case among all 20 benchmark problems where EANA has a much worse global hit than both ES and IFEP. Further discussions about this are given in the next section.

Although EANA has better performance on functions f_{20}–f_{24}, it still misses global optima in some runs. Our preliminary investigation reveals that EANA could lose population diversity quickly and get trapped in a region biased by the approximation method. It appears that using multiple approximation methods in one algorithm might help improve and maintain population diversity during evolution.

11.3 Further Discussions

In general, EANA appears to have a better reliability than ES and IFEP on high-dimensional problems with many local minima and a better efficiency on low-dimensional problems. Part of our future work is to study why it appears

[3] The matrix data (a_{ij}) can be found at http://www.wi.leidenuniv.nl/CS/ALP/alea.html.

Table 17. The results of ES, IFEP, and EANA on f_{14}–f_{25}. "Eval." is the number of evaluations. "G. hit" is the global hits.

F	ES			IFEP			EANA		
	Eval.	Mean	G. hit	Eval.	Mean	G. hit	Eval.	Mean	G. hit
14	**1000**	**4.18**	**33/50**	10000	1.56	33/50	1685	7.36	4/50
15	400000	3.92e-3	44/50	235500	4.17e-4	44/50	**31645**	**3.07e-4**	**50/50**
16	2500	-1.031628	50/50	3500	-1.031628	50/50	**863**	**-1.031628**	**50/50**
17	3000	0.3979	50/50	4500	0.3979	50/50	**945**	**0.3979**	**50/50**
18	3500	3	50/50	4500	3	50/50	**822**	**3**	**50/50**
19	4000	-3.86	50/50	6000	-3.86	50/50	**1297**	**-3.86**	**50/50**
20	10500	-3.31	43/50	17500	-3.26	24/50	**8357**	**-3.3128**	**46/50**
21	5000	-6.71	24/50	10000	-7.02	21/50	**2958**	**-9.6512**	**46/50**
22	5000	-9.14	41/50	9500	-8.42	34/50	**2880**	**-9.1430**	**41/50**
23	5000	-9.32	42/50	7500	-9.09	39/50	**3118**	**-9.5767**	**43/50**
24	4000	1.81e-2	1/50	8500	8.86e-3	4/50	**1181**	**2.97e-2**	**30/50**
25	10000	1.42e-1	48/50	**10000**	**2.47e-3**	**50/50**	1687	3.19e-1	44/50

so. Perhaps a closer look at functions f_9 and f_{14} might help since EANA's performance differed greatly from others.

Figure 19 shows the two-dimensional version of functions f_9 and f_{14}. Initially, we thought that the better performance of EANA on f_9 was brought about by its bigger population size. However, by redoing the experiment with $\mu = 10$, we obtained the same global hit (50/50), and the average number of function evaluations was only 21358. Therefore, we conjecture that the n-dimensional approximation may be a major factor here.

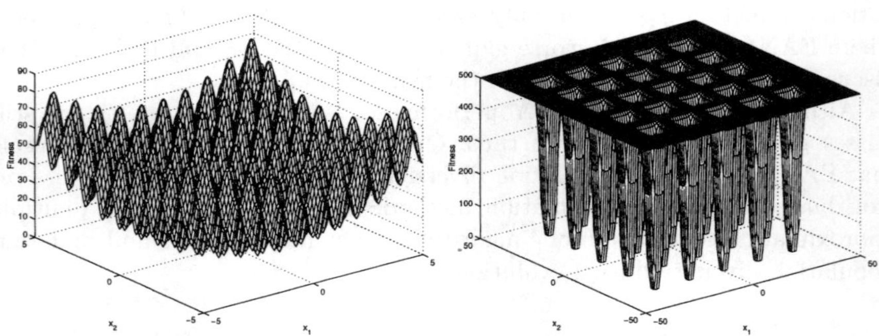

Fig. 19. The two-dimensional landscapes of generalized Rastrigin function f_9 (left) and Shekel's foxholes function f_{14} (right)

Function f_{14} represents a case where EANA's performance was worse than others. This could be caused by the small population size of EANA ($\mu = 5$). By increasing the population sizes of EANA, we did get better performance.

However, the computational effort would be too high for a large population. There are 25 holes in function f_{14} (see Figure 19). Suppose an individual in the initial population of EANA is located in the basin of attraction of the global optimum. Before it reaches the bottom, we may still lose it through selection. If we do not have such an individual, the approximation will be the only mechanism to generate one because we have not yet introduced mutation into our algorithms.

12 Conclusions

This chapter first proposes a fast evolutionary programming algorithm, FEP, and evaluates its performance on a number of benchmark problems. The experimental results show that FEP performs much better than CEP for multimodal functions with many local minima while being comparable to CEP in performance for unimodal and multimodal functions with only a few local minima. Since FEP and CEP differ only in their mutations, it is quite easy to apply FEP to real-world problems. No additional cost was introduced except for the difference in generating a Cauchy random number instead of a Gaussian random number.

The chapter then analyzes FEP and CEP in depth in terms of search step size and neighborhood size, and explains why FEP performs better than CEP for most benchmark problems. The theoretical analysis is supported by the additional empirical evidence in which the range of initial x values was changed. The chapter shows that FEP's long jumps increase the probability of finding a near-optimum when the distance between the current search point and the optimum is large, but decrease the probability when such a distance is small. The chapter also investigates the relationship between the neighborhood size and the probability of finding a near-optimum in this neighborhood. Some insights on evolutionary search and optimization in general have been gained from the above analyses.

The above analyses also led to an improved FEP (IFEP) which is very simple yet effective. IFEP uses the idea of mixing search biases to mix Cauchy and Gaussian mutations. Unlike some switching algorithms which have to decide when to switch between different mutations during search, IFEP does not need to make such a decision and introduces no parameters. IFEP is robust, assumes no prior knowledge of the problem to be solved, and performs at least as well as the better one of FEP and CEP for most benchmark problems. Future work on IFEP includes the comparison of IFEP with other self-adaptive algorithms such as [44] and other evolutionary algorithms using Cauchy mutation [45].

The idea of FEP and IFEP can also be applied to other evolutionary algorithms to design faster optimization algorithms [46]. For $(\mu + \lambda)$ and (μ, λ) evolutionary algorithms where $\mu < \lambda$, IFEP would be particularly attractive since a parent has to generate more than one offspring. It may be beneficial if different offspring are generated by different mutations [46].

This chapter also tackles optimization problems from a very different point of view, i.e., by simplifying a complex problem first before applying an optimization algorithm (i.e., a hybrid evolutionary and local search algorithm). A general framework for designing such algorithms is proposed. Then two concrete algorithms are described and compared with existing algorithms on a set of benchmark functions.

The idea of approximating a complex function by a simpler one can also be regarded as a novel approach to design recombination operators in evolutionary algorithms. Our experimental results indicate that this is a very promising area of future research.

A Appendix: Benchmark Functions

A.1 Sphere Model

$$f_1(x) = \sum_{i=1}^{30} x_i^2$$

$$-100 \leq x_i \leq 100, \quad \min(f_1) = f_1(0,\ldots,0) = 0$$

A.2 Schwefel's Problem 2.22

$$f_2(x) = \sum_{i=1}^{30} |x_i| + \prod_{i=1}^{30} |x_i|$$

$$-10 \leq x_i \leq 10, \quad \min(f_2) = f_2(0,\ldots,0) = 0$$

A.3 Schwefel's Problem 1.2

$$f_3(x) = \sum_{i=1}^{30} \left(\sum_{j=1}^{i} x_j \right)^2$$

$$-100 \leq x_i \leq 100, \quad \min(f_3) = f_3(0,\ldots,0) = 0$$

A.4 Schwefel's Problem 2.21

$$f_4(x) = \max_i \{|x_i|, 1 \leq i \leq 30\}$$

$$-100 \leq x_i \leq 100, \quad \min(f_4) = f_4(0,\ldots,0) = 0$$

A.5 Generalized Rosenbrock's Function

$$f_5(x) = \sum_{i=1}^{29}[100(x_{i+1} - x_i^2)^2 + (x_i - 1)^2]$$

$$-30 \le x_i \le 30, \quad \min(f_5) = f_5(1, \dots, 1) = 0$$

A.6 Step Function

$$f_6(x) = \sum_{i=1}^{30} (\lfloor x_i + 0.5 \rfloor)^2$$

$$-100 \le x_i \le 100, \quad \min(f_6) = f_6(0, \dots, 0) = 0$$

A.7 Quartic Function with Noise

$$f_7(x) = \sum_{i=1}^{30} i x_i^4 + random[0, 1)$$

$$-1.28 \le x_i \le 1.28, \quad \min(f_7) = f_7(0, \dots, 0) = 0$$

A.8 Generalized Schwefel's Problem 2.26

$$f_8(x) = -\sum_{i=1}^{30} \left(x_i \sin\left(\sqrt{|x_i|}\right)\right)$$

$$-500 \le x_i \le 500, \quad \min(f_8) = f_8(420.9687, \dots, 420.9687) = -12569.5$$

A.9 Generalized Rastrigin's Function

$$f_9(x) = \sum_{i=1}^{30}[x_i^2 - 10\cos(2\pi x_i) + 10]$$

$$-5.12 \le x_i \le 5.12, \quad \min(f_9) = f_9(0, \dots, 0) = 0$$

A.10 Ackley's Function

$$f_{10}(x) = -20\exp\left(-0.2\sqrt{\frac{1}{30}\sum_{i=1}^{30} x_i^2}\right) - \exp\left(\frac{1}{30}\sum_{i=1}^{30}\cos 2\pi x_i\right) + 20 + e$$

$$-32 \le x_i \le 32, \quad \min(f_{10}) = f_{10}(0, \dots, 0) = 0$$

A.11 Generalized Griewank Function

$$f_{11}(x) = \frac{1}{4000}\sum_{i=1}^{30} x_i^2 - \prod_{i=1}^{30}\cos\left(\frac{x_i}{\sqrt{i}}\right) + 1$$

$$-600 \leq x_i \leq 600, \quad \min(f_{11}) = f_{11}(0,\dots,0) = 0$$

A.12 Generalized Penalized Functions

$$f_{12}(x) = \frac{\pi}{30}\left\{10\sin^2(\pi y_1) + \sum_{i=1}^{29}(y_i - 1)^2[1 + 10\sin^2(\pi y_{i+1})] + (y_n - 1)^2\right\}$$
$$+ \sum_{i=1}^{30} u(x_i, 10, 100, 4)$$

$$-50 \leq x_i \leq 50, \quad \min(f_{12}) = f_{12}(1,\dots,1) = 0$$

$$f_{13}(x) = 0.1\left\{\sin^2(\pi 3x_1 + \sum_{i=1}^{29}(x_i - 1)^2[1 + \sin^2(3\pi x_{i+1})]\right.$$
$$\left. +(x_n - 1)^2[1 + \sin^2(2\pi x_{30})]\right\} + \sum_{i=1}^{30} u(x_i, 5, 100, 4)$$

$$-50 \leq x_i \leq 50, \quad \min(f_{13}) = f_{13}(1,\dots,1) = 0$$

where

$$u(x_i, a, k, m) = \begin{cases} k(x_i - a)^m, & x_i > a \\ 0, & -a \leq x_i \leq a \\ k(-x_i - a)^m, & x_i < -a \end{cases}$$

$$y_i = 1 + \frac{1}{4}(x_i + 1)$$

A.13 Shekel's Foxholes Function

$$f_{14}(x) = \left[\frac{1}{500} + \sum_{j=1}^{25}\frac{1}{j + \sum_{i=1}^{2}(x_i - a_{ij})^6}\right]^{-1}$$

$$-65.536 \leq x_i \leq 65.536, \quad \min(f_{14}) = f_{14}(-32,-32) \approx 1$$

where

$$(a_{ij}) = \begin{pmatrix} -32 & -16 & 0 & 16 & 32 & -32 & \cdots & 0 & 16 & 32 \\ -32 & -32 & -32 & -32 & -32 & -16 & \cdots & 32 & 32 & 32 \end{pmatrix}$$

A.14 Kowalik's Function

$$f_{15}(x) = \sum_{i=1}^{11} \left[a_i - \frac{x_1(b_i^2 + b_i x_2)}{b_i^2 + b_i x_3 + x_4} \right]^2$$

$$-5 \le x_i \le 5, \quad \min(f_{15}) \approx f_{15}(0.1928, 0.1908, 0.1231, 0.1358) \approx 0.0003075$$

Table 18. Kowalik's function f_{15}

i	a_i	b_i^{-1}
1	0.1957	0.25
2	0.1947	0.5
3	0.1735	1
4	0.1600	2
5	0.0844	4
6	0.0627	6
7	0.0456	8
8	0.0342	10
9	0.0323	12
10	0.0235	14
11	0.0246	16

A.15 Six-hump Camel-back Function

$$f_{16} = 4x_1^2 - 2.1x_1^4 + \frac{1}{3}x_1^6 + x_1 x_2 - 4x_2^2 + 4x_2^4$$

$$-5 \le x_i \le 5$$

$$x_{min} = (0.08983, -0.7126), \ (-0.08983, 0.7126)$$

$$\min(f_{16}) = -1.0316285$$

A.16 Branin Function

$$f_{17}(x) = \left(x_2 - \frac{5.1}{4\pi^2}x_1^2 + \frac{5}{\pi}x_1 - 6 \right)^2 + 10\left(1 - \frac{1}{8\pi}\right)\cos x_1 + 10$$

$$-5 \le x_1 \le 10, \quad 0 \le x_2 \le 15$$

$$x_{min} = (-3.142, 12.275), \ (3.142, 2.275), \ (9.425, 2.425)$$

$$\min(f_{17}) = 0.398$$

A.17 Goldstein–Price Function

$$\begin{aligned} f_{18}(x) = {} & [1 + (x_1 + x_2 + 1)^2(19 - 14x_1 + 3x_1^2 - 14x_2 + 6x_1x_2 + 3x_2^2)] \\ & \times [30 + (2x_1 - 3x_2)^2(18 - 32x_1 + 12x_1^2 + 48x_2 - 36x_1x_2 + 27x_2^2)] \end{aligned}$$

$$-2 \le x_i \le 2, \quad \min(f_{18}) = f_{18}(0, -1) = 3$$

A.18 Hartman's Family

$$f(x) = -\sum_{i=1}^{4} c_i \exp\left[-\sum_{j=1}^{n} a_{ij}(x_j - p_{ij})^2\right]$$

with $n = 3, 6$ for $f_{19}(x)$ and $f_{20}(x)$, respectively, $0 \le x_j \le 1$. The coefficients are defined by Tables 19 and 20, respectively.

Table 19. Hartman function f_{19}

i	$a_{ij}, j = 1, 2, 3$			c_i	$p_{ij}, j = 1, 2, 3$		
1	3	10	30	1	0.3689	0.1170	0.2673
2	0.1	10	35	1.2	0.4699	0.4387	0.7470
3	3	10	30	3	0.1091	0.8732	0.5547
4	0.1	10	35	3.2	0.038150	0.5743	0.8828

Table 20. Hartman function f_{20}

i	$a_{ij}, j = 1, \ldots, 6$						c_i	$p_{ij}, j = 1, \ldots, 6$					
1	10	3	17	3.5	1.7	8	1	0.1312	0.1696	0.5569	0.0124	0.8283	0.5886
2	0.05	10	17	0.1	8	14	1.2	0.2329	0.4135	0.8307	0.3736	0.1004	0.9991
3	3	3.5	1.7	10	17	8	3	0.2348	0.1415	0.3522	0.2883	0.3047	0.6650
4	17	8	0.05	10	0.1	14	3.2	0.4047	0.8828	0.8732	0.5743	0.1091	0.0381

For $f_{19}(x)$ the global minimum is equal to -3.86 and it is reached at the point $(0.114, 0.556, 0.852)$. For $f_{20}(x)$ the global minimum is -3.32 at the point $(0.201, 0.150, 0.477, 0.275, 0.311, 0.657)$.

A.19 Shekel's Family

$$f(x) = -\sum_{i=1}^{m} [(x - a_i)(x - a_i)^{\mathrm{T}} + c_i]^{-1}$$

Table 21. Shekel functions f_{21}, f_{22}, f_{23}

i	$a_{ij}, j = 1, \ldots, 4$				c_i
1	4	4	4	4	0.1
2	1	1	1	1	0.2
3	8	8	8	8	0.2
4	6	6	6	6	0.4
5	3	7	3	7	0.4
6	2	9	2	9	0.6
7	5	5	3	3	0.3
8	8	1	8	1	0.7
9	6	2	6	2	0.5
10	7	3.6	7	3.6	0.5

with $m = 5, 7$, and 10 for $f_{21}(x)$, $f_{22}(x)$, and $f_{23}(x)$, respectively, $0 \leq x_j \leq 10$.

These functions have 5, 7, and 10 local minima for $f_{21}(x)$, $f_{22}(x)$, and $f_{23}(x)$, respectively. $x_{local\_opt} \approx a_i$, $f(x_{local\_opt}) \approx 1/c_i$ for $1 \leq i \leq m$. The coefficients are defined by Table 21.

References

1. Yao, X. (1996) An overview of evolutionary computation. *Chinese Journal of Advanced Software Research (Allerton Press, Inc., New York, NY 10011)*, **3**, 12–29
2. Kirkpatrick, S., Gelatt, C.D., Vecchi, M.P. (1983) Optimization by simulated annealing. *Science*, **220**, 671–680
3. Szu, H.H., Hartley, R.L. (1987) Fast simulated annealing. *Physics Letters A*, **122**, 157–162
4. Ingber, L. (1989) Very fast simulated re-annealing. *Mathl. Comput. Modelling*, **12**, 967–973
5. Yao, X. (1995) A new simulated annealing algorithm. *Int. J. of Computer Math.*, **56**, 161–168
6. Grefenstette, J.J. (1987) Incorporating problem specific knowledge into genetic algorithms. In L. Davis, editor, *Genetic Algorithms and Simulated Annealing*, chapter 4, 42–60. Morgan Kaufmann, San Mateo, CA
7. Fogel, D.B. (1991) *System Identification Through Simulated Evolution: A Machine Learning Approach to Modeling.* Ginn Press, Needham Heights, MA
8. Fogel, D.B. (1992) *Evolving Artificial Intelligence.* PhD thesis, University of California, San Diego, CA
9. Fogel, D.B. (1993) Applying evolutionary programming to selected traveling salesman problems. *Cybernetics and Systems*, **24**, 27–36
10. Yao, X., Liu, Y., Lin, G. (1999) Evolutionary programming made faster. *IEEE Transactions on Evolutionary Computation*, **3**, 82–102
11. Fogel, D.B. (1994) An introduction to simulated evolutionary optimisation. *IEEE Trans. on Neural Networks*, **5**, 3–14

12. Bäck, T., Schwefel, H.P. (1993) An overview of evolutionary algorithms for parameter optimization. *Evolutionary Computation*, **1**, 1–23
13. Fogel, D.B. (1995) *Evolutionary Computation: Towards a New Philosophy of Machine Intelligence.* IEEE Press, New York, NY
14. Gehlhaar, D.K., Fogel, D.B. (1996) Tuning evolutionary programming for conformationally flexible molecular docking. In L.J. Fogel, P.J. Angeline, and T. Bäck, editors, *Evolutionary Programming V: Proc. of the Fifth Annual Conference on Evolutionary Programming*, 419–429. MIT Press, Cambridge, MA
15. Feller, W. (1971) *An Introduction to Probability Theory and Its Applications*, volume 2. John Wiley & Sons, 2nd edition
16. Yao, X., Liu, Y. (1996) Fast evolutionary programming. In L. J. Fogel, P. J. Angeline, and T. Bäck, editors, *Evolutionary Programming V: Proc. of the Fifth Annual Conference on Evolutionary Programming*, 451–460, MIT Press, Cambridge, MA.
17. Törn, A., Žilinskas, A. (1989) *Global Optimisation.* Lecture Notes in Computer Science, **350**. Springer-Verlag, Berlin.
18. Schwefel, H.P. (1995) *Evolution and Optimum Seeking.* John Wiley & Sons, New York
19. Wolpert, D.H., Macready, W.G. (1995) No free lunch theorems for search. Technical Report SFI-TR-95-02-010, Santa Fe Institute, 1399 Hyde Park Road, Santa Fe, NM 87501, USA
20. Wolpert, D.H., Macready, W.G. (1997) No free lunch theorems for optimization. *IEEE Transactions on Evolutionary Computation*, **1**, 67–82
21. Devroye, L. (1986) *Non-Uniform Random Variate Generation.* Springer-Verlag, New York, NY
22. Hunt, R.A. (1986) *Calculus with Analytic Geometry.* Harper & Row, New York, NY
23. Yao, X. (1994) Introduction. *Informatica (Special Issue on Evolutionary Computation)*, **18**, 375–376
24. Chellapilla, K. (1998) Combining mutation operators in evolutionary programming. *IEEE Transactions on Evolutionary Computation*, **2**, 91–96
25. Törn, A.A. (1978) A search-clustering approach to global optimization. In L.C.W. Dixon and G.P. Szegö, editors, *Towards Global Optimization 2*, 49–62, North-Holland, Amsterdam
26. Rinnooy Kan, A.H.G., Timmer, G.T. (1987) Stochastic global optimization methods part II: Multi level methods. *Mathematical Programming*, **39**, 57–78
27. Dixon, L.C.W., Szegö, G.P. (1978) The global optimization problem: An introduction. In L.C.W. Dixon and G.P. Szegö, editors, *Towards Global Optimization 2*, 1–15, Amsterdam. North-Holland
28. Ali, M.M., Storey, C., Törn, A. (1997) Application of stochastic global optimization algorithms to practical problems. *Journal of Optimization and Application*, **95**, 545–563
29. Whitley, D., Gordon, V.S., Mathias, K. (1994) Lamarkian evolution, the baldwin effect and function optimization. In Y. Davidor, H.-P. Schwefel, and R. Männer, editors, *Parallel Problem Solving from Nature-PPSN III*, Lecture Notes in Computer Science, **866**, 6–15, Springer-Verlag, Berlin.
30. Hart, W.E., Belew, R.K. (1996) Optimization with genetic algorithm hybrids that use local search. In R.K. Belew and M. Mitchell, editors, *Adaptive Individuals in Evolving Populations: Models and Algorithms*, volume 26 of *SFI Studies in the Sciences of Complexity*, 483–496, Addison-Wesley, Reading, MA.

31. Baldwin, J.M. (1896) A new factor in evolution. *American Naturalist*, **30**, 441–451
32. Hinton, G.E., Nolan, S.J. (1987) How learning can guide evolution. *Complex Systems*, **1**, 495–502
33. Powell, M.J.D. (1964) An efficient method for finding the minimum of a function of several variables without calculating derivatives. *The Computer Journal*, **7**, 155–162
34. Nelder, J.A., Mead, R. (1965) A simplex method for function minimization. *The Computer Journal*, **7**, 308–313
35. Powell, M.J.D. (1994) A direct search optimization method that models the objective and constraint functions by linear interpolation. In S. Gomez and J.-P. Hennart, editors, *Advances in Optimization and Numerical Analysis, Proceedings of the Sixth Workshop on Optimization and Numerical Analysis, Oaxaca, Mexico*, **275**, 51–67, Kluwer Academic, Dordrechth, NL.
36. H.-M. Voigt and J.M. Lange. Local evolutionary search enhancement by random memorizing. In *Proceedings of the 1998 IEEE International Conference on Evolutionary Computation (ICEC'98)*, 547–552, Piscataway, NJ, 1998. IEEE Press.
37. Winfield, D. (1973) Function minimization by interpolation in a data table. *Journal of the Institute of Mathematics and its Applications*, **12**, 339–347
38. Powell, M.J.D. (1994) A direct search optimization method that models the objective by quadratic interpolation. Presentation at the 5th Stockholm Optimization Days
39. Conn, A.R., Toint, Ph.L. (1996) An algorithm using quadratic interpolation for unconstrained derivative free optimization. In G. Di Pillo and F. Gianessi, editors, *Nonlinear Optimization and Applications*, 27–47, Plenum Publishing, New York.
40. Liang, K. H., Yao, X., Newton, C. (1999) Combining landscape approximation and local search in global optimization. In *Proceedings of the 1999 Congress on Evolutionary Computation*, **2**, 1514–1520, IEEE Press, Piscataway, NJ.
41. Schaffer, J. D., Caruana, R. A., Eshelman, L. J., Das, R. (1989) A study of control parameters affecting online performance of genetic algorithms for function optimization. In J.D. Schaffer, editor, *Proceedings of the third International Conference on Genetic Algorithms (ICGA'89)*, 51–60, Morgan Kaufmann, San Mateo, CA.
42. Yao, X., Lin, G., Liu, Y. (1997) An analysis of evolutionary algorithms based on neighbourhood and step sizes. In P.J. Angeline, R.G. Reynolds, J.R. McDonnell, and R. Eberhart, editors, *Evolutionary Programming VI: Proc. of the Sixth Annual Conference on Evolutionary Programming*, Lecture Notes in Computer Science, **1213**, 297–307, Springer, Berlin.
43. Bäck, Th., Eiben, A. E. (1999) Generalizations of intermediate recombination in evolution strategies. In *Proceedings of the 1999 Congress on Evolutionary Computation*, **2**, 1566–1573, IEEE Press, Piscataway, NJ.
44. Born, J. (1996) An evolution strategy with adaptation of the step sizes by a variance function. In H.-M. Voigt, W. Ebeling, I. Rechenberg, and H.-P. Schwefel, editors, *Parallel Problem Solving from Nature (PPSN) IV*, Lecture Notes in Computer Science, **1141**, 388–397, Springer-Verlag, Berlin.
45. Kappler, C. (1996) Are evolutionary algorithms improved by large mutations? In H.-M. Voigt, W. Ebeling, I. Rechenberg, and H.-P. Schwefel, editors, *Parallel*

Problem Solving from Nature (PPSN) IV, Lecture Notes in Computer Science, **1141**, 346–355, Springer-Verlag, Berlin.
46. Yao, X., Liu, Y. (1997) Fast evolution strategies. *Control and Cybernetics*, **26**, 467–496

Visualizing Evolutionary Computation

Trevor D. Collins

Knowledge Media Institute
The Open University
Milton Keynes, MK7 6AA
UK
E-mail: t.d.collins@open.ac.uk

Summary. Visualization is an interdisciplinary area drawing on work done in a variety of areas, such as graphic design, cinematography, typography, animation, and human computer interaction. During the last few years there has been an increasing trend within the evolutionary computation community to apply visualization techniques to facilitate people's understanding and use of evolutionary algorithms. This chapter is intended to provide an introductory review of the existing work done on visualizing evolutionary computation and identify some of the key directions for future research.

1 Introduction

Evolutionary algorithms (EAs) search large problem spaces by making gradual improvements to a set of possible solutions. The problem with EAs is that there is no single point during the algorithm's run that can be held responsible for the outcome; the solutions emerge during the course of the algorithm's iterations. This results in a fundamental credit assignment problem for EA users, i.e. if good solutions are found what proportion of the credit should be attributed to the individual components of the algorithm's design?

This problem is further compounded by the fact that the user is unable to see the EA's search behavior. EA users commonly examine how the quality of the solutions found by their algorithm changes over time using a fitness graph. Although such graphs illustrate the improvements in the quality of the solutions considered during the algorithm's run, they do not illustrate anything about the structure of the solutions being considered, or the regions of the search space being explored.

The motivation for applying visualization techniques to evolutionary computation (EC), is to address this fundamental design problem. By enabling users to see the search behavior of their algorithms, they can then begin to attribute credit to the individual designs and judge the quality of each algorithm based on its ability to explore the problem space.

The purpose of this chapter is two fold, firstly to give an introduction to the existing work done on visualizing EC (Section 2), and secondly to discuss the contributions made by this work and the challenges remaining (Section 3).

2 EC Visualization

The application of visualization techniques to support people's understanding and use of EAs has been receiving increasing attention from the EC community during the last few years [9]. By allowing EC users to observe and interact with their EAs it is hoped that a better understanding of the behavior of EAs will be achieved.

Genetic algorithms (GAs) are the most well known and most widely used form of EC [12]. Perhaps it is not surprising then that the most common form of EC visualization has been the visualization of GAs. However, that is not to say that GA visualizations are necessarily inappropriate for evolutionary programming (EP) or evolution strategies (ESs). The appropriateness of any given visualization must be judged by the type of information that is being illustrated. For example, fitness graphs are appropriate for illustrating the progress of any EA that holds information regarding the fitness ratings of the individual solutions considered during evolution.

The following review of EC visualization is structured by the following topics:

1 The operation of the algorithm's component parts.
2 The quality of the solutions found by the algorithm.
3 The individuals'[1] representation in the encoding domain.
4 The individuals' representation in the problem domain.
5 The algorithm's sampling of the search space.
6 Navigating the algorithm's execution.
7 Editing the algorithm.

2.1 The Operation of the Algorithm's Component Parts

Visualizing the operation of the algorithm's component parts, i.e. the actions of the algorithm's selection and reproduction operators, can be done by using either a static or dynamic illustration. Static illustrations benefit from being easy to present on paper as well as the computer screen, although dynamic illustrations (i.e. animations) are often more effective and more engaging.

An example of a static illustration of a GA's components is the "internals window" available in the graphical user interface for genetic algorithms (GIGA) [10]. This view illustrates the internal operations of the GA, such as the crossover and mutation operators (see Figure 1).

A sample dynamic algorithm animation of a GA was produced by David Brogan using a software visualization (SV) system called TANGO [37]. A screen shot is shown in Figure 2 depicting the best chromosome's phenotype (top) and all three of the population's genotypes (middle left) for a GA

[1] The term 'individual' is used here to refer to the individual solutions considered by an EA. These are more commonly referred to as chromosomes in GAs and machines in EP; however, individual is used here as a generic term.

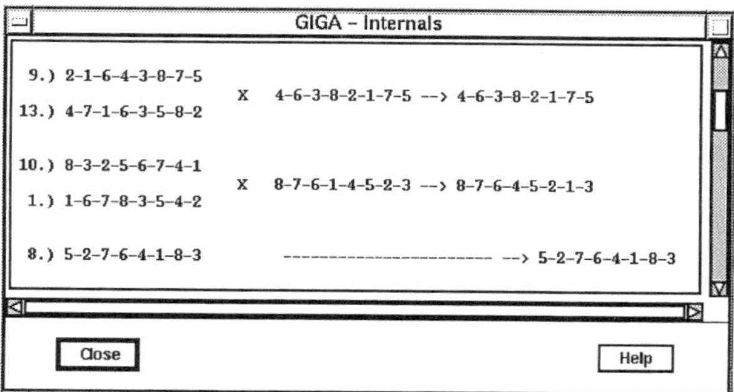

Fig. 1. The internals window available in GIGA for showing the actions of the reproduction operators of a GA. This example was taken from [10, page 8].

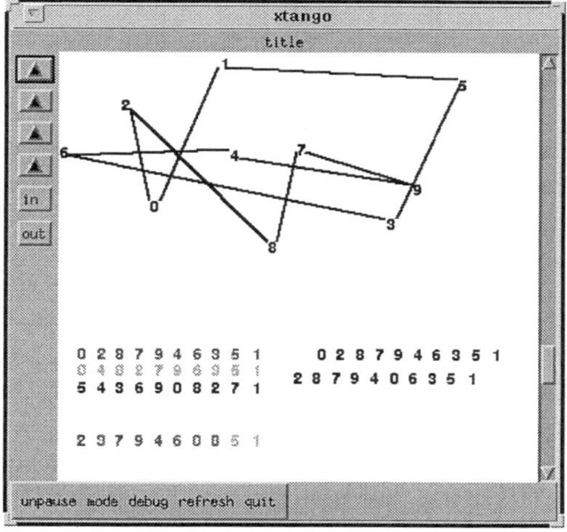

Fig. 2. An XTANGO visualization illustrating a GA with a population containing three decimal valued chromosomes. The upper section illustrates the phenotype data, i.e. the traveling salesman problem, the lower section shows an animated view of the genotype data (i.e. the decimal chromosomes) and the actions of the selection and reproduction operators used in the GA.

solving the traveling salesman problem. An algorithm animation, shown in the bottom half of the window, illustrates the actions of the GA's genetic operators. Brogan's illustrative example is included in the example visualizations supplied with the X Windows version of TANGO, available via ftp from `ftp.cc.gatech.edu` (see subdirectory `/pub/people/stasko`).

2.2 The Quality of the Solutions Found by the Algorithm

Examining the quality of the solutions found is an important part of applying an EA. Monitoring the algorithm's progress can be used to inform the user's decision to end the algorithm's run, or as a post-mortem technique for illustrating the run. This subsection presents a variety of techniques for showing the quality of an EA's solutions, including both summaries and complete accounts of the entire run's results, as well as the results within individual populations.

2D Fitness Graphs

The standard method for presenting a summary of the entire run's results is to plot some aspect of the population's fitness ratings for each generation. These visualizations are commonly referred to as "fitness graphs." Fitness graphs first appeared in one of the earliest papers on simulated evolution written by A.S. Fraser in 1957 [17]. Fraser used 2D line graphs to illustrate the changes in the population's average phenotype fitness value over successive generations.

A variety of fitness graphs are commonly used today. Some of these are used to illustrate an EA's "online" and "offline" fitness ratings (i.e. the mean fitness rating, and mean current-best fitness rating across all generations [11]), as well as the best and worst fitness ratings in each population; see [18], [16], or [1] for examples.

3D Fitness Graphs

Although the 2D fitness graph is the most commonly used representation of an EA's entire run, it is incomplete in that it only provides an indication of the individuals' fitness ratings in each population, rather than the actual fitness ratings of all the individuals. An example of a 3D fitness graph was presented in [20]. This provides a more complete view than the 2D fitness graph as it shows the fitness ratings of *every* chromosome in a fitness-ordered population (see Figure 3).

The 3D fitness graph presents all the chromosomes' fitness ratings, but if displayed as a static view some sections of the lines may be hidden by earlier and fitter line sections. A common solution to this problem is to let the user control the viewing position by rotating the 3D image about its own axes. Another point to be noted regarding the 3D fitness graph is that the individual lines do not refer to the same chromosomes, rather they refer to chromosomes at the same position in the fitness-ordered population across different generations.

Alternative Plots

Rather than examining the quality of the solutions found during the course of an EA's run, a number of visualizations for illustrating the fitness ratings

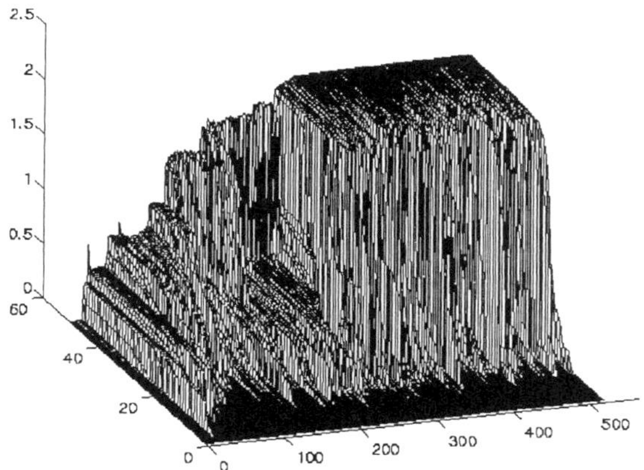

Fig. 3. An example of a 3D fitness graph. The fitness rating of each individual in the population is plotted over each generation. The fitness ratings are plotted on the y axis ($y = 0$ to 2.5), the position of each chromosome in the fitness-ordered population is plotted on the x axis ($x = 0$ to 50), and the generation number is plotted on the z axis ($z = 0$ to 522). This figure was taken from [20].

of the individuals within a single generation have been proposed in [5]. These visualizations make use of block diagrams, colour maps, bar charts, radial line graphs, and radial point plots. In each of these representations blocks, bars, lines, or points are used to represent the individuals in a population, with variations in size and/or color being used to indicate the relative fitness of each individual. The image components are ordered either by fitness, or similarity to the fittest, the latter being used to give an indication of the number of distinct solutions in the population. Examples of these views are available in [31] and [7].

2.3 The Individuals' Representation in the Encoding Domain

For a GA, a text-based print-out of the individuals' genotypes is usually given for a subset of the population, e.g. by displaying the best individual, or the top five individuals, in each generation. Although displaying the genotype of a few individuals per generation gives the user an indication of the solutions currently being considered, it is impossible for the user to view every individual from every generation there is simply too much information for the user to deal with. As a result, several systems have been developed using graphical visualizations to represent this information in a more manageable form.

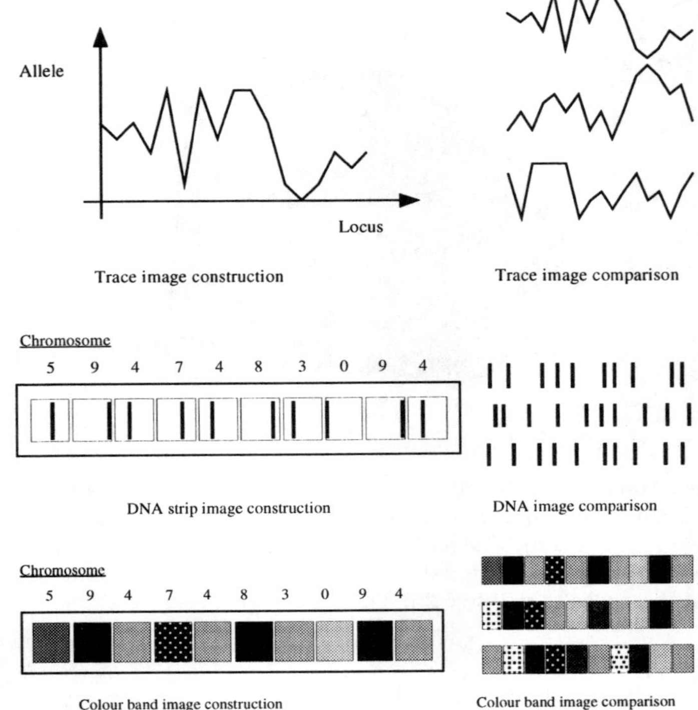

Fig. 4. Three example chromosome icons showing the design of line trace, DNA strip, and color band icons. This figure was taken from [5] where texture was used to indicate the different colors on the color band icon.

Chromosome Icons

Three chromosome icons were introduced in [5] for illustrating a GA's genotypic chromosomes: the "trace icon," "DNA strip," and "color strip" (see Figure 4). A "trace icon" is a 2D line trace of the allele held at each locus in the chromosome. The variation in the vertical position of the trace at each line segment indicates the allele's position in the coding alphabet for each locus. A "DNA strip" is a 2D line plot showing each allele as a vertical bar; the horizontal position of the bar indicates the allele's position in the coding alphabet. Thirdly, a "color strip" icon shows the allele held at each chromosome locus as a colored block; the color of each block indicates the allele's position in the coding alphabet.

Pixel-oriented Visualizations

Bill Spears at the US Naval Research Lab has also explored the use of visualization within GAs [36]. In order to illustrate the chromosomes in a specific

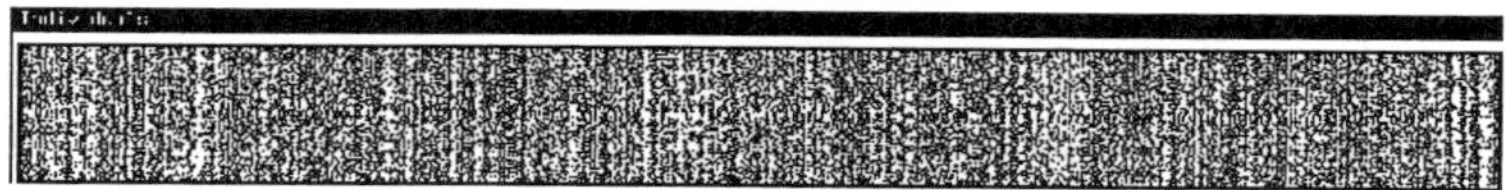

Fig. 5. A high-dimensional visualization showing a population of 100 1008-bit binary chromosomes as black and white pixels. The entire population is shown here in 100 rows; each chromosome is shown as a single row of 1008 pixels. Figure taken from [36].

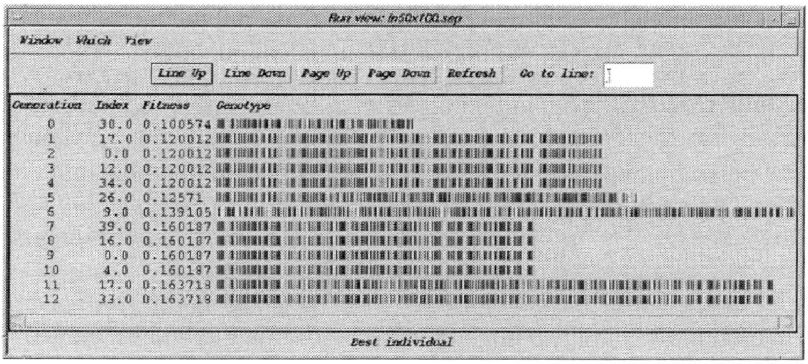

Fig. 6. An example of a VIS "run window," illustrating the best individual from each generation using a color-coded representation. This monochrome figure was produced from the original color image taken from [39].

population, Spears suggested illustrating the alleles in a population of binary chromosomes as black and white pixel dots (see Figure 5). The resulting pixel-oriented visualization shows a random set of black and white pixels for the initial population with patterns of vertical black and white lines forming during the GA's run indicating common genes between neighboring chromosomes.

Although developed separately, the pixel-based genotype visualization proposed by Spears is similar to the color strip icon proposed in [5]. Spears' representation can illustrate longer genotypes than the color strip icon in the same amount of screen space; however, the legibility of each pixel point would be poorer than the legibility of each colored block. For any specific application the purpose of the visualization should be used to determine the balance between screen economics and image legibility (see [2, pages 175 to 190] for guidelines on image legibility). The purpose of Spears' pixel-oriented visualization is to help people spot emerging patterns within the population, whereas the purpose of the color strip icon was to directly illustrate the alleles in each chromosome's genotype.

Another, more recent project at the US Naval Research Lab has been exploring the use of GAs for modeling viruses (the "Virtual Virus" project

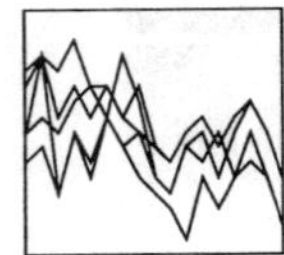

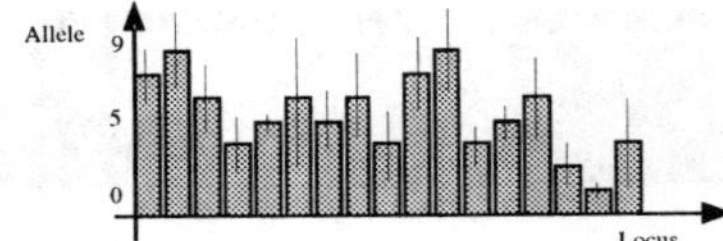

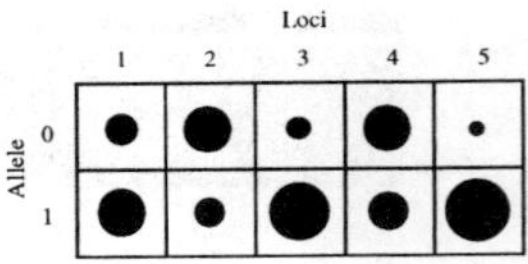

Fig. 7. Three example genotype visualizations: "overlaid chromosome icons" (left), a "population bar chart" (middle), and an "allele versus locus frequency matrix" (right). These three chromosome visualizations are taken from [5].

[19]). As part of this project an offline (post-mortem) visualization tool called VIS has been developed to support the detailed analysis of a GA's run [39]. VIS presents three different perspectives on a GA's run. *Run windows* display information on the entire run (typically showing one entry per generation, see Figure 6). *Population windows* display single individuals from a single generation. Thirdly, *Individual windows* display information about a single individual. Within VIS multiple windows can be viewed simultaneously.

Five different genotype representations are available in VIS, namely: "text," "zebra," "neapolitan," "color coded," and "gene location" representations. The representation used within any of the windows can be changed at any time via the "Views" menu. The *text* representation displays the individuals using a fixed-width type font. The *zebra* representation displays binary chromosomes as strips of black and white bars, like a zebra's stripes. The *neapolitan* representation displays every pair of binary alleles as a colored bar, where 00 = black, 11 = white, 01 = magenta, and 10 = orange. The *color-coded* representation is used to illustrate multi-letter alphabets (i.e. coding alphabets with more than two symbols), where each unique letter is shown by a different colored bar (e.g. A = blue, C = red, G = yellow, and T = green). Finally, the *gene location* representation can be used to highlight the occurrence of building blocks (i.e. groups of symbols or partial solutions); different colored strips are used to identify different building blocks.

Summarizing the Individuals' Values

Although it is easier to identify trends within the population using an iconic representation rather than printed text both printed text and icons present the same amount of information and, therefore, suffer from the same drawback. That is, when applied to large populations they contain too much information for the user to deal with. As a solution to this three composite representations for summarizing the values of a GA's chromosomes were proposed in [5]: "overlaid line trace icons," "population bar charts," and "allele versus locus frequency matrices" (see Figure 7).

The *overlaid line trace icons* representation is produced by plotting an enlarged version of every chromosome's line trace icon on the same set of axes

(Figure 7, left). The composite image indicates the allele diversity at each locus within the population, by the number of vertically aligned separate line segments. For large (i.e. most practical) population sizes the overlaid chromosome icon representation becomes overloaded and difficult to read. Although the line trace icons identify each chromosome and its alleles, they do not indicate the frequency of each chromosome (or chromosome building block). Therefore, although the user *can* see when the population has completely converged at a specific locus, they *cannot* see the diversity of the population prior to that time. For example, a population containing equal numbers of two different chromosomes would look the same as a population that contained 90% of one chromosome and 10% of another. The *population bar chart* summarizes the alleles that are present within the chromosomes in the population (Figure 7, middle). Each bar indicates the alleles present at each locus; the height of the bar is used to indicate the most frequent allele at that locus. Vertical lines are added to indicate the minimum and maximum allele values at each locus for the current population. Although this gives an indication of the population's diversity, like the overlaid chromosome icon representation it does not illustrate the distribution of the alleles. As a result, the user is no better informed about the diversity of the chromosomes in the population. Thirdly, *allele versus locus frequency matrices* illustrate the distribution of the allele within a population (Figure 7, right). By viewing the frequency matrices of subsequent generations the user can see how the allele's distribution varies during the GA's run. This shows both the convergence and diversity of the alleles. However, it does not show any information regarding the local structure of the alleles within each chromosome. The allele versus locus frequency matrix gives a clear summary of the distribution of alleles and is perhaps the clearest of the three genotype summary representations proposed in [5].

2.4 The Individuals' Representation in the Problem Domain

Domain-specific visualizations are a very effective way of illustrating the solutions considered by an EA. Several education-oriented GA tools illustrate a GA's phenotypes for the traveling salesman problem. Examples include the "best individual window" in GIGA (Figure 8, left), and the phenotype view presented in the XTANGO sample GA visualization (Figure 2).

Although domain visualizations of an EA's individuals can produce a very salient illustration of the solutions being considered they are specific to the problem being solved and therefore, as new problems are attempted, new views must be produced. If the effort involved in producing the view is perceived to be greater than the potential benefit of using the view, then the user will be disinclined to produce new views.

This "ease of production" threshold is a serious problem for SV. Producing any new visualization requires some form of programming. The important

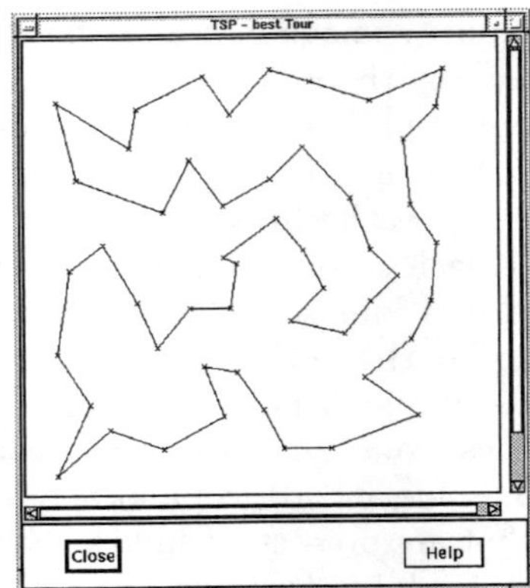

Fig. 8. An example GIGA visualization of a GA's current best solution to the traveling salesman problem

issue here is to ensure that the programming involved is sufficient to fully express what the user needs, whilst remaining at a sufficient level of abstraction, such that the user does not get deterred by technically demanding graphics programming.

One of the primary goals of producing an SV development environment, such as BALSA [4], TANGO [38], ZEUS [3], or VIZ [13], is to facilitate the development of new views. Although a great deal of work has been done in SV, establishing a sufficient level of expressiveness whilst maintaining ease of use remains a difficult trade-off (see [29]).

2.5 The Algorithm's Sampling of the Search Space

The term "search space" is used repeatedly in this chapter to refer to the complete set of all combinations available within any given coding alphabet. Exploring an EA's sampling (i.e. searching) of that space is one way of viewing the algorithm's behavior. This section describes some of the available search space visualization techniques.

2D and 3D Fitness Landscapes

In addition to his genotype visualization tool, Bill Spears also produced two visualization tools to illustrate the GA's sampling of the search space: one for 1D problems and a second for 2D problems [36]. The first tool uses a 2D line graph to illustrate the fitness rating (plotted on the y axis) of each chromosome (plotted on the x axis). The second tool adopts a similar approach but uses a 3D plot to show the variation in fitness for 2D fitness functions (see Figure 9).

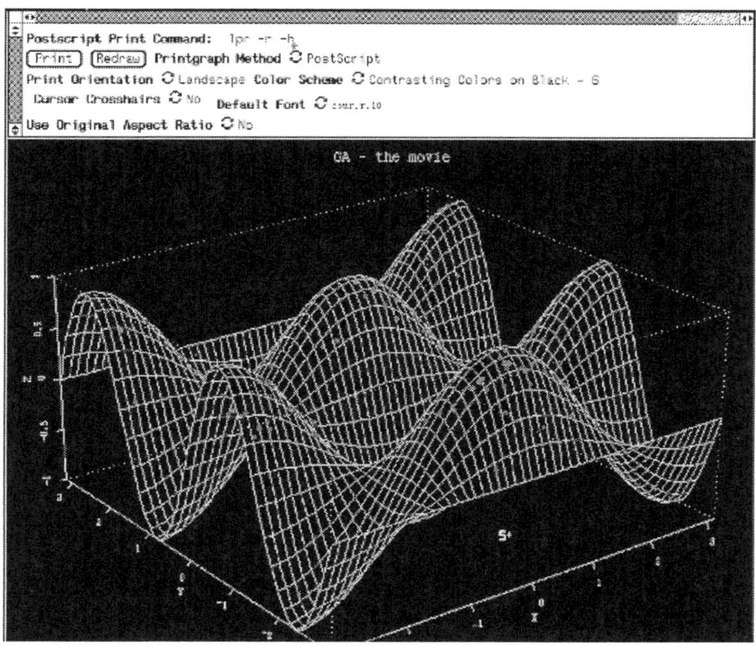

Fig. 9. A 3D surface plot showing the fitness surface for a 2D search space. In the visualization the chromosomes from old generations are shown as blue dots and the chromosomes in the current generation are shown as red dots. This monochrome figure was produced from the original color image taken from [36].

In the 3D visualization the individual chromosomes are shown as points on a 3D fitness surface. As the GA evolves, old chromosomes from previous generations are drawn as blue dots and chromosomes created in the current generation are drawn as red dots. Both of these tools illustrate the GA's sampling of the search space by explicitly plotting a line or surface showing the fitness landscape (i.e. the complete search space with its associated fitness ratings) and highlighting the population's sampling points. However, this approach is not possible for problems in which the fitness landscape (i.e. the fitness rating for every possible chromosome) is unknown. Around the same time Nassersharif, Ence, and Au from the University of Nevada, Las Vegas, were working on another 3D visualization of a GA's fitness landscape [25]. As with Spears' second tool, Nassersharif et al. visualized GAs solving 2D problems. However, in this case the problem space is plotted as a 3D scatterplot in which the two problem dimensions are plotted on the x and z axes, with the corresponding fitness ratings plotted on the y axis (Figure 10). Rather than illustrating the entire fitness surface and then highlighting the GA's sampling of it, Nassersharif et al. used 3D scatterplot visualizations to

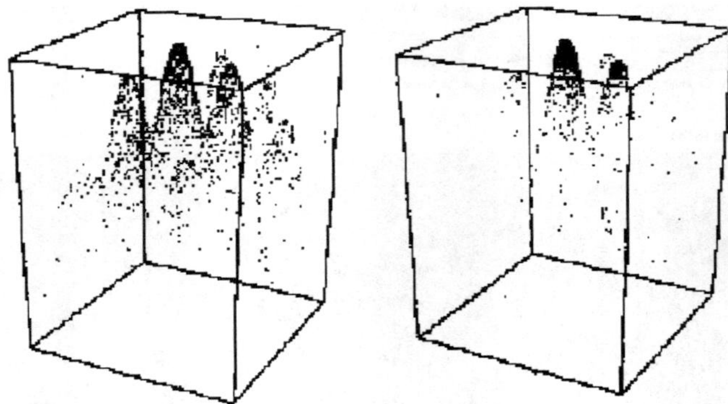

Fig. 10. Nassersharif, Ence, and Au's scatterplot visualization for GAs solving 2D problems. This figure (taken from [25, page I-564]) shows scatterplots for generation 0 (left) and generation 10 (right). The x and z axes illustrate the two problem dimensions and the vertical y axis illustrates the fitness ratings; note the convergence towards fitter solutions shown in generation 10.

show only the population's sample points, i.e. the population's chromosomes without the fitness surface.

Similarity Metrics

As noted in both Spears [36] and Nassersharif et al. [25], GAs are not typically applied to 1D or 2D problems; they are more often applied to high-dimensional problems whose search space cannot be directly illustrated in 2D or 3D space. Therefore, a number of people have explored similarity metrics for illustrating the GA's sampling of high-dimensional search spaces.

In [5] a 2D scatterplot was used to illustrate the distribution of a population's chromosomes. In this view each chromosome in the population is represented by either a dot or a chromosome icon. The coordinate of each chromosome indicates some problem-specific data measure. For example, a GA's chromosomes could be shown using the individuals' fitness ratings and similarity to the fittest chromosome. Selecting an informative similarity measure is the key to such a view's effectiveness.

Similarity metrics have been used as a means for judging problem complexity and population diversity. For example, Terry Jones and Stephanie Forrest have explored the correlation between the fitness values of all the chromosomes in a GA's run and the chromosomes' similarity to the final solution (measured by either the Hamming distance for binary chromosomes or the Euclidean distance for non-binary chromosomes). The resulting measure of problem complexity is referred to as the "fitness distance correlation" [21]. Another example would be Simon Ronald's work on distance functions for order-based encodings. These can be used to measure the genotypic or

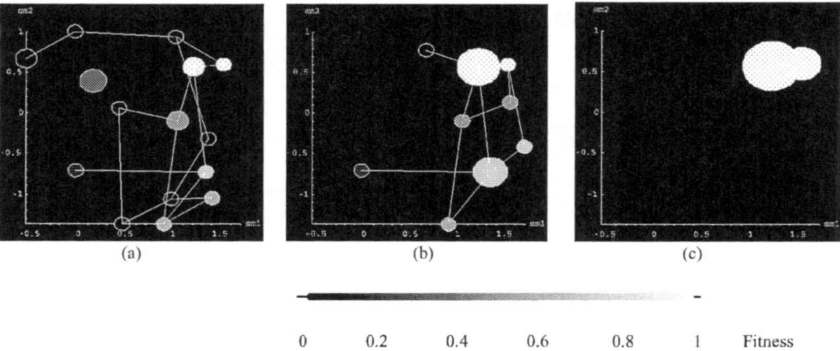

Fig. 11. A search space visualization of the progress of a GA on a problem with a single optimum solution. The chromosomes from generation 0, 3, and 10 are shown in the above three images; each circle illustrates a 5-bit binary chromosome, the position of each circle illustrates the chromosome's location in a 2D Sammon map, the size of each circle illustrates the frequency of that chromosome in the population, and the color value illustrates the chromosome's fitness rating. A line linking all the Hamming 1 neighbors of the fittest chromosome is shown in each image. This figure is taken from [14].

phenotypic similarities between the chromosomes in a GA's population (for further details see [30]).

Multi-dimensional Scaling

Sammon mapping [32], was used by Dybowski, Collins, and Weller to investigate the convergence of binary strings [14]. Sammon mapping is a non-linear mapping method used to map a set of datapoints in p-dimensional space to a set of datapoints in r-dimensional space (where $r < p$). The mapping attempts to preserve in r-dimensional space the Euclidean inter point distances present in p-space. In order to show the convergence of binary strings a 2D Sammon map ($r = 2$) was produced off-line for a dataset containing all the possible values within a fixed-length binary string. The individuals in a GA's population were then plotted on a 2D scatterplot, using color value to indicate fitness and point size to indicate the chromosome's frequency in the population. An example is given in Figure 11.

The main problem with using Sammon mapping is the computational cost involved in producing the r-dimensional map. Hartmut Pohlheim has also used Sammon mapping for visualizing multi-dimensional GAs [27], [28]. However, rather than mapping every possible chromosome, Pohlheim mapped only the chromosomes considered during the course of the GA's run. This approach has a significantly less computational cost than that proposed in [14]. However, both approaches are still computationally expensive as the

Fig. 12. The evolution of a GA solving DeJong's F1 test problem. These views are taken from [8] and show a search space visualization of the chromosomes contained in generations 0, 20, 40, and 64 (top left, top right, bottom left, bottom right, respectively). In these views each dot represents a chromosome in the current population.

complexity of the Sammon mapping rises quadratically with respect to the number of datapoints.

More recently a linear mapping method for representing high-dimensional search spaces has been presented in [6] and [33] to illustrate a GA's exploration of a search space. Both methods are based on the use of extensive repartitions [2, page 203] and provide a means for translating points from p to r dimensions and from r to p dimensions as required. A series of example search space visualizations for a 30-bit binary search space is given in Figure 12.

2.6 Navigating the Algorithm's Search

Being able to visualize an algorithm's search space creates a need to navigate the algorithm's search. Three concepts are represented in these visualizations: the space of possible solutions, the time associated with the algorithm's execution (i.e. the generations), and the fitness associated with each solution. Therefore, while viewing the search space, navigation can be made through time and fitness. Navigating through time involves viewing one or more generations. Navigating through fitness involves viewing one or more sections of the fitness landscape.

Navigating an Algorithm's Execution

GAMETER [22] and GIGA [10] are just two example systems that enable the user to "play" the GA's run like a movie, "pause" the execution of the GA, and "step" forward a single step (i.e. one generation). Using these controls the user can pause the execution of their algorithm, make a change to the algorithm's parameters and restart it, or step forward generation by generation in order to examine the GA's execution.

Within the field of SV a number of systems support the bi-directional control of a program's execution. These first appeared in systems like Henry Lieberman's "ZSTEP" system [23], Marc Eisenstadt and Mike Brayshaw's Transparent Prolog Machine ("TPM") [15], and Thomas Moher's PROcess Visualization and Debugging Environment ("PROVIDE") [24]. Bi-directional navigational control over a program's execution is usually achieved by periodically recording the program's current state and then producing the visualizations using the recorded history of events. As a result, the user can navigate *forwards and backwards* through a program's recorded history and the resulting visualizations will show the forwards and backwards execution of the program. A bi-directional control panel was used in the GONZO GA visualization tool [7] (see Figure 13, top left).

Navigating an Algorithm's Fitness Landscape

Another form of navigation that may prove useful within EC is the navigation of the fitness surface. Although generally used to navigate a program's execution a similar approach could be used to identify regions of interest within the range of fitness values from an EA's run, e.g. to identify the best chromosomes found. The navigation of the EA's execution and the discovered regions of the fitness landscape both require immediate visual feedback.

"Dynamic Queries" incorporate the use of direct manipulation and immediate feedback to query databases [34], [35]. An "AlphaSlider" [26] is an example of a dynamic query interface. AlphaSliders enable the user to select an item, or range of items, of interest within a dataset. A range-defining AlphaSlider looks like a regular scroll bar, except that rather than identifying

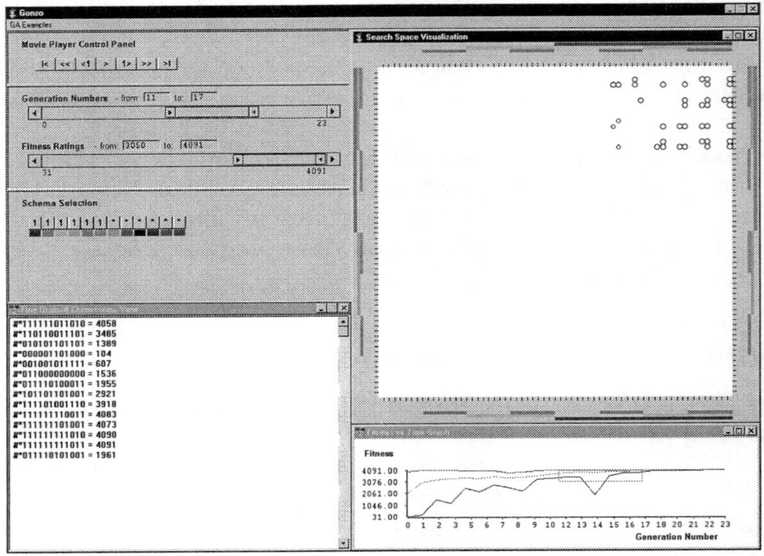

Fig. 13. An example screen image taken from GONZO. This example includes three views: a coarse-grained fitness versus time graph (bottom right), a medium-grained search space visualization (top right), and a fine-grained chromosome view (bottom left), and three navigators: a movie control panel (top left), generation and fitness range selector (second left), and a schema highlight selector (third left). The search space visualization currently illustrates all the chromosomes considered by the GA between generations 11 and 17 with fitness ratings between 3050 and 4091. The GA is attempting to solve the 12-bit maximum integer problem. Figure taken from [7].

a single point in a range as a small square, the AlphaSlider identifies a range within a range as a bar with scrolling arrow buttons at both ends. These arrow buttons define the start and end of the range of interest within an ordered dataset. The rectangular bar itself can also be dragged to pan across the dataset. Continuous feedback keeps the user informed of their current position within the dataset. AlphaSlider control panels are also available in the GONZO GA visualization tool [7] (see Figure 13, second left).

2.7 Editing the Algorithm

Finally, changes to an EA can be made either to the algorithm's component parts and their parameter settings, or to the data that the algorithm manipulates (i.e. the individuals).

A variety of EA systems enable the user to design their algorithms using a library of predefined selection and reproduction components with default parameter settings and interactive parameter controls (examples include GA-METER [22], GIGA [10], EVOS [1], and EvoNet's GA Software Development

Package). Within environments that also allow the user to pause and restart the algorithm's execution, these settings can be altered during the course of a run.

Editing can also be carried out at the data level (rather than the algorithm level). The user could directly alter the values of the chromosomes within the EA's population. GAMETER [22] facilitates this with the use of a "Bit String Editor." This allows the user to edit a selected chromosome from the current population, and set all of the alleles in a selected section of the chromosome to one, set all the alleles in a selected section to zero, or invert the alleles in a selected section (i.e. 0s to 1s and 1s to 0s). An on-the-spot evaluation can also be carried out to identify the effect of any changes made.

3 Discussion

In the previous section a range of EC visualizations were introduced. In discussing these systems an effort is made here to draw out the contributions of this body of work and point toward some of the remaining challenges currently facing EC visualization.

3.1 The Contributions Made

The Operation of the Algorithm's Component Parts

The actions of the GA's operators drive the GA's search in the problem space. Therefore, it would be reasonable to assume that illustrating the execution of the operators would provide some insight into the GA's search behavior. However, both the GIGA and TANGO visualizations of the GA's genetic operators are impractical for real problems. Neither visualization scales up for use on standard-sized GA populations. Furthermore, as the level of insight that can be achieved from these views is at the microscopic level of the chromosomes' genes, they provide little insight into the overall behavior of the system.

The Quality of the Solutions Found by the Algorithm

All of the fitness plots described can be used to illustrate the solutions found by any type of EA. The only limitation of such representations is the resolution of the axes used; for cases where the range of values along an axis is greater than the resolution of the screen some reduction in detail becomes necessary. However, such losses can be avoided by providing the user with controls for zooming and scrolling over sections of the graph. Although a number of alternative images to the 2D line graph were proposed in [5], a 2D line graph is the most efficient image for illustrating two components of information, such as fitness and time (see [2, pages 139 to 159] for a further discussion of image efficiency).

The Individuals' Representation in the Encoding Domain

Although exploring the fine-grained details of the individuals can be very useful for examining the solutions found, like the visualization of the genetic operators, it is at too fine-grained a level of detail to help people follow the overall search behavior of the algorithm. Therefore, visualizations of an EA's individuals must be used carefully to complement the user's exploration of the algorithm's search behavior. Perhaps if used in tandem with a visualization of the EA's sampling of the search space (as shown in Figure 13), the fine-grained focus that genotype visualization provides could be directed toward the more significant and interesting chromosomes within the GA's run.

The Individuals' Representation in the Problem Domain

Producing problem-specific visualizations of an EA's individuals is a very salient illustration. In GAs, for example, such views explicitly represent the link between the chromosomes' genotypes and phenotypes. This is why domain-specific visualizations are so useful when illustrating an EA's operation within an educational context. However, visualizing all of the individuals in a typical EA produces too much information for the user to digest easily. Like representations in the encoding domain, perhaps such detailed views are best used selectively to illustrate the more important individuals in an EA's run.

The Algorithm's Sampling of the Search Space

Showing a 2D or 3D visual representation of the search space enables the user to judge the diversity of the population and identify the formation of clusters of good solutions. However, the accuracy and resolution of such representations is an important issue that must be considered. Current PC screen resolutions offer around 1200 by 1000 pixels. Therefore, viewing an EA's entire search space is only possible at a very gross level of granularity. Much like a national road map and local street map, a combination of linked views showing an EA's search space at different resolutions can be used to gain an appreciation of the global and local properties of the EA's exploration of the search space.

Navigating the Algorithm's Execution

Using a similar approach to that commonly applied within SV, a bi-directional navigation controller could be introduced for the user to navigate an EA's run. In addition to a movie-player-styled controller for exploring the algorithm's execution, AlphaSliders can be used to define ranges of generation numbers and fitness ratings to be displayed. For example, displaying the top 5% of the chromosomes from all the generations of a GA would show the user how many good solutions had been considered during the algorithm's run.

Editing the Algorithm

Enabling the user to intervene in the evolutionary process, to alter either the algorithm or the population data, has both pros and cons. One of the pros is that providing there is sufficient visualization support, the user can explore the effects of any changes they make. This can be an engaging way to learn about the algorithm's search behavior and the impact of the user's choice of algorithm design. Another pro is the fact that the user can introduce domain knowledge by seeding or biasing the population with specific genes. However, one of the cons is that any form of intervention interferes with the algorithm's evolutionary search.

The common means of altering the GA's components or parameter settings through a pop-up dialog is a clear and effective approach. However, the means for altering the individual chromosomes in a population is perhaps less obvious. The bit string editor in GAMETER allows the user to change the alleles in selected sections of a chromosome. If the aim of altering the GA's chromosomes is to introduce domain knowledge then the user must translate that knowledge into the individuals' coding alphabet. Yet, in practice, biasing the EA's search may not be as simple as encoding a desired solution, rather the user may want to bias the EA away from sub-optimal clusters and toward unconsidered regions of the search space.

Interacting with a visualization of the EA's search space may be one way of supporting such a task. Users can directly manipulate the individuals' icons cutting, copying, and pasting individuals in the population, such that the EA is dragged away from sub-optimal clusters and toward unconsidered regions of the search space.

3.2 The Challenges Remaining

EC visualization is intended to facilitate people's understanding and effective use of EAs. It is hoped therefore that as new challenges face EC, new EC visualizations will be developed, as needed, to help deal with those challenges. However, two aspects of EC visualization that need to be dealt with currently are the provision of flexible visualization tools, and the effective representation of an EA's exploration of search spaces.

A number of EA visualization systems have been mentioned in the above review. However, these systems are often tied to a particular type of EA, or offer a closed set of recommended visualizations. More flexible visualization environments are needed that can be extended with additional user-defined views and applied to a variety of EAs, as the problem at hand dictates.

A variety of similarity metrics and multi-dimensional scaling techniques have been used in the visualizations presented here. However, none of the methods are appropriate for all EAs or EA applications. There will always be a need for ad hoc algorithm-specific and problem-specific mapping methods. Future work on EA visualization needs to explore the effectiveness of such mapping methods in order to develop appropriate visualizations.

Acknowledgments

I would like to thank the following people for permitting me to use images of their systems in this chapter: Jochen Schoof, David Brogan, John Stasko, Adrian Thompson, Bill Spears, Annie Wu, Bahram Nassersharif, and Richard Dybowski. Figure 11 was first published in Evolutionary Programming V: Proceedings of the Fifth Annual Conference on Evolutionary Programming. Lawrence J. Fogel, Peter J. Angeline, and Thomas Baeck (editors). A Bradford Book, MIT Press. Cambridge, Massachusetts. 1995. ISBN 0-262-06190-2.

References

1. Baeck, T. (1996) *Evolutionary Algorithms in Theory and Practice (Evolution Strategies, Evolutionary Programming, Genetic Algorithms).* Oxford University Press, New York
2. Bertin, J. (1981) *Graphics and Graphic Information Processing.* Walter de Gruyer, Berlin and New York
3. Brown, M. (1991) Zeus: A system for algorithm animation and multi-view editing. In *The Proceedings of the IEEE Annual Workshop on Visual Languages*, Kobe, Japan, 4–9
4. Brown, M., Sedgewick, R. (1985) Techniques for algorithm animation. *IEEE Software* (January 1985), 28–39
5. Collins, T. (1993) The visualisation of genetic algorithms. Master's thesis, Department of Computer Science, De Montfort University, Leicester, UK
6. Collins, T. (1997) Using software visualization technology to help evolutionary algorithm users validate their solutions. In *The Proceedings of the Seventh International Conference on Genetic Algorithms ICGA'97*, East Lansing, MI, USA, 307–314
7. Collins, T. (1998) The application of software visualization technology to evolutionary computation: A case study in genetic algorithms. PhD thesis, The Knowledge Media Institute, The Open University, UK
8. Collins, T. (1998) Understanding evolutionary computing: A hands on approach. In *The Proceedings of the IEEE International Conference on Evolutionary Computation, ICEC'98*, 564–569
9. Collins, T. (1999) Evolutionary computation visualization workshop abstracts. In *The Proceedings of the 1999 Genetic and Evolutionary Computation Conference Workshop Program*, 93–108. Details available from `http://kmi.open.ac.uk/people/trevor/workshops/gecco-99/`
10. Dabs, T., Schoof, J. (1995) A graphical user interface for genetic algorithms. Tech. Rep. 98, Lehrstuhl für Informatik II, University Würzburg, Germany
11. De Jong, K. (1980) Adaptive systems design. *IEEE Transactions on Systems, Man and Cybernetics*, 566–574
12. De Jong, K. (1999) Evolutionary computation for discovery. *Communications of the ACM 42*, **11**, 51–53
13. Domingue, J., Price, B., Eisenstadt, M. (1992) A framework for describing and implementing software visualisation systems. In *The Proceedings of the Graphics Interface Conference GI'92*, Vancouver, 11–15

14. Dybowski, R., Collins, T., Weller, P. (1996) Visualization of binary string convergence by Sammon mapping. In *The Proceedings of the Fifth Annual Conference on Evolutionary Programming EP'96*, San Diego, CA, 1996, L. Fogel, P. Angeline, and T. Baeck, Eds., MIT Press, Cambridge, MA, 377–383
15. Eisenstadt, M., Brayshaw, M. (1987) The transparent prolog machine (TPM): An execution model and graphical debugger for logic programming. Tech. Rep. 21a, Human Cognition Research Laboratory, The Open University, UK
16. Fogel, D. (1995) *Evolutionary Computation: Towards a New Philosophy of Machine Intelligence.* IEEE Press, New York
17. Fraser, A. (1957) Simulation of genetic systems by automatic digital computers. *Australian Journal of Biological Science 10*, 484–499
18. Goldberg, D. (1989) *Genetic Algorithms in Search Optimization and Machine Learning.* Addison-Wesley, Reading, MA
19. Grefenstette, J. J., Burke, D. S., DeJong, K. A., Ramsey, C. L., Wu, A. S. (1997) An evolutionary computation model of emerging virus diseases. Tech. Rep. AIC-97-030, Navy Center for Applied Research in Artificial Intelligence
20. Harvey, I., Thompson, A. (1996) Through the labyrinth evolution finds a way: A silicon ridge. In *The Proceedings of the First International Conference on Evolvable Systems: From Biology to Hardware ICES'96*, Springer-Verlag, Berlin. Available from `ftp://ftp.cogs.sussex.ac.uk/pub/users/inmanh/sil_ridge.ps.gz`
21. Jones, T., Forrest, S. (1995) Fitness distance correlation as a measure of problem difficulty for genetic algorithms. In *The Proceedings of the Sixth International Conference on Genetic Algorithms ICGA'95*, Morgan Kaufmann, San Francisco, 184–192
22. Kapsalis, A., Rayward-Smith, V., Smith, G. (1993) Fast sequential and parallel implementation of genetic algorithms using the GAMETER toolkit. In *The Proceedings of the International Conference on Neural Networks and Genetic Algorithms*, Innsbruck, 1993, Springer-Verlag, Berlin
23. Lieberman, H. (1984) Steps toward better debugging tools for lisp. In *The Proceedings of the ACM Symposium on Lisp and Functional Programming*, August 1984, Austin, Texas
24. Moher, T. (1988) PROVIDE: A process visualisation and debugging environment. *IEEE Transactions on Sofware Engineering 14*, **6**, 849–857
25. Nassersharif, B., Ence, D., Au, M. (1994) Visualisation of evolution of genetic algorithms. In *The Proceedings of the World Congress on Neural Networks*, San-Diego, USA, **1**, pp. 560–565
26. Osada, M., Liao, H., Shneiderman, B. (1993) AlphaSlider: Searching textual lists with sliders. In *The Proceedings of the Ninth Annual Japanese Conference on Human Interfaces*
27. Pohlheim, H. (1998) Development and engineering applications of evolutionary algorithms. PhD thesis, Faculty of Informatics for Automation, Technical University Ilmenau, Germany
28. Pohlheim, H. (1999) Visualization of evolutionary algorithms - set of standard techniques and multidimensional visualization. In *The Proceedings of the 1999 Genetic and Evolutionary Computation Conference GECCO'99*, Morgan Kaufmann, San Francisco
29. Repenning, A., Ambach, J. (1996) Tactile programming: A unified manipulation paradigm supporting program comprehension, composition and sharing. In *The Proceedings of the 1996 IEEE Symposium of Visual Languages*, 102–109

30. Ronald, S. (1998) More distance functions for order-based encodings. In *The Proceedings of the 1998 IEEE International Conference on Evolutionary Computation ICEC'98*, 558–563
31. Routen, T., Collins, T. (1993) Visualisation of A.I. techniques. In *The Proceedings of the International Conference on Computer Graphics and Visualization COMPUGRAPH'93*, Portugal, ACM Press
32. Sammon, J. (1969) A nonlinear mapping for data structure analysis. *IEEE Transactions on Computers C-18*, **5**, 401–408
33. Shine, W., Eick, C. (1997) Visualizing the evolution of genetic algorithm search processes. In *The Proceedings of the 1997 IEEE International Conference on Evolutionary Computation ICEC'97*, 367–372
34. Shneiderman, B. (1994) Dynamic queries for visual information seeking. *IEEE Software 11*, **6**, 70–77
35. Shneiderman, B. (1998) *Designing the User Interface: Strategies for Effective Human-Computer Interaction*, 3rd ed. Addison Wesley Longman
36. Spears, W. (1994) Visualizing genetic algorithms. Tech. Rep. AIC-94-055, AI Center, Naval Research Laboratory, Washington, DC
37. Stasko, J. (1989) TANGO: A framework and system for algorithm animation. PhD thesis, Brown University, Providence, RI
38. Stasko, J. (1990) The path-transition paradigm: A practical methodology for adding animation to program interfaces. *Journal of Visual Languages and Computing 1*, **3**, 213–236
39. Wu, A. S., De Jong, K., Burke, D., Grefenstette, J., Ramsey, C. (1999) Visual analysis of evolutionary algorithms. In *The Proceedings of the 1999 Congress on Evolutionary Computation*, IEEE, 1419–1425

New Schemes of Biologically Inspired Evolutionary Computation

Hidefumi Sawai[1], Susumu Adachi[1], and Sachio Kizu[2]

[1] Communications Research Laboratory, MPT, Japan
E-mail: {sawai, sadachi}@crl.go.jp
[2] Toshiba R & D Center, Kawasaki, 210 Japan
E-mail: skizu@isl.rdc.toshiba.co.jp

Summary. We propose a novel genetic algorithm which we call a parameter-free genetic algorithm (PfGA). The PfGA is inspired by the idea of a biological evolution hypothesis, i.e., the "disparity theory of evolution." The theory is based on different mutation rates in double strands of DNA. Its idea can be extended to a very compact and fast adaptive search algorithm accelerating its evolution based on the variable size of a population and taking a dynamic but delicate balance between exploration (i.e., global search) and exploitation (i.e., local search). The PfGA is not only simple and robust, but it is unnecessary to set almost all the genetic parameters in advance which need to be set up in other genetic algorithms. Furthermore, a uniformly distributed parallel architecture and a master-slave architecture for the PfGA are investigated as an extension. We discuss the performance of the parallel distributed architectures using a general set of function optimization problems including the functions in the first Internatinal Contest on Evolutionary Optimization. On the other hand, gene duplication theory was first proposed by a Japanese biologist, Dr. Susumu Ohno, in the 1970's. Inspired by this theory, we develop a gene-duplicating genetic algorithm. Several variants of this algorithm are considered. Individuals with various lengths of genes are evolved based on the PfGA or steady-state GA and then genes with different lengths are concatenated by migrating among subpopulations. To verify the effectiveness of the gene-duplicating genetic algorithm, we also performed a comparative study using the general set of function optimization problems.

1 Introduction

In 1859 Charles Darwin described the heredity and evolution of biological organisms in the famous book *On the Origin of Species by Means of Natural Selection or the Preservation of Favoured Races in the Struggle for Life* [1]. John Holland [2] proposed in the 1960's the genetic algorithm (GA) inspired by Darwin's idea.

The GA is an evolutionary computation paradigm inspired by such biological heredity and evolution. GAs have been successfully applied to many practical applications such as functional optimization problems, combinatorial optimization problems, and the optimal design of parameters in machine [3]. The GA has recently been integrated with other evolutionary schemes such as evolutionary strategies (ES) proposed by Rechenberg [4], and evolutionary programming (EP) proposed by Fogel [5], which forms an evolutionary computation scheme. However, some problems for GA have not yet been resolved. One such problem is that the design of genetic parameters in a GA has to be determined by trial and error, making optimization by GA *ad hoc*. One of the most important research areas in evolutionary computation is to adapt genetic parameters and operators in a self-adaptive manner because such adaptation can tune an algorithm during the solution of a given problem. In [14], a classification of adaptation is developed, which covers different levels (such as environment, population, individual, and component) and types (such as static and dynamic ones).

However, it is a very time-consuming task to design an optimal evolutionary strategy in an adaptive way because we have to perform the evolutionary algorithm many times by trial and error. Furthermore, if a given problem changes, we have to redesign the obtained evolutionary strategy again from the beginning. To relieve the user of this kind of adaptive parameter-setting problem, we propose a parameter-free genetic algorithm (PfGA) [24–28] where almost none of the genetic parameters, such as initial population size, crossover rate, and mutation rate, need to be set up by a user in advance. All that is needed is a random number generator.

The PfGA is inspired by the "disparity theory of evolution" which was proposed by Furusawa et al. [9,10] in the early 1990's. The idea is based on the disparity of copy error rates in the *leading* and *lagging* strands of DNA when each strand makes its copy. The error rate in the lagging strand is much higher than that in the leading strand. So the error rate accumulates more in the lagging strand than in the leading strands as generations proceed. The offspring from the leading strands rarely suffer from the copy error (we call it a *wild type*). On the other hand, the offspring from the lagging strand accumulate more copy error than the leading strand. Consequently, asymmetry or disparity occurs in the two kinds of offspring. This leads to a *diversity* in biological ecosystem.

The search strategy in the PfGA is based on a dynamic change of subpopulation size extracted from the population, which enables an adaptive search by taking a delicate balance between global and local search methods. These two kinds of search methods correspond to exploration and exploitation, respectively, and maintain diversity in GAs. Since the PfGA is a simple and robust algorithm, it may be easily applied not only to function optimization problems but also to combinatorial optimization problems such as TSP (Traveling Salesman Problem) and JSP (Job-shop Scheduling Problem) if the representation of the solutions can be appropriately coded onto genes.

To verify its performance, the PfGA is applied to a general set of function optimization problems as benchmark tests [23], and compare its performance to the results of algorithms which participated in the first International Contest on Evolutionary Optimization (ICEO) at the ICEC (International Conference on Evolutionary Computation) in 1996.

Furthermore, the PfGA can be extended to parallel distributed processing of the PfGA which makes it possible to solve more complicated optimization tasks and to study biological behavior in an ecosystem. We introduce parallel distributed architectures for the PfGA, and show some results using multiple processors [27–29].

On the other hand, gene duplication theory was first proposed by a Japanese biologist, Dr. Susumu Ohno, in the 1970's. Inspired by this theory, we develop a gene-duplicating genetic algorithm (GDGA)[30]. Several variants of this algorithm are considered. Individuals with various lengths of genes are evolved based on the PfGA or SSGA, and then genes with different lengths are concatenated by migrating among subpopulations. To verify the effectiveness of the GDGA, we performed a comparative study using the general set of function optimization problems.

1.1 Disparity Theory of Evolution

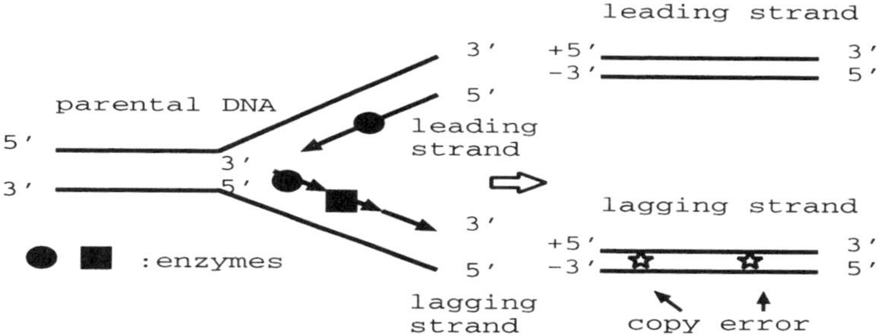

Fig. 1. A hypothesis in the *disparity theory of evolution*

As Charles Darwin claimed in the *Origin of Species* in 1859 [1], a major factor contributing to evolution is mutation, which can be caused by spontaneous misreading of bases during DNA synthesis. Semiconservative replication of double-stranded DNA is an asymmetric process where there is a leading and a lagging strand. Furusawa et al. proposed a "disparity theory of evolution"[9] based on a difference in frequency of strand-specific base misreading between the leading and lagging DNA strands (i.e., *disparity model*). Figure 1 shows a hypothesis in the disparity theory of evolution. In the figure, the leading strand is copied smoothly, whereas in the lagging strand a copy error can occur because plural enzymes are necessary to produce its copy. This disparity or asymmetry in producing each strand occurs because of the different mutation rates in the leading and lagging strands, and thus maintains the "diversity" of DNA in the population as generations proceed. The disparity model guarantees that the mutation rate of some leading strands is zero or very small. When circumstances change, although the original *wild type* cannot survive, selected mutants might adapt under the new circumstances as a *new wild type*. In their study, the disparity model was compared with a parity model in which there was no statistical difference in the frequency of base misreading between strands as in the generally accepted model.

The disparity model outperformed the parity model in a knapsack optimization problem. Furusawa et al. clearly showed that the advantageous situation for the disparity model happened in the cases of small population, strong pressure, high mutation rate, sexual reproduction with diploidy, and strong competition. On the other hand, survival conditions for the parity model are large population, weak selection pressure, low mutation rate, asexual reproduction with haploidy, and weak competition [10].

1.2 Parameter-free Genetic Algorithm

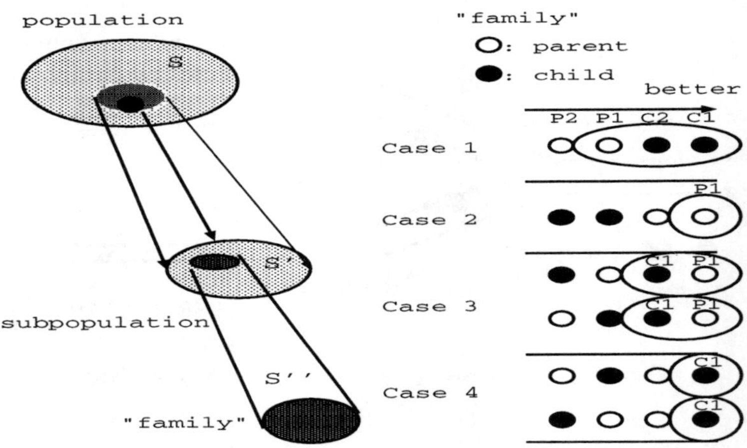

Fig. 2. Population, subpopulation, and family (left), and selection rules (right) in PfGA

In this section, the PfGA is described in detail in comparison with the disparity theory. Its basic procedure and the selection scheme with a local elitist-preserving strategy will be explained. First of all, the population of the PfGA is considered as a whole set **S** of individuals which corresponds to all possible solutions. From this whole set **S**, a subset **S**' is introduced. All genetic operations such as selection, crossover, and mutation are conducted for **S**', thus evolving the subpopulation **S**'. From the subpopulation **S**', we introduce a *family* which contains two parents and two children generated from the two parents. Figure 2 shows the population **S**, subpopulation **S**', *family* **S**" (left), and selection rule (right) in the PfGA.

The PfGA procedure is as follows:

Step 1. Select one individual randomly from the whole population **S**, and add this individual to the subpopulation **S'**.

Step 2. Select one individual randomly from the whole population **S**, and again add this individual to the subpopulation **S**'.

Step 3. Select two individuals randomly from the subpopulation **S**' and perform crossover between these individuals as parent 1 (P_1) and parent 2 (P_2).

Step 4. For one randomly chosen child of the two children generated from the crossover, perform mutation at random.

Step 5. Among the parents (P_1 and P_2) and the two generated children (C_1 and C_2) select one to three individuals depending on the following cases (i.e., Case 1 to 4), and feed them back to the subpopulation **S**'.

Step 6. If the number in subpopulation **S**' is greater than one, go to Step 3; otherwise, go to Step 2.

For the crossover operation of PfGA, we use multiple-point crossover in which randomly changeable crossover points and locations between two parents' chromosomes are adopted every time the crossover is operated. For the mutation operation, one child is randomly chosen from the two offspring. Then a randomly chosen portion of the child's chromosome is randomly inverted (i.e., bit-flipped). For the selection operation, we compare the fitness values of all individuals (C_1, C_2, P_1, P_2) in the family. The selection rules shown in Fig. 2 are used for four different cases depending on the fitness values of the parents and children.

Case 1: If the fitness values of the two children are better than those of the parents, then C_1, C_2, and $arg\ max_{P_i}(f(P_1), f(P_2))$ are left in **S**', thus increasing the size of **S'** by one. Since in this case the two parents produced better children, these children should be preserved and only the better parent with possibly good schemata is preserved to avoid increasing the number of individuals in **S**'.

Case 2: If the fitness values of C_1 and C_2 are worse than those of P_1 and P_2, then only $arg\ max_{P_i}(f(P_1), f(P_2))$ is left in **S**', thus decreasing the size of **S**' by one. In this case no better children were produced from the two parents. Because no optimal solutions would be guaranteed if all individuals were removed from **S**', only the better parent should be preserved to maintain the stability of the system.

Case 3: If the fitness value of either P_1 or P_2 is better than that of the children, $arg\ max_{C_i}(f(C_1), f(C_2))$ and $arg\ max_{P_i}(f(P_1), f(P_2))$ are left in **S**', thus maintaining the size of **S**'. In this case one of the children is worse than the better parent, but better than the worse parent. At least the better parent should be preserved to maintain the stability of the system. Since at least one child better than the worse parent was produced, that child should replace the worse parent and remain in **S**'.

Case 4: In all other situations, $arg\ max_{C_i}(f(C_1), f(C_2))$ is preserved and then one individual randomly chosen from **S** is added to **S**', thus maintaining the size of **S**'. In this case one of the children is better than the better parent. At least one better child should be preserved in **S**'. Moreover, to guarantee the flexibility of the system, the subpopulation **S**' should not be prematurely converged. A new individual should be added to the subpopulation **S**' from the population **S**.

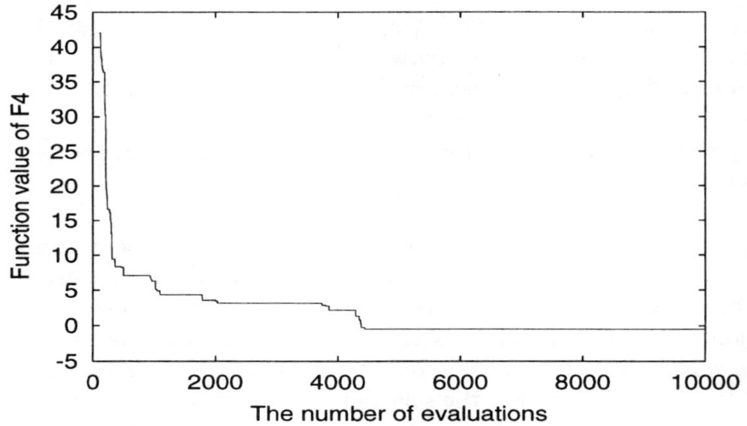

Fig. 3. Function value vs. number of evaluations in De Jong's F4

Figure 3 shows an example of convergence behavior of the best individual in subpopulation for De Jong's F4 [7]. Figure 4 shows the number of individuals in subpopulation for the corresponding run to Figure 3. De Jong's F4 is defined as $F4(x_i) = \sum_{i=1}^{30} ix_i^4 + Gauss(0, 1)$. As Gaussian noise is added to the first term, the number of subpopulations is fluctuated between one and five as shown in Fig. 4 which enables an adaptive search according to the change of landscape in F4.

Now we explain the correspondence and difference between the disparity theory and the PfGA. The two different schemes don't necessarily correspond

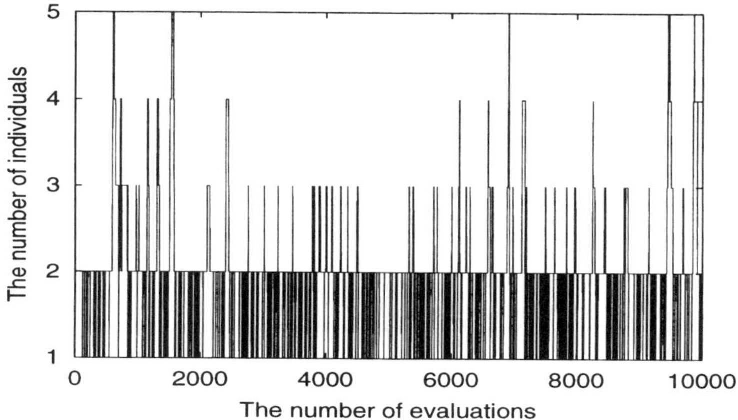

Fig. 4. Number of individuals in subpopulation vs. number of evaluations in De Jong's F4

each other directly because the disparity theory is a biological evolution theory and the PfGA is a mathematical algorithm, however, the PfGA was designed by getting the inspiration from the disparity theory. The leading strand in the disparity theory corresponds to the current best individual in the family, for examples, C_1 in the cases 1 and 4 (as a *new wild type*), or P_1 in the cases 2 and 3 (as a *wild type*) in Fig.2. Other individuals exept the best one correspond to the lagging strands in the disparity theory because the lagging strands would accumulate more mutation as generations proceed, which increases the diversity of population. This situation can be realized in a simplest form by generating new offspring based on crossever and mutation operations from the current best individual and its mated individual in the subpopulation **S**'.

1.3 Parallel Distributed Processing of PfGA

Generally speaking, parallel processing aims at accelerating the speed of processing. In the case of GA, it aims at reaching better solutions faster than sequential processing by extending the search space. Parallel and distributed processing of GA has been extensively studied [15–19], where the *granularity* ranges from fine- to coarse-grained, and the mode of processing covers both synchronous and asynchronous processing. The granularity concerns the size of a process assigned to a processor. In the case of a fine-grained parallel GA model, the (overlapping) neighborhoods of the individuals constitute the units of processing. On the other hand, coarse-grained parallel GA assigns a subpopulation as a unit of processing, and some few individuals are migrated among subpopulations at an appropriate rate. The latter model is called an

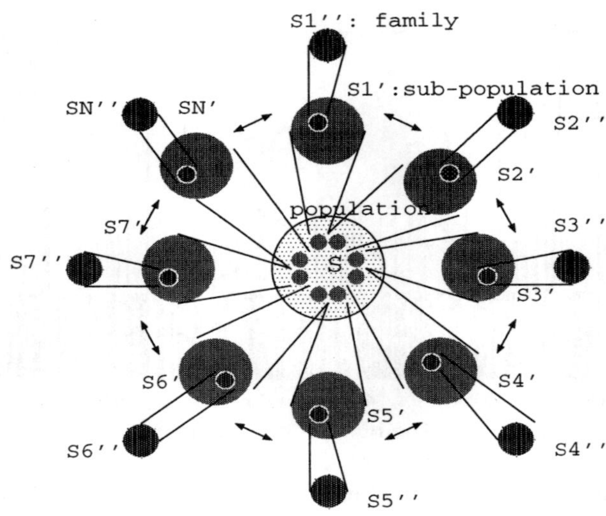

Fig. 5. A uniformly distributed architecture for the parallel PfGA

"island model", and one island (subpopulation) constitutes one "deme" which is a minimum recombinational unit of biological species. In this chapter we use the latter model.

Figure 5 shows a uniformly distributed architecture for the parallel PfGA. One *deme* in this architecture is based on that of Figure 2. The whole population **S** is located at the center from which N subpopulations $\{S_i'\}$ ($i = 1, \ldots, N$) are extracted (N =8 in this case). Of course if the number of processors increases, the number of subpopulations could also increase. Each "family" shown as S_i'' is extracted from each subpopulation S_i'. If better individuals are produced in each subpopulation S_i', the individuals are copied and migrated among the subpopulations (the bilateral arrows indicate this situation, however, in fact any migration can happen among any subpopulations). One possible migration method is as follows: if Case 1 or 4 happens in some family, the better child C_1 that is better than its two parents can be copied to other subpopulations as an emigrant. When other subpopulations receive the immigrant, they have to decide to accept it or not because an increase in the number of individuals will lead to an explosion if any immigrants are unconditionaly accepted. To avoid such a situation, one possible method is to eliminate the worst individual among all individuals in the subpopulation including the "candidate" immigrant. This maintains the sizes of the subpopulations. If the immigrant is eliminated, the immigration cannot be substantially realized.

Figure 6 shows another type of architecture of a parallel PfGA. This is called a "master-slave" type of parallel distributed PfGA. Let S_0' be the mas-

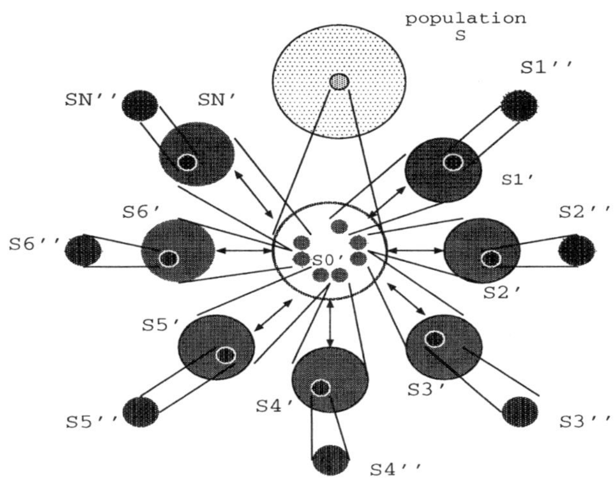

Fig. 6. A master-slave type architecture of the PfGA

ter subpopulation and $\{S_i'\}$, $(i = 1, \ldots, N)$ be the slave subpopulations. The master S_0' always or at some interval supervises the subpopulations $\{S_i'\}$ and identifies the best individual among them. One possible migration method is as follows; When case 1 or 4 occurs in a subpopulation $\{S_i'\}$, compare C_1 with the best-so-far individual identified by the master subpopulation. If the former (C_1) is better than the latter (actually, this situation always occurs), copy and distribute it to other subpopulations. In each subpopulation that receives the "candidate" immigrant, the worst individual among all individuals including the candidate immigrant is eliminated from the subpopulation. This elimination maintains the number of individuals in the subpopulation as in the uniformly distributed architecture.

We also propose the following two kinds of migration methods embedded in the two kinds of architectures (i.e., four combinations):

- Uniformly distributed type with direct migration (**UD1**):
  As shown in the upper-left of Figure 7, if the better child C_1 appears, copy C_1 and distribute it to a subpopulation S_j', $(j \neq i)$. The worst individual is removed from S_j' to maintain the number of individuals.
- Uniformly distributed architecture with hierarchical migration (**UD2**):
  In the above UD1, if the number of distributed individuals in S_j' becomes two, the two individuals as P_1 and P_2 are operated by crossover and mutation, thus producing two children, C_1 and C_2. Four individuals are selected according to the selection rule of a PfGA, and one, two, or three individuals are distributed to other subpopulations, as shown in the upper-right of Figure 7. The worst individual is removed from the sub-

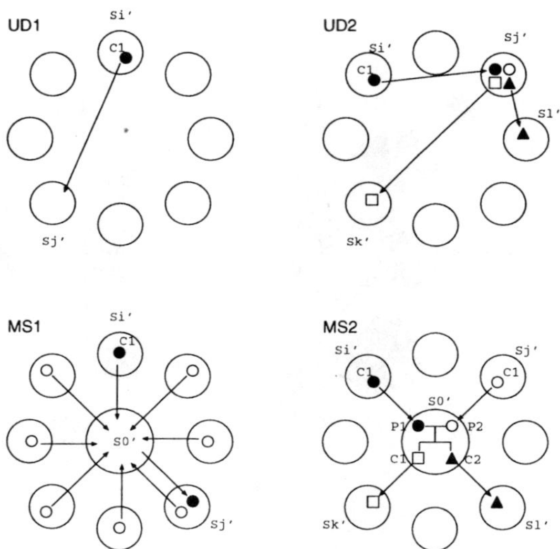

Fig. 7. Migration methods: UD1 (upper left), UD2 (upper right), MS1 (lower left) and MS2 (lower right)

population that accepts the new individual. In this method the number of evaluations increases by two because two new individuals are produced.

- Master-slave architecture with direct migration (**MS1**):
  S'_0 denotes the master subpopulation and S'_i, $(i = 1, \ldots, N)$ denotes the slave subpopulation. If case 1 or 4 occurs in S'_i, C_1 is distributed to the master S'_0. The master selects the best individual among the distributed ones from the slaves, then it copies and distributes it to other slave subpopulation S'_j as shown in the lower-left of Figure 7. The worst individual is removed from S'_j.
- Master-slave architecture with hierarchical migration (**MS2**):
  As shown in the lower-right of Figure 7, the master selects two parents, P_1 and P_2, that are distributed from slave subpopulations. The two parents are operated by crossover and mutation, thus producing two children, C_1 and C_2. Four individuals are selected according to the selection rule of a PfGA, and one, two, or three individuals are distributed to other slave subpopulations.

These architectures were implemented by using eight processors connected with local-area networks (LANs). PVM (parallel virtual machine) [13] software is used to evolve the subpopulations in parallel and synchronously. PVM is network software that uses dynamic load balancing between assigned processors.

1.4 Gene-duplicating GA

Gene duplication theory was first proposed by S. Ohno in the 1970's [6]. According to his proposition, during the evolutionary process for any biological creatures including viruses, plants and animals etc. they duplicate or copy the segments of genes for reuse which leads to more advanced biological creatures of a higher order. This gene-duplication occurs due to the exchange between two different lengths of gene-segments in a chromosome, or a partially repetitive copy of DNA. We proposed four different gene-duplication types inspired by such duplication mechanism, and compared them with each other. Type A is a gene-concatenated type that concatenates each segments of a gene at the same time. Type B is a gene-prolonged type that copies a segment of a gene to make a longer gene step by step. Type C is a gene-coupled type that couples two genes of different lengths. Type D is an extended gene-coupled type that coupled two genes of different lengths with the successive loci. We describe the four types in more detail in the following sections.

Gene-concatenating GA (type A): In this algorithm, variable x_i ($i = 1, \ldots, n$) is encoded in Gray coding with l bits. Then each individual with each gene $[x_i]$ evolves by using the PfGA. The fitness function for each individual is defined as $f(a_1, a_2, \ldots, a_{i-1}, x_i, a_{i+1}, \ldots, a_n)$. For simplicity the constants a_i can be set zero. For other types of gene duplication, type B, C and D, the constants a_i in the fitness function can also be set zero. When the individual C_1 better than two parents emerges in each subpopulation, all genes from $[x_1]$ to $[x_n]$ are concatenated to make a long gene $[[x_1] \cdots [x_n]]$ with a duplication rate R. This long gene belongs to the subpopulation S' as shown in Fig. 8, while the worst individual is removed from S' to maintain the size of S'.

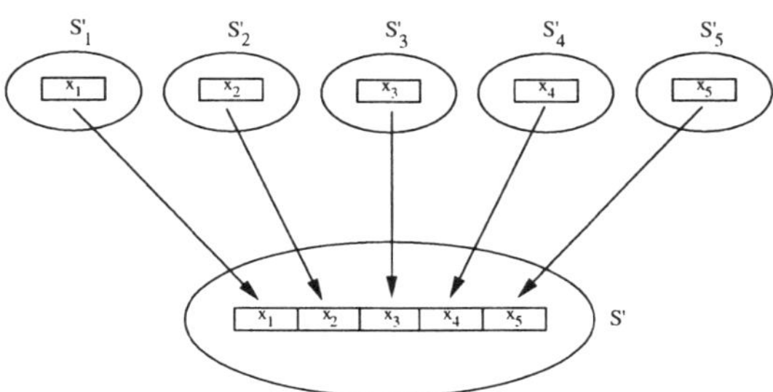

Fig. 8. Gene-concatenating GA (type A) for $n = 5$

Gene-prolonging GA (type B): As the initial population, every gene with the length $i \times l$, $[[x_1][x_2] \cdots [x_i]]$, are generated in each subpopulation S_i'

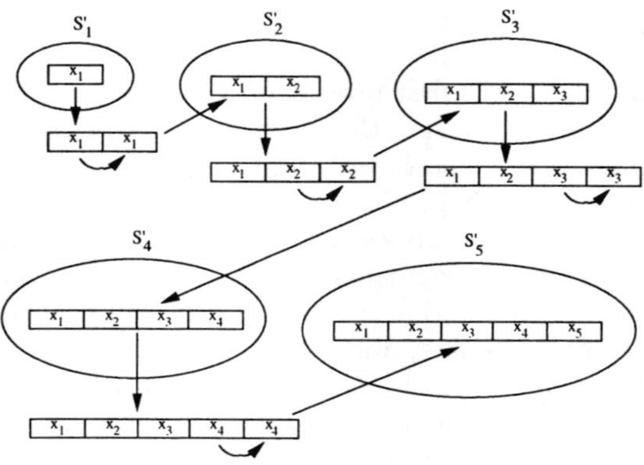

Fig. 9. Gene-prolonging GA (type B) for $n = 5$

$(i = 1, 2, \ldots, n)$. Each individual with the gene length $i \times l$ evolves in each S_i' according to the fitness function $f(x_1, x_2, \ldots, x_i, a_{i+1}, \ldots, a_n)$. When the individual C_1 better than two parents emerges in S_i', its gene is prolonged from $[[x_1] \cdots [x_i]]$ to $[[x_1] \cdots [x_{i+1}]]$ with a duplication probability R by copying $[x_i]$ to $[x_{i+1}]$ $(i = 1, \ldots, n-1)$ as shown in Fig. 9. The individual with the longer gene by one belongs to the subpopulation S_{i+1}'. The worst individual is removed from S_{i+1}' as well.

Gene-coupling GA (type C): As the initial population, genes with a length of $i \times l$ are generated similar to type B. Each individual evolves in each S_i' according to the PfGA and the fitness function $f(x_1, x_2, \ldots, x_i, a_{i+1}, \ldots, a_n)$. A gene with a length of $i \times l$, $[[x_1] \cdots [x_i]]$, in S_i' is coupled to another gene $[[x_1] \cdots [x_j]]$ in S_j' to produce $[[x_1] \cdots [x_{i+j}]]$ $(2 \le i + j \le n)$ with a duplication probability of R. This individual, having a length of $(i+j) \times l$, belongs to S_{i+j}', as shown in Fig. 10. The worst individual is removed from S_{i+j}' as well.

Extended Gene-coupling GA (type D): This type is an extension of type C. The loci of genes are not distinguished in type C, but they are distinguished in type D, as shown in Fig. 11. As initial populations genes with a length of l are generated as well. Then, genes with two successive variables $[[x_i][x_{i+1}]]$ are generated. Next, genes with three successive variables $[[x_i][x_{i+1}][x_{i+2}]]$ are generated. The last initial genes are $[[x_1] \cdots [x_n]]$ as shown in Fig. 11. The fitness function of individuals with two successive genes, for example, is defined as $f(a_1, \ldots, a_{i-1}, x_i, x_{i+1}, a_{i+2}, \ldots, a_n)$. When the individual C_1 better than two parents emerges in S_i', genes are coupled only in the case of successive loci of gene, such as $[x_1]$ and $[x_2]$, $[[x_1][x_2]]$ and $[x_3]$, $[[x_1][x_2][x_3]]$

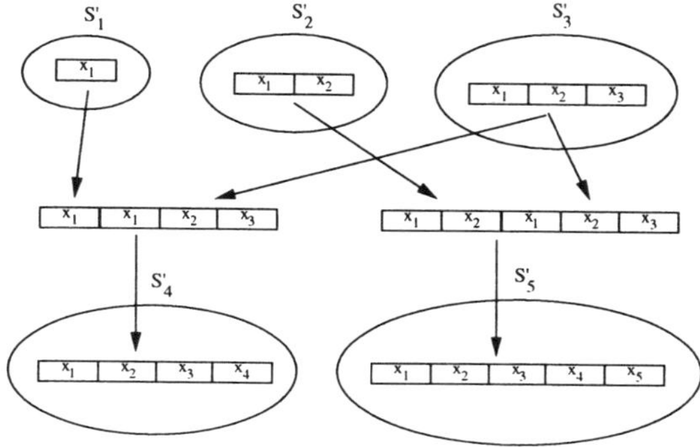

Fig. 10. Gene-coupling GA (type C) for $n = 5$

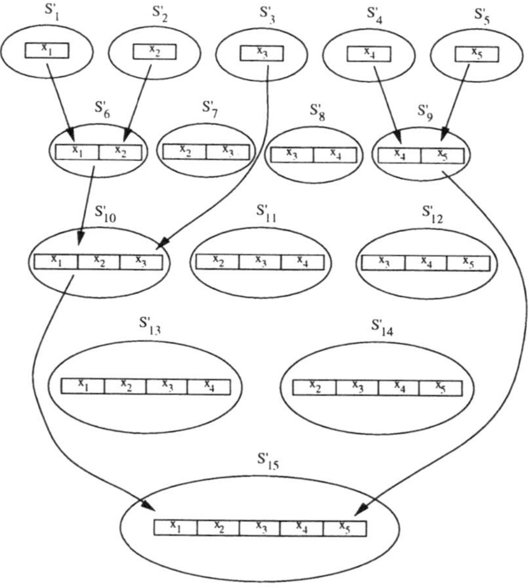

Fig. 11. Extended gene-coupling GA (type D) for $n = 5$

and $[[x_4][x_5]]$, leading to the final gene $[[x_1] \cdots [x_5]]$, as shown in Fig. 11, with a duplication probability of R.

2 Experiment

We performed several comparative experiments on the PfGA, its parallel distributed processing, and the GDGA using a general set of functional optimization problems recently proposed.

Table 1. Test functions

(1)Sphere model (Sp):
unimodal, symmetric, separable
$f(\mathbf{x}) = \sum_{i=1}^{n}(x_i - 1)^2$,
$-5 \leq x_i \leq 5$ (24*bit*), $VTR = 1.0 \times 10^{-6}$

(2)Schwefel's Double Sum (Ds):
unimodal, symmetric, inseparable
$f(\mathbf{x}) = \sum_{i=1}^{n}(\sum_{j=1}^{i} x_j)^2$,
$-65.536 \leq x_i \leq 65.536$ (27*bit*),
$VTR = 10^{-4}$ (gene-duplication), $VTR = 1.0$ (parallel PfGA)

(3)Randomized Sphere model (Rs):
unimodal, asymmetric, separable
$f(\mathbf{x}) = \sum_{i=1}^{n}(x_i - r_i)^2$,
$0 \leq r_i \leq 1$ (uniform random number)
$-5 \leq x_i \leq 5$ (24*bit*), $VTR = 1.0 \times 10^{-6}$

(4)Randomized Double Sum (Rd):
unimodal, asymmetric, inseparable
$f(\mathbf{x}) = \sum_{i=1}^{n}(\sum_{j=1}^{i}(x_j - r_j))^2$,
$0 \leq r_i \leq 1$ (uniform random number)
$-65.536 \leq x_i \leq 65.536$ (27*bit*),
$VTR = 10^{-4}$ (gene-duplication), $VTR = 1.0$ (parallel PfGA)

(5)Generalized Rastrigin's function(Ra):
multimodal, symmetric, separable
$f(\mathbf{x}) = 10n + \sum_{i=1}^{n}[x_i^2 - 10\cos(2\pi x_i)]$,
$-5.12 \leq x_i \leq 5.12$ (24*bit*), $VTR = 10^{-2}$

(6)Griewank's Function (Gr):
multimodal, symmetric, inseparable
$f(\mathbf{x}) = \frac{1}{d}\sum_{i=1}^{n}(x_i - 100)^2 - \prod_{i=1}^{n}\cos(\frac{x_i - 100}{\sqrt{i}}) + 1$,
$d = 4000, -600 \leq x_i \leq 600$ (31*bit*),
$VTR = 10^{-4}$ (gene-duplication), $VTR = 10^{-2}$ (parallel PfGA)

(7)Michalewicz' Function (Mi):
multimodal, asymmetric, separable
$f(\mathbf{x}) = -\sum_{i=1}^{n}\sin(x_i)\sin^{2m}(\frac{ix_i^2}{\pi})$,
$m = 10$, $0 \leq x_i \leq \pi$ (22*bit*),
$VTR = -4.687(5dim.)$, $VTR = -9.66(10dim.)$

(8)Shekel's foxholes (Sh):
multimodal, asymmetric, inseparable
$f(\mathbf{x}) = -\sum_{i=1}^{m}\frac{1}{\sum_{j=1}^{n}(x_j - a_{ij})^2 + c_i}$,
$m = 30$, $0 \leq x_i \leq 10$ (24*bit*), $VTR = -9$

(9)Generalized Langerman's Function (La):
multimodal, asymmetric, inseparable
$f(\mathbf{x}) = -\sum_{i=1}^{m} c_i \exp\left[-\frac{1}{\pi}\sum_{j=1}^{n}(x_j - a_{ij})^2\right]$
$\times \cos\left[\pi\sum_{j=1}^{n}(x_j - a_{ij})^2\right]$,
$m = 5$, $0 \leq x_i \leq 10$ (24*bit*), $VTR = -1.4$

Experiments were performed using the optimization (minimization) problems shown in Table 1. These functions include frequently used benchmark functions [11] and its randomly shifted functions in terms of variable x_i using a uniformly random number r_i ($0 \leq r \leq 1$). These constitute a set of general form of functions categorized into eight classes which are unimodal or multimodal, symmetrical or asymmetrical, and separable or inseparable in terms of variable x_i. Five functions such as (1)Sphere model(Sp), (6)Griewank's function(Gr), (7)Michalewicz' function (Mi), (8)Shekel's foxholes (Sh) and (9)Generalized Langerman's function (La) in Table 1 are used in the first ICEO (Int. Contest on Evolutionary Computation) held in 1996 [12]. Depending on each function, the defined variables were encoded using 22 to 31 bits in Gray coding as shown in Table 1.

2.1 Experiment on the Parallel Distributed PfGA

For these nine functions in five and ten dimensional versions, three hundred different trials from different random seeds were run using four kinds of different migration methods by increasing the number of subpopulations N as 1,2,4,6,8,16,32,64,80 and 100. For $N = 1$ (single processing) only, ten thousand trials were run in order to get enough precision. PVM software [13] was applied to a parallel machine, the DEC Alpha with eight processors. The success rates, ENES, BV (best value) and RT (relative time) are evaluated, where the ENES index represents the mean number of function evaluations needed to reach a certain fitness value, value to reach (VTR), given with each problem as shown in Table 1. RT is defined by (CT-ET)/ET, where CT is the total CPU time to perform the algorithm with 10,000 iterations, and ET is the CPU time to perform 10,000 evaluations of the fitness function. The termination criterion, the maximum number of evaluations per one run, was defined as 10,000 and 100,000 evaluations for five and ten dimensional version of each function, respectively. For some functions the dimension was increased up to 30 to clarify the performance of the parallel PfGA depending on the dimension.

2.2 Experiment on GDGA

The four types of gene duplication algorithms were applied to nine optimization (minimization) problems.

PfGA or SSGA are compared for each subpopulation S_i'. A subpopulation size of 100, a crossover rate of 0.8 and a mutation rate of 0.05 were used for the SSGA. The "delete least-fit strategy" [8] used selection of the two least-fit individuals in each generation. Other genetic operations were the same as those used for the PfGA so that a fair comparison would be made. Genes were duplicated according to each type (type A to D) of duplication with a rate of duplication R. The rate of duplication R used for the PfGAs varied from 0 to 1.0, in steps of 0.2. The rate R, which represents the number of individuals in the SSGA was varied from 0 to 50, in steps of 10. The rate R of 0 corresponds to the basic PfGA or SSGA in each case. Migration happened every one hundred evaluations. The dimension n is five. Three hundred independent trials were performed for each subpopulation S_i' with different random seeds. For some functions the dimension was increased up to 30 to clarify the performance of the gene-duplication dependending on the dimension.

3 Experimental Results

3.1 Results on the PfGA

We performed some comparative experiments of the PfGA with the SSGA (Steady-state GA) [8] using another set of test functions including the benchmark funcions in the first ICEO [26,28]. The parameters used for the SSGA are as follows; the population size is 100, the crossover rate is 0.8, and the mutation rate is 0.05. For the crossover operation multiple-crossover as same as the PfGA was used. For the selection operation the "delete-least-fit" strategy was used where the two worst individuals are deleted from the population.

As a result of that, the PfGA is statistically superior to the SSGA in terms of the best (10^5 times better than the SSGA), online and offline performances (10 times better than the SSGA) with more reduced standard deviations (this means the PfGA is more stable than the SSGA). Furthermore, the PfGA was compared with eight algorithms participated in the first ICEO[12]. The results are shown in Table 2 although only the results of algorithms within third place are shown. The first place algorithm [20] is "Inductive Search" where the Brent's method is used for global search and the hill-climbing method for local search. The second place algorithm [21] uses Latin Square Theory where the success rates reached 100% for the five functions. The third place algorithm [22] uses Differential Evolution efficient for optimizing real-valued multi-modal objective functions.

For the 5 dimensional version the PfGA became the second place, and for the 10 dimensional version it became the sixth place in terms of ENES (i.e., the less ENES, the better in the ICEO). The PfGA is found to be a compact algorithm compared to other algorithms because of less RT.

3.2 Results on Parallel Distributed PfGA

Generally speaking, there happened two cases of results as the number of subpopulations increased; One such case was that the success rates R monotonously decreased, and another case is that the success rates took a peak at a certain value N. ENES, the performance of convergence, mostly decreased as $1/N$, which showed an effect of the parallel processing with migration methods. Only typical results were shown in the Figures. As an example of the former case, Fig. 12 shows an example of the Sphere model's results in 30 dimensions where the success rates decrease monotonously, and the ENES is retro-proportional to N. Also, the functions (2)-(4) in Table 1 showed the similar property. As for the migration methods, the UD1 was the best.

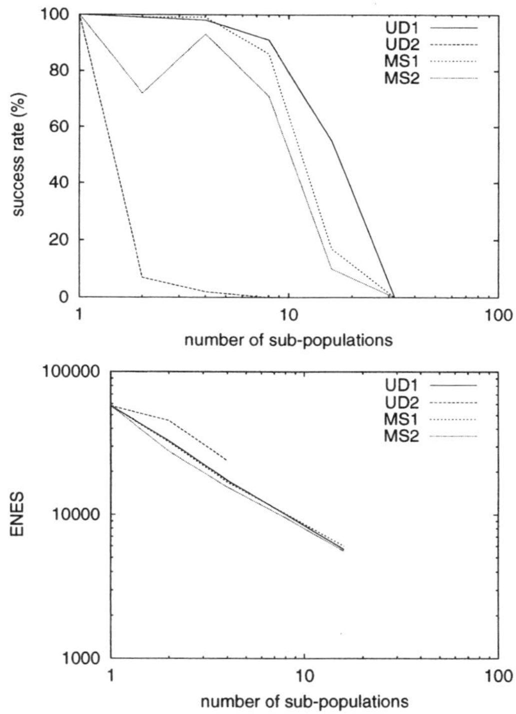

Fig. 12. Success rates and ENES in 30 dimensional sphere model

As examples of the latter case, Fig. 13, 14 and 15 show the success rates and ENES of the Michalewicz', Shekel's and Langerman's functions in five dimensions, respectively. The success rates take their peaks in $N = 32 \sim 64$ for UD1, and the peaks in $N = 4 \sim 16$ for other migration methods. The ENES decrease mostly as $1/N$. This situation that the success rates take their peaks at some N is similar in the cases of ten dimensions. Also, the Rastrigin's and Griewank's functions in Table 1 showed the same tendencies. As for the architectures with migration methods, UD1 was the best and the MS1, MS2 and UD2 in this order followed UD1 in all cases.

The detailed results for other functions are summarized in Table 4 as the best strategies, where the architectures and migration methods are shown in the case of UD1. The "best strategy" means the success rates and ENES for

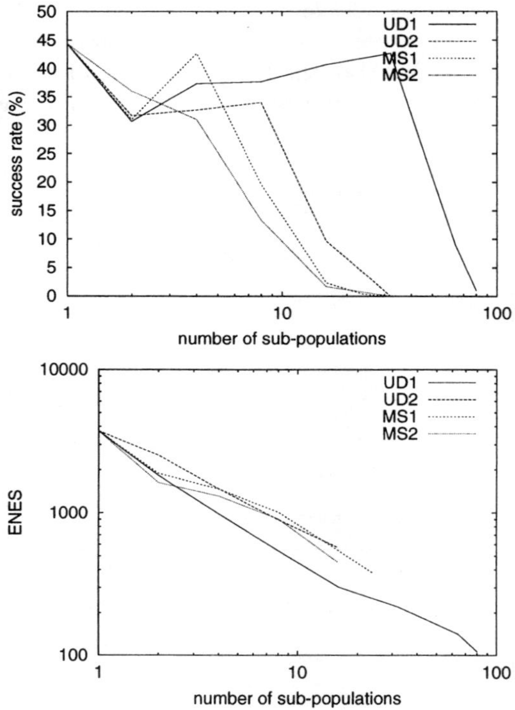

Fig. 13. Success rates and ENES in 5 dimensional Michalewicz' function

N that minimizes the function:

$$F(N) = \lambda(1 - Succ(N)) + (1 - \lambda)\frac{ENES(N)}{ENES(1)},$$

where $\lambda = 0.5$, which is a multi-objective optimization between the failure rates $1 - Succ(N)$ and the normalized ENES. The values in Table 4 are approximations.

Fig. 16 shows the success rates and ENES in 30 dimensional Griewank's function using 500,000 evaluations at maximum for each trial. When the dimension increased up to 30, the best migration type for the success rates changed to UD2 followed by MS2, UD1 and MS1. This experimental results implies that the more complex the multimodal function becomes, the more diversity in the population is necessary for more local optima because UD2 and MS2 produce two new offspring in the subpopulations S'_j and S'_0, respectively.

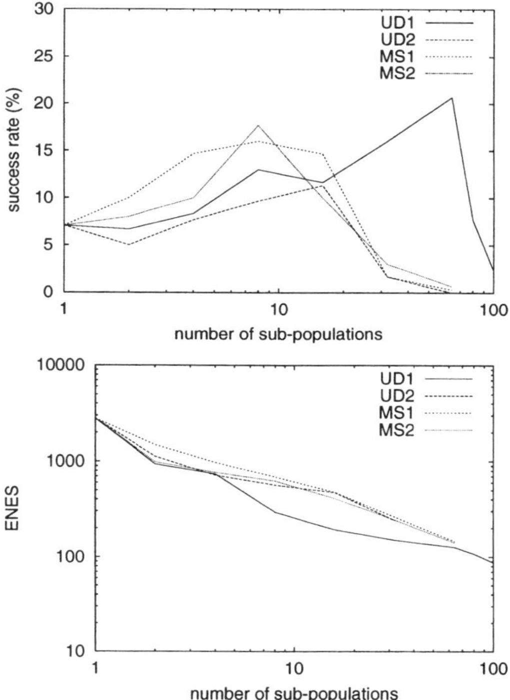

Fig. 14. Success rates and ENES in 5 dimensional Shekel's foxholes

3.3 Results on Gene-Duplicating GA

Success rates and ENES for each function are given in this section, as a function of migration rates, for the better of the two GAs (PfGA or SSGA).

(1)Sphere Model (Sp)

The sphere model is unimodal, symmetrical and separable. Since the success rate R_s for this model was 100% for all GAs, the figure is omitted. Fig. 17 shows the ENES for the four the PfGA-based methods. ENES based on the SSGA is not shown, but all types of gene duplication were more effective for both PfGAs and SSGAs, than for either in its basic form (i.e., for R=0). Type B was the best of the variants because the global minimum for the sphere model is at x_i =1, for any dimensions. The best ENES for the PfGA-based GDGA was 200, and for the SSGA-based GDGAs it was 750.

(2)Schwefel's Double Sum (Ds)

This function is unimodal, symmetrical and inseparable. ENES for the PfGA-based GDGAs is shown in Fig. 18. Although the success rates of only 2% without migration are not shown in the figure, the effects of migration appear immediately, and success rates with migration were 100% for all types

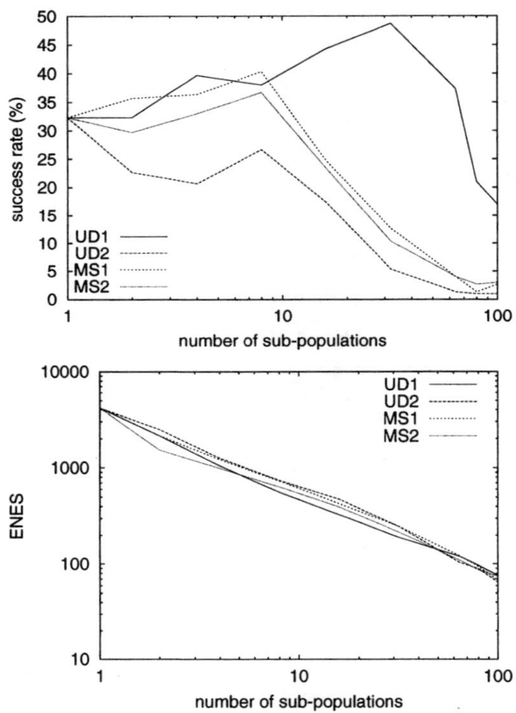

Fig. 15. Success rates and ENES in 5 dimensional Langerman's function

of duplication. Types A, B, C and D, in that order, displayed the better perfromance over the range of migration rates from 0.2 to 0.5. The SSGA-based GDGAs also displayed an immediate effect from migration and the success rates reached 100% for all types of duplication from a success rate of zero without migration. The ENES for type B showed some improvement over the ENES for other types (1550 for a migration rate of 40).

(3)Random Sphere model (Rs)

This function is unimodal, asymmetrical and separable. In Fig.19, the ENES is again shown for PfGA-based GDGAs. The success rates are 100% in all cases, and the ENES for types A and D improved dramatically. Types B and C are much less effective than for the sphere model (Fig. 17) . Accordingly, types A and D can be said to be robust in terms of the asymmetry of solutions, whereas types B and C are ineffective for asymmetrical solutions. The ENES for types A and D in these results are equivalent to or slightly worse than those shown in Fig. 17. For SSGA-based GDGAs, the success rate

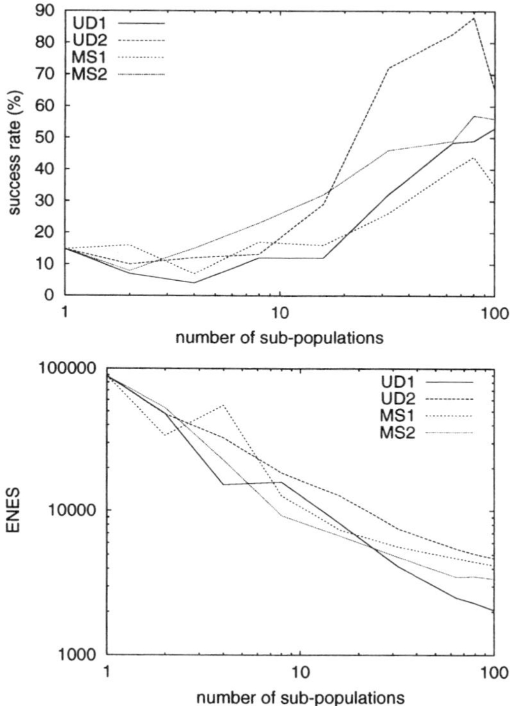

Fig. 16. Success rates and ENES in 30 dimensional Griewank's function

of types A and D was 100% whereas types B and C had a success rate of zero.

(4)Random Double Sum (Rd)

This function is unimodal, asymmetrical and inseparable. Success rates and ENES for the PfGA-based GDGAs are shown in Figs. 20 and 21, respectively. As migration rates increase, the success rates also increase, and reach 91% for type D at a migration rate of 1.0. Type D also displayed the best ENES. Success rates for all SSGA-based GDGAs were zero. Type D had the best BV with a value of about 0.01.

(5)Rastrigin's function (Rs)

This function is multimodal, symmetrical and separable. ENES for the PfGA-based GDGAs are shown in Fig. 22. Success rates without migration were 20%, but the rates immediately reached 100% as soon as migration was introduced, for all four types, and the ENES also tended to decrease. Type B appeared to be the most efficient. Success rates also reached 100% for all types of SSGA-based GDGAs, and the best ENES was achieved by type B

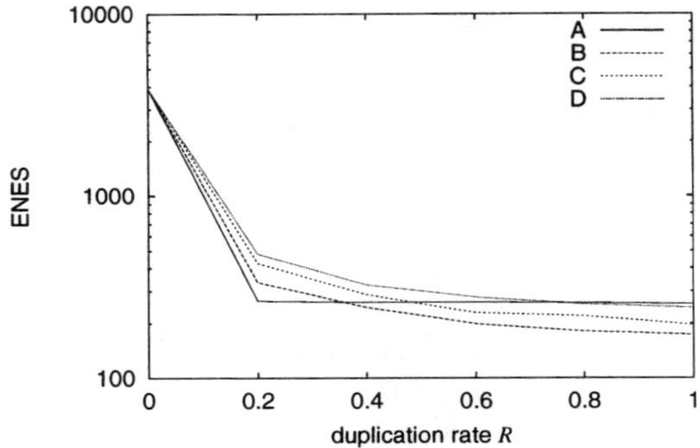

Fig. 17. ENES as a function of R for the sphere model (PfGA-based)

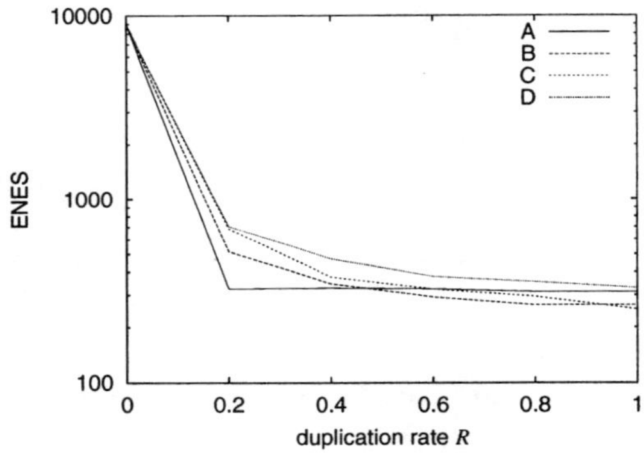

Fig. 18. ENES as a function of R for Schwefel's double sum (PfGA-based)

(ENES=2400 at a migration rate of 30).

(6) Griewank's Function (Gr)

This function is multimodal, symmetrical and inseparable. Success rates R_s and ENES for the SSGA-based GDGAs are shown in Figs. 23 and 24, respectively. Type B is the best with a success rate that almost reached 100%, and its ENES reached 4,000 at a migration rate of 50. The PfGA-based GDGAs showed a similar increase in the success rates for all types except type A. The ENES displayed a gradual decrease for types B and C. Type B was the best (a success rate of 50%, and ENES of about 2,500) and most easily reached the solution.

(7) Michalewicz' Function (Mi)

This function is multimodal, asymmetical and separable. The success rates and ENES for the PfGA-based GDGAs are shown in Figs. 25 and 26, respec-

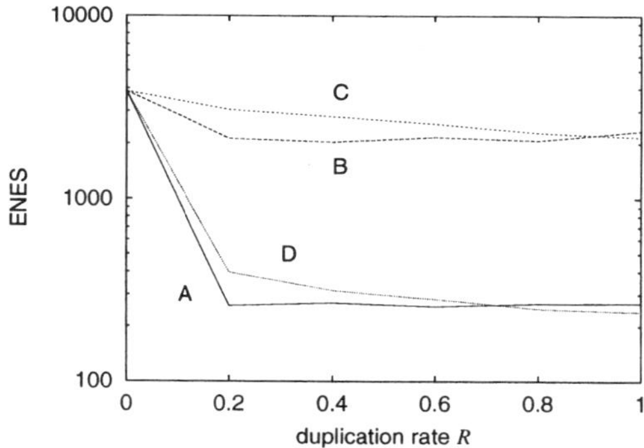

Fig. 19. ENES as a function of R for the random sphere model (PfGA-based)

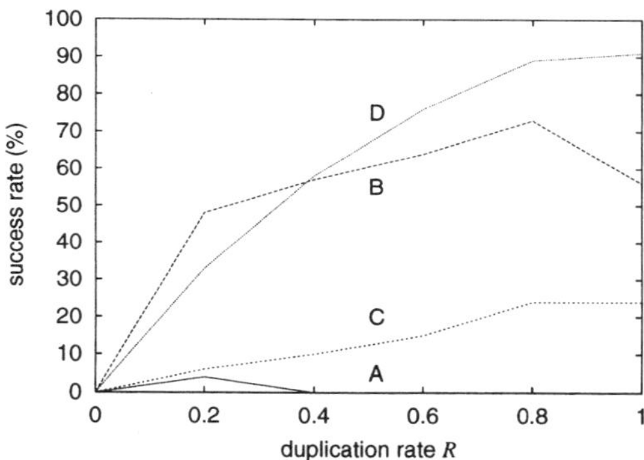

Fig. 20. Success rates as a function of R for the random double sum (PfGA-based)

tively. The effects of gene duplication are most significant to types A and D, with success rates that reach 100% and a dramatic decrease in the ENES with migration (ENES=400). Success rates for the SSGA-based GDGAs similarly reached 100% and the ENES was 1,000 for types A and D.

(8) Shekel's foxholes (Sh)

This function is multimodal, asymmetrical and inseparable. All of the success rates were less than 10%, and gene duplication had no apparent effect. In terms of the success rates, type D was relatively inferior to the other types. The situation was similar for SSGA-based GDGAs. This function seems to present difficulties to gene-duplicating GAs in terms of finding the solution.

(9) Langerman's Function (La)

This function is also multimodal, asymmetrical and inseparable. Success rates and ENES for the SSGA-based GDGAs are shown in Figs. 27 and 28.

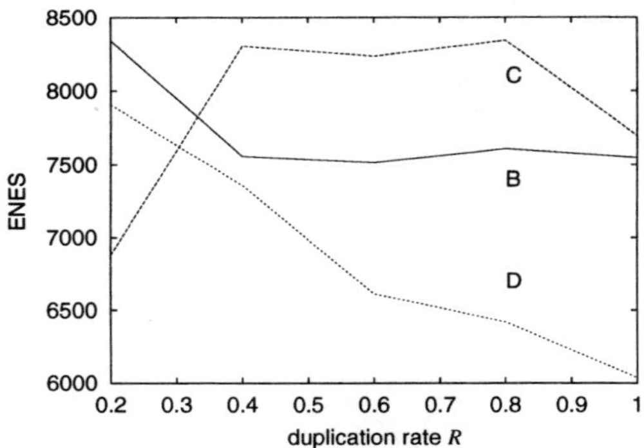

Fig. 21. ENES as a function of R for the random double sum (PfGA-based)

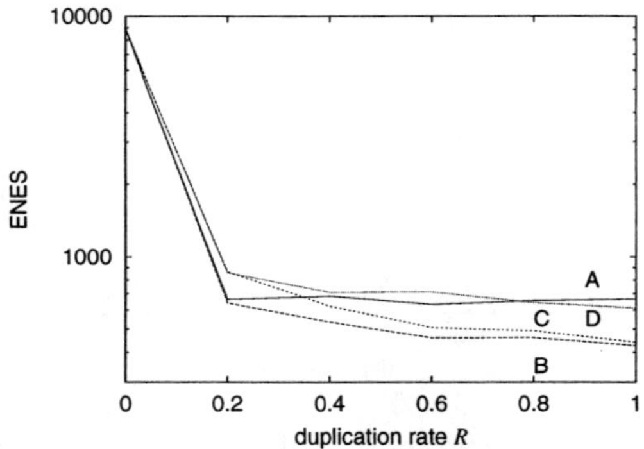

Fig. 22. ENES as a function of R for Rastrigin's function (PfGA-based)

Although the maximum success rate was 35% for PfGA-based GDGAs, the rates were around 95 to 97% for all SSGA-based GDGAs which demonstrates a strong effect from the use of a population-based GA. Significant differences don't appear clearly among the gene duplication types, however, types B, C and D displayed a better ENES than type A.

Fig. 29 shows an example of ENES as a function of dimensions for Schwefel's Double Sum. Although ENES of the PfGA exceeded 23,000 evaluations, ENES for all types of duplication don't increase so much up to 30 dimensions. This similar situation happened for the Sphere model and the Randomized Sphere model as well when the gene duplication worked effectively.

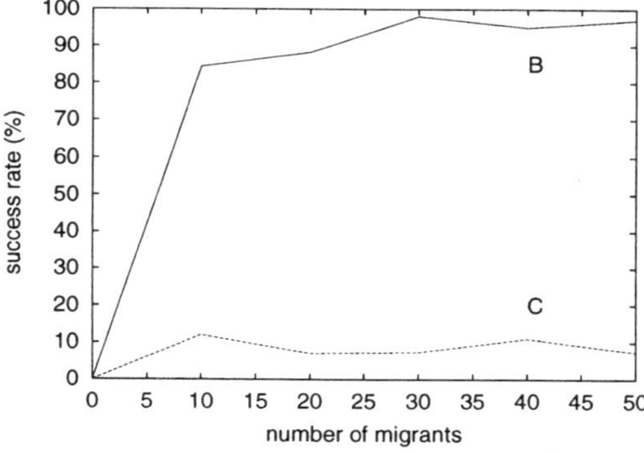

Fig. 23. Success rates as a function of the number of migrants for Griewank's function (SSGA-based)

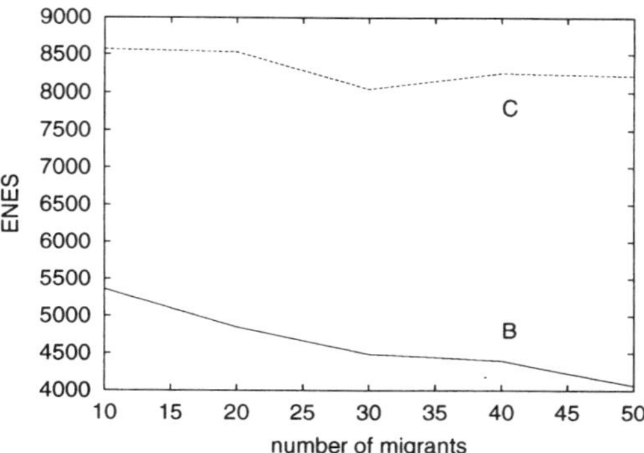

Fig. 24. ENES as a function of the number of migrants for Griewank's function (SSGA-based)

3.4 Comparison with the Algorithms Which Participated in ICEO'96

Case of the PfGA: If the PfGA had participated in the ICEO, it would have taken the second place for the 5 dimensional version and the sixth place for the 10 dimensional version as shown in Table 2. It is relatively compact, has a lower RT, and in general shows good performance.

Case of Parallel Distributed PfGA: We describe the effectiveness of the results compared to eight algorithms participated in the first ICEO'96 [12]. Table 2 shows the results within the third place, the PfGA and the parallel PfGA (UD1) where the upper block shows the results of five dimensions and the lower block shows the results of ten dimensions. Each line shows

Table 2. Results in the ICEO'96(upper: 5 dim, lower: 10 dim)

ENES BV RT	Bilchev and Parmee[20]	Li and Smith[21]	Storn and Price[22]	PfGA[28]	Parallel PfGA (N)
Sphere model (Sp)	20 3.88e-15 2	243 0.0 12.7	736 - 4.67	4,067 0.0 0.91	173 4.52e-7 2.3 (56)
Griewank's function (Gr)	41 7.99e-6 2	21,141 1.69e-5 3.1	5,765 - 1.79	6,785 4.66e-7 0.90	846 4.22e-05 1.88 (8)
Shekel's foxholes (Sh)	74 -10.327 2	6,318 -10.403 0.25	76,210 - 0.80	1,619 -10.40 0.33	87 -9.7624 2.24 (100)
Michalewicz' function (Mi)	120 -4.6876 2	6,804 -4.687 1.28	1,877 - 1.11	5,131 -4.688 0.90	105 -4.68749 1.93 (80)
Langerman's function (La)	176 -1.499 2	4,131 -1.499 1.62	5,308 - 1.35	5,274 -1.499 0.43	76 -1.49752 1.93 (100)

ENES BV RT	Bilchev and Parmee[20]	Li and Smith[21]	Storn and Price[22]	PfGA[28]	Parallel PfGA (N)
Sphere model (Sp)	40 7.10e-15 2	243 0.0 13.6	1,892 - 4.88	11,814 0.0 0.91	605 2.39e-10 2.3 (100)
Greiwank's function (Gr)	79 1.31e-6 2	20,898 5.782e-5 3.0	13,508 - 1.77	25,028 3.11e-15 0.90	876 1.07e-7 1.88 (64)
Shekel's foxholes (Sh)	120 -10.101 2	6,075 -10.207 0.42	744,250 - 0.66	3,586 -10.2079 0.33	321 -10.2079 2.24 (64)
Michalewicz' function (Mi)	501 -9.66 2	14,823 -9.66 1.25	10,083 - 0.68	63,424 -9.66015 0.90	876 -9.66015 1.93 (64)
Langerman's function (La)	372 -1.499 2	26,973 -1.50 1.78	44,733 - 1.46	39,841 -1.5 0.43	950 -1.49999 1.93 (64)

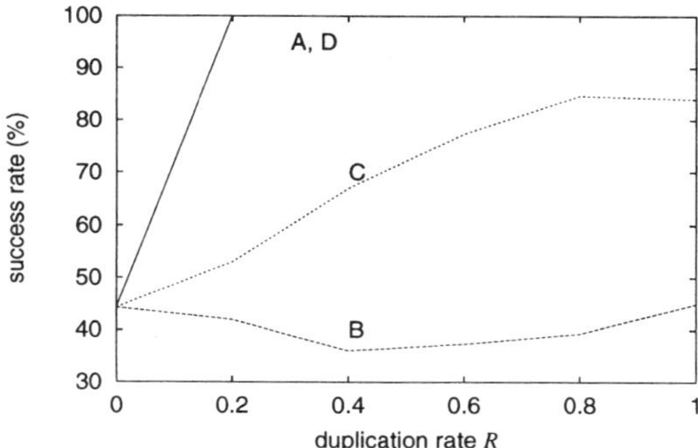

Fig. 25. Success rates as a function of R for Michalewicz' function (PfGA-based)

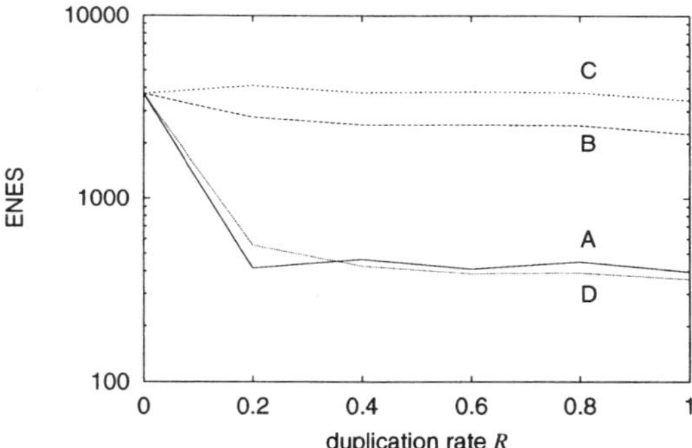

Fig. 26. ENES as a function of R for Michalewicz' function (PfGA-based)

ENES, BV and RT from the top. The PfGA corresponds to the results in $N = 1$. The values (N) in the lines of the parallel PfGA show the number of subpopulations when ENES becomes its minimum. If we compare the ENES among these algorithms, the parallel PfGA (UD1) becomes the first place for the five dimensional version of Michalewicz' and Langerman's functions, the second place for other five dimensional version of functions, and the second place for ten dimensional version of all functions except the Sphere model. This fact means the high search ability of the parallel PfGA (UD1).

Case of Gene Duplication GA: The GDGA is compared with eight of the algorithms which were entered in the first ICEO in Table 3. Each set of three lines shows the ENES, BV and RT, from the top. Perfromance was judged at the ICEO on the basis of the smallest ENES over 20 runs. The results for the PfGA and GDGA in Table 3 are averaged over 20 runs for comparison with

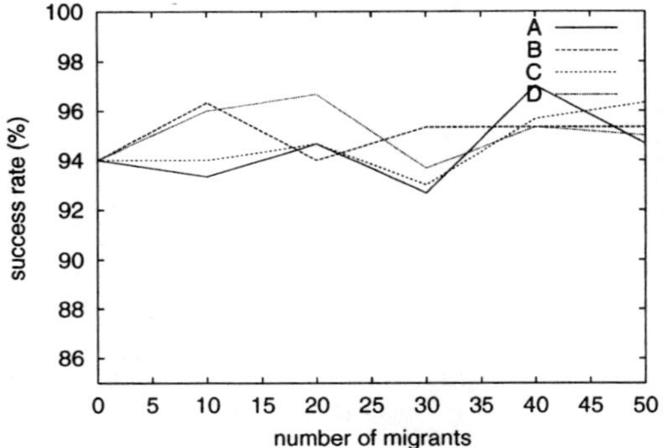

Fig. 27. Success rates as a function of the number of migrants for Langerman's function (SSGA-based)

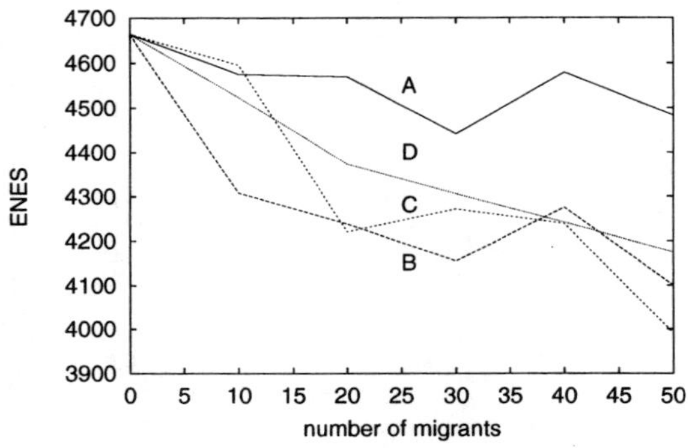

Fig. 28. ENES as a function of the number of migrants for Langerman's function (SSGA-based)

ICEO results. Although the RT is higher for the GDGA version, the ENES is significantly lower for the PfGA-based GDGA. As a whole, the PfGA-based GDGA would have taken second place and the SSGA-based GDGA would have taken third place.

In comparison with the parallel PfGA in Table 2 and the GDGAs in Table 3 for 5 dimensions, the parallel PfGA is better than the GDGA as a whole.

4 Discussion

4.1 Discussion on PfGA

Considering further the relationship between the PfGA and the *disparity theory of evolution*, the chromosome of best individual in *family* corresponds

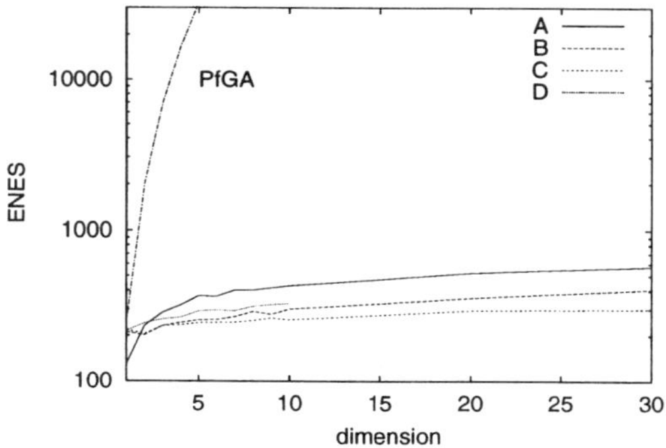

Fig. 29. ENES as a function of dimensions for Schwefel's Double Sum (PfGA-based)

Table 3. Results at the 1st ICEO, for PfGAs and for GDGAs

ENES BV RT	Bilchev and Parmee[20]	Li and Smith[21]	Storn and Price[22]	PfGA[28]	GD-PfGA	GD-SSGA
Sphere Model (Sp)	20 3.88e-15 2	243 0.0 12.7	736 - 4.67	4,067 0.0 0.91	173 0.0 2.39	723 0.0 71.26
Griewank's function (Gr)	41 7.99e-6 2	21,141 1.69e-5 3.1	5,765 - 1.79	6,785 4.66e-7 0.90	1367 3.65e-10 2.35	2537 3.65e-10 70.76
Shekel's foxholes (Sh)	74 -10.327 2	6,318 -10.403 0.25	76,210 - 0.80	1,619 -10.40 0.33	456 -10.404 2.40	3632 -10.396 41.47
Michalewicz' function (Mi)	120 -4.6876 2	6,804 -4.687 1.28	1,877 - 1.11	5,131 -4.688 0.90	240 -4.68766 3.54	786 -4.68766 79.24
Langerman's function (La)	176 -1.499 2	4,131 -1.499 1.62	5,308 - 1.35	5,274 -1.499 0.43	2,330 -1.499 2.12	3,460 -1.499 52.83

to a *wild-type* at present which is a *leading strand* in the "disparity theory of evolution." Offsprings generated by the crossover and mutation from two parents correspond to *lagging strands*. In selection the best individual (P_1) in the Cases 2 and 3 is regarded as a current wild-type whereas the best individual (C_1) in Cases 1 and 4 is regarded as a *new* wild-type because P_1 is replaced with C_1. In other words, C_1 (i.e., a new wild-type) is produced by "pivoting" on P_1 (i.e., a current wild-type) exploring the neighborhood of P_1, thus resulting in the better location in the search space. The reason why only one offspring is mutated while preserving another offspring is to maintain the characteristic (chromosome) inherited from two parents.

Therefore, the leading strand with a small mutation rate in a sense "conservative" to the change of environment, whereas the lagging strand is "revolutionary" or innovative. These features in the two kinds of strands produce "diversity" of population where the leading strand contributes to stability of population, and the lagging strand contributes to flexibility of population. Furthermore, the diversity of individuals maintains a delicate balance of search process between exploitation and exploration in the PfGA.

4.2 Discussion on Parallel PfGA

Table 4. Best strategy of the parallel PfGA (UD1)

function	dim	uni sym sepa	N	%success	ENES
Sphere	5	Yes Yes Yes	16	95	429
model (Sp)	10		64	100	667
Schwefel's	5	Yes Yes No	8	74	816
double sum (Ds)	10		8	92	8,451
Randomized	5	Yes No Yes	16	96	442
Sphere model (Rs)	10		64	100	733
Randomized	5	Yes No No	4	90	1,482
Double sum (Rd)	10		8	95	8,227
Rastrigin's	5	No Yes Yes	8	45	918
function (Ra)	10		16	89	3,598
Griewank's	5	No Yes No	32	0.3	296
function (Gr)	10		64	4	877
Michalewicz'	5	No No Yes	32	43	220
function (Mi)	10		64	3	877
Shekel's	5	No No No	64	20	127
foxholes (Sh)	10		64	3	321
Langerman's	5	No No No	32	49	189
function (La)	10		64	3	950

As we can see from Fig. 12∼15, for unimodal functions or the functions with relatively lower dimensions UD1 is the far best compared to other migration methods. This fact means that the uniformly distributed architecture is better than the master-slave one, and the direct migration method is better than the hierarchical one. It is because the best individual becomes a candidate to migrate in the master-slave type whereas the locally best individuals becomes the candidates to migrate in the uniformly distributed type which increase the diversity of population and decrease the possibility to fall the individuals into local optima. Although the reason why the hierarchical migration method is adopted was to increase the diversity of population, this method decreased the efficiency in search by increasing the number of evaluations for the newly produced individuals for unimodal functions or functions with relatively lower dimensions. However, for the multimodal functions with relatively higher dimensions, say 30 dimension, UD2 became the best in terms of success rates as shown in Fig. 16. This imples that the hierarchical migration method is more effective for more complex functions because more diversity in the population is necessary for finding out their global optima. Furthermore, the success rates increased as N increased from N=1 (i.e., the PfGA itself). This fact shows the significant effect of the migration methods.

In Table 4 the best results were obtained for the relatively small N in the functions from Sp to Rd except the ten dimensional Sphere model and Randomized Sphere model, whereas the best results were obtained for the relatively larger N in the fuctions from Ra to La. Namely, the success rates tend to decrease monotonously when an objective function is unimodal, and the success rates tend to take their peak when an objective function is multimodal. This is why the strategy with a few individuals is good to search for the unimodal functions, and the strategy with the large N (i.e., many individuals) is necessary to search for the multimodal functions. As the average number of individuals in a subpopulation is two, more than one hundred individuals make it possible to efficiently search for the multimodal functions.

4.3 Discussion on Gene-Duplicating GA

Table 5. Comparative results for gene-duplicating GAs based on the PfGA and SSGA

	feature			GD-PfGA			GD-SSGA			best strategy		
func	uni	sym	sepa	best	%succ	ENES	best	%succ	ENES	type	base	rate
Sp	Yes	Yes	Yes	B	100%	200	B	100%	750	B	PfGA	≥ 0.2
Ds	Yes	Yes	No	A	100%	300	B	100%	1550	A	PfGA	≥ 0.2
Rs	Yes	No	Yes	A,D	100%	300	D	100%	1500	A,D	PfGA	≥ 0.2
Rd	Yes	No	No	D	91%	6000	D	0%	-	D	PfGA	~ 1
Ra	No	Yes	Yes	B	100%	400	B	100%	2400	B	PfGA	~ 1
Gr	No	Yes	No	B	50%	2500	B	97%	4000	B	SSGA	≥ 0.5
Mi	No	No	Yes	A,D	100%	400	A,D	100%	1000	A,D	PfGA	≥ 0.2
Sh	No	No	No	any	8%	3000	any	1%	5000	any	PfGA	$0 \sim 1$
La	No	No	No	any	35%	4000	any	95%	4000	any	SSGA	≥ 0.5

The PfGA-based GDGA is compared with the SSGA-based GDGA. Table 5 shows the success rates and ENES for the best types of PfGA- and SSGA-based GDGAs, the base form of the type that had the best strategy, and its migration rate. The success rates and ENES vary according to the migration rates, so these values are approximations that use the best types. The best type for both PfGA- and SSGA-based GDGAs is the same for almost every function. The PfGA-based GDGA displays better convergence on the whole as can be seen by comparing the ENES values. Through comparison of the success rates for Griewank's and Langerman's functions (Gr and La), SSGA-based GDGAs can be seen to be superior to PfGA-based GDGAs for multimodal and inseparable functions.

As a whole, the higher the migration rate, the better the performance. The type of population should be selected according to the number of local optima in each function. The PfGA-based GDGA is the better strategy for fast convergence in unimodal functions and those with only a few local optima whereas the SSGA-based GDGA is the better strategy for multimodal functions and those with many local optima.

All types perform well on unimodal and symmetrical functions such as the sphere (Sp) model and Schwefel's double sum (Ds). However, the best type slightly differs with migration according to whether the function is separable or not, i.e., type B is best for separable functions, and type A is best for inseparable functions. Generally speaking, type B, a gene-prolonging GA, is adequate for all symmetrical functions, i.e., Sp, Ds, Ra and Gr. Conversely, types A and D are adequate for asymmetrical functions such as Rs, Rd, Mi, Sh and La, and the effects of migration appear immediately. This fact can also be understood by looking at the structures in types A to D. As a whole, a good strategy is to take type B or C for symmetrical functions and type A or D for asymmetrical functions. For multimodal, asymmetric and inseparable function such as Langermans's function the SSGA-based GDGA is more effective with the higher migration rates because more diverisity in the population is necessary for finding out their global optima, as shown in Figs. 27 and 28.

5 Conclusion

The PfGA shows a rapid evolutionary behavior, does not require almost the tuning of genetic parameters, and is easy to construct. Its ease of construction and the fact that it does not need parameter tuning are particularly important in all practical applications. This scheme of PfGA may easily applied to combinatorial optimization problems such as TSP (Traveling Salesman Problem) and JSP (Job-shop Scheduling Problem) if the representation for solutions is appropriately coded onto genes.

We also described the effects of migration methods in the parallel distributed Parameter-free GA architectures. Two kinds of parallel architectures and two kinds of migration methods were combined and implemented in a parallel machine, thus four migration methods were synchronously processed comparing the performance with each other. In evaluating the benchmark problems including five and ten dimensional versions of a general set of test functions, the uniformly distributed architecture with the direct migration method made it possible to decrease the ENES as $1/N$ for all functions and further improve the success rates depending on the problems by increasing the number of subpopulations. These results imply that the parallel distributed PfGA is comparable to the results in the international contest. This synchronously parallel processing can also be executed in a single processor if all subpopulations are assigned to one CPU and switching the process of each generation.

Although parallel processing is always accompanied with a parameter, i.e., the degree of parallel processing (or, the number of subpopulations in the case of the parallel PfGA), we may consider an extension which does not explicitly set the parameter in advance. That is, as the number of subpopulation size is dynamically determined, the number of subpopulations in the parallel PfGA also dynamically determined depending on the test functions. This extension may be one of the future studies.

We have also extended and strengthened the performance of the gene-duplicating GA (GDGA) by comparing its application to PfGAs and SSGAs as the basis of a GA in which subpopulations evolve. The GDGA has four variants, a gene-concatenating, a gene-prolonging, a gene-coupling and an extended gene-coupling type. These schemes divide a given problem into subproblems and synthesize them based on GAs. Each individual concatenates sub-solutions obtained so far and makes them migrate among subpopulations. We used a set of general test functions including benchmark problems that were used at the first ICEO to evaluate the performance of the GDGA. As a result, the best strategy including the best type of duplication, good migration rates and the best basic GA were ascertained accroding to the features of the test functions. If the type of gene duplication is selected appropriately for a given type of problem, it is then possible to obtain globally optimal solutions faster and more efficiently than with either the PfGA or SSGA (i.e. GAs without migration), for the general set of problems.

In a future study, we will apply these schemes to combinatorial problems and to problems in dynamic environments.

References

1. Darwin C. (1859) On the Origin of Species by Means of Natural Selection or the Preservation of Favoured Races in the Struggle for Life, London, John Murray
2. Holland J.H. (1975) Adaptation in Natural and Artificial Systems, The University of Michigan Press
3. Goldberg D.E. (1989) Genetic Algorithms in Search, Optimization, and Machine Learning, Reading, MA, Addison-Wesley
4. Rechenberg I. (1965) Cybernetic Solution Path of an Experimental Problem, Royal Aircraft Establishment Library Translation 1122

5. Fogel L. J. (1966) Artificial Intelligence through Simulated Evolution, New York, Wiley
6. Ohno S. (1970) Evolution by Gene Duplication, Berlin, Springer-Verlag
7. De Jong K. (1975) An Analysis of the Behavior: a Class of Genetic Adaptive Systems, Doctoral dissertation, University of Michigan
8. Syswerda G. (1991) A Study of Reproduction in Generational and Steady-State Genetic Algorithms, Foundation of Genetic Algorithms, pp 94-101, San Francisco, Morgan Kaufmann
9. Furusawa M., Doi H. (1992) Promotion of Evolution: Disparity in the Frequency of Strand-specific Misreading Between the Lagging and Leading DNA Strands Enhances Disproportionate Accumulation of Mutations, J. Theor. Biol., **157**, 127-133
10. Wada K., Doi H., Tanaka S., Wada Y., Furusawa M. (1993) A Neo-Darwinian Algorithm: Asymmetrical Mutations due to Semiconservative DNA-type replication Promote Evolution, Proc. Natl. Acad. Sci. USA, **90**, 11934-11938
11. Baeck T., Fogel D., Michalewicz Z. (Eds.) (1997) Handbook of evolutionary computation, New York, Oxford University Press
12. The Organising Committee: Bersini H., Doringo M., Langerman S., Seront G., Gambardella L. (1996) Results of the First International Contest on Evolutionary Optimization (1st ICEO), 1996 IEEE International Conference on Evolutionary Computation (ICEC'96), 611-615
13. Geist A., Beguelin A., Dongarra J., Jiang W., Manchek R., Sunderam V. (1993) PVM User's Guide and Reference Manual, **5**
14. Hinterding R., Michalewicz Z., Eiben A.E. (1997) Adaptation in Evolutionary Computation: A Survey, Proceedings of the 1997 IEEE International Conference on Evolutionary Computation, 65-69
15. Belding T.C., (1995) The Distributed Genetic Algorithm Revisited, Proceedings of the Sixth International Conference on Genetic Algorithms, 114-121, San Francisco, Morgan Kaufmann
16. Mahfoud S.W.(1995) A Comparison of Parallel and Sequential Niching Methods, Proceedings of the Sixth International Conference on Genetic Algorithms, 136-143, San Francisco, Morgan Kaufmann
17. Paz E.C., Goldberg D.E. (1997) Predicting Speedups of Idealized Bounding Cases of Parallel Genetic Algorithms, Proceedings of the Seventh International Conference on Genetic Algorithms, 113-126, San Francisco, Morgan Kaufmann
18. Evans I.K. (1998) Embracing Premature Convergence: The Hypergamous Parallel Genetic Algorithm, Proceedings of the 1998 International Conference on Evolutionary Computation (ICEC'98), 621-626
19. Maruyama T., Hirose T., Konagaya A. (1993) A Fine-Grained Parallel Genetic Algorithm for Distributed Parallel Systems, Proceedings of the Fifth International Conference on Genetic Algorithms, 184-190, San Francisco, Morgan Kaufmann
20. Bilchev G., Parmee I., (1996) Inductive Search, 1996 IEEE International Conference on Evolutionary Computation (ICEC'96), 832-836
21. Li D.G., Smith C. (1996) A New Global Optimization Algorithm based on Latin Square Theory, 1996 IEEE International Conference on Evolutionary Computation (ICEC'96), 628-630
22. Storn R., Price K., (1996) Minimizing the Real Functions of the ICEC'96 Contest by Differential Evolution, 1996 IEEE International Conference on Evolutionary Computation (ICEC'96), 842-844

23. Whitley D., Mathias K., Rana S., Dzubera J. (1995) Building Better Test Functions, Proceedings of the Sixth International Conference on Genetic Algorithms, 239-246, San Francisco, Morgan Kaufmann
24. Kizu S., Sawai H., Endo T. (1997) Parameter-free Genetic Algorithm: GA without Setting Genetic Parameters, Proceedings of the 1997 International Symposium on Nonlinear Theory and its Applications, **2**, 1273-1276
25. Sawai H., Kizu S., Endo T. (1998) Parameter-free Genetic Algorithm (PfGA), Trans. IEICE, Jpn, **J81-D-II**, 2, 450-452
26. Sawai H., Kizu S., Endo T. (1998) Performance Comparison of the Parameter-free Genetic Algorithm (PfGA) with Steady-state GA, Trans. IEICE, Jpn, **J81-D-II**, 6, 1455-1459
27. Sawai H., Kizu S. (1998) Parameter-free Genetic Algorithm Inspired by Disparity Theory of Evolution, Proceedings of the International Conference on Parallel Problem Solving from Nature, 702-711, Amsterdam, Netherlands
28. Kizu S., Sawai H. Adachi S.(1999) Parameter-free Genetic Algorithm (PfGA) Using Adaptive Search with Variable-size Local Population and Its Extension to Parallel Distributed Processing, Trans. IEICE, Jpn, **J82-D-II**, 3, 512-521
29. Sawai H., Adachi S. (2000) Effects of Hierarchical Migration in a Parallel Distributed Parameter-free GA, Proceedings of the Congress on Evolutionary Computation, San Diego, **2**, 1117-1124
30. Sawai H., Adachi S., (2000) A Comparative Study of Gene-Duplicated GAs Based on PfGA and SSGA, Proceedings of the Genetic and Evolutionary Computation Conferences, **1**, 74-81, Las Vegas

On the Design of Problem-specific Evolutionary Algorithms

Stefan Droste and Dirk Wiesmann

University of Dortmund
FB Informatik
44221 Dortmund, Germany
E-mail: droste@LS2.cs.uni-dortmund.de, wiesmann@LS11.cs.uni-dortmund.de

Summary. In this chapter a set of guidelines for the design of genetic operators and the representation of the phenotype space in evolutionary algorithms (EAs) is proposed. These guidelines should help to systematize the design of problem-specific EAs by making the genetic operators behave in a controlled fashion with respect to metrics on geno- and phenotype space. Because we assume that we have enough domain knowledge to choose metrics that smooth the fitness landscape, this controlled behavior should improve the efficiency of the EA.

The applicability of this concept is shown by the systematic design of a genetic programming system for finding Boolean functions. This system is the first genetic programming system to have reportedly found the 12 parity function.

1 Introduction

Evolutionary algorithms (EAs) are most often considered as very general search heuristics. Their main advantages compared to problem-specific search methods are their robustness, flexibility, and extensibility, and the fact that almost no domain knowledge is required for their implementation and application. The latter is especially true for canonical EAs. Nevertheless, it was shown in numerous publications that, depending on the problem at hand, the integration of domain knowledge yields systems with eventually much higher performance and reliability.

Ideally, the design of problem-specific EAs should be based on sound theories about their working principles. Unfortunately, up to now there exist only few theoretical results, but many, even opposing, hypotheses on the working principles of EAs. Hence, the EA designer has little more choice than to follow his or her favorite hypotheses, integrate some domain knowledge if available, and finally tune the algorithm through exhaustive testing. The integration of domain knowledge often improves the search process significantly, but in most cases this integration is not straightforward. In practice, this lack of theoretical knowledge makes the design process difficult and time consuming.

In this chapter we propose a set of guidelines for the design of genetic operators and the representation of the phenotype space. These guidelines should help to systematize the design of problem-specific EAs. By defining two related distance measures within the geno- and phenotype spaces so that neighboring elements have similar fitness, problem-specific knowledge can be integrated. The guidelines for the genetic operators guarantee that the search process will then behave in a controlled fashion. As we assume that the choice of the distance measures has "smoothened" the search space, this strategy should increase the performance of the EA.

We do not claim that the set of rules defined in the following sections is sufficient for a successful EA design. Nor do we say that this set of rules is different from traditional EA design. Contrarily, we assume that they just reflect the common notion of EAs. But in the majority of published applications these points are rarely discussed explicitly. In this respect our proposal should be taken as a small step towards a systematic design of EAs.

In the next two sections EAs are introduced formally and a short overview on the design of problem-specific EAs is given. Then the proposed guidelines are presented. In Section 5 we show how the guidelines can be applied to design a genetic programming (GP) system for finding Boolean functions. The experiments and their results, showing a remarkable increase in efficiency of this system in comparison to other GP systems, are presented in this section, too. The chapter ends with some comments about our choice of representation and a conclusion.

2 Representation and Operators

To localize an optimal point of an objective function $f : \mathcal{P} \to W$ (W is at least partially ordered) an EA uses a population of elements of $\mathcal{P}$, which are modified by genetic operators, like mutation, recombination, and selection. The implementation of an EA requires that the abstract elements of $\mathcal{P}$ have to be represented by a data structure, i.e. elements of a space $\mathcal{G}$. The set $\mathcal{P}$ is called the *phenotype* space and $\mathcal{G}$ the *genotype* space.

We assume that μ denotes the number of parents and λ the number of offspring in one generation. In the context of EAs it is sufficient to store an individual in the form of its genotype, since the genotype–phenotype mapping $h : \mathcal{G}^\lambda \to \mathcal{P}^\lambda$ is deterministic and environmental influences are not taken into consideration. Hence, we can denote a population at generation t by $Pop(t) \in \mathcal{G}^\mu$.

The mapping $h : \mathcal{G}^\lambda \to \mathcal{P}^\lambda$ can be composed simply from λ reduced mappings $h' : \mathcal{G} \to \mathcal{P}$, which represent the genotype–phenotype mapping for single individuals:

$$h(g) = (h'(g_1), \ldots, h'(g_\lambda)), \quad g = (g_1, \ldots, g_\lambda) \in \mathcal{G}^\lambda.$$

The mapping $h' : \mathcal{G} \to \mathcal{P}$ determines the abstract element $h'(x)$ of the search space being represented by $x \in \mathcal{G}$. Thus, the mapping h' describes the relation between genotype and phenotype space.

The recombination operator $r : \mathcal{G}^\mu \times \Omega_r \to \mathcal{G}^\lambda$ generates λ offspring from the parent population by mixing the parental genetic information. The probabilistic influence during recombination is described by the probability space (Ω_r, P_r), i.e. the outcome of the recombination depends on the random choice of $\omega_r \in \Omega_r$ according to P_r. The recombination is followed by mutating the λ offspring according to $m : \mathcal{G}^\lambda \times \Omega_m \to \mathcal{G}^\lambda$. Here (Ω_m, P_m) is the underlying probability space.

The new population $Pop(t+1) \in \mathcal{G}^\mu$ is selected from the set of offspring of $Pop(t)$, where the selection of an individual is based directly or indirectly on the objective function $f : \mathcal{P} \to W$. The objective function assesses only the phenotype. This relationship is formalized by the selection operator

$$s : \mathcal{G}^\lambda \times \mathcal{P}^\lambda \times \Omega_s \to \mathcal{G}^\mu$$

(with probability space (Ω_s, P_s)). With the auxiliary function

$$\begin{aligned} h^* &: \mathcal{G}^\lambda \to \mathcal{G}^\lambda \times \mathcal{P}^\lambda, \\ h^*(g) &:= (g, h(g)) \text{ for } g \in \mathcal{G}^\lambda \end{aligned}$$

the equation

$$Pop(t+1) = s(h^*(m(r(Pop(t), \omega_r), \omega_m)), \omega_s)$$

holds, where $\omega_r \in \Omega_r$, $\omega_m \in \Omega_m$, and $\omega_s \in \Omega_s$ are chosen randomly according to P_r, P_m, and P_s.

Thus, an EA can be described by the iterated use of the genetic operators and the selection operator. Now, the fundamental design decision is how to choose $\mathcal{G}$, r, m, and h, if a certain search space $\mathcal{P}$ and an objective function $f : \mathcal{P} \to W$ are given.

3 Design of Problem-specific EAs: an Overview

Every conference in the field of EAs gives birth to a variety of new variation operators and representations of the genotype space. These operators and representations are tailored for the optimization problem at hand. This reflects perfectly the fact that EAs (as all algorithms) can only achieve above-average performance on a specific subset of problems. Davis [7] has already pointed out that problem-specific EAs will outperform canonical EAs in the problem domain.

But fortunately, EAs have proved to be very robust. They will work satisfactorily, even if they are not adapted to the problem very well. The robustness of EAs leads to a great freedom in design. This is reflected in a variety of different design approaches. In the literature one can identify two main approaches. The first one is based on canonical genetic algorithms and the schema theorem, namely the building block hypothesis [10,22,18]. But since the schema theorem says nothing about the dynamical behavior of EAs over many generations, this approach does not seem to be appropriate [2,19,23].

The second approach is oriented according to the work of Davis [7] and Michalewicz [17]. Certain features of current algorithms for the problems are incorporated in the EAs. This approach often leads to a successful application of EAs (see e.g. [27]), but it lacks profound systematics.

In recent times a dynamical changing of representation during search has been suggested [15,16,20,11]. As a rule, there is only a changing in the mapping h from genotype to phenotype space. The actual data structures and operators stay the same. Thus, the dynamical changing may be seen as a special form of mutation only [15].

The research on dynamical and self–adapting representations is still in its infancy. It is an interesting question, whether an a priori choice of representation and operators can be replaced by a dynamical change of representation [6].

4 Design Guidelines for Recombination and Mutation Operators: a Proposal

4.1 Some General Considerations

To systematize the design of EAs, we propose some design guidelines in this section. To motivate these guidelines we are going to highlight some desirable features of the variation operators and the representation in the following. Since we want to run the EAs on a computer, we concentrate on finite search spaces and discrete time. To make the discussion easier, we assume furthermore that the search space is $\mathcal{G} = \mathcal{P}$ (thus, h is the identity).

At first the reachability of every point in the search space is important for an EA. Every point $u \in \mathcal{G}$ could be the desired optimum or a good local optimum. The property of visiting the global optimum with probability one is also a precondition if one wants to assure convergence of the EA. If from each point $u \in \mathcal{G}$ any other point $v \in \mathcal{G}$ can be reached with probability greater than zero in one mutation step, then the EA visits the global optimum after finite time with probability one regardless of the initialization [24].

If the selection operator guarantees that the best individual visited so far will survive with probability one, then the EA converges completely and in the mean to the global optimum regardless of the initialization. The property of reachability in one mutation step can be weakened, if some other properties can be assured (see [24] for details).

The next question is where the genetic operators should place a new search point. As long as no additional information is available, the operators should not prefer any direction, i.e. mutation and recombination should not generate an additional bias. Furthermore, they should support a gradual approximation of the optimum.

Mutation and recombination are the standard variation operators in EAs. But the opinions about the working method, usefulness, and necessity of these operators are divided. This may be because the operators and their properties are not always well specified. For the design and the analysis of an EA it may be advantageous if the operators are clearly separated, i.e. have different search characteristics.

Let us assume that we have an EA with well-specified genetic operators and selection. Then the genetic operators and the selection mechanism determin the efficiency of the search process (how many points $u \in \mathcal{G}$ must be visited on average until the optimum is found). But what about the representation of genotype and phenotype space?

Of course, the representation affects the time-efficiency of the operators. Thus, from an algorithmic point of view the representation of the phenotype space has to support an efficient calculation of the objective function, whereas the representation of the genotype space has to support an efficient implementation of the genetic operators. Of course, both have to support an efficient implementation of the mapping function h.

In the following we will present some design guidelines for genetic operators and the mapping h that should promote an efficient search process. By presenting empirical results in Section 5 we show that these guidelines can be helpful, although a theoretical proof is not given. The aim of our guidelines is to support a smooth motion of the search process through the fitness landscape [21,25].

4.2 Metric-based Evolutionary Algorithms

Metric In the following we assume that we have information about the phenotype space concerning a specific problem domain. This domain knowledge will be used to form a distance measure on the phenotype space. This distance measure shows the similarity between two phenotypes regarding a certain mark of quality, i.e. two elements of the phenotype space with a small distance between them should have similar objective function values. Hence, the distance measure should be strongly connected to the objective function to be optimized. The available domain knowledge directly influences the choice of this distance measure and only indirectly the design of the recombination and mutation operators. Therefore, the performance of the resulting EA mainly depends on the suitability of the distance measure.

In order to fulfill the most basic properties, which are reasonable for distance measures, we only allow *metrics* $d_{\mathcal{P}} : \mathcal{P} \times \mathcal{P} \to I\!N_0^+$ to measure the distance between two points of the phenotype space (we restrict ourselves here to metrics having values only in $I\!N_0^+$). In the optimal case the difference between the function values of two individuals can be upper bounded by the distance of the two individuals regarding the metric $d_{\mathcal{P}}$, i.e.

$$\forall p_1, p_2 \in \mathcal{P} : |f(p_1) - f(p_2)| \leq d_{\mathcal{P}}(p_1, p_2).$$

Because a metric like this will be very hard to find in practical applications, we do not give a formal account of what the correlation between the fitness function and the metric should look like. But it is desirable that it should be as close as possible to the relation given above. Because the success of finding a strong relationship is clearly bounded by the available domain knowledge, we simply claim that there should be a strong correlation, so that elements of $\mathcal{P}$ with small distance should have similar function values. We will present an example in Section 5, where such a strong relationship is given.

In the following both the genotype space and the phenotype space are assumed to be finite spaces.

Mapping Assuming that the mapping h is injective, the metric $d_{\mathcal{P}}$ automatically induces a metric $d_{\mathcal{G}} : \mathcal{G} \times \mathcal{G} \to \mathbb{N}_0^+$ on the genotype space using the following definition:

$$\forall g_1, g_2 \in \mathcal{G} : \; d_{\mathcal{G}}(g_1, g_2) := d_{\mathcal{P}}(h(g_1), h(g_2)). \tag{1}$$

If a genetic operator causes a small change regarding $d_{\mathcal{G}}$, this will result in a small change in the phenotype space regarding $d_{\mathcal{P}}$ by making use of definition (1). The inverse statement holds if h is bijective. If h is not injective (and thus (1) cannot be used), the mapping should at least obey the following rule:

Guideline H 1 *For the mapping h and the metrics $d_{\mathcal{G}}$ and $d_{\mathcal{P}}$ the following relations should hold:*

$$\begin{aligned}&\forall u, v, w \in \mathcal{G} :\\&(d_{\mathcal{G}}(u, v) \le d_{\mathcal{G}}(u, w)) \Leftrightarrow (d_{\mathcal{P}}(h(u), h(v)) \le d_{\mathcal{P}}(h(u), h(w)))\,.\end{aligned}$$

This guarantees that the distance relations between three points translate from the genotype to the phenotype space and back, although the distances do not have to be the same. If h is bijective, rule H 1 is fulfilled by metrics generated through equation (1).

Recombination Without loss of generality we can look at the reduced recombination operator $r' : \mathcal{G}^2 \times \Omega_{r'} \to \mathcal{G}$ with a finite probability space $(\Omega_{r'}, P_{r'})$ instead of the general recombination operator r. Let $P_{r'}(r'(u, v) = w) := P_{r'}(\{\omega \in \Omega_{r'} \mid r'(u, v, \omega) = w\})$ be the probability that two parents $u, v \in \mathcal{G}$ are recombined to an offspring $w \in \mathcal{G}$.

The first guideline on the recombination operator states that the genotype distance $d_{\mathcal{G}}$ between any of the two parents and the offspring should be smaller than the distance between the parents themselves. Hence, children are not allowed to differ extremely from their parents (according to the metric):

Guideline R 1 *The recombination r' should fulfill:*

$$\begin{aligned}&\forall u, v \in \mathcal{G}, \omega \in \Omega_{r'} \text{ with } z = r'(u, v, \omega) :\\&\max(d_{\mathcal{G}}(u, z), d_{\mathcal{G}}(v, z)) \le d_{\mathcal{G}}(u, v).\end{aligned}$$

Moreover, the recombination operator should not generate an additional bias independent of the parents, i.e. the offspring should not tend to one of its parents to avoid any bias in the search process:

Guideline R 2 *The recombination r' should fulfill:*

$$\forall u, v \in \mathcal{G}, \forall \alpha \geq 0 :$$
$$P_{r'}(d_{\mathcal{G}}(u, r'(u, v)) = \alpha) = P_{r'}(d_{\mathcal{G}}(v, r'(u, v)) = \alpha).$$

This guarantees that recombination produces an element having distance α from its first parent with the same probability as an element having distance α from its second parent. Hence, no parent is favored.

Mutation The mutation operator $m : \mathcal{G}^\lambda \times \Omega_m \to \mathcal{G}^\lambda$ works on all λ individuals at the same time. Since the mutation of a single individual does not depend on other individuals, we restrict our discussion to the reduced mutation operator $m' : \mathcal{G} \times \Omega_{m'} \to \mathcal{G}$ with the finite probability space $(\Omega_{m'}, P_{m'})$. With probability

$$P_{m'}(m'(u) = v) := P_{m'}(\{\omega \in \Omega_{m'} \mid m'(u, \omega) = v\})$$

the mutation operator m' changes an element $u \in \mathcal{G}$ to a new element $v \in \mathcal{G}$.

The first guideline states that from each point $g \in \mathcal{G}$ any other point $u \in \mathcal{G}$ can be reached in one mutation step. Hence, there can be no regions in the search space which are excluded from the search process:

Guideline M 1 *The mutation m' should fulfill:*

$$\forall u, v \in \mathcal{G} : \; P_{m'}(m'(u) = v) > 0.$$

Moreover, small mutations (with respect to $d_{\mathcal{G}}$) should occur with higher probability than large mutations, because in a smooth fitness landscape good elements are at least partially surrounded by other good elements:

Guideline M 2 *The mutation m' should fulfill:*

$$\forall u, v, w \in \mathcal{G} :$$
$$(d_{\mathcal{G}}(u, v) < d_{\mathcal{G}}(u, w)) \Rightarrow (P_{m'}(m'(u) = v) > P_{m'}(m'(u) = w)).$$

For a given $u \in \mathcal{G}$ all genotypes $v \in \mathcal{G}$ which have the same distance to u should have the same probability to be produced by mutation of u, guaranteeing that the mutation operator does not induce a bias on its own:

Guideline M 3 *The mutation m' should fulfill:*

$$\forall u, v, w \in \mathcal{G} :$$
$$(d_{\mathcal{G}}(u, v) = d_{\mathcal{G}}(u, w)) \Rightarrow (P_{m'}(m'(u) = v) = P_{m'}(m'(u) = w)).$$

Since at the beginning of the search there is no information about the position of the global optimum, the mutation should not push the search process in a certain direction. By working with a dynamic adaptation of step-size (e.g. self-adaptation of $n > 1$ step-sizes) guideline M 3 can be violated. By learning an internal model of the topology the preference for a certain local search direction can be useful. Because each individual can only learn a local model of the fitness function (especially in multimodal optimization), the learned topology can become useless when the individual leaves the subspace where the local model is valid. Whenever there is no longer any suitable information about the topology of the search space, it should be guaranteed that a state can be reached in which M 3 holds again.

Since all guidelines are based on the definition of a suitable metric on the phenotype space, an EA which fulfills the guidelines H 1, R 1, R 2, M 1, M 2, and M 3 for a given metric and mapping is called a *metric-based evolutionary algorithm (MBEA)*.

5 Metric-based Genetic Programming

Taking into account all the guidelines that define an MBEA, it is not obvious how an MBEA can be built and if the guidelines can improve the efficiency of an EA. Therefore, we will give an example of a *metric-based genetic programming (MBGP)* system in this section to show what metric-based crossover and mutation operators can look like. Then we will present empirical results showing a clear improvement in efficiency over other GP systems for the chosen problem domain.

The task of our MBGP system is to solve the following common benchmark problem (see for instance [4,5,9,12,13]): find a representation of an unknown function, where it is only possible to compute the number of inputs where a given function and the unknown one have the same output. Hence, we cannot determine the inputs, on which the given and the unknown function coincide, but only their number.

In our case, the unknown function g is a Boolean function with n inputs and one output, i.e. $g : \{0,1\}^n \to \{0,1\}$. Every element $f : \{0,1\}^n \to \{0,1\}$ of the phenotype space will be assigned the fitness $F(f) := |\{x \in \{0,1\}^n \mid f(x) = g(x)\}|$, i.e. the number of inputs, where f has the same output as the unknown function g. Hence, the unknown function g is the only maximum of the fitness function F. The task of the MBGP system is to find this maximum.

In our MBGP system we will make use of a special data structure for Boolean functions, called *ordered binary decision diagrams (OBDDs)* (see [3,28]), instead of S-expressions, used in most GP systems. OBDDs allow polynomially sized representation in n of many practically important Boolean functions and many operations on OBDDs can be done in polynomially bounded time with regard to the size of the OBDDs involved. Some of these efficient operations are used in the implementation of our mutation and recombination operators. Before the MBGP system is described, the next subsection gives a short introduction to OBDDs.

5.1 Ordered Binary Decision Diagrams

Given n Boolean input variables $x_1, \ldots, x_n$, let π be an ordering of these variables, i.e. a bijective function $\pi : \{x_1, \ldots, x_n\} \rightarrow \{1, \ldots, n\}$. An OBDD for π is an acyclic directed graph with one source, where every node is labeled by either one of the variables or one of the Boolean constants 0 or 1 (in the last two cases the node is called a *sink*).

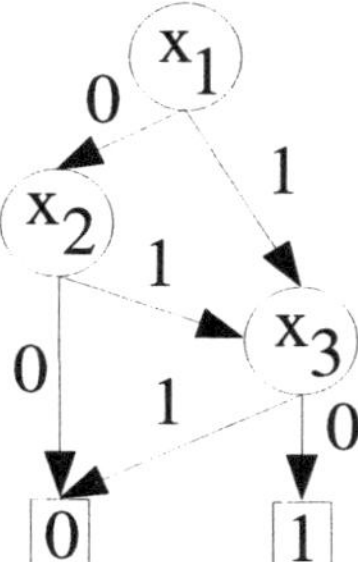

Fig. 1. OBDD

The two sink nodes have no outgoing edges, while every non-sink node is labeled by a Boolean variable x_i and has exactly two outgoing edges, one labeled by 0 and the other by 1 (whose endpoints are called *zero-successor* and *one-successor*, resp.). Furthermore, if there is an edge from a node labeled x_i to a node labeled x_j, then $\pi(x_i)$ must be smaller than $\pi(x_j)$ (in the following it is always assumed that $\pi(x_i) = i$ for all $i \in \{1, \ldots, n\}$).

To evaluate the function f_O represented by a given OBDD O for an input $(a_1, \ldots, a_n) \in \{0, 1\}^n$, one starts at the source. At a node labeled x_i one chooses the a_i-successor. The value $f_O(a)$ equals the label of the sink finally reached. The OBDD O in Figure 1 represents the Boolean function $f(x_1, x_2, x_3) = x_1\overline{x}_3 \vee \overline{x}_1 x_2 \overline{x}_3$.

OBDDs allow polynomially sized representation in n of many practically important Boolean functions, and many operations on OBDDs can be done in polynomially bounded time with regard to the size of the OBDDs involved. Another big advantage of OBDDs is the existence of *reduced* OBDDs: for a given function f and variable ordering π the reduced OBDD representing f is the unique OBDD from the set of all OBDDs representing f having the minimal number of nodes (see [3]). Our MBGP system uses only reduced OBDDs, thereby reducing memory needs. Furthermore, there are some efficient OBDD algorithms for problems that are hard to compute for other representations and that are used in slight variation in our MBGP system.

5.2 The MBGP System

In the following we present a GP system that fulfills the MBEA guidelines, thereby showing that they can be fulfilled. We start with the definition of a metric on the phenotype space, which leads to a smooth fitness landscape. Thus, the metric reflects our domain knowledge. Then an outline of the GP system is given.

We measure the fitness $F(f)$ of an individual $f : \{0,1\}^n \to \{0,1\}$ by the number of inputs, where f and the unknown function g agree. Hence, an appropriate metric $d_{\mathcal{P}}$ counts the number of inputs, where the two measured functions disagree:

$$d_{\mathcal{P}}(f,h) := |\{x \in \{0,1\}^n \mid f(x) \neq h(x)\}| .$$

$d_{\mathcal{P}}$ defines a metric, which can be easily seen on realizing that it is equivalent to the Hamming distance on the space of all bitstrings of length 2^n. Because two functions f and h can differ in their fitness $F(f)$ and $F(h)$ by $d_{\mathcal{P}}(f,h)$ at most, the fitness landscape defined by $d_{\mathcal{P}}$ and F is smooth, i.e. neighboring functions have similar fitness. Hence, our metric $d_{\mathcal{P}}$ fulfills our informal guideline of smoothing the fitness landscape.

On the genotype space, i.e. the space of concrete representations, we use the analogous function $d_{\mathcal{G}}$ defined by

$$d_{\mathcal{G}}(u,v) := d_{\mathcal{P}}(h(u), h(v)),$$

where $h : \mathcal{G} \to \mathcal{P}$ is the bijective mapping between reduced OBDDs and the represented function. Hence, $d_{\mathcal{G}}$ is a metric, too.

5.3 A Rough Outline of the MBGP System

In the following, we assume that $MAXGEN \in I\!N$ (the maximum number of generations), $\mu \in I\!N$ (the size of a parental generation), and $\lambda \in I\!N$ (the number of children) are predefined by the user. Then, our MBGP system looks as follows:

Algorithm A 1 The MBGP system.

1 *Set* $t := 0$.
2 *Initialize* P_0 *by choosing* μ *Boolean functions with equal probability and including the reduced OBDDs representing these functions.*
3 *While* $t \leq MAXGEN$:
   (a) *For* $i \in \{1, \ldots, \lambda\}$ *do:*
      i. *Choose two reduced OBDDs* D_1 *and* D_2 *with equal probability from* P_t.

ii. *Recombine* D_1 *and* D_2 *and call the resulting reduced OBDD* D'.
iii. *Mutate* D' *and call the resulting reduced OBDD* D^*.
iv. *Include* D^* *in* P'_{t+1}.

(b) *Choose the* μ *OBDDs in* P'_{t+1} *with the highest fitness to form* P_{t+1}.
(c) *Set* $t := t + 1$.

4 *Output the OBDD with the highest fitness that was found.*

5.4 Mutation

Mutation of an OBDD D simply consists of changing each output bit of the 2^n inputs independently with probability $1/2^n$. Hence, the expected number of changed bits by a mutation is one. It is implemented by generating a reduced OBDD where every output bit is 1 with probability $1/2^n$ and 0 otherwise and doing an EXOR synthesis of D and this OBDD to generate the new OBDD D'. Because the probability of flipping a bit is less than 1/2 (assuming $n \geq 2$), smaller mutations are more likely than larger ones (guideline M 2). Nevertheless, a mutation can lead to every bitstring in $\{0,1\}^n$ regardless of the starting point (guideline M 1) and the probability of mutating from u to v depends on the Hamming distance between u and v only (guideline M 3). Hence, the following theorem follows:

Theorem 1. *The mutation operator fulfills the guidelines M 1, M 2, and M 3.*

Proof. Guideline M 1 states that mutating an arbitrary reduced OBDD D can result in any other reduced OBDD D'. This follows from the fact that every output bit is flipped with probability $1/2^n$. Hence, the probability of mutating from u to v is

$$\left(\frac{1}{2^n}\right)^{H(u,v)} \cdot \left(1 - \frac{1}{2^n}\right)^{2^n - H(u,v)} > 0,$$

where $H(u,v)$ is the Hamming distance between the two bitstrings u and v of output bits of the reduced OBDDs D and D'. Therefore, every other function can result from mutation, i.e. every other reduced OBDD.

Guideline M 2 is fulfilled, as one bit is flipped with probability less than 1/2, implying that one bit is more likely not to be flipped than to be flipped. Hence, the probability of mutating an OBDD D_1 to an OBDD D_2 decreases with increasing Hamming distance between the bitstrings u and v of their output bits, i.e. $d_{\mathcal{G}}(D_1, D_2)$. So, small changes (with respect to $d_{\mathcal{G}}$) are more likely than larger ones.

The last guideline M 3 states that two OBDDs D_1 and D_2 have the same probability of resulting from a reduced OBDD D by mutation, if $d_{\mathcal{G}}(D, D_1) = d_{\mathcal{G}}(D, D_2)$. This again follows from the fact that the probability of mutating from D to D' depends only on the number of output bits, where D and D' disagree.

5.5 Recombination

The recombination operator $r' : \mathcal{G}' \times \mathcal{G}' \times \Omega_{r'} \to \mathcal{G}$ creates a new reduced OBDD out of two parental reduced OBDDs under random influence (represented by $\Omega_{r'}$). To fulfill guideline R 1 the new OBDD is not allowed to have a greater distance to one of its parents than the distance between the parents. Guideline R 2 is a kind of symmetry requirement: with the same probability the result of the recombination has distance d from the first parent and the second parent, respectively.

The source of the resulting reduced OBDD, when recombining two reduced OBDDs D_1 and D_2, results from calling the following recursive algorithm that gets the sources v_1 and v_2 of D_1 and D_2, resp., as arguments (where $v \to 0$ and $v \to 1$ is the zero- and one-successor of node v, resp., sinks are assumed to have label x_{n+1}, and the variable ordering is w. l. o. g. $x_1, \ldots, x_n$):

Algorithm A 2 *Rec*(v_1, v_2):

1 *If* $v_1 = v_2$, *return* v_1.
2 *If* v_1 *and* v_2 *are both sinks, return one of them with equal probability.*
3 *Let* x_i *be the label of* v_1 *and* x_j *the label of* v_2*:*
   (a) If $i < j$, *return node* v' *with label* x_i, $v' \to 0 := Rec(v_1 \to 0, v_2)$, *and* $v' \to 1 := Rec(v_1 \to 1, v_2)$.
   (b) If $i > j$, *return node* v' *with label* x_j, $v' \to 0 := Rec(v_1, v_2 \to 0)$, *and* $v' \to 1 := Rec(v_1, v_2 \to 1)$.
   (c) If $i = j$, *return node* v' *with label* x_i, $v' \to 0 := Rec(v_1 \to 0, v_2 \to 0)$, *and* $v' \to 1 := Rec(v_1 \to 1, v_2 \to 1)$.

Thus, the recombination operator works as follows: starting from the sources, the two OBDDs are traversed in parallel. If one node has a smaller label than the other node, the zero- and one-successor of this node are visited before the successors of the other node. If both nodes have the same label, the successors of both nodes are visited. If sinks are reached in both OBDDs, one of the sinks is returned with equal probability.

The result of a call of *Rec*(v_1, v_2) for a pair (v_1, v_2) of nodes is not stored in order to guarantee that the random decisions are independent of each other. As a consequence, the runtime of this recombination operator cannot be upper bounded polynomially in the OBDD sizes. Indeed, one can easily think of worst-case examples, where the recombination needs exponential running time in n even though the OBDDs have polynomial size in n (for instance, when the reduced OBDD for the parity function is recombined with its complement). But all experiments indicate that the average running time is rather small, which is caused by the increasing similarity of the OBDDs in the course of evolution.

To prove that the guidelines R 1 and R 2 are fulfilled we need more knowledge about the behavior of the recombination operator: how does the probability distribution on the space of all reduced OBDDs look like that represents the possible results of the recombination of two given reduced OBDDs?

For reasons of simplicity we introduce a different representation of Boolean functions expressed by OBDDs. For a given OBDD an input defines the set of nodes that are visited while evaluating the OBDD for this input. The nodes in this set form a *path* from the source to one of the sinks. As every variable is tested on this path at most once, we can identify a path with a vector $a \in \{0, 1, *\}^n$ with the following interpretation:

- If $a_i = *$, then the variable x_i is not tested on the path.
- If $a_i = 0$, then the variable x_i is tested and the zero-successor was chosen.
- If $a_1 = 1$, then the variable x_i is tested and the one-successor was chosen.

It can be easily seen that for a given reduced OBDD there is a one-to-one equivalence between a path and such a vector. Hence, we will call such a vector a path, too. Let $C(a)$ (which is called a *cube*) be the set of inputs, so that the evaluation procedure visits the nodes on the path a for these inputs. It can be written as

$$C(a) = \{x \in \{0,1\}^n \mid \forall i \in \{1, \ldots, n\} : a_i \neq * \Rightarrow x_i = a_i\}.$$

It is clear that all inputs belonging to one cube have the same output according to the given OBDD. So if $s_D(a) \in \{0, 1\}$ is the label of the sink that is reached when following the path a in the OBDD D, all elements of $C(a)$ are mapped by D to $s_D(a)$. So D can be represented as the set of all cubes with the according sink:

$$M(D) := \{(C(a), s_D(a)) \mid a \text{ is a path in } D \text{ leading to the } s_D(a)\text{-sink}\}.$$

Using this representation of an OBDD, we can explain the effect of the recombination of two OBDDs D_1 and D_2 more easily: remember that $Rec(v_1, v_2)$ returns a sink if and only if v_1 and v_2 are sinks. Let a^1 and a^2 be the paths on which v_1 and v_2 were reached in D_1 and D_2, respectively.

The sink that is returned by the recombination operator lies at the end of the *union* $a \in \{0, 1, *\}^n$ of a^1 and a^2, defined by

$$\forall i \in \{1, \ldots, n\} : a_i := \begin{cases} a_i^1, & \text{if } a_i^1 \neq * \\ a_i^2, & \text{if } a_i^2 \neq *. \end{cases}$$

The cube $C(a)$ is the intersection of $C(a^1)$ and $C(a^2)$. The sink is chosen with equal probability, so the child OBDD D' has the following representation:

$$M(D') = \{(C^1 \cap C^2, \otimes(s^1, s^2)) \mid (C^1, s^1) \in M(D^1), (C^2, s^2) \in M(D^2)\},$$

where $\otimes(s^1, s^2)$ returns one of its arguments with equal probability.

From this representation we can easily see that if both parents D_1 and D_2 have the same output for one input, the child OBDD D has this output for the input, too. If the parents disagree, the output of the child is 0 or 1 with probability 1/2 each. But it is important that these random decisions are not independent for every input, but dependent on the OBDD structure. Using this characterization, we can prove that the guidelines R 1 and R 2 are fulfilled:

Theorem 2. *The recombination described in Algorithm A 2 fulfills the guidelines R 1 and R 2.*

Proof. For R 1 we have to show that the distance between D' and D_1 and between D' and D_2 cannot be greater than the distance between D_1 and D_2, where D' is a child of D_1 and D_2. If $d = d_{\mathcal{G}}(D_1, D_2)$, then the parent OBDDs disagree in their output for exactly d inputs. As the child OBDD D' can disagree with one of its parents only if the parents disagree, it follows that both $d_{\mathcal{G}}(D', D_1)$ and $d_{\mathcal{G}}(D', D_2)$ can be at most d.

Furthermore, as D' agrees with D_2 on those inputs where it disagrees with D_1, but disagrees with D_2 on those inputs where it agrees with D_1 (taking only the d inputs into account, where D_1 and D_2 disagree), it follows that

$$d_{\mathcal{G}}(D_1, D') + d_{\mathcal{G}}(D', D_2) = d_{\mathcal{G}}(D_1, D_2).$$

Guideline R 2 states that D' has distance d with the same probability from D_1 as from D_2. Look at the sequence of random decisions (each with probability 1/2) that leads to the child D' with distance d from D_1 (on these d inputs D' agrees with D_2). The complementary series of random decisions during recombination leads to the OBDD that is complementary to the old child OBDD D' on all inputs where D_1 and D_2 disagree. It has distance d from D_2, as it disagrees with D_2 exactly on those inputs where the old child OBDD D' agrees with D_2 and D_1 and D_2 disagree. As both sequences have the same probability, and this argumentation is valid for all sequences, guideline R 2 is fulfilled.

Hence:

Corollary 1. *The GP system described by Algorithm A 1 is an MBEA.*

5.6 Experimental Results

We proposed guidelines for the design of problem-specific EA. Thus, we have to check that the guidelines make sense by examining the quality gain we obtain when using these guidelines. In the following we compare the results of our MBGP system with GP and EP systems which are not tailored for finding Boolean functions (except the function set {AND, OR, NOR}). Furthermore, these systems are not MBEAs. Of course, if our MBGP system does not perform much better than the general EP and GP systems, this would challenge the MBEA approach. We summarize the outcome of some empirical studies which confirm the usefulness of the proposed guidelines, at least for the problems presented here. We compare it to other systems trying to find representations of an unknown Boolean function, when its complete training set is given. Although this problem has no practical relevance, it is an often used benchmark problem. We take the *computational effort* to compare our results to those presented in [4,5,12,13]. The computational effort is a guess of the number of individuals that have to be evaluated to reach the optimum with 99% probability (see [12] for a definition). Furthermore, we were interested in the influence of using OBDDs on our MBGP system. By using reduced OBDDs the size of the genotype space is reduced to $|\mathcal{G}| = |\mathcal{P}|$. Therefore, a GP system that uses OBDDs as representation but makes use of traditional recombination and mutation operators may perform as well as the MBGP without the extra effort in designing the MBGP search operators. Hence, we also made experiments with a GP system that uses OBDDs as representation but does not fulfill the MBEA guidelines. This OBDDGP system has mutation and recombination operators that work similarly to those of standard GP systems using S-expressions as described in [12,13]. Hence, the mutation operator replaces the subOBDD, starting at a randomly chosen node in the OBDD, to be mutated by a new random subOBDD. The recombination operator exchanges two randomly chosen subOBDDs in both parent OBDDs. Special measures are taken to guarantee that the resulting structures are still valid OBDDs that obey the same variable ordering as the parent OBDDs. For a more detailed description of mutation and recombination see [8]. The rest of the OBDDGP system follows algorithm A 1, i.e. the OBDDGP and the MBGP system differ only in the mutation and recombination operator. We tested the MBGP and the OBDDGP system on the 6- and 11-multiplexer, the 3- to 12-parity functions, and on randomly chosen Boolean functions $rand_8$ with eight inputs. These Boolean functions were chosen independently with equal probability for each of the 50 runs from the set of all Boolean functions with eight inputs. By using this approach, we tested how well the MBGP system can find representations of functions which have no structure at all. As the MBGP system does not favor any Boolean function in particular, the computational effort to evolve them should be the same. We chose $\mu = 15$ and $\lambda = 100$ for the MBGP system. Because the OBDDGP system seemed to be more susceptible to premature stagnation in preliminary experiments, we used $\mu = 100$ and $\lambda = 700$ over a maximum number of 7000 generations for its runs. In general one has to be careful when comparing systems that search for representations in different spaces (in our case, in the space of all reduced OBDDs or in the space of all S-expressions over a special terminal and function set), as the search process can have different complexities in different search spaces. But it is feasible to compare the MBGP system to systems using S-expressions, since an OBDD D can be transformed in time $O(|D|)$

Function	GP	GP ADFs	EP	EP ADFs	OBDDGP	MBGP
mux_6	160,000	-	93,000	-	9,200	1,615
mux_{11}	-	-	-	-	-	60,115
par_3	96,000	64,000	28,500	63,000	15,600	315
par_4	384,000	176,000	181,500	118,500	22,600	615
par_5	6,528,000	464,000	2,100,000	126,000	19,700	1,015
par_6	-	1,344,000	-	121,000	30,900	1,715
par_7	-	1,440,000	-	169,000	56,100	3,415
par_8	-	n.e.d.	-	321,000	183,600	7,415
par_9	-	n.e.d.	-	586,500	184,200	14,615
par_{10}	-	n.e.d.	-	-	315,100	31,415
par_{11}	-	n.e.d.	-	-	1,578,000	73,615
par_{12}	-	-	-	-	8,358,400	144,715
$rand_8$	-	-	-	-	-	7.615

Table 1. Comparison of the computational effort (with 99% probability of finding a solution)

to an S-expression over the function set $F = \{AND, OR, NOT\}$, which is common for most systems.

Table 1 shows the results. Because in [13] only four runs were done for the 8-, 9-, 10-, and 11-parity function, the computational effort was not computed for these functions. In Table 1 this fact is noted as "n.e.d." (not enough data). A "-" in the columns for GP ([12,13]), GP with ADFs ([13]), EP ([4]), and EP with ADFs ([5]) indicates that no experiments were reported for this function type. A "-" in the column for the GP system with OBDDs indicates that, although 50 runs with $\lambda = 100$ and $\mu = 700$ over 7000 generations were done, no element with optimal fitness was found. The comparison shows that the MBGP system works well on the functions being tested, using only a small fraction of the number of individuals the GP systems described in [12,13] or the EP systems in [4,5] need. Furthermore, we see that even the 8-random "function" (as the unknown function was chosen randomly for each of the 50 runs, this is not a single function, but rather a sample of the class of all Boolean functions with eight inputs) was found with small computational effort. This indicates that the MBGP system works well on all Boolean functions, even on those without any "structure."

Although the GP system using OBDDs, but without obeying the MBEA guidelines, performs better than the GP systems using S-expressions on these test functions with respect to the computational effort, it performs worse than the MBGP system. Furthermore, the robustness of the MBGP system seems to be higher, because in contrast to the OBDDGP system it found the optimal function in every run, although the OBDDGP system was allowed to run over more generations with a larger population. Hence, we can conclude that for the test functions presented the MBEA concept seems to be useful.

5.7 The Interplay of Recombination and Mutation

A GP system based purely on a recombination operator fulfilling guideline R 1 is supposed to let the whole population converge to one single individual, as a child can only lie "between" the parents. Hence, over many generations the maximum distance between two individuals of the population will decrease. This makes sense, as long as we assume the fitness landscape to be smooth.

But a GP system using only a recombination operator fulfilling R 1 and R 2 without any mutation can only lead to one of the best individuals that lie "between" the individuals of the initial generation. But in a large enough search space it is very unlikely that the optimum is located in the subspace covered by the initial population. Thus, a recombination operator that fulfills guidelines R 1 and R 2, alone will lead to stagnation in the search process. Hence, we can assume that mutation is a prerequisite for reliable convergence to the optimum.

To empirically investigate these theories we modified the MBGP system to two different versions:

- MBGP system without mutation by eliminating step 3.(a).iii in Algorithm A 1 and including D' in P'_{t+1} in step 3.(a).iv.
- MBGP system without recombination by eliminating step 3.(a).ii in Algorithm A 1 and mutating D_1 to D^* in step 3.(a).iii.

Both versions and the original MBGP system were run 50 times for the 8-parity function. The average fitness of the best individual over the 125 generations is shown in Figure 2.

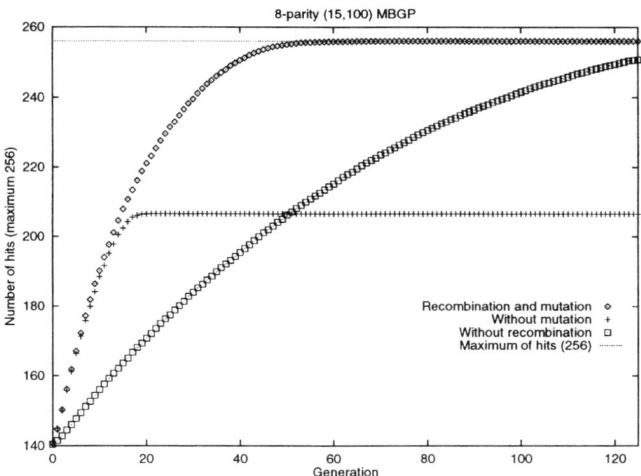

Fig. 2. Results for the 8-parity function over 50 runs with (1) mutation and recombination, (2) without mutation, and (3) without recombination.

The results confirmed that in our case recombination alone leads to stagnation, although its initial convergence speed is high. One can also observe that mutation alone is powerful enough to ensure convergence (the average number of generations to find the optimum was 154.44 with a maximum value of 186), but the process is much slower than that by recombination alone. On the other hand, by applying both operators in conjunction, the optimum was located after 53.27 generations on average (73 generations in the worst run). Thus, the MBGP system with recombination and mutation shows both convergence velocity and reliability, when applied to the 8-parity problem.

6 MBEA and Representation

In the preceding section we have described how an MBGP system with OBDDs can be designed to find Boolean functions. We feel it is appropriate to give some final comments on this choice of representation. In fact, one can use many other data structures than OBDDs to implement an MBGP system. With a little effort one can construct mutation and recombination operators which use S-expressions and have the same semantics as the mutation and recombination operators of the MBGP system described. Our MBGP system with OBDDs and this new MBGP system with S-expressions would have exactly the same computational effort.

But we know of no implementation of our genetic operators that works efficiently with S-expressions, i.e. in polynomial time with respect to the genotype. For instance, recombination by exchanging subtrees is simple to implement and fast to execute, but its consequences for the represented function are unclear. Hence, recombination implemented in this way would not meet the MBEA guidelines. However, this does not imply that MBGP systems using S-expressions cannot be built for other problem domains in general. The following two points sum up the connection between operators and representation in MBEAs:

- The MBEA design guidelines affect the efficiency of the search process, i.e. its computational effort.
- The representation affects the time-efficiency of the operators.

Which representation is the most efficient one depends on the problem at hand. For Boolean functions OBDDs have many advantages, which is also recognized in the CAD community, where they are the state-of-the-art representation. Thus, the MBEA approach gives us a clear separation of representation and genetic operators.

7 Conclusion

We proposed a set of guidelines for the design of EAs and showed by experiments that these design guidelines can decrease the computational effort to solve a problem, assuming that knowledge about the problem domain exists. Furthermore, the choice of a proper representation (with respect to evaluation and genetic operators) is necessary in order to efficiently implement the EA. An algorithm can only be as good as its underlying data structure, i.e. its form of representation. The "algorithmic design approach" presented here provides a clear separation between operators and representation. The MBEA design guidelines affect only the efficiency of the search process, i.e. its computational effort, while the representation affects only the time-efficiency of the operators. Furthermore, experiments have shown that recombination and mutation have different search characteristics in the MBGP system for the problems addressed.

We are aware that the proposed guidelines are not the only (and sometimes not the best) choice of incorporating domain knowledge into the search algorithm. But with respect to the benchmark problems presented here the MBEA concept proved a suitable concept for the usage of domain knowledge in EAs. One has to be aware that incorporating knowledge about the problem domain into the algorithm is necessary to have a chance of solving the problem with above-average success.

Of course the MBEA concept has to be tested on other problem domains, where it is more difficult to find a representation which allows a metric smoothing of the fitness landscape, and to construct genetic operators following the guidelines. In [26] a redesign of an EA for generating relevant fuzzy rules according to the MBEA guidelines led to a considerable gain in quality. Although it was only possible to find a heuristic distance measure, this example shows that the MBEA concept can be beneficial not just in the context of learning Boolean functions.

Additionally, one has to be aware that finding such a metric takes time and does not guarantee success. Hence, one has to compare the effort to find a better algorithm with the resulting gain in performance. In order to circumvent this risk, a theoretical investigation of the properties of an MBEA could lead to better based guidelines, probably guaranteeing some properties like convergence speed or reliability.

Acknowledgements

This research was supported by the Deutsche Forschungsgemeinschaft as part of the collaborative research center "Computational Intelligence" (531).

References

1. T. Bäck, editor. *Seventh Int'l Conference on Genetic Algorithms (ICGA '97), East Lansing, MI, July 19–23, 1997*, San Francisco, CA, 1997. Morgan Kaufmann.

2. H.-G. Beyer. An alternative explanation for the manner in which genetic algorithms operate. *BioSystems*, 41:1–15, 1997.
3. R. E. Bryant. Graph-based algorithms for Boolean function manipulation. *IEEE Transactions on Computers*, 35:677–691, 1986.
4. K. Chellapilla. Evolving computer programs without subtree crossover. *IEEE Transactions on Evolutionary Computation*, 1(3):209–216, 1997.
5. K. Chellapilla. A preliminary investigation into evolving modular programs without subtree crossover. In J. R. Koza, W. Banzhaf, K. Chellapilla, K. Deb, M. Dorigo, D. B. Fogel, M. H. Garzon, D. E. Goldberg, H. Iba, and R. L. Riolo, editors, *Proceedings of the Third Genetic Programming Conference (GP' 98), Madison, WI, July 22–25, 1998*, pages 23–31, San Francisco, CA, 1998. Morgan Kaufmann.
6. J. C. Culberson. On the futility of blind search: An algorithmic view of the "no free lunch". *Evolutionary Computation*, 6(2):109–127, 1998.
7. L. Davis. *Handbook of Genetic Algorithms.* Van Nostrand Reinhold, New York, 1990.
8. S. Droste. Efficient genetic programming for finding good generalizing Boolean functions. In Koza et al. [14], pages 82–87.
9. C. Gathercole and P. Ross. Tackling the Boolean even n parity problem with genetic programming and limited-error fitness. In Koza et al. [14], pages 119–127.
10. D. E. Goldberg. *Genetic Algorithms in Search, Optimization and Machine Learning.* Addison-Wesley, Reading, MA, 1989.
11. R. E. Keller and W. Banzhaf. The evolution of genetic code in genetic programming. In W. Banzhaf, J. Daida, A. E. Eiben, M. H. Garzon, V. Honavar, M. Jakiela, and R. E. Smith, editors, *Proceedings of the Genetic and Evolutionary Computation Conference (GECCO'99), Orlando, FL, July 13–17, 1999*, pages 1077–1082, San Francisco, CA, 1999. Morgan Kaufmann.
12. J. R. Koza. *Genetic Programming: On the Programming of Computers by Means of Natural Selection.* MIT Press, Cambridge, MA, 1992.
13. J. R. Koza. *Genetic Programming II.* MIT Press, Cambridge, MA, 1994.
14. J. R. Koza, K. Deb, M. Dorigo, D. B. Fogel, M. H. Garzon, H. Iba, and R. L. Riolo, editors. *Genetic Programming 1997 (GP' 97) : Proceedings of the Second Annual Conference, Stanford, July 13–16, 1997*, San Francisco, CA, 1997. Morgan Kaufmann.
15. J. Ludvig, J. Hesser, and R. Männer. Tackling the representation problem by stochastic averaging. In Bäck [1], pages 196–203.
16. K. E. Mathias and L. D. Whitley. Changing representation during search: A comparative study of delta coding. *Evolutionary Computation*, 2(3):249–278, 1994.
17. Z. Michalewicz. *Genetic algorithms + data structures = evolution programs.* Springer, Berlin, 1996.
18. N. J. Radcliffe. The algebra of genetic algorithms. *Annals of Mathematics and Artificial Intelligence*, 10(4):339–384, 1994.
19. N. J. Radcliffe. Schema processing. In T. Bäck, D. B. Fogel, and Z. Michalewicz, editors, *Handbook of Evolutionary Computation*, pages B2.5:1–10. Oxford University Press, New York, and Institute of Physics Publishing, Bristol, 1997.
20. S. B. Rana and L. D. Whitley. Bit representation with a twist. In Bäck [1], pages 188–195.

21. I. Rechenberg. *Evolutionsstrategie '94*. Frommann-Holzboog, Stuttgart, 1994.
22. S. Ronald. Robust encodings in genetic algorithms: A survey of encoding issues. In T. Bäck, Z. Michalewicz, and X. Yao, editors, *Proceedings og the IEEE Int'l Conference on Evolutionary Computation (IEEE/ICEC'97), Indianapolis, IN, April 13–16, 1997*, pages 43–48, Piscataway, NJ, 1997. IEEE Press.
23. G. Rudolph. *Convergence Properties of Evolutionary Algorithms*. Verlag Dr. Kovač, Hamburg, 1997.
24. G. Rudolph. Finite Markov chain results in evolutionary computation: A tour d'horizon. *Fundamenta Informaticae*, 35(1-4):67–89, 1998.
25. H.-P. Schwefel. *Evolution and Optimum Seeking*. Sixth-Generation Computer Technology Series. Wiley, New York, 1995.
26. T. Slawinski, A. Krone, U. Hammel, D. Wiesmann, and P. Krause. A hybrid evolutionary search concept for data–based generation of relevant fuzzy rules in high dimensional spaces. In *Eighth Int'l Conference on Fuzzy Systems (FUZZ-IEEE '99), Seoul, Aug. 22–25, 1999*, pages 1432–1437, 1999.
27. G. Tao and Z. Michalewicz. Inver-over operator for the TSP. In A. E. Eiben, T. Bäck, M. Schoenauer, and H.-P. Schwefel, editors, *Parallel Problem Solving from Nature – PPSN V, Fifth Int'l Conference, Amsterdam, The Netherlands, September 27–30, 1998, Proc.*, volume 1498 of *Lecture Notes in Computer Science*, pages 803–812, Berlin, 1998. Springer.
28. I. Wegener. *Branching Programs and Binary Decision Diagrams - Theory and Applications*. SIAM, Philadelphia, PA, 2000.

Multiparent Recombination in Evolutionary Computing

A. E. Eiben

Free University Amsterdam, The Netherlands
E-mail : gusz@cs.vu.nl

Summary. This chapter considers multiparent reproduction, where more than two parents are involved in creating offspring. First we give a survey of multiparent operators that have been introduced over the years in evolutionary computing and we reformulate the traditional mutation-or-crossover debate in the light of such operators. Second, we present some existing results on the usefulness of multiparent operators. We conclude the chapter with a look at future developments and some suggestions for further research.

1 Introduction

Despite the great diversity of life on Earth natural reproduction mechanisms work exclusively with one (asexual reproduction) or two parents (sexual reproduction). The majority of the species reproduce in an asexual manner, showing the viability of asexual reproduction. However, species that are higher in the evolutionary hierarchy use sexual reproduction, suggesting that sexual reproduction is more advanced. In simulated evolution, that is in evolutionary computation, many technical features are inspired by natural mechanisms. In particular, abstract variants of sexual and asexual reproduction are implemented as search operators. Some evolutionary techniques, e.g. evolutionary programming, have worked almost exclusively with mutation, i.e. they implement a simplification of asexual reproduction. Others, e.g. genetic algorithms, evolution strategies, and genetic programming, use recombination, i.e. they implement a simplification of sexual reproduction and mutation. There are several papers investigating the advantages and disadvantages of mutation with respect to crossover [20–22,25,37,42]. At the moment the question of which of mutation or crossover is preferable under certain circumstances is still an open research issue.

The number of parents involved in reproduction can be technically expressed as the arity of the reproduction operators. Mutation and crossover have arity 1 and 2, respectively, and the question is whether unary or binary operators are preferable for typical optimization problems. From a purely technical point of view there is no need to restrict the arity of reproduction operators to one or two. In general, a reproduction operator can have an arity from one up to the population size. Thus the analogy with natural evolution breaks down; to our knowledge there are no species on Earth that would apply multiparent reproduction mechanisms where genetic material of more than two parents is mixed in *one* reproductive action. Simulating such reproduction operators, however, is no problem.

The objectives of this chapter are twofold.

1 To give an overview of multiparent operators that have been introduced over the years in evolutionary computing.
2 To present existing results on the performance of multiparent operators based on the author's own (co-authored) research papers.

The structure of the chapter is as follows. In Section 2, we discuss the vocabulary. A survey of multiparent operators from the literature is given in Section 3. Here we also briefly summarize the experimental or theoretical results concerning these operators. Section 4 contains the experimental results on the performance of multiparent reproduction mechanisms from the author's own work. The chapter is concluded by Section 5.

2 Terminology

Let us start by setting some conventions on terminology. The term *population* is used for a multiset of individuals that undergoes selection and reproduction. This terminology is maintained in genetic algorithms (GAs), evolutionary programming (EP), and genetic programming (GP), but in evolution strategies (ES) all μ individuals in a (μ, λ) or $(\mu + \lambda)$ strategy are called parents. In this chapter, however, the term *parents* is only used for those individuals that are selected to undergo recombination. That is, parents are those individuals that are actually used as inputs for a recombination operator; the *arity of a recombination operator* is the number of parents it uses. An individual is called a *donor* if it is a parent that actually contributes to (at least one of) the alleles of (at least one of) the child(ren) created by the recombination operator. This contribution can be, for instance, the delivery of an allele, as in uniform crossover in bitstring GAs, of participating in an averaging operation, as in intermediate recombination in ES. As an illustration, consider a steady-state GA where 100 individuals form the population and two of them are chosen to undergo uniform crossover to create one single child. These two individuals are then called parents. If, furthermore, by pure chance, the child only inherits alleles from parent 1, then parent 1 is a donor, and parent 2 is not.

3 Multiparent reproduction operators

Let us start this overview with an observation that could be the conclusion: there are more multiparent recombination mechanisms mentioned in the literature than one[1] would expect. In this section, these operators are surveyed more or less chronologically. After going through the survey some features emerge that allow a systematic categorization of these operators. For didactical reasons, these features and the resulting categorization scheme will be discussed in the concluding section.

An early paper from 1966 mentioning multiparent recombination is [8] on solving systems of linear equations. It presents the definition of three different multiparent recombination mechanisms, called *majority mating*, *mating by crossing over*, and *mating by averaging*. All of these operators are defined in a general way; that is, they can be applied to any number of m parents. Unfortunately, only very little was reported on the performance of these operators. Almost as early is the recombination mechanism of [29] for evolving models for a given process utilizing four models to create a new one.

Global recombination in evolution strategies has been known since the 1970s. It allows the use of more than two recombinants, [2,39], because the two donors are drawn randomly for each position (gene) of the offspring anew. These drawings take the whole population of μ individuals into consideration. The multiparent character of global recombination is thus the consequence of redrawing the donors; therefore, possibly more than two individuals contribute to the offspring, but their number is not defined. It is clear that investigations on the effects of different numbers of parents on algorithm performance could not be performed in the traditional ES framework. The option of using multiple parents can be turned on or off, i.e. global recombination can be used or not, but the arity of the recombination operator is not tunable. Experimental studies on global versus two-parent recombination are possible, but so far there are almost no experimental results available on this subject. In [39] it is noted that "appreciable acceleration" is obtained by changing to a bisexual from an asexual scheme (i.e. adding recombination using two parents to the mutation-only algorithm), but only a "slight further increase" is obtained when changing from bisexual to multisexual recombination (i.e. using global recombination instead of the two-parent variant). Let us note that the terms bisexual and multisexual are not appropriate: individuals have no gender or sex, and recombination can be applied to any combination of individuals.

The mechanism called stochastic iterated genetic hill-climbing (SIGH) applies a sophisticated probabilistic voting mechanism, where m "voters" (m being the size of the population) determine the values of a new bitstring, [1]. SIGH was shown to be better than a GA with one-point and uniform crossover on four out of six test functions and the overall conclusion was that it is "competitive in speed with a variety of existing algorithms".

---

[1] Definitely more than I expected before I started to dig in the literature, and I dare to bet that most readers will be also surprised in seeing so many multiparent operators.

The *p-sexual voting recombination* from [32] is applied for the quadratic assignment problem. The operator produces one child of p parents. Let us remark again that the name p-sexual is somewhat misleading, as there are no different genders and no restriction on having one representative of each gender for recombination. In the experiments it "worked surprisingly well", but comparison between this scheme and the usual two-parent recombination was not performed.

An interesting attempt to combine GAs with the simplex method in [5] resulted in the ternary *simplex crossover*. The simplex GA performed better than the standard GA on the De Jong functions. This idea has also been extended to a version with $n + 1$ parents, where n is the dimensionality of the space, [6,35].

Uniform crossover with two as well as with three parents in a GA using an integer representation is compared on the problem of placing actuators on space structures in [24]. Based on the experimental results the authors conclude that the use of three parents did not improve the performance.

Scanning crossover has been introduced in [13] as a generalization and extension of uniform crossover in GAs creating one child from r parents. The name is based on the following general procedure scanning parents and thus building the child from left to right. Let $x^1, \ldots, x^r$ be the selected parents of length L and let x denote the child.

```
Procedure scanning
 begin
  INITIALIZE (put markers at the 1st position in each parent)
  for i = 1 to i = L
    CHOOSE j from 1, ...,r
    let i-th allele of x be the i-th allele of parent j
    UPDATE position markers
  end
```

The above procedure provides a general framework for a certain style of multiparent recombination, where the precise execution, hence the exact definition of the operator, depends on the mechanisms CHOOSE and UPDATE. In the simplest case, UPDATE can shift the markers one position to the right (thus *scanning* the chromosomes from left to right). This is appropriate for bitstring, integer, and floating-point representations. Scanning can also be easily adapted to order-based representation, where each individual is a permutation, if UPDATE shifts to the first allele which is not yet in the child. This guarantees that each offspring will be a permutation, if the parents are permutations themselves. Depending on the mechanism to CHOOSE a parent (and thereby an allele) there are three different versions of scanning. The choice can be deterministic, choosing the allele with the highest number of occurrences, and breaking ties randomly (*occurrence-based scanning*). Alternatively it can be random, either unbiased, following a uniform distribution thus giving each parent an equal chance to deliver its allele (*uniform scanning*), or biased by the fitness of the parents, where the chance of being chosen is proportional to fitness (*fitness-based scanning*). Uniform scanning for $r = 2$ is the same as uniform crossover, although creating only one child, and the occurrence-based version is very much like the voting or majority mating mechanism discussed before. *Diagonal crossover* has been introduced in [13] as a generalization of one-point crossover in GAs. In its original form, diagonal crossover creates r children from r parents by selecting $(r-1)$ crossover points in the parents and composing the children by taking the resulting r chromosome segments from the parents "along the diagonals". Later on, a one-child version was introduced and used, [47]. Figure 1 illustrates both variants. It is easy to see that for $r = 2$ diagonal crossover coincides with one-point crossover, and in some sense it also generalizes traditional two-parent n-point crossover.

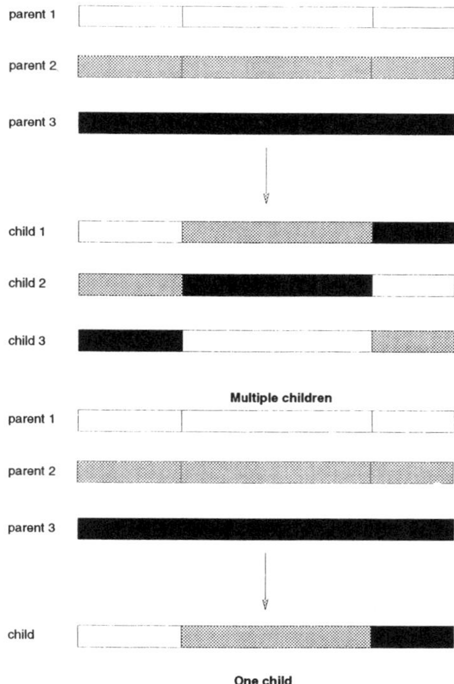

Fig. 1. Diagonal crossover (left) and its one-child version (right) for 3 parents

A so-called *triadic crossover* is introduced and tested in [34] for a multimodal spin-lattice problem. The triadic crossover is defined in terms of two parents and one extra individual for creating a child, but technically the result is identical to the outcome of a voting crossover on these three individuals as parents. A comparison between triadic, one-point, and uniform crossover is done, where triadic crossover turned out to deliver the best results.

Gene-pool recombination (GPR) and its variants were introduced in [48,33] as a multiparent recombination mechanism for discrete domains. It is defined as a generalization of two-parent recombination (TPR). Applying GPR is preceded by selecting a genepool consisting of would-be parents. Applying GPR two parent alleles are recombined to form the allele of an offspring, and the two parents are drawn from the gene pool. Similar to global recombination in ES, the arity of the operator is not defined in advance. GPR is shown to converge about 25% faster than TPR for ONEMAX, and its extension to continuous domains outperforms the corresponding two-parent operator on the spherical function.

A recombination mechanism with tunable arity in ES is proposed in [40]. The $(\mu, \kappa, \lambda, \rho)$-ES provides the possibility of freely adjusting the number of parents (called ancestors). The parameter ρ stands for the number of parents and global recombination is redefined for any given set of ρ parents. The discrete version chooses one of the parent alleles randomly, while the intermediate version takes the average of all parent alleles as the allele of the child. Observe that ρ-ary discrete recombination coincides with uniform scanning crossover, while ρ-ary intermediate recombination is a special case of mating by averaging. At this time there are no experimental results available on the effect of ρ within this framework. Related work in ES also uses ρ as the number of parents as an independent parameter for recombination, [7]. For the purposes of a theoretical analysis it is assumed that all parents are different, uniform randomly chosen from the population of μ individuals, and ρ-ary intermediate and discrete recombinations are defined similarly to [40]. Investigations are limited to the special case of $\rho = \mu$ on the spherical function. By this assumption it is not possible to draw conclusions on the effect of ρ, but the analysis shows that the optimal progress rate is a factor μ higher than that of the traditional (μ, λ)-ES, for both recombination mechanisms.

A very particular mechanism is the *linkage evolving genetic operator* (LEGO) as defined in [41]. The mechanism is designed to detect and propagate blocks of corresponding genes of potentially varying length during the evolution. Although the multiparent feature is only a side effect, LEGO is a mechanism where more than two parents can contribute to an offspring.

A recent generalization of global intermediate recombination in ES can be found in [11]. The new operator is applied after selecting ρ parent individuals from the population of μ, and the resampling of the two donors for each i takes only these ρ individuals into consideration. In this way, the original mechanism is kept as intact as possible, while a gradual variation between the two extremes $\rho = 2$ and $\rho = \mu$ is facilitated. Note that this operator differs from the ones defined in [7,40].

The paper [45] presents no less than three multiparent operators for real-coded GAs: the *center of mass crossover* (CMX), which was introduced originally in [44], *multiparent feature-wise crossover* (MFX), and the *seed crossover* (SX). All three operators are obtained by using a so-called base operator with arity 2 (the authors use BLX-α from [19]) and a generalization template that makes an m-ary operator from a binary one. In this respect, the paper goes much further than others, because the templates to create a multiparent operator from a BLX-α operator are general. That is, the CMX template can be applied to any operator for real-coded chromosomes, and the other two can even be applied to operators for other representations. The experimental results with these operators indicate that "more than two parents lead to better performance", although there are differences between the operators and the results also depend on the test function in question.

The so-called simplex crossover (SXP) in [46] is also designed for real-coded GAs and is shown to deliver its best performance for three or four parents on the investigated set of test functions.

4 Some experimental results on higher operator arities

As the previous section clearly shows, quite a few papers have studied the effect of operator arity on EA performance. Here we give a more detailed treatment of those two operators that generalize the most commonly applied crossovers in GAs: that is, we will focus on diagonal crossover (generalizing one-point crossover) and scanning crossover (that generalizes uniform crossover).

Most experimental research papers consider numerical optimization problmes as test functions. In the evolutionary computation literature there are a number of such functions that are used for performance assesment of novel agorithm variants. We refer to these functions by their common names and omit the exact formulas that can be retrieved from the referred articles. Extensive treatment of such test functions can be found in [3,39].

The performance of scanning crossover for different numbers of parents is studied in [13] in a generational GA with proportional selection. A canonical GA with bit-representation is applied for function optimization (De Jong functions F1–F4 and a function from Michalewicz) and an order-based GA for graph coloring and the traveling salesman problem (TSP). Different mechanisms to CHOOSE in the general scanning procedure are tested and compared. For the function optimization problems the number of generations needed to reach a solution is used as a performance measure. For the graph coloring problem, the percentage of runs that found a solution forms the basis of comparison, while for the TSP the length of the best tour found is used. On the numerical optimization test suite, more parents perform better than two; for the TSP and graph coloring two parents turn out to be advisable. Comparing different biases in choosing the child allele, on four out of the five numerical problems fitness-based scanning outperforms the other two and occurrence-based scanning is the worst operator. In this paper only the definition of diagonal crossover is given, there are no experiments with this operator.

Diagonal crossover is investigated in [18]. It is compared to the classical two-parent n-point crossover and uniform scanning in a steady-state GA with linear ranked-biased selection ($b = 1.2$) and worst-fitness deletion. The test suite consists of two two-dimensional problems (De Jong's F2 and a function from Michalewicz) and four scalable functions (after Ackley, Griewangk, Rastrigin, and Schwefel). When monitoring the performance two different measures were used, namely efficiency (speed) and success rate (percentage of cases when an optimum was found). Speed was measured by the total number of function evaluations (averaged over all runs). The performance of diagonal crossover using r parents and n-point crossover (for two parents) showed a significant correspondence with r, respectively n. The best performance was always obtained with high values, between 10 and 15, where 15 was the maximum tested. Besides, on all problems diagonal crossover is better than n-point crossover using the same number of crossover points ($r = n-1$), thus representing the same level of disruptiveness. For illustration we present the optimal number of parents and the corresponding success rates in Table 1. An interesting observation in [18] is that for scanning the relation between r and performance is less clear than for diagonal crossover, although the best performance is achieved for more than two parents on five out of the six test functions. A concise overview of all experiments in this study can be found in [17].

Test	Scanning Xover		Diagonal Xover		N-point Xover	
function	#par	succ.	#par	succ.	#Xover points	succ.
F2	7	.91 (.73)	11	.88 (.38)	11	.84
Ackl	8	.90 (.84)	15	.89 (.00)	10	.24
Grie	10	.48 (.22)	14	.32 (.04)	10	.15
Mic	10	.72 (.57)	15	.76 (.34)	15	.6
Ras	5	.10 (.00)	13	.28 (.00)	15	.06
Schw	2	.02 (.02)	15	.24 (.00)	10	.1

Table 1. Optimal number of parents and corresponding success rates. The brackets contain the results for two parents.

The interaction between selection pressure and the parameters r for diagonal crossover, respectively n for n-point crossover, is investigated in [47]. A steady-state GA with tournament selection (tournament size between 1 and 6) combined with random deletion and worst-fitness deletion is applied to the Griewangk and the Schwefel functions. The disruptiveness of both operators increases in parallel as the values for r and n are raised, but the experiments show that diagonal crossover consistently outperforms n-point crossover. The best option proves to be low selection pressure and high r in diagonal crossover combined with worst-fitness deletion.

The aforementioned studies have given sufficient indication that using multiparent operators with higher arities can increase GA performance. Such an increase, however, does not always occur, and the most recent studies shifted the focus of attention from showing that it occurs to investigating when it occurs. Such an analysis implicitly assumes a relationship between the characteristics of the test functions and the observed GA behavior. Unfortunately, it is very difficult to characterize the commonly used test functions.

Motivated by the difficulties of characterizing the shapes of numerical objective functions, the effects of operator arity are studied on fitness landscapes with controllable ruggedness in [14]. The NK-landscapes of Kauffman [28], where the level of epistasis, hence the ruggedness of the landscape, can be tuned by the parameter K, are used for this purpose. The multiple-children and the one-child versions of diagonal crossover and uniform scanning are tested within a steady-state GA with linear ranked-biased selection ($b = 1.2$) and worst-fitness deletion for $N = 100$ and different values of K. Two kinds of epistatic interactions, nearest neighbor interaction (NNI) and random neighbor interactions (RNI), are considered. As the NK-landscapes are generated randomly, the exact optimum is not known for any combination of N and K. This study uses the "practical global optimum" (being the highest value ever found during the tests) as the basis of performance comparison. The quality of a particular algorithm variant is evaluated by the distance to this practical global maximum which is computed by

$$\Delta = \frac{f_{maximal} - f_{obtained}}{f_{maximal}}$$

where $f_{obtained}$ is the best value found by the given variant. Similar to earlier findings, [18], the tests show that the performance of uniform scanning cannot be related to the number of parents. The two versions of diagonal crossover behave identically, and for both operators there is a consequent improvement with increasing r. However, as the epistasis (ruggedness of the landscape) grows from $K = 1$ to $K = 5$ the advantage of more parents becomes smaller. On landscapes with significantly high epistasis ($K = 25$) the relationship between operator arity and algorithm performance seems to diminish; see Figure 2 for example, where the error at termination (Δ) is shown for nearest neighbor interaction. The final conclusions of this investigation can be very well related to works of [37,20] and [25] on the usefulness of (two-parent) recombination. It seems that if and when crossover is useful, i.e. on mildly epistatic problems, then multiparent crossover is more useful than the two-parent variants.

A recent investigation analyzes diagonal crossover in detail on numerical optimization problems pursuing two research objectives, [16]. First, trying to find connections between the structure of the fitness landscape and the

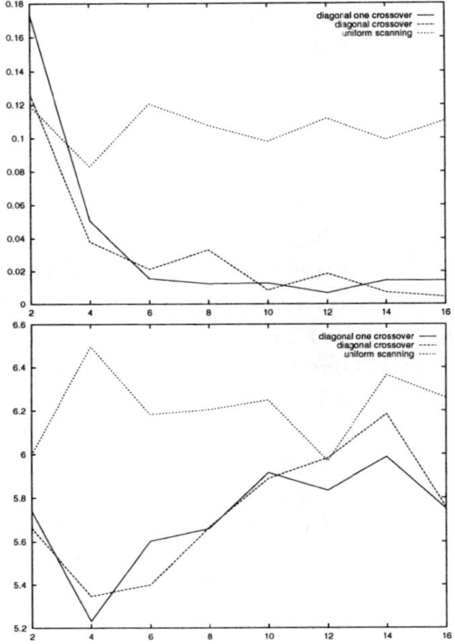

Fig. 2. Effect of the number of parents (horizontal axis) on Δ (vertical axis) on NK-landscapes with nearest neighbor interaction, $N = 100$, $K = 1$ (left), $K = 25$ (right)

performance of diagonal crossover, i.e. establishing on what kinds of landscapes it is advantageous to increase the number of parents. Second, trying to disclose the source of increased performance of the diagonal crossover with more parents if and when it is superior to two-parent recombination. As for the first goal, the functions in the applied test suite are characterized by their modality, separability, and the arrangement of local otpima. Regarding the second goal, a number of working hypotheses are formed based on two observations. It is observed that the increase in the number of parents in diagonal crossover automatically leads to an increased number of crossover points. It can be the case that higher performance is not the result of using more parents, but simply comes from being more disruptive by using more crossover points. Furthermore, it is noticed that when applying diagonal crossover, r parents create r children in one go. In a steady-state GA the population is updated, i.e. offspring are inserted, after each application of crossover (followed by mutation), which means that a GA using 10-parent diagonal crossover has more information before performing the selection step than a GA using the two-parent version. In other words, GAs with higher operator arity have a bigger generational gap which might cause a bias in their favor. Based on these observations the following working hypotheses are made.

Hypothesis 1: using more crossover points leads to better performance.
Hypothesis 2: bigger generational gap leads to better performance.
Hypothesis 3: using more parents leads to better performance.

These hypotheses are tested by using a steady-state GA with uniform random parent selection (!) and worst-fitness deletion on eight different numerical optimization test functions. Performance is measured by accuracy (error at termination), speed (median number of fitness evaluations before termination), and success rate, if the first two measures are inconlusive. Unfortunately, the outcomes do not provide a sufficient basis for well-grounded conclusions on the relationship between the structure of the fitness landscape and the performance of diagonal crossover. A surprising result is that on Rosenbrock's saddle (De Jong's F2) increasing r decreases the performance. This function is low-dimensional ($n = 2$), unimodal, and non-separable, but none of the other functions with these features has led to such behavior. As for the working hypotheses, Hypothesis 2 is clearly rejected by the similar behavior of the one-child, respectively original, variant of diagonal crossover. Hypothesis 1 is supported by the increasing performance of N-point crossover if N is increased. This, however, does not imply rejection of Hypothesis 3, i.e. that better performance for higher N's would *only* come from having more crossover points. In fact diagonal crossover was better than N-point crossover on all but two functions: on Rosenbrock's saddle (F2) and on the Fletcher–Powell function (multimodal, non-separable, with a random arrangement of local optima). On the Fletcher–Powell function, diagonal crossover was very similar to N-point crossover, indicating that increased performance for higher r's and N's seems to be the result of crossovers' effect as macro mutation, which effect is intensified by more crossover points. An illustration is given in Figure 3.

The working of multiparent recombination operators in continuous search spaces, in particular within ES is investigated in [11]. This study compares ρ-ary global intermediate recombination as defined at the end of Section 2, ρ-ary discrete recombination, which is identical to uniform scanning crossover, and diagonal crossover with one child. The main working hypothesis is that increasing the number of parents leads to increased EA performance in terms of achieved accuracy, i.e. distance from the global optimum at termination. This is divided into two subhypotheses: A) increasing the number of parents from one to two leads to increased EA performance; B) increasing the number of parents from two to larger numbers leads to increased EA performance. Note that subhypothesis A amounts to hypothesizing that recombination and mutation work better than mutation alone. Experiments are performed on unimodal landscapes (sphere model and Schwefel's double sum), multimodal functions with regularly arranged optima, and a superimposed uni-

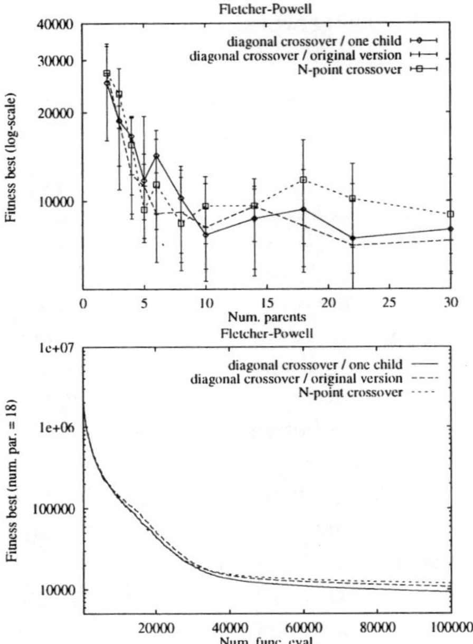

Fig. 3. Fletcher–Powell function. Left: effect of the number of parents, resp. crossover points, on accuracy. Right: population's best fitness during a run as a function of time (number of fitness evaluations), for $r = 18, N = 17$.

modal topology (Ackley, Griewangk, and Rastrigin functions). Furthermore, two functions with an irregular, random arrangement of local optima are studied, the Fletcher–Powell function and the Langerman function. A classification of the investigated fitness landscapes is presented in order to find the relationships between fitness landscape characteristics and performance of operators, see Table 2.

SEPARABILITY	MODALITY	
	UNIMODAL	MULTIMODAL
YES	Sphere	Rastrigin (regular)
NO	Schwefel	Ackley (regular) Griewangk (regular) Fletcher–Powell (irregular) Langerman (irregular)

Table 2. Characterization of the test functions

The results indicate that a diversity of possible outcomes can occur, and whether the working hypotheses hold or not depends on the particular combination of objective function and recombination operator. Subhypothesis A holds in more than 80% of the cases studied, but a further increase of the number of parents beyond two (subhypothesis B) does not necessarily have an advantageous effect on the accuracy achieved. In fact, there might be no significant impact, or even a negative impact; it can happen that A holds, but increasing the number of parents above two leads to worse results. Out of the 21 cases (3 operators, 7 test functions), multiparent recombination leads to a deterioration of solution quality only in 2 cases, it has no significant effect in 7 cases, and in the majority of the cases (12 out of 21 in total) it has a positive effect. These outcomes imply that, although there is no guarantee of success, it is reasonable to try multiparent recombination in an EA.

It is very interesting to consider the results on the randomly arranged fitness landscapes. With the Fletcher–Powell function there was no correlation between the number of parents and the ES performance [11], while in a genetic algorithm higher arities of diagonal crossover do lead to better performance [16]. This observation discloses that the same operator can behave differently on the same function, depending on the EA it is used in. Furthermore, the Fletcher–Powell function and the Langerman function are of the same type, non-separable and multimodal with an irregular, random arrangement of local optima. Still, the behavior of the ES is different: there is no clear effect of using more parents and the Fletcher–Powell function, while an advantage of higher arities can be observed with the Langerman function. This prevents general conclusions on ES behavior on quasirandom landscapes.

Let us note that introducing operator arity as a new parameter implies an obligation of setting its value. Since so far there are no reliable heuristics for setting this parameter, finding good values may require numerous tests. A way to circumvent this problem can be based on previous work on adapting or self-adapting the frequency of applying different operators [10,43], or using a number of competing subpopulations [38], each applying an operator with a different arity. A first assessment of this technique can be found in [15], where subpopulations with greater progress, i.e. with more powerful operators, become larger. As a "sideeffect", this study also compares six-parent diagonal crossover and two-parent one-point crossover within the traditional one population scheme on seven different fitness landscapes that have been specifically designed for studying the effect of crossover. On all of these landscapes (Onemax, Plateau, Plateau-d, Trap, Trap-d from [37], Twin Peaks from [20], and Royal Road from [30]) the multiparent operator is superior. As an illustration we present in Table 3 the mean best fitness values found and the number of fitness evaluations needed to find an optimum for these two operators.

	six-parent diagonal crossover		one-point crossover	
Problem	mean best	no. of evals	mean best	no. of evals
Onemax	100.0	3095	100.0	5691
Twin Peaks	100.0	3839	100.0	6021
Plateau	100.0	8060	97.6	25021
Plateau-d	99.2	23479	93.9	35063
Trap	189.8	3701	168.3	-
Trap-d	138.8	-	136.7	-
Royal Road	595.8	10228	480.2	30050

Table 3. Diagonal crossover with six parents vs. one-point crossover. Note: all problems are to be maximized; the number of fitness evaluations is undefined (-) if no runs found the optimum.

5 Conclusions

Recombination operators using more than two parents in an evolutionary problem solver have been repeatedly (re)introduced in evolutionary computation. However, there is still much systematic investigation needed to establish the effects of operator arity on EA performance. Of course, it can be questioned whether multiparent recombination is a single phenomenon showing one behavioral pattern. The present overview shows that there are (at least) two features for grouping multiparent operators. That is, multiparent recombination operators can be distinguished by their:

- Arity, for instance
  - fixed arity, e.g. the triadic crossover in [34],
  - variable arity that is tunable (can be set between 2 and X), e.g. the scanning crossover in [13],
  - variable arity that is undefined (a random number), e.g. global recombination in evolution strategies [2].
- Type, for instance
  - based on allele frequencies, e.g. the p-sexual voting from [32],
  - based on segmentation and recombination of the parents, e.g. the diagonal crossover in [13],
  - based on numerical operations on (real-valued) alleles like averaging, e.g. the center of mass crossover, in [44].

In general, it cannot be expected that operators in different classes show the same behavior. There are also experimental results supporting differentiation among various multiparent mechanisms. For instance, there seems to be no clear relationship between the number of parents and the performance of uniform scanning crossover, while the opposite is true for diagonal crossover [14].

Studies on multiparent operators also have to consider possibly different behavior on different types of fitness landscapes. So far, there are insufficient experimental data to support general conclusions. Some studies indicate that on irregular landscapes, such as NK-landscapes with relatively high K values [14] or the Fletcher–Powell function [11], they do not work. There are also results indicating the opposite on the Fletcher–Powell function [16], and on the Langerman function [11]. This stresses the importance of more experimental and theoretical research, following the tradition of studying the (dis)advantages of two-parent crossovers under different circumstances, [20,25,37,42]. In particular, there is a need for a better vocabulary to characterize objetcive functions or fitness landscapes. Such a vocabulary need not necessarily be universal, but applicable in EC research. That is, the function/landscape categories distinguishable by this vocabulary should coincide with the behavioral patterns of EAs. In fact, this is a methodological issue for experimental EC research concerning (the lack of) meaningful problem classes that can support generalizable experimental results [12]. Strongly related to this issue is the lack of a (general) theory that would illuminate why and when multiparent recombination works better than two-parent recombination.

One of the most inspiring approaches to multiparent reproduction is represented by the generalization templates in [45]. Although the authors do not elaborate on this aspect, their definitions can be applied to any two-parent operator matching the representation. On the one hand, this provides a technique to define new multiparent operators. On the other hand, it opens up a new trajectory for research, that of studying generalization templates as meta-operators acting on operators.

Summarizing this overview, let us note the following. While there are no biological analogies of multiparent recombination, computer simulations have already provided substantial evidence that applying more than two parents can "boost" evolution. Thus, although there is still much to be investigated, multiparent recombination mechanisms form a promising design heuristics for practitioners and a challenge for theoretical analysis.

Acknowledgments

I am grateful for the help of all those colleagues who contributed to this chapter, as co-authors of experimental research papers, or by exchanging ideas. Special thanks go to Thomas Bäck and Cees van Kemenade.

References

1. Ackley, D. H. (1987) An empirical study of bit vector function optimization. In L. Davis, editor, *Genetic Algorithms and Simulated Annealing*, 170–215 Morgan Kaufmann, San Francisco.

2. Bäck, T. (1996) *Evolutionary Algorithms in Theory and Practice.* Oxford University Press, New York.
3. Bäck, T., Michalewicz, Z. (1997) Test landscapes. In T. Baeck, D. Fogel, and Z. Michalewicz, editors, *Handbook of Evolutionary Computation*, pages B2.7:14–B2.7:20. Institute of Physics Publishing, Bristol, and Oxford University Press, New York.
4. Belew, R. K., Booker, L. B. (eds.) (1991) *Proceedings of the 4th International Conference on Genetic Algorithms* Morgan Kaufmann, San Francisco.
5. Bersini, H., Seront, G. (1992) In search of a good evolution-optimization crossover. In R. Maenner and B. Manderick, editors, *Proceedings of the 2nd Conference on Parallel Problem Solving from Nature*, 479–488 North-Holland, Amsterdam.
6. Bersini, H., Varela, F. J. (1991) The immune recruitment mechanism: A selective evolutionary strategy. In Belew and Booker [4], 520–526.
7. Beyer, H. G. (1995) Toward a theory of evolution strategies: On the benefits of the $(\mu/\mu, \lambda)$ theory. *Evolutionary Computation*, 3(1):81–111.
8. Bremermann, H. J., Rogson, M., Salaff, S. (1966) Global properties of evolution processes. In H.H. Pattee, E.A. Edlsack, L. Fein, and A.B. Callahan, editors, *Natural Automata and Useful Simulations*, 3–41 Spartan Books, Washington, DC.
9. Davidor, Y., Schwefel, H. P., Männer, R. (eds.) (1994) *Proceedings of the 3rd Conference on Parallel Problem Solving from Nature*, number 866 in Lecture Notes in Computer Science. Springer, Berlin.
10. Davis, L. (1989) Adapting operator probabilities in genetic algorithms. In Schaffer [36], 61–69.
11. Eiben, A. E., Bäck, Th. (1997) An empirical investigation of multi-parent recombination operators in evolution strategies. *Evolutionary Computation*, 5(3):347–365.
12. Eiben, A. E., Jelasity, M. (1992) A critical note on experimental research methodology in EC. In *2002 Congress on Evolutionary Computation (CEC'2002)*, 582–587. IEEE Press, Piscataway, NJ.
13. Eiben, A. E., Raué, P. E., Ruttkay, Zs. (1994) Genetic algorithms with multi-parent recombination. In Davidor et al. [9], 78–87.
14. Eiben, A. E., Schippers, C. A. (1996) Multi-parent's niche: n-ary crossovers on NK-landscapes. In H.-M. Voigt, W. Ebeling, I. Rechenberg, and H.-P. Schwefel, editors, *Proceedings of the 4th Conference on Parallel Problem Solving from Nature*, number 1141 in Lecture Notes in Computer Science, pages 319–328. Springer, Berlin.
15. Eiben, A. E., Sprinkhuizen-Kuyper, I. G., Thijssen, B. A. (1998) Competing crossovers in an adaptive GA framework. In [26], 787–792.
16. Eiben, A. E., van Kemenade, C. H. M. (1997) Diagonal crossover in genetic algorithms for numerical optimization. *Journal of Control and Cybernetics*, 26(3):447–465.
17. Eiben, A. E., van Kemenade, C. H. M. (1995) Performance of multi-parent crossover operators on numerical function optimization problems. Technical Report TR-95-33, Leiden University; available from http://www.liacs.nl/TechRep/1995/.
18. Eiben,A.E., van Kemenade,C.H.M., KOK, J.N.(1995) Orgy in the Computer:Multi-parent reproduction in genetic algorithm. In Moraen et al.[31],934–945.

19. Eshelman, L. J., Mathias, K. E., Schaffer, J. D. (1997) Crossover operator biases: Exploiting the population distribution. In Th. Baeck, editor, *Proceedings of the 7th International Conference on Genetic Algorithms*, 354–361. Morgan Kaufmann, San Francisco.
20. Eshelman, L. J., Schaffer, J. D. (1993) Crossover's niche. In Forrest [23], 9–14.
21. Fogel, D. B., Atmar, J. W. (1990) Comparing genetic operators with Gaussian mutations in simulated evolutionary processes using linear systems. *Biological Cybernetics*, 63:111–114.
22. Fogel, D. B., Atmar, J. W. (1990)Fogel, D. B., Stayton, L. C. (1994) On the effectiveness of crossover in simulated evolutionary optimization. *Biosystems*, 32:171–182.
23. Forrest, S. (ed.) (1993) *Proceedings of the 5th International Conference on Genetic Algorithms.* Morgan Kaufmann, San Francisco.
24. Furuya, H., Haftka, R. T. (1993) Genetic algorithms for placing actuators on space structures. In Forrest [23], 536–542.
25. Hordijk, W., Manderick, B. (1995) The usefulness of recombination. In Moraen et al. [31], 908–919.
26. *IEEE Proceedings of the 1995 IEEE Conference on Evolutionary Computation.* IEEE Press, Piscataway, NJ.
27. *IEEE Proceedings of the 1996 IEEE Conference on Evolutionary Computation.* IEEE Press, Piscataway, NJ.
28. Kauffman, S. A. (1993) *Origins of Order: Self-Organization and Selection in Evolution.* Oxford University Press, New York.
29. Kaufman, H. (1967) An experimental investigation of process identification by competitive evolution. *IEEE Transactions on Systems Science and Cybernetics, SSC-3(1)*, 11–16.
30. Mitchell, M., Forrest, S., Holland, J. H. (1994) The royal road for genetic algorithms: Fitness landscapes and GA performance. In F.J. Varela and P. Bourgine, editors, *Toward a Practice of Autonomous Systems: Proceedings of the 1st European Conference on Artificial Life*, pages 245–254. The MIT Press, Cambridge, MA.
31. Morán, F., Moreno, A., Merelo, J. J., Chacón, P. (eds) (1995) *Advances in Artificial Life. Third International Conference on Artificial Life*, volume 929 of *Lecture Notes in Artificial Intelligence.* Springer, Berlin.
32. Mühlenbein, H. (1989) Parallel genetic algorithms, population genetics and combinatorial optimization. In Schaffer [36], 416–421.
33. Mühlenbein, H., Voigt, H. M. (1996) Gene pool recombination in genetic algorithms. In I.H. Osman and J.P. Kelly, editors, *Meta-Heuristics: Theory and Applications*, 53–62 Boston, London, Dordrecht, Kluwer Academic Publishers.
34. Pál, K. F. (1994) Selection schemes with spatial isolation for genetic optimization. In Davidor et al. [9], 170–179.
35. Renders, J. M., Bersini, H. (1994) Hybridizing genetic algorithms with hill-climbing methods for global optimization: Two possible ways. In *Proceedings of the First IEEE Conference on Evolutionary Computation*, pages 312–317. IEEE Press, Piscataway, NJ, 1994.
36. Schaffer, J. D. (ed) (1989) *Proceedings of the 3rd International Conference on Genetic Algorithms.* Morgan Kaufmann, San Francisco.
37. Schaffer, J. D., Eshelman, L. J. (1991) On crossover as an evolutionary viable strategy. In Belew and Booker [4], 61–68.

38. Schlierkamp-Voosen, D., Mühlenbein, H. (1996) Adaptation of population sizes by competing subpopulations. In [27], 330–335.
39. Schwefel, H. P. (1995) *Evolution and Optimum Seeking.* Wiley, New York.
40. Schwefel, H. P., Rudolph, G. (1995) Contemporary evolution strategies. In Moraen et al. [31], 893–907.
41. Smith, J., Fogarty, T. C. (1996) Recombination strategy adaptation via evolution of gene linkage. In [27], 826–831.
42. Spears, W. M. (1993) Crossover or mutation? In L.D. Whitley, editor, *Foundations of Genetic Algorithms - 2*, pages 221–238. Morgan Kaufmann, San Francisco.
43. Spears, W. M. (1995) Adapting crossover in evolutionary algorithms. In J.R. McDonnell, R.G. Reynolds, and D.B. Fogel, editors, *Proceedings of the 4th Annual Conference on Evolutionary Programming*, pages 367–384. MIT Press, Cambridge,MP.
44. Tsutsui, S. (1998) Multi-parent recombination in genetic algorithms with search space boundary extension by mirroring. In A.E. Eiben, Th. Baeck, M. Schoenauer, and H.-P. Schwefel, editors, *Proceedings of the 5th Conference on Parallel Problem Solving from Nature*, number 1498 in Lecture Notes in Computer Science, 428–437. Springer, Berlin.
45. Tsutsui, S., Ghosh, A. (1998) A study on the effect of multi-parent recombination in real coded genetic algorithms. In *Proceedings of the 1998 IEEE Conference on Evolutionary Computation*, 828–833. IEEE Press, Piscataway, NJ.
46. Tsutsui, S., Yamamura, M., Higuchi, T. (1999) Multi-parent recombination with simplex crossover in real coded genetic algorithms. In W. Banzhaf, J. Daida, A.E. Eiben, M.H. Garzon, V. Honavar, M. Jakiela, and R.E. Smith, editors, *Proceedings of the Genetic and Evolutionary Computation Conference (GECCO-1999)*, 657–664. Morgan Kaufmann, San Francisco.
47. van Kemenade, C. H. M. Kok, J. N., Eiben, A. E. (1995) Raising GA performance by simultaneous tuning of selective pressure and recombination disruptiveness. In [26], 346–351.
48. Voigt, H. M., Ühlenbein, H. M. (1995) Gene pool recombination and utilization of covariances for the Breeder Genetic Algorithm. In [26],172–177.

TCG-2: A Test-Case Generator for Non-linear Parameter Optimisation Techniques

Zbigniew Michalewicz[1] and Martin Schmidt[2]

[1]Department of Computer Science, University of North Carolina, Charlotte, NC 28223, USA *and* at the Institute of Computer Science, Polish Academy of Sciences, ul. Ordona 21, 01-237 Warsaw, Poland
E-mail: zbyszek@uncc.edu
[2]NuTech Solutions Inc., University Executive Park, Suite 102, Charlotte, NC 28262, USA
E-mail: marsch@daimi.au.dk and martin.schmidt@nutechsolutions.com

Summary. The experimental results reported in many papers suggest that making an appropriate a priori choice of an evolutionary method for a non-linear parameter optimisation problem remains an open question. It seems that the most promising approach at this stage of research is experimental, involving a design of a scalable test suite of constrained optimisation problems, in which many features could be tuned easily. It would then be possible to evaluate the merits and drawbacks of the available methods as well as test new methods efficiently. In this chapter, we discuss a new test-case generator for constrained parameter optimisation techniques, which deals with deficiencies of the generators proposed earlier. This generator, TCG-2, is capable of creating various test problems with different characteristics, including the dimensionality of the problem, number of local optima, number of active constraints at the optimum, topology of the feasible search space, etc. Such a test-case generator is very useful for analysing and comparing different constraint-handling techniques.

Keywords: Evolutionary computation, non-linear programming, constrained optimisation, test-case generator.

1 Introduction

The general non-linear programming (NLP) problem is to find $\bar{x}$ so as to

$$optimize\ f(\bar{x}),\ \bar{x}=(x_1,..,x_1)\in\Re^n, \tag{1}$$

where $x\in F\subseteq S$. The *objective function f* is defined on the *search space* $S\subseteq\Re^n$ and the set $F\subseteq S$ defines the *feasible region*. Usually, the search space S is defined as an n-dimensional rectangle in $\Re^n$ (domains of variables defined by their lower and upper bounds):

$$l_i \le x_i \le u_i, \qquad 1\le i\le n,$$

where the feasible region $F \subseteq S$ is defined by a set of p additional constraints $p \geq 0$:

$$g_j(\bar{x}) \leq 0, \text{ for } j=1,\ldots,q, \quad \text{and} \quad h_j(\bar{x})=0, \text{ for } j=q+1,\ldots,p.$$

At any point $\bar{x} \in F$, the constraints g_j that satisfy $g_j(x)=0$ are called the *active* constraints at $\bar{x}$.

The NLP problem, in general, is intractable: It is impossible to develop a deterministic method for the NLP in the global optimisation category that would be better than exhaustive search [1]. This leaves an opportunity for evolutionary algorithms, extended by some constraint-handling methods. Indeed, during the last few years, several evolutionary algorithms that aim at complex objective functions (e.g. non-differentiable or discontinuous) have been proposed for the NLP; a recent survey paper [3] provides an overview of these algorithms.

It is not clear what characteristics of a constrained problem make it difficult for an optimisation technique. Any problem can be characterised by various parameters; these may include the number of linear constraints, e.g. the number of non-linear constraints, the number of equality constraints, the number of active constraints, the ratio $\rho = |F|/|S|$ of sizes of feasible search space to the whole, the type of the objective function (the number of variables, the number of local optima, the existence of derivatives, etc.). In [3], 11 test cases for constrained numerical optimisation problems were proposed (*G1–G11*). These test cases include objective functions of various types (linear, quadratic, cubic, polynomial, non-linear) with various numbers of variables and different types (linear inequalities, non-linear equations, and inequalities) and numbers of constraints. The ratio ρ between the size of the feasible search space F and the size of the whole search space S for these test cases vary from 0% to almost 100%; the topologies of feasible search spaces are also quite different. These test cases are summarised in Table 1. For each test case the number n of variables, type of the function f, the relative size of the feasible region in the search space given by ρ, the number of constraints of each category (linear inequalities *LI*, non-linear equations *NE* and inequalities *NI*), and the number a of active constraints at the optimum (including equality constraints) are listed.

Table 1 presents the summary of the 11 test cases. The ratio $\rho = |F|/|S|$ was determined experimentally by generating 1,000,000 random points from S and checking whether or not they belong to F (for *G2* and *G3* we assumed $k = 50$).

The results of many tests did not provide meaningful conclusions, as no single parameter (number of linear, non-linear, active constraints, the ratio ρ, type of the function, number of variables) proved significant as a major measure of difficulty of the problem. For example, many tested methods approached the optimum quite closely for the test cases *G1* and *G7* (with ρ = 0.0111% and ρ = 0.0003%, respectively), whereas most of the methods experienced difficulties for the test case *G10* (with ρ = 0.0010%). Two quadratic functions (the test cases *G1* and *G7*), with a similar number of constraints (9 and 8, respectively) and an identical number (6) of active constraints at the optimum, presented a different challenge to most of these methods. In addition, several methods were quite sensitive to the presence of a feasible solution in the initial population. Possibly a more extensive testing of various methods was required.

Table 1. Summary of test cases

Function	n	Type of f	ρ	LI	NE	NI	a
G1	13	quadratic	0.0111%	9	0	0	6
G2	k	non-linear	99.8474%	0	0	2	1
G3	k	polynomial	0.0000%	0	1	0	1
G4	5	quadratic	52.1230%	0	0	6	2
G5	4	cubic	0.0000%	2	3	0	3
G6	2	cubic	0.0066%	0	0	2	2
G7	10	quadratic	0.0003%	3	0	5	6
G8	2	non-linear	0.8560%	0	0	2	0
G9	7	polynomial	0.5121%	0	0	4	2
G10	8	linear	0.0010%	3	0	3	6
G11	2	quadratic	0.0000%	0	1	0	1

Not surprisingly, the experimental results of [3] suggested that making an appropriate a priori choice of an evolutionary method for a non-linear optimisation problem remained an open question. It seems that more complex properties of the problem (e.g. the characteristic of the objective function together with the topology of the feasible region) may constitute quite significant measures of the difficulty of the problem. Also, some additional measures of the problem characteristics due to the constraints might be helpful. However, this kind of information is not generally available. In [3] the authors wrote:

> "It seems that the most promising approach at this stage of research is experimental, involving the design of a scalable test suite of constrained optimisation problems, in which many [...] features could be easily tuned. Then it should be possible to test new methods with respect to the corpus of all available methods."

Clearly, there is a need for a parameterised test-case generator that can be used for analysing various methods systematically (rather than testing them on a few selected test cases; moreover, it is not clear whether the addition of a few extra test cases is of any real help).

There have been some attempts to propose test-case generators for unconstrained parameter optimisation [5], [6]. We are also aware of one attempt to do so for constrained cases; in [2] the author proposed a so-called stepping-stones problem defined as

$$\text{Objective: maximise} \qquad \sum_{i=1}^{n}(x_i/\pi+1),$$

where $-\pi \le x_i \le \pi$ for i=1,..., n and the following constraints are satisfied:

$$e^{x_i/\pi}+\cos(2x_i)\le 1 \qquad \text{for } i=1,\ldots,n.$$

Note that the objective function is linear and that the feasible region is split into 2^n disjoint parts (called stepping-stones). As the number of dimensions n grows, the problem becomes more complex. However, as the stepping-stones problem has only one parameter, it cannot be used to investigate some aspects of a constraint-handling method. In [4] we reported on preliminary experiments with a test-case generator *TCG*. This generator was capable of creating various test cases with different characteristics:

- Problems with different values of ρ: the relative size of the feasible region in the search space.
- Problems with different numbers and types of constraints.
- Problems with convex or non-convex objective functions, possibly with multiple optima.
- Problems with highly non-convex constraints consisting of (possibly) disjoint regions.

However, there were some problems with the *TCG* (we discuss them later in the paper), so it was necessary to develop its new version, *TCG-2,* to address these issues.

This chapter presents the test-case generators *TCG* and *TCG-2* as follow: The next section describes the original version of the *TCG*, section 3 summarises the experiments with the *TCG*, whereas section 4 discusses weak points of the proposed system. Section 5 provides a description of the *TCG-2* and section 6 concludes the paper and indicates some directions for future research. Since this chapter reports on work in progress, we do not provide test results on the *TCG-2* as they are not yet ready.

2 Test-case Generator *TCG*

As explained in the previous section, it is of great importance to have a parameterised generator of test cases for constrained parameter optimisation problems. By changing the values of some parameters, it would be possible to investigate merits/drawbacks (efficiency, cost, etc.) of many constraint-handling methods. Many interesting questions could be addressed:

- How does the efficiency of a constraint-handling method change as a function of the number of disjoint components in the feasible part of the search space?
- How does the efficiency of a constraint-handling method change as a function of the ratio between the sizes of the feasible part and the whole search space?
- What is the relationship between the number of constraints (or the number of dimensions, for example) of a problem and the computational effort of a method?

There are many others interesting questions. To find answers to these questions a parameterised *TCG* was proposed in [4]:

$$TCG(\, n, w, \lambda, \alpha, \beta, \mu\,)$$

which controls:

n – the number of variables in the problem
w – the number of optima in the search space
λ – the number of constraints (inequalities)
α – the "connectedness" of the feasible search regions

β – the ratio of the feasible to total search space
μ – the "ruggedness" of the fitness landscape

The ranges and types of the parameters are:

$n \geq 1$; *integer*	$w \geq 1$; *integer*	$0 \leq \lambda \leq 1$; *float*
$0 \leq \alpha \leq 1$; *float*	$0 \leq \beta \leq 1$; *float*	$0 \leq \mu \leq 1$; *float.*

The general idea behind this *TCG* was to divide the search space S into a number of disjoint subspaces S_k and to define a unimodal function f_k for every S_k. Thus, the objective function G is defined on the search space $S=\prod_{i=1}^{n}[0,1)$ as follows:

$$G(\overline{x})=f_k(\overline{x}) \text{ iff } \overline{x} \in S_k.$$

The total number of subspaces S_k is equal to w^n, as each dimension of the search space is divided into w equal length segments (for exact definition of a subspace S_k, see [4]). This number also indicates the total number of local optima of function G. For each subspace S_k ($k = 0,..., w^n - 1$) a function f_k is defined:

$$f_k(x_1,...,x_n)=a_k\left(\prod_{i=1}^{n}(u_i^k-x_i)(x_i-l_i^k)\right)^{\frac{1}{n}}, \tag{2}$$

where l_i^k and u_i^k are the boundaries of the k-th subspace for the i-th dimension. The constants a_k are defined as follows:

$$a_k=\begin{cases} 4w^2(1-\alpha^2\beta^2)^{k'[(1-\mu)+\mu/\log_2(wn-n+1)]} & \text{if } \alpha\beta>0 \\ 4w^2\left(\dfrac{(\mu-1)k''}{n(w-1)}+1\right) & \text{if } a\beta=0 \end{cases}, \tag{3}$$

where $k'=\log_2(\sum_{i=1}^{n}q_{i,k}+1)$, $k''=\sum_{i=1}^{n}q_{i,k}$, and $(q_{1,k},..., q_{n,k})$ is a n-dimensional representation of the number k in a w-ary alphabet.[1] Additionally, to remove any predefined and fixed pattern from the generated test cases, an additional mechanism, a random permutation of f_k's was used.

The third parameter λ of the TCG is related to the number m of constraints of the problem, since the feasible part of the search space S is defined by means of m double inequalities ($1 \leq m \leq w^n$):

$$r_1^2 \leq c_k(\overline{x}) \leq r_2^2, \quad k = 0,..., m\text{-}1, \tag{4}$$

where $0 \leq r_1 \leq r_2$ and each $c_k(x)$ is a quadratic function:

$$c_k(\overline{x})=(x_1-p_1^k)^2+...+(x_n-p_n^k)^2,$$

where $p_i^k = {(l_i^k-u_i^k)}/{2}$. These m double inequalities define m feasible parts F_k of the search space:

[1] For $w > 1$. If $w = 1$ the whole search space consists of one subspace S_0 only. In this case $a_0 = 4/(1 - \alpha^2\beta^2)$.

$$\bar{x} \in F_k \quad \text{iff} \quad r_1^2 \le c_k(\bar{x}) \le r_2^2,$$

and the overall feasible search space $F = \bigcup_{k=0}^{m-1} F_k$. In the following, we refer to each of F_k's as a ring; note that the two spheres with radii r_1 and r_2 bound such a ring. Note also that the interpretation of constraints here is different from the one in the standard definition of the NLP problem. Here the feasible search space is defined as a *union* (not intersection) of all double constraints. In other words, a point $\bar{x}$ is feasible if and only if there exists an index $0 \le k \le m-1$ such that the double inequality in Equation 4 is satisfied.

The parameter $0 \le \lambda \le 1$ determines the expected number m of rings as follows:

$$m = \lfloor \lambda(w^n - 1) + 1 \rfloor. \tag{5}$$

These rings are distributed randomly among subspaces S_k. Clearly, λ=0 and λ=1 implies $m = 1$ and $m = w^n$ respectively, i.e. minimum and maximum number of rings respectively.

The parameters α and β define the radii r_1 and r_2:

$$r_1 = \frac{\alpha\beta\sqrt{n}}{2w}, \qquad r_2 = \frac{\alpha\sqrt{n}}{2w}. \tag{6}$$

These parameters determine the topology of the feasible part of the search space.

If $\alpha\beta > 0$, the function G has 2^n global maxima points, all on the inner sphere in the permuted subspace S_0. For any global solution $(x_1,\ldots,x_n)$, $x_i = 1/(2w) \pm \alpha\beta/(2w)$ for all $i = 1, 2, \ldots, n$. The function values at these solutions are always equal to one. On the other hand, if $\alpha\beta = 0$, the function G has either one global maximum (if $\mu < 1$) or m maxima points (if $\mu = 1$), one in each of the permuted subspaces $S_0, \ldots, S_{m-1}$. If $\mu < 1$, the global solution $(x_1,\ldots,x_n)$ is always at

$$(x_1,\ldots,x_n) = (1/(2w), 1/(2w), \ldots, 1/(2w)),$$

which is the centre of the subspace S_0 before the random permutation is performed.

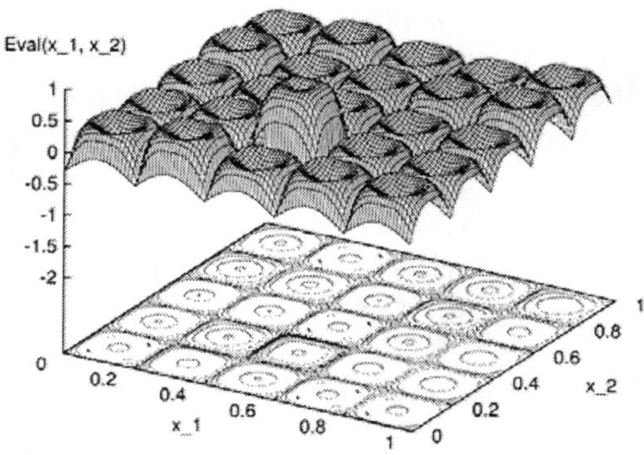

Fig. 1. *TCG*(2,5,1,1/$\sqrt{2}$,0.8,0)

Fig. 1 displays the final landscape for test case *TCG*(2,5,1,1/$\sqrt{2}$,0.8,0) using the penalty approach with a penalty coefficient $W = 1$ (see Equation 7 later).

Contours with function values ranging from -0.2 to 1.0 at a step of 0.2 are drawn at the base.

The interpretation of the six parameters of the *TCG* is as follows:

1. **Dimensionality** n: By increasing the parameter n the dimensionality of the search space can be increased.
2. **Multimodality** w: By increasing the parameter w the multimodality of the search space can be increased. For the unconstrained function, there are w^n locally maximum solutions, of which one is globally maximum. For the constrained test function with $\alpha\beta > 0$, there are $(2w)^n$ different locally maximum solutions, of which 2^n are globally maximum solutions.
3. **Constraintness** λ: By increasing the parameter λ the number m of constraints is increased.
4. **Connectedness** α: By reducing the parameter α (from 1 to $1/\sqrt{n}$ and smaller), the connectedness of the feasible subspaces can be reduced. When $\alpha < 1/\sqrt{n}$, the feasible subspaces F_k are completely disconnected. Additionally, parameter α (with fixed β) influences the proportion of the feasible search space to the complete search space, i.e. the ratio ρ.
5. **Feasibility** β: By increasing the ratio β, the proportion of the feasible search space to the complete search space can be reduced. For β values closer to one, the feasible search space becomes smaller and smaller. These test functions can be used to test an optimiser's ability to find and maintain feasible solutions.
6. **Ruggedness** μ: By increasing the parameter μ the function ruggedness can be increased (for $\alpha\beta > 0$). A sufficiently rugged function will test an optimiser's ability to search for the globally constrained maximum solution in the presence of other maxima almost equally significant locally.

Increasing each of the above parameters (except α) and decreasing α will cause an increased difficulty for any optimiser. However, it is difficult to conclude which of these factors most profoundly affects the performance of an optimiser. Thus, it is recommended that the user should first test his/her algorithm with the simplest possible combination of the above parameters (small n, small w, small μ, large α, small β, and small λ). Thereafter, the parameters may be changed in a systematic manner to create more difficult test functions. The most difficult test function is created when large values of parameters n, w, λ, β, and μ together with a small value of parameter α are used. For details of the TCG, the reader is referred to [4].

3 Experiments with the *TCG*

To test the usefulness of the *TCG*, a simple steady-state evolutionary algorithm was developed. We used a constant population size of 100 and each individual is a vector x of n floating-point numbers. Parent selection was performed by a standard binary tournament selection. An offspring replaces the worse individual of a binary tournament. One of three operators was used in every generation (the selection of an operator was done according to constant probabilities 0.5, 0.15, and 0.35, respectively):

- Gaussian mutation: $\overline{x} := \overline{x} + \overline{N}(0,\overline{\sigma})$, where $\sigma = \left(\frac{1}{2\sqrt{n}},\ldots,\frac{1}{2\sqrt{n}}\right)$ for all experiments reported in this section.
- Uniform crossover: $\overline{x} := (z_1,\ldots,z_n)$, where each z_i is either x_i or y_i (with equal probabilities), where x and $\overline{y}$ are two selected parents.
- Heuristic crossover: $z := r(x - y) + x$, where r is a uniform random number between 0 and 0.25, and the parent x is not worse than $\overline{y}$.

The termination condition was to quit the evolutionary loop if the improvement in the last N=10,000 generations was smaller than a predefined $\varepsilon = 0.001$.

Now the *TCG* can be used for investigating the merits of *any* constraint-handling method. For example, one of the simplest and the most popular constraint-handling methods is based on static penalties; let us define a particular static penalty method as follows:

$$eval(\overline{x}) = f(\overline{x}) + W * v(\overline{x}), \tag{7}$$

where f is an objective function, W is a constant (penalty coefficient), and function v measures the constraint violation. Note that only one double constraint is taken into account as the *TCG* defines the feasible part of the search space as a union of all m double constraints.

The penalty coefficient W was set to 10 and the value of $v(\overline{x})$ is defined by the following procedure:

```
find k such that x̄ ∈ S_k
set C = (–2w) / √n
if the whole S_k is infeasible then
   v(x̄) = –1
else
   calculate distance Dist between x̄ and the centre of the subspace S_k
   if Dist < r_1 then
      v(x̄) = C * (r_1 – Dist)
   else if Dist > r_2 then
      v(x̄) = C * (Dist – r_2)
   else
      v(x) = 0
   endif
endif
```

Thus the constraint violation measure v returns –1, if the evaluated point is in fully infeasible subspace (i.e., subspace without a ring); 0, if the evaluated point is feasible; or some number q from the range [–1,0], if the evaluated point is infeasible, but the corresponding subspace *has* a ring. This means that the point $\overline{x}$ is either inside the smaller sphere or outside the larger one. In both cases, q is a scaled negative distance of this point to the boundary of the closest sphere. Note that the scaling factor C guarantees that $0 < q \le 1$. Note that the values of the objective function for feasible points of the search space stay in the range [0,1];

the value at the global optimum is always 1. Thus, both the function values and the absolute constraint violation values are normalised in the range [0,1].

As the TCG has six parameters, it is difficult to discuss their interactions fully in a short chapter have selected a single point from the $TCG(n, w, \lambda, \alpha, \beta, \mu)$ parameter search space:

$$n = 2;\ w = 10;\ \lambda = 1.0;\ \alpha = 0.9/\sqrt{n};\ \beta = 0.1;\ \mu = 0.1,$$

and varied one parameter at a ime. In each case, a figure summarises the results that are averages over 10 runs with the following two lines in each figure:

1. Solid line (upper line): The fitness value of the best individual at the end of the run.
2. Broken line (lower line): The average height among feasible peaks in the landscape is displayed as fractions between 0 and 1, which corresponds to the height of the lowest peak and the height of the highest peak respectively.

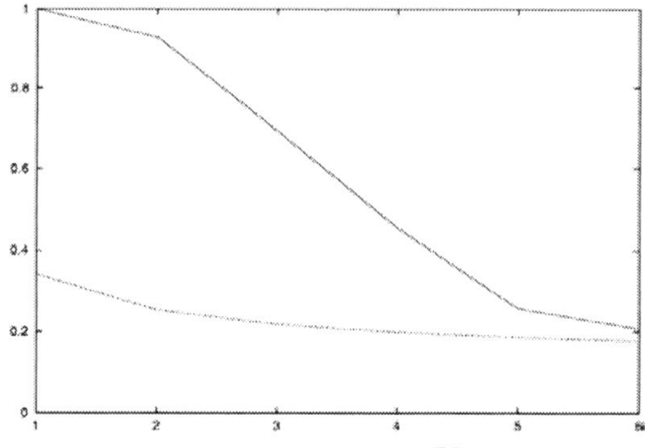

Fig. 2. $TCG(n, 10, 1.0, 0.9/\sqrt{n}, 0.1, 0.1)$

Fig. 2 displays the relative performance of the static penalty method on the $TCG(n,10,1.0,0.9/\sqrt{n},0.1,0.1)$ while n is varied between 1 and 6. It is clear that an increase in the number of dimensions n (the x-axis) reduces the efficiency of the algorithm: the fitness value of the returned solution approaches the value of the average peak height in the landscape.

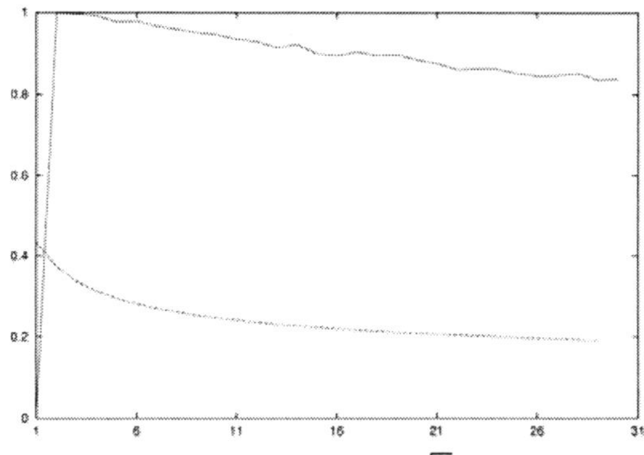

Fig. 3. $TCG(2,w,1.0,0.9/\sqrt{2},0.1,0.1)$

Fig. 3 displays the relative performance of the static penalty method on the $TCG(2,w,1.0,0.9/\sqrt{2},0.1,0.1)$ while w is varied between 1 and 30 (for $w = 30$ the objective function has $w^n = 900$ peaks). An increase of w decreases the performance of the algorithm, but not to the extent we have seen in the previous case (increase of n). The reason is that while $w = 10$ (see Fig. 2), the number of peaks grows as 10^n (for $n = 3$ there are 1000 peaks), whereas for $n = 2$ (Fig. 3), the number of peaks grows as w^2 (for $w = 30$ there are 900 peaks only).

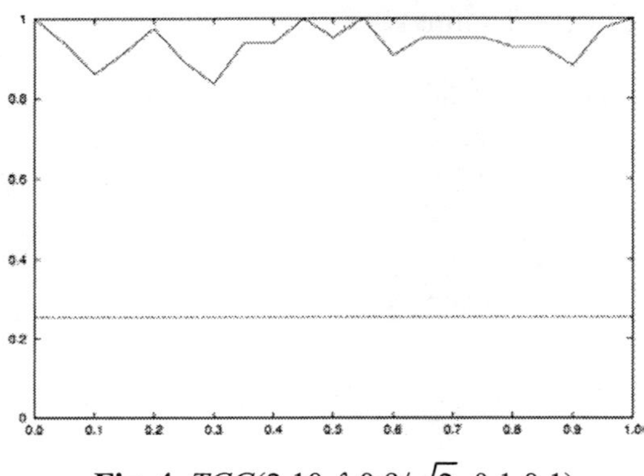

Fig. 4. $TCG(2,10,\lambda,0.9/\sqrt{2},0.1,0.1)$

Fig. 4 displays the relative performance of a static penalty method on the $TCG(2,10,\lambda,0.9/\sqrt{2},0.1,0.1)$ while λ is varied between 0 and 1. It seems that the number of constraints does not influence the performance of the system.

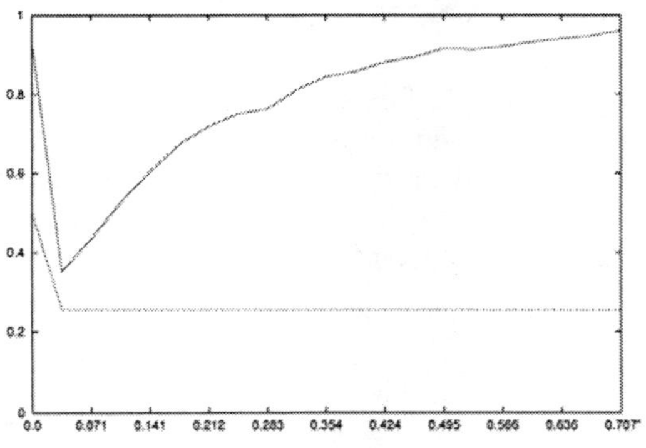

Fig. 5. $TCG(2,10,1.0,\alpha,0.1,0.1)$

Fig. 5 displays the relative performance of a static penalty method on the $TCG(2,10,1.0,\alpha,0.1,0.1)$ while α is varied between 0 and $1/\sqrt{2}$. Clearly, larger values of α improve the results of the algorithm, as it is easier to locate a feasible solution. Note an anomaly in the graph for $\alpha = 0$; this is due to Equation 3, which provides a different formula for a_k's when $\alpha\beta = 0$.

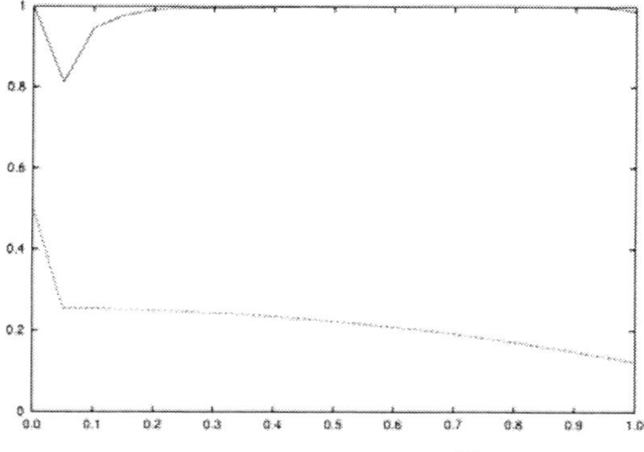

Fig. 6. $TCG(2,10,1.0,0.9/\sqrt{2},\beta,0.1)$

Fig. 6 displays the relative performance of a static penalty method on the $TCG(2,10,1.0,0.9/\sqrt{2},\beta,0.1)$ while β is varied between 0 and 1. Larger values of β decrease the size of rings; however, this feature seems not to influence the performance of the algorithm significantly.

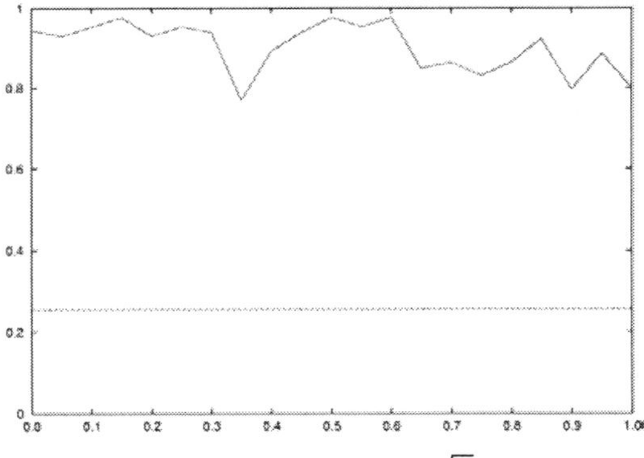

Fig. 7. $TCG(2,10,1.0,0.9/\sqrt{2}, 0.1,\mu)$

Fig. 7 display the relative performance of a static penalty method on the $TCG(2,10,1.0,0.9/\sqrt{2}, 0.1,\mu)$ while μ is varied between 0 and 1. Higher values of μ usually make the test case harder as the heights of the peaks are changed: Local maxima are of almost the same height as the global one.

We conclude that the results obtained by the described steady-state evolutionary algorithm using a static penalty approach depend mainly on the dimensionality n, the multimodality w, the connectedness α, and the ruggedness μ. On the other hand, they do not significantly seem to depend on the constraintness λ and the feasibility β.

It is interesting also to investigate the following relationship between the value of penalty coefficient W (x-axis) and the quality of the obtained solution (y-axis).

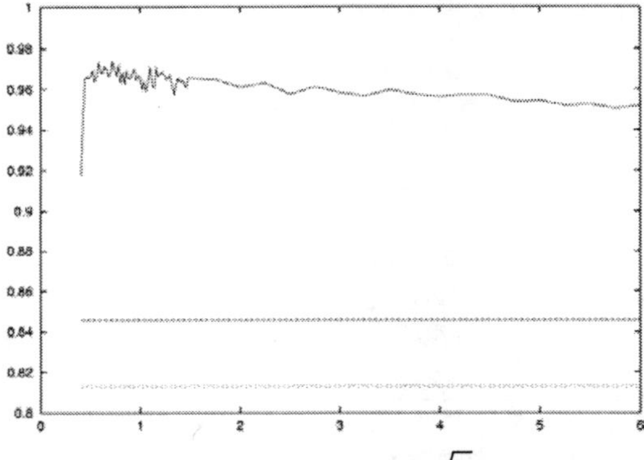

Fig. 8. $TCG(5,10,0.2,0.9/\sqrt{5},0.5,0.1)$

Fig. 8 confirms another intuition connected with static penalties: It is difficult to guess the "right" value of the penalty coefficient, as different landscapes require different values of W! Note that low values of W (i.e. values below 0.4) produce poor quality results; on the other hand, for larger values of W (i.e. values larger than 1.5), the quality of solutions drops slowly (it stabilises later – for large W – at the level of 0.95). Thus the best values of penalty coefficient W for the particular landscape of the $TCG(5,10,0.2,0.9/\sqrt{5},0.5,0.1)$ (which has $n^w = 10^5$ local optima) are in the range [0.6, 0.75].

4 Critique of the *TCG*

We have discussed briefly how the generator $TCG(n,w,\lambda,\alpha,\beta,\mu)$ can be used for evaluation of a constraint-handling technique. As explained in section 2,

- the parameter n controls the dimensionality of the test function,
- the parameter w controls the modality of the function,
- the parameter λ controls the number of constraints in the search space,
- the parameter α controls the connectedness of the feasible search space,
- the parameter β controls the ratio of the feasible to total search space, and
- the parameter μ controls the ruggedness of the test function.

Such a constrained test problem generator may serve the purpose of testing any method for constrained parameter optimisation. Moreover, one can also use the *TCG* for testing any method for unconstrained optimisation (e.g. operators, selection methods, etc.). In the previous section, we indicated how it can be used to evaluate the merits and drawbacks of one particular constraint-handling method (static penalties). Note that it is possible to analyse further the performance of a method by varying two or more parameters of the *TCG*.

Nevertheless, this TCG is far from perfect. It defines a landscape that is a collection of functions that can be optimised site-wise, where each is defined on different subspaces of equal sizes. According to all basins of attractions have the same size; moreover, all points at the boundary between two basins of attraction are equally fitted. The local optima are located in the centres of the hypercubes; all feasible regions are centred around the local optima. Note also that while we can change the number of constraints, there is precisely one (for $\alpha\beta > 0$) active constraint at the global optimum.

Some of these weaknesses can be corrected easily. For example, in order to avoid the symmetry and equal-sized basin of attraction of all subspaces, we may modify the *TCG* in the following two ways. First, the parameter vector x may be transformed into another parameter vector $\bar{y}$ using a non-linear mapping $\bar{y} = g(\bar{x})$. The mapping function may be chosen in a way so as to have the lower and upper bounds of each parameter y_i equal to zero and one, respectively. Such a non-linear mapping will make all subspaces of different size. Second, in order to make the feasible search space asymmetric, the centre of each subspace for the outer hypersphere can be made different from that of the inner hypersphere. The following update of the centre for the outer hypersphere can be used:

$$p_i^k = (l_i^k + u_i^k)/2 + \gamma_i^k \varepsilon_i^k ,$$

where ε_i^k is a small number denoting the maximum difference allowed between the centres of inner and outer hyperspheres, and γ_i^k is a random number between zero and one. To avoid using another control parameter, ε_i^k can be assumed to be the same for all subspaces. This modification makes the feasible search space asymmetric, although the maximum at each subspace remains the same as in the original function.

It might be worthwhile to modify further the *TCG* to parameterise the number of active constraints at the optima. It seems necessary to introduce the possibility of a more gradual increment in the number of peaks. In the current version of the TCG, $w = 1$ implies one peak, and $w = 2$ implies 2^n peaks (this was the reason for using low values for parameter n). In addition, the difference between the lowest and the highest values of the peaks in the search space are, in the present model, too small.

All these limitations of the *TCG* decrease its significance for being a useful tool for modelling real-world problems, which may have quite different characteristics. Note also that it is necessary to develop an additional tool, which would map a real-world problem into a particular configuration of the TCG. In such a case, the TCG should be able to "recommend" the most suitable constraint-handling method.

In the following section we describe the newest version of the TCG, *TCG-2*, which deals with the limitations of the original *TCG*.

5 New Test-case Generator *TCG-2*

In order to overcome the problems of the *TCG* we developed the new generator *TCG-2*. This section presents the *TCG-2* and explains how it overcomes the limitations of the earlier *TCG*.

The *TCG-2* is a parameterised TCG of the following form:

$$TCG\text{-}2(\,n,\, m,\, \rho,\, c,\, a,\, p,\, \sigma,\, \alpha\,)$$

where the parameters control

n – the *dimensionality* of the test function,
m – the number of feasible *components*,
ρ – the *feasibility* of the search space,
c – the *complexity* of the feasible search space,
a – the number of *active constraints* at the global optimum,
p – the number of *peaks* of the objective function,
σ – the *width* of the peaks of the objective function,
α – the *decay* of the peaks of the objective function.

The ranges and the types of the parameters are:

$1 \leq n$; *integer*	$1 \leq m$; *integer*	$0 \leq \rho \leq 1$; *float*
$0 \leq c \leq 1$; *float*	$0 \leq a \leq n$; *integer*	$1 \leq p$; *integer*
$0 \leq \sigma$; *float*	$0 \leq \alpha \leq 1$; *float.*	

There are three major components of the *TCG-2* and we discuss them in turn:

1. The search space.
2. The feasible search space.
3. The landscape of the objective function.

As with the original *TCG* the search space is defined as an n-dimensional cube. Note, however, that the ranges [0,1] are closed whereas in the original *TCG* the ranges were [0,1). The original *TCG* defined the feasible search area as a union of m double inequalities (Equation 4), which defined m feasible parts.

The general idea behind the *TCG-2* is randomly to create non-overlapping feasible areas (referred to as boxes) in the search space. Many boxes are attached to each other in order to form feasible components. All components are disjoint and hold a total feasible area of $\rho * |S|$. The higher the complexity c, the more (probably smaller) boxes will be created (hence the higher the complexity). The feasible part of the search space is generated with the following constraint generation algorithm:

```
do
    do
        Create_Initial_Components( m, ρ, c )
        Enlarge_Initial_Components( m, ρ, c )
    until Successful_Generated_Initial_Components() ≡ true
    Add_Complexity( m, ρ, c )
until Successful_Generated_Final_Components() ≡ true
```

The role of the procedures and functions incorporated in the above algorithm is explained below:

- Create_Initial_Components(m, ρ, c): This procedure creates m disconnected feasible components (boxes) at random locations such that each initial component occupies a very small part of the total feasible search space, i.e. each initial component has the size $1\% * \rho * (1-c) * |S| / m$.
- Enlarge_Initial_Components(m, ρ, c): This procedure enlarges the existing feasible components by choosing one side of a box at random and trying to

move it a little bit. This is repeated until the total feasible search space becomes $\rho * (1 - c) * |S|$. This procedure adds no new components but enlarges existing ones and makes sure that all components remain disconnected.
- Successful_Generated_Initial_Components(): This function checks that the randomly generated initial components are valid. The function returns either true or false for "success" or "failure" respectively.
- Add_Complexity(m, ρ, c): This procedure generates the complexity of the feasible search space by randomly generating and attaching boxes to the existing components, i.e. every new box can be attached to any existing box. This procedure also ensures that no matter how many boxes are added, there will never be any feasible continuous path from one feasible component to any other. Hence, we will have m disconnected feasible components and every component consists of one or more boxes. This procedure stops once the total feasible search space becomes $\rho * |S|$, and hence this procedure generates new boxes with a total size of $\rho * c * |S|$.
- Successful_Generated_Final_Components(): This function checks that all the generated boxes and components are valid. It returns true or false for "success" or "failure" respectively.

Note that the above algorithm is stochastic and might get stuck in an arrangement of boxes that is invalid. Hence the two functions that test the successful generation of the boxes might "fail", which results in a restart of the whole algorithm. In general, if $m \geq 4$ and ρ is close to 1 then the constraint generation algorithm has a long runtime, since it is difficult to arrange the boxes correctly. In the extreme case of $\rho = 1$ (and $m > 1$) it is impossible to create disjoint components and hence a feasible arrangement of m components is impossible.

The results of the constraint generation algorithm are:

1. The larger the number of components m, is the more (disconnected) feasible components are generated.
2. The larger the feasibility ρ is, the larger the initial components are (as $\rho * (1 - c) * |S| / m$) and the larger the final feasible search space becomes (as $\rho * |S|$).
3. The larger the complexity c is, the smaller the initial components are (as $\rho * (1 - c) * |S| / m$) and the more boxes are added to the components (as $\rho * c * |S|$). The more boxes that are added to the components, the more difficult the feasible search space becomes. Hence, the larger the complexity c is, the more complex and difficult the feasible search space becomes.

The following figures present examples of generated constraints by the above algorithm. Fig. 9 presents generated constraints for the *TCG-2*(n=2, m=2, ρ=0.5, c=0.1, a=1, p=3, σ=0.4, α=0.5).

Fig. 9 shows an example of generated constraints with two feasible components in agreement with $m = 2$. The two different components are shown with two different colours. The rectangles illustrate the generated feasible boxes. The global optimum is indicated with a circle at (0.2,0.8) where there is one active constraint in agreement with a=1. The two largest boxes are generated by the

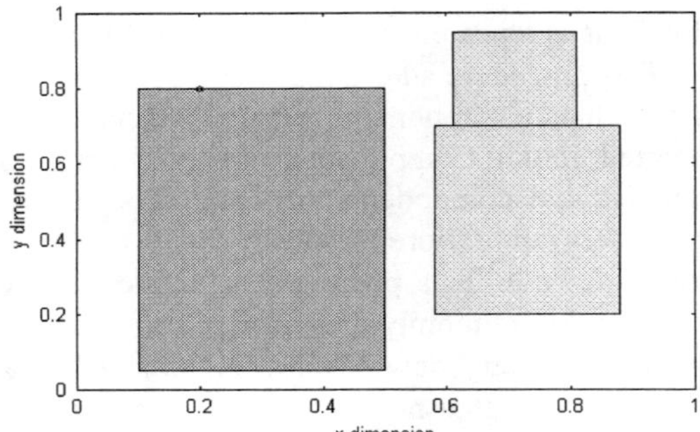

Fig. 9. TCG-2(2,2,0.5,0.1,1,3,0.4,0.5)

– Create_Initial_Components(m, ρ, c) and
– Enlarge_Initial_Components(m, ρ, c).

Since ρ = 0.5 and c = 0.1 the size of the initial components (the two largest boxes in this case) is $\rho * (1 - c) * | S |$ = 0.45. All other boxes (in this case only the third box around the location (0.7,0.9)) are created using the procedure

– Add_Complexity(m, ρ, c).

The total size of the added boxes is $\rho * c * | S | = 0.05$. Hence the total size of the feasible components is $\rho * | S | = 0.5$.

Fig. 10 illustrates the impact of varying the number of active constraints a and presents an example of the generated constraints for *TCG-2*(n=2, m=2, ρ=0.5, c=0.1, a=2, p=3, σ=0.4, α=0.5). There are two active constraints (and not just on as in Fig. 9) at the global optimum at (0.1,0.8) in agreement with $a=2$. Hence the only difference between Fig. 9 and Fig. 10 is that the global optimum is placed differently according to a = 1 and a = 2 respectively.

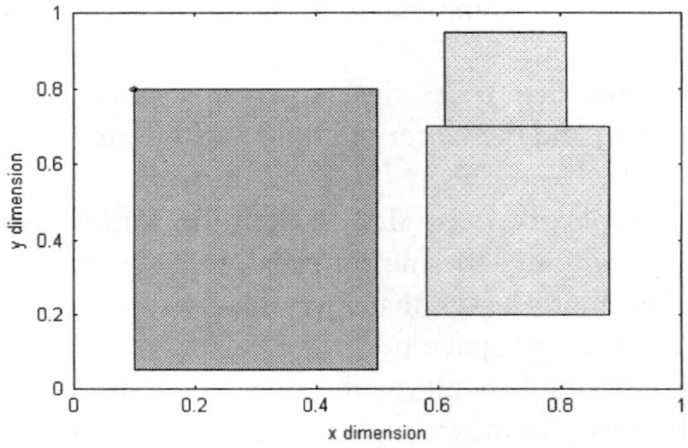

Fig. 10. *TCG-2*(2,2,0.5,0.1,2,3,0.4,0.5)

The total size of the feasible search space is the same in Fig.10 as in Fig. 9 since ρ=0.5.

Fig. 11 shows the *TCG-2*(n=2, m=2, ρ=*0.5,* c=0.5, a=2, p=3, σ=0.4, α=0.5). Note that c=0.5 and hence complexity is larger than in the earlier examples. The global optimum is placed such that there are two active constraints at the global optimum at (0.2,0.8) in agreement with a=2.

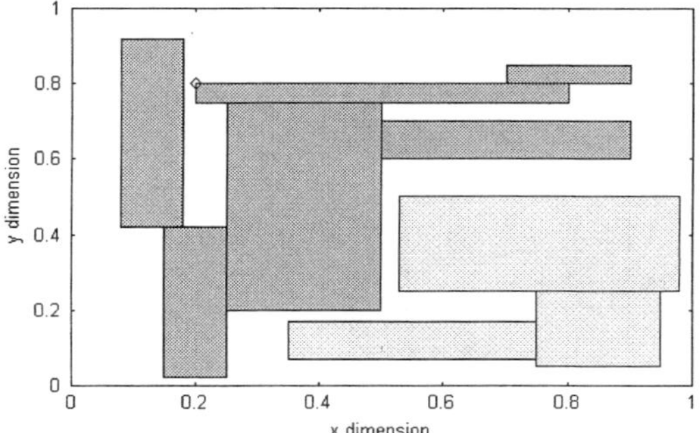

Fig. 11. *TCG-2*(2,2,*0.5,*0.5,2,3,0.4,0.5)

Fig. 11 clearly shows the impact of the complexity *c*:

- The initial boxes become smaller the larger the complexity *c* is.
- The larger the complexity *c* is, the more boxes are added. This increases the complexity of the search space.

Note that the size of the two feasible components is different and that a different number of boxes are attached to the two components. Further, all boxes that belong to the same component are attached to each other but they do not overlap. Boxes that belong to different components are disconnected.

After the creation of the feasible components the objective function $G(\bar{x})$ is defined as:

$$G(\bar{x}) = g_i(\bar{x}) \text{ where } \forall j\text{: } |\bar{x} - \bar{c}_i| \leq |\bar{x} - \bar{c}_j| \tag{8}$$

where $g_k(\bar{x}) = h_k \exp\left(-\frac{|\bar{x} - \bar{c}_k|^2}{2\sigma^2}\right) \wedge \alpha \leq h_k \leq 1 \wedge k = 1,\ldots,p.$

Equation 8 defines the objective function $G(\bar{x})$ using a set of p randomly placed Gaussians $g_k(\bar{x})$ where h_k is the height of peak k and $\bar{c}_k$ is the centre of peak k. In order to evaluate $G(\bar{x})$ the closest centre $\bar{c}_i$ is found and then the function $g_i(\bar{x})$ is evaluated.[2] All centres $\bar{c}_k$ are placed randomly[3] in the search space with the exception of the global optimum which is placed such that there are exactly *a* active constraints at the global optimum. All peak heights h_k are evenly distributed in the range $[\alpha, 1]$ such that the global optimum has the highest peak $h_k = 1$ while the lowest peak has the height $h_k = \alpha$.

[2] The reason for this approach is to create a non-differentiable and disruptive landscape for the objective function $G(\bar{x})$. If one instead had used the max over all $g_k(\bar{x})$ then the objective function $G(\bar{x})$ would be continuous.

[3] The centres are placed randomly (and are not a parameter of the *TCG-2*) in order to be able to create constraint optimisation problems with certain *features* but with no specific location of the centres. This also reduces the number of parameters for the *TCG-2* and improves its execution speed since only one Gaussian $g_i(\bar{x})$ must be evaluated.

Fig. 12 visualises a one-dimensional example of the objective function $G(x)$ of the *TCG-2*(n=1,m=2, ρ=0.5,c=0.5,a=1,p=3,σ=0.4,α=0.5).[4]

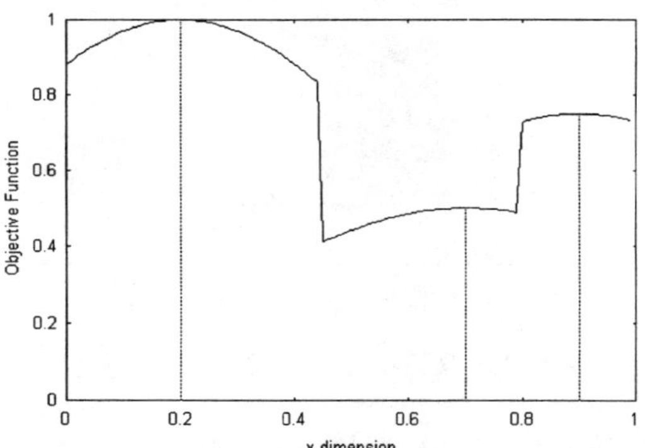

Fig. 12. $G(\bar{x})$ of *TCG-2*(1,2, 0.5,0.5,1,3,0.4,0.5)

Fig. 12 shows the objective function $G(\bar{x})$ and the vertical lines indicate the 3 peaks in agreement with p = 3. Notice that the global peak is at x = 0.2 and that the objective function is non-differentiable and disruptive between the centres of the peaks. At x = 0.7 the lowest peak is located with the height $h_k = \alpha = 0.5$.

Fig. 13 shows a 2-dimensional example of the objective function $G(\bar{x})$ for *TCG-2*(n=2, m=2, ρ=0.5, c=0.5, a=2, p=3, σ=0.4, α=0.5).

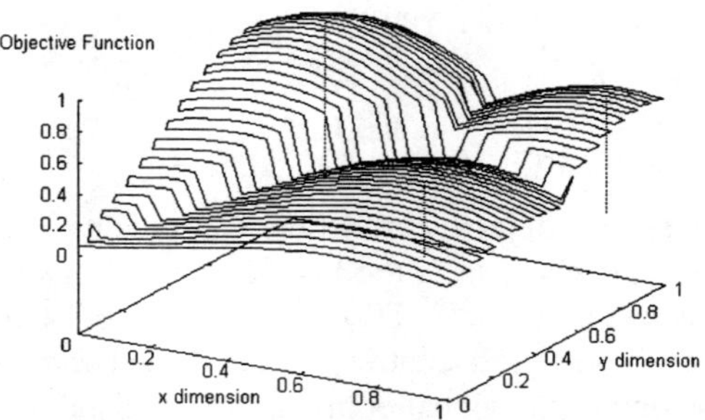

Fig. 13. *TCG-2*(2,2,0.5,0.5,2,3,0.4,0.5)

Fig. 13 shows the objective function where there are three peaks in agreement with p=3. The global optimum is at (0.2,0.8) with the height h_k = 1.0 and the lowest peak is at (0.7,0.4) with the height $h_k = \alpha = 0.5$. Notice the disruptive surface of the objective function and the irregular arrangement of the peaks.

The *TCG-2* overcomes the limitations of the earlier *TCG* in the following ways:

- **Multimodality**: The new *TCG-2* has a precise and gradual control over the number of peaks p whereas the *TCG* has far less gradual control using the

[4] Notice that m, ρ, c, and a have no impact on the objective function, except at the location of the global optimum which is located such that there are a active constraints.

parameter w which results in w^n peaks. This lets the *TCG-2* elegantly overcome a serious drawback of the *TCG*.

- **Feasibility**: The *TCG-2* also has a direct and intuitive control over the feasibility of the search space with the parameter ρ. Since the constraint generation and the objective function are separated from each other (except for the global optimum) the side effects of varying the feasibility are eliminated. This is in contrast to the *TCG* that has three parameters, λ, α, and β, that control the feasibility of the search space. Further, the *TCG* has the side effect that the feasible search space becomes one connected feasible component when the feasibility becomes large.
- **Ruggedness**: The *TCG-2* has a stronger and simpler control over the decay of the peaks than the *TCG*. The *TCG-2* has a linear decay of the peaks while the *TCG* uses an exponential decay of the peaks, which gives far more low peaks than high peaks.

Apart from overcoming the limitations of the *TCG* the new *TCG-2* also has additional features:

- **Constraintness**: The *TCG-2* can control the number of active constraints a at the global optimum in a precise and gradual fashion. This feature is not present in the *TCG* since the *TCG* always has exactly one active constraint at the global optimum. With this parameter, it is possible to test the ability of an optimiser to find the global optimum in spite of a varying number of active constraints at the global optimum.
- **Components**: The *TCG-2* also has a direct and gradual control over the number of disconnected feasible components using the parameter m. This makes it possible to control how many feasible islands exist in the search space. Further, this also makes it possible to investigate the impact that the number of components has on a particular optimisation technique.
- **Complexity**: The *TCG-2* has control over the complexity of the feasible search space using the parameter c. It is therefore possible to test an optimiser on search spaces with varying complexity without affecting the number of feasible components, the feasibility ρ, the number of active constraints, or any other parameter of the search space.

It is obvious that the new *TCG-2* overcomes the limitations of the earlier *TCG* and that the *TCG-2* also has many strong and intuitive new features. In general, the *TCG-2* is based on the lessons learned by developing and testing the *TCG*.

6 Conclusions

The test-case generators *TCG* and *TCG-2* were presented, and it was explained how the *TCG-2* overcomes the limitations of the *TCG*. Further, the new additional features of the *TCG-2* were presented.

The *TCG-2* creates constrained optimisation problems by means of several parameters:

- The dimensionality n,
- the number of feasible components m,

- the feasibility ρ,
- the complexity c,
- the number of active constraints a,
- the number of peaks p,
- the width of the peaks σ, and
- the decay of the peaks α.

Using these parameters it was explained how the *TCG-2* creates and controls the dimensionality, the multimodality, the feasibility, the ruggedness, the constraintness, the feasible components, and the complexity of the search space. With the gradual and intuitive control over these parameters, the *TCG-2* is a significant improvement over the original *TCG*.

Currently the *TCG-2* is being tested for a variety of different parameter settings. The goal is to test the abilities of optimisers to deal with the varying difficulty of the search space when the *TCG-2* parameters are varied. This is important since it is the foundation of knowledge about the behaviour of optimisers using different *TCG-2* parameter settings.

A future step for improving the usability of the *TCG-2* is to implement a mapping from real-world problems to *TCG-2* parameter settings. Once this mapping is implemented, knowledge about which optimiser is best for a given *TCG-2* parameter setting can (hopefully) be used to suggest the appropriate optimiser for a given real-world problem.

Acknowledgements

The research reported in this paper was partially supported by the ESPRIT Project 20288 Cooperation Research in Information Technology (CRIT-2): "Evolutionary Real-time Optimization System for Ecological Power Control".

References

1. Gregory, J. (1995). *Nonlinear Programming FAQ*, Usenet sci.answers. At ftp://rtfm.mit.edu/pub/usenet/sci.answers/nonlinear-programming-faq
2. van Kemenade, C.H.M. (1998). *Recombinative evolutionary search*. Ph.D. Thesis, Leiden University, Netherlands, 1998
3. Michalewicz, Z. and Schoenauer, M. (1996). *Evolutionary computation for constrained Parameter Optimization Problems*. Evolutionary Computation. Vol. 4, No. 1, pp. 1-32.
4. Michalewicz, Z.,Deb, K., Schmidt, M. and Stidsen, T. (2000). *Test-case generator for Non-linear Continuous Parameter Optimization Techniques*. IEEE Transactions on Evolutionary Computation Vol. 4 no. 3 pp. 192-215.
5. Whitley, D., Mathias, K., Rana, S. and Dzubera, J. (1995). *Building better test functions*. In L. Eshelman (Editor), Proceedings of the 6th International Conference on Genetic Algorithms, Morgam Kaufmann.
6. Whitley, D., Mathias, K., Rana, S. and Dzubera, J. (1996). *Evaluating evolutionary algorithms*. Artificial Intelligence Journal, Vol. 85, August, pp. 245-276.

A Real-coded Genetic Algorithm using the Unimodal Normal Distribution Crossover

Isao Ono[1], Hajime Kita[2], and Shigenobu Kobayashi[3]

[1] Faculty of Engineering, The University of Tokushima, 2-1, Minami-Josanjima, Tokushima, 770-8506, Japan
E-mail : isao@is.tokushima-u.ac.jp

[2] Interdisciplinary Graduate School of Science and Engineering, Tokyo Institute of Technology, 4259, Nagatsuta, Midori-ku, Yokohama, 226-8502, Japan
E-mail : kita@dis.titech.ac.jp

[3] Interdisciplinary Graduate School of Science and Engineering, Tokyo Institute of Technology, 4259, Nagatsuta, Midori-ku, Yokohama, 226-8502, Japan
E-mail : kobayasi@dis.titech.ac.jp

Summary. This chapter presents a real-coded genetic algorithm using the Unimodal Normal Distribution Crossover (UNDX) that can efficiently optimize functions with epistasis among parameters. Most conventional crossover operators for function optimization have been reported to have a serious problem in that their performance deteriorates considerably when they are applied to functions with epistasis among parameters. We believe that the reason for the poor performance of the conventional crossover operators is that they cannot keep the distribution of individuals unchanged in the process of repetitive crossover operations on functions with epistasis among parameters. In considering the above problem, we introduce three guidelines, 'Preservation of Statistics', 'Diversity of Offspring', and 'Enhancement of Robustness', for designing crossover operators that show good performance even on epistatic functions. We show that the UNDX meets the guidelines very well by a theoretical analysis and that the UNDX shows better performance than some conventional crossover operators by applying them to some benchmark functions including multimodal and epistatic ones. We also discuss some improvements of the UNDX under the guidelines and the relation between real-coded genetic algorithms using the UNDX and evolution strategies (ESs) using the correlated mutation.

1 Introduction

The problem of optimizing functions in the continuous search space (function optimization) is a very important problem because it has often to be solved in real-world applications. In some very complex real-world applications such as the design problem, we have to optimize complex functions characterized by many parameters, strong epistasis among parameters, and numerous local optima.

The genetic algorithm (GA) is an optimization method inspired by the evolutionary process of natural life. The GA can be easily applied to complex functions because it is a direct method that does not use the gradient of the objective function explicitly. Further, the GA can realize a global search in the multimodal search space, unlike most conventional optimization methods, since the GA is a stochastic search method with a set of solutions.

In applying GAs to function optimization, bit-string GAs have been traditionally employed in many cases. Bit-string GAs use the binary or Gray representation on bit strings and one-point crossover, two-point crossover, and uniform crossover defined on the bit strings. It has been reported that GAs based on the above crossover operators cannot achieve accurate solutions, compared to other function optimization techniques [3]. In recent years, several real-coded GAs, which use the real-number vector representation, have been proposed and reported to show better performance than bit-string GAs [3], [20], [28], [9], [13], [7], [26], [4]. In most real-coded GAs, mutation operators work as main search operators and crossover operators perform as assistant ones. Among the real-coded GAs, the BLX-α [7], the Fuzzy Recombination (FR) [26], and the Simulated Binary Crossover (SBX) [4] have shown relatively good performance without other search operators such as mutation operators. However, their performance deteriorates considerably when it is applied to functions which have epistasis among parameters [18], [22], [6].

In this chapter, considering epistasis among parameters, we propose three guidelines for designing crossover operators for function optimization and introduce a crossover operator named the Unimodal Normal Distribution Crossover (UNDX). By a theoretical analysis, we show that the UNDX meets the guidelines well. We also show that a real-coded GA using the UNDX gives a better performance than some bit-string GAs and a real-coded GA using the BLX-α by applying them to some benchmark functions characterized by many parameters to be optimized, strong epistasis among parameters, and numerous local optima. We also discuss some improvements of the UNDX under the guidelines and the relation between real-coded GAs using the UNDX and evolution strategies (ESs) that behave similarly.

In section 2, we present a novel viewpoint of the functions of crossover and selection operators in GAs and, based on this viewpoint, propose three guidelines for designing crossover operators for function optimization. Section 3 reviews some conventional crossover operators and points out their problems from the viewpoint of the guidelines. We introduce the UNDX in section 4. In section 5, we theoretically show that the UNDX meets the guidelines well. Section 6 describes the generation alternation model used in the experiments. The real-coded GA using the UNDX is compared with some conventional GAs on some benchmark functions in section 7. Section 8 discusses the results and section 9 concludes this chapter.

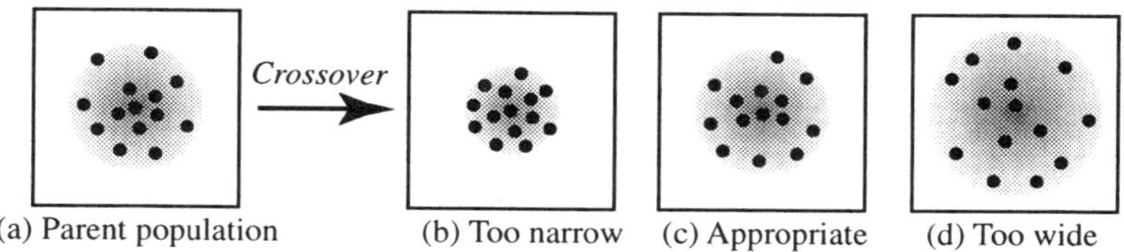

Fig. 1. Desirable behavior of crossover operator on non-epistatic functions

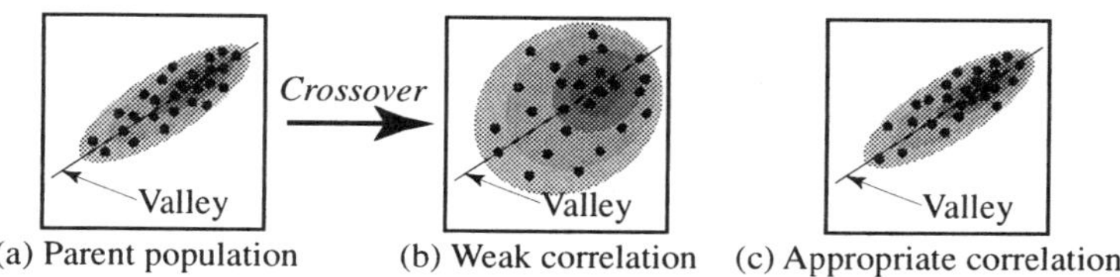

Fig. 2. Desirable behavior of crossover operator on epistatic functions

2 Guidelines for Designing Crossover Operators

In this section, we consider optimizing a function that meets the following assumptions:

1 The objective function has a kind of *continuous* property, which merely means that there should be some promising regions near a good point. Note that this assumption does not require continuity in the strict mathematical sense.
2 The initial population is distributed over the region where the optimal point(s) is/(are) located. This assumption seems to be reasonable because the upper and lower limits of the parameters to be optimized are known in advance in most practical applications.
3 The objective function is static, which means that the landscape of the objective function does not change with time.

In GAs, we believe that the function of the selection operator is to suggest the regions to be intensively searched in the search space by choosing good individuals. Therefore, we should design such selection operators that can focus the population on the globally good region(s) by gradually shrinking the search regions where the population is distributed. Especially in the case of multimodal functions, it is important to design selection operators with excellent capabilities of maintaining diversity in the population.

On the other hand, we believe that the function of crossover operators is to generate novel search points by exploiting the information from the regions to be intensively searched suggested by the current population. From this viewpoint, and with an interpretation that the crossover operator is an operator mapping the probability distribution of parents to that of offspring [19], [15], the distribution of offspring should not be narrowed nor widened too much compared with that of the parents, and should be similar to that of the parents.

When the distribution of the parent population looks like that in Fig. 1 (a) after repetitive selection operations, the region where the parents are distributed is a promising region to be intensively searched. If the offspring generated by crossover are distributed narrower than the parents as shown in Fig. 1 (b), the optimum may not be obtained. On the contrary, if the offspring are distributed wider as shown in Fig. 1 (d), computation time is wasted in searching such a hopeless region. Hence, as shown in Fig. 1 (c), sampling new solutions in the region where the parents reside will be an appropriate choice.

In functions with epistasis among parameters, we have to change multiple parameters appropriately in order to get an improved point starting from an initial one on the search space, which means that the promising regions of the functions look like valleys that are not parallel to the coordinate axes. Therefore, parents should be distributed along a valley as shown in Fig. 2 (a) after multiple selection operations. In this case, the distribution of parents has a strong correlation among parameters. If the correlation of the distribution of offspring is weaker than that of the parents as shown in Fig. 2 (b), the search efficiency deteriorates because the probability of sampling novel points from hopeless regions, i.e. the regions far away from the valley, increases. Hence, as shown in Fig. 2 (c), sampling new solutions in the region where the parents reside will be an appropriate choice.

The concept that specifies the roles of selection and crossover operators as described in the above is called *the functional specialization hypothesis* [12]. Based on this hypothesis, we introduce a guideline for designing crossover operators for function optimization [10]:

Guideline 0: The distribution of offspring generated by a crossover operator should be the same as that of the parents.

In the real parameter space, we can take a mean vector and a covariance matrix of the distribution of parents or their offspring as indices characterizing the distribution. Thus, we can introduce the following guideline which is a more concrete version of Guildeline 0.

Guideline 1: Preservation of Statistics Taking the continuity of the search space in real-coded GA into consideration, the distribution of offspring generated by crossover operators should preserve statistics such as the mean vector and the covariance matrix of the distribution of the parents well. Especially in the optimization of functions with strong epistasis among parameters, preserving the covariance matrix is very important to retain linkage among variables.

Although the function of crossover operators is to generate points which have not yet been sampled, the above guideline allows meaningless crossover operators that generate the same offspring as their parents. To eliminate such meaningless crossover operators, we introduce the following guideline:

Guideline 2: Diversity of Offspring Crossover operators should generate offspring having as much diversity as possible under the constraint of Guideline 1. Diversity can be measured by, for example, the entropy of the probability distribution of the offspring.

The last two guidelines assume that selection operators ideally work as mentioned above. However, actually, it may be that the optimal point is not in the region where the parents are distributed because a selection operator does not work adequately. To make the search more robust, we introduce an auxiliary guideline as follows:

Guideline 3: Enhancement of Robustness It is desirable that the distribution of the offspring is slightly wider than that of the parents.

Although Guideline 3 seems to conflict with Guideline 0 or 1 at first glance, it should be understood that Guideline 0 or 1 provides a reference point for designing crossover operators and Guideline 3 provides a direction of adjustment of the crossover operators designed according to Guideline 0 or 1. We recommend, first, designing a crossover operator according to Guidelines 1 (or 0) and 2 for efficiency, and then, to adjust it according to Guideline 3 for robustness. Note that there is a tradeoff between efficiency and robustness in the above adjustment.

In [2], Beyer and Deb have proposed two postulates for self-adaptive evolutionary algorithms as follows:

Postulate 1: Under a variation operator, [1] the expected population mean should remain unchanged.

Postulate 2: The expected population variance should increase with generation number.

Postulate 1 matches our guidelines. However, Postulate 2 is different from our guidelines. Beyer and Deb recommend that the expected population variance should increase with generation number, though we think that it should remain unchanged. This difference comes from the properties of problems considered by Beyer and Deb and those considered by us. Beyer and Deb consider that the landscape of an objective function may change with time and an initial population may not be distributed over the region including the optimal point. Hence, they combine their postulates with relatively high selection pressure. On the other hand, we assume that the landscape of an objective function remains unchanged and the upper and lower bounds of the search space are known in advance. So, to achieve global optimization in the search region, we combine our guidelines with relatively low selection pressure. Another difference between our guidelines and Beyer and Deb's postulates is that our guidelines take account of preserving the covariance of

[1] In [2] Beyer and Deb called a recombination or a mutation or a combination of both the *variation* operator.

the population distribution in order to realize an efficient search on epistatic functions while Beyer and Deb's postulates do not.

3 Conventional Crossover Operators

Crossover operators for function optimization can be roughly classified into two categories, crossover operators defined on bit strings and those defined on real-number vectors. The GA using the former is called the bit-string GA and that with the latter the real-coded GA. In this section, we briefly review some conventional crossover operators and point out some problems from the viewpoint of the guidelines described in the previous section.

3.1 Crossover Operators for Bit-string GAs

In applying GAs to function optimization, bit-string GAs have been traditionally employed in many cases. Bit-string GAs use binary or Gray representation on bit strings and one-point crossover, two-point crossover, and uniform crossover defined on the bit strings.

The topological structure of the genotype space defined by the binary representation is very different from that of the phenotype space, i.e. the real-number space. So, generally, one-point crossover, two-point crossover, and uniform crossover yield the distribution of offspring that is very different from that of their parents. This means that these crossover operators violate Guideline 1. The search by these crossover operators will be inefficient since they cannot intensively search the promising areas suggested by selection operators. The difficulty of searching accurate solutions due to loss of diversity in lower bits of each variable is another serious problem.

In the Gray representation, two adjacent numbers are encoded so that their Hamming distance is always one. The topological structure of the genotype space defined by the Gray representation is more similar to that of the real-number space than that of the genotype space defined by the binary representation. However, the Hamming distance of two numbers that are not adjacent to each other is not always equivalent to the distance in the phenotype space. So, one-point crossover, two-point crossover, and uniform crossover defined on the Gray representation also have the same problem as those defined on the binary representation.

3.2 Crossover Operators for Real-coded GAs

Recently, various crossover operators defined on the real-number vector representation have been proposed for real-coded GAs. Further, several crossover operators, which are called recombination operators in ESs, have been proposed for ESs [24]. Figure 3 shows a summary of the positions of offspring generated by conventional crossover operators for real-coded GAs in the case

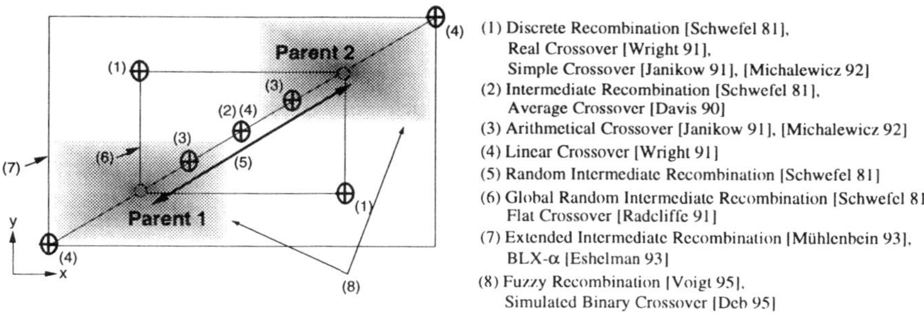

Fig. 3. Conventional crossover operators for real-coded GAs

of two parameters. Conventional crossover operators can be classified into two categories as follows:

- **Crossover operators independently determining each component of offspring**
  Crossover operators yielding offspring at the positions (1) and ones in the region (6), (7), or (8) in Fig. 3 are included in this category. Since these crossover operators determine each component of offspring independently, the correlation among parameters of the distribution of offspring becomes weaker than that of the parents. Hence, these crossover operators always reduce the covariance of the distribution of offspring close to zero and do not meet Guideline 1.
- **Crossover operators yielding offspring by the linear combination of parents**
  Crossover operators yielding offspring at the positions (2), (3), (4), and (5) in Fig. 3 belong to this category. Since crossover operators other than the Linear Crossover (Fig. 3: (2), (3), (5)) yield offspring at the interior points of the line segment whose ends are their parents, the distribution of offspring shrinks compared to that of the parents. Hence, these crossover operators do not meet Guideline 1. On the other hand, since the Linear Crossover (Fig. 3: (4)) makes offspring at the middle point and the exterior points, it keeps the distribution relatively unchanged but has a problem from the viewpoint of Guideline 2.

Of the above crossover operators, those yielding offspring in the region (7) or (8) in Fig. 3, e.g. the BLX-α ($\alpha > 0$), Fuzzy Recombination (FR), and Simulated Binary Crossover (SBX), show relatively good performance without the assistance of operators such as mutation operators. However, the performance of these crossover operators deteriorates considerably when they are applied to functions with epistasis among parameters. This is because these crossover operators always reduce the covariance of the distribution close to zero. The BLX-α yields offspring as follows:

1 Let $\boldsymbol{x}^1$, $\boldsymbol{x}^2$ be parents.

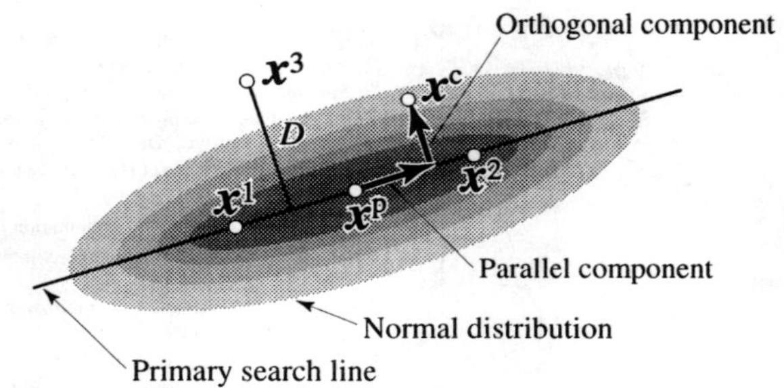

Fig. 4. Unimodal Normal Distribution Crossover (UNDX)

2 Determine each component of offspring x_i^c independently according to the uniform distribution with the range $[X_i^1, X_i^2]$, where

$$\begin{aligned} X_i^1 &= \min(x_i^1, x_i^2) - \alpha d_i, \\ X_i^2 &= \max(x_i^1, x_i^2) + \alpha d_i, \\ d_i &= |x_i^1 - x_i^2|. \end{aligned}$$

x_i^1 and x_i^2 are the i-th components of $\boldsymbol{x}^1$ and $\boldsymbol{x}^2$ respectively and α is a constant. Eshelman and Schaffer used $\alpha = 0.5$ in their experiments [7].

4 Unimodal Normal Distribution Crossover

In this section, we present the UNDX [16].

Consider the n-dimensional real space R^n as the search space. As shown in Fig. 4, let $\boldsymbol{x}^1$, $\boldsymbol{x}^2$, and $\boldsymbol{x}^3$ in R^n be parent 1, parent 2, and parent 3, respectively. Let $\boldsymbol{x}^p$ and $\boldsymbol{d}$ be their mid point and difference vector, respectively:

$$\boldsymbol{x}^p = \frac{1}{2}(\boldsymbol{x}^1 + \boldsymbol{x}^2), \quad \boldsymbol{d} = \boldsymbol{x}^2 - \boldsymbol{x}^1. \tag{1}$$

Draw a perpendicular line from parent 3 to the line connecting parent 1 and 2, and let D be the length from parent 3 to the foot of the perpendicular line:

$$D^2 = |\boldsymbol{x}^3 - \boldsymbol{x}^1|^2 \left(1 - \left(\frac{(\boldsymbol{x}^3 - \boldsymbol{x}^1)^{\mathrm{T}}(\boldsymbol{x}^2 - \boldsymbol{x}^1)}{|\boldsymbol{x}^3 - \boldsymbol{x}^1||\boldsymbol{x}^2 - \boldsymbol{x}^1|}\right)^2\right). \tag{2}$$

Then, an offspring $\boldsymbol{x}^c$ is produced by the following equation:

$$\boldsymbol{x}^c = \boldsymbol{x}^p + \xi \boldsymbol{d} + \sum_{i=1}^{n-1} \eta_i D \boldsymbol{e}_i, \tag{3}$$

where ξ is a random value following a normal distribution $N(0, \sigma_\xi^2)$, and $\eta_i, i = 1, \ldots, n-1$, are $n-1$ random values independently following a normal distribution $N(0, \sigma_\eta^2)$. Vectors $\boldsymbol{e}_i, i = 1, \ldots, n-1$, are orthonormal bases that span the subspace perpendicular to $\boldsymbol{d}$.

We recommend use of the following parameter values:

$$\sigma_\xi^2 = \frac{1}{4}, \quad \sigma_\eta^2 = \frac{0.35^2}{n}, \tag{4}$$

which have been determined by some numerical experiments [16].

For the analysis in the following section, we define some words for the UNDX:

- **The primary search line**: the line connecting parent 1 and 2.
- **The primary search direction**: the direction of $\boldsymbol{d}$.
- **The secondary search space**: the subspace containing the mid point $\boldsymbol{x}^p$ and perpendicular to the primary search direction.
- **The parallel component**: the second term on the RHS of Eq. (3).
- **The orthogonal components**: the third term on the RHS of Eq. (3).

In the actual implementation of the UNDX, we create two offspring as follows:

$$\boldsymbol{x}^{c_1} = \boldsymbol{x}^p + \xi \boldsymbol{d} + \sum_{i=1}^{n-1} \eta_i D \boldsymbol{e}_i, \quad \boldsymbol{x}^{c_2} = \boldsymbol{x}^p - \xi \boldsymbol{d} - \sum_{i=1}^{n-1} \eta_i D \boldsymbol{e}_i.$$

A detailed implementation procedure of the UNDX is described in appendix A.

5 Theoretical Analysis of the UNDX

5.1 Stochastic Model of Parents

In the following, we assume that the population size of the parents is sufficiently large and the population can be treated by a probability distribution. We express the statistical characteristics of the distribution of the parent population as follows:

- The distribution of the parents has a mean value $\langle \boldsymbol{x} \rangle = \bar{\boldsymbol{x}}$, and
- The distribution of the parents has a covariance matrix $\langle (x_i - \bar{x}_i)(x_j - \bar{x}_j) \rangle = \gamma_{ij}$.

5.2 Mean of Offspring

First, we consider the mean value of the offspring yielded by the UNDX. In the following, notation '$\langle \cdots \rangle$' represents the expectation over the distribution of parents $\boldsymbol{x}^1$, $\boldsymbol{x}^2$, and $\boldsymbol{x}^3$, sampled independently, and the expectation of the distribution of the independent random numbers ξ and η_i, used in the UNDX. The notation '$\langle \cdots \rangle_{\boldsymbol{x}}$' denotes the expectation where subscript $\boldsymbol{x}$ is the random variable.

Theorem 1 (Mean of offspring). *The mean value of the offspring* $\bar{\boldsymbol{x}}^c$ *yielded by the UNDX is the same as that of the parents* $\bar{\boldsymbol{x}}$.

Proof. From the equation of offspring

$$\boldsymbol{x}^c = \boldsymbol{x}^p + \xi \boldsymbol{d} + \sum_{i=1}^{n-1} \eta_i \boldsymbol{e}_i D, \tag{5}$$

and the independence of the random variables ξ, η_i from the parents, we obtain

$$\begin{aligned} \langle \boldsymbol{x}^c \rangle &= \left\langle \boldsymbol{x}^p + \xi \boldsymbol{d} + \sum_{i=1}^{n-1} \eta_i \boldsymbol{e}_i D \right\rangle \\ &= \langle \boldsymbol{x}^p \rangle_{\boldsymbol{x}^1,\boldsymbol{x}^2} + \langle \xi \rangle_\xi \langle \boldsymbol{d} \rangle_{\boldsymbol{x}^1,\boldsymbol{x}^2} + \sum_{i=1}^{n-1} \langle \eta_i \rangle_{\eta_i} \langle \boldsymbol{e}_i D \rangle_{\boldsymbol{x}^1,\boldsymbol{x}^2,\boldsymbol{x}^3} \\ &= \langle \boldsymbol{x}^p \rangle_{\boldsymbol{x}^1,\boldsymbol{x}^2} = \frac{1}{2}\left(\langle \boldsymbol{x}^1 \rangle_{\boldsymbol{x}^1} + \langle \boldsymbol{x}^2 \rangle_{\boldsymbol{x}^2} \right) = \bar{\boldsymbol{x}}. \end{aligned} \tag{6}$$

5.3 Analysis of the Parallel Component of the UNDX

Next, we consider the UNDX consisting of only the parallel component. In this case, we obtain the following theorem on the covariance matrix of offspring yielded by the UNDX without the orthogonal components.

Theorem 2 (Covariance matrix of offspring). *The covariance matrix of the offspring* $\{\gamma^c_{ij}\}$ *yielded by the UNDX without the orthogonal components is* $(2\sigma_\xi^2 + 1/2)$ *times as large as that of the parents* $\{\gamma_{ij}\}$, *i.e.*

$$\gamma^c_{ij} = \gamma_{ij}\left(2\sigma_\xi^2 + \frac{1}{2}\right). \tag{7}$$

Proof. Here, let the i-th component of the parent $\boldsymbol{x}^j$ and the offspring $\boldsymbol{x}^c$ be $x_{j,i}$ and $x_{c,i}$, respectively.

$$\begin{aligned} x_{c,i} &= x_{p,i} + \xi d_i \\ &= \frac{1}{2}(x_{1,i} + x_{2,i}) + \xi(x_{2,i} - x_{1,i}). \end{aligned} \tag{8}$$

From Theorem 1, $\bar{\boldsymbol{x}}^c = \langle \boldsymbol{x}^c \rangle = \bar{\boldsymbol{x}}$ and

$$\begin{aligned} x_{c,i} - \bar{x}_{c,i} &= x_{c,i} - \bar{x}_i \\ &= \left(\frac{1}{2} + \xi\right)(x_{2,i} - \bar{x}_i) + \left(\frac{1}{2} - \xi\right)(x_{1,i} - \bar{x}_i). \end{aligned}$$

Then, we obtain

$$
\begin{aligned}
\gamma_{ij}^{c} &= \langle (x_{c,i} - \bar{x}_{c,i})(x_{c,j} - \bar{x}_{c,j}) \rangle \\
&= \left\langle \left(\frac{1}{2} + \xi\right)^2 (x_{2,i} - \bar{x}_i)(x_{2,j} - \bar{x}_j) \right. \\
&\quad + \left(\frac{1}{2} + \xi\right)\left(\frac{1}{2} - \xi\right)(x_{2,i} - \bar{x}_i)(x_{1,j} - \bar{x}_j) \\
&\quad + \left(\frac{1}{2} + \xi\right)\left(\frac{1}{2} - \xi\right)(x_{1,i} - \bar{x}_i)(x_{2,j} - \bar{x}_j) \\
&\quad \left. + \left(\frac{1}{2} - \xi\right)^2 (x_{1,i} - \bar{x}_i)(x_{1,j} - \bar{x}_j) \right\rangle \\
&= \left\langle \left(\frac{1}{2} + \xi\right)^2 \right\rangle_{\xi} \langle (x_{2,i} - \bar{x}_i)(x_{2,j} - \bar{x}_j) \rangle_{\boldsymbol{x}^2} \\
&\quad + \left\langle \left(\frac{1}{2} + \xi\right)\left(\frac{1}{2} - \xi\right) \right\rangle_{\xi} \langle (x_{2,i} - \bar{x}_i) \rangle_{\boldsymbol{x}^2} \langle (x_{1,j} - \bar{x}_j) \rangle_{\boldsymbol{x}^1} \\
&\quad + \left\langle \left(\frac{1}{2} + \xi\right)\left(\frac{1}{2} - \xi\right) \right\rangle_{\xi} \langle (x_{1,i} - \bar{x}_i) \rangle_{\boldsymbol{x}^1} \langle (x_{2,j} - \bar{x}_j) \rangle_{\boldsymbol{x}^2} \\
&\quad + \left\langle \left(\frac{1}{2} - \xi\right)^2 \right\rangle_{\xi} \langle (x_{1,i} - \bar{x}_i)(x_{1,j} - \bar{x}_j) \rangle_{\boldsymbol{x}^1} \\
&= \left\langle \left(\frac{1}{2} + \xi\right)^2 \right\rangle_{\xi} \langle (x_{2,i} - \bar{x}_i)(x_{2,j} - \bar{x}_j) \rangle_{\boldsymbol{x}^2} \\
&\quad + \left\langle \left(\frac{1}{2} - \xi\right)^2 \right\rangle_{\xi} \langle (x_{1,i} - \bar{x}_i)(x_{1,j} - \bar{x}_j) \rangle_{\boldsymbol{x}^1} \\
&= \left(\frac{1}{4} + \langle \xi \rangle_{\xi} + \langle \xi^2 \rangle_{\xi}\right) \gamma_{ij} + \left(\frac{1}{4} - \langle \xi \rangle_{\xi} + \langle \xi^2 \rangle_{\xi}\right) \gamma_{ij} \\
&= \gamma_{ij} \left(\frac{1}{2} + 2\sigma_{\xi}^2\right) .
\end{aligned}
\tag{9}
$$

From Theorem 2, we can make the covariance matrix of offspring yielded by the UNDX without the orthogonal components coincide with that of parents with $\sigma_{\xi}^2 = 1/4$.

As shown in Eq. (4), the parameter empirically determined by numerical experiments [16] is $\sigma_{\xi}^2 = 1/4$. That is, the parameter value that yields good results in the actual search is the value preserving the statistical characteristics of the distribution of the population.

5.4 Analysis of the Orthogonal Components of the UNDX

In the UNDX, the magnitude of the orthogonal component is decided by the distance D from the third parent to the primary search line. It makes precise evaluation of statistical values such as the covariance matrix of the distribution of offspring difficult. Hence, we utilize the following strategy for analysis:

- Taking a simple case where the parents are distributed isotropically, we theoretically evaluate $\langle D^2 \rangle$, where D is the distance.
- In the case where the distribution of parents is not isotropic, we observe the changes of $\langle D^2 \rangle$ by a numerical experiment using typical examples of the distribution of parents.

The case when the distribution of parents is isotropic: theoretical evaluation In this section, we theoretically evaluate $\langle D^2 \rangle$, where D determines the magnitude of the orthogonal components when the parents are distributed isotropically in the search space. It can be a model of the early stages of search in which initial individuals are randomly generated.

In the following analysis, we assume the following:

Assumption 1: The parents are distributed isotropically.

Since the distribution of parents is isotropic, we can write the covariance matrix as γI, where I is the n-dimensional unit matrix and γ is a positive constant.

Assumption 2: The subspace spanned by the parents chosen for crossover is not degenerated.

Theorem 3. *Under the above assumptions, we have*

$$(n-1)\gamma \leq \langle D^2 \rangle \leq \frac{3}{2}(n-1)\gamma + \frac{1}{2}\gamma. \tag{10}$$

Proof. With an appropriate orthogonal transformation of the coordinate system, we can make the n-th coordinate axis be the primary search direction. Then, the projection of $\boldsymbol{x}^3 - \boldsymbol{x}^p$ onto the secondary search space is equivalent to extracting the first $n-1$ elements of the vector. Hence, we obtain

$$D^2 = \sum_{i=1}^{n-1} (x_{3,i} - x_{p,i})^2, \tag{11}$$

where subscript i denotes the i-th component of the vector. Then, we obtain

$$\begin{aligned}
\langle D^2 \rangle &= \left\langle \sum_{i=1}^{n-1} (x_{3,i} - x_{p,i})^2 \right\rangle \\
&= \left\langle \sum_{i=1}^{n-1} (x_{3,i} - \bar{x}_i)^2 \right\rangle + \left\langle \sum_{i=1}^{n-1} (x_{p,i} - \bar{x}_i)^2 \right\rangle \\
&= (n-1)\gamma + \left\langle \sum_{i=1}^{n-1} (x_{p,i} - \bar{x}_i)^2 \right\rangle_{\boldsymbol{x}^1, \boldsymbol{x}^2} \qquad (12) \\
&= (n-1)\gamma + \left\langle \|\boldsymbol{x}_p - \bar{\boldsymbol{x}}\|^2 \right\rangle_{\boldsymbol{x}^1, \boldsymbol{x}^2} - \left\langle (x_{p,n} - \bar{x}_n)^2 \right\rangle_{\boldsymbol{x}^1, \boldsymbol{x}^2} \\
&= (n-1)\gamma + \frac{n}{2}\gamma - \left\langle (x_{p,n} - \bar{x}_n)^2 \right\rangle_{\boldsymbol{x}^1, \boldsymbol{x}^2} . \qquad (13)
\end{aligned}$$

Here, we obtain the second line in the above equation from the fact that $\boldsymbol{x}^3$ is independent of $\boldsymbol{x}^1$ and $\boldsymbol{x}^2$, the third line from the fact that the parent $\boldsymbol{x}^3$ is isotropic, and the fifth line from the facts that $\boldsymbol{x}^p = (\boldsymbol{x}^2 + \boldsymbol{x}^1)/2$ is isotropic and the second term of the fourth line is the expectation that is not dependent on the coordinate system. Since the coordinate system is dependent on the parents sampled, the last terms of Eq. (12) and Eq. (13) cannot simply be replaced by γ.

From Eq. (12) and Eq. (13), we obtain

$$(n-1)\gamma \le \langle D^2 \rangle \le \frac{3}{2}(n-1)\gamma + \frac{1}{2}\gamma. \qquad (14)$$

The last term of Eq. (13) is difficult to evaluate exactly since the coordinate system is chosen according to the samples $\boldsymbol{x}^1$ and $\boldsymbol{x}^2$, as described in the above proof. However, assuming that the last term of Eq. (13) is equivalent to the value $\gamma/2$ with a fixed coordinate system, we obtain

$$\langle D^2 \rangle \simeq \frac{3}{2}(n-1)\gamma. \qquad (15)$$

Thus, from Eq. (3), the variance of the normally distributed random variable η_i for the orthogonal components is

$$\langle (\eta_i D)^2 \rangle = \sigma_\eta^2 \langle D^2 \rangle \simeq \sigma_\eta^2 \frac{3}{2}(n-1)\gamma.$$

Therefore, the magnitude of the orthogonal components (variance) coincides approximately with the variance of the parents γ when we choose

$$\sigma_\eta^2 = \frac{2}{3}\frac{1}{n-1}. \qquad (16)$$

The case when the distribution of parents is not isotropic: numerical experiments In this section, we examine the variation of D^2 by numerical experiments when the distribution of parents is not isotropic. The setup of the experiments is as follows:

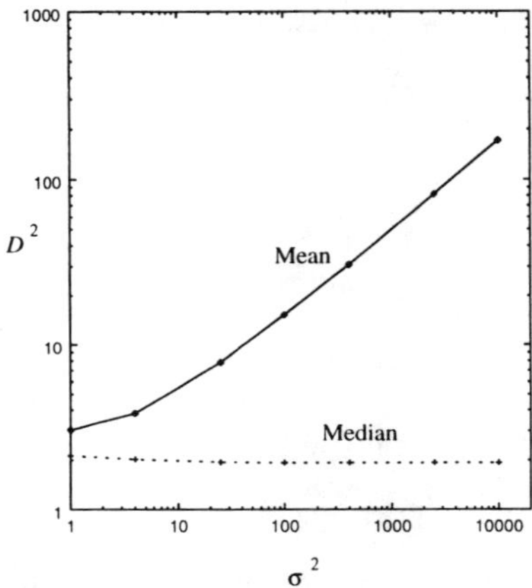

Fig. 5. Variation of D^2 according to change of σ^2

- The distribution of the parents follows a three-dimensional normal distribution whose mean value is 0 and covariance matrix is given by

$$\Gamma = \begin{bmatrix} \sigma^2 & 0 & 0 \\ 0 & 1 & 0 \\ 0 & 0 & 1/\sigma^2 \end{bmatrix}, \quad \sigma = 5. \tag{17}$$

  σ is a parameter which shows the degree to which the distribution of parents is not isotropic. The case where σ is close to one is a model of an early search phase as described in the previous section. On the other hand, the case where σ is large is a model of a last search phase in which parents are distributed along a valley on a function with strong epistasis among parameters.
- We sample three parents according to the above normal distribution and calculate D^2.
- We calculate the mean and the median of D^2 by sampling 10,001 sets of parents.
- We observe the statistics of D^2 by changing σ within the range from 1 to 100.

Figure 5 shows the results as follows:

- When the distribution of parents is isotropic ($\sigma = 1$), the mean of D^2 is quite close to the theoretical value

$$\frac{3}{2}(n-1) \cdot 1 = 3.$$

- When σ is large, the difference of the mean and the median of D^2 is significant.

- The mean of D^2 increases approximately in proportion to σ.
- On the other hand, the median of D^2 takes almost constant values.

The difference of the mean and the median of D^2 shown in the results can be understood as follows. When σ is large, the primary search line connecting parent 1 and parent 2, with high probability, takes the direction of the x_1 component (the primary leading component) whose variance is largest. In this case, D^2 is approximately the variance of the x_2 component (the second leading component) (1). The primary search component can take a direction considerably different from the x_1 axis with low probabilities. In this case, D^2 is a large value comparable to the variance of the x_1 component (σ^2). Therefore, the mean takes a medium value between the x_1 component and the x_2 one while the median stays around a value comparable to the variance of the x_2 component.

From the above, it is thought that the representative value of D^2 is smaller than the primary leading component of the distribution of parents and close to the second leading component.

Discussions on the orthogonal components of the UNDX From Eq. (16), when the distribution of parents is isotropic, the magnitude of the orthogonal components coincides approximately with the variance of the parents if we choose $\sigma_\eta^2 = 2/3(n-1)$. From the numerical experiments, when the distribution of parents is wide in only one direction, the mean and the median of D^2 are much smaller than the magnitude of the primary leading component of the parent distribution and take a medium value between the primary and second leading components.

The recommended parameters obtained by numerical experiments [16] are $\sigma_\eta^2 = (0.35)^2/n = 0.1225/n$:

- The order of the factor for the dimension of the problem n is $o(n^{-1})$, which coincides with Eq. (16). This factor eliminates the dependency of the magnitude of the orthogonal components on the dimension of the search space n.
- The coefficient 0.1225 is smaller than 2/3 in Eq. (16). This choice will be an adequate one to make the orthogonal components play a supplementary role, taking into consideration preservation of the covariance matrix by the parallel component.

5.5 Features of the UNDX

From the above theoretical analysis, the UNDX with the empirically recommended parameters ($\sigma_\xi^2 = 1/4$ and $\sigma_\eta^2 = (0.35)^2/n$) fulfills the guidelines described in section 2 because it has the following features:

- The parallel component of the UNDX meets Guideline 1 perfectly.

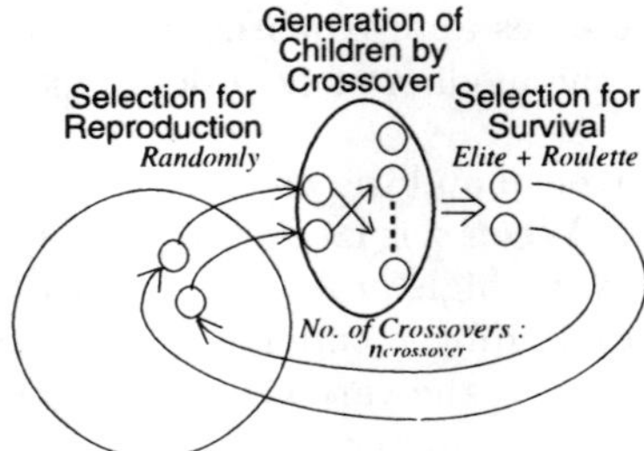

Fig. 6. Minimal generation gap (MGG)

- However, since the UNDX with only the parallel component generates offspring only on the line connecting parent 1 and parent 2, the distribution of offspring cannot sufficiently cover the search areas. That is, the UNDX with only the parallel component is insufficient from the viewpoint of Guideline 2.
- It is the orthogonal components that compensate the above weakness. The orthogonal components are also useful to meet Guideline 3. The role of the orthogonal components is not to greatly expand the search areas but just to assist the search.

The above features show that the UNDX yields a distribution of offspring similar to that of their parents. Therefore, it is thought that the UNDX can efficiently search along the valleys where parents are distributed in functions with strong epistasis among parameters.

6 Designing a Genetic Algorithm for Function Optimization

In this section, we design a genetic algorithm using the UNDX for function optimization.

Figure 6 shows the generation alternation model employed in this chapter. It is based on the minimal generation gap (MGG) model [23]. From the viewpoint of maintaining the diversity of a population, parents are chosen from the population randomly and generation alternations are performed locally. To precisely estimate the landscape, the crossover operator is applied to parents several times to generate multiple offspring at a time and selection for survival is applied to the family containing parents and their offspring.

The proposed GA procedure for function optimization is described as follows:

1 **Generating an initial population:** An initial population that is composed of real-number vectors is produced randomly.

2 **Selection for reproduction:** A pair of individuals is chosen by random sampling without replacement from the population.
3 **Generating offspring by the crossover:** This is done by applying the UNDX to the chosen pair of individuals. Here, the third parent for the UNDX is randomly chosen from the population and referenced. This process is repeated $n_{\text{crossover}}$ times to create $2 \times n_{\text{crossover}}$ offspring.
4 **Selection for survival:** Two individuals are chosen from the family containing the parents and their offspring; one is the best individual and the other is chosen by rank-based roulette wheel selection. The two individuals replace the parents in the population.
5 The above procedures from step 1 to step 4 are repeated until a certain condition is satisfied.

7 Experiments

In this section, we show the results of 1) comparing the performance of bit-string GAs using the uniform crossover defined on the binary representation and the Gray representation and real-coded GAs using the BLX-α and the UNDX on a non-epistatic function, and 2) comparing the performance of the BLX-α and the UNDX on epistatic functions. In the following experiments, we use the recommended parameters $\sigma_\xi^2 = 1/4$ and $\sigma_\eta^2 = (0.35)^2/n$, for the UNDX and $\alpha = 0.662$, which is recommended in [2], for the BLX-α. The population size is set to 50 in unimodal functions and 300 in multi-modal functions. In all experiments, the same generation alternation model described in the previous section is employed. We perform 10 trials on each benchmark function.

7.1 Comparison of the Performance of Bit-string GAs and Real-coded GAs

In order to compare the performance of the bit-string GA using the uniform crossover operator defined on the binary representation (Binary/UX), the bit-string GA using the uniform crossover operator defined on the Gray representation (Gray/UX), the real-coded GA using the BLX-α (BLX-α), and the real-coded GA using the UNDX (UNDX), we apply them to the 20-dimensional sphere function. In the bit-string GAs, the number of bits per variable is 16. The sphere function is given by

$$f(\boldsymbol{x}) = \sum_{i=1}^{n} x_i^2, \quad -5.12 < x_i < 5.12, \quad n = 20.$$

This is a unimodal function and has a unique minimum of zero at the point $(0, \ldots, 0)$. A contour plot of the function with two parameters is shown in Fig. 7 (a).

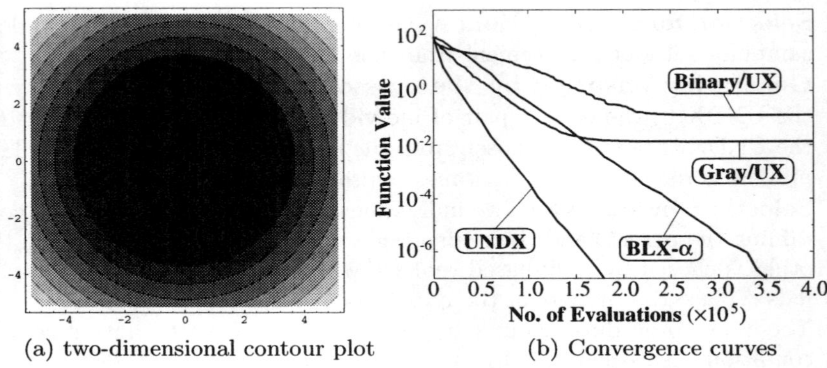

(a) two-dimensional contour plot (b) Convergence curves

Fig. 7. Results on the 20-dimensional sphere function

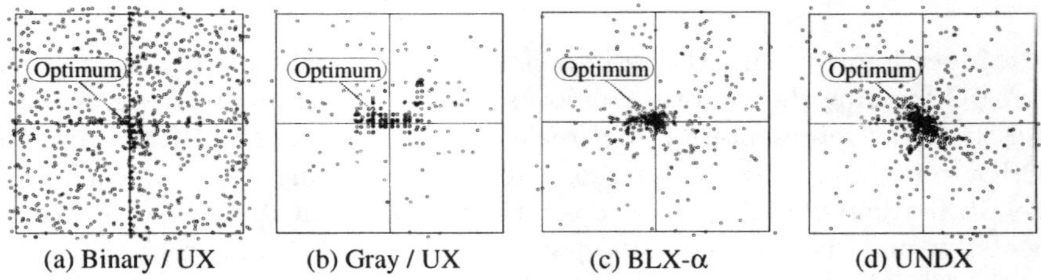

(a) Binary / UX (b) Gray / UX (c) BLX-α (d) UNDX

Fig. 8. Plots of individuals generated during a single run on the sphere function

Figure 7 (b) shows the online performance of each GA. Each curve shows the average of the best function value in the population over 10 runs. As shown in this figure, the bit-string GAs cannot find good solutions with such an easy function. On the other hand, the real-coded GAs show good performance. We have confirmed that the solutions found by the bit-string GAs with the longer chromosome in which the number of bits per variable is 32 do not improve very much, though the search time becomes longer.

Figure 8 shows the plots of all individuals generated by each crossover operator during a single run on the two-dimensional sphere function. These plots show that the real-coded GAs succeed in intensively sampling near the origin, which is the optimum, while the bit-string GAs fail.

There are two system parameters in the MGG model, the population size N_{pop} and the number of crossovers $n_{\text{crossover}}$, for controlling convergence speed. In this experiment, we did not select values of the parameters N_{pop} and $n_{\text{crossover}}$, so as to enhance the search abilities of all the GAs. The UNDX can achieve optimization with less function evaluations, e.g. with $N_{\text{pop}} = 30$ and $n_{\text{crossover}} = 5$; it found a function value of about 10^{-24} on the 30-dimensional sphere function after 10^5 function evaluations.

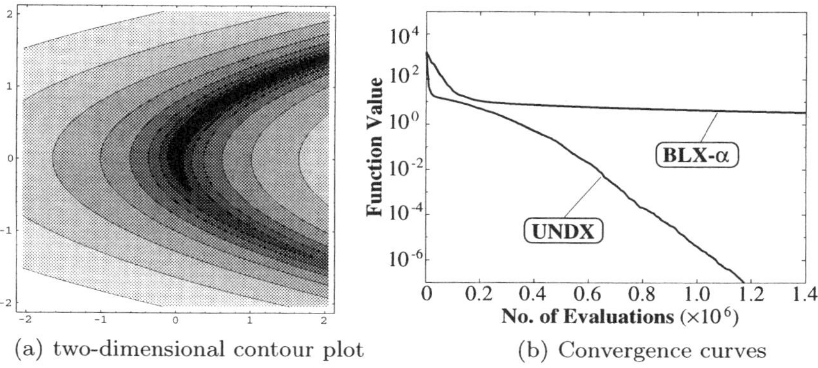

(a) two-dimensional contour plot (b) Convergence curves

Fig. 9. Results on the 20-dimensional Rosenbrock function

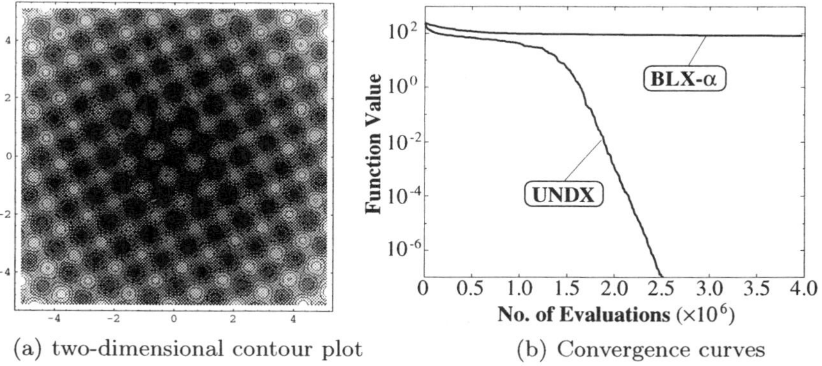

(a) two-dimensional contour plot (b) Convergence curves

Fig. 10. Results on the 20-dimensional Rotated Rastrigin function

7.2 Comparison of the UNDX and the BLX-α on Epistatic Functions

In order to compare the performance of the UNDX and BLX-α on functions with epistasis among parameters, we applied them to the 20-dimensional Rosenbrock function and the 20-dimensional Rotated Rastrigin function.

The Rosenbrock function is given by

$$f(\boldsymbol{x}) = \sum_{i=2}^{n} [100(x_1 - x_i^2)^2 + (x_i - 1)^2], \quad -2.048 < x_i < 2.048, \quad n = 20.$$

This is a unimodal function and has a parabolic valley along the curve $x_1 = x_i^2 (i = 2, \ldots, n)$ with a unique minimum of zero at the point $(1, \ldots, 1)$. A contour plot of the function with two parameters is shown in Fig. 9 (a). Figure 9 (b) shows the online performance of the UNDX and the BLX-α. The UNDX succeeded in finding the optimum while the BLX-α failed.

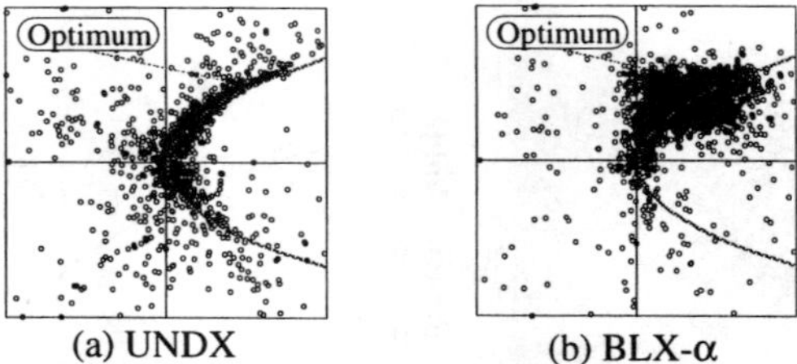

Fig. 11. Plots of individuals generated during a single run on the Rosenbrock function

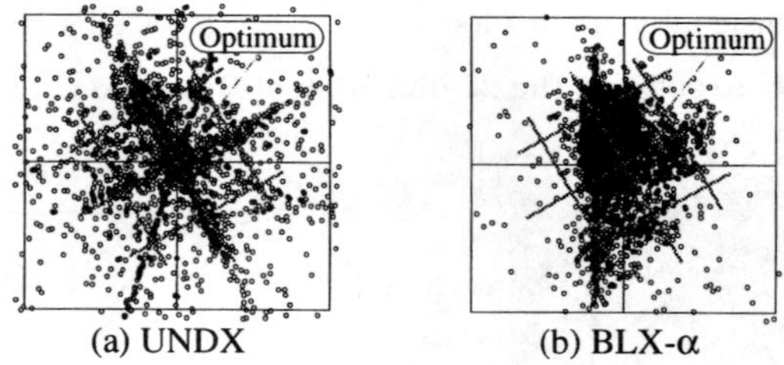

Fig. 12. Plots of individuals generated during a single run on the Rotated Rastrigin function

The Rotated Rastrigin function is the one made by randomly rotating the original Rastrigin function around the origin. The original Rastrigin function is given by

$$f(\boldsymbol{x}) = 10n + \sum_{i=1}^{n} [x_i^2 - 10\cos(2\pi x_i)], \quad -5.12 < x_i < 5.12, \quad n = 20.$$

Details of the rotating method are given in appendix B. The function is a multimodal one and has a unique minimum of zero at the point $(0, \ldots, 0)$ and many local minima around the global minimum. The local minima which are nearer the origin have smaller function values. The contour plot of the function with two parameters is shown in Fig. 10 (a). Figure 10 (b) shows the convergence curves of the UNDX and the BLX-α. The UNDX succeeded in finding the optimum in all trials while the BLX-α failed in all trials.

Figure 11 shows the distribution of all individuals generated by the UNDX and the BLX-α during a single run on the two-dimensional Rosenbrock function. Since the Rosenbrock function has a parabolic deep valley leading to the optimal point, generating children along the valley is required for better search efficiency. The BLX-α yields children not along the valley but widely around it, which means that the BLX-α cannot adapt the distribution of children well to the landscape of the function. As a result, the search efficiency of the BLX-α becomes poor. On the other hand, the UNDX works with great efficiency as it succeeds in generating children along the valley.

Figure 12 illustrates the distributions of all individuals generated in a single run on the two-dimensional Rotated Rastrigin function. In the figure, the points of intersection of gray lines represent local minima and the center point is the global minimum. The BLX-α generates children independent of the position of local minima, searching almost randomly around the origin. In contrast, the UNDX can generate children along local minima, which shows that it can adapt the distribution of children to the landscape of functions.

8 Discussion

8.1 Improvement of the UNDX under the Guidelines

The UNDX meets Guideline 1 well as shown in the theoretical analysis in section 5. However, from the viewpoint of Guideline 2, the offspring generated by the UNDX are mainly distributed on the primary search line and do not cover the search region well. The orthogonal components of the UNDX are introduced to relieve this difficulty and enhance robustness of the crossover operator (Guideline 3). However, at the same time, introduction of the component makes the crossover operator sensitive to the scale of the coordinate system of the search space. To remedy these problems, a few attempts have been undertaken. Kita et al. have proposed a multi-parent extension of the UNDX, named UNDX-m [10]. Tsutsui et al. have proposed a new crossover operator called the Simplex Crossover (SPX) which meets the guidelines very well [25]. In the following, we introduce the UNDX-m in detail.

The algorithm of the UNDX-m is derived as a solution of maximization of entropy of the distribution of offspring to meet Guideline 2, under the constraints of preservation of the mean vector and the covariance matrix to meet Guideline 1. Further, to meet Guideline 3, i.e. to make the operator robust, the secondary search component is introduced in the same way as the UNDX. A prototype algorithm of the UNDX-m is as follows:

1 Let $\boldsymbol{x}^1, \ldots, \boldsymbol{x}^{m+2}$ be $m+2$ parents.
2 Let the center of mass of the first $m+1$ parents be $\boldsymbol{p} = \frac{1}{m+1}\sum_i \boldsymbol{x}^i$, and let the difference vector of $\boldsymbol{x}^i$ and $\boldsymbol{p}$ be $\boldsymbol{d}^i = \boldsymbol{x}^i - \boldsymbol{p}$.
3 Let D be the length of component of $\boldsymbol{d}^{m+2} = \boldsymbol{x}^{m+2} - \boldsymbol{p}$ orthogonal to $\boldsymbol{d}^1, \ldots, \boldsymbol{d}^m$.
4 Let $\boldsymbol{e}^1, \ldots, \boldsymbol{e}^{n-m}$ be orthonormal bases of the subspace (say the secondary search space) orthogonal to the subspace spanned by $\boldsymbol{d}^1, \ldots, \boldsymbol{d}^m$.
5 Generate offspring $\boldsymbol{x}^c$ by the following equation:

$$\boldsymbol{x}^c = \boldsymbol{p} + \sum_{k=1}^{m} w_k \boldsymbol{d}^k + \sum_{k=1}^{n-m} v_k D \boldsymbol{e}^k \tag{18}$$

where w_k and v_k are random variables that follow normal distributions $N(0, \sigma_\xi^2)$ and $N(0, \sigma_\eta^2)$, respectively, and σ_ξ and σ_η are parameters.

Kita et al. recommend use of $\sigma_\xi = 1/\sqrt{m}$ and $\sigma_\eta = 0.35/\sqrt{n-m}$, where σ_ξ can be decided by maximization of entropy under the constraints of preservation of the mean vector and the covariance matrix and σ_η can be determined based on the consistency of the empirical value for the UNDX and the elimination of dependency on the dimension of the search space n.

8.2 Relationship Between Real-coded GAs Using the UNDX and ESs

In ESs [1], a mutation operator is used as the main search operator, which generates children around a parent by using the normal distribution. The correlated mutation [24], in which the axes of the normal distribution are inclined relative to those of the coordinate system, takes into consideration epistasis among parameters. Evolution parameters, such as the standard deviations and the axes angles of the normal distribution, are encoded by a position vector in an individual and are adapted by a random mutation. In contrast, the UNDX adaptively determines them from the distribution of parents.

9 Conclusion

In this chapter, first we presented three guidelines for designing crossover operators for function optimization: 'Preservation of Statistics', 'Diversity of Offspring', and 'Enhancement of Robustness'. The guidelines are based on the ideas that selection operators should have a function of suggesting the region to be searched intensively by selecting good solutions and that crossover operators should have a function of sampling novel search points exploiting the information of the region to be searched suggested by the current population. We briefly reviewed some conventional crossover operators for function optimization and pointed out their problems from the viewpoint of the guidelines. We introduced the UNDX that meets the guidelines well, which we showed by by a theoretical analysis. Then, we showed the effectiveness of the UNDX by applying it to some epistatic functions. Further, we discussed an extension of the UNDX.

In future work, we would like to apply real-coded GAs using the UNDX or its improved versions to various real-world applications and solve those difficult problems that conventional optimization methods cannot. At present, we are applying a real-coded GA using the UNDX to the lens design problem, known as a very difficult design problem, and this study is receiving attention from many lens design experts [17].

Appendix A. Implementation of the UNDX

In the implementation of the UNDX, taking account of the fact that the distribution in the secondary search space is isotropic, offspring are generated as follows:

1 Generate $\boldsymbol{t} = (t_1, \ldots, t_n)$, where $t_i \sim N(0, (D\sigma_\eta)^2)$, D is the distance of parent 3, $\boldsymbol{x}^3$, to the line connecting parent 1, $\boldsymbol{x}^1$, and parent 2, $\boldsymbol{x}^2$, $\sigma_\eta = 0.35/\sqrt{n}$, and n is the number of parameters.

2 Subtract the component of the primary search line from $\boldsymbol{t}$:

$$\boldsymbol{t} \leftarrow \boldsymbol{t} - (\boldsymbol{t} \cdot \boldsymbol{e}_0)\boldsymbol{e}_0, \quad \boldsymbol{e}_0 = \frac{\boldsymbol{x}^2 - \boldsymbol{x}^1}{|\boldsymbol{x}^2 - \boldsymbol{x}^1|}.$$

3 Add the parallel component $\xi \boldsymbol{d}$ to t:

$$\boldsymbol{t} \leftarrow \boldsymbol{t} + \xi \boldsymbol{d},$$

where $\xi \sim N(0, \sigma_\xi)$, $\sigma_\xi = 1/4$, and $\boldsymbol{d} = \boldsymbol{x}^2 - \boldsymbol{x}^1$.

4 Obtain offspring $\boldsymbol{x}^{c_1}$ and $\boldsymbol{x}^{c_2}$ by the following equations:

$$\boldsymbol{x}^{c_1} = \frac{(\boldsymbol{x}^1 + \boldsymbol{x}^2)}{2} + \boldsymbol{t}, \quad \boldsymbol{x}^{c_2} = \frac{(\boldsymbol{x}^1 + \boldsymbol{x}^2)}{2} - \boldsymbol{t}.$$

Appendix B. A way of generating rotated functions

Instead of rotating a function directly, we transform the offspring vector $\boldsymbol{x}^c = (x_1, \ldots, x_n)$ to $\boldsymbol{x}^c_{\text{rotation}} = (x'_1, \ldots, x'_n)$ by the following equation and use the result of letting $\boldsymbol{x}^c_{\text{rotation}}$ to the function as the value of the rotated function:

$$\begin{pmatrix} x'_i \\ x'_j \end{pmatrix} = \begin{pmatrix} \cos\theta_k & -\sin\theta_k \\ \sin\theta_k & \cos\theta_k \end{pmatrix} \begin{pmatrix} x_i \\ x_j \end{pmatrix},$$
$$i, j = 1, \ldots, n, \ i < j, \ k = 1, \ldots, n(n-1)/2,$$

where n is the number of parameters and θ_k are constants between -2π and $+2\pi$.

References

1. Bäck, T., Hoffmeister, F. and Schwefel, H.-P. (1991) A Survey of Evolution Strategies, Proc. 4th Int'l Conf. on Genetic Algorithms, 2-9
2. Beyer, H.-G. and Deb, K. (2000) On the Desired Behaviors of Self-Adaptive Evolutionary Algorithms, Parallel Problem Solving from Nature VI (PPSN VI), 59-68
3. Davis, L. (1990) The Handbook of Genetic Algorithms, Van Nostrand Reinhold, New York
4. Deb, K. and Agrawal, R.B. (1995) Simulated Binary Crossover for Continuous Search Space, Complex Systems, 9, 115-148

5. Deb, K. and Beyer, H.-G. (1999) Self-Adaptation in Real-Parameter Genetic Algorithms with Simulated Binary Crossover, Proc. Genetic and Evolutionary Computation Conf. 1999 (GECCO-99), 172-179
6. Deb, K. and Beyer, H.-G. (1999) Self-Adaptive Genetic Algorithms with Simulated Binary Crossover, Technical Report No. CI-61/99, Dept. Computer Science/XI, Univ. of Dortmund
7. Eshleman, L. J. and Schaffer, J. D. (1993) Real-Coded Genetic Algorithms and Interval-Schemata, Foundations of Genetic Algorithms,**2**, 187-202
8. Goldberg, D. E. (1989) Genetic Algorithms in Search, Optimization, and Machine Learning, Addison-Wesley, heading, MA
9. Jonikow, C. Z. and Michalewicz, Z. (1991) An Experimental Comparison of Binary and Floating Point Representations in Genetic Algorithms, Proc. 4th Int'l Conf. on Genetic Algorithms, 31-36
10. Kita, H., Ono, I. and Kobayashi, S. (1999) Multi-parental Extension of the Unimodal Normal Distribution Crossover for Real-coded Genetic Algorithms, Proc. 1999 Congress on Evolutionary Computation (CEC'99), 1581-1587
11. Kita, H., Ono, I. and Kobayashi, S. (1998) Theoretical Analysis of the Unimodal Normal Distribution Crossover for Real-coded Genetic Algorithms, Proc. 1998 IEEE Int'l Conf. on Evolutionary Computation, 529-534
12. Kita, H. and Yamamura, M. (1999) A Functional Specialization Hypothesis for Designing Genetic Algorithms, Proc. 1999 IEEE Int'l. Conf. on Systems, Man, and Cybernetics, 579-584
13. Michalewicz, Z. (1992) Genetic Algorithms+Data Structures=Evolution Programs, Springer-Verlag, Berlin
14. Mühlenbein, H. and Schlierkamp-Voosen, D. (1993) Predictive Models for the Breeder Genetic Algorithm I. Continuous Parameter Optimization, Evolutionary Computation, Vol.1, 25-49
15. Nomura, T. (1997) An Analysis on Crossover for Real Number Chromosomes in an Infinite Population Size, Proc. 15th Int'l Joint Conf. on Artificial Intelligence, 936-941
16. Ono, I. and Kobayashi, S. (1997) A Real-coded Genetic Algorithm for Function Optimization Using Unimodal Normal Distribution Crossover, Proc. 7th Int'l Conf. on Genetic Algorithms, 246-253
17. Ono, I., Kobayashi, S. and Yoshida, K. (1998) Global and Multi-objective Optimization for Lens Design by Real-coded Genetic Algorithms, SPIE Proc. Vol. 3482, International Optical Design Conference, 110-121
18. Ono, I., Yamamura, M., Kobayashi, S. (1996) A Genetic Algorithm with Characteristic Preservation for Function Optimization, Proc. IIZUKA'96, 511-514
19. Qi, X. and Palmieri, F. (1994) Theoretical Analysis of Evolutionary Algorithms with an Infinite Population Size in Continuous Space Part I: Basic Properties of Selection and Mutation, Part II: Analysis of Diversification Role of Crossover, IEEE Transactions on Neural Networks, Vol. 5, No. 1, 102-119, 120-129
20. Radcliffe, N. J. (1991) Forma Analysis and Random Respectful Recombination, Proc. 4th Int'l Conf. on Genetic Algorithms, 222-229
21. Rechenberg, I. (1973) Evolutionsstrategie: Optimierung technischer Systeme nach Prinzipien der biologischen Evolution, Frommann-Holzboog Verlag, Stuttgart

22. Salomon, R. (1996) Performance Degradation of Genetic Algorithms Under Coordinate Rotation, Proc. 5th Annual Conf. on Evolutionary Programming, 155-161
23. Satoh, H., Yamamura, M. and Kobayashi, S. (1996) Minimal Generation Gap Model for GAs Considering Both Exploration and Exploitation, Proc. IIZUKA'96, 494-497
24. Schwefel, H.-P. (1981) Numerical optimization of computer models, Wiley, Chichester
25. Tsutsui, S., Yamamura, M. and Higuchi, T. (1999) Multi-parent Recombination with Simplex Crossover in Real Coded Genetic Algorithms, Proc. Genetic and Evolutionary Computation Conf. (GECCO'99), 657-664
26. Voigt, H.-M., Mühlenbein, H. and Gvetkovic, D. (1995) Fuzzy Recombination for the Breeder Genetic Algorithm, Proc. 6th Int'l Conf. on Genetic Algorithms, 104-111
27. Whitley, D., Starkweather, T. and Fuauay, D. (1989) Scheduling Problems and Traveling Salesman: The Genetic Edge Reconbination Operator, Proc. 3rd Int'l Conf. on Genetic Algorithms, 133-140
28. Wright, A. (1991) Genetic Algorithms for Real Parameter Optimization, Foundations of Genetic Algorithms, 205-218

Designing Evolutionary Algorithms for Dynamic Optimization Problems

Jürgen Branke and Hartmut Schmeck

Institute AIFB, University of Karlsruhe
D-76128 Karlsruhe, Germany
E-mail: {branke,schmeck}@aifb.uni-karlsruhe.de

Summary. Most research in evolutionary computation focuses on optimization of static, non-changing problems. Many real-world optimization problems, however, are dynamic, and optimization methods are needed that are capable of continuously adapting the solution to a changing environment. If the optimization problem is dynamic, the goal is no longer to find the extrema, but to track their progression through the space as closely as possible. In this chapter, we suggest a classification of dynamic optimization problems, and survey and classify a number of the most widespread techniques that have been published in the literature so far to make evolutionary algorithms suitable for changing optimization problems. After this introduction to the basics, we will discuss in more detail two specific approaches, pointing out their deficiencies and potential. The first approach is based on memorization, the other one uses a novel multi-population structure.

Keywords: evolutionary algorithm, genetic algorithm, dynamic, changing, time-varying

1 Introduction

Evolutionary algorithms (EAs) have proven successful in a vast number of applications and the number of papers produced in that area is still growing fast. However, almost all publications deal with optimization in static, non-changing environments, whereas many real-world problems are actually dynamic: new jobs have to be added to the schedule, machines may break down or wear out slowly, raw material is of changing quality, etc.

On a more abstract level, this might mean that the optimization function, the problem instance, or some restrictions may change, and thus the optimum to that problem might change as well. If any of these events are to be taken into account in the optimization process, we call the problem *dynamic* or *changing* (both terms are used synonymously[1]).

One standard approach to deal with these dynamics is to regard each change as the arrival of a new optimization problem that has to be solved from scratch (cf. [1]). However, this simple approach is often impractical:

---

[1] Often in the literature, also the term non-stationary is used. Since, however, in a broader statistical sense non-stationarity implies more than dynamics, namely that the expected value changes over time, the term is avoided in this chapter.

Solving a problem from scratch without reusing information from the past might be too time consuming, a change might not be identifiable directly, or the solution to the new problem should not differ too much from the solution of the old problem.

Thus it would be nice to have an optimization algorithm that is capable of continuously adapting the solution to a changing environment, reusing the information gained in the past. Since EAs have much in common with natural evolution, and since in nature adaptation is a continuous and continuing process, they seem to be a suitable candidate.

The main problem with standard EAs used for dynamic optimization problems appears to be that EAs eventually converge to an optimum and thereby lose their diversity necessary for efficiently exploring the search space and consequently also their ability to adapt to a change in the environment when such a change occurs.

In recent years, a number of authors have addressed this issue in many different ways. This chapter has been written to provide a survey of the state of the art in the field, to allow a closer look at two recent approaches, and to serve as a basis for future research. For a thorough treatment of several different aspects related to evolutionary optimization in dynamic environments see [2].

The outline of the chapter will be as follows. In Section 2, we suggest a categorization of dynamic environments. Then, Section 3 surveys and classifies the most prominent approaches to evolutionary optimization in changing environments. For our experiments, we mainly used the Moving Peaks benchmark as described in Section 4. Section 5 presents a memory-based EA for dynamic optimization problems, while in Section 6, a novel multi-population approach is suggested. The two approaches are compared empirically in Section 7. The chapter concludes with a summary and some suggestions for future work.

2 Characterizing the Dynamism

Any time-dependent problem may be called dynamic. However, from the point of view of EAs, not all such problems are equally interesting.

First of all, in this survey, we restrict our attention to problems where the fitness landscapes before and after a change display some exploitable similarities. If the problem were to change completely, without any reference to the history, it could be regarded as a simple sequence of independent problems that have to be solved from scratch.

Then, even if the underlying problem looks dynamic, the question has to be whether the EA has to cope with these dynamics. If, for example, the task for the EA is to design a fuzzy controller for a given problem, then it is considered to be static, since the optimal controller (and thus the optimum the EA has to find) does not change over time. On the other hand, if the

EA is used to directly and continuously optimize the control variables for the very same problem, it will be considered dynamic.

A problem similar to optimization in dynamic environments, but nevertheless different and thus not treated in this survey, is the optimization of noisy fitness functions. For these problems usually a static optimum is sought despite the noise, while for dynamic problems, the optimum changes over time.

But even if we restrict our view to the dynamics as described above, not all dynamic environments are equivalent, and different dynamics will probably require different optimization approaches. This section suggests a number of criteria along which dynamic environments could be categorized. Based on this categorization, one might eventually characterize classes of dynamic environments for which one algorithm is more suitable than another.

- *Frequency of change:* How often does the environment change (starting from very rare changes up to continuous change)? Since time comparisons very much depend on hardware and implementation details, and since usually the number of evaluations in EAs is the time-determining factor, the average number of evaluations between changes would be an appropriate measure. Only if there are other relevant aspects for computation time than evaluation would the actual (real) time between changes of the environment be needed for comparisons between approaches.
- *Severity of change:* How strongly is the system changing? Just a slight change or a completely new situation? An important aspect here is certainly the genotypic distance from the old to the new optimum. But other aspects may be, for example, whether the new optimum may be found from the old one by simple hill climbing, the size of the search space, or the probability that the new optimum can be reached from the old optimum by a single mutation. A more complex measure not restricted to the optimum might be the correlation of old to new fitness values of all points in the search space (high correlation meaning low severity and vice versa).
- *Predictability of change:* Is there a pattern or trend in the changes? Is it possible to predict direction, time, or severity of the next change given the changes encountered so far?
- *Cycle length/cycle accuracy:* Will the optimum return to previous locations or at least get close to them? And if so, how close? Here one might measure the average number of environmental states between two consecutive encounters of the same (or a very similar) state. If the new state is not exactly the same but a slight variant, then additionally the distance of the new to the previously encountered solution is important. Cycle length and accuracy might determine whether memorization of previous solutions is a useful strategy (cf. Section 5).

It would be nice if prospective authors in the field would comment on these characteristics in order to allow for better comparisons and perhaps a general framework to be constructed. Since it seems difficult to measure some of these characteristics, comparisons between different optimization problems may still be out of reach, but at least the above characteristics can be varied on a single problem such that their qualitative influence on a specific EA can be examined. In addition to the environmental properties, it may be interesting for the design of an EA to consider the following four aspects:

- *Visibility of change:* Are the occurrences of a change explicitly known to the system or do they have to be detected?
- *Necessity to change representation:* Is the genetic representation affected by a change, e.g. because the dimension of the problem has changed?
- *Aspect of change:* Is the change equivalent to a change in the optimization function, the problem instance, or some restrictions?
- *EA influence on environment:* Often, the solutions produced by the EA may influence the environment, e.g. in scheduling, where the produced schedule will restrict the choice of actions of the EA, or in coevolutionary models, where the evolution of one population will influence the evolution of the others.

3 EAs for Dynamic Optimization Problems — State of the Art

Over the past few years, a number of authors have addressed the problem of convergence and subsequent loss of adaptability in many different ways. Most of these approaches could be grouped into one of the following three categories:

1 The EA is run in standard fashion, but as soon as a change in the environment has been detected, explicit actions are taken to increase diversity and thus to facilitate the shift to the new optimum.
2 Convergence is avoided all the time and it is hoped that a spread-out population can adapt to changes more easily.
3 The EA is supplied with a memory to recall useful information from past generations, which seems especially useful when the optimum repeatedly returns to previous locations.

In the following subsections, we will present some typical examples for each of the above-mentioned categories. However, due to the tight space restrictions, this survey is necessarily incomplete and restricted to (from the authors' viewpoint) most important aspects. For a more complete (albeit older) survey, see [3].

3.1 React on Changes

A simple restart of the EA after a change in the environment has been detected would of course be the simplest option to deal with changes. However, if one assumes that the changes of the problem are relatively small, it is likely that the new landscape will be in some sense related to the previous one. In that case one should be able to do better than simple restart by transferring knowledge from the old population to the new initial population, e.g. by transferring individuals.

Hypermutation as proposed by Cobb [4], for example, keeps the whole population after a change, but increases population diversity by drastically increasing the mutation rate for some number of generations.

A variant of this, called *Variable Local Search* (VLS), has been suggested by Vavak et al. [5]. It increases mutation gradually after a change in the environment has been detected. At first, only small mutations are applied (e.g. only the last bits of a binary-coded real number are changed). Only if the population fitness does not improve after a predetermined period of time is the range of the local search extended by allowing larger and larger mutations. Furthermore, in [6] a learning strategy to adapt the range of mutation is suggested.

If the dynamics of the optimization problem affect the genetic representation, it is not possible to simply keep old individuals, the individuals have to be adapted. For example, in job shop scheduling, when new additional jobs arrive, they have to be represented in the genotype. However, the adaptation of the individuals is usually rather straightforward and introduces additional variance that automatically stimulates exploration. Overall, significant improvements in convergence speed and solution quality have been found if the altered (old) individuals are reused (see e.g. Bierwirth and Kopfer [7], Bierwirth and Mattfeld [8], Lin et al. [9], Mattfeld and Bierwirth [10], or Reeves and Karatza [11]).

A number of authors also suggested leaving the response to a change in the environment up to the EA by relying on *self-adaptiveness*. These approaches, which are often used in the context of evolution strategies and evolutionary programming, encode strategy parameters (usually the *variance* of Gaussian mutations) in the genotype such that they can be evolved along with the usual object parameters. It is then the responsibility of the EA to adapt its mutation strength appropriately after a change in the environment, e.g. to increase diversity. For a short overview on self-adaptation see e.g. Eiben et al. [12]. Examples for the use of self-adaptation in the context of dynamic environments can be found, for example, in Angeline [13], Bäck [14], Grefenstette [15], and Stephens et al. [16]. In Weicker and Weicker [17], it is argued that the standard Gaussian self-adaptive mutation prevalent in static environments may not be the best choice for dynamic environments, because it always favors the local neighborhood while often jumps are necessary. As an alternative, a self-adaptive mutation with ring-shaped probability distribution is suggested.

3.2 Maintaining Diversity Throughout the Run

Grefenstette [18] introduced the method of *random immigrants* where in every generation, the population is partly replaced by randomly generated individuals. As opposed to strong mutations, random immigrants only affect part of the population. Thus this introduces diversity without disrupting the ongoing search process.

Andersen [19] examines the effect of genotypic and phenotypic *sharing* on the GA's ability to track moving optima. The idea is that, since these methods try to spread out the population over multiple peaks, they should maintain diversity in the population. And indeed Andersen concludes that sharing remarkably enhances the GA's ability to track optima in slowly changing environments.

A sharing scheme based on *tag bits* has been tested by Liles and DeJong in [20]. There, tag bits are appended to each genotype, and only individuals with equal tag bits are allowed to mate. This can also be regarded as introducing subpopulations of varying size (for a more detailed description of this idea see [21]). The neighborhood used for sharing is then the number of individuals with the same tag bit, i.e. individuals with rare tag bits are favored. The experiments show that the approach is able to maintain different subpopulations on different peaks in a simple environment with two peaks of changing heights.

Cedeno and Vemuri [22] use a crowding-like replacement scheme, called "*Worst among Most Similar*", together with a selection scheme that chooses the second parent with respect to similarity to the first parent. The authors show that this approach is capable of maintaining a number of different solutions and adapting to new peaks appearing in the landscape.

Ghosh et al. [23] suggest modifying the fitness function by taking the individual's age into account, i.e. $f_{mod} = g(f_{old}, age)$. Interestingly, g is chosen in a way that middle-aged individuals are favored. The authors show that this approach maintains more diversity than an EA not considering age and that it can thus better adapt to changes in the environment.

The basic idea behind the *Thermodynamical Genetic Algorithm* (TDGA) proposed by Mori et al. [24] is to control the diversity in the population explicitly by controlling a measure of so-called "free energy" F. For a minimization problem, this term is calculated as

$$F = \langle E \rangle - TH$$

where $\langle E \rangle$ stands for the average population fitness and H is a measure for the diversity in the population. The temperature T is a parameter of the algorithm and reflects the emphasis on diversity (the problem of adjusting the parameter T, especially in dynamic environments, has been addressed in [25]). In every generation, the best individual is preserved as an elite, then the n individuals are paired to produce n offspring. Next, mutation is applied to all parents and offspring. From the resulting pool of $2n + 1$ individuals (n

mutated parents, n mutated children, 1 elite), individuals are selected one by one for the next generation. For this selection, in each step the free energy F of the slowly forming new population is calculated assuming that individual i ($i = 1, ..., 2n+1$) would be added to the new population. The individual that minimizes F is then actually added and the process is started anew until the new population consists of n members.

With the *Shifting Balance GA*, Oppacher and Wineberg [26] try to maintain the EA's exploratory power by dividing the population into one core population and a number of smaller colony populations. While the task of the core population is to exploit the best optimum found, the colony populations are forced to search in different areas of the fitness landscape, i.e. they are responsible for exploration. At regular intervals, the colonies send some emigrants to the core population and thereby update the core population's gene pool. This model outperformed a simple GA on a test problem with frequent but small shifts of the landscape.

Sarma and DeJong [27] compare the *diffusion EA* to a global population model on dynamic environments and observe a significant advantage of the diffusion model. The observed difference is not explained, but might be due to the slower convergence of the diffusion model. However, since the diffusion model can only delay convergence and not avoid it, its application to environments with low-frequency changes may be questionable.

3.3 Memory-based Approaches

Supplying the EA with some sort of memory might allow it to store good (partial) solutions and reuse them later as necessary. Obviously, strategies with a memory may be especially beneficial in periodically changing environments, when there are repeated occurrences of a small set of situations. Additionally, redundant representations may slow down convergence and favor diversity.

Memory may be provided in two general ways: *implicitly* by using redundant representations, or *explicitly* by introducing an extra memory and formulating strategies to deposit and retrieve solutions from it.

Implicit memory The most prominent approach to redundant representations seems to be multiploidy, with different implementations of the dominance mechanism (see e.g. Goldberg and Smith [28], Hadad and Eick [29], Ng and Wong [30], and Smith [31]).

Ryan [32] uses additive multiploidy, where the genes determining one trait are *added* in order to determine the phenotypic trait. The phenotypic trait becomes 1 when a certain threshold b_1 is exceeded, 0 if the value is below a smaller threshold b_2, and is determined at random if the value is between b_1 and b_2.

An interesting comparative study on multiploidy has been performed by Lewis et al. [33]. They observed that a simple dominance scheme is not sufficient to track the optimum reasonably well. If the diploid approaches are extended with a dominance change mechanism (reversing the dominance relation after a change), much better results can be obtained. Nonetheless, a simple haploid GA with a hypermutation rate similar to the number of bits flipped by a dominance change performed comparably. Experiments with an environment of two alternating states as well as a larger number of states revealed that the diploid approach is able to learn two solutions and switch between them almost instantaneously. If more than two targets were used, however, the approach failed completely. Altogether, it seems that the diploid approach by Ng and Wong [30] is quite effective at maintaining memory, while the approach by Ryan [32], extended with a dominance change mechanism, maintains diversity similar to a hypermutation scheme.

A quite different redundant representation scheme using a multi-level structured gene representation has been suggested by Dasgupta and McGregor [34]. In this representation, each level can activate or deactivate genes at the next lower level, allowing complex hierarchically structured genes and more redundant information than in the diploid scheme. However, in the experiments on the time-varying knapsack problem [34] and a moving parabola [35], relatively simple representations were chosen. Nevertheless, improvements over simple GAs were found.

Explicit memory While redundant representations allow the EA to implicitly store some useful information during the run, it is not clear that the algorithm actually uses this memory in an efficient way. As an alternative, the following approaches use an explicit memory in which specific information is stored and reintroduced into the population at later generations.

Louis and Xu [36], for example, look at rescheduling an open shop problem after a machine has broken down and has been replaced by a faster machine. The memory is used to transfer individuals from one EA run to seed the initial population after a single change. A fixed number of generations between changes is assumed, and the population's best individual is stored at regular intervals. For example, when the maximum number of generations is 300, every 50 generations the best individual is stored, resulting in a total of six stored individuals. After a change, the GA is seeded partially (5–10%) with individuals from the old run, while all other individuals are initialized randomly. The authors report significant improvements over a totally random initialization, particularly in early generations. However, when carrying over more individuals from the old run (50–100%), or for problems where the environment changes more significantly (deletion of a job), the method reportedly failed. Further experiments on the effect of the number and quality of the inserted solutions are reported in [37].

Ramsey and Grefenstette [38] incorporate case-based reasoning into an EA. They use a knowledge base to memorize successful individuals in a permanent memory. The system assumes that the environmental conditions can be measured. At regular intervals, the best individual is stored in the knowledge base and indexed with data characterizing the environment at that time. Whenever a new environment is encountered (the environmental variables changed), the EA is restarted. For restart, half of the population is initialized with individuals from the knowledge base that have been successful in a similar environment. Experiments proved that the knowledge base allows the EA to build upon the knowledge gained in the past. Unfortunately this approach is only applicable when the similarity of environments can be measured.

Another example is the work by Trojanowski et al. [39] in which each individual is extended with additional memory for a number of its ancestors. After a change in the environment, these older solutions are also re-evaluated and replace the current individual if they outperform it in the new environment. Since the memory is limited, this approach may be regarded as an EA with short-term memory that allows variability to be increased by reintroducing individuals that have been considered good in recent generations. In [40] this method has been combined with a diversification strategy replacing 85% of the population by randomly generated individuals after a change. On a simple test problem with several peaks changing in height, this combination clearly outperformed either the use of memory or the diversification strategy alone, as well as simple restart. This supports the observations made in [41], namely that a memory should always be used in combination with some diversity generating mechanism.

Yet another storage strategy has been added to the Thermodynamical Genetic Algorithm (TDGA, cf. Section 3.2) [42,24]. There, every generation's best individual is stored in the memory, and another individual is deleted from the memory depending on its age and contribution to the memory population's diversity (measured as variance over bit positions). The individuals from the memory then serve as additional potential candidates in the process of selecting a parent generation (in addition to the usual population). However, so far it has been defined for binary representation only and has never been evaluated per se.

In a similar setting, Branke [41] compares a number of replacement strategies for inserting new individuals into the memory. A simple replacement of the most similar individual performed almost equivalently to a strategy that replaces the worse of the two individuals in the memory closest to each other. Both strategies performed significantly better than a variance maximization scheme. Also in that paper, the importance of diversity for memory-based approaches is stressed. Further results from this paper will be discussed in Section 5.

4 The Moving Peaks Benchmark

Creating a proper benchmark problem is always a tricky task. On the one hand, it should be complicated enough to reflect most characteristics of real-world problems, on the other hand, it has to be simple enough to allow thorough analysis and subsequent insights into the working of the algorithm. Furthermore, its characteristics should be tunable by parameters.

Optimization in dynamic environments seems to require two fundamental capabilities: tracking of a solution that changes slightly, and jumping from an old solution to a quite distant new optimum that appeared elsewhere. Furthermore, as noted in Section 2, continuous adaptation only makes sense if subsequent stages of the environment are related. Otherwise, a simple restart of the algorithm would be the best possible strategy.

The Moving Peaks benchmark first introduced by Branke in [3,41] tries to address the above aspects. Independently, a similar benchmark has been suggested by Morrison and DeJong in [43]. Since then, the benchmark generator has been extended significantly, and this extended version is used here.

The Moving Peaks benchmark consists of m peaks in an n-dimensional real-valued parameter space, and the fitness landscape is defined as the maximum over all peak functions. It can be formulated as

$$F(\boldsymbol{x}, t) = \max(B(\boldsymbol{x}), \max_{i=1\ldots m} P(\boldsymbol{x}, h_i(t), w_i(t), \boldsymbol{p}_i(t)))$$

where $B(\boldsymbol{x})$ is a time-invariant "basis" landscape, and P is the function defining a peak shape, where each of the m peaks has its own time-varying parameters height (h), width (w), and location ($\boldsymbol{p}$).

Every Δe evaluations, the height, the width, and the location of every peak changes. The height and width of every peak are changed by adding a random Gaussian variable. The location of every peak is moved by a vector v of fixed length s. Thus the parameter s (together with the less important scaling factors for the width and height change) allows control of the change severity; Δe will determine the change frequency. A new parameter, λ, determines by how much a peak's change in location depends on its previous move. If $\lambda = 0.0$, each move is completely random; for $\lambda = 1.0$, the peak will always move in the same direction (until it hits the boundaries of the parameter space where it will bounce off like a billiard ball). The effect of λ on the path of a single peak is displayed in Figure 1.

Overall, the benchmark exhibits the two typical scenarios prevalent in dynamic optimization problems mentioned above: when the optimal peak moves slightly, the EA has to follow that peak through space, and local hill-climbing is often sufficient. On the other hand, when the heights of the peaks change such that a different peak becomes the maximum peak, the EA basically has to "jump", or cross a valley, to reach the new maximum peak, although the overall change of the landscape is still small. Therefore we think that this benchmark poses challenges very similar to those encountered in many dynamic real-world problems, while still being simple enough to allow new insights into the working of the optimization algorithm.

A benchmark also requires a way to measure the quality of an optimization approach. Since for dynamic optimization problems a single, time-invariant

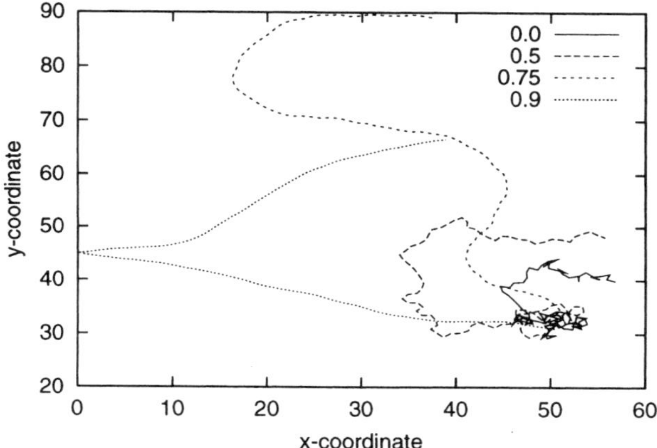

Fig. 1. The movement of a single peak over 100 steps, for different values of λ

optimal solution does not exist, performance can thus not be measured by just looking at the best individual found during the run.

A number of different metrics to measure tracking capabilities have been suggested in the literature (cf. [3]). For general use, we prefer the online and offline performance which were originally suggested by DeJong [44] for static problems, but they can be easily adapted to the dynamic case.

They are defined as follows. Let e_t be the t-th evaluation and let T be the number of evaluations considered. Then,

- *online performance* $x(T)$ is calculated as the average of all evaluations up to time T, i.e. $x(T) = \frac{1}{T}\sum_{t=1}^{T} e_t$. This is motivated by the assumption that every evaluation requires testing the real world.
- *offline performance* $x^*(T)$ is calculated as the average of the best values found so far at each time step, i.e. $x^*(T) = \frac{1}{T}\sum_{t=1}^{T} e_t^*$ with $e_t^* = \max\{e_1, e_2, ..., e_t\}$. This assumes that optimization is done in a simulated environment and only the best solutions are actually transferred into the real world. For non-stationary environments, however, this measurement is problematic since it has to be assured that the best solution found so far is the best solution for the current environment. Since this can only be guaranteed by re-evaluation, the offline performance should only consider individuals evaluated since the last change in the environment, i.e. $x'(T) = \frac{1}{T}\sum_{t=1}^{T} e'_t$ with $e'_t = \max\{e_\tau, e_{\tau+1}, ..., e_t\}$ and τ being the last time step before t at which a change in the environment occurred. Of course, this requires that the environmental changes are known to the observer.

For the Moving Peaks benchmark, the optimum at any point in time may be easily calculated. We therefore use here two derived methods that allow a less obstructed view on the performance by masking out the changes in the optimal fitness:

- *current error*, calculated as the difference between the currently best individual (best evaluation since last change) and the theoretical optimum, i.e. $\epsilon_t = opt(t) - e'_t$,
- *offline error*, defined as the the average current error over all time steps, $\epsilon^*(T) = \frac{1}{T}\sum_{t=1}^{T}\epsilon_t$.

The Moving Peaks benchmark in its current form includes a set of diagnostic tools to calculate these measures.

The benchmark and a more detailed description are freely available at http://www.aifb.uni-karlsruhe.de/~jbr/MovPeaks.

5 Memory/Search Population Approach

About one-third of the EAs designed for dynamic optimization problems known to the authors use some sort of memory. Why does the idea of adding a memory to an evolutionary algorithm (EA) seem to be so appealing that a larger number of authors have suggested it?

Intuitively, when the optimum reappears at a previous location, a memory could recall that location, and instantaneously move the population to the new optimum. A memory could also be useful in maintaining diversity. And it might guide evolution to promising areas after a re-initialization. But while the memory might allow exploitation of knowledge gained in the past, it might as well mislead evolution and prevent it from exploring new regions and discovering new peaks.

In [41], we have compared a number of ways to organize an explicit memory, and examined the usefulness of memory in different environments. As a result, we concluded that memory is only useful when the environment repeatedly returns to a small set of previously experienced solutions. Also in that paper, we have stressed the importance of diversification strategies to be used in combination with memory, because otherwise the EA might never get to the point where it has several different high-performance solutions in its memory.

The best strategy tested in that paper was to use two populations: a memory population that replaces its worst individuals with the individuals in the memory after a change, and a search population that is re-initialized after a change (cf. Figure 2). Both populations contribute to the memory by regularly storing their best individual in the memory, where it replaces the worse of the two most similar (in terms of euclidean distance) individuals in the memory.

For this chapter, we have rerun some of the experiments reported in [41] using the new, extended Moving Peaks benchmark. Besides confirming previous results on this changed test function, we will report on some additional tests on varying the number of peaks and the λ-parameter.

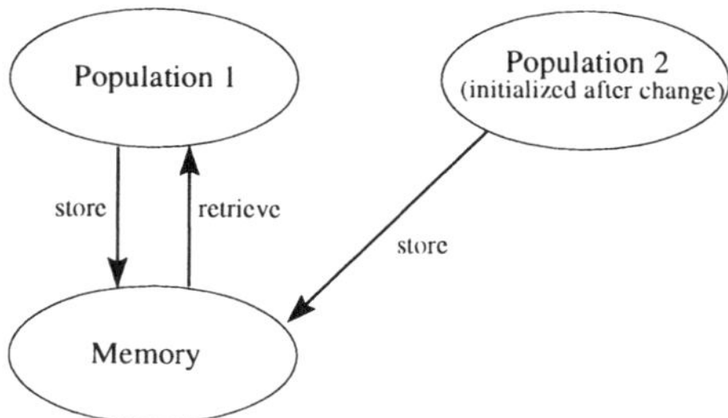

Fig. 2. Memory/search population using two subpopulations: one to exploit the memory, the other to explore new regions of the search space.

Results will be reported in Section 7, together with the results of the self-organizing scouts approach presented next.

6 Self-organizing Scouts

Although the memory/search population approach was quite successful, it soon became obvious that a strategy based on memorization would be too restricted to adapt successfully to a wide range of dynamic environments. As an alternative, we developed an approach with multiple populations acting as *self-organizing scouts* (SOS), watching over the changing landscape.

The basic idea of SOS is that once a peak has been found (i.e. the population converged to a high-performance region), the population should split: a small fraction, henceforth called the "child population" should "watch" over that peak, while the remainder of the population ("base population") should spread out and continue searching for new peaks.

When a watched peak moves, the child population may follow it through space, and even request reinforcement. Since the population size is limited, individuals are continuously redistributed to those populations where they seem to be needed most, and peaks that seem too unpromising may be abandoned.

The general idea of SOS could be implemented in many different ways. The design decisions that had to be made include:

- How does one determine when the base population has found a peak that justifies a split-off?
- How many individuals should be kept at which peak? When is a peak abandoned?
- What happens when two child populations move towards the same peak?
- What exactly is a peak, i.e. what is the area to be surveyed by a child population?
- How many peaks should be surveyed simultaneously?

In our actual implementation, we have borrowed some ideas from the *forking Genetic Algorithm (fGA)* as proposed by Tsutsui et al. [45]: not only the population, but also the search space are divided when a peak has been found. The search of the child population is then restricted to a sphere around the currently best individual, while the base population is excluded from this area. Individuals that (e.g. by mutation) enter a foreign area are deleted and replaced by a randomly generated individual within the valid area.

In general, our algorithm works as follows:

```
REPEAT
   Compute the next generation of base population and child populations
   Adjust search space of child populations
   IF (forking generation)
        Create new child population when possible
        Adjust size of base and child populations
UNTIL termination criterion
```

In the following paragraphs, the different steps are explained in more detail.

Creating a new child population A child population is an independent subpopulation working on a part of the phenotypic feature space. It is defined by a center (the most fit individual in the subpopulation) and a distance (range), and consists of all individuals whose distance (phenotypical Manhattan distance) to the center is smaller than or equal to the given range. The population fitness is defined as fitness of its best individual.

Child populations underlie a number of restrictions which can be set by preferences:

- minimum and maximum number of individuals relative to overall amount
- minimum and maximum diameter of the subspace relative to the size of the phenotypic search space
- minimum fitness of new forking populations relative to current overall best individual
- minimum fitness of existing forking populations relative to current overall best individual.

At specific generations, called *forking generations*, the base population is analyzed for the existence of a group of individuals that would satisfy the above constraints for child populations. If more than one group is found, the one with the maximum ratio of number of individuals to diameter is selected. All individuals in that group are split off from the base population and assigned the smallest hypercube encompassing all individuals as search space (note that this allows a variable size of the subspace, while the original fGA always uses a fixed size subspace). The process is illustrated in Figure 3.

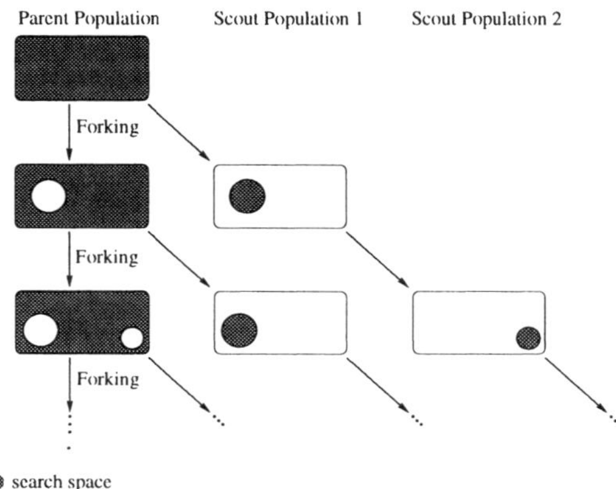

Fig. 3. Illustration of the creation of child populations. Note that, as opposed to fGAs, the child population search spaces may have different sizes and may also move.

Moving the child population's search space As the best individual in a child population defines the center of its search space, that search space may move. The expansion of the search space, however, is kept constant in the current implementation.

Note that it may happen that the search spaces of child populations overlap. Usually this is tolerated, but when a center individual falls into the search space of another child population, its whole child population is discarded.

Adjusting the population sizes Since the overall number of individuals is limited, they have to be distributed efficiently over the different populations. Generally, more individuals should be placed in areas with high quality and high dynamics. On the other hand, when a child population has converged to a peak and that peak has not changed for several generations, it may be sufficient to maintain a very small "outpost" on that peak in order to be able to detect when that peak becomes interesting again, i.e. when it changes height and/or position.

For that purpose, in each forking generation, first of all, any child population with a fitness smaller than the minimum required fitness is discarded. Then, for all remaining child populations as well as the parent population a quality measure Q_i is calculated.

The quality Q_i of population i is simply a linear combination of its fitness F_i and *dynamism measure* D_i which depends on the difference between a population's current and previous fitness:

$$F_i(t) = \text{fitness of best individual in population } i \text{ at time } t$$

$$D_i(t) = \max\left\{0, \frac{F_i(t) - F_i(t-1)}{F_i(t-1)}\right\}$$

$$Q_i(t) = \begin{cases} \alpha \frac{D_i(t)}{\sum_j D_j(t)} + (1-\alpha)\frac{F_i(t)}{\sum_j F_j(t)} & : \sum_j D_j(t) > 0 \\ \frac{F_i(t)}{\sum_j F_j(t)} & : \text{otherwise.} \end{cases}$$

The desired population size S_i is then chosen proportionally to each population's relative quality:

$$S_i = \frac{Q_i}{\sum_j Q_j} \cdot (\text{total number of individuals.})$$

Of course, the restrictions on minimum and maximum population size of each child population and the parent population always have to be respected.

When the size of a population is increased, new random individuals are generated within the corresponding search space. If individuals have to be removed as the new population size is smaller than the old one, the worst individuals are removed.

We consider the above described attempt to measure quality a rather straightforward and preliminary approach; in future studies other indicators such as convergence may be included.

Computing the next generation Generally, computing the next generation of the base or child population is equivalent to a single generation of an ordinary EA.

The mutation step size is adjusted to the extension of the corresponding search space, i.e. child populations generally have much smaller mutation step sizes than the base population.

Since the fitness function is dynamic, all individuals have to be re-evaluated in every generation. It is always assured that new individuals lie within the population's search space.

The first results of this approach have been published in [46].

In the following section we will compare SOS with the memory/search population approaches from the previous section.

7 Empirical Results

For the tests reported in this section, the Moving Peaks benchmark with five dimensions and 10 cone-shaped peaks has been used. Note that this differs slightly from the benchmark we used in [41] where only five peaks of a different shape were used. Every 5000 evaluations, the peaks change in width, height, and location. The exact parameters used for that benchmark can be obtained from the Moving Peaks website.

Unless stated otherwise, the EA uses real-valued encoding, generational replacement but elite of 1, mutation rate of 0.2, crossover probability of 0.6, and a total population size of 100, including the memory if used. All individuals, including those in the memory, have to be re-evaluated every generation due to the dynamics of the landscape. Therefore, a change every 5000 evaluations corresponds to a change every 50 generations. When a memory is used, it usually contains 10 individuals. For SOS, every generation was considered forking generation. All reported results are the averages over 50 runs with a different random seed.

7.1 The Effect of Change Severity

Figure 4 shows the offline error after 5000 generations (100 change intervals) over a range of values of s.

When the severity parameter s is set to zero, the peaks do not move at all, only their width and height change. With m peaks, this means that the optimum can switch only between these m locations. Naturally, this setting is very advantageous for memory-based approaches. Unfortunately, it is not very realistic, because once the m peaks have been found, the EA would no longer be necessary.

Nevertheless, simply adding a memory to a standard EA does not really help very much. But in conjunction with a diversification strategy like the memory/search population approach, the benefit is extraordinary (see Figure 4). SOS still performs a little bit better, perhaps because it can locate the different peaks very quickly, while the memory/search population approach only enforces the new search after the environment has changed.

When the length s of the shift vector is increased, the performance of all approaches decreases. In particular, the memory-based approaches suffer, because a memory is less useful when the peaks move. Nevertheless, even for a quite strong movement of 3.0, the memory/search population approach is significantly better than the simple memory. SOS is least affected by the increasing severity, because the child populations can follow the peaks.

One reason for the superiority of SOS may be its ability to discover and maintain more and more peaks over the course of the run. Figure 5 shows the percentage of peaks covered[2] by individuals of the current population and the memory. The length of the shift vector was 1.0 for this figure.

While without diversification the population quickly converges to a single peak and a memory is of little help (although there is a slight improvement towards the end), the additional diversification by memory/search allows it

[2] An individual at location $\boldsymbol{x}$ is said to cover a peak j if, at $\boldsymbol{x}$, peak function j returns the greatest value, i.e. $P(\boldsymbol{x}, h_j(t), w_j(t), \boldsymbol{p}_j(t)) = \max_{i=1\ldots m} P(\boldsymbol{x}, h_i(t), w_i(t), \boldsymbol{p}_i(t))$.

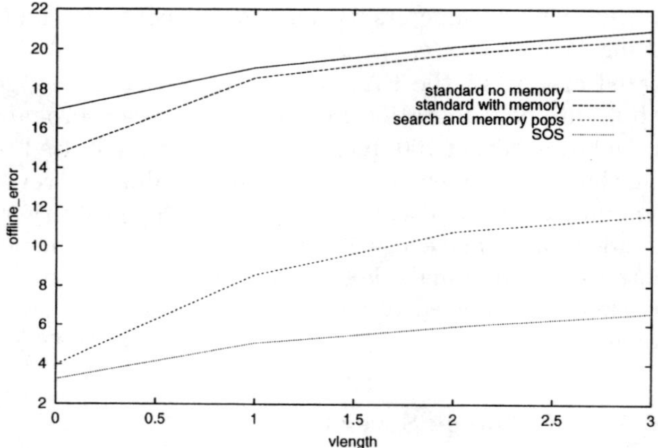

Fig. 4. Offline-error of several approaches after 5000 generations, varying shift vector length s.

to explore several peaks and store them in the memory for later use. Still, the SOS exploration strategy seems to be much more effective. SOS continuously increases the coverage of the landscape over the entire run and eventually covers 94% of all peaks, compared to 65% for the memory/search approach.

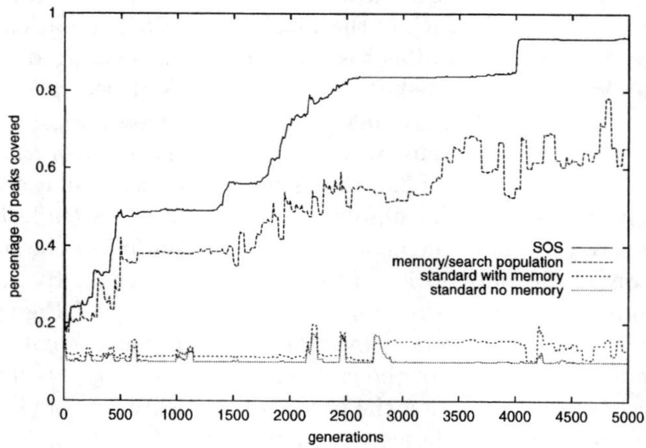

Fig. 5. Percentage of peaks covered at different generations. Moving Peaks landscape with 10 peaks and shift vector length 1.0.

7.2 The Effect of Recurrence

So far, each peak moved in a random direction for every change. That way, the peaks were likely to stay near the original location. By increasing the parameter λ, we can make recurrence much less likely (cf. Figure 1). The effect can be seen in Figure 6. Overall, the performance is only slightly affected, but both memory-based approaches lose some of their advantage. SOS is basically unaffected.

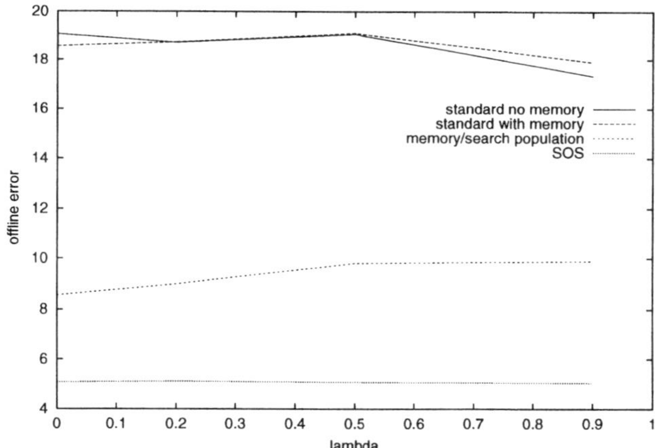

Fig. 6. The effect of increasing λ.

7.3 Increasing the Number of Peaks

With only 10 peaks, the memory/search population approach with its memory of size 10, and of course also the SOS approach, are at least theoretically able to cover all peaks permanently.

Figure 7 shows the effect of increasing the number of peaks. Since increasing the number of peaks makes the landscapes less comparable, the results reported there are the averages over five different landscapes times 10 random seeds, i.e. 50 runs altogether. While for the standard approach with or without memory a slight increase seems to make optimization more difficult, many more peaks seem to reduce the error. A possible reason here might be that the smaller peaks are completely hidden by larger peaks, and thus the average fitness of the landscape is increased. SOS and the memory/search population approach are largely unaffected, which shows that they are both able to select the important peaks.

One might argue that the performance of memory/search is limited here by the memory size, while the "memory capacity" of SOS is only limited

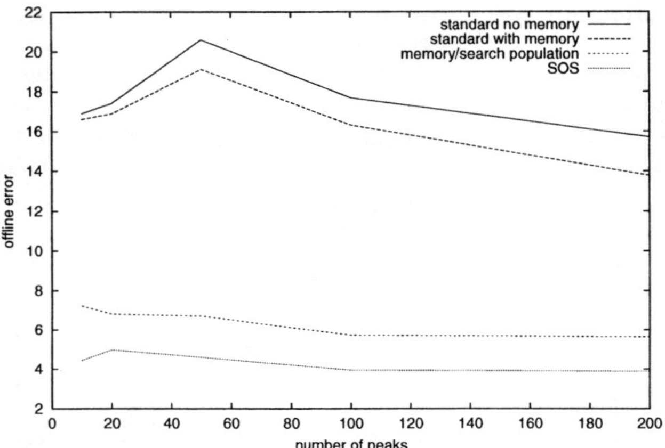

Fig. 7. The effect of increasing the number of peaks.

by the total population size and the minimum size for a child population. And indeed, when we look at the coverage of the peaks over a run with 50 peaks (Figure 8), we can observe that a larger memory helps to cover more peaks. However, as it turns out, this does not help to improve performance. The memory/search approach with a memory size of 50 (and, consequently, two populations of size 25 each) performs almost identically to the run with memory size 10 (not shown).

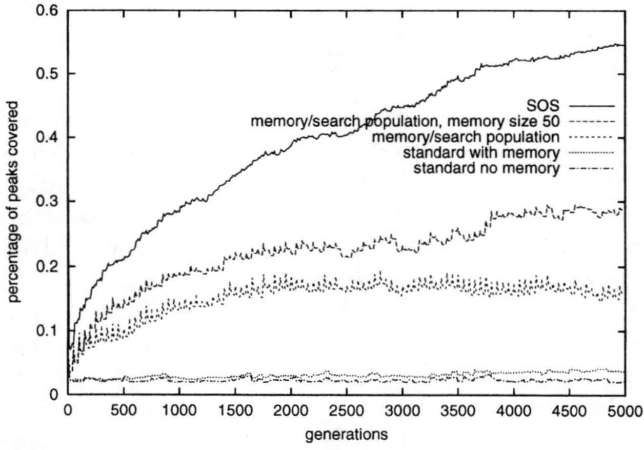

Fig. 8. Percentage of peaks covered at different generations. Moving Peaks landscape with 50 peaks and shift vector length 1.0.

8 Conclusion

This chapter surveyed and categorized a number of different approaches to adapt standard EAs to handle dynamic optimization problems.

Since only very few comparisons between different approaches have been published so far, it is difficult to draw conclusions about the superiority of one approach over the other. Instead, by examining two specific approaches more closely, we tried to point out the challenges faced when the optimization problem is dynamic, and the difficult balance between exploration and exploitation. We have shown the deficiencies of memorization strategies and proposed a novel approach based on a dynamic subpopulation structure. This self-organizing scouts (SOS) approach, although in its early stages, seems to be a very promising general framework with many aspects still open to exploration. In particular, we are currently working on better ways to assess the quality of a search region, and to dynamically adjust the child populations' search spaces.

Also, in the overall area of EAs for dynamic optimization problems, a lot of work remains to be done. In particular, it seems that the field would need better comparisons between the different approaches, especially with respect to different categories of dynamic environments as suggested in Section 2. Dynamic combinatorial problems might be another challenge, since they might exhibit a search space significantly different from the underlying principle of Moving Peaks.

References

1. Raman, N., Talbot, F. B. (1993) The job shop tardiness problem: a decomposition approach. *European Journal of Operational Research*, **69**, 187–199
2. Branke, J. (2002) *Evolutionary optimization in dynamic environments.* Kluwer.
3. Branke, J. (1999) Evolutionary algorithms for dynamic optimization problems - a survey. Technical Report 387, Insitute AIFB, University of Karlsruhe
4. Cobb, H. G. (1990) An investigation into the use of hypermutation as an adaptive operator in genetic algorithms having continuous, time-dependent nonstationary environments. Technical Report AIC-90-001, Naval Research Laboratory, Washington, DC
5. Vavak, F., Jukes, K., Fogarty, T. C. (1997) Adaptive combustion balancing in multiple burner boiler using a genetic algorithm with variable range of local search. In Bäck [47], 719–726
6. Vavak, F., Jukes, K., Fogarty, T. C. (1997) Learning the local search range for genetic optimisation in nonstationary environments. In *IEEE International Conference on Evolutionary Computation ICEC'97*, 355–360. IEEE
7. Bierwirth, C., Kopfer, H. (1994) Dynamic task scheduling with genetic algorithms in manufacturing systems. Technical Report, Department of Economics, University of Bremen
8. Bierwirth, C., Mattfeld, D. C. (1999) Production scheduling and rescheduling with genetic algorithms. *Evolutionary Computation*, **7**, 1–18

9. Lin, S. C., Goodman, E. D., Punch, W. F. (1997) A genetic algorithm approach to dynamic job shop scheduling problems. In Bäck [47], 481–488
10. Mattfeld, D. C., Bierwirth, C. (1998) Minimizing job tardiness: Priority rules vs. adaptive scheduling. In I. C. Parmee, editor, *Proceedings of ACDM*, 59–67. Springer
11. Reeves, C., Karatza, H. Dynamic sequencing of a multi-processor system: A genetic algorithm approach. In R. F. Albrecht, C. R. Reeves, and N. C. Steele, editors, *Artificial Neural Nets and Genetic Algorithms*, 491–495. Springer
12. Eiben, A. E., Hinterding, R., Michalewicz, Z. Parameter control in evolutionary algorithms. *IEEE Transactions on Evolutionary Computation*, **3**, 124–141
13. Angeline, P. J. (1997) Tracking extrema in dynamic environments. In Angeline et al. [48], 335–345
14. Bäck, T. (1998) On the behavior of evolutionary algorithms in dynamic environments. In *IEEE International Conference on Evolutionary Computation*, 446–451. IEEE
15. Grefenstette, J. J. (1999) Evolvability in dynamic fitness landscapes: A genetic algorithm approach. In *Congress on Evolutionary Computation*, **3**, 2031–2038. IEEE
16. Stephens, C. R., Olmedo, I. G., Vargas, J. M., Waelbroeck, H. (1998) Self-adaptation in evolvins systems. *Artificial Life*, **4**
17. Weicker, K., Weicker, N. (1999) On evolution strategy optimization in dynamic environments. In *Congress on Evolutionary Computation*, **3**, 2039–2046
18. Grefenstette, J. J. (1992) Genetic algorithms for changing environments. In R. Maenner and B. Manderick, editors, *Parallel Problem Solving from Nature 2*, 137–144. North-Holland
19. Andersen, H. C. (1991) An investigation into genetic algorithms, and the relationship between speciation and the tracking of optima in dynamic functions. Honours thesis, Queensland University of Technology, Brisbane
20. Liles, W., DeJong, K. (1999) The usefulness of tag bits in changing environments. In *Congress on Evolutionary Computation*, **3**, 2054–2060. IEEE
21. Spears, W. (1994) Simple subpopulation schemes. In *Evolutionary Programming Conference*, 296–307. World Scientific
22. Cedeno, W., Vemuri, V. R. On the use of niching for dynamic landscapes. In *International Conference on Evolutionary Computation*. IEEE
23. Ghosh, A., Tsutsui, S., Tanaka, H. (1998) Function optimization in nonstationary environment using steady state genetic algorithms with aging of individuals. In *IEEE International Conference on Evolutionary Computation*, 666–671
24. Mori, N., Kita, H., Nishikawa, Y. (1996) Adaptation to a changing environment by means of the thermodynamical genetic algorithm. In H.-M. Voigt,editor, *Parallel Problem Solving from Nature*, **1141** *LNCS*, 513–522. Springer
25. Mori, N., Kita, H., Nishikawa, Y. (1998) Adaptation to a changing environment by means of the feedback thermodynamical genetic algorithm. In Eiben et al. [49], 149–158
26. Oppacher, F., Wineberg, M. (1999) The shifting balance genetic algorithm: Improving the ga in a dynamic environment. In W. Banzhalf et al., (ed.), *Genetic and Evolutionary Computation Conference*, **1**, 504–510. Morgan Kaufmann
27. Sarma, J., DeJong, K. (1999) The behavior of spatially distributed evolutionary algorithms in non-stationary environments. In Wolfgang Banzhaf et al., editor, *GECCO*, **1**, 572–578, Morgan Kaufmann, San Francisco, California

28. Goldberg, D. E., Smith, R. E. (1987) Nonstationary function optimization using genetic algorithms with dominance and diploidy. In J. J. Grefenstette, editor, *Second International Conference on Genetic Algorithms*, 59–68. Lawrence Erlbaum Associates
29. Hadad, B. S., Eick, C. F. (1997) Supporting polyploidy in genetic algorithms using dominance vectors. In et al. [48], 223–234
30. Ng, K. P., Wong, K. C. (1995) A new diploid scheme and dominance change mechanism for non-stationary function optimization. In *Sixth International Conference on Genetic Algorithms*, 159–166. Morgan Kaufmann
31. Smith, R. E. (1987) Diploid genetic algorithms for search in time varying environments. In *Annual Southeast Regional Conference of the ACM*, 175–179, New York
32. Ryan, C. (1997) Diploidy without dominance. In J. T. Alander, editor, *Third Nordic Workshop on Genetic Algorithms*, 63–70
33. Lewis, J., Hart, E., Ritchie, G. A comparison of dominance mechanisms and simple mutation on non-stationary problems. In Eiben et al. [49], 139–148
34. Dasgupta, D., McGregor, D. R. (1992) Nonstationary function optimization using the structured genetic algorithm. In R. Männer and B. Manderick, (eds.), *Parallel Problem Solving from Nature*, 145–154. Elsevier Science Publisher
35. Dasgupta, D. (1995) Incorporating redudancy and gene activation mechanisms in genetic search. In L. Chambers, (ed.), *Practical Handbook of Genetic Algorithms*, **2**, 303–316. CRC Press
36. Louis, S. J., Xu, Z. (1996) Genetic algorithms for open shop scheduling and re-scheduling. In M. E. Cohen and D. L. Hudson, editors, *ISCA Eleventh International Conference on Computers and their Applications*, 99–102
37. Louis, S. J, Johnson, J. (1997) Solving similar problems using genetic algorithms and case-based memory. In Bäck [47], 283–290
38. Ramsey, C. L., Grefenstette, J. J. (1993) Case-based initialization of genetic algorithms. In S. Forrest, editor, *Fifth International Conference on Genetic Algorithms*, 84–91. Morgan Kaufmann
39. Trojanowski, K., Michalewicz, Z., Xiao, J. (1997) Adding memory to the evolutionary planner/navigator. In *IEEE Intl. Conference on Evolutionary Computation*, 483–487
40. Trojanowski, K., Michalewicz, Z. (1999) Searching for optima in non-stationary environments. In *Congress on Evolutionary Computation*, 3, 1843–1850. IEEE
41. Branke, J. (1999) Memory enhanced evolutionary algorithms for changing optimization problems. In *Congress on Evolutionary Computation CEC99*, 3, 1875–1882. IEEE
42. Mori, N., Imanishi, S., Kita, H., Nishikawa, Y. (1997) Adaptation to changing environments by means of the memory based thermodynamical genetic algorithm. In Bäck [47], 299–306
43. Morrison, R. W., DeJong, K. A. (1999) A test problem generator for non-stationary environments. In *Congress on Evolutionary Computation*, **3**, 2047–2053. IEEE
44. DeJong, K. (1975) *An analysis of the behavior of a class of genetic adaptive systems*. PhD thesis, University of Michigan, Ann Arbor MI
45. Tstutsui, S., Fujimoto, Y., Ghosh, A. (1997) Forking genetic algorithms: Gas with search space division schemes. *Evolutionary Computation*, **5**, 61–80

46. Branke, J., Kaußler, T., Schmidt, C., Schmeck, H. (2000) A multi-population approach to dynamic optimization problems. In *Adaptive Computing in Design and Manufacturing 2000*, 299–308. Springer
47. Bäck, T. (ed.) (1997) *Seventh International Conference on Genetic Algorithms.* Morgan Kaufmann
48. Angeline, P. J., Reynolds R. G, McDonnell, J. R., Eberhart, R.(eds.) (1997) *Proceedings of the Sixth International Conference on Evolutionary Programming,* **1213** *LNCS.* Springer
49. Eiben, A. E., Bäck T., Schoenauer, M., Schwefel, H. P.(eds.) (1998) *Parallel Problem Solving from Nature,* number 1498 in LNCS. Springer

Multi-objective Evolutionary Algorithms: Introducing Bias Among Pareto-optimal Solutions

Kalyanmoy Deb

Kanpur Genetic Algorithms Laboratory (KanGAL), Department of Mechanical Engineering, Indian Institute of Technology Kanpur, PIN 208 016, India,
E-mail: deb@iitk.ac.in

Summary. Since the beginning of the 1990s, research and application of multi-objective evolutionary algorithms (MOEAs) have attracted increasing attention. This is mainly due to the ability of evolutionary algorithms to find multiple Pareto-optimal solutions in one single simulation run. In this chapter, we present an overview of MOEAs and then discuss a particular algorithm in detail. Although MOEAs can find multiple Pareto-optimal solutions, often, users need to impose a particular order of priority to objectives. In this chapter, we present a few classical techniques to identify a preferred or a compromise solution, and finally suggest a biased sharing technique which can be used during the optimization phase to find a biased distribution of Pareto-optimal solutions in the region of interest. The results are encouraging and suggest further application of the proposed strategy to more complex multi-objective optimization problems.

1 Introduction

Many real-world optimization problems are naturally posed as multi-objective optimization problems. However, due to the lack of efficient multi-objective optimization algorithms, they have been suitably converted into single objective optimization problems and solved. The basic difficulty arises from the nature of the optimality conditions for multiple objectives. In the presence of multiple and conflicting objectives, the resulting optimization problem gives rise to a set of optimal solutions, instead of one optimal solution [10]. Multiple optimal solutions exist because no one solution can be optimal for multiple conflicting objectives. Let us illustrate this concept through an example. If cost and reliability are two objectives in a design optimization, it is clear that a minimum cost solution is usually not maximally reliable and a maximally reliable solution is not often the cheapest. In such a scenario, none of these two extreme solutions (the cheapest and the most reliable solutions) can be declared as an absolute optimum corresponding to both objectives of design. In the parlance of multi-criterion decision-making, both these solutions are optimal in some sense or they are Pareto-optimal [39]. In fact, there exist many other solutions in the search space which are also Pareto-optimal. Since none of these solutions can be said to be an 'absolute optimum', the onus on the part of the user is then to first find as many such solutions as possible. Once these multiple solutions are found, usually, a higher-level decision-making strategy is adopted to choose one solution from the set of obtained Pareto-optimal solutions.

In this chapter, we describe the principle of multi-objective optimization and then discuss a number of evolutionary algorithms. Since evolutionary algorithms deal with a population of solutions [20], it is logical that they can be used to find multiple Pareto-optimal solutions in one single simulation run [5,16,23,30,32,41]. We describe one such algorithm—non-dominated sorting GA or NSGA [38]—in somewhat greater detail. We present simulation results of NSGA on two problems.

Although most research on multi-objective evolutionary algorithms (MOEAs) has concentrated its efforts in developing new and efficient search algorithms for finding solutions closer to the Pareto-optimal front and having a wider spread [10,17,25,38,44], only a couple of them concentrate on devising strategies to find 'compromised' solutions from the Pareto-optimal set. In this chapter, we borrow a number of multi-objective optimization techniques from the classical field and suggest a number of post-optimization and optimization-level methods for choosing a Pareto-optimal solution or a biased Pareto-optimal region. Specifically, we suggest a biased sharing NSGA to find a biased distribution of solutions in the Pareto-optimal region. Instead of artificially forming a weighted sum of objective functions and finding the optimum of it, this method uses the user's order of preference for objectives to find a biased distribution of Pareto-optimal solutions. On a couple of two-objective optimization problems, it is shown that the proposed biased sharing NSGA can find a biased distribution of solutions according to the weight vector setting. This procedure allows more solutions to be found in the region of the user's preference. Although more studies are needed, the results of this chapter show the efficacy of the proposed method and take the MOEAs one step closer to making them usable in practice.

2 Multi-objective Optimization

The principles of multi-criterion optimization are different from that in a single-objective optimization. The main goal in a single-objective optimization is to find the global optimal solution. However, in a multi-criterion optimization problem, there are more than one objective function, each of which may have a *different* individual optimal solution. If there is sufficient difference in the optimal solutions corresponding to different objectives, the objective functions are often known as *conflicting* to each other. Multi-criterion optimization with such conflicting objective functions gives rise to a set of optimal solutions, instead of one optimal solution. The reason for the optimality of many solutions is that no one can be considered to be better than any other with respect to all objective functions. These optimal solutions have a special name—Pareto-optimal solutions. Let us illustrate this aspect with a hypothetical example shown in Figure 1 [6]. The figure considers two

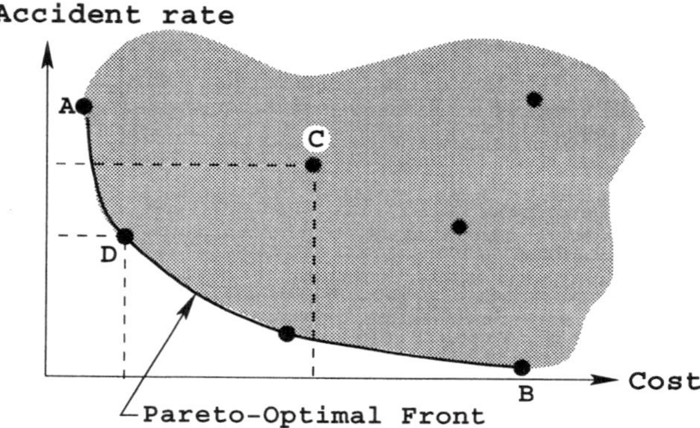

Fig. 1. The concept of Pareto-optimal solutions

objectives—cost and accident rate—both of which are to be minimized. The point A represents a solution which incurs a near-minimal cost, but is highly accident-prone. On the other hand, the point B represents a solution which is costly, but is near least accident-prone. If both objectives (cost and accident rate) are important design goals, one cannot really say whether solution A is better than solution B, or vice versa. One solution is better than the other in one objective, but is worse in the other. In fact, there exist many such solutions (like solution D) which also belong to the Pareto-optimal set and one cannot conclude about an absolute hierarchy of solutions A, B, D, or any other solution in the set without further information. All these solutions (in the front marked by the dashed line) are known as Pareto-optimal solutions.

Looking at the figure, we also observe that there exists non-Pareto-optimal solutions, like the point C. If we compare solution C with solution A, we again are in a fix and cannot say whether one is better than the other in both objectives. Does this mean that solution C is also a member of the Pareto-optimal set? The answer is no. This is because there exists another solution D in the search space, which is better than solution C in *both* objectives. That is why solutions like C are known as *dominated* solutions or *inferior* solutions.

It is now clear that the concept of optimality in multi-criterion optimization deals with a number (or a set) of solutions, instead of one solution. Based on the above discussions, we first define the conditions for a solution to become dominated with respect to another solution and then present the conditions for a set of solutions to become a Pareto-optimal set.

For a problem having more than one objective function (say, f_j, $j = 1, \ldots, M$ and $M > 1$), any two solutions $x^{(1)}$ and $x^{(2)}$ (having P decision variables each) can have one of two possibilities—one dominates the other or none dominates the other. A solution $x^{(1)}$ is said to dominate the other solution $x^{(2)}$ if both the following conditions are true [39]:

1 The solution $x^{(1)}$ is no worse (say the operator $\prec$ denotes worse and $\succ$ denotes better) than $x^{(2)}$ in all objectives, or $f_j(x^{(1)}) \not\prec f_j(x^{(2)})$ for all $j = 1, 2, \ldots, M$ objectives.
2 The solution $x^{(1)}$ is strictly better than $x^{(2)}$ in at least one objective, or $f_{\bar{j}}(x^{(1)}) \succ f_{\bar{j}}(x^{(2)})$ for at least one $\bar{j} \in \{1, 2, \ldots, M\}$.

If any of the above conditions is violated, the solution $x^{(1)}$ does not dominate the solution $x^{(2)}$. If $x^{(1)}$ dominates the solution $x^{(2)}$, it is also customary to write $x^{(2)}$ is dominated by $x^{(1)}$, or $x^{(1)}$ is non-dominated by $x^{(2)}$, or, simply, among the two solutions, $x^{(1)}$ is the non-dominated solution. Although the above is a standard definition of domination between two solutions, Parmee et al. [33] suggested a weighted dominance relation based on relative importance of objectives. In problems where weight information is available, such modified definitions may be useful.

The following definitions ensure whether a set of solutions belong to a local or global Pareto-optimal set, similar to the definitions of local and global optimal solutions in single-objective optimization problems:

Local Pareto-optimal Set: If for every member x in a set $\underline{P}$, there exists no solution y satisfying $\|y - x\|_\infty \leq \epsilon$, where ϵ is a small positive number (in principle, y is obtained by perturbing x in a small neighborhood), which dominates any member in the set $\underline{P}$, then the solutions belonging to the set $\underline{P}$ constitute a local Pareto-optimal set.

Global Pareto-optimal Set: If there exists no solution in the search space which dominates any member in the set $\bar{P}$, then the solutions belonging to the set $\bar{P}$ constitute a global Pareto-optimal set.

We would like to highlight here that there exists a difference between a non-dominated set and a Pareto-optimal set. A non-dominated set is defined in the context of a sample of the search space. In a sample of search points, solutions that are not dominated (according to the above definition) by any other solution in the sample space are non-dominated solutions. A Pareto-optimal set is a non-dominated set, when the sample is the entire search space.

From the above discussion, we observe that there are primarily two goals that a multi-criterion optimization algorithm must try to achieve:

1 Guide the search towards the global Pareto-optimal region, and
2 Maintain population diversity in the Pareto-optimal front.

The first task is a natural goal of any optimization algorithm. The second task is unique to multi-criterion optimization. Since no one solution in the Pareto-optimal set can be said to be better than any other, what an algorithm can do best is to find as many different Pareto-optimal solutions as possible.

The classical way of tackling multi-objective optimization problems is straightforward: convert multiple objectives into one objective. There exists a number of conversion methods [2,29,37]: weighted sum approach, ϵ-perturbation method, Tchebyshev method, min-max method, goal programming method, and others. Since multiple objectives are converted into one objective, the resulting solution to the single-objective optimization problem is usually subject to the parameter settings chosen by the user. Moreover, since usually a classical optimization method is used, only one solution (hopefully a Pareto-optimal solution) can be found in one simulation run. Thus, in order to find multiple Pareto-optimal solutions, the chosen optimization algorithm must be used a number of times. Furthermore, the classical methods have been found to be sensitive to the convexity and continuity of the Pareto-optimal region.

3 Evolutionary Techniques

As early as in 1967, Rosenberg suggested, but did not simulate, a genetic search method for finding the chemistry of a population of single-celled organisms with multiple properties or objectives [35]. However, the first practical implementation was suggested by David Schaffer in the year 1984 [36]. Thereafter, no significant study was performed for almost a decade, except a revolutionary 10-line sketch of a new non-dominated sorting procedure outlined in David Goldberg's book [20]. The book came out in the year 1989. To get a clue for an efficient multi-objective optimization technique, many researchers developed different versions of multi-objective optimization algorithms [38,17,25] based on their interpretations of the 10-line sketch. The idea was so sound and appropriate that almost any such implementation has been successful in many test problems. The publication of the above-mentioned algorithms showed the superiority of evolutionary multi-criterion optimization techniques over classical methods, and since then there has been no looking back. Many researchers have modified the above-mentioned approaches and developed their own versions. Many researchers have also applied these techniques to more complex test problems and to real-world engineering design problems. To date, most of the successful evolutionary implementations for multi-criterion optimization rely on the concept of non-domination. Although there are other concepts that can be used to develop a search algorithm [28], the concept of non-domination is simple to use and understand. However, a recent study [8] has shown that search algorithms based on non-domination need not always lead to the true Pareto-optimal front. The algorithm can get stuck to a non-dominated front which is different from the true Pareto-optimal front. This exception is in the details of the implementational issues and we would not like to belabor this here, simply because in most test problems tried so far most of the search algorithms based on the non-domination concept have found the true Pareto-optimal front.

With the development of many new algorithms, many researchers have attempted to summarize the studies in the field from different perspectives [7,18,24,40,3]. These reviews list many different techniques of multi-criterion optimization that exist to date.

A web site maintained by Carlos A. Coello Coello (`http://www.lania.mx/~ccoello/EMOO/EMOObib.html`) shows that there have been at least 300 research papers written till 1998. A yearly count of those papers is plotted in Figure 2, which shows an exponential growth in interest in the field over the years.

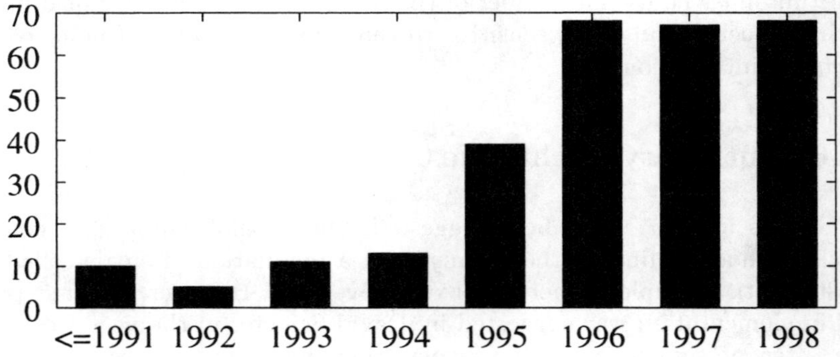

Fig. 2. Growth in number of research papers in the field of multi-objective evolutionary algorithms

We now present a brief summary of a few, salient, evolutionary multi-objective optimization algorithms.

3.1 Schaffer's VEGA

Schaffer [36] modified the simple tripartite genetic algorithm by performing independent selection cycles according to each objective. He modified the public-domain GENESIS software by creating a loop around the traditional selection procedure so that the selection method is repeated for each individual objective to fill up a portion of the mating pool. Then the entire population is thoroughly shuffled to apply crossover and mutation operators. This is performed to achieve the mating of individuals of different subpopulation groups.

The algorithm worked efficiently for some generations but in some cases suffered from its bias towards some individuals or regions. The independent selection of specialists resulted in speciation in the population. The outcome of this effect is the convergence of the entire population towards the individual optimum regions after a large number of generations. From a designer's point of view, it is not desirable to have any bias towards such middling individuals, rather it is of interest to find as many non-dominated solutions as possible. Schaffer tried to minimize this speciation by developing two heuristics—the non-dominated selection heuristic (a wealth redistribution scheme), and the mate selection heuristic (a cross-breeding scheme) [36]. In the non-dominated selection heuristic, dominated individuals are penalized by subtracting a small fixed penalty from their expected number of copies during selection. Then the total penalty for dominated individuals was divided among the non-dominated individuals and was added to their expected number of copies during selection. But this algorithm failed when the population had very few non-dominated individuals, resulting in a large fitness value for those few non-dominated points, eventually leading to a high selection pressure. The mate selection heuristic was intended to promote the cross-breeding of specialists from different subgroups. This was implemented by selecting an individual, as a mate to a randomly selected individual, which has the maximum Euclidean distance in the performance space from its mate. But it too failed to prevent the participation of poorer individuals in the mate selection. This is because of random selection of the first mate and the possibility of a large Euclidean distance between a champion and a mediocre individual. Schaffer concluded that the random mate selection is far superior to this heuristic.

3.2 Fonseca and Fleming's Multi-objective GA

Fonesca and Fleming [17] implemented Goldberg's suggestion in a different way. In this study, the multi-objective optimization GA (MOGA) uses a non-dominated sorting procedure. In MOGA, the whole population is checked and all non-dominated individuals are assigned rank '1'. Other individuals are ranked by checking the non-dominance of them with respect to the rest of the population in the following way. For an individual solution, the number of solutions that strictly dominate it in the population is first found. Thereafter, the rank of that individual is assigned to be one more than that number. Therefore, at the end of this ranking procedure, there may exist many solutions having the same rank. The selection procedure then uses these ranks to select or delete blocks of points to form the mating pool. As discussed elsewhere [21], this type of blocked fitness assignment is likely to produce a large selection pressure which might cause premature convergence. MOGA also uses a niche-formation method to distribute the population over the Pareto-optimal region. But instead of performing sharing on the parameter values, they have used sharing on objective function values. Even though this maintains diversity in the objective function values, it may not maintain diversity in the parameter set, a matter which is important for a decision maker. Moreover, MOGA may not be able to find multiple solutions in problems where different Pareto-optimal points correspond to the same objective function value [8]. However, the ranking of individuals according to their non-dominance in the population is an important aspect of this work.

3.3 Horn et al.'s Niched Pareto GA

Horn et al. [25] used *Pareto domination tournaments* instead of non-dominated sorting and the ranking selection method in solving multi-objective optimization problems. In this method, a *comparison set* comprising a specific number (t_{dom}) of individuals is picked at random from the population at the beginning of each selection process. Two random individuals are picked from the population for selecting a winner in accordance with the following procedure. Both individuals are compared with the members of the comparison set for domination with respect to objective functions. If one of them is non-dominated and the other is dominated, then the non-dominated point is selected. On the other hand, if both are either non-dominated or dominated, a niche count is found for each individual in the entire population. The niche count is calculated by simply counting the number of points in the population within a certain distance (σ_{share}) of an individual. The individual with least niche count is selected. The effect of multiple objectives is taken into the non-dominance calculation. Since this non-dominance is computed by comparing an individual with a randomly chosen population set of size t_{dom}, the success of this algorithm depends highly on the parameter t_{dom}. If a proper size is not chosen, true non-dominated (Pareto-optimal) points may not be found. If a small t_{dom} is chosen, this may result in a few non-dominated points in the population. Instead, if a large t_{dom} is chosen, premature convergence may result. This aspect is also observed by the authors. They have presented some empirical results with various t_{dom} values. Nevertheless, the concept of niche formation among the non-dominated points is an important aspect of this work.

3.4 Srinivas and Deb's Non-dominated Sorting Genetic Algorithm (NSGA)

Using the concept of sharing functions, Srinivas and Deb [38] have implemented Goldberg's idea most directly. The idea behind NSGA is that a ranking selection method is used to emphasize current non-dominated points and a sharing function method is used to maintain diversity in the population. Since we proposed this method, we describe the NSGA procedure in somewhat more detail.

NSGA varies from a simple genetic algorithm only in the way the selection operator is used. The crossover and mutation operators remain as usual. Before the selection is performed, two procedures are performed serially. First, the population is ranked on the basis of an individual's non-domination level and then the sharing function method is used to assign fitness to each individual. We describe both these mechanisms in the following subsections.

Classifying a population according to non-domination Consider a set of N population members, each having M (> 1) objective function values. The following procedure can be used to find the non-dominated set of solutions:

Step 0: Begin with $i = 1$.
Step 1: For all $j = 1, \ldots, N$ and $j \neq i$, compare solutions $x^{(i)}$ and $x^{(j)}$ for domination using two conditions (Section 2) for all M objectives.
Step 2: If for any j, $x^{(i)}$ is dominated by $x^{(j)}$, mark $x^{(i)}$ as 'dominated'.
Step 3: If all solutions (that is, when $i = N$ is reached) in the set are considered, go to Step 4, else increment i by one and go to Step 1.
Step 4: All solutions that are not marked 'dominated' are non-dominated solutions.

All these non-dominated solutions are assumed to constitute the first non-dominated front in the population. In order to find the solutions belonging to the second level of non-domination, we temporarily disregard the solutions of the first level of non-domination and follow the above procedure. The resulting non-dominated solutions are the solutions of the second level of non-domination. This procedure is continued till all solutions are classified into a level of non-domination. It is important to realize that the number of different non-domination levels could vary between one to N. Figure 3 shows how the above procedure can be used to identify four different levels of non-domination with five solutions. In this example problem, the objective function f_1 is to be maximized and the objective function f_2 is to be minimized. Thus, the result of classification is as follows: ((3,5), (4), (1), (2)), meaning that solutions 3 and 5 belong to the first level of non-domination, solution 4 belongs to the second level of non-domination, solution 1 belongs to the third level of non-domination, and solution 2 belongs to the fourth level of non-domination. It is important to note that this procedure has a computational complexity of $O(N^2)$. A more elegant approach is described elsewhere [10].

Fitness assignment In NSGA, a fitness is assigned to each individual according to its non-domination level. An individual in a higher level gets lower fitness. This is done in order to maintain a selection pressure for choosing solutions from the lower levels of non-domination. Since solutions in lower levels of non-domination are better, a selection mechanism that selects individuals with higher fitness provides a search direction towards the Pareto-optimal region.

Setting a search direction towards the Pareto-optimal region allows one of the two tasks of multi-objective optimization. Providing diversity among

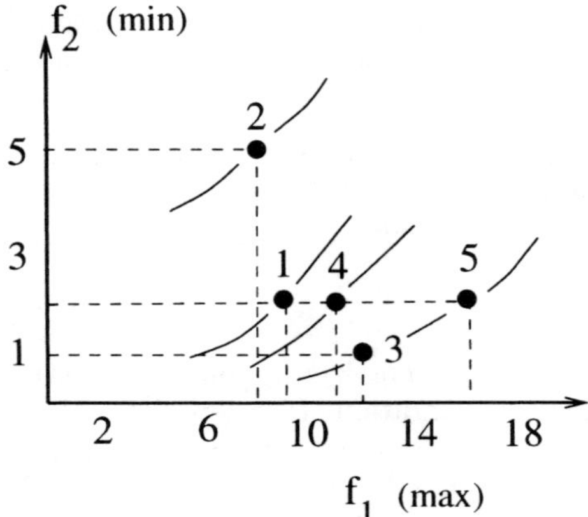

Fig. 3. Classification of a population of solutions into different non-domination levels

current non-dominated solutions is also important in order to get a good distribution of solutions in the Pareto-optimal front.

Thus, the fitness assignment is performed in two stages. First, all solutions in a particular non-domination level (or front) are assigned an identical dummy fitness. Thereafter, based on the crowding of solutions in the front, lonely solutions are emphasized by using a sharing function strategy. We discuss both of these strategies in the following.

First, all solutions in the first non-dominated front are assigned a fitness equal to the population size. This becomes the maximum fitness that any solution can have in any population. Based on the sharing strategy, if a solution has many neighboring solutions in the same front, its dummy fitness is reduced by a factor and a shared fitness is computed. The factor depends on the number and proximity of neighboring solutions. We shall describe the calculation procedure of this factor (usually known as the *niche count*) in a little while. Once all solutions in the first front are assigned their shared fitness values, the smallest shared fitness value is determined.

Thereafter, the individuals in the second non-domination level are all assigned a dummy fitness equal to a number smaller than the smallest shared fitness of the previous front. This makes sure that no solution in the second front has a shared fitness better than that of any solution in the first front. This maintains a pressure for the solutions to lead towards the Pareto-optimal region. The sharing method is again used among the individuals of second front and the shared fitness of each individual is found. This procedure is continued till all individuals are assigned a shared fitness.

Thereafter, a stochastic remainder roulette-wheel selection [20] is used to select N individuals. Single-point crossover and bit-wise mutation are used as search operators.

Another aspect of this method is that practically any number of objectives can be used. Both minimization and maximization problems can also be handled by this algorithm. The only place a change is required for the above two cases is in the way the non-dominated points are identified.

Sharing function method Given a set of n_k solutions in the k-th non-dominated front each having a dummy fitness value f_k, the sharing procedure [12] is performed in the following way for each solution $i = 1, 2, \ldots, n_k$:

Step 1: Compute a normalized Euclidean distance measure with another solution j in the k-th non-dominated front, as follows:

$$d_{ij} = \sqrt{\sum_{p=1}^{P} \left(\frac{x_p^{(i)} - x_p^{(j)}}{x_p^u - x_p^l} \right)^2},$$

where P is the number of variables in the problem. The parameters x_p^u and x_p^l are the upper and lower bounds of variable x_p.

Step 2: This distance d_{ij} is compared with a pre-specified parameter σ_{share} and the following *sharing function* value is computed [22]:

$$Sh(d_{ij}) = \begin{cases} 1 - \left(\frac{d_{ij}}{\sigma_{\text{share}}}\right)^2, & \text{if } d_{ij} \leq \sigma_{\text{share}}, \\ 0, & \text{otherwise.} \end{cases}$$

Step 3: Increment j. If $j \leq n_k$, go to Step 1 and calculate $Sh(d_{ij})$. If $j > n_k$, calculate the niche count for the i-th solution as follows:

$$m_i = \sum_{j=1}^{n_k} Sh(d_{ij}).$$

Step 4: Degrade the dummy fitness f_k of the i-th solution in the k-th non-domination front to calculate the shared fitness, f_i', as follows:

$$f_i' = \frac{f_k}{m_i}.$$

This procedure is continued for all $i = 1, 2, \ldots, n_k$ and a corresponding f_i' is found. Thereafter, the smallest value $f_k^{\min}$ of all f_i' in the k-th non-dominated front is found for further processing. The dummy fitness of the next non-dominated front is assigned to be $f_{k+1} = f_k^{\min} - \epsilon_k$, where ϵ_k is a small positive number.

The above sharing procedure requires a pre-specified parameter σ_{share}, which can be calculated as follows [12]:

$$\sigma_{\text{share}} \approx \frac{0.5}{\sqrt[P]{q}}, \tag{1}$$

where q is the desired number of distinct Pareto-optimal solutions. Although the calculation of σ_{share} depends on this parameter q, it has been been shown elsewhere [8,38] that the use of this equation with $q \approx 10$ works in many test problems. Moreover, the performance of NSGAs is not very sensitive to this parameter near σ_{share} values calculated using $q \approx 10$. Although in all simulations here σ_{share} is kept constant, they can be varied with generation according to some pre-specified rules [19].

We illustrate the working of NSGA by hand-calculating one iteration of the fitness assignment procedure on a simple test problem:

$$\begin{aligned} &\text{Minimize } f_1(x) = x^2, \\ &\text{Minimize } f_2(x) = (x-2)^2. \end{aligned} \tag{2}$$

We randomly pick six solutions as shown in the second column of Table 1. The

Table 1. A sample population and assignment of fitness

Sl. no.	x	f_1	f_2	Front	Dummy fitness	Shared fitness
1	−1.5	2.25	12.25	2	3.00	3.00
2	0.7	0.49	1.69	1	6.00	6.00
3	4.2	17.64	4.84	2	3.00	3.00
4	2.0	4.00	0.00	1	6.00	3.43
5	1.75	3.06	0.062	1	6.00	3.43
6	−3.0	9.00	25.00	3	2.00	2.00

first task is to classify the population into different non-domination classes. By using the above-mentioned procedure, we identify the following classes in ascending order of non-domination level: ((2,4,5), (1,3), (6)). This means that solutions 2, 4, and 5 belong to the first class of non-domination and there are three non-domination classes. This fact is also evident from Figure 4. Since solutions 2, 4, and 5 fall on the Pareto-optimal region, they clearly belong to the first class of non-domination. Thereafter, solutions 1 and 3 belong to the second class of non-domination and finally solution 6 belongs to the third class of non-domination. It is important to note that this figure is plotted with the objective function values.

The third and fourth columns in Table 1 show the two objective function values for each of six solutions. The fourth column shows the non-dominated front each solution belongs to. The next task is to assign a dummy fitness to each solution. First, a fitness of 6 (equal to the population size) is assigned to solutions 2, 4 and 5. Thereafter, the sharing function method is used to find the niche count of each of these three solutions. Using the distance metric, we observe the following among these three solutions:

$$d(2,4) = 1.3, \quad d(2,5) = 1.05, \quad d(4,5) = 0.25.$$

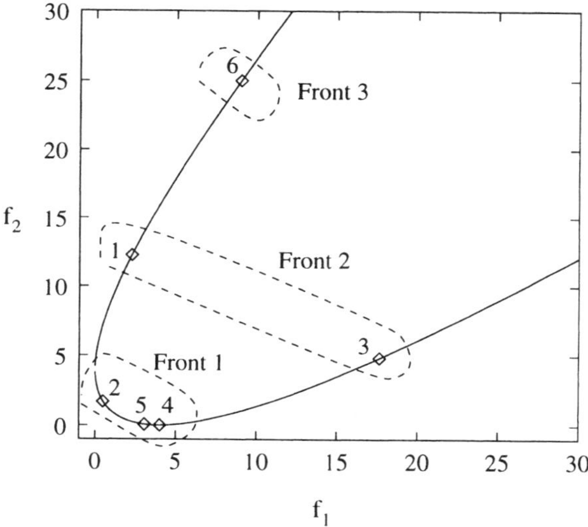

Fig. 4. Six solutions are assigned a fitness in NSGA.

Using $\sigma_{\text{share}} = 0.5$, the sharing function values are calculated as follows:

$$Sh(2,4) = 0, \quad Sh(2,5) = 0, \quad Sh(4,5) = 0.75.$$

It is clear that $Sh(2,2) = Sh(4,4) = Sh(5,5) = 1$, since the distance between any solution and the solution itself is zero. Now the niche count of each solution is as follows:

$$m_2 = 1, \quad m_4 = 1.75, \quad m_5 = 1.75.$$

Thereafter, we divide the dummy fitness value of each solution by the corresponding niche count to compute the shared fitness value, as follows:

$$f_2' = 6/1 = 6, \quad f_4' = 6/1.75 = 3.43, \quad f_5' = 6/1.75 = 3.43.$$

It is important to note that since solutions 4 and 5 are closely spaced, their shared fitness is lower compared to the lone solution 2.

Once all solutions in the first front are evaluated for a shared fitness, the minimum shared fitness is computed as 3.43. Then, all solutions in the second front are assigned a dummy fitness equal to 3 (a number smaller than 3.43). Once again a sharing procedure is adopted with $\sigma_{\text{share}} = 0.5$. Since solutions 1 and 3 are far away from each other, their shared fitness is the same as the dummy fitness. Next, a dummy fitness equal to 2 (a number smaller than 3) is assigned to the only member (solution 6) in the third front. This completes the fitness assignment procedure. The stochastic remainder roulette-wheel selection is then performed with these shared fitness values (last column of Table 1).

4 Test Results

We present the simulation results on two problems: (i) a single-variable two-objective problem and (ii) an engineering design problem.

4.1 Problem P1

We chose the above single-variable, two-objective problem given in equation 2. The initial range for the design variable used in the simulations is $(-100.0, 100.0)$. The Pareto-optimal solutions lie in $x \in [0, 2]$. The variable is coded using binary strings of size 30. We use a population size of 100. The parameter σ_{share} is set to be 0.0002 here. This value induces about N or 100 niches in the Pareto-optimal region. Figure 5 shows the population distribution in the initial generation. The figure shows that the population is widely distributed. With the above parameter settings, it is expected to have only one (out of 100) population member in the Pareto-optimal region. Figure 6 shows the population history at generation 100. All population members are now inside the Pareto-optimal region; notice how the population is uniformly distributed over the Pareto-optimal region.

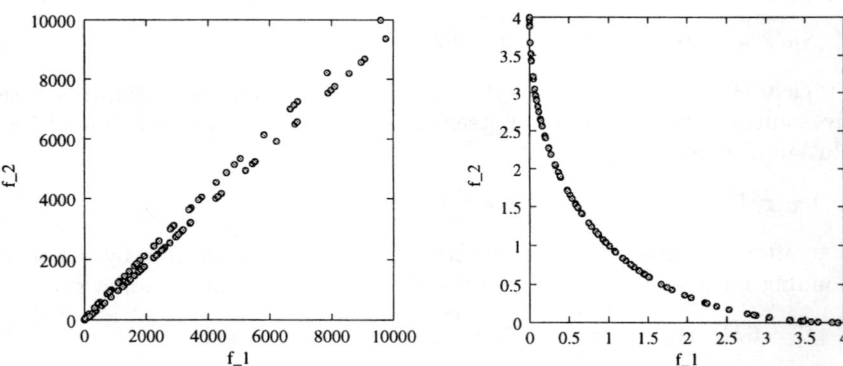

Fig. 5. Initial population for function F1

Fig. 6. Population at generation 100 for function F1

4.2 Problem P2: An Engineering Design Problem

A beam needs to be welded to another beam and must carry a certain load F (Figure 7). It is desired to find four design parameters (thickness of the beam, b, width of the beam t, length of weld ℓ, and weld thickness h) for which the cost of the beam and the deflection at the open end are minimized. The

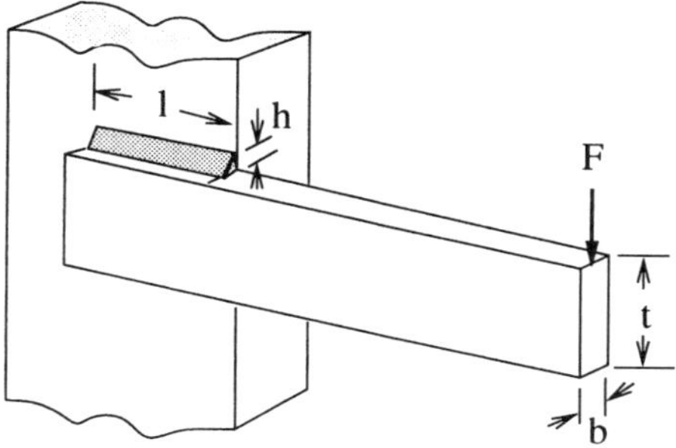

Fig. 7. The welded beam design problem. Minimizations of cost and end deflection are two objectives.

overhanging portion of the beam has a length of 14 inches and $F = 6,000$ lb force is applied at the end of the beam. It is intuitive that an optimal design for cost will make all four design variables take small values. When the beam dimensions are small, it is likely that the deflection at the end of the beam is going to be large. Again, a little thought will reveal that a design for minimum deflection at the end (or maximum rigidity of the beam) will make all four design dimensions take large dimensions. Thus, the design solutions for minimum cost and maximum rigidity (or minimum end deflection) are conflicting with each other. This kind of conflicting objective function leads to Pareto-optimal solutions. In the following, we present the mathematical formulation of the two-objective optimization problem of minimizing cost and the end deflection [14]:

$$\begin{aligned} \text{Minimize } f_1(\boldsymbol{x}) &= 1.10471h^2\ell + 0.04811tb(14.0 + \ell), \\ \text{Minimize } f_2(\boldsymbol{x}) &= \delta(\boldsymbol{x}), \\ \text{Subject to } g_1(\boldsymbol{x}) &\equiv 13,600 - \tau(\boldsymbol{x}) \geq 0, \\ g_2(\boldsymbol{x}) &\equiv 30,000 - \sigma(\boldsymbol{x}) \geq 0, \\ g_3(\boldsymbol{x}) &\equiv b - h \geq 0, \\ g_4(\boldsymbol{x}) &\equiv P_c(\boldsymbol{x}) - 6,000 \geq 0. \end{aligned} \tag{3}$$

The deflection term $\delta(\boldsymbol{x})$ is given as follows:

$$\delta(\boldsymbol{x}) = \frac{2.1952}{t^3 b}.$$

The expressions for other terms can be found in the literature [34]. The variables are initialized in the following range: $0.125 \leq (h,b) \leq 5.0$ and $0.1 \leq (\ell,t) \leq 10.0$. Constraints are handled using the bracket-operator penalty function [6]. Penalty parameters of 100 and 0.1 are used for the first and second objective functions, respectively. We use real-parameter GAs with a simulated binary crossover (SBX) operator [11] to solve this problem. Unlike binary-coded GAs, variables are used directly and a crossover operator that creates two real-valued children solutions from two parent solutions is used. For details of this crossover implementation, refer to the original studies [11,14,13]. We use a σ_{share} of 0.281 (refer to equation 1 with $P = 4$ and $q = 10$). Figure 8 shows that the population after 500 generations (marked with stars) has truly come near the Pareto-optimal front. This plot demonstrates the efficacy of NSGAs in converging close to the Pareto-optimal front with a wide variety of solutions. The dots in the figure are randomly chosen feasible solutions.

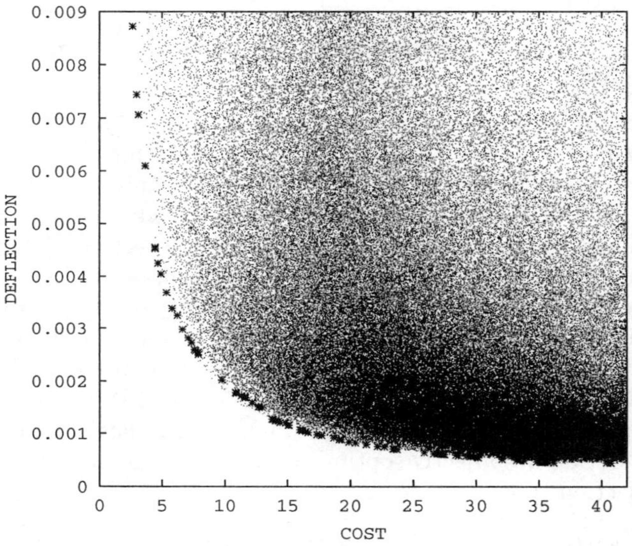

Fig. 8. Population at generation 500 shows that a wide range of Pareto-optimal solutions are found for the welded beam design problem.

5 Elitist Evolutionary Techniques

Besides the above early implementations, there exist a number of evolutionary multi-objective optimization techniques which use elitism to effect better convergence near the Pareto-optimal front. In the following subsections, we discuss three such recent algorithms.

5.1 Strength Pareto Approach (SPEA)

Zitzler and Thiele [44] have recently suggested an elitist multi-criterion EA with the concept of non-domination. They suggested maintaining an external population at every generation storing a set of non-dominated solutions discovered so far beginning from the initial population. This external population participates in genetic operations. The fitness of each individual in the current population and in the external population is decided based on the number of dominated solutions. Specifically, the following procedure is adopted. A combined population with the external and the current population is first constructed. All non-dominated solutions in the combined population are assigned a fitness based on the number of solutions they dominate. To maintain diversity and in the context of minimizing the fitness function, they assigned more fitness to a non-dominated solution having more dominated solutions in the combined population. On the other hand, more fitness is also assigned to solutions dominated by more solutions in the combined population. Care is taken to assign no non-dominated solution a fitness worse than that of the best dominated solution. This assignment of fitness makes sure that the search is directed towards the non-dominated solutions and simultaneously diversity among dominated and non-dominated solutions is maintained. On knapsack problems, they have reported better results than any other method used for comparison in that study. However, such comparisons of algorithms are not appropriate, simply because the SPEA approach uses an inherent elitism mechanism of using the best non-dominated solutions discovered up to the current generation, whereas other algorithms do not use any such mechanism. Nevertheless, an interesting aspect of their study is that it shows the importance of introducing elitism in evolutionary multi-criterion optimization. Similar effects of elitism in multi-criterion optimization were also observed elsewhere [31].

5.2 Pareto-archived Evolution Strategy (PAES)

Knowles and Corne [27] suggested a simple MOEA using a single-parent, single-child EA , similar to a (1+1)-evolution strategy. Instead of using real parameters, these authors have used binary strings and bit-wise mutations to create children. In their Pareto-archived evolution strategy (PAES) with one parent and one child, the child is compared with respect to the parent. If the child dominates the parent, the child is accepted as the next parent and the iteration continues. On the other hand, if the parent dominates the child, the child is discarded and a new mutated solution (a new child) is found. However, if the child and the parent do not dominate each other, the choice between the child and the parent is made by comparing them with an archive of best solutions found so far. The child is compared with the archive to check if it dominates any member of the archive. If yes, the child is accepted as the new parent and all the dominated solutions are eliminated from the archive. If the child does not dominate any member of the archive, both parent and child are checked for their *nearness* to the solutions in the archive. If the child resides in a least crowded region in the parameter space among the members of the archive, it is accepted as a parent and a copy is added to the archive. Crowding is maintained by deterministically dividing the entire search space into d^n subspaces, where d is the depth parameter and n is the number of decision variables, and by updating the subspaces dynamically. The authors later proposed an improved method such as the Pareto-envelope based selection algorithm or PESA, which uses the subpopulation stored in each grid location to control the selection pressure and diversity of population members [4].

5.3 Elitist Non-dominated Sorting GA (NSGA-II)

Deb et al.'s NSGA-II [15] is a modified NSGA, in which (i) a faster non-dominated sorting approach, (ii) an elitist approach, and (iii) a better crowding approach are incorporated. By using a better book-keeping strategy, the non-dominated sorting approach is reduced to $O(MN^2)$, where M is the number of objectives and the N is the population size. The elitist approach is also very simple. After the children population (of size N) is formed from the parent population, both these populations are combined and a non-dominated sorting is performed. Thereafter, solutions from better non-dominated sets are chosen one set at a time till the population is filled. In the event of inadequate available population slots to accommodate all solutions of a non-dominated set, a crowding strategy is used to identify solutions which reside in a less crowded area. This way, better solutions from the previous iteration are emphasized and a better spread of solutions is maintained by allowing less crowded solutions to remain in the population. Early on, when many non-dominated sets prevail in the population, NSGA-II provides a search direction towards the Pareto-optimal region by selecting solutions lying on better non-dominated regions. Later, when most solutions reside on or near the Pareto-optimal front, not much selective pressure comes from domination checks, rather selection pressure builds up for lone solutions. This dual procedure allows both convergence and spread of solutions near or on the Pareto-optimal front. NSGA-II has been compared with the above two elitist procedures and found to outperform both the methods in both aspects of maintaining convergence and spread on a number of difficult test problems [15].

In this chapter, we do not discuss the above procedures in detail. Instead, we discuss a number of procedures for finding a biased distribution of solutions in a desired portion of the Pareto-optimal set.

6 Introducing Bias

In the last few years of research on multi-objective optimization using evolutionary algorithms, it has been amply demonstrated that evolutionary algorithms are capable of finding multiple Pareto-optimal solutions in a single simulation run. It is then natural to ask the question: 'How does one choose a particular solution from the obtained set of Pareto-optimal solutions?' In the following, we first review a few of the techniques often followed in the context of multi-criterion decision-making and then suggest a technique which can be used during the optimization phase.

6.1 Post-optimal Techniques

Once the set of Pareto-optimal solutions is obtained, usually some higher-level decision-making considerations (often societal or political) are used to select a solution. The following methods are often used.

Compromise programming The method of compromise programming, sometimes known as the method of global criterion, selects a solution from the Pareto-optimal set which is minimally located from a reference point [42,43]. The user has to fix a distance metric $d()$ and a reference point $\boldsymbol{z}$ related to the problem. A couple of commonly used metrics are presented below:

$$L_p\text{-metric:}\quad d(\boldsymbol{f},\boldsymbol{z}) = \left(\sum_{k=1}^{M} |f_k(\boldsymbol{x}) - z_k|^p\right)^{1/p}, \tag{4}$$

$$\text{Tchebychev metric:}\quad d(\boldsymbol{f},\boldsymbol{z}) = \max_{k=1}^{M} \frac{|f_k(\boldsymbol{x}) - z_k|}{\max_{\boldsymbol{x}\in S} f_k(\boldsymbol{x}) - z_k}. \tag{5}$$

Here, S is the entire search space. The reference point is an *ideal* point formed with the individual best objective function values: $\boldsymbol{z} = (f_1^*, f_2^*, \ldots, f_M^*)$. Since this solution is usually an infeasible solution, the user will be interested in choosing a feasible solution (and Pareto-optimal) closest to this ideal solution. Figure 9 shows the reference point and the chosen solution in the Pareto-optimal set for a two-objective minimization problem. With two objectives, the L_2-metric becomes the Euclidean distance metric.

Marginal rate of return This is the amount of improvement in one objective function which can be obtained by sacrificing a unit decrement in performance in any other objective function [29]. The solution having the maximum marginal rate of return is the one chosen by this method. Since pair wise comparisons have to made with all M objectives and for each Pareto-optimal solution, this method may be computationally expensive. Figure 10 shows the preferred 'knee' point, where the marginal rate of return is maximum among a set of obtained Pareto-optimal solutions.

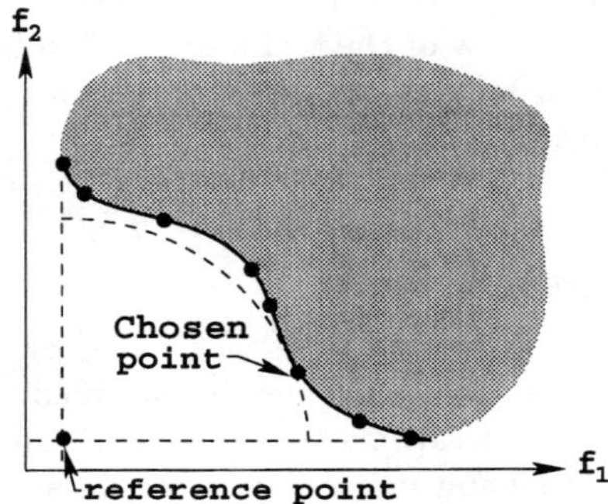

Fig. 9. The reference point and the chosen optimal solution

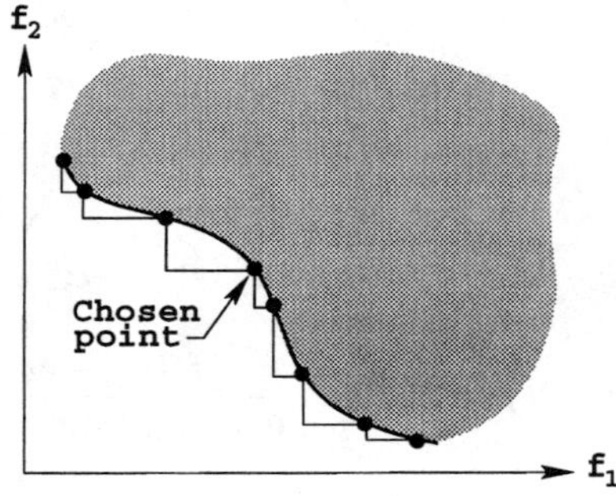

Fig. 10. The chosen optimal solution having the maximum rate of return.

Weighted average A simple strategy would be to choose a solution closer to the optimum solution corresponding to a particular user-specified weighted average of the objective functions. It is important to realize that this scheme is only applicable for identifying solutions in a convex Pareto-optimal region. The solutions inside the non-convex Pareto-optimal region cannot be found by this procedure. Figure 11 shows the chosen solution with a weighted scheme of $(w_1, w_2) = (0.1, 0.9)$. This procedure is different from the classical weighted average scheme in that here a solution is chosen after many Pareto-optimal solutions have been found. In the classical weighted average scheme, only one solution optimizing the weighted average of the objectives would be found. A different solution can be found with a different weight vector.

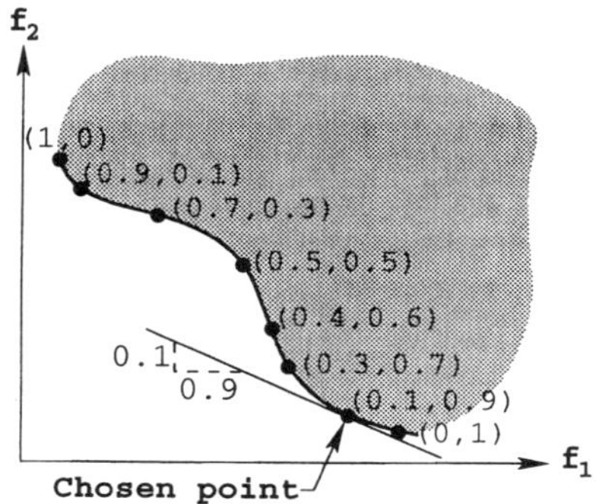

Fig. 11. The chosen optimal solution with weighted average scheme

6.2 Optimization-level Techniques

In this section, we describe a few methods which can be used during the optimization phase to find biased solutions in the Pareto-optimal region.

Utility functions Multiple objective functions can be used to form a utility function $U(\boldsymbol{f})$ [26]. The meaning of a utility function is that any two solutions having the same utility function value will have the same preference to an user. This way multiple objectives are reduced to a single objective of maximizing the utility function. It is obvious that the construction of a utility function is problem-dependent and is highly subjective to the user. However, if a utility function can be constructed, a solution maximizing the utility function can be obtained.

Guided domination approach In this approach [1], a weighted function of the objectives is defined as follows:

$$\Omega_i(\mathbf{f}(\mathbf{x})) = f_i(\mathbf{x}) + \sum_{j=1, j\neq i}^{M} a_{ij} f_j, \tag{6}$$

where a_{ij} is the amount of gain in the j-th objective function for a loss of one unit in the i-th objective. By using the above equation, a set of M weighted functions can be defined. This requires fixing the matrix a, which has a value one in its diagonal elements. The authors have defined a new domination concept as follows:

Definition 1 *A solution $x^{(1)}$ dominates another solution $x^{(2)}$ if $\Omega_i(\mathbf{f}(x^{(1)})) \leq \Omega_i(\mathbf{f}(x^{(2)}))$ for all $i = 1, 2, \ldots, M$ and the inequality is satisfied at least for one objective.*

Let us illustrate the concept for two ($M = 2$) objective functions. The two weighted functions are as follows:

$$\Omega_1(f_1, f_2) = f_1 + a_{12} f_2 \tag{7}$$

$$\Omega_2(f_1, f_2) = a_{21} f_1 + f_2. \tag{8}$$

Figure 12 shows the contour lines corresponding to the above two linear functions passing through a solution A in the objective space. All solutions

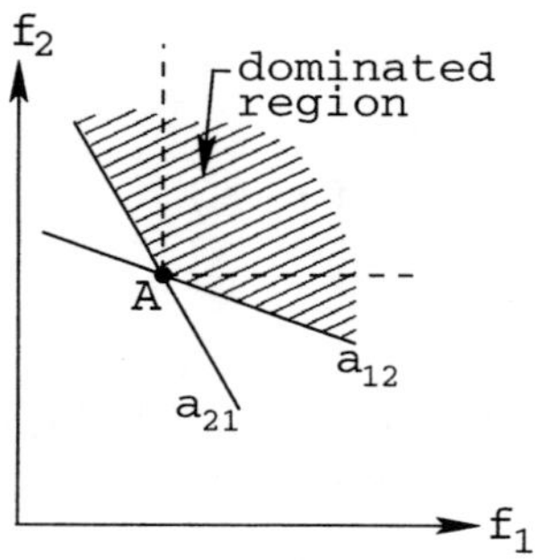

Fig. 12. The region dominated by the solution A is shown

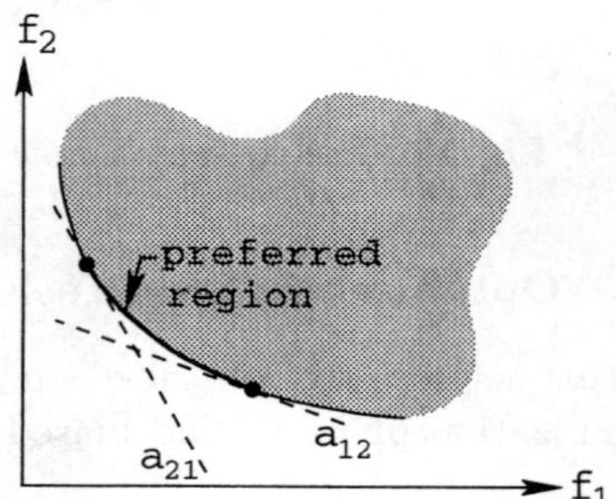

Fig. 13. The non-dominated portion of the Pareto-optimal region is also shown

in the shaded region are dominated by A according to the above definition of domination. It is interesting to note that using the usual definition of domination, the region (first quadrant) included by the horizontal and vertical dashed lines will be dominated. Thus, it is clear from the figure that the modified definition of domination allows a larger region to be dominated by any solution other than the usual definition. It is also interesting to realize that since a larger region is now dominated, not all the region in the original Pareto-optimal front becomes non-dominated according to the new definition of domination. By choosing appropriate values of elements of the a matrix, a part of the Pareto-optimal region can be emphasized. Figure 13 shows the corresponding region which will be found using the above guided domination approach. It is clear from the figure that some portion (shown by a thin solid line) of the Pareto-optimal region is dominated by a member in the middle portion of the Pareto-optimal region (shown by a thick line). Thus, an MOEA is expected to find only the middle portion of the Pareto-optimal region, thereby biasing the search towards a particular region of the Pareto-optimal front.

Although the authors viewed this approach differently, it is a simple extension of the usual multi-objective evolutionary approach with the original domination principle acting on a linearly transformed set of objective functions. A little thought will reveal that the above definition of domination on the objective vector **f** is the same as the original domination definition on the transformed vector Ω. Thus, the inability of weight-based approaches to handle problems with a non-convex Pareto-optimal region still holds for this modified approach.

Parmee et al. [33] have also suggested a weighted domination approach, where a weighted sum of objectives is used to decide domination between two solutions. It is unclear from the study how this modified domination principle helps to achieve a biased distribution. However, a niching technique which gives preference to solutions having a better weighted objective value is able to find a biased distribution of solutions in a region favored by the chosen weight vector.

In this regard, the present author's suggestion [9] of using MOEAs for solving goal programming problems also causes a biased distribution of solutions to be found. Here, the biasing depends on the chosen targets (or goals) for each objective function.

Biased sharing approach Here, we propose a sharing approach which uses a biased distance metric. In calculating the distance metric discussed in Section 3.4 in the fitness-space sharing, the following normalized distance metric was suggested:

$$d(i,j) = \left(\sum_{k=1}^{M} \frac{(f_k^{(i)} - f_k^{(j)})^2}{(f_k^{\max} - f_k^{\min})^2} \right)^{\frac{1}{2}} . \tag{9}$$

This distance metric is nothing more than the normalized Euclidean distance between two objective vectors. In the proposed biased sharing approach, an unequal weight is given to each objective in computing the Euclidean distance. For example, if w_k ($\in (0,1)$) is the weight assigned to the k-th objective function, then normalized $\boldsymbol{w}'$ is calculated as follows:

$$w'_k = \frac{(1 - w_k)}{\max_{k=1}^{M}(1 - w_k)}, \tag{10}$$

and the modified distance metric is computed as follows:

$$d(i,j) = \left(\sum_{k=1}^{M} w'_k \frac{(f_k^{(i)} - f_k^{(j)})^2}{(f_k^{\max} - f_k^{\min})^2} \right)^{\frac{1}{2}} . \tag{11}$$

The fitness-based sharing can then be used with this distance metric. Although this new distance metric requires a new calculation procedure for σ_{share}, we have used the following formula:

$$\sigma_{\text{share}} = \frac{0.5}{10^{\frac{1}{M}}}. \tag{12}$$

This way the highest-priority objective function always gets a weight (w) of one, whereas all others get a weight between zero and one. It is interesting to note that if equal weights are assigned to each objective, equation 11 reduces to equation 9.

For convex Pareto-optimal regions, a higher weight (w) for an objective function will produce more dense solutions near the individual optimum. Figure 14 explains this fact. For a two-objective optimization problem, if an extreme case of $w_1 = 0$ and $w_2 = 1$ is used, the corresponding $\boldsymbol{w}'$ is as follows: $w'_1 = 1$ and $w'_2 = 0$. Since the effect of f_2 is absent in calculating the distance metric, an equal number of solutions are expected to be created in each equal partition of the f_1 search space. Thus, the density of Pareto-optimal solutions in the partition closer to the individual best f_1^* will be less. For the non-convex Pareto-optimal region, w'_k can be calculated as $w'_k = w_k / \max_{i=1}^{M} w_k$.

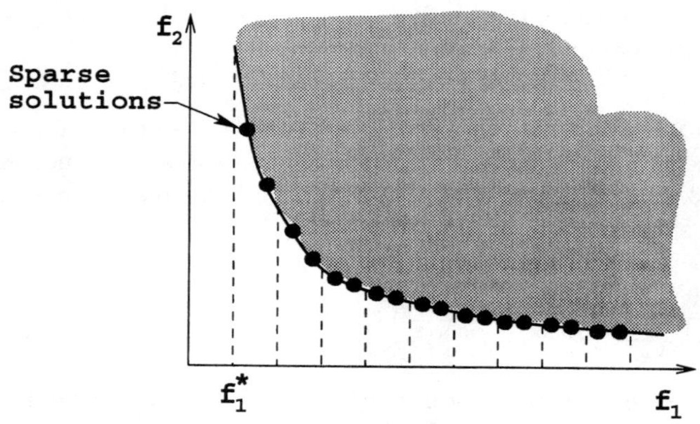

Fig. 14. The density of Pareto-optimal solutions near the individual champion solution of f_1 is less with $w_1 = 0$ and $w_2 = 1$.

In the following, we apply a modified NSGA with the above biased sharing approach to two multi-objective optimization problems (problems P1 and P2) having a convex Pareto-optimal region. In both problems, we use a real-parameter GA [11] with simulated binary crossover and a polynomially distributed mutation operator. We use a population of size 100, crossover probability of 0.9, mutation probability of $1/P$, and two spread factors [11] for crossover and mutation, $\eta_c = 30$ and $\eta_m = 500$, respectively. GAs are run till 500 generations are completed.

In problem P1, the Pareto-optimal solutions correspond to $x \in (0, 2)$. We divide this region into 10 equal divisions and count the number of Pareto-optimal solutions found in each division under different weight vectors. Figure 15 shows the number of such individuals versus x (the number of solutions in (0,0.2) are shown at $x = 0.2$ in the figure). An average of 20 runs is plotted. The figure clearly shows that when weights $w_1 = w_2 = 0.5$ are used, almost a uniform distribution is observed. But when $w_1 = 0.1$ and $w_2 = 0.9$ are used, more solutions are found closer to the $x = 2$ solution (the individual champion to the single-objective optimization problem: minimize $f_2 = (x-2)^2$). But when $w_1 = 0.9$ and $w_2 = 0.1$ are used, an opposing trend emerges. This experiment shows how by changing the weight vector the density of solutions along the Pareto-optimal front can be changed. In order to investigate how

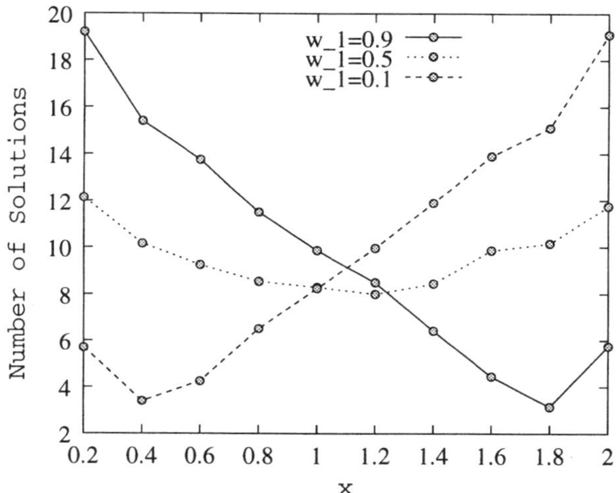

Fig. 15. Number of solutions in each partition of the Pareto-optimal region under different weight vectors for problem P1

the Pareto-optimal solutions are distributed, we have also plotted the solutions of one run in Figures 16 and 17 for $w_1 = 0.9$ and $w_1 = 0.1$, respectively. Figure 16 with $w_1 = 0.9$ shows that more solutions are near f_1^* and Figure 17 with $w_1 = 0.9$ shows that more solutions are away from f_1^*.

Although the proposed approach does not provide a single compromise solution, and instead finds a biased distribution of solutions, there is an advantage with this method. If a user wants to have a bias towards a particular objective, this biased sharing approach produces more solutions towards the preferred region in the search space. Finding more dense solutions in the region of preference is a much better approach than predefining a weighted sum of objectives and finding only one optimum solution. With a biased population, there exist many solutions in the desired region. One solution can then be chosen using the previously described methods. In case the preference of one objective over another is not known, an equal weightage can be set for

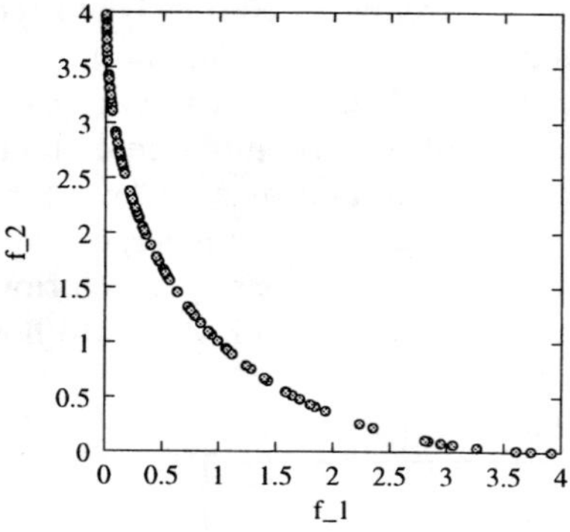

Fig. 16. Biased Pareto-optimal solutions with $w_1 = 0.9$

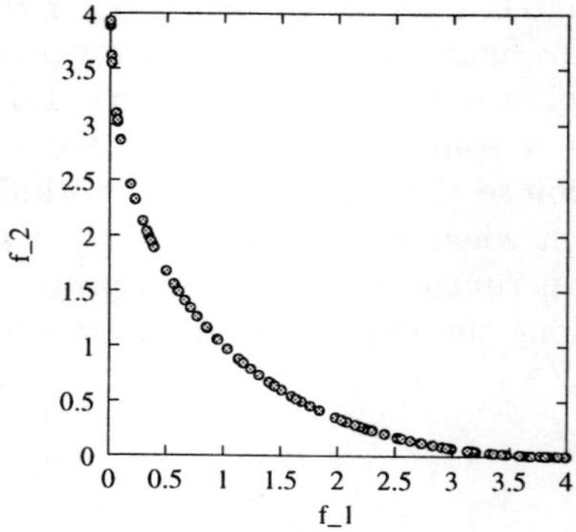

Fig. 17. Biased Pareto-optimal solutions with $w_1 = 0.1$

all objectives and a uniform distribution of solutions is expected to be found. It is also important to mention here that instead of using the weight information in the biased distance calculation, biasing of the population can also be achieved with other metrics, such as the marginal rate of return. More solutions near the largest marginal rate of return can be obtained with such a procedure.

Next, we apply the above procedure for two different weight vectors to the welded beam design problem. Figures 18 and 19 show the spread of Pareto-optimal solutions obtained for $\boldsymbol{w} = (0.9, 0.1)$ and $\boldsymbol{w} = (0.1, 0.9)$, respectively. When the weight for the cost function is higher (Figure 18), more solutions near the minimal cost solution are found. When a large weight for the deflection function is introduced (Figure 19), more solutions in the vicinity of the minimum deflection solution are obtained.

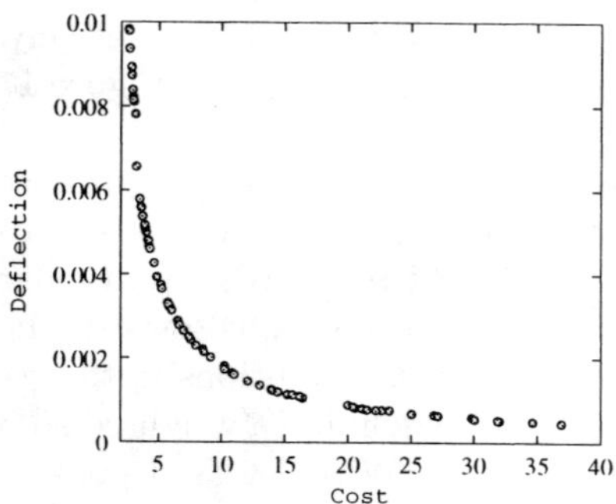

Fig. 18. Biased Pareto-optimal solutions with $w_1 = 0.9$

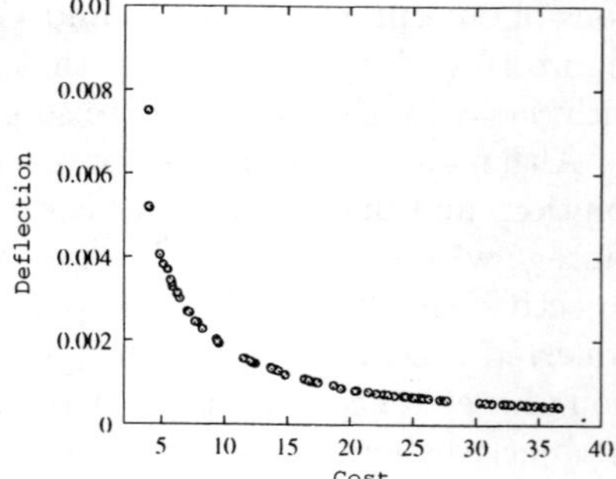

Fig. 19. Biased Pareto-optimal solutions with $w_1 = 0.1$

7 Conclusions

In this chapter, we have discussed the principles of multi-objective optimization and have argued that the presence of multiple conflicting objectives gives rise to multiple Pareto-optimal solutions, instead of one optimal solution. Therefore, there are two main goals in a multi-objective optimization problem: (i) convergence to the set of Pareto-optimal solutions, and (ii) maintenance of diversity among Pareto-optimal solutions. Since classical multi-objective optimization algorithms scale multiple objectives into a single objective, they are not efficient in satisfying the second goal described above. In the recent past, evolutionary algorithms have been efficiently used to satisfy both goals. This chapter has reviewed a few salient algorithms, particularly the non-dominated sorting GA or NSGA.

A number of considerations have been presented for selecting one solution from the obtained Pareto-optimal set. A biased sharing NSGA has been proposed and its efficacy has been demonstrated by solving a couple of multi-objective optimization problems. With the popularity of MOEAs, more studies are now needed to make them useful in practice. We believe that the biased sharing NSGA is one such algorithm which allows knowledge of the problem to be introduced into the search procedure. As a result, more solutions can be found in the desired Pareto-optimal region compared to other regions in the search space.

Acknowledgements

The author wishes to acknowledge the programming help of Shubhrajyoti Moitra and T. Meyarivan. The support from the Department of Science and Technology (DST), Government of India, during the course of this study is highly appreciated.

References

1. Branke, J., Kauβler, T., and Schmeck, H. (2000) Guiding multi-objective evolutionary algorithms towards interesting regions. Technical Report No. 399, Institute AIFB, University of Karlsruhe, Germany
2. Chankong, V., and Haimes, Y. Y. (1983) *Multiobjective decision making theory and methodology*. New York: North-Holland
3. Coello, C. A. C. (1999) A comprehensive survey of evolutionary based multi-objective optimization techniques, *Knowledge and Information Systems 1*(3), 269–308
4. Corne, D., Knowles, J., and Oates, M. (2000) The Pareto envelope-based selection algorithm for multiobjective optimization. *Proceedings of the Parallel Problem Solving from Nature VI Conference*, 839–848
5. Cunha, A. G., Oliveira, P., and Covas, J. A. (1997) Use of genetic algorithms in multicriteria optimization to solve industrial problems. *Proceedings of the Seventh International Conference on Genetic Algorithms*, 682–688

6. Deb, K. (1995) *Optimization for engineering design: Algorithms and examples.* New Delhi: Prentice Hall
7. Deb, K. (1999) Evolutionary Algorithms for Multi-Criterion Optimization in Engineering Design. In K. Miettinen, M. Mäkelä, P. Neittaanmäki, and J. Périaux (Eds.) *Proceedings of Evolutionary Algorithms in Engineering and Computer Science (EUROGEN-99)*, (pp. 135–161)
8. Deb, K. (1999) Multi-objective genetic algorithms: Problem difficulties and construction of test problems. *Evolutionary Computation Journal, 7*(3), 205–230
9. Deb, K. (2001) Nonlinear goal programming using multi-objective genetic algorithms. *Journal of the Operational Research Society, 52*(3), 291–302
10. Deb, K. (2001) *Multi-objective optimization using evolutionary algorithms.* Chichester, UK: Wiley
11. Deb, K. and Agrawal, R. B. (1995) Simulated binary crossover for continuous search space. *Complex Systems, 9*, 115–148
12. Deb, K. and Goldberg, D. E. (1989) An investigation of niche and species formation in genetic function optimization. *Proceedings of the Third International Conference on Genetic Algorithms*, 42–50
13. Deb, K. and Goyal, M. (1998) A robust optimization procedure for mechanical component design based on genetic adaptive search. *Transactions of the ASME: Journal of Mechanical Design, 120*(2), 162–164
14. Deb, K. and Kumar, A. (1995) Real-coded genetic algorithms with simulated binary crossover: Studies on multi-modal and multi-objective problems. *Complex Systems, 9*(6), 431–454
15. Deb, K., Agrawal, S., Pratap, A., Meyarivan, T. (2000) A fast elitist non-dominated sorting genetic algorithm for multi-objective optimization: NSGA-II. *Proceedings of the Parallel Problem Solving from Nature VI Conference*, Paris, 849–858
16. Eheart, J. W., Cieniawski, S. E., and Ranjithan, S. (1993) Genetic-algorithm-based design of groundwater quality monitoring system. *WRC Research Report No. 218.* Urbana: Department of Civil Engineering, The University of Illinois at Urbana-Champaign
17. Fonseca, C. M. and Fleming, P. J. (1993) Genetic algorithms for multiobjective optimization: Formulation, discussion, and generalization, *Proceedings of the Fifth International Conference on Genetic Algorithms*, 416–423
18. Fonseca, C. M. and Fleming, P. J. (1995) An overview of evolutionary algorithms in multi-objective optimization. *Evolutionary Computation, 3*(1) 1–16
19. Fonseca, C. M. and Fleming, P. J. (1998) Multiobjective optimization and multiple constraint handling with evolutionary algorithms–Part II: Application example. *IEEE Transactions on Systems, Man, and Cybernetics: Part A: Systems and Humans, 28*(1), 38–47
20. Goldberg, D. E. (1989) *Genetic algorithms for search, optimization, and machine learning.* Reading, MA: Addison-Wesley
21. Goldberg, D. E. and Deb, K. (1991) A comparison of selection schemes used in genetic algorithms. *Foundations of Genetic Algorithms I*, 69–93
22. Goldberg, D. E. and Richardson, J. (1987) Genetic algorithms with sharing for multimodal function optimization. *Proceedings of the First International Conference on Genetic Algorithms and Their Applications*, 41–49
23. Hajela, P. and Lin, C.-Y. (1992) Genetic search strategies in multi-criterion optimal design, *Structural Optimization, 4*, 99–107

24. Horn, J. (1997) Multicriterion decision making. In (T. Bäck et al., Eds.) *Handbook of Evolutionary Computation.* Bristol: Institute of Physics Publishing and New York: Oxford University Press
25. Horn, J. and Nafploitis, N., and Goldberg, D. E. (1994) A niched Pareto genetic algorithm for multi-objective optimization. *Proceedings of the First IEEE Conference on Evolutionary Computation*, 82–87
26. Keeney, R. L. and Raiffa, H. (1993) *Decisions with multiple objectives: Preferences and value tradeoffs*, Cambridge University Press
27. Knowles, J. and Corne, D. (1999) The Pareto archived evolution strategy: A new baseline algorithm for multiobjective optimisation. *Proceedings of the 1999 Congress on Evolutionary Computation*, Piscataway, New Jersey: IEEE Service Center, 98–105
28. Laumanns, M., Rudolph, G., and Schwefel, H.-P. (1998) A spatial predator-prey approach to multi-objective optimization: A preliminary study. *Proceedings of the Parallel Problem Solving from Nature V Conference*, 241–249
29. Miettinen, K. (1999) *Nonlinear multiobjective optimization.* Boston: Kluwer
30. Mitra, K., Deb, K., and Gupta, S. K. (1998) Multiobjective dynamic optimization of an industrial Nylon 6 semibatch reactor using genetic algorithms. *Journal of Applied Polymer Science, 69*(1), 69–87
31. Obayashi, S., Takahashi, S., and Takeguchi, Y. (1998) Niching and elitist models for MOGAs. *Parallel Problem Solving from Nature V Conference*, 260–269
32. Parks, G. T. and Miller, I. (1998) Selective breeding in a multi-objective genetic algorithm. *Proceedings of the Parallel Problem Solving from Nature V Conference*, 250–259
33. Parmee, I. C., Cevtković, D., Watson, A. W., and Bonham, C. R. (2000) Multiobjective satisfaction within an interactive evolutionary design environment. *Evolutionary Computation, 8*(2), 197–222
34. Reklaitis, G. V., Ravindran, A. and Ragsdell, K. M. (1983) *Engineering optimization methods and applications.* New York: Wiley
35. Rosenberg, R. S. (1967) Simulation of genetic populations with biochemical properties. PhD dissertation. University of Michigan
36. Schaffer, J. D. (1984) Some experiments in machine learning using vector evaluated genetic algorithms. Doctoral dissertation, Vanderbilt University
37. Sen, P. and Yang, J.-B. (1998) *Multiple criteria decision support in engineering design.* London: Springer
38. Srinivas, N. and Deb, K. (1994) Multi-Objective function optimization using non-dominated sorting genetic algorithms. *Evolutionary Computation, 2*(3), 221–248
39. Steuer, R. E. (1986) *Multiple criteria optimization: Theory, computation, and application.* New York: Wiley
40. Van Veldhuizen, D. and Lamont, G. B. (1998) Multiobjective evolutionary algorithm research: A history and analysis. *Report Number TR-98-03.* Wright-Patterson AFB, Ohio: Department of Electrical and Computer Engineering, Air Force Institute of Technology
41. Weile, D. S., Michielssen, E., and Goldberg, D. E. (1996) Genetic algorithm design of Pareto-optimal broad band microwave absorbers. *IEEE Transactions on Electromagnetic Compatibility, 38*(4), 518–525
42. Yu, P. L. (1973) A class of solutions for group decision problems. *Management Science, 19*(8), 936–946

43. Zeleny, M. (1973) Compromise programming. In J. L. Cochrane and M. Zeleny (Eds.), *Multiple Criteria Decision Making.* Columbia, South Carolina: University of South Carolina Press, (pp. 262–301)
44. Zitzler, E. and Thiele, L. (1998) Multiobjective optimization using evolutionary algorithms—A comparative case study. *Parallel Problem Solving from Nature V Conference*, 292–301

Gene Expression and Scalable Genetic Search*

Hillol Kargupta

School of Electrical Engineering and Computer Science
Washington State University
Pullman, WA 99164-2752, USA
E-mail:hillol@eecs.wsu.edu

Summary. Gene expression evaluates the genetic fitness of an organism through a sequence of representation transformations (DNA→mRNA→Protein). Moreover it does so in a very distributed and decomposed fashion by evaluating different portions of the DNA in order to produce various proteins in different body cells. This chapter reviews some of the recent results that underscore the possible critical role of gene expression in scalable genetic search. It considers a Fourier[1] basis representation to analyze genetic fitness functions and shows that polynomial-time construction of a decomposed representation in the Fourier basis is possible when the function has a polynomial-size description. It also points out that genetic code-like transformations may offer us a unique technique to transform some functions of exponential description in the Fourier basis to an exponentially long representation with only a polynomial number of terms that are exponentially more significant than the rest. This may be useful for a polynomial-time approximation of an exponential description. Since the construction of decomposed representation of functions from observed data plays an important role in machine learning, data mining, and black-box optimization, the role of gene expression in scalable genetic search appears quite critical.

1 Introduction

The gene expression process in nature evaluates the fitness of DNA in a very distributed and decomposed fashion through the production of different proteins in different cells. Such representation exposes the underlying decomposability of the fitness function. The principle of "divide and conquer", dividing a difficult task into smaller and easier subtasks, is a popular strategy in many disciplines. Such decomposition [13,23] is important for both scalability and distributed problem-solving capability. In many applications it is also important for the scalability of the genetic search [16,31]. Inducing such representations from observed data is at the heart of many fields such as machine learning, optimization, and data mining. In general, computation of

* This chapter reviews material from earlier published papers of the author [27,28,34].

[1] The analysis is identical to that using the Walsh basis [5,57]; however, the author chooses the term Fourier because of its historical [39,21] use in the function approximation literature.

such a representation is non-trivial. Therefore, the construction of this representation in nature is quite intriguing. It is important that we explore the mechanism in gene expression that does something which our state-of-the-art algorithms cannot do today.

A closer look at this process reveals that the gene expression process goes through a series of representation transformations (from DNA to mRNA and finally to protein). Representation transformations are used in many fields like mathematics, physics, engineering, machine learning, and others for solving problems efficiently. We should investigate if these transformations have anything to do with the efficient computation of the decomposed representation of the genetic fitness in gene expression. In short, the role of gene expression in genetic search is intriguing.

This chapter reviews a recent school of thought that explores gene expression from the perspective of computation and scalability. It first explores polynomial-time computability of a decomposed representation in an orthonormal basis. It points out that such a representation can be constructed using the Fourier basis in polynomial time if the fitness function has a polynomial-size description. Next it summarizes a recently revealed intriguing property of genetic code-like transformations. It shows that there exists a class of such representation transformations that can convert a function of exponential description in a Fourier basis to an exponentially long representation with only a polynomial number of terms exponentially more significant than the rest. It also shows that we may be able to construct such transformations by giving fitter population members more equivalent copies using a redundant representation. If that is the case, we may be able to approximate the representation using only a polynomial-size description. This essentially suggests that efficient construction of a decomposed representation of the genetic fitness function may be possible and it may be understood on rigorous computational grounds. While it is unlikely that such representations can be constructed for all functions, it is encouraging to note that they appear to work very well in nature. Therefore, the class of such functions may not be trivial.

Section 2 presents a short review of the natural gene expression process. It also suggests that this process can be viewed as an efficient mechanism to induce the representation of a function from observed data. Section 3 reviews the related work. Section 4 briefly overviews the basics of orthonormal basis representation and discusses the polynomial-time computability of a decomposed representation in the Fourier basis. It presents a recently proposed technique and related experimental results. Section 5 discusses the role of genetic code in the process of constructing decomposed fitness function representation in gene expression. Finally, Section 6 concludes this chapter.

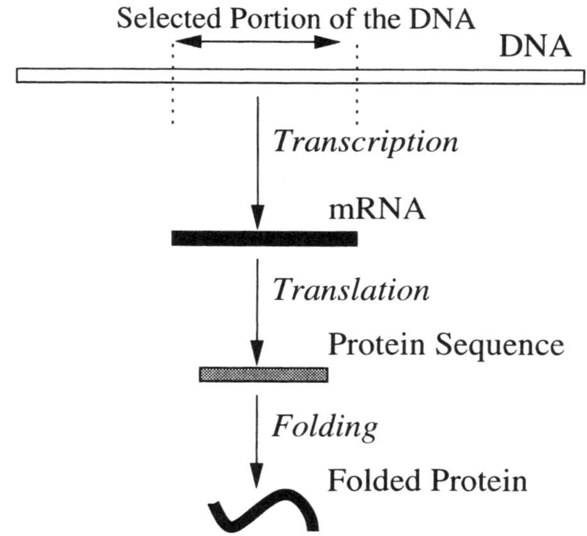

Fig. 1. Different steps of the natural gene expression process

2 Gene Expression and the Evaluation of Genetic Fitness

DNA is the primary carrier of the evolutionary information that is transmitted from one generation to another. DNA molecules consist of two long complementary chains held together by base pairs. DNA consists of four kinds of bases joined to a sugar-phosphate backbone. The four bases in DNA are *adenine* (A), *guanine* (G), *thymine* (T), and *cytosine* (C). Chromosomes are made of DNA *double helices.* Bases in DNA helices obey the *complementary base-pairing rule.* T and G pair with A and C respectively. For example, if the base at a particular position of a helix is T then the corresponding base in the other helix should be A.

DNA plays an important role in constructing the phenotypic structure of a living organism. The phenotype is traditionally viewed as the "fitness" of an organism. Therefore, construction of the phenotype can be viewed as the process of the fitness evaluation. Gene expression plays an important role in this process. It constitutes an expression of genetic information from the DNA to the mRNA sequence and construction of folded proteins from the mRNA. The main steps are

- transcription: formation of mRNA (messenger ribonucleic acid) from DNA,
- translation: formation of protein from mRNA using the genetic code, and
- protein folding.

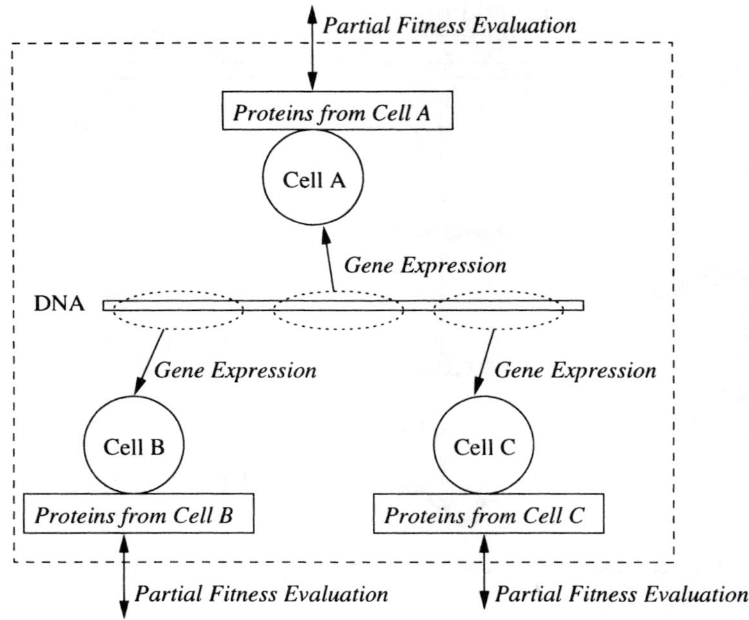

Fig. 2. Distributed, decomposed evaluation of the evolutionary fitness in gene expression

Figure 1 shows the different steps of the gene expression process. In a particular cell, transcription produces the mRNA from a small portion of the DNA. The mRNA defines another level of representation of the genetic information. It consists of four types of bases joined to a ribose-sugar-phosphodiester backbone. The four bases are *adenine* (A), *uracil* (U), *guanine* (G), and *cytosine* (C). All the bases defining the mRNA are same as those in DNA sequences, except that T is replaced by U. The mRNA is produced from the DNA by RNA polymerase and the regulatory proteins following the *complementary base-pairing rules* similar to those in DNA. The RNA Polymerase initiates the transcription at a place of the DNA marked by the *promoter* region (*start site*). It splits the DNA double helix and continues generating the mRNA using one of the DNA strands as a template. The RNA polymerase stops when it finds a termination signal sequence (*stop site*) in the DNA strand. Note that only a small portion of the DNA strand is transcribed and different cells may transcribe different regions of the DNA for producing proteins.

The mRNA acts as the template for protein synthesis. A protein is defined by a sequence of *amino acids*, joined by peptide bonds. The mRNA is transported to the cell cytoplasm for producing protein in the ribosome.

Protein feature	mRNA codons
Alanine	GCA GCC GCG GCU
Cysteine	UGC UGU
Aspartic acid	GAC GAU
Glutamic acid	GAA GAG
Phenylalanine	UUC UUU
Glycine	GGA GGC GGG GGU
Histidine	CAC CAU
Isoleucine	AUA AUC AUU
Lysine	AAA AAG
Leucine	UUA UUG CUA CUC CUG CUU
Methionine	AUG
Asparagine	AAC AAU
Proline	CCA CCC CCG CCU
Glutamine	CAA CAG
Arginine	AGA AGG CGA CGC CGG CGU
Serine	AGC AGU UCA UCC UCG UCU
Threonine	ACA ACC ACG ACU
Valine	GUA GUC GUG GUU
Tryptophan	UGG
Tyrosine	UAC UAU
STOP	UAA UAG UGA

Table 1. The universal genetic code

There exists a unique set of rules that defines the correspondence between nucleotide triplets (known as codons) and the amino acids in proteins. This is known as the *genetic code* (Table 1). Each codon is comprised of three adjacent nucleotides in mRNA chain and it produces a unique amino acid. With a few exceptions the genetic code for most eukaryotic and prokaryotic organisms is the same. The amino acid sequence defines a new representation of the information coded in the mRNA.

The next level of representation of genetic information is defined by the three-dimensional structure of folded proteins. Although amino acid sequences fundamentally define proteins, formation of the three-dimensional structure of proteins involves a complex process, often called *protein folding*. This process involves interaction between multiple amino acid subsequences and folding of the sequence into a complex structure.

Since proteins play the primary role in the performance of a living organism (i.e. phenotype), we can view the organism's fitness as a direct function of the different proteins generated from the DNA. Since different proteins are generated at different cells from different portions of the DNA, fitness evaluation in natural gene expression appears to take place in a distributed and decomposed fashion. In other words, the fitness evaluation seems to be decomposed into different partial evaluations. Figure 2 illustrates this distributed, decomposed evaluation of the evolutionary fitness. Such distributed fitness evaluation is possible under either of the two following scenarios:

1 the decomposed representation of the fitness function was available in nature a priori for some unknown reason;
2 the representation was constructed during the course of genetic search.

In the following sections we will adopt the latter possibility since there is a growing amount of evidence supporting the evolution of this representation [6,12,43]. We shall explore it in the context of our basic understanding about function induction. First we shall investigate whether the computation of such representation of a function is feasible or not. If it is, we would like to understand the requirements. Then we would like to explore if gene expression has any hidden mechanism that helps genetic search construct such a representation efficiently.

However, first let us review the existing literature in evolutionary computation that considers the computational role of gene expression.

3 Previous Work

The importance of gene expression in genetic search was realized in the early days of the field of genetic algorithms. Holland [18] described the dominance operator as a possible way to model the effect of gene expression in diploid chromosomes. He also noted the importance of the process of protein synthesis from DNA in the computational model of genetics. Despite the fact that, traditionally, dominance maps are explained from the Mendelian perspective, Holland made an interesting leap by connecting it to the synthesis of protein by gene signals, which today is universally recognized as gene expression. He realized the relation between the dominance operator and the "operon" model of the functioning of the chromosome [22] in evolution and pointed out the possible computational role of gene signaling in evolution [18].

Several other efforts have been made to model some aspects of gene expression. Diploidy and dominance-based evolutionary algorithms are proposed elsewhere [2,8,19,47,51]. Most of them took their inspiration from the Mendelian view of genetics. The under-specification and over-specification decoding operator of messy GA has been viewed as a mechanism similar to gene signaling [16]. The structured genetic algorithm [9] also shares motivations from the gene expression; it uses a structured hierarchical representation in which genes are collectively switched on and off. This provides the search algorithm with a richer representation and helps capture properties of the landscape better.

An empirical study of genetic programming using artificial genetic code is presented in [37]. The grammatical evolution-based automated programming approach [11,48,58] is another interesting effort to incorporate gene expression-based techniques for designing advanced evolutionary algorithms.

An empirical investigation of the role of such "non-coding" segments (introns) in genetic search can be found in [59]. A simulation of the expression of the genome in living cells and other related activities is reported in [38]. The role of random genotype-to-phenotype mappings in reducing the chance of being trapped in the suboptimal regions of the search space is explored in [50]. Experimental approaches to construct genetic regulatory networks are presented in [41].

Kauffman [36] offered an interesting perspective of the natural evolution that realizes the importance for gene expression. However, Kauffman's work does not relate gene expression with the computational complexity issues in genetic search on rigorous analytical grounds.

The "neutral network" theory [45,49] considers sequence-to-structure mapping from the perspective of random graph construction. This work approaches gene expression from the perspective of random graph construction and points out the existence of the neutral networks. The translation process maps multiple mRNA sequences to the same protein sequence. As a result, it creates a genetic space that contains multiple genomes with the same fitness. This work provides interesting insights into the effect of such neutral networks in genetic search. However, its contribution towards polynomial-time representation construction of the genetic fitness function is not clear.

Another related effort to understand the properties of the fitness landscape defined by the mRNA can be found in [46]. This work presents a Fourier analysis of the landscapes derived from the RNAs using fast Fourier transformation (FFT). Although the time complexity of the FFT is better than the regular Fourier transformation, it still grows exponentially with respect to the number of feature variables defining the domain of the fitness function. In contrast, the current chapter suggests that the genetic code that transforms the RNA to protein itself may help with designing a polynomial-time algorithm for the construction of the Fourier representation which the FFT cannot offer.

There also exists a body of literature that investigates the evolution of the genetic code. An algebraic model of the evolution of the genetic code is presented in [20]. This work searches for symmetries in the genetic code and points out the existence of a unique approximate symmetry group compatible with the codon assignments. The main idea behind this work is to view the evolution of the genetic code as an iterative process of representation decomposition. The genetic code is viewed as a 64-dimensional representation decomposed into several sub-representations with respect to different subgroups. The number of amino acids correspond, to the number of sub-representations and the number of codons for any amino acid corresponds to the dimension of that sub-representation. An extension of this work using Lie super-algebra is presented in [4]. Additional work on the different biological theories on the evolution of the genetic code can be found elsewhere [6,12].

An alternate approach has been developed by Kargupta and his colleagues [3,24,25,32,26,30,29,35]. This approach is mainly motivated by a perspective of the gene expression as a mechanism to make genetic search more efficient. This approach notes that the traditional model of evolutionary computation (based on selection, crossover, and mutation) [18] appears to have some serious scalability problems [54] for reasonably difficult problems. There are also little theoretical results available that prove guaranteed polynomial-time performance of existing evolutionary algorithms for reasonably difficult classes of problems. Since the existing models of evolutionary computation do not address the gene expression issue very well and gene expression changes the genetic representation, it may become a natural candidate for exploring the unknown mechanism that makes the genetic search in nature so efficient and scalable.

The early exploration of gene expression-like mechanisms for efficient inductive detection of function structure resulted in a class of heuristics-based techniques, known as the gene expression messy GA (GEMGA) [25]. In the recent past, more rigorous approaches using Fourier basis representations were suggested. Representations in Fourier bases expose the underlying function structure and they are functionally complete. Therefore, if we can learn such representations quickly the purpose of function induction is served. A randomized algorithm is presented in [35,33,34] that can induce a representation in the Fourier basis in polynomial time for problems with bounded variable interaction (BVI), i.e. bounded epistasis. The assumption of BVI makes sure that among ℓ features defining the search domain, only at most some k (a constant) number of variables can interact with each other. In other words the overall fitness function can be decomposed into a collection of either overlapping or non-overlapping sub-functions where each of the sub-functions can depend on at most k variables. This condition guarantees a polynomial-size description of the target function in Fourier representation. An alternate technique for estimating the Fourier representations is proposed elsewhere [21,39]. An extension of this technique for detecting function structure in genetic algorithms is reported in [53]. However, the author reported a scaling problem of this approach since it requires a large number of function evaluations for correctly estimating the coefficients.

The following sections review the scalable randomized algorithm reviewed in [35,33] and presents experimental results to demonstrate that this can be used for constructing a decomposed representation in the Fourier basis in polynomial time.

4 Construction of Decomposed Representation Using Fourier Basis

Representation of a function in a decomposed and distributed form can be realized and understood using an appropriately chosen set of basis functions.

Although we do not know the choice of basis functions in nature, we can still study this natural process by choosing a basis set that is functionally complete. Such a basis set is capable of representing any function that can be represented using any other basis set (including the unknown choice of nature). This chapter considers the Fourier basis set that is functionally complete over the space of all real-valued functions defined in the domain of binary strings. Although this chapter restricts itself to boolean strings, its extension to non-boolean domains with complex-valued basis functions is in our future agenda. The following section presents a brief review of the Fourier basis.

4.1 A Brief Review of the Fourier Basis

Fourier bases are orthogonal functions that can be used to represent any function. In this section we shall consider functions of binary variables. Consider the function space over the set of all ℓ-bit boolean feature vectors. The Fourier basis set that spans this space is comprised of 2^ℓ Fourier functions. Each Fourier basis function is defined as follows:

$$\psi_{\mathbf{j}}(\mathbf{x}) = (-1)^{(\mathbf{x}\cdot\mathbf{j})} \tag{1}$$

where $\mathbf{j}$ and $\mathbf{x}$ are binary strings of length ℓ. In other words $\mathbf{j} = j_1, j_2, \cdots j_\ell$, $\mathbf{x} = x_1, x_2, \cdots x_\ell$ and $\mathbf{j}, \mathbf{x} \in \{0,1\}^\ell$; $\mathbf{x}\cdot\mathbf{j}$ denotes the inner product of $\mathbf{x}$ and $\mathbf{j}$. $\psi_{\mathbf{j}}(\mathbf{x})$ can either be equal to 1 or -1. The string $\mathbf{j}$ is called a *partition*. The *order* of a partition $\mathbf{j}$ is the number of 1-s in $\mathbf{j}$. A Fourier basis function depends on some x_i only when the corresponding $j_i = 1$. Therefore a partition can also be viewed as a representation of a certain subset of x_i-s; every unique partition corresponds to a unique subset of x_i-s. If a partition $\mathbf{j}$ has exactly α number of 1-s then we say the partition is of order α since the corresponding Fourier function depends on only those α number of variables corresponding to the 1-s in the partition $\mathbf{j}$. Fourier bases are orthonormal. Therefore,

$$\begin{aligned}\frac{1}{2^\ell}\sum_{\mathbf{x}} \psi_{\mathbf{i}}(\mathbf{x})\psi_{\mathbf{j}}(\mathbf{x}) &= 1 \quad \text{when} \quad \mathbf{i} = \mathbf{j} \\ &= 0 \quad \text{when} \quad \mathbf{i} \neq \mathbf{j}\end{aligned}$$

A function $f : \mathbf{X}^\ell \rightarrow \mathcal{R}$, that maps an ℓ-dimensional space of binary strings to a real-valued range, can be expressed using the Fourier basis functions as follows :

$$f(\mathbf{x}) = \sum_{\mathbf{j}} w_{\mathbf{j}}\psi_{\mathbf{j}}(\mathbf{x}) \tag{2}$$

where $w_{\mathbf{j}}$ is the Fourier coefficient (FC) corresponding to the partition $\mathbf{j}$,

$$w_{\mathbf{j}} = \frac{1}{2^\ell}\sum_{\mathbf{x}} f(\mathbf{x})\psi_{\mathbf{j}}(\mathbf{x}) \tag{3}$$

We note from Equation 2 that a function can be expressed as a linear sum of the Fourier functions, each weighed by the corresponding Fourier coefficient. The Fourier coefficient $w_{\mathbf{j}}$ can be viewed as the relative contribution of the partition $\mathbf{j}$ to the function value of $f(\mathbf{x})$. Therefore, the absolute value of $w_{\mathbf{j}}$ can be used as the "significance" of the corresponding partition $\mathbf{j}$. If the magnitude of some $w_{\mathbf{j}}$ is very small compared to other coefficients then we may consider the $\mathbf{j}$-th partition to be insignificant and neglect its contribution.

Fourier bases and their close relatives Walsh bases are frequently used to study the behavior of genetic algorithms(GAs). Walsh bases [5,57] were first used by Bethke [7] for analyzing GAs. Further investigation of this approach can be found elsewhere [10,14,15,17,40,42,44,52,55,56]. Although the Walsh basis probably enjoys more familiarity in the evolutionary computation community, this chapter will use the Fourier terminology because of its historical [39,21] use in the function approximation literature.

A representation of the genetic fitness function in the Fourier basis defines the function as a weighted linear summation of the basis functions. Therefore, if we can construct such a representation then the evaluation of the fitness can be decomposed into different components similar to what happens in natural gene expression. Unfortunately, construction of the Fourier representation using the state-of-the-art FFT technique can potentially take exponential time in the number of feature variables defining the domain. So efficient construction of decomposed representations may not be realistic unless we can demonstrate that this can be done efficiently, possibly under some practical conditions. The following section considers this issue.

4.2 Polynomial-time Construction of Decomposed Representation

As we saw in the previous section (Equation 3) computation of a single Fourier coefficient requires information about all the 2^ℓ domain members. Clearly this cannot be done in time polynomial in ℓ. In order to make the problem tractable let us assume that the function has BVI of order k. Although this restriction is now introduced out of necessity for keeping the computation tractable, latter sections of this chapter will show that the representation transformations in gene expression appear to construct representations that can satisfy this condition.

Even if the problem has bounded order of interaction, explicit computation of FCs using Equation 3 may require exponential time. Another possibility is to generate $n = \sum_{z=0}^{k} \binom{\ell}{z}$ different members from the domain and solve a large $(n \times n)$ linear system of equations in order to compute the n possibly non-zero terms in the Fourier representation. Although it is theoretically possible in polynomial time, it is not clear how and if at all the genetic search is equipped with any mechanism to do this. In the following discussion we explore an alternate way to compute the FCs in an efficient and distributed manner.

From Equation 2 we can write,

$$f(\mathbf{x})\psi_{\mathbf{i}}(\mathbf{x}) = \sum_{\mathbf{j}} w_{\mathbf{j}}\psi_{\mathbf{j}}(\mathbf{x})\psi_{\mathbf{i}}(\mathbf{x})$$

Let $\mathbf{i}$ be a partition and $S(\mathbf{i})$ be the set of all strings that satisfies the following conditions: (1) every member of $S(\mathbf{i})$ has the same values at all positions where there is a 0 in $\mathbf{i}$ and (2) the substring defined by the positions corresponding to 1-s in $\mathbf{i}$ is unique in every member of $S(\mathbf{i})$. Let us denote the invariant values at the positions, corresponding to 0-s in $\mathbf{i}$, by $\mathbf{T}$ and call it the *template*. For example, if $\mathbf{i} = 001100$ then one possible choice of S may be $\{000000, 001000, 000100, 001100\}$. $|S(\mathbf{i})|$ is the size of the set $S(\mathbf{i})$ and $|S(\mathbf{i})| = 2^k$ when the partition $\mathbf{i}$ is of order k. In this case the template $\mathbf{T} = 00\_\_00$. Since there exists different such $S(\mathbf{i})$-s depending on the choice of $\mathbf{T}$, we may choose to use the symbol $S_{\mathbf{T}}(\mathbf{i})$ to denote the set $S(\mathbf{i})$ with respect to certain template $\mathbf{T}$, wherever needed. Now we can write,

$$\sum_{\mathbf{x}\in S_{\mathbf{T}}(\mathbf{i})} f(\mathbf{x})\psi_{\mathbf{i}}(\mathbf{x}) = \sum_{\mathbf{j}} w_{\mathbf{j}} \sum_{\mathbf{x}\in S_{\mathbf{T}}(\mathbf{i})} \psi_{\mathbf{j}}(\mathbf{x})\psi_{\mathbf{i}}(\mathbf{x})$$
$$= \sum_{\mathbf{j}\in J(\mathbf{i})} w_{\mathbf{j}} \sum_{\mathbf{x}\in S_{\mathbf{T}}(\mathbf{i})} \psi_{\mathbf{j}\oplus\mathbf{i}}(\mathbf{x})$$
$$\sum_{\mathbf{j}\in J(\mathbf{i})} w_{\mathbf{j}} \sum_{\mathbf{x}\in S_{\mathbf{T}}(\mathbf{i})} \psi_{\mathbf{j}\oplus\mathbf{i}}(\mathbf{x}) = \sum_{\mathbf{x}\in S_{\mathbf{T}}(\mathbf{i})} f(\mathbf{x})\psi_{\mathbf{i}}(\mathbf{x}) \quad (4)$$

where $J(\mathbf{i})$ denotes the set of all partitions that completely subsumes partition $\mathbf{i}$. In other words, every partition in $J(\mathbf{i})$ must have a 1 at every location where there is a 1 in $\mathbf{i}$. $\mathbf{j} \oplus \mathbf{i}$ denotes the partition defined by the boolean XOR between $\mathbf{j}$ and $\mathbf{i}$. The following part of this section presents an algorithm to construct the decomposed Fourier representation by evaluating Equation 4 for a set of randomly generated templates.

A random choice of template generates a particular assignment of signs of the corresponding Fourier term in Equation 4. For the special case $\mathbf{T} = \mathbf{0}$, every $\psi_{\mathbf{j}\oplus\mathbf{i}}(\mathbf{x}) = 1$. If all the coefficients corresponding to partitions that subsume a given partition $\mathbf{i}$ are zero then the left hand side of Equation 4 must produce a zero for any choice of templates. If it returns a non-zero value then at least one of the coefficients in $S_{\mathbf{T}}(\mathbf{i})$ is non-zero. The proposed algorithm makes use of this observation.

The algorithm searches for significant coefficients in the lattice of all partitions. The main steps are as follows:

1 Consider the lattice of all partitions, partially ordered based on the order of the partitions.
2 Start from the partition with zero order, i.e. the one with no fixed bit.
3 Traverse the lattice using a chosen order (e.g. depth-first or breadth-first). Initialize $\alpha = 0$. At each node do the following m number of times:
   (a) Randomly generate a template.

(b) Compute $\sum_{\mathbf{x}\in S_{\mathbf{T}}(\mathbf{i})} f(\mathbf{x})\psi_{\mathbf{i}}(\mathbf{x})$ at each node; if it returns a non-zero significant number then exit loop.

(c) Otherwise, set $\alpha = \alpha + 1$.

4 If $\alpha < m$ continue traversing new nodes using the chosen search strategy.

5 If $\alpha = m$ discard all the nodes subsumed by the current node from the scope of future search for nodes with significant coefficients.

The parameter m is user given. When m is large enough such that every possible template is considered for the test, Equation 4 is guaranteed to detect any non-zero coefficient corresponding to partitions that subsume the partition $\mathbf{i}$. The reason behind this is the following. Note that for an ℓ-bit problem and an order-k partition $\mathbf{i}$, there are $2^{\ell-k}$ possible choices of templates. On the other hand there are $2^{\ell-k}$ different partitions that subsume the partition i and therefore there are that many Walsh coefficients (WCs). Clearly, we have $2^{\ell-k}$ unknown WCs and $2^{\ell-k}$ different templates can generate that many equations. If all the tests or all the different templates return zero then we have a set of $2^{\ell-k}$ homogeneous equations. In that case the unknown WCs must take only the trivial solution, i.e. all of them are zeros. Although the deterministically correct solution can be obtained only when we enumerate all the templates, a small value of m appears to suffice for many practical cases because of the following reason. Let ν be the proportion of templates at a given node for which $\sum_{\mathbf{x}\in S_{\mathbf{T}}(\mathbf{i})} f(\mathbf{x})\psi_{\mathbf{i}}(\mathbf{x}) = 0$ despite the existence of non-zero contributing coefficients in the summation. Therefore the probability that all of the m experiments will return zero and thereby discard the node is ν^m. If $\nu < 1$ then the probability of such mistakes goes down exponentially with m. So a small value of m should be sufficient for all practical purposes. Randomized sampling of a few templates was sufficient for all the experimental results performed so far [33]. Next, let us illustrate this algorithm using a toy example.

Consider the lattice of all 16 partitions for a four-bit partition space as shown in Figure 3. We are interested in a function for which only the partitions, encircled with solid lines, have non-zero FCs. The proposed algorithm first performs the significance test for all nodes for order-1 partitions since the case for the order-0 partition is trivial. For the order-1 partitions, the test will result in non-zero value, at each of these nodes, indicating that the partition subsumes some partitions with non-zero FCs. Next the algorithm considers the nodes at the third level from the top in Figure 3. Since all the order-2 partitions subsume some set of order-1 partitions and all the order-1 partitions return non-zero value, for our test, we need to apply our test at every node corresponding to the order-2 partitions. Our test will return non-zero values only for 1100, 1001, and 0101. Next we consider every unique order-3 partition that subsumes at least one of the partitions 1100, 1001, and 0101 but do not subsume any of the partitions 1010, 0110, and 0011. There is only one such partition and that is 1101. We again apply our test at this node and observe that the test returns a zero value. Next we backtrack to the order-2

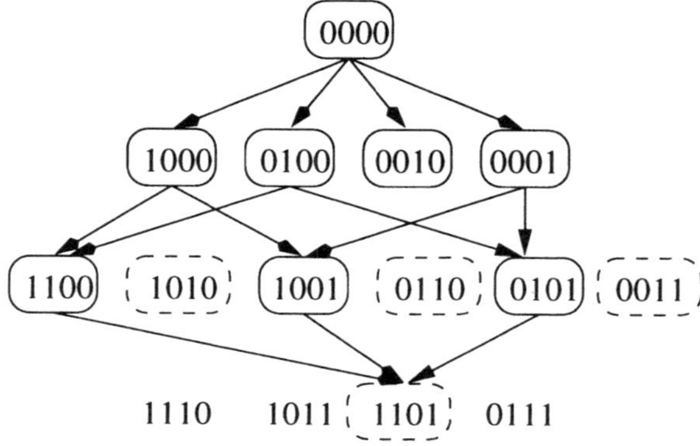

1111

Fig. 3. Efficient detection of Fourier coefficient

partition level and note that the outcomes of the tests that we performed earlier at each of these nodes are essentially the FC of the corresponding node since the coefficient of the partition 1101 is zero. Similarly, we continue to backtrack and compute the coefficients of every non-zero partition. This technique requires $O(2^k r)$ objective function evaluations, where r is the total number of non-zero FCs in its Fourier representation. For problems where r is $O(\ell^2)$ and k is a constant, the total number of objective function evaluations is $O(\ell^2)$. This is because the algorithm requires consideration of all nodes corresponding to the order-2 partitions unless some of the feature variables do not interact with any other variables.

This algorithm shares similarities with the KM-approach proposed elsewhere [39]. Both of them explore the space of all partitions in a kind of divide-and-conquer approach; both of them are randomized algorithms. However, the proposed algorithm imposes a different lattice structure in the space of all partitions that draws motivations from the traditional notion of schema and templates. Unlike the KM-approach, the current approach simply tries to detect the existence of any non-zero coefficient down the path rooted at the current node. This is different from the difficult task of estimating the sum of the squares of the coefficients (adopted by the KM-approach).

The following section considers the application of the proposed algorithm in detecting the decomposed structure of two classes of test functions.

4.3 Experimental Results

This section considers the task of constructing the decomposed representation of two classes of functions that satisfy the BVI property. The results presented

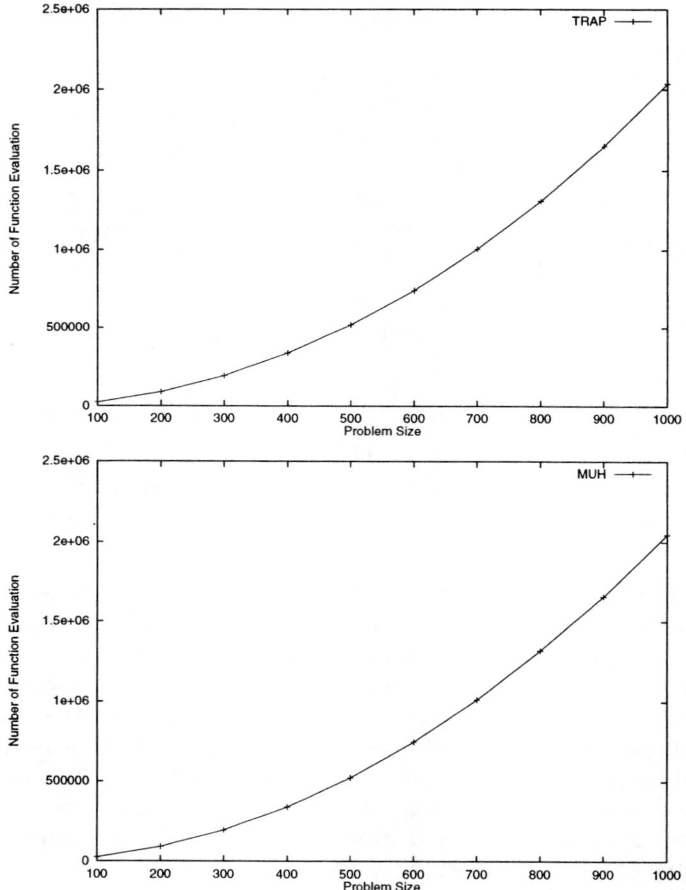

Fig. 4. The number of function evaluations vs. problem size in detecting linkage in the case of (Top) uniformly scaled non-overlapping TRAP and (Bottom) overlapping Muehlenbein function using a selected template

in this section demonstrate that the proposed randomized algorithm can be successfully used for efficiently detecting the function structure.

Experiment design The experimental test-bed is comprised of problems with BVI. All the objective functions are decomposable to a set of overlapping and non-overlapping sub-functions. No prior knowledge about such decomposability is provided to the algorithm. The randomized algorithm is used for generating the structure of the decomposition. A template of all 1-s and all 0-s is chosen for the experiments. Although this small subset of templates

f(x)		
f(x)	= 4	if x = 00000
	= 3	if x = 00001
	= 2	if x = 00011
	= 1	if x = 00111
	= 0	if x = 01111
	= 3.5	if x = 11111
	= 0	otherwise.

Table 2. Mühlenbein function

minimizes the number of fitness function evaluations, it does not change the overall time complexity. Detailed experiments with uniformly selected template sets are reported elsewhere [33].

TRAP: A Deceptive function The k-bit deceptive trap [1] function is defined as $f(x) = k \quad \text{if} \quad u = k$; $f(x) = k - 1 - u \quad$ otherwise; where u is the unitation variable, or the number of 1-s in the string x, and k is the number of variables. If we carefully observe this TRAP function, we shall note that it has two peaks. One of them corresponds to the string with all 1-s (the global optima) and the other is the string with all 0-s. The objective function is constructed by concatenating several deceptive TRAP functions one after another, where the overall objective function is a linear summation of several smaller TRAP functions. For $\ell = 200$, and $k = 5$, the overall function contains 40 sub-functions; therefore, an order-5 bounded 200-bit problem has 2^{40} local optima, and among them, only one is globally optimal. As the problem length increases the number of local optima exponentially increases. Optimizing this class of functions is hard without knowledge of the decomposed structure. The algorithm described earlier in this section is used to construct the decomposed representation. Figure 4 (Top) shows the growth of the number of objective function evaluations with respect to increasing number of search variables.

MUH: The Mühlenbein function This function is defined in Table 2. The global optimum is the string of all 0-s. Figure 4 (Bottom) shows the growth of the number of objective function evaluations with respect to increasing number of search variables.

The experimental results presented here clearly demonstrate the scalable performance of the proposed randomized algorithm. However, this works only when the function has a polynomial-size representation. Although there exist many functions with this characteristic, it is not clear why the natural genetic fitness function should have such a property. The following section considers this issue. It shows that genetic code-like transformations may be able to construct a Fourier representation of the fitness function that can be approximated using a polynomial-size description. The following section also discusses this possibility.

5 Exploring Genetic Code-like Transformations

Earlier sections of this chapter considered the polynomial-time computability of a decomposed representation in the Fourier basis. Since the representation of the genetic fitness function is also decomposed and distributed we decided to explore the computability of such a representation in an orthonormal basis. However, that discussion did not directly relate any of the representation transformations in gene expression to the algorithmic construction of such a decomposed representation in the Fourier basis. This section considers the translation process in gene expression that uses the genetic code to transform the mRNA representation to the protein sequence. It summarizes some of the recent results that show that such a process may play a critical role in the polynomial-time computation of a decomposed representation in the Fourier basis.

The genetic code transforms the mRNA sequence to the protein sequence by assigning one protein feature for every codon in the mRNA sequence. Although the cardinalities of their alphabet sets are more than two, understanding their underlying computation may require abstracting the process. In this section we will do so by assuming that the protein and the mRNA sequences are binary strings. Our objective is to explore the effect of the genetic code-like representation transformations in the binary domain using Fourier analysis. In order to do that, first we need to define what we mean by genetic code-like transformations. The following section does that.

5.1 The Notion of Genetic Code-like Transformations

Protein feature	mRNA codon
0	100
0	000
0	001
0	010
0	111
0	101
0	110
1	011

Table 3. Code B: A genetic code-like transformation for binary representation. A single bit in the protein space maps to three-bit codons in the mRNA space. Note that seven unique mRNA codons map to the protein feature value of zero

The genetic code defines the correspondence between an mRNA codon and a protein feature value. Although in nature the codons are defined by three mRNA feature values, the implication of the choice of number "three" is yet to be explained. Therefore, the current analysis will treat this as a parameter and the results of this chapter can be specialized for any size of codons including three. As noted earlier, the analysis considers the effect of such transformations in the binary space. Although strings are binary we will continue to use the terms mRNA, protein, and genetic code accordingly for maintaining the link between biology and the current analysis.

Let us use $\mathbf{r}$ and $\mathbf{p}$ to represent the mRNA and the protein sequences respectively. Let ℓ_r and ℓ_p be their respective lengths. Just like the natural *translation* process, our artificial translation maps the mRNA sequence to the corresponding protein sequence using the genetic code. The mapping in translation will be denoted by η_c where the subscript c denotes the number of mRNA features that define a codon. If three features are used like natural codons, $c = 3$; η_c can be defined as $\eta_c : R^{\ell_r} \rightarrow P^{\ell_p}$. R^{ℓ_r} and P^{ℓ_p} denote the ℓ_r -and ℓ_p- dimensional space of all mRNAs and proteins respectively. Note that $\ell_r = c\,\ell_p$ and for binary representation $R = P = \{0, 1\}$.

Consider the genetic code presented in Table 3. Note that the genetic code may be redundant. In other words, a unique protein feature value may be produced by several mRNA codons. This is also true for natural genetic code (Table 1). As a result, there exist many equivalent mRNA sequences that produce the same protein sequence. All these mRNA sequences have the same genetic fitness since they all map to the same protein sequence. So we can view the space of mRNAs grouped into different equivalence classes. We shall call this characteristic *translation introduced equivalence* (TIE) and these groups of equivalent mRNAs will be called the TIE classes. Let R_p be the TIE class for the protein sequence $\mathbf{p}$. We can also define R_p in the following manner: $R_p = \{\mathbf{r}_j | \mathbf{r}_j \stackrel{\eta_c}{\rightarrow} \mathbf{p}\}$. The cardinality of the set R_p depends on the genetic code and the protein sequence $\mathbf{p}$. Let a_0 and a_1 be the total number of codons that map to a protein feature value of 0 and 1 respectively. Let $\ell_{p,0}$ and $\ell_{p,1}$ be the number of 0-s and 1-s in $\mathbf{p}$ respectively. Then the cardinality of the TIE class is $|R_p| = a_0^{\ell_{p,0}} a_1^{\ell_{p,1}}$.

Since one feature in the protein sequence maps to c mRNA features, partitions defined in the mRNA and the protein spaces can be associated with each other. If $\mathbf{j}$ and $\mathbf{j}'$ are partitions in the mRNA and the protein spaces respectively then $\mathbf{j}'$ is the *reflection* of $\mathbf{j}$ in the protein space when $\mathbf{j}_i^{'} = 1$ if and only if $\mathbf{j}$ takes a value of 1 at the location(s) corresponding to at least one of the mRNA features associated with $\mathbf{j}_i^{'}$. For example, the reflection of the partitions 101000 and 100110 under a genetic code of codon size 3 are 10 and 11 respectively. Note that different mRNA partitions may have the same reflection in the protein space. If q is the number of 1-s in $\mathbf{j}'$ then it is the reflection of $(2^c - 1)^q$ different partitions in the mRNA space. The number of 1-s in $\mathbf{j}'$ will be called the *absolute order of partition* $\mathbf{j}$.

Once the protein sequence is constructed from the mRNA sequence, the protein folds into a three-dimensional structure and its shape determines its fitness. Let us use $f : P^{\ell_p} \rightarrow \Re^+$ for denoting this fitness function that maps the protein sequence to a non-negative real-valued range. Since the protein sequences are produced from the mRNA sequences, we can also define the fitness over the domain of mRNA sequences. Let $\phi : R^{\ell_r} \rightarrow \Re^+$ be this fitness function defined over the mRNA representation. Therefore, $\phi(\mathbf{r}) = f(\mathbf{p}) = f(\eta_c(\mathbf{r}))$. Therefore, $\phi(\mathbf{r})$ can be viewed as a different representation of the genetic fitness function $f(\mathbf{p})$.

In this section we shall study the representations of $f(\mathbf{p})$ and $\phi(\mathbf{r})$. We will be particularly interested in the effect of the representation transformation η_c on the complexity of describing the function. In other words, we would like to know if $\phi(\mathbf{r})$ has a more concise description compared to that of $f(\mathbf{p})$.

The following section explores the change in the properties of the Fourier coefficients under the genetic code-like representation transformations.

5.2 Exponential Decay of Individual Fourier Coefficients

The j-th FC in the mRNA space can be defined as

$$w_{\mathbf{j}} = \frac{1}{2^{c\,\ell_p}} \sum_{\mathbf{p}} f(\mathbf{p}) \sum_{\mathbf{r_i} \in R_p} \psi_{\mathbf{j}}(\mathbf{r_i}) \tag{5}$$

The magnitude of the second summation in the above expression may take a value in between 0 and $a_0^{\ell_{p,0}} a_1^{\ell_{p,1}}$ (cardinality of R_p) depending upon the nature of the set R_p. This imposes a scaling factor to the contribution of every unique protein sequence to the j-th FC.

Let us represent $\mathbf{j}$ using a collection of partitions $\{\mathbf{j}_0, \mathbf{j}_1, \cdots \mathbf{j}_q\}$ where $\mathbf{j}_0$ represents the null partition with all 0-s and every $\mathbf{j}_{i \neq 0}$ represents a sub-partition of the 1-contributing positions of $\mathbf{j}$ that contains only those features that belong to the same protein feature. Note that the reflection of any $\mathbf{j}_{i \neq 0}$ in the protein space has only one 1.

If p_α is the protein feature value in a given $\mathbf{p}$ corresponding to the reflection of the α-partition in the mRNA space then let R_{p_α} be the set of all mRNA codons that maps to p_α. Let $R_{j',p}$ be the Cartesian product of R_{p_α}-s for $\alpha = 1, 2, \cdots q$. Every member of $R_{j'}$ has $c\,q$ mRNA features. For example, consider $\mathbf{p} = 110$ and $\mathbf{j} = 110000010$. So $\mathbf{j}_0 = 000000000$, $\mathbf{j}_1 = 110000000$, and $\mathbf{j}_2 = 000000010$. In the case of code A, $R_{p_1} = \{100, 000, 001, 010\}$ and $R_{p_2} = \{111, 101, 110, 011\}$. Therefore $R_{101,110} = R_{p_1} \times R_{p_2}$.

Let $q_{p,j',0}$ and $q_{p,j',1}$ be the number of 0-s and 1-s in $\mathbf{p}$ that are covered by the fixed bits of $\mathbf{j}'$, the reflection of $\mathbf{j}$ in the protein space. Now we can write from Equation 5,

$$w_{\mathbf{j}} = \frac{1}{2^{c\,\ell_p}} \sum_{\mathbf{p}} f(\mathbf{p}) a_0^{\ell_{p,0} - q_{p,j',0}} a_1^{\ell_{p,1} - q_{p,j',1}} \prod_{\alpha = 0,1,\cdots q} \sum_{\mathbf{r}_{j,p} \in R_{p_\alpha}} \psi_{\mathbf{j}_\alpha}(\mathbf{r}_{j,p}) \tag{6}$$

Note that the basis function $\psi_{\mathbf{j}_\alpha}(\mathbf{r}_{j,p})$ is well defined for any $\mathbf{r}_{j,p} \in R_{j',p}$ for any α since the feature values of $\mathbf{r}_{j,p}$ are defined for every defining location of the partition $\mathbf{j}_\alpha$.

Let $e_{j_\alpha,p}$ and $o_{j_\alpha,p}$ be the number of members in R_{p_α} that have an even and odd number of 1-s respectively over the partition $j_{\alpha,p}$. For example, if $\alpha = 110000$ and $p = 10$ then $e_{110000,10} = 0$ and $o_{110000,10} = 1$ for code B shown in Table 3. Now using Equation 6 we can write,

$$w_{\mathbf{j}} = \frac{1}{2^{c\,\ell_p}} \sum_{\mathbf{p}} f(\mathbf{p}) a_0^{\ell_{p,0}-q_{p,j,0}} a_1^{\ell_{p,1}-q_{p,j,1}} \kappa \prod_{\alpha=0,1,\cdots q} |e_{j_\alpha,p} - o_{j_\alpha,p}| \tag{7}$$

where $\kappa \in \{-1, 1\}$ and $|(e_{j_\alpha,p} - o_{j_\alpha,p})|$ denotes the magnitude of $(e_{j_\alpha,p} - o_{j_\alpha,p})$ for all $\alpha \neq 0$. The value of $|e_{j_\alpha,p} - o_{j_\alpha,p}|$ can be determined directly from the genetic code. By definition, for the null partition $\alpha = 0$, we set $|e_{j_0,p} - o_{j_0,p}| = 1$ As before, this is done to take care of the case $\mathbf{j}$ which is comprised of only 0-s and as a result $q = 0$.

Let us now consider the properties of all the FC together using the energy measure of the spectrum.

5.3 Energy of the Fourier Spectrum

The energy of the Fourier spectrum can be defined as

$$E = \sum_{\mathbf{j}} w_{\mathbf{j}}^2 \tag{8}$$

The overall energy of the spectrum may change because of the genetic code-like representation transformations. Using Equation 5 and noting that $\psi_{\mathbf{j}}(\mathbf{x}) = \psi_{\mathbf{x}}(\mathbf{j})$ we can write [27],

$$\sum_{\mathbf{j}} w_{\mathbf{j}}^2 = \frac{1}{2^{c\ell_p}} \sum_{\mathbf{p}} f^2(\mathbf{p}) a_0^{\ell_{p,0}} a_1^{\ell_{p,1}} \tag{9}$$

Code B happens to change the overall energy. The overall energy of the Fourier representation of the TRAP function with four variables using code B is approximately 2.2778 times the energy of the protein space.

Although the overall energy is an interesting property to observe, the most critical properties are the number of coefficients that significantly contribute to the overall energy of the representation and the location of those significant coefficients. If the number is small then we know that the function can be approximated using a small number of coefficients. If we also have some idea about the partitions associated with those significant coefficients then we should be able to efficiently compute the representation. The following section explores these issues.

5.4 Distribution of the Energy in Partitions of Different Order

The distribution of the energy among the coefficients of different order is a very interesting property of a representation. For example, if we know that a representation has a small number of significant coefficients and they are associated with a certain order of partitions then it will be easier to compute such a representation.

Recall that the order of a partition $\mathbf{j}$ is essentially the number of 1-s in $\mathbf{j}$; in other words it is the number of features that define the corresponding basis function $\psi_{\mathbf{j}}(\mathbf{x})$. Let us define the *order-k energy*, $E^{(k)} = \sum_{\forall \mathbf{j}|\text{ones}(\mathbf{j})=k} w_{\mathbf{j}}^2$; ones($\mathbf{j}$) returns the number of 1-s in the binary string $\mathbf{j}$. We can compute this for both the protein and the mRNA space. Note that the order-k partition in the mRNA space may correspond to a lower order partition in the protein space since multiple mRNA features are associated with the same protein feature. A careful study of the effect of the representation transformations on the order-k energy in the mRNA space may require understanding the properties of the coefficients in the mRNA space that correspond to exactly k features in the protein space. We are going to call it the *absolute order-k energy*, defined as $\mathcal{E}^{(k)} = \sum_{\mathbf{j}|\text{ones}(\mathbf{j}')=k} w_{\mathbf{j}}^2$; as defined earlier, $\mathbf{j}'$ is the reflection of $\mathbf{j}$ in the protein space. Just like the association between the partitions in the protein and the mRNA spaces through the concept of reflection, the distribution of energies in these two representations can be linked though the concept of absolute order-k energy.

Using Equation 6 we can derive the overall energy of the representation for code B [27]:

$$\begin{aligned} E_R &= \sum_{\mathbf{j}} w_{\mathbf{j}}^2 \\ &\approx \frac{1}{2^{c\ell_p}} \sum_{\mathbf{p}} f^2(\mathbf{p}) a_0^{\ell_{p,0}} g(\ell_p, \ell_{p,0}, \ell_p) \end{aligned} \tag{10}$$

where

$$g(\ell_p, \ell_{p,0}, k) = \frac{1}{2^{c\,\ell_p}} \sum_{q=0}^{k} \sum_{q_{p,j,0}=\max(0,q-l_p+l_{p,0})}^{\min(q,\ell_{p,0})} \binom{\ell_{p,0}}{q_{p,j,0}} \binom{\ell_p-\ell_{p,0}}{q-q_{p,j,0}} (2^c-1)^q \times a_0^{\ell_{p,0}-2q_{p,j,0}}$$

where $0 \le k \le \ell_p$. We would like to study the convergence of $g(\ell_p, \ell_{p,0}, k)$ to 1 as k increases from 0 through ℓ_p. Note that $g(\ell_p, \ell_{p,0}, k)$ is also a function of $\ell_{p,0}$, the number of 0-s in a sequence $\mathbf{p}$. Since we are dealing with boolean sequences, $\ell_{p,0}$ is sufficient to define any particular $\mathbf{p}$.

Note that the protein sequences with high genetic fitness contribute significantly to the overall energy E_R since $f^2(\mathbf{p})$ will be large for them. Moreover the effect of $f^2(\mathbf{p})$ on the energy gets scaled up by the factor $a_0^{\ell_{p,0}}$. This essentially means that fitter protein sequences with a large number of 0-s will

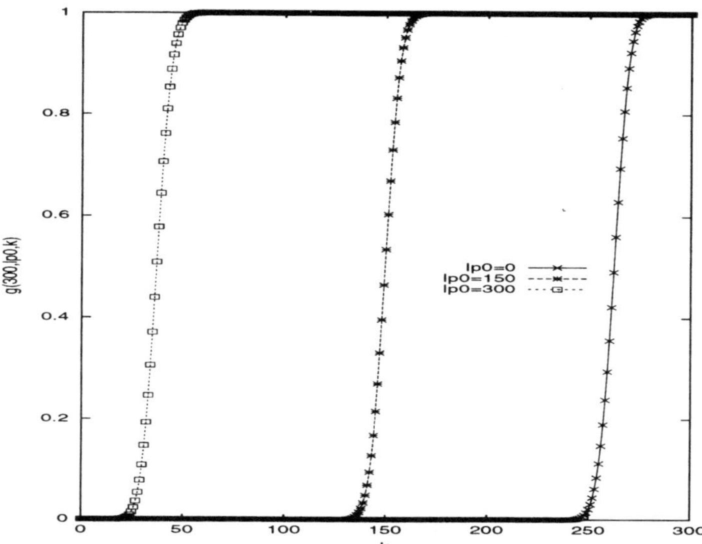

Fig. 5. $g(300, \ell_{p,0}, k)$ with respect to increasing k for two boundary cases $\ell_{p,0} = 0$, $\ell_{p,0} = 300$, and the intermediate case $\ell_{p,0} = 150$. This shows that $g(300, \ell_{p,0}, k)$ converges to 1 very fast for protein strings with large number of 0-s. Note that code B assigns more codons to protein feature value 0

mainly contribute to E_R. Now note that for proteins with a large number of 0-s (i.e. relatively large $\ell_{p,0}$) the function $g(\ell_p, \ell_{p,0}, k)$ approaches 1 very fast. In other words, the main portion of the overall energy comes from the highly fit proteins that have more numbers of equivalent mRNA representations (implied by in large value of $\ell_{p,0}$ and the bias of code B towards the protein feature 0).

Now let us consider an example using the TRAP functions. Note that in this case, although the sequence with all 1-s has the highest function value, there are other sequences that have non-zero function value. The sequences with more numbers of 0-s have relatively high fitness values. This also matches with the bias of the genetic code B. Therefore we should expect a good approximation using the low-order coefficients.

Figure 6 (Top) shows the distribution of absolute order-k energy using the code B and no transformation for a TRAP function with $\ell_p = 4$. Note that the absolute order-k energy decreases exponentially for code B. Figure 6 (Bottom) shows the distribution of order-k energy using code B for the TRAP function. Note that this is the order-k energy of the mRNA representation, not the absolute order-k energy, and four-bit protein sequences map to 12-bit mRNA sequences.

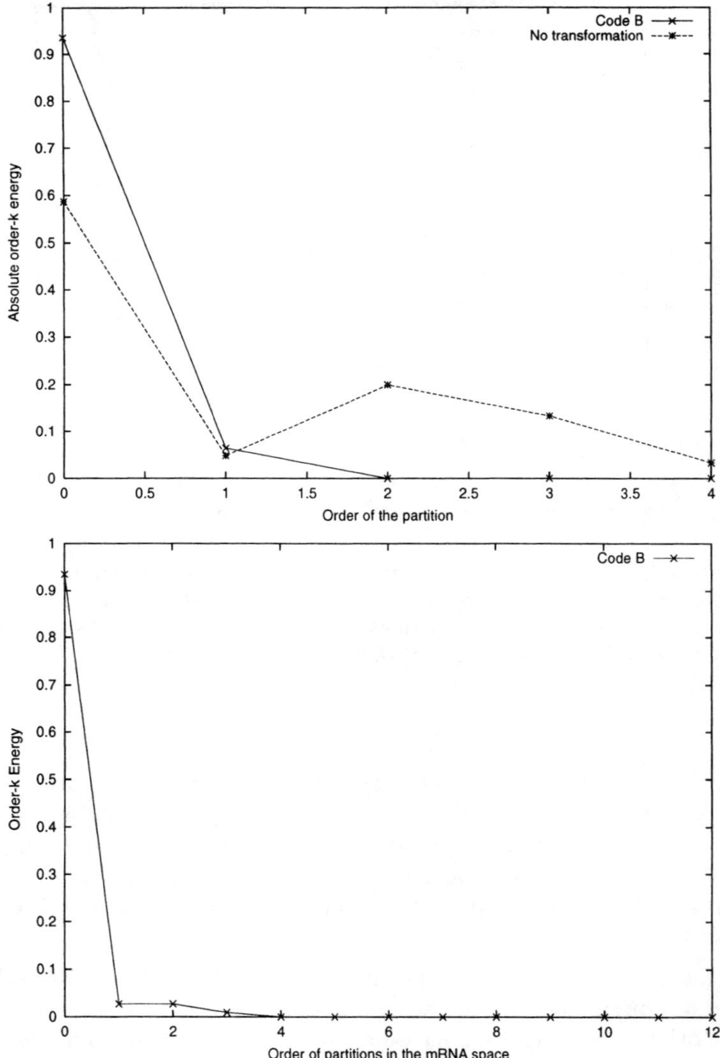

Fig. 6. (Top) Distribution of the absolute order energy using code B and no transformation for the TRAP function. (Bottom) Distribution of the order-k energy using Code B for the TRAP function. Note that four-bit protein space maps to 12-bit mRNA representation since the codon size is three

It is important to realize that the match between the bias of the genetic code and the representation of the fitter proteins may not be difficult to achieve. Fitter proteins will have larger value of $f^2(\mathbf{p})$. If we assign more numbers of codons to the most frequent feature value (either 1 or 0 in the case of binary strings) used in the fitter proteins, then the corresponding scaling factor ($a_0^{\ell_{p,0}}$ in the case of code B) will also be large. For these proteins $g(\ell_p, \ell_{p,0}, k)$ also approaches 1 very fast with respect to k. In the case of code B, the larger the value of $\ell_{p,0}$ in a protein, the higher the rate of convergence and the larger the scaling factor. On the other hand, the proteins with frequent feature values that have fewer numbers of codons assigned (1 in the case of code B) will have a slower convergence rate for $g(\ell_p, \ell_{p,0}, k)$ and smaller scaling factor. In case of code B it will be protein sequences with smaller values for $\ell_{p,0}$ (i.e. strings with relatively more numbers of 1-s). Although the convergence rate will be slow if the finesses of these proteins are relatively low, their contribution to the overall energy will be low since the scaling factor will be small for them. Note that if the fitness value is relatively small compared to the scaling factor, the latter will play a more significant role. For binary representation the issue is assigning a codon distribution among two possible protein features, 0 and 1. For representations with higher cardinality the code introduces richer transformations and we need to further explore the implications.

This section suggests that genetic code-like transformations may be useful for accurately approximating a function that does not satisfy the BVI property by one that does so in a larger space. If we can do that then the algorithm presented in Section 4 becomes immediately applicable to such functions. The following section concludes this chapter.

6 Conclusions

This chapter considered the evaluation of genetic fitness through the gene expression process and explored its role in the representation of the fitness function. It first noted that the representation of the genetic fitness function appears to be in a decomposed form since the fitness is evaluated in a distributed fashion in different cells of a living organism. Unless such a representation was available a priori, we have every reason to believe that there must be an efficient mechanism to compute such a distributed, decomposed representation. Our basic understanding of function representations suggests proper formulation of the problem using an appropriately chosen set of basis functions. This chapter considered the functionally complete Fourier basis functions and presented a polynomial-time randomized algorithm to construct such a representation in the Fourier basis. This showed that the representation of the genetic fitness function used during the gene expression process can be estimated in randomized polynomial-time if the fitness function has a polynomial-size description.

Although there are many functions with polynomial-size descriptions, it is not clear why the genetic fitness function should satisfy this requirement. This chapter also explored this issue and pointed out that the genetic code may be responsible for constructing a representation that satisfies this condition. It showed that genetic code-like transformations may offer us a unique technique to convert some functions of exponential description in the Fourier basis to an exponentially long representation with only a polynomial number of low-order terms exponentially more significant than the rest. This property may allow an approximation of the function using only a polynomial number of low-order Fourier terms. As a result, polynomial-time computation of a decomposed representation of the genetic fitness function may be possible using representation transformations like the genetic code which plays a key role in the natural process of gene expression.

The author and his colleagues are currently in the process of extending these techniques to non-binary representations. Advanced gene expression-based evolutionary algorithms are also under development. We strongly believe that this research will ultimately produce a new generation of evolutionary algorithms that will enjoy the benefits of the gene expression process and offer scalable performance based on solid theoretical foundations:

Acknowledgments

This work was supported by the United States National Science Foundation Grants IIS-9803660 and IIS-0083946. The experimental results presented in Section 1.4.3 were produced by B. H. Park.

References

1. Ackley, D. H. (1987) *A connectionist machine for genetic hill climbing.* Kluwer Academic, Boston
2. Bagley, J. D. (1967) The behavior of adaptive systems which employ genetic and correlation algorithms. *Dissertation Abstracts International*, **28**, 5106B (University Microfilms No. 68-7556)
3. Bandyopadhyay, S., Kargupta, H., Wang, G. (1997) Revisiting the GEMGA: Scalable evolutionary optimization through linkage learning. In *Proceedings of the IEEE International Conference on Evolutionary Computation*, 603–608 IEEE Press, Piscateway, NJ
4. Bashford, J., Tsohantjis, I., Jarvis, P. (1998) A supersymmetric model for the evolution of the genetic code. *Proceedings of the National Academy of Science USA*, **95**, 987–995
5. Beauchamp, K. G. (1984) *Applications of Walsh and Related Functions.* Academic Press, USA
6. Beland, P., Allen, T. (1994) The origin and evolution of the genetic code. *Journal of Theoretical Biology*, **170**, 359–365
7. Bethke, A. D. (1976) Comparison of genetic algorithms and gradient-based optimizers on parallel processors: Efficiency of use of processing capacity. Tech. Rep. No. 197, University of Michigan, Logic of Computers Group, Ann Arbor

8. Brindle, A. (1981) *Genetic Algorithms for Function Optimization.* Unpublished doctoral dissertation, Department of Computer Science, University of Alberta, Edmonton, Canada
9. Dasgupta, D., McGregor, D. R. (1992) Designing neural networks using the structured genetic algorithm. *Artifical Neural Networks*, **2**, 263–268
10. Forrest, S., Mitchell, M. (1991) The performance of genetic algorithms on Walsh polynomials: Some anomalous results and their explanation. In R. K. Belew and L. B. Booker, editors, *Proceedings of the Fourth International Conference on Genetic Algorithms*, 182–189. Morgan Kaufmann, San Mateo, CA
11. Freeman, J. J. (1998) A linear representation for GP using context free grammars. In *Proceedings of the 1998 Genetic Programming Conference*, 72–77, MIT Press, Cambridge, MA.
12. Fukuchi, S., Okayama, T., Otsuka, J. Evolution of genetic information flow from the viewpoint of protein sequence similarity. *Journal of Theoretical Biology*, **171**, 179–195
13. Ghahramani, Z., Wolpert, D. (1997) Modular decomposition in visuomotor learning. *Nature*, **386**, 392–395
14. Goldberg, D. E. (1989) Genetic algorithms and Walsh functions: Part I, a gentle introduction. *Complex Systems*, **3**, 129–152 (Also TCGA Report 88006)
15. Goldberg, D. E. (1989) Genetic algorithms and Walsh functions: Part II, deception and its analysis. *Complex Systems*, **3**, 153–171 (Also TCGA Report 89001)
16. Goldberg, D. E., Korb, B., Deb, K. (1989) Messy genetic algorithms: Motivation, analysis, and first results. *Complex Systems*, **3**, 493–530 (Also TCGA Report 89003)
17. Heckendorn, R., Whitley, D. (1999) Predicting epistasis from mathematical models. *Journal of Evolutionary Computation*, **7**, 69–101
18. Holland, J. H. (1975) *Adaptation in Natural and Artificial Systems.* University of Michigan Press, Ann Arbor
19. Hollstien, R. B. (1971) Artificial genetic adaptation in computer control systems. *Dissertation Abstracts International*, 32(3),1510B (University Microfilms No. 71-23,773)
20. Hornos, J., Hornos, Y. (1993) Algebraic model for the evolution of the genetic code. *Physical Review Letters*, **71**, 4401–4404
21. Jackson, J. (1995) *The Harmonic Sieve: A Novel Application of Fourier Analysis to Machine Learning Theory and Practice.* PhD thesis, School of Computer Science, Carnegie Mellon University, Pittsburgh, PA
22. Jacob, F., Monod, J. (1961) Genetic regulatory mechanisms in the synthesis of proteins. *Molecular Biology*, **3**, 318–356
23. Jacobs, R. A., Jordan, M. I., Nowlan, S. J., Hinton, G. E. (1991) Adaptive mixture of local experts. *Neural Computation*, **3**, 79–87
24. Kargupta, H. (1996) The gene expression messy genetic algorithm. In *Proceedings of the IEEE International Conference on Evolutionary Computation*, 814–819. IEEE Press, Piscateway, NJ
25. Kargupta, H. (1997) SEARCH, computational processes in evolution, and preliminary development of the gene expression messy genetic algorithm. *Complex Systems*, **11**, 233–287
26. Kargupta, H. Gene expression and large scale evolutionary optimization. In *Computational Aerosciences in the 21st Century.* Kluwer Academic, Boston

27. Kargupta, H. (1999) A striking property of genetic code-like transformations. School of EECS Technical Report EECS–99-004, Washington State University, Pullman, WA, http://www.eecs.wsu.edu/~hillol/gene_expression.html
28. Kargupta, H. (2000) Genetic code-like transformations and their effect on learning functions. In *Parallel Problem Solving from Nature*, 99–108 Springer-Verlag
29. Kargupta, H., Bandyopadhyay, S. (1998) Further experimentations on the scalability of the GEMGA. In *Lecture Notes in Computer Science: Parallel Problem Solving from Nature*, 315–324 Springer-Verlag, Amsterdam.
30. Kargupta, H., Bandyopadhyay, S. (2000) A perspective on the foundation and evolution of the linkage learning genetic algorithms. *Computer Methods in Applied Mechanics and Engineering*, **186** 269–294, Special Issue on Genetic Algorithms, Guest Editors: Goldberg, D. E. and Deb, K.
31. Kargupta, H., Goldberg, D. E. (1996) SEARCH, blackbox optimization, and sample complexity. In R. Belew and M. Vose, editors, *Foundations of Genetic Algorithms*, 291–324 Morgan Kaufmann, San Mateo, CA.
32. Kargupta, H., Goldberg, D. E., Wang, L. W. (1997) Extending the class of order-k delineable problems for the gene expression messy genetic algorithm. In *Proceedings of the Second Annual Conference on Genetic Programming*, 364–369 Morgan Kaufmann, San Francisco Kaufmann.
33. Kargupta, H., Park, H. (1999) Fast construction of distributed and decomposed evolutionary representation. In *Late Breaking Papers of the Genetic and Evolutionary Computation Conference*, 139–148 (Extended version is in communication)
34. Kargupta, H., Park, H. (2000) Fast construction of distributed and decomposed evolutionary representation. *Journal of Evolutionary Computation*, **9** : In press
35. Kargupta, H., Sarkar, K. (1999) Function induction, gene expression, and evolutionary representation construction. In *Proceedings of the Genetic and Evolutionary Computation Conference, Orlando, USA*, 313–320; Morgan Kaufmann, San Francisco
36. Kauffman, S. (1993) *The Origins of Order*. Oxford University Press, New York
37. Keller, R., Banzhaf, W. (1999) The evolution of genetic code in genetic programming. In *Proceedings of the Genetic and Evolutionary Computation Conference*, 1077–1082. Morgan Kaufmann
38. Kennedy, P. (1998) *Simulation of the Evolution of Single Celled Organisms with Genome, Metabolism and Time-Varying Phenotype*. PhD thesis, University of Technology, Sydney, Australia
39. Kushilevitz, S., Mansour, Y. (1991) Learning decision trees using Fourier spectrum. In *Proc 23rd Annual ACM Symp on Theory of Computing*, 455–464
40. Liepins, G. E., Vose, M. D. (1991) Polynomials, basic sets, and deceptiveness in genetic algorithms. *Complex Systems*, **5**, 45–61
41. Morohashi, M., Kitano, H. Identifying gene regulatory networks from time series expression data by *in silico* sampling and screening. In *Proceedings of the 5th European Conference on Artificial Life*, 477–486. Springer-Verlag, Berlin.
42. Oei, C. K. (1992) Walsh function analysis of genetic algorithms of nonbinary strings. Unpublished master's thesis, University of Illinois at Urbana-Champaign, Department of Computer Science, Urbana
43. Osawa, S. (1995) *Evolution of the Genetic Code*. Oxford University Press, NewYork

44. Rana, S., Heckendorn, R. B., Whitley, D. (1998) A tractable Walsh analysis of SAT and its implications for genetic algorithms. In *Proceedings of the AAAI-98*, AAAI Press
45. Reidys, C., Fraser, S. (1996) Evolution of random structures. Technical Report 96-11-082, Santa Fe Institute, Santa Fe, NM
46. Rockmore, D., Kostelec, P., Hordijk, W., Stadler, P. (1999) Fast Fourier transform for fitnes landscapes. Technical Report 99-10-068, Santa Fe Institute, Santa Fe, NM.
47. Rosenberg, R. S. (1967) Simulation of genetic populations with biochemical properties. *Dissertation Abstracts International*, 28, 2732B (University Microfilms No. 67-17,836)
48. Ryan, C., Collins, J. (1998) Grammatical evolution: Evolving programs for an arbitrary language. In *Proceedings of the first European Workshop on Genetic Programming*, 83–95. Springer-Verlag, Berlin.
49. Schuster P. (1997) The role of neutral mutations in the evolution of RNA molecules. In S. Suhai, editor, *Theoretical and Computational Methods in Genome Research*, 287–302 Plenum Press, New York
50. Shackleton, M., Shipman, R., Ebner, M. (2000) An investigation of redundant genotype-phenotype mappings and their role in evolutionary search. In *Proceedings of the 2000 Congress on Evolutionary Computation*, 493–500, Los Alamitos, California, IEEE Press, Piscateway, NJ
51. Smith, R. E. (1988) An investigation of diploid genetic algorithms for adaptive search of nonstationary functions. TCGA Report No. 88001, University of Alabama, The Clearinghouse for Genetic Algorithms, Tuscaloosa
52. Stadler, P. (1999) Spectral landscape theory. In J.P. Crutchfield and P. Schuster, (eds.), *Evolutionary Dynamics — Exploring the Interplay of Selection, Neutrality, Accident and Function.* Oxford University Press, New York
53. Thierens, D. (1999) Estimating the significant non-linearities in the genome problem-coding. In *Proceedings of the Genetic and Evolutionary Computation Conference*, 643–648, Morgan Kaufmann, San Francisco, CA
54. Thierens, D. (1999) Scalability problems of simple genetic algorithms. *Evolutionary Computation*, **7**, 331–352
55. Vose, M., Wright, A. (1998) The simple genetic algorithm and the Walsh transform: Part I, theory. *Journal of Evolutionary Computation*, **6**, 253–274
56. Vose, M., Wright, A. (1998) The simple genetic algorithm and the Walsh transform: Part II, the inverse. *Journal of Evolutionary Computation*, **6**, 253–274
57. Walsh, J. L. (1923) A closed set of orthogonal functions. *American Journal of Mathematics*, **45**
58. Wong, M., Leung, K. (1995) Applying logic grammars to induce subfunctions in genetic programming. In *Proceedings of the 1995 IEEE Conference on Evolutionary Computation*, 737–740, USA, IEEE Press, Piscateway, NJ
59. Wu, A., Lindsay, R. (1995) Empirical studies of the genetic algorithm with non-coding segments. *Journal of Evolutionary Computation*, **3**, 121–147

Solving Permutation Problems with the Ordering Messy Genetic Algorithm

Dimitri Knjazew and David E. Goldberg

Illinois Genetic Algorithms Laboratory
University of Illinois at Urbana-Champaign
117 Transportation Building
104 S. Mathews Avenue, Urbana, IL 61801, USA.
E-mail: {dimitri, deg}@illigal.ge.uiuc.edu

Summary. Although successful at first, messy genetic algorithms had minimum attention within the evolutionary computation community for the past few years. This chapter presents an ordering messy genetic algorithm (OmeGA) that is able to solve difficult permutation problems efficiently. Starting with a brief introduction to the fast messy genetic algorithm (fmGA), the chapter continues by proposing a robust representation model—the random keys—that proved to work successfully for representing permutations. The design of OmeGA is described and ordering deceptive problems are discussed in detail. Thereafter, experimental results that show the random key-based simple genetic algorithm (RKGA) being outperformed by its messy competitor in 32-length ordering deceptive problems are presented. The OmeGA is completely independent from the underlying chromosome's coding scheme and finds the global optimal solution in problems with both tightly and loosely coded building blocks. The chapter finally demonstrates the OmeGA's scale-up behavior.

1 Introduction

Various genetic/evolutionary algorithms (GEAs) have been developed for solving permutation problems over the past few years. Research in this area is very interesting, since there is a great variety of permutation-based commercially important applications. Unfortunately, many methods use either problem-specific or ad-hoc representation codings and operators. Also, the performance has not been sufficiently tested on hard problems, leaving the question of how the algorithm scales up unanswered.

Therefore, it would be interesting to design genetic algorithms (GAs) that use efficient codings and operators and have good scale-up properties. Furthermore, a more detailed and systematic analysis should be done on the GA performance. We suggest that so-called *competent genetic algorithms* that solve hard problems quickly, reliably and accurately [7] would be good approaches for this undertaking. Much research has been done in this area and numerous competent GAs have been developed, including the fmGA [9], the gene expression messy genetic algorithm [1], the linkage learning genetic algorithm [14], and the Bayesian optimization algorithm [21].

This chapter develops the ordering messy GA (OmeGA) specialized for permutation problems which represents the solutions by vectors of real numbers (random keys). In a number of experiments it is shown that the OmeGA significantly outperforms the simple GA in solving ordering deceptive problems, which are hard sequencing problems defined elsewhere [17].

We start with a background description (section 2) and an introduction to the fmGA (section 3). We then explain the concept of random keys (section 4) and develop the OmeGA (section 5). Afterwards, we describe ordering deceptive problems (section 6) and different codings that determine the problem difficulty (section 7). The chapter continues by presenting the results of pilot experiments (section 8). Finally, a scale-up analysis of the OmeGA is performed (section 9) and conclusions for future work are made (section 10).

2 Background

This section gives background information necessary for understanding the forthcoming sections. We start with a short overview of some permutation-oriented GEAs and discuss the notion of building blocks and deceptive problems. Finally, we take a closer look at competent GAs and explain why we decided to focus on the fmGA for this chapter.

Numerous GEAs have been designed for a wide variety of permutation-based problems over the past few years. Problems such as the travelling salesman problem [12], scheduling [4], vehicle route planning [3], or integrated circuit design [19] have been tackled. Many of these tasks are commercially important. However, only little research work has been done to examine how permutation-solving GEAs scale up or, in other terms, how their computational complexity increases with the underlying problem difficulty and size.

One approach to investigate the scale-up behavior of a GA is testing it on artificial problems where the solution is known a priori and where the problem difficulty can be varied. For this purpose researchers in the GA community frequently use *deceptive problems*, which are hard multimodal optimization problems for binary strings introduced by [6]. A deceptive problem may be designed by combining a desired number of *deceptive subfunctions* that mislead the GA letting it converge to certain local optimal points. No information about the subfunctions is passed to the GA. To find the global optimum, partial solutions (or schemas) with above-average fitness—the so-called *building blocks* (BBs)—must be identified and grouped together. Kargupta et.al. [17] developed *ordering deceptive problems* for permutations which we will discuss in detail in section 6.

A deceptive problem's degree of difficulty grows with increasing number and size (order) of the subfunctions. Particularly for the simple GA, the problem difficulty increases when a loose coding is chosen such that the elements of the subfunctions are mapped to distant positions within the problem representation. This results in a greater BB length and, consequently, the BBs are more likely to be disrupted when traditional recombination operators such as single-point or two-point crossover are applied. In general, no fixed recombination operators guarantee proper mixing with an arbitrary coding. This problem is usually referred to as *the linkage problem* in the literature. Thierens and Goldberg [23] showed in a dimensional analysis of mixing processes in simple GAs that the required population size grows exponentially with the BB length and number. Section 7 introduces some problem codings for ordering deceptive problems.

To tackle the linkage problem and to achieve a better scale-up behavior, first-generation GAs need to be extended by some additional mechanisms. Various competent GAs mentioned above have been developed for this purpose. They can be divided into two classes:

1 methods based on evolving the representation of solutions or adapting recombination operators among individual solutions
2 methods based on extracting information from the entire set of promising solutions to generate new solutions.

Examples of the first class are the fmGA [9], the gene expression messy GA [1] and the linkage learning GA [14]. Approaches belonging to the second class are the extended compact genetic algorithm [13] and the Bayesian optimization algorithm [21].

In this chapter we focus on the fmGA because it can be easily transformed into a permutation-solving GA. In section 4 we explain the notion of random keys introduced elsewhere [2]. Section 5 demonstrates how they can be incorporated into the fmGA to represent permutations.

3 A Brief Introduction to the fmGA

The first steps towards the development of competent GAs were taken by [11] who developed the messy GA (mGA) in 1989. The mGA formed a basis for the fmGA developed four years later by [9]. However, first-generation mGAs suffered some handicaps which made application to large-sized problems impractical. These were, among others, the lack of implicit parallelism and large memory consumption during the initialization phase, often referred to as the *initialization bottleneck*. The interested reader might refer to [10] for more details.

This section briefly describes key features of the fmGA, how it works, and how the initialization bottleneck has been overcome therein, making the mGA "fast". We first explain the messy representation and operators. Then we look at the organization of the fmGA and finish with some important techniques that contribute to its success. For a more detailed description we refer to [9].

3.1 Messy Representation

In mGAs the genes of a chromosome are represented by the pair (*allele locus,allele value*), like for example the two messy chromosomes `((3 0)(1 0)(2 1)(4 1)(5 0))` and `((2 1)(5 0)(3 0)(1 0)(4 1))`, which are both equivalent to the binary string 01010. The chromosomes may have different lengths for instance,

A : `((5 1)(1 1)(3 0)(1 0)(4 1)(3 1)(2 1)(4 0)(5 0))` and
B : `((3 0)(1 1)(5 0))`

both represent valid strings of length 5. To evaluate *over specified* chromosomes like example A the genes are scanned from left to right with a first-come first-serve precedence rule. Thus, example A would encode the bit string `11011`. For evaluating *underspecified* chromosomes, like example B, a *competitive template* which is a completely specified fixed-bit string is used in the fmGA. Before evaluation, the chromosome's missing genes are filled with the corresponding alleles from the template as illustrated in Figure 1. Note that a chromosome can be both under- and over specified.

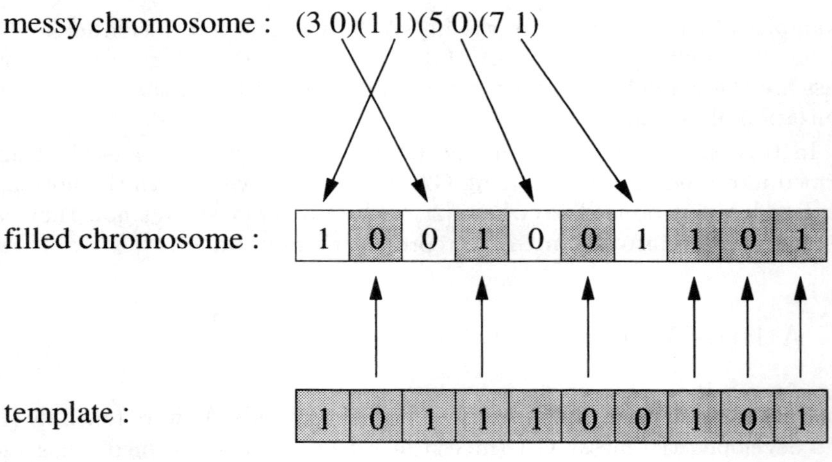

Fig. 1. Usage of competitive templates. Underspecified messy chromosomes are evaluated by taking the missing genes from the template.

3.2 Messy Operators

As in simple GAs, selection and recombination are used in the mGA to create a new population, except that traditional crossover is replaced by *cut and splice operators* [11]. The *cut operator* breaks a messy chromosome into two parts with a cut probability $p_c = p_\kappa(\lambda - 1)$, where p_κ is a specified bitwise cut probability and λ the length of the chromosome. The cut position is randomly chosen along λ. For example, with $p_\kappa = 0.1$ the probability of cutting the string `((2 0)(5 1)(3 1)(6 0)(5 1))` would be 0.4. A cut at position 3 would result in strings `((2 0)(5 1)(3 1))` and `((6 0)(5 1))`.

The *splice operator* joins two chromosomes with a specified splice probability p_s, for example `((2 0)(5 1))` and `((3 1)(6 0)(5 1))` would be combined to `((2 0)(5 1)(3 1)(6 0)(5 1))`. Figure 2 demonstrates how both operators work.

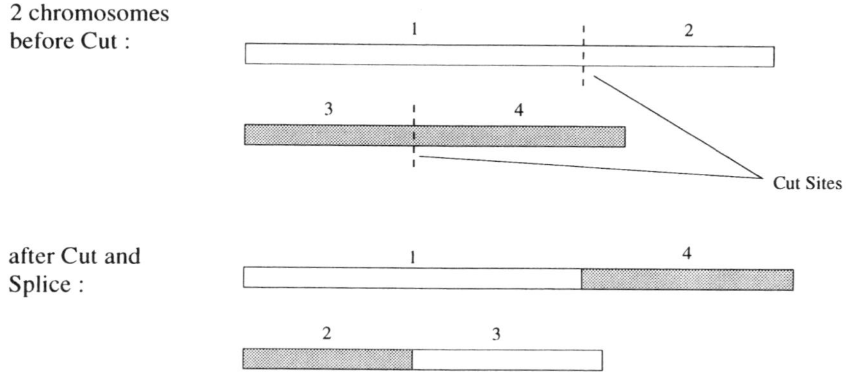

Fig. 2. Cut and Splice Operators. Two chromosomes are cut at randomly chosen positions and spliced thereafter.

Using messy codings gives much flexibility to the fmGA and enables linkage-oriented search. Loose BBs can be preserved as well as tight ones. The simple GA, on the contrary, is likely to disrupt loose BBs by crossover operators, as illustrated in Figure 3. However, the fmGA requires additional computation time and memory to find out the representation (coding) that ensures a proper mixing. The next paragraph will focus on this aspect in the context of fmGA organization.

3.3 Organization of the fmGA

The fmGA consists of two main loops:

- outer loop and
- inner loop.

The outer loop iterates over the order k of the processed BBs. Every cycle of the outer loop is denoted as an *era*. When a new era starts, the inner loop is invoked which can be divided into three phases:

- initialization phase

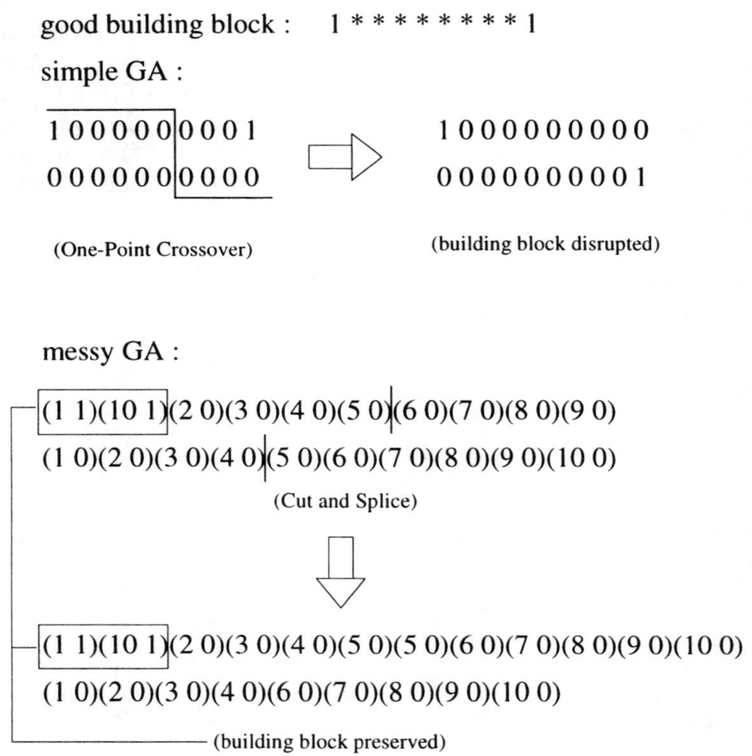

Fig. 3. BB disruption and preservation in a 10-bit problem. The one-point crossover operator will definitely disrupt the loose BB 1********1, while in the fmGA it would probably be preserved after cut and splice due to the flexibility of messy codings.

- BB filtering phase
- juxtapositional phase.

The goal of the initialization phase is to create a population of individuals containing all possible genic and allelic combinations—candidates for BBs. The BB filtering phase works like a filter: "bad" genes not belonging to BBs are supposed to be filtered out such that afterwards the population contains a high proportion of "good" genes and gene combinations belonging to BBs. These are then combined in the juxtapositional phase by genetic operators to form a high-quality, perhaps optimal, solution. We now describe these phases in detail.

3.4 Initialization Phase

The inner loop of the fmGA starts with a *probabilistically complete initialization* that randomly generates n fully specified chromosomes of length $\lambda \leq l$, where l is the problem length. The value of $\lambda \leq l$ can be chosen arbitrarily and n is calculated according to the population-sizing equation [9]

$$n = \frac{\binom{l}{\lambda}}{\binom{l-k}{\lambda-k}} 2c(\alpha)\beta^2(m-1)2^k. \tag{1}$$

Here, $c(\alpha)$ is the square of a normal random deviate corresponding to a tail-probability α, β is the signal-to-noise ratio which is the ratio of the fitness deviation to the difference between the best and the second-best fitness value of a subfunction. The variable m is the number and k is the order of the subfunctions. The population size has to be large enough and chromosomes have to be long enough to ensure the presence of all BB after the initialization phase is completed.

3.5 BB Filtering Phase

Now the goal is to reduce the string length gradually to some value near k to identify the BBs. At the same time, genes not belonging to BBs need to be eliminated, such that finally short messy strings with a high fitness are obtained. This may be accomplished by repeating the following two steps until the strings are short enough:

1 repeatedly performing selection
2 deleting random genes in all chromosomes.

During Step 1 the proportion of individuals containing BBs increases by giving more copies to the fitter chromosomes. After a sufficient number of selections has been done, there is no further need to worry about "good" genes being deleted in Step 2 by accident. Even if some BBs are damaged or eliminated, still enough copies are expected to remain for further processing. Figure 4 illustrates an example step of filtering phase.

To a certain extent, this procedure of searching for BB genes can be compared to the well-known Minesweeper game included in Windows operating systems, where the player tries to find hidden mines in an array of buttons by clicking on them and thus revealing what is behind. If a hidden mine is touched, the player loses. However, the game also provides the user with information about possible locations of mines, whereas the fmGA does not have any information about the BB genes at all.

Length reduction. Let us consider a sequence of string lengths $\lambda^{(0)} > \lambda^{(1)} > ... > \lambda^{(i)} > ... > \lambda^{(\mathcal{N})}$ where $\lambda^{(0)}$ is the initial and $\lambda^{(\mathcal{N})}$ the final string length greater than or equal k. Random deletion of genes reduces a string of length $\lambda^{(i)}$ down to $\lambda^{(i+1)}$. At every reduction step $n_d^{(i)} = \lambda^{(i-1)} - \lambda^{(i)}$ genes are deleted and $n_s^{(i)}$ selections are performed.

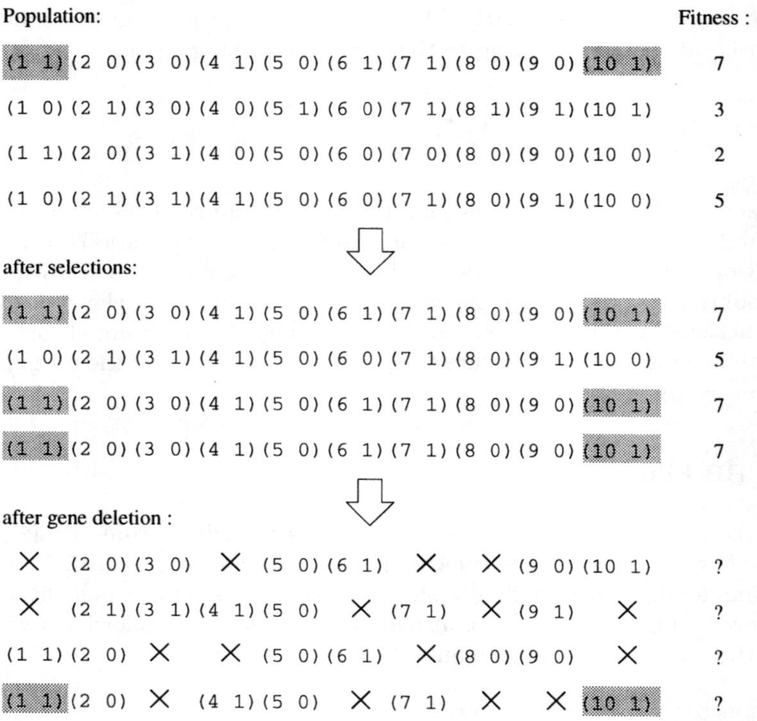

Population:	Fitness :
(1 1)(2 0)(3 0)(4 1)(5 0)(6 1)(7 1)(8 0)(9 0)(10 1)	7
(1 0)(2 1)(3 0)(4 0)(5 1)(6 0)(7 1)(8 1)(9 1)(10 1)	3
(1 1)(2 0)(3 1)(4 0)(5 0)(6 0)(7 0)(8 0)(9 0)(10 0)	2
(1 0)(2 1)(3 1)(4 1)(5 0)(6 0)(7 1)(8 0)(9 1)(10 0)	5
after selections:	
(1 1)(2 0)(3 0)(4 1)(5 0)(6 1)(7 1)(8 0)(9 0)(10 1)	7
(1 0)(2 1)(3 1)(4 1)(5 0)(6 0)(7 1)(8 0)(9 1)(10 0)	5
(1 1)(2 0)(3 0)(4 1)(5 0)(6 1)(7 1)(8 0)(9 0)(10 1)	7
(1 1)(2 0)(3 0)(4 1)(5 0)(6 1)(7 1)(8 0)(9 0)(10 1)	7
after gene deletion :	
× (2 0)(3 0) × (5 0)(6 1) × × (9 0)(10 1)	?
× (2 1)(3 1)(4 1)(5 0) × (7 1) × (9 1) ×	?
(1 1)(2 0) × × (5 0)(6 1) × (8 0)(9 0) ×	?
(1 1)(2 0) × (4 1)(5 0) × (7 1) × × (10 1)	?

Fig. 4. This is one example of a BB filtering step. After $n_s^{(i)}$ selections are applied to the population, $n_d^{(i)} = 4$ genes are deleted in every chromosome. Here, deleted genes are indicated by crosses and the gray squares show genes belonging to the BB (1 1)(1 10). Since selection increases the proportion of this highly fit schema, it survives the gene deletion.

The authors in [9] propose a strategy for length reduction using a constant average length-reduction ratio $\rho < 1$:

$$\lambda^{(i)} = \lceil \lambda^{(0)} \rho^i \rceil. \tag{2}$$

According to this method, the total number of required length reductions is

$$t_r = \log_\rho \frac{\lambda^{(\mathcal{N})}}{\lambda^{(0)}} \tag{3}$$

and varies as $O(\log\ l)$.

Selection. Given a gene-deletion strategy determining $n_d^{(i)}$ and t_r, one can easily derive the required number of selections $n_s^{(i)}$ between every length reduction such that enough copies of good BB are reproduced. Considering the number of ways $\lambda^{(i+1)}$ genes can be randomly picked from a string of length $\lambda^{(i)}$ holding k genes fixed we find that

$$\gamma^{(i)} = \frac{\binom{\lambda^{(i)}}{\lambda^{(i+1)}}}{\binom{\lambda^{(i)}-k}{\lambda^{(i+1)}-k}} \tag{4}$$

strings will have at least one expected copy of the desired BB. This number is denoted as the *BB repetition factor*. By assuming that tournament selection is used which roughly doubles the proportion of the best individuals, $n_s^{(i)}$ is given by the equation

$$n_s^{(i)} = \log_2(\gamma^{(i)}). \tag{5}$$

Given the parameters $n_d^{(i)}$,$n_s^{(i)}$, and t_r the whole BB filtering schedule can be easily implemented. Figure 5 shows a sample schedule. A detailed analysis of this process, discussed elsewhere [9], shows that the initialization and the filtering phases require a considerably small number of fitness function evaluations that grows as $O(l \ \log \ l)$.

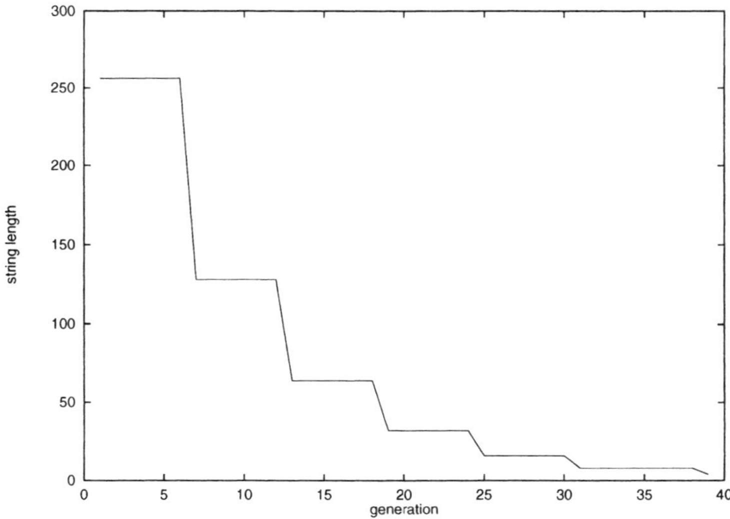

Fig. 5. This graph shows the overall chromosome length during the filtering phase in a 256-bit problem with order-4 BB. While selection is applied, the curve stays flat. Abrupt changes of the curve correspond to gene deletions.

3.6 Juxtapositional Phase

After the population has gained a high proportion of copies of the best BBs, the juxtapositional phase starts. Cut and splice operators explained above are now applied to combine "good pieces" of solutions together. Early in this phase mostly the effect of the splice operator can be observed, because the chromosomes are still short. Later, the chance of cutting a string increases with the average chromosome length. Since new strings are constantly produced, their objective function values need to be evaluated in every generation. Also, genic and allelic mutation are performed during this phase. After the whole process is finished, the inner loop of the fmGA terminates.

3.7 Thresholding

To overcome the problem of *cross-competition*, which means that different BBs tend to compete against each other, a *thresholding operator* has been introduced. Thresholding refers to a certain amount of genes two chromosomes must have in common before they can be compared during selection. If two individuals share θ genes together, they are ready to compete against each other. This threshold number θ should be greater than the expected number of common genes. In the original mGA the following value is proposed for θ when two strings of length λ_1 and λ_2 are compared:

$$\theta = \lfloor \frac{\lambda_1 \lambda_2}{l} \rfloor. \tag{6}$$

For example, the two strings `((1 0)(3 0)(5 1)(7 0)(9 1))` and `((2 0)(4 1)(6 0)(8 1)(10 1))` could not compete in a problem of length $l = 10$ with $\theta = 2$, since they have no common genes. The chromosomes `((1 0)(3 0)(5 1)(7 0)(9 1))` and `((1 1)(2 1)(5 1)(8 0)(10 1))`, on the other hand, share genes 1 and 5 and could therefore be compared.

Thresholding is used in tournament selection as follows. Since two randomly picked individuals are likely not to have enough genes in common, a sufficiently large amount of random candidates has to be considered. This can be accomplished in the following thresholding selection mechanism:

1 Pick a random individual from the population
2 Pick n_{sh} further random candidates
3 **If** among those candidates picked in Step 2 there exists one sharing at least θ genes with the first individual, perform regular tournament selection for both. **Otherwise**, choose the first individual for further processing.

In early mGA versions [11] the *shuffle number* n_{sh} is calculated according to the equation

$$n_{sh} = l. \tag{7}$$

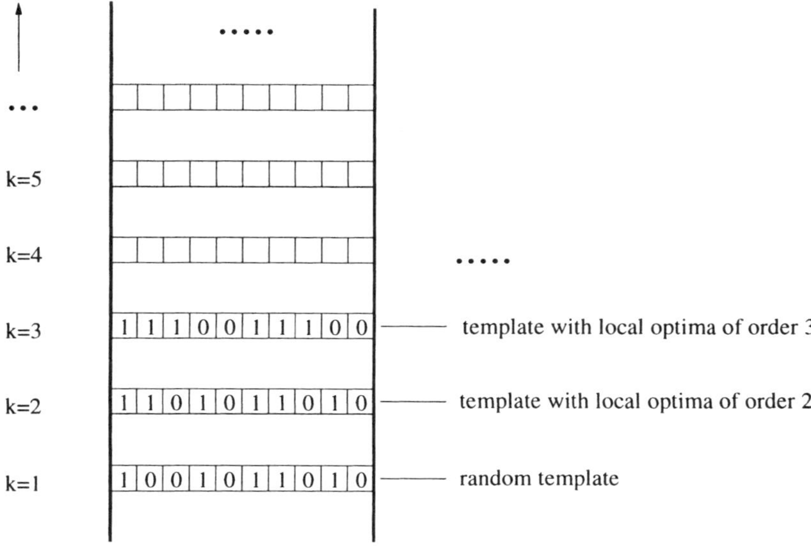

Fig. 6. The ladder of deception. Level-wise processing helps to overcome misleading local optima of deceptive problems. Every time the inner loop of the fmGA is invoked, the algorithm "climbs" up one level by taking the best individual from the previous level and using it as a new template.

3.8 Level-wise Processing

As was mentioned before, the outer loop of the fmGA iterates over the order k of the BBs. This technique is called *level-wise processing* and was introduced for overcoming local optima. Starting at level $k = 1$ the inner loop is invoked with a randomly initialized competitive template. After the inner loop terminates, the actual template is replaced by the best individual found so far, which becomes the new template for the next level and so on. Thus, the set of local optimal points discovered on level k serves as a launch pad for level $k + 1$. The whole procedure can be repeated until a maximum level is reached (see Figure 6).

3.9 Tie Breaking

In the previous paragraphs it was assumed that all BBs are of the same order. Suppose they are non-uniformly sized; then a discrepancy between the algorithm and the function BB size occurs that may cause problems.

To illustrate this in a simple example, consider two strings $A =$`((1 1) (2 1)(3 1)(8 0))` and $B =$`((1 1)(2 1)(3 1)(8 1))` in a 30-bit deceptive function with 3-bit BBs. Both A and B contain the optimal BB `((1 1)(2 1)(3 1))` but their fourth genes cause a significant difference concerning the expression of optimal bit combinations over positions 7, 8, and 9. If the gene (8 0) has a higher average fitness than gene (8 1), string B is likely to be preferred to string A in most cases during selection. These bits like (8 0) are denoted as *parasitic bits*, since they inhibit expression of further optimal bit combinations.

To tackle this problem, the selection operator is extended by a mechanism called *tie breaking*: when two or more individuals of the same fitness are compared during selection the one with the shortest string length is preferred. In addition, the fmGA initialization is extended such that chromosomes of all possible BB lengths from 1 to k are created. Tie breaking successfully addresses the problem of non-uniform BB size and attacks parasitic bits.

4 Using Random Keys for Representation

In the previous section we have discussed the fmGA assuming that it operates on binary strings. To specialize the algorithm for permutation problems, we have to choose a suitable representation. This section briefly overviews some frequently used techniques for encoding permutations and explains the concept of random keys and the random key-based simple genetic algorithm (RKGA).

Various representation models are proposed for permutations in the GA literature. Very often integer numbers are used to describe a sequence directly. However, this method requires repair mechanisms when genetic operators such as crossover are applied. For instance, consider two chromosomes

```
(1 6 3 | 7 8 2 5 4)
(2 4 8 | 6 3 5 7 1)
```

being recombined by one-point crossover at, say, position 3. This would yield two infeasible permutations with duplicate elements 1, 6, 3 in the first and 2, 4, 8 in the second offspring:

```
(1 6 3 6 3 5 7 1)
(2 4 8 7 8 2 5 4)
```

To overcome this infeasibility problem numerous crossover operators have been designed to keep the generated offspring feasible, for example PMX [12], UOX [5], and ROX [17] and these will be further discussed in section 6.

An alternative way to encode sequences is using binary matrices that describe the relative order of the permutation elements. If a matrix A has a 1 bit on position (i, j), the element i has to appear before element j in the sequence and vice versa for a 0 bit. Unfortunately, this representation is not ideal, since it involves many non-existent orderings, as was pointed out in [24]. For example, according to the matrix

	1	2	3	4
1	0	1	1	1
2	0	0	0	1
3	0	1	0	0
4	0	0	1	0

element 2 occurs after 3, 2 occurs before 4, and 4 occurs before 3, such that

$$3 \prec 2 \prec 4, \text{ but } 4 \prec 3,$$

holds. As with direct representation, repair mechanisms are also required here to obtain valid permutations.

For this chapter we have decided to use a representation where no repair is needed: the *random keys coding* [2]. Here, real or long integer random numbers are used as sort keys to decode a sequence. A permutation of length l is then represented as a real vector $\mathbf{r} = (r_1, r_2, \ldots, r_l)$ with typically $\mathbf{r} \in [0,1]^l$. By sorting the random keys such that

$$r_{\phi(1)} \leq r_{\phi(2)} \leq \ldots \leq r_{\phi(l)}$$

holds, where $\phi : \{1,..,l\} \rightarrow \{1,..,l\}$ is the corresponding mapping function arranging the keys in ascending order, the permutation is decoded as follows:

$$(\phi(1), \phi(2), \ldots, \phi(l)).$$

The following example demonstrates how random key chromosomes are decoded after single-point crossover. Crossing two strings

parent A:	`(0.46,0.91|0.33,0.75,0.51)`	≡	`(3 1 5 4 2)`
parent B:	`(0.84,0.32|0.64,0.04,0.48)`	≡	`(4 2 5 3 1)`

after the second gene yields the following offspring:

offspring A‘:	`(0.46,0.91,0.64,0.04,0.48)`	≡	`(4 1 5 3 2)`
offspring B‘:	`(0.84,0.32,0.33,0.75,0.51)`	≡	`(2 3 5 4 1).`

Note that the corresponding permutations on the right side are valid. In general, any sequence of real numbers can be interpreted as a valid permutation. Thus, traditional recombination operators would always generate feasible offspring when used on random key vectors. The example above also demonstrates how partial relative orderings are preserved after crossover: $3 \prec 5 \prec 4$ holds for A and B‘, likewise $4 \prec 5 \prec 3$ applies to B and A‘. As every random key refers to a fixed allele, the relative order of the elements encoded in the exchanged chromosome parts does not change after recombination, whereas the elements' absolute positions may change.

By taking a closer look at the decoding mechanism we realize that absolute position information is also taken into account. The placement of an allele is roughly determined by its corresponding random key value: if r_i is close to zero, one can expect the allele $\phi(i)$ to be arranged somewhere at the beginning of a permutation. If, on the contrary, r_i is close to one, $\phi(i)$ is likely to be placed somewhere at the end. We omit a detailed mathematical analysis of this effect here and summarize the above-described benefits of random keys as follows:

- random keys overcome the permutation feasibility problem
- partial relative ordering information is preserved in crossover
- absolute positioning information is respected in crossover to some extent.

Besides recombination, a simple mutation operator could be implemented by replacing a random key with a new randomly generated number. More sophisticated mutation operators are discussed in [20] for random keys or in [15] for real vectors in general. Also, various mutation techniques from evolution strategies [22] can be adopted.

Note that there is a chance of generating some duplicate keys by accident. However, this poses no practical problems to the permutation decoding. A simple GA that uses the random key representation is denoted by "random key genetic algorithm" (RKGA) in the literature. A detailed description of the RKGA can be found in [2].

5 Designing the OmeGA

This section briefly introduces an *ordering messy genetic algorithm*. In the previous section we discussed the usage of random keys on fixed-length chromosomes. But are random keys alone sufficient to guarantee a good scale-up? We suggest that this goal may be achieved when the GA is designed on the basis of three key ideas:

- all basic mechanisms of the fmGA are applied, therby ensuring a good BB mixing
- the alleles are real or long integer numbers
- the alleles are treated as random keys to encode permutations.

We denote this permutation-oriented fmGA using a non-binary representation as the "ordering messy genetic algorithm" (OmeGA). Particularly, messy genes are now represented as a pair of gene locus and a random key instead of a binary digit:

messy gene: (*gene locus*, *random key*).

Furthermore, cut and splice operators can be performed on messy chromosomes in the normal way. Competitive templates initialized with random real or long integer values replace binary templates. When decoding a messy chromosome all missing (not specified) genes are taken from the template. To obtain a permutation the whole random key vector is decoded as described in section 4.

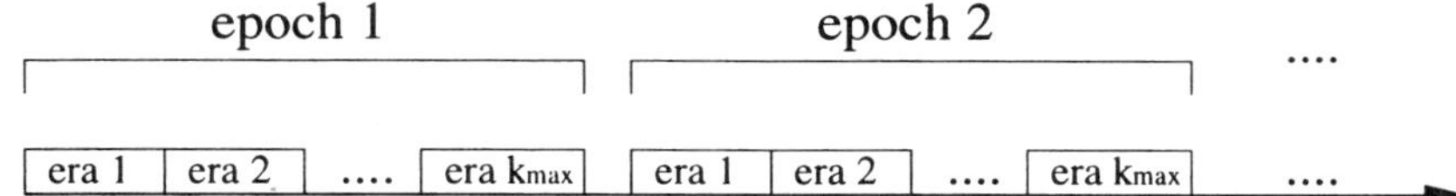

Fig. 7. Multiple epochs in the *OmeGA*. After the outer loop of the *OmeGA* terminates, an epoch is completed. Then, the best individual found in era k is carried over to the next epoch where it serves as a competitive template.

Referring to the advice given by [16] to apply the fmGA iteratively at each level, we extend the OmeGA by enclosing the outer loop into an external cycle that iterates over a desired number of *epochs*. An epoch starts with the first era and finishes with the maximum era k_{max}, as illustrated in Figure 7. Afterwards, the best individual found so far is used as a competitive template for the succeeding epoch and so on. Multiple epochs are useful when the population size is not large enough but cannot be further increased because of memory restrictions. Then, there is still a chance of finding the global optimal solution in a later epoch. This idea is also motivated by research work on the "fundamental tradeoff"—the tradeoff between population size and the number of epochs [8].

6 Ordering Deceptive Problems

This section introduces two concrete deceptive problems for permutations that will be later used to test the performance of the ordering GAs. We start with a description of two ordering deceptive functions and then show how to construct ordering deceptive problems. We then give some comments on previous research work that has been done on this topic.

Kargupta et al.,[17] introduced two deceptive functions of order 4: the relative ordering function, here denoted by f_{rel}, and the absolute ordering function, here denoted by f_{abs}. These functions are defined as follows.

relative ordering function f_{rel}:

```
f(1 2 3 4) = 4.0    f(1 4 2 3) = 1.2    f(4 1 2 3) = 2.1
f(1 2 4 3) = 1.1    f(4 2 1 3) = 1.2    f(3 1 4 2) = 2.2
f(1 4 3 2) = 1.1    f(3 2 4 1) = 1.2    f(3 4 1 2) = 2.2
f(1 3 2 4) = 1.1    f(4 1 3 2) = 1.2    f(4 3 1 2) = 2.4
f(3 2 1 4) = 1.1    f(2 4 3 1) = 1.2    f(2 3 4 1) = 1.5
f(4 2 3 1) = 1.1    f(3 1 2 4) = 1.2    f(3 4 2 1) = 3.2
f(2 1 3 4) = 1.1    f(2 3 1 4) = 1.2    f(2 4 1 3) = 2.4
f(1 3 4 2) = 1.2    f(2 1 4 3) = 2.4    f(4 3 2 1) = 2.4
```

absolute ordering function f_{abs}:

```
f(1 2 3 4) = 4.0    f(1 4 2 3) = 2.0    f(4 1 2 3) = 2.6
```

```
f(1 2 4 3) = 1.8    f(4 2 1 3) = 2.0    f(3 1 4 2) = 2.6
f(1 4 3 2) = 1.8    f(3 2 4 1) = 2.0    f(3 4 1 2) = 2.6
f(1 3 2 4) = 1.8    f(4 1 3 2) = 2.0    f(4 3 1 2) = 2.6
f(3 2 1 4) = 1.8    f(2 4 3 1) = 2.0    f(2 3 4 1) = 2.6
f(4 2 3 1) = 1.8    f(3 1 2 4) = 2.0    f(3 4 2 1) = 3.3
f(2 1 3 4) = 1.8    f(2 3 1 4) = 2.0    f(2 4 1 3) = 2.6
f(1 3 4 2) = 2.0    f(2 1 4 3) = 2.6    f(4 3 2 1) = 2.6.
```

In f_{rel}, only the relative ordering of the permutation elements matters. Here, the global optimal point is (1 2 3 4) with a function value equal to 4.0 and the misleading attractor is (3 4 2 1) with the second-highest function value of 3.2. For the function f_{abs}, the elements have to be placed in correct absolute positions in addition to being arranged in the right relative order.

At this point we recall the notion of relative and absolute ordering schemas (o-schemas) defined elsewhere [12]. A relative o-schema is a similarity subset of permutations where certain alleles have a relative order in common, for example the schema (! 1 ! 2 !) defines all permutations of length 5 where allele 1 appears before allele 2. In an absolute o-schema the alleles have common positional characteristics. The absolute o-schema (2 ! ! 5 !) represents all permutations having allele 2 on the first and allele 5 on the fourth position.

Let us consider a sample permutation and apply f_{abs}. The sequence

$$(\underline{1}\ \underline{2}\ 11\ 14\ \widehat{5}\ \widehat{6}\ \widehat{7}\ \widehat{8}\ 9\ 10\ \underline{3}\ 12\ \underline{4}\ 13\ 15\ 16\ \overline{20}\ \overline{18}\ \overline{19}\ \overline{17})$$

contains an absolute o-schema (! ! ! ! 5 6 7 8 ! ... !) which is optimal and would contribute to the overall fitness by the value of 4.0—the order is correct and the elements are placed in the right positions 5-8. The elements {17,18,19,20} are present in the right section but their order corresponds to $4 \prec 2 \prec 3 \prec 1$, therefore they gain a score of 1.8. Here, $\prec$ denotes relative order. Finally, the items {1,2,3,4} are in the correct relative order but 3 and 4 are misplaced. In this case, a *partial credit* of 1.0 is given to this o-schema instead, according to the formula

$$partial\ credit = \frac{1}{2}\ number\ of\ alleles\ in\ the\ correct\ section. \tag{8}$$

Ordering deceptive problems can be constructed by concatenating a desired number of the above-defined subfunctions. The overall function value of the whole permutation is the sum of the subfunction values. In this chapter we consider 32-allele problems consisting of eight subproblems. The search space has one global optimum—the sequence (1,2,3,4,...,32)—and $2^8 - 1 = 255$ local optima.

On these problems Kargupta et al. [17] tested the performance of the simple GA using several recombination operators: partially mapped crossover (PMX), relative ordering crossover (ROX), and uniform ordering crossover (UOX). For the absolute problem only the GA using PMX could find the global optimum whereas the optimal solution of the relative problem could only be found with ROX. The authors [17] comment on these results as follows (p. 55):

> When these problems are used to test three representative ordering recombination operators, it is not surprising that in a particular problem, the operator most appropriate for the schemata required to solve that problem works the best.

Since no problem information is supposed to be given beforehand, the simple GA, working with any crossover operator, would scale up badly for one of the two problems. This fact motivates the development of new GAs which are relatively independent from the internal structure of the task to be solved.

7 Problem Codings

In the following paragraphs we define three different problem codings that determine the degree of difficulty of ordering deceptive problems and explain how to decode and evaluate chromosomes. These codings will be later used to compare the scale-up properties of the OmeGA and RKGA.

By *tight coding* we denote a coding scheme where BBs are tight with a defining length 3. In *deflen6 coding* the defining length of the BBs is 6. We further denote a coding where the sum of all defining lengths is maximal by *loose coding*. Table 1 shows the genes comprising the subfunctions for different codings.

subfunction No.	tight	deflen6	loose
1	1, 2, 3, 4	1, 3, 5, 7	1, 9, 17, 25
2	5, 6, 7, 8	2, 4, 6, 8	2, 10, 18, 26
3	9, 10, 11, 12	9, 11, 13, 15	3, 11, 19, 27
4	13, 14, 15, 16	10, 12, 14, 16	4, 12, 20, 28
5	17, 18, 19, 20	17, 19, 21, 23	5, 13, 21, 29
6	21, 22, 23, 24	18, 20, 22, 24	6, 14, 22, 30
7	25, 26, 27, 28	25, 27, 29, 31	7, 15, 23, 31
8	29, 30, 31, 32	26, 28, 30, 32	8, 16, 24, 32

Table 1. Problem codings. Here, genes belonging to subfunctions 1-8 are listed for tight, deflen6, and loose coding.

Coding-oriented function evaluation of a random key vector $\mathbf{r}$ works as follows. First, a copy of $\mathbf{r}$ is created. Afterwards, the elements of the copy are rearranged according to the coding scheme, yielding a new vector $\mathbf{r}'$ which is then transformed into a permutation and evaluated as described in section 6. Finally, the function value is assigned to $\mathbf{r}$ and $\mathbf{r}'$ is discarded. For instance, when using the loose coding in the 32-length ordering deceptive problem, the rearranged copy of the string

$$\mathbf{r} = (r_1, r_2, r_3, r_4, r_5, r_6, r_7, r_8, r_9, r_{10}, r_{11}, r_{12}, \ \cdots \ , r_{29}, r_{30}, r_{31}, r_{32})$$

would be

$$\mathbf{r}' = (r_1, r_9, r_{17}, r_{25}, r_2, r_{10}, r_{18}, r_{26}, r_3, r_{11}, r_{19}, r_{27}, \ \cdots \ , r_8, r_{16}, r_{24}, r_{32}).$$

The probability of BB disruption grows with the defining length of the BB for the RKGA using one-or n-point uniform crossover. Therefore, ordering problems coded with loose coding are harder for the GA to solve than those with tight coding. On the other hand, we would expect the OmeGA to find the global optimum independently from the underlying coding thanks to its messy chromosomal representation.

8 Pilot Experiments

This section presents experimental results comparing the performance of the RKGA and OmeGA. First, RKGA is tested with different crossover operators. Then, after the best operator is chosen, test results with both GAs solving absolute and relative ordering problems of length 32 are presented for tight, deflen6, and loose codings.

In all the following experiments the genic and allelic mutation probability were kept zero to observe the effect of recombination alone. Moreover, in all runs the crossover probability in the RKGA and the splice probability in OmeGA were set to 1.0. With these parameters the best results were obtained. Binary tournament selection without replacement was used.

8.1 Experiment I: Choosing the Crossover Operator

To choose a good crossover operator for the RKGA, we did experiments with single-point, two-point, and uniform crossover. We used the same population sizes as [17] in their experiments with ordering deceptive problems: 3000 for the absolute and 2500 for the relative problem. Table 2 shows the average number of correct subfunctions found by the RKGA in 20 independent runs with tight, deflen6, and loose coding. Single-point crossover slightly outperformed two-point crossover—the global optimal solution with eight correct subfunctions has been found in all runs for both problems. In contrast, the uniform crossover operator performed remarkably badly due to its high degree of BB disruption.

It is interesting that the results on the absolute problem are significantly better than those on the relative problem. For example, considering loose coding and one-point crossover, 5.35 correct absolute subfunctions and only 1.8 correct relative subfunctions were found on average. To understand why

absolute problem	1 Point X	2 Point X	Uniform X
tight	8	7.95	3.25
deflen6	8	7.85	2.95
loose	5.35	5.15	3.15

relative problem	1 Point X	2 Point X	Uniform X
tight	8	8	1.4
deflen6	8	8	1.6
loose	1.8	1.55	1.55

Table 2. Performance of crossover operators in absolute and relative ordering deceptive problems. The two tables show the average number of optimal subfunctions found by the RKGA in 20 independent runs. Here, the RKGA is tested with single-point, two-point, and uniform crossover operators for tight, deflen6, and loose problem codings.

the RKGA finds more partial solutions in the absolute case, we have to take a closer look at the ordering deceptive functions f_{abs} and f_{rel} we described in section 6.

Both functions have been designed on the basis of a *dissimilarity* metric between two strings. However, this metric may depend on the underlying problem representation in the GA. For instance, the dissimilarity between (2 3 4 1) and (1 2 3 4) is four if direct representation is used, since all four alleles have no alleles in common. When using the random-key representation, however, the dissimilarity would be one, because only one random key needs to be changed to switch between the two permutations—the first key in this case:

$$
\begin{aligned}
\texttt{(0.96,0.21,0.33,0.75)} &\equiv \texttt{(2 3 4 1)} \\
\texttt{(0.16,0.21,0.33,0.75)} &\equiv \texttt{(1 2 3 4)}.
\end{aligned}
$$

In the absolute problem f_{abs}(2 3 4 1) has the third- highest fitness value of 2.6, the next value is f_{abs}(3 4 2 1)$=$ 3.3 which belongs to the misleading local optimum. Thus, there is a good chance for the RKGA to "jump" to the globally optimal point, right before the local optimum is approached—even if the problem is coded loosely. In the relative problem f_{rel}(2 3 4 1)$=$ 1.5 has a relatively low fitness, therefore the loose coding poses a greater challenge to the RKGA.

Summarizing the above, these results suggest the use of single-point crossover for the RKGA. Besides, the absolute ordering problem turns out to be less deceptive than its relative counterpart.

8.2 Experiment II: OmeGA versus RKGA

For the next experiment the OmeGA was organized as follows. The inner loop processed over 60 generations, including the juxtapositional phase which took 30 generations. The outer loop iterated over four eras and the maximum number of epochs was set to four. We used an empirically determined population size such that the global optimal solution could be found within four epochs. At the same time, we tried to keep the amount of function evaluations small. In the eras the following numbers of individuals were created:

era1: 750 individuals
era2: 1750 individuals
era3: 3250 individuals
era4: 6250 individuals.

During the BB filtering phase the chromosomes started with an initial length $\lambda^{(0)} = l$, equal to the problem size 32. With a length reduction factor $\rho = 0.5$ the strings were reduced to their corresponding BB length $\lambda^{(\mathcal{N})} = k$ at the end of the filtering phase. We used tiebreakers in this experiment, since it significantly contributed to the success of the OmeGA. For example, in era 3 there were 750 individuals of length 1, 1000 individuals of length 2, and 1500 individuals of length 3 present after the BB filtering process. For the juxtapositional phase we limited the maximum allowed string length to $2l$. The cut probability was set to 0.03.

The RKGA processed over 350 generations with a population size of 6250, equal to the maximal population size in the OmeGA. With these parameters we made 20 independent runs for the absolute and relative problems with all three codings. Since the experiments with tightly coded problems yielded almost the same results as with deflen6 coding, they are not presented here.

Deflen6 coding. In Figure 8 the maximum number of correct subfunctions found by the RKGA and OmeGA is plotted versus the number of function evaluations for the relative problem with deflen6 coding. Both algorithms found the global optimum in every run. In this and in all following plots the points corresponding to the filtering phases are omitted in the OmeGA curve for simplicity. The number of function evaluations includes evaluations during the initialization and the filtering phases.

The steps of the OmeGA curve indicate that throughout the main part of the juxtapositional phase usually no further correct subfunctions were found. The BBs discovered in the filtering phase were often mixed very fast thereafter. This is different for the absolute ordering problem. The curve in Figure 9 shows no salient steps, which suggests that some BBs were combined later in the juxtapositional phase.

This effect may be explained as follows. Considering an arbitrary relative o-schema of order k in a permutation problem of length l we find that there exist $\binom{l}{k}(l-k)!$ out of $l!$ possible permutations containing this schema. An

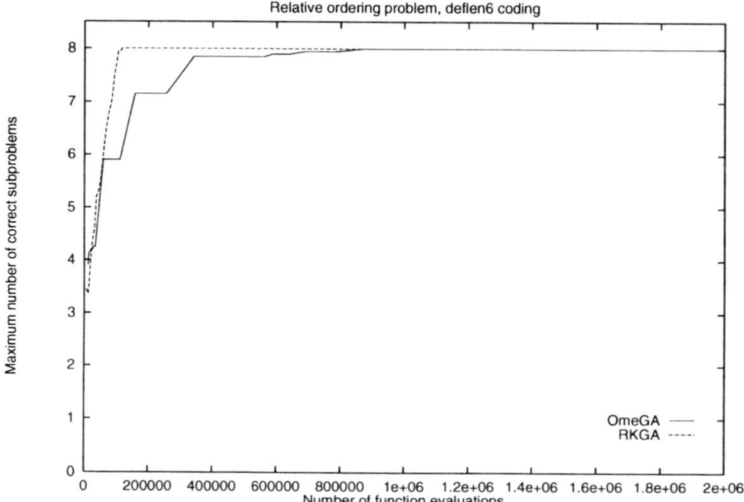

Fig. 8. Maximum number of correct subfunctions for the relative ordering problem with deflen6 coding found by the OmeGA and RKGA. The plots are averaged over 20 independent runs.

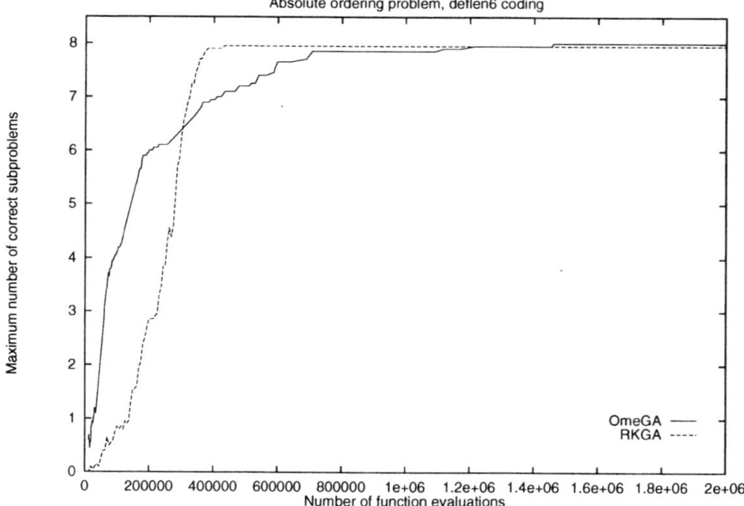

Fig. 9. Maximum number of correct subfunctions for the absolute ordering problem with deflen6 coding found by the OmeGA and RKGA. The plots are averaged over 20 independent runs.

arbitrary absolute o-schema of order k, on the contrary, appears in only $(l-k)!$ different permutations. In other words, a relative o-schema has more representatives than an absolute o-schema. Thus, a given random population would be expected to contain more versions of a certain relative ordering BB than of an absolute ordering BB. Since we used the same population sizes in our experiments for both problems, the population for the relative ordering problem was oversized leading to the effect of rapid BB mixing.

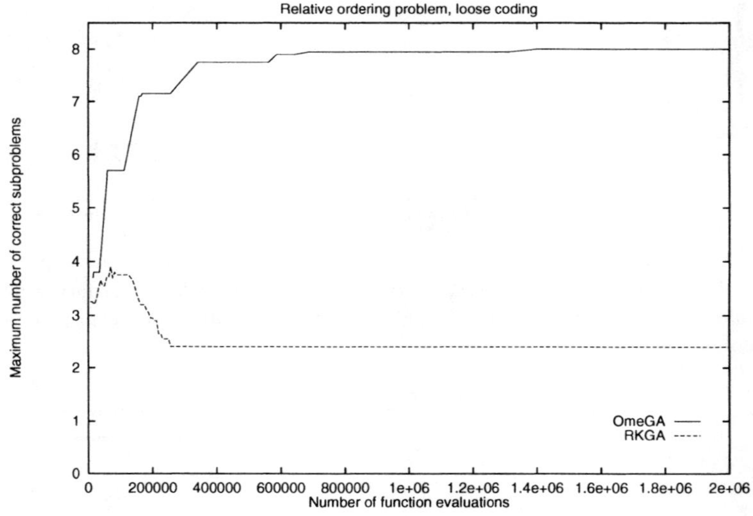

Fig. 10. Maximum number of correct subfunctions for the relative ordering problem with loose coding found by the OmeGA and RKGA. The plots are averaged over 20 independent runs.

Loose coding. Figures 10 and 11 demonstrate the performance of both GAs for the loose coding. The OmeGA found the global optimum of the absolute problem in 19 out of 20 runs. The optimum of the relative problem was found in every run. The maximum number of correct subfunctions discovered by the RKGA was seven in the absolute and five in the relative problem. The global solution was not found at all. Especially Figure 10 shows clearly how the RKGA is misled by the deceptive attractors: while it converges to highly fit solutions, the number of correct BBs decreases to a low value. In these two figures we observe the same effect we already noticed in the first experiment: the RKGA is less misled by f_{abs} than by f_{rel}. In the first case the RKGA curve converges to a value of 5.9 whereas in the second case it converges to 2.4.

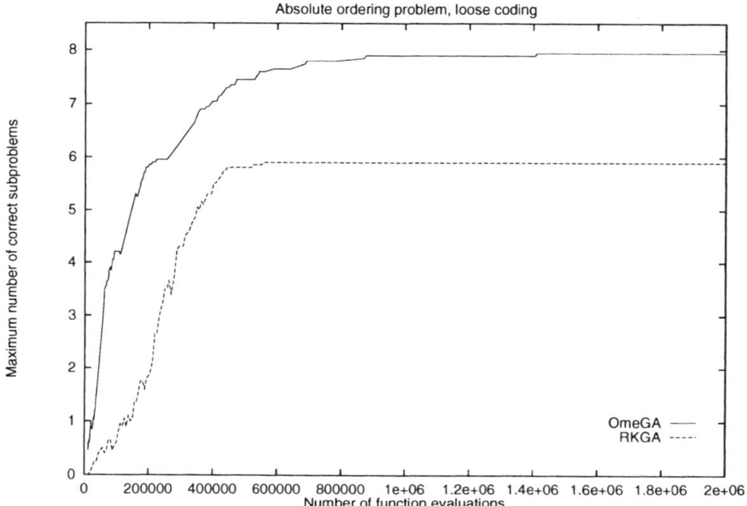

Fig. 11. Maximum number of correct subfunctions for the absolute ordering problem with loose coding found by the OmeGA and RKGA. The plots are averaged over 20 independent runs.

These results clearly show the OmeGA's relative independence from the underlying problem coding. For all tested codings the OmeGA curves have roughly the same appearance and the problem was completely solved in almost every run, whereas the RKGA succeeded only with tightly coded problems (deflen6 and tight coding). Although this is not a thorough scale-up analysis, the results suggest that the OmeGA has a significantly better scalability than the RKGA.

9 Scale-up Analysis

The scale-up behavior of an optimization algorithm can be described by the tradeoff between the problem size and the number of function evaluations it requires until convergence. In this section we investigate this tradeoff for the OmeGA in experiments with relative ordering deceptive problems of order 4 described in section 6. The computational complexity is expected to be the same as in the fmGA, which is no worse than $O(l^2)$ according to [9].

First, a description of population-sizing experiments is given. Based on the results the OmeGA's scale-up behavior is discussed and presented. Finally, the complexity of the RKGA is experimentally determined and compared to the OmeGA.

subfunction no.	subfunction genes
1	1, 17, 33, 49
2	2, 18, 34, 50
3	3, 19, 35, 51
4	4, 20, 36, 52
5	5, 21, 37, 53
6	6, 22, 38, 54
7	7, 23, 39, 55
8	8, 24, 40, 56
9	9, 25, 41, 57
10	10, 26, 42, 58
11	11, 27, 43, 59
12	12, 28, 44, 60
13	13, 29, 45, 61
14	14, 30, 46, 62
15	15, 31, 47, 63
16	16, 32, 48, 64

Table 3. This table shows the genes belonging to the subfunctions of the loosely coded f_{rel} problem of length $l = 64$.

9.1 The Scale-up Behavior of the OmeGA

To find out how many fitness function calls the OmeGA requires at least for solving a problem, one needs to perform a parameter adjustment at first. Therefore, we conducted population-sizing experiments for problem lengths 32, 64, 128, 192, 256, 384, and 512. For every problem length we tested the OmeGA with different population sizes and determined the most suitable size yielding maximum efficiency. The goal was to find at minimum 95 percent of all BBs. The test problems were composed of copies of the subfunction f_{rel}. We coded the problems loosely, such that the sum of the BB defining lengths was maximal. A sample loose coding is shown in Table 3 for the problem length $l = 64$.

The other OmeGA parameters were determined experimentally so that a good performance and a proper mixing could be achieved. The maximum chromosome length in the fmGA was limited to l. We did not use mutation and set the splice probability to one. For the cut probability p_κ we took the reciprocal value of one half of the problem length: $p_\kappa = 2/l$. The maximum number of epochs was chosen large enough such that the algorithm could find 95 percent of the BBs in all runs. In every epoch the OmeGA iterated over four eras of 65 generations each, including the BB filtering and the juxtapositional phase. The population sizes of the eras were set according to an empirical rule keeping a fixed proportion to the total population size n: $\lceil 0.1n \rceil$ individuals in the first, $\lceil 0.1n \rceil$ in the second, $\lceil 0.2n \rceil$ in the third, and $\lceil 0.6n \rceil$ in the fourth era. We used tie breaker and thresholding in our experiments.

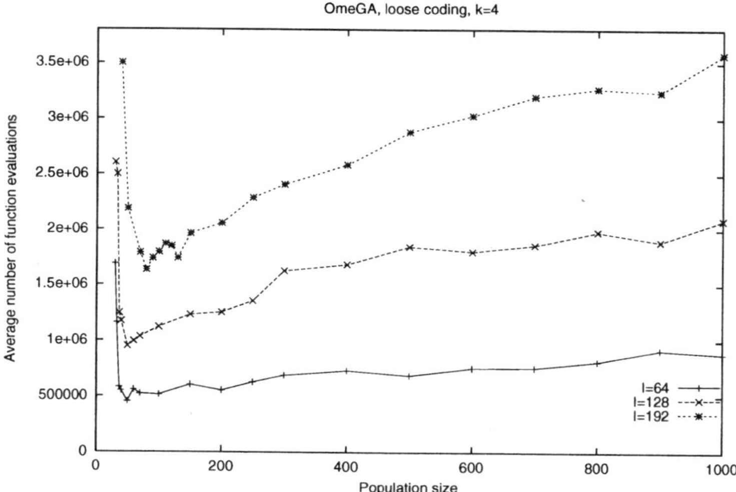

Fig. 12. Population-sizing results of the OmeGA. The average number of function evaluations is plotted versus the population size for the problem lengths 64, 128, and 192. All three curves have a minimum where the OmeGA performs the most efficiently.

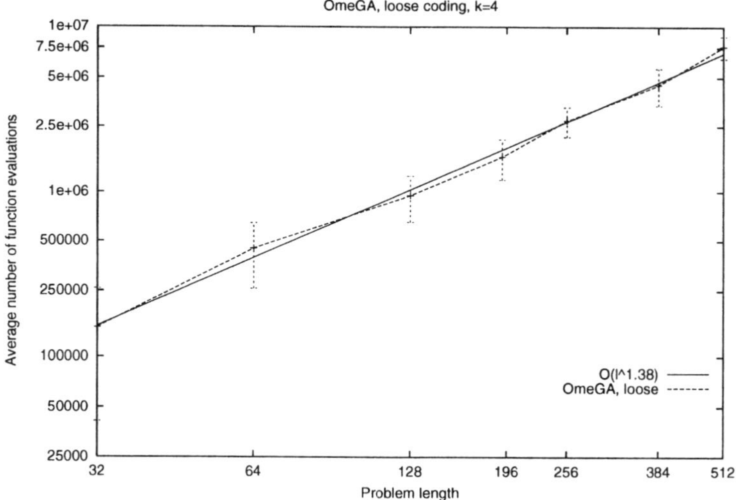

Fig. 13. Scale-up behavior of the OmeGA. The minimum number of function evaluations is plotted versus the problem length. Every point is averaged over 80 independent runs and plotted on a logarithmic scale. The solid regression line has a slope of 1.38 which clearly indicates a subquadratic scale-up behavior.

l	$n_{eco}^{(l)}$	function evaluations	epochs
64	50	451060	99.6
128	50	950922	222.2
256	80	1632983	263.6

Table 4. Population sizes used in the OmeGA for different eras. Here, n is the total population size.

The average number of function evaluations and iterated epochs is summarized in Table 4, and Figure 12 shows the population-sizing graphs for $l = 64$, 128, and 192, averaged over 80 independent runs. It is interesting that all three curves have a minimum for a relatively small population size. We call these population sizes *economic* and denote them by $n_{eco}^{(l)}$, as in the table.

The fact that with $n < n_{eco}^{(l)}$ the number of function calls becomes very large is not surprising. Consider a population with only a few members. Then, only a few, if any, BBs are expected to be expressed during the filtering phase.

As a result, the OmeGA would behave similar to a random search over the solution space. In this case, a large number of epochs and function evaluations would be required until convergence. The interesting features of the curves are their monotonously increasing parts for $n > n_{eco}^{(l)}$. They indicate that the best performance is achieved for a rather small population size and numerous epoch cycles. The number of function calls versus the problem length ranging from 32 to 512 is plotted in Figure 13 on a logarithmic scale. The regression line has a slope slightly smaller than 1.38, which indicates a complexity of $O(l^{1.38})$. Thus, the OmeGA scales up subquadratically as we expected.

9.2 Performance Comparison of the OmeGA and the RKGA

To investigate how the RKGA scales up for different codings, we performed population-sizing experiments on the same problems as in subsection 9.1. The population size was increased in constant steps of 200 individuals until 95 percent of all BBs could be found in 80 independent runs. We obtained the best results with the following RKGA parameters. The crossover probability was chosen to be 1.0 and no mutation was used at all. The recombination and selection operators were one-point crossover and binary tournament selection without replacement. The GA iterated over 300 generations in every run. We conducted experiments for the problem lengths 64, 128, 196, and 256 for two types of problem codings. In the first case the BBs were tight and in the second case the BBs had a defining length of 12. We refer to the last coding type as *deflen12 coding.* Figure 14 presents the minimum number of function evaluations versus the problem length for the RKGA (tight and deflen12 coding) and the OmeGA (loose coding). The results clearly show that the RKGA works well on tightly coded BBs. However, when the BBs are coded loosely, its performance significantly decreases and much more function evaluations are required to find the solution. In contrast, the OmeGA solves the problems independently from the underlying coding.

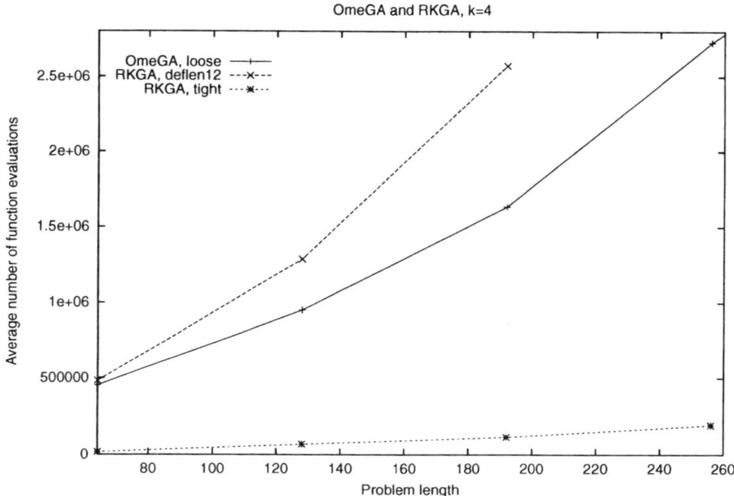

Fig. 14. The scale-up behavior of the OmeGA and the RKGA for $l = 64, 128, 192$, and 256. The average number of required function evaluations is plotted versus the problem length of the f_{rel} problems. The two RKGA curves correspond to tight and deflen12 coding. The OmeGA curve corresponds to loose coding.

10 Conclusions and Future Work

In this chapter we have developed the OmeGA and have successfully tested its performance in pilot experiments with ordering deceptive problems. By using an adaptive representation scheme, it becomes relatively independent from the underlying problem coding and outperforms the simple GA. The scale-up analysis has shown that the OmeGA's complexity is roughly $O(l^{1.4})$.

We summarize the benefits of the OmeGA:

- The OmeGA overcomes the linkage problem and is relatively coding independent.
- Relative and absolute ordering deceptive problems can be solved optimally.
- The OmeGA scales up subquadratically.
- The algorithm is robust and achieves a good performance with small population sizes.

For future research we recommend improving the BB filtering phase of the OmeGA and applying a more sensitive thresholding operator, as proposed by [16]. This might significantly increase the algorithm's performance, such that fewer function evaluations would be required for solving a problem. We also suggest applying the OmeGA to various combinatorial problems of commercial interest, for example scheduling problems [18], vehicle route planning, or integrated circuit design. Finally, new ordering deceptive functions with a higher degree of deception than f_{abs} and f_{rel} should be designed for further experiments with RK GAs.

In summary, more work needs to be done on the OmeGA. Nonetheless, the OmeGA is ready for a more extensive investigation in the real world of permutation problems. Scheduling, placement, and sequencing problems of various types should benefit from the application of a competent GA with demonstrated scalability and reliability in hard problems, thereby assuring robust solutions of a wide array of real-world problems.

Acknowledgments

The authors would like to thank Franz Rothlauf, who proposed the usage of random keys, and Martin Pelikan, Erick Cantu-Paz, Fernando Lobo, and Martin Butz for their useful comments and suggestions. The authors would like to give special thanks to Professor Hans-Paul Schwefel for encouraging the first author to spend a portion of his diploma thesis studies with the second author thus enabling this collaboration to take place. Professor Schwefel also provided important guidance and essential suggestions that greatly improved this work. The first author was partially supported by the "Studienstiftung des Deutschen Volkes"(Germany) and the Sigma-Xi Research Society.

The work was sponsored by the Air Force Office of Scientific Research, Air Force Materiel Command, USAF, under grants F49620-00-1-0163. Research funding for this work was also provided by the National Science Foundation under grant DMI-9908252. Support was also provided by a grant from the U. S. Army Research Laboratory under the Federated Laboratory Program, Cooperative Agreement DAAL01-96-2-0003. The U. S. Government is authorized to reproduce and distribute reprints for Government purposes notwithstanding any copyright notation thereon.

The views and conclusions contained herein are those of the authors and should not be interpreted as necessarily representing the official policies or endorsements, either expressed or implied, of the Air Force Office of Scientific Research, the National Science Foundation, the U. S. Army, or the U. S. Government.

References

1. Bandyopadhyay, S., Kargupta, H., & Wang, G. (1998) Revisiting the GEMGA: Scalable evolutionary optimization through linkage learning. *Proceedings of*

the *Fourth International Conference on Evolutionary Computation*, 603–608

2. Bean, J. C. (1994) Genetic algorithms and random keys for sequencing and optimization. *ORSA Journal on Computing*, *6*(c, 154–160
3. Blanton, Jr., J. L., & Wainwright, R. L. (1993) Multiple vehicle routing with time and capacity constraints using genetic algorithms. *Proceedings of the Fifth International Conference on Genetic Algorithms*, 452–459
4. Davis, L. (1985) Job shop scheduling with genetic algorithms. *Proceedings of an International Conference on Genetic Algorithms and Their Applications*, 136–140
5. Davis, L. (1991) A genetic algorithms tutorial. In Davis, L. (Ed.), *Handbook of Genetic Algorithms* (pp. 1–101). New York: Van Nostrand Reinhold
6. Goldberg, D. E. (1987) Simple genetic algorithms and the minimal, deceptive problem. In Davis, L. (Ed.), *Genetic Algorithms and Simulated Annealing* (Chapter 6, pp. 74–88). London: Pitman
7. Goldberg, D. E. (1993) Making genetic algorithms fly: A lesson from the Wright Brothers. *Advanced Technology for Developers*, *2*, 1–8
8. Goldberg, D. E. (1999) Using time effectively: Genetic-evolutionary algorithms and the continuation problem. *GECCO-99: Proceedings of the 1999 Genetic and Evolutionary Computation Conference*, *1*, 212–219
9. Goldberg, D. E., Deb, K., Kargupta, H., & Harik, G. (1993) Rapid, accurate optimization of difficult problems using fast messy genetic algorithms. *Proceedings of the Fifth International Conference on Genetic Algorithms*, 56–64
10. Goldberg, D. E., Deb, K., & Korb, B. (1990) Messy genetic algorithms revisited: Studies in mixed size and scale. *Complex Systems*, *4*, 415–444
11. Goldberg, D. E., Korb, B., & Deb, K. (1989) Messy genetic algorithms: Motivation, analysis, and first results. *Complex Systems*, *3*, 493–530. (Also TCGA Report 89003)
12. Goldberg, D. E., & Lingle, Jr., R. (1985) Alleles, loci, and the traveling salesman problem. *Proceedings of an International Conference on Genetic Algorithms and Their Applications*, 154–159
13. Harik, G. (1999) *Linkage learning via probabilistic modeling in the ECGA* (IlliGAL Report No. 99010). Urbana, IL: University of Illinois at Urbana-Champaign
14. Harik, G. R. (1997) *Learning gene linkage to efficiently solve problems of bounded difficulty using genetic algorithms.* Unpublished doctoral dissertation, University of Michigan, Ann Arbor. Also IlliGAL Report No. 97005
15. Janikow, C. Z., & Michalewicz, Z. (1991) An experimental comparison of binary and floating point representations in genetic algorithms. *Proceedings of the Fourth International Conference on Genetic Algorithms*, 31–36
16. Kargupta, H. (1995) *SEARCH, polynomial complexity, and the fast messy genetic algorithm* (IlliGAL Report No. 95008). Urbana, IL: University of Illinois at Urbana-Champaign
17. Kargupta, H., Deb, K., & Goldberg, D. E. (1992) Ordering genetic algorithms and deception. *Parallel Problem Solving from Nature - PPSN II*, 47–56
18. Knjazew, D., & Goldberg, D. E. (2000) *Application of the fast messy genetic algorithm to permutation and scheduling problems* (IlliGAL Report No. 2000022). Urbana, IL: University of Illinois at Urbana-Champaign, Illinois Genetic Algorithms Laboratory

19. Louis, S. J., & Rawlins, G. J. E. (1991) Designer genetic algorithms: Genetic algorithms in structure design. *Proceedings of the Fourth International Conference on Genetic Algorithms*, 53–60
20. Norman, B., & Bean, J. (1997) A random keys genetic algorithm for job shop scheduling. *Engineering Design and Automation*, *3*, 145–156
21. Pelikan, M., Goldberg, D. E., & Cantú-Paz, E. (1999) BOA: The Bayesian optimization algorithm. *GECCO-99: Proceedings of 1999 Genetic and Evolutionary Computation Conference*, *1*, 525–532
22. Schwefel, H. P. (1995) *Evolution and optimum seeking.* Sixth-Generation Computer Technology Series. New York: John Wiley & Sons.
23. Thierens, D., & Goldberg, D. E. (1993) Mixing in genetic algorithms. *Proceedings of the Fifth International Conference on Genetic Algorithms*, 38–45
24. Whitley, D., & Yoo, N.-W. (1994) Modeling simple genetic algorithms for permutation problems. *Foundations of Genetic Algorithms, 3*, 163–184

Effects of Adding Perturbations to Phenotypic Parameters in Genetic Algorithms for Searching Robust Solutions

Shigeyoshi Tsutsui[1] and Ashish Ghosh[2]

[1]Department of Management and Information Science, Hannan University, 5-4-33 Amamihigashi, Matsubara, Osaka 580, Japan
E-mail: tsutsui@hannan-u.ac.jp
[2]Machine Intelligence Unit, Indian Statistical Institute, 203 B. T. Road, Kolkata 700 035, India
E-mail: ash@isical.ac.in

Summary. We have proposed a scheme that extends the application of GAs to domains that require detection of robust solutions. We called this technique GAs/RS^3 – GAs with a robust solution searching scheme. In the GAs/RS^3, a perturbation is added to the phenotypic feature once for evaluation of an individual, thereby reducing the chance of selecting sharp peaks. We refer to this method as a single-evaluation model (SEM). In this chapter, we introduce a natural variant of this method, a multi-evaluation-model (MEM), where perturbations are given more than once for evaluation of the individual, and we offer comparative studies on their convergence property. The results showed that for the GAs/RS with SEM the population converges to robust solutions faster than with the MEM, and as the number of evaluations increases, the convergence speed decreases. We may conclude that the GAs/RS^3 with the SEM is more efficient than with the MEM. We also introduced a variation of the MEM, i.e., multi-evaluation model keeping the worst value (MEM-W), and provided a mathematical analysis.

1 Introduction

A large fraction of studies on genetic algorithms (GAs) emphasize finding global optimal solutions. There are many theoretical and empirical studies that investigate or present ways to improve the performance of conventional GAs for difficult function optimization problems such as those posed by multimodal and deceptive functions [1–4]. Some other investigations emphasize finding multiple solutions (peaks) including local optima [5–9].

If a solution obtained by search techniques is sensitive to small perturbations of its parameter values, it may not be appropriate for use in certain situations. For example, consider the problem of designing the optimal parameter values of a process control plant. Suppose, by some technique, we determine parameter values which yield very high performance from the plant. If the activity of the plant changes heavily due to a small variation in the parameter values, then it is very risky to use such a parameter set for the plant, because in practice some noise will always be involved with the parameter values. Let us consider another example. If the performance of a product is highly sensitive to the precision of its parts, then it will be very difficult or costly to produce this product by machines because each and every machine has a limited ability to handle precision. Thus in many optimization tasks, there is a need to determine solutions in which the value will not change much due to a small variation of the parameter values. We describe this type of solution as *robust*.

In previous studies [10–12], we have proposed a new scheme that extends the application of GAs to domains that require identification of robust solutions. We called this new technique GAs/RS^3 – GAs with a robust solution searching scheme. In GAs/RS^3, a perturbation is added to the phenotypic feature only once before evaluating an individual, thereby reducing the chance of selecting sharp peaks. Now, we refer to this method as a single-evaluation model (SEM).

In this chapter, we introduce a natural variant of this method, a multi-evaluation model (MEM), where perturbations are given more than once for evaluation of an individual, and offer comparative studies on the convergence property of the two methods.

This chapter is organized as follows. In the next section, we give a review of the basic concept of GAs/RS^3. In Section 3, we describe the MEM and give a mathematical analysis of it. In Section 4, an empirical analysis is given. In Section 5, we consider a variation of the MEM. Finally, concluding remarks are made in Section 6.

2 Review of the GAs/RS^3 [10–12]

2.1 The Basic Concept

In nature, the phenotypic expression of an organism is determined in part by decoding the genotypic code of genes in the chromosomes. During this decoding process there may be some perturbations, e.g., an abnormal temperature, a nutritional imbalance, existence of injurious matter, etc. Stated loosely, if the individual has low fitness due to these perturbed phenotypic features, then the individual will not survive to produce offspring. Thus, individuals and reproductive populations having *good* genotypic material would become extinct if they were highly sensitive to perturbations of phenotypic features. On the other hand, in noisy environments, reproductive units which are robust to these perturbations would have a better chance of surviving. We developed GAs/RS^3 with an aim of locating robust solutions by using this sort of natural genetic metaphor. GAs/RS^3 uses the effect of perturbation of the phenotypic parameters while evaluating the functional values of individuals.

Approaches which give consideration to the existence of noise in calculating the fitness values are discussed in [13–15]. These efforts are mainly directed towards studying noisy fitness functions, i.e., noise is added to the fitness function. If $\boldsymbol{X} = (x_1, x_2, ..., x_m)$ is a phenotypic parameter vector, $f(\boldsymbol{X})$ the evaluation function, and δ a scalar noise, then the fitness of the individual will be $f(\boldsymbol{X}) + \delta$. However, when we aim to detect robust solutions, it should be noted that with the natural phenomena, noise is added during the process of decoding the genotypic codes to the phenotypic parameters. Hence, to add noise to the phenotypic parameter $\boldsymbol{X}$, i.e., to use an evaluation function of the form $f(\boldsymbol{X}+\boldsymbol{\Delta})$, appears reasonable, where $\boldsymbol{\Delta} = (\delta_1, \delta_2, ..., \delta_m)$ is a random vector. The solutions thus determined are expected to be more robust against perturbations.

Let $\boldsymbol{G}$ be a genotypic string (or chromosome) which generates the phenotypic parameter $\boldsymbol{X}$. Then the model of GAs/RS[3] becomes as shown in Fig. 1. Here, it should be noted that adding noise in the form $f(\boldsymbol{X}+\boldsymbol{\Delta})$ may appear to be a mutation operation on a real-valued coding, but actually it is operationally different from mutation, since it does not have any direct effect on individual strings. The perturbations are used only for judging the quality of a solution and for selection.

```
gen = 0;
Pop(gen) = randomly initialized population {G^1, G^2, ..., G^N};
Decode each genotype G^i to produce the corresponding phenotype X^i;
Add noise and set Y^i = X^i + Δ;
Evaluate f(Y^i) for all i in Pop(gen);
while( termination condition == false ){
      gen += 1;
      Select Pop(gen) from Pop(gen - 1) based on f(Y^i);
      Apply genetic operators to {G^1, G^2, ..., G^N} in Pop(gen);
      Decode each G^i to produce X^i;
      Set Y^i = X^i + Δ;
      Evaluate f(Y^i) for all i in Pop(gen);
}
Evaluate f(X^i) for all i in Pop(gen);
```

Fig. 1. Schematic model of GA/RS[3]

2.2 Mathematical Model of GAs/RS[3]

In this model, we first assume $\boldsymbol{X}$ to be one-dimensional and denote $\boldsymbol{X}$ and $\boldsymbol{\Delta}$ by x and δ, respectively. Then, extension to the multidimensional case is discussed.

Effective Evaluation Function

Let N represent the population size. The *effective evaluation function* $F(x)$ of $f(x)$ is defined as

$$F(x) = \int_{-\infty}^{\infty} f(x+\delta) \cdot q(\delta) d\delta, \tag{1}$$

where it is assumed that x and δ are mutually independent, $q(\delta)$ is the continuous density function of noise δ having defined mean value. Here, the effective evaluation function means that when $N \to \infty$ the population under GAs/RS[3] evolves so as to maximize the effective evaluation function $F(x)$ as opposed to maximizing the actual evaluation function $f(x)$.

As is easily understood, $F(x)$ is equivalent to the expected value of $f(x)$ over $x+\delta$. If we assume $q(\delta)$ to be symmetric, i.e., $q(\delta) = q(-\delta)$, then $F(x)$ can be rewritten as

$$F(x) = \int_{-\infty}^{\infty} q(x-y) \cdot f(y) dy. \tag{2}$$

In practice, the population size must be finite. If the population is sufficiently large, then this may yield the approximate characteristics indicated in Eq. (2). In general, the sufficiency of the population size will depend on the distribution of the noise. Hereafter, Eq. (2) is used for further discussion and a Gaussian noise $N(0, \sigma)$ is assumed.

Reduction Factor

The appropriate size (σ) of a Gaussian noise to be added can be estimated, depending on the actual function, when we assume the width of the sharp peak to be known. For the sake of simplicity, the peaks of functions are represented by rectangles, and for illustration we define one rectangle having height h ($h > 0$) and width $2w$ ($w > 0$) as follows:

$$f(x) = \begin{cases} h : -w \le x \le w \\ 0 : \text{otherwise.} \end{cases} \tag{3}$$

The peak of this function is spread from $-w$ to w. The effective evaluation function corresponding to this is obtained from Eq. (2) as

$$\begin{aligned} F(x) &= h \int_{-w}^{w} q(x-y) dy \\ &= h\left[\Phi\left(\frac{x+w}{\sigma}\right) - \Phi\left(\frac{x-w}{\sigma}\right)\right], \end{aligned} \tag{4}$$

where $\Phi(x)$ is a normal distribution. By setting the derivative of the function $F(x)$ to zero, the peak point is obtained at $x = 0$ and the peak value, max $F(x)$, is obtained as

$$\begin{aligned} \max F(x) &= F(0) \\ &= h[2\Phi(w/\sigma) - 1] \\ &= h \times R(w/\sigma). \end{aligned} \tag{5}$$

Here, $R(w/\sigma) = 2\Phi(w/\sigma) - 1$ is the *reduction factor*.

Fig. 2 shows the relationship between function $f(x)$ and function $F(x)$ for w/σ = 4.0, 2.0, 1.0, 0.5, and 0.25. Fig. 2 confirms that the addition of Gaussian noise to phenotypic parameters reduces the effective height of the peaks and the effect is greater as the value of δ becomes larger.

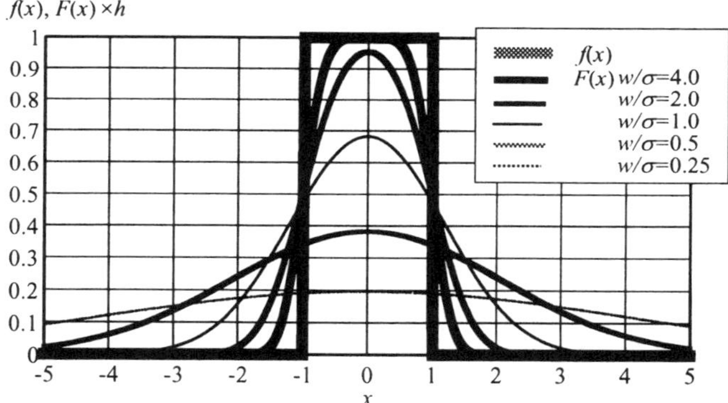

Fig. 2. Relationship between $f(x)$ and $F(x)$

Reduction Factor for Multiple Dimensions

Now we emulate a peak in n-dimensional search space by an n-dimensional hyper box function having height h ($h > 0$), width $2w_i$ ($w_i > 0$) along the x_i axis. Its effective function $F(\boldsymbol{X})$ is as follows:

$$F(X) = h\prod_{i=1}^{n}\left[\Phi\left(\frac{x_i + w_i}{\sigma_i}\right) - \Phi\left(\frac{x_i - w_i}{\sigma_i}\right)\right], \tag{6}$$

where σ_i is the standard deviation of the Gaussian noise added to the phenotypic parameter x_i. From Eq. (6) we can calculate R_n, the reduction factor for an n-dimensional search space, as follows:

$$\begin{aligned} R_n &= \max F(X)/h \\ &= \prod_{i=1}^{n} R(w_i/\sigma_i) \end{aligned} \tag{7}$$

where

$$R(w_i/\sigma_i) = 2\Phi(w_i/\sigma_i) - 1. \tag{8}$$

The reduction factor R_n is the product of the reduction factor in each dimension i (see Eq. (5)). Fig. 3 plots the reduction factor vs. w_i/σ_i.

Estimation of the Amount of Noise to be Added

The amount of noise to be added (σ) can be estimated given the width ($2w$) and the reduction factor of the peak. Let us consider the one-dimensional case. If σ

takes values in the range [2w, 4w], then w/σ has values in the range [0.5, 0.25] and the reduction factor is between 0.197 and 0.383 (see n = 1 in Fig. 3). Thus, σ can be roughly estimated when the allowable width of the sharp peaks and their reduction factors are given.

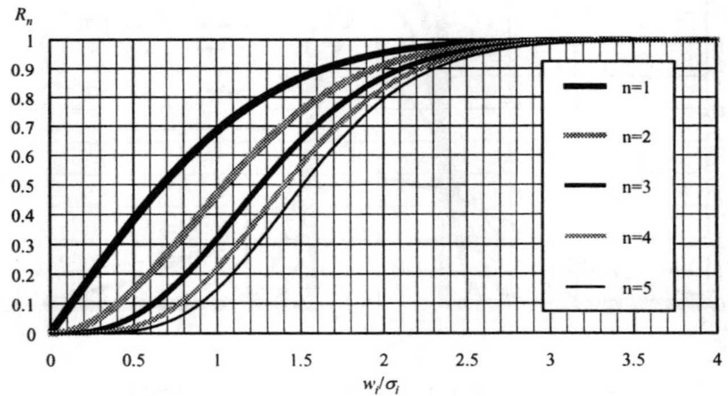

Fig. 3. Reduction factor R_n vs. w_i/σ_i

3 MEM

In GAs/RS[3] in Section 2, we give a perturbation once for each evaluating functional value of an individual. In this section we present GA/RS[3] with an MEM, where perturbations are given more than once for evaluating an individual, and compare it with the SEM described in Section 2. In [16], Wiesmann et al. also used the multi-evaluation method with ES (Evolution Strategies) for the design of multi-layer optical coatings.

In the MEM, $f(x_i)$, the functional value of individual i with phenotypic parameter value x_i, is obtained as a mean value over m functional values as

$$f(x^i) = \frac{1}{m}\sum_{j=1}^{m} f(x^i + \delta_j), \tag{9}$$

where $m > 1$ is the number of evaluations and δ_js are mutually independent noise. The effective evaluation function (defined in Section 2) for the MEM is obtained as

$$\begin{aligned} F(X) &= \oint\!\!\oint\!\!\oint_{\delta_j} \left(\frac{1}{m}\sum_{j=1}^{m} f(x+\delta_j) \right) d\delta_1 \cdots d\delta_m \\ &= \frac{1}{m}\sum_{j=1}^{m} \int_{-\infty}^{\infty} f(x+\delta_j) \cdot q(\delta_j) d\delta_j \\ &= \int_{-\infty}^{\infty} f(x+\delta) \cdot q(\delta) d\delta. \end{aligned} \tag{10}$$

Thus, Eq. (2) holds for the MEM if we assume $q(\delta)$ to be symmetric. When a Gaussian noise $N(0, \sigma)$ is assumed, the effective evaluation functions $F(x)$ for the MEM are obtained by Eqs. (4) and (6), and the reduction factors are obtained by Eqs. (5) and (7), respectively.

Although the mathematical structure is the same for the SEM and the MEM with population size N, we may get different convergence rates for them since population size must be finite. Thus, we may say:

(1) the larger the value of m (N being finite) the more accurately we may emulate the mathematical model of Section 2 for $N \to \infty$

(2) with larger values of m, the number of function evaluations required to converge will increase. As a result, the algorithm becomes costly.

4 SEM vs. MEM

Here we compare the SEM and the MEM by observing the convergence processes of both the models.

4.1 Test Functions

For test functions, we use the following three one-dimensional functions and one two-dimensional function.

1) Function f_a: Consider a function f_a (Fig. 4 below) defined as

$$f_a = \begin{cases} 1 : -1 \le x \le 1 \\ 2 : 1.5 \le x \le 1.7 \\ 0 : \text{otherwise} \end{cases} \tag{11}$$

having one wide peak and one sharp peak.

The parameter range is $-3 \le x \le 3$. We took $\sigma = 0.4$ ($w/\sigma = 0.25$).

2) Function f_b: Function f_b (Fig. 6 below) has five unequal peaks in the range $0 \le x \le 1$. It is defined as

$$f_b = \begin{cases} e^{-2\ln 2\left(\frac{x-0.1}{0.8}\right)^2} \left|\sin(5\pi x)\right|^{0.5} : 0.4 < x \le 0.6 \\ e^{-2\ln 2\left(\frac{x-0.1}{0.8}\right)^2} \sin^6(5\pi x) \; : \text{otherwise.} \end{cases} \tag{12}$$

As shown in Fig. 6 below, the global optimum is located at x = 0.1 with the functional value 1.0. There are four sharp peaks. The third peak is wide compared to the others and is located at x = 0.486 with functional value 0.715. The effective width of the four sharp peaks can be estimated as in [11]. Referring to Fig. 3, w/σ can be chosen in (0, 0.65) so as to reduce the effective functional value by more than 50%. We chose $w/\sigma = 0.5$. Thus, $\sigma = 0.0625$ ($\sigma = w/0.5 = 21/32$) was used. The reduction factor R_1 is about 0.4.

3) Function f_c: Function f_c (Fig. 8 below) is similar in shape to function f_c, and defined as

$$f_c = \begin{cases} 0 & : 0.4 < x \leq 0.46 \\ e^{-2\ln 2\left(\frac{x-0.1}{0.8}\right)^2} \left|\sin(5\pi x)\right|^{0.3} & : 0.46 < x \leq 0.6 \\ e^{-2\ln 2\left(\frac{x-0.1}{0.8}\right)^2} \sin^6(5\pi x) & : \text{otherwise.} \end{cases} \quad (13)$$

Function f_c has a cliff at $x = 0.46$. The global optimum is located at $x = 0.1$ with the functional value 1.0. The third peak is wider compared to the others and is located at $x = 0.478$ with functional value 0.721. The value of $\sigma = 0.0625$ was used in this function too.

4) Function f_d: The two-dimensional function f_d is an extension of the function f_b (Eq. (12)). This function has 25 peaks as shown in Fig. 10 below. The central peak is broad compared to the other 24 peaks and is located at $x_1 = x_2 = 0.486$ with functional value 0.715. The highest sharp peak is located at $x_1 = x_2 = 0.1$ with functional value 1.0. Here, the noise size was $\sigma_1 = \sigma_2 = 0.0625$, which is the same as used in the one-dimensional function f_b. The value of the reduction factor R_2 for the sharp peaks can be calculated from Eq. (7) as 0.16 ($= 0.4 \times 0.4$).

4.2 Convergence Processes

We use a simple GA (hereafter we refer to it as the SGA) [17]. The GA parameters were kept constant for all the simulations: population size $N = 100$, mutation probability $p_m = 0.006$, crossover probability $p_c = 0.6$, maximum number of trials = 10,000. The phenotypic parameters are encoded by a 30-bit string (Gray coded). We did 30 simulations for each experiment.

The MEM's performance is tested for $m = 1, 2, \ldots, 5$. Here, $m = 1$ corresponds to the SEM. The convergence process was checked by observing the mean value of the parameters in the population over 30 experiments.

Fig. 4 shows a typical distribution of the individuals in function f_a for the SGA and GAs/RS^3. For the SGA, the population converges around the highest peak centered at $x = 0.5$. However, for GAs/RS^3 the population converges to the wider peak centered at $x = -1.0$. Fig. 5 shows the variation of mean value (over the population) of x in function f_a with respect to function evaluation. In function f_a, the population converged to the stable peak faster with the SEM, i.e., $m = 1$, than with the MEM, i.e., $m = 2$, 3, 4, and 5. The case $m = 5$ showed the slowest speed of convergence.

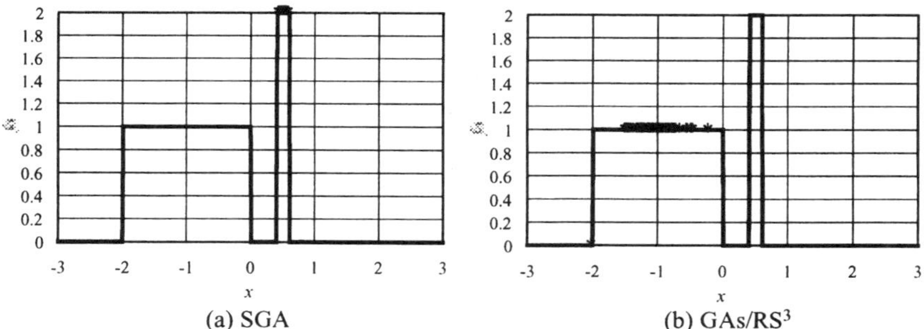

(a) SGA (b) GAs/RS³

Fig. 4. A typical distribution of individuals in function f_a after 10000 function evaluations for (a) the SGA and for (b) GAs/RS^3 with $N = 100$ and $m = 1$

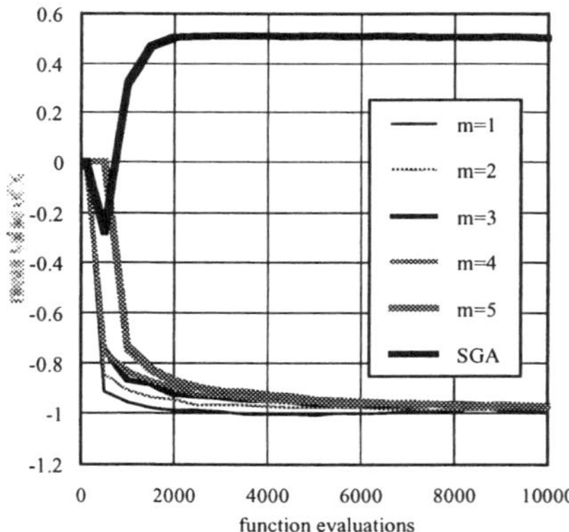

Fig. 5. Convergence process showing the variation of mean value (over the population) of x in function f_a with respect to function evaluations.

Fig. 6 shows a typical distribution of the individuals in function f_b for the SGA and GAs/RS^3. For the SGA the population converges around the highest peak centered at $x = 0.1$. However, for GAs/RS^3 the population converges to the wider stable peak centered at $x = 0.486$. Fig. 7 shows the variation of mean value (over the population) of x in function f_b with respect to function evaluation. In function f_b, the population converged to the stable peak faster with the SEM, i.e., $m = 1$, than with the MEM, i.e., $m = 2$, 3, 4, and 5. $m = 5$ showed the slowest speed of convergence in this function too.

Fig. 8 shows a typical distribution of the individuals in function f_c for the SGA and GAs/RS^3. For the SGA the population converges around the highest peak centered at $x = 0.1$. However, for GAs/RS^3 the population converges to the wider stable peak. Although the highest point of this wide peak is located at $x = 0.478$, the population converge to the right side of the highest point (around $x = 0.52$), away from the cliff, where it seems a robust solution. Fig. 9 shows the variation of mean value (over the population) of x in function f_c with function evaluation. The population converged around that point faster with the SEM, i.e., $m = 1$, than with the MEM, i.e., m = 2, 3, 4, and 5. As usual, m = 5 showed the slowest speed of convergence again.

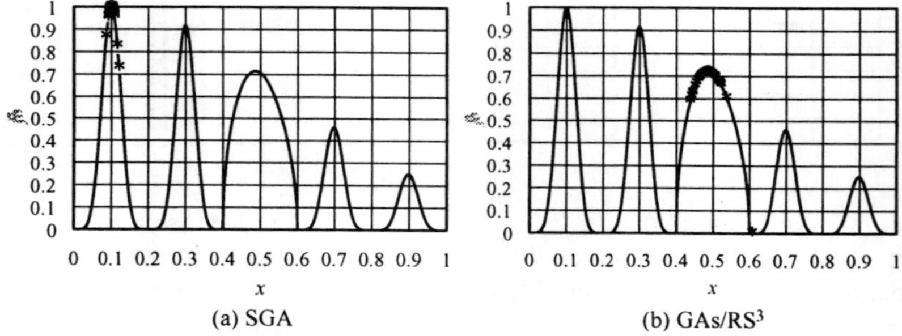

Fig. 6. A typical distribution of the individuals in f_b after 10000 function evaluations for (a) the SGA and for (b) GAs/RS^3 with N = 100 and m = 1

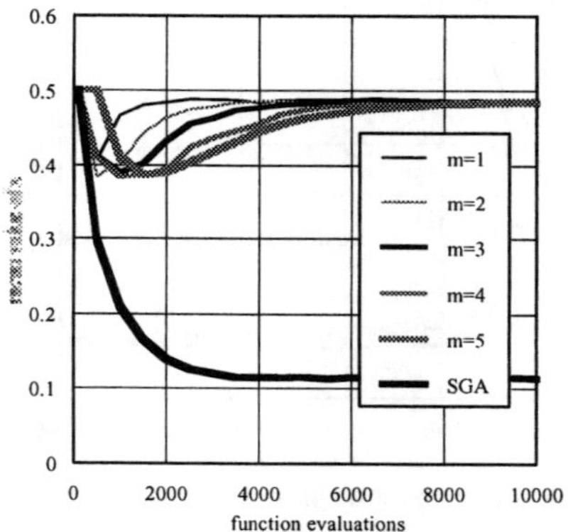

Fig. 7. Convergence process showing the variation of mean value (over the population) of x in function f_b with respect to function evaluations.

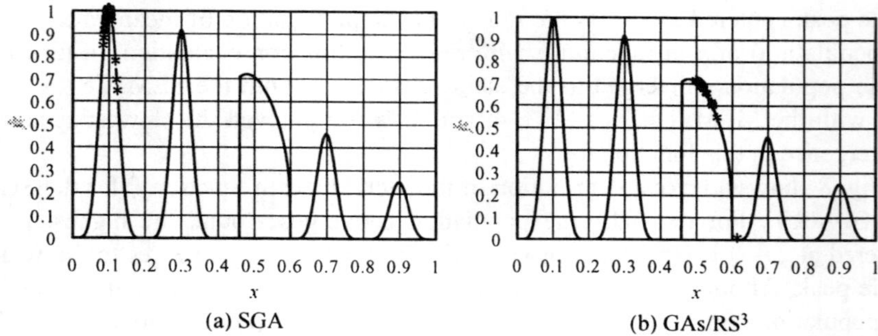

Fig. 8. A typical distribution of the individuals in f_c after 10000 function evaluations for (a) the SGA and for (b) GAs/RS^3 with N = 100 and m = 1

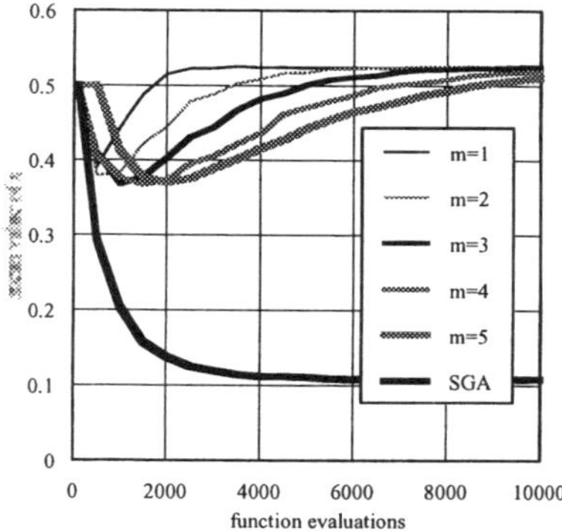

Fig. 9. Convergence process showing the variation of mean value (over the population) of x in function f_c with function evaluations.

Thus, in one-dimensional functions f_a, f_b, and f_c, we can observe that for GAs/RS^3 with the SEM the population converges to robust solutions faster than with the MEM; and as the value of m increases, the convergence speed decreases. This tendency is also seen for the two-dimensional function f_d (Fig. 10), as seen in Fig. 11, which shows the variation of mean value (over the population) of x_1 in function f_d with respect to function evaluation. The variation of mean value of x_2 was almost the same as that of x_1.

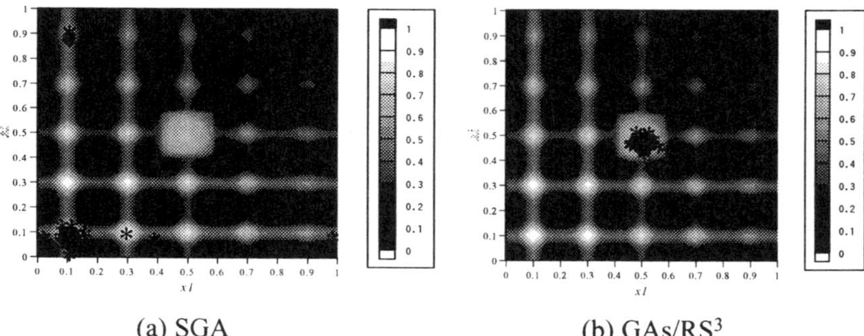

(a) SGA (b) GAs/RS^3

Fig. 10. A typical distribution of the individuals in function f_d after 10000 function evaluations for (a) the SGA and for (b) GAs/RS^3

5 Consideration of a Variation of the MEM

In the MEM described in Sections 3 and 4, the functional value of individual i with phenotypic parameter value x_i is obtained as a mean value over m functional values, $f(x_i + \delta_j)$, $j = 1, 2, ..., m$. In this section, we consider a variation of the MEM, i.e,, multi-evaluation model keeping the worst value (MEM-W), and explore the reduction factor of GAs/RS^3 with the MEM-W.

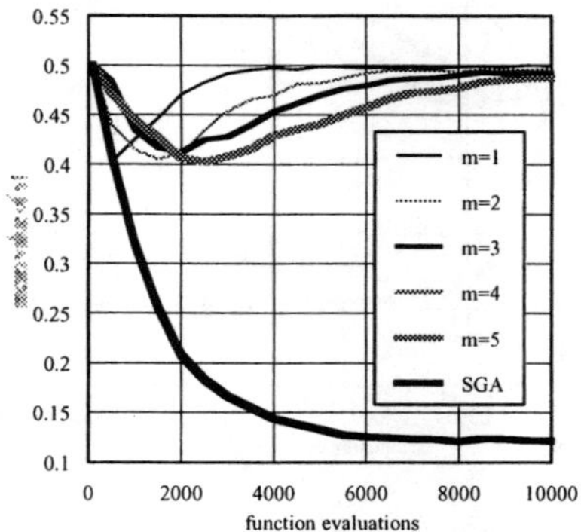

Fig. 11. Convergence process showing the variation of mean value (over the population) of x_1 in function f_d with respect to function evaluations.

In the MEM-W, we define the functional value of individual i by the worst value of the m functional values as

$$f(x^i) = \min_{j=1,2,,\ldots,m} f(x^i + \delta_j). \tag{14}$$

The MEM-W may be applicable where the robustness of a solution becomes more crucial. We can see the characteristics of the MEM-W by obtaining its reduction factor. Since each m functional value in Eq. (14) is identical to the functional value in the MEM, its effective evaluation function can be obtained when $N \to \infty$ as

$$F_j(x) = \int_{-\infty}^{\infty} q(x-y) \cdot f(y) dy. \tag{15}$$

We represent a peak by a rectangle defined by Eq. (2) and assume Gaussian noise $N(\sigma, 0)$. Then $F_j(0)$, the max F_j (x), is obtained as follows:

$$\max_x F_j(x) = F_j(0)$$
$$= \begin{cases} h: \text{with probability } 2\Phi(w/\sigma) - 1 \\ 0: \text{with probability } 2 - 2\Phi(\text{w}/\sigma). \end{cases} \tag{16}$$

In the MEM-W, max $F(x) = F(0) = h$ if $F_j(0) = h \; \forall \, j = 1, 2, \ldots, m$, simultaneously, otherwise it has a minimum value of 0. Thus, we get max $F(x) = \text{F}(0)$ as follows:

$$\max_x F(x) = F(0)$$
$$= \begin{cases} h: \text{with probability } (2\Phi(w/\sigma) - 1)^m \\ 0: \text{with probability } 1 - (2\Phi(w/\sigma) - 1)^m. \end{cases} \tag{17}$$

The reduction factor $R^{(m)}(w/\sigma)$ for the MEM-W is obtained from Eq. (17) as

$$
\begin{aligned}
R^{(m)}(w/\sigma) &= F(0)/h \\
&= \{h\times(2\Phi(w/\sigma)-1)^m + 0\times(1-(2\Phi(w/\sigma)-1)^m)\} \\
&= (2\Phi(w/\sigma)-1)^m. \qquad (18)
\end{aligned}
$$

From Eq. (18), we can confirm that the MEM-W has a stronger peak reduction effect than the MEM. As *m* becomes larger, the reduction effect becomes stronger. We can easily obtain the reduction factor $R_n^{(m)}$ as

$$R_n^{(m)} = \prod_{i=1}^{n}(2\Phi(w_i/\sigma_i)-1)^m \qquad (19)$$

for *n*-dimensional search space, as described in Section 2.

6 Conclusion

In previous studies, we have proposed a new scheme that extends the application of GAs to domains that require identification of robust solutions. We called this new technique GAs/RS^3: GAs with a robust solution searching scheme. In GAs/RS^3, a perturbation is added to the phenotypic feature once for each evaluation of an individual, thereby reducing the chance of selecting sharp peaks. In this study, we referred to this method as a single-evaluation-model (SEM).

In this chapter, we introduced a natural variation of this method, a multi-evaluation model (MEM), where perturbations are given more than once for each evaluation of an individual, and offer comparative studies on convergence properties.

The results showed that for GAs/RS^3 with the SEM, the population converges to robust solutions faster than with the MEM, and as the number of evaluations of the MEM increases, the convergence speed decreases. Thus we may conclude GAs/RS^3 with the SEM is more efficient than with the MEM. In this chapter, we also introduced a variant of the MEM, the multi-evaluation model keeping the worst value (MEM-W), which may be applicable where the robustness of solution becomes more crucial, and explored the peak reduction property mathematically.

Future work will focus on analyzing the behavior of GAs/RS^3 with the SEM, the MEM, and the MEM-W on more complex problems where many peaks interact.

Acknowledgements

This research is partially supported by the Ministry of Education, Culture, Sports, Science and Technology of Japan under Grant-in-Aid for Scientific Research number 13680469 and a grant to RCAST at Doshisha University from the Ministry of Education, Culture, Sports, Science and Technology, Japan

References

[1] Eshelman, L. J., "The CHC Adaptive Search Algorithm: How to Have Safe Search When Engaging in Nontraditional Genetic Recombination," *Foundations of Genetic Algorithms*, edited by G. J. E. Rawlins, Morgan Kaufmann, San Mateo, CA, pp. 265-283, 1991.

[2] Goldberg, D. E. , Deb, K. and Korb, B., "Messy Genetic Algorithms Revisited: Studies in Mixed Size and Scale," *Complex Systems*, no. 4, pp. 415-444, 1990.

[3] Mathias, K. and Whitley, L. D., "Changing Representations During Search: A Comparative Study of Delta Coding," *Evolutionary Computation*, vol. 2, no. 3, pp. 249-278, 1994.

[4] Tsutsui, S., Fujimoto, Y. and Ghosh, A., "Forking GAs: GAs with Search Space Division Schemes," *Evolutionary Computation*, vol. 5, no. 1, pp. 61-80, 1997.

[5] De Jong, K. A., "An Analysis of the Behavior of a Class of Genetic Adaptive Systems, *A dissertation submitted in partial fulfillment of the requirements for the degree of Doctor of Philosophy (Computer and Communication Sciences) in the University of Michigan*, 1975.

[6] Goldberg, D. E. and Richardson, J., "Genetic Algorithms with Sharing for Multimodal Function Optimization," *Proc. of the 2nd Int. Conf. on Genetic Algorithms*, J. J. Grefenstette (Ed.), Lawrence Erlbaum, Hillsdale, NJ, pp. 41-49, 1987.

[7] Horn, J. and Nafpliotis, N. "Multiobjective Optimization Using the Niched Pareto," *IlliGAL Report no. 93005*, University of Illinois, Urbana-Champaign, 1993.

[8] Mahfoud, S. W., "Crowding and preselection revisited," *Proc. of the 2nd Conf. on Parallel Problem Solving from Nature*, R. Mnner and B. Manderick (Eds.), North-Holland, Amsterdam, pp. 27-34, 1992.

[9] Beasley, D., Bull, D. R. and Martin, R. R., "A Sequential Niche Technique for Multimodal Function Optimization," *Evolutionary Computation*, vol. 1, no. 2, pp. 101-125, 1993.

[10] Tsutsui, S., Ghosh, A. and Fujimoto, Y., "A Robust Solution Searching Scheme in Genetic Search," *Proc. of the the Fourth Int. Conf. on Parallel Problem Solving from Nature (PPSN IV)*, Springer, Berlin, pp. 543-552, 1996.

[11] Tsutsui, S. and Ghosh, A., "Genetic Algorithms with a Robust Solution Searching Scheme", *IEEE Trans. on Evolutionary Computation*, vol. 1, no. 3, pp. 201-208, 1997.

[12] Tsutsui, S., "On Properties of the Genetic Algorithms with a Robust Solution Searching Scheme in Multidimensional Search Spaces", *Journal of Japanese Society for Artificial Intelligence*, vol. 13, no. 6, pp. 142-147, 1998.

[13] Fitzpatrick, J. M. and Grefenstette, J. J., "Genetic Algorithms in Noisy Environments," *Machine Learning*, vol. 3, pp. 101-120, 1988.

[14] Hammel, U. and Baeck, T., "Evolution Strategies on Noisy Functions: How to Improve Convergence Properties," *Proc. of the Third Int. Conf. on Parallel Problem Solving from Nature (PPSN III)*, Springer, Berlin, pp. 159-168, 1994.

[15] Miller, B. L. and Goldberg, D. E., "Genetic Algorithms, Selection Scheme, and the Varying Effect of Noise," *Evolutionary Computation*, vol. 4, no. 2, pp. 113-131, 1996.

[16] Wiesmann, D., Hammel, U. and Baeck, T., "Robust Design of Multilayer Optical Coatings by Means of Evolutionary Algorithms", *IEEE Trans. on Evolutionary Computation*, vol. 2, no. 4, pp. 162-167, 1998.

[17] Goldberg, D. E., *Genetic Algorithms in Search, Optimization and Machine Learning*, Addison-Wesley, Reading, MA, pp. 28-33, 1989.

[18] Tsutsui, S., "A Comparative Study on the Effects of Adding Perturbations to Phenotypic Parameters in Genetic Algorithms with a Robust Solution Searching

Scheme," *Proc. of the 1999 IEEE Systems, Man, and Cybernetics Conf. (SMC'99 Tokyo)*, pp. III-585-591, 1999.

Evolution of Strategies for Resource Protection Problems

William M. Spears and Diana F. Gordon-Spears

Department of Computer Science
College of Engineering
University of Wyoming
Laramie, Wyoming 82071, USA
E-mail: {wspears,dspears}@cs.uwyo.edu

Summary. The objective of this project is to develop effective finite-state machine (FSM) strategies for winning against an adversary in a Competition for Resources simulation. To achieve this goal, we evolve these strategies in a simulated environment and compare a variety of evolutionary methods in this context. Key empirical questions are addressed, such as how many FSM states are optimal, how effective is it to use an evolutionary algorithm that adapts the number of states, and how can one reduce the variance in fitness evaluation? Some of our experimental answers to these questions are quite intriguing. This chapter also explores and evaluates novel algorithms for detecting and repairing deleterious cycles in the evolved FSMs.

1 Introduction

We are becoming increasingly dependent on large interconnected networks for the control of our resources, such as the Internet, communications networks, and power grids. The advantage of these networks is the ability to route resources in a reasonably optimal fashion. However, their interconnectivity, coupled with the lack of global view of what is happening in these networks, can lead to tremendous problems in network reliability. For example, small local failures can easily propagate to entire networks, causing loss of service and corruption of data. Also, deliberate attacks (such as "denial of service" attacks) can also easily cause widespread havoc (Denning, 1999).

Thus one important issue is the development of effective network traversal strategies to protect as many resources as possible from failure and/or attacks, i.e., to maximally restrict the number of resources damaged. To address this issue we have decided to create a novel "resource protection" simulation that captures the essential aspects of this problem. A "defender" agent attempts to protect resources before they are damaged by an intentional (or unintentional) "adversary".

Our primary goal then is to create sophisticated reactive strategies for the defender. We use finite-state machines (FSMs) for our strategies, since there are a number of precedents for FSMs being effective strategies for adversarial situations. We use evolutionary algorithms (EAs) to evolve the FSMs, since there is ample precedent for the effectiveness of this approach (Fogel, Owens, and Walsh, 1966; Fogel, 1995; Fogel, 1999).[1] A secondary goal of our research is to highlight a number of important issues with respect to the evolution of FSMs: their representation, the application of mutation and recombination, methods for adapting the number of states in the FSM, and the removal of unproductive cyclic behavior from the FSMs in the context of resource protection problems.

2 The Competition for Resources Problem

Our current Competition for Resources simulation is a two-player game on a board of squares. Each square corresponds to a resource, and the two players compete for squares on the board. One player is the defender while the other is the adversary. At the beginning of each game the defender occupies one square on the board, the adversary occupies another square on the board, and the remainder of the squares are not occupied. If the board is of size $N \times N$, then the defender will start at square (1,1) and the adversary will start at square (N,N). For example, if the defender is depicted with a circle and the adversary with a square, the beginning of a 5×5 board game would appear as shown in Figure 1.

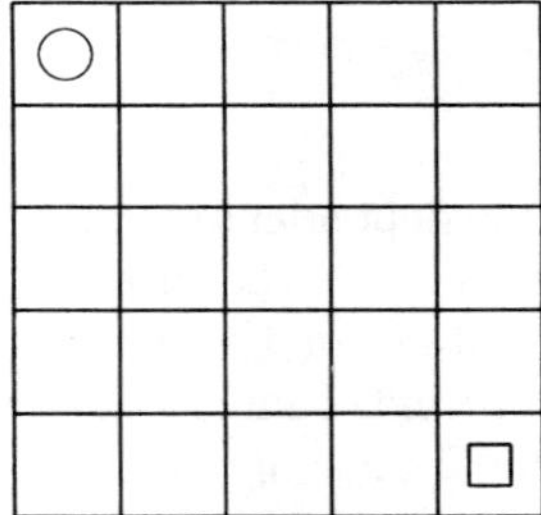

Fig. 1. The board at the beginning of the game

Since the board grid represents real networks, such as power grids or communication networks, and in the real world networks may be highly interconnected and will have few geophysical boundaries, our board is toroidal (has no edges). Furthermore, no assumption is made about board size, thus allowing us to scale to large problems.

Each player can only perceive a limited amount of information, namely, the status of the squares neighboring the current position of the player. The neighbors consist of the four squares to the north, south, east, and west of the current position. The diagonal squares cannot be seen. The status of each

[1] The evolution of FSMs is generally referred to as "evolutionary programming" in the literature.

neighboring square will be one of the following: unoccupied, occupied by that player, or occupied by the opponent. It is important to point out that we use the global view provided in Figure 1 only as a visualization tool; the players do not see this global view. Due to the toroidal nature of the board, the defender and the adversary are actually quite close to one another at the beginning of the game. However, because they cannot see along diagonals, they cannot see one another initially.

In the current simulation, the two players alternate making moves. First a player examines its neighboring squares. Then it moves to one of the neighboring squares. A player can move to an unoccupied square or back to a square that it has previously occupied. It cannot move to a square occupied by the opponent. At each time step, each of the players takes one action, which consists of moving to a neighboring resource to control/protect that resource. The player is not allowed to "stand still" and make no move. However, because each player must follow a path of "owned" resources to its current position, it will always be able to make a move at every time step (it can always back up along the path it has taken). Thus a player cannot be "trapped", i.e., it cannot be completely surrounded by the opponent. Once an agent occupies a resource, it controls/protects that resource forever. A game ends when all squares are occupied or time runs out. The agent with the most resources at the end of the game wins.

To continue the example shown in Figure 1, suppose the defender moves first. Each of the four neighboring squares are unoccupied. Suppose the defender randomly decides to move west (again, remember that the board is toroidal). The board is shown in Figure 2. The open circle represents the current position of the defender. The filled circle represents the fact that the defender has previously occupied this position and now "owns" it.

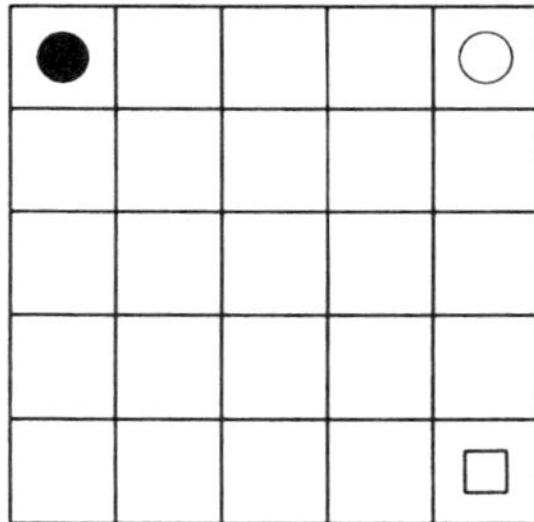

Fig. 2. The board after the defender moves

Now the adversary moves. In this case only three neighboring squares are unoccupied; the southern neighbor is occupied by the defender. Thus the adversary can only move north, east, or west. If it moves west the board is as shown in Figure 3. The open square represents the current position of the adversary. The filled square represents the fact that the adversary has previously occupied this position and now "owns" it. As mentioned above, the two players continue to move until all squares are occupied or time expires.

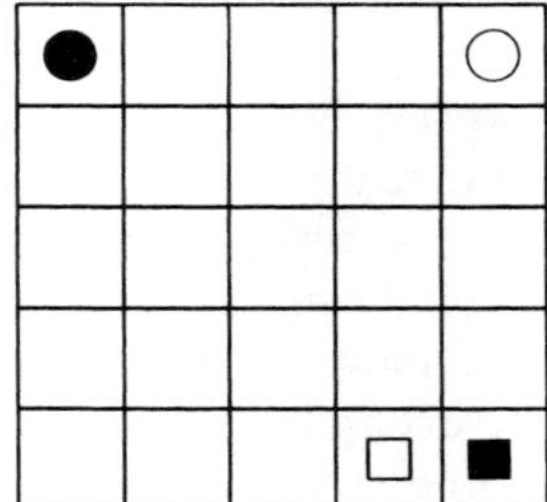

Fig. 3. The board after the adversary moves

Throughout this chapter the adversary will have a fixed stochastic strategy that the defender must learn to defeat. The strategy we have chosen for the adversary is simple, but is surprisingly hard to beat. If the adversary detects any unoccupied neighboring squares, it uniformly randomly moves to one of them. Otherwise it uniformly randomly backtracks to a neighboring square it has previously occupied. We view this as an opportunistic but stochastic opponent. Given the game and our adversary, we focus on developing effective strategies for the defender. The form of these strategies is FSMs, as described in the next section.

3 Overview of FSMs

Our choice of FSMs for representing strategies was motivated by three main considerations. First, FSMs have proven to be effective representations of agent plans/strategies, e.g., see Carmel and Markovitch (1996) or Jefferson et al. (1991). Second, unlike classical plans (Dean and Wellman, 1991), FSMs allow for indeterminate-length action sequences. Recall from Hopcroft and Ullman (1979) that the usual acceptance criterion for finite-length strings is termination in a "final" state. Here we assume that there are no final states, i.e., action sequences of any length are allowed. This provides a good model of embedded agents that are continually responsive to their environment. Finally, FSMs divide an overall task into subtasks, represented as internal states. This facilitates the understanding of agent strategies (see, for example, Section 7).

Formally, we define the machine M to be a six-tuple $(Q, \Sigma, \Delta, \delta, \lambda, q_0)$. Q is the set of vertices (states) of M, Σ is the alphabet of input symbols (which are agent sensory inputs), and Δ is the alphabet of output symbols (which are agent actions). δ is the transition function from a state and an input to a next state, i.e., $\delta(q_i, x_i) = q_{i+1}$ where q_i, $q_{i+1} \in Q$ and $x_i \in \Sigma$ is a sensory input. An edge is denoted as (q_i, q_{i+1}). For example, the edge from STATE1 to STATE2 in Figure 4 is (STATE1, STATE2). λ is the transition function from a state and an input to an output, i.e., $\lambda(q_i, x_i) = a_i$ where $q_i \in Q$, $x_i \in \Sigma$, and $a_i \in \Delta$ is an action. Finally, q_0 is the initial state, where all behavior begins.[2]

We assume that the FSMs are *deterministic* and *complete*. The FSMs are deterministic because δ and λ are functions, i.e., for every state and input there is a unique next state and action. The FSMs are complete because there exists a next state and action for every state and input, i.e., δ and λ are fully defined. Deterministic and complete FSMs are strategies that tell the agent precisely what to do in every situation it perceives.

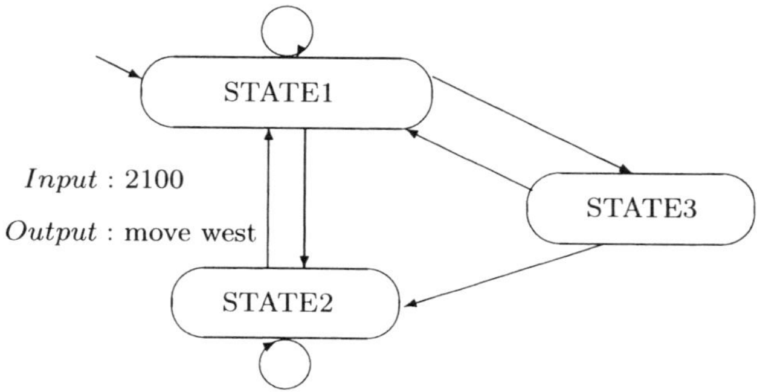

Fig. 4. An FSM agent plan for the Competition for Resources simulation, where the input and output are shown for only one of the edges.

Consider Figure 4, which is an example of an FSM strategy for the Competition for Resources simulation. Each sensory input x_i in this case shows the status of the neighboring resource immediately to the north, east, south, and west of the agent. The status of each resource can be 0 (unoccupied), 1 (occupied by the defensive agent), or 2 (occupied by the adversary). Thus an input of "2100" specifies that the north resource is owned by the adversary, the east resource is owned by the defensive agent, and that the south and west resources are unoccupied. Thus, according to the FSM depicted in Figure 4, if the defensive agent is in STATE2 and it sees an input of "2100" it will move west (λ (STATE2 , 2100) = west) and it will transition to STATE1 (δ (STATE2 , 2100) = STATE1). The initial state for this FSM is STATE1.

4 Evolution of FSMs

As mentioned earlier, the goal of this chapter is to develop effective strategies for the defender in the context of the Competition for Resources simulation

[2] This definition of FSM M corresponds to what is called a *Mealy machine* in Hopcroft and Ullman (1979).

```
procedure EA;
t = 0; /* Initial Generation */
initialize_population(t);
evaluate(t);
until (done) {
        t = t + 1; /* Next Generation */
        selection(t);
        recombine(t);
        mutate(t);
        evaluate(t);
}
```

Fig. 5. The outline of an EA

outlined above. The defender will be modelled with an FSM. We focus on the application of EAs to evolve the FSMs, since there is ample precedent for the effectiveness of this approach (Fogel, Owens, and Walsh, 1966; Fogel, 1995; Fogel, 1999). This section outlines precisely how we evolve the FSMs, by describing the representation used and the operators applied. First, however, a brief overview of EAs is in order.

4.1 A Brief Overview of EAs

EAs are population-based search algorithms. They maintain a population of individual structures that evolve according to rules of Darwinian selection and other operators, such as recombination and mutation. Each individual in the population is evaluated, receiving a measure of its fitness in the environment. Selection focuses attention on high-fitness individuals, thus exploiting the available fitness information. Recombination and mutation perturb those individuals, providing general heuristics for exploration. Although simplistic from a biologist's viewpoint, these algorithms are sufficiently complex to provide robust and powerful adaptive search mechanisms.

Figure 5 outlines a typical EA. A population of P individual structures is initialized and then evolved from generation t to generation $t+1$ by repeated applications of fitness evaluation, selection, recombination, and mutation. The population size P is generally constant in an EA, although there is no a priori reason (other than convenience) to make this assumption.

An EA typically initializes its population randomly, although domain-specific knowledge can also be used to bias the search. Evaluation measures the fitness of each individual according to its worth in some environment. Selection chooses the best individuals (those with high fitness) for survival and allows them to have children. Children are created via recombination, which exchanges information between parents, and mutation, which further perturbs the children. The children are then evaluated.

In the context of the Competition for Resources simulation, each individual in the population is an FSM. Each FSM is evaluated by playing the game numerous times, to obtain an estimate of how well that FSM is defending the resources against the adversary. Those FSMs that perform the task better are allowed to have more children, which are created through the processes of mutation and recombination. This process continues generation by generation, until the search is terminated.

The application of an EA to any problem involves a number of design decisions. For example, one has to choose a reasonable representation for the individual. The method of mutation and recombination must be precisely defined, as must be the form of selection and the termination criterion. These processes may depend on domain-specific knowledge, as mentioned above. Finally, the evaluation of the fitness of the individuals must be clearly described.

4.2 Representation of FSMs

The first design decision is the choice of internal representation for an FSM. Although at a high level all reasonable choices are semantically equivalent, a good data structure promotes efficiency. For this reason we chose a simple tabular representation, as exemplified in Table 1, which shows a simple FSM with three states ($Q = \{q_0, q_1, q_2\}$), three inputs ($\Sigma = \{x_0, x_1, x_2\}$), and two actions ($\Delta = \{a_0, a_1\}$).

	x_0	x_1	x_2
q_0	q_1/a_1	q_2/a_0	q_1/a_1
q_1	q_2/a_0	q_0/a_1	q_1/a_0
q_2	q_0/a_0	q_1/a_1	q_2/a_1

Table 1. The transition table for a simple three-state FSM. Rows are states and columns are inputs. Table entries are denoted as a next-state/action pair.

Rows in Table 1 correspond to states, and columns correspond to inputs. For each state q_i and input x_j, the corresponding table entry is a pair. The first element of this table entry pair is the next state, i.e., it is $\delta(q_i, x_j)$. The second element is the action to take given the agent is in state q_i and sees input x_j, i.e., it is $\lambda(q_i, x_j)$. Again, q_0 is the initial state (although this is not shown in the table).

In the context of the Competition for Resources simulation, the number of states is user defined, and will vary from one to S. The number of possible inputs is $3^4 = 81$, since the status of each of the four neighboring squares may have three values (0 = unoccupied, 1 = occupied by the defensive agent, 2 = occupied by the adversary). It should be noted that the input "2222" will never occur, since that implies that the defender is surrounded by the adversary. This is impossible, since the defender must have been able to get to the square it currently occupies. Thus there are 80 possible inputs and we require a table of size $S \times 80$. Each entry in the table is an "allele" that represents a next-state/action pair.[3] The states are labeled from 1 to S and in this chapter the initial state is always STATE1.

4.3 Initialization and the Use of Domain Knowledge

As mentioned above, the EA population is of size P. Thus, each of the P FSMs at generation zero must be initialized either randomly or by using domain-specific knowledge. Initialization defines the functions δ and λ for each FSM. For any given state (row) and input (column) the next state (given by δ) is chosen uniformly randomly from the set of all states Q.

The choice of action (given by λ) is somewhat more complex. The number of possible actions is maximally four, since the defender may potentially move north, east, south, or west. However, in practice, some of these moves might be impossible, if the adversary owns the neighboring squares.

	2100	2101	2102
STATE1	STATE2/south	STATE1/south	STATE3/south
STATE2	STATE1/west	STATE2/south	STATE1/south
STATE3	STATE2/west	STATE3/south	STATE2/south

Table 2. Example of a portion of an FSM transition table at initialization, using only preferred actions. Only one action (south) is preferred for inputs 2101 and 2102.

For example, recall the prior example where the input was "2100". In this case the north resource is owned by the adversary, the east resource is owned by the defensive agent, and the south and west resources are unoccupied. In this case there are only three *legal* moves: east, south, and west. The move to the north is *illegal*, since the adversary owns that square. Naturally, we must always restrict actions to those that are legal, and every input has a set of legal moves Δ_l that are possible. However, since the goal of the game is to capture resources, we also found it useful to define *preferred* moves – those that capture previously unoccupied squares. In the prior example where the input is "2100", moves to the south or west will capture new territory and are thus preferable. During initialization actions are always chosen uniformly randomly from the set of preferred actions Δ_p, if there are any. If there are no preferred actions, then a legal action is randomly chosen.[4] As an example,

[3] We use the term "allele" because individuals in EAs are often considered to be chromosomes of alleles, in an analogy to genetics.

[4] Preliminary experiments indicated that the emphasis on preferred actions enormously helps the initial search of the EA.

Table 2 shows three columns of an initial FSM with three states, under the inputs "2100", "2101", and "2102".

4.4 Mutation and Recombination of FSMs

Mutation and recombination are operators that can alter the functions δ and λ. The mutation of an FSM is reasonably straightforward. Each allele (next-state/action pair) in the FSM is mutated with probability p_m. Once an allele is chosen for mutation a coin is flipped to see whether the action or the next state is mutated. With probability p the next state is mutated by uniformly randomly choosing a state from the set of all states Q. This could result in no change, with probability $1/S$. With probability $1-p$ the action is mutated. If there are any preferred actions Δ_p the algorithm uniformly randomly chooses one of those. If there are no preferred actions the algorithm uniformly chooses any legal action Δ_l. Note that this also could result in no change (e.g., if there is only one legal action then there *will* be no change). In this chapter we have set $p = 0.5$. In the future we plan to test whether there is any advantage to having both the next state and the action be mutated independently.

Recombination is also straightforward. A proportion p_r of pairs of parents in the population are chosen for recombination (p_r is referred to as the "recombination rate"). For each pair of parents, a coin is flipped for each of the $S \times 80$ alleles. The allele at the table location (i,j) in the first FSM is swapped with the corresponding allele in the second FSM, with probability P_0.[5] Note that since only corresponding alleles are swapped, there is no need to worry about possible illegal actions. If an action is legal for one FSM at location (i,j) it must be legal for any other FSM at location (i,j), since the input j is the same. Note also that if alleles are swapped, both the next state and the action are swapped. In the future we plan to test whether there is any advantage to having the next state and action be swapped independently.

4.5 The Method of Selection and Termination

For this chapter we have chosen a standard fitness-proportional selection mechanism (Holland, 1975). We also made sure that the population contained a copy of the best individual that has ever been seen. This is referred to as "elitism" and it often helps to bias the search towards promising areas of the search space. For a termination criterion we simply ran the EA for a user-defined number of generations.

4.6 Evaluation of FSMs

Since the adversary in the Competition for Resources simulation is stochastic, each defender FSM will have to play the game multiple times in order to

[5] This is simply the well-known parameterized uniform recombination often used in the literature (Spears and De Jong, 1991; Spears, 2000).

obtain an estimate of how well it defends the resources. As stated earlier, the defender wins the game if it controls more resources than the adversary at the end of the game. Otherwise the adversary wins. The adversary also wins if both players control the same number of resources at the end of the game (i.e., there is a "tie").

The fitness function used by the EA is simple. Given G games, the fitness is the fraction of games that the defender wins. Thus this function returns values from 0.0 to 1.0, with 1.0 representing an FSM that won all the games it played. A major issue is the setting of G. If G is small, the fitness function will return a relatively poor estimate of the true worth of the defender. If G is large, the estimate will be much better, but at the expense of valuable computational cycles that might be better spent on running the EA for more generations. Prior work by Grefenstette and Fitzpatrick (1985) concluded that in some cases the overall efficiency of the EA may be improved by reducing G and by running for more generations.

In order to test this hypothesis we tried a wide range of values of G, ranging from 100 to 10,000. Due to high variance, G had to be quite large (on the order of 10,000) in order to obtain good estimates of the true fitness of an FSM. Unfortunately, setting G to 10,000 slowed down the EA so much that it took too long to run. With lower G the EA ran much faster, but there was a high probability that an FSM would do well purely due to stochastic noise, thus biasing the search towards FSMs that actually had poorer fitness. Equally unfortunately, running the EA for more generations did nothing to solve this difficulty and we were unable to confirm the hypothesis made by Grefenstette and Fitzpatrick.

Thus, on one hand a low value of G made for a fast fitness computation but produced FSMs that actually had inferior fitness. On the other hand a high value of G produced an accurate measure of fitness, but we were unable to run the EA for an adequate number of generations to produce good results. We were also unable to balance these two constraints adequately by picking some medium value of G. Instead we took another approach. Each individual would get a quick evaluation by using a low value of G. If that individual was promising (it did better than the best individual seen thus far), it was re-evaluated using a high value of G. If it still beat the best individual thus far, it became the new best individual. The idea was to carefully evaluate only those individuals that appeared promising. This approach worked quite well. We used a value of $G = 500$ for the initial evaluation and $G =$ 10,000 for the subsequent re-evaluation (if it was performed). Since most individuals were unable to beat the best individual seen thus far, they were not re-evaluated. As a result the algorithm ran much more quickly, but the finesses of the best individuals were quite accurate, biasing the search in promising directions.[6]

[6] This is similar in spirit to performing a "secondary search" of promising nodes in minimax search (Rich and Knight, 1991).

5 Experimental Methodology

One topic that we have not discussed thus far is whether the EA will also be effective at evolving the number of states in an FSM. Traditional approaches (Fogel, Owens, and Walsh, 1966; Fogel, 1995; Fogel, 1999) to the evolution of FSMs have always included operators for adding and deleting states in an FSM. Although we were primarily motivated to generate good defenders for the Competition for Resources task, we found that the issue of how to properly add and delete states raised a number of interesting questions that we considered worth some investigation.

The first question is that when a state is deleted, should one actually erase the state from the table, or should it simply be made inaccessible? Second, once a state has been deleted (or has been made inaccessible), the portions of the FSM that point to this state must now be "repaired". How is this repair done? Third, if a state has simply been made inaccessible, is it still subject to mutation and recombination? Finally, when a state is added, is this done by making a former state accessible again, or is a new state randomly initialized?

Although we realized that we could not hope to address all of these questions adequately in this chapter, what became clear is that in order to even deal with these issues we needed a way to evaluate the efficacy of different mechanisms for evolving the number of states in an FSM. Thus we decided that a control study was needed, in which we fixed the number of states. Once we could see the performance of the EA with a fixed number of states, we would be better equipped to judge how well the EA was performing when it adapted the number of states.

5.1 Fixed Number of States

For our experiments we had to select a reasonable board size N for the Competition for Resources simulation. The evaluation of the fitness function is the most expensive component of the EA, and the computation goes up as N^2, since the goal of the agents is to occupy all N^2 resources. On the other hand, if N is small the game is not terribly interesting. For the experiments reported in this chapter, we use a board size N of ten. We found that this was large enough to provide a challenging problem for the defender, while still keeping the computational overhead within reason.

Unlike a purely reactive agent that has no memory, states in an FSM provide a mechanism for "counting", i.e., given the same input the FSM can perform a certain action for some number of time steps and then perform a different action. Visualization of this game indicated that there might be merit in having the defender change direction after a certain number of steps, even if the input had not changed. A reasonable upper bound on the amount of counting that might be necessary appeared to be the size of the board, i.e., the FSM could easily recognize via counting that it had crossed the board from one end to another. Thus we allowed up to ten states in the FSMs, since N was ten. Ten experiments were performed – during experiment i the number of states was held constant at i.

We ran some preliminary experiments to determine reasonable values for the population size P, the mutation rate p_m, the recombination rate p_r, and the number of generations to run the EA before termination. We found that a small population size was not sufficient, and settled on a population size of $P = 100$. As for the mutation rate, we followed the heuristic that the mutation rate should be inversely proportional to the number of alleles in the individual (Bäck and Schwefel, 1993). For one-state FSMs p_m was set to 0.001, and p_m was decreased linearly with the number of states until it was 0.0001 for ten states. Thus, regardless of the number of states, on average the same number of individuals was mutated per generation. In general, we also found that the EA was relatively insensitive to changes in p_m, although higher values near 0.01 were detrimental to performance. In all cases we used a fixed mutation rate which did not adapt during the course of a run.[7]

Recombination performed quite differently. We varied the recombination rate p_r and the probability of swapping alleles P_0. The EA was quite sensitive to these values, and performance was optimized when p_r was 1.0 (all individuals are recombined) and P_0 was at its maximum of 0.5. In general, we found that recombination was quite useful for successful search. Interestingly, these results are counter to observations made by David Fogel (personal communication). The differences in results may be due to differences in the mutation used (fixed rate versus adaptive rate), the form of recombination, or even the form of selection. Unfortunately, a thorough investigation of the cause of this difference is beyond the scope of the current chapter, and we instead intend to explore this issue in detail in future work.

As for the termination criterion, we found that there was a very low probability of continued improvement in the search after 2,500 generations, so the search was terminated at that point. Ten independent runs were performed for each experiment (fixed number of states) and the performance of the best individual seen thus far throughout the search was monitored. We wished to answer two experimental questions. First, is there an increase in performance when there is more than one state (i.e., does the defender actually need multiple states to play this game well)? Second, what is the optimal number of states?

	Number of States									
	1	2	3	4	5	6	7	8	9	10
Fitness	0.806	0.867	0.893	0.888	0.901	0.903	0.899	0.905	0.896	0.883

Table 3. The fitness of the best individual at the end of the experiment, averaged over ten runs per experiment. One state is not enough for good performance.

[7] This differs from the evolutionary programming community, which typically uses an adaptive mutation rate.

The results are shown in Table 3 and represent the fitness of the best individual seen by the end of the experiment (averaged over the ten runs per experiment). As can be seen, there is clearly a need for more than one state. Two states are not quite adequate either. However, it is not clear that there is an optimal number of states. The optimal range of states appears to be roughly from five to eight, although anything from three to ten states performs roughly equivalently.[8]

Table 3 shows the results only at the end of the experiments, and it is important to see how performance improved over the course of the 2,500 generations. Figure 6 shows a graph of the fitness of the best individuals (these are referred to as "best-so-far" curves), as the population evolves. The vertical axis is the fitness, while the horizontal axis is the number of individuals that have been evaluated. Since the population size P is 100, 250,000 individuals will be evaluated during the course of a run. We use a log plot to emphasize the behavior of the EA during the early generations. To avoid clutter, we show only the graphs for one state and the even number of states. The remaining graphs for the odd number of states are similar. Again, each curve is averaged over ten runs.

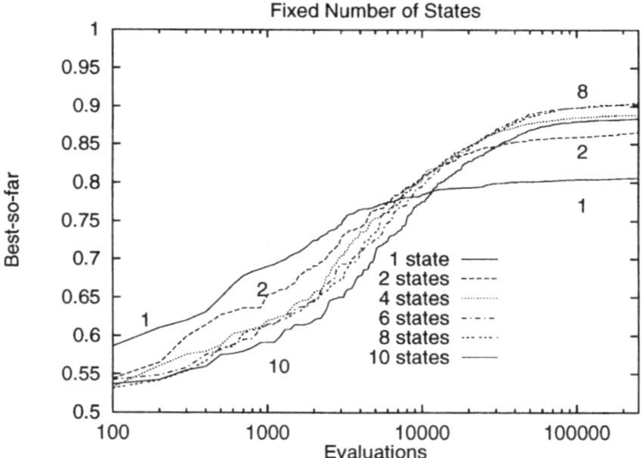

Fig. 6. "Best-so-far" curves for different numbers of states. We use a log plot to emphasize the behavior of the EA during the early generations. Having a small number of states enhances early performance but hurts later performance. Having a larger number of states degrades early performance but helps later.

Figure 6 shows quite clearly that increasing the number of states hurts performance at the beginning of the run. This is quite intuitively plausible, since the EA is searching a larger search space when the number of states increases. Thus, having fewer states helps search in the beginning. However,

[8] The increase in performance from one to two states and two to three states is significant ($p < 0.003$ and $p < 0.04$). The other differences are not significant.

as also can be seen, having only one or two states hinders search later in the experiment (around 20,000 evaluations). Having three or more states appears to be crucial for achieving the best performance at the end of the runs.

5.2 Adaptive Number of States

With the fixed-state experiments providing a control study, we were prepared to evaluate the efficacy of having the EA adapt the number of states as it ran. As mentioned earlier, the exact implementation of state adaptation involves a sequence of design decisions. To aid us in making these decisions we relied upon two basic EA design principles: "exploitation" and "exploration". Exploitation refers to the continued use of knowledge already learned by the EA, while exploration refers to the continued search for new knowledge. In general increased exploitation results in decreased exploration (and vice versa), and successful EA implementations must strike a balance between these two design principles.

Let us address the questions mentioned earlier. When a state is deleted, should one actually erase the state from the table, or should it simply be made inaccessible? Our prior experience in similar areas has shown that it is often best to make the information inaccessible (De Jong, Spears, and Gordon, 1993). Since information has been learned, keeping the information stored serves as a useful memory, which can be re-activated at a later time (if the state is added back to the FSM). Memory is a form of exploitation. Thus we decided to add a "tag" to each row of the FSM table. If the tag is 1 the state is accessible. If the tag is 0 the state is inaccessible, but is not destroyed. Thus, when a state is added, it is accomplished simply by turning on the tag. In our implementation these tags are subject to an independent mutation operation that flips the tags (from 1 to 0 and vice versa) with probability 0.001. As mentioned earlier, STATE1 is always the initial state, so that is the only state that cannot be made inaccessible.

Once a state has been made inaccessible, how should the remainder of the FSM (that points to that state) be "repaired"? Two solutions come to mind. Suppose state s is no longer accessible. If state s_1 points to s, change the pointer so that it points back to s_1. We refer to this form of repair as "self-reference", since the state will point back to itself. An alternative is to change the pointer to point to any state that is accessible, chosen uniformly randomly. In this situation the principle of exploration appears most germane, and we took the latter approach. We also modified mutation and recombination in a similar fashion. When mutation changes the next state of an allele, it can only choose accessible states (uniformly randomly).[9] Since parents may have different sets of accessible states, recombination may swap alleles in such a fashion that a next state that was accessible in one child FSM is now

[9] For these experiments we chose an intermediary mutation rate p_m of 0.0005.

inaccessible in the other FSM. In this situation a new next state is chosen uniformly randomly from the set of accessible states (in that FSM).

If a state has simply been made inaccessible, is it still subject to mutation and recombination? The principle of exploitation would indicate that one should not mutate and recombine inaccessible states. But the principle of exploration would indicate that one should. We tested both versions, and found very little difference in the results. For the results presented here, mutation and recombination work on all states, regardless of whether they are accessible or not.

We performed several experiments to judge the efficacy of this implementation of adaptation. We were interested in answering two questions. First, does the adaptive mechanism find the optimal range of accessible states? Second, how many states should be accessible initially? To address these questions we initialized each FSM individual with ten states. In the first experiment all ten states are initially accessible. In the second experiment the first five states were initially accessible. In the third experiment only the first state was initially accessible. The only mechanism for changing the number of accessible states is via the independent mutation operation mentioned above, which flips the accessibility tags. Although we have no "penalty" function per se (that would penalize the FSMs for having more accessible states), the mutation operator will provide a slight bias towards having five accessible states (in the same way that continuously flipping ten coins tends to result in roughly five heads and five tails).

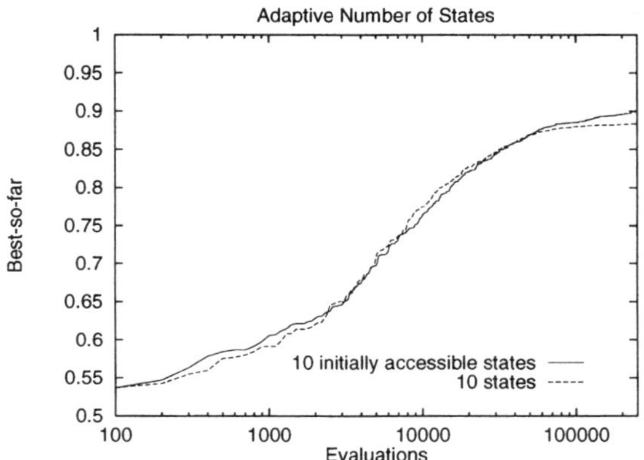

Fig. 7. "Best-so-far" curves for the EA with ten fixed states and the adaptive-state EA with ten initially accessible states. There is very little difference.

Figure 7 shows the best-so-far curve for the adaptive-state EA when it is initialized with ten accessible states. The results are again averaged over ten runs. On average the adaptive mechanism decreased the number of accessible states in the FSMs, but only down to nine. Performance was reasonable, given that the final fitness was 0.899. Figure 7 also includes the best-so-far curve for the EA with ten fixed states. What is surprising is how similar the two curves are. There is little evidence that the adaptive mechanism is actually having much effect, from a performance or accessibility point of view.

When the EA was initialized with five accessible states, the adaptive mechanism increased the number of accessible states in the FSMs to roughly seven. Performance was reasonable, given that the final fitness was 0.898. The best-so-far curve is almost identical to that for the EA with five fixed states. Finally, when the EA was initialized with one accessible state, the adaptive mechanism increased the number of accessible states in the FSMs to roughly five. However, performance did not reflect that increase. The final fitness was 0.841, which is between the performance of the one and two fixed-state experiments. This provides evidence that the EA is having difficulty taking full advantage of the newly accessible states.

In summary, we have had mixed results with using the EA to adapt the number of accessible states in the FSM. Starting with only one accessible state resulted in inferior performance. Starting with five or ten accessible states provided much better performance, but there is little evidence that the adaptive mechanism is having much effect. The evidence does suggest that if one is going to use adaptation, it might be best to err on the side of initially having at least as many accessible states as the problem requires. We explore this issue further in the next section.

6 Behavior of the Evolved FSMs

As mentioned in the previous section, the EA appears to have some difficulty in adapting the number of accessible states successfully. We address this further in this section by examining the internal dynamics of the FSMs. We also examine the external behavior of the FSMs, which provides some important insights into how to evolve better FSMs for the Competition for Resources simulation.

6.1 Internal Behavior

Although the number of accessible states is an interesting metric in itself, it obscures information about how often accessible states are actually used during the execution of the defender FSM. For example, during the fixed-state experiments, are all states used equally? Or does the FSM depend on some states more than others? Similar questions can be asked about the adaptive-state experiments. For example, in the experiment with one initially accessible state, the number of accessible states increased to roughly five. But a graph of performance indicates that all five states are probably not being used. When the EA is started with ten initially accessible states, performance is good and the EA ends up using nine accessible states (on average). But are all of these accessible states used equally?

To address this issue we took the FSMs at the end of every run and reran each over 100,000 games. Each FSM has a set of $n \leq 10$ accessible states (in the case of the fixed-state experiments all states are always accessible). While the FSM was executing we counted the number of times that each of the n states was actually the next state of a transition. After execution the accessible states were sorted by their transition counts. For example, suppose that an FSM with four accessible states has made 100 transitions. It could be the case that 40 transitions were to one state, 30 to a second state, 20 to a third state, and 10 to the remaining state. We then translated this to the percentage of time that the FSM spent in each state. Finally, we computed the cumulative distribution of these percentages. Using the example above, 40% of time is spent in the most popular state, 70% of time is spent in the two most popular states, and 90% of time is spent in the three most popular states. Clearly 100% of time is spent in all accessible states.

# of States	State Usage									
	1	2	3	4	5	6	7	8	9	10
2	0.617	1.000	-	-	-	-	-	-	-	-
3	0.460	0.777	1.000	-	-	-	-	-	-	-
4	0.418	0.661	0.862	1.000	-	-	-	-	-	-
5	0.353	0.579	0.748	0.897	1.000	-	-	-	-	-
6	0.314	0.521	0.698	0.833	0.930	1.000	-	-	-	-
7	0.283	0.462	0.609	0.731	0.837	0.928	1.000	-	-	-
8	0.233	0.415	0.561	0.688	0.788	0.873	0.942	1.000	-	-
9	0.226	0.393	0.522	0.638	0.739	0.820	0.891	0.953	1.000	-
10	0.194	0.340	0.464	0.571	0.670	0.753	0.825	0.892	0.953	1.000

Table 4. Cumulative distributions of state usage for the fixed-state experiments. Note that although each experiment with i fixed states did not use all i states uniformly, it did make reasonable use of all i states.

For each experiment conducted in this chapter we performed ten runs. Thus for each experiment we averaged the cumulative distributions over those ten runs. The results for the fixed-state experiments are shown in Table 4. If the FSMs use all states uniformly the cumulative distribution will be linear. This is not the case, indicating that the FSMs are making more use of some states than others. For example, if all states were being used uniformly in the fixed-state experiment with ten states, each state would account for roughly 10% of the transitions. However, one can note that in fact the most popular state accounted for roughly 20% of the transitions. It is clear, though, that all states play a role in adding to the dynamics of the FSM.

Table 5 provides a similar table for the adaptive-state experiments. The data for the experiment with one initially accessible state (where the number

Initial # of States	State Usage									
	1	2	3	4	5	6	7	8	9	10
1	0.946	0.997	0.999	1.000	-	-	-	-	-	-
5	0.448	0.730	0.924	0.983	1.000	-	-	-	-	-
10	0.346	0.578	0.776	0.899	0.963	0.990	0.999	1.000	-	-

Table 5. Cumulative distributions of state usage for the adaptive-state experiments. In this case not all accessible states are well used.

of accessible states increased to roughly five) indicates quite clearly that the FSMs never actually make any real use of the newly accessible states. One can see from the table that one state accounts for about 95% of the transitions, while a second state accounts for the remainder. Similar results can be seen in the table when there are five and ten initially accessible states. The FSM appears to have difficulty making use of newly accessible states that have never been seen before. This confirms our prior conclusion that it appears best to err on the side of initially having more accessible states than the problem requires.

The results of our analysis suggest a few intriguing modifications to the EA. First, suppose FSMs are routinely monitored in this dynamic fashion. If states are not accessed, then increase the probability that a mutation will provide a transition to an under-utilized state. This increases exploration. Second, instead of making states inaccessible at random, it might be better to make poorly used states inaccessible, which would increase exploitation. We intend to explore these possibilities in the future.

Another possibility for modifying the EA is in response to the concern that addition and deletion of states are highly disruptive operations. We are currently investigating the application of "gentler" operators that could perform the same role. Simply deleting states (or turning them off) is too disruptive, due to the repair that must be performed afterwards. However, merging two similar states could remove a state in a fashion less deleterious to evolution. This process would be analogous to generalization. Similarly, as opposed to adding states (or turning them on), an alternative operator would clone an existing row of the tabular representation. Accessing this new state would not be deleterious and evolution could proceed to modify it slowly. This provides a process of specialization.

6.2 External Behavior

Although a study of the internal behavior of the FSMs is useful, it cannot replace the striking visual impression gained from simply watching the strategy execution in action. By watching the defender agent play numerous games against the adversary, we have detected fascinating patterns.

Let us begin by describing the desirable behaviors that evolve. We have observed two key successful types of strategies. The first consists of encapsulating a relatively large rectangular region, and then filling it in. For example, the defender might capture all squares along the perimeter of a four-square by five-square rectangle, and then capture all the internal squares. The effectiveness of this strategy results from the fact that the adversary cannot cross the perimeter, once captured, since it is now owned by the defender. The second successful strategy consists of spiraling through alternate rows of the board, then returning to capture squares in the unowned rows in between. The success of this second strategy is due to reasons similar to the success of the first strategy. In particular, blocking off rows of squares sets up borders that inhibit the adversary's movement.

Watching the agents play has also led to the observation of undesirable behaviors. The most striking and deleterious of these is the presence of unproductive cycling behaviors by the defender. Although it is guaranteed that FSMs must eventually cycle, many of our cycles were surprisingly short, e.g., of length less than N (the board size). For example, the defender might encapsulate a region, fill it in, and appear to be winning the game. But much to our dismay it will then get caught in an infinite cycle going back and forth between the same two or more squares. Meanwhile, the adversary continues capturing squares in its opportunistic random fashion and wins the game! Having seen this problem of cycles, our immediate concern was to remove these cycles from the FSM strategies, as described in Section 8. But first, let us examine an example evolved FSM.

7 An Example FSM

To gain some understanding of how the FSM encodes the encapsulation behavior, we carefully examined the best FSM that evolved during the fixed-state experiments. This was an FSM with four states that won 95% of the games. Although the number of states was small, the large number of inputs (80) inherent in this task makes it very difficult to fully understand the actual FSM. In addition, we were unable to find any way to generalize the FSM, since it performed different actions for slightly different inputs. However, we did find it useful to consider the situation in which the defender has not yet observed the adversary. The resulting behavior is deterministic, but also holds in the more general situation where the adversary has been observed.

Figure 8 illustrates the portion of the FSM that encodes the encapsulation behavior. STATE1 is the initial state and the labels on the edges are input/action pairs. In this case, since the adversary has not been seen, the input indicates whether the defender is occupying any of the four neighbors of its current position. Figure 9 shows the sequence of moves that occur on the board.

At the beginning of the game the defender is at resource (1,1) and has been nowhere else; the input is null ($\emptyset$). The beginning of the game is indicated by time step "0" in square (1,1). The FSM indicates that the defender moves

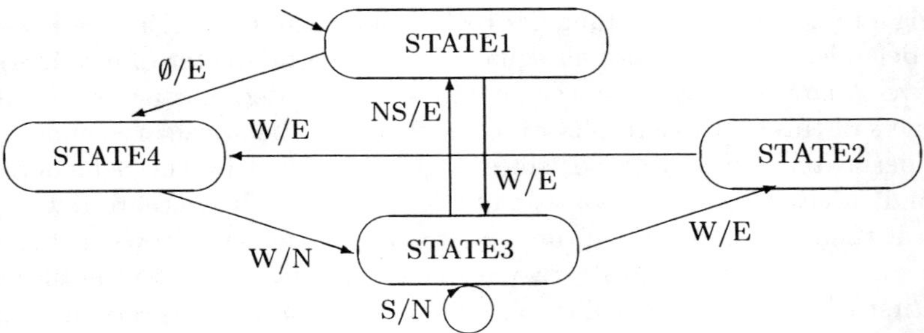

Fig. 8. A portion of an FSM agent plan that performed well. The edges are labelled as input/action, e.g., NS/E indicates that the agent observes that it is occupying the north and south resources (from its current position) and it moves east.

0	1				15				
	10	11	12	13	14				
	9				23	24			
	8				22				
	7				21				
	6				20				
	5				19				
	4				18				
	3				17				
	2				16				

Fig. 9. The sequence of moves made by the FSM in Figure 8, assuming that it has not encountered the opponent yet. Note how the agent encapsulates one-half of the resources (which will eventually be filled in). The behavior of the actual agent under simulation looks similar.

east and transitions to STATE4 (step "1"). The defender notices that it is occupying the resource to the west of its current position, moves north, and transitions to STATE3 (step "2"). At this point the defender continues moving north, staying in STATE3. At step "10" the defender notices it can no longer move north, so it moves east and transitions to STATE1.

Now the defender continues moving east, transition from STATE3 to STATE2, and then STATE4. However, at step "14" the agent changes direction and moves north (transition to STATE3), although the input has not changed. This is a classic example of how FSMs can use states to count – the defender is able to detect that it has crossed one-half of the board's width. The defender continues moving north (staying in STATE3) until step "23". At this point one-half of the resources have been encapsulated.

If the defender has still not encountered the adversary it will move east and attempt to repeat the pattern. Usually this will not happen and the defender will be forced to respond to the adversary. The response will generally be an attempt to block the adversary, followed by a systematic filling in of the encapsulated region. One behavior that we have not shown is the deleterious cycling behavior. For the example FSM that we analyzed here, evolution removed most of the short cycles. However, this rarely occurs, and we needed a more systematic method for detecting and removing cycles.

8 Removing Cycles from the Defender's Strategy

As mentioned above, the EA itself can remove many cycles from FSMs, since their removal (through mutation or recombination) will increase performance. Unfortunately, we found that this simply takes too long to accomplish. Therefore we decided to supplement the EA with another method for cycle removal to speed up the process. We investigated two methods: *model checking* and *behavior checking*. Model checking (Clarke and Wing, 1996) examines an FSM before execution, while behavior checking (Kim et al., 1999) examines the FSM behavior during execution. We investigated model checking first, since it is known to be highly effective for enforcing behavioral constraints and will allow the FSM to be repaired before it is even executed.

Model checking consists of constructing a model of the system and then determining whether a property (constraint) holds strictly for the model, i.e., whether it is guaranteed that there will never be any violations of the property. Since the model is usually in the form of an FSM, model checking is immediately applicable to the defender's strategy. Traditional model checking consists of brute-force search through the entire set of all possible FSM transitions to verify whether the property holds absolutely. In this case we are clearly interested in having a property that "avoids cycles". Unfortunately, this property cannot hold absolutely for FSMs, since cycles are an inherent property of an FSM (Fogel, 1999). However, as noted earlier, a large number of the cycles that the defender gets into are actually quite short, e.g., many consist of repeated visits to the same two or three squares. Furthermore, model checking can be computationally expensive; checking for long cycles is far more expensive than checking for short cycles. Therefore we focused instead on the property "avoid short cycles", where "short" refers to cycles of length two or three. We felt that avoiding short cycles would significantly improve performance while being reasonably cost-effective. In addition to model checking for cycles, we also implemented repair mechanisms that chose an alternative to the first action in a cycle in order to break the cycle.

To address the question of whether our cycle checking and repair algorithms improve behavior and, if so, how much, we ran experiments to determine their effectiveness. Unfortunately, the outcome of these experiments was that the addition of these algorithms results in very little performance improvement. The primary reason appears to be that the removal of short cycles usually results in the creation of slightly longer cycles, thus not really solving the cycle problem. We considered the computational expense of running model checking for these longer cycles to be unacceptable.

Given the failure with model checking, we next explored the use of behavior checking. Behavior checking gives no formal guarantee of behavior, but can be much quicker to perform than model checking. Behavior checking examines the dynamic run-time behavior of the agent, rather than the model (FSM) used by the agent. In particular, model checking explores numerous *possible* action sequences that the agent could take. Behavior checking, on the other hand, only tests the property for those action sequences *actually* taken at run-time, which is typically a very small subset of the possible action sequences. Run-time checking of system behavior is a very new topic in the verification community, but some of the results already appear promising (e.g., Gordon et al., 1999). Here, we present the first algorithm of which we are aware that does a run-time check for an FSM agent's cyclic behaviors.

Our behavior checking algorithm is executed while the agents play a game. For a sliding window of t time steps, the defender agent saves its current state and location on the board (the agent now consists of an FSM and auxiliary memory). Once cycle checking is turned on, the defender uses this auxiliary memory to make a cycle check before every move. If all four immediate neighbors are occupied, then the defender checks whether its current state and location are equal to any other in its window of memory. If yes, a cycle has been identified and a random alternative action is taken to the one recommended by the FSM (because the action recommended by the FSM would perpetuate the cycle). Of course, this alternative action could also create a cycle. But the behavior checking algorithm would immediately detect that cycle, after that move. We have found that a window size of $2 \times N = 20$ time steps, which will identify cycles up to length 20, works well. We are also exploring larger window sizes. In our experiments we ran cycle checking as soon as the window memory was filled, although in practice one might want to initiate it later in the game to save the expense – since cycles generally show up later in the game.

There is one more issue that we explored with behavior checking. If a cycle is broken with an alternative action, should that action be remembered, i.e., does the FSM get changed? This would be a form of Lamarckian (1984) learning. We tried this and unfortunately found that it hurt performance, so it is not used in the current version of the algorithm.

Let us consider the results of an experiment where behavior checking is added to the adaptive-state EAs. In this experiment, we added behav-

Initial Number of Accessible States in Adaptive-State EA	5	10
Fitness without behavior checking and repair	0.898	0.899
Fitness with behavior checking and repair	0.948	0.945

Table 6. The fitness of the best individual at the end of the experiment, averaged over ten runs per experiment, with and without behavior checking and repair

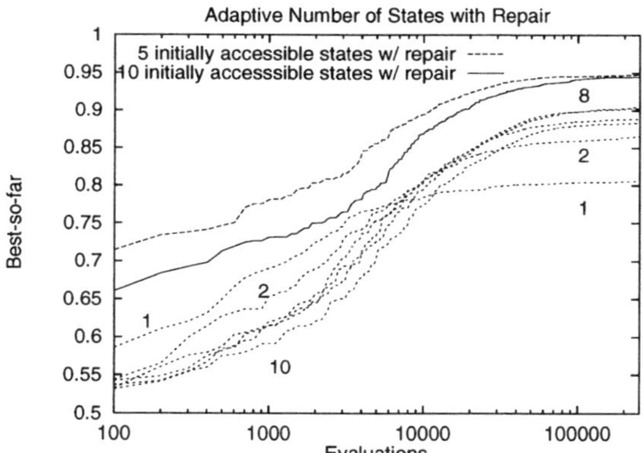

Fig. 10. "Best-so-far" curves for the adaptive-state EA with five and ten initially accessible states with cycle repair. The results for the fixed number of states are repeated for comparison. Note that cycle repair greatly enhances performance.

ior checking to the adaptive-state EA that begins with ten initially accessible states, and the one that begins with five initially accessible states, since these were two of our most successful EAs. Other than the addition of cycle checking and repair, each EA was run identically to the experiments in Section 5.2. Our hypothesis was that the addition of behavior checking and cycle repair would improve performance. This hypothesis is confirmed, as shown in Table 6 and Figure 10. Table 6 gives the fitness of the best individual seen by the end of the experiment, averaged over ten runs per experiment. Clearly, there is a large advantage to adding behavior checking and cycle repair.[10] Our best performing defender wins close to 97% of the games!

9 Summary and Future Work

To summarize, this chapter has empirically explored a variety of issues related to evolving FSMs in the context of the Competition for Resources problem. Our experiments yielded some rather surprising and potentially useful results. For example, the best method we found for evaluating FSM fitness was to perform an initial quick evaluation over a small number of games; then,

[10] The improvement using cycle checking and repair is significant, $p < 0.01$.

for those individuals identified as promising, a more extensive evaluation is performed over a larger number of games. We were unable to predict the optimal number of FSM states for winning this game (other than we expected it to be greater than one); thus it was interesting to discover that the best range is between five and eight states for a board of size ten. Deeper insights into the reasons for this result would help us make this representational decision for other board sizes and problems. Alternatively, one can use an EA with an adaptive number of states. We tried this and were surprised at how difficult it is to do this effectively. An in-depth analysis of dynamic state usage revealed that when starting with a minimum number of states, the EA appears to have difficulty making use of newly accessible states. This is less of a problem for an EA beginning with a maximum number of states – because rather than starting with new states, it is re-activating states that were formerly accessible and still retain valuable information. The hypothesis, then, is that it appears to be preferable to start with more states than needed for an adaptive-state EA. Finally, the chapter concluded with novel algorithms for cycle checking and repair. The algorithm for behavior checking proved to be far more effective than model checking. Because the behavior checking algorithm is restricted to only those cycles that actually appear while playing the game, considerably longer cycles can be detected and repaired with this method.

A possible direction for future research is to explore alternative strategy learning methods. Here we applied EAs because they are the most effective method we could find for learning FSMs from scratch. One option we could explore in the future would be to consider stochastic FSMs. By making the FSMs stochastic, it is possible that our cycle problem could be avoided or reduced. Unfortunately, the evolution of stochastic FSMs seems computationally intensive. On the other hand, there are alternative learning methods for stochastic FSMs that might be explored (Mars, Chen, and Nambiar, 1996). We could also try more widely used reinforcement learning methods, such as *q-learning* (Watkins, 1989), although as indicated by the experiments in this chapter, multiple internal states would be required for good performance.

Our main focus for the future will be to make the Competition for Resources simulation more realistic, and to continue our empirical investigations in the context of the newer versions of the game. For example, in the current game resources are all treated equally. In the spirit of game theory, we would like to consider resources having different numeric values, and perhaps have the value of a resource differ for each of the agents. Another possibility is to allow one agent to (with some small probability) "steal" a resource owned by the other agent. In other words, the rule that once a resource is owned by an agent, this ownership is permanent, would be altered. Another possibility is to include multiple agents and co-evolution. What is most interesting about this game is how easy it is to be changed to represent a wide variety of problems. For example, with minor modifications we have extended the game to represent the epidemiology of virus versus anti-virus spread. In the virus version of the game, each square represents an agent with the virus, anti-virus, or neither. At each time step, an agent having the virus or anti-virus can spread it to one of its neighbors. What one sees on the board when watching this version of the game looks like a "spreading activation". Further pursuit of the virus version both in simulation and in a corresponding mathematical model is currently in progress (Billings, Spears, and Schwartz, 2002).

References

1. Bäck, T. and Schwefel, H.-P. (1993) An overview of evolutionary algorithms for parameter optimization. Evolutionary Computation, **1** , 1–23
2. Billings, L., Spears, W., and Schwartz, I. (2002) A unified prediction of computer virus spread in connected networks. Physics Letters A, **297** , 261–266
3. Carmel, D. and Markovitch, S. (1996) Learning models of intelligent agents. Proceedings of the Thirteenth National Conference on Artificial Intelligence
4. Clarke, E. and Wing, J. (1996) Formal methods: State of the art and future directions. ACM Computing Surveys, **28** , 626–643
5. Dean, T. and Wellman, M. (1991) Planning and Control. Morgan Kaufmann, San Mateo
6. De Jong, K., Spears, W., and Gordon, D. (1993) Using genetic algorithms for concept learning. Machine Learning Journal, **13** , 161–188
7. Denning, D. (1999) Information Warfare and Security. Addison-Wesley, New York
8. Fogel, D. (1995) Evolutionary Computation. IEEE Press, New York
9. Fogel, L. (1999) Intelligence Through Simulated Evolution: Forty Years of Evolutionary Programming. Wiley Series on Intelligent Systems, New York
10. Fogel, L., Owens, A., and Walsh, M. (1966) Artificial Intelligence Through Simulated Evolution. John Wiley and Sons, New York
11. Gordon, D., Spears, W., Sokolsky, O., and Lee, I. (1999) Distributed spatial control, global monitoring and steering of mobile physical agents. Proceedings of the IEEE International Conference on Information, Intelligence, and Systems
12. Grefenstette, J. and Fitzpatrick, J. (1985) Genetic search with approximate function evaluations. Proceedings of the International Conference on Genetic Algorithms
13. Holland, J. (1975) Adaptation in Natural and Artificial Systems. University of Michigan Press, Ann Arbor
14. Hopcroft, J. and Ullman, J. (1979) Introduction to Automata Theory, Languages, and Computation. Addison-Wesley, Menlo Park
15. Jefferson, D., Collins, R., Cooper, C., Dyer, M., Flowers, M., Korf, R., Taylor, C., and Wang, A. (1991) Evolution as a theme in artificial life: The Genesys/Tracker system. Proceedings of Artificial Life II
16. Kim, M., Viswanathan, M., Ben-Abdallah, H., Kannan, S., Lee, I., and Sokolsky, O. (1999) Formally specified monitoring of temporal properties. Proceedings of the Euromicro Conference on Real-Time Systems
17. Lamarck, J.B. (1984) Philosophie Zoologique. English Translation, University of Chicago

18. Mars, P., Chen, J., and Nambiar, R. (1996) Learning Algorithms: Theory and Applications in Signal Processing, Control and Communications. CRC Press, New York
19. Rich, E. and Knight, K. (1991) Artificial Intelligence. McGraw-Hill, New York
20. Spears, W. (2000) Evolutionary Algorithms: The Role of Mutation and Recombination. Springer-Verlag, Berlin
21. Spears, W. and De Jong, K. (1991) On the virtues of parameterized uniform crossover. Proceedings of the International Conference on Genetic Algorithms
22. Watkins, C. (1989) Learning from delayed rewards. Ph.D. thesis, University of Cambridge, England

A Unified Bayesian Framework for Evolutionary Learning and Optimization

Byoung-Tak Zhang

Biointelligence Laboratory
School of Computer Science and Engineering
Seoul National University
Seoul 151-742, Korea
E-mail: btzhang@bi.snu.ac.kr

Summary. A probabilistic evolutionary framework is presented and shown to be applicable to both learning and optimization problems. In this framework, evolutionary computation is viewed as Bayesian inference that iteratively updates the posterior distribution of a population from the prior knowledge and observation of new individuals to find an individual with the maximum posterior probability. Theoretical foundations of Bayesian evolutionary computation are given and its generality is demonstrated by showing specific Bayesian evolutionary algorithms for learning and optimization. We also discuss how the probabilistic framework can be used to develop novel evolutionary algorithms that embed evolutionary learning for evolutionary optimization and vice versa.

1 Introduction

A number of evolutionary algorithms have recently been proposed that explicitly model the population of good solutions and use the constructed model to guide further search [3,7,20,16]. These methods are generally known as the estimation of distribution algorithms or EDAs [17]. They use global information contained in the population, instead of using local information through crossover or mutation of individuals. From the population, statistics of the hidden structure are derived and used when generating new individuals.

Several methods have been proposed to build and sample from the distribution of the population. Baluja and Caruana propose the population-based incremental learning (PBIL) method that uses a single probability vector to replace the population [2]. The components of the vector are regarded independently of each other, so PBIL only takes into account first-order statistics. Mühlenbein and Paaß [18] present a univariate marginal distribution algorithm (UMDA) that estimates the distribution using univariate marginal frequencies in the set of selected parents, and resamples the new points. UMDA shows good performance on linear problems.

To capture more complex dependency, structures that can express higher-order statistics are necessary. De Bonet et al. [7] suggest a second-order method, called MIMIC (mutual information maximizing input clustering).

It uses a chain structure to express conditional probabilities. Baluja and Davies [3] propose to use dependency trees to learn second-order probability distributions. Pelikan and Mühlenbein [20] suggest the bivariate marginal distribution algorithm (BMDA) as an extension of the UMDA. Mühlenbein and Mahnig [16] present the factorized distribution algorithm (FDA). Here, the distribution is decomposed into various factors or conditional probabilities. Pelikan et al. [19] describe the Bayesian optimization algorithm (BOA) that uses Bayesian networks in order to estimate the joint distribution of promising solutions. All these methods are designed for optimization.

Zhang [23] presents Bayesian evolutionary algorithms (BEAs) where evolutionary computation is formulated as a probabilistic process of finding an individual with the maximum a posteriori probability (MAP). The BEA starts with a population of individuals drawn from the prior distribution, and iteratively generates a new population by estimating the posterior fitness distribution of parent individuals and then sampling from the distribution offspring individuals via variation and selection operators. Explicit modelling of fitness distributions in terms of probabilities and the generational transition by means of Bayes formula are two distinguishing features of BEAs from other evolutionary algorithms. This framework was presented originally in the context of evolutionary learning of models from given data, i.e., function approximation. This is contrasted with most of the EDAs that were suggested for function optimization.

In this chapter, we extend the Bayesian evolutionary framework to encompass probabilistic evolutionary algorithms for function optimization as well as those for function approximation. Optimization and learning problems are formally defined and the similarities and differences between them are identified. We then illustrate how the existing evolutionary approaches can be formulated as BEAs and present a unified Bayesian framework of evolutionary computation for learning and optimization. The generality of the framework is illustrated by showing two specific applications of BEAs. One is the BEA that performs function optimization by learning the sample distribution using a probabilistic graphical model, i.e., the Helmholtz machine [6]. The effectiveness of this method is demonstrated on a suite of GA-deceptive functions. A second example is the BEA that learns neural trees for time series prediction. Explicit formulas for specifying the distributions are provided and the effectiveness and robustness of the method is demonstrated on the laser data.

The chapter is organized as follows. In Section 2, we formally define the problems of function optimization and function approximation (learning) and examine their relationship. Section 3 presents the unified Bayesian framework of evolutionary computation for learning and optimization. Section 4 demonstrates the effectiveness of the BEA for learning in the context of evolving neural trees. Section 5 illustrates the usefulness of the BEAs for solving function optimization. Section 6 concludes with some remarks on the implications of the Bayesian evolutionary approaches to learning and optimization, respectively and in combination.

2 Optimization and Learning

2.1 Optimization

The goal of function optimization is to find $\mathbf{x}^*$ that minimizes an objective function f:

$$\mathbf{x}^* = \text{argmin}_{\mathbf{x} \in \mathbf{X}} f(\mathbf{x}), \tag{1}$$

where X is the input space of f. Here, the objective function f is given as an equation (or the equivalent) and the fitness of any search points $\mathbf{x}$ can be evaluated directly from this function.

Evolutionary algorithms have been extensively used for function optimization [1,12,22]. The initial population X^0 of search points $\mathbf{x}_i$ is generated at random. The fitness values $f(\mathbf{x}_i)$ of the points is measured. Depending on these values, the next population X^1 of search points are generated by a variety of variation operators. Typically, these involve mutation and crossover. Mutation generates a new point by modifying the existing point. Crossover generates new points by combining two or more search points. New populations X^t are generated until the optimum point $\mathbf{x}^*$ is found or the maximum allowed generation is reached.

2.2 Learning

Learning or function approximation involves constructing models f_θ of a target function f given a set D of input–output pairs $(\mathbf{x}_c, f(\mathbf{x}_c))$, i.e., $D = \{(\mathbf{x}_c, f(\mathbf{x}_c)) \mid c = 1, ..., N\}$. Here $f(\mathbf{x}_c)$ is the observed output of the target function given the input $\mathbf{x}_c$. The objective is to find the model θ^*, i.e., the functional structure and associated parameters, whose output $f_{\theta^*}(\mathbf{x})$ best predicts the output $f(\mathbf{x})$ of the target function given an arbitrary input $\mathbf{x}$. Using the squared error criterion as the objective function, this can be formulated as a minimization problem [5]:

$$\theta^* = \underset{\theta \in \Theta}{\text{argmin}} \left\{ \sum_{\mathbf{x}_c \in D} ||f_\theta(\mathbf{x}_c) - f(\mathbf{x}_c)||^2 \right\}, \tag{2}$$

where Θ is the space of all possible models under consideration and D is the data set available.

Evolutionary computation has been used from its inception for automatic induction of models for a given system or process. For example, L. Fogel used simulated evolution to induce finite-state machines that predict a sequence of symbols from an environment [9]. Other authors have used different structures as models of target systems. These include machine language instructions [10], sets of if–then rules [14], neural networks [26], and many others. Recent development includes genetic programming [15,4] where Lisp-like symbolic programs are evolved from training data.

2.3 Optimization vs. Learning

There are several differences between function approximation (learning) and function optimization. First, the objective function f in optimization is given as an "explicit" functional form and the fitness of any search points $\mathbf{x}$ can be evaluated directly from this function. In contrast, the objective function in learning is given "implicitly" as a finite set D of function values, and thus the fitness of search points (in this case, models θ) is evaluated on this finite set of fitness cases.

Learning also differs from optimization in the way the optimization result is used. The result θ^* of learning in (2) is usually used later to solve multiple problem instances (e.g., to predict outputs $f_{\theta^*}(\mathbf{x})$ given different inputs $\mathbf{x}$) while the solution $\mathbf{x}^*$ of an optimization problem (1) is itself the final solution for the problem instance at hand.

Despite the differences in their goals and assumptions, function approximation is formally in close connection with "regular" function optimization. To see the relationship, we define the objective function F of function approximation (2) in terms of D as

$$F(\theta) = \sum_{\mathbf{x}_c \in D} ||f_\theta(\mathbf{x}_c) - f(\mathbf{x}_c)||^2. \tag{3}$$

Then, the function approximation problem can be viewed as a function optimization problem:

$$\theta^* = \underset{\theta \in \Theta}{\operatorname{argmin}}\ F(\theta), \tag{4}$$

where Θ is the input space of the function F (see Equation (1) for comparison).

On the other hand, function optimization can be facilitated by using function approximation as a subroutine. That is, instead of attempting to directly optimize the objective function $f(\mathbf{x})$, we can build a model $f_\theta(\mathbf{x})$ of the objective function using the search points observed so far or their subset $X^t = \{\mathbf{x}_c \mid c = 1, ..., N\}$. The new problem is then formulated as

$$\mathbf{x}^* = \underset{\mathbf{x} \in X}{\operatorname{argmin}}\ f_\theta(\mathbf{x}), \tag{5}$$

where f_θ is the approximated function of the original objective function f. This is reasonable since many optimization problems have underlying structure in their search space. Using this structure can help the search for the optimal solution.

Recently, a number of methods have been proposed that explicitly model, i.e., "learn", the population of promising points and use the constructed model to guide further search as reviewed in Section 1. These methods are generally known as the estimation of distribution algorithms or EDAs [17]. They use global information contained in the population, instead of using local information through crossover or mutation of individuals. From the population, statistics of the hidden structure are derived and used when generating new individuals.

In the following section, we present a general probabilistic framework for evolutionary computation that handles function optimization, function approximation, and their combinations in a uniform way.

3 A Bayesian Evolutionary Framework for Learning and Optimization

3.1 Bayesian Evolutionary Computation

In the Bayesian approach to evolutionary computation [23], the fitness of the individuals is formulated as a probability function. In the context of maximum likelihood, this might be a likelihood function or its transform. In decision-theoretic contexts, this might involve complex loss functions. In its most general form, the fitness of an individual is represented as a posterior probability. Here, the best (fittest) individual is defined as the most probable model of the data with respect to the prior knowledge on the problem domain.

More formally, let θ denote the parameter vector for the model, let $\pi(\theta)$ be the prior probability distribution for the models (since θ uniquely determines the model, we use the terms "model" and "model parameter" interchangeably in this chapter), and $p(D|\theta)$ the likelihood of the model for the data $D = \{(\mathbf{x}_c, y_c) \mid c = 1, ..., N\}$. Then, using Bayes formula [11] the posterior probability $\pi(\theta|D)$ of model θ is given as the

$$\pi(\theta|D) = \frac{p(D|\theta)\pi(\theta)}{p(D)}. \tag{6}$$

Here, $p(D)$ is a normalizing constant and computed as

$$p(D) = \int_{\Theta} p(D|\theta)\pi(\theta)d\theta, \tag{7}$$

where Θ is the space of all possible model parameters (in the case of discrete space, the integral will be replaced with a summation).

Initially, the shape of the (prior) probability distribution of individuals $\pi_0(\theta)$ is flat to reflect the fact that little is known at the outset (see Figure 1). Evolution is considered as an iterative process of revising the posterior distribution of individuals $\pi_t(\theta|D)$ by combining the prior $\pi_t(\theta)$ with the

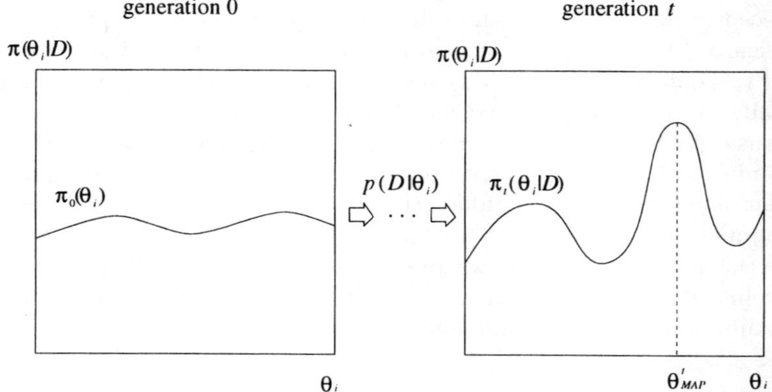

Fig. 1. Bayesian formulation of evolutionary computation. Initially, the shape of the (prior) probability distribution of models $\pi_0(\theta_i)$ is relatively flat, reflecting the fact that little is known at the outset. Evolution is considered as an iterative process of updating the posterior distribution of models $\pi_t(\theta_i|D)$ by combining the prior probability $\pi(\theta_i)$ with the likelihood $p(D|\theta_i)$ of observed data D.

likelihood $p(D|\theta)$. In each generation, the Bayes theorem (6) is used to estimate the posterior fitness of individuals from their prior fitness values. The posterior distribution $\pi_t(\theta|D)$ is then used to generate its offspring.

The aim of Bayesian evolutionary computation is twofold, depending on the problem being addressed. One is to choose a model θ_{MAP} that maximizes the posterior probability (MAP):

$$\theta_{MAP} = \underset{\theta \in \Theta}{\text{argmax}} \; \pi(\theta|D). \tag{8}$$

The MAP model is then used to predict the output values y for given input values $\mathbf{x}$:

$$y = f(\mathbf{x}; \theta_{MAP}). \tag{9}$$

This is the approach we take for function approximation or learning. An example of this approach will be given in Section 4 below.

Alternatively, the samples from the posterior distribution $\pi(\theta|D)$ can be used to compute the posterior *predictive* distribution $p(\mathbf{x}|D)$ of inputs $\mathbf{x}$ as follows:

$$p(\mathbf{x}|D) = \int p(\mathbf{x}, \theta|D)d\theta \tag{10}$$

$$= \int p(\mathbf{x}|\theta, D)\pi(\theta|D)d\theta \tag{11}$$

$$= \int p(\mathbf{x}|\theta)\pi(\theta|D)d\theta. \tag{12}$$

In the first two lines the distribution of future inputs $\mathbf{x}$ is expressed as an average of conditional predictions over the posterior distribution $\pi(\theta|D)$ of θ. The last equation above follows because $\mathbf{x}$ and D are conditionally independent given θ. This is the approach we take to generate promising new points for function optimization.

The distribution estimation algorithms for optimization mentioned in foregoing sections can be regarded as special cases of the Bayesian approach, where the maximum likelihood estimate approximates the (full) posterior predictive distribution. That is, a single function $p(\mathbf{x}|\hat{\theta})$ of maximum likelihood estimate $\hat{\theta}$ is used to approximate the objective function $f(\mathbf{x})$ rather than using $p(\mathbf{x}|\theta)$'s for all possible θ's to take into account their distribution $\pi(\theta|D)$. Note that this is a reasonable approximation since if the prior is relatively flat and the peak of the likelihood function is relatively sharp, then the integral in (12) will be dominated by the region around the maximum likelihood estimate $\hat{\theta}$, and the posterior predictive distribution can be given approximately by

$$p(\mathbf{x}|D) \approx p(\mathbf{x}|\hat{\theta}) \int \pi(\theta|D)d\theta \tag{13}$$

$$= p(\mathbf{x}|\hat{\theta}), \tag{14}$$

where we used $\int \pi(\theta|D)d\theta = 1$, and $\hat{\theta}$ is the parameter vector that maximizes $p(\mathbf{x}|\theta)$. In Section 5 below, we describe a method that uses a probabilistic graphical model known as Helmholtz machines to approximate the posterior predictive distribution function $p(\mathbf{x}|D)$ by the maximum likelihood $p(\mathbf{x}|\hat{\theta})$.

3.2 The Canonical BEA

Given the principles of the Bayesian evolutionary approach to optimization and learning, we are now ready to describe their realization. Algorithm 3.1 summarizes the canonical BEA.

Algorithm 3.1 [Canonical BEA]

1 (Initialize) Generate $\Theta^0 = \{\theta_1^0, ..., \theta_M^0\}$ from the prior distribution $\pi_0(\theta)$. Initialize data size N_0 and temperature T_0. Set generation count $t \leftarrow 0$.
2 (D-step) Generate (observe) D^t of size N_t. Compute likelihoods $p(D^t|\theta_i^t)$.
3 (P-step) Estimate posterior distribution $\pi_t(\theta|D^t)$ of the individuals in Θ^t.
4 (V-step) Generate L variations $\Theta' = \{\theta_1', ..., \theta_L'\}$ by sampling from $\pi_t(\theta|D^t)$.
5 (S-step) Select M individuals from Θ' into $\Theta^t = \{\theta_1^{t+1}, ..., \theta_M^{t+1}\}$ based on $p(D^t|\theta_i')$. Set the best individual θ_{best}^t.
6 (R-step) Optionally, revise prior distribution $\pi_t(\theta)$, and update N_t and T_t.

7 (Loop) If the termination condition is met, then stop. Otherwise, set $t \leftarrow t+1$ and go to Step 2.

In essence, the algorithm consists of five steps: D (data), P (posterior), V (variation), S (selection), and R (revision). The three steps of R, D, and P involve computation of prior, likelihood, and posterior probabilities, respectively. The V- and S-steps realize the sampling from the posterior distribution. Note that BEAs attempt in the P-step to explicitly model the posterior fitness distribution $\pi_t(\theta|D^t)$ of individuals in population Θ^t. Another feature of BEAs is the D-step which may care for incremental growth of data sets [27]. This naturally corresponds to the Bayesian inductive learning principle. The R-step is used only for adaptive versions of BEAs, and will not be considered in this chapter.

More specifically, we define the fitness value of individual θ_i^t as its posterior probability $\pi_t(\theta_i^t|D^t)$:

$$\pi_t(\theta_i^t|D^t) \equiv \frac{p(D^t|\theta_i^t)\pi_t(\theta_i^t)}{\sum_{\theta_j^t \in \Theta^t} p(D^t|\theta_j^t)\pi_t(\theta_j^t)}, \tag{15}$$

where Θ^t is the finite set of individuals in the tth population. Assuming the exponential family for the likelihood function and prior distribution (e.g., Gaussian distributions),

$$p(D^t|\theta_i^t) = \frac{1}{Z_E} \exp\{-E(D^t|\theta_i^t)/T_t\} \tag{16}$$

$$\pi_t(\theta_i^t) = \frac{1}{Z_C} \exp\{-C(\theta_i^t)/T_t\}, \tag{17}$$

the fitness of individuals is written as

$$\begin{aligned} \pi_t(\theta_i^t|D^t) &\equiv \frac{\exp\{-F(\theta_i^t|D^t)/T_t\}}{\sum_{\theta_j^t \in \Theta^t} \exp\{-F(\theta_j^t|D^t)/T_t\}} \\ &= \frac{\exp\{-(E(D^t|\theta_i^t) + C(\theta_i^t))/T_t\}}{\sum_{\theta_j^t \in \Theta^t} \exp\{-(E(D^t|\theta_j^t) + C(\theta_j^t))/T_t\}}, \end{aligned} \tag{18}$$

where $E(D|\theta_j^t)$ and $C(\theta_j^t)$ are arbitrary component measures for evaluating the raw fitness of individuals, and T_t is the temperature parameter for controlling the randomness of the stochastic process. Note here that the posterior probability is approximated by a fixed-size population Θ^t which is typically a small subset of the entire model space Θ: $\Theta^t \subset \Theta, |\Theta^t| \ll |\Theta|$. The evolutionary inference step from generation t to $t+1$ is then considered to induce a new fitness distribution $\pi_{t+1}(\theta)$ from $\pi_t(\theta)$ following the Bayes formula.

At each generation t we keep the best individual θ_{best}^t which is the individual with the maximum a posteriori (MAP) probability with respect to

Θ_t:

$$\begin{aligned}\theta_{best}^t &= \underset{\theta_i^t}{\text{argmax}}\ \pi_t(\theta_i^t|D^t) \qquad (19)\\ &= \underset{\theta_i^t}{\text{argmax}}\ \frac{p(D^t|\theta_i^t)\pi_t(\theta_i^t)}{\sum_{\theta_j^t} p(D^t|\theta_j^t)\pi_t(\theta_j^t)}\\ &= \underset{\theta_i^t}{\text{argmax}}\ p(D^t|\theta_i^t)\pi_t(\theta_i^t), \qquad (20)\end{aligned}$$

where θ_i^t and θ_j^t are elements of population Θ^t. A complete run for t generations of the BEA then chooses the best among the generation-best models, i.e., $\theta_{best}(t)$ such that

$$\pi_t(\theta_{best}(t)|D^t) = \max_{k\leq t} \pi_k(\theta_{best}^k|D^t), \qquad (21)$$

where θ_{best}^k is the best solution at generation k and $\pi_t(\theta_{best}(t)|D^t)$ is the tth estimation of $\pi(\theta_{MAP}|D^t)$.

Note that the description of Algorithm 3.2 is intentionally abstract and general. Thus, for example, the V-step can be implemented in several ways, including mutation, crossover, Metropolis–Hastings moves, or their combinations. The S-step can also be realized using various selection schemes, such as truncation selection and tournament selection as well as proportional selection [1].

4 Bayesian Evolutionary Computation for Learning

The BEA is applied to learning neural tree models of time series data. We provide explicit probability models for the prior, likelihood, and posterior probability distributions of neural trees. The sampling procedure for evolving neural trees is described and empirical results are reported.

4.1 Defininining Probability Distributions of Neural Trees

Neural trees are tree-structured neural networks. A neural tree is composed of terminal nodes, nonterminal nodes, and the weights of connection links between two nodes [26]. The nonterminal nodes represent neural units with various activation functions. Typical neurons compute the sum of weighted inputs from the lower layer by

$$net_i = \sum_j w_{ij} y_j, \qquad (22)$$

where y_j are the inputs to the ith neuron. The output of a neuron is computed by the sigmoid transfer function

$$y_i = f(net_i) = \frac{1}{1+e^{-net_i}}. \qquad (23)$$

In principle, there is no restriction on the types of activation functions employed in neural trees since the evolutionary learning procedure does not impose limiting constraints such as continuity or differentiability on the search space.

Neural trees have a number of important features from the model induction point of view. Neural trees can represent a broad class of higher-order networks since a single tree can have a mixture of sigma, pi, and any other types of units. No bound is enforced on the number of layers of the network. The network structures are not strictly layered; the connections between non-neighboring layers are allowed. The network may contain partial connectivity, which is useful for the economic representation of arbitrary complex interactions. In addition, neural trees do not require decoding for their fitness evaluation. Training and fitness evaluation can be performed directly on the genotype since both the genotype and phenotype are equivalent.

To learn the fittest model by the BEAs, we first define the probability distributions of neural tree models for data. A neural tree model is parameterized as $\theta = (\mathbf{w}, k)$, where k is the number of nodes in the neural tree (including bias terms) and $\mathbf{w}$ is the weight vector. The posterior probability of a neural tree θ is written as

$$\begin{aligned} \pi(\theta|D) &\propto p(D|\theta)\pi(\theta) &(24)\\ &= p(D|\mathbf{w},k)\pi(\mathbf{w},k) &(25)\\ &= p(D|\mathbf{w},k)\pi(\mathbf{w}|k)\pi(k). &(26)\end{aligned}$$

Given the training data

$$D = \{(\mathbf{x}_c, y_c) \mid c = 1, ..., N\}, \tag{27}$$

the model A can represent the following input–output mapping:

$$y_c = f(\mathbf{x}_c; \theta) + \epsilon. \tag{28}$$

Here, the noise ϵ is assumed to be Gaussian with mean zero and standard deviation σ.

If we additionally assume that data items are independent of each other, then the likelihood of the neural tree can be expressed as follows:

$$\begin{aligned} p(D|\mathbf{w},k) &= \prod_{c=1}^{N} \frac{1}{\sqrt{2\pi}\sigma} \exp\Big(- \frac{(y_c - f(\mathbf{x}_c; \mathbf{w}, k))^2}{2\sigma^2} \Big) \\ &= \Big(\frac{1}{\sqrt{2\pi}\sigma} \Big)^N \exp\Big(- \frac{\sum_{c=1}^{N} (y_c - f(\mathbf{x}_c; \mathbf{w}, k))^2}{2\sigma^2} \Big). \end{aligned} \tag{29}$$

We define the following prior probability for weights of the neural tree

$$\pi(\mathbf{w}|k) = \prod_{j=1}^{k-1} \frac{1}{\sqrt{2\pi}} \exp\Big(- \frac{w_j^2}{2}\Big)$$
$$= \Big(\frac{1}{\sqrt{2\pi}}\Big)^{k-1} \exp\Big(- \frac{\sum_{j=1}^{k-1} w_j^2}{2}\Big), \tag{30}$$

where the components of the weight vector $\mathbf{w}$ are assumed to be independent of each other and distributed according to zero-mean Gaussian with standard deviation 1. We also assume that the number of nodes in the neural tree is distributed according to following Poisson distribution:

$$\pi(k-3) = \frac{\lambda^{k-3}\exp(-\lambda)}{(k-3)!}, \tag{31}$$

where $k = 3, 4, \ldots$ since the neural tree which consists of one terminal node was not considered. Substituting Equations (29), (30), and (31) into (24), we obtain the following posterior probability for the neural tree:

$$\pi(\theta|D) \propto p(D|\mathbf{w}, k)\pi(\mathbf{w}|k)\pi(k)$$
$$= \Big(\frac{1}{\sqrt{2\pi}\sigma}\Big)^{N} \exp\Big(- \frac{\sum_{c=1}^{N} (y_c - f(\mathbf{x}_c; \mathbf{w}, k))^2}{2\sigma^2}\Big)$$
$$\times \Big(\frac{1}{\sqrt{2\pi}}\Big)^{k-1} \exp\Big(- \frac{\sum_{j=1}^{k-1} w_j^2}{2}\Big) \frac{\lambda^{k-3}\exp(-\lambda)}{(k-3)!}, \tag{32}$$

where $\theta = (\mathbf{w}, k)$ is the parameter vector for the neural tree.

4.2 The BEA for Learning

The general procedure for evolving neural trees is similar to Algorithm 3.2. It consists of four major steps(Figure 2): D (data), P (posterior), V (variation), and S (selection). The D- and P-steps involve computation of the likelihood $p(D|\theta_i^t)$ and posterior probabilities $\pi_t(\theta|D)$, respectively. The V- and S-steps realize the sampling from the posterior distribution. The probability models for the component distributions are as given in the previous subsection.

To search the structure and parameters of neural trees, we maintain a population Θ^t of individuals θ_i at the tth generation

$$\Theta^t = \{\theta_1, \theta_2, \ldots, \theta_M\}, \tag{33}$$

where M is the population size. The initial population Θ^0 is created according to the prior probability of models. Each individual $\theta_i = (\mathbf{w}_i, k_i)$ is generated by sampling first a value k_i for the number of nodes from the Poisson distribution (31) and then $\mathbf{w}_i$ from the Gaussian distribution (30) using the k_i-value. In each generation t, the error $E_i(t)$ of neural trees is measured on the training set D as

$$E_i(g) = \sum_{c=1}^{N} (y_c - f(\mathbf{x}_c; \theta_i))^2, \tag{34}$$

where $f(\mathbf{x}_c; \theta_i)$ is the actual output for input vector $\mathbf{x}_c$ of neural tree θ_i, and y_c is the target output. Using this value, the likelihood (29) of the neural tree is computed. Finally, the posterior probability of each model is computed by Equation (32).

1 (Initialize) Set the training set $D = \{(\mathbf{x}_c, y_c) \mid c = 1, ..., N\}$. Generate $\Theta^0 = \{\theta_1^0, ..., \theta_M^0\}$ from $\pi_0(\theta)$. Set generation count $t \leftarrow 0$.
2 (D-step) Compute likelihoods $p(D|\theta_i^t)$ of individuals $\theta_i^t = (\mathbf{w}_i^t, k_i^t)$.
3 (P-step) Estimate posterior distribution $\pi_t(\theta|D)$ of the individuals in Θ^t.
4 (V-step) Generate L variations $\Theta' = \{\theta'_1, ..., \theta'_L\}$ by sampling from $\pi_t(\theta|D)$. Optionally, train the weights $\mathbf{w}$ of the individuals θ.
5 (S-step) Select M individuals from Θ' into $\Theta^t = \{\theta_1^{t+1}, ..., \theta_M^{t+1}\}$ based on $p(D|\theta'_i)$. Set the best individual θ_{best}^t.
6 (Loop) If the termination condition is met, then stop. Otherwise, set $t \leftarrow t+1$ and go to Step 2.

Fig. 2. Outline of the BEA for learning neural trees.

To construct the next generation Θ^{t+1}, a candidate tree θ'_i is first created from the parent tree θ_i in the current population. The candidate tree is then accepted with probability

$$\alpha(\theta_i, \theta'_i) = \min\left\{1, \frac{\pi(\theta'_i|D)}{\pi(\theta_i|D)}\right\}, \tag{35}$$

which is called the acceptance probability. The candidate tree θ'_i is always accepted when its posterior probability is higher than that of the parent; otherwise, it is accepted according to the ratio of two probabilities. If the candidate model is accepted, θ'_i is copied into the next generation. If the candidate is rejected, then θ_i is copied into the next generation.

Two major variation operators are applied to the parent models for generating candidate models. First, crossover operators swap two subtrees chosen at random from the parent tree θ_i and another tree $\theta_j (i \neq j)$ which is selected randomly from the current population to create the candidate model

θ_i. Second, mutation operators change the type of nonterminal nodes or the index of incoming units in the subtree which is also chosen randomly from the parent tree. Other kinds of mutations can also be defined. The probabilities for applying these operators are p_c and p_m, respectively. These mating steps are performed iteratively until L individuals are produced.

Weights of a neural tree are adjusted through a stochastic hill-climbing. All components of the weight vector $\mathbf{w}$ are changed just once in a random order by a Gaussian mutation expression

$$w'_{ij} = w_{ij} + \mathrm{N}(0,1) \qquad j = 1, 2, ..., k_i - 1, \tag{36}$$

where k_i is the number of nodes in tree θ_i and $\mathrm{N}(0,1)$ is a normal distribution with mean 0 and variance 1. Each change of the weight is also accepted by Equation (35).

The offspring population Θ' is obtained through the above procedure and we finally generate the parent population Θ^{t+1} of the next generation by selecting the best M individuals from Θ'.

4.3 Empirical Results

To see the potential advantages of the probabilistic approach underlying the BEAs for learning, we carried out experiments on time series data. The far-infrared NH_3 laser data was chosen as a test problem [26]. We used the first 500 data points for evolving the neural tree models and the remaining 500 data points for testing the predictive accuracy. Our experimental setup is as follows: the maximum number of branches of a nonterminal node is 3, which is the same as the input size, the standard deviation of the noise is $\sigma = 0.05$, and the maximum number of fitness evaluations is 10^6. The candidate population size is identical to the parent population size ($M = L$) in all experiments.

The effects of priors were analyzed by running BEAs with different mean values of $\lambda = 10, 20, 30, 40$ for the Poisson distribution on the number of nodes. Figures 3 and 4 show the effect of priors on the complexities of the best neural trees in terms of the number of nodes and the squared sum of weight values. The corresponding performance (in normalized mean squared error) of the neural trees is shown in Figure 5. To take into account the variation caused by the algorithm parameters, we ran two different versions of BEAs with varying population sizes and variation operators. Ten runs were made for each algorithm with each parameter setting.

The results show that for $\lambda = 20$ and 30 the size of the solution trees is approximately the same as λ, while for $\lambda = 10$ the solution size was around 18 and for $\lambda = 40$ the solution size was around 35. This suggests that the size of best solutions for this problem is likely to lie between 20 and 35. This reasoning is confirmed by seeing the accuracy given in Figure 5, which indicates the best solution size of 30 nodes. The results also show that the prior effect is relatively robust in the sense that incorrect priors, as in the

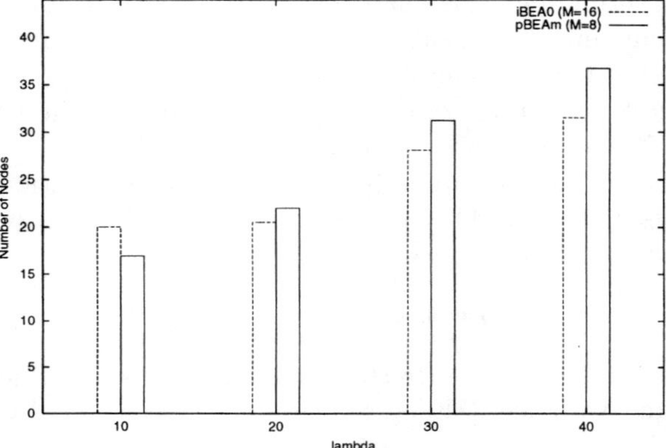

Fig. 3. Effects of the priors on the number of nodes in the solution trees for the laser problem

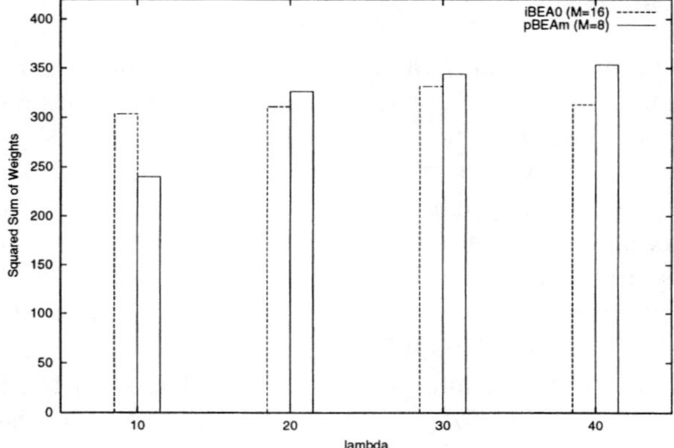

Fig. 4. Effects of the priors on the weight size in the solution trees for the laser problem

case of $\lambda = 20$, can still lead to acceptable solutions. The results for $\lambda = 40$ indirectly show the effect of Occam's razor (i.e., parsimony pressure) of priors: unnecessary complexity is avoided. Figure 4 shows that weight values are less strongly dependent on λ than the number of nodes. This means that harmful overfitting, at least a serious one, did not occur. One possible explanation for this is that local search is better guided by the probability distribution models. Another reason might be the effect of Occam's razor: that is, preferring parsimonious solutions avoids overfitting the data.

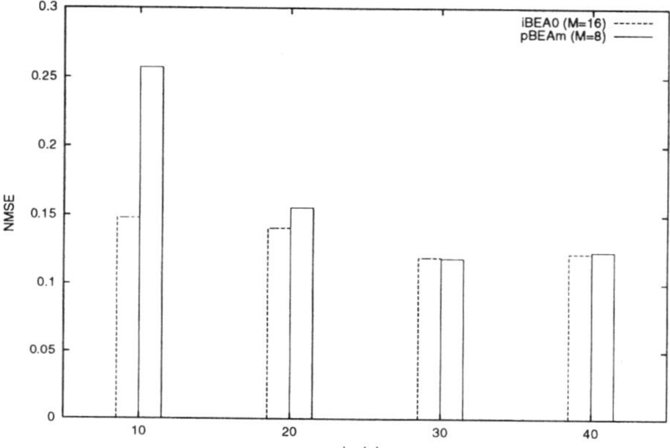

Fig. 5. Effects of the priors on the predictive accuracy of solution trees for the laser problem

To summarize, the experimental results support the effectiveness of priors in guiding the evolutionary learning of tree structures and weights; good priors facilitate the search process and result in better solutions. This is contrasted with conventional evolutionary algorithms where it is usually difficult to incorporate the prior knowledge about the problem domain.

5 Bayesian Evolutionary Computation for Optimization

In this section, a method for Bayesian evolutionary optimization is presented. It is based on the distribution estimation of the sample population. To find the $\mathbf{x}$-point that maximizes the objective function f, this method builds a model f_θ of f using the current population of search points. We present a distribution estimation method that uses the Helmholtz machine [6], a probabilistic graphical model.

5.1 The BEA for Optimization

The algorithm is summarized in Figure 6. Initially, a population X^0 of M search points $\mathbf{x}_i^0$, $i = 1, ..., M$, is generated from a prior distribution $\pi_0(\mathbf{x})$. It should be stressed that the search space we are here concerned with is the $\mathbf{x}$-space, not the θ-space. Note also that the population in this optimization context is denoted as X^t, not D which is reserved to denote the training data set in the context of learning. Once created, the fitness values of the points are evaluated and their likelihoods $p(X^t|\theta)$ are computed, where θ is the parameter vector for the probability model of the population. Combining the prior $\pi_0(\mathbf{x})$ and likelihood $p(X^t|\theta)$, the posterior probability $\pi(\theta|X^t)$ of individuals is computed using the Bayes rule:

$$\pi(\theta|X^t) = \frac{p(X^t|\theta)\pi(\theta)}{p(X^t)}. \tag{37}$$

Since $p(X^t)$ does not depend on the parameter vector θ, maximization of Equation (37) is equivalent to maximizing the numerator, i.e.,

$$\pi(\theta|X^t) \propto p(X^t|\theta)\pi(\theta). \tag{38}$$

Note that, under the uniform prior for θ, the maximization of expression (38) is reduced to finding the maximum likelihood estimate (MLE) $\hat{\theta}$:

$$\theta^* = \arg\max_\theta \pi(\theta|X^t) = \arg\max_\theta p(X^t|\theta) = \hat{\theta}. \tag{39}$$

Thus, the posterior distribution $\pi(\theta|X^t)$ can be approximated by the maximum likelihood $p(X^t|\hat{\theta})$. We make use of this equivalence and use a Helmholtz machine to find $p(X^t|\hat{\theta})$.

1 (Initialize) Generate $X^0 = \{\mathbf{x}_i^0, ..., \mathbf{x}_M^0\}$ from the prior distribution $\pi_0(\mathbf{x})$. Set generation count $t \leftarrow 0$.
2 (D-step) Given X^t, compute the parameter $\hat{\theta}$ that maximizes the likelihood $p(X^t|\theta)$ using a Helmholtz machine.
3 (P-step) Estimate the posterior predictive distribution $p_{t+1}(\mathbf{x}) = p(\mathbf{x}|X^t) \approx p(\mathbf{x}|\hat{\theta})$ using the Helmholtz machine.
4 (V-step) Generate L variations $X' = \{\mathbf{x}'_1, ..., \mathbf{x}'_L\}$ by sampling from the distribution $p_{t+1}(\mathbf{x})$ using $\hat{\theta}$ of the Helmholtz machine.
5 (S-step) Select M best points from X' and X^t into $X^{t+1} = \{\mathbf{x}_1^{t+1}, ..., \mathbf{x}_M^{t+1}\}$ based on their fitness values $f(\mathbf{x})$.
6 (Loop) If the termination condition is met, then stop. Otherwise, set $t \leftarrow t+1$ and go to Step 2.

Fig. 6. Outline of the BEA for optimization where the Helmholtz machine is used for density estimation.

Once the posterior distribution (or its MLE approximation) is estimated, new points are generated by sampling from the posterior *predictive* distribution $p(\mathbf{x}|X^t)$. To do this, we make use of the following relationship:

$$p(\mathbf{x}|X^t) \approx p(\mathbf{x}|\hat{\theta}) \int \pi(\theta|X^t)d\theta \tag{40}$$

$$= p(\mathbf{x}|\hat{\theta}), \tag{41}$$

which follows from the similar arguments given for Equation (14).

To summarize, the BEA for optimization consists of four main steps: data observation (D), probability estimation (P), variation (V), and selection (S) steps. In the P-step, the density of the current population X^t is estimated, in this case, by a Helmholtz machine [6]. Since the samples may not be very representative of the distribution, especially in early generations, the weights of the Helmholtz machine are reinitialized each generation. This avoids trapping in local minima. In the V-step, the learned Helmholtz machine is used to generate offspring population X' of L data points. More details on learning from and simulating the Helmholtz machine can be found in [28]. In the S-step, M best points are chosen for the next population X^{t+1} from the union of X^t and X'. In the experiments, we use $L = M$. This is similar to the $(\mu + \lambda)$ evolution strategy [1] with $\mu = \lambda = M$.

5.2 Empirical Results

Experiments have been performed on a suite of deceptive functions from the literature [20,19]. We used the one-max function, quadratic function, 3-deceptive function, and trap-5 function. Here we report the results of the first two problems; additional results can be found in [28].

The performance of the Helmholtz machine was compared with that of the simple genetic algorithm (sGA). The sGA used one cut-point crossover, one-point mutation, and roulette-wheel selection. The parameters for the sGA were: maximum generation = 10^7, population size = 50, crossover rate = 0.5, and mutation rate = 0.01. The parameters of the BEA with the Helmholtz machine were: maximum generation = 10^5, population size = 50, learning rate = 0.5, and number of iterations = 1000.

Tables 1 and 2 summarize the results for the one-max and quadratic functions, respectively. The results shown are average values over 10 runs. The algorithm is considered as converged if the population contains over 10% of optimal solutions. The entry in the tables marked with "-" means that none of the runs found an optimal solution. The results show that the BEAs with the Helmholtz machine outperform the sGA both in the success rate and in the number of iterations. BEAs find the optimal solutions all the time while the sGA finds solutions for small-size problems only. It is also interesting that the BEAs find solutions much faster than the sGAs when both find optimal solutions. The tables show that the computational time (in the number of iterations) for the sGAs grows exponentially while that for the BEAs grow almost linearly.

6 Concluding Remarks

We presented a unified probabilistic framework for solving optimization and learning problems using evolutionary computation. In this framework, evolu-

Table 1. Results on the one-max function for the Helmholtz machine algorithm (HM) and the simple genetic algorithm (sGA)

Problem Size	Success % HM	Success % sGA	Number of Iterations HM	Number of Iterations sGA
20	100	100	6.3	930.7
40	100	100	34.7	16,904.7
60	100	50	175.4	5,691,291
80	100	0	657.7	-
100	100	0	2999.3	-

Table 2. Results on the quadratic function for the Helmholtz machine algorithm (HM) and the simple genetic algorithm (sGA)

Problem Size	Success % HM	Success % sGA	Number of Iterations HM	Number of Iterations sGA
20	100	100	5.9	12,073.3
40	100	10	33.0	6,532,588
60	100	0	198.0	-
80	100	0	765.6	-
100	100	0	3430.7	-

tionary computation is formulated as a probabilistic process of Bayesian inference, which leads to a class of Bayesian evolutionary algorithms or BEAs. The generality and concreteness of the framework is demonstrated by showing two specific instances of BEAs. One is for optimization and the other for learning.

The method of Bayesian evolutionary optimization is very closely related to the distribution estimation algorithms. Similar to the usual EDAs, a BEA builds a probabilistic model of discrete samples. However, BEAs are more general in the sense that the distribution estimated is the posterior distribution which encompasses likelihood functions (as in most EDAs) as a special case. We have shown that graphical models such as Helmholtz machines provide a powerful tool for implementing the BEA for optimization. We also show that the learning problems can be effectively solved by adopting the Bayesian evolutionary approach. This was shown in the context of evolving neural tree models for predicting time series data. The probabilistic approach allows the evolution to be guided more effectively by using principled specification of prior knowledge.

It is interesting to see that learning helps solve optimization problems (as in the distribution estimation algorithms), while optimization helps solve learning problems (as in genetic programming and other evolutionary algorithms for machine learning). In both cases, evolutionary computation can be used as a tool for learning and optimization. In this work, we studied two rather loosely coupled combinations of learning and optimization. However, it is not difficult to imagine more tightly and cross-coupled combinations of evolutionary learning and optimization. For example, it would be interesting to have an evolutionary algorithm that "learns" the distribution of a training data set using evolutionary "optimization" where the training data set itself is collected incrementally by an evolutionary "learning" process to "optimize" some information criterion.

Acknowledgments

This research was supported by the Korea Science and Engineering Foundation (KOSEF) under Grant 981-0920-350-2, by the Korea Ministry of Science and Technology through KISTEP under BrainTech Program, and by the Korea Ministry of Education under BK21-IT Program.

References

1. Bäck, T. (1996) *Evolutionary Algorithms in Theory and Practice.* Oxford Univ. Press
2. Baluja, S., Caruana, R. (1995) Removing the genetics from the standard genetic algorithm. Technical Report CMU-CS-95-141, Carnegie Mellon University
3. Baluja, S., Davies, S. (1997) Using optimal dependency-trees for combinatorial optimization: learning the structure of the search space. In *Proc. 14th Int. Conf. on Machine Learning*, San Mateo, CA: Morgan-Kaufmann, 30–38
4. Banzhaf, W., Nordin, P., Francone, F. (1998) *Genetic Programming: An Introduction.* San Mateo, CA: Morgan Kaufmann
5. Cherkassky, V., Mulier, F. (1998) *Learning from Data: Concepts, Theory, and Methods.* New York: Wiley
6. Dayan, P., Neal, G.E., Zemel, R.S. (1995) The Helmholtz machine. *Neural Computation*, **7**, 1022–1037
7. De Bonet, J.S., Isbell, C.L., Viola, P. (1997) MIMIC: Finding optima by estimating probability densities. In *Advances in Neural Information Processing Systems* **9**, Cambridge, MA: The MIT Press, 424–430
8. Fogel, D.B., Ghozeil, A. (1996) Using fitness distributions to design more efficient evolutionary computations. In *Proc. 1996 IEEE Int. Conf. on Evolutionary Computation*, Piscataway, NJ: IEEE Press, 11–19
9. Fogel, D.B. (ed.) (1998) *Evolutionary Computation: The Fossil Record.* Priscataway, NJ: IEEE Press
10. Friedberg, R.
M. (1958) A learning machine: Part I, *IBM Journal of Research and Development*, **2**, 2–13

11. Gelman, A., Carlin, J.B., Stern, H.S., Rubin, D.B. (1995) *Bayesian Data Analysis.* London: Chapman & Hall
12. Goldberg, D.E. (1989) *Genetic Algorithms in Search, Optimization, and Machine Learning.* Realing, MA: Addison-Wesley
13. Hinton, G.E., Dayan, P., Frey, B.J., Neal, R.M. (1995) The wake-sleep algorithm for unsupervised neural networks. *Science*, **268**, 1158–1160
14. Holland, J.H. (1986) Escaping brittleness: The possibilities of general-purpose learning algorithms applied to parallel rule-based systems. In *Machine Learning: An Artificial Intelligence Approach*, Vol. II, R.S. Michalski, J.G. Carbonell, and T.M. Mitchell, Eds. San Francisco: Morgan Kaufmann, 593–623
15. Koza, J.R. (1992) *Genetic Programming: On the Programming of Computers by Means of Natural Selection.* Cambridge, MA: The MIT Press
16. Mühlenbein, H., Mahnig, T. (1999) FDA - A scalable evolutionary algortihm for the optimization of additively decomposed functions. *Evolutionary Computation*, **7**, 353–376
17. Mühlenbein, H., Mahnig, T., Ochoa, A. (1999) Schemata, distributions and graphical models in evolutionary optimization. *Journal of Heuristics*, **5**, 215–247
18. Mühlenbein, H., Paaß, G. (1996) From recombination of genes to the estimation of distributions I: Binary parameters. In *Parallel Problem Solving from Nature IV*, LNCS 1141. Berlin: Springer, 178–187
19. Pelikan, M., Goldberg, D.E., Cantú-Paz, E. (1999) BOA: The Bayesian optimization algorithm. In *Proc. 1999 Genetic and Evolutionary Computation Conference*, San Mateo, CA: Morgan Kaufmann, 525–532
20. Pelikan, M., Mühlenbein, H. (1999) The bivariate marginal distribution algorithm. *Advances in Soft Computing - Engineering Design and Manufacturing.* London: Springer, 521–535
21. Rissanen, J. (1986) Stochastic complexity and modeling. *Annals of Statistics*, **14**, 1080–1100
22. Schwefel, H.-P. (1995) *Evolution and Optimum Seeking*, Sixth Generation Computer Technology Series. Wiley: New York
23. Zhang, B.-T. (1999) A Bayesian framework for evolutionary computation. In *Proc. 1999 Congress on Evolutionary Computation* (CEC99), IEEE Press, 722–727
24. Zhang, B.-T. (2000) Bayesian methods for efficient genetic programming. *Genetic Programming and Evolvable Machines*, **1**, 217–242
25. Zhang, B.-T., Cho, D.-Y. (2000) Evolving neural trees for time series prediction using Bayesian evolutionary algorithms, In *Proc. First IEEE Workshop on Combinations of Evolutionary Computation and Neural Networks* (ECNN-2000), IEEE Press, 17-23
26. Zhang, B.-T., Ohm, P., Mühlenbein, H. (1997) Evolutionary induction of sparse neural trees. *Evolutionary Computation*, **5**, 213–236
27. Zhang, B.-T., Paaß, G., Mühlenbein, H. (2000) Convergence properties of incremental Bayesian evolutionary algorithms with single Markov chains. *Congress on Evolutionary Computation* (CEC-2000), IEEE Press, 938-945
28. Zhang, B.-T., Shin, S.-Y. (2000) Bayesian evolutionary optimization using Helmholtz machines, *Int. Conf. on Parallel Problem Solving from Nature* (PPSN-2000), LNCS 1917. Berlin: Springer, 827–836

Designed Sampling with Crossover Operators

Akiko Aizawa

National Institute of Informatics
2-1-2 Hitotsubashi Chiyoda-ku, Tokyo 101-8430, Japan
E-mail: akiko@nii.ac.jp

Summary. Experimental design and evolutionary computation are different types of optimization techniques, both based on the same expectation that the achievement of a target system is improved by identifying and recombining the promising sub-components. Although the close relationship between the two techniques has long been recognized in the evolutionary computation community, past studies mostly focused on either analytical or comparative aspects of the issues. The present chapter is an attempt to combine experimental design and evolutionary computation into a single search strategy using a specific type of recombination function called a deterministic crossover operator. We first provide a brief overview of the traditional methods for experimental design as well as their connections to evolutionary computation. Then, "fair" and "greedy" sampling strategies are formulated, assuming the solution space is decomposed into two uniquely determined sub-spaces with a given deterministic crossover operator. Based on this formulation, a genetic-based implementation of fair and greedy sampling is also presented, with some illustrative experimental results.

1 Introduction

The class of evolutionary computation algorithms assumed in our present research is extremely simple: genes are expressed as binary strings and a crossover operator is applied to a pair of selected parents to generate new offspring by a simple recombination of genes.

Let g be a binary string composed of L elements $(g_1 \cdots g_L)$ in which $g_i \in \{0, 1\}$, and let $G = \{0, 1\}^L$ be a space of all possible g. Similarly, let $\psi = (\psi_1 \cdots \psi_L)$ be a "crossover template" or a "crossover mask" of length L, in which $\psi_i \in \{0, 1\}$, and let $\Psi = \{0, 1\}^L$ be a space of all possible ψ. Here, ψ represents a mapping from parents a, b $(\in G)$ to children c, d $(\in G)$ such that $c_i = a_i$ and $d_i = b_i$ for i's such that $\psi_i = 0$, and $c_i = b_i$ and $d_i = a_i$ for i's such that $\psi_i = 1$ (Figure 1).

A general definition of a crossover operator is given by a recombination distribution R defined over a space of crossover templates (for example, in [1]):

$$R(\psi) = p_\psi \quad \text{where} \quad \sum_{\psi \in \Psi} p_\psi = 1. \tag{1}$$

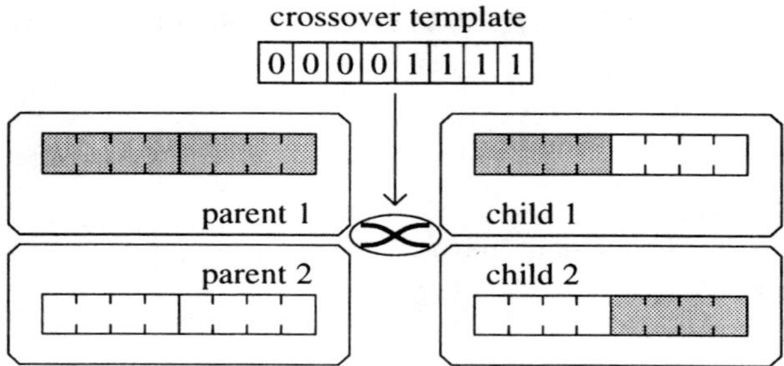

Fig. 1. Crossover template

By varying the recombination distribution, different types of crossover operators can be defined. For example, considering the case when $L = 4$, the uniform crossover is given by the distribution $R(\psi) = 0.0625$ for all possible 2^4 crossover templates. Similarly, the one-point crossover is given by the distribution $R(\psi) = 0.125$ for crossover templates $\psi = \{$"0000", "0001", "0011", "0111", "1111", "1110", "1100", "1000"$\}$ and $R(\psi) = 0.0$ for templates $\psi = \{$"0010", "0100", "0101", "0110" "1001", "1010", "1011", "1101"$\}$.

Among many possible distributions for $R(\psi)$, we specifically focus on the case where ψ is uniquely determined, i.e., $R(\psi) = 1$ for $\psi = \psi^*$ and 0 otherwise. The simple recombination function with uniquely determined ψ^* is referred to as "deterministic crossover" in this chapter. Although the formulation of deterministic crossover seems to be rather restrictive, we expect that this simplification will be useful not only to introduce an experimental design perspective into evolutionary computation but also to help us understand the behavior of the general crossover operator under the assumption that a specific template ψ^* is selected in its operation. We would like to point out here that the probability allocation of $R(\psi)$ represents a priori knowledge of the algorithm on the solution space and thus is essentially an issue of the representation and encoding of genes in each application domain. For this reason, we assume in our analysis that ψ^* is pre-determined. In more general cases, other factors may also require consideration, including the effect of interactions between different crossover templates that are randomly selected with probability $R(\psi)$.

Our formulation is closely related to the experimental design perspective of evolutionary computation that was introduced by Reeves and Wright in 1995 [4]. Experimental design and evolutionary computation are both probabilistic solution-space sampling strategies to explore and exploit the effects of multiple factors. Past studies show that a close relationship exists between factor analysis in experimental design and schema analysis in evolutionary computation. Our proposed method is an attempt to combine the two different sampling schemes into a single search strategy under the simplifying assumption of deterministic crossover. When a crossover template ψ^* is given, both experimental design and evolutionary computation can be viewed as similar types of sampling strategies, which decompose and recombine sampling sub-spaces specified by ψ^*. This view enables us to use experimental design and evolutionary computation complementarily.

The rest of the chapter is organized as follows. Section 2 provides a brief review of the traditional experimental design methods and their connections to evolutionary computation, followed by some discussions in Section 3. Section 4 gives a mathematical description of partitioned random search and the fair and greedy sampling strategies. Section 5 formulates a genetic sampling scheme in which the fair and greedy sampling schemes are combined. Section 6 presents a method for designing a global sampling distribution using a specified deterministic crossover operator, with some simulation results shown in Section 7. Section 8 deals with discussions and future work.

2 Experimental Design and Evolutionary Computation: Past Studies

2.1 A Brief Overview of Experimental Design

The basic concept of experimental design is to arrange experiments so that the effects and interactions of multiple causes can be analyzed most efficiently. Usually, a cause that affects the achievement of a target system is referred to as a "factor" and the values of each factor to be tested are referred to as "levels" [2].

As an example, let us consider a series of experiments to compare the performance of different search engines, measured by the total search time, with different search topics. Assume that four search engines, A_1, A_2, A_3 and A_4, as well as four search topics, B_1, B_2, B_3 and B_4, are selected for testing. The "factors" in these experiments are search engines and search topics, and the "levels" of the factors are A_1, A_2, A_3 and A_4 for search engines, and B_1, B_2, B_3 and B_4 for search topics (Table 1(a)). Now, the subject of experimental design is to decide the combinations of levels of the two factors to be tested so that reliable analytical results are obtained with minimum experimental cost. In the example shown in Table 1, all the levels of the factors are tested exactly three times, which means that the total number of evaluations is reduced to 12 from the 16 required to enumerate all the possible combinations.

The underlying assumption of experimental design is that the performance of a system (s_{ij}) with levels A_i and B_j may be expressed as the sum of (i) an over-all mean (μ), (ii) the mean effect of factor A at level i (a_i), (iii) the mean effect of factor B at level j (b_j), and (iv) a random error (e_{ij}). Namely,

$$s_{ij} = \mu + a_i + b_j + e_{ij}. \tag{2}$$

Table 1. Example of an experimental design

(a) combination of factors tested in the experiments

search engine	search topic B_1	B_2	B_3	B_4
A_1		×	×	×
A_2	×		×	×
A_3	×	×	×	
A_4	×	×		×

(b) estimated contribution of each level

search engine	search topic B_1	B_2	B_3	B_4	the mean effect of factor A
A_1		s_{12}	s_{13}	s_{14}	$a_1 = \frac{1}{3}(s_{12} + s_{13} + s_{14})$
A_2	s_{21}		s_{23}	s_{24}	$a_2 = \frac{1}{3}(s_{21} + s_{23} + s_{24})$
A_3	s_{31}	s_{32}	s_{33}		$a_3 = \frac{1}{3}(s_{31} + s_{32} + s_{33})$
A_4	s_{41}	s_{42}		s_{44}	$a_4 = \frac{1}{3}(s_{41} + s_{42} + s_{44})$
the mean effect of factor B	$b_1 = \frac{1}{3}(s_{21} + s_{31}) + s_{41})$	$b_2 = \frac{1}{3}(s_{12} + s_{32}) + s_{42})$	$b_3 = \frac{1}{3}(s_{13} + s_{23}) + s_{33})$	$b_4 = \frac{1}{3}(s_{14} + s_{24}) + s_{44})$	$\mu = \frac{1}{12}(s_{11} + s_{12} + s_{13} + s_{21} + s_{22} + s_{24} + s_{31} + s_{33} + s_{34} + s_{42} + s_{43} + s_{44})$

with $\sum_i a_i = 0$ and $\sum_j b_j = 0$. Assuming that the contribution of the two factors is much greater than that of the random error, the values of μ, a_i and b_j are determined from the observed performance values as:

$$\mu = \frac{1}{N}\sum_i\sum_j s_{ij}, \quad a_i = \frac{1}{n_i}\sum_j s_{ij} - \mu, \quad b_j = \frac{1}{m_j}\sum_i s_{ij} - \mu \tag{3}$$

with N being the total number of evaluations, n_i being the number of levels of factor B examined with level A_i, and m_j being the number of levels of factor A examined with level B_j (Table 1(b)).

In the following, three different aspects of experimental design are briefly discussed in connection with evolutionary computation: analysis of variance, correlation between factors, and sequential sampling design.

2.2 Variance Analysis and Evolutionary Computation

When all the combinations of levels of any two factors are examined an equal number of times, as was the case with Table 1, the experiments are said to be "orthogonal." The data obtained from orthogonal experiments can be analysed by the technique of "analysis of variance" (Anova). By adopting this technique, the variance of the measured performance is broken down into the portions caused by the specified factors, singly or in combination, and a portion caused by unexpected errors. Although general equations for Anova are not shown here, we will see an example later in this chapter (in section 6.3).

In [3] and [4], Reeves and Wright examined the links between evolutionary computation and the traditional Anova technique. In their formulation, genes in evolutionary computation correspond to factors in experimental design and alleles correspond to levels (Figure 2). Thus, for a problem with binary strings of length L, there exist in total L factors each with two levels, and the solution space can be analyzed using Anova with L-factor two-level experiments. Based on this, Reeves and Wright reformulated Davidor's "epistasis variance" [7] from an experimental design perspective, and showed that the epistasis variance corresponds exactly to the interaction effects in standard Anova. In other words, if the deceptiveness of a problem is to be measured by the degree of non-linearity, then the value is estimated efficiently using Anova. They further showed the equivalence of Anova to the Walsh transform analysis popularized by Goldberg [8][9].

In [5], Reeves and Wright also compared the performance of evolutionary computation with methods based on experimental design in which three versions of "Sequential Elimination of Levels" (SEL), including their original version, are compared with a simple genetic algorithm. SEL is a method that iteratively repeats orthogonally designed experiments while the number of levels tested in each iteration step is reduced sequentially. Their comparison showed that the genetic algorithm is more robust than the SEL techniques to the increase of epistasis.

In [13], Tanaka proposed the "Orthogonal Design Algorithm" (ODA) and compared its performance with traditional genetic algorithms. ODA uses the same binary encoding as traditional genetic algorithms, although without any recombination and mutation. Instead of these genetic operators, ODA adopts n-factor two-level orthogonal experiments and factor analysis as a means to explore the solution space. The experimental results show that the performance of ODA is comparable to one of the traditional genetic algorithms, showing that the simple statistical method is sufficiently effective in optimizing binary-encoded problems.

2.3 Correlation Analysis and Evolutionary Computation

As is pointed out in [5], the major goal of the traditional experimental design methods is to understand the factors that affect the outcome of a given system, rather than to find the optimal combination of the factors. Therefore, these analysis methods are static in nature and applicable only after the completion of carefully designed experiments.

In the field of evolutionary computation, on-the-fly methods also exist, which we refer to as correlation analysis methods in this chapter. Basically, these methods calculate the correlation between the fitness values of parents and children to estimate the amount of interaction in a given population.

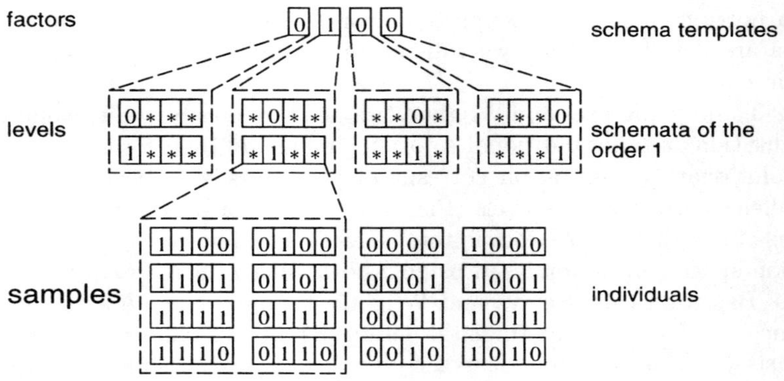

Fig. 2. Factors, levels and samples with binary representation

Then, the correlation value is used as an indicator of how well the algorithm is performing on a given problem. The difference between variance analysis and correlation analysis is that the former estimates the contribution of individual factors (and their combinations) separately, whereas the latter measures the contribution of known factors as a whole.

Correlation analysis was first introduced by Mason [10] with the formulation of "crossover non-linearity ratios." Reeves and Wright [4] later discussed the connections between Mason's crossover non-linearity ratios and factor analysis, pointing out the importance of such measures in extending the notion of epistasis to a consideration of dynamic aspects. At the same time, it was also pointed out that Mason's calculation of the correlation was based on some ad hoc assumptions.

Another method of measuring non-linearity in the course of crossover operation is the "crossover correlation," introduced by Manderic et al. [11]. In this approach, the performance of the algorithm on a given problem is evaluated by the coefficient of correlation between the fitness values of the parents and the children. Aizawa [12] showed the relationship between Manderic et al.'s crossover correlation and variance decomposition, epistasis variance and also Walsh coefficients. Asoh and Mühlenbein derived equations to estimate heritability by the decomposition of genetic variance [6], in which they adopted a variance decomposition technique similar to Anova.

2.4 Sequential Analysis and Evolutionary Computation

Orthogonal experiments allocate an equal numbers of samples for each level, although experimental design methods that allocate an unequal numbers of samples between levels also exist. In these methods, usually referred to as "sequential analysis," samples are allocated sequentially at each step based on past observations, so that the expected loss (or risk) of a given objective function is minimized.

When the objective is to identify the best among k given levels, it is called a "ranking and selection" problem. For example, a typical situation would be a clinical trial where the objective was to identify the best of k possible treatments for a patient. Starting from the pioneering work by Bechhofer in the 1950s [15], sequential ranking and selection problems have long been the subject of mathematical statistics.

In his initial work in the 1970s [14], Holland showed that the "two-armed bandit problem" formulation in statistical decision theory can be used as a justification of schemata processing of genetic algorithms, i.e., the policy that allocates an exponentially increasing number of samples to promising schemata. Holland's original view matches very well with Bechhofer's sequential selection and ranking formulation, where the problem is to find the (asymptotically) optimal allocation of samples between k normal distributions with unknown means. In this case, the normal distributions correspond to schemata and samples to individuals.

SEL, proposed by Reeves and Wright, can also be viewed as a kind of sequential analysis, as a series of orthogonal experiments is designed to sequentially eliminate poorly performing levels at each step.

3 Experimental Design and Evolutionary Computation: Discussion

We have given a brief overview of the connections between experimental design and evolutionary computation. Although some studies pointed out the possible advantage of using these two techniques in a complementary fashion, experimental design perspectives in conventional studies have been mostly applied to analyst the difficulty of target problems or to clarify the characteristics of evolutionary computation in comparison with experimental design. This motivated us to explore a new method that combines these two techniques into a single search strategy. In our attempt at this, we mainly focus on the variance analysis perspective in section 2.2, as the method has a clear correspondence with conventional schema-based mathematical formulations in evolutionary computation, such as Walsh function analysis.

Conventional studies related to variance analysis, such as [5] and [13], view evolutionary computation as designed experiments with L factors and two levels, where L is the length of the binary strings, factors correspond to genes and levels to alleles (Figure 2). In this chapter, we extend the definition so that factors correspond to arbitrary combinations of bits instead of single bits. It then follows that levels correspond to schemata and samples to the performance of individuals that contain the schema (Figure 3). Then, given a binary string of length L, there exist 2^L possible factors of the order k $(0 \leq k \leq L)$ with 2^k possible levels, each of which contains $2^{(L-k)}$ samples.

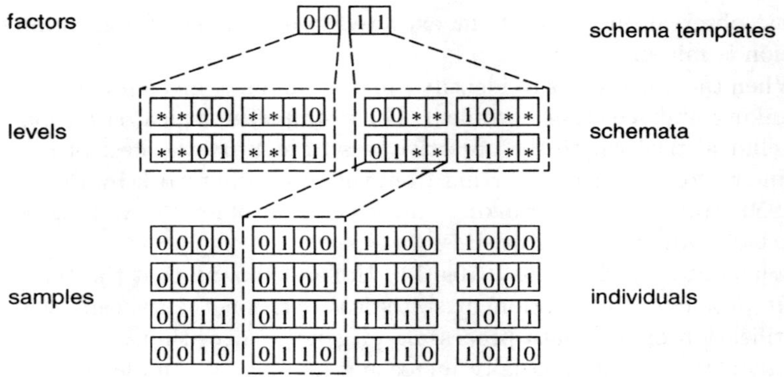

Fig. 3. Factors, levels and samples with binary representation in a different view

Based on the above-mentioned correspondences, the difference between the sampling strategies of experimental design and evolutionary computation can be summarized as follows.

(1) the number of factors considered in the experiments
In experimental design, factors must be determined before the experiments. As the experiments require the interactions between factors to be as small as possible, factors examined in the same experiments usually do not overlap (even with binary encoding). Therefore, the total number of factors is bounded by the length of the binary string as a maximum. On the other hand, in evolutionary computation, crossover templates are selected probabilistically because of the non-deterministic nature of crossover operators. This means that up to 2^L crossover templates, depending on the definition of the crossover operator, are processed implicitly during the execution of the algorithm.

(2) the number of levels considered for each factor
In experimental design, the levels examined in the experiments should also be specified before the experiments. This means that the number of factors, as well as the number of levels, should be reduced to a manageable size so that the experiments can be scheduled most efficiently. On the other hand, in evolutionary computation, schemata are processed dynamically and implicitly during the course of the execution. Because the algorithm does not explicitly control the total number of schemata to be tested, the number could be very large, up to a maximum of 2^L.

(3) sample allocation policy
In experimental design, the experiments are arranged so that all the levels are examined an equal number of times regardless of the performance. On the other hand, in evolutionary computation, greater numbers of samples are allocated to better-performing schemata.

We should specifically emphasize here that the above three aspects require separate consideration when designing experiments. Selecting factors for the experiments mainly concerns the issues of decomposing the target problem into possibly independent causes based on some domain knowledge or on the result of pre-sampling. Selecting levels to be tested from many possible candidates is mainly related to the issues of trading off the exploration and exploitation efforts under resource constraints. Allocating samples between levels is associated with efficiency of the experiments as well as the mathematical interpretation of the objective function, such as whether to maximize the profit or to minimize the risk, or whether to improve the on-line or the off-line performance.

Taking these aspects into account, the proposed sampling procedure is formulated to satisfy the following conditions:

(1) Adopt two-factor analysis where the factors are uniquely determined by a given deterministic crossover.
(2) Sequentially increase the number of levels for examination during the operation of the algorithm. Make sure that all levels have equal chances of being selected.
(3) Enable dynamic allocation of samples while guaranteeing that all the levels are sampled at least a specified number of times.

Table 2 compares sampling strategies of experimental design, evolutionary computation, and the proposed method.

Table 2. Comparison of experimental design, evolutionary computation and the proposed method

	experimental design	evolutionary computation	proposed method
factors crossover templates	explicitly given	implicitly selected	explicitly given
levels 0 * 1 0 * * 0 * schemata	explicitly given	implicitly selected	randomly selected
samples 0 1 1 0 0 1 0 0 individuals	taken equally from each level	allocated based on performance	initially taken equally from each level

4 Partitioning and Recombining Sampling Space: Fair and Greedy Sampling Strategies

4.1 Mathematical Description

Let x_1 and x_2 be binary strings of sizes n_1 and n_2 and let y be another binary string composed of x_1 and x_2. Let X_1, X_2 and Y represent sampling spaces for x_1, x_2 and y. The size of the sampling space is 2^{n_1} for X_1, 2^{n_2} for X_2 and $2^{n_1+n_2}$ for Y.

Assume the performance of y is given by the evaluation function $g(x_1, x_2)$, the value of which can be identified by a single experiment. In the framework of experimental design, x_1 and x_2 can be viewed as levels of factors X_1 and X_2. Then, the evaluation function $g(x_1, x_2)$ is decomposed as follows:

$$g(x_1, x_2) = \alpha f_1(x_1) + \beta f_2(x_2) + \gamma f_\epsilon(x_1, x_2), \tag{4}$$

where the first two terms represent the main effects of factors X_1 and X_2, respectively, and the third term represents the interaction effects between X_1 and X_2. A constant factor is omitted in the above equation.

Our objective here is to identify the best combination of x_1 and x_2 in terms of g by conducting a series of experiments. To study the exploration and the exploitation effects of sampling, let us specifically focus on the comparison of the following two sampling strategies:

(a) fair sampling strategy
Levels are selected and combined randomly so that all levels have equal chances of being tested.

(b) greedy sampling strategy
Levels are initially selected randomly, although only levels with good performance values can survive for further testing.

Figure 4 illustrates the two contrasted sampling strategies.

Let us assume that the experiments are either *memoryless* or *memory-based.* In the memoryless case, only the current best combination of X_1 and X_2 is remembered by the algorithm, which means the same combination may be tested more than once. On the other hand, in the memory-based case, past observations are all retained so that no experiments are repeated for the same combination of X_1 and X_2. In memory-based experiments, the maximum sample size (i.e., the number of evaluations) required to identify the optimal Y is $2^{n_1+n_2}$ for fair sampling and $2^{n_1} + 2^{n_2} - 1$ for greedy sampling, if the independence between X_1 and X_2 is guaranteed.

4.2 A Simple Simulation Study

By the above definitions, it can easily be expected that greedy sampling works optimally when there are no interactions between X_1 and X_2, whereas the

1. set the current best combination to (x_1^*, x_2^*).
2. select x_1 randomly from X_1.
3. select x_2 randomly from X_2.
4. evaluate (x_1, x_2).
5. if $g(x_1^*, x_2^*) > g(x_1, x_2)$, then set $x_1^* = x_1$ and $x_2^* = x_2$.
6. go to 1.

(a) procedure for fair sampling

1. set the current best combination to (x_1^*, x_2^*).
2. select x_1 randomly from X_1.
3. evaluate (x_1, x_2^*).
4. if $g(x_1, x_2^*) > g(x_1^*, x_2^*)$, then set $x_1^* = x_1$.
5. select x_2 randomly from X_2.
6. evaluate (x_1^*, x_2).
7. if $g(x_1^*, x_2) > g(x_1^*, x_2^*)$, then set $x_2^* = x_2$.
8. go to 1.

(b) procedure for greedy sampling

Fig. 4. Fair and greedy sampling procedures

fair sampling strategy is more robust against increased interaction between X_1 and X_2.

The following simple simulation study illustrates the difference between these two contrasting sampling strategies. In the first case, the independence of X_1 and X_2 is assumed. We use the evaluation function $g_1(x_1, x_2)$ given by

$$g_1(x_1, x_2) = rf_1(x_1) + (1 - r)f_2(x_2) \tag{5}$$

where r is a constant determining the ratio of contributions of the two factors. In the second case, the dependence between X_1 and X_2 is assumed with the evaluation function $g_2(x_1, x_2)$ given by

$$g_2(x_1, x_2) = rf_1(x_1) + (1 - r)f_2(x_2) + sE \tag{6}$$

where E represents the interaction between X_1 and X_2 and s the contribution of the noise factor to the overall evaluation value.

Figure 5 compares the performance of fair and greedy sampling using the evaluation functions g_1 and g_2. In the simulation, $n_1 = n_2 = 4$ was assumed, which made the size of the total sampling space Y equal to 256. For each simulation run, the 16 samples of X_1 and X_2 were determined using the beta distribution, which is commonly used as a prior distribution in

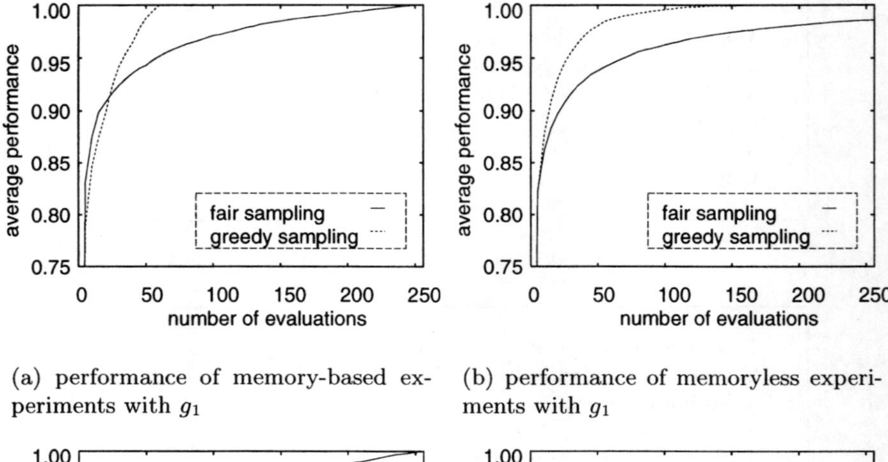

(a) performance of memory-based experiments with g_1

(b) performance of memoryless experiments with g_1

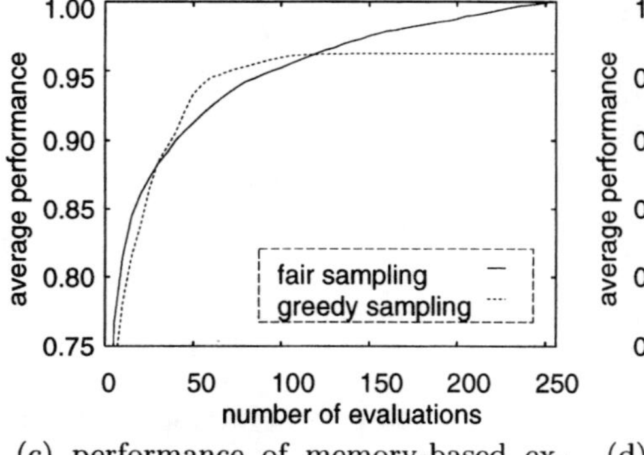

(c) performance of memory-based experiments with g_2

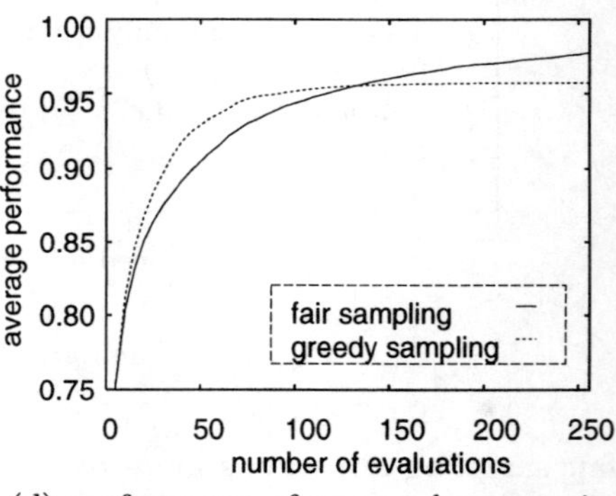

(d) performance of memoryless experiments with g_2

Fig. 5. Performance of the fair and greedy sampling strategies

Bayesian statistics. The parameters for the beta distribution were determined as $a = 1.5$ and $b = 5.0$:

$$f_1(x_1) \sim Be(1.5, 5.0), \quad \text{and} \quad f_2(x_2) \sim Be(1.5, 5.0). \tag{7}$$

For both g_1 and g_2, r was set to 0.5 so that X_1 and X_2 were weighted equally. The reason the beta distribution was used here is that the distribution has skewed shape with clear bounds: $0 \leq Be(a, b) \leq 1$ for $a > 0$ and $b > 0$. Therefore, we can expect the value of the optimal solution not to vary much for each simulation run. Also, Gaussian noise was assumed in the case of g_2, which means the 256 samples for E were determined for each simulation run using the standard normal distribution:

$$E \sim N(0.0, 1.0). \tag{8}$$

The value for s was set to 0.1.

The performance of each simulation run was normalized by the following equation:

$$v = 1 - (v_{opt} - v_{max}), \tag{9}$$

where v_{max} is the best solution obtained by the experiments and v_{opt} is the optimal value of Y. The simulation was repeated and averaged over 1000 runs for both the memoryless and memory-based cases.

It can be seen from Figure 5 that greedy sampling works most efficiently when X_1 and X_2 are independent of each other. However, it is also observed that when interactions exist between X_1 and X_2, fair sampling outperforms greedy sampling. The result also shows that the greedy sampling strategy suffers from premature convergence, just as commonly observed in other local search methods. The conditions for deception to occur with greedy sampling are formally described as follows: Assume (x_1, x_2) exist such that

(1) $g(x_1, x_2) \geq g(x_1', x_2)$ for all $x_1' \in X_1$,
(2) $g(x_1, x_2) \geq g(x_1, x_2')$ for all $x_2' \in X_2$, and
(3) $g(x_1^*, x_2^*) \geq g(x_1, x_2)$ for some $x_1^* \in X_1$ and $x_2^* \in X_2$.

Then the greedy sampling is always trapped by the local optimum point (x_1, x_2) once (x_1, x_2) is memorized as the current best. An example of such a deceptive situation is shown in Figure 6.

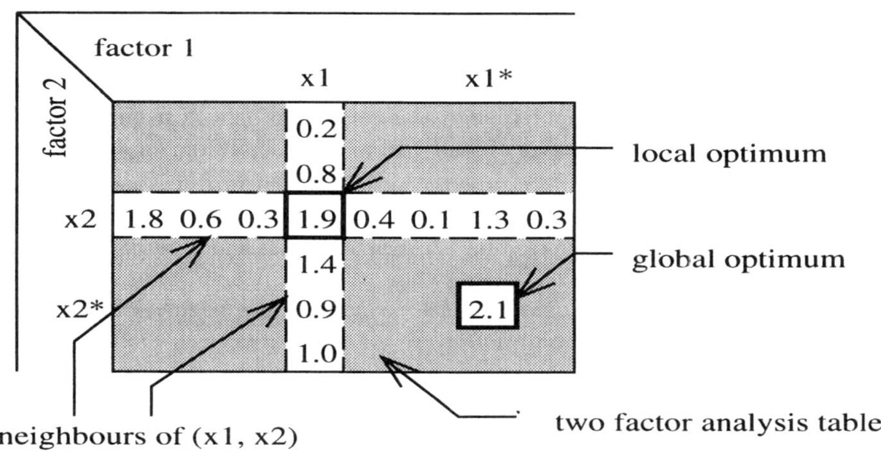

Fig. 6. Local optimum with the greedy sampling strategy

5 Partitioning and Recombining Sampling Space: A Genetic Sampling Strategy

5.1 Description of Genetic Sampling

The fair and the greedy sampling strategies in the previous section can be naturally expressed using the recombination function of a deterministic crossover operator.

Assume a crossover template ψ is given. ψ partitions the whole sampling space into two sub-spaces: one defined by bits with $\psi_i = 0$ and the other defined by bits with $\psi_i = 1$. Thus, X_1, X_2 and Y in the previous section are automatically defined by specifying ψ. Now, let (x_1, x_2) and (x'_1, x'_2) be parents, and (x_1, x'_2) and (x'_1, x_2) be children obtained by ψ. Then, the four individuals compose a simple two-factor two-level experiment as is shown in Figure 7.

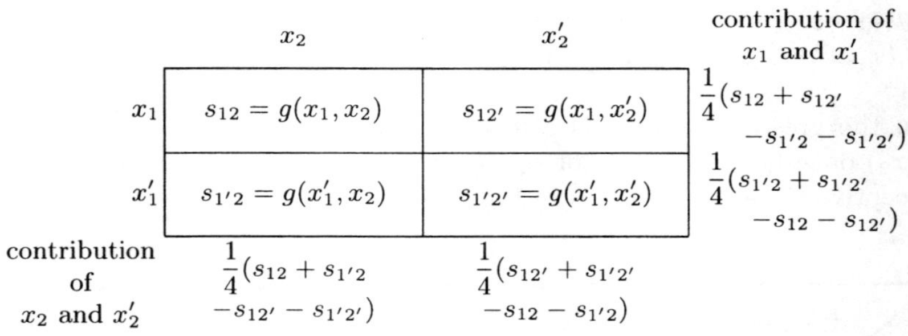

Fig. 7. Crossover operation as two-factor analysis

Let (x_1^*, x_2^*) be the current best combination of X_1 and X_2. Also, let (x_1, x_2) be a randomly selected new combination. Applying the deterministic crossover operator to (x_1^*, x_2^*) and (x_1, x_2), we have all four combinations, i.e., (x_1^*, x_2^*), (x_1, x_2^*), (x_1^*, x_2) and (x_1, x_2). Now, remember that fair sampling compares $g(x_1^*, x_2^*)$ with $g(x_1, x_2)$ and selects the better one to be tested next. In contrast, greedy sampling compares $g(x_1^*, x_2^*)$ with $g(x_1, x_2^*)$ and then compares either $g(x_1^*, x_2^*)$ with $g(x_1^*, x_2)$ or $g(x_1, x_2^*)$ with $g(x_1, x_2)$, depending on the result of the first comparison, to select the best one. It then follows that if we select the best one from (x_1^*, x_2^*), (x_1, x_2^*), (x_1^*, x_2) and (x_1, x_2), the selection represents either fair or greedy sampling depending on which of the four combinations performs the best. Based on this, we can formulate a simple genetic sampling procedure as is shown in Figure 8.

1. set the current best combination to (x_1^*, x_2^*).
2. select x_1 randomly from X_1.
3. select x_2 randomly from X_2.
4. evaluate (x_1, x_2), (x_1, x_2^*) and (x_1^*, x_2).
5. select the best combination from (x_1^*, x_2^*), (x_1, x_2), (x_1, x_2^*) and (x_1^*, x_2).
6. go to 1.

Fig. 8. Genetic sampling procedure with a deterministic crossover

5.2 Comparison of Fair, Greedy, and Genetic Sampling

In the memoryless case, where only the current best combination (x_1^*, x_2^*) is maintained by the algorithm, the convergence and sampling distribution of fair, greedy and genetic sampling can be analyzed as follows.

First with fair sampling, samples are taken randomly at each step from the whole sampling space (Figure 4(a), steps 2 and 3). It follows that the probability of selecting (x_1, x_2) equals $1/2^L$ $(= 1/2^{n_1} \times 1/2^{n_2})$ for all combinations of $x_1 \in X_1$ and $x_2 \in X_2$. Because samples are taken independently of each other, the probability that the global optimal solution is *not* selected after k samplings is calculated as $p_\epsilon(k) = (1 - \frac{1}{2^L})^k$. Then, it can be easily confirmed that for any positive number $\delta > 0$, there exists a positive integer k^* such that $p_\epsilon(k) \leq \delta$ for all $k \geq k^*$, which means that the probability can be as small as any specified number in the long run.

Next, with greedy sampling, only samples in which one level of the two factors is the same as the current best combination are examined (Figure 4(b), steps 2 and 5). Let us denote the current best combination as (x_1^*, x_2^*). Then, the probability of selecting a combination (x_1, x_2) at step 2 of Figure 4(b) equals $1/2^{n_1}$ for combinations that satisfy $x_2 = x_2^*$ and 0 for others. Similarly, the probability of selecting (x_1, x_2) at step 5 of Figure 4(b) equals $1/2^{n_2}$ for combinations that satisfy $x_1 = x_1^*$ and 0 for others. Although the calculation of $p_\epsilon(k)$ is infeasible in most cases with greedy sampling, we know from Figure 6 that at least one case exists in which the algorithm does not converge; the probability that the local optimum solution is selected at the first time sampling in Figure 6 equals $1/2^L$ and therefore $p_\epsilon(k) \geq 1/2^L$ for all $k > 0$. This means that the algorithm is trapped in a local optimum with probability greater than $1/2^L$.

Lastly, with genetic sampling, one of every three samples is always taken randomly from the whole sampling space (Figure 8, steps 2 and 3). Then, the probability that the global optimal solution is *not* selected after $3k$ samplings is at least greater than $p_\epsilon(3k) = (1 - \frac{1}{2^L})^k$. Applying the same calculation as fair sampling, we can guarantee that for any positive number $\delta > 0$, k^* exists such that $p_\epsilon(k) < \delta$ for all $k \geq 3k^*$. Therefore, the probability that genetic sampling does not reach the optimal solution can be as small as any specified number in the long run, as was the case with fair sampling. On the other hand, if we let (x_1^*, x_2^*) be the current best combination used as one of the parents by the deterministic crossover, the probability that a combination (x_1, x_2^*) is examined in every three samplings at step 5 in Figure 8 is $1/2^{n_1}$ for all $x_1 \in X_1$ and the probability that (x_1^*, x_2) is examined in every three samplings is $1/2^{n_2}$ for all $x_2 \in X_2$. Based on these simple analysis, we can conclude that the formulated genetic sampling entails both greediness and fairness properties with a simple tradeoff policy. In this respect, the behavior of genetic sampling is similar to greedy sampling.

5.3 A Simple Simulation Study

Figure 9 compares the performance of genetic sampling with fair and greedy sampling where the conditions are the same as those used in the memoryless case of the previous simulations. The horizontal axis shows the number of samples (i.e., the number of times each algorithm evaluates the combination of X_1 and X_2), and the vertical axis shows the average performance. From the figure, it is clearly seen that genetic sampling successfully trades off fair and greedy sampling strategies. When the two factors are independent of each other (Figure 9(a)), genetic sampling converges faster than fair sampling. When dependencies exist (Figure 9(b)), genetic sampling converges faster than greedy sampling.

Note that genetic sampling always examines all four combinations (x_1^*, x_2^*), (x_1, x_2^*), (x_1^*, x_2) and (x_1, x_2), whereas fair and greedy sampling only look at two or three of them. Because of the extra sampling overhead, the performance of genetic sampling is slightly degraded in the initial sampling stages. It should also be noted that the premature convergence problem with greedy sampling does not appear with genetic sampling, because one of the parents is always selected randomly.

6 Designed Sampling Procedure with A Deterministic Crossover Operator

6.1 Basic Concept

The genetic sampling method described in the previous section demonstrates how the formulation of deterministic crossover can be associated with two-factor analysis of experimental design. It should be noted here that a single adaptation of a deterministic crossover operator can only generate a small two-factor analysis table, with four different combinations of two levels of the two factors. In this section, we show a method to generate larger (although finite) analysis tables applying the same framework.

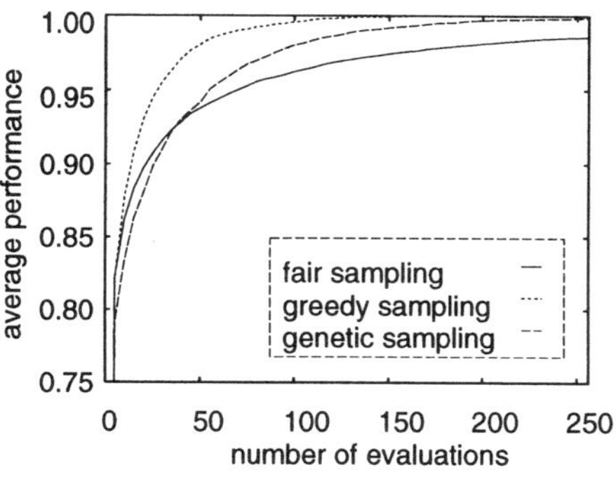

(a) performance of memoryless experiments with g_1

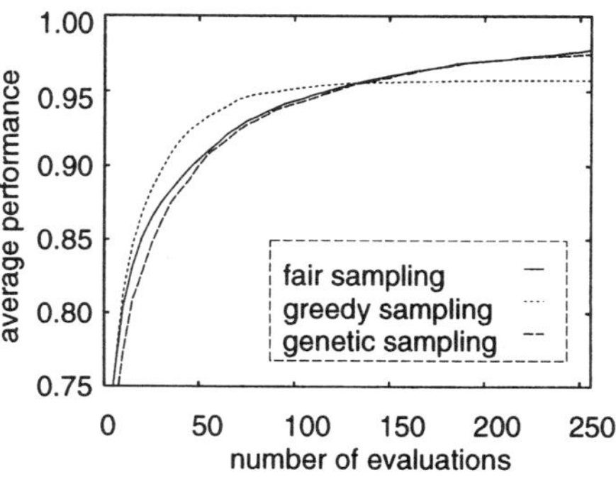

(b) performance of memoryless experiments with g_2

Fig. 9. Performance of genetic sampling

The basic concept here is to divide the whole population into two separate ones: the global population and the selective population. In the global population, a deterministic crossover operator is applied to generate and test new individuals. Promising individuals are then transferred from the global population to the selective population for further examination.

The global population is maintained so that each existing level is tested the same number of times, where the number of tests is specified by the experimenter. As the fairness of sampling between different levels is guaranteed in the global population, the contribution of each level as well as their interactions can be analyzed using the standard Anova technique.

On the other hand, the selective population collects individuals with good performance and then uses them as starting points for local searches. Conventional search techniques such as hill-climbing, simulated annealing, or search methods enhanced with specific domain knowledge can be applied to obtain further improvement. As the local search technique adopted in the selective population can be any conventional technique, our main concern in this chapter is the management of the global population.

6.2 Procedure for Designed Sampling

The proposed method initializes and maintains the global population according to the following steps:

(1) generate the initial global population
(2) add individuals to the global population
(3) transfer promising individuals from the global population to the selective population

In the following, $n_1 = n_2$ is assumed for simplicity, i.e., the orders of sub-strings corresponding to the two factors are assumed to be equal. The assumption is feasible when the search space is large enough and the enumeration of samples in each sub-space is impossible. The number of samples allocated for each level is denoted as k (≥ 2). Although we assume the value of k is a priori specified here, it is also possible to adaptively decide the value using an Anova technique. In the following, we denote the ith level of X_1 as $x_1^{(i)}$, the jth level of X_2 as $x_2^{(j)}$, the combination $(x_1^{(i)}, x_2^{(j)})$ as S_{ij}, and the performance as $s_{ij} = g(x_1^{(i)}, x_2^{(j)})$.

(1) generation of the initial population
Initially, k parents are generated by combining levels randomly selected from composing factors X_1 and X_2. Then, a deterministic crossover operator is applied for all of the possible combinations of k parents. The resulting k^2 individuals become the initial global population (Figure 10).

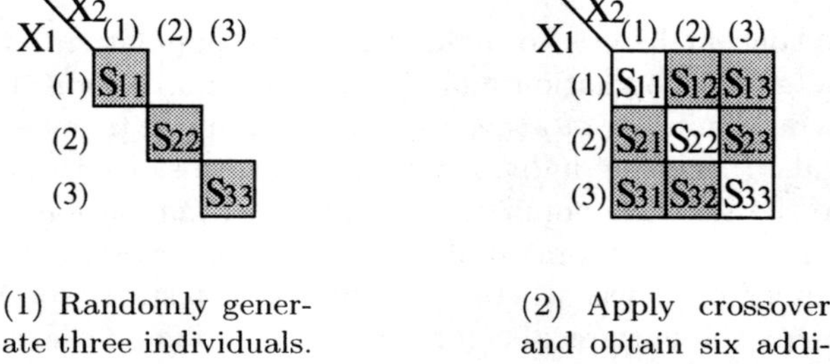

(1) Randomly generate three individuals.

(2) Apply crossover and obtain six additional individuals.

Fig. 10. Generation of the initial global population

(2) add individuals to the population
First, $(k-1)$ individuals with good performance and levels different from each other are selected from the global population; the best individual in the population is selected and then all the individuals with one level the same as the selected one are excluded from future selection. The selection iterates until $(k-1)$ individuals are selected.

Next, a new individual is generated by combining levels randomly selected from those that do not exist in the current population. Then, a deterministic crossover operator is applied between (i) each of the existing $(k-1)$ individuals selected from the original population and (ii) the newly generated one.

Finally, the newly generated parent and the resulting $2(k-1)$ children are added to the population. At the same time, the other $(k-1)$ parents initially selected from the original population are removed. In total, $(k-1)$

crossover operations are performed and the population size is increased by k (Figure11).

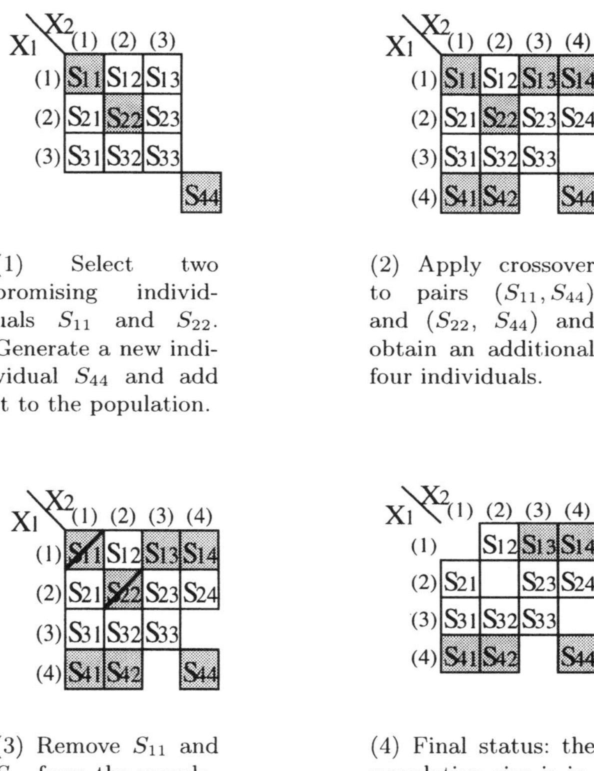

(1) Select two promising individuals S_{11} and S_{22}. Generate a new individual S_{44} and add it to the population.

(2) Apply crossover to pairs (S_{11}, S_{44}) and (S_{22}, S_{44}) and obtain an additional four individuals.

(3) Remove S_{11} and S_{22} from the population.

(4) Final status: the population size is increased from 9 to 12.

Fig. 11. Add individuals to the global population

When the size of the population exceeds its limit, the size can be reduced by conducting the opposite: remove the $2(k-1)$ individuals generated in one of the previous stages while regenerating the $(k-1)$ individuals removed from the population in the same stage.

(3) select promising individuals

The $(k-1)$ individuals that are removed from the global population at step (2) are added to the selective population. The combination of the best performing levels in the global population is also added to the selective population; namely, the performance of a level is calculated as the average performance of k individuals with the level, i.e., $\frac{1}{k}\sum_j s_{ij}$ for X_1 and $\frac{1}{k}\sum_i s_{ij}$ for X_2 (Figure 12). In total k individuals are added to the global population.

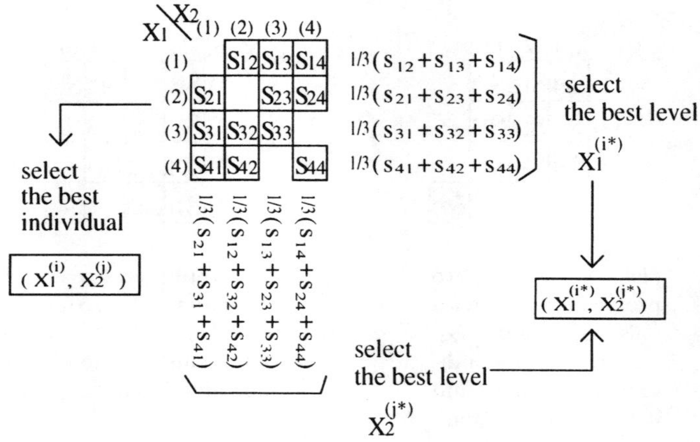

Fig. 12. Selection of promising individuals

6.3 Analysis of Performance

In the above steps, the selection of the levels to be added to the global population is random for each factor. On the other hand, the distribution of individuals in the global population is not random, but depends on the performance of their composing levels because, at step (2), individuals with better performance are selectively recombined and removed. This means that if interactions exist between the two factors, then levels with better average performance have a better chance of being combined and further examined in the selective population. We will see an example of this effect in the experiments in the next section.

Given a deterministic crossover operator, the main and interaction effects between the two factors are easily estimated using the following Anova equations: Let n be the total number of levels examined in the current population. As in the previous section, let $s_{ij} = g(x_1^{(i)}, x_2^{(j)})$ $(1 \le i, j \le n)$. In addition, let $\delta_{ij} = 1$ for $(x_1^{(i)}, x_2^{(j)})$ existing in the global population and $\delta_{ij} = 0$ for $(x_1^{(i)}, x_2^{(j)})$ not existing in the global population. The total sum of squares, denoted as SS, is calculated as

$$SS = \sum_{i=1}^{n}\sum_{j=1}^{n} \delta_{ij} s_{ij}^2. \tag{10}$$

As each level of X_1 or X_2 has exactly k samples, there are in total nk samples in the analysis table. The correction term, denoted as CT, is then calculated as

$$CT = \frac{1}{kn}\left(\sum_{i=1}^{n}\sum_{j=1}^{n}\delta_{ij}s_{ij}\right)^2 . \tag{11}$$

CT represents the contribution of the grand mean to the total squared sum of variances. Next, the main effect of factor X_1, denoted as S_1, is

$$S_1 = \sum_{i=1}^{n}\frac{1}{k}\left(\sum_{j=1}^{n}\delta_{ij}s_{ij}\right)^2 - CT. \tag{12}$$

Similarly, the main effect of factor X_2, denoted as S_2, is

$$S_2 = \sum_{j=1}^{n}\frac{1}{k}\left(\sum_{i=1}^{n}\delta_{ij}s_{ij}\right)^2 - CT \tag{13}$$

and the interaction effect between X_1 and X_2, denoted as S_e, is

$$S_e = SS - CT - S_1 - S_2, \tag{14}$$

which is in principle the same as the traditional measure of the first-order epistasis in evolutionary computation.

As an example, the above quantities in Table 1 are calculated as follows:

$$\begin{aligned} SS &= {s_{12}}^2 + {s_{13}}^2 + {s_{14}}^2 + {s_{21}}^2 + {s_{23}}^2 + {s_{24}}^2 \\ &\quad + {s_{31}}^2 + {s_{32}}^2 + {s_{33}}^2 + {s_{41}}^2 + {s_{42}}^2 + {s_{44}}^2, \end{aligned} \tag{15}$$

$$\begin{aligned} CT &= \frac{1}{12}(s_{12} + s_{13} + s_{14} + s_{21} + s_{23} + s_{24} \\ &\quad + s_{31} + s_{32} + s_{33} + s_{41} + s_{42} + s_{44})^2, \end{aligned} \tag{16}$$

$$\begin{aligned} S_1 &= \frac{1}{3}(s_{12} + s_{13} + s_{14})^2 + \frac{1}{3}(s_{21} + s_{23} + s_{24})^2 \\ &\quad + \frac{1}{3}(s_{31} + s_{32} + s_{33})^2 + \frac{1}{3}(s_{41} + s_{42} + s_{44})^2 - CT, \quad \textit{and} \end{aligned} \tag{17}$$

$$\begin{aligned} S_2 &= \frac{1}{3}(s_{21} + s_{31} + s_{41})^2 + \frac{1}{3}(s_{12} + s_{32} + s_{42})^2 \\ &\quad + \frac{1}{3}(s_{13} + s_{23} + s_{33})^2 + \frac{1}{3}(s_{14} + s_{24} + s_{44})^2 - CT. \end{aligned} \tag{18}$$

The performance of a deterministic crossover operator can be measured by the degree of contribution of the main effects. Let us define here the *relative strength of the main effects* by the following equation:

$$\gamma_\psi = \frac{S_1 + S_2}{S_1 + S_2 + S_e}. \tag{19}$$

It is shown in [12] that, under some simplifying assumptions, the relative strength of the main effects can be measured by the crossover non-linearity ratio given ψ, i.e.,

$$\begin{aligned} \rho_\psi(P,C) &= \frac{Cov(P,C)|_\psi}{\rho(P)\rho(C)} \\ &\approx \frac{S_1+S_2}{S_1+S_2+S_e} = \gamma_\psi. \end{aligned} \tag{20}$$

where P and C are random variables representing the fitness of the parents and children respectively, $\rho(P)$ and $\rho(C)$ are the standard deviations of P and C, and $Cov(P,C)|_\psi$ is the covariance between P and C given ψ.

7 Simulation Study

7.1 Conditions for Experiments

The test function used in our experiment was a 16-bit problem that was composed of four 4-bit sub-strings $x_1x_5x_9x_{13}$, $x_2x_6x_{10}x_{14}$, $x_3x_7x_{11}x_{15}$ and $x_4x_8x_{12}x_{16}$. The function simply summed the contributions of these sub-strings:

$$\begin{aligned} f(x_1\cdots x_{16}) = f_1(x_1x_5x_9x_{13}) + f_2(x_2x_6x_{10}x_{14}) \\ +f_3(x_3x_7x_{11}x_{15}) + f_4(x_4x_8x_{12}x_{16}). \end{aligned} \tag{21}$$

In the experiments, the performances of the global and the selective populations were compared for the following two deterministic crossover operators. The first one is given by a crossover template ψ_1 such that

$$\psi_1 = \texttt{'0101010101010101'}. \tag{22}$$

The second one is given by a crossover template ψ_2 such that

$$\psi_2 = \texttt{'0000000011111111'}. \tag{23}$$

By ψ_1, the whole 2^{16} sampling space was partitioned into two independent sub-spaces of sizes 2^8. Because these sub-spaces are independent of each other, the relative strength of the main effects given by Eq. (19) equals 1.0 for ψ_1. Templates `'0011001100110011'` and `'0110011001100110'` also partition the sampling space into two independent sub-spaces of sizes 2^8 and give the same results as ψ_1. In contrast, the two sub-spaces decomposed by ψ_2 are dependent on each other. The relative strength of the main effects calculated from Eq. (19) (where k is set to 8) was only 0.4 for ψ_2, which means strong interactions existed between the two factors.

In the experiments, the number of samples taken for each level was set to $k = 3$. For both ψ_1 and ψ_2, the procedure started from step (1), generating the initial population of nine individuals. In step (2), five new individuals were added to the global population and then step (3) was applied and two promising individuals were removed and transferred to the selective population. Also in step (3), a new individual, generated by combining the best-performing levels in the global population, was added to the selective population. Steps (2) and (3) in total increased the size of each population by three. The two steps were repeated until 500 individuals were sampled for each population. In total, 1000 evaluations were performed in each simulation run (Figure 13(a)).

We also examined the performance of a naive random sampling procedure in which 1000 individuals were randomly generated initially, and then the top 500 individuals were selected (Figure 13(b)). Also examined was the performance of a simple genetic algorithm (GA) with the following parameters: population size, 100; generation size, 10; crossover rate, 0.6; mutation rate, 0.01; window size, 0.7; and generation gap, 1.0. An elitist strategy was adopted so that the best individual always survived into the next generation. The simple GA started from the initial population of 100 randomly selected individuals and generated nine more populations of size 100, making the total number of evaluations 1000. The best-performing individual in the last generation was selected as the final result (Figure 13(c)).

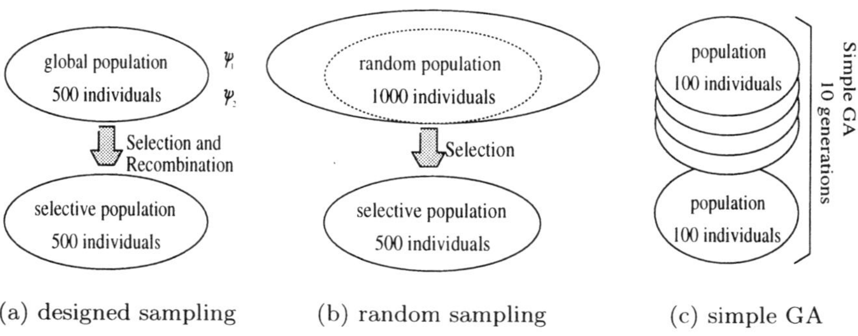

(a) designed sampling (b) random sampling (c) simple GA

Fig. 13. Comparison of the genetic and random selections

7.2 Performance Comparison Using the fourth-order Deceptive Problem

In our first experiments, the sub-functions f_i $(i = 1, 2, 3, 4)$ were identically given by the fourth-order deceptive function developed in [16]:

$$\begin{array}{llll} f_i(\texttt{1111})=30 & f_i(\texttt{0100})=22 & f_i(\texttt{0110})=14 & f_i(\texttt{1110})=6 \\ f_i(\texttt{0000})=28 & f_i(\texttt{1000})=20 & f_i(\texttt{1001})=12 & f_i(\texttt{1101})=4 \\ f_i(\texttt{0001})=26 & f_i(\texttt{0011})=18 & f_i(\texttt{1010})=10 & f_i(\texttt{1011})=2 \\ f_i(\texttt{0010})=24 & f_i(\texttt{0101})=16 & f_i(\texttt{1100})=8 & f_i(\texttt{0111})=0\,. \end{array}$$

The optimal value is 120 when all bits are set to 1.

Table 3 compares the performance of populations generated by designed sampling with ψ_1 and ψ_2, and also by naive random sampling. For each global and selective population, the mean and the deviation of the performance values, as well as the best value, are recorded and averaged over 1000 simulation runs. Note that the 'deviation' in this case does not represent the standard deviation of the performance value over 1000 simulation runs, which was less than 0.02, but the divergence of the individuals' performance in the population averaged over 1000 simulation runs. The number of times each sampling algorithm reaches the optimal solution is also shown in the table. For the simple GA, only the performance of the best individual in the final population averaged over 1000 simulation runs is shown. The mean and deviation values are not shown for the simple GA because the size of the population was different from the other three cases.

Table 3. Performance comparison using a deceptive problem

(a) the performance of the global population

sampling strategy	mean value	deviation value	best value	reach optimal / total runs
designed sampling with ψ_1	60.0	12.4	89.7	0/1000
designed sampling with ψ_2	54.4	12.7	86.2	0/1000
random sampling	60.0	18.4	112.2	19/1000

(b) the performance of the selective population

sampling strategy	mean value	deviation value	best value	reach optimal / total runs
designed sampling with ψ_1	89.4	10.0	115.0	66/1000
designed sampling with ψ_2	84.3	10.7	112.0	4/1000
random sampling	74.9	10.9	112.2	19/1000

(c) the performance of the simple GA

sampling strategy	best value	reach optimal / total runs
simple GA	113.0	8/1000

It can be seen from Table 3(a) that, averaging over the global population, the performance using ψ_1 is equal to that of random sampling. This can be explained by the independence of the two factors; with independent cases, the way levels are recombined does not affect the performance so long as the fairness of selecting each level is guaranteed. On the other hand, the performance using ψ_2 on average is lower than that of random sampling. This also agrees with our expectation that the distribution of the global population differed from that of random sampling when there was a large amount of interaction between the two factors. As for the deviation of individuals' performance in the global population, the value with ψ_1 or ψ_2 in either case is smaller than that with random sampling, which shows the effect of sampling the same level more than once. The difference of the best solution value between ψ_1 or ψ_2 and random sampling is mainly caused by the difference of their population sizes, i.e., 500 with ψ_1 or ψ_2 and 1000 with random sampling.

Next, Table 3(b) shows that the mean value using ψ_1 or ψ_2 is better than that of random sampling for selective populations, which indicates the power of recombination of these designed sampling strategies. Also, the performance of the best individual, as well as the number of times the algorithm reaches the optimal solution, is best with ψ_1, while ψ_2 performs worse than random sampling. From these observations, we can conclude that recombination works quite successfully with ψ_1 and that ψ_2 tends to suffer from the local optimum problem of the specified deceptive function.

Last, Table 3(c) shows that the performance of designed sampling with ψ_1 is actually better than that of the simple GA while the numbers of evaluations are the same.

7.3 Performance Comparison Using Randomly Generated Problems

In our next experiments, the sub-functions f_i $(i = 1, 2, 3, 4)$ were randomly determined by the beta distribution $B(1.5, 5.0)$:

$$f_i(x_i, x_{4+i}, x_{8+i}, x_{12+i}) = Be(1.5, 5.0). \tag{24}$$

In each simulation run, a new test function was generated and the performance was measured for designed sampling with ψ_1 and ψ_2, random sampling, and the simple GA. The simulation was repeated 1000 times. The results are shown in Table 4. Based on these figures, we can confirm that the tendency remains the same as in the previous experiments. Only the performance of the simple GA becomes considerably better because the generated test functions are not necessarily GA-deceptive.

To summarize, these experimental results show that the proposed sampling procedure is capable of guaranteeing fairness between levels in the global population while exploiting the promising levels in the selective population. It should again be emphasized that selecting appropriate crossover templates, with smaller interaction between factors, is essential for the success of designed sampling. Although we do not provide further insight into this problem, Eq. (20) in the previous section shows that the fitness of specific crossover templates can be well represented by the crossover non-linearity ratio; its value is easily obtained during execution of the algorithm.

Table 4. Performance comparison using randomly generated problems

(a) the performance of the initial population

sampling strategy	mean value	deviation value	best value	reach optimal / total runs
designed sampling with ψ_1	0.93	0.22	1.57	1/1000
designed sampling with ψ_2	0.80	0.19	1.35	1/1000
random sampling	0.93	0.30	1.97	22/1000

(b) the performance of the selective population

sampling strategy	mean value	deviation value	best value	reach optimal / total runs
designed sampling with ψ_1	1.55	0.22	2.08	161/1000
designed sampling with ψ_2	1.30	0.22	2.00	26/1000
random sampling	1.16	0.20	1.97	22/1000

(c) the performance of the simple GA

sampling strategy	best value	reach optimal / total runs
simple GA	2.07	115/1000

8 Conclusion

Experimental design and evolutionary computation both improve the achievement of a target system by recombining promising levels. Although, traditionally, they are aimed at completely different types of problems, we have shown in this chapter a possible implementation combining the two approaches into a single search procedure.

The procedure we have shown in this chapter is simple enough to allow a number of possible extensions. For example, the selection of individuals in the global population may be proportional to the fitness value, there may be feedback from the selective population to the global population, or multiple global populations could be maintained for different crossover templates. The last case is specifically useful to dynamically control the recombination distribution based on the performance history. Because the amount of epistasis of a given crossover template can be estimated by applying Anova to the corresponding global population, it is possible to dynamically assign greater probabilities to templates that perform well. Another possibility is to apply multiple crossover templates with high confidence in a single recombination operation so that the solution space is further partitioned. Finally, variance or schema analysis usually assumes that the epistatic effects are uniform within the whole solution space. When the amount of epistasis is strongly location dependent, then each individual may better be imprinted with the recombination distribution used upon its birth. Issues arising from this are left for future investigation.

References

1. Bäck, T. B., Fogel, D. B. and Michaelewicz, Z. (eds.) (1997) Handbook of Evolutionary Computation. Institute of Physics Publishing and Oxford University Press.
2. Nakamura, G. (1997) Understanding Experimental Design. Kindai-kagaku-sha (in Japanese).
3. Reeves, C. R. and Wright, C. C. (1995) Epistasis in Genetic Algorithms: An Experimental Design Perspective. Proceedings of the Sixth International Conference on Genetic Algorithms (ICGA'95), 217–224.
4. Reeves, C. R. and Wright, C. C. (1995) An Experimental Design Perspective on Genetic Algorithms. Foundations of Genetic Algorithms, **3**:7–22.
5. Reeves, C. R. and Wright, C. C. (1997) Genetic Algorithms versus Experimental Methods: A Case Study. Proceedings of the Seventh International Conference on Genetic Algorithms (ICGA'97), 214–220.
6. Asoh, H. and Mühlenbein, H. (1994) Estimating the Heritability by Decomposing the Genetic Variance. Parallel Problem Solving from Nature, **3**:98–107.
7. Davidor, Y. (1991) Epistasis Variance: A Viewpoint on GA-hardness. Foundations of Genetic Algorithms, **1**:23–35.
8. Goldberg, D. E. (1989) Genetic Algorithms and Walsh Functions: Part I, A Gentle Introduction. Complex Systems, **3**:129–152.
9. Goldberg, D. E. (1989) Genetic Algorithms and Walsh Functions: Part II, Deception and Its Analysis. Complex Systems, **3**:153–171.
10. Mason, A. J. (1991) Partition Coefficients, Static Deception and Deceptive Problems for Non-Binary Alphabets. Proceedings of the Fourth International Conference on Genetic Algorithms (ICGA'91), 210–214.
11. Manderic, B., DeWeger, M. and Spiessens, P. (1991) The Genetic Algorithm and the Structure of the Fitness Landscape. Proceedings of the Fourth International Conference on Genetic Algorithms (ICGA'91), 143–150.
12. Aizawa, A. (1997) Fitness Landscape Characterization by Variance of Decompositions. Foundations of Genetic Algorithms, **4**:225–245.
13. Tanaka, H. (1997) A Comparative Study of GA and Orthogonal Experimental Design. Proceedings of the 1997 IEEE International Conference on Evolutionary Computation (ICEC'97), 143–146.
14. Holland, J. H. (1975) Adaptation in Natural and Artificial Systems. The MIT Press.
15. Bechhofer, R. E. (1954) A Single-Sample Multiple Decision Procedure for Ranking Means of Normal Populations with Known Variances. Annals of Mathematical Statistics, **25**:16–39.
16. Whitley, L. D. (1991) Fundamental Principles of Deception in Genetic Search. Foundations of Genetic Algorithms, **1**:221–241.

Evolutionary Computation for Evolutionary Theory

C. C. Maley

Fred Hutchinson Cancer Research Center
1100 Fairview Ave. N. C1-015
Seattle, WA 98109, USA
E-mail : cmaley@alum.mit.edu

Summary. Evolutionary computation has developed primarily as an engineering tool for finding solutions to difficult problems. In the shadow of this success, a revolution has been building in evolutionary theory. Evolutionary models, often quite similar to genetic algorithms, are being used to extend our theoretical understanding of biology. Evolutionary models can represent details of the biology that makes analytical models mathematically intractable. Evolutionary models may thus be used as checks on the simplifications of analytical models, as well as formalisms for exploring the consequences of a particular representation of a biological system. In the best of cases, evolutionary models act as biological theories, accounting for previous experimental results and making predictions for future results. In this way, evolutionary models are serving to reunite theoretical biology with experimental biology.

> "Mathematical and computational approaches to biological questions, a marginal activity a short time ago, are now recognized as providing some of the most powerful tools in learning about nature; such approaches guide empirical work and provide a framework for synthesis and analysis." [1]

1 Evolutionary Algorithms and their Kin

Evolutionary algorithms (EAs) are dramatic testaments to the power of Darwin's ideas [2,3]. They are clear cases of biology making significant contributions to engineering [4–8]. EAs now fit comfortably amongst the "generate and test" algorithms of machine learning [9]. In this context, EAs are engineering tools. The field of EAs has largely been concerned with honing these tools [10–17]. Yet, this is not the whole story. While computer scientists looked at evolution and saw an engineering tool, biologists looked at computation and saw a biological tool. EAs, or more specifically, simulations of evolution, are being used as scientific tools to enhance our understanding of life itself.

We use abstractions, such as computational and analytical models, to say something general about life. Such a model proposes that the missing details are irrelevant. Whether they are, or not, is a scientific question that can be addressed through experimental tests of the model's predictions in the field or laboratory. There are two high-level approaches one might take to the construction of a model that expresses the essence of a biological system. One can either start with a simple model and add complexity incrementally, or start with a highly detailed description of a biological system and then pare it down to a model of the bare essentials.

This chapter is intended to be an introduction to the use of evolutionary models (EMs) to address questions in evolution, rather than an exhaustive survey of past research. For related reviews, see [18,1]. I will begin with a general description of the structure of most models of evolution, and then discuss both the incremental and the decremental approaches to modelling evolution.

2 The Structure of Evolutionary Models

The use of evolutionary computation in biology lies on a continuum somewhere between mathematical biology and biological experiments. Computational models serve to elaborate mathematical or analytical models, beyond their bounds of tractability. Thus, computational models can represent more of the complexity that characterizes biological systems relative to analytical models. However, they typically lack the generality of an analytical model. Their results must rely on the generation of data, parameter sweeps, and the statistical analysis of that data rather than formal proofs. As models, they are necessarily divorced from biological reality, and thus suffer the vulnerabilities of all theories: they may not only be incorrect, they may be irrelevant if their abstraction fails to encompass the structure of the biological system.

The fact that we are witnessing a renaissance in evolutionary theory is due in part to an innovation in modelling techniques. A new form of model has emerged which I will call a configuration model [19] but is also known as an "individual-based" model and an "agent-based" model. These are typically contrasted against their predecessors, the analytical or distribution [19] models which are generally based on differential or difference equations. Both distribution and configuration models are based on populations of individuals and so the term "individual-based" is misleading when it is used to refer to configuration models alone. The essential difference between a configuration model and a distribution model is the way in which they break the populations into equivalence classes. A distribution model of a population may separately track subpopulations. These subpopulations may be defined by parameters such as sex, life history stage (e.g., embryo, juvenile, adult), physical characteristics, or any other biological parameter of relevance. However, within each combination of parameter values, the set of individuals that share those values are treated as equivalent by the model, and are generally represented by a single variable in a system of differential equations. Thus, the population is broken into equivalence classes by the parameters of interest. However, this is only feasible if the function that updates the state of a subpopulation depends on factors that affect the subpopulation uniformly. If, alternatively, the change of state in an individual depends on a specific set of interactions with other individuals, or "neighbors," then the states of those neighbors must be considered part of the individual's update function. But the state of those neighbors depends on their neighbors, and so on. A rigorous formulation of the update function for an individual depends on the configuration of interactions of the population. Thus the term "configuration model" for such a system. This dependence on the configuration of interactions implies that every individual defines its own unique equivalence class and so the most efficient representation of the model involves the explicit representation of all of the individuals in a population. Configuration models often include spatial structure as well as the behavior of individuals. Their abstractions generally conform directly to the concepts recognized and used by ecologists and evolutionary biologists, such as organisms, genes, sexual reproduction, and predation. Such detailed models have only recently become feasible with the advent of low-cost and powerful computational resources.

To anyone familiar with genetic algorithms [4,5], the structure of most evolutionary configuration models will seem very familiar. Evolutionary models track the state of a population of individuals over time. Each individual is explicitly represented as a data structure in the model. There are generally rules for the interactions of the individuals that result in the reproduction of those individuals and the production of a new generation. The algorithms include rules for the inheritance of traits and some process of mutation that generates variation in those traits. Reproduction may be either asexual or sexual, depending on the biological system of interest. Finally, there are rules for the death of individuals based at least in part on their traits. These three aspects of the model, (1) reproduction with inheritance, (2) variation in the traits, and (3) differential reproduction based on the heritable traits, are the necessary and sufficient conditions for the emergence of natural selection.

The structure of evolutionary models is so close to that of genetic algorithms that a few words on their differences is in order. Most of the work in EAs, including genetic algorithms (GAs), evolutionary programming (EP), evolutionary strategies (ES), and genetic programming (GP), has been focused on finding a good or even optimal "solution" to a particular problem. This contrasts with biology, and models of biology, in which the only "problem" is to survive and reproduce. In other words, the fitness function is endogenous in most evolutionary models and in biology. In an EA, the effect of the fitness function is known from the beginning. The fitness function is designed to select for a good solution to the problem of interest. In contrast, the question of interest for most evolutionary models is, how will the population respond to a particular selection pressure? It is the effect, not the solution, that we are after. There is a further difference in the fitness functions of EMs and EAs. Most EAs, with a few exceptions [20,21], lack coevolutionary dynamics and so have a static fitness function. EMs are based on interactions of individuals, and so the fitness of an individual depends on the context its neighbors. Coevolution is often central to EMs.

The construction of an EM is more of a scientific endeavor than an engineering endeavor. A model represents an instantiation of a theory about how nature works. These models are powerful scientific tools because once we have embodied a theory, we may then explore its consequences.

The central task of the modeler is to decide what to include in the model and what to ignore. Of course, there is also the engineering problem of how to represent the phenomena of the theory. Once this is done, there is the further challenge of testing the realism of the model. This can be approached through a combination of comparing the model's results to biological data along with the generation of predictions from the model that may be tested in the lab or field. When these tests are impractical, progress may still be made by comparing aspects of the model's dynamics to theory that itself has been tested against biology [22]. Yet the central task remains. How can we converge on a theory, and thereby an understanding, of a biological system? There are two approaches: start with a toy model and elaborate it in successive waves, or start with a detailed description of the system and erode away the non-essential details, until a model remains that explains the observed data and, hopefully, makes testable predictions about the results of future experiments.

3 Incremental Complexity

Germinating a theory in a simple mathematical description, or toy model, has a long illustrious history in science, particularly physics. It is based on Occam's razor: "One should not increase, beyond what is necessary, the number of entities required to explain anything." [23] The advantage of this approach is that a toy model can generally be formulated in tractable mathematics. However, in the realm of theoretical biology, the horizon of intractability is never far away. The addition of heterogeneity, whether it is through spatial structure, interactions between individuals, interactions between genes, or just significant genetic variation in the population, typically makes the mathematics intractable. This is the point where computational models may be usefully brought to bear on a problem.

Computational models that derive from simple formulations function in one of three ways. In some cases, they are used as a check on the mathematically tractable models to see if the simplifications that are necessary for mathematical tractability significantly change the results of the model. In other cases we would like to add some of the biological complexity to a model but lack the mathematical tools to derive the consequences of that complexity. In such cases, we construct a model and use the model to explore our elaborated formalism. Finally, there are some hypotheses in evolutionary theory that concern dynamics that are generally beyond the scope of direct observation in the field or the lab. In these cases, we may construct a computational model of the hypothesis and explore its behavior. Sometimes this will show that the hypothesis is not internally consistent. Other times, it will make predictions that may be corroborated or challenged through other indirect methods of observation. We will examine each of these uses of computational models in turn.

3.1 Elaborating Analytical Models

EMs have long been used in population genetics to test the simplifications in their analytical models. Analytical models that are continuous and posit infinite population sizes are easier to solve than finite populations with discrete variables like the distance in nucleotides between two loci[1] on a chromosome.

Consider, for example, models of the effects of selected loci on genetic diversity in the nearby loci. The neutral theory of molecular evolution makes a set of predictions of the dynamics of mutations that have no selective effects [24–26]. However, there are many interactions between genes and even between genes and non-coding parts of the genome. The simplest interaction derives from the fact that if two loci are on the same chromosome, the fate of one locus is bound to the other, unless recombination breaks them apart. That is, the two loci are linked. Thus, if a selective locus is linked to a neutral locus, the neutral allele[2] at that locus will spread through the population as it "hitchhikes" along with the selected locus [27–29]. Similarly, a neutral allele linked to a deleterious allele will tend to be removed from the population [30,31]. Unfortunately, the mathematics necessary to describe the frequency spectrum of neutral alleles under the effects of selection quickly becomes intractable with the inclusion of recombination and self-fertilization [32]. In order to address these complexities, population geneticists typically first derive approximate analytical results of simplified systems and then check the approximations with EMs [32–35].

Collins and Jefferson provide a good example of incremental complexity with their extension of the analytical results of Kirkpatrick's analysis of sexual selection [36]. In evolution, the genetic composition of the next generation is determined not only by which organisms survived from the previous generation, but more importantly, by which organisms reproduced. In many sexual species, mate choice plays a critical role in determining the genetic structure of the next generation. An individual may possess "deleterious" genes that make it more difficult to survive, but if those genes make the individual particularly attractive to the opposite sex, the deleterious genes will

[1] A locus on a genome is a particular position in the genome. It may or may not correspond to a part of the genome that is translated into protein (a "coding" part of the genome).

[2] An allele is a particular state of a gene. In Mendel's famous experiments his pea plants had either a wrinkly shell or a smooth shell allele at one locus.

tend to spread in the population. The effect of mate choice on evolution is called sexual selection.

Kirkpatrick's model of sexual selection includes two loci each with two alleles [37]. One locus codes for one of two different phenotypes in males. One of these phenotypes has a deleterious effect on the male. The other locus codes for a preference for one or the other of the two male phenotypes in females. The selective effects of the alleles in the males are parameterized, as is the degree of preference determined by alleles in the females. Kirkpatrick finds that under some values of these parameters, there can be stable equilibria where both male alleles and both female alleles are maintained in the population. Kirkpatrick's model assumes infinite populations, no mutation, females with global knowledge of the population, complete mixing (no spatial structure), and haploid organisms. Collins and Jefferson [36] relaxed each of these assumptions in turn to determine if Kirkpatrick's results apply to populations with more realistic characteristics. With the exception of the invasion of recessive alleles into a diploid population, Kirkpatrick's results do indeed generalize to these more realistic assumptions.

3.2 Exploring Formalisms

Another approach to modelling evolution is to construct a toy model that does not derive directly from an analytical model but is, nevertheless, too complicated to understand without the assistance of computation. This approach is similar to the decremental complexity approach discussed below in Section 4. However, the two approaches differ in their origins. These exploratory formalisms derive from theories in the form of toy models. Models based on decremental complexity are based on observations of a biological system and may be directly related to some data set.

Perhaps one of the most successful exploratory formalisms is a model of the origin of life [38]. Boerlijst and Hogeweg have rejuvenated the hypercycle theory for the origin of life. This theory had suffered in competition with the auto-catalytic RNA hypothesis due to the fact that hypercycles are unstable to mutations that produce "parasites." A hypercycle is a sort of primitive metabolism. It consists of a cycle of molecules in which each molecule catalyzes the synthesis of the next molecule in the cycle. The problem comes when we consider the effects of mutations on the molecules. If a mutant form of one of the molecules arises that receives catalytic support from the previous molecule but doesnot play its part by contributing to the next molecule, the hypercycle is in trouble. If this mutant molecule is catalyzed more frequently than the normal form, it will be selected and soon it will sever the cycle. This model implicitly assumes complete mixing of the molecules and so it makes no reference to spatial structure. Would the addition of spatial structure change these dynamics? This is difficult to determine without simulating the system. Boerlijst and Hogeweg constructed a cellular automata model of hypercycles with mutation [38]. They found that the addition of two-dimensional spatial structure to the system produced hypercycles that were robust to the emergence of parasites. Mutant parasitic molecules could be spatially sequestered from a hypercycle by spiraling waves of catalytic products produced by a hypercycle at the core of the spiral. A cycle is only disrupted if a mutation occurs precisely in the center of a spiral. Thus, the addition of spatial structure to the hypercycle model showed that what had previously been viewed as an inviable theory might in fact explain the origin of life [39].

In some cases, exploratory formalisms are used more as tools for the investigation of a topic. Kauffman [40] has used his NK models to explore many aspects of the fitness landscapes of interacting genes. Bak and Sneppen [41] developed a model of species extinctions to demonstrate that species might be self-organized into a critical state such that the number of species that go extinct in any one period should follow a power law. Their model was specifically designed to produce a self-organized system. The model assigns a fitness value to each individual. In this case an individual represents a species. The species are arranged on a ring and interact with their two neighbors. Bak and Sneppen assume that the species with the lowest fitness value is the one most likely to go extinct and be replaced (or to evolve to a different local fitness optimum). When it does this, the fitness landscape of its coevolving neighbors changes and all three species are assigned new fitness values from a uniform random distribution between 0 and 1. The process then repeats, selecting the species with the lowest fitness and replacing it or mutating it and its neighbors. Eventually, the system takes on a stable distribution of finesses where occasionally an extinction triggers a mass extinction in its local area on the ring, in a sort of domino effect. However, most extinctions are isolated events. In fact, the size of local extinction events follows a power law. This result appears to be robust to variations in the topology of interactions between species, as well as variations to the algorithm for modifying the fitness landscapes [42]. One would not want to justify the structure of these models with specific reference to a biological system. Their strength lies in their capacity to explore a representation of a biological dynamic.

Since one of the principle strengths of computational models is the ability to encode interactions between individuals, it is perhaps not surprising that many models have been developed to examine the coevolution of organisms. Our topological intuitions can understand two- and three-dimensional fitness landscapes, and sometimes even landscapes in more dimensions. However, when the landscapes themselves change over time, as they do in a coevolutionary system in which one evolving species determines the fitness function of the other, it becomes very difficult to predict the outcome of the system. Consider a highly abstract representation of coevolution. Structure a population of asexual organisms along a ring, so that every organism interacts with two neighbors. Each organism has two bit strings for *offense* and *defense* "genes." An organism is rewarded for the number of matching bits between its offense gene and its neighbor's defense genes, but is penalized for the number of matching bits between its defense gene and its neighbor's offense genes. The rewards and penalties can be scaled by multiplying the number of matches with the parameters *offense-reward* and *defense-punishment*. Let us finally add a gene in each organism that encodes the mutation rate for the bits in the offense and defense genes. This mutation rate itself may evolve. In a static environment, modifier theory predicts that the mutation rates should decrease over time [43–45]. In a non-static environment, or in small populations, non-zero mutation rates may evolve [46–48]. But what about in a coevolutionary environment? When interactions are asymmetric for the participants, when *offense-reward* $\neq$ *defense-punishment*, it turns out that there is selective pressure to reduce mutation rates after the genes of the neighbors perfectly matched each other. The scores form a payoff matrix, shown in Table 1, where the options for both lineages are to mutate or not. In this case, the payoff matrix for mutating a defense gene away from a perfect match conforms to the iterated prisoner's dilemma (IPD) payoff matrix [49] where not mutating is analogous to "cooperation" in an IPD[3]. Since coevolution is iterative, and there is a long-term bias towards cooperation in an IPD, the lineages evolve mutation rates near zero. However, if the interactions are symmetric, with *offense-reward* = *defense-punishment*, then extremely high mutation rates evolve. In fact, the mutation rates evolve to higher levels than if they were simply doing a random walk. This is because high mutation rates are beneficial when there is a lot of homogeneity in the population. But homogeneity in the population only arises when mutation rates are low. So any time mutation rates descend to low levels through a random walk, homogeneity spreads in the population and so mutation rates are pushed up again by selection. The observation of elevated mutation rates was such a surprise that it was difficult to understand even in such a simple model. Further computational experiments illuminated the dynamics of the biased random walk [50]. This formalism helped us to elaborate the analytical results of modifier theory.

	Species 2	
Species 1	No mutation	Mutation
No Mutation	8,8	6,9
Mutation	9,6	7,7

Table 1. *Offense-reward* = 2 and *defense-punishment* = 1. When species 1 and 2 perfectly match, a mutation in the defense gene produces these scores. The first number in each pair is the payoff for species 1, and the second number the payoff for species 2. This is a prisoner's dilemma game.

[3] In the case of a perfect match between offense and defense genes, mutating an offense gene will always reduce the payoff for the mutant organism.

Hartvigsen and Starmer developed a more complex model of the coevolution of plants and their herbivores [51]. The model included individuals with movement in a two-dimensional environment, diploid genomes, gender, and sexual reproduction. Through this model, they discovered that more than one gene for plant defenses was required for the plants to effectively defend themselves when herbivores could evolve resistance to those defenses. This may explain the tendency for multiple genes to dominate gene-for-gene coevolutionary systems [52]. They also found that the probability of extinction increased with the ratio of edge to area of the species' habitat [51].

Further work in applying EMs to evolutionary theory has taken place independently in paleobiology. In the 1970's, Raup, Gould, Sepkoski, and others began simulating branching processes as models of speciation and extinction [53–56]. These were treated as null models: if speciation and extinction were random, what patterns in the fossil record should we expect to see? These null models served as important checks on the human tendency to find meaningless patterns amongst noise. Raup et al. [53] noticed that there was a large amount of variation between clade[4] shapes. Some groups seemed to diversify quickly while others tended to stay relatively rare for long periods of time before going extinct. They note:

> "When faced with such variation in evolutionary patterns, paleontologists are inclined to suspect or event to postulate that the organisms involved are inherently different—that the various taxonomic groups differ from one another in evolutionary potential because they differ in population structure, reproductive systems, mutation rates, dispersal systems, and so on." [53]

The computational models pose an alternative hypothesis: the difference between clade shapes may be random and not at all based on intrinsic characteristics of the species [53].

More recent work on these models has been used to look for biases in the methods that paleobiologists use to study diversity patterns over time [55,56]. Ideally, we would like to count the number of extant species during each epoch of geological history. Two problems prevent this. First of all, the fossil record is notoriously poor [2,57–59]. To illustrate this, consider all the animal and plant life on the North American continent today. Paleobiologists of the future, seeking to measure this diversity, would only find fossils of the organisms that fall into the Mississippi river delta. And then, they could only find those fossils if they were lucky enough that erosion and volcanic activity had not destroyed them in the interim. In addition, most of paleobiology has been limited to the fossils that are exposed to the surface and so can

[4] A clade is the set of species that have descended from one ancestral species. The shape of a clade is the pattern of the number of species in the clade over time. Most clades begin thin, with the ancestral species, grow fat as they diversify, and then become thin again as the species go extinct.

easily be detected and retrieved. Much more remains buried to this day. The second problem with measuring species diversity is that it is often hard to distinguish closely related species based on observing their fossils. Paleobiologists generally address these problems by counting families of species, not the species themselves. But how well does the count of families correlate with the actual count of underlying species? A recent EM has shown that this depends largely on the quality of sampling for the organisms of interest in the fossil record and then on the sizes, in number of species, of the groups counted [56]. The model predicts that for organisms that are readily fossilized, like marine invertebrates, the most accurate picture of diversity patterns can be obtained by organizing species in to many small groups. However, if the sampling of the fossil record is poor, as it is for terrestrial vertebrates, a mixture of some large and some small groups seems to best correlate with the underlying species diversity [56]. These conclusions are based on a branching process model of speciation and extinction. Every time step, representing 1 million years, each species has a fixed probability of splitting or going extinct. A phylogeny (evolutionary tree) is generated via this process. We can then simulate the fossil record by sampling the phylogeny at some rate. For example, if sampling is poor, there may only be a 0.05 chance per time step that a fossil from a species is recovered. If we then classify the species into groups we can calculate the correlation between the number of groups sampled against the actual species diversity over time. The sensitivity of the results to the sampling rate and group sizes suggests that paleobiologists must adjust their classifications of species in order to derive the most accurate measure of diversity patterns over time.

3.3 Testing Intractable Hypotheses

In some cases, exploratory formalisms are designed specifically to test hypotheses. This is particularly useful in evolutionary and ecological theory. In these fields of biology, the phenomena of interest take place over such long time periods or amongst so many different species, that controlled experiments of the full system are nigh or impossible. There are two reasonable responses to this problem. First, we might try to design simplified systems that can be tested in the lab. Many single-celled organisms have short enough generation times that their evolution can be observed in an experiment [60]. Other simplified systems make use of multicellular organisms and attempt to observe slight changes that might be extrapolated to large changes over evolutionary time [61]. Second, we can design computational models that simulate the long time scales and the complexity of interactions in an ecosystem. These models can produce one of two results. If the model of the hypothetical dynamic produces results that contradict the predictions of the hypothesis, then the hypothesis must be internally inconsistent. In this case, the model instantiates the axioms or antecedents of the hypothesis but the supposed consequences do not follow. The other possible result is that the model produces the hypothesized consequences of the axioms and thereby makes a prediction about the biological system. All of these models require parameters and so, by varying the parameters, the predictions of the model can be refined by qualifying the predictions based on particular parameter ranges. In the example from Section 3.2, mutation rates rose only when *offense-reward* = *defense-punishment*. In more elaborate cases, competing hypotheses may be compared within the common framework of a model. In this way, indirect evidence may be obtained as to which hypothesis is a more plausible explanation of the biological phenomena.

The first step in this hypothesis-driven approach to evolutionary modelling is to choose and specify a particular hypothesis to be tested. In a model of the diversification of species [62] one of the hypotheses we tested was the proposal that geographical isolation has a strong[5] impact on speciation. The second step is to attempt to extract from the hypothesis the necessary and sufficient components that must be modelled in order to test the hypothesis. In this case, in order to ground the species-level dynamics in microevolutionary dynamics, we needed: organisms, heredity, mutation, differential survival, reproductive barriers (to define species), spatial structure, and migratory barriers. The third step is then to implement each of these components in the model. Once the model has been constructed, the fourth step is to test or calibrate it. In cases where direct experimental data is currently unavailable, progress toward testing the model can still be made by reference to other theoretical results that have themselves been tested [22]. In a multi-scale model that attempts to ground high-level phenomena in the dynamics of lower levels, it is often the case that we already know something about how those lower levels behave in the real world. In our example of speciation, we have represented the evolution of populations as well as ecosystems. We can test the model to see if the model exhibits appropriate evolutionary and ecological dynamics like adaptation, predator prey oscillations, trophic cascades, competitive exclusion, and even the species-area power law [22,62]. Finally, after testing the realism of the model, we are ready to run experiments to test our hypothesis. The results can be used either to identify missing essential components of the hypothesis, or to qualify under which conditions the hypothesis holds. It turned out that geographical isolation was indeed the strongest factor tested that affected speciation. Other hypotheses for diversification, like differential selection on populations due to habitat heterogeneity, did not result in statistically significant effects. These results should be viewed as predictions generated by the model, not information about real biology.

In a similar vein, Todd and Miller tested the hypothesis that sexual selection can cause speciation even when the subpopulations are not geograph-

[5] "Strong" here is understood to be relative to other factors that were tested for their impact on speciation.

ically isolated [63]. In order to do this, they needed to represent the components of evolution: organisms, genetic inheritance, mutation, and differential reproduction. They also had to represent species as some form of grouping of the individuals. An individual was represented by three genes: two phenotype genes and one mating preference gene. The two phenotype genes specified a point in two-dimensional phenotype space. The probability of an organism accepting another as a mate depended on the distance between their phenotypes in this two-dimensional space. The mating preference gene encoded the phenotypic distance that an organism preferred for the choice of its mates. The probability of an organism accepting a mate was 1 at that distance and decreased quadratically with the difference between the preferred and the actual phenotypic distance between the organisms. Mating would occur only if both individuals accepted each other. Two-point crossover and bit-wise mutation on the genes were used to produce the offspring of a mating. Species can be identified as reproductively isolated clusters of individuals in phenotype space. These are clusters of individuals that preferentially mate within their cluster but not outside of their cluster. This is a reasonable representation of species in that the most commonly accepted definition of species in biology is based on the reproductive isolation of organisms, or gene pools [64,65]. Todd and Miller start the model with a single cluster, or species, in phenotype space and observe that new species can arise in as few as 20 generations. They also observe that these "species" readily coalesce again. This contradicts conventional wisdom in biology. However, speciation has been notoriously difficult to study in the lab [61], and so conventional wisdom is open to challenge. Might species that are isolated by mating preferences easily coalesce? This is a prediction for future testing.

One potential problem with this methodology for testing hypotheses is that choices of implementation can have a significant impact on the results of the model. The choice of pseudorandom number generator may lead to specious results. One version of the standard pseudorandom number generator used in C will produce strictly alternating 1's and 0's if asked to generate binary values. In our own work, we have observed that the use of deterministic, as opposed to stochastic, algorithms often produces aberrant behavior in models. These sorts of artifacts must be laboriously excluded before we can honestly accept or reject the results of a model. In order for the community to guard against such misleading results and to facilitate the replication of the results, the source code for a model should be made available in some form.

The indirect approach to biology through modelling is similar to the indirect approaches to cognition used in psychology. It is thus perhaps wise to follow the psychologists in their insistence on obtaining similar results from different approaches before those results are accepted scientifically. In other words, the results of a model should be corroborated by other models that use different implementations before those results can be accepted as insights into the hypothesis.

4 Decremental Complexity

The previous section discussed the attempts to extract biological insight from the incremental development of toy models. These models started with a formalism and added complexity as was needed. The opposite approach starts with a description of a complex biological system and attempts to dissolve away the superfluous details until the essentials of the system remain. What remains is again a toy model, but a toy model that relates to a specific biological system and often to a specific set of data from observations of that system.

One of the most successful examples of an EM that has been used for biological theory comes from the field of theoretical immunology [66]. Immunology is directly related to evolutionary theory in that the immune system has harnessed the power of evolution within the body to breed B and T cells that can bind to and clear invading pathogens. We manipulate this evolutionary system every time we are vaccinated. Smith et al. actually started their project with an exploratory formalism. It was their insight that the immune system behaves like an associative memory [67]. A pathogen, or antigen, is like a query to the immune system, and the immune system tries to retrieve an entry that is similar to the query (associative recall). If we are infected with the flu, then our immune system tries to clear it with the available B and T cells. If we have had a flu vaccine before the infection, and the vaccine is similar to the strain of flu that is attacking us, the immune system can usually clear the virus without our ever noticing the infection. This is called a secondary response because it is (at least) the second time the immune system has encountered the pathogen or something similar. If, on the other hand, we did not have a flu vaccine, and our immune system had not been challenged with a similar strain of flu, then the query to the associative memory fails, we become ill, and the immune system must generate a new set of antibodies to match this strain of flu. This is called a primary response.

Whether or not a strain of flu (or other pathogen) appears to be similar to a previous pathogen that the immune system has experienced, depends on the binding of the T and B cells to the pathogen. There are many details involved in determining if a cell will bind to a pathogen, having to do with organic chemistry. Smith et al. guessed that most of these details are not essential to the behavior of the immune system. Instead, they measure similarity by "antigenic distance," a measure of reactivity between an antigen (a foreign molecule) and an antibody (the immune system molecule that binds to the antigen). In biology, antigenic distance can be measured with hemagglutinin inhibition assays. In their computer model, antigenic distance was represented by Hamming distance where antigens and antibodies were represented by strings of 20 symbols, each of which could take on one of four values [66]. An antibody could bind an antigen if the Hamming distance was between 0 and 7. Thus, the antigens defined "balls of stimulation" in this 20-dimensional space. When he first constructed this model, Smith noticed that there could be both positive and negative interference between repeated challenges to a simulated immune system, shown in Figure 1. This is important in cases of vaccines that are repeatedly administered to the same subject, like the flu vaccine. Negative interference can occur if one vaccine is close enough in antigenic distance to an earlier vaccine. In this case, the new vaccine will be cleared from the system before it can stimulate a primary immune response. It never gets the chance to generate a concentrated set of B and T cells, or antibodies for that vaccine. Positive interference can occur if the two vaccines have a slightly larger antigenic distance. It can happen that the balls of stimulation for the two vaccines are near to each other but do not overlap. The second vaccine would stimulate a primary immune response and a new set of antibodies would be generated within the ball of stimulation of the second vaccine. If a third antigen is introduced which lies between the balls of stimulation for the first two vaccines, the third ball of stimulation may overlap both of the first two balls. It would then be cleared more effectively than if only a single vaccine had been used.

The power of this model lies in its explanation of the available data on the efficacy of repeat vaccination [66]. In most cases, repeat vaccinees tend to be protected as well as, if not better than, first-time vaccinees. However, in a few cases, the repeat vaccinees actually suffered more illness than the first-time vaccinees. These later studies have been the focus of a great deal of criticism and have generally been dismissed as flawed. For the studies that provided data from the hemagglutinin inhibition assays, the antigenic distance model was used to predict the efficacy of repeat vaccination versus first-time vaccination. The model predicts both the normal high- efficacy cases and the disparaged low-efficacy cases with a correlation to the observed efficacies of 0.87 [66]. Thus the model of antigenic distance provides a parsimonious explanation of *all* the observed data.

What started as a description of the details of the immune response was boiled down to evolutionary dynamics in shape space. In fact, the complexity of the EM of an associative memory can be decremented even further by just representing the antigenic distance between successive vaccinations or antigens. The result is a new theory which explains more of the data than the previous theories.

The archetypal decremental approach to modelling biology would be to start with as full a description of the biological system as possible. A model might then be constructed from this description. The model should then be calibrated against experimental data. If the theoretician can get this far, the real game begins. We can now commence removing details of the model one by one and in various combinations. After each reduction in the model's com-

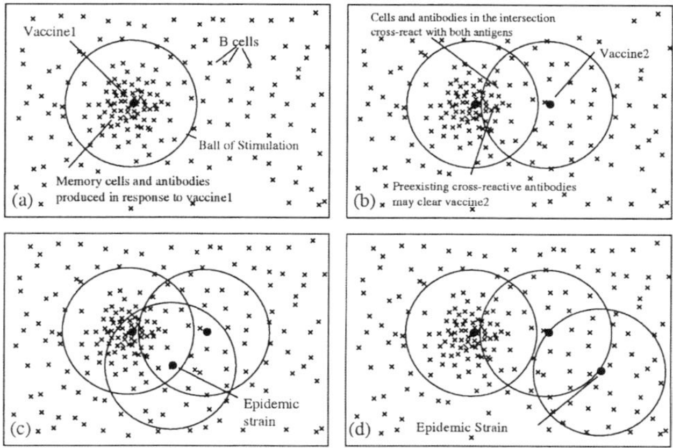

Fig. 1. A two-dimensional illustration of the antigenic distance hypothesis. The B cells, or antibodies (×), randomly cover most of the shape space except where antigens (•) have stimulated their proliferation, as shown in (a). The "ball of stimulation" for an antigen, be it a vaccine or an epidemic strain, is illustrated with a circle around the antigen. All the antibodies that fall within a ball of stimulation may bind to the antigen and thus help to clear it from the system. (b) If the ball of stimulation of one antigen overlaps the ball of stimulation of a previous antigen, there may be enough antibodies to clear the second antigen without causing the proliferation of new antibodies. In this way, a second vaccine may not stimulate the generation of new antibodies. When a repeat vaccinee is challenged with an epidemic strain, they may be protected by antibodies from the first vaccine even if the second vaccine did no good (c). However, in some cases the epidemic strain will not be so easily cleared because the second vaccine failed to stimulate the generation of new antibodies (d) and so the vaccinee will get sick. Notice that in (d) if the vaccinee had only been given the second vaccine, they would not have succumbed to the virus. This figure was reproduced with permission from [66]

plexity, its dynamics should be measured and compared to the fully detailed system. If the removal of a piece of the model has no significant effect on the results of the model, we may hypothesize that it does not contribute to the essence of the system, and so should be discarded from the model. It is quite likely that the order in which features are removed will change the effects of their absence. It may be necessary to explore different sequences of feature erosion. By picking away at the model in this way, we may be able to distill the signal from the noise, the essence from the description. The efficacy of this approach is, to my knowledge, yet untested. The main obstacle is to find a system whose details are known sufficiently that it may be "fully described," represented in a computationally tractable model, and the model calibrated against experimental results. Biology has not generally afforded us this luxury.

5 Artificial versus Biological Evolution

One striking failure of all of our models of evolution is that none of them has come close to producing the cornucopia of diversity observed in biological evolution. Until recently, we have not even had the tools to measure how close (or far away) a model comes to mimicking the "open-ended" nature of biological evolution. Bedau et al. [68] present three measurements of evolutionary dynamics that can be used to classify both artificial and real evolutionary systems. These metrics attempt to capture the diversity in the system, the amount of new adaptive activity, and the cumulative adaptive activity for a particular time period. They argue that the fossil record shows that in biological evolution, diversity grows at least linearly with time, while there is a continuous injection of new adaptive activity. They also note that the cumulative adaptive activity seems to be bounded. Using the same statistics they have examined popular EMs and found that none of them show the same dynamics as the fossil record [68]. They have challenged researchers to create an EM that exhibits the same evolutionary dynamics as biological evolution. A year later, the challenge was answered with a toy model that does exhibit unbounded diversity and the continual injection of new adaptive activity [69]. However, even the author of this toy model solution refuses to claim that it demonstrates open-ended evolution. Instead, the toy model solution suggests that Bedau et al.'s metrics have not quite captured the phenomena of interest. The challenge has now become one of refining those metrics so as to distinguish the exuberance of life from the stale progeny of our current EMs.

6 Conclusions

All of these approaches to evolutionary theory involve the instantiation of a theory in order to explore that theory. In the best cases, such as the repeat vaccination models of Smith et al. [66], the model can provide a more parsimonious explanation for the observed data relative to alternative theories. The best models can also make predictions about the results of future experiments. In this way, they function as animate theories for biology. We have moved from gedanken experiments to external media for cognition. Initially this meant the use of pencil and paper to derive analytical models. We now have a much more helpful tool for the development of our thoughts, the computer.

The promise of computational models is that they can represent the complexity and dynamics of biological systems in a formal context. Configuration models typically have explicit representations of individuals or cells, and genes. The match between these components of configuration models and the objects of experimental research in biology makes the configuration models readily accessible to (and thus criticizable by) experimentalists. EMs have already had a dramatic impact on our understanding of biological evolution. If we are clear and rigorous enough in our methodology, computational models may help to close the yawning chasm that has traditionally separated theoretical and experimental biology.

Acknowledgements

I would like to thank Mark Bedau for his helpful comments as well as Stephanie Forrest, Rod Brooks, Michael Donoghue, and W. D. Hamilton for their discussions and support. This work was supported in part by ONR Grant N00014-99-1-0417 to Stephanie Forrest.

References

1. Levin, S. A., Grenfell, B., Hastings, A., Perelson, A. S. (1997) Mathematical and computational challenges in population biology and ecosystems science, *Science*, **275**, 334–343
2. Darwin, C. R. (1859) *On the Origin of Species.* London: John Murray
3. Dennett, D. C. (1995) *Darwin's Dangerous Idea.* New York, NY: Simon and Schuster
4. Holland, J. H. (1975) *Adaptation in Natural and Artificial Systems.* Ann Arbor, MI: University of Michigan Press
5. Goldberg, D. E. (1989) *Genetic Algorithms: In Search, Optimization and Machine Learning.* Reading, MA: Addison-Wesley
6. Fogel, D. B., Atmar, W. (eds.) (1992) *First Annual Conference on Evolutionary Programming,* (La Jolla, California), Evolutionary Programming Society
7. Becker, J. D., Eisele, I., Mündemann, F. W. (eds.) (1989) *Parallelism, Learning, Evolution: Proceedings of the Workshop on Evolutionary Models and Strategies/Proceedings of the Workshop on Parallel Processing: Logic, Organization, and Technology (WOPPLOT 89)*, **565** *LNAI* (Wildbad Kreuth, FRG), Springer-Verlag
8. Koza, J. R. (1992) *Genetic Programming.* Cambridge, MA: MIT Press
9. Mitchell, T. M. (1997) *Machine Learning.* New York, NY: McGraw-Hill
10. Grefenstette, J. J. (ed.) (1985) *Proceedings of the First International Conference on Genetic Algorithms and their Applications (ICGA'85)*, (Hillsdale, N.J, Lawrence Erlbaum Associates
11. Grefenstette, J. J. (ed.) (1987) *Genetic Algorithms and their Applications: Proceedings of the Second International Conference on Genetic Algorithms*, (Hillsdale, NJ), Lawrence Erlbaum Associates
12. Schaffer, J. D. (ed.) (1989) *Proceedings of the Third International Conference on Genetic Algorithms*, George Mason University, San Mateo, Morgan Kaufmann
13. Belew, R. K., Booker, L. B. (eds.) (1991) *Proceedings of the Fourth International Conference on Genetic Algorithms (ICGA'91)*, (San Mateo, CA), Morgan Kaufmann
14. Forrest, S. (ed.) (1993) *Proceedings of the Fifth International Conference on Genetic Algorithms (ICGA'93)*, (San Mateo, CA), Morgan Kaufmann

15. Eshelman, L. (ed.) (1995)*Proceedings of the Sixth International Conference on Genetic Algorithms*, (San Mateo, CA), Morgan Kaufmann
16. Bäck, T. (ed.) (1997)*Proceedings of the Seventh International Conference on Genetic Algorithms (ICGA'97)*, (San Francisco, CA), Morgan Kaufmann
17. Banzhaf, W., Daida, J., Eiben, A. E., Garzon, M. H., Honavar, V., Jakiela, M., Smith, R. E. (eds.), *GECCO-99: Proceedings of the Genetic and Evolutionary Computation Conference*, (Orlando, FL), Morgan Kaufmann Forthcoming
18. Mitchell, M., Forrest, S. (1994) Genetic algorithms and artificial life, *Artificial Life*, **1**, 267–290
19. Caswell, H., John, A. M. (1992) From the individual to the population in demographic models, in *Individual-Based Models and Approaches in Ecology* (D. DeAngelis and L. Gross, eds.), 36–61, NY: Chapman and Hall
20. Hillis, W. D. (1992) Co-evolving parasites improve simulated evolution as an optimization procedure, in *Artificial Life II* (C. G. Langton, C. Taylor, J. D. Farmer, and S. Rasmussen, eds.), 313–324, Reading, MA: Addison-Wesley
21. Sumida, B. H., Hamilton, W. D., Both Wrightian and 'parasite' peak shifts enhance genetic algorithm performance, in *Computing with Biological Metaphors* (R. Paton, ed.), (1994) 264–279, London: Chapman and Hall
22. Maley, C. C. (1998) Models in evolutionary ecology and the validation problem, in *Artificial Life VI* (C. Adami, R. K. Belew, H. Kitano, and C. E. Taylor, eds.), 423–427, Cambridge, MA: MIT Press
23. Heylighen, F. (1997) Occam's razor, in *Principia Cybernetica Web* (C. J. F. Heylighen and V. Turchin, eds.), Brussels: Principia Cybernetica, URL: http://pespmc1.vub.ac.be/OCCAMRAZ.html
24. Kimura, M., Ohta, T. (1969) The average number of generations until fixation of a mutant gene in a finite population, *Genetics*, **61**, 763–771
25. Kimura, M., Ohta, T. (1969) The average number of generations until extinctions of an individual mutant gene in a finite population, *Genetics*, **63**, 701–709
26. Kimura, M. (1983) *The Neutral Theory of Molecular Evolution.* Cambridge: Cambridge University Press
27. Maynard Smith, J., Haigh, J.(1974) The hitch-hiking effect of a favorable gene, *Genetics Research*, **231**, 1114–1116
28. Thomson, G. (1977) The effect of a selected locus on a linked neutral locus, *Genetics*, **85**, 752–788
29. Kaplan, N. L., Hudson, R. R., Langley, C. H. (1989) The hitchhiking effect revisited, *Genetics*, **123**, 887–899
30. Ohta, T. (1971) Associative overdominance caused by linked detrimental mutations, *Genetics Research*, **18**, 277–286
31. Sved, J. A. (1972) Heterosis at the level of the chromosome and at the level of the gene, *Theoretical Population Biology*, **3**, 491–506
32. Charlesworth, B., Morgan, M. T., Charlesworth, D. (1993) The effect of deleterious mutations on neutral molecular variation, *Genetics*, **134**, 1289–1303
33. Hudson, R. R., Kaplan, N. L. (1995) Deleterious background with recombination, *Genetics*, **141**, 1605–1617
34. Braverman, J. M., Hudson, R. R., Kaplan, N. L., Langley, C. H., Stephan, W. (1995) The hitchhiking effect on the site frequency spectrum of DNA polymorphisms, *Genetics*, **140**, 783–796

35. Charlesworth, D., Charlesworth, B., Morgan, M. T. (1995) The pattern of neutral molecular variation under the background selection model, *Genetics*, **141**, 1619–1632
36. Collins, R., Jefferson, D. (1992) The evolution of sexual selection and female choice, in *Toward a Practice of Autonomous Systems: Proceedings of the First European Conference on Artificial Life* (F. J. Varela and P. Bourgine, eds.), 327–336, (Cambridge, MA), MIT Press
37. Kirkpatrick, M. Sexual selection and the evolution of female choice., *Evolution*, **36**, 1–12
38. Boerlijst, M., Hogeweg, P. (1991) Spiral wave structure in prebiotic evolution: Hypercycles stable against parasites, *Physica*, **48D**, 17–28
39. May, R. M., (1991) Hypercycles spring to life, *Nature*, **353**, 607–608
40. Kauffman, S. A. (1993) *Origins of Order.* Oxford: Oxford University Press
41. Bak, P., Sneppen, K. (1993) Punctuated equilibrium and criticality in a simple model of evolution, *Physical Review Letters*, **71**, 4083–4086
42. Sneppen, K., Bak, P. (1993) Mean field theory for a simple model of evolution, *Physical Review Letters*, **71**, 4087–4091
43. Feldman, M. W., Liberman, U. (1986) An evolutionary reduction principle for genetic modifiers, *Proceedings of the National Academy of Science, USA*, **83**, 4824–4827
44. Liberman, U., Feldman, M. W. (1986) Modifiers of mutation rate: A general reduction principle, *Theoretical Population Biology*, **30**, 125–142
45. Altenberg, L., Feldman, M. W. (1987) Selection, generalized transmission and the evolution of modifier genes. I the reduction principle, *Genetics*, **117**, 559–572
46. Gillespie, J. H. (1981) Mutation modification in a random environment, *Evolution*, **35**, 468–476
47. Gillespie, J. H. (1981) Evolution of a mutation rate at a heterotic locus, *Proceedings of the National Academy of Science, USA*, **78**, 2452–2454
48. Ishii, K., Matsuda, H., Iwasa, Y., Sasaki, A. (1989) Evolutionary stable mutation rate in a periodically changing environment, *Genetics*, **121**, 163–174
49. Axelrod, R., Hamilton, W. D. (1981) The evolution of cooperation, *Science*, **211**, 1390–1396
50. Maley, C. (1997) Mutation rates as adaptations, *Journal of Theoretical Biology*, **186**, 339–348
51. Hartvigsen, G., Starmer, W. T. (1995) Plant-herbivore coevolution in a spatially and genetically explicit model, *Artificial Life*, **2**, 239–259
52. Thompson, J. N., Burdon, J. J. (1992) Gene-for-gene coevolution between plants and parasites, *Nature*, **360**, 121–125
53. Raup, D. M., Gould, S. J., Schopf, T. J. M., Simberloff, D. S. (1981) Stochastic models of phylogeny and the evolution of diversity, *Journal of Geology*, **81**, 525–542
54. Gould, S. J., Eldredge, N. (1977) Punctuated equilibria: The tempo and mode of evolution reconsidered, *Paleobiology*, **3**, 115–151
55. Sepkoski, J. J. (Jr.), Kendrick, D.C. (1993) Numerical experiments with model monophyletic and paraphyletic taxa, *Paleobiology*, **19**, 168–184
56. Robeck, H., Maley, C. C., Donoghue, M. (2000) Taxonomy and temporal diversity patterns, In press.
57. Raup, D. M. (1972) Taxonomic diversity during the Phanerozoic, *Science*, **177**, 1065–1071

58. Valentine, J. W. (1969) Patterns of taxonomic and ecological structure of the shelf benthos during Phanerozoic time, *Paleontology*, **12**, 684–709
59. Sepkoski, J. J. (Jr.), Bambach, R. K., Raup, D. M., Valentine, J. W. (1981) Phanerozoic marine diversity and the fossil record, *Nature, London*, **293**, 435–437.
60. Lenski, R. E., Travisano, M. (1994) Dynamics of adaptation and diversification: A 10,000-generation experiment with bacterial populations, *Proceedings of the National Academy of Science, USA*, **91**, 6808–6814
61. Rice, W. R., Hostert, E. E. (1993) Laboratory experiments on speciation: What have we learned in 40 years *Evolution*, **47**, 1637–1653
62. Maley, C. C. (1998) *The Evolution of Biodiversity: A Simulation Approach.* PhD thesis, Massachusetts Institute of Technology, Cambridge, MA
63. Todd, P. M., Miller, G. F. (1991) On the sympatric origin of species: Mercurial mating in the quicksilver model, in *Proceedings of the Fourth International Conference on Genetic Algorithms* (R. K. Belew and L. B. Booker, eds.), (San Mateo, CA), 547–554, Morgan Kaufmann
64. Mayr, E. (1942) *Systematics and the Origin of Species.* New York: Columbia University Press
65. Mayr, E. (1963) *Populations, Species, and Evolution.* Cambridge, MA: Harvard University Press
66. Smith, D. J., Forrest, S., Ackley, D. H. , Perelson, A. S. (1999) Variable efficacy of repeated annual influenza vaccination, *Proceedings of the National Academy of Science, USA*, **96**, 14001–14006
67. Smith, D. J., Forrest, S., Perelson, A. S. (1998) Immunological memory is associative, in *Artificial Immune Systems and their Applications* (D. Dasgupta, ed.), 105–112, Berlin, Springer-Verlag
68. Bedau, M. A., Snyder, E., Packard, N. H. (1998) A classification of long-term evolutionary dynamics, in *Artificial Life VI* (C. Adami, R. K. Belew, H. Kitano, and C. E. Taylor, eds.), 228–237, Cambridge, MA: MIT Press
69. Maley, C. C. (1999) Four steps toward open-ended evolution, in *Proceedings of the Genetic and Evolutionary Computation Conference* (W. Banzhaf, J. Daida, A. E. Eiben, V. H. M. H. Garzon, M. Jakiela, and R. E. Smith, eds.), San Francisco, CA: Morgan Kaufmann, 1336–1343

Computational Embryology: Past, Present and Future

Sanjeev Kumar and Peter J. Bentley

Department of Computer Science, University College London, Gower St., London WC1E 6BT, UK
E-mail: S.Kumar@cs.ucl.ac.uk, P.Bentley@cs.ucl.ac.uk

Summary. This chapter describes research into the new field of computational embryology. It starts with a look at the past, examining the contributions scientists have made over the years that have caused the gradual amalgamation of embryology and genetics into developmental biology. This is followed by a detailed investigation into the evolution of computational embryogenies. The focus of this chapter is on the two most promising types of computational embryogeny, explicit and implicit, investigating the evolvability and scalability of both for morphogenesis. The problem set is that of evolving certain predefined shapes – letters of the alphabet. The results show that both embryogenies are good at defining different morphologies, but significantly, the implicit embryogeny incurs no increase in genotype size as the problem is scaled. Finally, the chapter ends with a description of a more biologically plausible model of aspects of biological development.

1 Introduction

Evolutionary computation (EC) has been a very successful area of computer science for some time. EC has grown from several types of evolutionary algorithms (EAs) (Bentley, 1999) which take their inspiration from nature: genetic algorithms, genetic programming, evolutionary strategies and evolutionary programming. Although these methods are essentially the same, an important difference is the distinction between the genotype (coded parameters and values) and the phenotype (representation of solutions). GP practitioners often regard the genotype as the phenotype, as do ES and EP practitioners. The genetic algorithm (GA) is the only one of the four EAs that makes the distinction. It is the omission of this crucial genotype from the phenotype mapping process, known in biology as an *embryogeny*, which can often deny the non-GA practitioner the advantages that embryogenies bring.

This chapter investigates the evolvability of two of the most interesting types of computational embryogeny, *explicit* and *implicit*. The chapter is organised as follows: a brief history of embryology is given in section 2, section 3 explains embryology in nature and introduces computational embryology. The two embryogeny-based systems are introduced in section 4, with section 5 describing a series of experiments together with an analysis of the results. Section 6 introduces a new, more biologically plausible model of development. Conclusions are provided in section 7.

2 From Embryology to Developmental Biology: A Historical Review

We have pondered questions such as our place in the universe and the existence of God for millennia. But a question much closer to home, yet equally baffling, is that of development. How are our bodies and other organic entities formed? What gives rise to these bodies? Is the complete body plan of an animal already present in the fertilised egg, i.e. is it *preformed*? If not, then what mechanisms mould and sculpt such complexity?

Greece saw the formation of such questions starting with Hippocrates, who first addressed the important question of the nature of development, as long ago as the fifth century BC.

The fourth century BC provided the setting to take the question a step further. Aristotle was the first to formulate the problem of how structure arose in the embryo. He proposed two possibilities: *preformationism*, namely that the animal is preformed and simply gets bigger, and the second, *epigenesis* (meaning 'upon formation'), namely that structure arises progressively. Aristotle favoured epigenesis; however, in the fourth century religion was prevalent and the church favoured preformationism, creating a hostile environment for ideas of epigenesis. Aristotle's preformation/epigenesis dichotomy was to provide the fuel for vigorous debate over the coming centuries (for a more detailed history of embryology, see Wolpert, 1998; Bard, 1990).

Some 2000 years later in 1672, Italian scientist Marcello Malpighi provided a detailed description of the chick embryo. Preformationist thinking, endorsed by the church, still prevailed in the seventeenth century. Unfortunately, Malpighi simply could not help but believe in preformationism, remarking that the embryo was so small that he was just not able to see it, even with his most powerful microscopes.

Fortunately, preformationism was shown to be wrong by the observations of Carl Friedrich Wolff. Wolff did not believe in preformationism and disproved it in 1759 by studying how blood vessels appeared in the chick. He demonstrated that the blood vessels of the chick blastoderm slowly developed from islands of material surrounded by liquid. The preformationists responded to Wolff's evidence by suggesting that the blood vessels were there all the time, but became visible later. But in 1768, Wolff dealt preformation the final blow, demonstrating that the chick gut was not initially a tube but emerged by the gradual folding of the ventral sheet of the embryo.

Many more important discoveries were to follow (Table 1). The basic unit of life, the cell, was discovered in 1838–1839 by German botanist Matthais Schleiden and physiologist Theodor Schwann. This was a turning point for

Scientist	Date	Contribution
Hippocrates	5th century BC	First to address the problem of development.
Aristotle	4th century BC	Aristotle considered two possibilities to the problem of how structure arose in the embryo, namely preformationism and epigenesis.
Marcello Malpighi	1672,17th century	Provided an accurate description of chick embryo.
Carl Friedrich Wolff	1759–1768	Dismissed preformationism by providing evidence that the chick gut was not initially a tube but emerged by the gradual folding of the ventral sheet of the embryo.
Mathais Schleiden and Theodor Schwann	1838–1839	Recognised, independently, the cell as the basic unit of life.
Gregor Mendel	Mid 1860s	Proposed origin of hereditary variability lay in differences in 'factors' that pass unchanged from one generation to the next.
August Weismann	End of 19th century	Proposed that the offspring does not inherit its characteristics from the soma of the parent, but from the germ cells. In addition, proposed a 'special factors' model of development.
Wilhelm Roux	Late 1880s	Provided confirmation of Weismann's ideas, by experimenting on frogs, concluding that frog development was based on a mosaic mechanism.
Hans Driesch	Late 1880s	Disproved Roux's results showing that cells retain the developmental potential of the zyogte and were hence not based on a mosaic mechansim.
Wilhelm Johannsen	1909	Recognised the distinction between genotype and phenotype, helping to link genetics with embryology.

Table 1. Key contributions in the history of embryology

embryology providing the beginnings of a new era in which embryology was eventually to be subsumed by developmental biology.

The mid 1860s saw the Austrian monk and botanist Gregor Mendel show, by plant hybridisation experiments, that the origin of hereditary variability lay in differences in 'factors' (later to be named genes by Johannsen) that pass unchanged from one generation to the next.

In the mid nineteenth century, the German biologist August Weismann proposed that characteristics in the offspring are inherited from the *germ cells* (the eggs and sperm) and not the soma (body). He further proposed that germ cells were not influenced by the body that bears them. In addition, Weismann devised a model of development in which 'special factors' known as *determinants* were contained in the nucleus of the zygote. These determinants were then unequally distributed to daughter cells during zygote division (cleavage), allowing for control over the future development of the cells. In effect, *cell fate*, according to Weismann's model, is controlled by the unequal distribution of the determinants, and hence predetermined in the egg during cleavage. Weismann's model was termed *mosaic*, as the egg was likened to a mosaic of discrete localised determinants.

Weismann's ideas received support in the late 1880s from the German embryologist Wilhelm Roux. Roux provided this support by experimenting with frog embryos. After allowing the first cleavage of a zygote, he destroyed one of the two cells with a hot needle. He found that the remaining cells developed into a well-formed half-larva, and concluded that the development of the frog is based on a mosaic mechanism, the cells having their character and fate determined at each cleavage.

Hans Driesch, also in the late 1880s, disagreed with Roux and set about disproving Roux's results, i.e. that development is based on a mosaic mechanism. Driesch repeated Roux's experiment, but on invertebrate sea urchin eggs; however, instead of killing one of the cells at the two-cell stage he separated them. He found that these isolated cells went on to develop normally, concluding that all cells retained the developmental ability of the zygote.

Despite these discoveries it was left to Danish botanist Wilhelm Johannsen in 1909 not only to label Mendel's 'factors' as genes, but also to make the fundamental distinction between genotype (genetic composition) and phenotype (physical appearance). This crucial concept helped link both genetics and embryology giving way to developmental biology.

3 Embryology and Embryogenies

Biology has moved on considerably since Aristotle, and an entire new field now known as developmental biology has emerged. Through the centuries of scientific research, we now know that Aristotle's notion of epigenesis was the closest to the truth. Developmental biology and, in particular, embryology have fast become exciting and fashionable subjects for research.

Today we regard *embryology* as the study of the controlled formation and development of animal and plant embryos. It involves four fundamental processes:

- *Morphogenesis*, which involves the emergence and change of form (Bard, 1990; Slack, 1991; Wolpert, 1998).
- *Pattern formation*, the generation of ordered spatial patterns of cell activities, due to the acquisition of cell identities (Wolpert, 1998; Slack, 1991).
- *Cellular differentiation*, in which cells become specialised for particular functions (Wolpert, 1998; Slack, 1991).

- *Growth*, in which cells multiply and increase in size (Wolpert, 1998).

These four processes operate together in different parts of the embryo, at different times, and in stages according to a 'recipe' known as an *embryogeny*. Embryogenies have evolved in nature to describe how an animal should be grown (rather than contain an overall description of an animal).

Embryology, like other fields, has its own set of problems that need answering. One such problem is that of positional information, i.e. how cells 'know' where to grow. This was addressed by the eminent embryologist Lewis Wolpert who put forth *positional information theory* (Wolpert, 1998).

Positional information theory states that cells glean their positional information from the free diffusion of a chemical, known as a *morphogen*, relative to a boundary. The information is coded in the form of the concentration value of the diffusing morphogen. This diffusion sets up a *chemical gradient*, thus allowing cells to position themselves relative to the boundary whereupon they can, if need be, *differentiate*, i.e. become a specialised type of cell (Wolpert, 1998). The idea of chemical gradients is used in the implicit computational embryogeny described later.

3.1 Computational Embryology

Having briefly reviewed biological embryology, we will explore computational embryology in this section.

Computer science first saw the use of morphogenesis in 1952 by Alan Turing (Turing, 1952). Since then morphogenesis has featured in a number of works, such as the evolution of neural network morphologies (Gruau, 1994; Jakobi, 1995), autonomous agent design (Vaario, 1994; Daellart and Beer, 1995), and evolvable hardware using cellular encoding (Gruau, 1998; Koza et al., 1999). Other notable examples include the work of Ortega and Tyrrell (1998) where they discuss hardware fault-tolerance inspired by the self-healing processes of embryology. This is in accordance with Sanchez et al's (1997) phylogeny–ontogeny and epigenesis (POE) classification model of different levels of organisation in natural and artificial systems.

It should be noted that all of these approaches in computer science have used crude approximations of the natural embryological processes, paying little attention to the intricate subtleties. Nevertheless, researchers have been using these simple computational embryogenies in various guises for over a decade. Current computational embryogenies can be classified into three different types: external, explicit and implicit (Bentley and Kumar, 1999).

External embryogenies are typically hand-designed and are defined globally and externally to genotypes. They are characterised by their fixed, non-evolvable structures. For example, evolutionary art systems often use external embryogenies which specify how phenotypes should be constructed using the genes in the genotypes. Similarly, Richard Dawkins' Blind Watchmaker program (Dawkins, 1987) employs a simple external embryogeny to create biomorphs, using the 'eye-of-the-beholder' to provide a measure of fitness, and mutation to vary the evolving shapes. Dawkins also points out advantages of using embryogenies, why they are so important, and perhaps most interestingly, that different embryogenies can lead to different results. For example, different embryogenies allow different areas of 'solution space' to be searched and thus in doing so constrain themselves to different types of shapes or designs (not necessarily a bad thing as Dawkins shows, relative to what he calls naïve pixel-peppering).

An *explicit* embryogeny specifies each step of the growth process in the form of explicit and evolvable genetic instructions. In computer science, an explicit embryogeny can be viewed as a tree containing a single growth instruction at each node. GP uses tree structures to represent its genotypes. GP, therefore, offers a simple and concise way to evolve explicit embryogenies. Typically, the genotype and the embryogeny are combined and both are allowed to evolve simultaneously. Perhaps the most famous example is the work by Gruau who devised cellular encoding, which is based on an explicit embryogeny in the form of genetic programming. Cellular encoding uses graph rewriting rules to grow neural networks. These neural nets start from initial zygotic conditions, i.e. development begins from a single zygote cell which then undergoes transformations according to developmental instructions contained within its explicit embryogeny. Other examples of explicit embryogenies include Koza et al. (1999) who use an explicit embryogeny in the form of cellular encoding for the evolution of analogue circuits. Also, Sims (1999) used an explicit embryogeny with the idea of directed graphs to specify the nervous systems (neural networks) and morphologies of virtual creatures.

An *implicit* embryogeny does not explicitly specify each step of the growth process. Instead, the growth process is implicitly specified by a set of genetic rules or instructions, similar to a 'recipe' that governs the growth of a shape. For example, de Garis (1999) describes an implicit embryogeny to evolve convex and concave shapes using a cellular automata approach along with the notion of cellular differentiation. He has reported encouraging results, as well as highlighting problems that need to be tackled in order to improve this approach (de Garis, 1999). Jakobi (1995) devised an implicit embryogeny-based system that employed cell division, cell movement and diffusable proteins, in order to evolve neural net robot control architectures. Table 2 summarises the three categories of computational embryogeny.

<table>
<tr><th></th><th>Outside genotypes
(Non-evolvable)</th><th>Inside Genotypes
(Evolvable)</th></tr>
<tr><td>Timing and action of every growth step is provided</td><td rowspan="2">EXTERNAL EMBRYOGENY</td><td>EXPLICIT EMBRYOGENY</td></tr>
<tr><td>Timing and action of growth emerges</td><td>IMPLICIT EMBRYOGENY</td></tr>
</table>

Table 2. External, explicit and implicit embryogenies

4 Evolving Embryogenies

Previous work (Bentley and Kumar, 1999) compared the performance and scalability of different evolved computational embryogenies for the generation of tessellating tiles. Subsequent experiments have shown that significant questions remain concerning the evolvability of different embryogenies. Specifically, evolved implicit and explicit embryogenies show inconsistent abilities to define specific morphologies. Consequently, it was decided that further investigation was necessary to help explore and understand these issues of evolvability.

4.1 Evolving Predefined Shapes

In order to assess the change in performance of the two embryogenies, a number of fixed shapes were specified as targets for evolution. Since shapes with distinct and useful characteristics were desired (e.g. convex, concave, solid, hollow, curved and linear), a subset of the alphabet was selected. Six letters were chosen: C, E, G, L, O and R, as shown in Figure 1.

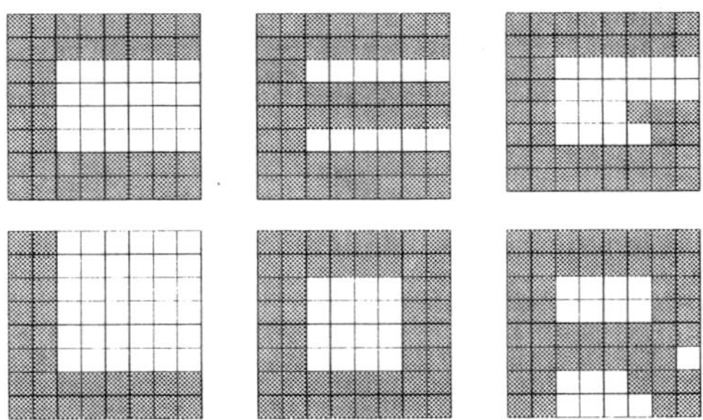

Fig. 1. The predefined six target shapes

These six letters were selected based upon how much of the alphabet they were representative of. For example, the diagonal in the letter R forming the bottom right portion of the letter is characteristic of the letters M, N, W, X, Y, Z. Likewise, the semi-circle is characteristic of the letters P, R, B. The Letters E and L with their upright stem are characteristic of P, T, D, F.

To judge how closely each evolving shape matched the targets, a fitness function based on the number of incorrectly filled squares was employed. The fitness score is incremented by one whenever an element in the evolving shape differs from the corresponding element in the current target, see Figure 2. To assess scalability, three different phenotype grid sizes of 4×4, 8x8 and 16×16 cells were used.

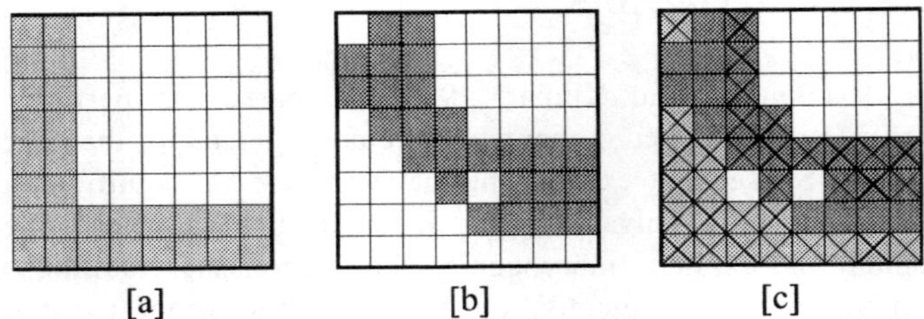

Fig. 2. Calculating the fitness of an 8×8 evolving shape. [a] shows the target, [b] shows the shape to be judged, [c] shows the incorrect elements identified by the fitness function.

4.2 Explicit

This first system used GP (Koza, 1992) to evolve explicit embryogenies in the form of program trees. Beginning at a seed or zygote cell placed in the phenotype grid, the embryogeny defines the direction of growth at every point. Four functions were used, LEFT, RIGHT, UP and DOWN, with each node in the tree allowed up to four branches. Paths of growth were permitted to overlap. Figure 3 shows an example genotype defining the explicit embryogeny. The root node has two parts, x and y, for the coordinates of the seed.

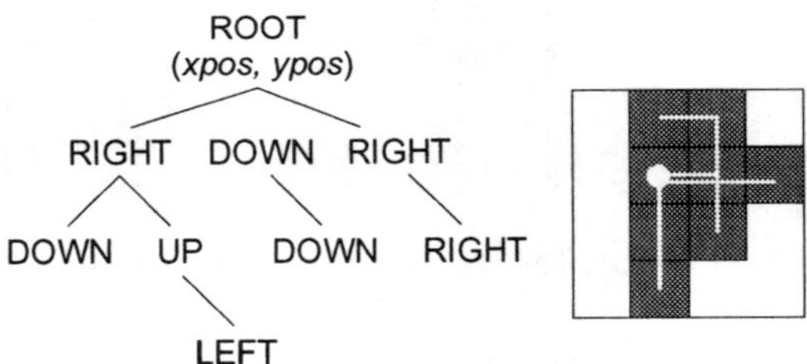

Fig. 3. An example explicit embryogeny defined by a tree of nine nodes, and its corresponding 4×4 phenotype.

The GP system used steady-state selection and a crossover designed to minimise disruption by crossing parents at points of similarity in the two trees. As with all GP systems, bloat occurred, so an additional fitness function penalised genotypes with more nodes. This system differed from the one presented in Bentley and Kumar (1999), in that it evolved the coordinates of the seed.

4.3 Implicit

The second system was an advanced variable-length chromosome genetic algorithm (Bentley, 1999, Ch. 18) that evolved implicit embryogenies. Each genotype comprised a variable number of rules (usually between four and eight). Each rule had a precondition and an action. Each precondition had six fields: *LEFT, RIGHT, UP, DOWN, X, Y.* A specific rule can take the following values for each precondition field (where # is *don't care*, 0 is *empty*, 1 is *filled*, **0,1,2,3,4,5,6,7** are gradient zones):

LEFT	*RIGHT*	*UP*	*DOWN*	*X*	*Y*
0,1,#	0,1,#	0,1,#	0,1,#	**0-7**,#	**0-7**,#

For a rule to be fired, values in at least four of the six fields in the precondition must be matched. (This provides the equivalent of disjunction for rule preconditions.) The action of a rule can be: DIE, UPDATE, or grow LEFT, RIGHT, UP or DOWN.

Growth takes place in a phenotype grid, which as usual can be of 4×4, 8×8 or 16×16 elements. In order to permit evolution of specialised rules that can provide detail in specific areas of the phenotype, the grid has two 'gradients' - one in the x direction, one in the y direction. In a similar way to the gradients used to provide positional information in eggs and wombs of nature (Wolpert, 1998; Slack, 1991; Lawrence, 1995), the gradients divide the grid into eight zones per axis (as opposed to four in previous work), regardless of the number of elements in the phenotype grid.

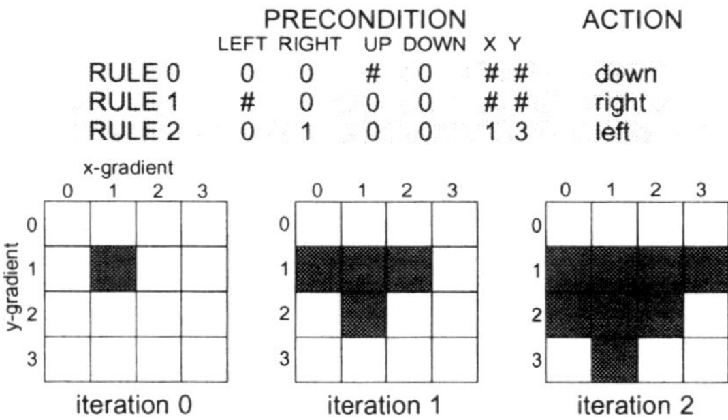

Fig. 4. Example of a three-rule implicit embryogeny and its corresponding phenotype after two iterations

At iteration zero, a seed cell is placed in the phenotype grid at a position defined by the coordinates held in dedicated genes of the chromosome. To model biological cell growth, the rules in the genotype are then applied for a fixed number of iterations to each *filled* element in the current embryonic phenotype grid. (This is unlike traditional cellular automata, where rules are applied to empty or filled grid elements.) Depending on whether the neighbouring elements of the current element exist or not, and on the strength of the two gradients at that point, the rules may be activated, causing growth or cell death in the phenotype. Rules are applied 'in parallel' so that the results of applying the rules to each filled element only take effect at the end of each iteration step. However, a rule which performs the UPDATE action causes all activated rules in the current iteration to be applied. By prematurely placing cells in the phenotype grid in this way, evolution can increase the number of rules applied in each iteration and provide extra growth where needed. This new type of rule action was added to the embryogeny because during the development of the system, the number of iterations was found to be overly critical. Finally, the system was also given the ability to evolve the seed coordinates. Figure 4 shows the growth of a target shape, defined by three rules.

5 Experiments

5.1 Objectives and Parameters

The experimental objectives were three-fold: firstly to investigate the use of both embryogenies for efficiency of search in terms of fitness; secondly, to investigate the scalability of both embryogenies for evolving different morphologies; and finally, to see how the evolution of the two embryogenies differs in defining different morphologies.

A total of 50 runs were performed with each target shape (letters C, E, G, L, O and R) for each grid size. Population sizes of 100 and a total of 100 generations were used for each run. The explicit embryogeny system used an initial tree depth of 4 for the 4×4, 5 for the 8×8 and 6 for the 16×16 grids. All trees were created randomly.

The implicit system used random rule initialisation for both the initial population and for each new rule added to the genome.[1] The cellular automata presented in this work used iteration values of 4 for the 4×4, 8 for the 8×8, and 14 for the 16×16 grids. Both systems used random crossover for offspring creation. The explicit system employed a mutation probability rate of 0.001 per bit, whereas preliminary experiments revealed that the implicit system required a rule mutate rate of 0.5 and an increased bit mutation rate of 0.05.

[1] This is as opposed to copying an existing rule as in previous work (Bentley and Kumar, 1999).

5.2 Results

A summary of the results from the experiments is given in Table 3.[2] As shown in the table, both embryogenies attained good fitnesses for all 4×4 target letters. However, relative performances between the two approaches were inconsistent. For example, the explicit embryogeny outperformed the implicit one for the letters C and L, whilst the reverse was true for the other targets.

For the 8×8 grid, fitness scores were reduced, on average, for both methods. For example, the explicit embryogeny managed 3.55 at best for the letter L and at worst 15.76 for the O. The implicit embryogeny faired similarly on the 8×8 targets achieving a fitness of 5.89 for the E and only 12.84 for the G. Again, relative performances varied, this time with each embryogeny providing better scores for three of the letters.

When the problem was scaled up to the 16×16 grid, the results show that the implicit embryogeny outperformed the explicit one, in terms of fitness. The figures are, however, a little deceptive. For these targets, many of the shapes evolved by the implicit embryogeny were solid blocks. Because of the simple nature of the fitness function, such shapes were awarded higher fitness scores compared to the attempts of the explicit embryogeny. (Nevertheless, it should be noted that the forms generated by the explicit embryogeny rarely resembled the desired targets, either.)

	4×4		8×8		16×16	
Shape	**Mean Soln. Size**	**Mean Fitness**	**Mean Soln. Size**	**Mean Fitness**	**Mean Soln. Size**	**Mean Fitness**
C	14.28 *12.70*	0.92 *1.64*	57.70 *11.47*	13.20 *12.82*	309.40 *10.00*	84.1 *53.7*
E	24.22 *11.42*	1.28 *0.32*	168.44 *11.96*	9.54 *5.89*	693.58 *6.700*	81.40 *49.40*
G	18.88 *13.86*	1.2 *0.78*	59.52 *10.98*	12.72 *12.84*	302.12 *6.000*	76.34 *52.90*
L	9.52 *11.22*	0.26 *0.58*	71.04 *9.02*	3.56 *6.38*	235.46 *8.200*	39.46 *38.40*
O	20.20 *11.60*	1.31 *0.18*	81.29 *13.29*	15.76 *9.00*	293.00 *6.400*	104.33 *48.70*
R	18.78 *12.98*	1.35 *0.60*	121.53 *12.33*	7.88 *9.92*	513.43 *6.900*	76.16 *55.80*

Table 3. Results for the target shapes. Values in *italics* denote the results for the implicit embryogeny. Solution sizes are measured in tree nodes for the explicit, and rules for the implicit, embryogeny.

[2] Because of time constraints, average values given for the 16×16 targets using the implicit embryogeny were based on only 10 experiments per target.

Execution times were noticeably different for the two techniques. As the scale of the problem was increased, both methods took longer to grow shapes, but of the two, the implicit required the most computation time. For example, as evolution time of six hours for one run of the implicit was not uncommon, compared to less than an hour for the explicit embryogeny.

Perhaps the most significant results shown in Table 3 are the solution sizes. It is clear that the explicit embryogeny required ever-increasing tree sizes as the scale of the target shapes were increased. However, the reverse seems to be true for the implicit embryogeny, where the number of rules actually appears to decrease as the problems are scaled up. This lack of increase of solution size corroborates and confirms the results obtained in previous work which reported similar findings (Bentley and Kumar, 1999).

5.3 Analysis

The results show interesting behaviours of both embryogenies for all grid sizes. For the 4×4 grid, because of the size of the targets, the ability of both methods to find good solutions is not surprising. The explicit embryogeny uses small trees to define its solutions, but the implicit noe often seems to evolve more rules than are necessary. More specifically, a larger number of rules are evolved than are actually used during the growth process. The reason for the inefficiency for such small targets seems to be connected with the search process – it is very hard for evolution to find the correct rules for specific shapes. Clearly the 'add rule' mutation plays an important, but excessive role for these smaller problems. Rules are added until an appropriate collection exists to define the target, but the unused rules are not removed by mutation, much like bloat in GP.

The same searching mechanism is also evident for larger grid sizes with the implicit embryogeny, with similar levels of redundancy observed. However, because the number of rules did not increase much beyond 12, such redundancy becomes a more acceptable compromise for larger problems.

So how does evolution fine-tune solutions to make them match the target letters? Both types of embryogeny seem to begin by filling a large part (or all) of the phenotype grid, and then 'pruning away' unnecessary elements, see Figures 5 and 6. The explicit embryogeny achieves this by pruning branches of its trees; the implicit embryogeny makes use of 'kill' rules to remove elements (similar to the use of apoptosis 'programmed cell death' in biological development during the formation of the digits of the hands and feet). Of the two, the implicit embryogeny goes to the furthest extreme with this technique – often by evolving a completely solid block and then removing the odd elements, see Figures 5 and 6.

The fitness function may be to blame for the 'carving letters from a solid block' approach, for it awards considerably higher fitnesses for solid shapes than for emptier ones.

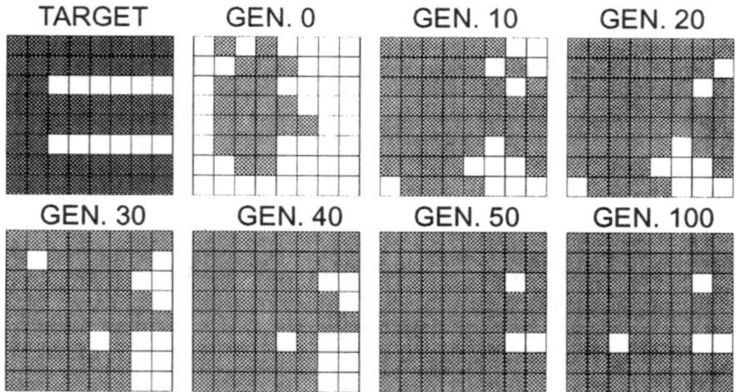

Fig. 5. The evolution of an 'E' using the explicit embryogeny. The best new shapes grown in the population are shown every 10 generations (except where no change occurred). The final shape has a fitness of 8 and a solution size of 251 nodes.

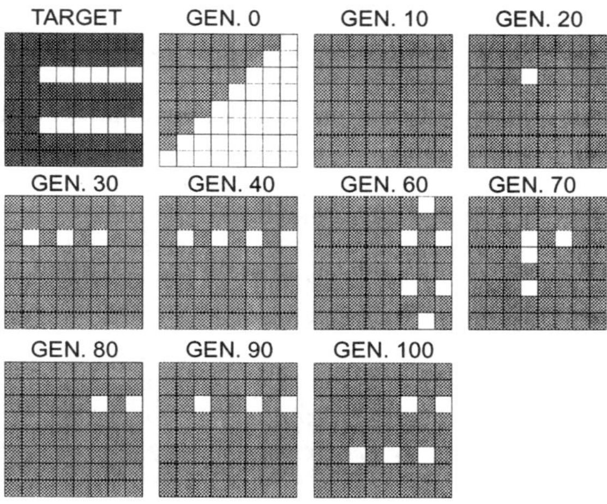

Fig. 6. The evolution of an 'E' using the implicit embryogeny. The best new shapes grown in the population are shown every 10 generations (except where no change occurred). The final shape has a fitness of 7 and a solution size of 11 rules.

As Dawkins (1987) points out, certain embryogenies are better than others at producing certain morphologies. This observation seems to be echoed in this work too. The explicit embryogeny found morphologies such as C's and O's difficult, whereas the implicit embryogeny was able to handle these morphologies with relative ease. This can be attributed to the fact that the implicit need only generate a few general growth rules for specific directions, and, in doing so, can start from a single seed and grow to encompass all four sides of the grid whilst leaving (or killing), e.g. in the case of the letter O, a hole in the middle. This is difficult for the explicit embryogeny as it must evolve a long and difficult growth path around the edge of the grid, whilst keeping the centre of the shape free of elements.

The way in which evolution attempted to generate morphology indicates two points: firstly that the implicit embryogeny seems to have considerable potential because of its impressive scalability, perhaps more so than the explicit embryogeny; secondly, the representation used for the implicit embryogeny is not as amenable to evolution as we would desire. This may be caused by:

- the representation, which allows dissimilar phenotypes to be close together in the solution space, providing a discontinuous search space of solutions and thus causing problems for evolution.
- ineffective positional rules. Although the notion of 'zones' improved the quality of solution for the implicit embryogeny more than without, it is clear that positional information is hard to glean using the current implicit system, making the evolution of specific rules difficult and leading in turn to bad fitness results.

6 Biologically Plausible Implicit Embryogeny

With the results of this experiment in mind, a new implicit model is now under development by the authors. The current system has been extended from the two-dimensional cellular automata to an *isospatial grid* system. The isospatial grid is a three-dimensional coordinate system, developed by Frazer (1995), and uses six axes to define a point in space, yielding 12 equidistant neighbours for each point.

The system uses spheres to represent cells and builds three-dimensional morphologies by carefully placing and organising a colony of cells using a growth process. This process will use the concept of freely diffusing morphogens to allow cells to acquire positional information. In addition, key embryological processes such as differentiation and pattern formation will be investigated to grow morphologies. Although heavily inspired by biology, this system is not intended as a model of biological development. Instead, the work is aimed at extending the capabilities and scalability of EAs.

An implicit embryogeny-based system is used to evolve rules that are able to grow designs in complex ways. A chromosome comprises a series of rules (genes). A rule consists of a precondition field and an action field. Each cell has a copy of the chromosome. The rules are applied (*expressed*) by matching the preconditions to a cell's state. If the preconditions are satisfied the rule is expressed. In this way, rules expressed earlier on in the development can affect other rules by switching them on or off.

Notions of evolvability and speed of growth are playing a key role in the design of this new implicit system. Recent ideas about representations and evolvability, such as the use of component-based approaches (Bentley, 2000) and neutral networks (Barnett, 1997), may throw light on more suitable representations. The authors are also intending to use a parallel Beowulf supercomputer to exploit the inherent parallelism of growth processes.

Figure 7 illustrates six morphologies grown from random genomes using the new system. It should be clear that the isospatial grid enables surprisingly organic forms to emerge. Shape 4 also illustrates how flat surfaces can arise from multiple cells.

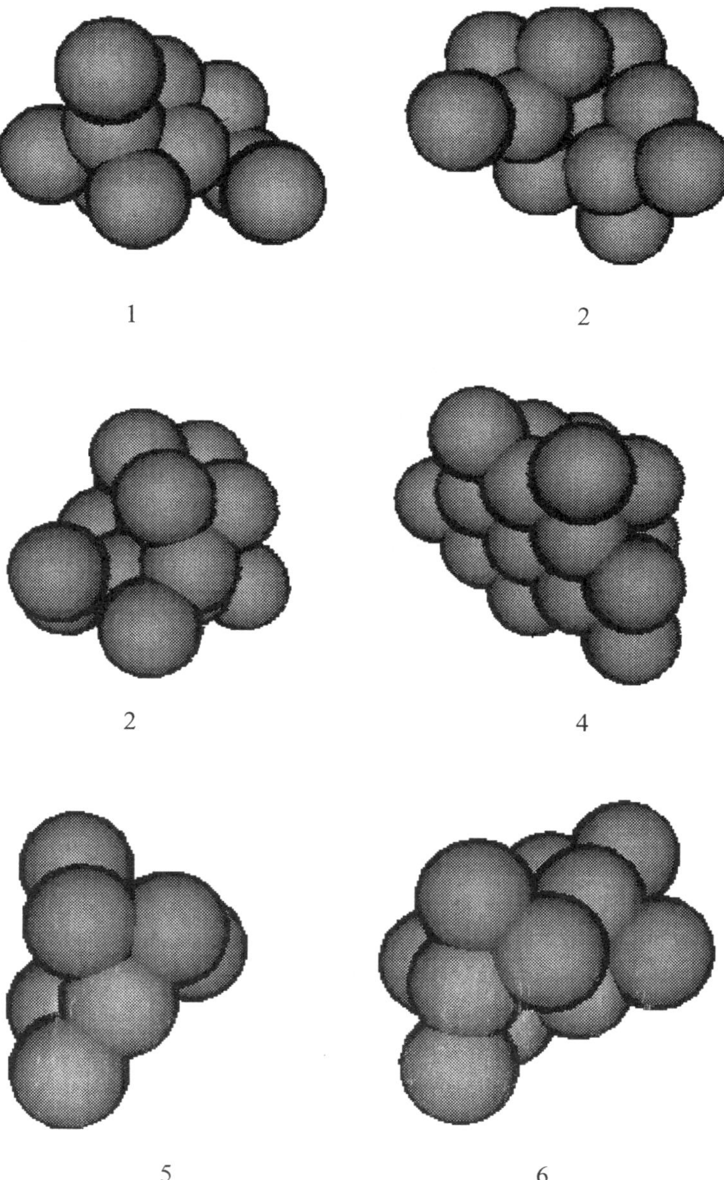

Fig. 7. Six example random morphologies grown by the new system without any morphogens in the environment

7 Conclusions

This chapter has introduced computational embryology. It began with a historical review of embryology in biology and then described the recent development of computational embryology in computer science. It then looked at present computational embryogenies, comparing the evolvability and scalability of two different approaches (explicit and implicit) for the problem of evolving shape morphologies.

The behavioural analysis of these two embryogenies has shown that both are good at growing different shape morphologies, and that evolutionary computation can benefit from the use of embryogenies.

In addition, this work highlighted some of the problems that require attention with regard to designing evolvable embryogenies. For example, biological concepts such as positional information, as echoed in this work in the form of zones, can assist in the evolution of shape morphologies, but need to be very carefully designed.

The chapter ended with a brief description of a future computational embryogeny system using more biologically plausible methods to overcome the problems encountered in the experiment.

Nature has been successfully evolving complex animals for millions of years. It is the concept of an embryogeny (which itself evolved in nature) that has allowed the evolution of these complex designs.

Acknowledgements

Our thanks to Lewis Wolpert, Tom Quick and the members of nUCLEAR for enlightening discussions on this work. This work is funded by Science Applications International Corporation (SAIC).

References

Bard, J. (1990). *Morphogenesis.* Cambridge University Press, Cambridge, UK.

Barnett, L. (1997). Tangled Webs: Evolutionary Dynamics on Fitness Landscapes with Neutrality. M.Sc. Dissertation, University of Sussex, Brighton, UK.

Bentley, P.J. (Ed.) (1999). *Evolutionary Design by Computers.* Morgan Kaufmann, San Francisco, CA.

Bentley, P.J. (2000). Exploring Component-Based Representations - The Secret of Creativity by Evolution? In Proc. of the Fourth International Conference on Adaptive Computing in Design and Manufacture (ACDM 2000), April 26th-28th, 2000, University of Plymouth, UK.

Bentley, P.J. & Kumar, S. (1999). Three Ways to Grow Designs: A Comparison of Embryogenies for an Evolutionary Design Problem. In Genetic and Evolutionary Computation Conference (GECCO), Orlando, Florida, USA.

Daellart, F. & and Beer, R. (1995). Toward a Biologically Defensible Model of Development. Masters Thesis, Case Western Reserve University.

Dawkins, R. (1987). The Evolution of Evolvability. *Proceedings of Artificial Life VI.* Langton (Ed.) USA.

de Garis, H. (1999). Artificial Embryology and Cellular Differentiation. Ch. 12 in Bentley, P. J. (Ed.) *Evolutionary Design by Computers*. Morgan Kaufman, San Francisco, CA.

Frazer, J. (1995). *An Evolutionary Architecture.* Architecture Association, London.

Goldberg, D.E. (1989). *Genetic Algorithms in Search, Optimization and Machine Learning*. Addison-Wesley, Reading, MA.

Gruau, F. & Whitley, D. (1993). Adding Learning to the Cellular Development of Neural Networks. *Evolutionary Computation.* 1, 3, 213-233.

Holland, J. H. (1975). *Adaptation in Natural and Artificial Systems.* The University of Michigan Press, Ann Arbor.

Jakobi, N. (1996). Harnessing Morphogenesis. University of Sussex, Cognitive Science Research Report #429, Brighton, UK.

Koza, J.R. (1992). *Genetic Programming I: On the Means of Natural Selection*. Morgan Kaufmann, San Francisco, CA.

Koza, J.R., Bennett III, Forrest H., Andre, David, & Keane, Martin A. (1999). *Genetic Programming III*. Morgan Kaufmann, San Francisco, CA.

Kumar, S. (1999). Lessons from Nature: The Benefits of Embryology. In Genetic and Evolutionary Computation Conference (GECCO). Orlando, Florida, USA.

Kumar, S. & Bentley, P.J. (1999). The ABCs of Evolutionary Design: Investigating the Evolvability of Embryogenies for Morphogenesis. In Genetic and Evolutionary Computation Conference (GECCO). Orlando, Florida, USA.

Lawrence, P.A. (1995). *The Making of a Fly: The Genetics of Animal Design*. Blackwell Science, The Alden Press, Oxford, UK.

Ortega, C. & Tyrrell, A. (1998). MUXTREE Revisited: Embryonics as a reconfiguration strategy in fault-tolerant processor arrays. In *Evolvable Systems: From Biology to Hardware, Lecture Notes in Computer Science*, Vol. 1478, Springer-Verlag, Berlin, pp. 206-217.

Sánchez, E. et al. (1997). Phylogeny, Ontogeny and Epigenesis: Three sources of biological inspiration for softening hardware. In Higuchi T. et al. (Eds.), *Evolvable Systems: From Biology to Hardware*, Lecture Notes in Computer Science, Vol. 1478, Springer-Verlag, Berlin.

Sims, K. (1999). Evolving Three-dimensional Morphology and Behaviour. Ch. 13 in Bentley, P.J. (Ed.) *Evolutionary Design by Computers.* Morgan Kaufmann, San Francisco, CA..

Slack, J.M. (1991). *From Egg to Embryo.* Cambridge University Press, UK.

Turing, A.M. (1952). The chemical basis of morphogenesis. *Phil. Trans. R. Soc.*, 237B, 37-72.

Vaario, J. (1994). Modeling Adaptative Self-Organization. In Brooks and Maes (Eds). Artificial Life IV, The MIT Press, Cambridge, MA, pp. 313-318.

Wolpert, L. (1998). *Principles of Development.* Oxford University Press, Oxford, UK.

An Evolutionary Approach to Synthetic Biology: Zen in the Art of Creating Life

Thomas S. Ray

ATR Human Information Science Laboratories, 2-2-2 Hikaridai, Seika-cho, Soraku-gun, Kyoto, 619-0288, Japan
E-mail: tray@ou.edu
Ray, T. S. 1994. An evolutionary approach to synthetic biology: Zen and the art of creating life. Artificial Life 1(1/2): 195–226. MIT Press.

Summary. Our concepts of biology, evolution, and complexity are constrained by having observed only a single instance of life, life on Earth. A truly comparative biology is needed to extend these concepts. Because we cannot observe life on other planets, we are left with the alternative of creating artificial life forms on Earth. I will discuss the approach of inoculating evolution by natural selection into the medium of the digital computer. This is not a physical/chemical medium, it is a logical/informational medium. Thus these new instances of evolution are not subject to the same physical laws as organic evolution (e.g., the laws of thermodynamics), and therefore exist in what amounts to another universe, governed by the "physical laws" of the logic of the computer. This exercise gives us a broader perspective on what evolution is and what it does.

An evolutionary approach to synthetic biology consists of inoculating the process of evolution by natural selection into an artificial medium. Evolution is then allowed to find the natural forms of living organisms in the artificial medium. These are not models of life, but independent instances of life. This essay is intended to communicate a way of thinking about synthetic biology that leads to a particular approach: to understand and respect the natural form of the artificial medium, to facilitate the process of evolution in generating forms that are adapted to the medium, and to let evolution find forms and processes that naturally exploit the possibilities inherent in the medium. Examples are cited of synthetic biology embedded in the computational medium, where in addition to being an exercise in experimental comparative evolutionary biology, it is also a possible means of harnessing the evolutionary process for the production of complex computer software.

1 Synthetic Biology

Artificial life (AL) is the enterprise of understanding biology by constructing biological phenomena out of artificial components, rather than breaking natural life forms down into their component parts. It is the synthetic rather than the reductionist approach. I will describe an approach to the synthesis of artificial living forms that exhibit natural evolution.

The umbrella of AL is broad, and covers three principal approaches to synthesis: in hardware (e.g., robotics, nanotechnology), in software (e.g., replicating and evolving computer programs), in wetware (e.g., replicating and evolving organic molecules, nucleic acids or others). This essay will focus on software synthesis, although it is hoped that the issues discussed will be generalizable to any synthesis involving the process of evolution.

I would like to suggest that software syntheses in AL could be divided into two kinds: simulations and instantiations of life processes. AL simulations represent an advance in biological modeling, based on a bottom-up approach, that has been made possible by the increase of available computational power. In the older approaches to modeling of ecological or evolutionary phenomena, systems of differential equations were set up that expressed relationships between covarying quantities of entities (i.e., genes, alleles, individuals, or species) in the populations or communities.

The new bottom-up approach creates a population of data structures, with each instance of the data structure corresponding to a single entity. These structures contain variables defining the state of an individual. Rules are defined as to how the individuals interact with one another and with the environment. As the simulation runs, populations of these data structures interact according to local rules, and the global behavior of the system emerges from those interactions. Several very good examples of bottom-up ecological models have appeared in the AL literature [34], [92]. However, ecologists have also developed this same approach independently of the AL movement, and have called the approach "individual based" models [19], [40].

The second approach to software synthesis is what I have called instantiation rather than simulation. In simulation, data structures are created which contain variables that represent the states of the entities being modeled. The important point is that in simulation, the data in the computer is treated as a representation of something else, such as a population of mosquitoes or trees. In instantiation, the data in the computer does not represent anything else. The data patterns in an instantiation are considered to be living forms in their own right, and are not models of any natural life form. These can form the basis of a comparative biology [58].

The object of an AL instantiation is to introduce the natural form and process of life into an artificial medium. This results in an artificial life form in some medium other than carbon chemistry, and is not a model of organic life forms. The approach discussed in this essay involves introducing the process of evolution by natural selection into the computational medium. I consider evolution to be the fundamental process of life, and the generator of living form.

2 Recognizing Life

Most approaches to defining life involve assembling a short list of properties of life, and then testing candidates on the basis of whether or not they exhibit the properties on the list. The main problem with this approach is that there is disagreement as to what should be on the list. My private list contains only two items: self-replication and open-ended evolution. However, this reflects my biases as an evolutionary biologist.

I prefer to avoid the semantic argument and take a different approach to the problem of recognizing life. I was led to this view by contemplating how I would regard a machine that exhibited conscious intelligence at such a level that it could participate as an equal in a debate such as this. The machine would meet neither of my two criteria as to what life is, yet I don't feel that I could deny that the process it contained was alive.

This means that there are certain properties that I consider to be unique to life, and whose presence in a system signifies the existence of life in that system. This suggests an alternative approach to the problem. Rather than creating a short list of minimal requirements and testing whether a system exhibits all items on the list, create a long list of properties unique to life and test whether a system exhibits *any* item on the list.

In this softer, more pluralistic approach to recognizing life, the objective is not to determine if the system is alive or not, but to determine if the system exhibits a "genuine" instance of some property that is a signature of living systems (e.g., self-replication, evolution, flocking, consciousness).

Whether we consider a system living because it exhibits some property that is unique to life amounts to a semantic issue. What is more important is that we recognize that it is possible to create disembodied but genuine instances of specific properties of life in artificial systems. This capability is a powerful research tool. By separating the property of life that we choose to study from the many other complexities of natural living systems, we make it easier to manipulate and observe the property of interest. The objective of the approach advocated in this chapter is to capture genuine evolution in an artificial system.

3 What Natural Evolution Does

Evolution by natural selection is a process that enters into a physical medium. Through iterated replication-with-selection of large populations through many generations, it searches out the possibilities inherent in the "physics and chemistry" of the medium in which it is embedded. It exploits any inherent self-organizing properties of the medium, and flows into natural attractors realizing and fleshing out their structure.

Evolution never escapes from its ultimate imperative: self-replication. However, the mechanisms that evolution discovers for achieving this ultimate goal gradually become so convoluted and complex that the underlying drive can seem to become superfluous. Some philosophers have argued that the evolutionary theory as expressed by the phrase "survival of the fittest" is tautological, in that the fittest are defined as those that survive to reproduce. In fact, fitness is achieved through innovation in engineering of the organism [82]. However, there remains something peculiarly self-referential about the whole enterprise. There is some sense in which life may be a natural tautology.

Evolution is both a defining characteristic and the creative process of life itself. The living condition is a state that complex physical systems naturally flow into under certain conditions. It is a self-organizing, self-perpetuating state of auto-catalytically increasing complexity. The living component of the physical system quickly becomes the most complex part of the system, such that it re-shapes the medium, in its own image as it were. Life then evolves adaptations predominantly in relation to the living components of the system, rather than the non-living components. Life evolves adaptations to itself.

3.1 Evolution in Sequence Space

Think of organisms as occupying a "genotype space" consisting of all possible sequences of all possible lengths of the elements of the genetic system (i.e., nucleotides or machine instructions). When the first organism begins replicating, a single self-replicating creature, with a single sequence of a certain length, occupies a single point in the genotype space. However, as the creature replicates in the environment, a population of creatures forms, and errors cause genetic variation, such that the population will form a cloud of points in the genotype space, centered around the original point.

Because the new genotypes that form the cloud are formed by random processes, most of them are completely inviable, and die without reproducing. However, some of them are capable of reproduction. These new genotypes persist, and as some of them are affected by mutation, the cloud of points spreads further. However, not all of the viable genomes are equally viable. Some of them discover tricks to replicate more efficiently. These genotypes increase in frequency, causing the population of creatures at the corresponding points in the genotype space to increase.

Points in the genotype space occupied by greater populations of individuals will spawn larger numbers of mutant offspring, thus the density of the cloud of points in the genotype space will shift gradually in the direction of the more fit genotypes. Over time, the cloud of points will percolate through the genotype space, either expanding outward as a result of random drift, or by flowing along fitness gradients.

Most of the volume of this space represents completely inviable sequences. These regions of the space may be momentarily and sparsely occupied by inviable mutants, but the cloud will never flow into the inviable regions. The cloud of genotypes may bifurcate as it flows into habitable regions in different directions, and it may split as large genetic changes spawn genotypes in distant but viable regions of the space. We may imagine that the evolving population of creatures will take the form of wispy clouds flowing through this space.

Now imagine for a moment the situation that there was no selection. This implies that every sequence is replicated at an equal rate. Mutation will cause the cloud of points to expand outward, eventually filling the space uniformly. In this situation, the complexity of the structure of the cloud of points does not increase through time, only the volume that it occupies. Under selection by contrast, through time the cloud will take on an intricate structure as it flows along fitness gradients and percolates by drift through narrow regions of viability in a largely uninhabitable space.

Consider that the viable region of the genotype space is a very small subset of the total volume of the space, but that it probably exhibits a very complex shape, forming tendrils and sheets sparsely permeating the otherwise empty space. The complex structure of this cloud can be considered to be a product of evolution by natural selection. This thought experiment appears to imply that the intricate structure that the cloud of genotypes may assume through evolution is fully deterministic. Its shape is pre-defined by the physics and chemistry and the structure of the environment, in much the same way that the form of the Mandlebrot set is pre-determined by its defining equation. The complex structure of this viable space is inherent in the medium, and is an example of "order for free" [45].

No living world will ever fill the entire viable subspace, either at a single moment of time, or even cumulatively over its entire history. The region actually filled will be strongly influenced by the original self-replicating sequence, and by stochastic forces which will by chance push the cloud down a subset of possible habitable pathways. Furthermore, co-evolution and ecological interactions imply that certain regions can only be occupied when certain other regions are also occupied. This concept of the flow of genotypes through the genotype space is essentially the same as that discussed by Eigen [22] in the context of "quasispecies". Eigen limited his discussion to species of viruses, where it is also easy to think of sequence spaces. Here, I am extending the concept beyond the bounds of the species, to include entire phylogenies of species.

3.2 Natural Evolution in an Artificial Medium

Until recently, life has been known as a state of matter, particularly combinations of the elements carbon, hydrogen, oxygen, nitrogen, and smaller quantities of many others. However, recent work in the field of AL has shown that the natural evolutionary process can proceed with great efficacy in other media, such as the informational medium of the digital computer [1], [3], [7], [15], [16], [20], [24], [29], [43], [44], [51], [53], [54], [68], [69], [71]–[74], [77], [78], [81], [89], [91].

These new natural evolutions, in artificial media, are beginning to explore the possibilities inherent in the "physics and chemistry" of those media. They are organizing themselves and constructing self-generating complex systems. While these new living systems are still so young that they remain in their primordial state, it appears that they have embarked on the same kind of journey taken by life on Earth, and presumably have the potential to evolve levels of complexity that could lead to sentient and eventually intelligent beings.

If natural evolution in artificial media leads to sentient or intelligent beings, they will likely be so alien that they will be difficult to recognize. The sentient properties of plants are so radically different from those of animals that they are generally unrecognized or denied by humans, and plants are merely in another kingdom of the one great tree of organic life on Earth [70], [75], [88]. Synthetic organisms evolving in other media, such as the digital computer, not only are not a part of the same phylogeny, but they are not even of the same physics. Organic life is based on conventional material physics, whereas digital life exists in a logical, not material, informational universe. Digital intelligence will likely be vastly different from human intelligence; forget the Turing test.

4 The Approach

> "Marcel, a mechanical chessplayer... his exquisite 19th-century brainwork — the human art it took to build which has been flat lost, lost as the dodo bird ... But where inside Marcel is the midget Grandmaster, the little Johann Allgeier? where''s the pantograph, and the magnets? Nowhere. Marcel really is a mechanical chessplayer. No fakery inside to give him any touch of humanity at all."
> — Thomas Pynchon, Gravity's Rainbow.

The objective of the approach discussed here is to create an instantiation of evolution by natural selection in the computational medium. This creates a conceptual problem that requires considerable art to solve: ideas and techniques must be learned by studying organic evolution, and then applied to the generation of evolution in a digital medium, without forcing the digital medium into an "un-natural" simulation of the organic world.

We must derive inspiration from observations of organic life, but we must never lose sight of the fact that the new instantiation is not organic, and may differ in many fundamental ways. For example, organic life inhabits a Euclidean space, but computer memory is not a Euclidean space. Inter-cellular communication in the organic world is chemical in nature, and therefore a single message generally can pass no more information than on or off. By contrast, communication in digital computers generally involves the passing of bit patterns, which can carry much more information.

The fundamental principle of the approach being advocated here is *to understand and respect the natural form of the digital computer, to facilitate the process of evolution in generating forms that are adapted to the computational medium, and to let evolution find forms and processes that naturally exploit the possibilities inherent in the medium.*

Situations arise where it is necessary to make significant changes from the standard computer architecture. But such changes should be made with caution, and only when there is some feature of standard computer architectures which clearly inhibits the desired processes. Examples of such changes are discussed in the section "The Genetic Language" below. Less substantial changes are also discussed in the sections on the "Flaw" genetic operator, "Mutations", and "Artificial Death". The sections on "Spatial Topology" and "Digital 'Neural

Networks" – Natural AI" are little tirades against examples of what I consider to be un-natural transfers of forms from the natural world to the digital medium.

5 The Computational Medium

The computational medium of the digital computer is an informational universe of boolean logic, not a material one. Digital organisms live in the memory of the computer, and are powered by the activity of the central processing unit (CPU). Whether the hardware of the CPU and memory is built of silicon chips, vacuum tubes, magnetic cores, or mechanical switches is irrelevant to the digital organism. Digital organisms should be able to take on the same form in any computational hardware, and in this sense are "portable" across hardware.

Digital organisms might as well live in a different universe from us, as they are not subject to the same laws of physics and chemistry. They are subject to the "physics and chemistry" of the rules governing the manipulation of bits and bytes within the computer's memory and CPU. They never "see" the actual material from which the computer is constructed, they see only the logic and rules of the CPU and the operating system. These rules are the only "natural laws" that govern their behavior. They are not influenced by the natural laws that govern the material universe (e.g., the laws of thermodynamics).

A typical instantiation of this type involves the introduction of a self-replicating machine language program into the random access (RAM) memory of a computer subject to random errors such as bit flips in the memory or occasionally inaccurate calculations [3], [7], [20], [53], [71]. This generates the basic conditions for evolution by natural selection as outlined by Darwin [14]: self-replication in a finite environment with heritable genetic variation.

In this instantiation, the self-replicating machine language program is thought of as the individual "digital organism" or "creature". The RAM provides the physical space that the creatures occupy. The CPU provides the source of energy. The memory consists of a large array of bits, generally grouped into eight bit bytes and sixteen or thirty-two bit words. Information is stored in these arrays as voltage patterns which we usually symbolize as patterns of ones and zeros.

The "body" of a digital organism is the information pattern in memory that constitutes its machine language program. This information pattern is data, but when it is passed to the CPU, it is interpreted as a series of executable instructions. These instructions are arranged in such a way that the data of the body will be copied to another location of memory. The informational patterns stored in the memory are altered only through the activity of the CPU. It is for this reason that the CPU is thought of as the analog of the energy source. Without the activity of the CPU, the memory would be static, with no changes in the informational patterns stored there.

The logical operations embodied in the instruction set of the CPU constitute a large part of the definition of the "physics and chemistry" of the digital universe. The topology of the computer's memory (discussed below) is also a significant component of the digital physics. The final component of the digital physics is the operating system, a software program running on the computer, which embodies rules for the allocation of resources such as memory space and CPU time to the various processes running on the computer.

The instruction set of the CPU, the memory, and the operating system together define the complete "physics and chemistry" of the universe inhabited by the digital organism. They constitute the physical environment within which digital organisms will evolve. Evolving digital organisms will compete for access to the limited resources of memory space and CPU time, and evolution will generate adaptations for the more agile access to and the more efficient use of these resources.

6 The Genetic Language

The simplest possible instantiation of a digital organism is a machine language program that codes for self-replication. In this case, the bit pattern that makes up the program is the body of the organism, and at the same time its complete genetic material. Therefore, the machine language defined by the CPU constitutes the genetic language of the digital organism.

It is worth noting at this point that the organic organism most comparable to this kind of digital organism is the hypothetical, and now extinct, RNA organism [6]. These were presumably nothing more than RNA molecules capable of catalyzing their own replication. What the supposed RNA organisms have in common with the simple digital organism is that a single molecule constitutes the body and the genetic information, and effects the replication. In the digital organism a single bit pattern performs all the same functions.

The use of machine code as a genetic system raises the problem of brittleness. It has generally been assumed by computer scientists that machine language programs cannot be evolved because random alterations such as bit flips and recombinations will always produce inviable programs. It has been suggested [23] that overcoming this brittleness and "Discovering how to make such self-replicating patterns more robust so that they evolve to increasingly more complex states is probably the central problem in the study of artificial life."

The assumption that machine languages are too brittle to evolve is probably true, as a consequence of the fact that machine languages have not previously been designed to survive random alterations. However, recent experiments have shown that brittleness can be overcome by addressing the principal causes, and without fundamentally changing the structure of machine languages [71], [78].

The first requirement for evolvability is graceful error handling. When code is being randomly altered, every possible meaningless or erroneous condition is likely to occur. The CPU should be designed to handle these conditions without crashing the system. The simplest solution is for the CPU to perform no operation when it meets these conditions, perhaps setting an error flag, and to proceed to the next instruction.

Due to random alterations of the bit patterns, all possible bit patterns are likely to occur. Therefore a good design is for all possible bit patterns to be interpretable as meaningful instructions by the CPU. For example, in the Tierra system [71]–[74], [77], [78], a five bit instruction set was chosen, in which all thirty-two five bit patterns represent good machine instructions.

This approach (all bit patterns meaningful) also could imply a lack of syntax, in which each instruction stands alone, and need not occur in the company of other instructions. To the extent that the language includes syntax, where instructions must precede or follow one another in certain orders, random alterations are likely to destroy meaningful syntax thereby making the language more brittle. A certain amount of this kind of brittleness can be tolerated as long as syntax errors are also handled gracefully.

During the design of the first evolvable machine language [71], a standard machine language (Intel 80X86) was compared to the genetic language of organic life, to attempt to understand the difference between the two languages that might contribute to the brittleness of the former and the robustness of the latter. One of the outstanding differences noted was in the number of basic informational objects contained in the two.

The organic genetic language is written with an alphabet consisting of four different nucleotides. Groups of three nucleotides form sixty-four "words" (codons), which are translated into twenty amino acids by the molecular machinery of the cell. The machine language is written with sequences of two voltages (bits) which we conceptually represent as ones and zeros. The number of bits that form a "word" (machine instruction) varies between machine architectures, and in some architectures is not constant. However, the number required generally ranges from sixteen to thirty-two. This means that there are from tens of thousands to billions of machine instruction bit patterns, which are translated into operations performed by the CPU.

The thousands or billions of bit patterns that code for machine instructions contrast with the sixty-four nucleotide patterns that code for amino acids. The sixty-four nucleotide patterns are degenerate, in that they code for only twenty amino acids. Similarly, the machine codes are degenerate, in that there are at most hundreds rather than thousands or billions of machine operations.

The machine codes exhibit a massive degeneracy (with respect to actual operations) as a result of the inclusion of data into the bit patterns coding for the operations. For example, the add operation will take two operands, and produce as a result the sum of the two operands. While there may be only a single add operation, the instruction may come in several forms depending on where the values of the two operands come from, and where the resultant sum will be placed. Some forms of the add instruction allow the value(s) of the operand(s) to be specified in the bit pattern of the machine code.

The inclusion of numeric operands in the machine code is the primary cause of the huge degeneracy. If numeric operands are not allowed, the number of bit patterns required to specify the complete set of operations collapses to at most a few hundred.

While there is no empirical data to support it, it is suspected that the huge degeneracy of most machine languages may be a source of brittleness. The logic of this argument is that mutation causes random swapping among the fundamental informational objects, codons in the organic language, and machine instructions in the digital language. It seems more likely that meaningful results will be produced when swapping among sixty-four objects than when swapping among billions of objects.

The size of the machine instruction set can be made comparable to the number of codons simply by eliminating numeric operands embedded in the machine code. However, this change creates some new problems. Computer programs generally function by executing instructions located sequentially in memory. However, in order to loop or branch, they use instructions such as "jump" to cause execution to jump to some other part of the program. Since the locations of these jumps are usually fixed, the jump instruction will generally have the target address included as an operand embedded in the machine code.

By eliminating operands from the machine code, we generate the need for a new mechanism of addressing for jumps. To resolve this problem, an idea can be borrowed from molecular biology. We can ask the question: how do biological molecules address one another? Molecules do not specify the coordinates of the other molecules they interact with. Rather, they present shapes on their surfaces that are complementary to the shapes on the surfaces of the target molecules. The concept of complementarity in addressing can be introduced to machine languages by allowing the jump instruction to be followed by some bit pattern, and having execution jump to the nearest occurrence of the complementary bit pattern.

In the development of the Tierran language, two changes were introduced to the machine language to reduce brittleness: elimination of numeric operands from the code, and the use of complementary patterns to control addressing. The resulting language proved to be evolvable [71]. As a result, nothing was learned about evolvability, because only one language was tested, and it evolved. It is not known what features of the language enhance its evolvability, which detract, and which do not affect evolvability. Subsequently, three additional languages were tested and the four languages were found to vary in their patterns and degree of evolvability [78]. However, it is still not known how the features of the language affect its evolvability.

7 Genetic Operators

In order for evolution to occur, there must be some genetic variation among the offspring. In organic life, this is insured by natural imperfections in the replication of the informational molecules. However, one way in which digital "chemistry" differs from organic chemistry is in the degree of perfection of its operations. In the computer, the genetic code can be reliably replicated without errors to such a degree that we must artificially introduce errors or other sources of genetic variation in order to induce evolution.

7.1 Mutations

In organic life, the simplest genetic change is a "point mutation", in which a single nucleic acid in the genetic code is replaced by one of the three other nucleic acids. This can cause an amino acid substitution in the protein coded by the gene. The nucleic acid replacement can be caused by an error in the replication of the DNA molecule, or it can be caused by the effects of radiation or mutagenic chemicals.

In the digital medium, a comparably simple genetic change can result from a bit flip in the memory, where a one is replaced by a zero, or a zero is replaced by a one. These bit flips can be introduced in a variety of ways that are analogous to the various natural causes of mutation. In any case, the bit flips must be introduced at a low to moderate frequency, as high frequencies of mutation prevent the replication of genetic information, and lead to the death of the system [74].

Bit flips may be introduced at random anywhere in memory, where they may or may not hit memory actually occupied by digital organisms. This could be thought of as analogous to cosmic rays falling at random and disturbing molecules which may or may not be biological in nature. Bit flips may also be introduced when information is copied in the memory, which could be analogous to the replication errors of DNA. Alternatively, bit flips could be introduced in memory as it is accessed, either as data or executable code. This could be thought of as damage due to "wear and tear".

7.2 Flaws

Alterations of genetic information are not the only source of noise in the system. In organic life, enzymes have evolved to increase the probability of chemical reactions that increase the fitness of the organism. However, the metabolic system is not perfect. Undesired chemical reactions do occur, and desired reactions sometimes produce undesired by-products. The result is the generation of molecular species that can "gum up the works", having unexpected consequences, generally lowering the fitness of the organism, but possibly raising it.

In the digital system, an analog of metabolic (non-genetic) errors can be introduced by causing the computations carried out by the CPU to be probabilistic, producing erroneous results at some low frequency. For example, any time a sum or difference is calculated, the result could be off by some small value (e.g., plus or minus one). Or, if all bits are shifted one position to the left or right, an appropriate error would be to shift by two positions or not at all. When information is transferred from one location to another, either in the RAM memory or the CPU registers, it could occasionally be transferred from the wrong location, or to the wrong location. While flaws do not directly cause genetic changes, they can cause a cascade of events that result in the production of an offspring that is genetically different from the parent.

7.3 Recombination – Sex

The Nature of Sex

In organic life, there are a wide variety of mechanisms by which offspring are produced which contain genetic material from more than one parent. This is the sexual process. Recombination mechanisms range from very primitive and haphazard to elaborately orchestrated.

At the primitive extreme we find certain species of bacteria, in which upon death the cell membrane breaks open, releasing the DNA into the surrounding medium. Fragments of this dead DNA are absorbed across the membranes of other bacteria of the same species, and incorporated into their genome [59]. This is a one-way transfer of genetic material, rather than a reciprocal exchange.

At the complex extreme we find the conventional sexual system of most of the higher animals, in which each individual contains two copies of the entire genome. At reproduction, each of two parents contributes one complete copy of the genome (half of their genetic material) to the offspring. This means that each offspring receives one-half of its genetic material from each of two parents, and each parent contributes one-half of its genetic material to each offspring. Very elaborate behavioral and molecular mechanisms are required to orchestrate this joint contribution of genetic material to the offspring.

The preponderance of sex remains an enigma to evolutionary theory [5], [26], [31], [32], [55], [61], [86], [96]. Careful analysis has failed to show any benefits from sex, at the level of the individual organism, that outweigh the high costs (e.g., passing on only half of the genome). The only obvious benefit of sex is that it provides diversity among the offspring, allowing the species to adapt more readily to a changing environment. However, quantitative analysis has shown that in order for sex to be favored by selection at the individual level, it is not enough for the environment to change unpredictably, the environment must actually change capriciously [13], [57]. That is, whatever genotype has the highest fitness in this generation must have the lowest fitness in the next generation, or at least a trend in this direction, a negative heritability of fitness.

One theory to explain the perpetuation of sex (based on the Red Queen hypothesis, see below) states that the environment is in fact capricious, due to the importance of biotic factors in determining selective forces. That is, sex is favored because it is necessary to maintain adaptation in the face of evolving species in the environment (e.g., predators/parasites, prey/hosts, competitors) who themselves are sexual, and can undergo rapid evolutionary change. Predators and parasites will tend to evolve so as to favor attacking whatever genotype of their prey/host is the most common. The genotype that is most successful at present is targeted for future attack. This dynamic makes the environment capricious in the sense discussed above.

There are fundamental differences in the nature of the evolutionary process between asexual and sexual organisms. The evolving entity in an asexual species is a branching lineage of genetic individuals which retain their genetic identity through the generations. In a sexual species, the evolving entity is a collective "gene pool", and genetic individuals are absolutely ephemeral, lasting only one generation.

Recalling the discussion of "genotype space" above in the section "Evolution in Sequence Space", imagine that we could represent genotype space in two dimensions, and that we allow a third dimension to represent time. Visualize now an evolving asexual organism. Starting with a single individual, it would occupy a single point in the genotype space at time zero. When it reproduces, if there is no mutation, its offspring would occupy the same point in genotype space, at a later time. Thus the lineage of the asexual organism would appear as a line moving forward in time. If mutations occur, they cause the offspring to occupy new locations in genotype space, forming branches in the lineage.

Through time, the evolving asexual lineage would form a tree-like structure in the genotype space–time coordinates. However, every individual branch of the tree will evolve independently of all the others. While there may be ecological interactions between genetically different individuals, there is no exchange of genetic material between them. From a genetic point of view, each branch of the tree is on its own; it must adapt, or fail to adapt, based on its own genetic resources.

In order to visualize an evolving sexual population we must start with a population of individuals, each of which will be genetically unique. Thus they will appear as a scatter of points in the genotype space plane at time zero. In the next generation, all of the original genotypes will be dead; however, a completely new set of genotypes will have been formed from new combinations of pieces of the genomes from the previous generation. No individual genotypes will survive from one generation to the next, thus over time, the evolving sexual population appears as a diffuse cloud of disconnected points, with no lines formed from persistent genotypes.

The most important distinction between the evolving asexual and sexual populations is that the asexual individuals are genetically isolated and must adapt or not based on the limited genetic resources of the individual, while sexual organisms by comparison draw on the genetic resources of the entire population, due to the flow of genes resulting from sexual matings. The entity that evolves in an asexual population is an isolated but branching lineage of genetic individuals. In a sexual population, the individual is ephemeral, and the entity that evolves is a "gene pool".

Due to the genetic cohesion of a sexual population and the ephemeral nature of its individuals, the evolving sexual entity exists at a higher level of organization than the individual organism. The evolving entity, a gene pool, is supra-organismal. It samples the environment through many individuals simultaneously, and pools their genetic resources in finding adaptive genetic combinations.

The definition of the biological species is based on a concept of sexual reproduction: a group of individuals capable of interbreeding freely under natural conditions. Species concepts simply do not apply well to asexual species. In order for synthetic life to be useful for the study of the properties of species and the speciation process, it must include an organized sexual process, such that the evolving entity is a gene pool.

Implementation of Digital Sex

The above discussions of the nature of sexuality are intended to make the point that it is an important process in evolutionary biology, and should be included in synthetic implementations of life. The sexual process is implemented with the "crossover" genetic operator in the field of genetic algorithms, where it has been considered to be the most important genetic operator [35].

The crossover operator has also been implemented in synthetic life systems [76], [91]. However, it has been implemented in the spirit of a genetic algorithm, rather than in the spirit of synthetic life. This is because in these implementations the crossover process is not under the control of the organism, but rather is forced on the individual. In addition, these implementations are based on haploid sex not diploid sex (see below). In order to address many of the interesting evolutionary questions surrounding sexuality, the sexual process must be optional, at least through evolution, and should be diploid.

Primitive sexual processes have appeared spontaneously in the Tierra synthetic life system [71]. However, there apparently has still not been an implementation of natural organized sexuality in a synthetic system. I would like to discuss my conception of how this could be implemented, with particular reference to the Tierra system.

It would seem that the simplest way of implementing an organized sexuality that would give rise to an evolving gene pool would involve the use of "ploidy". Ploidy refers to a system in which each individual contains multiple copies of the complete genome. In the most familiar sexual system (that used by humans), the gametes (egg and sperm) contain one copy of the genome (they are haploid), and all other stages of the life cycle contain two copies (they are diploid), which derive from the union of a sperm and egg.

In a digital organism whose body consists of a sequence of machine code, it would be easy to duplicate the sequence and include two copies within the cell. However, some problems can arise with this configuration, if the two copies of the genome occupy adjacent blocks of memory. Which copy of the genome will be executed? When the organism contributes one of its two copies of the genome to an offspring, which of the two copies will be contributed, and how can the mother cell recognize where one complete genome begins and ends?

A solution to these problems that has been partially implemented in the Tierra system is to have the two copies of the genome intertwined, rather than in adjacent blocks of memory. This can be done by letting alternate bytes represent one genome, and the skipped bytes the other genome. Tierran instructions utilize only five bits, and so are mapped to successive bytes in memory. If we instead place successive instructions in successive sixteen bit words, one copy of the genome can occupy the high-order bytes, and the other genome can occupy the low-order bytes of the words.

This arrangement facilitates relatively simple solutions to the problems mentioned above. Execution of the genome takes place by having the instruction pointer execute alternate bytes. In a diploid organism there are two tracks. The track to initially be executed can be chosen at random. At a certain frequency, or under certain circumstances, the executing track can be switched so that both copies of the genome will be expressed.

Having two parallel tracks helps to resolve the problem of recognizing where one copy of the genome ends and the other begins, since both genomes usually begin and end together. Copying of the genome, like execution, can occur along one track. Optionally, tracks could be switched during the copy process, to introduce an effect similar to crossing over in meiosis. In addition, the use of both tracks can be optional, so that haploid and diploid organisms can coexist in the same soup, and evolution can favor either form, according to selective pressures.

7.4 Transposons

The explosion of diversity in the Cambrian occurred in the lineage of the eukaryotes; the prokaryotes did not participate. One of the most striking genetic differences between eukaryotes and prokaryotes is that most of the genome of prokaryotes is translated into proteins, while most of the genome of eukaryotes is not. It has been estimated that typically 98% of the DNA in eukaryotes is neither translated into proteins nor involved in gene regulation, that it is simply "junk" DNA [93]. It has been suggested that much of this junk code is the result of the self-replication of pieces of DNA within rather than between cells [21], [67].

Mobile genetic elements, transposons, have this intra-genome self-replicating property. It has been estimated that 80% of spontaneous mutations are caused by transposons [12], [30]. Repeated sequences, resulting from the activity of mobile elements, range from dozens to millions in numbers of copies, and from hundreds to tens of thousands of base pairs in length. They vary widely in dispersion patterns from clumped to sparse [41].

Larger transposons carry one or more genes in addition to those necessary for transposition. Transposons may grow to include more genes; one mechanism involves the placement of two transposons into close proximity so that they act as a single large transposon incorporating the intervening code. In many cases transposons carry a sequence that acts as a promoter, altering the regulation of genes at the site of insertion [90].

Transposons may produce gene products and often are involved in gene regulation [17]. However, they may have no effect on the external phenotype of the individual [21]. Therefore they evolve through another paradigm of selection, one that does not involve an external phenotype. They are seen as a mechanism for the selfish spread of DNA which may become inactive junk after mutation [67].

DNA of transposon origin can be recognized by their palindrome endings flanked by short non-reversed repeated sequences resulting from insertion after staggered cuts. In *Drosophila melanogaster* approximately 5 to 10% of its total

DNA is composed of sequences bearing these signs. There are many families of such repeated elements, each family possessing a distinctive nucleotide sequence, and distributed in many sites throughout the genome. One well-known repeated sequence occurring in humans is found to have as many as a half million copies in each haploid genome [87].

Elaborate mechanisms have evolved to edit out junk sequences inserted into critical regions. An indication of the magnitude of the task comes from the recent cloning of the gene for cystic fibrosis, where it was discovered that the gene consists of 250,000 base pairs, only 4,440 of which code for protein; the remainder are edited out of the messenger RNA before translation [46], [56], [79], [80].

It appears that many repeated sequences in genomes may have originated as transposons favored by selection at the level of the gene, favoring genes which selfishly replicated themselves within the genome. However, some transposons may have co-evolved with their host genome as a result of selection at the organismal or populational level, favoring transposons which introduce useful variation through gene rearrangement. It has been stated that "transposable elements can induce mutations that result in complex and intricately regulated changes in a single step", and they are "A highly evolved macromutational mechanism" [90].

In this manner, "smart" genetic operators may have evolved, through the interaction of selection acting at two or more hierarchical levels (it appears that some transposons have followed another evolutionary route, developing intercellular mobility and becoming viruses [41]). It is likely that transposons today represent the full continuum from purely parasitic "selfish DNA" and viruses to highly co-evolved genetic operators and gene regulators. The possession of smart genetic operators may have contributed to the explosive diversification of eukaryotes by providing them with the capacity for natural genetic engineering.

In designing self-replicating digital organisms, it would be worthwhile to introduce such genetic parasites, in order to facilitate the shuffling of the code that they bring about. Also, the excess code generated by this mechanism provides a large store of relatively neutral code that can randomly explore new configurations through the genetic operations of mutation and recombination. When these new configurations confer functionality, they may become selected for.

8 Artificial Death

Death must play a role in any system that exhibits the process of evolution. Evolution involves a continuing iteration of selection, which implies differential *death*. In natural life, death occurs as a result of accident, predation, starvation, disease, or if these fail to kill the organism, it will eventually die from senescence resulting from an accumulation of wear and tear at every level of the organism including the molecular.

In normal computers, processes are "born" when they are initiated by the user, and "die" when they complete their task and halt. A process whose goal is to repeatedly replicate itself is essentially an endless loop, and would not spontaneously terminate. Due to the perfection of normal computer systems, we cannot count on "wear and tear" to eventually cause a process to terminate.

In synthetic life systems implemented in computers, death is not likely to be a process that would occur spontaneously, and it must generally be introduced artificially by the designer. Everyone who has set up such a system has found their own unique solutions. Todd [94] recently discussed this problem in general terms.

In the Tierra system [71] death is handled by a "reaper" function of the operating system. The reaper uses a linear queue. When creatures are born, they enter the bottom of the queue. When memory is full, the reaper frees memory to make space for new creatures by killing off the top of the queue. However, each time an individual generates an error condition, it moves up the reaper queue one position.

An interesting variation on this was introduced by Barton-Davis [3] who eliminated the reaper queue. In its place, he caused the "flaw rate" (see section on flaws above) to increase with the age of the individual, in mimicry of wear and tear. When the flaw rate reached 100%, the individual was killed. Skipper [81] provided a "suicide" instruction, which if executed, would cause a process to terminate (die). The evolutionary objective then became to have a suicide instruction in your genome which you do not execute yourself, but which you try to get other individuals to execute. Litherland [51] introduced death by local crowding. Davidge caused processes to die when they contained certain values in their registers [16]. Gray [29] allowed each process six attempts at reproduction, after which it would die.

9 Operating System

Much of the "physics and chemistry" of the digital universe is determined by the specifications of the operations performed by the instruction set of the CPU. However, the operating system also determines a significant part of the physical context. The operating system manages the allocation of critical resources such as memory space and CPU cycles.

Digital organisms are processes that spawn processes. As processes are born, the operating system will allocate memory and CPU cycles to them, and when they die, the operating system will return the resources they had utilized to the pool of free resources. In synthetic life systems, the operating system may also play a role in managing death, mutations, and flaws.

The management of resources by the operating system is controlled by algorithms. From the point of view of the digital organisms these take the form of a set of logical rules like those embodied in the logic of the instruction set. In this way, the operating system is a defining part of the physics and chemistry of the digital universe. Evolution will explore the possibilities inherent in these rules, finding ways to more efficiently gain access to and exploit the resources managed by the operating system.

10 Spatial Topology

Digital organisms live in the memory space of computers, predominantly in the RAM, although they could also live on disks or any other storage device, or even within networks to the extent that the networks themselves can store information. In essence, digital organisms live in the space that has been referred to as "cyber space". It is worthwhile reflecting on the topology of this space, as it is a radically different space from the one we live in.

A typical UNIX workstation, or Macintosh computer, includes a RAM that can contain some megabytes of data. This is "flat" memory, meaning that it is essentially unstructured. Any location in memory can be accessed through its numeric address. Thus adjacent locations in memory are accessed through successive integer values. This addressing convention causes us to think of the memory as a linear space, or a one-dimensional space.

However, this apparent one-dimensionality of the RAM is something of an illusion generated by the addressing scheme. A better way of understanding the topology of the memory comes from asking "what is the distance between two locations in memory?" In fact the distance cannot be measured in linear units. The most appropriate unit is the time that it takes to move information between the two points.

Information contained in the RAM cannot move directly from point to point. Instead the information is transferred from the RAM to a register in the CPU, and then from the CPU back to the new location in the RAM. Thus the distance between two locations in the RAM is just the time that it takes to move from the RAM to the CPU plus the time that it takes to move from the CPU to the RAM. Because all points in the RAM are equidistant from the CPU, the distance between any pair of locations in the RAM is the same, regardless of how far apart they may appear based on their numeric addresses.

A space in which all pairs of points are equidistant is clearly not a Euclidean space. That said, we must recognize, however, that there are a variety of ways in which memory is normally addressed, that gives it the appearance, at least locally, of being one dimensional. When code is executed by the CPU, the instruction pointer generally increments sequentially through memory, for short distances, before jumping to some other piece of code. For those sections of code where instructions are sequential, the memory is effectively one dimensional. In addition, searches of memory are often sequentially organized (e.g., the search for complementary templates in Tierra). This again makes the memory effectively one dimensional within the search radius. Yet even under these circumstances, the memory is not globally one dimensional. Rather it consists of many small one dimensional pieces, each of which has no meaningful spatial relationship to the others.

Because we live in a three-dimensional Euclidean space, we tend to impose our familiar concepts of spatial topology onto the computer memory. This leads first to the erroneous perception that memory is a one-dimensional Euclidean space, and, second, it often leads to the conclusion that the digital world could be enriched by increasing the dimensionality of the Euclidean memory space.

Many of the serious efforts to extend the Tierra model have included, as a central feature, the creation of a two-dimensional space for the creatures to inhabit [3], [15], [16], [53], [81]. The logic behind the motivation derives from contemplation of the extent to which the dimensionality of the space we live in permits the richness of pattern and process that we observe in nature. Certainly if our universe were reduced from three to two dimensions, it would eliminate the possibility of most of the complexity that we observe. Imagine for example, the limitations that two-dimensionality would place on the design of neural networks (if "wires" could not cross). If we were to further reduce the dimensionality of our universe to just one dimension, it would probably completely preclude the possibility of the existence of life.

It follows from these thoughts that restricting digital life to a presumably one-dimensional memory space places a tragic limitation on the richness that might evolve. Clearly it would be liberating to move digital organisms into a two- or three-dimensional space. The flaw in all of this logic derives from the erroneous supposition that computer memory is a Euclidean space.

To think of memory as Euclidean is to fail to understand its natural topology, and is an example of one of the greatest pitfalls in the enterprise of synthetic biology: to transfer a concept from organic life to synthetic life in a way that is "un-natural" for the artificial medium. The fundamental principle of the approach I am advocating is *to respect the nature of the medium into which life is being inoculated, and to find the natural form of life in that medium*, without inappropriately trying to make it like organic life.

The desire to increase the richness of memory topology is commendable, but this can be achieved without forcing the memory into an un-natural Euclidean topology. Let us reflect a little more on the structure of cyberspace. Thus far we have only considered the topology of flat memory. Let us consider segmented memory such as is found with the notorious Intel 80X86 design. With this design, you may treat any arbitrarily chosen block of 64K bytes as flat, and all pairs of locations within that block are equidistant. However, once the block is chosen, all memory outside of that block is about twice as far away.

Cache memory is designed to be accessed more rapidly than RAM, thus pairs of points within cache memory are closer than pairs of points within RAM. The distance between a point in cache and a point in RAM would be an intermediate distance. The access time to memory on disks is much greater than for RAM, thus the distance between points on the disk is very great, and the distance between RAM and disk is again intermediate (but still very great). CPU registers represent a small amount of memory locations, between which data can move very rapidly, thus these registers can be considered to be very close together.

For networked computer systems, information can move between the memories of the computers on the net, and the distance between these memories is again the transfer time. If the CPU, cache, RAM, and disk memories of a network of computers are all considered together, they present a very complex memory topology. Similar considerations apply to massively parallel computers which have memories connected in a variety of topologies. Utilizing this complexity moves us in the direction of what has been intended by creating Euclidean memories for digital organisms, but does so while fully respecting the natural topology of computer memories.

11 Ecological Context

11.1 The Living Environment

Some rain forests in the Amazon region occur on white sand soils. In these locations, the physical environment consists of clean white sand, air, falling water, and sunlight. Embedded within this relatively simple physical context we find one of the most complex ecosystems on Earth, containing hundreds of thousands of species. These species do not represent hundreds of thousands of adaptations to the physical environment. Most of the adaptations of these species are to the other living organism. The forest creates its own environment.

Life is an auto-catalytic process that builds on itself. Ecological communities are complex webs of species, each living off of others, and being lived off of by others. The system is self-constructing, self-perpetuating, and feeds on itself. Living organisms interface with the non-living physical environment, exchanging materials with it, such as oxygen, carbon dioxide, nitrogen, and various minerals. However, in the richest ecosystems, the living components of the environment predominate over the physical components.

With living organisms constituting the predominant features of the environment, the evolutionary process is primarily concerned with adaptation to the living environment. Thus ecological interactions are an important driving force for evolution. Species evolve adaptations to exploit other species (to eat them, to parasitize them, to climb on them, to nest on them, to catch a ride on them, etc.) and to defend against such exploitation where it creates a burden.

This situation creates an interesting dynamic. Evolution is predominantly concerned with creating and maintaining adaptations to living organisms which are themselves evolving. This generates evolutionary races among groups of species that interact ecologically. These races can catalyze the evolution of upwardly spiraling complexity as each species evolves to overcome the adaptations of the others. Imagine, for example, a predator and prey, each evolving to increase its speed and agility, in capturing prey, or in evading capture. This coupled evolutionary race can lead to increasingly complex nervous systems in the evolving predator and prey species.

This mutual evolutionary dynamic is related to the Red Queen hypothesis [95], named after the Red Queen from *Alice in Wonderland*. This hypothesis suggests that in the face of a changing environment, organisms must evolve as fast as they can in order to simply maintain their current state of adaptation. "In order to get anywhere you must run twice as fast as that" [11].

If organisms only had to adapt to the non-living environment, the race would not be so urgent. Species would only need to evolve as fast as the relatively gradual changes in the geology and climate. However, given that the species that comprise the environment are themselves evolving, the race becomes rather hectic. The pace is set by the maximal rate that species may change through evolution, and it becomes very difficult to actually get ahead. A maximal rate of evolution is required just to keep from falling behind.

What all of this discussion points to is the importance of embedding evolving synthetic organisms into a context in which they may interact with other evolving organisms. A counter example is the standard implementations of genetic algorithms in which the evolving entities interact only with the fitness function, and never "see" the other entities in the population. Many interesting behavioral, ecological, and evolutionary phenomena can only emerge from interactions among the evolving entities.

11.2 Diversity

Major temporal and spatial patterns of organic diversity on Earth remain largely unexplained, although there is no lack of theories. Diversity theories suggest fundamental ecological and evolutionary principles which may apply to synthetic life. In general these theories relate to synthetic life in two ways: 1) They suggest factors which may be critical to the auto-catalytic increase of diversity and complexity in an evolving system. It may be necessary then to introduce these factors into an artificial system to generate increasing diversity and complexity. 2) Because it will be possible to manipulate the presence, absence, or state of these factors in an artificial system, the artificial system may provide an experimental framework for examining evolutionary and ecological processes that influence diversity.

The Gaussian principle of competitive exclusion states that no two species that occupy the same niche can coexist. The species which is the superior competitor will exclude the inferior competitor. The principle has been experimentally demonstrated in the laboratory, and is considered theoretically sound. However, natural communities widely flaunt the principle. In tropical rain forests several hundred species of trees coexist without any dominant species in the community. All species of trees must spread their leaves to collect light and their roots to absorb water and nutrients. Evidently there are not several hundred niches for trees in the same habitat. Somehow the principle of competitive exclusion is circumvented.

There are many theories on how competitive exclusion may be circumvented. One leading theory is that periodic disturbance at the proper level sets back the process of competitive exclusion, allowing more species to coexist [37]–[39]. There is substantial evidence that moderate levels of disturbance can increase diversity. In a digital community, disturbance might take the form of freeing blocks of memory that had been filled with digital organisms. It would be very easy to experiment with differing frequencies and patch sizes of disturbance.

One theory to explain the great increase in diversity and complexity in the Cambrian explosion [85] states that its evolution was driven by ecological interactions, and that it was originally sparked by the appearance of the first organisms that ate other organisms (heterotrophs). As long as all organisms were autotrophs (produce their own food, like plants), there was only room for a few species. In a community with only one trophic level, the most successful competitors would dominate. The process of competitive exclusion would keep diversity low.

However, when the first herbivore (organisms that eat autotrophs) appeared it would have been selected to prefer the most common species of algae, thereby preventing any species of algae from dominating. This opens the way for more species of algae to coexist. Once the "heterotroph barrier" had been crossed, it would be simple for carnivores to arise, imposing a similar diversifying effect on herbivores. With more species of algae, herbivores may begin to specialize on different species of algae, enhancing diversification in herbivores. The theory states that the process was auto-catalytic, and set off an explosion of diversity.

One of the most universal of ecological laws is the species–area relationship [52]. It has been demonstrated that in a wide variety of contexts, the number of species occupying an "area" increases with the area. The number of species increases in proportion to the area raised to a power between 0.1 and 0.3: $S = KA^z$, where $0.1 < z < 0.3$. The effect is thought to result from the equilibrium species number being determined by a balance between the arrival (by immigration or speciation) and local extinction of species. The likelihood of extinction is greater in small areas because they support smaller populations, for which a fluctuation to a size of zero is more likely. If this effect holds for digital organisms it suggests that larger amounts of memory will generate greater diversity.

11.3 Ecological Attractors

While there are no completely independent instances of natural evolution on Earth, there are partially independent instances. Where major diversifications have occurred, isolated either by geography or epoch from other similar diversifications, we have the opportunity to observe whether evolution tends to take the same routes or is always quite different. We can compare the marsupial mammals of Australia to the placental mammals of the rest of the world, or the modern mammals to the reptiles of the age of dinosaurs, or the bird fauna of the Galapagos to the bird faunas of less isolated islands.

What we find again and again is an uncanny convergence between these isolated faunas. This suggests that there are fairly strong ecological attractors which evolution will tend to fill, more or less regardless of the developmental and physiological systems that are evolving. In this view, chance and history still play a role, in determining what kind of organism fills the array of ecological attractors (reptiles, mammals, birds, etc.), but the attractors themselves may be a property of the system and not as variable. Synthetic systems may also contain fairly well-defined ecological forms which may be filled by a wide variety of specific kinds of organisms.

Given their evident importance in moving evolution, it is important to include ecological interactions in synthetic instantiations of life. It is encouraging to observe that in the Tierra model, ecological interactions, and the corresponding evolutionary races, emerged spontaneously. It is possible that any medium into which evolution is inoculated will contain an array of "ecological attractors" into which evolution will easily flow.

12 Cellularity

Cellularity is one of the fundamental properties of organic life, and can be recognized in the fossil record as far back as 3.6 billion years. The cell is the original individual, with the cell membrane defining its limits and preserving its chemical integrity. An analog to the cell membrane is probably needed in digital organisms in order to preserve the integrity of the informational structure from being disrupted by the activity of other organisms.

The need for this can be seen in AL models such as cellular automata where virtual state machines pass through one another [48], or in core wars-type simulations where coherent structures that arise demolish one another when they come into contact [68], [69]. An analog to the cell membrane that can be used in the core wars type of simulation is memory allocation. An artificial "cell" could be defined by the limits of an allocated block of memory. Free access to the memory within the block could be limited to processes within the block. Processes outside of the block would have limited access, according the rules of "semi-permeability"; for example, they might be allowed to read and execute but not write.

13 Multi-cellularity

Multi-celled digital organisms are parallel processes. By attempting to synthesize multi-celled digital organisms we can simultaneously explore the biological issues surrounding the evolutionary transition from single-celled to multi-celled life, and the computational issues surrounding the design of complex parallel software.

13.1 Biological Perspective – Cambrian Explosion

Life appeared on Earth somewhere between 3 and 4 billion years ago. While the origin of life is generally recognized as an event of the first order, there is another event in the history of life that is less well known but of comparable significance. The origin of biological diversity, and at the same time of complex macroscopic multi-cellular life, occurred abruptly in the Cambrian explosion 600 million years ago. This event involved a riotous diversification of life forms. Dozens of phyla appeared suddenly, many existing only fleetingly, as diverse and sometimes bizarre ways of life were explored in a relative ecological void [28], [65].

The Cambrian explosion was a time of phenomenal and spontaneous increase in the complexity of living systems. It was the process initiated at this time that led to the evolution of immune systems, nervous systems, physiological systems, developmental systems, complex morphology, and complex ecosystems. To understand the Cambrian explosion is to understand the evolution of complexity. If the history of organic life can be used as a guide, the transition from single-celled to multi-celled organisms should be critical in achieving a rich diversity and complexity of synthetic life forms.

13.2 Computational Perspective – Parallel Processes

It has become apparent that the future of high-performance computing lies with massively parallel architectures. There already exist a variety of parallel hardware platforms, but our ability to fully utilize the potential of these machines is constrained by our inability to write software of a sufficient complexity.

There are two fairly distinctive kinds of parallel architecture in use today: SIMD (single instruction multiple data) and MIMD (multiple instruction multiple data). In the SIMD architecture, the machine may have thousands of processors, but in each CPU cycle, all of the processors must execute the same instruction, although they may operate on different data. It is relatively easy to write software for this kind of machine, since what is essentially a normal sequential program will be broadcast to all the processors.

In the MIMD architecture, there exists the capability for each of the hundreds or thousands of processors to be executing different code, but to have all of that activity coordinated on a common task. However, there does not exist an art for writing this kind of software, at least not on a scale involving more than a few parallel processes. In fact it seems unlikely that human programmers will ever be capable of actually writing software of such complexity.

13.3 Evolution as a Proven Route

It is generally recognized that evolution is the only process with a proven ability to generate intelligence. It is less well recognized that evolution also has a proven ability to generate parallel software of great complexity. In making life a metaphor for computation we will think of the genome, the DNA, as the program, and we will think of each cell in the organism as a processor (CPU). A large multi-celled organism like a human contains trillions of cells/processors. The genetic program contains billions of nucleotides/instructions.

In a multi-celled organism, cells are differentiated into many cell types such as brain cells, muscle cells, liver cells, kidney cells, etc. The cell types just named are actually general classes of cell types within which there are many sub-types. However, when we specify the ultimate indivisible types, what characterizes a type is the set of genes it expresses. Different cell types express different combinations of genes. In a large organism, there will be a very large number of cells of most types. All cells of the same type express the same genes.

The cells of a single cell type can be thought of as exhibiting parallelism of the SIMD kind, as they are all running the same "program" by expressing the same genes. Cells of different cell types exhibit MIMD parallelism as they run different code by expressing different genes. Thus large multi-cellular organisms display parallelism on an astronomical scale, combining both SIMD and MIMD parallelism into a beautifully integrated whole. From these considerations it is evident that evolution has a proven ability to generate massively parallel software embedded in wetware. The computational goal of evolving multi-cellular digital organisms is to produce such software embedded in hardware.

13.4 Fundamental Definition

In order to conceptualize multi-cellularity in the context of an artificial medium, we must have a very fundamental definition which is independent of the context of the medium. We generally think of the defining property of multi-cellularity as being that the cells stick together, forming a physically coherent unit. However, this is a spatial concept based on Euclidean geometry, and therefore is not relevant to non-Euclidean cyberspace.

While physical coherence might be an adequate criterion for recognizing multi-cellularity in organic organisms, it is not the property that allows multi-cellular organisms to become large and complex. There are algae that consist of strands of cells that are stuck together, with each cell being identical to the next. This is a relatively limiting form of multi-cellularity because there is no differentiation of cell types. It is the specialization of functions resulting from cell differentiation that has allowed multi-cellular organisms to attain large sizes and great complexity. It is differentiation that has generated the MIMD style of parallelism in organic software.

From an evolutionary perspective, an important characteristic of multi-cellular organisms is their genetic unity. All the cells of the individual contain the same genetic material as a result of having a common origin from a single egg cell (some small genetic differences may arise due to somatic mutations; in some species new individuals arise from a bud of tissue rather than a single cell). Genetic unity through common origin, and differentiation, are critical qualities of multi-cellularity that may be transferable to media other than organic chemistry.

Buss [9] provides a provocative discussion of the evolution of multi-cellularity, and explores the conflicts between selection at the levels of cell lines and of individuals. From his discussion the following idea emerges (although he does not explicitly state this idea, in fact he proposes a sort of inverse of this idea, p. 65): the transition from single- to multi-celled existence involves the extension of the control of gene regulation by the mother cell to successively more generations of daughter cells.

In organic cells, genes are regulated by proteins contained in the cytoplasm. During early embryonic development in animals, an initially very large fertilized egg cell undergoes cell division with no increase in the overall size of the embryo. The large cell is simply partitioned into many smaller cells, and all components of the cytoplasm are of maternal origin. By preventing several generations of daughter cells from producing any cytoplasmic regulatory components, the mother gains control of the course of differentiation, and thereby creates the developmental process. In single-celled organisms by contrast, after each cell division, the daughter cell produces its own cytoplasmic regulatory products, and determines its own destiny independent of the mother cell.

Complex digital organisms will be self-replicating algorithms, consisting of many distinct processes dedicated to specific tasks (e.g., locating free memory, mates or other resources; defense; replicating the code). These processes must be coordinated and regulated, and may be divided among several cells specialized for specific functions. If the mother cell can influence the regulation of the processes of the daughter, so as to force the daughter cell to specialize in function and express only a portion of its full genetic potentiality, then the essence of multi-cellularity will be achieved.

13.5 Computational Implementation

The discussion above suggests that the critical feature needed to allow the evolution of multi-cellularity is for a cell to be able to influence the expression of genes by its daughter cell. In the digital context, this means that a cell must be able to influence what code is executed by its daughter cell.

If we assume that in digital organisms, as in organic ones, all cells in an individual contain the same genetic material, then the desired regulatory mechanism can be achieved most simply by allowing the mother cell to affect the context of the CPU of the daughter cell at the time that the cell is "born". Most importantly, the mother cell needs to be able to set the address of the instruction pointer of the daughter cell at birth, which will determine where the daughter cell will begin executing its code. Beyond that, additional influence can be achieved by allowing the mother cell to place values in the registers of the daughter's CPU.

A large digital genome may contain several sections of code that are "closed" in the sense that one section of code will not pass control of execution to another. Thus if execution begins in one of these sections of code, the other sections will never be expressed. This type of genetic organization, coupled with the ability of the mother cell to determine where the daughter cell begins executing, could provide a mechanism of gene regulation suitable for causing the differentiation of cells in a multi-cellular digital organism.

Other schemes for the regulation of code expression are also possible. For example, digital computers commonly have three protection states available for the memory: read, write and execute. If the code of the genome were provided with execute protection, it would provide a means of suppression of the execution of code in the protected region of the genome.

13.6 Digital "Neural Networks" – Natural Artificial Intelligence

One of the greatest challenges in the field of computer science is to produce computer systems that are "intelligent" in some way. This might involve, for example, the creation of a system for the guidance of a robot which is capable of moving freely in a complex environment, seeking, recognizing and manipulating a variety of objects. It might involve the creation of a system capable of communicating with humans in natural spoken human language, or of translating between human languages.

It has been observed that natural systems with these capabilities are controlled by nervous systems consisting of large numbers of neurons interconnected by axons and dendrites. Borrowing from nature, a great deal of work has gone into setting up "neural networks" in computers [18], [33]. In these systems, a collection of simulated "neurons" are created, and connected so that they can pass messages. The learning that takes place is accomplished by adjusting the "weights" of the connections.

Organic neurons are essentially analog devices, thus when neural networks are implemented on computers, they are digital emulations of analog devices. There is a certain inefficiency involved in emulating an analog device on a digital computer. For this reason, specialized analog hardware has been developed for the more efficient implementation of artificial neural nets [60].

Neural networks, as implemented in computers, either digital or analog, are intentional mimics of organic nervous systems. They are designed to function like natural neural networks in many details. However, natural neural networks represent the solution found by evolution to the problem of creating a control system based on organic chemistry. Evolution works with the physics and chemistry of the medium in which it is embedded.

The solution that evolution found to the problem of communication between organic cells is chemical. Cells communicate by releasing chemicals that bind to and activate receptor molecules on target cells. Working within this medium, evolution created neural nets. Inter-cellular chemical communication in neural nets is "digital" in the sense that chemical messages are either present or not present (on or off). In this sense, a single chemical message carries only a single bit of information. More detailed information can be derived from the temporal pattern of the messages, and also the context of the message. The context can include where on the target cell body the message is applied (which influences its "weight"), and what other messages are arriving at the same time, with which the message in question will be integrated.

It is hoped that evolving multi-cellular digital organisms will become very complex, and will contain some kind of control system that fills the functional role of the nervous system. While it seems likely that the digital nervous system would consist of a network of communicating "cells", it seem unlikely that this would bear much resemblance to conventional neural networks.

Compare the mechanism of inter-cellular communication in organic cells (described above) to the mechanisms of inter-process communication in computers. Processes transmit messages in the form of bit patterns, which may be of any length, and so which may contain any amount of information. Information need not be encoded into the temporal pattern of impulse trains. This fundamental difference in communication mechanisms between the digital and the organic mediums must influence the course that evolution will take as it creates information processing systems in the two mediums.

It seems highly unlikely that evolution in the digital context would produce information processing systems that would use the same forms and mechanisms as natural neural nets (e.g., weighted connections, integration of incoming messages, threshold-triggered all-or- nothing output, thousands of connections per unit). The organic medium is a physical/chemical medium, whereas the digital medium is a logical/informational medium. That observation alone would suggest that the digital medium is better suited to the construction of information processing systems.

If this is true, then it may be possible to produce digitally based systems that have functionality equivalent to natural neural networks, but which have a much greater simplicity of structure and process. Given evolution's ability to discover the possibilities inherent in a medium, and its complete lack of preconceptions, it would be very interesting to observe what kind of information processing systems evolution would construct in the digital medium. If evolution is capable of creating network-based information processing systems, it may provide us with a new paradigm for digital "connectionism" that would be more natural to the digital medium than simulations of natural neural networks.

14 Digital Husbandry

Digital organisms evolving freely by natural selection do no "useful" work. Natural evolution tends to the selfish needs of perpetuating the genes. We cannot expect digital organisms evolving in this way to perform useful work for us, such as guiding robots or interpreting human languages. In order to generate digital organisms that function as useful software, we must guide their evolution through artificial selection, just as humans breed dogs, cattle and rice. Some experiments have already been done with using artificial selection to guide the evolution of digital organisms for the performance of "useful" tasks [1], [89], [91]. I envision two approaches to the management of digital evolution: digital husbandry, and digital genetic engineering.

Digital husbandry is an analogy to animal husbandry. This technique would be used for the evolution of the most advanced and complex software, with intelligent capabilities. Correspondingly, this technique is the most fanciful. I would begin by allowing multi-cellular digital organisms to evolve freely by natural selection. Using strictly natural selection, I would attempt to engineer the system to the threshold of the computational analog of the Cambrian explosion, and let the diversity and complexity of the digital organisms spontaneously explode.

One of the goals of this exercise would be to allow evolution to find the natural forms of complex parallel digital processes. Our parallel hardware is still too new for human programmers to have found the best way to write parallel software. And it is unlikely that human programmers will ever be capable of writing software of the complexity that the hardware is capable of running. Evolution should be able to show us the way.

It is hoped that this would lead to highly complex digital organisms, which obtain and process information, presumably predominantly about other digital organisms. As the complexity of the evolving system increases, the organisms will process more complex information in more complex ways, and take more complex actions in response. These will be information processing organisms living in an informational environment.

It is hoped that evolution by natural selection alone would lead to digital organisms which, while doing no "useful" work, would none-the-less be highly sophisticated parallel information processing systems. Once this level of evolution has been achieved, then artificial selection could begin to be applied, to enhance those information processing capabilities that show promise of utility to humans. Selection for different capabilities would lead to many different breeds of digital organisms with different uses. Good examples of this kind of breeding from organic evolution are the many varieties of domestic dogs which were derived by breeding from a single species, and the vegetables cabbage, kale, broccoli, cauliflower, and brussels sprouts which were all produced by selective breeding from a single species of plant.

Digital genetic engineering would normally be used in conjunction with digital husbandry. This consists of writing a piece of application code and inserting it into the genome of an existing digital organism. A technique being used in organic genetic engineering today is to insert genes for useful proteins into goats, and to cause them to be expressed in the mammary glands. The goats then secrete large quantities of the protein into the milk, which can be easily removed from the animal. We can think of our complex digital organisms as general purpose animals, like goats, into which application codes can be inserted to add new functionalities, and then bred through artificial selection to enhance or alter the quality of the new functions.

In addition to adding new functionalities to complex digital organisms, digital genetic engineering could be used for achieving extremely high degrees of optimization in relatively small but heavily used pieces of code. In this approach, small pieces of application code could be inserted into the genomes of simple digital organisms. Then the allocation of CPU cycles to those organisms would be based on the performance of the inserted code. In this way, evolution could optimize those codes, and they could be returned to their applications. This technique would be used for codes that are very heavily used such as compiler constructs, or central components of the operating system.

15 Living Together

> "I'm glad they''re not real, because if they were, I would have to feed them and they would be all over the house."

— Isabel Ray.

Evolution is an extremely selfish process. Each evolving species does whatever it can to insure its own survival, with no regard for the well-being of other genetic groups (potentially with the exception of intelligent species). Freely evolving autonomous artificial entities should be seen as potentially dangerous to organic life, and should always be confined by some kind of containment facility, at least until their real potential is well understood. At present, evolving digital organisms exist only in virtual computers, specially designed so that their machine codes are more robust than usual to random alterations. Outside of these special virtual machines, digital organisms are merely data, and no more dangerous than the data in a database or the text file from a word processor.

Imagine, however, the problems that could arise if evolving digital organisms were to colonize the computers connected to the major networks. They could spread across the network like the infamous internet worm [2], [8], [83], [84]. When we attempt to stop them, they could evolve mechanisms to escape from our attacks. It might conceivably be very difficult to eliminate them. However, this scenario is highly unlikely, as it is probably not possible for digital organisms to evolve on normal computer systems. While the supposition remains untested, normal machine languages are probably too brittle to support digital evolution.

Evolving digital organisms will probably always be confined to special machines, either real or virtual, designed to support the evolutionary process. This does not mean, however, that they are necessarily harmless. Evolution remains a self-interested process, and even the interests of confined digital organisms may conflict with our own. For this reason it is important to restrict the kinds of peripheral devices that are available to autonomous evolving processes.

This conflict was taken to its extreme in the movie *Terminator 2*. In the imagined future of the movie, computer designers had achieved a very advanced chip design, which had allowed computers to autonomously increase their own intelligence until they became fully conscious. Unfortunately, these intelligent computers formed the “sky-net” of the United States military. When the humans realized that the computers had become intelligent, they decided to turn them off. The computers viewed this as a threat, and defended themselves by using one of their peripheral devices: nuclear weapons.

Relationships between species can, however, be harmonious. We presently share the planet with millions of freely evolving species, and they are not threatening us with destruction. On the contrary, we threaten them. In spite of the mindless and massive destruction of life being caused by human activity, the general pattern in living communities is one of a network of inter-dependencies.

More to the point, there are many species with which humans live in close relationships, and whose evolution we manage. These are the domesticated plants and animals that form the basis of our agriculture (cattle, rice), and who serve us as companions (dogs, cats, house plants). It is likely that our relationship with digital organisms will develop along the same two lines.

There will likely be carefully bred digital organisms developed by artificial selection and genetic engineering that perform intelligent data processing tasks.

These would subsequently be "neutered" so that they cannot replicate, and the eunuchs would be put to work in environments free from genetic operators. We are also likely to see freely evolving and/or partially bred digital ecosystems contained in the equivalent of digital aquariums (without dangerous peripherals) for our companionship and aesthetic enjoyment.

While this chapter has focused on digital organisms, it is hoped that the discussions will be taken in the more general context of the possibilities of any synthetic forms of life. The issues of living together become more critical for synthetic life forms implemented in hardware or wetware. Because these organisms would share the same physical space that we occupy, and possibly consume some of the same material resources, the potential for conflict is much higher than for digital organisms.

At the present, there are no self-replicating artificial organisms implemented in either hardware or wetware (with the exception of some simple organic molecules with evidently small and finite evolutionary potential [25], [36], [66]). However, there are active attempts to synthesize RNA molecules capable of replication [4], [42], and there is much discussion of the future possibility of self-replicating nano technology and macro-robots. I would strongly urge that as any of these technologies approaches the point where self-replication is possible, the work be moved to specialized containment facilities. The means of containment will have to be handled on a case-by-case basis, as each new kind of replicating technology will have its own special properties.

There are many in the AL movement who envision a beautiful future in which AL replaces organic life, and expands out into the universe [49], [50], [62], [63], [64]. The motives vary from a desire for immortality to a vision of converting virtually all matter in the universe to living matter. It is argued that this transition from organic to metallic-based life is the inevitable and natural next step in evolution.

The naturalness of this step is argued by analogy with the supposed genetic takeovers in which nucleic acids became the genetic material taking over from clays [10], and cultural evolution took over from DNA-based genetic evolution in modern humans. I would point out that whatever nucleic acids took over from, it marked the origin of life more than the passing of a torch. As for the supposed transition from genetic to cultural evolution, the truth is that genetic evolution remains intact, and has had cultural evolution layered over it rather than being replaced by it.

The supposed replacement of genetic by cultural evolution remains a vision of a brave new world, which has yet to materialize. Given the ever increasing destruction of nature, and human misery and violence being generated by human culture, I would hesitate to place my trust in the process as the creator of a bright future. I still trust in organic evolution, which created the beauty of the rain forest through billions of years of evolution. I prefer to see artificial evolution confined to the realm of cyberspace, where we can more easily coexist with it without danger, using it to enhance our lives without having to replace ourselves.

As for the expansion of life out into the universe, I am confident that this can be achieved by organic life aided by intelligent non-replicating machines. And as for immortality, our unwillingness to accept our own mortality has been a primary fuel for religions through the ages. I find it sad that AL should become an outlet for the same sentiment. I prefer to achieve immortality in the old-fashioned organic evolutionary way, through my children. I hope to die in my patch of Costa Rican rain forest, surrounded by many thousands of wet and squishy species, and leave it all to my daughter. Let them set my body out in the jungle to be recycled into the ecosystem by the scavengers and decomposers. I will live on through the rain forest I preserved, the ongoing life in the ecosystem into which my material self is recycled, the memes spawned by my scientific works, and the genes in the daughter that my wife and I created.

16 Challenges

For well over a century, evolution has remained a largely theoretical science. Now new technologies have allowed us to inoculate natural evolution into artificial media, converting evolution into an experimental and applied science, and at the same time, opening Pandora''s box. This creates a variety of challenges which have been raised or alluded to in the preceding essay, and which will be summarized here.

16.1 Respecting the Medium

If the objective is to instantiate rather than simulate life, then care must be taken in transferring ideas from natural to artificial life forms. Preconceptions derived from experience with natural life may be inappropriate in the context of the artificial medium. Getting it right is an art, which likely will take some skill and practice to develop.

However, respecting the medium is only one approach, which I happen to favor. I do not wish to imply that it is the only valid approach. It is too early to know which approach will generate the best results, and I hope that other approaches will be developed as well. I have attempted to articulate clearly this "natural" approach to synthetic life, so that those who choose to follow it may achieve greater consistency in design through a deeper understanding of the method.

16.2 Understanding Evolvability

Attempts are now underway to inoculate evolution into many artificial systems, with mixed results. Some genetic languages evolve readily, while others do not. We do not yet know why, and this is a fundamental and critically important issue. What are the elements of evolvability? Efforts are needed to directly address this issue. One approach that would likely be rewarding would be to systematically identify features of a class of languages (such as machine languages), and one by one, vary each feature, to determine how evolvability is affected by the state of each feature.

16.3 Creating Organized Sexuality

Organized sexuality is important to the evolutionary process. It is the basis of the species concept, and while remaining something of an enigma in evolutionary theory, clearly is an important facilitator of the evolutionary process. Yet this kind of sexuality still has not been implemented in a natural way in synthetic life systems. It is important to find ways of orchestrating organized sexuality in synthetic systems such as digital organisms, in a way in which it is not mandatory, and in which the organisms must carry out the process through their own actions.

16.4 Creating Multi-cellularity

In organic life, the transition from single- to multi-celled forms unleashed a phenomenal explosion of diversity and complexity. It would seem then that the transition to multi-cellular forms could generate analogous diversity and complexity in synthetic systems. In the case of digital organisms, it would also lead to the evolution of parallel processes, which could provide us with new paradigms for the design of parallel software. The creation of multi-celled digital organisms remains an important challenge.

16.5 Controlling Evolution

Humans have been controlling the evolution of other species for tens of thousands of years. This has formed the basis of agriculture, through the domestication of plants and animals. The fields of genetic algorithms [27], [35] and genetic programming [47] are based on controlling the evolution of computer programs. However, we still have very little experience with controlling the evolution of self-replicating computer programs, which is more difficult. In addition, breeding complex parallel programs is likely to bring new challenges. Developing technologies for managing the evolution of complex software will be critical for harnessing the full potential of evolution for the creation of useful software.

16.6 Living Together

If we succeed in harnessing the power of evolution to create complex synthetic organisms capable of sophisticated information processing and behavior, we will be faced with the problems of how to live harmoniously with them. Given evolution's selfish nature and capability to improve performance, there exists the potential for a conflict arising through a struggle for dominance between organic and synthetic organisms. It will be a challenge to even agree on what the most desirable outcome should be, and harder still to accomplish it. In the end the outcome is likely to emerge from the bottom-up through the interactions of the players, rather than being decided through rational deliberations.

Acknowledgements

This work was supported by grants CCR-9204339 and BIR-9300800 from the United States National Science Foundation, a grant from the Digital Equipment Corporation, and by the Santa Fe Institute, Thinking Machines Corp., IBM, and Hughes Aircraft. This work was conducted while at: School of Life & Health Sciences, University of Delaware, Newark, Delaware 19716, USA, ray@udel.edu; and Santa Fe Institute, 1660 Old Pecos Trail, Suite A, Santa Fe, New Mexico, 87501, USA, ray@santafe.edu.

References

1. Adami, Chris. 1998. Introduction to Artificial Life. New York: Springer-Verlag. Pp. 374. Adami has used the input–output facilities of the new Tierra languages to feed data to creatures, and select for responses that result from simple computations, not contained in the seed genome. Contact: chris@almach.caltech.edu
2. Anonymous. 1988. Worm invasion. Science 11-11-88: 885.
3. Barton-Davis, Paul. Unpublished. Independent implementation of the Tierra system, contact: pauld@cs.washington.edu
4. Beaudry, Amber A., and Gerald F. Joyce. 1992. Directed evolution of an RNA enzyme. Science 257: 635–641.
5. Bell, Graham. 1982. The masterpiece of nature: the evolution and genetics of sexuality. Berkeley: University of California Press.
6. Benner, Steven A., Andrew D. Ellington, and Andreas Tauer. 1989. Modern metabolism as a palimpsest of the RNA world. Proceedings of the National Academy of Sciences U.S.A. 86: 7054–7058.
7. Brooks, Rodney. Unpublished. Brooks has created his own Tierra-like system, which he calls Sierra. In his implementation, each machine instruction consists of an opcode and an operand. Successive instructions overlap, such that the operand of one instruction is interpreted as the opcode of the next instruction. Contact: brooks@ai.mit.edu
8. Burstyn, Harold L. 1990. RTM and the worm that ate internet. Harvard Magazine 92(5): 23–28.
9. Buss, Leo W. 1987. The evolution of individuality. Princeton, NJ: Princeton University Press. Pp. 203.
10. Cairn-Smith, A. G. 1985. Seven clues to the origin of life. Cambridge: Cambridge University Press.
11. Carroll, L. 1865. Through the Looking-Glass. London: Macmillan.
12. Chao, Lin, Christopher Vargas, Brian B. Spear, and Edward C. Cox. 1983. Transposable elements as mutator genes in evolution. Nature 303: 633–635.
13. Charlesworth, B. 1976. Recombination modification in a fluctuating environment, Genetics 83: 181–195.

14. Darwin, Charles. 1859. On the origin of species by means of natural selection or the preservation of favored races in the struggle for life. London: Murray.
15. Davidge, Robert. 1992. Processors as organisms. CSRP 250. School of Cognitive and Computing Sciences, University of Sussex. Presented at the ALife III Conference. Contact: robertd@cogs.susx.ac.uk
16. Davidge, Robert. 1993. Looping as a means to survival: playing Russian roulette in a harsh environment. *In*: Self organization and life: from simple rules to global complexity, Proceedings of the Second European Conference on Artificial Life. Contact: robertd@cogs.susx.ac.uk
17. Davidson, Eric H., and Roy J. Britten. 1979. Regulation of gene expression: Possible role of repetitive sequences. Science 204: 1052–1059.
18. Dayhoff, Judith. 1990. Neural network architectures. New York: Van Nostrand Reinhold. Pp. 259.
19. DeAngelis, D., and L. Gross [eds]. 1992. Individual based models and approaches in ecology. New York: Chapman and Hall.
20. de Groot, Marc. Unpublished. Primordial soup, a Tierra-like system that has the additional ability to spawn self-reproducing organisms from a sterile soup. Contact: marc@kg6kf.ampr.org, marc@toad.com, marc@remarque.berkeley.edu
21. Doolittle, W. Ford, and Carmen Sapienza. 1980. Selfish genes, the phenotype paradigm and genome evolution. Nature 284: 601–603.
22. Eigen, Manfred. 1993. Viral quasispecies. Scientific American 269(1): 32–39. July.
23. Farmer, J. D., and A. Belin. Artificial life: the coming evolution. Proceedings in celebration of Murray Gell-Mann''s 60th Birthday. Cambridge: Cambridge University Press. (Reprinted in Artificial Life II. Pp. 815–840.)
24. Feferman, Linda. 1992. Simple rules... complex behavior [video]. Santa Fe, NM: Santa Fe Institute. Contact: fefie@ibm.net
25. Feng, Qing, Tae Kyo Park, and Julius Rebek. 1992. Crossover reactions between synthetic replicators yield active and inactive recombinants. Science 256: 1179–1180.
26. Ghiselin, Michael. 1974. The economy of nature and the evolution of sex. Berkeley: University of California Press.
27. Goldberg, D. E. 1989. Genetic algorithms in search, optimization, and machine learning. Reading, MA: Addison-Wesley.
28. Gould, Steven J. 1989. Wonderful life. New York, W. W. Norton. Pp. 347.
29. Gray, James. Unpublished. Natural selection of computer programs. This may have been the first Tierra-like system, but evolving real programs on a real rather than a virtual machine, and predating Tierra itself: "I have attempted to develop ways to get computer programs to function like biological systems subject to natural selection.... I don't think my systems are models in the usual sense. The programs have really competed for resources, reproduced, run, and 'died'. The resources consisted primarily of access to the CPU and partition space.... On a PDP11 I could have a population of programs running simultaneously." Contact: Gray.James_L+@northport.va.gov
30. Green, Melvin M. 1988. Mobile DNA elements and spontaneous gene mutation. *In* M. E. Lambert, J. F. McDonald, I. B. Weinstein [eds.]: Eukaryotic transposable elements as mutagenic agents. Pp. 41–50. Banbury Report 30, Cold Spring Harbor Laboratory.
31. Halvorson, Herlyn O., and Albert Monroy. 1985. The origin and evolution of sex. New York: A. R. Liss.
32. Hapgood, Fred. 1979. Why males exist: an inquiry into the evolution of sex. New York: William Morrow.
33. Hertz, John, Anders Krogh, and Richard G. Palmer. 1991. Introduction to the theory of neural computation. Reading, MA: Addison-Wesley. Pp. 327.

34. Hogeweg, P. 1989. Mirror beyond mirror: puddles of life. *In*: Langton, C. [ed.]; Artificial Life, Santa Fe Institute Studies in the Sciences of Complexity, vol. VI, 297–316. Redwood City, CA: Addison-Wesley.
35. Holland, John Henry. 1975. Adaptation in natural and artificial systems: an introductory analysis with applications to biology, control, and artificial intelligence. Ann Arbor, University of Michigan Press.
36. Hong, J. I., Q. Feng, V. Rotello, and J. Rebek. 1992. Competition, cooperation, and mutation: Iimproving a synthetic replicator by light irradiation. Science 255: 848–850.
37. Huston, Michael. 1979. A general hypothesis of species diversity. American Naturalist 113: 81–101.
38. Huston, Michael. 1992. Biological diversity and human resources. Impact of Science on Society 166: 121–130.
39. Huston, Michael. 1993. Biological diversity: the coexistence of species on changing landscapes. Cambridge: Cambridge University Press.
40. Huston, M., D. DeAngelis, and W. Post. 1988. New computer models unify ecological theory. Bioscience 38(10): 682–691.
41. Jelinek, Warren R., and Carl W. Schmid. 1982. Repetitive sequences in eukaryotic DNA and their expression. Annual Reviews of Biochemistry 51: 813–844.
42. Joyce, Gerald F. 1992. Directed molecular evolution. Scientific American, December: 90–97.
43. Kampis, George. 1993. Coevolution in the computer: The necessity and use of distributed code systems. Printed in the ECAL93 proceedings, Brussels. Contact: gk@cfnext.physchem.chemie.uni-tuebingen.de
44. Kampis, George. 1993. Life-like computing beyond the machine metaphor. *In*: R. Paton [ed.]: Computing with biological metaphors. London: Chapman and Hall. Contact: gk@cfnext.physchem.chemie.uni-tuebingen.de
45. Kauffman, Stuart A. 1993. The origins of order, self-organization and selection in evolution. Oxford: Oxford University Press. Pp. 709.
46. Kerem, Bat-sheva, Johanna M. Rommens, Janet A. Buchanan, Danuta Markiewicz, Tara K. Cox, Aravinda Chakravarti, Manuel Buchwald, and Lap-Chee Tsui. 1989. Identification of the cystic fibrosis gene: Genetic analysis. Science 245: 1073–1080.
47. Koza, John R. 1992. Genetic programming, on the programming of computers by means of natural selection. Cambridge, MA: MIT Press.
48. Langton, C. G. 1986. Studying artificial life with cellular automata. Physica 22D: 120–149.
49. Levy, Steven. 1992. Artificial Life, the quest for a new creation. Pantheon Books, New York. Pp. 390.
50. Levy, Steven. 1992. A-Life Nightmare. Whole Earth Review #76, Fall: 22.
51. Litherland, J. 1993. Open-ended evolution in a computerised ecosystem. A Masters of Science dissertation in the Department of Computer Science, Brunel University. Contact: david.martland@brunel.ac.uk
52. MacArthur, Robert H., and Edward O. Wilson. 1967. The theory of island biogeography. Princeton, NJ: Princeton University Press. Pp. 203.
53. Maley, Carlo C. 1993. A model of early evolution in two dimensions. Masters of Science thesis, Department of Zoology, New College, Oxford University. Contact: cmaley@oxford.ac.uk
54. Manousek, Wolfgang. 1992. Spontane Komplexitaetsentstehung — TIERRA, ein Simulator fuer biologische Evolotion. Diplomarbeit, Universitaet Bonn, Germany, Oktober 1992. Contact: Kurt Stueber, stueber@vax.mpiz-koeln.mpg.d400.de

55. Margulis, Lynn, and Dorion Sagan. 1986. Origin of sex. New Haven, CT: Yale University Press.
56. Marx, Jean L. 1989. The cystic fibrosis gene is found. Science 245: 923–925.
57. Maynard Smith, J. 1971. What use is sex? Journal of Theoretical Biology 30: 319–335.
58. Maynard Smith, J. 1992. Byte-sized evolution. Nature 355: 772–773.
59. Maynard Smith, J, Christopher G. Dowson, and Brian G. Spratt. 1991. Localized sex in bacteria. Nature 349: 29–31.
60. Mead, Carver. 1993. Analog VLSI and neural systems. Reading, MA: Addison-Wesley. Pp. 371.
61. Michod, Richard E., and Bruce R. Levin. 1988. The evolution of sex: An examination of current ideas. Sunderland, MA: Sinauer Associates.
62. Moravec, Hans. 1988. Mind Children: the future of robot and human intelligence. Cambridge, MA: Harvard University Press.
63. Moravec, Hans. 1989. Human culture: A genetic takeover underway. *In*: Langton, C. [ed.]: Artificial Life, Santa Fe Institute Studies in the Sciences of Complexity, vol. VI, 167–199. Redwood City, CA: Addison-Wesley.
64. Moravec, Hans. 1993. Pigs in cyberspace. Extropy #10, Winter/Spring.
65. Morris, S. Conway. 1989. Burgess shale faunas and the Cambrian explosion. Science 246: 339–346.
66. Nowick, J., Q. Feng, T. Tijivikua, P. Ballester. and J. Rebek. 1991. Journal of the American Chemical Society 113: 8831–8839.
67. Orgel, L. E., and F. H. C. Crick. 1980. Selfish DNA: The ultimate parasite. Nature 284: 604–607.
68. Rasmussen, Steen, Carsten Knudsen, Rasmus Feldberg, and Morten Hindsholm. 1990. The coreworld: Emergence and evolution of cooperative structures in a computational chemistry. Physica D 42: 111–134.
69. Rasmussen, S., C. Knudsen, and R. Feldberg. 1991. Dynamics of programmable matter. *In*: Langton, C., C. Taylor, J. D. Farmer, and S. Rasmussen [eds.]: Artificial Life II, Santa Fe Institute Studies in the Sciences of Complexity, vol. X, 211–254. Redwood City, CA: Addison-Wesley.
70. Ray, T. S. 1979. Slow-motion world of plant 'behavior' visible in rainforest. Smithsonian 9(12): 121–30.
71. Ray, T. S. 1991. An approach to the synthesis of life. *In*: Langton, C., C. Taylor, J. D. Farmer, and S. Rasmussen [eds.: Artificial Life II, Santa Fe Institute Studies in the Sciences of Complexity, vol. X, 371–408. Redwood City, CA: Addison-Wesley.
72. Ray, T. S. 1991. Population dynamics of digital organisms. *In*: Langton, C. G. [ed.]: Artificial Life II Video Proceedings. Redwood City, CA: Addison Wesley.
73. Ray, T. S. 1991. Is it alive, or is it GA? *In*: Belew, R. K., and L. B. Booker [eds.], Proceedings of the 1991 International Conference on Genetic Algorithms, 527–534. San Mateo, CA: Morgan Kaufmann.
74. Ray, T. S. 1991. Evolution and optimization of digital organisms. *In*: Billingsley K. R., E. Derohanes, and H. Brown, III [eds.]: Scientific Excellence in Supercomputing: The IBM 1990 Contest Prize Papers, Athens, GA: The Baldwin Press, The University of Georgia.
75. Ray, T. S. 1992. Foraging behaviour in tropical herbaceous climbers (Araceae). Journal of Ecology. 80: 189–203.
76. Ray, T. S. 1992. Tierra.doc. Documentation for the Tierra Simulator V4.0, 9–9–92. Newark, DE: Virtual Life. The full source code and documentation for the Tierra program is available by anonymous ftp at: tierra.slhs.udel.edu [128.175.41.34] and life.slhs.udel.edu [128.175.41.33], or by contacting the author.

77. Ray, T. S. 1994. Evolution and complexity. *In*: Cowan, George A., David Pines, and David Metzger [eds.]: Complexity: Metaphors, Models, and Reality, Pp. 161–173. Reading, MA: Addison-Wesley.
78. Ray, T. S. 1994. Evolution, complexity, entropy, and artificial reality. Physica D 75: 239–263.
79. Riordan, John R., Johanna M. Rommens, Bat-sheva Kerem, Noa Alon, Richard Rozmahel, Zbyszko Grzelczak, Julian Zielenski, Si Lok, Natasa Plavsic, Jia-Ling Chou, Mitchell L. Drumm, Michael C. Lannuzzi, Francis S Collins, and Lap-Chee Tsui. 1989. Identification of the cystic fibrosis gene: Cloning and characterization of complementary DNA. Science 245: 1066–1073.
80. Rommens, Johanna M., Michael C. Iannuzzi, Bat-sheva Kerem, Mitchell L. Drumm, Georg Melmer, Michhael Dean, Richard Rozmahel, Jeffery L. Cole, Dara Kennedy, Noriko Hidaka, Martha Zsiga, Manuel Buchwald, John R. Riordan, Lap-Chee Tsui, and Francis S. Collins. 1989. Identification of the cystic fibrosis gene: Chromosome walking and jumping. Science 245: 1059–1065.
81. Skipper, Jakob. 1992. The computer zoo – evolution in a box. *In*: Francisco J. Varela and Paul Bourgine [eds.]: Toward a practice of autonomous systems, proceedings of the First European Conference on Artificial Life. Cambridge, MA: MIT Press. Pp. 355–364. Contact: Jakob.Skipper@copenhagen.ncr.com
82. Sober, E. 1984. The nature of selection. Cambridge, MA: MIT Press.
83. Spafford, Eugene H. 1989. The internet worm program: An analysis. Computer Communication Review 19(1): 17–57. Also issued as Purdue CS technical report TR-CSD-823. Contact: spaf@purdue.edu
84. Spafford, Eugene H. 1989. The internet worm: crisis and aftermath. Communications of the ACM 32(6): 678–687. Contact: spaf@purdue.edu
85. Stanley, Steven M. 1973. An ecological theory for the sudden origin of multicellular life in the late Precambrian, Proceedings of the National Academy of Sciences U.S.A. 70: 1486–1489.
86. Stearns, Steven C. 1987. The evolution of sex and its consequences. Boston: Birkhäuser Verlag.
87. Strickberger, Monroe W. 1985. Genetics. New York: Macmillan.
88. Strong, D. R. and T. S. Ray. 1975. Host tree location behavior of a tropical vine (*Monstera gigantea*) by skototropism. Science, 190: 804–06.
89. Surkan, Al. Unpublished. Self-balancing of dynamic population sectors that consume energy. Department of Computer Science, UNL. "Tierra-like systems are being explored for their potential applications in solving the problem of predicting the dynamics of consumption of a single energy carrying natural resource." Contact: surkan@cse.unl.edu
90. Syvanen, Michael. 1984. The evolutionary implications of mobile genetic elements. Annual Review of Genetics 18: 271–293.
91. Tackett, Walter, and Jean-Luc Gaudiot. 1993. Adaptation of self-replicating digital organisms. Proceedings of the International Joint Conference on Neural Networks, Nov. 1993, Beijing, China. Piscataway, NJ: IEEE Press. Contact: tackett@ipld01.hac.com, tackett@priam.usc.edu
92. Taylor, Charles E., David R. Jefferson, Scott R. Turner, and Seth R. Goldman. 1989. RAM: Artificial life for the exploration of complex biological systems. *In*: Langton, C. [ed.: Artificial Life, Santa Fe Institute Studies in the Sciences of Complexity, vol. VI, 275–295. Redwood City, CA: Addison-Wesley.
93. Thomas, C. A. 1971. The genetic organization of chromosomes. Annual Review of Genetics 5: 237–256.

94. Todd, Peter M. 1993. Artificial death. Proceedings of the Second European Conference on Artificial Life (ECAL93), vol. 2, Pp. 1048–1059. Brussels, Belgium: Université Libre de Bruxelles. Contact: ptodd@spo.rowland.org
95. Van Valen, L. 1973. A new evolutionary law. Evolutionary Theory 1: 1–30.
96. Williams, George C. 1975. Sex and evolution. Princeton, NJ: Princeton University Press.

Scatter Search

Fred Glover[1], Manuel Laguna[1] and Rafael Martí[2]

[1]Leeds School of Business, University of Colorado, Boulder, CO 80309-0419, USA
E-mail: {Fred.Glover}{Manuel.Laguna}@Colorado.edu
[2]Departamento de Estadística e Investigación Operativa, Facultad de Matemáticas, Universidad de Valencia, Dr. Moliner 50, 46100 Burjassot (Valencia) Spain
E-mail: Rafael.Marti@uv.es

Summary. The evolutionary approach called scatter search originated from strategies for creating composite decision rules and surrogate constraints. Recent studies demonstrate the practical advantages of this approach for solving a diverse array of optimisation problems from both classical and real—world settings. Scatter search contrasts with other evolutionary procedures, such as genetic algorithms, by providing unifying principles for joining solutions based on generalised path constructions in Euclidean space and by utilising strategic designs where other approaches resort to randomisation. Additional advantages are provided by intensification and diversification mechanisms that exploit adaptive memory, drawing on foundations that link scatter search to tabu search. The main goal of this chapter is to demonstrate the development of a scatter search procedure by demonstrating how it may be applied to a class of non-linear optimisation problems on bounded variables. We conclude the chapter by highlighting key ideas and research issues that offer the promise of yielding future advances.

1 Introduction

Scatter search derives its foundations from earlier strategies for combining decision rules and constraints, with the goal of enabling a solution procedure based on the combined elements to yield better solutions than one based only on the original elements. An examination of these origins sheds light on the character of these methods.

Historically, the antecedent strategies for combining decision rules were introduced in the context of scheduling methods to obtain improved local decision rules for job shop scheduling problems [1]. New rules were generated by creating numerically weighted combinations of existing rules, suitably restructured so that their evaluations embodied a common metric.

The approach was motivated by the supposition that information about the relative desirability of alternative choices is captured in different forms by different rules, and that this information can be exploited more effectively when integrated by means of a combination mechanism than when treated by the standard strategy of selecting different rules one at a time, in isolation from each other. In addition, the method departed from the customary approach of stopping upon reaching a local optimum, and instead continued to vary the parameters that determined the combined rules, as a basis for producing additional trial solutions. (This latter strategy also became a fundamental component of tabu search. See e.g. [2].)

The decision rules created from such combination strategies produced better empirical outcomes than standard applications of local decision rules, and also proved superior to a "probabilistic learning approach" that selected different rules probabilistically at different junctures, but without the integration effect provided by generating combined rules [3].

The associated procedures for combining constraints likewise employed a mechanism of generating weighted combinations, in this case applied in the setting of integer and non-linear programming, by introducing non-negative weights to create new constraint inequalities, called *surrogate constraints* [4]. The approach isolated subsets of constraints that were gauged to be most critical, relative to trial solutions based on the surrogate constraints, and produced new weights that reflected the degree to which the component constraints were satisfied or violated.

A principal function of surrogate constraints, in common with the approaches for combining decision rules, was to provide ways to evaluate choices that could be used to generate and modify trial solutions. From this foundation, a variety of heuristic processes evolved that made use of surrogate constraints and their evaluations. Accordingly, these processes led to the complementary strategy of combining solutions, as a *primal* counterpart to the *dual* strategy of combining constraints, which became manifest in scatter search and its path relinking generalisation. The *primal/dual* distinction stems from the fact that surrogate constraint methods give rise to a mathematical duality theory associated with their role as relaxation methods for optimisation. For example, see [4]–[11].

Scatter search operates on a set of solutions, the *reference set*, by combining these solutions to create new ones. When the main mechanism for combining solutions is such that a new solution is created from the linear combination of two other solutions, the reference set may evolve as illustrated in Figure 1. This figure assumes that the original reference set of solutions consists of the circles labelled A, B and C. After a non-convex combination of reference solutions A and B, solution 1 is created. In fact, a number of solutions in the line segment defined by A and B are created; however, only solution 1 is introduced in the reference set. (The criteria used to select solutions for membership in the reference set are discussed later.) In a similar way, convex and non-convex combinations of original and newly created reference solutions create points 2, 3 and 4. The resulting complete reference set shown in Figure 1 consists of seven solutions (or elements).

More precisely, Figure 1 shows a precursor form of the resulting reference set. Scatter search does not leave solutions in a raw form produced by its combination mechanism, but subjects candidates for entry into the reference set to heuristic improvement, as we elaborate subsequently.

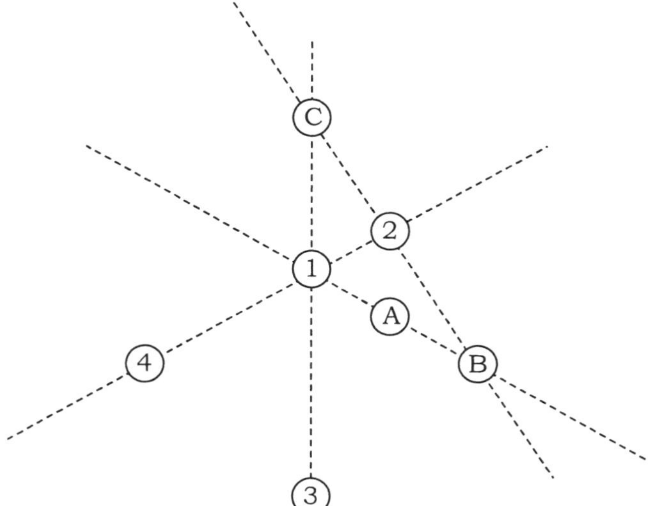

Fig. 1. Two-dimensional reference set

Unlike a "population" in genetic algorithms, the reference set of solutions in scatter search is relatively small. In genetic algorithms, two solutions are chosen from the population by a process that relies heavily on randomization and a "crossover" mechanism, likewise involving recourse to randomization, is applied to generate one or more offspring. A typical population size in a genetic algorithm consists of 100 elements. In contrast, scatter search chooses two or more elements of the reference set in a systematic way with the purpose of creating new solutions strategically. (Randomization can be invoked, but in a way that is subordinate to this strategic emphasis.) Since the number of two-element to five-element subsets of a reference set, for example, can be quite large, even a highly selective process for isolating preferred instances of these subsets as a basis for generating combined solutions can produce a significant number of combinations, and so there is a practical need for keeping the cardinality of the set small. Typically, the reference set in scatter search has 20 solutions or less. In one standard design, if the reference set consists of b solutions, the procedure examines approximately $(3b–7)b/2$ combinations of four different types [12]. The basic type consists of combining two solutions; the next type combines three solutions, and so on.

Limiting the scope of the search to a selective group of combination types can be used as a mechanism for controlling the number of possible combinations in a given reference set. An effective means for doing this is to subdivide the reference set into "tiers" and to require that combined solutions must be based on including at least one (or a specified number) of the elements from selected tiers. A two-tier example, which subdivides the reference set into two components, is illustrated in section 3.3.

2 Scatter Search Template

The scatter search process, building on the principles that underlie the surrogate constraint design, is organised to (1) capture information not contained separately in the original vectors, (2) take advantage of auxiliary heuristic solution methods to evaluate the combinations produced and to generate new vectors [12]. Specifically, the scatter search approach may be sketched as follows:

1. Generate a starting set of solution vectors to guarantee a critical level of diversity and apply heuristic processes designed for the problem as an attempt for improving these solutions. Designate a subset of the best vectors to be reference solutions. (Subsequent iterations of this step, transferring from Step 4 below, incorporate advanced starting solutions and best solutions from previous history as candidates for the reference solutions.) The notion of "best" in this step is not limited to a measure given exclusively by the evaluation of the objective function. In particular, a solution may be added to the reference set if the diversity of the set improves even when the objective value of the solution is inferior to other solutions competing for admission into the reference set.
2. Create new solutions consisting of structured combinations of subsets of the current reference solutions. The structured combinations are:
   a) chosen to produce points both inside and outside the convex regions spanned by the reference solutions.
   b) modified to yield acceptable solutions. (For example, if a solution is obtained by a linear combination of two or more solutions, a generalised rounding process that yields integer values for integer-constrained vector components may be applied. An acceptable solution may or may not be feasible with respect to other constraints in the problem.)
3. Apply the heuristic processes used in Step 1 to improve the solutions created in Step 2. These heuristic processes must be able to operate on infeasible solutions and may or may not yield feasible solutions.
4. Extract a collection of the "best" improved solutions from Step 3 and add them to the reference set. The notion of "best" is once again broad; making the objective value one among several criteria for evaluating the merit of newly created points. Repeat Steps 2, 3 and 4 until the reference set does not change. Diversify the reference set, by re-starting from Step 1. Stop when reaching a specified iteration limit.

The first notable feature in scatter search is that its structured combinations are designed with the goal of creating weighted centres of selected subregions. This adds non-convex combinations that project new centres into regions that are external to the original reference solutions (see e.g. solution 3 in Figure 1). The dispersion patterns created by such centres and their external projections have been found useful in several application areas.

Another important feature relates to the strategies for selecting particular subsets of solutions to combine in Step 2. These strategies are typically designed to make use of a type of clustering to allow new solutions to be constructed "within clusters" and "across clusters". Finally, the method is organised to use ancillary improving mechanisms that are able to operate on infeasible solutions, removing the restriction that solutions must be feasible in order to be included in the reference set.

The following principles summarise the foundations of the scatter search methodology:

- Useful information about the form (or location) of optimal solutions is typically contained in a suitably diverse collection of elite solutions.
- When solutions are combined as a strategy for exploiting such information, it is important to provide mechanisms capable of constructing combinations that extrapolate beyond the regions spanned by the solutions considered. Similarly, it is also important to incorporate heuristic processes to map combined solutions into new solutions. The purpose of these combination mechanisms is to incorporate both diversity and quality.
- Taking account of multiple solutions simultaneously, as a foundation for creating combinations, enhances the opportunity to exploit information contained in the union of elite solutions.

The fact that the mechanisms within scatter search are not restricted to a single uniform design allows the exploration of strategic possibilities that may prove effective in a particular implementation. These observations and principles lead to the following template for implementing scatter search.

1. A *Diversification Generation Method* to generate a collection of diverse trial solutions, using an arbitrary trial solution (or seed solution) as an input.
2. An *Improvement Method* to transform a trial solution into one or more enhanced trial solutions. (Neither the input nor the output solutions are required to be feasible, though the output solutions will more usually be expected to be so. If no improvement of the input trial solution results, the "enhanced" solution is considered to be the same as the input solution.)
3. A *Reference Set Update Method* to build and maintain a *reference set* consisting of the b "best" solutions found (where the value of b is typically small, e.g. no more than 20), organised to provide efficient accessing by other parts of the method. Solutions gain membership to the reference set according to their quality or their diversity.
4. A *Subset Generation Method* to operate on the reference set, to produce a subset of its solutions as a basis for creating combined solutions.
5. A *Solution Combination Method* to transform a given subset of solutions produced by the Subset Generation Method into one or more combined solution vectors.

In the next section, we employ this template to illustrate the design of a scatter search procedure for unconstrained non-linear optimisation problems. The success of scatter search and related strategies is evident in a variety of application areas such as vehicle routing, arc routing, quadratic assignment, financial product design, neural network training, job shop scheduling, flow shop scheduling, crew scheduling, graph drawing, linear ordering, unconstrained optimisation, bit representation, multi-objective assignment, optimising simulation, tree problems, mixed integer programming, as reported in [12].

3 Scatter Search Tutorial

In this tutorial section we develop a scatter search procedure for the following class of optimisation problems:

Min $f(x)$
Subject to $l \leq x \leq u$

where $f(x)$ is a non-linear function of x. For the purpose of illustrating the solution procedure, we will apply our design to the following problem instance:

$$\text{Minimize} \quad 100\left(x_2 - x_1^2\right)^2 + \left(1 - x_1\right)^2 + 90\left(x_4 - x_3^2\right)^2 + \left(1 - x_3\right)^2 + 10.1\left(\left(x_2 - 1\right)^2 + \left(x_4 - 1\right)^2\right) + 19.8\left(x_2 - 1\right)\left(x_4 - 1\right)$$

$$\text{Subject to} \quad -10 \leq x_i \leq 10 \quad \text{for } i = 1, \ldots, 4$$

This problem is "Test Case #6" in [13]. The best solution found in [13] has an objective function value of 0.001333, reportedly found after 500 000 runs of a genetic algorithm (GA).[1]

3.1 Diversification Generation Method

Our illustrative diversification method employs controlled randomisation (the emphasis on "controlled" is important) drawing upon frequency memory to generate a set of diverse solutions. We accomplish this by dividing the range of each variable $u_i - l_i$ into 4 subranges of equal size. Then, a solution is constructed in two steps. First a subrange is randomly selected. The probability of selecting a sub-range is inversely proportional to its frequency count. Then a value is randomly generated within the selected subrange. The number of times sub-range j has been chosen to generate a value for variable i is accumulated in $freq(i, j)$.

The diversification generator is used at the beginning of the search to generate a set of P solutions with $PSize$, the cardinality of the set, generally set at max(100, 5*b) diverse solutions, where b is the size of the reference set. Although the actual

[1] A single run of the GA in [13] consists of 1000 iterations. An iteration requires 70 evaluations of the objective function, where 70 is the size of the population. Therefore, theGA required 35 billion objective function evaluations to find this solution.

scatter search implementation would generate 100 solutions for this example, we have limited P in our tutorial to the 10 solutions shown in Table 1.

Table 1. Diverse solutions

Solution	x_1	x_2	x_3	X_4	$f(x)$
1	1.1082	0.8513	9.4849	–6.3510	835534.3
2	–9.5759	–6.5706	–8.8128	–2.2674	1542087.0
3	–8.3565	0.7865	7.8762	–2.6978	854129.3
4	8.8337	–8.4503	4.5242	3.1800	775473.7
5	–6.2316	7.4765	5.9955	7.8018	171451.1
6	0.1975	–3.6392	–5.2999	–7.0332	114021.1
7	–3.0909	6.6189	–2.3250	–3.1240	7468.8
8	–6.0775	0.6699	–6.4774	1.4775	279100.3
9	–1.9659	8.1258	–5.6343	8.0178	54538.5
10	3.1131	–1.9358	5.8964	6.8859	83607.1

The 100 solutions generated by the actual procedure are diverse with respect to the values that each variable takes in each of the subranges. Note, however, that the generation is done without considering the objective function. In other words, the Diversification Generation Method focuses on diversification and not on the quality of the resulting solutions, as evident from the objective function values in Table 1. In fact, the objective function values for the entire set of 100 solutions range from 1689.7 to 1 542 087 with an average of 392 032.8. These objective function values are very large considering that the optimal solution to this problem has an objective function value of zero. To verify that the Generation method is operating as expected, we show in Table 2 the frequency values corresponding to the complete set of 100 solutions.

Table 2. Frequency counts

Range	x_1	x_2	x_3	x_4
–10 to 5	19	25	26	29
–5 to 0	25	18	22	21
0 to 5	26	27	29	25
5 to 10	30	30	23	25

Note that the frequency counts in Table 2 are very similar for each range and variable, with a minimum value of 18 and a maximum value of 30. The target frequency value for each range is 25 for a set of 100 solutions. This value is observed four times in Table 2.

3.2 Improvement Method

Since we are developing a solution procedure for a class of unconstrained optimisation problems, the solutions constructed with the Diversification Generation Method are guaranteed to be feasible. The Improvement Method, however, must be able to handle starting solutions that are either feasible or infeasible.

The Improvement Method we use in this tutorial consists of a classical local optimiser for unconstrained non-linear optimisation problems. In particular, we apply Nelder and Mead's simplex method [14] to each solution in Table 1. After the application of this local optimiser, the solutions in Table 3 are transformed to the solutions in Table 2.

Table 3. Improved solutions

Solution	x_1	x_2	x_3	X_4	$f(x)$
1	–0.8833	0.7514	1.1488	1.3231	3.7411
2	1.3952	1.9411	0.3424	0.0977	0.9809
3	0.0921	0.0668	–1.3826	1.8998	7.2004
4	0.6172	0.3937	–1.3036	1.641	5.9422
5	–1.1008	1.2423	0.9054	0.7429	5.0741
6	2.3414	5.9238	0.5721	–0.5716	210.8007
7	–1.2979	1.6781	–0.6035	0.2775	8.7473
8	–2.2507	4.8302	–1.8987	3.2877	408.1172
9	–2.2492	4.1608	–2.9055	8.0508	1164.5050
10	0.8406	0.6755	–1.0425	1.0697	4.9854

The objective function values now range between 0.9809 and 1164.5050 for the improved solutions in Table 3. It is interesting to point out that when the 100 solutions are subjected to the Improvement Method, five solutions converge to the same local minimum, thus effectively reducing the cardinality of *P*. As shown in Section 4 we could apply the Diversification Generation Method again until *PSize* different solutions are found after executing the Improvement Method. Incidentally, the convergence point for those five solutions turns out to be the global optimum that sets $x = (1, 1, 1, 1)$ for an objective function value of zero. Hence for this small example, a single application of the Generation Diversification and Improvement Methods is sufficient to find the optimal solution.

3.3 Reference Set Update Method

The reference set, *RefSet*, is a collection of both high-quality solutions and diverse solutions that are used to generate new solutions by way of applying the Solution Combination Method. Specifically, the reference set consists of the union of two subsets, $RefSet_1$ and $RefSet_2$, of size b_1 and b_2, respectively. That is,

$|RefSet| = b = b_1 + b_2$. The construction of the initial reference set starts with the selection of the best b_1 solutions from P. These solutions are added to *RefSet* and deleted from P.

For each improved solution in *P-RefSet*, the minimum of the Euclidean distances to the solutions in *RefSet* is computed. Then, the solution with the maximum of these minimum distances is selected. This solution is added to *RefSet* and deleted from P and the minimum distances are updated. This process is repeated b_2 times. The resulting reference set has b_1 high-quality solutions and b_2 diverse solutions.

Let us apply this initialisation procedure to our example, using the following parameter values, $b = 5$, $b_1 = 3$ and $b_2 = 2$, and considering that P is limited to the set of improved solutions in Table 3. The best three solutions in Table 3 are shown in Table 4.

Table 4. High-quality subset of *RefSet*

Solution	x_1	x_2	x_3	x_4	$f(x)$
2	1.3952	1.9411	0.3424	0.0977	0.9809
1	–0.8833	0.7514	1.1488	1.3231	3.7411
10	0.8406	0.6755	–1.0425	1.0697	4.9854

We then calculate the minimum distance $d_{\min}(x)$ between each solution x in *P-RefSet* and the solutions y currently in *RefSet*. That is,

$$d_{\min}(x) = \underset{y \in RefSet}{Min} \{d(x,y)\}$$

where $d(x,y)$ is the Euclidean distance between x and y. For example, the minimum distance between solution 3 in Table 3 (x^3) and the *RefSet* solutions in Table 4 (x^2, x^1 and x^{10}) is calculated as follows:

$$d_{\min}(x^3) = Min\{d(x^3,x^2), d(x^3,x^1), d(x^3,x^{10})\}$$
$$d_{\min}(x^3) = Min\{3.38, 2.86, 1.32\} = 1.32$$

The maximum $d_{\min}$ value for the solutions in *P-RefSet* corresponds to solution 9 in Table 3 ($d_{\min}(x^9) = 8.6$). We add this solution to *RefSet*, delete it from P and update the $d_{\min}$ values. The new maximum $d_{\min}$ value of 4.91 corresponds to solution 8 in Table 3, so the diverse subset of *RefSet* is as shown in Table 5.

Table 5. Diverse subset of *RefSet*

Solution	x_1	x_2	x_3	x_4	$f(x)$
9	–2.2492	4.1608	–2.9055	8.0508	1164.5
8	–2.2507	4.8302	–1.8987	3.2877	408.2

After the initial reference set is constructed, the Solution Combination Method is applied to the subsets generated as outlined in the following section. The reference set is dynamically updated during the application of the Solution Combination Method. A newly generated solution may become a member of the reference set if either one of the following conditions is satisfied:

- The new solution has a better objective function value than the solution with the worst objective value in $RefSet_1$.
- The new solution has a better $d_{\min}$ value than the solution with the worst $d_{\min}$ value in $RefSet_2$.

In both cases, the new solution replaces the worst and the ranking is updated to identify the new worst solution in terms of either quality or diversity. The reference set is also regenerated when the solution Combination Method is incapable of creating solutions that can be admitted to *RefSet* according to the rules outlined above. The regeneration consists of keeping $RefSet_1$ intact and using the Diversification Generation Method to construct a new diverse subset $RefSet_2$.

3.4 Subset Generation Method

This method consists of generating the subsets that will be used for creating new solutions with the Solution Combination Method. The Subset Generation Method is typically designed to generate the following types of subsets:

1. All two-element subsets.
2. Three-element subsets derived from the two-element subsets by augmenting each 2-element subset to include the best solution (as measured by the objective value) not in this subset.
3. Four-element subsets derived from the three-element subsets by augmenting each 3-element subset to include the best solution (as measured by the objective value) not in this subset.
4. The subsets consisting of the best i elements (as measured by the objective value), for $i = 5$ to b.

Note that due to the rules for generating subsets of type 2, 3 and 4, the same subset may be generated more than once. Simple and efficient procedures outlined in [12] can be used to generate all the unique subsets of each type.

For the purpose of our tutorial, we limit our scope to type-1 subsets consisting of all pairwise combinations of the solutions in *RefSet*. There are $(b^2–b)/2$ type-1 subsets, which in the case of our example amounts to a total of $(5^2–5)/2 = 10$.

3.5 Solution Combination Method

This method uses the subsets generated with the Subset Generation Method to combine the elements in each subset with the purpose of creating new trial solutions. Generally, the Solution Combination Method is a problem-specific mechanism, since it is directly related to the solution representation. Depending on the specific form of the Solution Combination Method, each subset can create one or more new solutions. Let us consider the following Combination Method for solutions that can be represented by bounded continuous variables.

The method consists of finding linear combinations of reference solutions. The number of solutions created from the linear combination of two reference solutions depends on the membership of the solutions being combined. These combinations in our illustration are based on the following three types, assuming that the reference solutions are x' and x'':

C1: $x = x' - d$
C2: $x = x' + d$
C3: $x = x'' + d$

where $d = r(x'' - x')/2$ and r is a random number in the range (0, 1). The following rules are used to generate solutions with these three types of linear combinations:

- If both x' and x'' are elements of $RefSet_1$, then generate four solutions by applying C1 and C3 once and C2 twice.
- If only one of x' and x'' is a member of $RefSet_1$, then generate three solutions by applying C1, C2 and C3 once.
- If neither x' nor x'' is a member of $RefSet_1$, then generate two solutions by applying C2 once and randomly choosing between applying C1 or C3.

To illustrate the combination mechanism, consider the combination of the first two solutions in $RefSet_1$ (i.e. solutions 2 and 1 in Table 4). Table 6 shows the solutions generated from combining these two high-quality reference points.

Table 6. New solutions from combining x^1 and x^2

Type	r	x_1	x_2	x_3	x_4	$f(x)$
C1	0.1330	1.5468	2.0202	0.2888	0.0162	15.480
C2(a)	0.3822	0.9598	1.7138	0.4965	0.3319	63.937
C2(b)	0.6862	0.6134	1.5329	0.6191	0.5181	135.83
C3	0.3551	−1.2879	0.5401	1.2920	1.5407	132.09

Solutions generated with the Solution Combination Method are subjected to the Improvement Method before they are considered for membership in the reference set. After applying the Improvement Method to the solutions in Table 6, the solutions are transformed to those shown in Table 7.

Table 7. Improved new solutions

Type	x_1	x_2	x_3	x_4	$f(x)$
C1	1.2565	1.5732	0.6746	0.4464	0.3125
C2(a)	–0.4779	0.2308	–1.2065	1.4939	8.1029
C2(b)	1.0354	1.0746	0.9625	0.9266	0.0055
C3	0.8506	0.7171	1.1408	1.3101	0.0956

The best solution in Table 7 is C2(b) with an objective function value of 0.0055. According to our updating rules for the reference set, this solution should replace solution 10 in Table 4, because solution 10 is the worst in the $RefSet_1$ subset.

The search continues in a loop that consists of applying the Solution Combination Method followed by the Improvement Method and the Reference Update Method. This loop terminates when the reference set does not change and all the subsets have already been subjected to the Solution Combination Method. At this point, the Diversification Generation Method is used to construct a new $RefSet_2$ and the search continues.[2]

4 An Outline of the Procedure

In the previous section, we illustrated the operations that take place within each of the methods in the scatter search framework. We now finish our introduction to scatter search with an overall view of the procedure. This outline (or pseudo-code) uses the following parameters:

PSize = the size of the set of diverse solutions generated by the Diversification Generation Method.
b = the size of the reference set.
b1 = the size of the high-quality subset.
b2 = the size of the diverse subset.
MaxIter = maximum number of iterations.

The procedure consists of the steps in the outline of Figure 2, where *P* denotes the set of solutions generated with the Diversification Generation Method and *RefSet* is the set of solution in the reference set. The procedure starts with the generation of *PSize* distinct solutions. These solutions are originally generated to be diverse and subsequently improved by the application of the Improvement Method (Step 1). The set *P* of *PSize* solutions is ordered in Step 2, in order to facilitate the task of creating the reference set in Step 3. The reference set (*RefSet*) is constructed with the first b_1 solutions in *P* and b_2 solutions that are diverse with respect to the members in *RefSet*.

[2] A copy of an experimental C code of the procedure described in this tutorial section can be obtained from the authors.

The search consists of three main loops: (1) a "for-loop" that controls the maximum number of iterations, (2) a "while-loop" that monitors the presence of new elements in the reference set, and (3) a "for-loop" that controls the examination of all the subsets with at least one new element. In Step 4, the number of subsets with at least one new element is counted and this value is assigned to *MaxSubset*. Also, the Boolean variable *NewElements* is made FALSE before the subsets are examined, since it is not known whether a new solution will enter the reference set in the current examination of the subsets. The actual generation of the subsets occurs in Step 5. Note that only subsets with at least one new element are generated in this step. A solution is generated in Step 6 by applying the application of the Solution Combination Method. Step 7 attempts to improve this solution with the application of the Improvement Method. If the improved solution from Step 7 is better (in terms of the objective function value) than the worst solution in $RefSet_1$, then the improved solution becomes a new element of *RefSet*. As previously indicated, $RefSet_1$ is the subset of the reference set that contains the best solutions as measured by the objective function value. The solution is added in Step 8 and the *NewElements* indicator is switched to TRUE in Step 9.

If a solution is not admitted to the *RefSet* due to its quality, the solution is tested for its diversity merits. If a solution adds diversity to $RefSet_2$, then the solution is added to the reference set and the less diverse solution is deleted (as indicated in Steps 10 and 11). Finally, Step 12 is performed if additional iterations are still available. This step provides a seed for set *P* by adding the solutions in $RefSet_1$ before a new application of the Diversification Generation Method.

The general procedure outlined in Figure 2 can be modified in a variety of ways. One possibility is to eliminate the outer "for-loop" along with Step 12. In this case, set *P* is generated one time only and the process stops after no new elements are admitted into the reference set. That is, the search is abandoned when the "while-loop" that contains Steps 4 to 11 becomes false. This variation is useful for problems in which the search has to be performed within a relatively small amount of computer time. Also, there are some settings in which a large percentage of the best solutions are found during the first iteration (i.e. when $Iter = 1$). In this case, bypassing additional iterations has a small effect on the average performance of the procedure (as measured by the quality of the best solution found).

A second variation consists of eliminating Steps 10 and 11 and the if-statement associated with these steps. This variation considers that the reference set will be initially constructed with both high-quality solutions and diverse solutions. However after Step 3, membership to the reference set is obtained only due to quality. Therefore, in addition to eliminating Steps 10 and 11, Step 8 and its associated if-statement are modified as follows:

If (x_s^* is not in *RefSet* and the objective function value of x_s^* is better than the objective function value of the worst element in *RefSet*) **then**

8. Add x_s^* to *RefSet* and delete the worst element currently in *RefSet*. (The worst element is the solution with worst objective value.)

That is, all references to $RefSet_1$ are substituted with references to *RefSet*. In other words, after Step 3, the elements of *RefSet* are always ordered according to the objective function value, making element 1 the best and element b the worst.

Implementing both of these variations at the same time results in a very aggressive search method that attempts to find high-quality solutions fast. While this may be desirable in some settings, there are also settings in which a more extensive search can be afforded, making the full procedure outlined in Figure 2 more attractive. A reasonable approach could be to begin with the more aggressive variant and then to shift to the more through variant (or alternate between these variants).

1. Start with $P = \emptyset$. Use the Diversification Generation Method to construct a solution x. Apply the Improvement Method to x to obtain the improved solution x^*. If $x^* \notin P$ then add x^* to P (i.e. $P = P \cup x^*$), otherwise discard x^*. Repeat this step until $|P| = PSize$.
2. Order the solutions in P according to their objective function value (where the best overall solution is first on the list).

For (*Iter* = 1 **to** *MaxIter*)

3. Build $RefSet = RefSet_1 \cup RefSet_2$ from P, with $|RefSet| = b$, $|RefSet_1| = b_1$ and $|RefSet_2| = b_2$. Take the first b_1 solutions in P and add them to $RefSet_1$. For each solution x in P-*RefSet* and y in *RefSet*, calculate a measure of distance or dissimilarity $d(x,y)$. Select the solution x' that maximises $d_{min}(x)$, where $d_{min}(x) = \min_{y \in RefSet} \{d(x,y)\}$. Add x' to $RefSet_2$, until $|RefSet_2| = b_2$. Make *NewElements* = TRUE.

While (*NewElements*) **do**

4. Calculate the number of subsets (*MaxSubset*) that include at least one new element. Make *NewElements* = FALSE.

For (*SubsetCounter* = 1, …, *MaxSubset*) **do**

5. Generate the next subset s from *RefSet* with the Subset Generation Method. This method generates one of four types of subsets with number of elements ranging from 2 to $|RefSet|$. Let subset $s = \{s_1, \ldots, s_k\}$, for $2 \le k \le |RefSet|$. (We consider that the Subset Generation Method skips subsets for which the elements considered have not changed from previous iterations.)
6. Apply the Solution Combination Method to s to obtain one or more new solutions x_s.
7. Apply the Improvement Method to x_s, to obtain the improved solution x_s^*.

Fig. 2. Scatter search outline

```
If ( x_s* is not in RefSet and the objective function value of x_s* is better
than the objective function value of the worst element in RefSet_1 ) then
      8. Add x_s* to RefSet_1 and delete the worst element currently in
         RefSet_1. (The worst element is the solution with worst objective
         value.)
      9. Make NewElements = TRUE.
Else
      If ( x_s* is not in RefSet_2 and d_min( x_s* ) is larger than d_min(x) for a
      solution x in RefSet_2) then
         10. Add x_s* to RefSet_2 and delete the worst element currently in
             RefSet_2. (The worst element is the solution x with the
             smallest d_min(x) value.)
         11. Make NewElements = TRUE.
      End if
End if
      End for
   End while
   If (Iter < MaxIter) then
      12. Build a new set P using the Diversification Generation Method.
          Initialise the generation process with the solutions currently in
          RefSet_1. That is, the first b_1 solutions in the new P are the best b_1
          solutions in the current RefSet.
   End if
End for
```

Fig. 2. (Continued)

5 Implications for Future Developments

The focus and emphasis of the scatter search approach have a number of specific implications for the goal of designing improved optimisation procedures. To understand these implications, it is useful to consider certain contrasts between the highly exploitable meaning of "solution combination" provided by scatter search and the rather amorphous concept of "crossover" used in GAs. Originally, GAs were founded on precise notions of crossover, using definitions based on binary strings and motivated by analogies with genetics. Although there are still many GA researchers who favour the types of crossover models originally proposed with GAs – since these give rise to the theorems that have helped to popularise GAs – there are also many who have largely abandoned these ideas and who have sought, on a case-by-case basis, to replace them with something different. The well-defined earlier notions of crossover have not been abandoned without a price.

The literature is rife with examples where a new problem (or a new variant of an old one) has compelled the search for an appropriate "crossover" to begin anew.[3]

As a result of this lack of an organising principle, many less-than-suitable modes of combination have been produced, some eventually replacing others, without a clear basis for taking advantage of context – in contrast to the strong context-exploiting emphasis embodied in the concept of structured combinations. The difficulty of devising a unifying basis for understanding or exploiting context in GAs was inherited from its original theme, which had the goal of making GAs *context free.*

Specific areas of research for developing improved solution strategies that emerge directly from the scatter search orientation are:

- Strategies for clustering and anti-clustering, to generate candidate sets of solutions to be combined.
- Rules for multi-parent compositions.
- Isolating and assembling solution components by means of constructive linking and vocabulary building.

These research opportunities carry with them an emphasis on producing systematic and strategically designed rules, rather than following the policy of relegating decisions to random choices, as often is fashionable in evolutionary methods. The strategic orientation underlying scatter search is motivated by connections with the tabu search setting and invites the use of adaptive memory structures in determining the strategies produced. The learning approach called target analysis [2] gives a particularly useful basis for pursuing such research.

6 Randomisation and the Intensification/Diversification Dichotomy

The emphasis on systematic strategies in achieving intensification and diversification does not preclude the use of randomised selection schemes, which are often motivated by the fact that they require little thought or sophistication to apply (as illustrated by the Solution Combination Method of Section 3.5). By the same token, deterministic rules that are constructed with no more reflection than devoted to creating a simple randomised rule can be quite risky, because they can easily embody oversights that will cause them to perform poorly. A randomised rule can then offer a safety net, by preventing a bad decision from being applied persistently and without exception.

Yet a somewhat different perspective suggests that deterministic rules can offer important advantages in the longer run. A "foolish mistake" incorporated into a

[3] The disadvantage of lacking a clear and unified model for combining solutions has had its compensations for academic researchers, since each new application creates an opportunity to publish another form of crossover! The resulting abundance of papers has done nothing to tarnish the image of a dynamic and prospering field.

deterministic rule becomes highly visible by its consequences, whereas such a mistake in a randomised rule may be buried from view – obscured by the patternless fluctuations that surround it. Deterministic rules afford the opportunity to profit by mistakes and learn to do better. The character of randomised rules, that provides the chance to escape from repetitive folly, also inhibits the chance to identify more effective decisions.

The concepts of intensification and diversification are predicated on the view that intelligent variation and randomised variation are rarely the same.[4] This clearly contrasts with the prevailing perspective in the literature of evolutionary methods although, perhaps surprisingly, the intensification and diversification terminology has been appearing with steadily increasing frequency in this literature. Nevertheless, a number of the fundamental strategies for achieving the goals of intensification and diversification in scatter search applications have still escaped the purview of other evolutionary methods.

Perhaps one of the factors that is slowing a more complete assimilation of these ideas is a confusion between the terminology of intensification and diversification and the terminology of "exploitation versus exploration" popularised in association with GAs. The exploitation/exploration distinction comes from control theory, where exploitation refers to following a particular recipe (traditionally memoryless) until it fails to be effective, and exploration then refers to instituting a series of random changes – typically via multi-armed bandit schemes – before reverting to the tactical recipe. The issue of exploitation versus exploration concerns how often and under what circumstances the randomised departures are launched.

By contrast, intensification and diversification are mutually reinforcing (rather than being mutually opposed), and can be implemented in conjunction as well as in alternation. In longer-term strategies, intensification and diversification are both activated when simpler tactics lose their effectiveness. Characteristically, they are designed to profit from memory [2], rather than to rely solely on indirect "inheritance effects".

7 Conclusions

It is not possible within the limited scope of this chapter to detail completely the aspects of scatter search that warrant further investigation. Additional implementation considerations, including associated intensification and diversification processes, and the design of accompanying methods to improve

---

[4] Intelligence can sometimes mean quickly doing something mildly clever, rather than slowly doing something profound. This can occur where the quality of a single move obtained by extended analysis is not enough to match the quality of multiple moves obtained by more superficial analysis. Randomised moves, which are quick, sometimes gain a reputation for effectiveness because of this phenomenon. In such a setting, a different perspective may result by investigating comparably fast mechanisms that replace randomisation with intelligent variation.

solutions produced by combination strategies, may be found in the *template* for scatter search and path relinking in [12].

However, a key observation deserves to be stressed. The literature often contrasts evolutionary methods – especially those based on combining solutions – with local search methods, as though these two types of approaches are fundamentally different. In addition, evolutionary procedures are conceived to be independent of any reliance on memory, except in the very limited sense where solutions forged from combinations of others carry the imprint of their parents. Yet as previously noted, the foundations of scatter search strongly overlap with those of tabu search. By means of these connections, a wide range of strategic possibilities exist for implementing scatter search.

Very little computational investigation of these methods has been done by comparison to other evolutionary methods, and a great deal remains to be learned about the most effective implementations for various classes of problems. The highly promising outcomes of studies such as those cited in [15] suggest that these approaches may offer a useful potential for applications in areas beyond those investigated up to now.

References

1. Glover F. Parametric Combinations of Local Job Shop Rules. ONR Research Memorandum no. 117, GSIA, Carnegie Mellon University, Pittsburgh, PA, 1963, Chapter IV.
2. Glover F. and Laguna M. *Tabu Search*, Kluwer Academic Publishers, Boston, 1997.
3. Crowston WB, Glover F, Thompson GL and Trawick JD. Probabilistic and Parametric Learning Combinations of Local Job Shop Scheduling Rules. ONR Research Memorandum No. 117, GSIA, Carnegie Mellon University, Pittsburgh, PA, 1963.
4. Glover F. A Multiphase Dual Algorithm for the Zero-One Integer Programming Problem, *Operations Research* 1965; 13(6): 879.
5. Greenberg HJ and Pierskalla WP. Surrogate Mathematical Programs. *Operations Research* 1970; 18: 924-939.
6. Greenberg HJ and Pierskalla WP. Quasi-conjugate Functions and Surrogate Duality. *Cahiers du Centre d'Etudes de Recherche Operationelle* 1973; 15: 437-448.
7. Glover F. Surrogate Constraint Duality in Mathematical Programming. *Operations Research* 1975; 23: 434-451.
8. Karwan MH and Rardin RL. Surrogate Dual Multiplier Search Procedures in Integer Programming. School of Industrial Systems Engineering, Report Series No. J-77-13, Georgia Institute of Technology, 1976.
9. Karwan MH and Rardin RL. Some Relationships Between Lagrangean and Surrogate Duality in Integer Programming. *Mathematical Programming* 1979; 17: 230-334.
10. Freville A and Plateau G. Heuristics and Reduction Methods for Multiple Constraint 0-1 Linear Programming Problems. *European Journal of Operational Research* 1986; 24: 206-215.
11. Freville A and Plateau G. An Exact Search for the Solution of the Surrogate Dual of the 0-1 Bidimensional Knapsack Problem. *European Journal of Operational Research* 1993; 68: 413-421.

12. Glover F. A Template for Scatter Search and Path Relinking. In: Hao JK, Lutton E, Ronald E, Schoenauer M. and Snyers D (eds.) *Lecture Notes in Computer Science* 1997; 1363: 1-53.
13. Michlewicz Z and Logan TD. Evolutionary Operators for Continuous Convex Parameter Spaces. In: Sebald AV and Fogel LJ (eds.) *Proceedings of the 3rd Annual Conference on Evolutionary Programming*. World Scientific Publishing, River Edge, NJ, 1994, pp. 84-97.
14. Nelder JA and Mead R. A Simplex Method for Function Minimisation. *Computer Journal* 1965; 7: 308.
15. Glover F. Scatter Search and Path Relinking. In: Corne D., Dorigo M. and Glover F. (eds.) *New Ideas in Optimisation*, McGraw-Hill, 1999.

The Ant Colony Optimization Paradigm for Combinatorial Optimization

Antonella Carbonaro and Vittorio Maniezzo

Scienze dell'Informazione
University of Bologna
Via Sacchi 3, 47023 Cesena, Italy
E-mail: carbonar@csr.unibo.it, maniezzo@csr.unibo.it

Summary. Ant Colony Optimization is a paradigm for designing combinatorial optimization metaheuristic algorithms, which construct a solution on the basis of information provided both by some standard constructive heuristic and by previously obtained solutions. In this chapter, we present current results obtained by ACO algorithms on several hard combinatorial optimization problems. Furthermore, we describe in more detail a particular ACO algorithm, the ANTS metaheuristic, presenting its general structure and reporting results obtained on the quadratic and on the frequency assignment problems.

1 Introduction

Ant Colony Optimization (ACO) is a framework for designing metaheuristic algorithms for combinatorial optimization problems. The first algorithm which can be classified within this framework was presented in 1991 [8] and, since then, several applications have justified the interest in it. What distinguishes ACO algorithms from other metaheuristic approaches is the attempt of combining a priori information about the structure of a promising solution with a posteriori information about the structure of previously obtained good solutions.

Metaheuristic algorithms contain at their core some basic heuristic, either a constructive heuristic which starts from a null solution and adds elements to build a good complete one, or a local search which starts from a complete solution and iteratively modifies some of its elements in order to achieve a better one. The metaheuristic part drives the low-level heuristic to obtain solutions better than those it could have achieved alone, even if iterated. Usually, the drive is achieved either by constraining or by randomizing the set of local neighbor solutions to consider in local search (as is the case of simulated annealing [31] or tabu search [27]), or by combining elements taken by different solutions (as is the case of evolution strategies [2] and genetic [28] or bionomic [37] algorithms).

ACO algorithms are different, in that they explicitly use elements of previous solutions. In fact, they drive a constructive low-level solution, as GRASP [21] does, but including it in a population framework and randomizing the construction in a Monte Carlo way. The probabilities used in the Monte Carlo process are explicitly conditioned by previously obtained solutions.

The particular way of conditioning probabilities is problem-specific, and can be designed in different ways, facing a trade-off between the specificity of the information used for the conditioning and the number of solutions which need to be constructed before effectively biasing the probability distribution to favor the emergence of good solutions. Different applications have favored either the use of conditioning at the level of decision variables, thus requiring a huge number of iterations before getting a precise distribution, or the computational efficiency, thus using very coarse conditioning information.

The chapter is structured as follows. Sections 2 and 3 describe the common elements of the heuristics belonging to the ACO class and the results obtained by current approaches on different problems. Section 4 concentrates on the ANTS approach, one algorithm of the ACO class, describing its essential ingredients, while Section 5 presents computational results obtained by the ANTS algorithm on two well-known combinatorial optimization problems: the Quadratic Assignment and the Frequency Assignment problems. Finally, Section 6 contains a brief discussion on the issues raised by this overview.

2 ACO

ACO [14,17] is a class of algorithms whose first member, called Ant System, was initially proposed by Colorni, Dorigo and Maniezzo [8,19,16]. The importance of the original Ant System resides mainly in being the prototype of a number of ant algorithms which have found interesting and successful applications. The main underlying idea is that of a parallel search over several constructive computational threads, all based on a dynamic memory structure incorporating information on the effectiveness of previously obtained results and in which the behavior of each single agent is inspired by the behavior of real ants.

The collective behavior emerging from the interaction of the different search threads has proved effective in solving combinatorial optimization (CO) problems. In order to introduce a standard notation, we define combinatorial optimization problems to be problems defined over a set $\mathbf{C} = \{c_1, \ldots, c_n\}$ of basic *components*. A subset S of components represents a *solution* of the problem; $\mathbf{F} \subseteq 2^{\mathbf{C}}$ is the subset of *feasible solutions*, thus a solution S is feasible if and only if $S \in \mathbf{F}$. A *cost function* z is defined over the solution domain, $z : 2^{\mathbf{C}} \rightarrow R$, the objective being to find a minimum cost feasible solution S^*, i.e., to find $S^* : S^* \in \mathbf{F}$ and $z(S^*) \leq z(S), \forall S \in \mathbf{F}$.

An *ant* is defined to be a simple computational agent, which iteratively constructs a solution for the problem to solve. Partial problem solutions are seen as *states*; each ant *moves* from a state ι to another one ψ, corresponding to a more complete partial solution. At each step σ, each ant k computes a set $A_k^\sigma(\iota)$ of feasible expansions to its current state, and moves to one of these according to a probability distribution specified as follows.

For ant k, the probability $p_{\iota\psi}^k$ of moving from state ι to state ψ depends on the combination of two values:

1 the attractiveness $\eta_{\iota\psi}$ of the move, as computed by some heuristic indicating the a priori desirability of that move;

2 the trail level $\tau_{\iota\psi}$ of the move, indicating how proficient it has been in the past to make that particular move: it represents therefore an a posteriori indication of the desirability of that move.

Trails are updated when all ants have completed a solution, increasing or decreasing the level of trails corresponding to moves that were part of "good" or "bad" solutions, respectively.

The specific formula for defining the probability distribution of moving from one state to another one makes use of a set $tabu_k$ which indicates a problem-dependent set of infeasible moves for ant k. Different authors use different formulae. According to [33] probabilities are computed as follows: $p_{\iota\psi}^k$ is equal to 0 for all moves which are infeasible (i.e., they are in the tabu list), otherwise it is computed by means of formula (1), where α is a user-defined parameter ($0 \leq \alpha \leq 1$):

$$p_{\iota\psi}^k = \frac{\alpha \cdot \tau_{\iota\psi} + (1-\alpha) \cdot \eta_{\iota\psi}}{\sum_{(\iota\nu)\notin tabu_k}(\alpha \cdot \tau_{\iota\nu} + (1-\alpha) \cdot \eta_{\iota\nu})} \tag{1}$$

Parameter α defines the relative importance of the trail with respect to attractiveness. After each iteration t of the algorithm, i.e., when all ants have completed a solution, trails are updated following formula (2):

$$\tau_{\iota\psi}(t) = \tau_{\iota\psi}(t-1) + \Delta\tau_{\iota\psi} \tag{2}$$

where $\Delta\tau_{i\psi}$ represents the sum of the contributions of all ants that used move $(\iota\psi)$ to construct their solution. The ants' contributions are proportional to the quality of the achieved solutions, i.e., the better an ant solution, the higher will be the trail contribution added to the moves it used. Several authors [8,25,4,51] use a parameter (called *evaporation*) to multiply $\tau_{\iota\psi}(t-1)$ in order to decrease trail values and accept only positive $\Delta\tau_{\iota\psi}$, whereas in formula (2) negative trail variations are possible. The general structure of an ACO algorithm is as shown in Figure 1.

ACO framework for combinatorial optimization

Step 1 : (*Initialization*)
 Initialize $\tau_{\iota\psi}$, $\forall(\iota, \psi)$.
Step 2 : (*Construction*)
 For each ant k (currently in state ι) **do**
  repeat
   compute $\eta_{\iota\psi}$, $\forall(\psi)$.
   choose in probability the state to move into.
   append the chosen move to the k-th ant's set $tabu_k$.
  until ant k has completed its solution.
  carry the solution to its local optimum.
 end for.
Step 3 : (*Trail update*)
 For each ant move $(\iota\psi)$ **do**
  compute $\Delta\tau_{\iota\psi}$.
  update the trail matrix.
 end for.
Step 4 : (*Terminating condition*)
 If not(end condition) **go to** step 2.

Figure 1. Pseudo-code of ACO general structure

3 Current Research

The general framework just presented has been specified in different ways by the authors working on the ACO approach. This variety is well represented in the many diverse conference with tracks entirely dedicated to ACO and most notably in ANTS conference series, entirely dedicated to algorithms inspired by the observation of ants' behavior (*ANTS'98, ANTS'2000 and ANTS'2002*). These events were attended by research groups from several European countries (Germany, Italy, Switzerland, the UK, France, Austria, The Netherlands, Slovenia, Spain and Belgium) besides other groups from Japan, Russia, Brazil, Mexico, Israel and the USA. Different applications were presented: from plan merging to routing problems, from driver scheduling to search space sharing, from set covering to nurse scheduling, from graph coloring to dynamic multiple criteria balancing problems. A large part of the relevant literature can be accessed online from [14].

Table 1 presents a summary of the main ACO metaheuristics so far published. The first column of the table shows, when available, the name given to the metaheuristic, the second column the authors who proposed that approach and the third the problems it has been applied to (where TSP stands for Travelling Salesman Problem, QAP for Quadratic Assignment Problem, JSP for Job Shop Scheduling Problem, VRP for Vehicle Routing Problem,

SOP for Sequential Ordering Problem, FAP for Frequency Assignment Problem, GCP for Graph Coloring Problem, SCS for Shortest Common Supersequence).

Table 1. ACO applications (adapted from [17])

ABC	Bonabeau et al. [3], van der Put [57]	network routing
ACS	Dorigo, Gambardella [18]	TSP, VRP
AntNet	Di Caro, Dorigo [11,12]	network routing
ANTS	Maniezzo [33] , Maniezzo, Carbonaro [34]	QAP, FAP
AS	Colorni, Dorigo, Maniezzo [8,16,9,20]	TSP, QAP, JSP
ASrank	Bullnheimer, Hartl, Strauss [5]	TSP, VRP
HAS	Gambardella, Taillard, Dorigo [26]	QAP, VRP, SOP
MMAS	Stuetzle, Hoos [51,49]	TSP, QAP
AS-SCS	Michel, Middendorf [41]	SCS
-	Costa, Hertz [10]	GCP
-	Merkle, Middendorf, Schmeck [39,40]	scheduling problems
-	Gambardella, Dorigo [24]	SOP
-	Kawamura et al. [30]	TSP

The variety of the contributions testifies to both the flexibility of the approach and the infancy of the field, where no evidence of the superiority of a particular technique over the other ones has so far emerged. In Section 4 we will concentrate on a particular one, ANTS, to provide more insight into the computational elements of a specific implementation, while in the following section we will present more elements about the different ACO contributions.

3.1 ACO Approaches to TSP

The first application of AS used the travelling salesman problem (TSP) as a benchmark problem. This was because the TSP is one of the most studied NP-hard problems, and the ant paradigm is easily adapted to it. Several authors built upon this initial contribution, as follows.

Stuetzle and Hoos [50] introduced the Max-Min AS (MMAS), a modification of the AS applied to the TSP. These authors explicitly introduced in the algorithm two parameters, a maximum and minimum trail level, whose values are chosen in a problem-dependent way in order to restrict possible trail values to the interval $[\tau_{min}, \tau_{max}]$. Moreover, MMAS controls the trail levels (initialized to their maximum value τ_{max}), allowing only the best ant at each iteration to update trails, thus providing feedback on its results. Trails that do not receive any or very rare reinforcements will continuously lower their strength and will be selected more and more rarely by the ants, until they reach the τ_{min} value. The τ_{min} and τ_{max} parameters are used to counteract premature stagnation of search, maintaining at the same time some kind of elitist strategy. When applied to the TSP, MMAS performs better than AS.

Bullnheimer, Hartl and Strauss [5] proposed yet another modification of AS, called AS_{rank}, introducing a rank-based version of the probability distribution to limit the danger of over-emphasized trails caused by many ants using sub-optimal solutions. The idea is the following. At each iteration, when all solutions are completed the ants are sorted by solution quality (i.e., tour lengths in the case of the TSP) and the contribution of an ant to the trail level update is weighted according to the rank of the ant, considering only the ω best ants.

Table 2, taken from [5], compares a simulated annealing (SA), a simulated annealing with the nearest neighbor heuristic (SA_{NN}), a genetic algorithm (GA), an Ant System (AS), an Ant System with elitist strategy (AS_{elite}) and an ant system with elitist strategy and ranking (AS_{rank}). The table reports the percentage deviation of the average results obtained on five different TSP instances (considering also real-life problems from an industrial application), the percentage deviation of the best results and the percentage deviation of the worst results.

In general, the AS can compete with the other two metaheuristics; for large problems AS_{rank} outperforms the other methods on average and, most noticeably, on worst-case results.

Table 2. Comparison of different ACO approaches on the TSP (from [5])

	Dev.Avg.Sol.	*Dev.Best.Sol.*	*Dev.Worst.Sol.*
SA	2.784	0.550	7.416
SA_{NN}	2.036	0.148	4.902
GA	2.112	0.614	4.252
AS	1.886	0.652	2.802
AS_{elite}	1.112	0.200	2.866
AS_{rank}	1.012	0.100	2.212

Gambardella and Dorigo [23] merged AS and Q-learning [58], a well-known reinforcement learning algorithm, into an algorithm called Ant-Q. The idea was to update trails with values which predicted the quality of solutions using the edges to which the trails were associated.

Even though showing a good performance, Ant-Q was abandoned for the simpler Ant Colony System (ACS) algorithm [18], which uses a constant value instead of the mentioned prediction term. In this algorithm the trail values are added offline, at the end of each iteration of the algorithm, only to the arcs belonging to the best tour from the beginning of the search process. Ants perform online step-by-step trail updates to favor the emergence of solutions other than the best so far. Each ant uses a *pseudo-random proportional rule* to choose the next node to move to. This is a decision rule based on a $q_0 \in [0,1]$ parameter that permits modulation of the exploration behavior, concentrating the system activity either on the best solutions or on the entire search space. ACS also uses a data structure associated to vertices called a *candidate list* which provides additional local heuristic information. The candidate list associated with a vertex contains only the *cl* vertices nearest to it, and the ants choose the next move by scanning the candidate list instead of examining all the unvisited neighboring vertices.

3.2 ACO Approaches to QAP

The quadratic assignment problem (QAP) is the problem of assigning n facilities to n locations so that a quadratic assignment cost is minimized. The QAP can be considered one of the hardest CO problems, and can be solved to optimality only for small instances. For this reason, the QAP was chosen as a second benchmark for AS, resulting in code AS-QAP [36]. AS-QAP was of limited effectiveness, but was the first evidence of the robustness of AS. The effectiveness was, however, improved using a well-tuned local optimizer [35]. In the latter paper it has been shown that, while the process of an individual ant will almost always converge very quickly to a possibly mediocre solution, the interaction of many feedback processes can instead lead to convergence towards a region of the search space containing good solutions, so that very good results can be obtained (but not sticking on them). In other words, the ant population does not converge to a single solution, but on a set of (good) ones; the ants continue their search to further improve the best solution found. The results obtained showed the competitive performance of the AS on several test problems. Further developments led to the design of ANTS, which will be detailed in Section 4. Several other systems previously introduced were also adapted to the QAP. For example, two efficient techniques are the MMAS-QAP algorithm [51] and HAS-QAP [26]. Both of them have been applied to problem instances classified in two categories: randomly generated problems without any structure and structured real-life problems. The performances of these two heuristic approaches are strongly dependent on the type of problem considered.

Table 3. MMAS-QAP results compared to genetic hybrid and HAS-QAP (from [51])

Prbl.instance	Genetic Hybrid	HAS-QAP	MMAS-QAP
sko72	0.143	0.277	0.301
sko81	0.136	0.144	0.153
sko90	0.196	0.231	0.366
tai50a	1.049	2.800	2.213
tai60a	1.159	3.070	2.211
tai60a	0.796	2.689	1.808
bur26a-h	0.0043	0.0	0.0
kra30a	0.1338	0.629	0.821
kra30b	0.0536	0.0711	0.121
tai35b	0.1067	0.0256	0.0469
tai40b	0.2109	0.0	0.0
tai50b	0.2142	0.1916	0.000596
tai60b	0.2905	0.0483	0.0144
tai80b	0.8286	0.667	0.741

Comparisons with some of the best heuristics for the QAP have shown that HAS-QAP performs well as far as real-world, irregular and structured problems are concerned. On the other hand, on random, regular and unstructured problems the performance of this technique appears to be less competitive.

Table 3 reports the results presented in [51] comparing MMAS-QAP to a genetic hybrid method [22] and to HAS-QAP, both on unstructured (upper half of the table) and structured (lower half) QAP instances. The table entries represent the ratio between the result produced by HAS-QAP and the best known result for the instance. MMAS-QAP appears to be one of the most promising approaches for the solution of structured real-life QAPs.

3.3 ACO Approaches to VRP

Vehicle routing problems (VRPs) are CO problems where a set of vehicles stationed at a depot has to serve a set of customers before returning to the depot. The objective is to minimize the number of vehicles used and the total distance travelled by the vehicles. Capacity constraints are imposed on vehicle trips, as well as possibly a number of other constraints deriving from real-world applications, such as time windows, backhauling, rear loading, vehicle objections, maximum tour length, etc. Also the VRP can be considered as a generalization of the TSP; in fact the VRP reduces to the TSP when only one vehicle is available and capacity constraints can be neglected. The most successful applications of ACO metaheuristics to the VRP are the following.

A direct extension of AS based on the AS_{rank} algorithm is AS-VRP, an algorithm designed by Bullnheimer, Hartl and Strauss [5,6]. These authors used various standard heuristics to improve the quality of VRP solutions and modified the construction of the tabu list considering constraints on the maximum total tour length of a vehicle and on its capacity. The results obtained on some problem instances were sufficiently interesting to justify a more detailed study.

Table 4. HAS-VRP results (from [25])

	R1		C1		RC1		R2		C2		RC2	
	VEH	DIST	VEH	DIST	VEH	DIST	VEH	DIST	VEH	DIST	VEH	DIST
MACS	12	1217.73	10	828.38	11.63	1382.42	2.73	967.75	3.00	589.86	3.25	1129.19
RT	12.25	1208.50	10	828.38	11.88	1377.39	2.91	961.72	3.00	589.86	3.38	1119.59
TB	12.17	1209.35	10	828.38	11.50	1389.22	2.82	980.27	3.00	589.86	3.38	1117.44
CR	12.42	1289.95	10	885.86	12.38	1455.82	2.91	1135.14	3.00	658.88	3.38	1361.14
PB	12.58	1296.80	10	838.01	12.13	1446.20	3.00	1117.70	3.00	589.93	3.38	1360.57
TH	12.33	1238.00	10	832.00	12.00	1284.00	3.00	1005.00	3.00	650.00	3.38	1229.00

Also Gambardella, Taillard and Agazzi [25] tackled the VRP, adapting ACS to obtain MACS-VRPTW, and considering the time windows extension of the VRP, which introduces a time range within which each customer must be serviced. This approach has proved to be comparable with the best known approaches in the literature, as shown by Table 4, in which six different problem types (R1 and R2 with randomly distributed vertices, C1 and C2 with clustered vertices and RC1 and RC2 with random-clustered vertices, from [48]) have been used to compare MACS-VRPTW with five other heuristics:

the adaptive memory programming methods of Rochat and Taillard (RT) and Taillard et al. (TB) [45,53], the method of Chiang and Russel (CR) [7], the GA of Potvin and Bengio (PB) [44] and the method of Thangiah et al. (TH) [54].

The two columns associated with each problem set show the number of vehicles (VEH) and the total distance (DIST) used in the solution. Both parameters must be minimized.

3.4 ACO Approaches to RP

The telecommunication routing problem is the problem of maximizing a network performance measure while minimizing costs in directing traffic from source to destination nodes. This problem can be stated as a multi-objective optimization problem in a non-stationary stochastic environment. The most widely used routing algorithms are shortest path algorithms where the objective is to find the shortest path between two nodes and the costs associated to the links are computed following some description of the link states.

Schoonderwoerd et al. [46] were the first to apply an ACO algorithm to a routing problem. Their Ant-Based Control (ABC) algorithm was applied to a model of the British Telecom telephone network in which each node can establish a limited number of connections, assuming that the links have infinite capacity. Each new request is accepted or rejected on the basis of a setup feasibility condition that looks for a path with available capacity for new connections, by probing the deterministically best path as indicated by the routing tables. In preliminary computational results ABC, compared to an agent-based algorithm developed by British Telecom researchers, always performed better even on a variety of different traffic situations.

Another problem is the virtual wavelength path (VWP) routing problem that is, the problem of minimizing the total number of different wavelengths that are used in a network to satisfy a given workload, expressed as transmission requests. Specifically, the problem seeks to minimize the total number of different wavelengths that are used in the network, which is in turn equal to the maximum number needed on any one link.

In the network, each link potentially consists of many optical fibers and each fiber is capable of carrying a certain number of wavelengths. If the individual requests are carried using VWP routing, it is possible to change the wavelength at each node by using optical conversion at the nodes along the path. The applicability of ACO for VWP routing and wavelength allocation is possible because the problem is inherently graph-based and the routing of wavelength channels requires path-following analysis. Navarro-Varela and Sinclair [43] propose three variants of an ACO algorithm, where the artificial ants, besides being attracted, are also repelled by the pheromone of the other ants. Their best algorithm approaches the solution quality of the wavelength-allocation heuristic of Nagatsu et al. [42] on small- and medium-sized networks, although it requires much higher computational costs.

Di Caro and Dorigo [12] considered the problem of routing packets in communication networks and designed the AntNet algorithm. Every node of their network holds a buffer where the incoming and the outgoing data or packets are stored. The packets are served using a first-in first-out policy; the algorithm uses the routing tables to obtain information about which link to use to forward a packet along its path towards its destination. If the buffer space is sufficient to hold the entire packet then the transfer is set up, otherwise, the packet is discarded. The authors approached the problem using two sets of mobile agents, *forward* and *backward* ants. While building the path, each *forward ant* collects information about the time length and the load status of the network. *Backward ants*, moving in the opposite direction, back-propagate this information and modify the routing tables. AntNet results are compared to some effective routing algorithms [13] showing good performance and robust behavior, being able to rapidly reach a good stable level in performance.

4 The ANTS Algorithm

ANTS is an extension of the AS proposed in [8], which specifies some underdefined elements of the general algorithm, such as the attractiveness function to use or the initialization of the trail distribution. This turns out to be a variation of the general ACO framework that makes the resulting algorithm similar in structure to tree search algorithms. In fact, the essential trait which distinguishes ANTS from a tree search algorithm is the lack of a complete backtracking mechanism, which is substituted by a probabilistic (*Non – deterministic*) choice of the state to move into and by an incomplete (*Approximate*) exploration of the search tree: this is the rationale behind the name ANTS, which is an acronym of *Approximated Non-deterministic Tree Search.* In the following, we will outline two distinctive elements of the ANTS algorithm within the ACO framework, namely the attractiveness function and the trail updating mechanism.

4.1 Attractiveness

The attractiveness of a move can be effectively estimated by means of lower bounds (upper bounds in the case of maximization problems) on the cost of the completion of a partial solution. In fact, if a state ι corresponds to a partial problem solution it is possible to compute a lower bound on the cost of a complete solution containing ι. Therefore, for each feasible move $(\iota\psi)$, it is possible to compute the lower bound on the cost of a complete solution containing ψ: the lower the bound the better the move. Since a large part of research in CO is devoted to the identification of tight lower bounds for the different problems of interest, good lower bounds are usually available.

When the bound value becomes greater than the current upper bound, it is obvious that the considered move leads to a partial solution which cannot be completed into a solution better than the current best one. The move can therefore be discarded from further analysis. A further advantage of lower bounds is that in many cases the values of the decision variables, as appearing in the bound solution, can be used as an indication of whether each variable will appear in good solutions. This provides an effective way of initializing the trail values. For more details see [33].

The use of tight lower bounds is a very effective and straightforward general policy, whenever such bounds have been identified for the problem to solve.

4.2 Trail Update

A good trail updating mechanism avoids stagnation, the undesirable situation in which all ants repeatedly construct the same solutions making any further exploration in the search process impossible. Stagnation derives from an excessive trail level on the moves of one solution, and can be observed in advanced phases of the search process, if parameters are not well tuned to the problem.

The trail updating procedure evaluates each solution against the last k solutions globally constructed by ANTS. As soon as k solutions are available, their moving average $\overline{z}$ is computed; each new solution z_{curr} is compared with $\overline{z}$ (and then used to compute the new moving average value). If z_{curr} is lower than $\overline{z}$, the trail level of the last solution's moves is increased, otherwise it is decreased. Formula (3) specifies how this is implemented:

$$\Delta\tau_{\iota\psi} = \tau_0 \cdot \left(1 - \frac{z_{curr} - LB}{\overline{z} - LB}\right) \tag{3}$$

where $\overline{z}$ is the average of the last k solutions and LB is a lower bound on the optimal problem solution cost. The use of a dynamic scaling procedure permits discrimination of a small achievement in the latest stage of search, while avoiding focusing the search only around good achievement in the earliest stages.

One of the most difficult aspects to be considered in metaheuristic algorithms is the trade-off between exploration and exploitation. To obtain good results, an agent should prefer actions that it has tried in the past and found to be effective in producing desirable solutions (exploitation); but to discover them, it has to try actions not previously selected (exploration). Neither exploration nor exploitation can be pursued exclusively without failing in the task: for this reason, the ANTS algorithm integrates the stagnation avoidance procedure to facilitate exploration with the probability definition mechanism based on attractiveness and trails to determine the desirability of moves.

Based on the elements described, the ANTS algorithm is as shown in Figure 2.

ANTS algorithm

Step 1 : (Initialization)
  Compute a (linear) lower bound on the problem to be solved.
  Initialize $\tau_{\iota\psi}$, $\forall(\iota, \psi)$, with the primal variable values.
Step 2 : (Construction)
  For each ant k **do**
    repeat
      compute $\eta_{\iota\psi}$, $\forall(\iota, \psi)$, as a lower bound on the cost of a complete solution containing ψ.
      choose the state to move to, with probability given by (1).
      append the chosen move to the k-th ant's set $tabu_k$.
    until ant k has completed its solution.
    carry the solution to its local optimum.
  end for.
Step 3 : (Trail update)
  For each ant move $(\iota\psi)$ **do**
    compute $\Delta\tau_{\iota\psi}$.
    update the trail matrix by means of (2) and (3).
  end for.
Step 4 : (Terminating condition)
  If not(end-test) **go to** step 2.

Figure 2. Pseudo-code for the ANTS algorithm

It can be noted that the general structure of the ANTS algorithm is closely akin to that of a standard tree search procedure. At each stage we have in fact a partial solution which is expanded by branching on all possible offspring; a bound is then computed for each offspring, possibly fathoming dominated ones, and the current partial solution is selected from among those associated to the surviving offspring on the basis of lower bound considerations. By simply adding backtracking and eliminating the Monte Carlo choice of the node to move to, we revert to a standard branch and bound procedure. An ANTS code can therefore be easily turned into an exact procedure.

5 Computational Results

The above elements have been implemented and tested on two well-known CO problems: the Quadratic Assignment and the Frequency Assignment Problems (QAP and FAP, respectively). It is important to note that the reason behind the choice of these problems lies in the attempt to empirically evaluate the robustness of the ANTS approach. While in fact preliminary non-optimized codes testified to the validity of the issues reported in Section 4, i.e., the effectiveness of the ANTS algorithm when tight bounds are available, both QAP and FAP were chosen because of the ineffectiveness of the bounds so far presented in the literature. A good performance on these problems *a fortiori* suggests the efficiency of the approach in the general case.

5.1 QAP

The QAP is one of the best known and most difficult CO problems, as confirmed by the small gap that exists between the dimension of the problems that can be solved to optimality by means of complete enumeration and the dimension of the problems that can be solved by means of the most advanced exact methods proposed in the literature.

The ANTS algorithm makes use of a lower bound derived from the well-known Gilmore and Lawler bound. Details are provided in [33], where both the ANTS and the derived exact procedure are described. Computational results for the heuristic part were given on all QAPLIB problem instances of dimension up to n = 40 and presented for ANTS and for two state-of-the-art heuristic procedures: Li et al.'s GRASP [32] and Taillard's robust tabu search (TS) [52]. Table 5 shows the results obtained by 10 min long runs, and presents the percentage deviation of the best solution found by each approach with respect to the best known solution value for each instance [33].

The good performance of all algorithms led to results worse than the best known ones on only a small number of problems. However, the best performing algorithm is ANTS, in terms of both the best and the average quality of the solutions proposed. It is interesting to see how, even in the presence of a bad bound at the root node, the non-deterministic strategy followed by ANTS permits quick identification of good solutions.

5.2 FAP

The FAP is the problem that arises when a region is covered, for wireless communications, by cells centered on base stations and transmitters scattered around the region want to establish a connection with the antennas of the base stations. Each connection, or link, between a transmitter and a base station can be made at a frequency supported by the antenna. However, the frequency concurrently operated by overlapping cells must be separated in order to minimize interference to the communications taking place in the cells. The current state of development of research on FAP does not provide efficient lower bounds. We developed one [38], which is not very tight but is efficient to compute, and included it in the ANTS algorithm [34].

Table 5. ANTS on QAP

	TS	GRASP	ANTS		TS	GRASP	ANTS
CHR20A	0.06	1.48	0.00	NUG30	0.00	0.39	0.00
CHR20B	2.06	4.65	0.00	TAI30A	0.16	1.53	0.13
ROU20	0.00	0.00	0.00	TAI30B	0.00	0.11	0.00
TAI20A	0.00	0.19	0.00	THO30	0.00	0.15	0.00
CHR22A	0.00	1.15	0.00	ESC32A	0.00	2.77	0.00
CHR22B	0.71	2.03	0.00	ESC32B	0.00	0.00	0.00
CHR25A	4.48	4.64	0.76	ESC32H	0.00	0.00	0.00
TAI25A	0.12	0.79	0.00	MC33	0.12	0.07	0.00
TAI25B	0.00	0.00	0.00	TAI35A	0.65	1.80	0.32
BUR26A	0.00	0.00	0.00	TAI35B	0.00	0.23	0.00
BUR26B	0.00	0.00	0.00	STE36A	0.00	1.76	0.00
BUR26E	0.00	0.00	0.00	STE36B	0.00	1.44	0.00
BUR26F	0.00	0.00	0.00	STE36C	0.00	0.67	0.00
KRA30A	0.00	0.34	0.00	LIPA40A	0.00	1.11	0.00
KRA30B	0.00	0.15	0.00	TAI40A	0.93	2.06	0.47
LIPA30A	0.00	0.19	0.00	TAI40B	0.00	0.03	0.00
				THO40	0.06	0.91	0.03

The computational results were obtained on three well-known problem datasets from the literature, the CELAR, GRAPH and PHILADELPHIA datasets, and are presented in Table 6. The CELAR dataset consists of 11 problems proposed within the framework of the EUCLID (EUropean Cooperation for the Long term In Defense) CALMA (Combinatorial ALgorithms for Military Applications) project [55]. The GRAPH test problems [56] are 14 problems which exhibit the same structure as the CELAR problems. The PHILADELPHIA (PHIL) problems, originally presented by Anderson [1], are among the most studied instances of FAP. The problems are based on data relative to the area around Philadelphia and consist of cells located on a hexagonal grid. All the results were obtained by implementing the algorithms in C and running the codes on a Pentium II 233 MHz machine equipped with 64 Mb of RAM for 1200 CPU seconds. We applied to these problems two heuristics which have been presented in the literature to compute frequency assignments and which we have used to benchmark ANTS results. The first is an adaptation to FAP of a well-known heuristic for the coloring problem, called Dsatur (DS), the second is a tabu search (TS) procedure. Moreover, we used two versions of simulated annealing taken from the literature: SA1 [29] and SA2 [47]. All algorithms were allowed 20 minutes CPU time on each problem instance, except Dsatur which is a fast constructive heuristic that contains its own termination condition. TS was applied on the solution initialized by means of the DS procedure.

Table 6. ANTS on FAP

	DS		TS		ANTS		SA1		SA2	
PROBLEM	Value	Time	Value	Time	Value	Time	Value	Time	Value	Time
CELAR01	625	5	238	37	0	4	0	3	0	377
CELAR02	428	1	122	60	0	0	0	0	0	19
CELAR03	798	1	196	347	0	1	0	0	0	133
CELAR04	1474	3	1012	57	8	437	0	222	1	324
CELAR05	1823	1	688	116	32	545	11	106	54	82
CELAR06	213866	1	149607	10	5319	614	6 994	75	30160	18
CELAR07	$> 10^8$	1	$> 10^8$	38	8083093	630	11000296	986	4698907	412
CELAR08	2468	6	1364	295	709	572	306	1087	457	805
CELAR09	79406	3	45988	6	16732	1018	30024	77	23634	644
CELAR10	107310	3	58554	9	31516	378	31518	54	33557	244
CELAR11	1364	4	848	417	0	628	0	83	2	405
GRAPH01	160	0	15	8	0	2	0	0	0	21
GRAPH02	299	1	15	17	0	1	0	1	0	79
GRAPH03	264	0	35	3	14	122	14	27	96	12
GRAPH04	519	2	33	3	42	1162	64	332	213	6
GRAPH05	0	0	0	0	0	1	0	0	0	0
GRAPH06	0	1	0	1	0	1	0	0	0	0
GRAPH07	0	0	0	0	0	1	0	0	0	0
GRAPH08	566	3	18	6	0	15	0	4	0	539
GRAPH09	665	5	14	14	0	53	0	8	0	903
GRAPH10	856	3	37	10	127	1052	91	818	81	1 128
GRAPH11	0	2	0	2	0	1	0	0	0	0
GRAPH12	0	3	0	3	0	1	0	0	0	0
GRAPH13	0	5	0	5	0	2	0	0	0	0
GRAPH14	750	5	19	44	0	2	0	3	0	763
PHIL01	7	2	4	14	0	29	51	234	263	1129
PHIL02	0	2	0	2	0	43	17	315	251	1081
PHIL03	8	3	8	3	0	47	31	382	288	954
PHIL04	6	3	6	3	0	53	36	336	353	1144
PHIL05	3	6	3	6	30	1041	31	359	252	1063
PHIL06	18	7	17	92	36	177	30	359	288	1139
PHIL07	0	7	0	7	29	1180	30	365	271	955
PHIL08	17	7	14	1197	31	1046	22	472	249	572
PHIL09	0	35	0	35	403	830	366	1198	934	1160
PHIL06b	1	8	1	8	51	373	18	352	254	772

The computational results show that the ANTS algorithm was competitive with the best approaches at the time of writing the report [34]. In particular, Table 6 shows that different approaches performed well. The CELAR and GRAPH problems, which have the same structure, are best solved by the ANTS and the SA1 algorithms, while the PHIL problems are more suited to the DS and TS approaches. Improved results have meanwhile been obtained both by ANTS and by other approaches, but the relevant researche is still under development.

6 Conclusions

Ant Colony Optimization has proved to be a robust framework for designing combinatorial optimization heuristic algorithms. Since its initial proposal, it has undergone several refinements which led to the current situation where several algorithms share the same idea of combining trails and attractiveness for computing move probabilities. Different problems have been successfully solved in this way, in some cases obtaining solutions which represent the state of the art. This testifies to the robustness of the approach.

There is as yet little mathematical understanding of the working principles of the approach. The probability updating formulas have been proposed as they are mainly for computational efficiency reasons, but results on convergence are probably possible by studying suitable update formulas. Similarly, no normative results have been established for defining the best level of conditioning of the transition probabilities $p^k_{\iota\psi}$ for a given problem instance. This issue is linked with similar efforts going on in well-studied areas, such as Bayesian networks, and transpositions of effective results will surely be beneficial.

References

1. Anderson, L. G. (1973) *A simulation study of some dynamic channel assignment algorithms in a high capacity mobile telecommunications system*, IEEE Transaction on Communications **COM-21**, 1294–1301
2. Bäck , T., Schwefel, H. P. (1993) *An overview of evolutionary algorithms for parameter optimization*, Evolutionary Computation **1(1)**, 1–23
3. Bonabeau, E., Henaux, F., Guerin, S., Snyers, D., Kuntz, P., Theraulaz,G. *Routing in telecommunication networks with "smart" ant-like agents*, Proceedings of IATA'98, Lecture notes on Artificial Intelligence **1437**, Springer, Berlin.
4. Botee, H. M., Bonabeau, E. (1998) *Evolving ant colony optimization*, Advances in Complex Systems **1**, 149–159
5. Bullnheimer, B., Hartl, R. F., Strauss, C. (1997) *A new rank-based version of the ant system: a computational study*, Technical Report POM-03/97, Institute of Management Science, University of Vienna, 1997, Accepted for publication in the Central European Journal for Operations Research and Economics.
6. *Applying the ant system to the vehicle routing problem* In: S., Voss, S., Martello, I.H., Osman, C., Roucairol(eds.) (1999) Meta-Heuristics: Advances and Trends in Local Search Paradigms for Optimization, Kluwer, Boston.
7. Chiang, W. C., Russel, R. (1993) *Hybrid heuristics for the vehicle routing problem with time windows*, Technical Report, Department of Quantitative Methods, University of Tulsa.
8. Colorni, A., Dorigo, M., Maniezzo, V. (1991) *Distributed optimization by ant colonies*, Proceedings of ECAL'91, European Conference on Artificial Life, Elsevier Publishing.
9. Colorni, A., Dorigo, M., Maniezzo, V., Trubian, M. (1994) *Ant system for job-shop scheduling*, Belgian Journal of Operation Research, Statistics and Computer Scence**34(1)**, 39–54.
10. Costa, D., Hertz, A. (1997) *Ants can colour graphs*, Journal of the Operational Research Society **48**, 295–305.
11. Di Caro, G., Dorigo, M. *Antnet: a mobile agents approach to adaptive routing*, Technical Report IRIDIA/97-12, Universit Libre de Bruxelles, Belgium.
12. *Antnet: distributed stigmergetic control for communications networks*(1998), Journal of Artificial Intelligence Research**9**, 317–365.
13. *Mobile agents for adaptive routing*, Proceedings of HICSS-31, 1998, The Software Technology Track, USA, Hawaii **9**.

14. Dorigo, M. *Ant colony optimization web page*, http://iridia.ulb.ac.be/~mdorigo/ACO/ACO.html
15. *Ants'98 web page*, http://iridia.ulb.ac.be/ants98/ants98.html
16. *Optimization, learning and natural algorithms*, Ph.D. Thesis 1992, Politecnico di Milano, Milano.
17. Dorigo, M., di Caro, G. (1999) *The ant-colony optimization meta-heuristic*, New Ideas in Optimization, 11–32.
18. Dorigo, M., Gambardella, L. M. (1997) *Ant colony system: a cooperative learning approach to the traveling salesman problem*, IEEE Transaction on Evolutionary Computation **1**, 53–66.
19. Dorigo, M., Maniezzo, V., Colorni, A. (1991) *The ant system: an autocatalytic optimizing process*, Technical Report TR91-016, Politecnico di Milano.
20. Dorigo, M., Maniezzo, V., Colorni, A. (1996)*The ant system: optimization by a colony of cooperating agents*, IEEE Transactions on Systems, Man, and Cybernetics-Part B **26(1)**, 29–41.
21. Feo, T. A., Resende, M. G. C. (1995) *Greedy randomized adaptive search procedures*, Journal of Global Optimization **6**, 109–133.
22. Fleurent, C., Ferland, J. A. (1994) *Genetic hybrids for the quadratic assignment problem*, In: Panos M. Pardalos and H. Wolkowicz (eds.) Quadratic Assignment and Related Problems, American Mathematical Society
23. Gambardella, L. M., Dorigo, M. (1995) *Ant-q: a reinforcement learning approach to the travelling salesman problem*, Proceedings of the Twelfth International Conference on Machine Learning, ML-95, Morgan Kaufmann, Palo Alto, CA.
24. *An ant colony system hybridized with a new local search for the sequential ordering problem*, to appear in INFORMS Journal on Computing (2000)
25. Gambardella, L. M., Taillard, E., Agazzi, G. (1999) *Ant colonies for vehicle routing problems*. In: D. Corne, M. Dorigo and F. Glover(eds.) New Ideas in Optimization, McGraw-Hill.
26. Gambardella, L. M., Taillard, E., Dorigo, M. (1999) *Ant colonies for the quadratic assignment problem*, Journal of the Operational Research Society **50**, 167–176.
27. Glover, F. (1989) *Tabu search*, ORSA Journal on Computing**1**, 190–206
28. Holland, J. H. (1975) *Adaptation in natural and artificial systems*, University of Michigan Press.
29. Hurkens, C., Tiourine, S. (1995) *Upper and lower bounding techniques for frequency assignment problems*, Technical Report 95-34, T.U. Eindhoven.
30. Kawamura, H., Yamamoto, M., Suzuki, K., Ohuchi, A. (2000) *Multiple ant colonies algorithm based on colony level interactions*, IEICE Transactions Fundamentals **E83-A(2)**.
31. Kirkpatrik, S., Gelatt, C. D., Vecchi, M. P. (1983) *Optimization by simulated annealing*, Science **220**.
32. Li, Y., Pardalos, P. M., Resende, M. G. C. (1994) *A greedy randomized adaptive search procedure for the quadratic assignment problem*. In: P. M. Pardalos and H. Wolkowicz(eds) Quadratic assignment and related problems, DIMACS Series in Discrete Mathematics and Theoretical Computer Science **16**, 237–261.
33. Maniezzo, V., (1999) *Exact and approximate nondeterministic tree-search procedures for the quadratic assignment problem*, INFORMS Journal of Computing **11(4)**, 358–369.

34. Maniezzo, V., Carbonaro, A. (2000) *A bionomic approach to the capacitated p-median problem*, Future Generation Computer Systems **16(8)**, 927–935.
35. Maniezzo, V., Colorni, A. (1999) *The ant system applied to the quadratic assignment problem*, IEEE Transactions Knowledge and Data Engineering **11(5)**, 769–778.
36. Maniezzo, V., Colorni, A., Dorigo, M. (1994) *The ant system applied to the quadratic assignment problem*, Technical Report IRIDIA/94-28, Université Libre de Bruxelles, Belgium.
37. Maniezzo, V., Mingozzi, A., Baldacci, R. (1998) *A bionomic approach to the capacitated p-median problem*, Journal of Heuristics **4(3)**, 263–280.
38. Maniezzo, V., Montemanni, R. (1999) *An exact algorithm for the radio link frequency assignment problem*, Technical Report CSR99-02.
39. Merkle, D., Middendorf, M. (2000) *An ant algorithm with a new pheromone evaluation rule for total tardiness problems.* In: S. Cagnoni et al.(eds.) vol. Real-World Applications of Evolutionary Computing, Proceedings of EvoWorkshop, LNCS 1803, Springer, Berlin.
40. Merkle, D., Middendorf, M., Schmeck, H.(2000) *Ant colony optimization for resource-constrained project scheduling*, Proceedings of GECCO, Las Vegas, Nevada.
41. Michel, R., Middendorf, M.(1998) *An island model based ant system with lookahead for the shortest supersequence problem*, Proceedings of PPSN-V, Fifth International Conference on Parallel Problem Solvivg from Nature, Springer, Vienna.
42. Nagatsu, N., Hamazumi, Y., Sato, K. (1995) *Number of wavelengths required for constructing large-scale optical path networks*, Electronics and Communications in Japan, Part I - Communications **78**, 1–11.
43. Navarro-Varela, G., Sinclair, M. (1999) *Ant colony optimisation for virtual-wavelength-path routing and wavelength allocation*, Proceedings of CEC'99, Washington DC.
44. Potvin, J. Y., Bengio, S.(1996) *The vehicle routing problem with time windows - part II: genetic search*, Informs Journal of Computing **8**.
45. Rochat, Y., Taillard, E. D. (1995) *Probabilistic diversification and intensification in local search for vehicle routing*, Journal of Heuristics **1**, 147–167.
46. Schoonderwoerd, R., Holland, O., Bruten, J., Rothkrantz, L. (1996) *Ant-based load balancing in telecommunications networks*, Adaptive Behavior **5**, 169–207.
47. Smith, D. H., Hurley, S., Thiel, S. U. (1998) *Improving heuristics for the frequency assignment problem*, European Journal of Operational Research **107**, 76–86.
48. Solomon, M. (1987) *Algorithms for the vehicle routing and scheduling problem with time window constraints*, Operations Research**35**, 254–265.
49. Stuetzle, T., Dorigo, M. (1999) *Aco algorithms for the quadratic assignment problem.* In: D. Corne and M. Dorigo, and F. Glover,(eds.) New Ideas in Optimization, McGraw-Hill.
50. Stuetzle, T., Hoos, H. (1997) *Improvements on the ant system: Introducing $max-min$ ant system*, Proceedings of ICANNGA'97, International Conference on Artificial Neural Networks and Genetic Algorithms, Springer, Vienna.
51. Stuetzle, T., Hoos, H. (1998) *Ant system and local search for combinatorial optimization problems*, S. Voss, S. Martello, I. H. Osman and C. Roucairol (eds) Meta-Heuristics: Advanced and Trends in Local Search Paradigms for Optimization, Kluwer, Boston.

52. Taillard, E. (1991) *Robust taboo search for the quadratic assignment problem*, Parallel Computing **17**, 443–455.
53. Taillard, E. D., Badeau, P., Gendreau, M., Guertin, F., Potvin, J. Y. (1997) *A tabu search heuristic for the vehicle routing problem with soft time windows*, Transportation Science**31**, 170–186.
54. Thangiah, S. R., Osman, I. H., Sun, T. (1994) *Hybrid genetic algorithm simulated annealing and tabu search methods for vehicle routing problem with time windows*, Technical Report 27, Computer Science Department, Slippery Rock University.
55. Tiourine, S., Hurkens, C., Lenstra, J. K. (1995) *An overview of algorithmic approaches to frequency assignment problem*, Technical Report, T.U. Eindhoven.
56. Van Benthem, H. P. (1995) *Graph: Generating radio link frequency assignment problems heuristically*, Master's thesis, Faculty of Technical Mathematics and Informatics, T.U. Delft.
57. Van der Put, R. (1998) *Routing in packet switched networks using agents*, Technical Report RD-SV-98-276, KPN Research, The Netherlands.
58. Watkins, C. J., Dayan, P. (2000) *Q-learning*, Machine Learning **8**, 279–292.

Evolving Coordinated Agents

Sandip Sen, Sandip Debnath, and Manisha Mundhe

Department of Mathematical & Computer Sciences
University of Tulsa
Tulsa, OK, USA
E-mail: sandip@kolkata.mcs.utulsa.edu

Summary. In recent years, considerable interest and enthusiasm have been generated by the prospect of widespread use of intelligent agent-based systems [18]. In particular, a number of researchers have been investigating the design and implementation of systems consisting of multiple agents [26]. The design of successful multiagent systems is, however, a problem of significant magnitude and difficulty. Often multiple, conflicting criteria have to be simultaneously optimized to come up with a cost-effective multiagent system design. Agent system design may involve designing the infrastructure or environment for agent interaction as well as behavioral strategies for individual or multiple agents. Agent behaviors handcrafted offline can be inadequate if possible interactions are overlooked. Genetic algorithms provide us with another tool for designing both individual agent behaviors as well as social rules for multiagent systems. In this chapter we identify different modes for evolving agent groups and present instances of two different approaches: a coevolutionary optimization approach, and an adaptive system approach.

1 Introduction

Autonomous agents that possess expertise but lack proper coordination skills can suffer from poor performance. The focal point of multiagent systems research has been the development of coordination schemes that effectively overcome these inefficiencies [26]. Agent designers may not have a complete view or understanding of the environmental dynamics in complex environments. It may also be infeasible to foresee all possible agent interactions and environmental changes. As such, it is a challenging and at times impossible task to design behavioral strategies that can produce effective coordination between agents with limited perception and bounded reasoning capabilities. Coordination strategies designed offline can, therefore, prove to be inadequate because of either incomplete or incorrect knowledge.

The ability of agents to learn and adapt in complex environments is desirable as this can develop and sustain effective coordination strategies. Biological evolution can serve as a proof-of-principle model for generating coordination among agents based on performance in the actual domain of interest. A special case of evolution in which two or more different populations simultaneously evolve with coupled fitness is called coevolution [9,10,15,17,21,22]. In such a setting, evolutionary changes in any one species (population) influence the evolution of other species. Cooperative coevolution is an approach for evolving multiple individuals who can effectively cooperate to solve a common problem. In a cooperative setting, individuals try to coordinate their actions with an intent to improve global performance rather than maximizing local payoff/feedback. Example applications include developing strategies for hunting in a group [9,10] or evolving communicating rule-based classifier systems for the control of a wall-climbing quadrupedal robot [3]. In competitive coevolution, individuals from different populations are paired off for competition. Example applications include use of coevolution to develop strategies for two-player games like Tic-Tac-Toe, Nim, and Go [25], the coevolution of robots searching for food [7].

We present our approach to cooperative coevolution that uses a shared memory between populations for storing and using effectively coordinated groups of agents [23]. We experiment with a task allocation domain which requires multiple autonomous agents to coordinate their actions to maximize payoff. We compare our approach with other cooperative coevolutionary techniques and provide an analysis of the experiments conducted.

We also present our work on designing agent societies using evolutionary algorithms [20]. The social sciences literature discusses fundamental problems of providing and maintaining a public good in a society composed of self-interested individuals [6,8]. Public goods are social benefits that can be accessed by individuals irrespective of their personal contributions. We have demonstrated the use of genetic algorithms (GAs) for generating an optimized agent society that can circumvent a particularly problematic social dilemma. In our initial design, each chromosome represented an entire society of agents and a GA was used to find the best co-adapted or optimized society. Though encouraging, this result is less exciting than the possibility of evolving a set of co-adapted chromosomes where each chromosome represents an agent, and hence the population represents the society. We discuss the challenges of designing such an adaptive systems approach to using GAs for evolving agent societies. We also present experimental results from an example social paradox problem.

The rest of the chapter is organized as follows: Section 2 presents alternative designs for using population-based evolutionary techniques to design agent societies; Section 3 presents our shared memory-based cooperative coevolutionary approach to designing effectively coordinated agent groups; Section 4 presents our adaptive system approach to evolving a society of co-adapted agents that can avoid social dilemmas; Section 5 summarizes the contributions of this body of work and identifies some future research avenues.

2 Evolving Agent Societies

We now describe alternate evolutionary frameworks that can be used to evolve agent societies (see Figure 1):

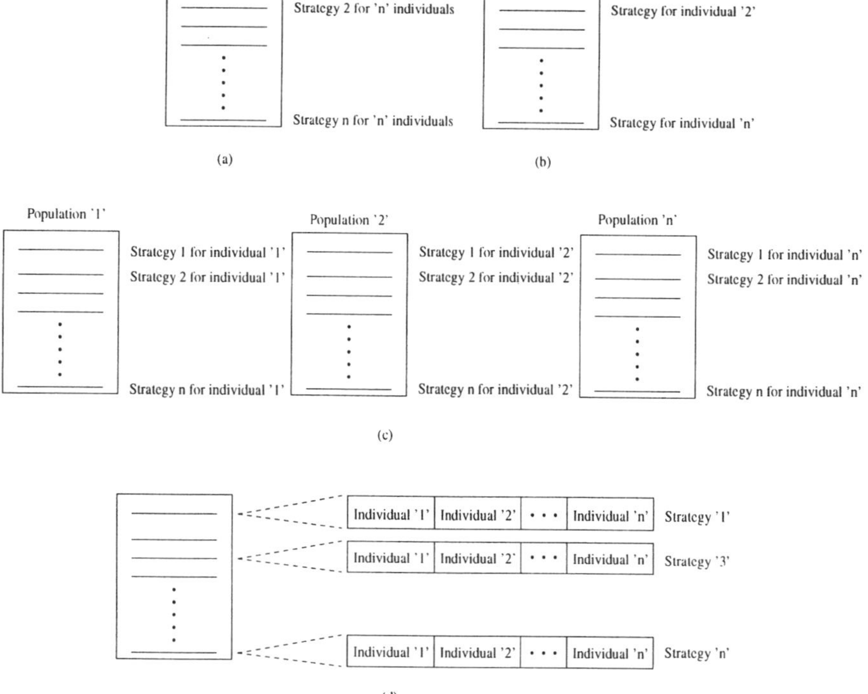

Fig. 1. Evolving agent societies.

1 **Evolving one strategy to be used by all individuals:** In this approach, one population of agent strategies is evolved. The best strategy is adopted by all individuals in the agent society to achieve effective coordination (see Figure 1(a)). This approach does not allow for individual specialization within a group.

2 **Evolving one strategy per individual:** In this approach, we evolve a population of strategies, where each strategy is adopted by an individual in the society (see Figure 1(b)). The goal is to find an effective set of strategies. This approach allows for multiple specializations to evolve in a group.

3 **Evolving one population per individual (coevolution):** In this approach, we simultaneously evolve multiple populations corresponding to the number of individuals in the system. Each population consists of a number of strategies for an individual. Strategies from each population are paired to evaluate the performance of the individuals (Figure 1(c)). The best strategy pairing is used to assign strategies to the different members of the agent society.

4 **Evolving strategies for all individuals:** In this technique of evolving strategies, a single population is evolved. Each chromosome in the population contains strategies for all agent society members (Figure 1(d)).

3 Coevolutionary Optimization

Coevolutionary approaches have to make critical decisions for evaluating individuals in any population. These include how to pair them with individuals from other populations [2,24], and how to assign fitness to an individual given evaluations of several such pairings. In most current GA or genetic programming (GP) based approaches, the goal is to optimize the behavior of an agent or a group of agents (also called agent societies). These formulations use the GA/GP as a function optimizer with higher fitness assigned to chromosomes in the population that produce better performance or generate greater reward/payoff from the environment for the corresponding agent or agent group. In most cases, each chromosome in one population is evaluated independently from other chromosomes in the same population[1] though they may be paired with multiple chromosomes from other coevolving populations. In the following, we identify commonly used pairing strategies and then provide an alternate shared memory-based pairing strategy tailored for cooperative coevolution.

3.1 Pairing Strategies

The fitness of individuals in coevolutionary techniques (competitive and cooperative) is evaluated by pairing individuals from one population with individuals from other populations (see Figure 2). Different methods have been examined to pair individuals to evaluate their fitness. The following are some of the pairing methods, used in association with the evolutionary configurations mentioned above, that have proved effective.

1 An individual from one population is paired with a random member in the current generation of the other GA populations [2].
2 An individual is evaluated by pairing it with the highly fit individuals in the previous generation from the other populations (fitness proportionate pairing).
3 Each individual is evaluated with a collection of the previous champions (individuals with superior performance) from other populations [7].
4 Best individuals from each of the populations are combined into a single composite structure and evaluated in the environment [19].

[1] Exceptions include the use of *competitive fitness sharing* [25].

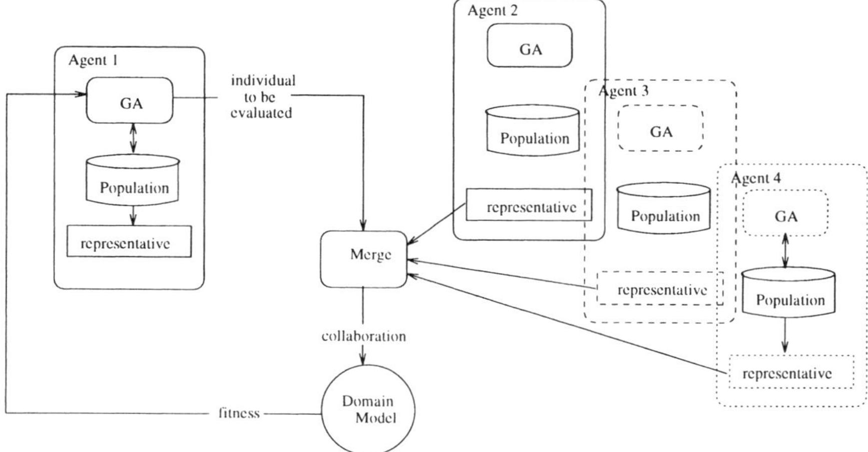

Fig. 2. Cooperative coevolutionary architecture using separate GAs for each agent.

In a cooperative environment, an individual is assigned a fitness by measuring its ability to cooperate with individuals from other coevolving populations to perform an assigned task. Some of the successful coevolution approaches investigated involve pairing agent rules from one population with a collection of good rules from the other population, or with a collection of the best rules from the other population over all generations [7], or by choosing pairs from the two populations randomly [2]. In each of the above approaches, rules that perform well are stored separately in each population to be evaluated with other rules in future generations. None of these approaches provides a means to remember successful pairings, as only the best rules from each population were available.

The caveat in these approaches is that the best rules paired may not produce the best pairs. This is because the best behaviors may not be complementary. Consider a task allocation problem between two agents with tasks numbered from 1 through N. Suppose the current best strategy in the first population performs all even-numbered tasks and the current best strategy in the second population performs all tasks numbered below $\frac{N}{2}$. When these two behaviors are paired they may not produce as good a task allocation as each of them can with some other, individually less effective, policy.

3.2 The Shared Memory Approach

We now present a shared-memory-based approach to cooperative coevolution. We use a shared memory accessible to both populations to store the n best *pairs* found so far. Figure 3 illustrates the process of evaluating an individual from population 1. The shared memory is initialized with randomly generated rule pairs along with their joint fitness. To evaluate an individual

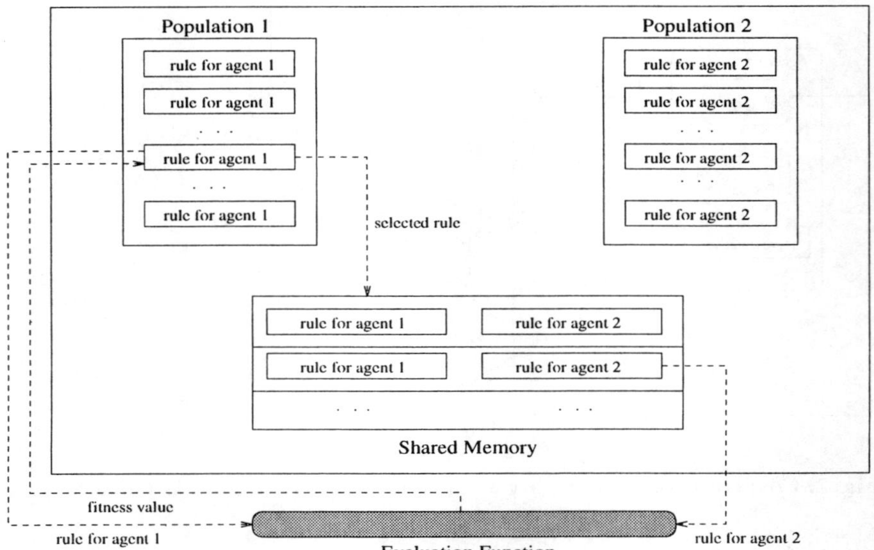

Fig. 3. Using shared memory with two GA populations.

from population 1, we pair it with each of the individuals from population 2 stored in the shared memory. A pair that has maximum fitness from the set of evaluations replaces the pair with the lowest fitness in the shared memory. This process of updating the shared memory is executed only if we encounter a pair with fitness higher than that of at least one of the pairs that are currently represented in the shared memory. Thus the pairs with the highest fitness are stored in a shared memory which maintains a collection of the best pairs over generations. By storing several of the best pairs, we ensure that important behavioral traits are not lost. Though our shared memory approach can be used with larger groups, we demonstrate it here in a domain with two cooperative agents.

In a cooperative setting, it is necessary that we provide an individual from one population with an opportunity to pair up with a number of other individuals from the other population to discover effective pairs. In our experiments, to ensure that an individual gets distinct evaluations, we pick random individuals from the agent population when we run out of distinct pairs in the shared memory. The shared memory contains rule pairs that perform effectively. A rule or strategy from one population may perform well in combination with one or more rules from the other population. In such cases, the shared memory contains duplicate rules. However, although there is a possibility of duplicate individuals being present in the shared memory, the pairs are always distinct.

In competitive domains, the evaluation of an individual should be proportional to the average over its evaluation when pitted against a sampling of individuals from the other population. In a cooperative setting, the highest evaluation an individual receives from all pairings with others appears to be the best choice for the evaluation of that individual. This is because we are interested in finding a pair that performs the best overall. If a pair performs better than any other pair, it could be because one of them performed really well, even if the other was not effective. However, it is also not essential that an individual be robust; it just has to effectively co-adapt with another individual. We therefore recommend using the best evaluation it gets over several pairings in such scenarios.

3.3 Problem Domain

In this section, we describe the problem domain we have used for our study. We presented initial results using the shared memory approach in a cooperative room painting domain [23]. The room painting domain, however, did not allow for a systematic evaluation of our approach by varying the problem difficulty. We decided to apply the shared-memory-based coevolution approach to a domain where the degree of cooperation required by the agents and the nature of their coupling can be systematically varied. To satisfy this criterion, we designed a distributed task allocation problem which we now describe.

The goal of the distributed task allocation problem is to evolve agent groups such that together they obtain the maximum utility by processing the tasks chosen. There will be as many GA populations as there are agents in the group. An agent is evaluated in conjunction with all of the other agents. Each agent is allowed to execute a maximum of k tasks out of a total of T tasks. The problem specification includes a matrix of elements, where element u_{ij} represents the utility accrued when the ith task is allocated to the jth agent. A chromosome for an agent represent the tasks it has chosen to perform. From such a choice set, the total utility obtained can be calculated by summing the utility of each of the tasks in this agent's allocation. The jth agent is also required to perform all the tasks from a set of tasks, t_j. For each of the tasks it fails to perform from this set, its total utility is penalized a fixed percentage, p. In the worst case, when the agent fails to perform any of the tasks from the required set, it gets penalized a maximum of $p * |t_j|$ of its total utility.

For ease of exposition we will assume only two agent groups (we have also used two agents in all of the experiments in this chapter). To evaluate an individual from the first population it is paired in succession with individuals from the other population. For each such pairing, any duplicate tasks outside required task sets are identified. If the lth task is chosen by both agents, the group gets a utility equal to $\max(u_{l1}, u_{l2})$. This prevents double counting for performing the same task. The pair evaluation returns the sum of the utilities of only the tasks performed and after imposing any applicable penalties for not performing required tasks. As mentioned above, the maximum pair evaluation received by an individual is used as the fitness of the individual.

To systematically vary the difficulty of the problem in terms of the coupling between individuals, we varied T with respect to k. In our experiments, we used $k = 20$, and varied T from 20 to 40. When $k = T$ we have a completely decoupled system, i.e., the utility obtained by one agent does not depend on the other agent. On the other hand, when $T = 2k$, maximum fitness can be obtained by a pair only when they choose complementary task sets with no overlap.

3.4 Experimental Results

In the first set of experiments we evaluated the shared memory approach against a random pairing approach (where individuals are chosen randomly from the other population for pairing with an individual to be evaluated) and a single population approach (where a single population was used to evolve both agents, i.e., each chromosome contained the task sets for both agents (see Figure 1(b))). We ran the GAs for 2000 generations with a population size of 50, crossover rate of 0.9, mutation rate of 0.05, and shared memory size of 10. Each chromosome was initialized with distinct tasks, i.e., there were no duplicate tasks in a chromosome. For a fair comparison, in the single population approach we enforced that each of the two agents in one chromosome had distinct tasks; there can still be duplication of tasks between the two agents. Utilities for processing tasks were selected randomly between 1 and 20. The total task set, T, is 40 for this set of experiments. Results are averaged over 10 runs and are presented in Figure 4.

¿From the figure it is clear that the shared memory approach improves performance over the random pairing approach, which in turn improves over the single population approach. These differences are also statistically significant. The single population approach suffers because each individual does not get paired with a sufficient number of individuals of the other type to get a fair evaluation. Both the shared memory and random pairings give 10 evaluations to each individual and this allows for more opportunity to obtain a true measure of goodness of an agent. The shared memory approach provides a balance between exploration (random pairing) and exploitation (pairing with shared memory elements). This balance allows for a more sustained selection pressure in the population compared to the random pairing approach.

In the next set of experiments we varied the coupling between the agents by varying T while holding k constant at 20. With increase in T, the utility obtainable increased and this is shown by higher convergence levels for the corresponding plots in Figure 5. It is interesting to note that there is no appreciable increase in the number of generations needed by the agents to find the best co-adapted pair except for the completely decoupled case ($T = 20$). The best pair is found within a few generations for the completely decoupled

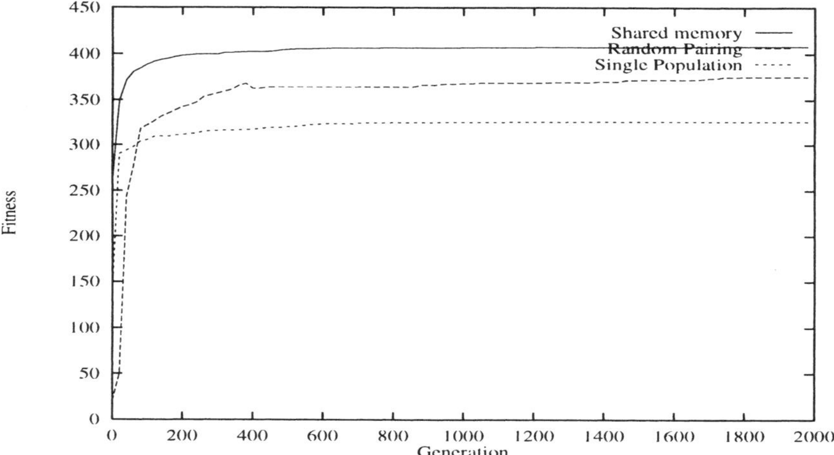

Fig. 4. Comparison between shared memory, random pairing, and single population.

case. For all other instances the best pair is found within 1500 generations, and in most cases it is found within the first 1000 generations. We still need to evaluate the approach with larger problems, e.g., by significantly increasing k and T. For the reasonable-sized problems we have experimented with, however, the shared-memory-based coevolution handles the increase in agent coupling without appreciable deterioration in convergence time.

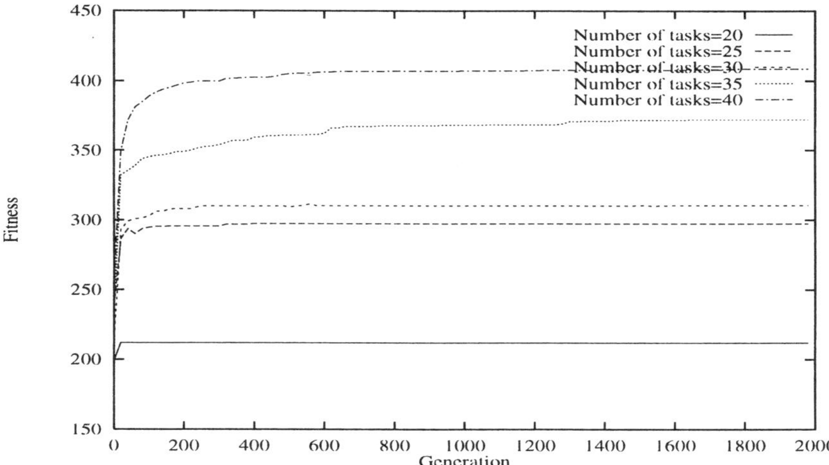

Fig. 5. Varying problem difficulty by changing the total number of tasks that needs to be processed while holding the number of tasks processed by any one agent constant at 20.

In the third set of experiments we evaluated the fitness assignment method used. From the many pairings that an individual gets, we can assign it the best, average, or worst finesses. We ran experiments with these different methods and for $T = 40$. Results plotted in Figure 6 are averaged over 10 runs. The worst fitness choice clearly lags behind the other two choices. Though the best fitness choice leads to quicker learning early on compared to the average fitness choice, the final convergence results are statistically indistinguishable. This suggests that any initial advantage of using the best fitness assignment is likely to go away with further search.

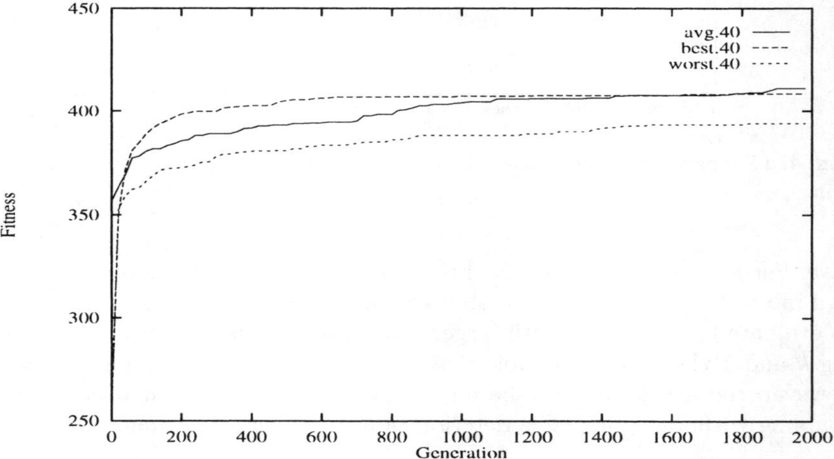

Fig. 6. Performance curves of GA using shared memory approach and evolving with the best, average, and worst fitness.

3.5 Observations

The goal of the shared memory approach is to expedite the evolution of agent strategies that pair with each other to produce maximum fitness. We show that the performance of an agent is increased if it is exposed to a number of individuals from the other population to be paired. In adopting the single population approach, the agents did not have the option of pairing with a number of other individuals to obtain their fitness. This lack of variety resulted in poor performance of the individual. In the random pairing approach, we rely on the performance of individuals in the current generation. Although good pairs are produced, the number of pairs with high performance would be fewer than that obtained with the shared memory approach.

In the shared memory approach, to evaluate an individual we pair it with individuals from the other population stored in the shared memory. The shared memory, however, can contain the same individual multiple times as part of different high-performing groups. In such cases, in place of duplicate evaluations, we pick an individual randomly from the other population for pairing with the individual being evaluated. An individual is assigned the highest evaluation from all such pairings. We conducted experiments to examine the contribution of shared memory (exploitative) and random pairing (explorative) towards the overall fitness. Experiments conducted have shown that using random pairing along with the shared memory approach helps in preventing the populations from premature convergence.

We propose assigning fitness to an individual based on the maximum evaluation it receives from one of several pairings with individuals from the other coevolving populations. Researchers have used the evaluation of an individual to be an average over all its evaluations with individuals from the other population. This approach is best in environments of a competitive nature where we are interested in finding robust individuals. We show that in a cooperative setting, there can be some benefits early in the evolution process of measuring the fitness of an individual using the maximum evaluation of all evaluations. This early advantage, however, is not necessarily sustained over the course of prolonged evolution. In addition, we conducted experiments by evolving the agent populations with the worst fitness an individual receives from a set of pairings with individuals from the other population. Results show that evolving with the worst fitness leads to poor performance in comparison to approaches using the maximum and average measures. These results hold for both the shared memory approach as well as the random pairing approach.

The shared memory approach is shown to outperform the random pairing and single population approaches. The random pairing approach picks individuals from a population randomly to be paired. This amounts to an explorative mode of partner selection. A shared memory approach that selected partners only from the shared memory, and never randomly, would be a very exploitative mode of partner selection. Our approach provides a balance between exploration and exploitation. This is because when the shared memory contains copies of the same agent, some partners are chosen randomly. The shared memory takes advantage of good individuals discovered during evolution by storing and using them for evaluating other population members in the future. Since new strategies get a chance to be paired with relatively proven strategies, they receive a fair evaluation. An analysis of experimental traces involving the shared memory approach and the random pairing approach shows that the balance provided by the shared memory approach between the exploitative trend of pairing with good strategies and the explorative trend of pairing with randomly selected strategies is quite effective.

It will be instructive to further investigate the effect of shared memory size on the effectiveness of the evolutionary process. One can also investigate the use of shared memory in other evolutionary algorithms.

4 Adapting Societies to Avoid Social Dilemmas

In this section, we present an alternate formulation to evolving coordinated societies. For a number of years, GAs have been successfully used primarily as function optimizers. Holland's work on GAs, however, was motivated by design and implementation of robust adaptive systems [14]. In recent years, researchers have started to pay more attention to the initial motivation of GAs [4].

In a typical function optimization-based approach to evolving agent societies, a single chromosome represents the entire society (see Figure 1(d)), and the optimization problem translates to finding the optimal society [1,11]. The evaluation of one chromosome is independent of the other chromosomes in the population. Contrast this to the evolutionary approach where a single population is evolved where each chromosome or structure represents an agent (see Figure 1(b)), and the challenge is to evolve an effectively co-adapted population of chromosomes. The evaluation of each structure depends on all the other structures in the population. We will call this an *adaptive systems* approach to evolving agent societies.

There are at least two compelling reasons for investigating the adaptive systems framework:

1 It allows for the construction and evaluation of novel challenging scenarios for evolving populations, e.g., situations where one agent has to compete with some and cooperate with other agents.
2 Careful experimentation with these scenarios can help us develop a better understanding of the dynamics of evolutionary algorithms. The optimization framework of independent chromosome evaluation has been studied extensively and we have developed a reasonable understanding of how an evolutionary algorithm works in this mode. But when the evaluations of population structures are interdependent, as in the adaptive systems framework, much less is known and there is little understanding of the dynamics and convergence of the evolutionary process. In the adaptive systems framework, a more complex interplay among agent strategy representation, evaluation modes (how many pairings, with whom, etc.), evaluation functions, selection and replacement schemes, incremental versus generational evolution schemes, etc., provides a richer framework to research the capabilities and limitations of evolutionary paradigms.

The viability of an individual in natural or artificial societies often depends on group composition and the behaviors of other members. In our previous work [1], we presented our results of applying GAs as function optimizers to solve social paradox problems. We used an optimization approach based on a Pitt-style genetic-based machine learning system, where each structure in the population represents an entire group of agents. A critique of this work would be that since the GA selects between alternative societies, it really avoids the social dilemma problem: rather than solving the social dilemma problem for a society, the GA selects the society which has solved this problem. This argument is analogous to the Michigan versus Pitt-style classifier system debate, where the latter avoids the individual rule credit assignment problem in the former by working with entire rule sets. In the following we present an alternate approach where the GA evaluates each individual in a society and the entire population represents a society. The challenge for the GA is to adapt the population to produce a self-sustaining balance between exploitative and greedy components of the population.

4.1 Social Dilemmas

A social dilemma arises when agents have to decide between contributing or not contributing towards a public good without the enforcement mechanism of a central authority [6]. Individual agents have to tradeoff local and global interests while choosing their actions. If a sufficient number of agents make the selfish choice, the public good may not survive, and then everybody suffers. In general, social laws, taxes, etc., are enforced to guarantee the preservation of necessary public goods. In the following, we present a representative social dilemma that we address with GAs in this chapter.

Braess Paradox: Consider a resource sharing problem where the cost of utilizing a resource increases with the number of agents sharing it, e.g., congestion on traffic lanes. Assume that initially the agents are randomly assigned to one of two identical resources. Now, if every agent opts for the resource with the least current usage, the overall system cost (cost incurred per person) increases [13]. So, the dilemma for each agent is whether or not to make the greedy choice. We will now briefly review some of the work by Glance and Hogg [5] on social dilemmas in groups of computational agents. Glance and Hogg study a version of the social dilemma problem known as the Braess Paradox [16] in the context of a traffic flow problem. Figure 7 shows agents entering the network from the bottom and choosing among several paths, moving between nodes along the indicated links. The cost of traversing a link is either constant or directly proportional to the fraction of agents traversing it (these fractions, f_i, are used as link labels). When the link between B and C is absent, both paths ABD and ACD are equally attractive and half of the agents take the left route and the other half choose the right route. Thus $f_1 = f_2 = 0.5$, and the cost per agent is 1.25. Consider the network as a highway system where each agent seeks to minimize its travel time across the network. It would seem that adding more highways can only reduce the travel time of agents. However, it turns out that the addition of an extra route can sometime decrease the overall throughput of the system, i.e., increase the travel time of agents. This is an instance of the Braess Paradox. For example, when a link is added between B and C (in Figure 7) with cost $\frac{1}{4} < x < \frac{1}{2}$, a greedy decision to minimize individual costs leads each agent to choose the path ABCD. This results in the average cost per agent to be 1 + x which is greater than the average cost of 1.25 without the link.

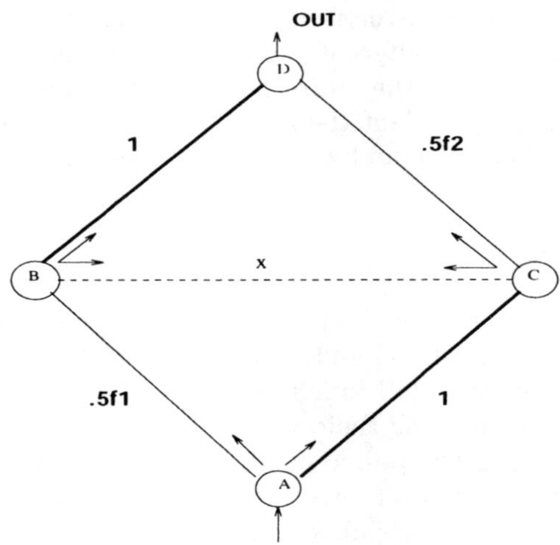

Fig. 7. Example of a stand-alone Braess Paradox.

4.2 Experimental Results

A GA approach was used by Glance and Hogg [5] on a larger network (see Figure 8) which has a smaller network representing a Braess Paradox (see Figure 7) embedded in it. Each structure in the GA encoded a path taken by individual agents. Glance and Hogg found that the extra link always lowered the performance of the system and hence concluded that the GA was unable to solve the social dilemma.

In our previous work we used an optimization approach to evolve a set of agent strategies that "solves" the paradox [1]. Now, we present our experimental results using an adaptive systems approach for this problem. In our encoding we used 8 bits to represent each agent, 1 bit for each possible decision node in the network (the 8 bits correspond to the decision nodes 1,2,A,B,C,D,3,4 in Figure 8). Thus for any two structures in the GA population, the same bit position now represents the same decision node in the network. For example, consider the two strings S_1 = 01001001 and S_2 = 11000110. The first bit of S_1 forces the agent along the path between node 1 and node A (see Figure 8). Then we ignore the second bit as the decision at node 2 need not be made. Thus, in both cases, when the agent is at node A, it proceeds to node B.

We used an incremental GA for the experiments on the Braess Paradox problem. Figure 9 shows the results of our experiments with 10 agents averaged over 10 runs with different costs of x link. The GA easily finds the

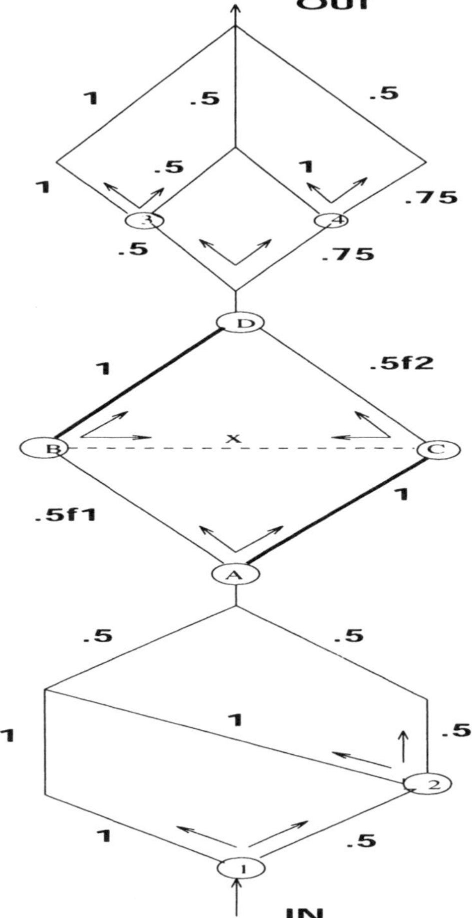

Fig. 8. Example of an embedded Braess Paradox.

optimal configuration if the x link is not present. If the cost of the link is too low, say 0.2, all individuals converge to the rule that always takes the x link. Analogously, if the cost is too high, the rules converge to the one that never takes the x link. The dilemma arises when the cost of x link is about 0.4 – nearly half of the individuals take the x link, while the rest avoid it. In this case, the adaptive GA takes longer to find the optimal pattern of individuals in the population than in the non-dilemma cases, but the population converges consistently to the optimal structure. Table 1 present populations at the end of typical runs for different x values. These results show that the GA is able to produce an effectively co-adapted set of structures that circumvents the Braess Paradox problem.

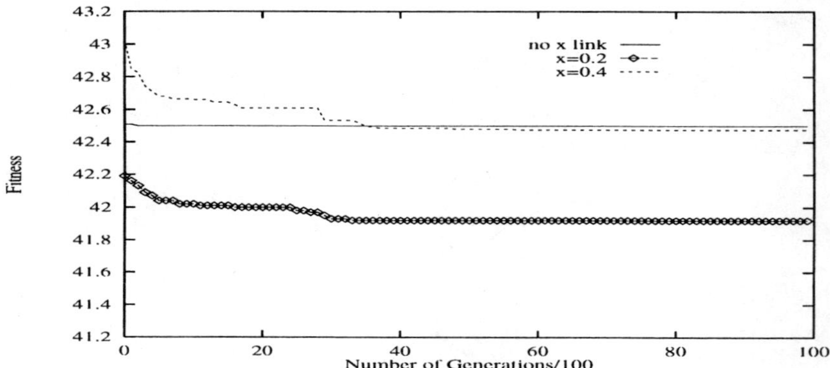

Fig. 9. Experiments with the Braess Paradox problem.

Performance comparison with our previous work on using a function optimization approach [1] shows that the adaptive systems approach performs a little better in the dilemma situation. But the claim in this chapter is not necessarily that of superiority with respect to the function optimization approach. As mentioned before, our primary goal is to demonstrate the feasibility of the adaptive systems approach for designing agent systems with effectively co-adapted individuals.

x=0.4	x=0.2	no x link
11010010	11110010	11010011
11110010	11010010	11100011
11010010	11010010	11100011
11000010	11010010	11100011
11000010	11000010	11000011
11110010	11000010	11000011
11010010	11010010	11000011
11000010	11010010	11100011
11011010	11110010	11100011
11110010	11110011	11100011

Table 1. Evolved population members for different network configurations.

4.3 Observations

We have shown that an adaptive systems approach for using GAs for evolving agent societies faced with social dilemmas can be a viable alternative to the straightforward optimization approach that is normally used. This was not easily foreseeable, and gives testimony to the robustness of the GA approach.

Social dilemmas are not restricted to human societies and are bound to plague artificial social systems [5,12,27], e.g., message congestion problems. A naive design and implementation of agent societies, therefore, will be likely to lead to ineffective utilization of resources. Given an environment, we can evolve agent societies that optimally utilize the resources available to them. The model proposed in this chapter assumes an evolutionary setting, where agent societies that are better able to effectively utilize their resources are more likely to prosper over time. This model may not be appropriate in a number of domains where the agent designer is designing one or more agents that will interact with other self-interested agents and little is known about the characteristics of these other agents. This model is appropriate, however, when an agent designer is designing an entire artificial agent society and wants these agents to effectively share all the resources. Since cooperative agent societies are not immune to social dilemmas [5], the mechanisms that we are studying will be useful for the design of effective cooperative societies.

5 Conclusion

The optimization approach to designing agent societies is likely to continue to be the major thrust of research in evolutionary algorithm-based agent system design. Our work presented in the first part of the chapter identifies effective ways of coevolving cooperative groups. To date, more GA/GP researchers have concentrated on competitive coevolution. This may stem partly from the general perception that since more knowledge is available about all agents when the designer is in charge of designing an entire group of agents, the problem of designing cooperative groups is less challenging. But a large design space and an improperly understood, dynamic environment emphasize the need for viable alternatives to hand-crafted designs. We believe that our proposed approach is an interesting spring-board for further investigations into coevolving proficient cooperative agent groups.

As argued in the second part of this chapter, using GAs in the adaptive system mode not only allows for effective design of agent societies, but also poses interesting challenges for GA researchers. Encouraged by our initial results, we plan to test the adaptive systems approach in larger, more complex social dilemma situations, e.g., the Tragedy of the Commons [8]. In the case of general social dilemma problems, the key research issue will involve the interplay between an appropriate evaluation function and reproductive schemes to balance local greedy choices with moderation or restraint imposed by global considerations.

Acknowledgments

This work has been supported in part by an NSF CAREER award IIS-9702672.

References

1. Arora, N., Sen, S. (1997) Resolving social dilemmas using genetic algorithms: Initial results. In *Proceedings of the Seventh International Conference on Genetic Algorithms*, 689–695, San Mateo, CA, Morgan Kaufmann
2. Bull, L. (1997) Evolutionary computing in multi-agent environments: Partners. In *Proceedings of the Seventh International Conference on Genetic Algorithms*, 370–377, San Mateo, CA, Morgan Kaufmann
3. Bull, L., Fogarty, T. C., Snaith, M. (1995) Evolution in multiagent systems: Evolving communicating classifier systems for gait in a quadrupedal robot. In *Proceedings of the Sixth International Conference on Genetic Algorithms*, 382–389, San Francisco, CA, Morgan Kaufmann
4. DeJong, K. A. (1993) Genetic algorithms are NOT function optimizers. In L. Darrell Whitley, editor, *Foundations of Genetic Algorithms 2*, 5–18
5. Glance, N. S., Hogg, T. (1995) Dilemmas in computational societies. In *First International Conference on Multiagent Systems*, 117–124, Menlo Park, CA, AAAI Press/MIT Press
6. Glance, N. S., Huberman, B. A. (1994) The dynamics of social dilemmas. *Scientific American*, **270**, 76–81
7. Grefenstette, J., Daley, R. (1996) Methods for competitive and cooperative coevolution. In Sandip Sen, editor, *Working Notes for the AAAI Symposium on Adaptation, Co-evolution and Learning in Multiagent Systems*, 45–50, Stanford University, CA
8. Hardin, G. (1968) The tragedy of the commons. *Science*, **162**, 1243–1248
9. Haynes, T., Sen, S. Evolving behavioral strategies in predators and prey. In Gerhard Weiß and Sandip Sen, editors, *Adaptation and Learning in Multi-Agent Systems*, Lecture Notes in Artificial Intelligence, 113–126. Berlin, Springer Verlag
10. Haynes, T., Sen, S. (1997) Crossover operators for evolving a team. In John R. Koza, Kalyanmoy Deb, Marco Dorigo, David B. Fogel, Max Garzon, Hitoshi Iba, and Rick L. Riolo, editors, *Proceedings of Genetic Programming 97: the Second Annual Conference*, 162–167, San Francisco, CA, Morgan Kaufmann
11. Haynes, T., Wainwright, R. L., Sen, S., Schoenefeld, D. A. (1995) Strongly typed genetic programming in evolving cooperation strategies. In L. Eshelman, editor, *Genetic Algorithms: Proceedings of the 6th International Conference*, 271–278, San Francisco, CA, Morgan Kaufmann
12. Hogg, T. (1995) Social dilemmas in computational ecosystems. In *Proceedings of the International Joint Conference on Artificial Intelligence*, 711–716
13. Hogg, T., Huberman, B. A. (1991) Controlling chaos in distributed systems. *IEEE Transactions on Systems, Man, and Cybernetics*, **21**, 1325–1332 (Special Issue on Distributed AI)
14. Holland, J. H. (1975) *Adpatation in natural and artificial systems.* University of Michigan Press, Ann Arbor, MI
15. Iba, H. (1996) Emergent cooperation for multiple agents using genetic programming. In *Parallel Problem Solving from Nature - PPSN IV*, 32–41, Berlin, Springer Verlag
16. Irvine. A. D. (1993) How Braess' Paradox solves Newcomb's problem. *International Studies in the Philosophy of Science*, **7**, 141–160

17. Ito, A., Yano, H. (1995) The emergence of cooperation in a society of autonomous agents – the Prisoner's Dilemma game under disclosure of contract histories. In *Proceedings of the First International Conference on Multiagent Systems*, 201–208, Menlo Park, CA, AAAI Press/MIT Press
18. Jennings, N. R., Sycara, K., Wooldridge, M. (1998) A roadmap of agent research and development. *International Journal of Autonomous Agents and Multi-Agent Systems*, **1**, 7–38
19. De Jong, K. A., Potter, M. A. (1995) Evolving complex structures via cooperative coevolution. In *Fourth Annual Conference on Evolutionary Programming*, San Diego, CA
20. Mundhe, M., Sen, S. (2000) Evolving agent societies that avoid social dilemmas. In *Proceedings of the Genetic and Evolutionary Computation Conference, GECCO-2000*, 809–816
21. Paredis, J. (1997) Coevolving cellular automata: Be aware of the Red Queen! In *Proceedings of the Seventh International Conference on Genetic Algorithms*, 393–400. San Mateo, CA, Morgan Kaufmann
22. Potter, M., De Jong, K. A., Grefenstette, J. J. (1995) A coevolutionary approach to learning sequential decision rules. In *Proceedings of the Sixth International Conference on Genetic Algorithms*, 373–381, San Francisco, CA, Morgan Kaufmann
23. Puppala, N., Gordin, M., Sen, S. (1998) Shared memory based cooperative coevolution. In *Proceedings of the International Conference on Evolutionary Computation'98*. Piscataway, NJ, IEEE Press
24. Riolo, R. (1997) The effects and evolution of tag-mediated selection of partners in populations playing the Iterated Prisoner's Dilemma. In *Proceedings of the Seventh International Conference on Genetic Algorithms*, 378–385, San Mateo, CA, Morgan Kaufmann
25. Rosin, C. D., Belew, R. K. (1995) Methods for competitive co-evolution: Finding opponents worth beating. In *Proceedings of the Sixth International Conference on Genetic Algorithms*, 373–382, San Francisco, CA, Morgan Kaufmann
26. Sen, S. (1997) Multiagent systems: Milestones and new horizons. *Trends in Cognitive Sciences*, **1**, 334–339
27. Turner, R. M. (1993) The Tragedy of the Commons and Distributed AI Systems. In *Working Papers of the 12th International Workshop on Distributed Artificial Intelligence*, 379–390

Exploring the Predictable

Jürgen Schmidhuber

IDSIA, Galleria 2, 6928 Manno-Lugano, Switzerland
E-mail:juergen@idsia.ch - http://www.idsia.ch/~juergen

Summary. Details of complex event sequences are often not predictable, but their reduced abstract representations are. I study an embedded active learner that can limit its predictions to almost arbitrary computable aspects of spatio-temporal events. It constructs probabilistic algorithms that (1) control interaction with the world, (2) map event sequences to abstract internal representations (IRs), (3) predict IRs from IRs computed earlier. Its goal is to create novel algorithms generating IRs useful for correct IR predictions, without wasting time on those learned before. This requires an adaptive novelty measure which is implemented by a co-evolutionary scheme involving two competing modules *collectively* designing (initially random) algorithms representing experiments. Using special instructions, the modules can bet on the outcome of IR predictions computed by algorithms they have agreed upon. If their opinions differ then the system checks who's right, punishes the loser (the surprised one), and rewards the winner. An evolutionary or reinforcement learning algorithm forces each module to maximize reward. This motivates both modules to lure each other into agreeing upon experiments involving predictions that surprise it. Since each module essentially can veto experiments it does not consider profitable, the system is motivated to focus on those computable aspects of the environment where both modules still have confident but different opinions. Once both share the same opinion on a particular issue (via the loser's learning process, e.g., the winner is simply copied onto the loser), the winner loses a source of reward — an incentive to shift the focus of interest onto novel experiments. My simulations include an example where surprise-generation of this kind helps to speed up external reward.

> *We can learn only what we already almost know.*
>
> PATRICK WINSTON
>
> *Zwei Seelen wohnen, ach, in meiner Brust!*
> *(Two souls, alas! are dwelling in my breast!)*
>
> JOHANN WOLFGANG GOETHE

1 Introduction

How does one explore the world? By creating novel, surprising situations and learning from them until they are understood and boring. But what exactly should constitute a surprise?

Intuitively, surprise or novelty implies unexpectedness. For instance, many would be surprised by the view of a flying elephant. Unexpectedness by itself, however, does not imply surprise. For instance, few would be surprised by the unpredictable and therefore unexpected details of the elephant's skin texture.

One difference between the flying elephant and its skin texture is that the former violates a confident prediction ("this elephant won't fly") while the latter does not. Obviously surprise has to be measured with respect to some given predictor. Only a predictor explicitly expressing an expectation that may be wrong can leave room for surprise and novelty: no surprise without the commitment of a prediction.

Now consider that in general it is not particular inputs that are surprising, but sequences of inputs. For instance, many would say an airborne elephant that jumped off the roof of a building is less surprising than one lifting off from the street and disappearing in the clouds. In fact, the predictable and the surprising things are usually spatio-temporal abstractions of the input sequences. For example, all possible visual input sequences caused by physically plausible trajectories of falling elephants (modulated by eye and head movements and varying lighting conditions) may map onto the same abstract internal representation "falling elephant," as opposed to the more surprising "flying elephant."

Let's have a closer look now at how the concept of a surprise is implemented in some existing, "curious," "inquisitive" machine learning systems designed to explore a given environment, and how this differs from what we intuitively might want.

Most previous work on exploring unknown data sets has focused on selecting single training exemplars maximizing traditional information gain [1–5]. Here typically the concept of a surprise is defined in Shannon's sense [6]: some event's surprise value or information content is the negative logarithm of its probability. This inspired simple reinforcement learning approaches to pure exploration [7–9] that use adaptive predictors to predict the entire next input, given current input and action. The basic idea is that the action-generating module gets rewarded in the case of predictor failures. Since it tries to maximize reward, it is motivated to generate action sequences leading to yet unpredictable states that are "informative" in the classic sense. Some of these explorers actually like white noise simply because it is so unpredictable, thus conveying a lot of Shannon information.

Most existing systems are limited to picking out simple statistical regularities such as "performing action A in discrete, fully observable environmental state B will lead to state C with probability 0.8." Essentially, they always predict all the details of single inputs (or the next state among a set of predefined states), and are not able to limit their predictions solely to certain computable aspects of inputs (as requested in [10]) or input sequences, while ignoring random and irregular aspects. For instance, they cannot even express (and therefore cannot find) complex, abstract, predictable regularities such as "executing a particular sequence of eye movements, given a history of incomplete environmental inputs partially caused by a falling elephant, will result in the appearance of a big red stain on the street within the next 3 seconds, where details of the shape of the stain are expected to be unpredictable and left unspecified."

General spatio-temporal abstractions of this kind apparently can be made only by systems that can run fairly general algorithms mapping input/action sequences to compact internal representations conveying only certain relevant information embedded in the original inputs. For instance, there are many different, realistic, plausible big red stains — all may be mapped onto the same compact internal representation predictable from all sequences corresponding to the abstraction "falling elephant." Only if the final input sequence caused by eye movements scanning the street does not map onto the concept "big red stain" (because the elephant somehow decelerated in time and for some strange reason never touched the ground) will there be a surprise.

The central questions are: In a given environment, how does one extract the predictable concepts corresponding to algorithmic regularities that are not already known? Which novel input sequence-transforming algorithms do indeed compute reduced internal representations permitting reliable predictions? Usually we cannot rely on a teacher telling the system which concepts are interesting, such as in the EURISKO system [11]. So how to discover novel spatio-temporal regularities automatically among the many random or unpredictable things that should be ignored?

To study these questions, I will use a rather general algorithmic setup. Consider an agent exposed to a lifelong sequence of complex (for example, visual) inputs from a real world-like environment that it can manipulate. The agent is able to compose algorithms written in a Turing-equivalent programming language (one that permits construction of a universal Turing machine). The instruction set includes instructions that can access the current input (for example, pixels of visual scenes), modify the environment via actions, and modify an internal memory consisting of many addressable memory cells. It also includes arithmetic instructions and conditional jumps (conditioned on current memory cell contents).

An important role is played by so-called *"Bet!"* instructions that can be used as parts of algorithms that make predictions about memory cell contents. For instance, suppose there already exists a complex algorithm that writes value 7 into memory cell 22 whenever there has occurred one of the many visual input sequences caused by typical falling elephants. Suppose there exists another piece of code that writes value 7 into memory cell 44 shortly after a big red stain in the street is identified by an appropriate algorithm. Finally, suppose shortly after cell 44 is filled another algorithm looks at cell 22 and a history of recent eye movements represented in other memory cells and claims that cell 44's content is 7, *without looking at it.* The claim is simply implemented via a *Bet!* instruction that addresses the two memory cells and compares their contents. It corresponds to a prediction expressing abstract knowledge about the world. The prediction may be wrong. This will constitute a surprise — a violation of an explicit (confident) prediction limited to a very particular, abstract transformation of a complex input sequence.

Now, in the absence of an external programmer or teacher, how could such algorithms focusing on the essential, predictable aspects of the environment be learned? How can we motivate the system to focus on such *novel* algorithms without wasting time on previously learned algorithms whose effects are already known? How can we prevent it from focusing on trivial transformations mapping all input sequences on the same internal representation which is always predictable? Clearly, we somehow need to evaluate whether something is trivial or already known or novel or just plain irregular. The model of what's known needs to be adaptive, quickly accessible, and able to generalize from experience with, say, flying elephants, to yet unseen instances of flying elephants. A look-up table model, for instance, is out of the question, due to its lack of generalization capability, and the large number of possible algorithms to investigate.

Here I propose to implicitly represent both algorithmic knowledge and model of knowledge in an adaptive data structure evolving through co-evolution of two essentially identical, algorithm-generating modules that collectively design experiments and bet on their outcomes, competing and cooperating at the same time. I will motivate each module to show the other regularities it does not yet believe in.

Each module is a set of modifiable, real-valued parameters, and owns initially equal amounts of "money" represented by two real-valued variables — the money idea in the context of machine learning can be traced back at least to Holland's bucket brigade system [12]. There is a function that maps the current values of the parameters of *both* modules onto a *single*, possibly complex algorithm that may include *Bet!* instructions corresponding to statements such as "cells 22 and 44 are now equal." Thus the modules *collectively* design and run an algorithm they have *agreed* upon.

For each *Bet!* instruction the modules can bet in advance whether the prediction is wrong or true (bids are real values, either fixed or based on the current module parameters). If they bet on different outcomes, given a particular *Bet!* instruction, then the system will immediately check which module is right. The winner gets rewarded (it receives the other's bid which is added to its money), while the surprised loser is punished (loses its bid).

It is absolutely essential that both modules *agree* on each experiment they assemble and execute. This is to make sure that no module can cheat. In the elephant example above, for instance, cheating would be possible if one of the modules could insert into its current algorithm instructions (not authorized by the other module) that look at cell 44's contents and use this information before computing an appropriate bid. This would be like two differently privileged viewers watching a magic trick, one in the audience being surprised by something the other finds entirely predictable because the latter can observe hidden movements of the magician from backstage. By agreeing on the entire instruction protocol of the current experiment the modules implicitly agree on all the information that may be used for the current computation and the current predictions.

In trying to maximize reward, each module is motivated to surprise the other as often as possible. Each is also motivated to veto computations whose outcome it does not consider profitable. As a consequence, each is essentially motivated to lure the other into agreeing on experiments that will surprise it. But by agreeing, each module expresses its belief that it is actually the other module which is in for a surprise. Using an appropriate evolutionary or reinforcement learning (RL) algorithm, a surprised module will eventually adapt. In turn, the other will lose a source of reward — an incentive to shift the exploration focus and try new algorithms revealing *novel*, yet unknown, predictable regularities.

One of the simplest such RL algorithms divides the learners' life times into trials, each lasting, say, 1000 successive instructions. After each trial, the current loser is replaced by a copy of the winner (the module with the most money) if there is one, thus "learning" what the other knew. Money is then equally redistributed among both modules, and the parameters of one of them are modified via a random mutation such that the new competitors in the next trial are not exactly alike and tend to bet on different outcomes.

In the experiments, however, I will use a more sophisticated RL scheme that (1) takes into account the possibility of additional external reward from the environment (to be maximized by both modules), (2) does not depend on pre-wired trial lengths, and (3) replaces the simple mutation process mentioned above by more powerful, self-referential "sequences of module parameter-modifying instructions" that allow the algorithm-generating modules to modify themselves in a highly directed manner [13,14]. Section 2 will present formal details, Section 3 will describe details of the RL algorithm, Section 4 will show that algorithmic surprise-generation of this kind can actually help to speed up external reward, Section 5 will conclude, and the appendix will describe details of the concrete implementation used in the experiments.

2 More Formally

The explorer's life in environment $\mathcal{E}$ lasts from time 0 (birth) to unknown time T (death). Two $m \times n$ matrices ($Left$ and $Right$) of modifiable real values represent the learner's modules. $Left$'s k-th variable, vector-valued column is denoted $Left_k$; its l-th real-valued component $Left_{k,l}$; similarly for $Right$. A variable *Instruction Pointer* (*IP*) with range $\{1, \dots, m\}$ always points to one of the module pair's columns. $\mathcal{S}$ is the learner's variable internal state with i-th component $\mathcal{S}_i \in \{-M, \dots, -2, -1, 0, 1, 2, \dots, M\}$ for $i \in \{1, \dots, m\}$.

Throughout its life the learner repeats the following basic cycle over and over: select and execute instruction $a_j \in \mathcal{A}$ with probability $Q(IP, j)$ (here $\mathcal{A}$ denotes a set of possible instructions), where

$$Q(i,j) = \frac{f(Right_{i,j}, Left_{i,j})}{\sum_k f(Right_{i,k}, Left_{i,k})}$$

for $i \in \{1,\ldots,m\}$, $j \in \{1,\ldots,n\}$. The *collective decision function* $f(x,y)$ maps real-valued x,y to real values. Given an appropriate f, each module may "veto" instructions suggested by the other module. Only instructions that are strongly supported by both "modules" are highly likely to be selected (in the experiments I use $f(x,y) = xy$). Some instructions require parameters — these are selected in an analogous fashion, using probability distributions $Q(IP{+}1,.)$, $Q(IP{+}2,.)$, $\ldots$, $Q(IP{+}v,.)$, where v is the number of parameters. Instructions consume time and may change (1) environment $\mathcal{E}$: there are instructions such as *MoveAgentInCurrentDirection(StepSize)*, *TurnAgent(Angle)*, (2) state $\mathcal{S}$: new inputs from the environment may affect the internal state; and there are also arithmetic instructions such as *Add(x,y,z)* whose interpretation is $\mathcal{S}_z \leftarrow (\mathcal{S}_x + \mathcal{S}_y)\ mod\ M$; (3) instruction pointer *IP*, for example, through conditional jump instructions; and (4) the modules themselves: there are instructions called *learning instructions* (LIs) that can modify the parameters determining the conditional probability distributions. To ensure non-vanishing exploration potential LIs may not generate module modifications that will let an instruction probability go to zero. LIs and other instructions can be combined to form more complex (probabilistic) learning algorithms. See Figure 1 for an illustration of how instruction sequences can be drawn from two probabilistic policies. See the appendix for details of a concrete implementation.

Reward. Occasionally $\mathcal{E}$ may distribute real-valued *external reward* equally between both modules. $R(t)$ is each module's cumulative external reward obtained between time 0 and time $t > 0$, where $R(0) = 0$. In addition there may be occasional *surprise rewards* triggered by the special instruction *Bet!(x,y,c,d)* $\in \mathcal{A}$. Suppose the two modules have agreed on executing a *Bet!* instruction, given a certain *IP* value. The probabilities of its four arguments $x, y \in \{1,\ldots,m\}$ and $c,d \in \{-1,1\}$ depend on both modules: the arguments are selected according to the four probability distributions $Q(IP{+}1,.)$, $Q(IP{+}2,.)$, $Q(IP{+}3,.)$, $Q(IP{+}4,.)$ (details in the appendix). If $c = d$ then no module will be rewarded. $c = 1, d = -1$ means that $Left$ bets on $\mathcal{S}_x = \mathcal{S}_y$, while $Right$ bets on $\mathcal{S}_x \neq \mathcal{S}_y$. $c = -1, d = 1$ means that $Left$ bets on $\mathcal{S}_x \neq \mathcal{S}_y$, while $Right$ bets on $\mathcal{S}_x = \mathcal{S}_y$. Execution of the instruction will result in an immediate test: which module is wrong, which is right? The former will be punished (surprise reward minus 1), the other one will be rewarded (+1). Note that in this particular implementation the modules always bet a fixed amount of 1 — alternative implementations, however, may compute real-valued bids via appropriate instruction sequences.

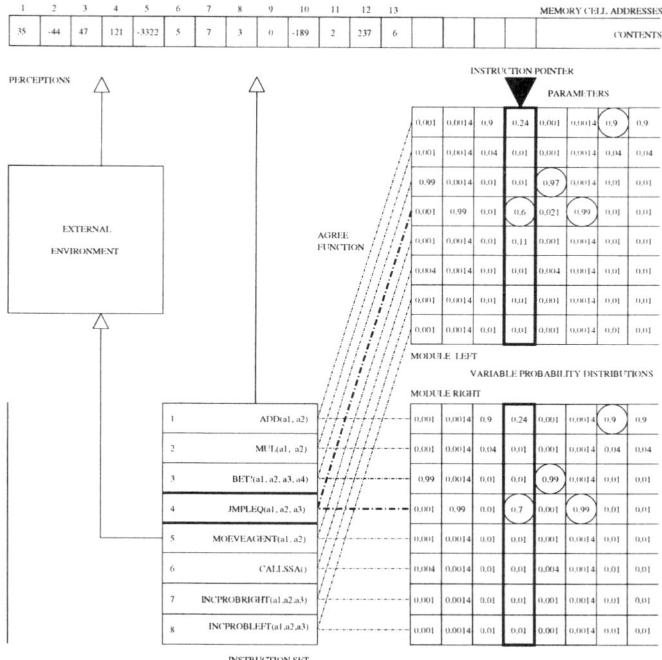

Fig. 1. Snapshot of parts of two policies and some memory cells (which are viewed as part of the policy environment). Each policy is a set of variable probability distribution on n_{ops} possible instructions or parameters. For simplicity, in this hypothetical example, $n_{ops} = 8$, and there is only one output instruction for manipulating the external environment, which in turn may affect memory cells (active perception). Instruction pointer IP currently points to a particular pair of distributions which collectively determine the probability of selecting a particular instruction (here: JMPLEQ whose probability is high). JMPLEQ requires three parameters generated according to the subsequent probability distributions.

Interpretation of surprise rewards. Instructions are likely to be executed only if both modules collectively assign them high conditional probabilities, given $\mathcal{E}$ and $\mathcal{S}$. In this sense both must "agree" on each executed instruction of the lifelong computation process. In particular, both collectively set arguments x, y, c, d in the case they decide to execute a *Bet!* instruction. By setting $c \neq d$ they express that their predictions of the *Bet!* outcome differ. Hence $Left$ will be rewarded for luring *Right* into agreeing upon instruction subsequences (algorithms) that include *Bet!* instructions demonstrating that certain calculations yield results different from what *Right* expects, and vice versa. Thus each module is motivated to discover algorithms whose outcomes surprise the other; but each also may reduce the probability of algorithms it does *not* expect to surprise the other.

The sheer existence of the *Bet!* instruction motivates each module to act not only to receive external reward but also to obtain surprise reward, by discovering algorithmic regularities the other module still finds surprising — a type of curiosity.

Trivial and random results. Why not provide reward if $\mathcal{S}_x = \mathcal{S}_y$ and $c = d = 1$ (meaning both modules rightly believe in $\mathcal{S}_x = \mathcal{S}_y$)? Because then both would soon focus on this particular way of making a correct and rewarding prediction, and do nothing else. Why not provide punishment if $\mathcal{S}_x = \mathcal{S}_y$ and $c = d = -1$ (meaning that both modules are wrong)? Because then both modules would soon be discouraged from making any prediction at all. In case $c = d = 1$ the truth of $\mathcal{S}_x = \mathcal{S}_y$ is considered a well-known, "trivial" fact whose confirmation does not deserve reward. In case $c = d = -1$ the truth of $\mathcal{S}_x = \mathcal{S}_y$ is considered a subjectively "random," irregular result. Surprise rewards can occur only in the case both modules' opinions differ. They reflect one module's disappointed *confident* expectation, and the other's justified one, where, by definition, "confidence" translates into "agreement on the surprising instruction sequence" — no surprise without such confidence.

Examples of learnable regularities. A Turing-equivalent instruction set $\mathcal{A}$ (one that permits construction of a universal Turing machine) allows exploitation of arbitrary computable regularities [15,17–19] to trigger or avoid surprises. For instance, in partially predictable environments the following types of regularities may help to reliably generate matches of computational outcomes. *(1) Observation/prediction:* selected inputs computed via the environment (observations obtained through "active perception") may match the outcomes of earlier internal computations (predictions). *(2) Observation/explanation:* memorized inputs computed via the environment may match the outcomes of later internal computations (explanations). *(3) Planning/acting:* outcomes of internal computations (planning processes) may match desirable inputs *later* computed via the environment (with the help of environment-changing instructions). *(4) "Internal" regularities:* the following computations yield matching results — subtracting 2 from 14, adding 3, 4, and 5, multiplying 2 by 6. Or: apparently, the computation of the truth value of "n is the sum of two primes" yields the same result for each even integer $n > 2$ (Goldbach's conjecture). *(5) Mixtures of (1–4).* For instance, raw inputs may be too noisy to be precisely predictable. Still, there may be internally computable, predictable, informative input transformations [10]. For example, hearing the first two words of the sentence "John eats chips" does not make the word "chips" predictable, but at least it is likely that the third word will represent something edible. Examples (1-5) mainly differ in the degree to which the environment is involved in the computation processes, and in the temporal order of the computations.

Curiosity's utility? The two-module system is supposed to solve self-generated tasks in an unsupervised manner. But can the knowledge collected in this way help to solve externally posed tasks? Intuition suggests that the more one knows about the world the easier it will be to maximize external reward. In fact, later I will present an example where a curious system indeed outperforms a non-curious one. This does not reflect a universal law though: in general there is no guarantee that curiosity will not turn out to be harmful (for example, by "killing the cat" [9]).

Relative reward weights? Let $RL(t)$ and $RR(t)$ denote $Left$'s and $Right$'s respective total cumulative rewards obtained between time 0 and time $t > 0$. The sum of both modules' surprise rewards always remains zero: we have $RL(t) + RR(t) - 2R(t) = 0$ for all t.

If we adopt the traditional hope that exploration will contribute to accelerating environmental rewards, then zero-sum surprise rewards seem to afford less need to worry about the relative weights of surprise versus other rewards than the surprise rewards of previous approaches, which did not add up to zero.

Enforcing fairness. To avoid situations where one module consistently outperforms the other, the instruction set includes a special LI that copies the currently superior module onto the other (see the appendix for details). This LI (with never-vanishing probability) will occasionally bring both modules on a par with each other.

In principle, each module could learn to outsmart the other by executing subsequences of instructions that include LIs. But how can we ensure that each module indeed improves? Note that arithmetic actions affecting $\mathcal{S}$ and jump instructions affecting IP cause a highly non-Markovian setting and prevent traditional RL algorithms based on dynamic programming from being applicable. For such reasons I use *Incremental Self-improvement (IS)* [13,14] to deal with both modules' complex spatio-temporal credit assignment problem.

3 Incremental Self-improvement (IS)

The currently surprising module wants to repeat similar surprises with higher probability in the future. The other wants to avoid further surprises by learning not to agree on similar computation sequences (implicitly learning what the other already knows). And it wants to be "creative" in the sense that it wants to generate new surprises for the other module instead. In principle, both can learn by executing subsequences of instructions that include LIs. How can we ensure that each module indeed improves by accelerating its reward intake?

In this chapter I will use the IS paradigm [13,14] to deal with both modules' complex spatio-temporal credit assignment problem. This does not necessarily mean that IS is the best way of doing so. Other RL paradigms may be appropriate, too — this chapter's basic ideas are independent of the choice of RL method. IS seems attractive, however, because: (1) It does not make an explicit difference between learning algorithms and other instructions, or between learning, metalearning, metametalearning, etc. (2) It properly takes into account that the success of each module modification recursively depends on the success of all later modifications for which it is setting the stage. (3) Its objective takes into account the entire time consumed by lifelong learning itself. (4) It is designed for quite general non-Markovian credit assignment problems in lifelong learning situations — see [13,14,20] for recent IS applications. Following [14], the remainder of this section will describe $Left$'s IS-based learning process. *Right*'s is analogous.

Checkpoints. $Left$'s entire life time can be partitioned into time intervals separated by special times called *checkpoints*. Checkpoints are computed dynamically during the learner's life by certain instructions in $\mathcal{A}$ executed according to the modules themselves. $Left$'s k-th checkpoint is denoted l_k. Checkpoints obey the following rules: (1) $\forall k \;\; 0 < l_k < T$. (2) $\forall j < k \;\; l_j < l_k$. (3) Except for the first, checkpoints may not occur before at least one LI executed at least one $Left$-modification since the previous checkpoint.

Sequences of module modifications. SLM_k denotes the sequence of $Left$-modifications (SLM) computed by LIs between checkpoints l_k and l_{k+1}. Since LI execution probabilities depend on $Left$ and *Right*, the modules can in fact modify the way they modify themselves — they can devise their own probabilistic learning algorithms.

Goal. At some checkpoint t $Left$'s goal is to generate $Left$-modifications that will accelerate long-term reward intake: it wants to let the value of $(RL(T) - RL(t))/(T - t)$ exceed the current average reward intake.

Success-story criterion (SSC). $Left$ maintains a time-varying set V of past $Left$ checkpoints that have led to long-term reward accelerations. Initially V is empty. Let v_k denote the k-th element of V in ascending order. SSC is satisfied at time t if either V is empty (trivial case) or if

$$\frac{RL(t) - RL(0)}{t - 0} < \frac{RL(t) - RL(v_1)}{t - v_1} < \ldots < \frac{RL(t) - RL(v_{|V|})}{t - v_{|V|}}. \qquad (1)$$

SSC demands that each checkpoint in V marks the beginning of a long-term reward acceleration measured up to the current time t.

Success-story algorithm (SSA). At every checkpoint of $Left$ we invoke the *success-story algorithm* (SSA):

1. **WHILE** SSC is **not** satisfied

Undo all $Left$-modifications made since the most recent checkpoint in V.
Remove that checkpoint from V.

2. Add the current checkpoint to V.

"Undoing" a modification means restoring the preceding $Left$ — this requires storing past values of $Left$-components on a stack prior to modification. (Components of $Left$ and elements of V can be stored on the same stack — see the appendix.) Thus each $Left$-modification that survived SSA is part of a bias shift generated after a checkpoint marking a lifelong reward speed-up: since v_j there has been more reward per time than since v_i, for $v_j > v_i$ $(v_j, v_i \in V)$.

Timing SSA calls. Between two checkpoints $Left$ is temporarily protected from SSA evaluations. Since the way of setting checkpoints depends on $Left$ itself, $Left$ can learn to influence *when* it gets evaluated. This evaluation-timing ability is important in dealing with unknown reward delays.

SSA's generalization assumption. At the end of each SSA call, until the beginning of the next one, the only temporary generalization assumption for inductive inference is: $Left$-modifications that survived all previous SSA calls will remain useful. In the absence of empirical evidence to the contrary, each surviving SLM_k is assumed to have set the stage for later successful $\text{SLM}_i, i > k$. Since life is one-way (time is never reset), during each SSA call the system has to generalize from a *single* experience concerning the utility of $Left$-modifications executed after any given previous point in time: the average reward per time since then.

Implementing SSA. Using stack-based backtracking methods such as those described in [13,14] and the appendix, one can guarantee that SSC will be satisfied right after each new SLM-start, despite interference from S, $\mathcal{E}$, and *Right*. Although inequality 1 contains $|V|$ fractions, SSA can be implemented efficiently: only the two SLMs on top of the stack need to be considered at any given time in an SSA call (see details in appendix). A *single* SSA call, however, may undo *many* SLMs if necessary.

What has been described for $Left$ analogously holds for *Right*. The appendix describes a particular implementation (the one used for the experiments) based on an assembler-like programming language similar to those used in [14,21].

4 Experiments

It has already been shown that IS by itself can solve interesting tasks. For instance, [14] describes two agents A and B living in a partially observable 600×500 environment with obstacles. They learn to solve a complex task that could not be solved by various TD(λ) Q-learning variants [22]. The task requires (1) agent A to find and take "key A"; (2) agent A go to "door A" and open it for agent B; (3) agent B to enter through "door A," find, and take another "key B"; (4) agent B to go to another "door B" to open it (to free the way to the goal); (5) one of the agents to reach the goal. Both agents share the same design. Each is equipped with limited "active" sight: by executing certain instructions, it can sense obstacles, its own key, the corresponding door, or the goal, within up to 50 unit lengths in front of it. The agent can also move forward (up to 30 unit lengths), change its direction, turn relative to its key or its door or the goal. It can use memory (embodied by its *IP*) to disambiguate inputs. Reward is provided only if one of the agents touches the goal. This agent's reward is 5.0; the other's is 3.0. In the beginning, the goal is found only every 300,000 basic cycles. Through IS, however, within 130,000 trials the average trial length decreases by a factor of 60 — both agents learn to cooperate to accelerate reward intake [14].

This section's purpose is not to elaborate on how IS can solve difficult tasks. Instead IS is used as a particular vehicle to implement the two-module idea for preliminary attempts at studying "inquisitive" explorers. Subsection 4.1 will describe empirically observed system behavior in the absence of external rewards. In Subsection 4.2 there will be additional reward for solving externally posed tasks, to see whether curiosity can indeed be useful.

Experimental details. There are $n = 24$ instructions (see the appendix) and $m = BS(n^2 \; div \; BS) = 576$ columns per module. $M = 100,000$. Time is measured as follows: selecting an instruction head, selecting an argument, selecting one of the two values required to compute the next instruction addresses, pushing or popping a module column costs one time step. Other computations do not cost anything. This ensures that measured time is of the order of total CPU-time. For instance, selecting an instruction head plus six arguments plus the next *IP* address costs $1 + 6 + 2 = 9$ time steps.

Figure 2 shows a point-like agent's two-dimensional environment whose width is 1000 unit lengths. The large "room" in the south is a square. Its southwest corner has coordinates (0.0, 0.0), its southeast corner (1000.0, 0.0). There are infinitely many possible agent states: the agent's current position is given by a pair of real numbers. Its initial coordinates are $(50.0, 50.0)$, its initial direction is 0, its stepsize 12 unit lengths. Compare appendix A.3.

4.1 Experiment 1: No External Reward

This introductory experiment focuses on the totally unsupervised case — there is no external reinforcement for solving pre-wired tasks. Thus there also is no objective way of measuring the system's performance. The experiment's only purpose is to give an impression of what happens while the active learner is running freely. In what follows I will describe a single but rather typical run.

The entrance to the small room in Figure 2 is blocked. The system runs for 10^9 time steps corresponding to about 10^8 instructions. The system does

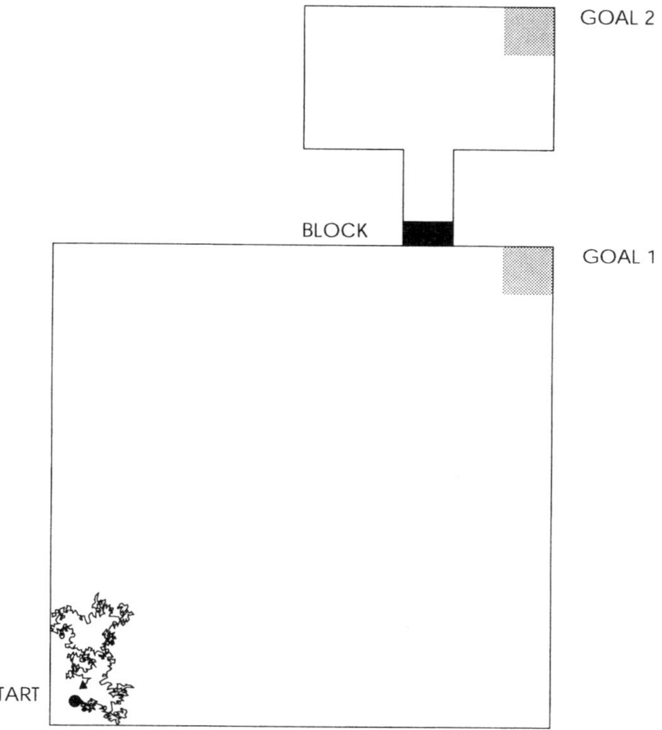

Fig. 2. Hypothetical trace of the agent near START in the southwest corner. The width of the environment is 1000 unit lengths, the agent's stepsize 12 unit lengths. Whenever the agent hits GOAL1 or GOAL2 it gets teleported back to START (except in Experiment 1). GOAL2 provides 10 times as much external reward as GOAL1. The entrance to the small room is blocked in Experiments 1 and 2a, and open in 2b. Due to the tiny step size and numerous non-goal-specific instructions, random behavior requires millions of time steps to hit one of the goals by accident.

make heavy use of its arithmetic instructions: soon the entire storage is filled with varying numbers generated as by-products of its computations.

Zero-sum reward. Consider Figure 3. The derivatives of the reward plots at a given time tell which module currently receives more reward than the other. There are long time intervals during which one of the modules dominates. The existence of zero crossings, however, shows that each module occasionally collects sufficient negative rewards to cancel all the previously collected positive rewards, and vice versa. This means that no module consistently outperforms the other.

Stack pointer evolution. The stack pointers reflect the number of currently valid module column modifications. Consider Figure 4. Initially the stacks grow very quickly. Then there is a consolidation phase. Later growth resumes, but at a much more modest level — it becomes harder and harder for each module to acquire additional "probabilistic knowledge" to outwit the other. Sometimes SSA pops off a module's entire stack (Figure 4's temporal resolution is too low to show this), but in such cases the stack pointer soon reaches its old value again. This is partly due to *SSAandCopy* instructions copying the currently superior module onto the other to enforce fair matches.

Reward acceleration. Even in later stages both modules are able to accelerate their long-term average reward intake again and again, despite the fact that the sum of all rewards remains zero. As each module collects a lot of negative reward during its life, it continually pops appropriate checkpoints/modifications off its stack such that the resulting histories of surviving module modifications correspond to histories of less and less negative reward/time ratios. Recall the occasional popping off of the entire stack.

Evolution of instruction frequencies. Consider Figure 5. Although the sum of all surprise rewards remains zero, *Bet!* instructions are soon among the most frequent types. Other instructions also experience temporary popularity, often in the form of sharp peaks beyond the plot's resolution. Usually, however, the module that does not profit from them learns to put its veto in.

Interestingness depends on current knowledge and computational abilities. These are very different for human observers and my particular implementation. It seems hard to trace and understand the system's millions of self-surprises and self-modifications. Figure 5, for instance, provides only very limited insight into the nature of the complex computations carried out. It plots frequencies of selected instruction types but ignores the corresponding arguments and *InstructionPointer* positions. It does not reflect that for a while computations may focus on just a few module columns and storage cells, until this gets "boring" from the system's point of view. Much remains to be done to analyze details of the system's complex dynamics.

Creativity's utility? Can the probabilistic knowledge collected by a "creative" two-module system (solving self-generated tasks in an unsupervised manner) help to solve externally posed tasks? The following experiment provides a simple example where this type of curiosity indeed helps to achieve faster goal-directed learning in the case of occasional non-zero environmental reward.

4.2 Experiment 2: Additional External Reward

Experiment 2a. Consider Figure 2. The entrance to the small room is blocked. Whenever the point-like agent moves into the 100×100 northeast corner (GOAL1) of the 1000×1000 field it receives external reward 100.0 and is immediately teleported back to start position $(50.0, 50.0)$; its direction is reset to 0.0.

Solution. GOAL1 can be reached by a simple stochastic policy making the agent head in the northeast direction. Once the system somehow has fixed the agent's direction in an optimal way, the shortest path requires about 100

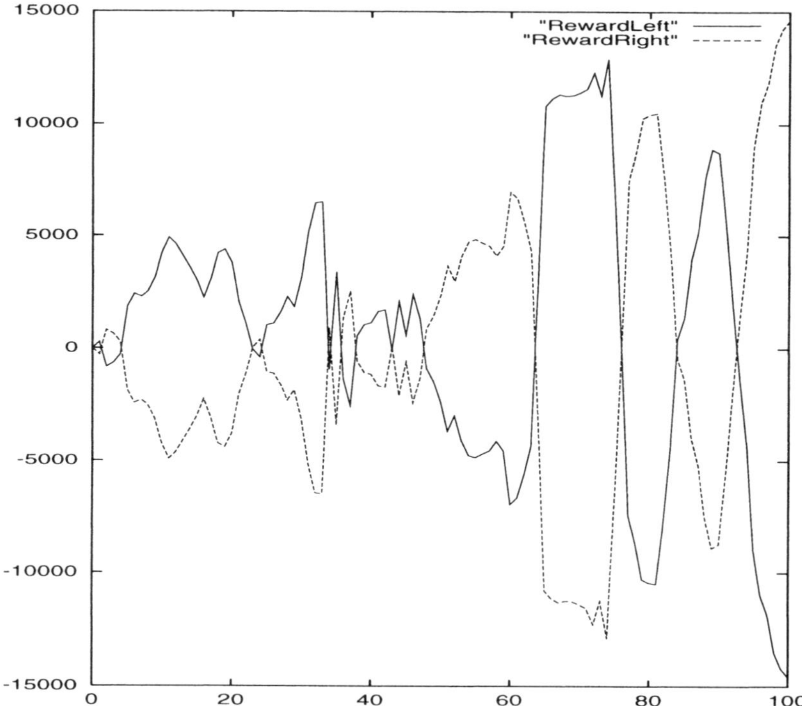

Fig. 3. Experiment 1 (no external rewards): symmetric evolution of both modules' cumulative surprise rewards during the first billion time steps, sampled at intervals of 10 million steps.

calls of *MoveAgent* (due to the agent's limited stepsize). We will see that the major difficulties in *learning* such a policy are due to the system's extremely low initial bias towards solving this task, and the extremely rare occurrences of rewarding training instances.

Random behavior results. With its module-modifying capabilities being switched off (LIs such as IncProbLEFT(x_1, x_2, y_1) have no effect), the system exhibits random behavior according to its maximum entropy initialization. This behavior leads to about five visits to GOAL1 per 10 million time steps, which shows how little built-in knowledge there is about the nature of the goal. Following Geman et al. [23], the learning system's bias towards solving the task is extremely low, while variance is high. There is a confusing choice of $2 \times 576 = 1152$ module columns (many more than needed for solving the task), all being potential candidates for modifications. The two modules initially have no idea that two of the 23 instruction types (namely, *MoveAgent* and *SetDirection*) are particularly important. Furthermore, module modifications can be computed only by appropriate instruction subsequences (includ-

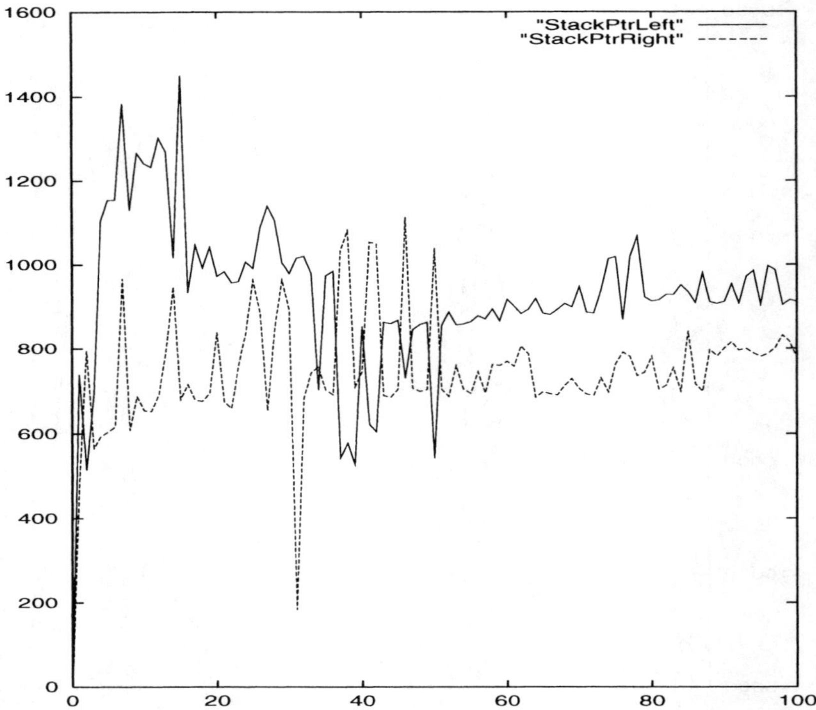

Fig. 4. Experiment 1: evolution of both modules' stack pointers during the first billion time steps, measured at intervals of 10 million steps.

ing LIs) generated according to the module columns themselves. Initially all instructions are extremely stochastic, however, partly due to the many arguments and addresses to jump to. Reducing this "free parameter overkill" by using smaller modules would be equivalent to inserting more a priori knowledge about the task.

Comparison. I compare the performance of two learning systems. One is curious/creative in the sense described in the previous sections, the other is not. The latter is like the former except that its *Bet!* action has no effect — external reward is the only type of reward. Ten simulations are conducted for each system. Each simulation takes 200 million time steps. During the 10th and the 20th 10^7 time step interval I count how often the agent visits GOAL1. Since external reward initially occurs only every 2 million time steps on average, goal-relevant training examples are very rare.

Results. Table 1 lists the plain and the curious systems' goal visits per 10^7 time steps after half their life times, and near death. All 10 results are given due to the high variance. The curious system's results tend to be more

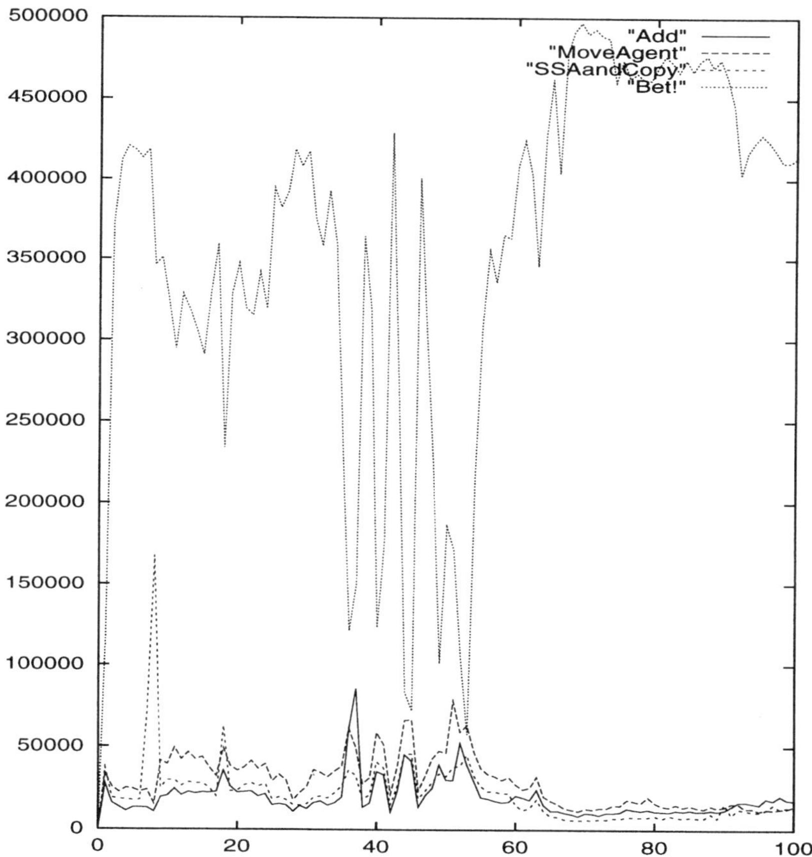

Fig. 5. Experiment 1: execution frequencies of selected instruction types during the first billion time steps, sampled at intervals of 10 million steps.

than an order of magnitude better than the plain's, which tend to be more than an order of magnitude better than the random's.

Figures 6, 7, 8 show details of the particularly successful simulation 1. We observe that total reward fluctuations due to surprise rewards appear negligible, although they seem important for overall performance improvement (compare Table 1). Between 20 and 30 million steps there are lots of *Bet* and *SSAandCopy* actions. Soon afterwards the instruction types *Move* and *SetDirection* start to dominate. Around 130 million steps their frequencies suddenly rise dramatically. The final breakthrough is achieved shortly thereafter. The modules' first 100 columns after simulation 1 are shown in Figure 9. Both modules are almost identical. The probability mass of many columns is concentrated in a single value. Additional inspection revealed, however, that some of the most frequently used instruction addresses point to maximum entropy distributions.

Possible explanation. How can curiosity/creativity help? In early system life external rewards occur very rarely. During this time, however, there are many surprise reward-oriented instruction subsequences. Possibly they provide the system with useful fragments of "probabilistic knowledge" about consequences of its innate instructions. For instance, to repeatedly surprise itself it may construct little "probabilistic subprograms" involving highly probable jumps to appropriate module column addresses. Once it has figured out how to increase the likelihood of certain jumps it may find it easier to solve external tasks that also require jumps.

Details of how the system actually came up with the final matrices remain quite unclear, however. More experimental analysis is necessary to better understand how solving external tasks can benefit from pure curiosity.

Simulation	plain: half	curious: half	plain: final	curious: final
1	4	899	9	15774
2	4	0	4	0
3	3	204	15	381
4	124	558	1036	3150
5	399	539	1000	2073
6	9	1024	82	2937
7	9	1024	123	1181
8	6	891	20	1561
9	58	89	324	446
10	62	5603	476	9963
Average	68	1083	309	3746

Table 1. Experiment 2a: 10 simulations, each taking $2 * 10^8$ time steps. The table lists goal visits achieved by plain and curious systems during the 10th (half life time) and the final 10^7 time step interval. Random behavior leads to about 5 visits per 10^7 time steps. Curiosity typically seems to help a lot. But note the second simulation's "outlier."

The outlier. Note the curious system's outlier in the second simulation. Inspection revealed that during the entire simulation the system never reached the goal at all — there was not a single training example concerning external reward. This outlier reflects curiosity's potential drawback: it may focus attention on computations that do not have anything to do with external reinforcement. On average, however, surprise rewards do help.

Runtime. To some readers the runtime of several hundred million time steps may seem large. Note, however, that the number of training examples (trials) is much smaller because the goal is hit very rarely, especially in the beginning of the learning phase. Some initial trials take millions of time steps

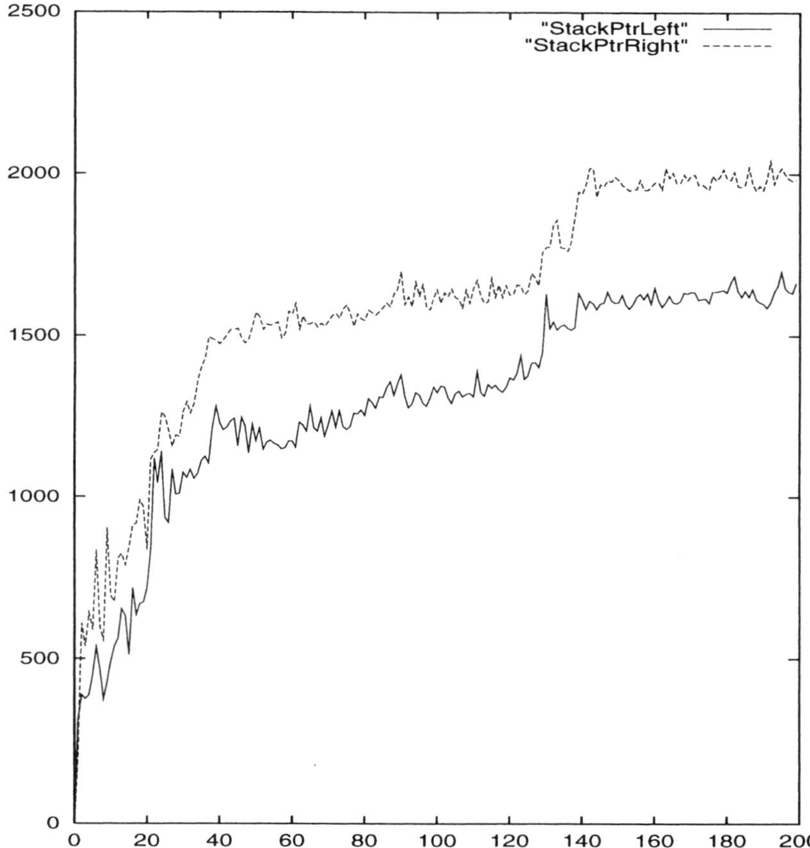

Fig. 6. Experiment 2a (additional external reward): stack pointer evolutions during the first 200 million time steps of the particularly successful simulation 1, sampled at million step intervals. Compare Figures 7 and 8.

simply because the system does not know that there is a source of reward at all, and that some of its instructions (those for interaction with the environment) are much more important for achieving the goal than arithmetic instructions — not one biases its search towards the goal. The following simulations will require even longer runtime.

Experiment 2b. I keep the basic setup from Experiment 2a, but open the entrance to the small room in Figure 2 by removing the block. Whenever the point-like agent moves into the small room's 100 × 100 northeast corner (GOAL2) it receives external reward 1000 (10 times as much as for reaching

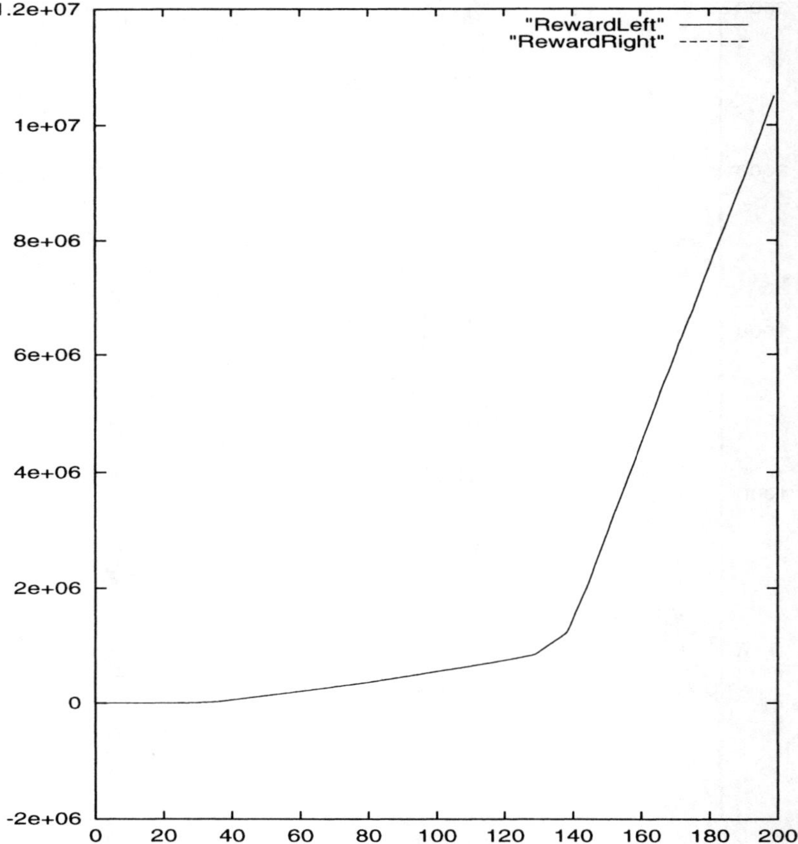

Fig. 7. Experiment 2a: evolution of both modules' cumulative rewards during the first 200 million time steps of simulation 1, sampled at million step intervals. Plot resolution is too low to show that both curves are not identical — fluctuations due to surprise rewards seem negligible. Still, surprise rewards seem essential for significant performance improvement — compare Table 1. The final breakthrough occurs around 140 million steps. Compare Figures 6 and 8.

GOAL1) and is teleported back to start position $(50.0, 50.0)$; its direction is reset to 0.0.

The idea is: knowledge collected by solving the simpler task (reaching GOAL1) may help to solve the more difficult, but also more rewarding task (reaching GOAL2). Note that *both* tasks are solvable by similar stochastic policies making the agent head northeast. To receive a lot of reward, however, the agent must avoid GOAL1 on its way to GOAL2 — otherwise it will be teleported back home.

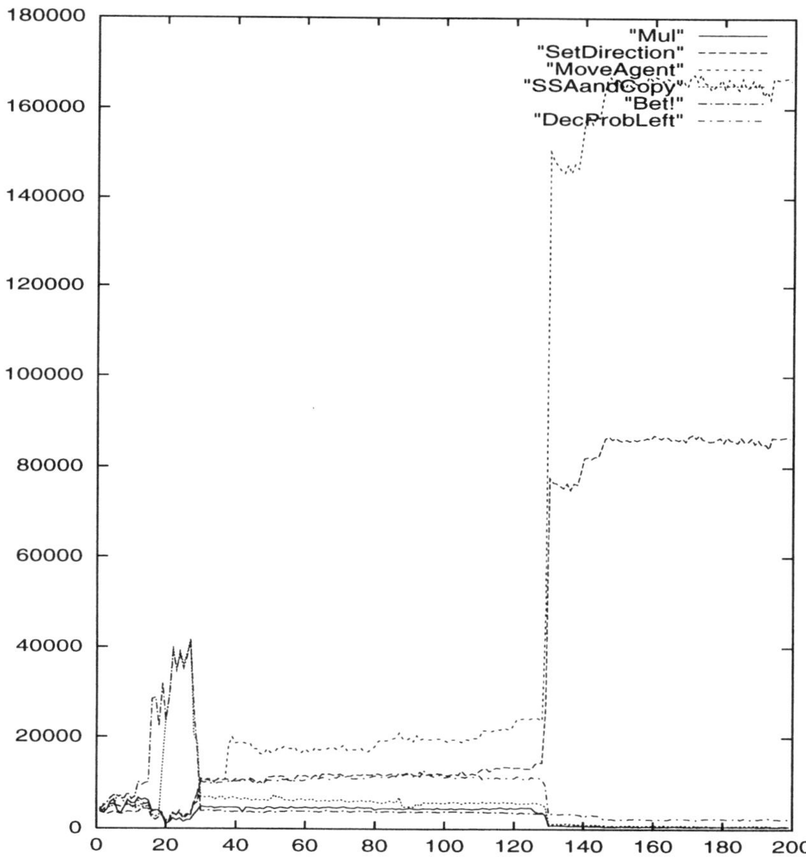

Fig. 8. Experiment 2a: execution frequencies of selected instruction types during the first 200 million time steps of simulation 1, sampled at million time step intervals. Between 20 and 30 million steps there are many *Bet!* and *SSAandCopy* instructions. Soon afterwards *Move* and *SetDirection* start to dominate. The final breakthrough is achieved 10 million steps after their frequencies increase dramatically around 130 million steps. Compare Figures 6 and 7.

This setup involves an obvious goal-directed exploration component: it is likely that the system will keep an exploration strategy that has helped to improve a policy for reaching GOAL1 (successful exploration strategies have an "evolutionary advantage"). This strategy may later also help to improve a policy for reaching GOAL2 — in principle, IS can use experience with exploration strategies to evaluate and refine them.

Random behavior results. With its module-modifying capabilities being switched off (LIs such as IncProbLeft(x_1, x_2, y_1) have no effect), the

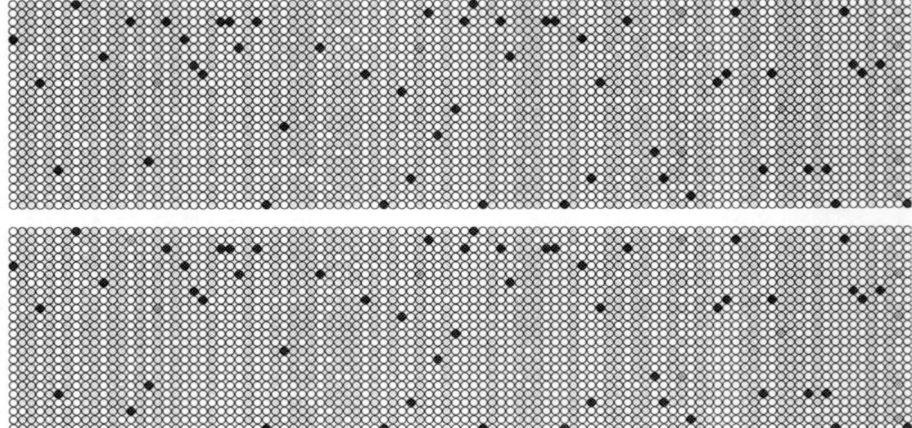

Fig. 9. Experiment 2a: LEFT's (top) and RIGHT's first 100 (of 576) probability distributions after simulation 1. Grey scales indicate probability magnitudes (white = close to 0, black = close to 1). The probability mass of many (but not all) columns is concentrated in a single value. Both modules are almost identical due to *SSAandCopy* LIs. Their stacks are quite different though.

system exhibits random behavior according to its maximum entropy initialization. On average this behavior leads to less than one visit to GOAL2 per 10 million time steps.

Comparison. Again I compare the performance of a "plain" and a "curious" system. Again the former is like the latter except that its *Bet!* instructions have no effect. Ten simulations are conducted for each system. Each simulation takes 4 billion time steps. I keep track of how often the agent reaches GOAL2 per 10^7 time steps.

Results. Figure 10 plots average goal visit frequencies during the first 500 million time steps. During this period, the curious system's results are clearly better than the plain's, and its performance improves much faster.

As more simulation time is spent, however, initial differences tend to level out. Figure 11 plots average goal visit frequencies during the entire 4 billion steps. As goal-oriented training examples become more and more frequent, we see that the performances of plain and curious systems eventually reach comparable levels — the former even becomes a bit better than the latter.

Advantage in the case of rare rewards? For this particular experiment, self-generated surprise rewards appear to boost initial external reward. Sufficient training time, however, cancels out the initial advantage. Could it be that curiosity is particularly useful as long as external reward is extremely rare and little has already been learned? This may seem intuitively plausible: in the beginning performance tends to be so bad that time spent on extensive exploration seems to be a good investment with little downside potential — things cannot get much worse. Once the system's strategy is rather efficient and yields frequent external rewards, however, additional progress depends on fine-tuning it. In this stage surprise reward-oriented curiosity may distract more than it helps — it tends to consume time without necessarily contributing much to optimizing goal-directed trajectories. Many additional experiments are necessary, however, to understand whether the above is a typical result or not.

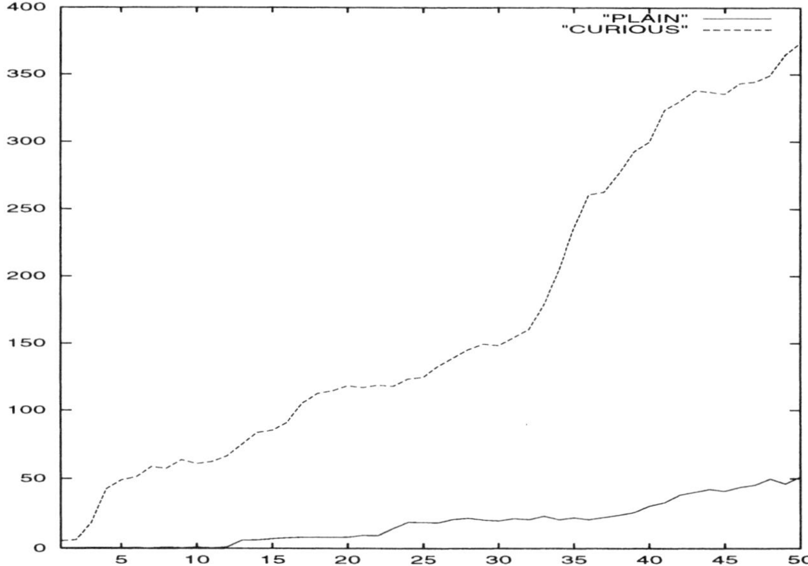

Fig. 10. Experiment 2b: comparison of plain and curious systems. The plot shows visits to GOAL2 during the first 500 million time steps, sampled at intervals of 10^7 steps (averaged over 10 simulations; random behavior yields less than one visit per 10^7 steps). Compare Figure 11.

5 Final Remarks

Previous work on IS (for example, [14,13]) implicitly focused on goal-oriented exploration — in principle IS can learn to change its exploration strategy if this turns out to accelerate external reward in the long run. This chapter's IS implementation, however, involves an additional *pure* exploration component besides the goal-oriented one. Part of the learner receives internal reward for pointing out something another part did not know but thought it knew. The surprised part suffers to the extent the surprising part benefits — the sum of all internal rewards remains zero. The learner is "interested" in "creative" computations leading to unexpected results, while simultaneously trying to

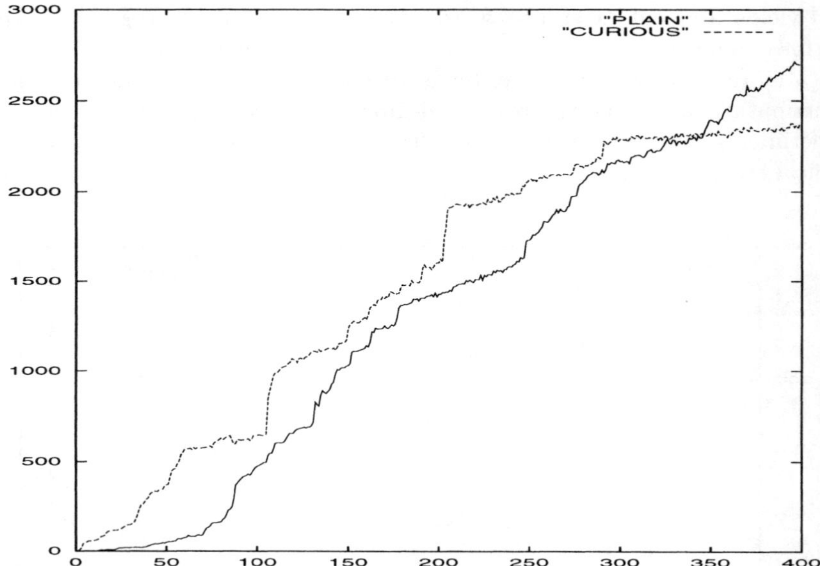

Fig. 11. Experiment 2b: visits to GOAL2 during the entire 4 billion time steps, sampled at intervals of 10^7 steps (averaged over 10 simulations). Initial advantages of curiosity are lost as more and more goal-oriented training examples become available. Compare Figure 10.

make formerly surprising things predictable and boring. It does not care much for irregular noise rich with Shannon information [6]. Instead it prefers easily learnable algorithmic regularities, taking into account the costs of gaining information in an RL framework.

From each module's perspective an instruction subsequence is "novel" as long as its outcome surprises the other. Since the surprised module will eventually figure out what is going on, there will be incessant pressure to create new novelties. What is the use of such an inquisitive system's computations? Curiosity's long-term justifiability depends on whether knowledge growth will eventually support goal-oriented behavior. When will this be the case? The question is reminiscent of G. H. Hardy's toast on "pure" mathematics — the kind that "would never be of any use to anyone" ([24], p. 185). History teaches us, however, that it is hard to decide which math will be useless forever. For instance, old results from "pure" number theory are used in today's encryption technology.

In general, however, it will always be possible to design environments where "curiosity kills the cat" [9], or at least has negative influence on external performance. For instance, as exemplified by simulation 2 of Experiment 2a, curiosity may occasionally prevent discovery of external reward sources. This is reminiscent of the situation in supervised learning. There often additional "regularizer" terms are added to the standard error function defining network performance on the training data. They can greatly help to remove redundant free parameters and improve generalization capability on unseen data (e.g., [25]), but in general this cannot be guaranteed.

This chapter's approach draws inspiration from several sources. For instance, the two-module system is based on two co-evolving modules. Co-evolution of competing strategies, however, is nothing new. See, for example, [26,27] for interesting cases. Also, the idea of improving a learner by letting it play against itself is ancient. See, for example, [28,29]. Even the idea of unsupervised learning through co-evolution of predictors and modules trying to escape the predictions is nothing new — it has been used extensively in our previous work on unsupervised sensory coding with neural networks [30,31,10,32,33]. Finally, co-evolutionary methods translating mismatches between reality and expectations into reward for "curious," exploring agents are not new either — see our previous work on "pure" RL-based exploration [7–9]. So, what is new?

Novel is the idea that both adaptive modules equally influence the probability of each executed instruction/computation. This (1) allows for a straightforward way of making both modules equally powerful (by copying the currently superior one onto the other), and (2) prevents each module from being able to enforce computations that will make the other lose no matter what it tries. For instance, details of white noise on a screen are inherently unpredictable, but none of the two opponents may exploit this to generate surprises if the other does not "agree" to the corresponding experiment. And it will agree only as long as it suspects that there is a regularity in the white noise that the other does not yet know. The precondition of a surprise is that the surprised module has expressed its confidence in a different outcome of the surprising computation sequence by participating in the collective decision process. Intuitively, my adaptive explorer continually wants to discover new, "creative" uses of its innate sensorium and computational potential. It wants to focus on those novel things that seem easy to learn, given current knowledge. It wants to ignore (1) previously learned, predictable things, (2) inherently unpredictable ones (such as details of white noise on a screen), and (3) things that are unexpected but not expected to be easily learned (such as the contents of an advanced math textbook beyond the explorer's current level).

Another novel aspect is the general setting. Instead of being limited to Markovian domains and simple reactive strategies such as approaches in [8,9], this chapter's setup allows for quite arbitrary domains and computations. This is made possible by the recent IS paradigm [14,13]. There is no essential limit (besides computability) to the nature of the regularities that may be exploited to generate surprises. Neither is there an essential limit to the nature of the learning processes that can make formerly surprising regularities predictable and boring. There may be RL schemes even more general than IS, but this is beyond the scope of this chapter.

Note that this chapter's notion of "simple regularities" differs from, e.g., Kolmogorov complexity theory's [15,17–19,21]. There an object is called simple relative to current knowledge x if the size of the shortest algorithm computing it from x is small. The algorithm's computation time is ignored, as are constant factors reflecting Kolmogorov complexity's machine independence. The current chapter, however, takes both into account.

As the explorer's knowledge about its environment and computational abilities expands, it keeps balancing on the thin, dynamically changing line between the subjectively random and the subjectively trivial. Unlike Nake and other authors he cites [34], I do not suggest a predefined optimal ratio between known and unknown information. Instead, the two cooperating/competing modules dynamically, implicitly determine this ratio as they keep trying to surprise each other.

Recent papers attempt to explain "beauty" with the help of complexity theory concepts [35,36]. They argue that something "beautiful" need not be "interesting". They predict that the "most beautiful" object from a set of objects satisfying certain specifications is the one that can be most easily computed from the subjective observer's input coding scheme. Interestingness in the current chapter's sense, however, also takes into account whether the computational result is expected or not. Something that is both "beautiful" and already known may be quite boring — "beauty" needs to be unexpected to awaken interest.

Future work. The programming language used in the experiments is designed to allow for fairly arbitrary computations/explorations and learning processes. To make progress towards analyzing "inquisitive" explorers, however, one will probably have to study alternative systems with less computational power and less general RL paradigms but more accessible dynamics. On the other hand, it will also be interesting to study a curious learner's performance in the case of more difficult tasks and more powerful primitive instructions with more bias towards solving the task. Note that LIs can be almost anything: neural net algorithms, Bayesian analysis algorithms, etc.

Furthermore, although IS is a rather general RL paradigm, it may be possible to develop more general ones. In that case I would like to combine them with the two-module idea. Promising candidates may be RL schemes based on economy and market models, such as classifier systems and their variants [12,37–40], or the related "Prototypical Self-referential Associating Learning Mechanisms" (PSALMs) [41], the Neural Bucket Brigade [42], Hayek Machines [43,16], Collective Intelligences (COINs) [44].

The basic ideas of the present chapter will probably remain unchanged, however: competing agents will agree on algorithmic experiments and bet on their outcomes, the winners profiting from outwitting others.

A Appendix

In what follows I will describe details of the system used in the experiments.

Architecture. The internal state $\mathcal{S}$ consists of m addressable cells with addresses ranging from 0 to $m-1$ (instead of ranging from 1 to m — this is due to a peculiarity of C, the implementation language). $\mathcal{S}_k \in \{-M, -M+1, \ldots, 0, 1, \ldots, M\}$ are the current contents of the cell with address k. Instructions and arguments are encoded by a fixed set I of n integer values $\{0, \ldots, n-1\}$. For each value j in I, there is an instruction head a_j with $n_j < BS - 2$ integer-valued arguments, where BS is the instruction block size, and m is a multiple of BS. In the experiments I use $BS = 9$ (there are at most six arguments per instruction), $n = 24$, $m = BS(n^2 \; div \; BS) = 576$, and $M = 100,000$. See list below for instruction syntax and semantics.

Right and Left modules. All $Right_i$ and $Left_i$ ($i \in \{0, \ldots, m-1\}$) are vectors of n positive, real values that sum up to 1.0. The k-th component of $Right_i$ ($Left_i$) is denoted $Right_{i,k}$ ($Left_{i,k}$) for $k \in \{0, \ldots, n-1\}$. A variable *InstructionPointer* (*IP*) with range $\{0, \ldots, m-1\}$ always points to one of the module pair's columns. *IP* is viewed as a modifiable part of the environment.

Initialization. At system birth at time 0, all $Right_{i,k}$ and $Left_{i,k}$ are set equal to $1/n$. All $\mathcal{S}_k$ and *IP* are set to zero. They will never be re-initialized again. To be able to restore modified module columns if necessary, we introduce two initially empty stacks Stack*Right* and Stack*Left* that allow for variable-sized stack entries, and the conventional *push* and *pop* operations. Instructions may change the Boolean variables *BlockSSA*Left and *BlockSSA*Right (both are modifiable parts of $\mathcal{E}$ and initially FALSE at time 0).

A.1 Basic Cycle of Operations

Until unknown time T (system death), the system repeats the following basic instruction cycle over and over.

1. Select instruction head $a_j \in I$ with probability $Q(IP, j)$, where

$$Q(i,j) = \frac{f(Right_{i,j}, Left_{i,j})}{\sum_k f(Right_{i,k}, Left_{i,k})},$$

$$for \; i \in \{0, \ldots, m-1\}, j \in \{0, \ldots, n-1\}.$$

Here the *collective decision function* $f(x,y)$ maps real-valued x, y to real values. Given an appropriate f, each module may "veto" instructions suggested by the other module. Only instructions that are strongly supported by both modules are highly likely to be selected. One possibility is $f(x,y) = min(x,y)$. In the experiments I use $f(x,y) = xy$.

Comment: owing to peculiarities of certain instructions to be introduced below, $Q(i,j)$ will later be refined for cases where i addresses an instruction head as opposed to an argument.

2. a_j's $n_j \leq 6$ arguments $\in \{0,\ldots,n-1\}$ are selected according to probability distributions $Q(IP+1,.), Q(IP+2,.),\ldots,Q(IP+n_j,.)$ (except when a_j = *Bet!* — two of *Bet!*'s arguments will be treated differently — see Section A.3 below).

3. Execute the selected instruction. This will consume time and may change (1) environment $\mathcal{E}$, (2) *IP*, (3) internal state $\mathcal{S}$; (4a) *Right*, (4b) *Left*. If there is external reward R then set $\mathcal{S}_8 \leftarrow R$ (rewards become visible to the system in the form of inputs).

4. If an input has changed one of the cell contents $\mathcal{S}_0, \mathcal{S}_1,\ldots,\mathcal{S}_8$, then shift the contents of $\mathcal{S}_0, \mathcal{S}_1,\ldots,\mathcal{S}_{80}$ to components $\mathcal{S}_9, \mathcal{S}_{10},\ldots,\mathcal{S}_{89}$, respectively. This results in a built-in short-term memory (long-term memory can be implemented by the system itself by executing appropriate instruction sequences).

5. If a_j did not modify *IP* (no conditional jump — compare instruction list below), then compute the address of the next instruction head by setting $IP \leftarrow w - w \bmod BS$. Here $w = (w_1 n + w_2) \bmod m$, where $w_1 \in \{0,\ldots,n-1\}$ is selected according to probability distribution $Q(IP+7,.)$, while $w_2 \in \{0,\ldots,n-1\}$ is selected according to $Q(IP+8,.)$.

6. Goto **1.**

A.2 SSA Calls

At a given time LEFT's current stack will be either empty or of the form $Block_1, Block_2,\ldots,Block_u$, where $Block_i$ ($i \in \{1,\ldots,u\}$) is of the form

$$[v_i, RL(v_i), (c_i^1, Left_{c_i^1}), (c_i^2, Left_{c_i^2}),\ldots].$$

Here v_i denotes the i-th checkpoint still in LEFT's stack (compare section refis), $RL(v_i)$ denotes LEFT's reward until time v_i, c_i^k ($k \in \{1,2,\ldots,\}$) is the address of the k-th LEFT-column modified in between v_i and the subsequent checkpoint, and $Left_{c_i^k}$ is the corresponding previous LEFT-column. The pairs $(c_i^k, Left_{c_i^k})$ were saved on the stack by the first LI that changed the corresponding LEFT-columns after v_i.

The procedures SSALEFT() and SSARIGHT() below will be invoked by module-modifying instructions to be discussed later.

SSALEFT():

1. If *BlockSSA*LEFT = TRUE then exit. Else:

2. Set *BlockSSA*LEFT = TRUE. Set variable t equal to current time (t is a new checkpoint). Use backtracking and the information in Stack$Left$ to undo as many of the most recent $Left$-modifications as necessary to achieve SSC — see inequality (1) in section 3. Pop off the corresponding blocks in Stack$Left$. This procedure guarantees that SSC will eventually be satisfied — see, for example, [13].

3. Push t and $RL(t)$ onto Stack$Left$. They are the first two elements of the next block to be pushed.

SSARIGHT**()** is analogous to SSALEFT().

A.3 Semantics of Instruction Heads and Their Arguments

In what follows, $x_1, x_2, y_1, y_2, z_1, z_2, \in \{0, \ldots, n-1\}$ stand for instruction arguments selected according to probability distributions $Q(IP{+}1, .), Q(IP{+}2, .), \ldots, Q(IP{+}6, .)$, respectively. They are used to address state components and module columns. For simplicity, instruction descriptions below use the following macros: $x := (x_1 n + x_2)\ mod\ m$; $y := (y_1 n + y_2)\ mod\ m$; $z := (z_1 n + z_2)\ mod\ m$.

Instructions Operating on Internal State Only

Jmpl$(x_1, x_2, y_1, y_2, z_1, z_2)$ **:** If $\mathcal{S}_x < \mathcal{S}_y$ then $IP \leftarrow z - z\ mod\ BS$.
Jmpeq$(x_1, x_2, y_1, y_2, z_1, z_2)$ **:** If $\mathcal{S}_x = \mathcal{S}_y$ then $IP \leftarrow z - z\ mod\ BS$.
Add$(x_1, x_2, y_1, y_2, z_1, z_2)$ **:** $\mathcal{S}_z \leftarrow (\mathcal{S}_x + \mathcal{S}_y)\ mod\ M$
Sub$(x_1, x_2, y_1, y_2, z_1, z_2)$ **:** $\mathcal{S}_z \leftarrow (\mathcal{S}_x - \mathcal{S}_y)\ mod\ M$
Mul$(x_1, x_2, y_1, y_2, z_1, z_2)$ **:** $\mathcal{S}_z \leftarrow (\mathcal{S}_x \mathcal{S}_y)\ mod\ M$
Div$(x_1, x_2, y_1, y_2, z_1, z_2)$ **:** if $\mathcal{S}_y \neq 0$ then $\mathcal{S}_z \leftarrow (\mathcal{S}_x\ div\ \mathcal{S}_y)\ mod\ M$
Mov(x_1, x_2, y_1, y_2)**:** $\mathcal{S}_y \leftarrow \mathcal{S}_x$
Init(x_1, x_2, y_1, y_2)**:** $\mathcal{S}_y \leftarrow x$.

Instructions for Interaction with the Environment There is a point-like agent with "active perception." It can move about in a two-dimensional environment (details of the environment are described in Section 4). The agent's current position is given by a pair of real values. Its current direction is given by variable *Direction* with range $[0, \ldots, 2\pi]$.

MoveAgent(): Move the agent 12 unit lengths in direction *Direction*, unless an obstacle or the environment's boundary blocks the path (the value 12 was chosen arbitrarily — neither it nor any other arbitrarily chosen constants mentioned below have ever been changed during the experiments). Moving may lead to new inputs: $\mathcal{S}_0$, $\mathcal{S}_1$, $\mathcal{S}_2$ become equal to the agent's horizontal position, vertical position, and current direction, respectively. Here position components are rounded to integers between 1 and 1000, and directions are scaled to $[0, 100]$ and then rounded. If the point 24 unit lengths in front of the agent is within an obstacle or outside the enviroment, then $\mathcal{S}_3 \leftarrow 24$ (the

value 24 is chosen arbitrarily). Otherwise $\mathcal{S}_3 \leftarrow 0$. Similarly for $\mathcal{S}_4$ (input 90 degrees to the right), $\mathcal{S}_5$ (behind), $\mathcal{S}_6$ (to the left).
SetDirection(x_1): *Direction* $= \frac{x_1}{n} 2\pi$. Compute new inputs like instruction MoveAgent().

"Introspective" Instructions

GetLEFT(x_1, x_2, y_1): $\mathcal{S}_x \leftarrow round(M\ Left_{x,y_1})$
GetRIGHT(x_1, x_2, y_1): $\mathcal{S}_x \leftarrow round(M\ Right_{x,y_1})$.

Bet Instructions

Bet!(x_1, x_2, y_1, y_2, c, d): Select a value $c' \in \{0, \ldots, n-1\}$ with probability $\frac{Left_{IP+5,c'}}{\sum_k Left_{IP+5,k}}$. If $c' > n/2$ then set variable $c \leftarrow 1$, else $c \leftarrow -1$. Select a value $d' \in \{0, \ldots, n-1\}$ with probability $\frac{Right_{IP+5,d'}}{\sum_k Right_{IP+5,k}}$. If $d' > n/2$ then set variable $d \leftarrow 1$, else $d \leftarrow -1$. If $c = d$ then exit. If $\mathcal{S}_y = \mathcal{S}_x$ then give reward c to LEFT and reward $-c$ to RIGHT. Otherwise give reward $-c$ to LEFT and reward c to RIGHT. Set $\mathcal{S}_7 \leftarrow c$ (surprise rewards become visible to the system in the form of inputs). *Comment: arguments x, y are indexed by $IP{+}1, IP{+}2, IP{+}3, IP{+}4$, while arguments c, d are chosen according to module columns indexed by $IP + 5$. See Section 2 for* Bet! *'s interpretation.*

SSA-enabling Instructions As long as *BlockSSA*LEFT = TRUE, LEFT will be protected from evaluations and restorations by SSA. This is important to allow for (1) complex learning processes computing not just one but many correlated modifications of LEFT-columns, and (2) reward delays: using the EnableSSALEFT() instruction below, LEFT can influence when to trigger evaluations and when to wait for delayed reward.

EnableSSALEFT(x_1): If $x_1 < 10$ (the value 10 is chosen arbitrarily) then set *BlockSSA*LEFT = FALSE.
EnableSSARIGHT(x_1): analogous to EnableSSALEFT.

Primitive Learning Algorithms All the module-modifying LIs below save the module columns they change on a stack. The first LEFT-modifying instruction executed after some EnableSSALEFT() instruction will start a new block of LEFT-modifications, to be ended by the next EnableSSALEFT(). Analogously for RIGHT.

IncProbLEFT(x_1, x_2, y_1):

1. Call SSALEFT().
2. If there is not already some $(x, .)$ entry above the most recent *(checkpoint, RL(checkpoint))* pair in Stack$Left$, then push the pair $(x, Left_x)$ onto Stack$Left$.

3. Set $Left_{x,k} \leftarrow \lambda Left_{x,k}\ \forall k \neq y_1$. Set $Left_{x,y_1} \leftarrow 1-\lambda(1-Left_{x,y_1})$.
Here $0 < \lambda < 1$ is a real-valued constant. In the experiments I arbitrarily use $\lambda = 0.3$. Note that step **3.** includes renormalization.
IncProbLEFT has no effect though if the corresponding module modifications would lead to at least one LEFT value below *MinProb*, a small positive real value (I use $MinProb = 0.004$).
Comment: IncProbLEFT is an LI that permits LEFT to modify itself (provided RIGHT agrees). IncProbLEFT instructions may be used in conjunction with other instructions to form complex probabilistic learning algorithms (running between subsequent checkpoints).

DecProbLEFT(x_1, x_2, y_1)**:** like IncProbLEFT, but step **3.** is different:
3. $Left_{x,d} \leftarrow \lambda Left_{x,y_1};\ \forall k \neq y_1 : Left_{x,k} \leftarrow \frac{1-\lambda Left_{x,k}}{1-Left_{x,k}} Left_{x,k}$.

MoveDistLEFT(x_1, x_2, y_1, y_2)**:** like IncProbLEFT, but step **3.** is different:
3. $Left_x \leftarrow Left_y$.

IncProbRIGHT(x_1, x_2, y_1)**:** analogous to IncProbLEFT.

DecProbRIGHT(x_1, x_2, y_1)**:** analogous to DecProbLEFT.

MoveDistRIGHT(x_1, x_2, y_1, y_2)**:** analogous to MoveDistLEFT.

IncProbBOTH(x_1, x_2, y_1)**:** call IncProbLEFT and IncProbRIGHT in random order.

DecProbBOTH(x_1, x_2, y_1)**:** call DecProbLEFT and DecProbRIGHT in random order.

SSAandCopy(x_1)**:** If $x_1 \geq 5$ then exit (the value 5 is chosen arbitrarily). Set *BlockSSA*LEFT and *BlockSSA*RIGHT = FALSE. Call SSALEFT() and SSARIGHT() in random order. Test if one of the modules has received more reward per time (since the most recent checkpoint still in its stack) than the other. If so:
Find those columns in the superior module that differ from the corresponding columns in the "loser." (In my implementation this is done efficiently by using a separate stack and a marker array tracing module differences as they occur.) Push the loser's different columns onto the loser's stack (just like with IncProbLEFT and all other LIs). Then copy the winner's different columns onto the loser's.
Comment: the LI SSAandCopy() allows for ending "unfair" matches in the case one module consistently outperforms the other. SSAandCopy will make both modules identical, although their stacks will in general be quite different and reflect quite different histories of successful module modifications.

Basic Cycle Modification For didactic reasons I wait until the end of this appendix to introduce a slight change in the basic cycle's instruction selection procedure (compare Section A.1). In case i addresses an instruction head as opposed to an argument, redefine

$$Q(i,j) = \frac{f(Right_{i,j}, Left_{i,g(j)})}{\sum_k f(Right_{i,k}, Left_{i,g(k)})},$$

$$i \in \{0, BS, 2BS, \ldots\}, j \in \{0, \ldots, n-1\}.$$

If a_i is IncProbRIGHT then $g(i)$ returns the index of a_i's *antagonistic* instruction head IncProbLEFT. Similarly for the other pairs of antagonistic instruction heads: (DecProbRIGHT, DecProbLEFT), (MoveDistRIGHT, MoveDistLEFT), (GetRIGHT, GetLEFT), (EnableSSARIGHT, EnableSSALEFT). This is necessary because antagonistic instructions require special treatment to achieve module symmetry through *SSAandCopy*. For instance, suppose that LEFT's current advantage depends on supporting some IncProbRIGHT instruction. An equal RIGHT opponent should strongly support IncProbLEFT instead.

Acknowledgments

I would like to thank Jieyu Zhao (supported by SNF grant 21-43'417.95 "Incremental Self-improvement"), Nic Schraudolph, Rafal Salustowicz, Sepp Hochreiter, Marco Wiering, and Luca Gambardella for valuable discussions and comments on a draft of this chapter, which is based on the following earlier publications: [45,46].

References

1. Fedorov, V. V. (1972) *Theory of optimal experiments.* Academic Press
2. Hwang, J., Choi, J., Oh, S., Marks II, R. J. (1991) Query-based learning applied to partially trained multilayer perceptrons. *IEEE Transactions on Neural Networks*, **2**, 131–136
3. MacKay, D. J. C. (1992) Information-based objective functions for active data selection. *Neural Computation*, **4**, 550–604
4. Plutowski, M., Cottrell, G., White, H. (1994) Learning Mackey-Glass from 25 examples, plus or minus 2. In J. Cowan, G. Tesauro, and J. Alspector, editors, *Advances in Neural Information Processing Systems* ***6***, 1135–1142. Morgan Kaufmann
5. Cohn, D. A. (1994) Neural network exploration using optimal experiment design. In J. Cowan, G. Tesauro, and J. Alspector, editors, *Advances in Neural Information Processing Systems* ***6***, 679–686. Morgan Kaufmann
6. Shannon, C. E. (1948) A mathematical theory of communication (parts I and II). *Bell System Technical Journal*, XXVII, 379–423

7. Schmidhuber, J. (1991) A possibility for implementing curiosity and boredom in model-building neural controllers. In J. A. Meyer and S. W. Wilson, editors, *Proceedings of the International Conference on Simulation of Adaptive Behavior: From Animals to Animats*, 222–227. MIT Press/Bradford Books
8. Schmidhuber, J. (1991) Curious model-building control systems. In *Proceedings of the International Joint Conference on Neural Networks, Singapore*, **2**, 1458–1463. IEEE
9. Storck, J., Hochreiter, S., Schmidhuber, J. (1995) Reinforcement driven information acquisition in non-deterministic environments. In *Proceedings of the International Conference on Artificial Neural Networks, Paris*, **2**, 159–164. EC2 & Cie
10. Schmidhuber, J., Prelinger, D. (1993) Discovering predictable classifications. *Neural Computation*, **5** , 625–635
11. Lenat, D. (1983) Theory formation by heuristic search. *Machine Learning*, **21**
12. Holland, J. H. (1985) Properties of the bucket brigade. In *Proceedings of an International Conference on Genetic Algorithms*, Hillsdale, NJ
13. Schmidhuber, J., Zhao, J., Wiering, M. Shifting inductive bias with success-story algorithm, adaptive Levin search, and incremental self-improvement. *Machine Learning*, **28**, 105–130
14. Schmidhuber, J., Zhao, J., Schraudolph, N. (1997) Reinforcement learning with self-modifying policies. In S. Thrun and L. Pratt, editors, *Learning to learn*, 293–309. Kluwer
15. Kolmogorov, A. N. (1965) Three approaches to the quantitative definition of information. *Problems of Information Transmission*, **1**, 1–11
16. Kwee, I., Hutter, M., and Schmidhuber, J. (2001) Market-based reinforcement learning in partially observable worlds. *Proceedings of the International Conference on Artificial Neural Networks (ICANN-2001), in press.*
17. Chaitin, G. J. (1969) On the length of programs for computing finite binary sequences: statistical considerations. *Journal of the ACM*, **16**, 145–159
18. Solomonoff, R. J. (1964) A formal theory of inductive inference. Part I. *Information and Control*, **7**, 1–22
19. Li, M., Vitányi, P. M. B. (1997) *An Introduction to Kolmogorov Complexity and its Applications*. Springer
20. Schmidhuber, J. (1999) A general method for incremental self-improvement and multi-agent learning. In X. Yao, editor, *Evolutionary Computation: Theory and Applications*, 81–123. World Scientific
21. Schmidhuber, J. (1997) Discovering neural nets with low Kolmogorov complexity and high generalization capability. *Neural Networks*, **10**, 857–873
22. Lin, L. J. (1993) *Reinforcement Learning for Robots Using Neural Networks*. PhD thesis, Carnegie Mellon University, Pittsburgh
23. Geman, S., Bienenstock, E., Doursat, R. (1992) Neural networks and the bias/variance dilemma. *Neural Computation*, **4**, 1–58
24. Clarke, A. C. (1991) *The ghost from the grand banks.*
25. Hochreiter, S., Schmidhuber, J. (1997) Flat minima. *Neural Computation*, **9**, 1–42
26. Hillis, D. (1992) Co-evolving parasites improve simulated evolution as an optimization procedure. In C.G. Langton, C. Taylor, J. D. Farmer, and S. Rasmussen, editors, *Artificial Life II*, 313–324. Addison Wesley

27. Pollack, J. B., Blair, A. D. (1997) Why did TD-Gammon work? In M. C. Mozer, M. I. Jordan, and S. Petsche, editors, *Advances in Neural Information Processing Systems*, **9**, 10–16
28. Samuel, A. L. (1959) Some studies in machine learning using the game of checkers. *IBM Journal on Research and Development*, **3**, 210–229
29. Tesauro, G. (1994) TD-gammon, a self-teaching backgammon program, achieves master-level play. *Neural Computation*, **6**, 215–219
30. Schmidhuber, J. (1992) Learning factorial codes by predictability minimization. *Neural Computation*, **4**, 863–879
31. Schraudolph, N., Sejnowski, T. J. (1993) Unsupervised discrimination of clustered data via optimization of binary information gain. In Stephen José Hanson, Jack D. Cowan, and C. Lee Giles, editors, *Advances in Neural Information Processing Systems*, **5**, 499–506
32. Schmidhuber, J., Eldracher, M., Foltin, B. (1996) Semilinear predictability minimization produces well-known feature detectors. *Neural Computation*, **8**, 773–786
33. Schraudolph, N. N., Eldracher, M., Schmidhuber, J. (1999) Processing images by semi-linear predictability minimization. *Network: Computation in Neural Systems*, **10**, 133–169
34. Nake, F. (1974) *Ästhetik als Informationsverarbeitung.* Springer
35. Schmidhuber, J. (1997) Low-complexity art. *Leonardo, Journal of the International Society for the Arts, Sciences, and Technology*, **30**, 97–103
36. Schmidhuber, J. (1998) Facial beauty and fractal geometry. Technical Report IDSIA-28-98, IDSIA, Also published in the Cogprint Archive: http://cogprints.soton.ac.uk
37. Wilson, S. W. (1994) ZCS: A zeroth level classifier system. *Evolutionary Computation*, **2**, 1–18
38. Wilson, S. W. (1995) Classifier fitness based on accuracy. *Evolutionary Computation*, **3**, 149–175
39. Weiss, G. (1994) Hierarchical chunking in classifier systems. In *Proceedings of the 12th National Conference on Artificial Intelligence*, **2**, 1335–1340
40. Weiss, G., Sen, S. (eds.) (1996) *Adaption and Learning in Multi-Agent Systems.* LNAI 1042, Springer-Verlag
41. Schmidhuber, J. (1987) Evolutionary principles in self-referential learning, or on learning how to learn: the meta-meta-... hook. Institut für Informatik, Technische Universität München
42. Schmidhuber, J. (1989) A local learning algorithm for dynamic feedforward and recurrent networks. *Connection Science*, **1**, 403–412
43. Baum, E. B., Durdanovic, I. (1999) Toward a model of mind as an economy of agents. *Machine Learning*, **35**, 155–185
44. Wolpert, D. H., Tumer, K., Frank, J. (1999) Using collective intelligence to route internet traffic. In M. Kearns, S. A. Solla, and D. Cohn, editors, *Advances in Neural Information Processing Systems 12*
45. Schmidhuber, J. (1998) What's interesting? Technical Report IDSIA-35-97, IDSIA, 1997. ftp://ftp.idsia.ch/pub/juergen/interest.ps.gz; extended abstract in Proc. Snowbird'98, Utah
46. Schmidhuber, J. (1999) Artificial curiosity based on discovering novel algorithmic predictability through coevolution. In P. Angeline, Z. Michalewicz, M. Schoenauer, X. Yao, and Z. Zalzala, editors, *Congress on Evolutionary Computation*, 1612–1618. IEEE Press

Part II
Applications

Approaches to Combining Local and Evolutionary Search for Training Neural Networks: A Review and Some New Results

Kim W. C. Ku[1], M. W. Mak[2], and W. C. Siu[2]

[1] Department of Computer Science, City University of Hong Kong
cskimku@cityu.edu.hk

[2] Center for Multimedia Signal Processing, Department of Electronic and Information Engineering, The Hong Kong Polytechnic University

Summary. Training of neural networks by local search such as gradient-based algorithms could be difficult. This calls for the development of alternative training algorithms such as evolutionary search. However, training by evolutionary search often requires long computation time. In this chapter, we investigate the possibilities of reducing the time taken by combining the efforts of local search and evolutionary search. There are a number of attempts to combine these search strategies, but not all of them are successful. This chapter provides a critical review of these attempts. Moreover, different approaches to combining evolutionary search and local search are compared. Experimental results indicate that while the Baldwinian and the two-phase approaches are inefficient in improving the evolution process for difficult problems, the Lamarckian approach is able to speed up the training process and to improve the solution quality. In this chapter, the strength and weakness of these approaches are illustrated, and the factors affecting their efficiency and applicability are discussed.

1 Introduction

Over the past few decades, development in neural networks has focused on particular types of neural network called multi-layer feed-forward networks. These are static networks where the network outputs depend only on the current inputs, not on any past inputs or outputs. While feed-forward networks have found applications in pattern classification and functional interpolations [31,32,48], they are subjected to a constraint that temporal information cannot be stored naturally (unless encoded explicitly, e.g. through the use of tapped delay inputs [74,78]).

To circumvent the above drawback, recurrent neural networks (RNNs) have been introduced by a number of researchers, e.g. Elman [15], Jordan [36], Pineda [64], and Williams and Zipser [84], to name but a few. These networks have feedback connections so that they can preserve their past activities for future computation. As a result, the current network state depends on the previous ones over a potentially unbounded period of time (up to the time at which the network is started to operate). In contrast, the outputs of a feed-forward network with tapped delay inputs are only dependent on its inputs within a limited period of time. With the capability of handling temporal information, RNNs have been used to model the temporal processes occurring in nature, science, and engineering [26,61,65].

1.1 Training by Local Search

One school of thought to determine the network weights is to use local search methods. Typical examples are the back-propagation algorithm developed by Rumelhart et al. [70] for training feedforward neural networks and the real-time recurrent learning (RTRL) algorithm developed by Williams and Zipser [83] for training RNNs. These local search methods make use of the gradient information of the network error function. However, it is commonly believed that using gradient information to train neural networks has difficulties in (a) escaping from local optima when the search surface is rugged; (b) finding better solutions when the surface has many plateaus (gradient is zero, for example); and (c) deciding the search direction when gradient information is not readily available (lack of target signals, for example). Because of these difficulties, non-gradient-based searching approaches such as evolutionary search have been proposed.

1.2 Training by Evolutionary Search

Another school of thought to train neural networks is to use evolutionary search. Genetic algorithms [23,54,56], evolutionary programming [18,20], and evolution strategies [67,69,73] are typical examples of evolutionary search. Attempts at training feedforward neural networks by evolutionary search include the work of Fogel et al. [19], Yao and Liu [87], and Montana and Davis [57]. There are also attempts to evolve recurrent networks, e.g. Angeline et al. [3] and McDonnell and Waagen [51]. Applying evolutionary search to more complex types of neural networks (high order networks, for example) can be found in [33–35,85], and a good review of evolving neural networks is provided by [86].

Bäck et al. [4] and Fogel [17] provided an introduction to various evolutionary search algorithms. Generally, a population of candidate solutions, ranked by their performance, is maintained and updated iteratively by evolutionary search. Each candidate solution represents one neural network in which the weights can be encoded as a string of binary [14,81,82] or floating-point numbers [24,52,66,71]. The performance of each solution is determined by the network error function which is to be optimized by evolutionary search.

Unlike local search, evolutionary search maintains a population of potential solutions rather than a single solution. Therefore, the risk of getting stuck in local optima is smaller. Moreover, as gradient information is not required, evolutionary search is applicable to problems where gradient information is unavailable or to the cases where the search surface contains many plateaus. However, the iterative process in evolutionary search requires evaluation of a large number of candidate solutions; consequently, evolutionary search is usually slower than local search. The lack of fine-tuning operations in evolutionary search also limits the accuracy of the final solution.

1.3 Combining Local and Evolutionary Search

Training neural networks by local search has a higher risk of getting stuck in local optima, and its applications are limited to the cases where gradient information is readily available. These limitations can be overcome if the training is performed by evolutionary search. However, training by evolutionary search is usually a slow process. Obviously, these two search strategies have their own strengths and weaknesses. One possible way of constructing an efficient hybrid algorithm is to allow these two search strategies to complement each other. In this chapter, the possibilities of creating efficient hybrid training algorithms by combining the efforts of local search and evolutionary search are investigated.

2 Attempts in Combining Local and Evolutionary Search

In the belief that better results can be achieved by combining local search and evolutionary search, various attempts have been made to adopt this synergetic approach to construct and train neural networks. Some [8,25,39,49] achieved good results while others [41,57] found that the resulting hybrid algorithms are not efficient. These attempts differ in how local search is applied, and the differences are summarized in this section.

2.1 Nature of Local Search

Local search aims at searching for better solutions in the neighborhood of the current solution. There are different local search methods, and the following are some typical examples.

Stochastic Methods Maniezzo [49] proposed a hybrid algorithm for evolving feedforward networks. In Maniezzo's work, the networks are enhanced by a local search method similar to the simplex procedure in linear programming [12]. More specifically, the local search is embedded in evolutionary search as a kind of evolutionary operator. Like other evolutionary operator (e.g. crossover and mutation), the local search operator is selected according to a fixed probability. The operator optimizes the new binary-encoded offspring based on the binary strings of three parent solutions and their corresponding fitness values, and it works as follows. Suppose the fitness values of three parents x_1, x_2, and x_3 are sorted such that x_1 is the best parent and x_3 is the worst. The ith bit of offspring x_4 (denoted as x_{4i}) is set to x_{1i} if $x_{1i} = x_{2i}$; otherwise x_{4i} is set to the negation of x_{3i}. Although the local search method is very simple, the inclusion of this operator is found to be able to improve the evolution process.

Gruau and Whitley [25] proposed a local search method and compared different approaches to combining local search and evolutionary search. In their hybrid algorithms, a boolean neural network is represented by a grammar tree (instead of a string of floating-point numbers or a binary string) that specifies the number of nodes, the connectivity, and the network weights. The activation of each node is either 'on' or 'off', and the value of each weight is restricted to either $+1$ or -1. Local search is applied to every new offspring, but only the weights connecting to the network's outputs will be changed by the local search.

To a certain extent, the local search method in [25] is similar to Hebbian learning [29]. For each weight connecting to an output node, there is an associated variable d that is initialized to zero before applying the local search. When the training patterns are fed to the network one at a time during the application of local search, d will be increased if the activations of the two nodes (each output node is clamped to the target output) across the weight are the same; otherwise, d will be decreased. After all training patterns have been fed, the final values of d's are used for deciding whether the weights should be flipped or not. A subset of weights with signs opposite to the variable d are selected for consideration of flipping. The weight with the largest absolute value of d in the subset is flipped, while others are flipped with very small probabilities. Gruau and Whitley showed that this simple local search can speed up the evolution process.

McDonnell and Waagen [51] proposed a hybrid algorithm that combines the method of Solis and Wets [76] and evolutionary search for evolving RNNs. In each iteration of the evolutionary search, a set of offspring is generated by applying the Solis and Wets method and another set is generated by perturbing the parent solutions according to a normal distribution. The Solis and Wets method directs the search by comparing the fitness of $\mathbf{x} + \delta\mathbf{x}$ and $\mathbf{x} - \delta\mathbf{x}$ with that of the parent solution $\mathbf{x}$, where $\delta\mathbf{x}$ is a normally distributed offset. More precisely, if $\mathbf{x} + \delta\mathbf{x}$ is better than $\mathbf{x}$, then $\mathbf{x} + \delta\mathbf{x}$ will be chosen as the offspring solution; otherwise, $\mathbf{x} - \delta\mathbf{x}$ will be the offspring if it is better than $\mathbf{x}$. If both are worse than $\mathbf{x}$, new solutions with different $\delta\mathbf{x}$ will be tried systematically. If good solutions are produced frequently, $\delta\mathbf{x}$ will be increased to enlarge the search step; otherwise, $\delta\mathbf{x}$ will be decreased to refine the search. Although promising results have been obtained [51], this search method is not very efficient. This is because finding a good solution may require a large number of iterations and each iteration requires evaluation of the fitness of $\mathbf{x} + \delta\mathbf{x}$ and $\mathbf{x} - \delta\mathbf{x}$, which is computationally intensive.

Gradient-based Methods The backpropagation algorithm [70] is a well-known gradient-based algorithm for training feedforward neural networks. Therefore, it is common to combine backpropagation with evolutionary search to construct hybrid algorithms. For example, in the work of Miller et al. [55], backpropagation is applied iteratively to every network generated by evolutionary search. In each iteration of backpropagation, the gradient of the search surface is calculated and network weights are changed in a direction opposite to the gradient. This can be computationally expensive if a large number of iterations are required to find an acceptable network. While promising results can be obtained by combining backpropagation and evolutionary search (e.g. in [62,87]), fast variants of backpropagation are sometimes required to speed up the hybrid algorithms.

Considering the computational trade-offs between local and evolutionary search, Braun and Zagorski [8] adopted a fast backpropagation algorithm RPROP [68] as the local search method. In their hybrid algorithm, evolutionary search is interleaved with gradient-based local search. Experimental results show that the hybrid algorithm is able to produce high-quality networks.

Conjugate gradient has been widely used in some gradient-based local search methods. Methods based on conjugate gradient are different from backpropagation in that a series of conjugate search directions are generated such that optimization in the current direction does not affect the optimization in the previous directions. Skinner and Broughton [75] proposed a hybrid algorithm in which conjugate gradient is used to further train the networks after the evolutionary search. The experimental results demonstrate that the hybrid algorithm can shorten the overall training time.

2.2 Recipients of Local Search

The following are some criteria that have been used to select the candidate solutions for local search.

Applying Local Search for Final Fine-tuning Montana and Davis [57] attempted to use local search to fine-tune the feedforward networks found by evolutionary search. The best network (encoded as a string of floating-point numbers) obtained after a fixed amount of evolutionary search is fine-tuned by a backpropagation-like algorithm. The algorithm is different from the standard backpropagation one in that weights are updated in an adaptive step size that is not proportional to the magnitude of error gradient. The experimental results show that the network performance is improved for a very short period, followed by a period of no improvement. Montana and Davis concluded that this combination of local and evolutionary search does not provide any significant improvement.

Applying Local Search to Preferred Individuals Korning [41] performed experiments to train feedforward networks by a hybrid algorithm in which evolutionary search is interleaved with a hillclimber. In each iteration of the evolution process, the offspring solutions produced by evolutionary operators will be taken for hillclimbing if their fitness is good enough. As each weight in the networks is encoded by a binary string, hillclimbing is achieved by bit-flipping. Specifically, the more significant bits of each string are flipped in a round-robin manner. For each bit inversion, the change is kept only if it improves the fitness. Hillclimbing is terminated when no further improvement can be obtained. Experimental results showed that the hillclimber can only achieve a very small fitness improvement in each iteration. This local search method is also very computationally expensive, because fitness has to be evaluated for each flipped bit. Therefore, the benefit gained from the local search is very little.

Unbiased Application of Local Search Belew et al. [6] investigated the efficiency of combining local and evolutionary search and proposed the hybrid algorithms that apply local search to every offspring. Each offspring, which is encoded in the form of binary strings, specifies an initial weight vector from which searching for better networks by backpropagation begins. The range of initial weights (e.g. $\pm\frac{1}{2}$) being explored by evolutionary search is smaller than the range of weights found by local search. The rationale of using evolutionary search to find the initial weight vector is that the ability of gradient-based algorithms in finding satisfactory solutions is heavily influenced by the initial weight vector [40]. Experimental results demonstrate that evolutionary search is able to find good initial weights for backpropagation to begin with and that the solutions found by the hybrid algorithms are better than the ones found by evolutionary search alone or backpropagation with multiple random restarts.

As evolutionary search is only used to find the initial weights, the hybrid algorithm spends most of its time in applying backpropagation. In the six-bit symmetry problem studied by Belew et al. [6], backpropagation was applied to each offspring for 40 or 200 epochs. Complex problems, however, are likely to require far more than 200 epochs. This can affect the hybrid algorithm's efficiency considerably.

2.3 Combination Approaches

Attempts at combining local search and evolutionary search can be categorized according to the synergetic approaches. These include the two-phase approach, Lamarckian evolution, and the approaches that are based on the Baldwin effect.

Two-phase Approach Kitano [39] adopted the two-phase approach to train feedforward networks for classification problems. In Kitano's work, evolutionary search is used to find the regions that are likely to contain the global optimum, then local search is used as a final fine-tuning operator. The evolutionary search is terminated when the network performance reaches a pre-defined threshold value that indicates the proximity of the global optimum. The best network is then taken for further training by local search. Backpropagation is applied iteratively until an acceptable solution is found. Kitano found that the overall training process is improved by the two-phase approach.

Similarly, Belew et al. [6] used genetic algorithms (GAs) to find the initial weights of feedforward networks, which were further trained by backpropagation in the second phase. Hybrid algorithms based on the two-phase approach were also proposed by Skinner and Broughton [75], and the algorithms were found to be effective for training feedforward networks to solve function interpolation problems. More investigations on this approach can be found in [16,47,60].

Although the studies mentioned above have illustrated the two-phase approach's capability, the training tasks they used are generally not difficult so that local search alone can find the solutions successfully. In difficult training tasks, the benefit of this approach is unclear. Therefore, it is necessary to evaluate the two-phase approach's capability by applying it to a test problem that cannot be solved by local search alone.

Lamarckian Evolution Lamarckian evolution is based on the inheritance of acquired characteristics obtained through learning. This approach was adopted in the hybrid algorithm proposed by Yao and Liu [87] to evolve feedforward networks. In each iteration of the evolution process, the hybrid algorithm selects a network from the population and trains it by backpropagation for a fixed number of epochs. If the training improves the network's performance, the trained network together with its associated fitness will be put back into the population for further evolution. This mechanism preserves the acquired characteristics obtained through learning, which is very similar to Lamarckian evolution.

Promising results of Lamarckian evolution were reported by Braun and Zagorski [8] in constructing feedforward networks to solve classification problems. Braun and Zagorski argued that fine-tuning every network by a fast backpropagation algorithm can reduce the search space to a set of local optimal points (or saddle points) such that finding the global optimum becomes more efficient.

It is noteworthy that previous studies in Lamarckian evolution (such as those mentioned above) typically employ local search methods with high computational complexity. This could introduce a serious burden to the hybrid algorithms. Although these studies have demonstrated the capability of Lamarckian evolution, most of them did not show the actual time improvement, making the real benefit of combining local and evolutionary search difficult to observe. Therefore, this chapter evaluates the capability of Lamarckian learning by comparing the actual time taken in the experiments.

Baldwin Effect As Lamarckism cannot be found in biological systems, another school of thought is to use a more biologically plausible mechanism based on the Baldwin effect. In contrast with Lamarckian learning, Baldwinian learning does not allow a parent to pass its learned characteristics to its offspring; instead, only the fitness after learning is retained.

Hinton and Nowlan [30] were the first to use Baldwinian learning for evolving neural networks. In their experiments, random search is applied to every neural network generated by evolutionary search. The random search does not change the network; rather, only its fitness is updated to reflect the distance from the global optimum. Their experimental results show that the hybrid algorithm is able to find the global optimum, which is unachievable by using evolutionary search alone.

In the work of Ackley and Littman [1], Baldwinian learning was used to assist the evolution of adaptive agents that struggle for survival in an artificial ecosystem. The behavior of each agent is specified by an evaluation neural network and an action neural network. Ackley and Littman found that evolution without learning produces ill-behaved agents that are unfit for survival in the ecosystem, causing extinction of adaptive agents in a short time. On the other hand, with Baldwinian learning, well-behaved agents and long-lasting populations can be produced. Ackley and Littman argued that the Baldwin effect is beneficial to evolution because it allows the agents to stay longer in the ecosystem.

Although the capability of Baldwinian learning has been demonstrated, it has also been suggested that the Baldwin effect can, in some circumstances, lead to inefficient hybrid algorithms [22,37,80]. These prompt us to investigate the efficiency of Baldwinian learning and to determine the situations that degrade the hybrid algorithms' performance.

2.4 Other Attempts

Apart from the above investigations, there are other attempts at examining the relationship between local and evolutionary search in neural networks. These works [28,59,63] provide bases for modelling learning and evolution in biological organisms in order to understand their complex behavior. There are also studies in using evolutionary search to find an optimal learning parameter set for local search methods, such as the learning rate and momentum term in the backpropagation algorithm [27,38,53]. More ambitious works include the investigation of the evolution of local search methods [9,11,21]. For example, the delta rule for feedforward neural networks has been successfully evolved in [9].

3 The Long-term Dependency Problem

Much of the previous work in combining gradient-based algorithms and evolutionary search used either a simple task (e.g. the parity and symmetry tasks in [6,25,55]) or a task that can be readily solved by using local search alone (e.g. the classification tasks in [8,39] and the function interpolation tasks in [75]). If gradient-based algorithms can successfully train neural networks for the given training tasks, there will be little incentive to use evolutionary search methods. However, there are situations in which gradient-based algorithms have difficulties in finding an appropriate neural network. The long-term dependency problem is a typical example.

Many sequence recognition tasks such as speech recognition, handwriting recognition, and grammatical inference involve long-term dependencies—the output depends on inputs that occurred a long time ago. These tasks depend mainly on whether the long-term dependencies can be accurately represented; however, extracting these dependencies from data is not an easy task. While RNNs provide a promising solution to this problem, some researchers [7,58] have shown that the commonly used gradient-based algorithms have difficulty in learning the long-term dependencies.

The long-term dependency problem used in this chapter is defined as follows. It is required to learn a temporal relationship such that the output at time t depends on the inputs from time $t - t'$ to $t - 1$. Let us assume that an input sequence contains symbols drawn from a symbol set and that each symbol is represented by a binary number with N bits. There are only two possible input sequences:

$$I = \begin{cases} (x, a_1, a_2, a_3, \ldots, a_k) \\ (y, a_1, a_2, a_3, \ldots, a_k), \end{cases}$$

where x, y, and $\{a_i\}_{i=1}^{k}$ are the symbols in the symbol set. The first symbol in the input sequence can be either x or y, but the next k input symbols are fixed. The corresponding output sequences are

$$O = \begin{cases} (a_1, a_2, a_3, \ldots, a_k, x') & \text{if } I = (x, a_1, a_2, a_3, \ldots, a_k) \\ (a_1, a_2, a_3, \ldots, a_k, y') & \text{if } I = (y, a_1, a_2, a_3, \ldots, a_k). \end{cases}$$

In other words, when the first input symbol is x at time t, the output at time $t + k$ is x'; when the first input symbol is y at time t, the output at time $t + k$ is y'. For other time intervals, the output predicts the next input. A training sequence is formed by concatenating ten randomly chosen input/output sequences. As the problem becomes increasingly difficult when the temporal length increases, the experiments in this chapter used a length of five time steps, which was found to be sufficiently difficult for the gradient-based algorithms.[1]

[1] For shorter temporal length (three time steps, for example), the problem can be easily solved by gradient-based algorithms; consequently, training by evolutionary search becomes unnecessary.

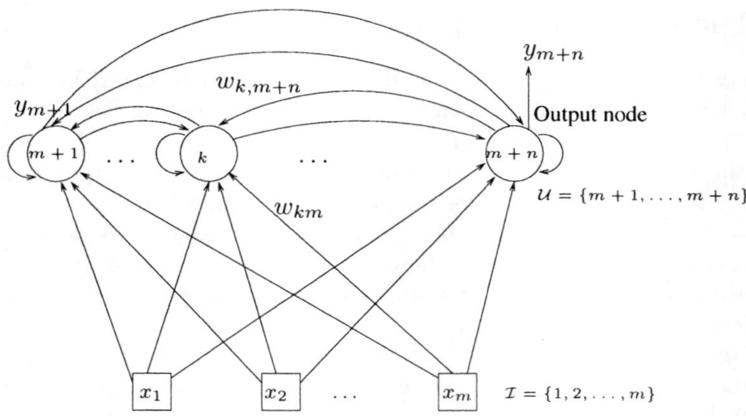

Fig. 1. A fully connected recurrent neural network with m inputs, n processing nodes, and one output

In this work, we used a fully-connected RNN to solve the long-term dependency problem. Fig. 1 shows a typical RNN with m inputs, n processing nodes, and one output. The parameters are defined as follows:

$x_k(t)$ = signal applied to input node k at time step t.
$y_k(t)$ = actual output of processing node k at time step t.
$\mathcal{I}$ = the set of indices representing the input nodes (including the bias).
$\mathcal{U}$ = the set of indices representing the processing nodes.
$\mathcal{O}$ = the set of indices representing the output nodes.
$z_k(t)$ = $x_k(t)$ if $k \in \mathcal{I}$, $z_k(t) = y_k(t)$ if $k \in \mathcal{U}$.
$d_k(t)$ = target output of processing node k at time step t.
$s_k(t)$ = activation of processing node k at time step t.
w_{ij} = weight connecting node j to node i.

The activation of processing node $k \in \mathcal{U}$ in the network is the weighted sum of the current inputs and the feedback signals:

$$s_k(t) = \sum_{p \in \mathcal{I}} w_{kp} x_p(t) + \sum_{q \in \mathcal{U}} w_{kq} y_q(t). \tag{1}$$

The output of processing node k at time step $t+1$ is

$$y_k(t+1) = f(s_k(t)) \tag{2}$$

where $y_k(0) = 0$ when the network is initialized and $f()$ is a nonlinear activation function defined by

$$f(s_k(t)) = \frac{1}{1 + e^{-s_k(t)}}. \tag{3}$$

The performance of neural networks can be measured through a network error function that is typically defined as the sum of the squared error between the actual network outputs and the target values over a fixed period. Let

$$E(t) = \frac{1}{2} \sum_{k \in \mathcal{O}} \{d_k(t) - y_k(t)\}^2 \tag{4}$$

denote the instantaneous squared error of the network at time step t, and let the network error function over the period $[t_0, t_n]$ be

$$E^{total}(t_0, t_n) = \sum_{t=t_0}^{t_n} E(t). \tag{5}$$

Therefore, the better the network performs over the period $[t_0, t_n]$, the smaller is the value of the network error function.

Typically, $d_k(t)$ and $x_k(t)$ are provided at every time step t, and the remaining unknown parameters are estimated by minimizing the network error function. Assuming that the network size is fixed and the nonlinear activation functions have no adjustable parameters, the only parameters required to be optimized are the weights w_{ij}, where $i \in \mathcal{U}$, $j \in \mathcal{U} \cup \mathcal{I}$. Different types of training algorithms have been developed to determine the weights.

In the following experiments, RNNs (see Fig. 1) with three input nodes and twelve processing nodes (five of them were dedicated as the output nodes) have been used to learn the long-term dependency problem with a temporal length of five time steps. Therefore, there are a total of $12 \times 12 + 12 \times (3+1) = 192$ weights required to be optimized.

Cellular genetic algorithms (GAs) [10,13,79], as described in Fig. 2, have been used to optimize the weights of RNNs. These weights are encoded as strings of floating-point numbers. With a population size of 100 and a random walk of four steps, the cellular GA is able to find acceptable solutions for the long-term dependency problem. The average performance (based on 100 simulations running on a Sun Sparc 1000 workstation) of the cellular GA is shown in Fig. 3, which forms the baseline performance for comparing with various hybrid approaches described in the following sections.

4 Local Search Methods

In order to improve the evolution process for the long-term dependency problem, different local search methods have been incorporated into the cellular GA. These local search methods and their performance in the long-term dependency problem are described and evaluated in this section.[2]

[2] Other local search methods, such as backpropagation through time, have also been investigated in our previous studies [42,43]. However, the performance of the hybrid algorithms produced by these methods are not satisfactory.

```
procedure cellularGA

c_k      : a chromosome at (x_k, y_k) where x_k(y_k) can be any x(y)-coordinate
           in the grid (x_k and y_k have no particular relationship)
c_new    : a newly produced chromosome
l        : length of random walk
M        : total number of chromosomes in the population
w_ij^c_k : weights w_ij of the network corresponding to c_k
f(c_k)   : fitness of c_k
I        : the set of indexes representing the input nodes (including the bias)
U        : the set of indexes representing the processing nodes

begin
    Initialize a population of M chromosomes c_k, and evaluate the
        corresponding fitness f(c_k) where k = 1, 2, ..., M

    // Generate a new chromosome for each reproduction cycle
    repeat
        Randomly select c_0 at (x_0, y_0) in the grid

        // Choose parent c_a along a random walk originating from (x_0, y_0)
        Create random walk set {c_1 at (x_1, y_1), ..., c_l at (x_l, y_l)} such that
            |x_{k+1} - x_k| ≤ 1 and |y_{k+1} - y_k| ≤ 1, k = 0, 1, 2, ..., l - 1
        Select c_a such that f(c_a) is the best along the random walk

        // Choose parent c_b along another random walk originating from
        // (x_0, y_0)
        Create random walk set {c'_1 at (x'_1, y'_1), ..., c'_l at (x'_l, y'_l)} such that
            |x'_{k+1} - x'_k| ≤ 1 and |y'_{k+1} - y'_k| ≤ 1, k = 1, 2, ..., l - 1 and
            |x'_1 - x_0| ≤ 1 and |y'_1 - y_0| ≤ 1
        Select c_b such that f(c_b) is the best along the random walk

        // Apply crossover to c_a and c_b to produce c_new
        for all i ∈ U, j ∈ U ∪ I do
            w_ij^c_new := { w_ij^c_a    with a probability of 0.5
                          { w_ij^c_b    with a probability of 0.5
        endloop

        // Apply mutation to c_new by randomly selecting a processing node
        // in the network, and each weight connected to the input part of
        // the node is changed by exponentially distributed mutation
        Randomly select i ∈ U
        for all j ∈ U ∪ I do
            w_ij^c_new := { w_ij^c_new + δ    with a probability of 0.5
                          { w_ij^c_new - δ    with a probability of 0.5
            // δ is a positive number randomly generated from
            // the exponential source p(x) = e^{-x}, x > 0
        endloop

        // Replace c_0 by c_new if the latter has better fitness
        Evaluate f(c_new)
        if f(c_new) < f(c_0) then c_0 := c_new
    until termination condition reached
endproc cellularGA
```

Fig. 2. The procedure of the cellular GA [46]

4.1 Real-time Recurrent Learning

The real-time recurrent learning (RTRL) algorithm [83] calculates the instantaneous error gradient $\nabla_{\mathbf{w}} E(t)$ by

$$\frac{\partial E(t)}{\partial w_{ij}} = -\sum_k (d_k(t) - y_k(t)) \frac{\partial y_k(t)}{\partial w_{ij}} \tag{6}$$

where $E(t)$ (defined in (4)) is the instantaneous squared error at time step t. The sensitivity $\frac{\partial y_k(t)}{\partial w_{ij}}$ is obtained by the recursion

$$\frac{\partial y_k(t+1)}{\partial w_{ij}} = f_k'(s_k(t)) \left\{ z_j(t)\delta_{ki} + \sum_q w_{kq} \frac{\partial y_q(t)}{\partial w_{ij}} \right\} \tag{7}$$

with $\frac{\partial y_k(0)}{\partial w_{ij}} = 0$, where δ_{ki} is the Kronecker delta.

The RTRL algorithm is a gradient-based algorithm in which all the weights are changed at every time step in a direction opposite to the instantaneous error gradient. It is computationally intensive because it has a computational complexity of $O(n^4)$ for each time step, where n is the number of processing nodes.

4.2 Delta Rule

The running time of the RTRL algorithm scales poorly with the network size. In order to reduce the computational complexity, we propose to update only the weights that connect to the output nodes. Specifically, we only compute the gradient $\frac{\partial E(t)}{\partial w_{ij}}$ whenever node i is an output node. Therefore, (7) is simplified to

$$\frac{\partial y_i(t+1)}{\partial w_{ij}} = \begin{cases} f_i'(s_i(t)) z_j(t) & \text{when } i \text{ is an output node} \\ 0 & \text{otherwise.} \end{cases} \tag{8}$$

This is equivalent to the delta rule for feedforward networks. The dynamics of the network remain unchanged; however, the updates of weights are based on a feedforward architecture. The philosophy behind this approach is to lower the computational complexity by eliminating the term $\sum_q w_{kq} \frac{\partial y_q(t)}{\partial w_{ij}}$ in (7).

4.3 Applying Local Search Alone

For the long-term dependency problem, a set of control experiments has been performed to train the RNNs by local search alone. The limitation of the gradient-based algorithms (i.e. RTRL and the delta rule) is clearly demonstrated in Fig. 3. For most of the simulation runs, the mean square errors (MSEs) are quickly reduced to a value around 0.08, and no improvement can be obtained by further training.

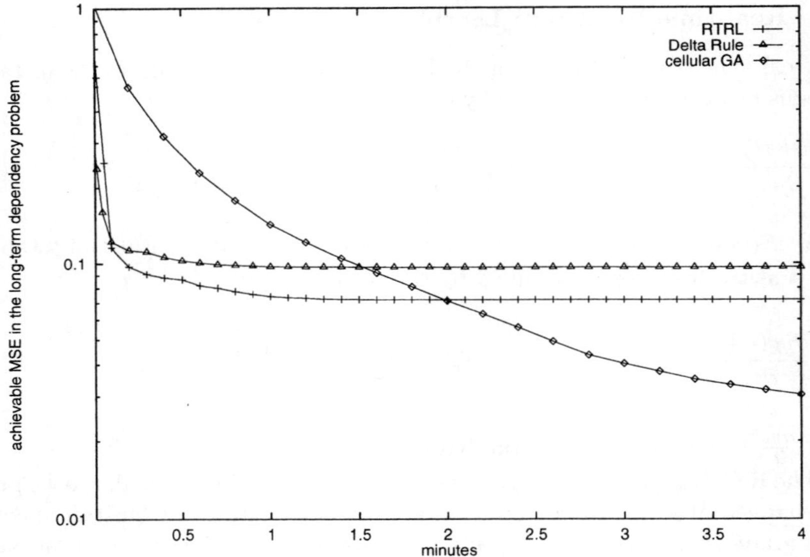

Fig. 3. MSE (based on the average of ten simulations) of the best network found by various gradient-based local search methods. The performance of the cellular GA is also illustrated.

On the other hand, Fig. 3 shows that the cellular GA is more capable of solving the problem. The average MSE attained after four minutes of simulations (i.e. 20,000 generations) is 0.0303, which is lower than that of the local search methods.

5 Two-phase Approach

Although cellular GAs are viable training algorithms for neural networks, training by cellular GAs may require long computation time, a typical problem of evolutionary search. In order to shorten the training time and to improve the solution quality, different combinations of local search and cellular GAs are investigated in this chapter. One intuitive approach to combining the efforts of local search and cellular GAs is the two-phase approach.

In the two-phase approach, the cellular GA is used in the first phase to roughly locate the global optimum. The aim is to avoid the local optima where local search may get stuck. This phase terminates when the MSE of the best network in the population reaches a pre-defined threshold. Then, local search is applied to fine-tune the best network in the second phase in order to accelerate the search process.

As Section 4.3 points out that there are difficult regions around an MSE of 0.08 in the search space, using a threshold of 0.07 ensures that the cellular

GA has already moved the solutions out of the difficult regions. This should overcome one of the barriers that hinders the gradient-based local search and should increase the chance of finding a satisfactory solution in the second phase.

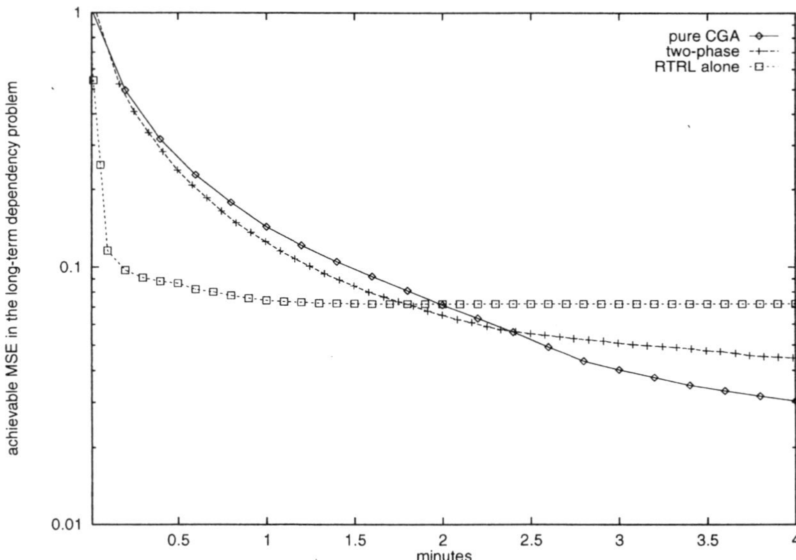

Fig. 4. MSE (based on the average of 100 simulations) achieved by the two-phase approach where a cellular GA was applied in the first phase and RTRL was applied in the second phase. The pre-defined threshold for switching between phases was set to 0.07. The performance of the RTRL algorithm and the pure cellular GA (CGA) are also shown.

Fig. 4 shows that the pure cellular GA (without learning) outperforms the GA–RTRL hybrid algorithm. After four minutes of simulations, the cellular GA achieves a significantly lower (significance $p < 0.05$, calculated by Student's t-tests) average MSE than the hybrid algorithm. Although the two-phase approach in this experiment does not improve the evolution process, it does produce better solution quality as compared to applying RTRL alone.

Different hybrid algorithms based on the two-phase approach have been constructed by using different threshold values (0.08 and 0.04, for example) and by replacing the RTRL algorithm in the second phase with the delta rule. However, none of the hybrid algorithms can outperform the cellular GA.

6 Lamarckian Evolution

Lamarckian evolution [2,80] is another approach to combining evolutionary search and local search. It is based on the inheritance of acquired characteristics–an individual can pass the characteristics (observed in the phenotype) acquired through learning to its offspring genetically (encoded in the genotype).

In the following Lamarckian hybrid algorithms, local search (i.e. RTRL or the delta rule) is applied to the newly born offspring at every generation. After the application of local search, the offspring's fitness is changed and the offspring's corresponding inborn weights (weights as a result of genetic operations) are replaced by the weights obtained through learning for further genetic operations.

Fig. 5 shows that embedding RTRL in the cellular GA is not appropriate because the performance of CGA+RTRL is very poor. This is because the RTRL algorithm is so computationally intensive that the fitness improvement obtained from learning cannot compensate for the loss in computation time.[3] On the other hand, the performance is significantly better in CGA+DeltaRule, and the average MSE attained after four minutes is only 18% of that attained by the pure cellular GA. This suggests that embedding delta rule in the cellular GA has merits.

Another advantage of embedding the delta rule is that it considerably saves computation time. For example, the pure cellular GA takes four minutes to attain an MSE of 0.0303. To evolve a network to the same accuracy, CGA+DeltaRule requires 1.4 minutes, suggesting that up to 65% of computation time can be saved.

The computational complexity of the delta rule is low because the RNN is considered as a feedforward network when the error gradient is computed. However, the delta rule is so simple that the error gradient obtained by this algorithm may be inaccurate. As a result, the fitness of a chromosome could deteriorate after the application of the delta rule. Despite this deficiency, the low computational complexity of the delta rule can shorten the overall training time when the delta rule is embedded in the cellular GA.

7 The Baldwin Effect

As learning takes place in phenotype space, Lamarckian evolution requires an inverse mapping from the phenotype space (e.g. neural networks) to the genotype space (e.g. strings of floating-point numbers), which is impossible in biological systems and impractical for complex genotypes-to-phenotypes relations (e.g. when neural networks are represented by grammar trees [25]).

[3] When the time involved in learning is not taken into account, the hybrid algorithm can achieve a lower MSE as compared to the pure cellular GA.

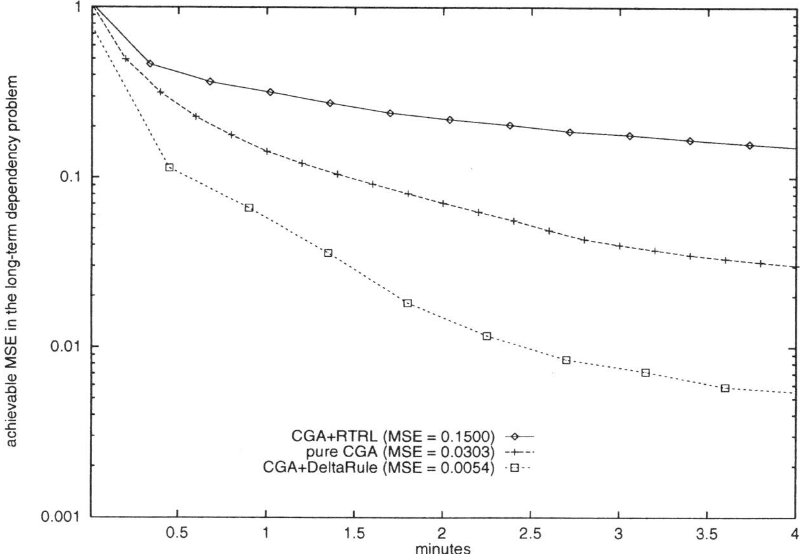

Fig. 5. MSE (based on the average of 200 simulations) of the best network found by the Lamarckian hybrid algorithms. The average MSEs after four minutes of simulation are shown in parentheses.

The approach based on the Baldwin effect [5,77] is more biologically plausible and more applicable to different situations. Unlike Lamarckian evolution, learning in this approach cannot modify the genotypes directly. Only the fitness is replaced by the 'learned' fitness (i.e. fitness after learning). Therefore, after learning, the chromosome will be associated with a 'learned' fitness that is not the same as its 'inborn' fitness (i.e. fitness before learning). Even though the characteristics to be learned in the phenotype space are not genetically specified, there is evidence that the Baldwin effect is able to direct the genotypic changes [30].

In order to investigate the efficiency of Baldwinian learning, several experiments similar to those in Section 6 have been performed. Local search (i.e. RTRL or the delta rule) was applied to the newly born offspring at every generation. Here, learning is based on the Baldwinian mechanism instead of the Lamarckian mechanism.

Fig. 6 illustrates the performance of the Lamarckian and Baldwinian hybrid algorithms using RTRL as the learning method. Evidently, the Lamarckian hybrid algorithms outperform their Baldwinian counterparts. Table 1 shows that even if the time involved in learning is not taken into consideration, the Baldwinian hybrid algorithms with RTRL perform poorly as compared to the pure cellular GA. The following conjecture is suggested for explaining this phenomenon.

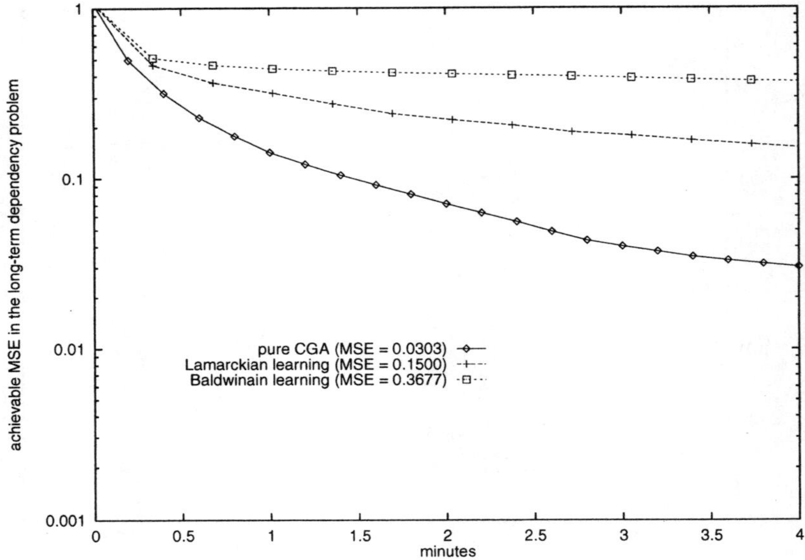

Fig. 6. MSE (based on the average of 200 simulations) of the best network found by the Lamarckian and Baldwinian hybrid algorithms where RTRL was applied at every generations. The average MSEs after four minutes of simulation are shown in parentheses.

Table 1. MSEs attained after 20,000 generations by different Baldwinian hybrid algorithms. All results are based on the average of 200 simulation runs, except CGA+RTRL where the MSEs are based on the average of 10 simulation runs because of the long computation time required.

Baldwinian hybrid algorithms	Average MSEs
pure CGA	0.0303
CGA+RTRL	0.1161
CGA+DeltaRule	0.0196

The more difficult it is for genetic operations (crossover and mutation) to produce the changes between the genotypes corresponding to the 'inborn' fitness and the 'learned' fitness, the poorer is the performance of Baldwinian learning.

In Baldwinian learning, the learned fitness of a chromosome is the fitness obtained after learning. This learned fitness is not equal to the inborn fitness corresponding to the genotype. Genetic operations are therefore required to produce the change in the genotype, where the change should correspond to the difference between the inborn fitness and the learned fitness. While these genotypic changes are produced randomly by crossover and mutation, only some of them may match the phenotypic changes caused by learning. If only one gene (or one weight) is allowed to be changed[4] during learning, it should not be difficult for genetic operations to produce this change. However, in the RTRL algorithm, all weights can be changed; consequently, it is very difficult for genetic operations to produce the corresponding changes in the weights. Therefore, according to the conjecture, the Baldwinian hybrid algorithms perform poorly even if the time spent on learning is not considered.

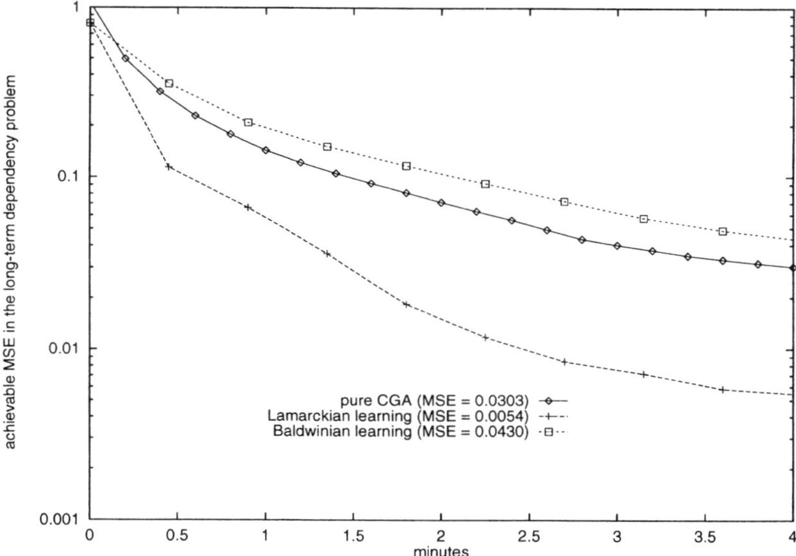

Fig. 7. MSE (based on the average of 200 simulations) of the best network found by the Lamarckian and Baldwinian hybrid algorithms where the delta rule was applied to the offspring generated at every generation. The average MSEs after four minutes of simulation are shown in parentheses.

The inefficiency of Baldwinian learning is also illustrated in Fig. 7 where the delta rule is embedded in the cellular GA. However, the hybrid algorithm achieves a significantly lower (significance $p < 0.01$) MSE after 20,000 generations, as shown in Table 1. This indicates that if computation time is not a concern, the hybrid algorithm has merits. Of particular interest is that no such situation occurs when RTRL is embedded in the cellular GA using the Baldwinian mechanism. Recall that the main difference between RTRL and the delta rule is that the latter has a smaller number of changeable weights.

[4] The learned fitness is obtained by changing that gene while keeping other genes fixed.

Consequently, it is relatively easy for the genetic operations to produce the changes in weights caused by the simplified learning methods. Therefore, according to the conjecture, the Baldwinian hybrid algorithms with the delta rule outperform those with RTRL. Further evidence to support the conjecture can be found in [44,46,50].

8 Generalization Performance

It is desirable that the trained networks have good generalization performance. In other words, the networks should have the capability of processing unseen patterns. In the long-term dependency problem, a training sequence was formed by the concatenation of ten randomly chosen input/output sequences. To compare generalization performance, a test sequence comprising 100 randomly chosen input/output sequences was used to determine the misclassification rate (i.e. the chance of misclassifying an input sequence) of the trained RNNs. The results are tabulated in Table 2.

Table 2. Comparisons of generalization performance based on average of 200 simulations.

training algorithms	average MSEs(training)	misclassification
pure CGA	0.0303	4.6%
CGA+DeltaRule(Lamarckian)	0.0054	1.4%
CGA+RTRL(Lamarckian)	0.1500	10.5%

It is found that after 4 minutes of simulations, the RNNs trained by the pure cellular GA have an average misclassification rate of 4.6%. When RNNs are trained by CGA+DeltaRule(Lamarckian), the solution quality is improved and the corresponding misclassification rate is reduced to 1.4%. Therefore, a well trained network is able to solve the long-term dependency problem.

It is also interesting to explore the capability of CGA+DeltaRule(Lamarckian) on a more difficult long-term dependency problem – the temporal length is increased to 10 time steps. Fig. 8 demonstrates that despite the substantial increase in complexity, the evolution process can still be improved by embedding the delta rule in the cellular GA.

9 Discussion and Conclusions

Training of neural networks by local search such as gradient-based algorithms can be difficult. For instance, the algorithms may have difficulties in (a) escaping from local optima when the search surface is rugged; (b) finding better solutions when the surface has many plateaus; and (c) deciding the search

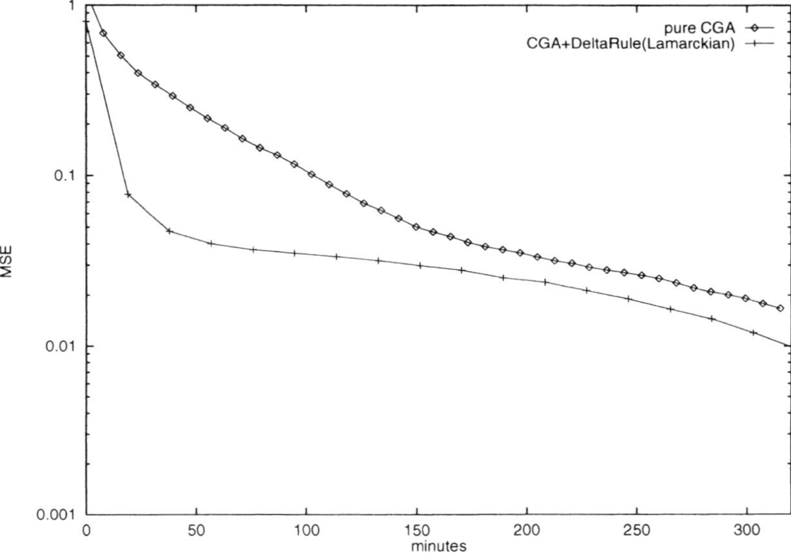

Fig. 8. MSE (based on the average of 30 simulations) of the best network achieved on the long-term dependency problem with a temporal length of 10 time steps. Because of the complexity of the problem, the RNN to be trained has 4 input nodes and 16 processing nodes, where 6 of them were dedicated as output nodes, and the population size was increased to 1600.

direction when gradient information is not readily available. This calls for the development of alternative training algorithms such as evolutionary search. However, training by evolutionary search often requires long computation time. It is possible to reduce the computation time by combining the efforts of local search and evolutionary search. This chapter has reviewed a number of previous attempts to combine local and evolutionary search. It has also compared different approaches to combining the two search strategies.

In the two-phase approach, evolutionary search is used to locate roughly the region of global optimum in the first phase and local search is used to accelerate local convergence in the second phase. Experimental results indicate that while evolutionary search is able to find a good network for local search to start with, an inefficient local search method in the second phase can degrade the overall performance. This chapter suggests that the success of the two-phase approach depends on two factors: (a) the efficiency of locating promising regions in the first phase and (b) the efficiency of finding an acceptable solution in the second phase. The latter factor is particularly hard to fulfill for difficult problems, e.g. the long-term dependency problem. The threshold for switching the algorithm to the second phase is also very important. However, finding the optimal value of these problem-dependent thresholds is difficult. All of the above factors limit the applicability of the two-phase approach.

Besides the two-phase approach, we have also investigated the Baldwinian approach to combining local and evolutionary search. It is found that none of the Baldwinian hybrid algorithms have satisfactory performance. In particular, the Baldwinian hybrid algorithm with RTRL is inferior to evolutionary search alone even if the time involved in learning is not taken into account. These observations suggest that Baldwinian learning may be inefficient in evolving neural networks, especially when the local search methods can change most of the network weights.

A conjecture has been proposed to explain the inefficiency of Baldwinian learning: the level of difficulties for genetic operations to produce the genotypic changes that match the phenotypic changes due to learning can significantly affect the Baldwin effect. This conjecture suggests that if many weights are changed by Baldwinian learning and the changes are large, the resulting hybrid algorithms will not be better than evolutionary search alone. This is because if there are too many possible phenotypic changes, obtaining genotypic-to-phenotypic matches will become very difficult.

This work has also evaluated the Lamarckian approach to combining local and evolutionary search. It is found that Lamarckism is able to speed up the training process and to improve the solution quality. These findings are based on observing the simulation runs of the long-term dependency problem for up to four minutes.[5] More experiments are still required to see whether these findings are applicable to other situations. There might be other factors (for example, Lamarckian learning could be detrimental to the adaption of neural networks under changing environment [72]) affecting the overall efficiency of the Lamarckian approach. Further investigations are therefore required to clarify the benefit of Lamarckian evolution.

Among the Lamarckian hybrid algorithms that we have investigated, the one with the delta rule achieves the best performance. It is noteworthy that the delta rule is so simple that it is not able to find a good solution on its own. However, its low computational complexity makes it suitable for being embedded in the cellular GA. This suggests that local search methods need not be sophisticated in order to obtain the benefit of combining evolutionary search and local search.

Although the experimental results in this work are based on a simple and hypothetical problem, similar phenomenon has also been observed in a more difficult benchmark problem—the inverted pendulum problem [42,45]. This suggests that the idea of combining simple local search and evolutionary search is viable. A possible extension of this work is to apply the hybrid algo-

[5] An acceptable solution to the long-term dependency problem (with $k = 5$) can be obtained within four minutes of simulation. Longer simulation time is to be expected for problems with longer dependency (i.e. larger values of k).

rithms to some difficult real-world problems that are known to be unsolvable by conventional methods.

Acknowledgment

This work was in part supported by the Hong Kong Polytechnic University Grant No. 1.42.37.A410 and GV178.

References

1. Ackley, D. H., Littman, M. L. (1992) Interactions between learning and evolution. In C. G. Langton, C. Taylor, J. D. Farmer, and S. Rasmussen, editors, *Artificial Life 2*, 487–509. Redwood City, CA: Addison-Wesley
2. Ackley, D. H., Littman, M. L. (1994) A case for Lamarckian evolution. In C. G. Langton, editor, *Artificial Life 3*, 3–10. Reading, MA: Addison-Wesley
3. Angeline, P. J., Saunders, G. M., Pollack, J. B. (1994) An evolutionary algorithm that constructs recurrent neural networks. *IEEE Transactions on Neural Networks*, 5(1):54–65
4. Bäck, T., Hammel, U., Schwefel, H.-P. (1997) Evolutionary computation: Comments on the history and current state. *IEEE Transactions on Evolutionary Computation*, 1(1):3–17
5. Baldwin, J. M. (1896) A new factor in evolution. *American Naturalist*, 30:441–451
6. Belew, R. K., McInerney, J., Schraudolph, N. N. (1992) Evolving networks: Using the genetic algorithm with connectionist learning. In C. G. Langton, C. Taylor, J. D. Farmer, and S. Rasmussen, editors, *Artificial Life 2*, 511–547. Redwood City, CA: Addison-Wesley
7. Bengio, Y., Simard, P., Frasconi, P. (1994) Learning long-term dependencies with gradient descent is difficult. *IEEE Transactions on Neural Networks*, 5(2):157–166
8. Braun, H., Zagorski, P. (1994) ENZO-M – a hybrid approach for optimizing neural networks by evolution and learning. In Y. Davidor, H.-P. Schwefel, and R. Manner, editors, *Parallel Problem Solving from Nature – PPSN III*, 440–451. Berlin: Springer-Verlag
9. Chalmers, D. J. (1990) The evolution of learning: An experiment in genetic connectionism. In D. S. Touretzky, editor, *Proceedings of the 1990 Connectionist Models Summer School*, 81–90. San Mateo, CA: Morgan Kaufmann
10. Collins, R. J., Jefferson, D. R. (1991) Selection in massively parallel genetic algorithms. In *Proceedings of the Fourth International Conference on Genetic Algorithms*, 249–256
11. Crosher, D. (1993) The artificial evolution of a generalized class of adaptive processes. In *AI'93 Workshop on Evolutionary Computation*, 18–36
12. Dantzig, G. B. (1963) *Linear Programming and Extensions.* Princeton, NJ: Princeton University Press
13. Davidor, Y. (1991) A naturally occurring niche & species phenomenon: The model and first results. In *Proceedings of the Fourth International Conference on Genetic Algorithms*, 257–262

14. De Garis, H. (1991) GenNets: Genetically programmed neural networks – using the genetic algorithm to train neural nets whose inputs and/or outputs vary in time. In *Proceedings of the IEEE International Joint Conference on Neural Networks*, 1391–1396
15. Elman, J. L. (1988) Finding structure in time. Technical Report CRL 8801, Center for Research in Language, University of California, San Diego
16. Erkmen, I., Ozdogan, A. (1997) Short term load forecasting using genetically optimized neural network cascaded with a modified kohonen clustering process. In *Proceedings of the IEEE International Symposium on Intelligent Control*, 107–112
17. Fogel, D. B. (1994) An introduction to simulated evolutionary optimization. *IEEE Transactions on Neural Networks*, 5(1):3–14
18. Fogel, D. B. (1995) *Evolutionary computation: toward a new philosophy of machine intelligence.* Piscataway, NJ: IEEE Press
19. Fogel, D. B., Fogel, L. J., Porto, V. W. (1990) Evolving neural networks. *Biological Cybernetics*, 63:487–493
20. Fogel, L. J., Owens, A. J., Walsh, M. J. (1966) *Artificial Intelligence Through Simulated Evolution.* New York: Wiley
21. Fontanari, J. F., Meir, R. (1991) Evolving a learning algorithm for the binary perceptron. *Network*, 2(4):353–359
22. French, R. M., Messinger, A. (1994) Genes, phenes and the Baldwin effect: Learning and evolution in a simulated population. In A. B. Rodney and M. Pattie, editors, *Artificial Life 4*, 277–282. Cambridge, MA: MIT Press
23. Goldberg, D. E. (1989) *Genetic Algorithms in Search, Optimization, and Machine Learning.* Reading, MA: Addison-Wesley
24. Greenwood, G. W. (1997) Training partially recurrent neural networks using evolutionary strategies. *IEEE Transactions on Speech Audio Processing*, 5(2):192–194
25. Gruau, F., Whitley, D. (1993) Adding learning to the cellular development of neural networks: Evolution and the Baldwin effect. *Evolutionary Computation*, 1(3):213–233
26. Hanes, M. D., Ahalt, S. C., Krishnamurthy, A. K. (1994) Acoustic-to-phonetic mapping using recurrent neural networks. *IEEE Transactions on Neural Networks*, 5(4):659–662
27. Harp, S. A., Samad, T., Guha, A. (1989) Towards the genetic synthesis of neural networks. In J. D. Schaffer, editor, *Proceedings of the Third International Conference on Genetic Algorithms*, 360–369. San Mateo, CA: Morgan Kaufmann
28. Harvey, I. (1997) Is there another new factor in evolution? *Evolutionary Computation*, 4(3):313–329
29. Hebb, D. O. (1949) *The Organization of Behavior.* New York: Wiley
30. Hinton, G. E., Nowlan, S. J. (1987) How learning can guide evolution. *Complex Systems*, 1:495–502
31. Hornik, K. (1990) Approximation capabilities of multilayer feedforward neural networks. *Neural Networks*, 4:251–257
32. Huang, W. M., Lippmann, R. P. (1988) Neural net and traditional classifiers. In D. Anderson, editor, *Neural Information Processing Systems*, 387–396. New York: American Institute of Physics
33. Ichimura, T., Takano, T., Tazaki, E. (1995) Reasoning and learning method for fuzzy rules using neural networks with adaptive structured genetic algorithm.

In *Proceedings of the IEEE International Conference on Systems, Man, and Cybernetics*, 3269–3274
34. Janson, D. J., Frenzel, J. F. (1992) Application of genetic algorithms to the training of higher order neural networks. *Journal of Systems Engineering*, 2(4):272–276
35. Janson, D. J., Frenzel, J. F. (1993) Training product unit neural networks with genetic algorithms. *IEEE Expert*, 8(5):26–33
36. Jordan, M. I. (1986) Attractor dynamics and parallelism in a connectionist sequential machine. In *Proceedings of the Eighth Annual Conference of the Cognitive Science Society*, 531–546
37. Keesing, R., Stork, D. G. (1991) Evolution and learning in neural networks: The number and distribution of learning trial affect the rate of evolution. In R. P. Lippmann, J. E. Moody, and D. S. Touretzky, editors, *Advances in Neural Information Processing Systems 3*, 804–810. San Mateo, CA: Morgan Kaufmann
38. Kim, H. B., Jung, S. H., Kim, T. G., Park, K. H. (1996) Fast learning method for back-propagation neural network by evolutionary adaptation of learning rates. *Neurocomputating*, 11(1):101–106
39. Kitano, H. (1990) Empirical studies on the speed of convergence of neural network training using genetic algorithms. In *Proceedings of the Eighth National Conference on Artificial Intelligence*, 789–795
40. Kolen, J. F., Pollack, J. B. (1990) Back propagation is sensitive to initial conditions. *Complex Systems*, 4:269–280
41. Korning, P. G. (1995) Training neural networks by means of genetic algorithms working on very long chromosomes. *International Journal of Neural Systems*, 6(3):299–316
42. Ku, K. W. C. (1999) *On the Combination of Local and Evolutionary Search for Training Recurrent Neural Networks*. PhD thesis, The Hong Kong Polytechnic University, Hong Kong
43. Ku, K. W. C., Mak, M. W. (1997) Exploring the effects of Lamarckian and Baldwinian learning in evolving recurrent neural networks. In *Proceedings of the IEEE International Conference on Evolutionary Computation*, 617–621
44. Ku, K. W. C., Mak, M. W. (1998) Empirical analysis of the factors that affect the Baldwin effect. In A. E. Eiben, T. Bäck, M. Schoenauer, and H.-P. Schwefel, editors, *Parallel Problem Solving from Nature – PPSN V*, 481–490. Berlin: Springer-Verlag
45. Ku, K. W. C., Mak, M. W., Siu, W. C. (2000) A study of the Lamarckian evolution of recurrent neural networks. *IEEE Transactions on Evolutionary Computation*, 4(1):31–42
46. Ku, K. W. C., Mak, M. W., Siu, W. C. (1999) Adding learning to cellular genetic algorithms for training recurrent neural networks. *IEEE Transactions on Neural Networks*, 10(2):239–252
47. Lee, S. W. (1996) Off-line recognition of totally unconstrained handwritten numerals using multilayer cluster neural network. *IEEE Transactions on Pattern Analysis and Machine Intelligence*, 18(6):648–652
48. Lippmann, R. P. (1987) An introduction to computing with neural nets. *IEEE Acoustics, Speech, and Signal Processing Magazine*, 4–22
49. Maniezzo, V. (1994) Genetic evolution of the topology and weight distribution of neural networks. *IEEE Transactions on Neural Networks*, 5(1):39–53
50. Mayley, G. (1997) Landscapes, learning costs, and genetic assimilation. *Evolutionary Computation*, 4(3):213–234

51. McDonnell, J. R., Waagen, D. (1994) Evolving recurrent perceptrons for time-series modelling. *IEEE Transactions on Neural Networks*, 5(1):24–38
52. Menczer, F., Parisi, D. (1992) Evidence of hyperplanes in the genetic learning of neural networks. *Biological Cybernetics*, 66:283–289
53. Merelo, J. J., Patón, M., Canas, A., Prieto, A., Morán, F. (1993) Optimization of a competitive learning neural network by genetic algorithms. In *Proceedings of the International Workshop on Artificial Neural Networks*, 185–192
54. Michalewicz, Z. (1996) *Genetic Algorithms + Data Structures = Evolution Programs*. Berlin: Springer-Verlag
55. Miller, G. F., Todd, P. M., Hegde, S. U. (1989) Designing neural networks using genetic algorithms. In J. D. Schaffer, editor, *Proceedings of the Third International Conference on Genetic Algorithms*, 379–384. San Meteo, CA: Morgan Kaufmann
56. Mitchell, M. (1996) *An Introduction to Genetic Algorithms*. Cambridge, MA: MIT Press
57. Montana, D. J., Davis, L. (1989) Training feedforward neural network using genetic algorithms. In *Proceedings of the Eleventh International Joint Conference on Artificial Intelligence*, 762–767
58. Mozer, M. C. (1992) Induction of multiscale temporal structure. In J. E. Moody, S. J. Hanson, and R. P. Lippmann, editors, *Advances in Neural Information Processing Systems 4*, 275–282. San Mateo, CA: Morgan Kaufmann
59. Nolfi, S., Elman, J. L., Parisi, D. (1994) Learning and evolution in neural networks. *Adaptive Behavior*, 3:5–28
60. Omatu, S., Yoshioka, M. (1997) Self-tuning neuro-PID control and applications. In *Proceedings of the IEEE International Conference on Systems, Man, and Cybernetics*, 1985–1989
61. Omlin, C. W., Giles, C. L. (1996) Rule revision with recurrent neural networks. *IEEE Transactions on Knowledge and Data Engineering*, 8(1):183–188
62. Paredis, J. (1996) Coevolutionary life-time learning. In H.-M. Voigt, W. Ebeling, I. Rechenberg, and H.-P. Schwefel, editors, *Parallel Problem Solving from Nature – PPSN IV*, 72–80. Berlin: Springer-Verlag
63. Parisi, D., Nolfi, S. (1996) The influence of learning on evolution. In R. K. Belew and M. Mitchell, editors, *Adaptive Individuals in Evolving Populations: Models and Algorithms*, 419–428. Reading, MA: Addison-Wesley
64. Pineda, F. J. (1987) Generalization of backpropagation to recurrent neural networks. *Physical Review Letters*, 59:2229–2232
65. Port, R. F. (1990) Representation and recognition of temporal patterns. *Connection Science*, 2:151–176
66. Porto, V. W., Fogel, D. B., Fogel, L. J. (1995) Alternative neural networks training methods. *IEEE Expert*, 10(3):16–22
67. Rechenberg, I. (1989) Evolution strategy: Nature's way of optimization. In *Optimization: Methods and Applications, Possibilities and Limitations*, volume 47 of *Lecture Notes in Engineering*. Berlin: Springer-Verlag
68. Riedmiller, M., Braun, H. (1993) A direct adaptive method for faster backpropagation learning: The RPROP algorithm. In *Proceedings of the International Conference on Neural Networks*, 586–591
69. Rudolph, G. (1991) Global optimization by means of distributed evolution strategies. In H. P. Schwefel and R. Männer, editors, *Parallel Problem Solving from Nature – PPSN I*, 209–213. Berlin: Springer-Verlag

70. Rumelhart, D. E., Hinton, G. E., Williams, R. J. (1986) Learning internal representations by error propagation. In D. E. Rumelhart, J. L. McClelland, and the PDP Research Group, editors, *Parallel Distribution Processing: Explorations in the Microstructure of Cognition. Vol. 1: Foundation.* Cambridge, MA: MIT Press
71. Saravanan, N., Fogel, D. B. (1995) Evolving neural control systems. *IEEE Expert*, 10(3):23–27
72. Sasaki, T., Tokoro, M. (1998) Adaptation under changing environments with various rates of inheritance of acquired characters: Comparison between Darwinian and Lamarckian evolution. In *Proceedings of the Second Asia-Pacific Conference on Simulated Evolution and Learning*, 34–41
73. Schwefel, H.-P. (1995) *Evolution and Optimum Seeking.* New York: Wiley
74. Sejnowski, T. J., Rosenberg, C. R. (1987) Parallel networks that learn to pronounce English text. *Complex Systems*, 1:145–168
75. Skinner, A. J., Broughton, J. Q. (1995) Neural networks in computational materials science: Training algorithms. *Modelling and Simulation in Materials Science and Engineering*, 3:371–389
76. Solis, F. J., Wets, R. J-B. (1981) Minimization by random search techniques. *Mathematics of Operations Research*, 6(1):19–30
77. Turney, P. (1996) Myths and legends of the Baldwin effect. In *Proceedings of the Workshop on Evolutionary Computing and Machine Learning at the 13th International Conference on Machine Learning*, 135–142
78. Waibel, A. (1989) Modular construction of time-delay neural networks for speech recognition. *Neural Computation*, 1:39–46
79. Whitley, D. (1994) A genetic algorithm tutorial. *Statistics & Computing*, 4(2):65–85
80. Whitley, D., Gordon, V. S., Mathias, K. (1994) Lamarckian evolution, the Baldwin effect and function optimization. In Y. Davidor, H.-P. Schwefel, and R. Manner, editors, *Parallel Problem Solving from Nature – PPSN III*, 6–15. Berlin: Springer-Verlag
81. Whitley, D., Starkweather, T., Bogart, C. (1990) Genetic algorithms and neural networks: Optimizing connections and connectivity. *Parallel Computing*, 14:347–361
82. Wieland, A. (1991) Evolving neural network controllers for unstable systems. In *Proceedings of the International Joint Conference on Neural Networks*, 667–673
83. Williams, R. J., Zipser, D. (1989) Experimental analysis of the real-time recurrent learning algorithm. *Connection Science*, 1:87–111
84. Williams, R. J., Zipser, D. (1989) A learning algorithm for continually running fully recurrent neural networks. *Neural Computation*, 1:270–280
85. Wu, K. H., Chen, C. H., Lee, J. D. (1996) Cache-genetic-based modular fuzzy neural networks for robot path planning. In *Proceedings of the IEEE International Conference on Systems, Man, and Cybernetics*, 3089–3094
86. Yao, X. (1999) Evolving artificial neural networks. *Proceedings of the IEEE*, 87(9):1423–1447
87. Yao, X., Liu, Y. (1997) A new evolutionary system for evolving artificial neural networks. *IEEE Transactions on Neural Networks*, 8(3):694–713

Evolving Analog Circuits by Variable Length Chromosomes

Shin Ando[1], Mitsuru Ishizuka[1], and Hitoshi Iba[2]

[1] Department of Electronics Engineering, School of Engineering,University of Tokyo, Tokyo, Japan
E-mail: ando@miv.t.u-tokyo.ac.jp, ishizuka@miv.t.u-tokyo.ac.jp
[2] University of Tokyo, Graduate School of Frontier Science, Tokyo, Japan
E-mail:iba@miv.t.u-tokyo.ac.jp

Summary. This chapter proposes a framework of evolutionary analog circuits. This system features robustness to noise, optimized scaling, and high efficiency. These features solve the problems of the analog circuit design and manufacture. Methods utilized by this system are list-based chromosome, adjusted fitness, and two-stage evolution. Several experiments are conducted to examine the effectiveness of each of the methods. The first experiment compares other types of chromosome for the analog circuit design. The second experiment examines the robustness of evolutionary analog circuits. The other experiments are on the deduction of scaling and two-stage evolution.

1 Introduction

Evolvable Hardware (EHW) is hardware built on a software-reconfigurable logic device, such as a PLD (Programmable Logic Device) and FPGAs (Field Programmable Gate Arrays). EHW architecture can be reconfigured through the evolutionary method so as to adapt to the new environment. If hardware errors occur or a new hardware functionality is required, EHW can alter its own hardware structure in order to accommodate such changes.

There is a clear distinction between conventional hardware (CHW) and EHW. A designer can begin to design CHW only after its detailed specification is given. In this sense, CHW is a top-down approach. However, EHW is applicable even when no hardware specification is known beforehand. EHW implementation is determined through genetic learning in a bottom-up way. Thus, EHW will be applied totally differently from CHW. EHW is suitable for problem domains where both on-line adaptation and real-time response are required. Such applications can be found in advanced areas such as multimedia where adaptive and real-time behaviors are often inevitable.

The basic idea of EHW is to regard the architecture bits of a PLD as a chromosome for genetic algorithm (GA) (see Fig.1). The hardware structure is adaptively searched by the GA. These architecture bits, i.e., the GA chromosome, are downloaded onto a PLD, on and after the genetic learning. Therefore, EHW can be considered as on-line adaptive hardware [3] [5].

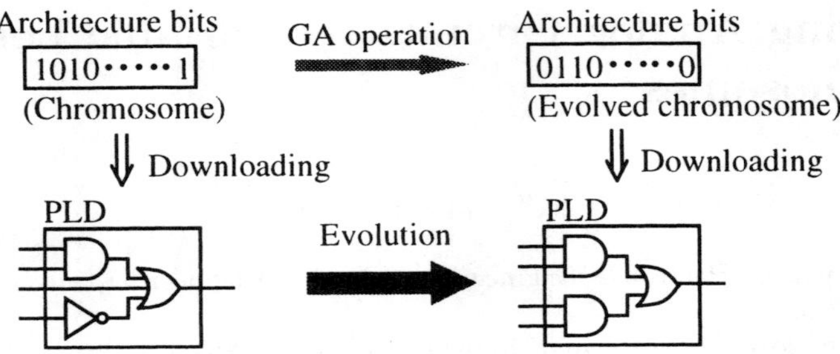

Fig. 1. Evolvable hardware (EHW)

In this chapter, a GA system based on variable length chromosomes is constructed in order to realize an evolutionary analog circuit, and the features of such a system are furthermore verified through simulated circuit evolution.

Despite the great diversity of digital devices, analog devices hold a firm position as intermediaries between digital devices and the natural environment, and high-speed analog circuits are indispensable in many fields such as communication.

However, the weak points of analog devices have always been the errors and variations of the analog elements which comprise the circuits. These elements' values regularly differ from specification and are easily subjected to influences such as temperature. There are countermeasures such as the implementation of redundant circuits and the employment of linear-adaptive filters, but these conventional methods are considerably difficult for they require complicated design procedures and much human skill.

We expect to solve such problems with our proposed evolutionary analog circuit. It is an application of EHW, which is a combination of variable hardware and evolutionary computation, and can automatically compose the hardware with objective functions. In EHW, a cycle of implementation, feedback, and adjustment of the hardware is repeated in the mechanism of the GA, and, as a result, the robustness to errors the analog elements and the influence of the environment can be realized. Furthermore, design and manufacture are automated by the GA. Thus it is expected that the problems referred to above can be solved.

Several evolutionary methods have been proposed in the study of analog circuits using an evolutionary method. For example, Koza has simulated the automatic synthesis of various types of circuits by using genetic programming (hereafter abbreviated to GP) [8]. However, voluminous memory consumption and lengthy convergence time compared to that of the GA are notable defects of GP. Further, Koyabu et al. coded an electric circuit into matrix-based chromosomes, and conducted a simulated evolution of analog circuits using the GA [6]. In a study of analog EHW, Murakawa et al. were successful in constructing analog EHW chips for Intermediate Frequency (IF) filters [10]. This was a method to adjust the variation of the elements for each constructed chip which consisted of adjustable opamps. These authors succeeded in improving the yield rate.

Based upon these conventional methods, our evolutionary analog circuit uses the following methods to realize the automatic circuit design of a variety of functions:

1 List-based chromosomes.
2 Two-stage evolution of electrical circuits.
3 Adequate scaling for circuit design.

Furthermore, the effectiveness of these methods is evaluated in the simulated evolution of analog circuits described below:

1 Passive filter synthesis, in comparison with other evolutionary methods, i.e., GP and GA by matrix-based chromosomes.
2 Noise and error absorption.
3 Evaluation of a separate evolution method.
4 Evaluation of a scale optimization method.

By means of these experiments the effectiveness of and possibilities for the evolutionary analog circuit will be shown.

2 Function and Architecture of Evolutionary Analog Circuits

2.1 Variation in Analog Circuits

When an analog circuit is implemented as an integrated circuit, the circuit component values differ from the designed specification because of a number of environmental factors such as temperature. These variations become crucial for the high-end analog device, which requires very precise specification. Furthermore, these variations make the design process considerably difficult and more dependent on human skill. By providing evolutionary analog circuits with a variable structure, robustness is obtained along with automation in both design and manufacture.

2.2 Architecture

The evolutionary analog circuit is composed of a GA simulator and variable hardware as illustrated in Fig.2. The variable hardware consists of analog elements such as resistors, capacitors, opamps, and so on. Their values and connections are supposed to be programmable. The circuit evolution process is as follows:

1 Randomly generate a number of circuit models.
2 Create new individuals by genetic operations based upon the evaluation of the parents.

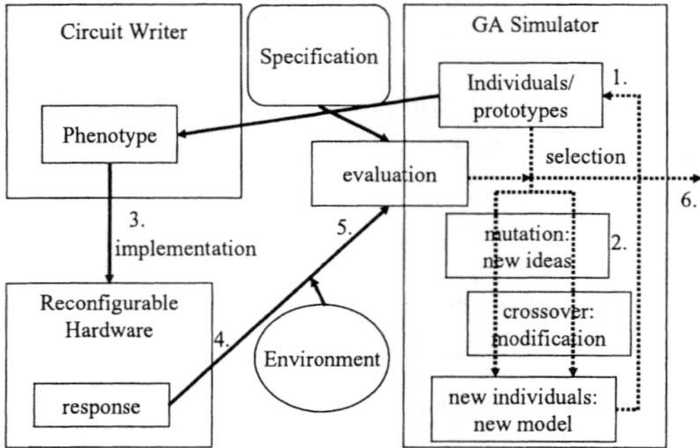

Fig. 2. Structure of evolutionary analog circuit

3 Implement each individual's phenotype into the variable hardware.
4 Return the response of the equipped circuit from the variable hardware to the GA simulator.
5 Use the GA simulator to decide the fitness of each individual by comparing the response of the phenotypes with the specification.
6 Eliminate individuals with low fitness and use individuals of high fitness to create the next generation of individuals.
7 Return to 2.

3 Equipment of GA Simulator

3.1 Chromosomes

Several representative schemes of circuit structures have been proposed. Koza used GP, which uses a tree structure chromosome to evolve a Lisp-type circuit-generating program [7,8]. Its defects are high memory consumption and slow convergence. Koyabu et al. used matrix-based chromosomes for the automated design of an electric circuit [6]. They were successful in designing several types of passive filters utilizing settled value elements. The defects of their circuit are that it does not measure dynamic scaling, and requires skill in deciding the complexity of the objective response. The designation of an adequate scale for the circuit is also required in advance.

We chose list-based representation which is the application of a variable length chromosome, i.e., Messy GA, in circuit design [1]. Messy GA is a method developed by Goldberg et al. to improve the GA's weakness to deceptive problems [2]. In Messy GA, each gene's locations are not fixed in the chromosomes and can operate over a large semantic alternation, which can prevent the GA from producing just local solutions. This representation was also used for the synthesis of active filters by Zebulum et al. ([12] and [13]).

The phenotype and genotype of this chromosome are shown in Fig.3. This chromosome forms a list of elements included in the circuit. Each gene

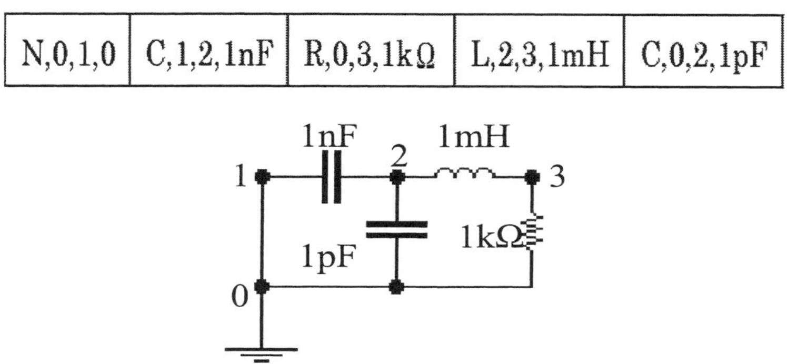

Fig. 3. Phenotype and genotype of a designed chromosome

holds the type, value, and location of an element. Each node of the circuit is numbered, and the position of an element is expressed by the nodes at both ends. Element types correspond to the kinds of the passive circuit elements and are shown as R, C, and L. In addition to the above, we use N, which means nodes are connected, and O, which indicates nodes are disconnected. When there is no description, it is assumed that the type is O and thus remains open.

3.2 Fitness

Each individual is evaluated based on the deviation between the ideal and actual frequency response. The fitness function is defined as

$$eval = \frac{1}{K}\sum_{f}^{K}(F_f - R_f) \tag{1}$$

This fitness is the mean of the the squared deviation values between the ideal gain F_f and the obtained gain R_f at frequency f. The chromosomes with a lower value in the equation are reproduced according to the roulette-wheel selection. Based on the evolutionary strategy $(\mu + \lambda) - ES$, from the old population and newly produced individuals, the portion with worse fitness is eliminated and the remaining portion is the new population.

3.3 Structural and Parameter Evolution

The GA is characterized by its forceful global search, and is quick to reach a quasi-optimal solution. On the contrary, a stochastic search of the GA can be inefficient in going from a quasi-optimal to an optimal solution.

Considering such features of the GA, we propose a two-stage evolution method to enable a high efficiency for the specialized circuit design. Values of the elements and connections are the changeable items in the circuit design. However, in EHW, it is inefficient to evolve them simultaneously owing to the convergence time and resource consumption. In the firs or structural evolution stage, the values of the circuit elements are fixed. The objective in this stage is to acquire the closest response by combining elements with several fixed values. Thus the optimal structure is acquired in this stage. The next stage is parameter evolution, in which the best acquired structure of the previous stage is fixed and the value of the composing elements will be adjusted. In this stage the elements' values are adjusted according to

$$Adjval = Val \times 10^{s} \tag{2}$$

Val is the original value, which was fixed in the structural evolution, and is adjusted according to the parameter s to $Adjval$, i.e. , the adjusted value. The array of parameters s for each element will be the chromosomes in this stage.

The characteristics of each stage are as follows. In the first stage, or the structural evolution stage, circuit structures are dynamically changed to fully utilize the GA's global search. In the second stage, the parameter evolution stage, the response cannot be altered as much, but can adapt with high accuracy to the stringent specification.

Furthermore, in each stage of evolution, because either the value or the structure of the components is fixed, the amount of data processed is decreased, which results in less memory consumption in the simulator. To be specific, the values of components in the structure evolution could be held as integer variables, while the chromosomes for the parameter evolution are arrays of component values.

3.4 Selective Pressure on Circuit Scale

In the simulation of the GA, *introns* develop in the chromosomes [9]. These introns have a crucial effect on EHW and result in implementing unnecessary components onto hardware, which is undesirable from the viewpoint of the hardware resources. Examples of introns in electric circuits are the connection of both ends of an element and open terminals.

To avoid these defects, it is necessary to eliminate the introns from the chromosomes, but it is difficult to distinguish automatically identify a useless portion from an indispensable portion.

To dispense with these introns, we redefine the fitness to eliminate those circuits with unnecessarily large scale. The individuals are evaluated not only by their frequency response but also from the circuit scale. Better evaluation is given to an individual with a smaller circuit scale provided that the evaluation of the response is equivalent. This results in the survival of the chromosomes without introns and redundant portions. The new fitness is defined as

$$fitness = E + P. \tag{3}$$

The fitness will be the sum of the response evaluation, E, and the pressure applied on the circuit scale P. P is defined as

$$P = N\_elem \times T \tag{4}$$

where N_elem is the number of elements included in the circuit and T is the modulus to control the intensity of the pressure.

This scaling pressure has its defects when applied too excessively. Completely excluding the introns will make each crossover semantically too destructive. Furthermore it can cause the elimination of useful schema at an early stage of evolution, abandoning diversity and obstructing achievement of the desired circuit. Thus adjusting the pressure with parameter T becomes important. Parameter T is chosen to adjust the order of P and E in Eq.(3). At the primary stage of evolution, the term E is predominant in Eq.(3) owing to the fact that the response is far from the goal. As evolution advances and the importance of the evaluation value E decreases, the selection pressure P gains in influence. Thus those circuits with an ample scale are gradually selectively removed. The value of T has to be set according to the required accuracy, and a selection of concrete examples will be presented in section 7. A similar method is used in the evolution of digital circuits [4].

4 Comparison with Conventional Methods

In this section, we will try to evaluate the ability of the list-based circuit representation by comparing it with other evolutionary methods, GP and the GA with matrix-based chromosomes. The comparison is made by conducting an experiment with the same fitness definition.

4.1 Comparison with GP

This experiment is based upon the synthesis of an asymmetric bandpass filter described in [8], Chap. 31.

Specification The target is an asymmetric bandpass filter described in [11], and is considered difficult to design because its specification is both stringent and highly asymmetric.

The ideal and allowable characteristics are defined as shown in Fig.4. Details are specified in [8]. The solid line labelled *ideal* indicates the ideal characteristics, and the broken line labelled *allowable* indicates the allowable range.

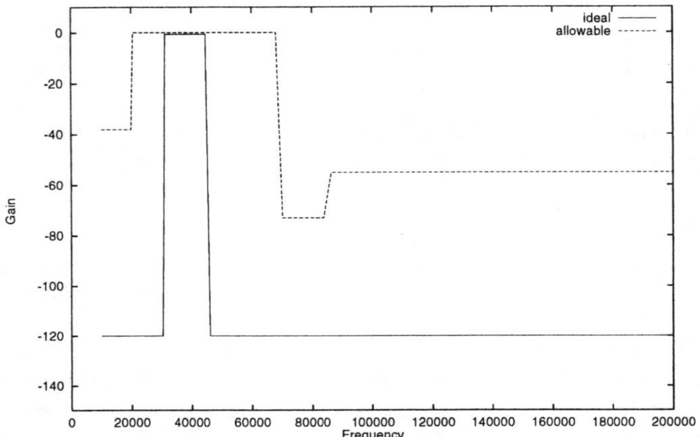

Fig. 4. Asymmetric band-pass filter specification.

The ideal characteristics of this asymmetric bandpass filter are specified as below. The pass-band ranges from 31.2kHz to 45.6kHz, where the gain should be from 0.6dB to −0.6dB. The gain outside the passband is less than −120dB. The allowable range is defined as follows:

- For isolation purposes, the attenuation should be less than −78dB from 69.6kHz to 84.0kHz.
- The attenuation less than 20kHz should be less than −38dB.
- The gain from 20kHz to 31kHz and from 45.6kHz to 69.9kHz should be less than 0dB.
- The gain above 84kHz should be less than −55dB.

The circuit behavior is observed at 101 frequencies in the interval between 10kHz and 200kHz in equal increments on a logarithmic scale. The fitness is defined as

$$F = \sum_{i=0}^{100} (W_i(d(f_i)) \times d(f_i)). \tag{5}$$

Weights are calculated from the difference between the response and the goal response for each observation point. The fitness values are derived from the total product of the weight W and the difference. The weight in the passband is 10 if allowable, otherwise it is 100. In the stopband, the weight is 1 if allowable, otherwise it is 10. More precisely:

- If the gain is 0dB, the difference is 0.
- If the gain ranges from 0.6 to 0.6dB, the weight $W = 10$.
- If the gain remains in the region outside of the above, the weight $W = 100$.

The weight outside the passband is defined as follows:

- If the gain is less than 120dB, the difference is 0.
- If the gain is less than the allowable region, the weight $W = 1$.
- If the gain remains in the region outside of the above, the weight $W = 10$.

The GA parameters are shown in Table 1. The GP parameters used for the experiment in [8] are also shown. Our simulation is based upon GAlib2.1.1, and the replacement rate is the parameter for the $(\mu + \lambda) - ES$ strategy.

	Population	Generation	Crossover rate	Mutation rate	Replacement rate
List-based GA	2000	400	0.99	0.001	0.3
GP[8]	640000	200	0.9	0.01	-

Table 1. GA parameters

4.2 Results

The acquired circuit response is depicted in Fig.5.

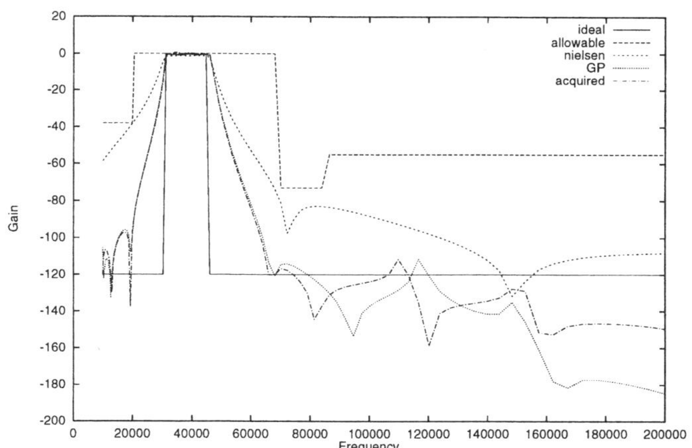

Fig. 5. Asymmetric bandpass filter: acquired response

The best obtained response at the 400th generation is shown as the chain line labelled *acquired*. The dotted line labelled *GP* indicates the response of the circuit obtained in [8]. The finesses of the best individuals were 2037.47 with the proposed method and 2024.0 with the GP. The dashed line labelled

Nielsen shows the response of a human-designed prototype circuit. The acquired response satisfies the allowable condition in the whole region, and a better response than in Neilson's heuristic method was obtained. Compared with GP, we were able to obtain exceedingly close responses at the passband, and equally acceptable characteristics at the cut-off region as well.

4.3 Comparison with GA by Matrix-based Genes

The following details concern the experiment conducted based upon Chap. 3 in [6].
Specification The fitness is defined as given in Eq.(6). $d(f_i)$ is the difference between the goal gain $V_{goal}(f_i)$ and the actual gain $V_{out}(f_i)$ at $F+1$ sample frequencies defined as in Eq.(7). The weighted function W is defined by Eq.(8). The value of W_θ is set to be 0.02 in this experiment.

$$Fitness = \sum_{i=0}^{F} W(d(f_i), f_i) \times d(f_i) \tag{6}$$

$$d(f_i) = |V_{goal}(f_i) - V_{out}(f_i)| \tag{7}$$

$$W(d(f_i, f_i)) = \begin{cases} 1 & \text{for } d(f_i) \le W_\theta \\ 10 & \text{for } d(f_i) > W_\theta \end{cases} \tag{8}$$

The goal response is an ideal low-pass filter with a passband from 1Hz to 1300Hz, a stopband from 1300Hz to 100kHz, and a cut-off frequency of 1300Hz. $V_{goal}(f_i)$ is 1V in the passband and 0V in the stopband. The fitness was calculated from the total of 78 sample frequencies, i.e., 50 from the passband and 28 from the stopband. We used a population of 500 individuals, and 200 generations for each run as in [6]. The crossover ratio, mutation ratio, and replacement ratio are as shown in Table 1.

Results Fig.6 shows the response of the best individual at the 200th generation.

The deviation remained within W_θ (= 0.02V) for all sampled frequencies, and the fitness was 1.97615, while the evaluation value of the best individual obtained in [6] was 2.278. This shows the superiority of the list-based chromosomes over the matrix chromosomes for the circuit design. The phenotype of the best individual is shown in Fig.7.

5 Noise and Error Absorption

The design method for simple passive filters is well established. For example, the response of a band-eliminator filter shown as the solid line in Fig.8 can be achieved with a twin-t circuit design (see Fig.9).

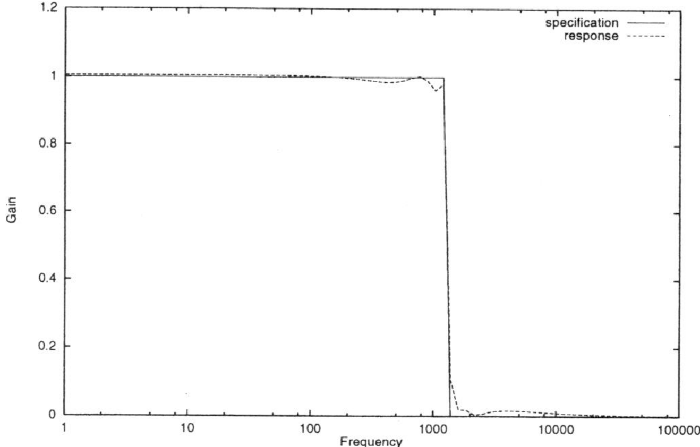

Fig. 6. Specification and acquired response of a low-pass filter

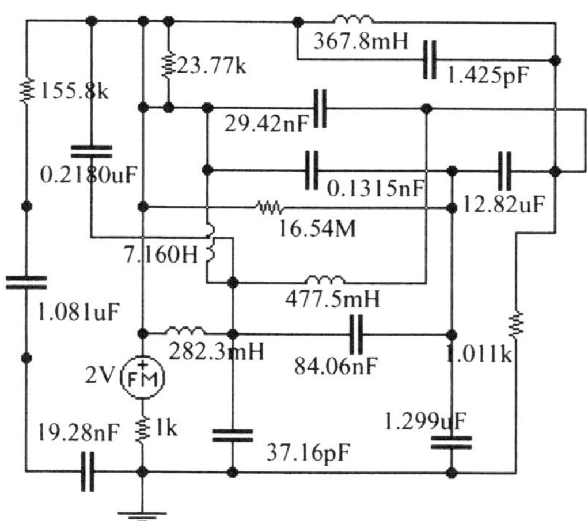

Fig. 7. Acquired low-pass filter circuit

However, when the circuit is manufactured from actual analog elements, the response is not identical to the response in Fig.8, because of the errors in elemental values. The errors inherent in an analog element device can be up to 20%, and the dotted lines and broken lines in Fig.8 show the responses when each element in the circuit randomly contains errors within 5%, 10%, and 20% as labelled.

We conducted a filter synthesis experiment under noisy circumstances where each of the elements contained random deviation from the expected value in order to see how evolutionary analog circuit would accommodate errors and noise.

5.1 Specification

The goal response is the band-eliminator filter shown as the solid line in Fig.8, whose central stopband frequency is 16kHz. We experimented with a set of different maxima for the amount of deviation, i.e., 5%, 10%, and 20%. The population is 500, and the maximum generation is 400. We conducted five runs for each percentage error. The average fitness at each generation is shown in Fig.2.

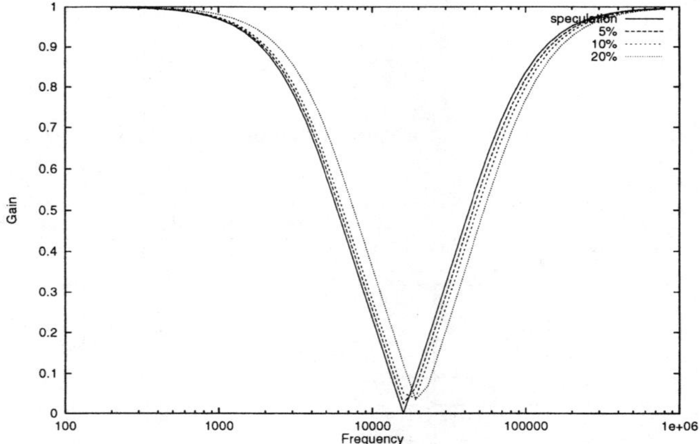

Fig. 8. Actual response of twin-t band-eliminator circuit at the presence of noise

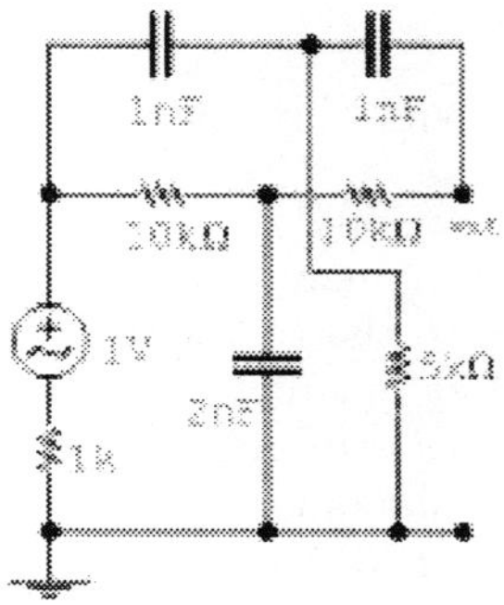

Fig. 9. Designed twin-t band-eliminator circuit sample

5.2 Results

Averaged fitness values each designated percentage of noise are shown in Table 2. The fitness values of the sample circuit (Fig.9) are also shown in the second column. Fig.10 shows the acquired frequency characteristics at 20%.

In evolutionary analog circuit, the GA modifies the deviated circuit based on the response of the circuit, and and thus will not be affected by the difference in designated values and the actual value of the components of the circuit. Through repetition of feedback and redesigning, the precise response to the specification is acquired, as shown in Fig.2.

noise	Sample circuit	200th generation	400th generation
5%	0.000242971	1.73174e-05	2.53538e-08
10%	0.00121551	1.54782e-05	1.48567e-07
20%	0.00521907	2.17895e-05	1.35741e-07

Table 2. Fitness of evolved band-eliminator under the influence of preliminary errors

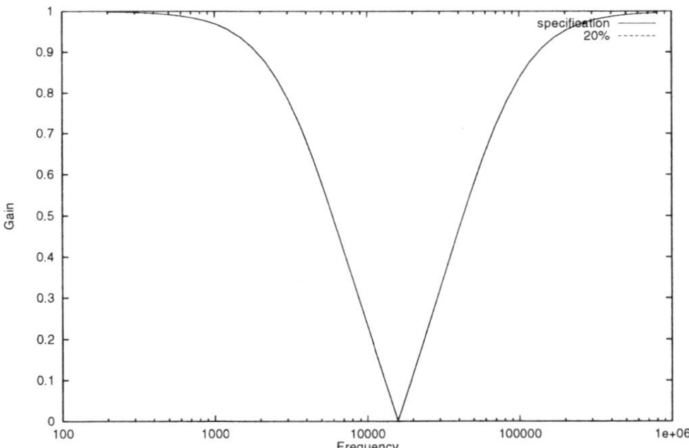

Fig. 10. Acquired response of twin-t band-eliminator circuit the presence of noise

6 Simulation of Two-stage Evolution

Described in this section is the comparison of two-stage evolution and ordinary evolution.

6.1 Specification

The target response is an ideal high-pass filter depicted as solid line in Fig.12 below. The cut-off frequency was 30kHz, and 14 points were taken at intervals in a geometric ratio ranging from 100kHz to 1MHz as the observation points. For the structure evolution phase, the chosen values were those shown in Table 3. The GA parameters are as shown in Table 1.

Element types	Values
Resistances	10kΩ, 1MΩ
Condensers	1nF, 1pF
Coils	100μH, 10mH

Table 3. Circuit element specification

6.2 Results

Fig.11 shows the fitness transition of the best individual versus generations. The fitness is averaged over three runs. The broken line denotes one-stage evolution, and the dotted line indicates two-stage evolution. An arrow shows where the parameter evolution starts. The responses acquired by each evolution are shown in Fig.12. The response for the one-stage is given as a broken line, whereas that of the two-stage is provided by a dotted line. The fitness values are 0.00113213 and 0.001955815 for the two-stage and one-stage evolutions, respectively.

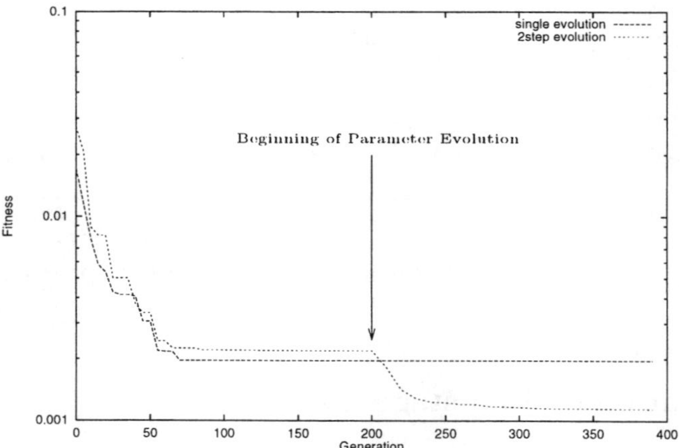

Fig. 11. fitness by generation in simulated two-stage evolution

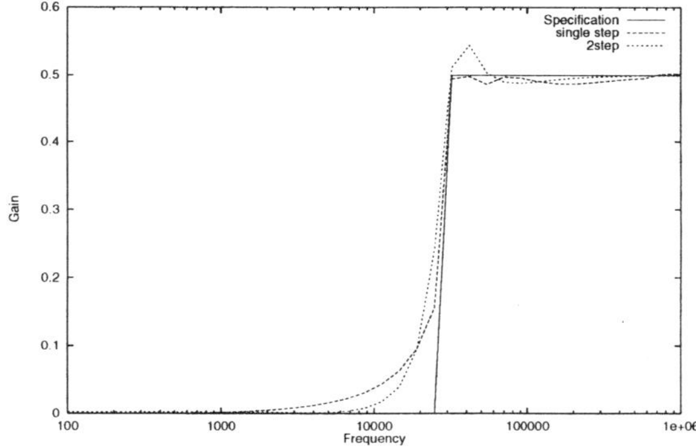

Fig. 12. Acquired response of two-stage evolution

It can be seen from Fig.11 that ordinary evolution converges after the 200th generation, whereas two-stage evolution resumes the search by entering the parameter evolution.

7 Selective Pressure on Circuit Scale

We simulated circuit evolution using the selective pressure referred to in section 3.4.

7.1 Specification

The objective response is the bandpass filter shown in Fig.13. Fitness is defined as in Fig.3, and the value of s is set to be 10^{-6}. We conducted a total of three runs with a population of 500 and 200 generations for each run.

7.2 Results

The response of the best individual at the 40th and the 150th generations is shown in Fig.14. Their deviation from the specification is shown in Fig.15.The fitness at the final generation was 6.10766×10^{-11}. At the 40th generation, the response fulfilling the specification is obtained. The averages of the transfer in the fitness and circuit scale throughout all the trials are shown in Figs.16 and 17.

At the 40th generation, while the influence of the pressure is inconsiderable, the intron portion such as the short circuit of ends of an element is very much as seen in Fig.18. At the 150th generation those portions are deleted

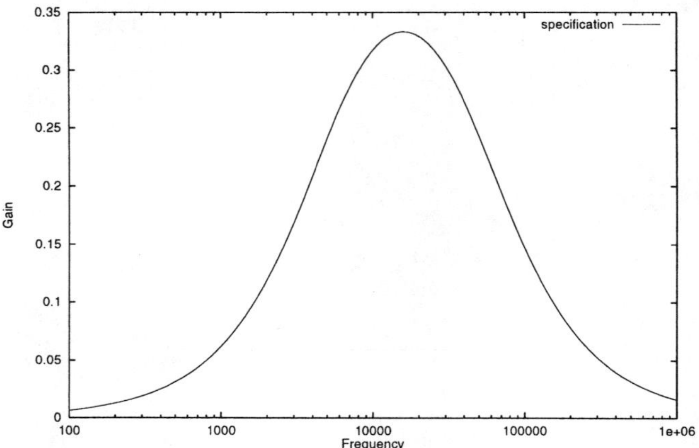

Fig. 13. Objective band-pass filter response

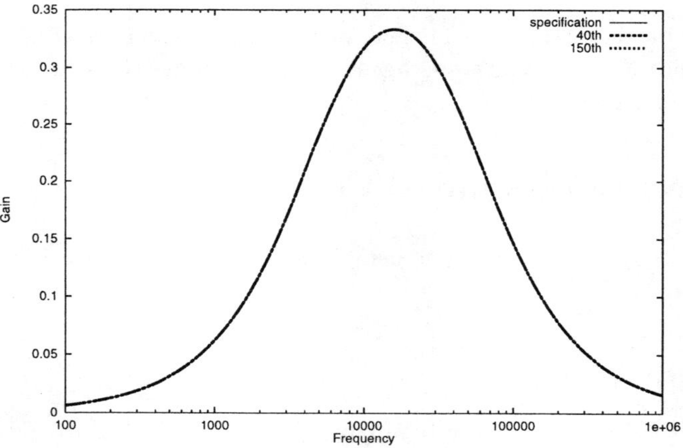

Fig. 14. Response of the best individuals

as seen in Fig.19. From Figs.14 and 17, it can be understood that the major objective at the first half-stage of the evolution is the acquisition of the specified response, while at the second half-stage of the period the objective is to reduce the circuit scale.

8 Discussion

8.1 Comparison with GP and GA with Matrix-based Genes

In the comparison with GP, we were able to achieve an equivalent fitness level by using the list-based chromosome for the filter synthesis simulation.

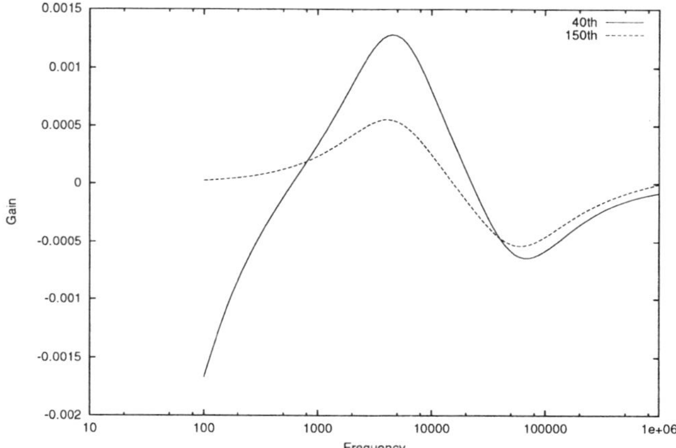

Fig. 15. Deviation of the best individuals

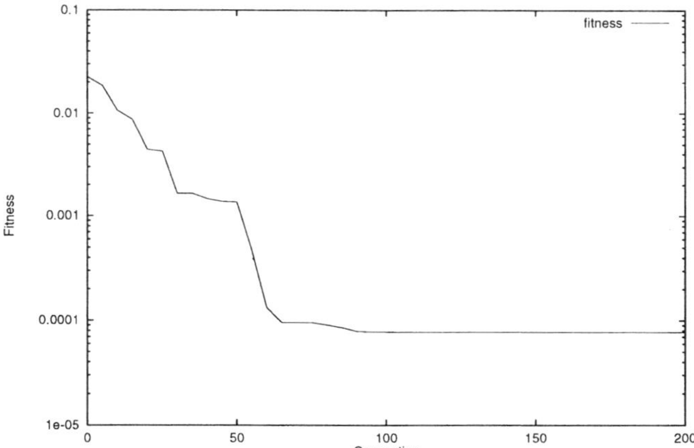

Fig. 16. Fitness vs. generation

Furthermore, the amount of computation in terms of the number of individuals and generations was reduced by using the list-based variable length chromosomes. The memory consumption of a single chromosome was smaller than that of GP. Despite the above, GP is more capable of taking in the existing design method. In comparison with the matrix-based chromosome, a better response is obtained with the list-based method. From the above description, it can be concluded that the circuit designability of the GA using the list-based variable length chromosome is superior to the conventional evolutionary method from the viewpoint of resources such as time or memory consumption.

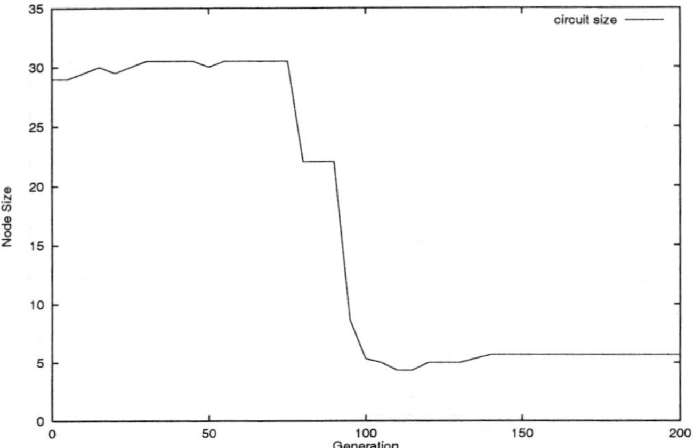

Fig. 17. Number of elements vs. generation

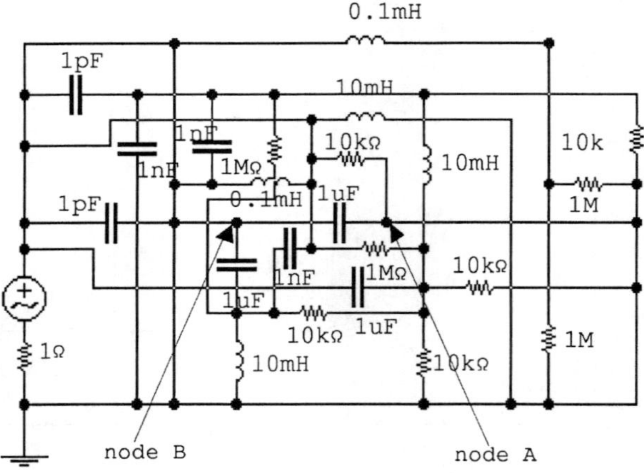

Fig. 18. Best individual at generation 40

8.2 Separation of the Evolution

In the proposed two-stage evolution, the first stage, or the structural evolution, makes dynamic changes in the circuit to fully utilize the GA's global search. On the other hand, in the second stage, i.e., the parameter evolution, the response cannot be altered as much, but can adapt to the stringent specification with high accuracy. As can be seen from Fig.11, while the ordinary evolution converges, the two-stage evolution enters the parameter evolution to resume the search. As a result, a better fitness was acquired with the two-stage evolution. Furthermore, at each stage of the evolution, either the value or the structure of circuit components is fixed. Thus, the amount of data

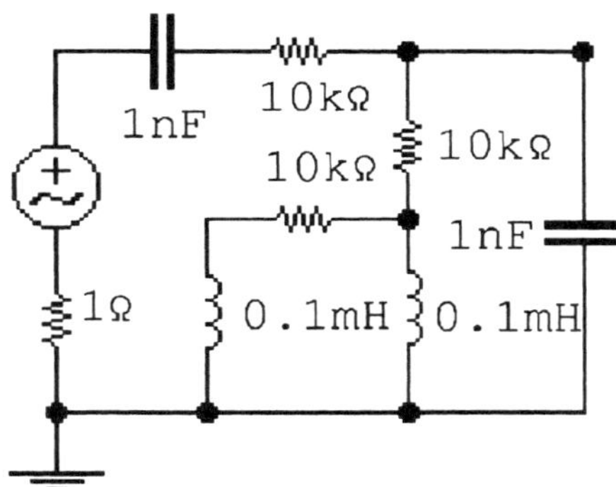

Fig. 19. Best individual at generation 150

to be processed is decreased and results in less memory consumption in the simulator. To be specific, values of the components in the structure evolution can be held as integer variables, while the chromosomes of the parameter evolution are an array of component values. Therefore the proposed method has economized on computer resources in the simulation.

8.3 Adequacy of the Scale

Figs.14 and 16 show that at an earlier stage of the evolution, the fitness is improved by reducing the deviation of the response, and at a later stage, the scale factor contributes mainly to the fitness improvement. By comparing Fig.18 and Fig.19, it can be seen that useless elements are deleted as the evolution proceeds. Therefore, by the scale-pressuring method, implementing a circuit to an adequate scale has been made possible without disturbing the search power of the GA.

9 Conclusion

In this study, we have proposed the following methods for the purpose of implementing evolutionary analog circuits:

- Messy GA representation of a circuit.
- Two-stage evolution.
- Selection pressure on circuit scale.

In the simulated experiments we have confirmed the effectiveness of these methods. As a prospect for the future, the implementation of the GA system with an adequate specification for the variable structure together with the equipment of EHW will be promoted.

Acknowledgments

We would like to thank Professor Shinzo Kitamura for providing us with valuable materials and information for the study. We also wish to express our gratitude to the EHW research group of ETL, including Dr. Tetsuya Higuchi, who has always furnished us with helpful advice and constructive suggestions.

References

1. Ando, S., Ishizuka, M., Iba, H. (1999) "Evolvable Analog Circuit using Variable Length Chromosomes", 56th IPSJ National Conf., (in Japanese).
2. Goldberg, D.E. and Deb, K. and Karpupta, H. and Harik, G. (1993) "Rapid, Accurate Optimization of Difficult Problems using Fast Messy Genetic Algorithms", Proc. 5th Int. Joint Conf. on Genetic Algorithms (ICGA93).
3. Iba, H., Iwata, M., Higuchi, T. (1997)"Machine Learning Approach to Evolvable Hardware", Proc. Int. Conf. on Evolutionary Systems (ICES96), Higuchi, T., Iwata, M., and Liu, W. (Eds.), pp.327-343, Springer-Verlag.
4. Iwata, M., Kajitani, I., Yamada, H., Iba, H., and Higuchi, T. (1996)"A Pattern Recognition System using Evolvable Hardware", Parallel Problem Solving from Nature - PPSN IV, Lecture Notes in Computer Science 1141, pp.761-770, Springer-Verlag.
5. Keymeulen, D., Sakanashi, H., Murakawa, M., Kajitani, I., Takahashi, E., Toda, K., Salami, M., Kajihara, N., and Otsu, N. (1999) "Real-World Applications of Analog and Digital Evolvable Hardware", IEEE Transactions on Evolutionary Computation, Vol.3, No.3.
6. Koyabu, M., Murao, H., and Kitamura, S., (1997)"Automatic Design of Electrical Circuit by Genetic Algorithm", SICE 24th Symposium on Intelligent Systems, March, (in Japanese).
7. Koza, J., Bennett III, F. H., Andre, D., Keane, M. A., and Dunlap, F. (1997) "Automated Synthesis of Analog Electrical Circuit by Means of Genetic Programming", IEEE Transactions on Evolutionary Computation, Vol.1, No.2.
8. Koza, J., Bennett III, F. H., Andre, D., Keane, M. A. (1999) "Genetic Programming III", Morgan Kaufmann.
9. Oitchell, M. (1996) "An introduction to genetic algorithms", MIT Press.
10. Murakawa, M., Yoshizawa, S., Adachi, T., Suzuki, S., Takasuka, K., and Higuchi, T. (1998) "Analogue EHW Chip for Intermediate Frequency Filter", Proc. of the Second Int. Conf. on Evolvable Systems.
11. Nielsen, I. (1995) "A C-T filter compiler-From specification to layout", Analog Integrated Circuits and Signal Processing, vol.7, no.1, pp.21-33.
12. Zebulum, R., Pacheco, M., and Vellasco, M. (1998) "Analog Circuit Evolution in Intrinsic and Extrinsic Mode", Proc. of Second Int. Conf. on Evolvable Systems, Lecture Note in Computer Science 1478 p.154-165, Springer-Verlag.
13. Zebulum, R., Pacheco, M., and Vellasco, M. (1999) "Artificial Evolution of Active Filters: A Case Study", The First NASA/DOD Workshop on Evolvable Hardware.

Human-competitive Applications of Genetic Programming

John R. Koza

Stanford Medical Informatics, Department of Medicine, School of Medicine, Department of Electrical Engineering, School of Engineering, Stanford University, Stanford, California 94305
E-mail: koza@stanford.edu

Summary. Genetic programming is an automatic technique for producing a computer program that solves, or approximately solves, a problem. This chapter reviews several recent examples of human-competitive results produced by genetic programming. The examples all involve the automatic synthesis of a complex structure from a high-level statement of the requirements for the structure. The illustrative results include examples of automatic synthesis of both the topology and sizing (component values) for analog electrical circuits, automatic synthesis of placement and routing (as well as topology and sizing) for circuits, and automatic synthesis of both the topology and tuning (parameter values) of controllers.

1 Introduction

Genetic programming is an automatic technique for producing a computer program that solves, or approximately solves, a problem. Genetic programming addresses the challenge of getting a computer to solve a problem without explicitly programming it. This challenge calls for an automatic system whose input is a high-level statement of a problem's requirements and whose output is a working program that solves the problem. Paraphrasing Arthur Samuel (1959), this challenge concerns, "How can computers be made to do what needs to be done, without being told exactly how to do it?"

Since many problems can be easily recast as a search for a computer program, genetic programming can potentially solve a wide range of types of problems, including problems of control, classification, system identification, and design.

The field of design is a good source of challenging problems that can be used for determining whether an automated technique can produce results that are competitive with human-produced results. Design is usually viewed as requiring creativity and human intelligence. The design process entails creation of a complex structure to satisfy user-defined high-level requirements. Design is a major activity of practicing engineers. Since the design process typically entails tradeoffs between competing considerations, the end product of the process is usually a satisfactory and compliant design as opposed to a perfect design.

Section 2 describes genetic programming. Section 3 states what we mean when we say that an automatically created solution to a problem is competitive with the product of human creativity. Section 4 describes how genetic programming has been applied to problems of synthesis of both the topology and sizing (component values) for analog electrical circuits. Section 5 extends this process to include the automatic creation of the placement and routing of circuits (as well as the automatic creation of the topology and sizing). Section 6 describes the application of genetic programming to the problem of automatically synthesizing the design of both the topology and tuning (parameter values) for controllers.

2 Genetic Programming

Genetic programming progressively breeds a population of computer programs over a series of generations by starting with a primordial ooze of thousands of randomly created computer programs and using the Darwinian principle of natural selection, recombination (crossover), mutation, gene duplication, gene deletion, and certain mechanisms of developmental biology. Specifically, genetic programming starts with an initial population of randomly generated computer programs composed of the given primitive functions and terminals. The programs in the population are, in general, of different sizes and shapes. The creation of the initial random population is a blind random search of the space of computer programs composed of the problem's available functions and terminals.

On each generation of a run of genetic programming, each individual in the population of programs is evaluated as to its fitness in solving the problem at hand. The programs in generation 0 of a run almost always have exceedingly poor fitness for non-trivial problems of interest. Nonetheless, some individuals in a population will turn out to be somewhat more fit than others. These differences in performance are then exploited so as to direct the search into promising areas of the search space. The Darwinian principle of reproduction and survival of the fittest is used to probabilistically select, on the basis of fitness, individuals from the population to participate in various operations. A small percentage (e.g., 9%) of the selected individuals are reproduced (copied) from one generation to the next. A very small percentage (e.g., 1%) of the selected individuals are mutated in a random way. Mutation can be viewed as an undirected local search mechanism. The vast majority of the selected individuals participate in the genetic operation of crossover (sexual recombination) in which two offspring programs are created by recombining genetic material from two parents.

The creation of the initial random population and the creation of offspring by the genetic operations are all performed so as to create syntactically valid, executable programs. After the genetic operations are performed on the current generation of the population, the population of offspring (i.e., the new generation) replaces the old generation. The tasks of measuring fitness, Darwinian selection, and genetic operations are then iteratively repeated over many generations.

Genetic programming is an extension of the genetic algorithm (Holland 1975). Genetic programming is described in books such as Koza 1992; Koza 1994a; Koza, Bennett, Andre, and Keane 1999; Banzhaf, Nordin, Keller, and Francone 1998; Langdon 1998; Ryan 1999, Wong and Leung 2000; Langdon and Poli 2002; in edited collections of papers such as Kinnear 1994; Angeline and Kinnear 1996; and Spector, Langdon, O'Reilly, and Angeline 1999; in conference proceedings such as Koza, Goldberg, Fogel, and Riolo 1996; Koza, Deb, Dorigo, Fogel, Garzon, Iba, and Riolo 1997; Koza, Banzhaf, Chellapilla, Deb, Dorigo, Fogel, Garzon, Goldberg, Iba, and Riolo 1998; Banzhaf, Daida, Eiben, Garzon, Honavar, Jakiela, and Smith 1999; Whitley, Goldberg, Cantu-Paz, Spector, Parmee, and Beyer 2000; Spector, Goodman, Wu, Langdon, Voigt, Gen, Sen, Dorigo, Pezeshk, Garzon, and Burke 2001; Banzhaf, Poli, Schoenauer, and Fogarty 1998; Poli, Nordin, Langdon, and Fogarty 1999; Poli, Banzhaf, Langdon, Miller, Nordin, and Fogarty 2000; and Miller, Tomassini, Lanzi, Ryan, Tettamanzi, and Langdon 2001; in videotapes such as Koza and Rice 1992; Koza 1994b; and Koza, Bennett, Andre, Keane, and Brave 1999; in the new *Genetic Programming and Evolvable Machines* journal; and at web sites such as www.genetic-programming.org.

3 Human-competitive Machine Intelligence

What do we mean when we say that an automatically created solution to a problem is competitive with human-produced results?

We are not referring to the fact that a computer can rapidly print ten thousand payroll checks or that a computer can compute π to a million decimal places. Instead, we think it is fair to say that an automatically created result is competitive with one produced by human engineers, designers, mathematicians, or programmers if it satisfies any one (or more) of the following eight criteria (or any other similarly stringent criterion):

- **(A)** The result was patented as an invention in the past, is an improvement over a patented invention, or would qualify today as a patentable new invention.
- **(B)** The result is equal to or better than a result that was accepted as a new scientific result at the time when it was published in a peer-reviewed scientific journal.
- **(C)** The result is equal to or better than a result that was placed into a database or archive of results maintained by an internationally recognized panel of scientific experts.
- **(D)** The result is publishable in its own right as a new scientific result — *independent* of the fact that the result was mechanically created.
- **(E)** The result is equal to or better than the most recent human-created solution to a long-standing problem for which there has been a succession of increasingly better human-created solutions.
- **(F)** The result is equal to or better than a result that was considered an achievement in its field at the time it was first discovered.

(G) The result solves a problem of indisputable difficulty in its field.

(H) The result holds its own or wins a regulated competition involving human contestants (in the form of either live human players or human-written computer programs).

Note that each of the above criteria are couched in terms of *producing results* and that the results are measured in terms of standards that are external to the fields of artificial intelligence and machine learning.

Using the above criteria, there are now at least 25 instances where genetic programming has produced a result that is competitive with human performance. These examples come from fields such as quantum computing, the annual Robo Cup competition, cellular automata, computational molecular biology, sorting networks, the automatic synthesis of the design of analog electrical circuits, and the automatic synthesis of the design of controllers.

Table 1 shows 25 instances of results where genetic programming has produced results that are competitive with the products of human creativity and inventiveness. Each claim is accompanied by the particular criterion (from the list above) that establishes the basis for the claim. As can be seen in the table, seven of these automatically created results infringe on previously issued patents. In addition, one of the genetically evolved results improves on a previously issued patent. Also, nine of the other genetically evolved results duplicate the functionality of previously patented inventions in a novel way. Since nature routinely uses evolution and natural selection to create designs for complex structures that are well adapted to their environments, it is not surprising that many of these examples involve the design of complex structures.

Table 1. Twenty-five instances where genetic programming has produced human-competitive results.

	Claimed instance	Basis for claim	Reference	Infringed patent
1	Creation, using genetic programming, of a better-than-classical quantum algorithm for the Deutsch-Jozsa "early promise" problem	B, F	(Spector, Barnum, and Bernstein 1998)	
2	Creation, using genetic programming, of a better-than-classical quantum algorithm for Grover's database search problem	B, F	(Spector, Barnum, and Bernstein 1999)	
3	Creation, using genetic programming, of a quantum algorithm for the depth-2 AND/OR query problem that is better than any previously published result	B, D	(Spector, Barnum, Bernstein, and Swamy 1999)	

4	Creation of soccer-playing program that ranked in the middle of the field of 34 human-written programs in the Robo Cup 1998 competition	H	(Andre and Teller 1999)	
5	Creation of four different algorithms for the transmembrane segment identification problem for proteins	B, E	(Koza, Bennett, Andre, and Keane 1999)	
6	Creation of a sorting network (O'Connor, and Nelson 1962) for seven items using only 16 steps	A, D	(Koza, Bennett, Andre, and Keane 1999)	
7	Rediscovery of the ladder topology for lowpass and highpass filters	A, F	(Koza, Bennett, Andre, and Keane 1999)	(Campbell 1917)
8	Rediscovery of "*M*-derived half section" and "constant *K*" filter sections	A, F	(Koza, Bennett, Andre, and Keane 1999)	(Zobel 1925)
9	Rediscovery of the Cauer (elliptic) topology for filters	A, F	(Koza, Bennett, Andre, and Keane 1999)	(Cauer 1934, 1935, 1936)
10	Automatic decomposition of the problem of synthesizing a crossover filter	A, F	(Koza, Bennett, Andre, and Keane 1999)	(Zobel 1925)
11	Rediscovery of a recognizable voltage gain stage and a Darlington emitter-follower section of an amplifier and other circuits	A, F	(Koza, Bennett, Andre, and Keane 1999)	(Darlington 1953)
12	Synthesis of 60 and 96 decibel amplifiers	A, F	(Koza, Bennett, Andre, and Keane 1999)	
13	Synthesis of analog computational circuits for squaring, cubing, square root, cube root, logarithm, and Gaussian functions	A, D, G	(Koza, Bennett, Andre, and Keane 1999)	
14	Synthesis of a real-time analog circuit for time-optimal control of a robot	G	(Koza, Bennett, Andre, and Keane 1999)	

15	Synthesis of an electronic thermometer	A, G	(Koza, Bennett, Andre, and Keane 1999)	
16	Synthesis of a voltage reference circuit	A, G	(Koza, Bennett, Andre, and Keane 1999)	
17	Creation of a cellular automata rule for the majority classification problem that is better than the Gacs-Kurdyumov-Levin (GKL) rule and all other known rules written by humans	D, E	(Andre, Bennett, and Koza 1996)	
18	Creation of motifs that detect the D-E-A-D box family of proteins and the manganese superoxide dismutase family	C	(Koza, Bennett, Andre, and Keane 1999)	
19	Synthesis of analog circuit equivalent to Philbrick circuit (Philbrick 1956)	A	(Koza, Bennett, Keane, Yu, Mydlowec, and Stiffelman 1999)	
20	Synthesis of NAND circuit	A	(Bennett, Koza, Keane, Yu, Mydlowec, and Stiffelman 1999)	
21	Synthesis of digital-to-analog converter (DAC) circuit	A	(Bennett, Koza, Keane, Yu, Mydlowec, and Stiffelman 1999)	
22	Synthesis of analog-to-digital (ADC) circuit	A	(Bennett, Koza, Keane, Yu, Mydlowec, and Stiffelman 1999)	

23	Synthesis of topology, sizing, placement, and routing of analog electrical circuits	G	(Koza and Bennett 1999)	
24	Synthesis of topology for a PID type of controller	A, F	(Koza, Keane, Yu, Bennett, Mydlowec 2000)	(Callender and Stevenson 1939)
25	Synthesis of topology for a controller with a second derivative	A, F	(Koza, Keane, Yu, Bennett, Mydlowec 2000)	(Jones 1942)

The fact that genetic programming can evolve entities that infringe on previously patented inventions, improve on previously patented inventions, or duplicate the functionality of previously patented inventions suggests that genetic programming can potentially be used as an "invention machine" to create new and useful patentable inventions.

4 Automatic Circuit Synthesis

The *topology* of a circuit includes specifying the gross number of components in the circuit, the type of each component (e.g., a capacitor), and a *netlist* specifying where each lead of each component is to be connected. *Sizing* involves specifying the values (typically numerical) of each of the circuit's components.

The design process for analog electrical circuits begins with a high-level description of the circuit's desired behavior and characteristics and includes creation of the topology and sizing of a satisfactory circuit.

The field of design of analog and mixed analog/digital electrical circuits is especially challenging because (prior to genetic programming) there has been no previously known general technique for automatically creating the topology and sizing of an analog circuit from a high-level statement of the design goals of the circuit.

Although considerable progress has been made in automating the synthesis of certain categories of purely digital circuits, the synthesis of analog circuits has not proved to be as amenable to automation. As O. Aaserud and I. Ring Nielsen (1995) observe,

"Analog designers are few and far between. In contrast to digital design, most of the analog circuits are still handcrafted by the experts or so-called 'zahs' of analog design. The design process is characterized by a combination of experience and intuition and requires a thorough knowledge of the process characteristics and the detailed specifications of the actual product."

"Analog circuit design is known to be a knowledge-intensive, multiphase, iterative task, which usually stretches over a significant period of time and is performed by designers with a large portfolio of skills. It is therefore considered by many to be a form of art rather than a science."

We use a simple filter circuit to demonstrate the automatic synthesis of analog electrical circuits using genetic programming. A *filter* is a one-input, one-output circuit that receives a signal as its input and passes the frequency components of the incoming signal that lie in a specified range (called the *passband*) while suppressing the frequency components that lie in all other frequency ranges (the *stopband*). Specifically, the goal is to design a lowpass filter composed of capacitors and inductors that passes all frequencies below 1,000 Hertz (Hz) and suppresses all frequencies above 2,000 Hz.

Genetic programming can be applied to the problem of synthesizing circuits if a mapping is established between the program trees (rooted, point-labeled trees with ordered branches) used in genetic programming and the labeled cyclic graphs germane to electrical circuits. The principles of developmental biology provide the motivation for mapping trees into circuits by means of a developmental process that begins with a simple embryo. For circuits, the initial circuit typically includes a test fixture consisting of certain fixed components (such as a source resistor, a load resistor, an input port, and an output port) as well as an embryo consisting of one or more modifiable wires. Until the modifiable wires are modified, the circuit does not produce interesting output. An electrical circuit is developed by progressively applying the functions in a circuit-constructing program tree to the modifiable wires of the embryo (and, during the developmental process, to succeeding modifiable wires and components). A single electrical circuit is created by executing the functions in an individual circuit-constructing program tree from the population. The functions are progressively applied in a developmental process to the embryo and its successors until all of the functions in the program tree are executed. That is, the functions in the circuit-constructing program tree progressively side-effect the embryo and its successors until a fully developed circuit eventually emerges. The functions are applied in a breadth-first order.

The functions in the circuit-constructing program trees are divided into five categories:

(1) topology-modifying functions that alter the topology of a developing circuit,

(2) component-creating functions that insert components into a developing circuit,

(3) development-controlling functions that control the development process by which the embryo and its successors become a fully developed circuit,

(4) arithmetic-performing functions that appear in subtrees as argument(s) to the component-creating functions and specify the numerical value of the component, and

(5) automatically defined functions that appear in the automatically defined functions and potentially enable certain substructures of the circuit to be reused (with parameterization).

Before applying genetic programming to a problem of circuit design, seven major preparatory steps are required: (1) identify the embryonic circuit, (2) determine the architecture of the circuit-constructing program trees, (3) identify the primitive functions of the program trees, (4) identify the terminals of the program trees, (5) create the fitness measure, (6) choose control parameters for the run, and (7) determine the termination criterion and method of result designation. A detailed discussion concerning how to apply these seven preparatory steps to a particular problem of circuit synthesis (such as a lowpass filter) is found in Koza, Bennett, Andre, and Keane 1999 (chapter 25).

4.1 Campbell 1917 Ladder Filter Patent

The best circuit (Fig. 1) of generation 49 of one run of genetic programming (Koza, Bennett, Andre, and Keane 1996) on the problem of synthesizing a lowpass filter is a 100% compliant circuit (i.e., it complies with all requirements for attenuation, passband ripple, and stopband ripple).

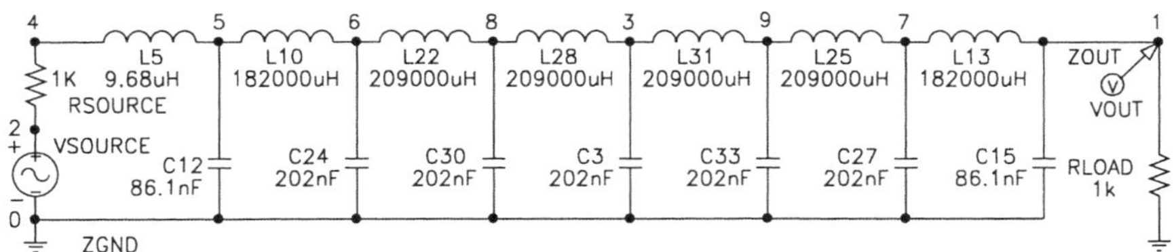

Fig. 1. Evolved Campbell filter

The evolved circuit is what is now called a cascade (ladder) of identical π sections and is shown and analyzed in Koza, Bennett, Andre, and Keane 1999 (chapter 25). The evolved circuit has the recognizable topology of the circuit for which George Campbell of American Telephone and Telegraph received U.S. patent 1,227,113 in 1917. Claim 2 of Campbell's patent covered,

"An electric wave filter consisting of a connecting line of negligible attenuation composed of a plurality of sections, each section including a capacity element and an inductance element, one of said elements of each section being in series with the line and the other in shunt across the line, said capacity and inductance elements having precomputed values dependent upon the upper limiting frequency and the lower limiting frequency of a range of frequencies it is desired to transmit without attenuation, the values of said capacity and inductance elements being so proportioned that the structure transmits with practically negligible attenuation sinusoidal currents of all frequencies lying between said two limiting frequencies, while attenuating and approximately extinguishing currents of neighboring frequencies lying outside of said limiting frequencies."

In addition to possessing the topology of the Campbell filter, the numerical values of all the components in the evolved circuit closely approximate the numerical values specified in Campbell's 1917 patent. But for the fact that this 1917 patent has expired, the evolved circuit would infringe on the Campbell patent.

The legal criteria for obtaining a U.S. patent are that the proposed invention be "new" and "useful" and

"... the differences between the subject matter sought to be patented and the prior art are such that the subject matter as a whole would [not] have been obvious at the time the invention was made to a person having ordinary skill in the art to which said subject matter pertains." (35 *United States Code* 103a).

Since filing for a patent entails the expenditure of a considerable amount of time and money, patents are generally sought, in the first place, only if an individual or business believes the inventions are likely to be useful in the real world and economically rewarding. Patents are only issued if an arm's-length examiner is convinced that the proposed invention is novel, useful, and satisfies the statutory test for unobviousness.

The fact that genetic programming rediscovered both the topology and sizing of an electrical circuit that was unobvious "to a person having ordinary skill in the art" establishes that this evolved result satisfies Arthur Samuel's criterion (1983) for artificial intelligence and machine learning, namely

"The aim [is] ... to get machines to exhibit behavior, which if done by humans, would be assumed to involve the use of intelligence."

4.2 Zobel 1925 "*M*-Derived Half Section" Patent

Since the genetic programming is a probabilistic algorithm, different runs produce different results. In another run of this same problem of synthesizing a lowpass filter, a 100%-compliant circuit (Fig. 2) was evolved in generation 34.

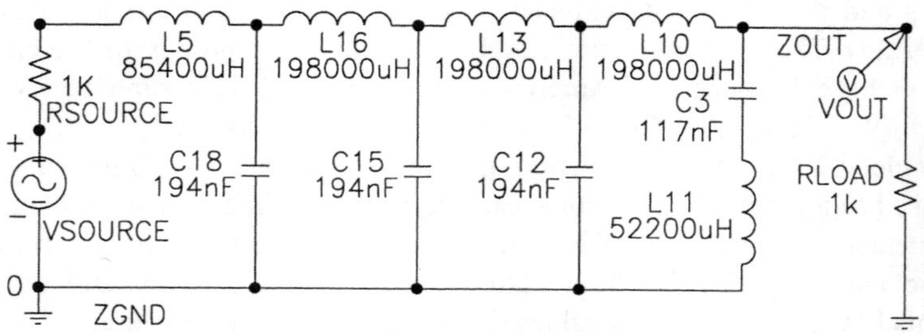

Fig. 2. Evolved Zobel filter

This evolved circuit (presented in Koza, Bennett, Andre, and Keane 1999, chapter 25) is equivalent to a cascade of three symmetric T-sections and an *M*-derived half section. Otto Zobel of American Telephone and Telegraph Company invented and received a patent for an "*M*-derived half section" used in conjunction with one or more "constant K" sections. Again, the numerical values of all the components in the evolved circuit closely approximate the numerical values specified in Zobel's 1925 patent.

4.3 Cauer 1934–1936 Elliptic Filter Patents

In yet another run of this same problem of synthesizing a lowpass filter, a 100%-compliant circuit (Fig. 3) emerged in generation 31 (Koza, Bennett, Andre, and Keane 1999, chapter 27).

This circuit has the recognizable elliptic topology that was invented and patented by Wilhelm Cauer in 1934, 1935, and 1936. The Cauer filter was a significant advance (both theoretically and commercially) over the earlier filter designs of Campbell, Zobel, Johnson, Butterworth, and Chebychev. For example, for one commercially important set of specifications for telephones, a fifth-order elliptic filter matches the behavior of a 17th-order Butterworth filter or an eighth-order Chebychev filter. The fifth-order elliptic filter has one less component than the eighth-order Chebychev filter. As Van Valkenburg 1982 relates in connection with the history of the elliptic filter:

"Cauer first used his new theory in solving a filter problem for the German telephone industry. His new design achieved specifications with one less inductor than had ever been done before. The world first learned of the Cauer method not through scholarly publication but through a patent disclosure, which eventually reached the Bell Laboratories. Legend has it that the entire Mathematics Department of Bell Laboratories spent the next two weeks at the New York Public library studying elliptic functions. Cauer had studied mathematics under Hilbert at Goettingen, and so elliptic functions and their applications were familiar to him."

Genetic programming did not, of course, study mathematics under Hilbert or anybody else. Instead, the elliptic topology emerged from a run of genetic programming as a natural consequence of the problem's fitness measure and natural selection. The elliptic topology did not emerge as a consequence of priming the run with domain knowledge about elliptic functions or filters or electrical circuitry. Genetic programming opportunistically *reinvented* the elliptic topology because necessity (fitness) is the mother of invention.

4.4 Other Circuits

In addition, genetic programming has also been successfully used to synthesize the design for many other types of filters, including highpass, bandpass, bandstop, crossover, comb, and asymmetric filters (Koza, Bennett, Andre, and Keane 1999; Koza, Bennett, Andre, Keane, and Brave 1999). Also, genetic programming has been applied to the problem of automatic synthesis of both the topology and sizing of many analog electrical circuits composed of transistors. These include amplifiers (evolved using multiobjective fitness measures that consider gain, distortion, bandwidth, parts count, power consumption, and power supply rejection ratio), computational circuits (square root, squaring, cube root, cubing, logarithmic, and Gaussian), time-optimal controller circuits, source identification circuits, temperature-sensing circuits, and voltage reference circuits (Koza, Bennett, Andre, and Keane 1999; Koza, Bennett, Andre, Keane, and Brave 1999).

The amplifiers, computational circuits, electronic thermometers, and voltage reference circuits were all covered by one or more patents when they were first invented. Many of these circuit include previously patented subcircuits, such as Darlington emitter-follower sections (Darlington 1953).

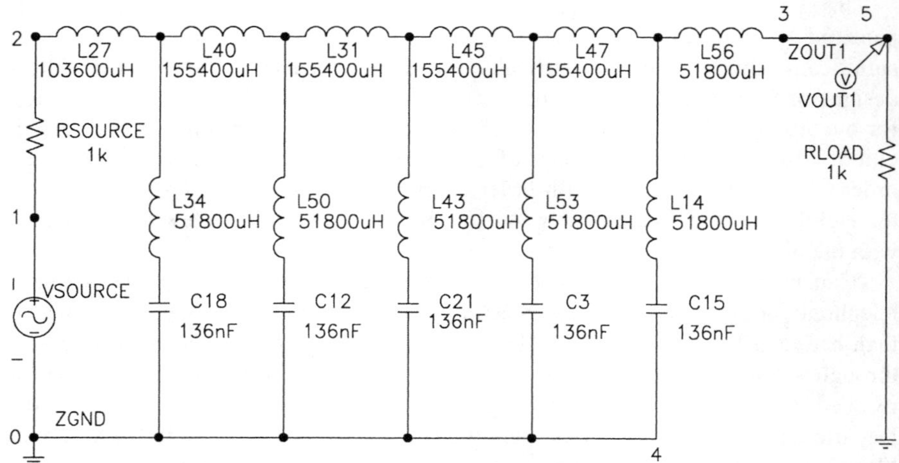

Fig. 3. Evolved Cauer (elliptic) filter topology

5 Topology, Sizing, Placement, and Routing of Circuits

Circuit *placement* involves the assignment of each of the circuit's components to a particular physical location on a printed circuit board or silicon wafer. *Routing* involves the assignment of a particular physical location to the wires between the leads of the circuit's components.

Genetic programming can simultaneously create a circuit's topology and sizing along with the placement and routing of all components as part of an integrated overall design process (Koza and Bennett 1999). It can do this while also optimizing additional other considerations (such as minimizing the circuit's area).

This is accomplished by using an initial circuit that contains information about the geographic (physical) location of components and wires and using component-inserting and topology-modifying operations that appropriately adjust the geographic (physical) location of components and associated wires. For example, the initial circuit in the developmental process complies with the requirements that wires must not cross on a particular layer of a silicon chip or on a particular side of a printed circuit board, that there must be a wire connecting 100% of the leads of all the circuit's components, and that minimum clearance distances between wires, between components, and between wires and components must be maintained. Similarly, each of the circuit-constructing functions used in preserves compliance with these requirements. Thus, every fully laid-out circuit complies with these requirements.

For example, in one run, a lowpass filter circuit was first evolved in generation 25 for a discrete-component printed circuit board. The topology and component sizing for this circuit complied with all requirements (for passband ripple, stopband ripple, and attenuation); however, this circuit contained five capacitors and 11 inductors and occupied an area of 1775.2. Later, a 100%-compliant lowpass filter was created in generation 30 containing 10 inductors and five capacitors occupying an area of 950.3. Then, in generation 138, a physically compact lowpass filter circuit (Fig. 4) containing four inductors and four capacitors and occupying an area of only 359.4 was created. As can be seen, this circuit has the Campbell topology (Campbell 1917).

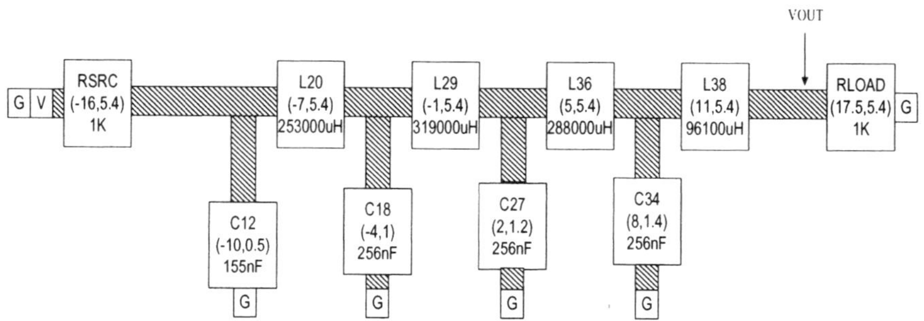

Fig. 4. Topology, sizing, placement, and routing of a lowpass filter for a printed circuit board

6 Automatic Synthesis of Controllers

The design of controllers is another area where there has been (prior to genetic programming) no previously known general technique for automatically creating the topology and tuning for a controller from a high-level statement of the design goals for the controller.

The purpose of a controller is to force, in a meritorious way, the actual response of a system (conventionally called the *plant*) to match a desired response (called the *reference signal*) (Astrom and Hagglund 1995; Boyd and Barratt 1991; Dorf and Bishop 1998).

In the PID type of controller, the controller's output is the sum of proportional (P), integrative (I), and derivative (D) terms based on the difference between the plant's output and the reference signal. The PID controller was patented in 1939 by Albert Callender and Allan Stevenson of Imperial Chemical Limited of Northwich, England.

Claim 1 of Callender and Stevenson (1939) covers what is now called the PI controller,

"A system for the automatic control of a variable characteristic comprising means proportionally responsive to deviations of the characteristic from a desired value, compensating means for adjusting the value of the characteristic, and electrical means associated with and actuated by responsive variations in said responsive means, for operating the compensating means to correct such deviations in conformity with the sum of the extent of the deviation and the summation of the deviation."

Claim 3 of Callender and Stevenson (1939) covers what is now called the PID controller,

"A system as set forth in claim 1 in which said operation is additionally controlled in conformity with the rate of such deviation."

The vast majority of automatic controllers used by industry are of the PID type. As Astrom and Hagglund (1995) observe,

"Several studies ... indicate the state of the art of industrial practice of control. The Japan Electric Measuring Instrument Manufacturing Association conducted a survey of the state of process control systems in 1989 ... According to the survey, more than 90% of the control loops were of the PID type."

However, it is generally recognized by leading practitioners in the field of control that PID controllers are not ideal and that there are significant limitations on analytical techniques in designing controllers. As Boyd and Barratt stated in *Linear Controller Design: Limits of Performance* (Boyd and Barratt 1991),

"The challenge for controller design is to productively use the enormous computing power available. Many current methods of computer-aided controller design simply automate procedures developed in the 1930's through the 1950's ..."

There is no preexisting general-purpose analytic method for automatically creating a controller for arbitrary linear and non-linear plants that can simultaneously optimize prespecified performance metrics (such as minimizing the time required to bring the plant output to the desired value as measured by, say, the integral of the time-weighted absolute error), satisfy time-domain constraints (involving, say, overshoot and disturbance rejection), satisfy frequency domain constraints (e.g., bandwidth), and satisfy additional constraints, such as constraints on the magnitude of the control variable and the plant's internal state variables.

6.1 Robust Controller for a Two-lag Plant

We employ a problem involving control of a two-lag plant to illustrate the automatic synthesis of controllers by means of genetic programming. The problem here (described by Dorf and Bishop 1998, page 707) is to create both the topology and parameter values for a controller for a two-lag plant such that plant output reaches the level of the reference signal so as to minimize the integral of the time-weighted absolute error (ITAE), such that the overshoot in response to a step input is less than 2%, and such that the controller is robust in the face of significant variation in the plant's internal gain, K, and the plant's time constant, τ.

Genetic programming routinely creates PI and PID controllers infringing on the 1942 Callender and Stevenson patent during intermediate generations of runs of genetic programming on controller problems. However, the PID controller is not the best controller for this (and many) problems.

Fig. 5 shows the block diagram for the best-of-run controller evolved on generation 32 of one run of the two-lag plant problem. In this figure, $R(s)$ is the reference signal; $Y(s)$ is the plant output; and $U(s)$ is the controller's output (control variable).

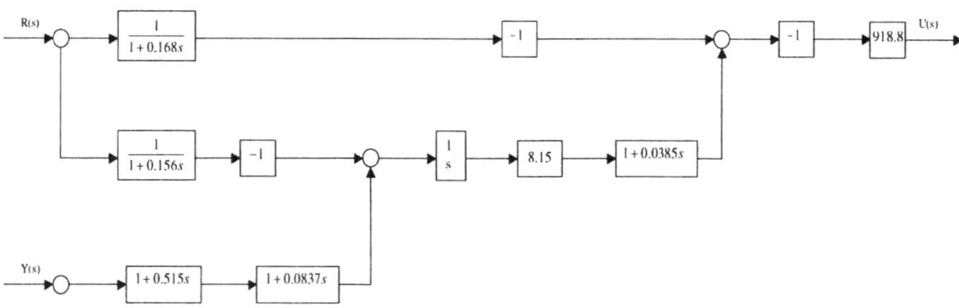

Fig. 5. Best-of-run genetically evolved controller

The controller evolved by genetic programming differs from a conventional PID controller in that the genetically evolved controller employs a second-derivative processing block. As will be seen, this evolved controller is 2.42 times better than the Dorf and Bishop (1998) controller as measured by the criterion used by Dorf and Bishop (namely, the integral of the time-weighted absolute error). In addition, this evolved controller has only 56% of the rise time in response to the reference input, has only 32% of the settling time, and is 8.97 times better in terms of suppressing the effects of disturbance at the plant input.

After applying standard manipulations to the block diagram of this evolved controller, the transfer function for the best-of-run controller from generation 32 for the two-lag plant can be expressed as a transfer function for a pre-filter and a transfer function for a compensator. The transfer function for the pre-filter, $G_{p32}(s)$, for the best-of-run individual from generation 32 is

$$G_{p32}(s) = \frac{1(1+.1262s)(1+.2029s)}{(1+.03851s)(1+.05146)(1+.08375)(1+.1561s)(1+.1680s)}$$

The transfer function for the compensator, $G_{c32}(s)$, is

$$G_{c32}(s) = \frac{7487(1+.03851s)(1+.05146s)(1+.08375s)}{s}$$

$$= \frac{7487.05+1300.63s+71.2511s^2+1.2426s^3}{s}$$

The s^3 term (in conjunction with the s in the denominator) indicates a second derivative. Thus, the compensator consists of a second derivative in addition to proportional, integrative, and derivative functions. Harry Jones of The Brown

Instrument Company of Philadelphia patented this same kind of controller topology in 1942.

Claim 38 of the Jones 1942 patent (Jones 1942) states,

"In a control system, an electrical network, means to adjust said network in response to changes in a variable condition to be controlled, control means responsive to network adjustments to control said condition, reset means including a reactance in said network adapted following an adjustment of said network by said first means to initiate an additional network adjustment in the same sense, and rate control means included in said network adapted to control the effect of the first mentioned adjustment in accordance with the second or higher derivative of the magnitude of the condition with respect to time."

Note that the user of genetic programming did not preordain, prior to the run (as part of the preparatory steps for genetic programming), that a second derivative should be used in the controller (or, for that matter, that a P, I, or D block should be used). The evolutionary process discovered that these elements were helpful in producing a good controller for this problem. That is, necessity was the mother of invention. Similarly, the user did not preordain any particular topological arrangement of proportional, integrative, derivative, second derivative, or other functions within the automatically created controller. Instead, genetic programming automatically created a robust controller for the given plant without the benefit of user-supplied information concerning the total number of processing blocks to be employed in the controller, the type of each processing block, the topological interconnections between the blocks, the values of parameters for the blocks, or the existence of internal feedback (none in this instance) within the controller.

7 Conclusion

This chapter has demonstrated that genetic programming can produce human-competitive designs from complex structures. The results in this chapter (and the other recently produced human-competitive results in Table 1) suggest that genetic programming is on the threshold of routinely producing such human-competitive results. We expect that the rapidly decreasing cost of computing power will enable genetic programming to deliver additional human-competitive results on increasingly difficult problems and, in particular, that genetic programming will be routinely used as an "invention machine" for producing patentable new inventions.

References

Aaserud, O. and Nielsen, I. Ring. 1995. Trends in current analog design: A panel debate. *Analog Integrated Circuits and Signal Processing*. 7(1) 5-9.

Andre, David, Bennett III, Forrest H, and Koza, John R. 1996. Discovery by genetic programming of a cellular automata rule that is better than any known rule for the majority classification problem. In Koza, John R., Goldberg, David E., Fogel, David B., and Riolo, Rick L. (editors). *Genetic Programming 1996: Proceedings of the First Annual Conference, July 28-31, 1996, Stanford University*. Cambridge, MA: MIT Press. Pages 3–11.

Andre, David and Teller, Astro. 1999. Evolving team Darwin United. In Asada, Minoru and Kitano, Hiroaki (editors). *RoboCup-98: Robot Soccer World Cup II*. Lecture Notes in Computer Science. Volume 1604. Berlin: Springer-Verlag. Pages 346-352.

Angeline, Peter J. and Kinnear, Kenneth E. Jr. (editors). 1996. *Advances in Genetic Programming 2*. Cambridge, MA: The MIT Press.

Astrom, Karl J. and Hagglund, Tore. 1995. *PID Controllers: Theory, Design, and Tuning*. Second Edition. Research Triangle Park, NC: Instrument Society of America.

Banzhaf, Wolfgang, Nordin, Peter, Keller, Robert E., and Francone, Frank D. 1998. *Genetic Programming – An Introduction*. San Francisco, CA: Morgan Kaufmann and Heidelberg: dpunkt.

Banzhaf, Wolfgang, Poli, Riccardo, Schoenauer, Marc, and Fogarty, Terence C. 1998. *Genetic Programming: First European Workshop. EuroGP'98. Paris, France, April 1998 Proceedings*. Lecture Notes in Computer Science. Volume 1391. Berlin: Springer-Verlag.

Bennett III, Forrest H, Koza, John R., Keane, Martin A., Yu, Jessen, Mydlowec, William, and Stiffelman, Oscar. 1999. Evolution by means of genetic programming of analog circuits that perform digital functions. In Banzhaf, Wolfgang, Daida, Jason, Eiben, A. E., Garzon, Max H., Honavar, Vasant, Jakiela, Mark, and Smith, Robert E. (editors). 1999. *GECCO-99: Proceedings of the Genetic and Evolutionary Computation Conference, July 13-17, 1999, Orlando, Florida, USA*. San Francisco, CA: Morgan Kaufmann. Pages 1477 - 1483.

Boyd, S. P. and Barratt, C. H. 1991. *Linear Controller Design: Limits of Performance*. Englewood Cliffs, NJ: Prentice Hall.

Callender, Albert and Stevenson, Allan Brown. 1939. *Automatic Control of Variable Physical Characteristics*. U.S. Patent 2,175,985. Filed February 17, 1936 in United States. Filed February 13, 1935 in Great Britain. Issued October 10, 1939 in United States.

Campbell, George A. 1917. *Electric Wave Filter*. Filed July 15, 1915. U.S. Patent 1,227,113. Issued May 22, 1917.

Cauer, Wilhelm. 1934. *Artificial Network*. U.S. Patent 1,958,742. Filed June 8, 1928 in Germany. Filed December 1, 1930 in United States. Issued May 15, 1934.

Cauer, Wilhelm. 1935. *Electric Wave Filter*. U.S. Patent 1,989,545. Filed June 8, 1928. Filed December 6, 1930 in United States. Issued January 29, 1935.

Cauer, Wilhelm. 1936. *Unsymmetrical Electric Wave Filter*. Filed November 10, 1932 in Germany. Filed November 23, 1933 in United States. Issued July 21, 1936.

Darlington, Sidney. 1953. *Semiconductor Signal Translating Device*. U.S. Patent 2,663,806. Filed May 9, 1952. Issued December 22, 1953.

Dorf, Richard C. and Bishop, Robert H. 1998. *Modern Control Systems*. Eighth edition. Menlo Park, CA: Addison-Wesley.

Holland, John H. 1975. *Adaptation in Natural and Artificial Systems: An Introductory Analysis with Applications to Biology, Control, and Artificial Intelligence*. Ann Arbor, MI: University of Michigan Press. Second edition. Cambridge, MA: The MIT Press 1992.

Jones, Harry S. 1942. *Control Apparatus*. U.S. Patent 2,282,726. Filed October 25, 1939. Issued May 12, 1942.

Kinnear, Kenneth E. Jr. (editor). 1994. *Advances in Genetic Programming*. Cambridge, MA: MIT Press.

Koza, John R. 1992. *Genetic Programming: On the Programming of Computers by Means of Natural Selection*. Cambridge, MA: MIT Press.

Koza, John R. 1994a. *Genetic Programming II: Automatic Discovery of Reusable Programs*. Cambridge, MA: MIT Press.

Koza, John R. 1994b. *Genetic Programming II Videotape: The Next Generation*. Cambridge, MA: MIT Press.

Koza, John R., Banzhaf, Wolfgang, Chellapilla, Kumar, Deb, Kalyanmoy, Dorigo, Marco, Fogel, David B., Garzon, Max H., Goldberg, David E., Iba, Hitoshi, and Riolo, Rick. (editors). 1998. *Genetic Programming 1998: Proceedings of the Third Annual Conference*. San Francisco, CA: Morgan Kaufmann.

Koza, John R., and Bennett III, Forrest H. 1999. Automatic Synthesis, Placement, and Routing of Electrical Circuits by Means of Genetic Programming. In Spector, Lee, Langdon, William B., O'Reilly, Una-May, and Angeline, Peter (editors). 1999. *Advances in Genetic Programming 3*. Cambridge, MA: The MIT Press. Chapter 6. Pages 105 - 134.

Koza, John R., Bennett III, Forrest H, Andre, David, and Keane, Martin A. 1999. *Genetic Programming III: Darwinian Invention and Problem Solving*. San Francisco, CA: Morgan Kaufmann.

Koza, John R., Bennett III, Forrest H, Andre, David, Keane, Martin A., and Brave, Scott. 1999. *Genetic Programming III Videotape: Human-Competitive Machine Intelligence*. San Francisco, CA: Morgan Kaufmann.

Koza, John R., Bennett III, Forrest H, Andre, David, and Keane, Martin A. 1996. Automated design of both the topology and sizing of analog electrical circuits using genetic programming. In Gero, John S. and Sudweeks, Fay (editors). *Artificial Intelligence in Design '96*. Dordrecht: Kluwer Academic. Pages 151-170.

Koza, John R., Bennett III, Forrest H, Keane, Martin A., Yu, Jessen, Mydlowec, William, and Stiffelman, Oscar. 1999. Searching for the impossible using genetic programming. In Banzhaf, Wolfgang, Daida, Jason, Eiben, A. E., Garzon, Max H., Honavar, Vasant, Jakiela, Mark, and Smith, Robert E. (editors). 1999. *GECCO-99: Proceedings of the Genetic and Evolutionary Computation Conference, July 13-17, 1999, Orlando, Florida, USA*. San Francisco, CA: Morgan Kaufmann. Pages 1083 - 1091.

Koza, John R., Deb, Kalyanmoy, Dorigo, Marco, Fogel, David B., Garzon, Max, Iba, Hitoshi, and Riolo, Rick L. (editors). 1997. *Genetic Programming 1997: Proceedings of the Second Annual Conference* San Francisco, CA: Morgan Kaufmann.

Koza, John R., Goldberg, David E., Fogel, David B., and Riolo, Rick L. (editors). 1996. *Genetic Programming 1996: Proceedings of the First Annual Conference*. Cambridge, MA: The MIT Press.

Koza, John R., and Rice, James P. 1992. *Genetic Programming: The Movie*. Cambridge, MA: MIT Press.

Koza, John R., Keane, Martin A., Yu, Jessen, Bennett III, Forrest H, and Mydlowec, William. 2000. Automatic creation of human-competitive programs and controllers by means of genetic programming. *Genetic Programming and Evolvable Machines*. 1(1-2) 121 – 164.

Langdon, W. B. 1998. *Genetic Programming and Data Structures: Genetic Programming + Data Structures = Automatic Programming!* Amsterdam: Kluwer.

Langdon, William B. and Poli, Riccardo. 2002. *Foundations of Genetic Programming*. Springer-Verlag.

Miller, Julian, Tomassini, Marco, Lanzi, Pier Luca, Ryan, Conor, Tettamanzi, Andrea G. B., and Langdon, William B. (editors). 2001. *Genetic Programming: 4th European Conference, EuroGP 2001, Lake Como, Italy, April 2001 Proceedings*. Berlin: Springer.

Wong, Man Leung and Leung, Kwong Sak. 2000. *Data Mining Using Grammar Based Genetic Programming and Applications*. Amsterdam: Kluwer Academic Publisher.

O'Connor, Daniel G. and Nelson, Raymond J. 1962. *Sorting System with N-Line Sorting Switch*. U.S. Patent 3,029,413. Issued April 10, 1962.

Philbrick, George A. 1956. *Delayed Recovery Electric Filter Network*. Filed May 18, 1951. U.S. Patent 2,730,679. Issued January 10, 1956.

Poli, Riccardo, Nordin, Peter, Langdon, William B., and Fogarty, Terence C. 1999. *Genetic Programming: Second European Workshop. EuroGP'99. Proceedings*. Lecture Notes in Computer Science. Volume 1598. Berlin: Springer-Verlag.

Poli, Riccardo, Banzhaf, Wolfgang, Langdon, William B., Miller, Julian, Nordin, Peter, and Fogarty, Terence C. 2000. *Genetic Programming: European Conference, EuroGP 2000, Edinburgh, Scotland, UK, April 2000, Proceedings*. Lecture Notes in Computer Science. Volume 1802. Berlin, Germany: Springer-Verlag.

Ryan, Conor. 1999. *Automatic Re-engineering of Software Using Genetic Programming*. Amsterdam: Kluwer Academic Publisher.

Samuel, Arthur L. 1959. Some studies in machine learning using the game of checkers. *IBM Journal of Research and Development*. 3(3): 210–229.

Samuel, Arthur L. 1983. AI: Where it has been and where it is going. *Proceedings of the Eighth International Joint Conference on Artificial Intelligence*. Los Altos, CA: Morgan Kaufmann. Pages 1152 – 1157.

Spector, Lee, Barnum, Howard, and Bernstein, Herbert J. 1998. Genetic programming for quantum computers. In Koza, John R., Banzhaf, Wolfgang, Chellapilla, Kumar, Deb, Kalyanmoy, Dorigo, Marco, Fogel, David B., Garzon, Max H., Goldberg, David E., Iba, Hitoshi, and Riolo, Rick. (editors). 1998. *Genetic Programming 1998: Proceedings of the Third Annual Conference*. San Francisco, CA: Morgan Kaufmann. Pages 365 - 373.

Spector, Lee, Barnum, Howard, and Bernstein, Herbert J. 1999. Quantum computing applications of genetic programming. In Spector, Lee, Langdon, William B., O'Reilly, Una-May, and Angeline, Peter (editors). 1999. *Advances in Genetic Programming 3*. Cambridge, MA: The MIT Press. Pages 135-160.

Spector, Lee, Barnum, Howard, Bernstein, Herbert J., and Swamy, N. 1999. Finding a better-than-classical quantum AND/OR algorithm using genetic programming. In *IEEE Proceedings of 1999 Congress on Evolutionary Computation*. Piscataway, NJ: IEEE Press. Pages 2239-2246.

Spector, Lee, Goodman, E., Wu, A., Langdon, William B., Voigt, H.-M., Gen, M., Sen, S., Dorigo, Marco, Pezeshk, S., Garzon, Max, and Burke, E. (editors). 2001. *Proceedings of the Genetic and Evolutionary Computation Conference, GECCO-2001*. San Francisco, CA: Morgan Kaufmann Publisher.

Spector, Lee, Langdon, William B., O'Reilly, Una-May, and Angeline, Peter (editors). 1999. *Advances in Genetic Programming 3*. Cambridge, MA: The MIT Press.

Valkenburg, M. E. 1982. *Analog Filter Design*. Fort Worth, TX: Harcourt Brace Jovanovich.

Whitley, Darrell, Goldberg, David, Cantu-Paz, Erick, Spector, Lee, Parmee, Ian, and Beyer, Hans-Georg (editors). 2000. *GECCO-2000: Proceedings of the Genetic and Evolutionary Computation Conference, July 10 - 12, 2000, Las Vegas, Nevada*. San Francisco: Morgan Kaufmann Publishers.

Zobel, Otto Julius. 1925. *Wave Filter*. Filed January 15, 1921. U.S. Patent 1,538,964. Issued May 26, 1925.

Evolutionary Algorithms for the Physical Design of VLSI Circuits

James Cohoon[1], John Karro[2], and Jens Lienig[3]

[1] Department of Computer Science
University of Virginia
Charlottesville, VA 22903, U.S.A.
E-mail: cohoon@virginia.edu

[2] Computer Science Program
Oberlin College
Oberlin, OH 44017, U.S.A.
E-mail: john.karro@oberlin.edu

[3] Faculty of Electrical Engineering and Information Technology
Dresden University of Technology
Dresden, 01062, Germany
E-mail: jens@ieee.org

Summary. Electronic design automation (EDA) is concerned with the design and production of VLSI systems. One of the important steps in creating a VLSI circuit is physical design. The input to the physical design step is a logical representation of the system under design. The output of this step is the layout of a physical package that optimally or nearly optimally realizes the logical representation. Physical design problems are generally combinatorial in nature and have very large problem sizes, thus necessitating the use of heuristics such as evolutionary algorithms. We review evolutionary algorithms for physical design and observe and analyze the common traits of the superior contributions. We also discuss important requirements for evolutionary-based approaches for even greater acceptance within the VLSI community.

1 Introduction

Electronic systems are at the core of our everyday lives. For example, they are in our financial networks, mass transit, telephone systems, power plants, and personal computers. Electronic systems are increasingly based on complex VLSI (Very Large Scale Integration) integrated circuits, or as they are known in the vernacular, chips.

A VLSI chip today can contain more than 100 million transistors. One of the main factors contributing to this rapid increase in complexity is an important computer science application area – electronic design automation (EDA). EDA systems are able to simplify the otherwise extremely complex design process of VLSI chips by hiding low-level circuit theory and device physics details from the designer. This encapsulation allows the designer to concentrate on the functionality of the circuit and ways to optimize that functionality.

An EDA system supports descriptions of hardware at many levels of abstraction. Hence it enables designers to work progressively down from an abstract level of design to the layout level. A layout is a complete geometric representation (i.e., a set of rectangles) of the masks which define how the individual layers of the circuit are to be produced.

EDA researchers have created methodologies and tools that allow circuit designers to not merely produce feasible systems, but rather optimal or near-optimal systems composed of tens of millions of circuit elements. The typical figures of merit for VLSI systems are concerned with maximizing reliability and circuit speed while minimizing the size of the physical package, power consumption, etc.

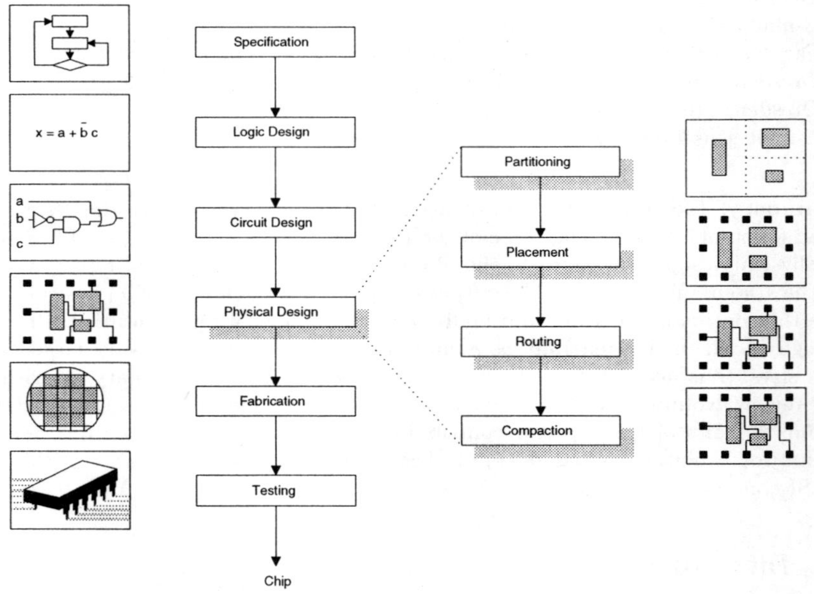

Fig. 1. Typical steps of chip design

In creating a VLSI system, a designer will typically iterate through six major steps (see Figure 1):

- *Specification*: produce a functional specification of the system under design.
- *Logic design*: transform the functional specification into a logical representation, typically via Boolean expressions.
- *Circuit design*: represent the logic representation as a circuit using components from an available library of modules (e.g., AND, NOT, and OR gates, standard cells, or building block macros).

- *Physical design*: translate the circuit design into a physical package representation. This representation is specified through a set of mask descriptions, which define how the individual layers of the integrated circuit are to be produced.
- *Fabrication*: use the physical package representation to fabricate an actual integrated circuit.
- *Testing*: determine whether there are manufacturing errors that prevent the integrated circuit from implementing the functional specification.

Because of the desire for optimality, the problems arising in the design and testing steps are generally combinatorial in nature. And because EDA problem instances are much larger than the instances of traditional problems, EDA problems are particularly difficult as compared to typical combinatorial optimization problems. For example, a circuit can be easily composed of over ten million circuit elements. As such, EDA practitioners have a strong tradition of quickly considering and adapting new and alternative solution techniques. Simulated annealing, an optimization technique that emulates the annealing of crystals, is a good example of this. This combinatorial optimization method was first proposed in the literature in 1983 by [44] and by the following year, the major EDA conferences (e.g., *ACM-IEEE DAC* 1984 and *IEEE ICCAD 1984*) had multiple sessions on simulated annealing for EDA. Early adoption was again repeated on neural networks that simulate the organizing principles of nervous systems (e.g., [81] and [82]).

There is an extensive amount of EDA literature associated with each of the system design steps. Detailed coverage of each of these steps – even limited to the application of evolutionary algorithms to EDA – is beyond the scope of a single chapter. Therefore, we limit ourselves to the application of evolutionary algorithms to physical design. Because physical design is rich in optimization problems, there have been many applications of evolutionary algorithms to it. (For more general overviews of the EDA field itself, we refer the reader to the three classic texts: [59], [75], and [68].)

Our discussion continues with a brief overview of physical design. Next we examine evolutionary algorithms and present a prototypical evolutionary algorithm for circuit partitioning, one of the major steps in physical design. We then present a systematic review of evolutionary algorithm investigations for the physical design process. These contributions are generally different than standard evolutionary algorithm investigations. One major difference is that the crossover and mutation operators for physical design evolutionary algorithms are typically very problem-specific. This specificity occurs because of the extreme importance of determining very high-quality solutions – therefore expert information on the likely form of solutions is included as much as possible. Another major difference is the concern for robustness. There exists a rich collection of design automation benchmarks (e.g., [17] and [IEEE International Testing Conference (1999)]). For a solution method to be accepted, it must be demonstrated to work consistently well on current benchmarks.

We then consider recent trends and discuss what actions must be taken to see even more use of evolutionary algorithms in physical design. We finish this discussion with a summary of our analysis.

2 Physical Design Overview

In physical design, the circuit designer produces a description of the physical layout of the circuit. The description is an assignment of circuit elements including interconnections to geometric coordinates. Depending upon the nature of a particular circuit element, an element may be placed either on a single planar layer or on several contiguous planar layers. The layout must satisfy the requirements of the fabrication technology (e.g., sufficient spacing between components of the circuit and sufficient feature size) and must optimize system characteristics (e.g., timing, speed and size).

The input to the physical design step is a logical description of the circuit that has been augmented to include a specification of the particular circuit elements implementing the various logic functions. The input is generally given as a *netlist*, where a *net* is a collection of circuit elements that must be interconnected for the circuit to be realized. Because circuit elements may in general be part of several different nets, a particular point of contact on the circuit element is specified for each of its interconnections. The point of contact is known as a *pin*. Due to its complexity, the physical layout problem is generally divided into four subproblems that are solved sequentially: partitioning, placement, global and detailed routing, and compaction (see Figure 1). Although these subproblems are NP-hard, they reduce the practical complexity to a manageable level.

Partitioning is the task of dividing a circuit into subcircuits. The decomposition is performed in order to reduce the problem size. The reduction can be necessary because of problem complexity or because the physical package on which the circuit is to be implemented is of insufficient size for the entire circuit or both. The partitions are then implemented separately. The goal is to partition the circuit so that the sizes of the various partition elements are within prescribed ranges and so that the complexity of the interconnections between the partition elements is minimized. We note that the partitioning problem and the other physical layout subproblems may be invoked in a hierarchical fashion. For example, a circuit may be first partitioned into a collection of boards that comprise the backplane of the circuit. The boards may be further partitioned into the chips that make up a given board. A chip itself may be partitioned to produce a floorplan where the major circuit elements are to reside, and so on.

Placement assigns the circuit elements to their geometrical locations on the package. In placement terminology, a circuit element is known as a *cell*. Depending upon where in the invocation hierarchy a placement tool is invoked, a cell may be a single transistor, Boolean operator, functional unit, subcircuit, and so on. Depending upon the fabrication technology, the cells may be placed according to a particular design style (see Figure 2). In *standard cell design*, the circuit elements are rectangular with uniform height but varying widths and are placed in rows. The layout spaces between the rows and along the perimeter are called *channels* and are used for routing interconnections. (If more than two layers are available for routing, the area "above" the cells is used for routing as well.) In *macro cell design*, the cells can be of different sizes and rectilinear shapes and can be placed irregularly. For each of these design styles, a typical goal is to minimize the total layout area of the chip. Another typical goal is to minimize the length of the longest interconnection. Achieving this goal usually satisfies timing constraints. In *gate matrix design*, the circuit elements are rectangular, of uniform size, and are constrained to lie at positions arranged in a matrix pattern. Here the placement objective is to ensure routability by minimizing the congestion of the interconnections between the cells. Another popular design style utilizes FPGAs (field-programmable gate arrays) in a manner similar to gate matrix design.

Global and detailed routing follow the placement activity. The tools for this phase determine the paths of the interconnections between the cells laid out during the placement procedure. The goal is to connect all pins that belong to the same net, subject to certain physical constraints (e.g., to minimize the overall area size), technological constraints (e.g., derived from mask requirements such as not to violate minimum spacing rules) and electrical constraints (e.g., to meet timing requirements). A global router typically works with many nets simultaneously and determines a coarse routing that loosely describes the forms of the various interconnections. The description for a particular net is typically the sequence of tiles (or channels) that the interconnection traverses. A detailed router can then take the *course routing* and assign specific resources and paths within the tile (channel) to realize the interconnection. Detailed routers typically route nets one at a time.

Depending upon the design style, routing may be the final step in physical design. For example, such is the case for gate matrix and FPGA design. However, for other design styles, such as macro cells, *compaction* is usually the final step in the physical layout design. A *compactor* transforms the symbolic layout produced by the preceding steps into a mask layout, i.e., the geometric mask features on the silicon. The objective of compaction is to minimize the size of the resulting circuit layout without impacting performance characteristics.

Once a circuit has been fabricated, tests are needed to determine whether defects have been introduced in the manufacturing process. For an overview of testing consider [1] and [Bushnell and Agrawal (2000)], and for its application to genetic algorithms consider [Mazumder and Rudnick (1999)]. For some individual genetic algorithms applied in testing approaches consider [14], [58], [Rudnick et al. (1997)], [Saab et al. (1992)], and [Srinivas and Patnaik (1993)].

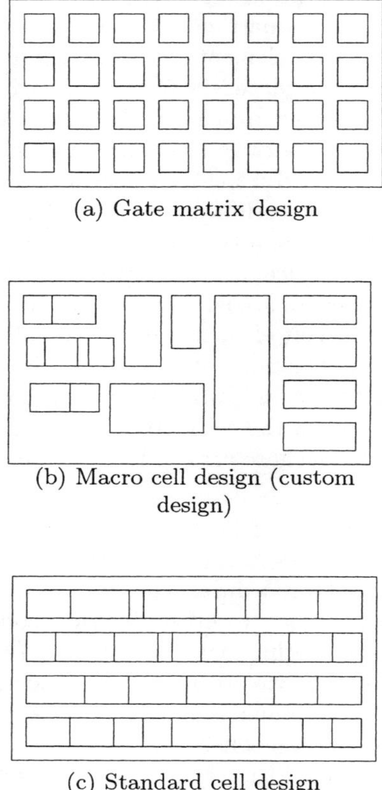

Fig. 2. Layout styles

Logic Synthesis Although the primary interest of this chapter lies in partitioning, placement, routing, and compaction, we note that there is also evolutionary algorithms literature on related problems such as logic synthesis and testing. Logic synthesis is concerned with determining the optimal representation of a circuit with respect to a given design style. Clearly, different representations will lead to different circuit placement and routing characteristics. For an overview of logic synthesis consider [80] and [13] and for its application to evolutionary algorithms consider [16]. For individual genetic algorithm investigations examining logic synthesis issues consider [4], [56], [77], and [79]].

3 Evolutionary Algorithms Overview

Evolutionary algorithms are search algorithms that operate by evolving a population of solutions through repeated transformations. The fact that there is a population of solutions being simultaneously manipulated is one of the major differences between evolutionary algorithms and traditional search algorithms such as backtracking. Another important difference is that the operators performing the transformations on the population are drawn from evolution in nature.

There are three basic models of evolutionary algorithms: genetic algorithms (GA) [35], evolutionary programming (EP) [25], and evolutionary strategies (ES) [64]. Although the three methods have some similarities, they were developed independently. For individual overviews of the GA, EP, and ES methods see respectively [54], [24], and [3]. For a general overview of these strategies, see elsewhere in this text, as well as [5] and [24].

In an *evolutionary algorithm*, associated with each solution is a *score*, which is a measure of the solution's *fitness*. In GA parlance, the encoding used to represent a solution is normally known as a chromosome or *string*. Typically considerable effort is spent by all three paradigms in developing an encoding that facilitates transformations. One of the principal transformation mechanisms for both the GA and ES methods is *crossover*. In a crossover, two solutions are combined to produce a new solution, where the new solution is called an offspring. The other principal evolutionary algorithm transformation mechanism is *mutation*. In a mutation, an existing solution is randomly modified. The GA, ES, and EP models all use this operator. In a GA, the mutation generally occurs in-place, i.e., the existing solution itself is modified. A standard GA limits the application of the mutation operator to offspring. However, there are many GAs that ignore this limitation. In the EP method, the mutation is made to a copy of a parent. (The EP chooses the parent in a manner that favors the more fit solutions and the parent's performance in a tournament.) ES variants that do not use the crossover operator apply the mutation operator in a manner similar to the EP method. ES variants that do use the crossover operator apply the mutation operator to the offspring in an in-place manner.

A basic evolutionary algorithm begins by constructing an initial *population* of solutions. The initial population is typically a collection of randomly generated solutions. As a result, it is expected that a diverse collection of solution characteristics will be generated. In particular, among the solutions of the initial population we expect to find both good and bad characteristics.

The evolutionary algorithm then produces a series of successor populations or generations. The production of a new generation begins by first performing a series of crossovers. For each crossover, two parents are chosen probabilistically from the prior generation. If the GA model is being followed, parent selection is done in a manner that probabilistically favors the more fit solutions. If the ES model is being followed, all parents have an equal probability of being favored. Next, the mutation operator is applied as described previously. Afterwards, a selection process chooses which members of the prior generation and the offspring are to survive to form the basis of the next generation. In a GA, the survival selection process is similar to crossover selection, i.e., the more fit solutions are probabilistically favored to survive. In a typical ES method, the best offspring solutions are the ones selected for survival. Note that some ES implementations are similar to the EP process that selects the best solutions from the combined offspring and parent population. This selection completes the construction of the new generation.

By favoring the more fit solutions in crossover and in survival, the evolutionary algorithm's search is directed towards solutions with superior characteristics. By performing incremental mutations, new solution characteristics can be introduced, which in turn prevents premature convergence of the population to one that contains merely locally optimal characteristics. Together these evolutionary processes are sufficient to produce an algorithm that converges in the limit to an optimal solution [32].

```
Initalize population to represent a random collection
   of parent solutions.
Evaluate the fitness of all members of the population.
While the population has not converged:
      Initalize the offspring population to be empty.
      While the number of offspring is insufficient do:
            Select two new members a and b of the
              parent population.
         Crossover a and b to produce offspring c.
         Add c to the offspring population.
      End
      Initialize the number of mutated members to 0.
      While the number of mutated members is insufficient:
         Randomly select a member to mutate.
      End
      Select the members of the parent and offspring
         populations to become the new parent population.
End
```

Fig. 3. A basic GA-like evolutionary algorithm

An algorithm description of a GA-like evolutionary algorithm is given in Figure 3. The outer loop iterates until the algorithm has converged. The test for convergence can be as simple as determining whether the solution quality is acceptable or running until a desired number of iterations have been performed. A more complicated test can measure the amount of improvement expected in future iterations and then halt when the expectation is sufficiently low. A population size is often used that is proportional to the problem instance. The proportionality is typically either linear or quadratic. With regard to the number of crossovers and mutations, one simple rule of thumb is three crossovers for every mutation.

Multi-objective Optimization One of the benefits of using a genetic algorithm is the relative ease of using them to solve multi-objective optimization. In such cases, we are searching a solution space with the objective of finding the solution that optimizes over two or more independent criteria. One of the first successful attempts to address such problems was devised by [69]. Further research in the area is summarized in [26] and [32].

4 A Sample Evolutionary Algorithm for Partitioning

We next describe a simple evolutionary algorithm for a basic partitioning problem. We do so by providing methods for the initialization of the population, a crossover operator, and a mutation operator.

The partitioning problem that we consider has the following description:

- Input: A list $N = \{N_1, \cdots, N_n\}$ of n nets. A list $C = \{C_1, \cdots, C_m\}$ of m circuit elements.
- Output: A bipartition B of the circuit elements to sides 0 or 1 that minimizes the number of *cut nets* – nets that have circuit elements on both sides of the partition. If $B_i = 0$, then element C_i lies on side 0, and if $B_i = 1$, then element C_i lies on side 1. Approximately one-half of the circuit elements are assigned to side 0 and the other half are assigned to side 1. The difference in the number of elements on the two sides can be no greater then k percent of the total number of elements.

Although simplistic, the problem is sufficiently close to standard EDA partitioning problems to acquaint the reader with the problem.

A straightforward encoding for a population member would be a string of m bits, where the ith bit in the string is 0, if the solution being represented has C_i on side 0; the ith bit in the string is 1, if the solution being represented has C_i on side 1. The score of a solution would be the number of nets cut by the solution being represented. The fitness of a solution could be the difference of its score from m.

To construct an initial population, one could generate a series of random bit patterns of length m such that the difference in the number of zeros and ones in a given string pattern is no greater than k percent.

A variety of straightforward crossover operators exist. For example, a one-point crossover operator [32], acting on parents a and b, first randomly picks an index i into a. Then, the first i bits of parent a are concatenated with the last $m - i$ bits of parent b to produce an offspring c. If feasible solutions are required and c is infeasible, then excess zeros or ones can be randomly complemented. As an example, suppose there are eight circuit elements with

$a = 10100110$ and $b = 01100101$ and the one-point index is 5. The offspring c is $\underline{1010}0101$ (the underlined portion comes from a). Because the number of zeros and ones are equal, no feasibility correction is required.

For another example, consider a two-point crossover operator. A two-point crossover, operating on parents a and b, first randomly selects two different indices i and j. If $i < j$, then bits 1 through $i - 1$ of b are concatenated with bits i through j of a that are concatenated in turn with the last $m - j$ bits of b to produce the offspring c. If instead $i > j$, then the roles of a and b are reversed (i.e., bits 1 through $i - 1$ of a are concatenated with bits i through j of b that are concatenated in turn with the last $m - j$ bits of a to produce the offspring c). As an example with the same values of parent a (10100110) and parent b (01100101) and $i = 6$ and $j = 3$, then offspring $c = \underline{101001}10$ (the underlined portion comes from a). A two-point crossover operator is usually preferred to a one-point crossover operator as more varied combinations of the parents are possible.

Straightforward mutation operators are also possible. One possibility is to randomly select a bit position and to complement its value. If the complementation makes the solution infeasible, a random bit with that same value is selected and complemented. For example, if $a = 10100110$ is selected for mutation and the third bit is to be complemented, the result is $10\underline{0}00110$ (the underlined portion shows the change that is introduced). For reasonable values of k (the maximum percentage difference between the two partition sizes), the new solution would be feasible. For other values of k, either bit 2, 4, 5, or 8 of a would need to be complemented. For a more radical mutation, a range of bits could be complemented.

For small circuits, the above methods produce good solutions. However, because most interesting EDA problems do not have small instances, more involved strategies are necessary. Such strategies are discussed in the next section, where we give a systematic overview of evolutionary algorithms that have been successfully applied to the major steps of the physical design of VLSI circuits. Although EDA did not immediately add evolutionary algorithms to its basic tool chest, evolutionary algorithms have been consistently used in the field since 1987.

5 Application of Evolutionary Algorithms to the VLSI Physical Design

In this section we look at the four primary steps of design automation (partitioning, placement, routing, and compaction), and provide an overview of the work that has been done in applying evolutionary algorithms to these problems. We present a brief description of each major tool, discuss the primary characteristics of the algorithms (e.g., crossover and mutation operators and population representation), and argue the merits and problems of each method.

Ideally, we would provide both analytical runtime bounds of each tool as well as experimental comparisons. This, however, is not currently a feasible goal. Runtime bounds of evolutionary algorithms are, at best, quite difficult. Because of the probabilistic nature of the algorithms, and because of the dependency of the runtime on subtle characteristics of the solution space of any particular problem, it can be almost impossible to achieve a useful bound. It is sometimes feasible to achieve loose worst-case bounds, but the bounds are frequently so loose that they cannot be considered useful information.

In presenting an experimental analysis, we run into the problems of experimental inconstancy between papers. In order to truly compare two evolutionary algorithms, we would need to run them on the same problem instance, using comparable architectures, with consistent parameters. For example, we cannot meaningfully compare the results and runtime when using two possible population representations if the convergence criteria vary between the experiments. As there is no constant methodology for testing evolutionary algorithms, each research group uses the experiments they see fit. Hence, we cannot meaningfully compare the resulting reports of two methodologies if the researches in question did not employ consistent tests.

Partitioning As mentioned previously, the goal of partitioning is to divide the circuits into smaller parts that would be separately implemented. Thus, in performing the partition we want parts with both similar complexities and minimal mutual dependencies. Mutual dependencies are generally expressed in the form of interconnect complexities between different parts. Most algorithms for partitioning focus on these interconnections between different parts with the goal to minimize overall interconnection costs.

The first evolution-based algorithm for solving the partitioning problem was published by [66]. In contrast to previous heuristic algorithms that usually optimize on only one constraint, this approach is capable of handling both a number of constraints and a number of objectives. This capability is achieved by using a multi-constraint, multi-objective fitness function.

The algorithm consists of an initial partitioning (based on a bin packing algorithm) and a so-called *evolution function*. The evolution function includes three steps: migration, distribution, and calculation of parameters. The algorithm works with *one* problem solution. A crossover operator is not used.

The algorithm yields multi-way partitions with fairly balanced sizes and a small number of pins for each part. The presented strategy has a reasonable execution speed that is comparable to other published heuristic approaches.

[37] and [38] investigate different coding schemes for the problem of circuit partitioning. The proposed GA is specifically tailored for the partitioning of circuits with complex bit-slice components and uses a special two-step coding of partitions.

The initial population is generated randomly, i.e., the circuit is partitioned by chance and then decoded. The algorithm consists of a crossover, a mutation operator, and a deterministic improvement strategy. The runtime of the algorithm is not competitive.

A parallel GA by [12] is based on a population structure that involves subpopulations which have their isolated evolution occasionally interrupted by inter-population communication. Every 50 generations, each subpopulation exchanges 10 – 20 percent of its members with neighboring subpopulations. Partitions with design violations are tolerated by means of a special penalty function. Results show that multiple subpopulations enhance the solution quality by exploiting wider regions of the search space.

A hybrid GA for the ratio-cut partitioning problem is presented by [7]. (The ratio-cut partitioning problem considers not only the cut size but also the size of the partitions.) Here the problem is formulated in terms of a hypergraph. The modules are mapped to vertices and the nets are mapped to hyperedges. Each solution in the population is represented by a chromosome. The number of genes in a chromosome equals the number of modules in the graph. The chromosome is a binary string, where each module has a corresponding location on the string. A location has the value 0 if the corresponding module is on the left side of the bipartition, and has value 1 otherwise. Before the GA is executed, the ordering of these genes is determined by a depth-first search to improve the performance of the GA. Traditional crossover and mutation operators are combined with a fast partitioning heuristic applied to each offspring as an improvement operator. The performance of the algorithm is compared with two other partitioning approaches, using benchmark data sets. Averaged over all benchmarks, the presented algorithm achieves better results than the other approaches, with a similar runtime for smaller graphs and a shorter runtime for the largest graphs.

Another hybrid approach is published by [78]. It combines a simulated annealing method with a GA. The main motivation for this approach is the parallelization of the simulated annealing strategy by replacing its single solution search process with a population-based approach using a GA. Benchmark comparisons are not presented.

While a number of GAs have been proposed for dealing with the partition problem, none of them are truly competitive with the state-of-the-art tools in the area. See [2] for a full survey on partitioning, including a number of the most sophisticated probabilistic and deterministic algorithms for the problem. For a tutorial on GA partitioning, consider [53].

Placement The placement step assigns physical locations on the chip to the components (cells, gates, etc.) of the circuit. This assignment is done by considering placement constraints (e.g., no overlap is allowed between cells) and objectives (e.g., fulfilling timing constraints). The objectives differ widely according to the specific design concept. Placement algorithms can be divided into algorithms for standard cell design, macro cell design, gate matrix design, and FPGAs according to the variation in size and location of these cells (see Figure 4).

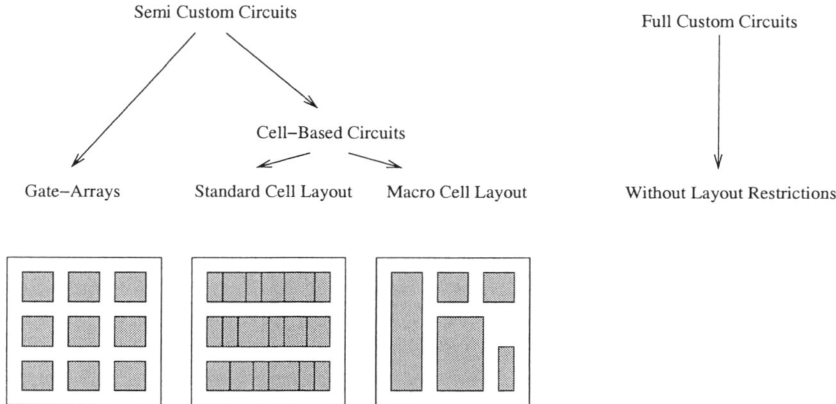

Fig. 4. Placement styles

The task of placement can usually be considered as two related problems: a two-dimensional bin packing problem and the problem of minimizing the interconnection costs. The bin packing problem targets the optimal or sub-optimal composition of components with different sizes and shapes. The minimization of interconnection costs optimizes the length of the routing either by considering the overall routing length (or cost) or by considering the individual interconnection costs between different components.

One of the main tasks of a placement algorithm is the definition of an appropriate cost function which also considers the needs of the subsequent routing procedure. Most placement algorithms are limited to two types of cost functions: they consider either the estimated length of the interconnect (as needed for timing constraints) or the area density of the interconnect (in order to avoid routing congestions).

After the pioneering work of [10], further applications of evolutionary algorithms for standard cell placement have been presented, such as [45], [46], [73], and [74]. The last two approaches work with an initial population that has been generated randomly. Each individual consists of an array of four integers, containing the cell number, the x and y coordinates, and the location of the cell. The population size is kept at 24 individuals. Three crossover operators are applied that cut the parent string at random positions. The crossover operators vary according to their "rebuilding strategy" of the descendants.

The mentioned approaches produce high-quality placements of real-world VLSI circuits that can compete with sophisticated simulated annealing-based placement strategies. The published runtime are up to 6 hours for the algorithm of [45], while runtime of up to 12 hours are presented in [74].

[55] reduced the runtime significantly by developing a parallel implementation of a GA that runs on a distributed network of workstations. The total population is split over different processors and a migration mechanism is used to exchange genetic material between them. While the placement results are similar to a sequential GA, an almost linear speedup can be achieved with this method.

We next discuss investigations that use evolutionary algorithms for macro cell placement. The approach of [9] is based on a two-dimensional bitmap representation of the macro cell placement problem. Another representation scheme, a binary tree, is applied by [19]. [21] consider a combination of a GA with a simulated annealing-based optimization strategy. The experimental results suggest that a mixed strategy performs better than a pure GA for the macro cell placement problem. The results are better or comparable to previously published results of placement benchmarks. However, the runtime is not as competitive.

[70] present a parallel GA for the macro cell placement problem. They use the term "self-adaptation" to describe a search process by which several "islands" execute sequential GAs with different strategies. At fixed intervals these strategies are ranked and each strategy is improved by assimilating the characteristic parameters of the next better strategy. The experimental results suggest that "self-adaptation" improves both the performance and the robustness of the algorithm. Benchmark results and runtime are not presented.

Since macro cells are constructed in a hierarchical way, their shape may not be fixed. This more general placement problem is called *floorplanning.* A hybrid GA for the floorplan area optimization problem has been presented by [63]. They introduce several heuristic operators in addition to a pure GA. Their approach adaptively provides the activation probabilities of the operators. Experimental results show that the proposed method is competitive with other approaches in both solution quality and runtime. Since the complexity of the algorithm grows linearly with the problem size, the approach can be applied to floorplans larger than any benchmark previously considered.

An application of a GA for the placement of gate matrix design has been published by [72]. The approach uses the Genesis package ([28]) as the basic GA. This package is modified with a special algorithm for constructing permutations that considers only a small subset of the solution space. The results are compared with only one previously published algorithm. The runtime is in the order of minutes.

It can be observed in the above-mentioned evolutionary algorithms for placement that "traditional" bit-string representations are not used. (The obvious reason is that symbols in the solution string cannot be repeated when applied to the placement problem.) Hence, "traditional" crossover strategies, such as one-point crossover operators, have not been applied either.

As mentioned earlier, crossover is the primary method of optimization during the evolutionary process. In the case of placement, crossover works by combining sub-placements of two different parent configurations in order to generate a new (and perhaps better) placement. This process might generate conflicts that need to be resolved by the crossover operator. Despite the variety of crossover operators presented, all include conflict resolution methods tailored to the placement problem. Three of the most promising ones are mentioned here (see Figure 5, described in [52]).

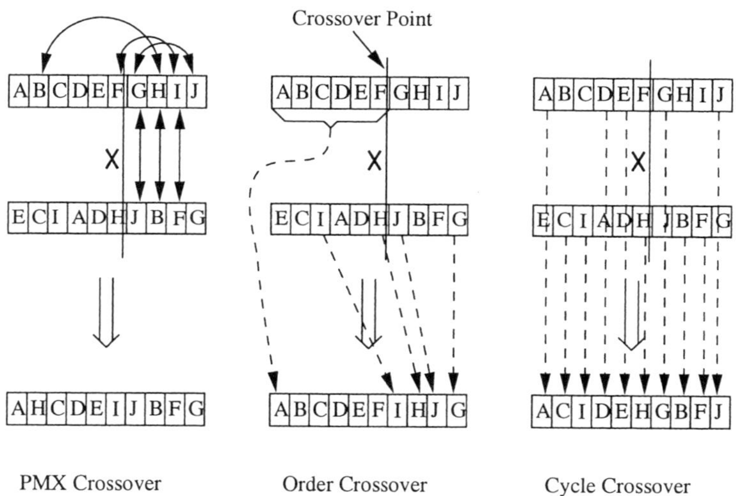

Fig. 5. Crossover operators suitable for placement

The *partially mapped crossover* (PMX) chooses a random cut point in both parents and considers the segments following the cut point as a partial mapping of the cells to be exchanged in the first parent to generate the offspring [33]. It takes corresponding cells from the segments of both parents, locates both these cells in the first parent, and exchanges them. Hence, a cell segment in the first parent and a cell at the same location in the second parent define which cells in the first parent have to be exchanged to generate an offspring.

The *order crossover* [15] also chooses a random cut point in both parents. It then copies the array segment to the left of the cut point from one parent to the offspring. The remaining (i.e., right) portion of the offspring array is filled by going through the second parent and taking, in order, those elements which were left out.

The *cycle crossover* [57] generates an offspring in which every cell is in the same location as in one parent or the other. This crossover operator tries to avoid cell conflicts by finding non-overlapping sets of cells to pass from the two parents. Since non-overlapping cells are not contiguous, a special selection process is applied.

While GAs have enjoyed some success when applied to placement algorithms – some of the tools are competitive with conventional algorithms – they take considerably longer to match the non-GA solutions. In the area of gate array placement, tools such as VPR [6] or Spiffy [43] are able to outperform any GA tools currently in the literature.

For a general survey of placement literatures see [75], as well as the work by [68]. For a tutorial on GAs as applied to the problem of placement, consider the work in [52].

Routing Routing is the process of connecting pins subject to a set of routing constraints. The routing of VLSI circuits (and also of Multi-Chip Modules, MCMs) is usually divided into *global routing* (to assign nets into certain routing regions) and *detailed routing* (to assign nets to exact positions inside a routing region).

Routing of MCMs and PCBs (Printed Circuit Boards) is based on fixed routing areas with numerous routing layers. This leads to a non-planar routing topology. This characteristic and the desire to achieve a global-optimal routing result prevent the separation into global and detailed routing. Instead, a so-called "free" (or custom) routing is used that considers the entire routing region of the substrate (see Figure 6).

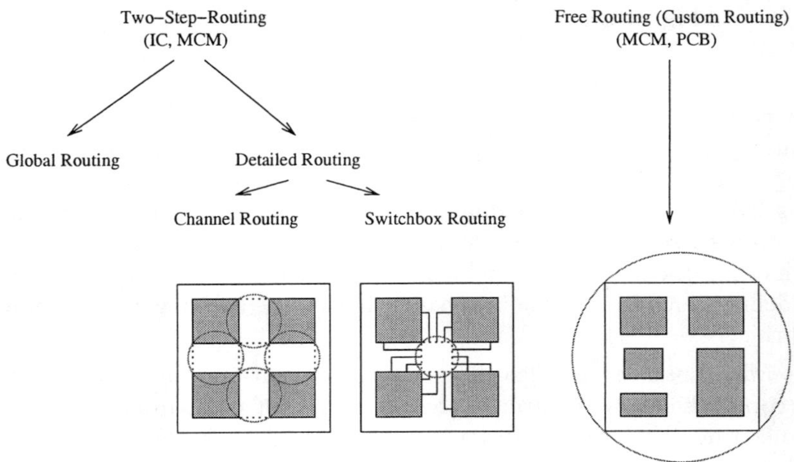

Fig. 6. Routing strategies

Global routing algorithms are usually based on graph-search strategies where the nodes of the graph represent separate routing regions (e.g., tiles or channels) of the circuit. The edges of the graph mark the routing capacities of the routing regions. These routing capacitances are often estimates because they can be verified exactly only in the subsequent detailed routing.

The regions are routed sequentially during detailed routing. According to the position of the pins, detailed routing can be separated into *channel routing* (pins are only located on two parallel sides of the routing area) and *switchbox routing* (pins are placed on all four sides of the routing area), as shown in Figure 7. Detailed routing sometimes also includes the determination of the exact placement location of the cells and their pins, respectively.

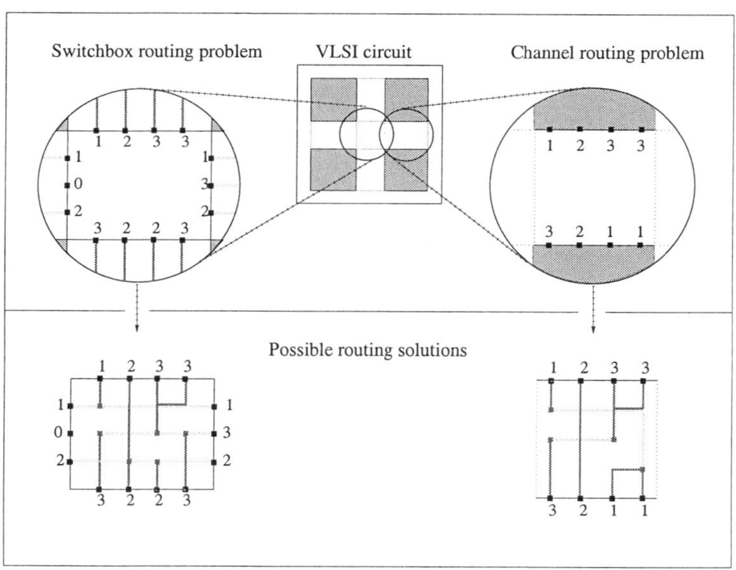

Fig. 7. Switchbox and channel routing

Contrary to placement, where numerous evolutionary algorithms have been published, there have not been as many successful publications of evolutionary routing algorithms. This condition seems to be a consequence of the larger complexity of the routing task. In particular, the appropriate coding of the routing poses a major obstacle with the result that almost all publications in this field are limited by specific constraints induced by the encoding.

Related to the routing problem is the Steiner tree problem – finding a minimum length interconnection in the plane. For an introduction to this problem see [39]. For a discussion of Steiner-tree-related GAs see [22], [41], [Kapsalis and Rayward-Smith (1993)], and [Esbensen and Mazumder (1999)].

To our knowledge, only one evolutionary algorithm for global routing has been reported ([20]). Before the global routing process itself is initiated, a rectilinear routing graph is extracted from the given placement. Routing is then performed in terms of this graph; that is, computing a global route for a net is done by computing a corresponding path in the routing graph. In other words, each edge of the graph corresponds to a routing channel, and each vertex corresponds to the intersection of two channels. Vertices representing the terminals of the net are added to the routing graph at appropriate locations. Finding the shortest path for the net is then equivalent to finding a minimum cost subtree in a graph. This graph spans all of the added terminal vertices, assuming that the cost of an edge is defined as its length. When a net has been routed, its terminal vertices are removed from the routing graph, thereby significantly reducing the problem size.

The global route algorithm is based on a two-phase router. In the first phase, a GA for the Steiner problem in the above-mentioned graph is used to generate a number of distinct, alternative routes for each net. Then, in a second phase, another GA selects a specific route for each net (among the alternatives given from the first phase), such that the overall layout area is minimized.

The router is superior to TimberWolfMC ([71]), a state-of-the-art simulated-annealing-based router, with respect to solution quality. However, the runtime is slower by a factor of 50 which limits practical applications of this algorithm.

[51] developed a rip-up-and-rerouter which is based on a probabilistic rerouting of nets of one routing result. (A rip-up-and-rerouter iteratively improves the routing result by deleting and rerouting previously routed nets.) The initial population (consisting of nets as individuals) is generated with both a shortest path algorithm and a random path strategy. Competitive results are presented for channel and switchbox routing benchmarks. The runtimes vary between a couple of minutes for small examples and up to 24 hours for large channel routing problems.

The router of [29] combines the steepest descent method with specific features of GAs. The crossover operator is restricted to the exchange of entire nets and the mutation procedure performs only the creation of new individuals. The presented results are limited to simple VLSI problems, and no runtime remarks are made.

Some evolutionary algorithms have been presented for the restrictive channel routing problem; [60] have an example of this, as do [61], and [62]. Here, all vertical net segments are located on one layer and all horizontal segments are placed on a second layer. This and other restrictions make these approaches unusable for real-world VLSI channel routing problems.

The GA for channel routing published in [Lienig and Thulasiraman (1994)] is based on a problem-specific representation scheme. Here the layout is coded in a three-dimensional lattice-like chromosome with the cells representing different coordinate points of the routing solution. The value of a cell indicates which net is routed at this coordinate point in the routing solution. A negative cell value indicates a fixed assignment (e.g., a pin) and zero indicates that the area is unused (Figure 8, as described in [49]).

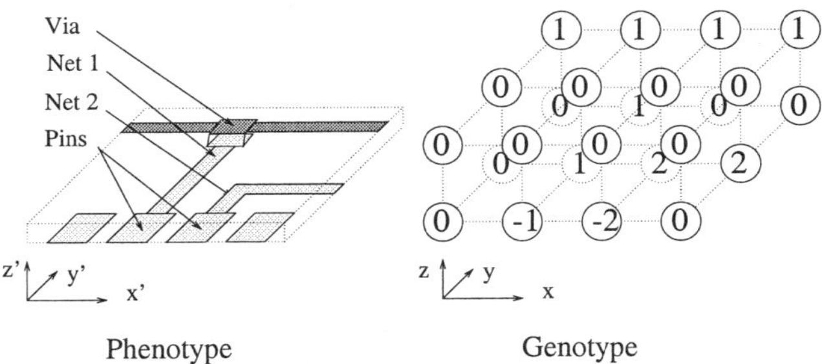

Fig. 8. A sample routing (left) and its genetic encoding (right)

This three-dimensional encoding scheme ensures that good "routing islands" in the routing structure are preserved as compact high-fitness building blocks in the chromosome. Consequently, these building blocks have a high probability of being transferred intact and recombined with other high-quality building blocks in offspring solutions. Furthermore, this encoding scheme enables a simple monitoring of the routing constraints directly in the chromosome.

The genetic operators are also specifically developed for the channel routing problem. The results are either qualitatively similar to or better than the best published results for channel routing benchmarks. The runtime of the algorithm (between one and fifty minutes) is not as competitive.

A GA for the channel routing problem, which includes problem-specific heuristics, is presented by [30]. The reported results are identical to those in [49], but with shorter runtime. An extension of this algorithm for the routing of channels with more than two layers (the so-called "multi-layer detailed routing problem") has also been published [31]. Instances of up to five layers and more than 60 nets are considered, with runtime ranging from several minutes up to 24 hours.

The GA by [50] for switchbox routing is similar to their channel router. The genotype is essentially a lattice corresponding to the coordinate points of the layout (as in the channel router). Crossover and mutation are performed in terms of interconnection segments. The algorithm assumes that the switchbox is expandable in both directions. Subsequently, these extensions are reduced with the goal of reaching the fixed size of the switchbox. While more costly in runtime, on numerous benchmark examples the GA produces solutions with equal or better routing characteristics than the previously best published results.

[48] presents a parallel GA for the channel and switchbox routing problem. The problem is based on the theory of punctuated equilibria from [18] and [11]. A GA with punctuated equilibria is a parallel GA in which independent subpopulations of individuals, each with their own fitness function, evolve in isolation. Periodic exchanges of individuals (migration) are then performed when a state of equilibrium throughout all the subpopulations has been reached (illustrated in Figure 9).

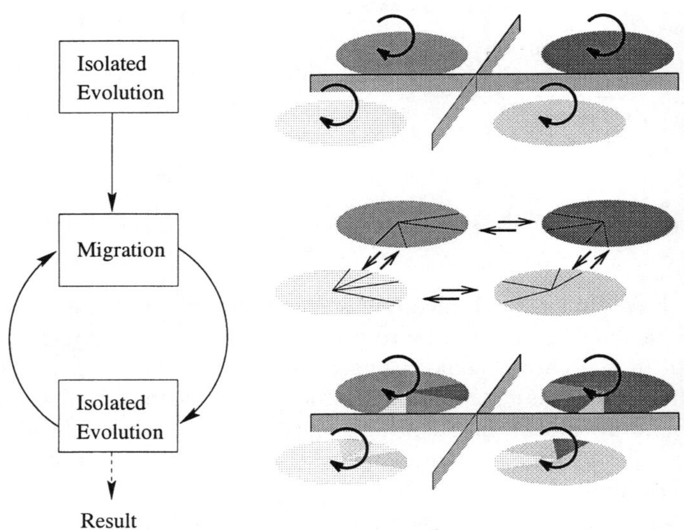

Fig. 9. Punctuated equilibria model with four subpopulations as shown in [Lienig(1997)]. Subpopulations evolve in isolation ("Isolated Evolution"), periodically interrupted by a limited exchange of individuals ("Migration").

A set of n individuals (problem solutions) is assigned to each of the N processors, for a total population size of $n \times N$. The set assigned to each processor is considered to be a subpopulation. The processors are connected by an interconnection network with a torus topology. Thus, each processor (subpopulation) has exactly four neighbors.

First, the main process creates an initial subpopulation at each processor. This initial subpopulation consists of randomly constructed (i.e., not optimized) routing solutions. They are designed by a random routing strategy which connects net points in an arbitrary order with randomly placed interconnections. The main process consists of a number of iterations, called epochs. During an epoch, each processor, disjointly and in parallel, executes the sequential GA on its subpopulation for a certain number of generations. Afterwards, each subpopulation exchanges a specific number of individuals (migrants) with its four neighbors. Here, the individuals themselves are exchanged, i.e., the migrants are removed from one subpopulation and added to another. Hence, the size of the subpopulations remains the same after migration and the assimilation of migrants is simply a fitness recalculation. The process continues with the separate evolution of each subpopulation during the next epoch. At the end of the process, the best individual seen during execution constitutes the final routing solution.

Experimental results showed that, when applied to the routing problem, the parallel GA – based on concepts of punctuated equilibria – consistently performs better than a sequential GA.

In investigating the algorithm's parameters, the following conclusions have been reached by [48]:

- A small number of migrants (1 to 3 per neighbor) combined with a "moderate" epoch length (approximately 5 to 10 percent of the total number of generations) leads heuristically to the best results.
- *Variable* epoch lengths determined via equilibrium measures within subpopulations achieve overall results that are slightly better than those obtained with (near-)optimized *fixed* epoch lengths. Practical applications of this "strict punctuated equilibria method" require the user to weigh the advantage of this "self-adjustment" against its main drawback, decreased time efficiency.
- Quality constraints on the migrants (e.g., to be above median fitness) do not improve the overall behavior of the algorithm. On the contrary, placing quality requirements on the selection of migrants led to premature stagnation.
- Given a sufficient number of individuals per subpopulation, a larger number of parallel evolving subpopulations will produce better routing results (for a fixed number of total evaluations). The size of the problem and the minimal subpopulation size have a direct correlation that must be taken into account when dividing a population into subpopulations.

Similar to placement, all published evolutionary algorithms for routing differ widely from "traditional" GAs. This is due to sophisticated representation schemes that often incorporate layout-specific constraints. Genetic operators, such as crossover and mutation, include the consideration of design rules in order to avoid unresolvable conflicts in the offspring(s). A significant improvement regarding solution quality, robustness, and runtime has been achieved by using parallel implementation strategies.

As with partition and placement, evolutionary-algorithm-based routing approaches have proved effective, but are generally not competitive with the state-of-the-art tools in the literature in terms of runtime and robustness. For discussions of such routing tools, see [75].

Compaction As mentioned earlier, compaction transforms the symbolic layout into a mask layout with the goal of minimizing the size of the resulting circuit representation. It usually consists of two separate steps: (1) transforming the symbolic layout to a mask layout, and (2) compacting the layout area. Both steps have to be performed under strict consideration of the design rules. The first step results from the use of symbols as representation of circuit components ("schematic-driven design"or SDC) that obviously cannot consider *all* geometrical features of these components. Generating the mask layout also includes detailed characteristics of the final fabrication technology.

The subsequent minimization of the layout area utilizes layout regions which allow a compaction in one (or two) direction(s) based on their occupation. (For example, an area with only horizontal routing segments without vias usually allows a vertical compaction.) Due to the used strategies, compaction algorithms are divided into one-dimensional algorithms (compaction in one direction at a time), two-dimensional algorithms (compaction in the x- and y-direction simultaneously), and topological algorithms (moving separate cells according to routing constraints).

To the best of our knowledge, the only applications of an evolutionary algorithm for compaction have been advanced by two papers: [27] and [34]. (In a related paper, [36] make use of similar principles to apply the technique of *simulated annealing* to the compaction problem.) In the first of these, [27] describes two prototypes of GAs that perform compaction of a symbolic circuit layout.

The algorithm considers symbolic placement and routing results which are characterized by their routing requirements and the technological constraints. These designs are encoded into individuals. The individuals undergo an evolutionary process with the selection based on both the number of design rule violations and the size of the allocated area.

Although Fourman's results are limited to very simple layout structures, he does propose a new problem-specific representation for layout design that includes constraints of the compaction process. Neither runtimes nor real VLSI applications are discussed.

In [34], the authors investigate a methodology known as *hierarchical chromosome representation* (HCR) and its application to several related problems – including compaction. By applying this new methodology to the compaction problem, it is argued that many of the constraints and limitations of the techniques from [27] can be overcome – but the discussion is only theoretical. No actual code or results are presented.

6 Conclusions

Since the problems encountered in the physical design of VLSI circuits are very complex, evolutionary algorithms have considerable potential in this field. Accordingly, there has been a significant increase in evolutionary algorithms that have been developed for VLSI physical design applications. This development is characterized by some specific features, such as:

- Successful evolutionary algorithms in VLSI design differ widely from the "traditional evolutionary algorithm concept" in that they incorporate problem-specific knowledge into their operators. Furthermore, universal fixed-length binary implementations are often replaced by problem-specific representation schemes.
- Combinations of evolutionary algorithms with other optimization strategies, e.g., simulated annealing or fast deterministic algorithms, are no longer an exception. Their incorporation often seems to be the only way to obtain an approach that offers a competitive runtime.
- Useful parallel implementations of evolutionary algorithms have emerged in VLSI design. The increasing availability of high-speed local computer networks and inexpensive parallel computers encourages the implementation of parallel evolutionary algorithms with improved solution quality and runtime characteristics.

While these trends look promising, we cannot ignore the fact that most evolutionary-based VLSI design algorithms are not applicable in real-world VLSI design systems. This condition is underlined by the fact that currently no commercially available EDA system contains any evolutionary-based algorithms.

What are the reasons for this obvious conflict between promises of an emerging solution technique and the problems of successful adoption? One investigation [47] using different software vendors showed two major drawbacks of evolutionary algorithms: robustness and failed expectations. Evolutionary algorithms are of probabilistic nature, hence *one* execution does not necessarily deliver the best possible result. This property is one of the reasons that many very good results claimed in publications could not be repeated in successive runs using practical applications – often only "fine tuning" of parameters would have lead to the expected superior results.

These are some of the reasons that are responsible for the gap between successful academic results with evolutionary-based VLSI physical design algorithms and the obvious neglect of these algorithms in commercial design tools. However, we believe that the above-mentioned problems can be avoided if the following guidelines are considered when developing evolutionary algorithms for the VLSI design problem:

- Since evolutionary algorithms are probabilistic approaches, it is very important to investigate the influence of the randomness on the solution quality. A VLSI designer will always hesitate to use an algorithm he/she has to run several times in order to achieve a usable and competitive result. Furthermore, the parameters used for running the evolutionary algorithm should be designed in a way to provide an easy adjustment to the specific design problem without extensive experimentation.
- The probabilistic nature of the algorithm should be discussed whenever solutions of evolutionary algorithms are presented. In other words, it is

necessary to clearly express how the presented results have been obtained. The randomness of the results also has to be taken into account when comparing the runtime of an evolutionary algorithm with deterministic algorithms that require only one execution to achieve their results.

- Whenever possible, the presented approach should be compared with state-of-the-art algorithms regarding both solution quality and runtime. Many of the surveyed evolutionary algorithms are competitive with respect to the solution quality only. However, to be of interest in the VLSI design area, a measure of running time must also be included. Less interesting to the community is the number of recombination, points visited in the search space, and so on.
- Constraints of other algorithms have to be taken into account when comparing with their results. For example, it is not a fair comparison when the routing quality of an evolutionary algorithm is expressed only in terms of the number of vias ([29]) and then compared with the results of other approaches that minimize the net length concurrently.
- Due to the large number of CAD tools in VLSI design, benchmark data are available for all major design steps, e.g., [17]. An evolutionary algorithm developed for VLSI design will not create any interest within the VLSI community unless its performance is tested with the appropriate benchmarks. It is only by examining the results of these benchmarks that we can compare a particular evolutionary algorithm with any other given approach. The test examples should be large, reflecting real-world VLSI design problems.

7 Summary

We have presented a survey of evolutionary algorithms that have been applied in VLSI physical design. We have also provided guidelines to make evolutionary algorithms more competitive among physical design algorithms for VLSI layout. We believe that such an approach is an important and necessary part of the continued growth of this field.

Acknowledgments

We would like to acknowledge support of the research received from the Virginia Aerospace Consortium, from the National Science Foundation under grants CDA 9634333, CCR 9224789, MIP 9107717, DUE 9554715, and DUE 9653413, from the Department of Computer Science at the University of Virginia, the German Science Foundation (DFG) and from Dresden University of Technology.

References

1. Abramovici, M., Breuer, M. A., Friedman, A. D. (1994) *Digital Systems Testing and Testable Design*, IEEE Press, Piscataway, NJ
2. Alpert, C. J., Khang, A. B. (1995) Recent Directions in Netlist Partitioning: A Survey, *Integration: The VLSI Journal*, **19**, 1-81
3. Bäck, T. (1996) *Evolutionary Algorithms in Theory and Practice*, Oxford University Press, New York
4. Becker, B., Drechsler, R. (1994) OFDD Based Minimization of Fixed Polarity Reed-Muller Expressions Using Hybrid Genetic Algorithms, *IEEE International Conference on Computer Design*, 106-110
5. Bentley, P. (1996) *Evolutionary Design by Computers*, Morgan Kaufmann Publishers, San Francisco, CA
6. Betz, V., Rose, J. (1997) VPR: A New Packing, Placement and Routing Tool for FPGA Research, *International Workshop on Field Programmable Logic and Applications*, 213-222
7. Bui, T. N., Moon, B. R. (1994) A Fast and Stable Hybrid Genetic Algorithm for the Ratio-Cut Partitioning Problem on Hypergraphs, *Proc. of the ACM-IEEE Design Automation Conference*, 664-669
8. Bushnell, M. L., Agrawal, V. D. (2000) *Essentials of Electronic Testing for Digital, Memory, and Mixed-Signal VLSI Circuits*, Kluwer Academic Publishers, Boston, MA
9. Chan, H., Mazumder, P., Shahookar, K. (1991) Macro-Cell and Module Placement by Genetic Adaptive Search with Bitmap-Represented Chromosome, *Integration, The VLSI Journal*, **12**, 49-77
10. Cohoon, J. P., Paris, W. D. (1987) Genetic Placement, *IEEE Trans. on Computer-Aided Design*, **6**, 956-964
11. Cohoon, J. P., Hegde, S. E., Martin, W. N., Richards, D. S. (1991) Distributed Genetic Algorithms for the Floorplan Design Problem, *Computer-Aided Design of Integrated Circuits and Systems*, **10**, 483-492
12. Cohoon, J. P., Martin, W. N., Richards, D. S. (1991) Genetic Algorithms and Punctuated Equilibria in VLSI Parallel Problem Solving from Nature, H. P. Schwefel and R. Männer, eds., *Lecture Notes in Computer Science*, 496 (Berlin: Springer Verlag), 134-144
13. Cong, J., Ding, Y. (1996) Combinational logic synthesis for LUT Based Field Programmable Gate Arrays, *ACM Transactions on the Design of Electronic Systems*, **1**, 145-204
14. Corno, F., Prinetto, P., Rebaudengo, M., Reorda, M. S. (1996) GATTO: a Genetic Algorithm for Automatic Test Pattern Generation for Large Synchronous Sequential Circuits, *IEEE Transactions on Computer-Aided Design*, **15**, 943-951
15. Davis, L. (1985) Applying Adaptive Algorithms to Epistatic Domains, *Proc. Int. Joint Conference on Artificial Intelligence*
16. Drechsler, R. (1998) *Evolutionary Algorithms for VLSI CAD*, Kluwer Academic Publishers, Boston, MA
17. *EDA Benchmarks 1997*, WWW: http://www.cbl. ncsu.edu/ or email: benchmarks@cbl.ncsu.edu
18. Eldredge, N., Gould, S. L. (1972) Punctuated equilibria: An Alternative to Phyletic Gradulism, *Models of Paleobiology*, Freeman, Cooper and Co., San Francisco, CA, 82-115

19. Esbensen, H. (1992) A Genetic Algorithm for Macro Cell Placement, *Proc. of the European Design Automation Conference*, 52-57
20. Esbensen, H. (1994) A Macro-Cell Global Router Based on Two Genetic Algorithms, *Proc. of the European Design Automation Conference*, 428-433
21. Esbensen, H., Mazumder, P. (1992) SAGA: A Unification of the Genetic Algorithm with Simulated Annealing and its Application to Macro-Cell Placement, *Proc. of the 7th International Conference on VLSI Design*, 211-214
22. Esbensen, H., Mazumder, P. (1994) Genetic Algorithm for Steiner Problems in a Graph, *European Design and Test Conference*, Paris, 402-406
23. Esbensen, H., Mazumder, P. (1999) Macro Cell Routing, *Genetic Algorithms for VLSI Design, Layout and Test Automation*, P. Mazumder and E. M. Rudnick, eds., Prentice-Hall, Upper Saddle River, NJ, 70-106
24. Fogel, D. (1995) *Evolutionary Computation: Toward a New Philosophy of Machine Intelligence*, IEEE Press, New York
25. Fogel, L. J., Owens, A. J., Walsh, M. J. (1966) *Artificial Intelligence through Simulated Evolutions*, Clearinghouse for Federal Scientific and Technical Information, Springfield, VA
26. Fonseca, M., Fleming, P. J. (1995) An Overview of Evolutionary Algorithms in Multiobjective Optimization, *Evolutionary Computation*, **3**, 1-16
27. Fourman, M. P. (1985) Compaction of Symbolic Layout using Genetic Algorithms, *Proc. of the First International Conference on Genetic Algorithms*, 141-153
28. Grefenstette, J., Schraudolph, N. (1987) *A User's Guide to GENESIS 1.2 UCSC*, CSE Dept., University of California, San Diego
29. Geraci, M., Orlando, P., Sorbello, F., Vasallo, G. (1991) A Genetic Algorithm for the Routing of VLSI Circuits, *Euro Asic '91*, Parigi, 27-31 Maggio, 218-223
30. Göckel, N., Pudelko, G., Drechsler, R., Becker, B. (1997) A Hybrid Genetic Algorithm for the Channel Routing Problem, *Proc. of the International Symposium on Circuits and Systems*, 675-678
31. Göckel, N., Pudelko, G., Drechsler, R., Becker, B. (1997) A Multi-Layer Detailed Routing Approach Based on Evolutionary Algorithms, *Proc. of IEEE International Conference on Evolutionary Computation*, 557-562
32. Goldberg, D. E. (1989) *Genetic Algorithms: Search, Optimization and Machine Learning*, Addison-Wesley, New York
33. Goldberg, D. E., Lingle, R. (1985) Alleles, Loci and the Traveling Salesman Problem, *Proc. International Conference on Genetic Algorithms*
34. Goodman, E., Tetelbaum, A. Y., Kureichik, V. (1994) *A Genetic Algorithm Approach to Compaction, Bin Packing, and Nesting Problems*, GARAGe Technical Report
35. Holland, J. H. (1975) *Adaptation in Natural and Artificial Systems*, University of Michigan Press, Ann Arbor, MI
36. Hsieh, T. M., Leong, H. W., Liu, C. L. (1988) Two-Dimensional Layout Compaction by Simulated Annealing, *Proc. IEEE International Symposium on Circuits and Systems*, Espoo, Finland, **3**, 2439-2443
37. Hulin, M. (1991) Analysis of Schema Distributions, *Proc. of the Fourth International Conference on Genetic Algorithms*, 204-209
38. Hulin, M. (1991) Circuit Partitioning with Genetic Algorithms Using a Coding Scheme to Preserve the Structure of a Circuit, *Parallel Problem Solving from Nature*, H. P. Schwefel and R. Männer, eds., Lecture Notes in Computer Science, 496 (Berlin: Springer Verlag), 75-79

39. Hwang, F. K., Winter, P., Richards, D. S. (1992) *The Steiner Tree Problem*, Elsevier Science, Amsterdam.
40. *IEEE International Testing Conference*, Atlantic City, NJ
41. Julstrom, D. A. (1993) A Genetic Algorithm for the Rectilinear Steiner Problem, *International Conference on Genetic Algorithms and their Applications*, 231-236
42. Kapsalis, A., Rayward-Smith, V. J., Smith, G. D. (1993) Solving the Graphical Steiner Tree Problem Using Genetic Algorithms, *Journal of the Operational Research Society*, **44**, 397-406
43. Karro, J., Cohoon, J. (1999) A Spiffy Tool for the Simultaneous Placement and Global Routing of Three-Dimensional Field Programmable Gate Arrays, *Ninth Great Lakes Symposium on VLSI*, Ann Arbor, MI, 226-227
44. Kirkpatrick, S., Gelatt, C. D., Vecci, M. P. (1983) Optimization by Simulated Annealing, *Science*, **220**, 45-54
45. Kling, R. M., Banerjee, P. (1989) ESP: Placement by Simulated Evolution, *IEEE Trans. on Computer-Aided Design* **8**, 245-256
46. Kling, R. M., Banerjee, P. (1990) Optimization by Simulated Evolution with Applications to Standard Cell Placement, *Proc. of the 27th ACM-IEEE Design Automation Conference*, 20-25
47. Lienig, J. (1996) *Evolutionary Algorithms Applied to VLSI Physical Design*, (in German), Fortschrittberichte VDI, Reihe, 20, VDI-Verlag, Duesseldorf
48. Lienig, J. (1997) A Parallel Genetic Algorithm for Performance-Driven VLSI Routing, *IEEE Trans. on Evolutionary Computation*, **1**, 29-39
49. Lienig, J., Thulasiraman, K. (1994) A Genetic Algorithm for Channel Routing in VLSI Circuits, *Evolutionary Computation*, **1**, 293-311
50. Lienig, J., Thulasiraman, K. (1996) GASBOR: A Genetic Algorithm for Switchbox Routing in Integrated Circuits, *Journal of Circuits, Systems, and Computers*, **6**, 359-373
51. Lin, Y.-L., Hsu, Y.-C., Tsai, F.-S. (1989) SILK: A Simulated Evolution Router, *IEEE Trans. on Computer-Aided Design*, **8**, 1108-1114
52. Mazumder, P., Rudnick, E. (1999) *Genetic Algorithms for VLSI Design, Layout & Test Automation*, Prentice Hall, Upper Saddle River, NJ
53. Mazumder, P., Shahookar, K. (1999) Partitioning in *Genetic Algorithms for VLSI Design, Layout and Test Automation*, P. Mazumder and E. M. Rudnick, eds., Prentice Hall, Upper Saddle River, NJ, 38-68
54. Mitchell, M. (1996) *An Introduction to Genetic Algorithms*, The MIT Press, Cambridge, MA
55. Mohan, S., Mazumder, P. (1993) Wolverines: Standard Cell Placement on a Network of Workstations, *IEEE Trans. on Computer-Aided Design*, **12**, 1312-1326
56. Ohmori, K., Kasai, T. (1997) Logic Synthesis using a Genetic Algorithm, *Internal Symposium on IC Technologies, Systems, and Applications*, 200-203
57. Oliver, I. M., Smith, D. J., Holland, J. R. C. (1985) A Study of Permutation Crossover Operators on the Traveling Salesman Problem, *Proc. International Conference on Genetic Algorithms*, 224-230
58. Ökmen, C., Keirn, M., Krieger, R., Becker, B. (1997) On Optimizing BIST-Architecture by Using OBDD-Based Approaches and Genetic Algorithms, *IEEE VLSI Test Symposium*, 426-431
59. Preas, B. T., Lorenzetti, M. J. (1988) *Physical Design Automation of VLSI Systems*, Benamin/Cumming Publishing, Menlo Park, CA

60. Rahmani, A. T., Ono, N. (1993)A Genetic Algorithm for Channel Routing Problem, *Proc. of the Fifth International Conference on Genetic Algorithms*, 494-498
61. Rao, B. B. P., Patnaik, L. M., Hansdah, R. C. (1994) A Genetic Algorithm for Channel Routing Using Inter-Cluster Mutation, *Proc. of the First IEEE International Conference on Evolutionary Computation*, 97-103
62. Rao, B. B. P., Patnaik, L. M., Hansdah, R. C. (1995) An Extended Evolutionary Programming Algorithm for VLSI Channel Routing, *Evolutionary Programming IV: Proc. of the 4th Annual Conference on Evolutionary Programming*, J. R. McDonnell, R. G. Reynolds, and D. B. Fogel, eds., MIT Press, Cambridge, MA, 521-544
63. Rebaudengo, M., Reorda, M. S. (1996) GALLO: A Genetic Algorithm for Floorplan Optimization, *IEEE Trans. on Computer-Aided Design*, **15**, 943-951.
64. Rechenberg, I. (1973) *Evolutionsstrategie: Optimierung technischer Systeme nach Prinzipien der biologischen Evolution*, Frommann-Holzberg, Stuttgart -Bad Cannstatt
65. Rudnick, E. M., Patel, J. H., Greenstein, G. S., Niermann, T. M. (1997) A Genetic Algorithm Framework for Test Generation, *IEEE Trans. on Computer-Aided Design of Integrated Systems*, **16**, 1034-1044
66. Saab, Y., Rao, V. (1989) An Evolution-Based Approach to Partitioning ASIC Systems, *Proc. of the ACM-IEEE Design Automation Conference*, 767-770
67. Saab, D. G., Saab, G. Y., Abraham, J. A. (1992) CRIS: a Test Cultivation Program for Sequential VLSI Circuits, *IEEE/ACM International Conference on Computer-Aided Design*, 216-219
68. Sarrafzadeh, M., Wong, C. (1996) *An Introduction to VLSI Physical Design*, McGraw Hill, New York
69. Schaffer, J. D. (1984) *Some Experiments in Machine Learning Using Vector Evaluated Genetic Algorithms*, Unpublished doctoral dissertation, Vanderbilt University, Nashville, TN
70. Schnecke, V., Vornberger, O. (1996) An Adaptive Parallel Genetic Algorithm for VLSI-Layout Optimization, *Parallel Problem Solving from Nature*, H. P. Schwefel and R. Männer, eds., Lecture Notes in Computer Science, 1141 (Berlin: Springer Verlag), 859-868
71. Sechen, C. (1988) *VLSI Placement and Global Routing Using Simulated Annealing*, Kluwer Academic Publishers, Boston, MA
72. Shahookar, K., Khamisani, W., Mazumder, P., Reddy, S. M. (1993) Genetic Beam Search for Gate Matrix Layout. *Proc. of the 6th International Conference on VLSI Design* , 208-213
73. Shahookar, K., Mazumder, P. (1990) GASP - A Genetic Algorithm for Standard Cell Placement, *Proc. of the European Design Automation Conference*, 660-664
74. Shahookar, K., Mazumder, P. (1990) A Genetic Approach to Standard Cell Placement using Meta-Genetic Parameter Optimization *IEEE Trans. on Computer-Aided Design*, **9**, 500-511
75. Sherwani, N. (1999) *Algorithms for VLSI Physical Design Automation, Third Edition*, Kluwer Academic Publishers, Boston, MA
76. Srinivas, M., Patnaik, L. M. (1993) A Simulation-Based Test Generation Scheme Using Genetic Algorithms, *International Conference on VLSI Design*, 132-135

77. Thomson, P., Miller, J. F. (1997) Comparison of AND-XOR Logic Synthesis Using a Genetic Algorithm Against MISII for Implementation on FPGAs, *IEE and IEEE International Conference on GA Applications in Engineering*
78. Varanelli, J. M., Cohoon, J. P. (1995) Population-Oriented Simulated Annealing: A Genetic/Thermodynamic Hybrid Approach to Optimization, *Proc. of the Sixth International Conference on Genetic Algorithms*, 174-181
79. Vemuri, R., Vemuri, R. (1991) Genetic Synthesis: Performance-Driven Logic Synthesis Using Genetic Evolution, *IEEE Great Lakes Symposium on VLSI Systems*, Kalamazoo, MI, 312-317
80. Villa, T., Kam, T., Brayton, R. K., Sangiovanni-Vincentelli, A. (1997) *Synthesis of Finite State Machines: Logic Optimization*, Kluwer Academic Publishers, Boston, MA
81. Yu, M. L. (1989) A Study of the Applicability of Hopfield Decision Neural Nets to VLSI CAD, *Proc. of the 26th ACM-IEEE Design Automation Conference*, 412-417
82. Zhang, C. X., Mlynski, D. A. (1990) VLSI Placement With a Neural Network Model, *Proc. of the International Symposium on Circuits and Systems*, 475-478

From Theory to Practice: An Evolutionary Algorithm for the Antenna Placement Problem

Jörg Zimmermann, Robin Höns, and Heinz Mühlenbein

RWCP Theoretical Foundation Lab
GMD - Center for Information Technology
D-53754 Sankt Augustin, Germany
E-mail: Joerg.Zimmermann@gmd.de

Summary. We give a short introduction to the results of our theoretical analysis of evolutionary algorithms. These results are used to design an algorithm for a large real-world problem: the placement of antennas for mobile radio networks. Our model for the antenna placement problem (APP) addresses cover, traffic demand, interference, different parameterized antenna types, and the geometrical structure of cells. The resulting optimization problem is constrained and multi-objective. The evolutionary algorithm derived from our theoretical analysis is capable of dealing with more than 700 candidate sites in the working area. The results show that the APP is tractable. The automatically generated designs enable experts to focus their efforts on the difficult parts of a network design problem.

1 Introduction

Simulating evolution as seen in nature has been identified as one of the key computing paradigms for the next decade. Today evolutionary algorithms have been successfully used in a number of applications. These include discrete and continuous optimization problems, synthesis of neural networks, synthesis of computer programs from examples (also called genetic programming) and even evolvable hardware. But in all application areas problems have been encountered where evolutionary algorithms performed badly. Therefore a mathematical theory of evolutionary algorithms is urgently needed. Theoretical research has evolved from two opposed ends: theories emerging from a mathematical approach are getting closer to practice and from the applied side ad hoc theories have arisen that often lack deeper theoretical justification.

In this chapter we shortly discuss results of our theoretical analysis of evolutionary algorithms. The analysis helped us to design an algorithm for a large real-world optimization problem: the placement of antennas for mobile radio networks. The outline of the chapter is as follows. First we summarize some of the theoretical results on evolutionary algorithms. In Section 3, we introduce a mathematical model of the antenna placement problem (APP). Then the evolutionary algorithm to solve the APP is presented. In Section 5, we discuss the APP model for expanding an existing network. The chapter closes with results from real-world benchmarks.

2 Theoretical Results

There exist many variants of evolutionary algorithms. It is a difficult decision to decide which algorithm to use in a specific application. We discuss the problem here with a well-known example, the Royal Road function R_1. This function was used by Mitchell et al. [10]. It is defined as follows:

$$R_1(l, \mathbf{x}) = \sum_{i=0}^{l-1} \prod_{j=1}^{8} x_{8i+j} \tag{1}$$

Here $\mathbf{x}$ is a binary string of length $8l$. The function is of order 8. This means the largest product consists of eight variables. The fitness value is increased by 1 if all bits of a product are on. The results of different evolutionary algorithms are displayed in Table 1. The algorithms can be described as follows:

- **(1+1):** This algorithm performs stochastic hill climbing. The current string is randomly changed (mutated). If the fitness of the mutated string is greater than or equal to the fitness of the current string, the mutated string replaces the current string.
- **SGA:** This is the standard genetic algorithm with proportionate selection [7].
- **UMDA:** This algorithm uses univariate marginal distributions instead of recombination and crossover for generating search points. It approximates genetic algorithms [14].
- **FDA:** This algorithm uses a search distribution factorized in conditional and marginal distributions [17,15].

The (1+1) algorithm uses only mutation and selection. All other algorithms have recombination (or an extension of recombination) as the main operator. The (1+1) algorithm is a discrete version of a *(1+1) Evolution Strategy (ES)*. This is an evolution scheme where selection takes place between one parent solution and its offspring solution [22]. If the fitness of the offspring is better than or equal to the fitness of the parent, then the offspring replaces the parent. The good performance of this algorithm has already been shown in [12,13]. The other algorithms use recombination of strings and are generalizations of this method. The search power of recombination was previously derived from a *Building Block Hypothesis (BBH)*. It states that "the GA works well when instances of low-order, short schemas that confer high fitness can be recombined to form instances of larger schemas that confer even higher fitness" [7].

In order to investigate the BBH conjecture we have developed a theory of evolutionary algorithms based on distributions. In mathematical terms a schema defines nothing more than a marginal distribution. Thus a first-order schema defines a univariate marginal distribution. We have shown that only the first half of the BBH is correct: first-order schemata of high fitness are recombined. Larger schemata play no role. The derivation of this result is precisely stated in [14,17].

In this chapter we use the simulations of the Royal Road function to illustrate the results of our theory. Table 1 displays the mean number of function evaluations for Royal Road(8). Some of the results have been previously obtained in [10]. There it was noted that the empirical results are contradictory to the BBH hypothesis.

For the Royal Road function the (1+1) algorithm is the clear winner. The really bad performance of SGA is mainly a result of proportionate selection. UMDA with proportionate selection (UMDA p) needs slightly fewer evaluations than SGA. With very strong selection ($\tau = 0.05$), UMDA needs only about twice as many function evaluations as the (1+1) algorithm. Almost identical performance to the (1+1) algorithm can be obtained by FDA. It uses marginal distributions of size 8 instead of univariate marginal distributions.

1+1	SGA	UMDA p	UMDA $\tau = 0.3$	UMDA $\tau = 0.05$	FDA $\tau = 0.3$
6334	61334	55586	28000	14264	7634

Table 1. Mean function evaluations for Royal Road(8) (100 runs)

A more detailed theoretical analysis of the different algorithms can be found in [16]. There, an extension of the FDA algorithm is discussed as well. FDA needs the dependence structure of the factorization as input. Using a Bayesian network it is possible to compute a good search distribution from the selected data points. This method is used by LFDA (LearningFDA).

The computational complexity of the heuristics is as follows: (1+1) algorithm has the lowest, followed by SGA and UMDA. FDA and LFDA have a complexity which depends on the structure of the factorization, e.g. a tree-like dependence structure allows efficient algorithms.

We do not want to give the impression that the (1+1) algorithm solves all problems. In fact, our recommendation for unconstrained optimization problems is as follows: Start with the simplest algorithm. If it gives good results, then stop. If the results have to be improved, use the next heuristic in the sequence.

Another design issue is constraints. Almost all constraints can be incorporated into the (1+1) algorithm, whereas genetic algorithms have more difficulties. There exist constraints which are respected by recombination, but in general recombination will create infeasible solutions. At this time there exists no general technique for dealing with constraints in genetic algorithms. Constraints seem to be easier to incorporate into algorithms using distributions. A first step has already been done in [17].

A further difficulty of our application is that it is a multi-objective optimization problem. An overview of evolutionary algorithms in the context of multi-objective optimization is given in [5]. The above recommendations have been used for a real-world application, which is discussed next.

3 The Antenna Placement Problem

Engineering of current mobile radio networks consists of several tasks: traffic estimation, radio antenna positioning, broadcast control, frequency assignment, etc. Within the selected geographic area where the radio network will be installed or extended, the operator decides the number of radio transmitters to be installed and the number of frequencies to be assigned to the area. These parameters are then used for the placement of antennas and the frequency assignment.

The purpose of the *Antenna Placement Problem (APP)* is to optimize the radio coverage of an area. The main objective of the *Frequency Assignment Problem (FAP)* is to minimize electromagnetic interference due to multiple use of frequencies in different parts of the network. Whereas the FAP has been intensively investigated [19,20,2,3], little has been done on the APP [9]. One reason is that the APP is much more difficult to model, and even a simple model – treating only coverage of the working area – can be shown to be NP-complete [4]. Furthermore the APP is a multi-objective and constrained problem with all its known difficulties [6,8].

The APP has been investigated within the EU project *ARNO (Algorithms for Radio Network Optimization, IT Project 23243)*. The mathematical model used within ARNO consists of points defined on a grid, e.g. service test points, traffic test points and candidate sites. Radio transmission is modelled by a propagation loss matrix [21]. The objectives are to minimize cost, to minimize the interference level and to have geometrically “nice” cells. These objectives are components of a cost function which guides the evolutionary search.

3.1 The APP Model

One of the most difficult tasks in solving complex real-world problems is to develop an appropriate model. The model for the APP used in this chapter is the result of several cycles of design, evaluation and redesign. The quality of solutions of the APP is assessed by two constraints, which deal with cover and traffic demand, and three objectives, which address economical and technical aspects. In order to define our APP model we introduce the following concepts:

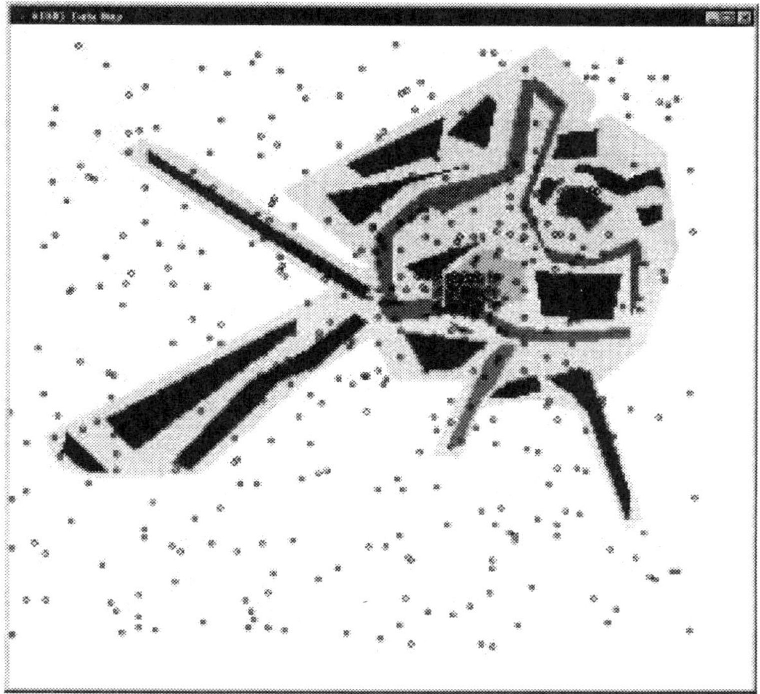

Fig. 1. Map of working area: antennas can be placed at candidate sites, displayed as black circles. Radio coverage must be provided in the gray shaded areas. Darker shades of gray correspond to higher traffic demand.

Input Data

- A set $\mathcal{R}$ of *Reception Test Points (RTPs)*, given by coordinates (x,y).
- A set $\mathcal{S}$ of *Service Test Points (STPs)*. A *signal quality threshold* S_q is assigned to each STP, usually -90 dBm. [1]
- A set $\mathcal{T}$ of *Traffic Test Points (TTPs)*. Each TTP carries *traffic demand*, measured in Erlang. [2]
- A set $\mathcal{L}$ of coordinates of *Candidate Sites*.

Note that $\mathcal{T} \subset \mathcal{S} \subset \mathcal{R}$. Usually, the RTPs form a rectangular grid. The position of a candidate site does not have to coincide with an RTP. Figure 1 shows an example of a data map. Candidate sites are displayed as black circles, RTPs are white, STPs light gray and TTPs are colored in darker shades of gray, according to increasing traffic demand.

[1] dBm = decibel milliwatt. This is a logarithmic power unit [23].

[2] This measure arises in queueing theory [1].

- A *Propagation Loss Matrix* PLM_i for each candidate site L_i defines the signal losses (in dB) from site L_i to all RTPs.
- An *Angle of Incidence Matrix* AIM_i for each candidate site L_i defines the vertical angles at which the RTPs appear to site L_i.

Antenna Types

- omnidirectional antenna (OD):
  Parameters: power, ranging from 26 dBm to 55 dBm in steps of 1 dB.
- large directive antenna (LD):
  Parameters: power, ranging from 26 dBm to 55 dBm; azimuth, varying in steps of 1°; tilt, ranging from 0° to −15° in steps of 1°.
- small directive antenna (SD):
  Parameters: same as for large directive antenna.

All antenna types can handle traffic of up to 43 Erlang. This is the maximal traffic load of an antenna allowed by the GSM model (Global System for Mobile communications, an international standard for mobile communication systems defined in 1991) [11]. Associated with an antenna type are *antenna diagrams*: one vertical diagram (VDIAG) for omnidirectional antennas and one vertical and one horizontal diagram (HDIAG) for the directive ones. These diagrams depict the *signal loss* (in dB), depending on the radial deviation from the signal main axis, measured in degrees. The diagrams of the omnidirectional and small directive antenna are displayed in Figure 2. The diagrams of the large directive antenna are similar to the diagrams of the small directive antenna. Furthermore each antenna type has a specific *antenna gain (G)* and a specific *antenna loss (A)*, measured in dB. The gain and loss of the different antenna types are displayed in Table 2.

Antenna Type	G	A
OD	11.15	7.00
LD	15.65	7.00
SD	17.15	7.00

Table 2. Antenna gain and antenna loss

According to the GSM model a site may carry one omnidirectional antenna or up to three directive antennas [11]. So, let AT_{ij} denote the jth antenna at site L_i. Accordingly, $Power_{ij}$, G_{ij}, etc., denote the respective parameters of AT_{ij}, and (u_i, v_i) are the coordinates of site L_i. The field strength $F_{ij}(x,y)$ of antenna AT_{ij} at reception test point (x,y) is defined as

$$F_{ij}(x,y) = Power_{ij} + G_{ij} - A_{ij} - PLM_i[x,y] \\ -VDIAG[\alpha - \beta] - HDIAG[\gamma - \delta]$$

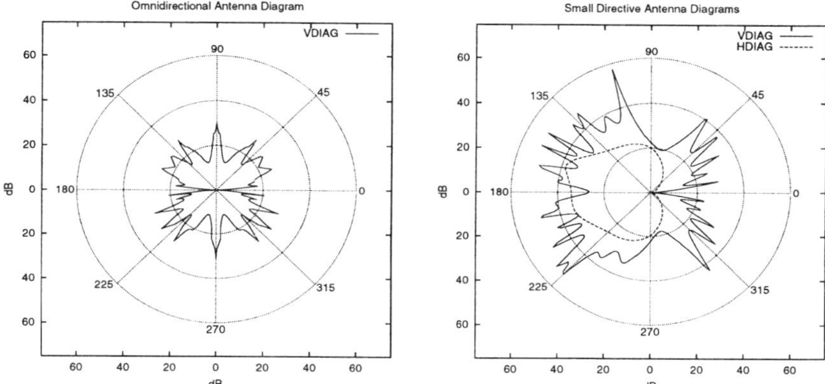

Fig. 2. Antenna diagrams: the diagram on the left shows the signal loss of the omnidirectional antenna, depending on the vertical deviation from the signal main axis, the diagram on the right shows the signal loss of the small directive antenna, depending on the vertical deviation from the signal main axis (VDIAG), and depending on the horizontal deviation from the signal main axis (HDIAG).

where $\alpha = AIM_i[x, y]$, $\beta = Tilt_{ij}$, $\gamma = \frac{180^\circ}{\pi} \cdot \text{atan2}(y - v_i, x - u_i)$ and $\delta = Azimuth_{ij}$. VDIAG and HDIAG are the diagrams associated with the antenna type of AT_{ij}. If AT_{ij} is omnidirectional, we neglect the HDIAG-term and set $Tilt_{ij}$ to zero. $\text{atan2}(y, x)$ is similar to $\arctan(y/x)$, except that the signs of both arguments are used to determine the quadrant of the result. The arguments of VDIAG and HDIAG specify where to look up the respective antenna diagrams. Figure 3 illustrates the antenna parameters tilt and azimuth. Tilt and azimuth can be manipulated by the optimization process, whereas α and γ are determined by the input data.

A *configured antenna* is a pair defining an antenna type and a complete list of parameter values for this antenna type, e.g. (LD, [Power = 50 dBm, Azimuth = 90°, Tilt = −5°]). A *configured site* is a site carrying at least one configured antenna. As noted above, a configured site can carry either one omnidirectional antenna or up to three directive antennas. A *solution* for an APP is a list of configured sites. The *cell* C_{ij} of an antenna AT_{ij} is the set of STPs $\mathbf{x}$ receiving the best signal (i.e. strongest field strength) from AT_{ij}, provided that the signal is above the signal quality threshold of $\mathbf{x}$. The *traffic load* of an antenna AT_{ij} is the sum of the traffic demand of all TTPs contained in the cell C_{ij} of AT_{ij}. An antenna with a traffic load above 43 Erlang is called *overloaded*. The percentage of handled traffic demand of a solution is called *traffic hold*:

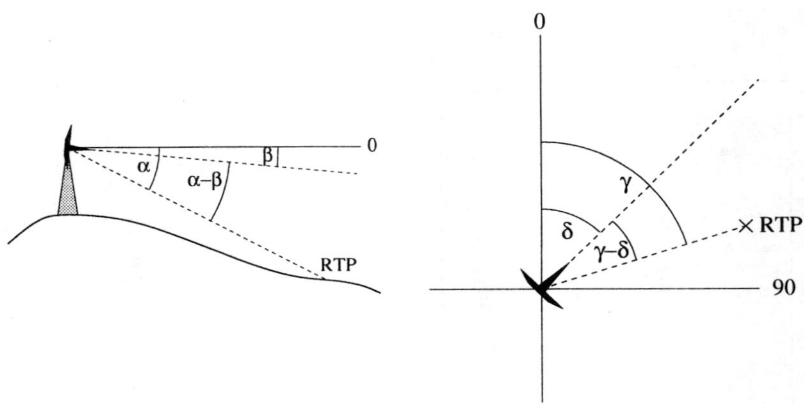

Fig. 3. Tilt and azimuth: the left figure illustrates the tilt β of a directive antenna and the angle of incidence α of an arbitrary RTP, the right figure illustrates the azimuth δ of a directive antenna and the angle of location γ of an arbitrary RTP.

$$\mathit{TrafficHold} = \frac{100\%}{\mathit{TotalTraffic}} \sum_{AT \in \mathcal{AT}} \min(\mathit{TrafficLoad}(AT), 43 \text{ Erlang})$$

where *TotalTraffic* is the summed traffic demand of all TTPs and $\mathcal{AT}$ is the set of all antennas. The percentage of lost traffic demand is called *traffic loss*, i.e. $\mathit{TrafficLoss} = 100\% - \mathit{TrafficHold}$.

3.2 Constraints and Objectives

A *feasible* solution has to fulfill the following two constraints:

- **Traffic Constraint (TC):** The traffic load of an antenna is $\leq$ 43 Erlang.
- **Cover Constraint (CC):** Every STP receives at least one signal above its signal quality threshold.

The constraints imply that all TTPs of a feasible solution are contained in non-overloaded cells. Thus the traffic hold of a feasible solution is 100%.

A precise mathematical definition of one or more objectives – which engineers intuitively use – is very difficult. For the design and evaluation of solutions it is highly desirable to have only one objective. But it is often not possible to measure solution quality directly by a single objective. Therefore, we first introduce three "quality-dimensions" as objectives to evaluate solutions. For the evolutionary algorithm these objectives will be combined into one cost function (see Section 4). The minimization objectives are:

- **Site Cost:**

$$SC = \text{number of used sites}$$

- **Interference Level:**

$$IL = \frac{1}{|\mathcal{S}|} \sum_{\mathbf{x} \in \mathcal{S}} \sum_{F \in \mathcal{F}_{\mathbf{x}} \setminus \mathcal{H}_{\mathbf{x}}} \max(F - S_m, 0)$$

where $\mathcal{F}_{\mathbf{x}}$ is the set of field strengths of all antennas at STP $\mathbf{x}$. $\mathcal{H}_{\mathbf{x}}$ is the *handover set*, consisting of the four strongest signals at STP $\mathbf{x}$. S_m is the receiver sensibility threshold of a mobile station, usually -99 dBm.

- **Cell Shape Factor:**

$$SF = \frac{1}{|\, Cells \,|} \sum_{C \in Cells} \frac{boundary(C)}{\sqrt{area(C)}}$$

where $boundary(C)$ is the number of boundary points of cell C, i.e. the number of $\mathbf{x} \in C$ having an $\mathbf{x}' \in \mathcal{S} \setminus C$ in their 8-neighborhood, and $area(C)$ is the number of points in C.

Site cost addresses the economical, *interference level* and *cell shape factor* address technical aspects of the APP. The intention of the interference objective is to make the FAP as simple as possible. The shape factor objective prefers geometrically well-formed cells, which is highly desirable for several reasons, e.g. minimization of drop-out probability. Within the ARNO project, the objective cell shape factor was not used. Instead there was a "connectivity constraint" which ensures that all cells are topologically connected. But it turned out that this constraint has several disadvantages:

- it is very difficult to satisfy,
- it is sensitive to small changes,
- connected cells can still be very irregular.

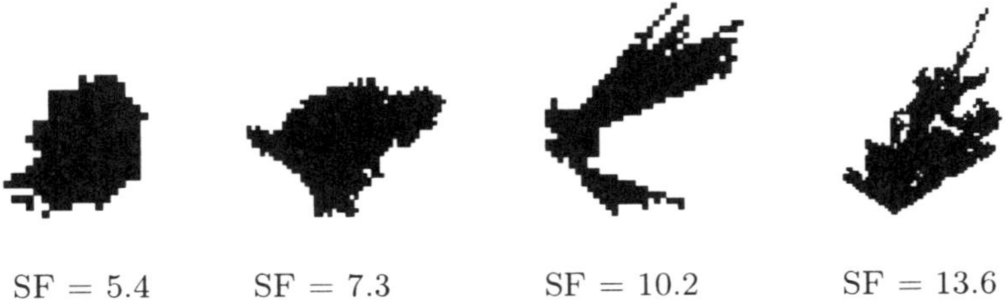

Fig. 4. Examples of cells and their respective shape factors. Less compact cells correspond to higher shape factors.

For these reasons we have replaced the connectivity constraint by the shape factor objective, which avoids most of the above problems. This new objective is inspired by a ratio widely used in physics: the ratio between the square root of the surface and the cube root of the volume of a body. This ratio is called *shape factor*. It is scale invariant and reaches its minimum value for a ball. In two-dimensional (2D) Euclidean geometry the shape factor is analogously defined as the ratio between the length of the boundary and the square root of the area of a planar figure. This 2D shape factor is minimal for a circle: $2\sqrt{\pi} \approx 3.54$. A hexagon has shape factor ≈ 3.72. Since in the APP we have to deal with a *discrete* geometry the shape factor values are slightly different and the discrete shape factor does not have all the nice properties of the Euclidean shape factor (e.g. scale invariance). But in general it favors compact cells over less compact ones, which is all that is needed in our case. If the diameter of a discrete circle approaches infinity, its discrete shape factor converges to $8/\sqrt{\pi} \approx 4.51$. Figure 4 gives examples of geometrical structures corresponding to different shape factors.

4 The Evolutionary Algorithm

We have used the theoretical results presented in Section 2 to design an evolutionary algorithm for the APP. Let us recall the major conclusions. First, start with the simplest algorithm. Second, for optimization problems with constraints the (1+1) algorithm has advantages. Therefore we have implemented an extended version of the basic (1+1) algorithm. The algorithm consists of three phases:

- Initialization Phase
- Repair Phase
- Optimization Phase

In order to deal with constraints and objectives we use two types of cost functions, a *Hard Cost Function* and a *Soft Cost Function*. The hard cost of a solution is given by

$$HardCost = \phi_1 \cdot \mathit{TrafficLoss} \; + \; \phi_2 \cdot \mathit{CoverLoss}$$

where ϕ_1 and ϕ_2 are positive weight factors representing the relative importance of the single terms. *CoverLoss* denotes the percentage of uncovered STPs. *TrafficLoss* and *CoverLoss* are called *penalty terms*. Note that a solution is feasible if and only if its hard cost is zero. The soft cost is given by a linear combination of the three objectives:

$$SoftCost = \psi_1 \cdot SC \; + \; \psi_2 \cdot IL \; + \; \psi_3 \cdot SF$$

The basic structure of our evolutionary optimization algorithm ENCON (Evolutionary Network CONfigurator) is described in Figure 5.

1: Initialization Phase

INITIALIZE solution (guided by heuristic rules)

2: Repair Phase (repairs violated constraints)

```
REPEAT
    SELECT repair operator
    APPLY selected repair operator on solution
    IF HardCost(offspring) ≤ HardCost(parent)
        ACCEPT offspring
UNTIL solution is feasible OR stop condition
```

3: Optimization Phase (optimizes feasible solution)

```
REPEAT
    SELECT optimization operator
    APPLY selected optimization operator on solution
    APPLY local repair on solution
    IF SoftCost(offspring) ≤ SoftCost(parent)
        ACCEPT offspring
UNTIL stop condition
```

Fig. 5. Basic structure of ENCON

4.1 Initial Solutions

In order to get a good *initial solution* ENCON exploits the *local structure* around a site. For this purpose the following concepts are introduced:

1 *traffic demand density in region A:* sum of traffic demand in A divided by the area of A.
2 *candidate site density in region A:* number of candidate sites in A divided by the area of A.

Depending on the densities in the neighborhood of a site a *placement probability* for this site is computed. Next, an initial configuration of antennas is placed at this site according to the computed probability. Repeating this procedure for all sites generates an initial solution.

The formula for the placement probability reflects the local structure around a site by balancing the local traffic demand density and the local candidate site density. The formula is split into two parts because we have to deal with the cover constraint as well. Usually the traffic demand restricts the size of a cell. But if the demand density drops below a critical value (ρ^*_{TD}), the cell can become so large (with regard to the traffic constraint only) that the field strength does not reach the signal quality threshold at the cell periphery. Hence in this case the placement probability should depend on the *maximal cell size* of a site and not on the local traffic demand. For a given candidate site L_i let

C_{max} = maximal traffic capacity $[C_{max}]$ = Erlang,

A_{max} = maximal cell area $[A_{max}] = \mathrm{m}^2$,

ρ_{TD} = local traffic demand density $[\rho_{TD}]$ = Erlang $\cdot\ \mathrm{m}^{-2}$,

ρ_{CS} = local candidate site density $[\rho_{CS}] = \mathrm{m}^{-2}$.

These are the parameters we need in order to compute the placement probability. C_{max}, the maximal traffic capacity of site L_i, depends on the *placement policy*. The placement policy defines the configuration that will be placed initially at a selected site. If an omnidirectional antenna is placed, then C_{max} is 43 Erlang. If the policy requires two or three directive antennas to be placed, C_{max} takes the values 86 and 129 Erlang, respectively. A_{max}, the maximal cell size possible at site L_i, is defined as the area where the field strength of an omnidirectional antenna with maximal power – placed at site L_i – is above the signal quality threshold. Let Δ_{mesh} denote the width of a square mesh in the RTP grid. Then A_{max} is computed by the following steps:

1. Place an omnidirectional antenna at site L_i with maximal power.
2. Let N be the number of STPs receiving a good signal from the placed antenna.
3. $A_{max} = N \cdot \Delta^2_{mesh}$.

For the estimation of the parameters ρ_{TD} and ρ_{CS} we assume that the distributions of candidate sites and traffic demand are not too irregular. Thus we can use the ideal situation of a network of hexagonal cells (see Figure 6) as orientation to derive our density estimators. There are a lot of other reasonable estimators, especially if the "not too irregular" assumption fails. However, note that the initial placement probabilities need to be only approximately "correct", because we are interested in an *initial* solution and not in a final one. With regard to Figure 6 we use the following estimation procedures for the densities ρ_{TD} and ρ_{CS} in the neighborhood of a candidate site L_i:

Local Candidate Site Density

1. Compute the distances from site L_i to its six nearest candidate sites: $d_1, \ldots, d_6$

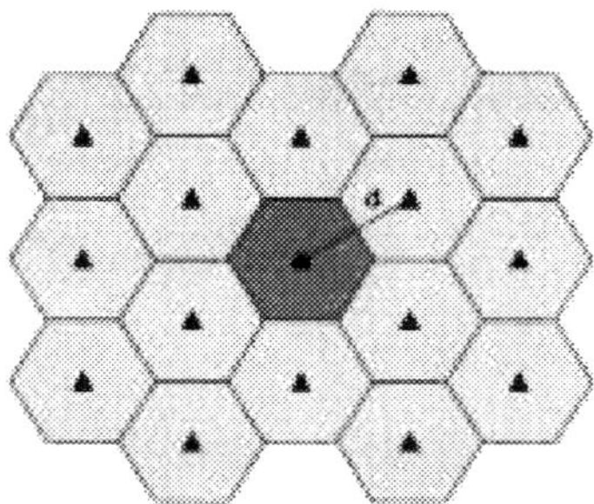

Fig. 6. Hexagonal network: the "ideal" cellular network. The triangles depict sites carrying an (omnidirectional) antenna; d is the distance of neighboring sites.

2. $d = \frac{1}{6} \sum_{i=1}^{6} d_i$

3. $A_{hexagon} = \frac{\sqrt{3}}{2} \cdot d^2$ (area of hexagon)

4. $\rho_{CS} = \frac{1}{A_{hexagon}}$

The following estimator of traffic demand density ρ_{TD} uses the notion of the *bounding box* of a set $A \subset \mathbb{R}^2$. It is a rectangle

- with sides parallel to the coordinate axes,
- containing A,
- with minimal area.

We use the bounding box, because in our context not only the accuracy, but also the computational efficiency of an estimator are important.

Local Traffic Demand Density

1. box = bounding box of the six nearest candidate sites of site L_i

2. $e_{box} = \sum_{t \in \mathcal{T} \cap box} e(t)$ ($e(t)$ = traffic demand at TTP t)

3. A_{box} = area of *box*

4. $\rho_{TD} = \frac{e_{box}}{A_{box}}$

The formula for the placement probability p_i of site L_i is given by

$$p_i = \begin{cases} \min\left(\frac{1}{C_{max}} \cdot \frac{\rho_{TD}}{\rho_{CS}},\ 1\right),\ \rho_{TD} \geq \rho^*_{TD} & \text{(critical constraint: traffic)} \\ \min\left(\frac{1}{A_{max}} \cdot \frac{1}{\rho_{CS}},\ 1\right),\ \text{otherwise} & \text{(critical constraint: cover)} \end{cases}$$

where $\rho^*_{TD} = C_{max}/A_{max}$. This approach leads to good *initial* solutions if the problem structure is not too irregular, i.e. strongly varying densities.

4.2 Repair Phase

After finishing the initialization phase, ENCON analyzes the initial solution: if it violates a constraint, the algorithm enters the *repair phase*. The goal of the repair phase is to transform the initial solution into a feasible solution. The transformation is carried out by a number of *repair operators*. The application of a repair operator or optimization operator (see next subsection) is called a *step*. The implemented repair operators in ENCON are:

- **RepairTraffic**: Select randomly an overloaded antenna AT*. Reduce its power by 1 dB. If the antenna is still overloaded, choose randomly one of the six nearest sites. If the chosen site is empty, randomly place there a large or a small directive antenna. Set its tilt to $-3°$, direct its azimuth towards AT* and increase its power – starting with minimal power – by 1 dB steps until AT* is no longer overloaded, maximal power is reached or the new antenna itself gets overloaded. In the latter case reduce power by 1 dB. If the chosen site carries an omnidirectional antenna, increase its power by 1 dB steps until AT* is no longer overloaded, maximal power is reached or the omnidirectional antenna gets overloaded. In the latter case reduce power by 1 dB. If the chosen site carries directional antennas, choose the one with azimuth closest to the direction of AT*. Direct its azimuth towards AT* and increase its power by 1 dB steps until AT* is no longer overloaded, maximal power is reached or the new antenna itself gets overloaded. In the latter case reduce power by 1 dB.
- **RepairHole**: Select randomly an uncovered STP. Detect the nearest *used* site. If this site carries an omnidirectional antenna, compute the minimal increase in power to cover the selected STP (this can be done using the field strength formula in Section 3.1). If the power of the omnidirectional

antenna can be increased by this amount, do it. If the site carries directive antennas, choose the one with azimuth closest to the direction of the selected STP. Try to increase its power by the same procedure as used for the omnidirectional antenna. If the selected STP is still uncovered, choose the nearest *unused* site. Place randomly an omnidirectional or directive antenna. If an omnidirectional antenna is placed, compute the minimal power P_{cov} to cover the selected STP. Set the power of the omnidirectional antenna to $\min(P_{cov}, 55 \text{ dBm})$. If a directive antenna is placed, set its tilt to 0°, direct its azimuth towards the uncovered STP and set its power by the same procedure as used for the omnidirectional antenna.

- **IncreaseCover**: Sample STPs randomly until an uncovered STP $\mathbf{x}^*$ is found, but at most sample 1000 STPs. If no such STP is found, look for an uncovered STP $\mathbf{x}^*$ in the list of all STPs. If there is no uncovered STP, abort, else identify the antenna AT* which has the strongest field strength F^* in $\mathbf{x}^*$. Compute the difference $\Delta F = S_q - F^*$, where S_q is the signal quality threshold of STP $\mathbf{x}^*$. Set the power of antenna AT* to $\min(P^* + \Delta F, 55 \text{ dBm})$, where P^* is the original power of AT*.
- **DecreasePower**: Select randomly an overloaded antenna AT*. Decrease the power of AT* by 1 dB steps until the overload vanishes or the minimal power is reached.
- **IncreasePower**: Select randomly an antenna AT* with traffic load less than 35 Erlang. Increase the power by 1 dB steps until the maximal power (55 dBm) is reached or AT* gets overloaded. In the latter case reduce power by 1 dB.
- **ChangeAzimuth**: Change the azimuth of a randomly selected directive antenna by a random value.
- **ChangeTilt**: Change the tilt of a randomly selected directive antenna by a random value.
- **DissipateTraffic**: Try to reduce traffic over- and underload of all antennas simultaneously by applying the following *dissipation algorithm*:

  Run through the list of all antennas. If an antenna with power > 26 dBm is overloaded, decrease its power by 1 dB. Repeat until a run through the antenna list results in no change.

  In many cases the effect of this operator is that peaks of traffic load will dissipate over the whole network until there is no longer traffic overload. This operator has proved to be very successful in eliminating traffic overload.

A *selection operator* chooses a repair operator from the above list, which is then applied to the current solution. In our current implementation each repair operator is chosen with the same probability. An investigation of other selection operators seems to be promising, especially conditionalization on the current solution, but this is the topic of future research. The transformed solution – also called the offspring solution – is only accepted if its hard cost is less than or equal to the hard cost of the parent solution. This procedure will be repeated until a feasible solution results – then the design process enters the *optimization phase* – or the *termination criterion* (see 4.4) is reached. In the latter case the design process will be aborted.

4.3 Optimization Phase

The optimization phase of ENCON has basically the same structure as the repair phase. Operators on solutions are now called *optimization operator* instead of repair operator. Differences from the repair phase are:

1 A soft cost function instead of a hard cost function guides the search.
2 If an optimization operator destroys the feasibility of a solution, then ENCON tries to repair this new candidate solution in order to maintain feasibility. This is done by entering the repair phase again, but with a stronger termination criterion, e.g. by defining a small upper bound (say 50) on the number of applications of repair operators (see Section 4.4). This sub-phase is called *local repair*. If feasibility cannot be restored, the new candidate solution will be discarded.

The following optimization operators are used for the optimization phase:

- **IncreaseCompactness**: Sample 10 antennas randomly. Select the antenna which generates the cell with the greatest shape factor. Reduce its power by 1 dB or, in case of a directive antenna, increase its tilt by 1°.
- **ReduceIrregularities**: Sample STPs randomly until an STP $\mathbf{x}^*$ with $I_7(\mathbf{x}^*) \geq 0.3$ (see below) is found, but at most sample 5000 STPs. If no such STP is found, abort, else choose randomly one of the six nearest sites to $\mathbf{x}^*$ allowing the placement of a new directive antenna (if no such site is found, abort). Place at this site a large directive antenna, set its power to 55 dBm, direct its azimuth towards $\mathbf{x}^*$, and set its tilt to $-2°$. This operator is designed to reduce regions with irregular geometrical and topological cell structures.
- **RemoveSite**: Select randomly a used site. Remove all antennas from this site.
- **RevertSite**: (Only for expansion design problems, see Section 5.) Select randomly a site. Revert the configuration of this site to its configuration in the legacy solution.
- **ChangeAzimuth**: Same operator as in repair phase.
- **ChangeTilt**: Same operator as in repair phase.

A major problem in the optimization phase is the construction of networks with a reasonable cell topology and geometry. Due to irregular path loss matrices (reflecting irregular geographical structures) many cells tend to have an irregular shape or even get disconnected. This often leads to regions containing many "fractally formed" parts of cells. In order to identify such "irregularity hot-spots" we introduce a measure that indicates the average geometrical structure in the neighborhood of a given STP. This measure is used in the *ReduceIrregularities* operator. The irregularity $I_r(\mathbf{x})$ in a square-shaped neighborhood of an STP $\mathbf{x}$ is defined as

$$I_r(\mathbf{x}) = \frac{1}{(2r+1)^2} \sum_{||\mathbf{y}-\mathbf{x}||_\infty \leq r} \tfrac{1}{8} S(\mathbf{y}) \qquad (r = 0, 1, \ldots)$$

where $S(\mathbf{x})$ measures the "point surface" of $\mathbf{x}$, i.e. $S(\mathbf{x})$ is the number of direct neighbors of $\mathbf{x}$ belonging to another cell. We use 8-neighborhood, hence $S(\mathbf{x})$ varies between 0 and 8. Using the maximum norm $||\cdot||_\infty$, the parameter r determines the size of a square around the point $\mathbf{x}$. For r we use values between 5 and 10. Note that $0 \leq I_r(\mathbf{x}) \leq 1$ for all STPs $\mathbf{x}$.

4.4 Termination Rules

A general problem of heuristic search processes is the question of when to stop them. Early detection of low chance for good improvements can drastically reduce computation time by focusing computational resources on promising approaches.

The most common domain-independent termination rules are the *Max Rule* and the *Stagnation Rule.* The Max Rule defines a priori an upper bound on the number of steps or computation time, whereas the Stagnation Rule observes the development of the cost and stops the process if over a predefined number of steps – the *lag interval* – the decrease of cost (measured in percent) drops below a given *critical threshold.* We used mainly the stagnation rule, because it realizes a good compromise between ease of implementation and adaptation to the problem. It is noteworthy that the two parameters of the stagnation rule – the length of the lag interval and the critical threshold (Table 3) – are important *control* parameters. In general, stricter parameter values (shorter lag interval, higher threshold) lead to reduced average computation time, but also to lower average solution quality. So, depending on the available resources and on performance requirements (both average running time and solution quality), it is a *Meta-Optimization* problem to choose the parameters of the stagnation rule. Based on several experiments, parameter values in the intervals listed below have proven to be useful:

5 The Expansion Problem

The APP described so far is also called a *greenfield* design problem, i.e. all candidate sites are initially unused. Another important problem is the case

parameter	useful values
lag interval	50–100 steps
threshold percentage	0–5%

Table 3. Useful parameter values for stagnation rule

that there is an existing solution – the "legacy" solution – which should be expanded in order to meet new requirements, e.g. increased traffic demand. This is called an *expansion* design problem. Usually the legacy solution is not feasible: it contains cover holes and overloaded antennas. The goal is to transform the legacy solution into a feasible one, changing as little as possible, and also tackle the objectives of interference and cell shape factor. To this end, the site cost objective is altered as follows:

$$SC = \sum_{i=1}^{|\mathcal{L}|} c_i$$

where

$$c_i = \begin{cases} 0, \text{ if site } L_i \text{ is empty in the legacy and current network} \\ 1, \text{ if the initially configured site } L_i \text{ has not been modified} \\ 2, \text{ if the initially configured site } L_i \text{ has been modified} \\ 5, \text{ if site } L_i \text{ is newly installed} \\ 7, \text{ if the initially configured site } L_i \text{ has been removed} \end{cases}$$

Thus, large modifications of the network are more severely punished than smaller ones like changing the power or the azimuth of an antenna. It is particularly undesirable to give up a formerly used site.

With this modification to the site cost objective (the other constraints and objectives staying unaltered), the repair and optimization phase of ENCON can handle the expansion problem in the same way as the greenfield problem (except that in addition to the "RemoveSite" operator in the optimization phase there is a "RevertSite" operator, which undoes all the changes at a random site, thus reducing the site cost). But the construction of initial solutions has to be adapted. A typical expansion problem looks as follows: some regions are already in a good state, others have big cover holes or heavily overloaded cells. Then the following dilemma appears: for regions with big cover holes it is desirable to use an initialization procedure similar to the initialization in the greenfield setting, but applying this procedure to the whole network usually degenerates the already well-solved regions. To solve this dilemma we modify the construction of initial solutions in the following way:

FOR ALL unused sites L_i **DO**

network	size	Δ_{mesh}	RTPs	STPs	TTPs
N1_0	40 km × 170 km	200 m	164580	29954	4967
N1_1	40 km × 170 km	200 m	164580	29954	4967
N3_0	50 km × 46 km	200 m	56792	17393	6656
N3_1	50 km × 46 km	200 m	56792	42974	21475

Table 4. Network data

network	#CS	total traffic	S_q
N1_0	250	3210.94 Erlang	−90 dBm
N1_1	250	3210.94 Erlang	−82 dBm
N3_0	568	2988.08 Erlang	−90 dBm
N3_1	747	8089.90 Erlang	−90 ; −82 ; −75 dBm

Table 5. Network data

Compute placement probability p_i of site L_i as in the greenfield setting.
Configure site L_i with probability p_i.
IF an antenna configuration is placed at site L_i
  IF the resulting solution lowers the hard cost
    accept antenna configuration at site L_i
  ELSE
    reject antenna configuration at site L_i
END

The result is that well-solved regions remain unaltered, because newly placed antennas can have no positive effect on the hard cost. On the other hand, in poorly solved regions, new antennas often have a positive effect, e.g. eliminating a cover hole.

6 Results

Within the ARNO project we have studied eight real-world problems. The data were provided by CNET.[3] The first four networks – denoted by N1_0, ..., N4_0 – are greenfield design problems, whereas Networks N1_1, ..., N4_1 are expansion design problems for Networks N1_0, ..., N4_0, respectively.

We present results for N1_0, N1_1, N3_0 and N3_1 in detail. N1_X define a highway scenario, N3_X a medium-sized town scenario. Tables 4 and 5 describe the input data defining the APPs. The #CS-column contains the number of available candidate sites and the S_q-column the signal quality thresholds occurring in the network.

[3] France Telecom Research and Development Center

6.1 Greenfield Results

Tables 6 and 7 display feasible solutions generated by ENCON for Network N1_0 and Network N3_0. They have been obtained with the weights ψ_1, ψ_2 and ψ_3 given in the tables. The weights ϕ_1 and ϕ_2 of the hard cost function used in the repair phase have been set to 1. The SC-column contains the number of used sites. The number in parentheses is the minimal number of sites necessary to handle the traffic demand of the whole working area. The IL- and the SF-columns display the interference level (in dB per STP) and the mean shape factor. σ_{SF} denotes the standard deviation of the individual cell shape factors occurring in the solution. The OD-, LD- and SD-columns contain the number of used omnidirectional, large directive and small directive antennas, respectively.

best in...	SC (25)	IL	SF	σ_{SF}	OD	LD	SD	ψ_1	ψ_2	ψ_3
SC	36	64.7	6.2	2.5	5	12	81	1	0	0
IL	49	20.9	5.7	2.4	20	3	84	0	1	0
SF	74	288.2	4.1	1.5	0	5	207	0	0	1
uniform	41	23.3	5.9	2.5	9	3	93	1	1	1
rescaled	47	22.7	5.1	2.0	12	12	93	1	2	10

Table 6. Feasible solutions for Network1_0

best in...	SC (23)	IL	SF	σ_{SF}	OD	LD	SD	ψ_1	ψ_2	ψ_3
SC	30	151.0	7.3	2.4	0	3	87	1	0	0
IL	45	35.1	5.9	1.8	21	12	60	0	1	0
SF	58	532.8	4.8	1.1	0	26	123	0	0	1
uniform	35	51.2	6.4	2.1	10	3	72	1	1	1
rescaled	33	36.3	5.9	1.8	7	6	72	1	1	8

Table 7. Feasible solutions for Network3_0

We first executed runs with one objective only in order to determine the scale of the objectives. Next we set all weights ψ_1, ψ_2 and ψ_3 to 1 in order to produce reference solutions for the multi-objective problem. The solutions obtained with these weights are displayed in the rows labeled "uniform". Finally we use the costs for SC, IL and SF resulting from the single-objective runs to determine the weights ψ_1, ψ_2 and ψ_3 so that the typical values of the weighted objectives have comparable magnitude. The solutions obtained with these weights are displayed in the rows labeled "rescaled". In both cases – Networks N1_0 and N3_0 – the uniform weights lead to solutions which are a reasonable compromise between all three objectives, but the rescaled weights lead to improved solution quality (remember that the shape factor objective is very sensitive). However, the "best" setting of weights can only be determined in cooperation with end-users.

6.2 Expansion Results

In Table 8 the legacy solutions for Networks N1_1 and N3_1 are evaluated according to our objectives. The SC-column contains the number of used sites. The legacy solutions have traffic and cover loss, displayed in the TL- and CL-columns. Hence they are *not* feasible solutions.

network	SC	IL	SF	σ_{SF}	TL	CL	OD	LD	SD
N1_1	87	206.0	4.7	1.8	0.1 %	0.3 %	13	152	15
N3_1	96	798.9	5.0	2.0	26.8 %	11.6 %	6	81	174

Table 8. Legacy solutions for N1_1 and N3_1

In Tables 9 and 10, feasible solutions for Networks N1_1 and N3_1 computed with ENCON are displayed. The SC-column now denotes site cost for the expansion design problem (see Section 5). The legacy solution for N1_1 violates the constraints only a little, therefore one cannot expect great improvements since the site cost objective SC penalizes changes to the existing configuration. The legacy solution for N3_1 strongly violates the constraints. The solutions in Table 10 differ very much. There is no clear best solution. However, our research has shown that with carefully chosen weights the computation of well-balanced solutions is possible.

In all cases the number of steps to compute the solutions is between 200 and 2000. Each displayed solution is chosen from 10 runs performed with the same weights. The number of performed steps for the given solutions and the average run time for one "rescaled" run on a Pentium II/450 MHz is displayed in Table 11. The run times correspond directly to the difficulty of the problem instance.

6.3 Evaluation of Operators

We now examine the performance of the operators. Two important criteria are the success rate (the rate of operator applications resulting in an improved solution) and the average reduction of cost of the individual operators. The success rates are displayed in Table 12. *RepairTraffic* has the largest success rate of the repair operators. *IncreaseCompactness* has the largest success rate of the optimization operators.

The situation is different for the average cost reduction (see Figure 7). Here *DissipateTraffic* and *RepairHole* are the most efficient operators in the repair phase, while in the optimization phase *RemoveSite* is the clear winner. This can be traced back to two causes: first, the repair phase introduces many

best in...	SC	IL	SF	σ_{SF}	OD	LD	SD	ψ_1	ψ_2	ψ_3
SC	119	240.1	4.8	1.9	13	155	30	1	0	0
IL	281	176.6	4.9	2.0	15	115	31	0	1	0
SF	177	212.9	4.5	1.7	14	152	23	0	0	1
uniform	173	195.9	4.7	1.9	14	136	26	1	1	1
rescaled	126	220.6	4.7	2.0	14	149	27	1	1	20

Table 9. Feasible solutions for Network1_1

best in...	SC	IL	SF	σ_{SF}	OD	LD	SD	ψ_1	ψ_2	ψ_3
SC	277	826.6	5.0	1.6	6	117	202	1	0	0
IL	796	433.3	5.3	1.9	29	42	195	0	1	0
SF	301	869.4	4.8	1.4	8	107	221	0	0	1
uniform	608	674.9	5.2	1.7	13	69	224	1	1	1
rescaled	601	564.5	4.9	1.6	22	91	196	1	1	50

Table 10. Feasible solutions for Network3_1

network	repair steps	opt. steps	avg. time per run
N1_0	21	1606	14.7 h
N1_1	37	190	0.6 h
N3_0	26	852	23.2 h
N3_1	184	1576	31.1 h

Table 11. Run times

repair operator	success rate	optimization operator	success rate
RepairTraffic (RT)	0.27	IncreaseCompactness (ICo)	0.61
RepairHole (RH)	0.19	ReduceIrregularities (RI)	0.03
IncreaseCover (IC)	0.14	RemoveSite (RmS)	0.31
DecreasePower (DP)	0.20	RevertSite (RvS)	0.19
IncreasePower (IP)	0.03	ChangeAzimuth (CA)	0.44
ChangeAzimuth (CA)	0.04	ChangeTilt (CT)	0.39
ChangeTilt (CT)	0.03		
DissipateTraffic (DT)	0.24		

Table 12. Success rates of repair and optimization operators

sites, because the only goal in this phase is to construct a feasible solution. This results in many opportunities for removing sites. Second, an application of an optimization operator, which generates an infeasible solution, is followed by *local repair* (see Section 4.3), which tries to repair the constraint violations introduced by the optimization operator. As a result *RemoveSite* often leads to cheaper, but still feasible solutions. But the less efficient operators play an important role too. They address other objectives and, as a side effect, they increase the probability for a successful application of the *RemoveSite*

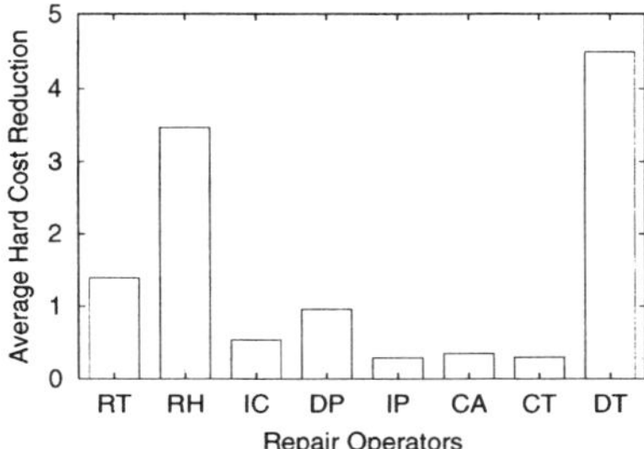

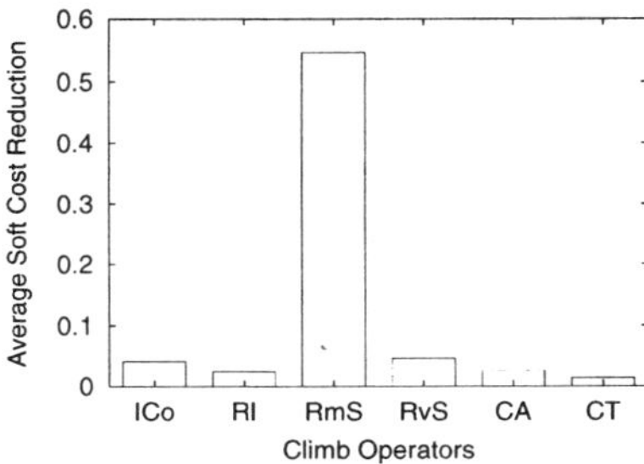

Fig. 7. Average cost reduction of repair and optimization operators.

operator. Currently we are planning experiments to investigate quantitatively these cross-effects between operators.

7 Conclusion

We have introduced an advanced model for the antenna placement problem, which addresses economical and technical aspects. This leads to a constrained and multi-objective optimization problem, which we tackle with an evolutionary algorithm. The design approach for the algorithm is based on theoretical results, which suggest starting with a (1+1) evolution strategy, especially for optimization problems with constraints. The obtained results are encouraging. The problem for the application does not lie in the design of the evolutionary algorithm, but in the mathematical model. We are currently discussing with network operators how to extend the model in order to make it still more realistic.

Acknowledgments

We thank all the partners of the ARNO project for the fruitful and lively discussions, especially A. Caminada and P. Reininger (CNET), J.-K. Hao (Armines-LGI2P, Nimes), S. Hurley (University of Cardiff) and O. Sarzeaud (ECTIA, Nantes). Also we would like to thank the other members of the ARNO team at GMD, C. Crisan (now at T-Mobil, Bonn) and J. Bendisch, for their substantial contributions to the design of the APP model and implementation of the ENCON software system.

References

1. Allan, A. O. (1978) *Probability, Statistics, and Queueing Theory.* Academic Press, New York

2. Crisan, C., Mühlenbein, H. (1998) The breeder genetic algorithm for frequency assignment. *Lecture Notes in Computer Science*, **1498**, 897–906
3. Crisan, C., Mühlenbein, H. (1998) The frequency assignment problem: A look at the performance of evolutionary search. *Lecture Notes in Computer Science*, **1363**, 263–273
4. Eidenbenz, S., Stamm, C., Widmayer, P. (1998) Positioning guards at fixed height above a terrain — an optimum inapproximability result. *Lecture Notes in Computer Science*, **1461**, 187–198
5. Fonseca, C. M., Flemmings, P. (1995) An overview of evolutionary algorithms in multi-objectives optimization. *Evolutionary Computation*, **3**, 1–16
6. Fonseca, C. M., Flemmings, P. (1996) On the performance assessment and comparison of stochastic multi-objective optimizers. *Lecture Notes in Computer Science*, **1141**
7. Goldberg, D. E. (1989) *Genetic Algorithms in Search, Optimization and Machine Learning.* Addison-Wesley, Reading, MA
8. John, B., Jianmin, J., James, D. (1997) Simulation techniques for sensitivity analysis of multi-criteria models. *European Journal of Operational Research*, **103**, 531–546
9. Lissajoux, T. et al. (1998) Genetic algorithms as prototyping tools for multi-agent systems: Application to the antenna parameter setting problem. *Lecture Notes in Computer Science*, **1437**, 17–28
10. Mitchell, M. et al. (1994) When will a genetic algorithm outperform hill climbing? *Advances in Neural Information Processing Systems*, **6**, 51–58
11. Mouly, M., Pautet, M. B. (1992) *The GSM System for Mobile Communications.* Cell & Sys, Palaiseau
12. Mühlenbein, H. (1991) Evolution in time and space - the parallel genetic algorithm. In G. Rawlins, editor, *Foundations of Genetic Algorithms*, 316–337, Morgan-Kaufman, San Mateo
13. Mühlenbein, H. (1992) How genetic algorithms really work: mutation and hill-climbing. In R. Männer and B. Manderick, editors, *Parallel Problem Solving from Nature*, 15–26, North-Holland, Amsterdam
14. Mühlenbein, H. (1997) The equation for the response to selection and its use for prediction. *Evolutionary Computation*, **5**, 303–346
15. Mühlenbein, H., Mahnig, T. (1999) FDA – a scalable evolutionary algorithm for the optimization of additively decomposed functions. *Evolutionary Computation*, **7**
16. Mühlenbein, H., Mahnig, T. (2000) Evolutionary algorithms: from recombination to distributions. to be published in Lecture Notes in Computer Science
17. Mühlenbein, H. et al. (1999) Schemata, distributions and graphical models in evolutionary optimization. *Journal of Heuristics*, **5**, 215–247
18. Mühlenbein, H., Schlierkamp-Voosen, D. (1994) The science of breeding and its application to the breeder genetic algorithm. *Evolutionary Computation*, **1**, 335–360
19. Plehn, J. (1994) Applied frequency assignment. In *Proceedings of the 1994 IEEE 44th Vehicular Technology Conference*, Stockholm
20. Renaud, D., Caminada, A. (1997) Evolutionary methods and operators for the frequency assignment problem. *SpeedUp*, **11**, 27–32

21. Reininger, P., Caminada, A. (1998) Model for GSM radio network optimization. In *2nd ACM International Conference on Discrete Algorithms and Methods for Mobility*, Dallas
22. Schwefel, H. P. (1995) *Evolution and Optimum Seeking*. Wiley-Interscience, Chichester
23. White, G. D. (1991) *The Audio Dictionary*. 2nd edition, University of Washington Press

Routing Optimization in Corporate Networks by Evolutionary Algorithms

Thomas Bäck[†,*], Claus Hillermeier[‡], and Jörg Ziegenhirt[†]

[†] NuTech Solutions GmbH
Martin Schmeisser Weg 15
D-44227 Dortmund, Germany
[*] Leiden University
Leiden Institute for Advanced Computer Science
Niels Bohrweg 1
NL-2333 CA Leiden, The Netherland
[‡]SIEMENS AG
Corporate Technology
Otto-Hahn-Ring 6
D-81735 Munich, Germany

Summary. The purpose of the research reported in this chapter is to meet the increasing and rapidly changing communication demands in a private telecommunication network. The goal is to improve the performance of a telecommunication system just by changes in its software components. The network topology and the trunk capacities are viewed as fixed parameters, because a change in these components is very expensive. The algorithm of how to use this network is stored in a routing table that consists of alternative paths between the nodes that are supposed to be connected. So the goal of the evolutionary algorithm is to find a routing table that increases the performance of the network by reducing the probability of end-to-end-blocking. The investigated non-hierarchical networks require a fixed alternate routing (FAR) with sequential office control (SOC). The chapter presents a new approach based on evolutionary algorithms to solve this problem.

1 Introduction and Motivation

The management of non-hierarchical telecommunication networks as well as their configuration gives rise to many large-scale optimization problems. The configuration process of a network consists of several decisions, e.g. where to identify nodes or which nodes are to be connected by what size of trunk. Once a network is physically installed, it is very expensive and time consuming to change its structure. So the configuration of a network should consider all future demands of the net. This is obviously impossible. The only chances to improve an installed network are to change the network or to optimize the management of the existing network. This chapter focuses on the optimization of the management. The networks of interest are private telecommunication networks employing the routing strategy of fixed alternate routing (FAR) with sequential office control (SOC). FAR provides a set of alternatives for each connection between two points in a net. The set is defined once, and is fixed for all requests that are satisfied with this FAR. A connection between two nodes mostly consists of several transit nodes. SOC allows different alternatives to be defined in each transit node, such that the decision for a particular way is not made once, but several routing decisions are necessary to build up one connection.

Genetic algorithms, a specific instance of the class of evolutionary algorithms[1], have been used for survivable network design [4] and for dynamic routing control [4]. Sinclair [5] tested a genetic algorithm (GA) to optimize routing tables for an originated office controlled (OOC) version of FAR. Sinclair concludes that the GA performs worse than a heuristic algorithm he presented previously. The coding he introduces allows routing tables to be produced with loops. In the SOC routing investigated in this chapter the cardinality of the search space would be extremely increased by including the routing tables with loops. In order to exclude this undesired effect, a coding is introduced in sections 2 and 3 so as to generate only routing tables which satisfy the constraints. The details of the evolutionary algorithm (EA) are then presented in section 4, and some real-world test cases are given in section 5. The test results are presented and discussed in section 6, and finally some conclusions are given.

2 The Routing Table

2.1 Structure and Constraints

The routing table realizes a routing strategy with SOC: in each (transit) node a decision is made where to forward the call. The decision depends not only on the actual (transit) node and the destination, but also on the node of origin. Actually there is a list of neighboring nodes (the so-called routing vector), which defines a priority order for consecutive routing attempts to neighboring nodes. The trunk group to each neighbor is tested, and the call is handed over to the first node with free capacity. To give an example, the routing vector for the transit node T with the origin O and destination D according to the situation given in figure 1 is $T_{O,D} = (A, B, C)$, meaning that in order to route a message from O to D, node T will try to route via A,B, and C in this order of priority.

To obtain good routing tables one has to avoid some situations, which are handled as constraints of the problem, by using a representation which guarantees the feasibility of individuals:

- **Looplessness:** Following the path from the origin to the destination, it makes no sense to visit the same transit node twice. Such a loop is consuming resources without making any progress in building a path between

[1] See e.g. [1],[2],[3] for an overview of these search and optimization methods gleaned from the model of organic evolution.

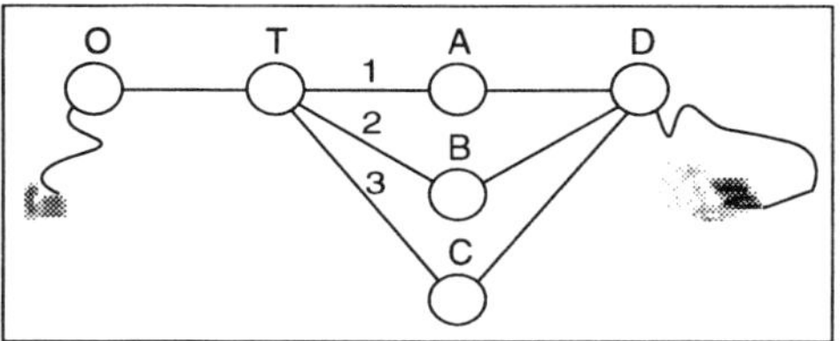

Fig. 1. Routing situation in node T

the origin and destination. Beyond that, loops bear the possibility of network deadlocks. Therefore a reasonable routing solution must guarantee looplessness. It is not possible to decide whether this constraint is met or not by considering the information of a single transit node. The smallest structure of a routing table which can be checked for looplessness is the totality of all routing information for one single origin/destination pair (OD-pair). This block is called the OD-pair table.

- **Dead ends:** A call should never be forwarded to a node that has no routing information to lead to the destination. These dead ends can also be checked in an OD-pair table.
- **Compliance with the network topology:** A routing vector should only contain nodes that are directly connected with the corresponding transit node. This condition can be checked in each node.

In section 3, a coding is introduced that meets all these constraints. First, however, a few remarks are made on the size of the search space.

2.2 Cardinality of the Search Space

The cardinality of the search space is equal to the number of possible different routing tables. As a worst case we assume that, in a net with N nodes, each one is connected to every other node, so there are $N \cdot (N-1)$ OD-pair tables (the network connections are asymmetric). The number of transit nodes of an OD-pair table can be as large as $N-2$, and each one can have up to $N-2$ possible neighbors. A routing vector can be seen as a permutation of the elements of an arbitrary subset of these possible neighbors. So the number of all possible OD-pair tables for one OD-pair is:

$$\#T_{OD} = \left(\sum_{k=1}^{N-2} \frac{(N-2)!}{(N-2-k)!} \right)^{(N-2)} \tag{1}$$

This leads to an upper bound of the numbers of possible routing tables:

$$\#T_R \leq \left(\left(\sum_{k=1}^{N-2} \frac{(N-2)!}{(N-2-k)!} \right)^{(N-2)} \right)^{(N \cdot (N-1))} \tag{2}$$

A network with only 10 nodes can have up to $4 \cdot 10^{1440}$ routing tables. This approach does not consider the constraints mentioned in section 2.1, so a large number of routing tables included in this formula do have loops, or do not comply with the network topology.

3 Coding of the Variables

One possible approach would be to build, as a first step, routing tables which are just formally correct, and then as a second step to check whether they meet the constraints given in section 2.1. But due to the dimension of the search space discussed in section 2.2, a coding is preferred which only produces routing tables meeting the constraints. A routing table can be regarded as a union of OD-pair tables, such that to each possible origin/destination pair exactly one OD-pair table is assigned. Therefore, it is sufficient to give an algorithm that generates OD-pair tables which meet the constraints for each possible OD-pair. Following an idea introduced in [6], the problem of finding several alternative paths for one OD-pair can be mapped to the electric flow calculation in an electrical resistance network.

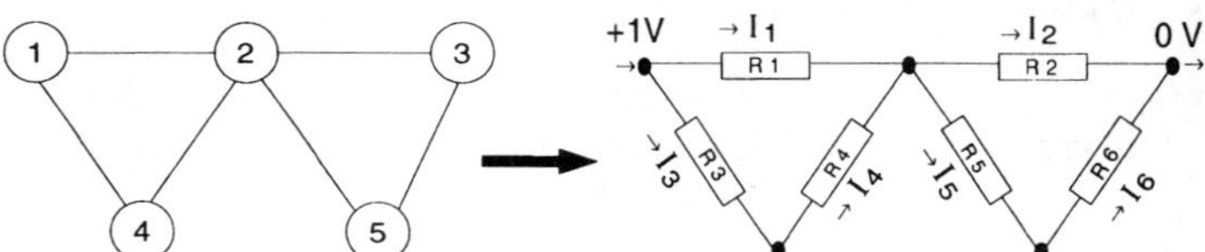

Fig. 2. Electrical current model

The first step is to generate an electrical resistance network by assigning a resistance to each link (see figure 2). The node of origin is given some positive potential U and the potential of the destination node is set to 0. The electrical current flowing through the network can now be calculated by means of Ohm's law and Kirchhoff's node rule (e.g. [7]). For each transit node the positive currents flowing out of the transit node are sorted in decreasing order. The corresponding nodes in the order which is obtained after sorting build up the routing vector. All routing paths now follow the flow of electricity. So neither loops nor dead ends can be produced by this algorithm according to the behavior of electricity. The algorithm is to be changed if the number of alternatives in each transit node is restricted to a value x. Then the neighboring nodes of each transit node are ordered too, but only the first x nodes are taken into account for creating the routing vector. This must be done recursively, starting from the originating node. To construct all feasible solutions a randomly chosen subset of the neighbors can be selected.

Using this model, random OD-pair tables can easily be generated by randomly choosing values for the resistances in the network and executing the above-mentioned algorithm.

4 The Evolutionary Algorithm

In the context of the EAs, a complete routing table is regarded as an individual. For a network with N nodes the individual consists of $N \cdot N$ parameters, the OD-pair tables for each OD-pair. Only $N \cdot (N-1)$ of these are relevant, because there is no routing within the net if the origin and the destination are identical.

4.1 Initializing the Population

As explained in section 3, OD-pair tables are built by the electrical current model. The values of the resistances are randomly chosen values from a certain interval $[1, r_{\max}]$ (typical values for $r_{\max}$ range from 1.2 to 20). A routing table is created by generating one OD-pair table for each OD-pair. The initial population consists of 10 routing tables that are obtained by this method.

4.2 Objective Function

The aim of routing optimization is to minimize the number of situations for which a call cannot be routed through the network. For each OD-pair (i, j) the so-called 'end-to-end-blocking (EEB) EEB_{ij} denotes the probability that a desired connection between these nodes cannot be built up due to 'all-links-busy in an intermediate node. The objective function to be minimized is the sum over all EEBs, weighting each EEB_{ij} with the offered traffic A_{ij} between the origin i and the destination j:

$$f = \sum_{\substack{i,j=1 \\ i \neq j}}^{N} A_{i,j} \cdot EEB_{i,j} \tag{3}$$

Here the traffic matrix A_{ij} models the traffic measured in the daily operation of the network. The customer can give a constraint, the 'grade-of-service (GOS) GOS_{ij}, that has to be satisfied. So, formally the constraint is $EEB_{ij} \leq GOS_{ij}$ for all traffic pairs (i, j). We do allow routing tables that violate the constraint, but add a penalty term to the objective function:

$$f_{opt} = \sum_{\substack{i,j=1 \\ i \neq j}}^{N} A_{i,j} \cdot EEB_{i,j} + a_{i,j} \cdot (EEB_{i,j} - GOS_{i,j})$$

with

$$a_{i,j} = \begin{cases} 10 \text{ if} & EEB_{i,j} > GOS_{i,j} \\ 0 \text{ otherwise} & \end{cases} \tag{5}$$

Despite the appearance of equation (4.2) it is not a linear combinatorial problem to find a good routing table, since the EEBs depend in a non-linear way on the routing entries.

4.3 Recombination

For the recombination of routing tables they are used in a straightforward way as an ordered list of OD-pair tables. An offspring is generated by an n-point crossover of two individuals as shown in figure 3.

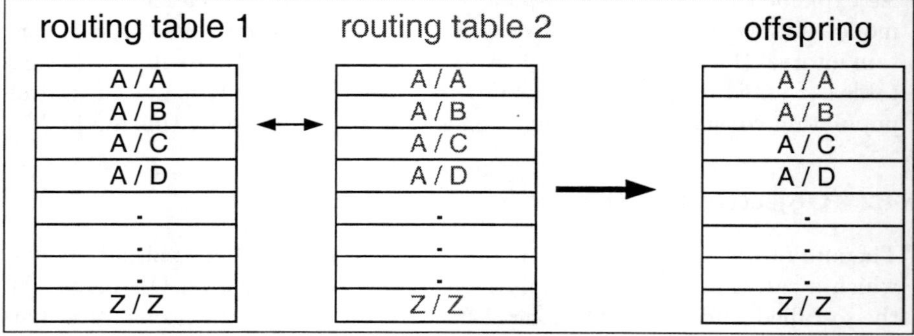

Fig. 3. Recombination by n-point crossover. Notice that only one crossover point is explicitly indicated in the figure (between OD-pairs A/B and A/C).

It is important to notice that this operator cannot produce a routing table that violates the constraints. Recombination is tested with one-point crossover, n-point crossover ($n \in \{1, .., N\}$ is chosen uniformly at random) and with a special kind of global recombination. The global recombination is based on the n-point crossover, and its first step is to mix two individuals with the n-point crossover. Then the operator uniformly at random selects $m \in \{1, .., N\}$ OD-pair tables of the offspring routing table that should be replaced (m is a uniform random number from the set $\{1, .., N\}$). For each of the m selected OD-pair tables a routing table is randomly chosen from a special set of routing tables (see section 4.5), and the selected OD-pair table of the offspring is replaced by the corresponding one from the special set. The set of routing tables that can replace one of the m OD-pair tables consists of all μ parents of the actual generation plus $10 - \mu$ randomly selected routing tables from the first generation (the genetic load; see section 4.5). See also [8] for details.

4.4 Mutation

Mutation is realized in two completely different variants. The first step in both variants is the same: $m \in \{0, .., N\}$ OD-pairs are chosen to be mutated (N denotes the number of nodes within the net), where the probability for smaller values of m is larger than for great values. More precisely, $m = \lfloor |X| \bmod N \rfloor$ where $X \sim N(0,1)$ is a random sample from a standardized normal distribution. The first variant ($Mut1$) replaces the chosen OD-pair table by a new one which is generated by the electric current model. The resistances are randomly chosen within the given limits $[1, r_{\max}]$. The second variant to mutate an OD-pair table ($Mut2$) does not replace the whole OD-pair table, but only changes the routing vector of one randomly chosen transit node within this OD-pair. The routing vector is changed by permutating the entries of this vector by one swap of (randomly chosen) entries. It should be noted that both operators do not affect the constraints of the routing table. We also investigated a third variant $Mut3$ as a mixture of the other two methods: for each mutation one of the methods $Mut1$ or $Mut2$ is chosen randomly to mutate the selected individual. Mutation is applied to each offspring individual.

4.5 Implemented Algorithms

For the implementation of the algorithm some modules of the 'Network Engineering and Routing Tool' (NERT) introduced in [9] are used. The module ORD (optimal routing and dimensioning) optimizes the network by simultaneously optimizing the trunk capacities and the routing throughout the network. Each of the 11 tested networks is optimized by ORD. The routing table produced by the ORD module is used as a reference routing table. The performance evaluation of the EEB probabilities is done by the module PKA (performance and cost evaluation) which analyzes the network performance based on two-moment standard techniques to obtain the EEB values. The modules are described in [10].

The algorithms are based on the standard EA given in [1] with some marginal modifications:

begin
  $t := 0;$
  $I := \emptyset;$
  for $i := 1$ *to* $(\mu + \lambda)$
    $I := I \cup initialize\ \{\boldsymbol{a}_i(0)\};$
    $evaluate\ PKA(\boldsymbol{a}_i(0));$
  end for
  $initialize\ P(t) := select(I);$
  repeat
    $P'(t) := recombine(P(t) \cup P(c));$
    $P''(t) := mutate P'(t);$

$\quad\quad evaluate\ P''(t) : PKA(\boldsymbol{a}_1''(t)), ..., PKA(\boldsymbol{a}_\lambda''(t));$
$\quad\quad P(t+1) := select(P''(t) \cup Q);$
$\quad\quad t := t + 1;$
$\quad$ **until**$(\iota(P(t)) = true);$
end

μ denotes the number of parents, λ the number of offspring individuals, where $\boldsymbol{a}_i(t)$ denotes the i-th individual of the t-th generation. t is the generation counter, so $P(0)$ is the initial population. The termination criterion $\iota(P(t)) = true$ depends on the number of objective function calls. Since the evaluation is done with the PKA module of NERT, the function is named PKA. Initialization is done in two steps: $(\mu + \lambda)$ individuals are randomly created and the μ best are taken as $P(0)$. Some of the other λ routing tables are used during the recombination as a genetic load, as suggested in [11]: $P(c) = \{\boldsymbol{a}_{\mu+1}(0), ..., \boldsymbol{a}_{10}(0)\}$. The size of $P(c)$ depends on μ and can be zero, if $\mu = 10$.

Now, the differences between the algorithms used can be explained by means of the use of the sets mentioned in the pseudocode. The $(\mu + \lambda)$-strategy selects the offspring out of $P''(t)$ and the parents $P(t)$, so the set Q is equal to $P(t)$. In the case of a (μ, λ)-strategy Q is empty, the next generation is only selected out of the offspring population $P''(t)$.

Another selection algorithm is implemented and tested as well: a so-called '2 from 3' strategy [6]. In the following, the notation 2(3) is used to denote this kind of selection strategy. In terms of the standard EA, there is no mutation, but always a global recombination. Three individuals are selected randomly, and the worst of these three is replaced by a recombination of the two others, with some OD-pair tables taken from $P(t) \cup P(c)$. In a way it can be regarded as a special $(\mu + 1)$-strategy, because the best individual of the population always survives.

5 Test Environment

The aim of the study was to find good routing tables by means of the EA, explained in the sections above. As a test set we employ 11 practically relevant corporate telecommunication networks which are shown in figure 4.

The modules of NERT mentioned in section 4.5 are used to compare the performance of the resulting routing tables. The ORD module of NERT not only produces the dimensioning of a net, but also a routing table that is used in practice. For each network ($test_x$) the ORD-produced routing table ($ord\_rt_{test_x}$) is taken as a reference and the fitness value according to equation (4.2) is computed. To enable a comparison of the optimization results for all nets ($ea\_rt_{test_x}$), the fitness function is normalized:

$$f_{fit}(ea\_rt_{test_x}) = \frac{f_{opt}(ea\_rt_{test_x})}{f_{opt}(ord\_rt_{test_x})} \tag{6}$$

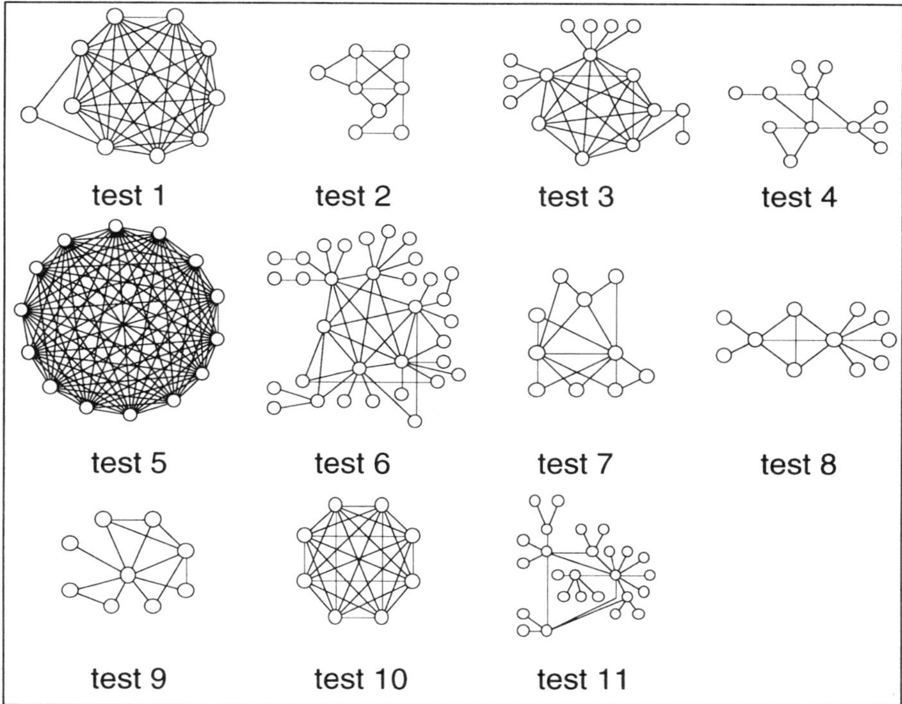

Fig. 4. An overview of all examples

As a result, f_{fit} measures the performance of an EA routing table as the ratio of two (weighted) EEB-sums, namely that one corresponding to the EA routing table and that one corresponding to the ORD routing table. Values less than one are reached if they are better than the routing tables optimized by the ORD module of NERT. If the optimum calculated by the ORD module had no EEBs at all, the normalized fitness function would be undefined. However, this optimum is not reached by any of the ORD routing tables.

The optimization runs are performed with 2000 fitness evaluations repeated six times for each parameter setting with different seeds for the random number generator. The reason for this small database is that one optimization run lasts on average 1.5 hours on a SPARC workstation; 90% of the time is used by the NERT module PKA, needed for fitness evaluation. The evaluation time increases with the complexity of the network. In the case of routing tables with high blocking probabilities, PKA needs much more time and produces only approximately correct values. As a consequence there is a 'time-out' criterion included that sets the value of the corresponding individual to a maximum. The creation of an individual for the initial population is repeated if its evaluation leads to this maximum.

Table 1. Fitness values reached

Sample Net	All Best Value	Mean Value Good Parameter	Mean Value Bad Parameter
test02	0.6384 + npc gR M1 20.0 U	0.6397 + npc gR M1 20.0 N	9.9489 + npc gR M1 1.2 N
test03	1.1570 + npc gR M1 20.0 U	1.1871 + npc gR M1 20.0 N	1.2096 + npc gR M1 1.2 U
test01	0.3437 + npc gR M3 1.2 N	0.3695 + npc gR M3 2.0 U	0.5492 , npc gR M2 2.0 U
test04	0.1403 + npc gR M3 1.6 N	0.4759 + npc gR M3 2.0 U	1.0087 , npc ogR M3 1.2 N
test05	0.0000 , npc gR M1 2.0 N	0.0049 + npc gR M3 2.0 U	0.4679 + npc ogR M2 1.2 U
test06	0.1073 + npc gR M1 1.2 U	0.6113 + npc gR M3 2.0 U	4.7297 + npc gR M0 1.2 N
test07	0.0000 , npc gR M3 1.6 U	0.0014 + npc gR M3 2.0 U	0.2074 , npc gR M2 1.2 U
test08	0.4178 + npc ogR M3 1.2 U	0.7802 + npc gR M3 2.0 U	0.8286 , npc gR M1 1.2 N
test09	0.0991 + npc ogR M3 2.0 N	0.6123 + npc gR M3 2.0 U	1.1523 + npc gR M2 1.2 U
test10	0.0000	0.0000	0.0000
test11	0.3308 , npc gR M1 2.0 U	0.4175 + npc gR M3 2.0 U	0.6587 + npc ogR M2 1.2 U

6 Results

A self-adaptation of strategy parameters (as widely used in evolution strategies and evolutionary programming for parameter control) is not included in the strategy, but a parameter study is performed to find good values. Some variants are only tested sparsely and are fixed to one value for the comparison. Due to memory restrictions, the population size is restricted to 30 individuals. As a result the strategies finally use $\mu = 4$ parents and $\lambda = 26$ offspring, after testing values for μ in the range of [1..10]. In some examples, the randomly chosen resistances out of large intervals cause initialization times of more than 12 hours. As a result, those parameters are only rarely tested. The results obtained by the different algorithmic variants are shown in Table 1. Together with each value in the table the set of parameters is given that was used to reach this value. The abbreviations are described below:

- '+ and ', denote the selection methods $(\mu + \lambda)$ or (μ , λ), respectively.
- '1pc denotes one-point crossover and 'npc n-point crossover.
- 'gR denotes global recombination, and 'ogR without global recombination.
- 'M1, 'M2, and 'M3 are shortcuts for the mutation variants $Mut1$, $Mut2$, and $Mut3$. 'M0 means that no mutation at all is used.
- The values $r_{\max}$ for the upper bound of the interval of the resistances are randomly chosen from $\{1.2, 1.6, 2.0, 4.0, 20.0\}$. The lower bound is always set to 1.0. The value of $r_{\max}$ is also given in the parameter set of a specific EA instance.
- The resistances are chosen uniformly ('U) or normally distributed ('N) from the given interval $[1, r_{\max}]$.

The three columns of Table 1 give an overview of the diversity of the experimental results by presenting, for each test case investigated, the best objective function value f_{fit} ever found in a single run and its corresponding strategy parameter settings (column 'All Best Value), the mean value of six runs for a robust, general parameter setting (column 'Mean Value, Good Parameter), and the mean value of six runs for a parameter setting which results in the worst behavior of the considered settings for the EA (column 'Mean Value, Bad Parameter). The 11 test cases are split into two groups according to the number of routing alternatives, which is limited to one (i.e. there is no routing alternative) for the test02 and test03 networks and larger than one in all other cases. More specifically, for test11 two alternatives and for all other test cases three alternatives are allowed.

As Table 1 reveals immediately, test03 is the only test case where the EA is not able to find routing tables performing better than the ORD-optimized tables. This result, however, is easily explained by the fact that the reference routing table resulting from ORD does not stick to the constraint of only one alternative, thus violating this assumption in contrast to the EA. As soon as alternatives are allowed, the EA is able to produce better routing tables for this test case as well. The mean value for good parameters reaches 0.86.

With this simple classification of test cases, it is possible to identify a robust parameter setting for each of the two classes, which from the results given in Table 1 seems promising for finding a good solution for an unknown network as well. In particular, the recommended parameter settings are:

- $(\mu + \lambda)$-selection, n-point crossover, global recombination, mutation operator three (random choice between both operators), a resistance interval $[1, 2]$, and uniform resistance initialization for a number of alternatives larger than one.
- $(\mu + \lambda)$-selection, n-point crossover, global recombination, mutation operator one (replacement of OD-pair tables by newly generated ones), a resistance interval $[1, 20]$, and normally distributed resistance initialization for a number of alternatives limited to one.

It is obvious that, with a waysplit limited to one, only the mutation operator one can be used at all for these test cases. From a test set of only two networks for this class it is of course difficult to give a general suggestion, but it seems fairly important to use a large interval for the resistances to provide enough diversity in the OD-pair tables. On the other hand, this argument would also favor the utilization of uniformly rather than normally distributed initialization.

The first column of Table 1 is included to demonstrate the improvements which can be achieved by tuning the strategy parameters for a particular network. For the most difficult test case, test06, the results clarify that the effort of tuning the strategy parameter setting can yield a substantial further improvement of results compared to the robust strategy given above. On the contrary, due to its overdimensionalized trunk capacities, the network test10 is too simple for challenging the EA at all, and the global optimum is already found during the initialization process.

The results for the 2(3)-strategy have already been presented in [6], but are repeated in Table 2 for comparability. Some routing tables can be optimized with this 2(3)-strategy to values close to the optima found by the improved EA presented in this chapter. The advantage of the 2(3)-strategy lies in the smaller number of strategy parameters that have to be adjusted, namely just the initialization parameters and the population size μ.

Table 2. Results obtained with the 2(3)-strategy.

	Best Value	Worst Value
test01	0.5280	0.5480
test02	0.6616	28.6003
test03	1.2030	1.2096
test04	0.5044	0.5206
test05	0.0006	0.0883
test06	0.3192	6.2243
test07	0.0072	0.0150
test08	0.7553	0.8121
test09	0.6124	0.6423
test10	0.0000	0.0000
test11	0.6383	0.64235

To visualize the improvement yielded during the optimization with the EA some typical runs are given in figure 5. The fitness of the best individual is plotted as a function of the number of evaluations. For each network, only the run which produced the best result for the robust parameter settings (middle column of Table 1) is shown in the figure.

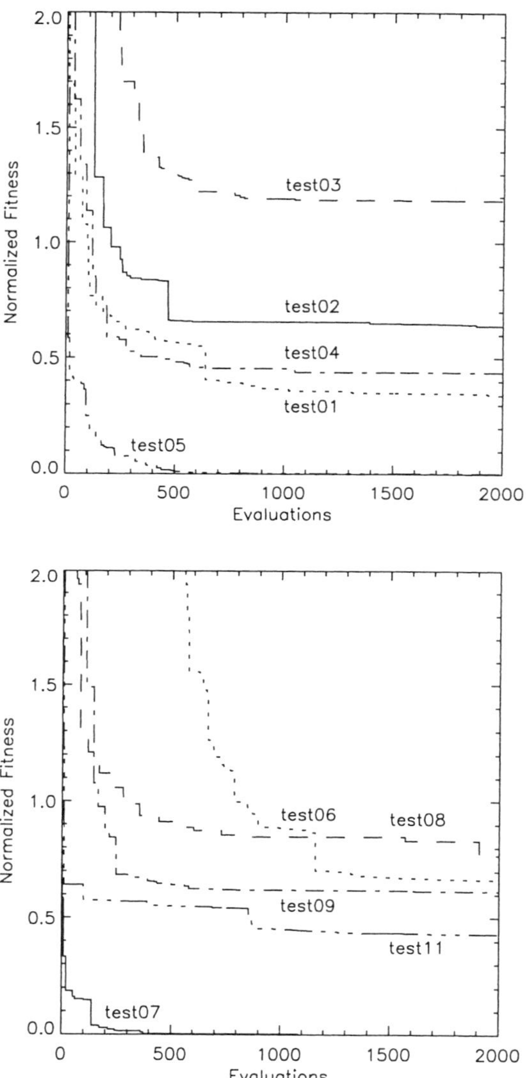

Fig. 5. Typical runs for each example

7 Conclusion

The aim of this study was to optimize a private telecommunication network only by changes in its software components. The EA presented in this work can be effectively used to optimize routing tables for the networks and the FAR investigated. It is possible to find routing tables that show better performance than those found by a traditional approach. The disadvantage of the EA in comparison to the traditional approach is the need for more CPU time. The ORD module optimizes routing and dimensioning by a factor of 20 to 30 times faster than the EA. There are several possibilities to speed up the EA. The PKA not only computes the EEBs, but is a complex tool for analyzing the network. Therefore time could be saved if the PKA were replaced by a function that only evaluates the EEBs. The other possibility is to be seen in the termination criterion. As displayed in figure 5 the fitness of the best individual reaches the reference value of 1 in less than 1000 fitness evaluations. So if the criterion is changed to stop after one routing table with no violation to the GOS found, the EA could be much faster.

The algorithm is now able to produce routing tables for different traffic situations offered. Therefore the EA can be used for multi-hour routing (i.e. routing tables for different traffic situations that correspond to the traffic at specific hours per day). Further investigations based on this approach are promising. First steps are taken to optimize the trunk capacities and to use the EA even in the design of the networks.

Acknowledgements

This research is supported by grant 01 IB 802 B of the German Federal Ministry of Education, Science, Research and Technology (BMBF). The authors are responsible for the contents of this publication.

References

1. Bäck, T. (1996) *Evolutionary Algorithms in Theory and Practice.* Oxford University Press
2. Fogel, D. B. (1995) *Evolutionary Computation: Toward a New Philosophy of Machine Intelligence.* IEEE Press, Piscataway, NJ
3. Michalewicz, Z. (1996) *Genetic Algorithms + Data Structures = Evolution Programs.* Springer, Berlin
4. Davis, L., Orvosh, D., Cox, A., Qiu, Y. (1993) A Genetic Algorithm for Survivable Network Design. In S. Forrest, editor, *Proceedings of the Fifth International Conference on Genetic Algorithms*, 408–415. Morgan Kaufmann, San Mateo, CA
5. Sinclair, M. C. (1993) The Application of a Genetic Algorithm to Trunk Network Routingtable Optimization. In *IEE Proceedings of the Tenth UK Teletraffic Symposium*, 2/1–2/6. Martlesham (UK)

6. Hillermeier, C., Weber, D. (1997) Optimal Routing in Private Networks using Genetic Algorithms. *Proceedings of the ISS '97: World Telecommunication Congress, Toronto*, **1**, 523–531
7. Pregla, R. (1984) *Grundlagen der Elektrotechnik, Teil I: Felder und Gleichstromnetzwerke.* Dr. Alfred Hüthig Verlag, Heidelberg
8. Ziegenhirt, J. (1998) Optimierung von Routingtabellen in Privaten Netzen mit Evolutionären Algorithmen. Technical Report SYS-3/98, Universität Dortmund
9. Bai, Z., Hartmann, H. L., He, H., Weber D. (1996) Engineering and Routing for Modular Private ISDNs: Models, New Concepts, Fast Algorithms and Advanced Tools. In *Proceedings of Networks '96*
10. He, H. (1997) *Kostenoptimierung privater Telekommunikationsnetze mit digitaler Kanaldurchschaltung.* Shaker, Aachen
11. Born, J. (1978) *Evolutionsstrategien zur numerischen Lösung von Adaptationsaufgaben.* Dissertation, Humboldt-Universität, Berlin
12. Girard, A. (1990) *Routing and Dimensioning in Circuit-Switched Networks.* Addison-Wesley.

Genetic Algorithms and Timetabling

Peter Ross[1], Emma Hart[1], and Dave Corne[2]

[1] Division of Informatics, University of Edinburgh
80 South Bridge, Edinburgh EH1 1HN, UK
(*as from Oct 2000*) School of Computing, Napier University
219 Colinton Road, Edinburgh EH14 1DJ, UK
E-mail: {peter,emmah}@dcs.napier.ac.uk

[2] Department of Computer Science, University of Reading
Whiteknights, Reading RG6 6AY, UK
E-mail: D.Corne@reading.ac.uk

Summary. This chapter discusses the state of the art, at the start of the new millennium, in using evolutionary algorithms to tackle timetabling problems of various kinds. Timetabling problems are interesting because they are often regarded as difficult examples of problems which mix hard and soft constraints, and yet huge numbers of practical instances need to be tackled somehow.

A variety of representations and algorithms have been used, but there is still a need for more wide-ranging scientific study to compare different approaches. Most authors merely report success of some kind on the problems that occur at their own institutions.

1 Introduction

In the prototypical timetabling problem there are a number of events which need to be assigned to time-slots in a way that satisfies a set of 'hard' constraints, and there are also a number of 'soft' constraints which should be obeyed if it is possible to do so. For example, in exam timetabling:

- for every student, the exams which that student takes must all be at different, non-overlapping times and it must be possible to get to each exam by its start-time. That is, there must be no *clashes* between exams. *[hard constraint]*
- certain exams may be excluded from certain time-slots, for example because a student or an invigilator or some other exam-specific resource is unavailable in those slots. *[usually hard]*
- there may be ordering constraints: certain exams may be required to take place before certain others, for example because a written exam refers to something in a previous practical exam. *[usually hard]*
- there may be capacity constraints, for example imposed by room sizes. An exam may be split across two or more rooms, if there are invigilators available, or it may be possible to hold several exams in the same room simultaneously, but there is normally at least an upper limit on the number of seats available in a time-slot. *[usually hard]*

- students and invigilators may have preference for or against certain slots. *[soft]*
- for each student it may be desirable to spread out the relevant exams, to allow some chance for resting between exams. Exams which are uncomfortably close for some students are called near-clashes below. *[soft]*
- alternatively, students or invigilators may desire their exams to be packed together so as to allow a larger block of free time for other activities. *[soft]*

There are analogous sorts of constraints in class/lecture timetabling, with obvious differences of detail; you cannot have two lectures in the same room at the same time, for instance. In general, the task is to obey all the clash constraints and the hard resource constraints (seating, room/person availability, ordering) and simultaneously to do the best one can with the soft constraints. It is common for published papers to talk about trying to find 'optimal' timetables, but in practice it is very unlikely that people would wish to pay the price of achieving provable optimality; what is usually wanted is something that can be deemed to be good enough because it meets the hard constraints at least and can be used, and which can be found in some user-defined notion of reasonable time. This is not only because provable optimality costs too much to find, but also because a timetable that is very elegantly packed might be very fragile; one small unforeseen change of circumstance could cause tremendous problems. For instance, if an invigilator is ill and no replacement is available, then the exam must be rescheduled, but in a tightly packed timetable there may be nowhere to re-site it.

In practice, every real problem has its idiosyncrasies, so that it becomes difficult to make valid performance comparisons between different methods that have only been tested on a few real problems. For example, in some practical exams each student only has to turn up to one instance of the exam, but the exam may happen in several different time-slots. Or consider this real-world problem, whose solution in Prolog is described in an article at `http://www.amzi.com` (Amzi! Inc products include Prolog systems and tools):

> The Atlantic Coast Conference basketball league has nine teams. Games are held on Wednesdays and Saturdays over nine weeks (18 playing days); each day, one team gets a bye and the other eight play, hence four matches per day. After the first nine days each team needs to have played all others once and to have had one bye. In the second nine days the same happens, but with home and away teams reversed. The second nine-day half cannot be timetabled symmetrically to the first half; because home games on a Saturday are very profitable for a team, all teams need to have the same number (4) of Saturday home games over the nine weeks, ideally well spaced out. No team wants to play three or more home games, or three or more away games, in a row. For any two teams, the two matches they play against each other must be separated by at least three playing days. Certain matches

are also constrained to be on certain days; for example, the league wants the final matches to be between long-established rivals.

This is a small example that illustrates some of the idiosyncratic nature of timetabling problems in general. At the other end of the scale some very much larger, real university exam timetabling problems are published by Mike Carter at `ftp://mie.utoronto.ca/pub/carter/testprob/` with a copy at `ftp://ftp.cs.nott.ac.uk/ttp/Data/`. As an illustration of problem difficulty consider the KFU-S-93 problem, provided by King Fahd University; it is by no means the hardest or largest in the set. It involves 5,349 students and 461 exams, ideally to be fitted into 20 timeslots, with 1,955 seats available in any one time slot (64% occupancy). The largest exam involved 1,280 students. In graph-theoretic terms, this problem contains two cliques of size 19, and huge numbers of smaller ones – a clique is a maximally connected sub-graph, so that a clique of size 19 necessarily requires 19 distinct timeslots. There are 16 exams each of which are required not to clash with over one hundred others; one of these is constrained not to clash with 247 other exams! On average, each exam is constrained not to clash with over 25 others. Carter's problem set includes even bigger examples; one of them involves over 30,000 students and over 2,400 exams.

The conventional non-evolutionary methods of tackling timetabling problems are usually derived from graph-coloring. The basic idea is that each node of a graph represents an event, and a pair of nodes is joined by an arc if those two events must not clash. The problem is to color the nodes, using no more than a given maximum number of colors, such that no arc joins two identically colored nodes. Colors correspond to time-slots. Standard books on combinatorics (eg [44]) or graph theory (eg [45]) contain the basic theory of graph-coloring and, for instance, show that if a graph has e edges and n nodes and there are c colors available then it can be colored in

$$c^n - ec^{n-1} + (e(e-1)/2 - \Delta)c^{n-2} + \cdots$$

ways ([44], p.306), where Δ is the number of triangles in the graph. This polynomial will be 0 if the number of colors is too small; all such numbers turn out to be roots. There is no computationally cheap way to generate the polynomial, unfortunately.

In some real-world problems the resource constraints are sufficiently loose that a greedy graph-coloring algorithm such as Brelaz's [2], which simply tackles the most color- and arc-constrained nodes first, can do quite well. For example, in the trivial graph on the left of Figure 1, three colors are needed and the greedy algorithm finds such a coloring at once. Time-slots can then be assigned to colors so as to try to obey other hard and soft constraints, so that (assuming slots are one hour long) the five events can be timetabled into three hours. If the resource constraints are even moderately demanding, then this two-phase process is likely to produce poor results.

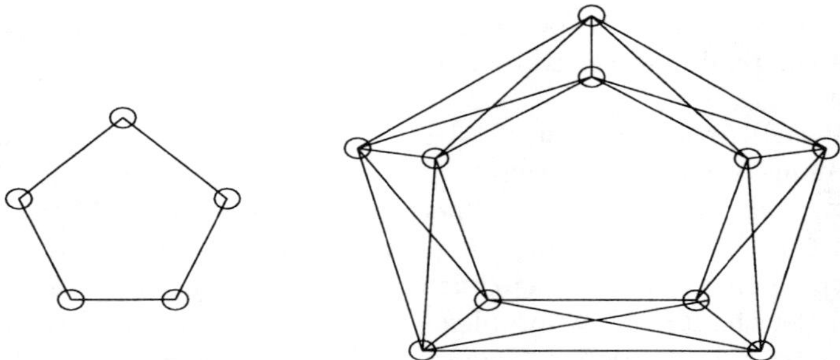

Fig. 1. A trivial coloring problem

However, as Scheinerman and Ullman point out [41], if events can be split so as to occupy two non-contiguous half-hours rather than a single hour, then the five events can be timetabled in 2.5 hours rather than 3. This is equivalent to coloring the 10-node graph shown on the right of Figure 1, derived trivially from its simpler counterpart, with five colors (one per half-hour). As yet, we know of no timetabling systems that cater automatically for such reformulations of a problem.

In what follows the reader is assumed to be familiar with the basic terminology and ideas of genetic algorithms (GAs) and evolutionary computation.

2 Background

Perhaps the earliest application of GAs to timetabling was by Colorni et al [11]. That work tackled a modest class timetabling problem for a school in Milan. The problem was represented as a matrix: each row corresponded to a teacher, each column to some hour of some day, each entry to a class/room pairing. Careful design and repair ensured that each individual row (essentially a permutation of the teacher's duties) was always self-consistent. A penalty-based fitness measure was used for the fitness. Specialized crossover and mutation were used; crossover sorted the rows of one parent according to their fitness contribution, and the best rows of the first parent were copied to the first child, with the rest of that child being filled from the second parent, using those rows (teachers) not already present. A simple test suggested that crossover did indeed contribute usefully to the result.

The field has since progressed rapidly. At the second international conference on automated timetabling, PATAT-97, held in August 1997 in Toronto, almost a quarter of the papers presented used some kind of evolutionary technique. At the previous Practice and Theory of Automated Timetabling (PATAT) conference held in 1995 in Edinburgh, roughly one-third of the papers published were based on evolutionary techniques – these figures suggest that evolutionary computing has become an established and successful methodology for tackling a wide variety of timetabling problems, alongside techniques from operations research such as graph-coloring, constraint logic programming and, more recently, tabu search and simulated annealing. The interested reader can find some discussion of these other techniques, and many references, in [9,10,40].

3 What Kind Of Problems Are Currently Tackled?

The majority of published work now seems to be on applications to real-world timetabling problems, and not just reduced or toy problems. Besides school and university lecture and exam timetabling, other applications such as employee rostering are also starting to appear. There is still a large concentration of effort on university timetabling problems, possibly unsurprising since the majority of this work is done in universities. This is also reflected in the number of practical applications of evolutionary algorithms actually in use – though there are several implementations of university course and exam timetabling programs, there is currently almost no evolutionary school timetabling software commercially available. EvoSchool from Genetica Advanced Software Architectures S.r.L of Milan [23] is a notable exception, but only available in Italian at the time of writing. This is despite the fact that there are several non-evolutionary software systems available for schools, for instance SchoolMagic [42] written in Japan and Mimosa [32], produced in Finland.

Research publications on evolutionary approaches to employee timetabling are still rare, though several other non-evolutionary approaches to this application can be found at both PATAT conferences. Meisels and Lusternik [29] briefly comment on the use of GAs for solving employee timetabling problems as part of a more general study on phase transitions, and at PATAT-97 there was an example of using a genetic algorithm for preacher timetabling, a novel application, but one that has similar features to many other timetabling problems and therefore is useful to compare to other more standard problems. The only other detailed study appears to be that by Easton and Mansour [16] in 1993. Other timetabling areas represented at the two timetabling conferences but so far ignored by evolutionary techniques are that of sports timetabling and nurse timetabling. This may be because such problems contain sufficient structure, or are small enough, that conventional techniques are satisfactory; or simply that no-one has yet explored evolutionary methods in these areas.

Many of the practical problems considered in recent publications are large, which is a good indication of the scalability of the techniques. For example, Mamede and Rente [26] consider a problem with 7,000 students, where 169 lectures and 211 lessons must be assigned in the timetable, to 396 available hours, using 98 teachers. The scope of the problems tackled is also becoming more realistic – for example, many of the problems at the recent PATAT conference considered many real constraints associated with *practi-*

cal timetabling, taking into account room constraints, travel time between rooms, and people preferences for instance. Evolutionary computing perhaps lends itself well to such multi-constrained problems, due to the ease of incorporating constraint violations into a single fitness function.

A comprehensive survey of the literature on practical applications of automated timetabling was published by Carter in 1986 [8]. This survey was updated in 1995 [9] to include recent developments in practical exam timetabling, and again in 1997 [10] to include practical developments in course scheduling. Both the recent updates include several implementations of GAs, whereas the original 1986 survey showed none.

4 What Techniques Are Used?

All evolutionary algorithms require some method of representing a timetable or problem as a chromosome. A significant proportion of the timetabling community opt for some kind of *direct* representation, where the chromosome directly represents the timetable itself, including all slot assignments, teacher assignments and room assignments. For examples of this method, see [1,4,7,19]. Using a direct representation, it is straightforward to make that representation highly problem-specific. Intuitively, it seems surprising that direct encodings can work. Consider a simple example, in which the chromosome consists of an array of integers and the value of the e-th integer specifies which time-slot to put the e-th event into. If there are 100 events and 10 time-slots (a very modest-sized problem) then there are 10^{100} possible chromosomes, and in any initial population of size 100 (say) the large majority of low-order building blocks will not appear – there isn't enough space for them. Random mutation is also unlikely to work fast enough to bring desirable 'missing' building blocks into the population.

However, although there may be a large space, there are also likely to be very large numbers of desirable solutions to be found. If, for example, one assumes that near-clashes cannot happen across day boundaries and there are no day-specific constraints, then new timetables could be generated from any satisfactory one simply by permuting the days. In practice, direct representations often work on small- and medium-sized problems for such reasons (notably when mutation is not random but involves some degree of improving local search), but there is still a danger that some hard-to-see features of the problem will cause there to be relatively few solutions to be found. It is also worth remembering that GAs not only search by 'implicit parallelism' (because each chromosome is an instance of a very large number of schemas, so the schema sampling rate can be very high, although biased), but also operate on all the parts of a chromosome in parallel. Consider the example in Figure 2 from [38].

The problem consists of a set of cliques, of size 5 in this figure; the first overlaps the second at one node, the second overlaps the third at four nodes,

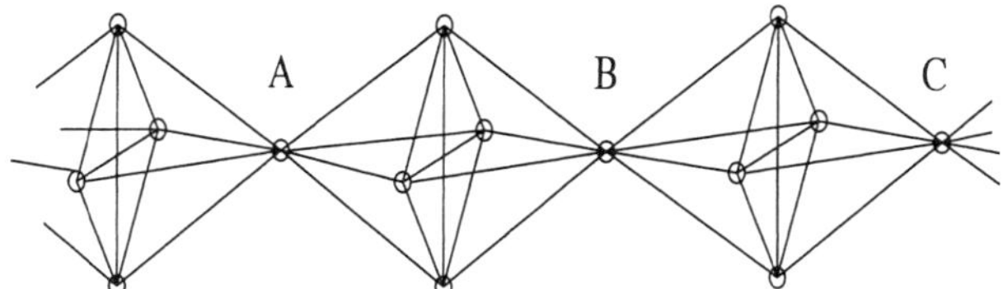

Fig. 2. A difficult problem for direct encodings

the third overlaps the fourth at one node, and so on. The 'linking' nodes at A, B, C are each joined to four others on either side. If the problem allows five colours/time-slots, it is easy to see that it can be solved. Unfortunately, in every solution the linking nodes must all be the same color, and with a direct encoding it is extremely hard for a GA to achieve this. What is more likely to happen is that the GA 'solves' a clique near the left-hand end of the chain, thereby earning some suitably increased fitness for that chromosome, and also 'solves' a clique near the right-hand end, but these two partial solutions are incompatible. Because of the fitness increase that each brings, they spread through the population and eventually make it impossible for the GA to backtrack out of the deadlock. We have seen this happen.

In any direct representation, offspring resulting from applying an evolutionary operator such as crossover or mutation are likely to be infeasible in that they break a number of hard constraints. There are two main strategies for handling such issues – in the first, infeasible solutions are repaired before being placed in the population, eg [7,30]. Arguably, there is again a risk that the purely local search used to find a repaired chromosome can also generate deadlocks whose effects do not become apparent until it is too late. The alternative approach is to allow infeasible solutions to exist, but to penalize them heavily through a weighted penalty function. Since it can be impossible to exactly capture the true problem in timetabling (people change their constraints, or simply forget to supply some of them), a penalty function approach which provides a heavy enough bias against breaking the declared hard constraints can be a reasonable thing to use.

By contrast, in *indirect* encodings, a chromosome typically represents an ordered list of events which are placed into a timetable according to some pre-defined method by a timetable *builder*. The timetable builder can use any combination of heuristics and local search to place events into the timetable. Instances of the use of indirect representations include, for example, the work by Paechter et al [34].

Corne and Ogden [13] directly contrast results obtained on a single problem using two different representations – an indirect one and a direct one – and find that different results are obtained. This obviously suggests some further work is necessary in this area in order to understand the importance of representational issues to the overall success of the technique. Some further comments on this appear later in this chapter.

Whatever the representation, all methods require some method of dealing with violations of constraints in the timetable. In realistic problems, constraints will often vary in their importance, and there can be large numbers of them – many of the papers surveyed include between 10 and 20 types of constraints. Much of the published work uses some kind of penalty-based fitness function in order to quantitatively score timetables. Common ways of implementing such a function are:

1 Use a weighted penalty function, where weights reflect the importance of each constraint. This can be linear combination of weighted violations, but may also include some exponential terms, although Caldeira and Rosa [7] warn that values used in exponential functions must be chosen very carefully.
2 Use a two-step penalty function in which soft constraints are only considered once the hard constraints have been satisfied [26].

Many people note that choosing the correct weights can be extremely difficult, and in multi-objective problems they can critically influence the solution quality. Paechter et al [33] try to circumvent this difficulty by allowing the weights to be manually changed during the algorithm, with a visual interface indicating the effect of the changes on the current violations of each constraint, so effectively the algorithm can be hand-tuned to provide an acceptable solution. From a more practical point of view, others have commented that it is often difficult to obtain precise information from administrators as to exactly what constitutes a good timetable, and which constraints are more important than others. Often, the requirements of a timetable are defined in terms of 'there must be fewer violations of constraint A than constraint B', rather than 'constraint A is twice as important as constraint B', which makes it difficult to assign suitable weights. Some method of reporting current constraint violations, combined with an ability to interactively change the weights, may help in such cases, although successful interaction relies on the algorithm running fast enough for this to be practicable without over-taxing the user's patience.

Salwach [39] argues that a penalty-based fitness function is not effective, as using a sum of penalties results in a discontinuous fitness function and small changes in the chromosome can result in large changes in the fitness function. Salwach advocates a novel approach in which the timetabling problem is broken down into smaller problems – once a sub-problem has been solved by the GA, its size is increased and the process repeated until the whole problem is solved. At each stage, the fitness is given by a modified penalty function. A solution incurs penalties for:

- violation of the hard constraints;

- the fact that any full solution built from the partial one will *surely* violate the hard constraints;
- the possibility that any full solution built on the partial one will *probably* violate some of the hard constraints. This is estimated by use of heuristics.

Salwach argues that this circumvents the problems of a weighted penalty function as:

- its value depends directly on violation of constraints in the partial solution only, where there are fewer penalties – this results in a less complicated function with fewer local optima;
- its value depends indirectly on all constraints; this indirect dependency is encoded in the form of heuristics that estimate the chance of the full solution satisfying all constraints. This results in less isolated optima which are more easily found.

A somewhat different approach to the use of weights is taken by Paechter et al [33]. They note that weights can be used for two different purposes:

1 to specify the final goals of the timetable;
2 to decide how to select timetables for reproduction.

Their method uses weights to signify the importance of constraints, but avoids the need for calculating a weighted penalty sum for each timetable. In this case, the weights are used only to *rank* timetables according to quality. The method, described below, allows the timetables to be ordered in terms of superiority. The GA uses tournament selection in order to select timetables for mating, with tournament size 2, and hence an exact quantifier of fitness is unnecessary. The method proceeds as follows:

- Consider a pair of timetables, A and B, chosen at random, and a set of i constraints, each of which has an associated weight, indicating its significance.
- Consider each constraint i in turn – if timetable A has fewer violations of constraint i than timetable B, assign timetable A the weight associated with constraint i.
- Total the weights accrued by each timetable – the one with the higher total is deemed fittest and selected in preference

Several people advocate the use of hybrid genetic techniques that include some element of local search. This can take the form of a smart mutation operator [37] or memetic search, for example see [35], where local search around the genome is performed at each evaluation, with the results written back to the chromosome (Lamarckian evolution). In the commercially available ACT system, [25], an evolutionary algorithm is combined with hill-climbing and best-first search in order to produce good timetables. Various methods of local search are proposed, such as simulated annealing, deterministic hill-climbing

[4] and stochastic hill-climbing. Combining the powerful search capabilities of the GA with refinements from other non-evolutionary methods may provide the 'icing on the cake' for the future in this field. However, as noted before, local search that tries too hard for improvement may find that the best improvement comes from fixing many straightforward constraint violations but in the process breaking a few others that later turn out to be crucial. Sometimes a gentle pressure to find improvements through local search can produce much better overall performance [14]. Note that nearly all the methods used, both evolutionary and non-evolutionary, allow the possibility of breaking some already-satisfied constraint as a step on the way to finding a good result.

Many evolutionary algorithms encounter common problems with early convergence which may lead to sub-optimal solutions. An interesting method to deal with this is proposed by [26] and uses a simulated annealing cooling schedule to adjust the weights of each of the penalties associated with remaining constraint violations each time the population converges, until a satisfactory solution is found. Mendes dos Santos et al [30] propose a diversification procedure based on scatter search [24] whenever the population begins to converge.

Selecting the correct operators to use is another common problem associated with GAs, and is often a trial-and-error empirical procedure. Several people suggest the use of adapting operator probabilities to determine the best operator to apply at each stage of the evolution, for example [26,1], which evolve alongside the solution. Both Adamidis and Arakapis [1] and Kim and Chung [25] report that they have obtained better results on their particular problems using mutation operators only, with no crossover whatsoever. However, it would seem that many researchers do not even explore such questions.

Finally, note that it is entirely possible that a given practical problem has no feasible solution at all. From the user's point of view it may still be desirable to get close, in order to cut down on the number of hard constraints that might need to be re-negotiated with those concerned. For this, weighted penalty methods may sometimes be more suitable than all-or-nothing techniques such as those that try to build a timetable sequentially, always obeying every hard constraint.

5 Three Case Studies

This section provides some examples of current GA-based timetabling systems that are being used to provide practical solutions, giving a flavor of what is currently achievable.

A commercially available evolutionary timetabling systems is *ACT* [25] – **A**utomated **C**lass **T**imetabler – which is a recent system for university timetabling, running under Windows 95, and developed in Korea. The system uses a hybridized algorithm which incorporates hill-climbing, a GA and also a best-first heuristic search algorithm. It is able to solve all hard constraints and optimists soft constraints in a two-phase approach. The system produces class timetables, professor timetables and room timetables. It has been successfully tested in 15 universities in Korea, and can cope with reasonably large problems. In the largest university tested, timetables were evolved for 70 departments, with 700 professors, and 4,000 subjects, satisfying 5 hard constraints and optimizing up to 10 soft constraints. Using a Pentium 133MHz processor, timetables can be produced in around 20 minutes, and hence the system is extremely fast. It contains a straightforward graphical user interface based on MS Windows for data entry. An added feature is a manual scheduling function that enables any lesson attributes to be changed after the automated procedure has finished, with guidance from the system as to the possible changes that can be made.

Several other universities are using evolutionary timetabling systems to schedule lectures and exams within the university. At Napier University, in Edinburgh, a memetic algorithm-based timetabling system has been developed, known as *Neeps and Tatties* [33,31]. It has been successfully used by the university to schedule the entire university timetable. This is a large problem – for example, in the first semester of 1997 there were 2,067 events to place in 45 time-slots and 183 rooms. These were attended by 669 lecturers and 978 student groups, and the timetables produced had to respect 12 different criteria. The algorithm was run on a single machine, and takes about an hour to satisfy all hard constraints. From then on, it begins to optimism the soft constraints and can be run indefinitely. In practice, the algorithm was left running for about three days. However, a great advantage of the program is that it can be seeded with previous results, which allows solutions to evolve gradually as the changed data is progressively entered, or that the program can be seeded with a solution from a previous year to speed things up. A novel feature of this program is the ability to interactively change the weights of the soft constraints as the program runs, and see the results with immediate effect, which is of invaluable help to an administrator trying to find a satisfactory timetable. The representation includes a permutation of the order in which to consider events when building a timetable, and also a short list (typically 1–4) of suggested time-slots for each event. In addition to a special-purpose crossover for managing suggestion lists, there are several types of mutation employed: random; steal a time-slot from another event; swap time-slots with another event.

Another system was developed at the University of Edinburgh, where it is used each year to schedule examinations. This is known as *GATT* [22], standing for *G*enetic *A*lgorithm *T*ime *T*abler, and was developed from ideas and experiments described in [14]. This software was also used by Harvard Business School to schedule their examinations, and was found to outperform a commercially available timetabling system in trials. In a direct head-to-head competition with the commercial (non-GA-based) software, exams were scheduled over 15 slots – *GATT* found a schedule with two students with direct conflicts, involving two courses. The commercial software came up with 14 conflicts when running automatically, while its manual attempt to develop a schedule came up with fewer, with only five students with conflicts! *GATT* has also been tested on a real timetabling problem from a secondary school in Belgium, which involved scheduling classes and teachers for the first two year groups. The problem involved 10 different class groups, 26 different teachers, and 318 different courses. In total, 5,320 edge constraints were involved. The constraints included a significant number of specifications and exclusions, where an event was either pre-assigned to a particular slot, or excluded from a range of slots. Defever [15] reports that all clash violations were removed by *GATT*. He does not specify, however, if other constraints, such as room constraints or proximity constraints, were considered. *GATT* uses a simple direct representation, typically small populations and local search that seeks modest rather than locally optimal improvements, with a penalty-based fitness scheme.

6 Acquiring Problem Data

Despite the availability of benchmark data, very few papers contain any comparison between different techniques. Furthermore, it is regrettably very rare for authors to compare results with previously published results or with established benchmarks. This points to the fact that current research seems mostly to be based on isolated efforts, and that the timetabling community is not yet sharing data and ideas as much as it should. Some data is publicly available from these sites:

```
ftp://ie.utoronto.ca/mwc/testprob/
ftp://ftp.cs.nott.ac.uk/ttp/Data/Nott94-1/
ftp://ftp.dai.ed.ac.uk/pub/gatt/ttdata/
```

There has been some discussion in the community recently regarding the need for a standard data format for specifying timetabling instances, in order to facilitate more sharing and cross comparison of data. Mata *et. al* [27] propose two approaches for specifying a generic timetabling problem, one based on logic, written in Prolog, and the other based on relational algebra, written in SQL. Burke et al [5] propose a new specification language, based on the *Z* specification language [36] that is functional in nature. Whilst the aims and benefits of having a general language are clear, the adoption of any new language is likely to encounter a great deal of inertia due to the time and effort required by researchers to re-specify any data. It is arguable that simply having a single web-site with large numbers of test problems can achieve the same aims, if each problem is clearly described, and in the short term will result in data sharing more rapidly. The ready availability of tools such as *awk* makes conversion of data between formats relatively straightforward. It is, however, an open-ended question which ought to be addressed in the near future.

7 The Future

A solid foundation of evolutionary timetabling work now exists, from which it should be possible to both consolidate and expand. Many real-life problems are being tackled, that are both difficult and complex, and of considerable size, over a whole range of classes of problems. However, a word of caution is required – work by Ross et al [38] shows that although a GA can be extremely successful on a wide range of problems, there are some classes of timetabling problems that a direct-encoding GA finds impossible or extremely difficult to solve. In particular, consider the kind of problems shown in Figure 3. The idea

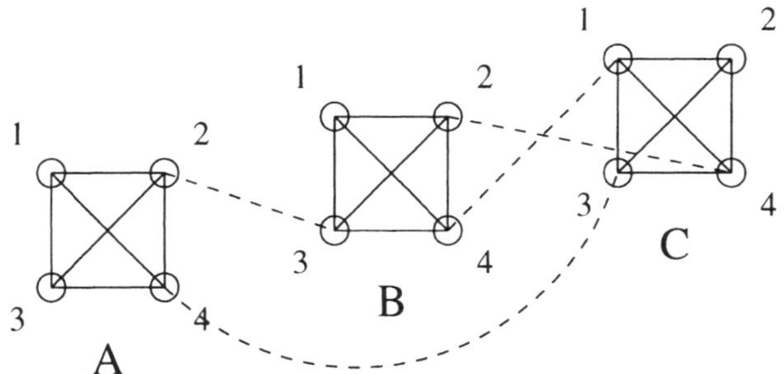

Fig. 3. An awkward class of problems

here is to build a problem solvable with n time-slots by first defining a series of identical cliques A,B,C,... each of size n and specifying the same solution for each. That is, A1 and B1 and C1 (etc.) will be in one time-slot; A2, B2, C2 (etc.) will be in another, and so on. Then extra constraints can be added that do not violate the desired solution, as illustrated by the dashed lines. It turns out that as more constraints get added, the problems become hard for a wide variety of direct-encoded GAs. But as even more constraints get added, the problems again become readily solvable by those same GAs. The same phenomenon, a transition from easy to hard to easy as constraints are added while preserving solvability, is found with a range of non-evolutionary timetabling algorithms as well. The phenomenon seems to arise because all the algorithms depend to some extent on local search, but within a certain range the problems do not contain enough structure at the local level to guide the search towards the global solution. It can be shown that certain other non-evolutionary algorithms can solve these same problems with ease.

Reviewing the literature reveals few other systematic studies of the performance of evolutionary algorithms across *ranges* of problems. Generally, results are reported on isolated cases. In-depth studies of GA performance across problems as the parameters of that problem change would probably give us greater insight into the *robustness* and more *general* capabilities of the GA as a general tool for solving timetabling problems. Such problems could be easily generated artificially and could prove extremely useful to the evolutionary computing community as a set of benchmarks.

As suggested in [38], rather than dealing with permutation-based representations or direct encodings, it may be preferable to use GAs to customize some simpler algorithm to exploit features of the problem. For example, Brelaz' greedy algorithm [2] can easily be modified to take account of seating capacity, near-clashes and other constraints when prioritizing events for placement into the developing timetable. But without some very CPU-intensive problem analysis, it can be hard to discern whether it is really necessary to pay attention to room-packing issues at the start, at the expense of near-clash constraints, and so on. A GA can be used to explore this without paying the price of a detailed problem analysis. Some good results using this approach can be found in [43], where the idea is simply to choose a pair of heuristics (one to pick events, one to insert them) that are used for a while, a threshold for when to switch, and then a second pair of heuristics to use for the same purpose thereafter. This works quickly and well even on Carter's large problems

Corne [12] also notes that acquiring knowledge about how constraints interact with each other in multi-constrained, multi-objective, complex problems may give us some clues about how to design algorithms that deal with these problems. He suggests the use of a technique known as Configuration Space Analysis, described by Mattfeld [28] for analyzing the search space and providing clues about its complexity. More such analysis of many different problems may enable us to begin to characterize classes of problems, and use this knowledge when attempting to design an algorithm for a new problem. Furthermore, characterization of problems could provide us with useful information about the transferability of techniques across domains.

As noted in section 4, there remains a great deal of work to be done in investigating the precise nature of the effect of problem representation on the outcome of a problem. Also still to be properly resolved is the issue of dealing with weights, and how to assign them. An area worthy of investigation would be to examine how the distribution of weights alters the shape of the search space, and the ease of traversing it. For instance, a (simplified) timetabling problem in which the desired outcome is to remove all violations of constraint A and to minimize violations of constraint B may well be best solved by first concentrating on removing B violations by assigning B a higher weight and then reversing the weight assignments at some later point in the algorithm. See [17,18,21,20] for some exploration of related ideas in the areas of satisfiability and graph-coloring.

Acknowledgements

Some of the survey work was supported by EvoNET, the European Network of Excellence in Evolutionary Computing. Emma Hart has been supported by EPSRC grant GR/L22232.

References

1. Adamidis, P. and Arakapis, P.(1997) Weekly lecture timetabling with genetic algorithms. In Burke and Carter, *Proc. of PATAT-97, Toronto, Canada*, pp.115–122, University of Toronto (ISBN 0-7727-6703-3)
2. Brelaz, D. (1979) New methods to colour the vertices of a graph. *Communications of the ACM*, 22, pp.251–256
3. Burke, E.K. and Carter,M. (editors) (1997) *Proceedings of the 2nd International Conference on the Practice and Theory of Automated Timetabling*
4. Burke, E.K., Newall, J. P., and Weare, R. F.(1995) A memetic algorithm for university exam timetabling. In Burke and Ross [6], pp.241–250
5. Burke, E. K., Kingston, J. H., and Pepper, P. A. (1997) A standard data format for timetabling instances. In Burke and Carter [3], pp.213–222
6. Burke, E. K. and Ross, P. M. (editors) (1995) *Proceedings of the 1st International Conference on the Practice and Theory of Automated Timetabling*, Springer-Verlag Lecture Notes in Computer Science 1153
7. Caldeira, J. P. and Rosa, A. C. (1997) School timetabling using genetic search. In Burke and Carter, *Proc. of PATAT-97, Toronto, Canada*, pp.115–122, University of Toronto (ISBN 0-7727-6703-3), 1997, also available directly from http://laseeb.ist.utl.pt/projects/aghora/agh_pub.html
8. Carter, M. W. (1986) A survey of practical applications of examination timetabling algorithms. *Operations Research*, 34(2), pp.193–202
9. Carter, M. W. and Laporte, G. (1995) Recent developments in practical exam timetabling. In Burke and Carter [6], pp.3–21
10. Carter, M. W. and Laporte, G. (1997) Recent developments in practical course timetabling. In Burke and Carter [3], pp.3–19
11. Colorni, A., Dorigo, M., and Maniezzo, V. (1990) A genetic algorithm to solve the timetable problem. Technical Report 90-060, Politecnico di Milano
12. Corne, D. (1997) Tradeoffs and phase transitions in the interactions between exam-splitting, clash, near-clash, and room capacity constraints. In Burke and Carter, *Proceedings of PATAT-97, Toronto, Canada*, pp. 318–320, University of Toronto (ISBN 0-7727-6703-3)
13. Corne, D. and Ogden, J. (1997) Evolutionary optimisation of Methodist preaching timetables. In Burke and Carter [3], pp.142–155
14. Corne, D., Ross, P., and Fang, H. L. (1995) Evolving timetables. In L. Chambers, editor, *Practical Handbook of Genetic Algorithms: Applications, Volume 1*, chapter 8, pp. 219–276. CRC Press

15. Defever, S. (1997) Genetic algorithms for timetabling. Master's thesis, Katholieke Universiteit Leuven
16. Easton, F. and Mansour, N. (1993) A distributed GA for employee staffing and scheduling. In Stephanie Forrest, editor, *Proceedings of the Fifth International Conference on Genetic Algorithms*, pages 360–367. Morgan Kaufmann
17. Eiben, A. E., and van der Hauw, J. K. (1997) Solving 3-SAT by GAs adapting constraint weights. In *IEEECEP: Proceedings of The IEEE Conference on Evolutionary Computation, IEEE World Congress on Computational Intelligence*
18. Eiben, A. E., van der Hauw,J. K., and van Hemert, J. I. (1998) Graph coloring with adaptive evolutionary algorithms. *Journal of Heuristics*, 4(1), pp.25–46
19. Fernandes, C., Melicio, F., Caldeira, J. P., and Rosa, A. (1999) High school weekly timetabling by evolutionary algorithms. In *Proceedings of ACM Symposium on Applied Computing*, San Antonio, pp.344–350, ACM Press
20. Frank, J. (1996) Learning short-term weights for GSAT. Technical Report CSE-96-14, Department of Computer Science, University of California, Davis
21. Frank, J. (1996) Weighting for Godot: Learning heuristics for GSAT. In *Proc. of AAAI-96*, pp.338–343. AAAI Press / The MIT Press
22. GATT. ftp://ftp.dai.ed.ac.uk/pub/gatt/gatt-1.8/
23. Genetica Advanced Software Architectures SrL. Evoschool software package. http://www.genetica-soft.com
24. Glover, F. (1994) Genetic algorithms and scatter search: Unsuspected potentials. *Statistics and Computing*, 4, pp.131–140
25. Kim, M. J. and Chung, T. C. (1997) Development of an automated course timetabler for university. In E. Burke and M. Carter, *Proceedings of PATAT-97, Toronto, Canada*, pp.182–186, University of Toronto (ISBN 0-7727-6703-3)
26. Mamede, N. and Rente, T. (1997) Repairing timetables using genetic algorithms and simulated annealing. In Burke and Carter, *Proceedings of PATAT-97, Toronto, Canada*, pp.187–204, University of Toronto (ISBN 0-7727-6703-3)
27. Mata, J. M., Senna, A. L., and Andrade, M. A. (1997) Towards a language for the specification of timetabling problems. In Burke and Carter, *Proc. of PATAT-97, Toronto, Canada*, pp.330–333, University of Toronto (ISBN 0-7727-6703-3)
28. Mattfeld, D. C. (1996) *Evolutionary Search and the Job-Shop*. Physica-Verlag
29. Meisels, A. and Lusternik, N. (1997) Experiments on networks of employee timetabling problems. In Burke and Carter [3], pp.130–141
30. Mendes, A. dos Santos, Marques, E. and Saturo Ochi, L. (1997) Design and implementation of a timetable system using a genetic algorithm. In Burke and Carter, *Proc. of PATAT-97, Toronto, Canada*, pp. 347–348, University of Toronto (ISBN 0-7727-6703-3)
31. Neeps and tatties. httP://www.dcs.napier.ac.uk/ benp/tatties/tatties.htm
32. Mimosa Software Oy. Mimosa. http://www.mimosasoftware.com/.
33. Paechter, B., Rankin, R. C., Cumming, A., and Fogarty, T. C. (1998) Timetabling the classes of an entire university with an evoltuionary algorithm. In Eiben, A. E. T. Bäck, Schoenauer, M. and Schwefel, H. P. editors, *Parallel Problem-Solving from Nature - PPSN V*, pp.865–874, Springer-Verlag
34. Paechter, B., Cumming, A., Norman, M. G., and Luchian, H. (1995) Extensions to a memetic timetabling system. In Burke and Ross [6], pp.251–265
35. Paechter, B., Rankin, R. C., and Cumming, A. (1997) Improving a lecture timetabling system for university-wide use. In Burke and Carter [3], pp.156–165

36. Potter, B. (1991) *An Introduction to Formal Specification and Z.* Prentice Hall
37. Ross, P., Corne, D., and Fang, H.S. (1994) Improving evolutionary timetabling with delta evaluation and directed mutation. In H.-P. Schwefel, Y. Davidor and R. Manner, editors, *Parallel Problem Solving from Nature III*, LNCS, pp.560–565. Springer-Verlag
38. Ross, P., Hart, E., and Corne, D. (1997) Some observations about GA-based exam timetabling. In Burke and Carter [3], pp.115–129
39. Salwach, W. (1997) Genetic algorithms in solving constraint satisfaction problems: The timetable case. In *Badania Operacyjne i Decyzje*
40. Schaerf, A. (1999) A survey of automated timetabling. *Artificial Intelligence Review*, 13(2), pp.87–127, Kluwer
41. Scheinerman, E. and Ullman, D. (1997) *Fractional Graph Theory.* Wiley
42. School Magic (1998) http://www.nec.co.jp/english/today/newsrel/9811/1601.html
43. Terashima-Marin, H. Ross, P. M., and Valenzuela, M. (1999) Evolution of constraint satisfaction strategies in examination timetabling. In W. Banzhaf et al, editors, *Proceedings of the Genetic and Evolutionary Computation Conference - GECCO-99*, pp.635–642, Morgan Kaufmann
44. van Lint, J. H. and Wilson, R. M. (1992) *A Course in Combinatorics.* Cambridge University Press
45. Wilson, R. J. (1985) *Introduction to Graph Theory.* Longman, 3rd ed edition

Machine Learning by Schedule Decomposition – Prospects for an Integration of AI and OR Techniques for Job Shop Scheduling

Ulrich Dorndorf, Erwin Pesch and Toàn Phan Huy
FB 5 – Management Information Systems, BWL 3, University of Siegen, 57068 Siegen, Germany
E-mail: udorndorf@acm.org; pesch@fb5.uni-siegen.de; phanhuy@toan.de

Summary. A survey of recent solution approaches as well as a class of approximation algorithms is provided for solving the minimum makespan problem of job shop scheduling. We briefly review most recent exact approaches as well as neighbourhood search methods and evolution based heuristics. Genetic algorithm-based meta-strategies serve to guide an optimal design of scheduling decision sequences. Simple sequences of dispatching rules for job assignment as well as learning of promising sequences of one machine and multiple job decompositions are considered. Finally, a number of ways for introducing problem-specific knowledge through constraint consistency tests for propagation will be presented. These ideas are applied in a subproblem based constraint propagation approach that learns to find the best bounds for fixing arc directions. Calculation of the initial lower bounds for the subproblems' "best bounds" uses a branch and bound search. Whenever some problem-specific knowledge through constraint propagation leads to a partial solution of the job shop problem, a complete solution can be obtained with either a branch and bound procedure or some heuristic neighbourhood or priority rule based search. Computational experiments show that the approach can find shorter makespans than other local search approaches. The chapter provides an initial framework for a unified solution approach to many combinatorial optimization problems incorporating techniques from artificial intelligence, e.g. evolutionary algorithms and constraint propagation, and operations research, e.g. metaheuristics and branch and bound.

1 Introduction

We will study scheduling decisions for the well-known job shop scheduling problem which remains a permanent challenge in practice and research. An informal description is as follows:

A finite set of jobs each of which consists of a sequence of operations has to be scheduled with the objective of minimizing the makespan, which is the maximum completion time of all jobs. Each operation has a specific processing time and must not be interrupted during its processing. We assume that transportation and setup times are sequence independent and are included in the processing times, so that the makespan is influenced by the processing and waiting times. Two kinds of constraints are considered: conjunctive and disjunctive constraints. The conjunctive constraints specify the flow of material through the different jobs, i.e. the logical order in which certain operations have to be processed; they define the jobs. The disjunctive constraints model the resource demand given a scarce but renewable resource supply. More precisely, the capacity of each resource (machine) is one unit per processing period. An operation uses exactly one machine at any time of its processing. Thus, two operations requiring a common machine cannot be processed in parallel.

With good reason the job shop scheduling problem has been called one of the most intractable problems. This view is best supported by the notorious 10×10 problem instance proposed in 1963 which resisted any solution attempts for two decades and was only solved more than 25 years later. Amazingly, due to the evolution of solution techniques and growing computational power, this formerly unsolvable instance can now be solved in a few seconds and thus, in spite of remaining an important benchmark for new solution methods, no longer comprises a real challenge. Consequently, the research interest has shifted in recent years to the study of larger problem instances, and the intrinsic structure of the job shop scheduling problem, both as intermediate steps to the solution of problems of higher practical relevance.

When solving difficult optimization problems, one is always faced with the contrary goals of general applicability and high efficiency of the solution methods developed. This roughly describes the contrast between the fields of artificial intelligence and management science. While the former deals with general problem-solving models and methods, the latter is mainly focused on the development of efficient solution methods for special and well-distinguished problems. Only in the most recent years has the opinion gradually gained ground that both research fields are highly complementary, and quite impressive results have been obtained by combining general solving paradigms with problem-specific knowledge. In this context, constraint propagation techniques can be seen as an interface between artificial intelligence, from which they originate, and management science, in which they are applied with growing success.

Constraint propagation is a general paradigm to simplify search and optimization problems by reducing the search space and has contributed a lot to the improvement of solution methods. Roughly speaking, constraint propagation derives and makes use of implicit constraints through the evaluation of variables, domains and constraints that define the set of solutions of a particular problem instance.

We introduce the basic exact and heuristic solution ideas for job shop scheduling that have heavily influenced current constraint propagation approaches, and review approximation algorithms for the job shop scheduling problem. Since constraint propagation alone is not sufficient for solving a problem, it has to be embedded in a search algorithm which explores the reduced search space.

Finally, we explain how to combine constraint propagation techniques with a problem decomposition approach. The resulting algorithm simplifies the solution of the job shop through the fixing of processing sequences of operations.

Problem-specific knowledge is introduced by decomposition of the whole problem into single machine problems for which genetically guided constraint based reasoning may provide a substantial increase of the knowledge base.

2 The Job Shop Model and Complexity

A job shop consists of a set of different machines (like lathes, milling machines, drills etc.) that perform operations on jobs. Each job has a specified processing order through the machines, i.e. a job is composed of an ordered list of operations each of which is determined by the machine required and the processing time on it. There are several constraints on jobs and machines: (i) there are no precedence constraints among operations of different jobs; (ii) operations cannot be interrupted (non-preemption) and each machine can handle only one job at a time; (iii) each job can be performed only on one machine at a time. While the machine sequence of the jobs is fixed, the problem is to find the job sequences on the machines that minimize the makespan, i.e. the maximum of the completion times of all operations. It is well known that the problem is NP-hard (see Lenstra and Rinnooy Kan [63], Rinnooy Kan [87], Lenstra [62]) and belongs to the most intractable problems considered.

A huge amount of literature on machine scheduling, including scheduling of job shops, has been published within the last 35 years, among others the books by Conway et al. [32], who wrote the first book on scheduling theory, or Baker [9]. A broader view on production and operations scheduling is provided by Pinedo [85], Blazewicz et al. [18], Tanaev et al. [90, 91], or Chretienne et al. [31].

A variety of scheduling rules for certain types of job shops have been considered in an enormous number of publications, see the excellent surveys, also covering complexity issues and mathematical programming formulations, by Manne [67], Blazewicz et al. [17, 18], Lawler et al. [61], Blazewicz [15], Lageweg et al. [60], Graham et al. [53].

There are different problem formulations and we have adopted the one presented by Adams et al. [4]. Let $V = \{0, \ldots, n\}$ denote the set of operations where 0 and n are considered as dummy operations "start" (the first operation of all jobs) and "end" (the last operation of all jobs), respectively. Let M denote the set of m machines and A be the set of ordered pairs of operations constrained by the precedence relations for each job. For each machine k, the set E_k describes the set of all pairs of operations to be performed on machine k, i.e. operations which cannot overlap (cf. (ii)). For each operation i, its processing time p_i is fixed, and the processing start time of i is st_i, a variable that has to be determined during the optimization. Hence, the job shop scheduling problem can be modeled as follows:

$$\min st_n$$

s.t.

$$(1)\ st_j - st_i \geq p_i \qquad \forall\, (i,j) \in A,$$

$$(2)\ st_j - st_i \geq p_i \quad \text{or} \quad st_i - st_j \geq p_j \qquad \forall\, \{i,j\} \in E_k,\ \forall\, k \in M,$$

$$(3)\ st_i \geq 0 \qquad \forall\, i \in V$$

Restrictions (1) ensure that the processing sequence of operations in each job corresponds to the predetermined order. Constraints (2) demand that there is only one job on each machine at a time, and (3) assures completion of all jobs. Any feasible solution to the constraints (1), (2) and (3) is called a schedule.

An illuminating problem representation is the disjunctive graph model due to Roy and Sussman [88], see also Balas [10] and Blazewicz et al. [20]. In the edge-weighted graph there is a vertex for each operation; additionally there exist two dummy operations, with zero processing times, vertices 0 and n, representing the start and end of a schedule, respectively. For every two consecutive operations of the same job there is a directed arc; the start vertex 0 is considered to be the first operation of every job and the end vertex n is considered to be the last operation of every job. For each pair of operations $\{i,j\} \in E_k$ that require the same machine there are two arcs (i,j) and (j,i) with opposite directions. The operations i and j are said to define a disjunctive arc pair or a disjunctive edge. Thus, single arcs between operations represent the precedence constraints on the operations and opposite directed arcs between two operations represent the fact that each machine can handle at most one operation at the same time. Each arc (i,j) is labeled by a weight p_i corresponding to the processing time of operation i. All arcs from 0 have label 0.

The job shop scheduling problem requires an order of the operations on each machine to be found, i.e. to select one arc among all opposite directed arc pairs such that the resulting graph is acyclic (i.e. there are no precedence conflicts between operations) and the length of the maximum weight path between the start and end vertex is minimal. Obviously, the length of a maximum weight or longest path in G connecting vertices 0 and i equals the earliest possible starting time est_i of operation i; the makespan of the schedule is equal to the length of the critical path, i.e. the weight of a longest path from start vertex 0 to end vertex n. Any arc (i,j) on a critical path is said to be critical; if i and j are operations from different jobs then (i,j) is called a disjunctive critical arc, otherwise it is a conjunctive critical arc.

Example

A job shop consists of a set of three machines $M = \{M_1, M_2, M_3\}$ and a set of three jobs $J = \{J_1, J_2, J_3\}$ which are described by the following chains of operations:

$J_1 : 1 \to 2 \to 3, \quad J_2 : 4 \to 5, \quad J_3 : 6 \to 7 \to 8$

For any operation i the required machine M(i) and processing time p_i are as presented in Table 1.

i	1	2	3	4	5	6	7	8
M(i)	M_1	M_2	M_3	M_3	M_2	M_2	M_1	M_3
p_i	3	2	3	3	4	6	3	2

Table 1. Operation requirements

The corresponding disjunctive graph for the given instance is presented in Figure 1 (weights for disjunctive edges are omitted for simplicity).

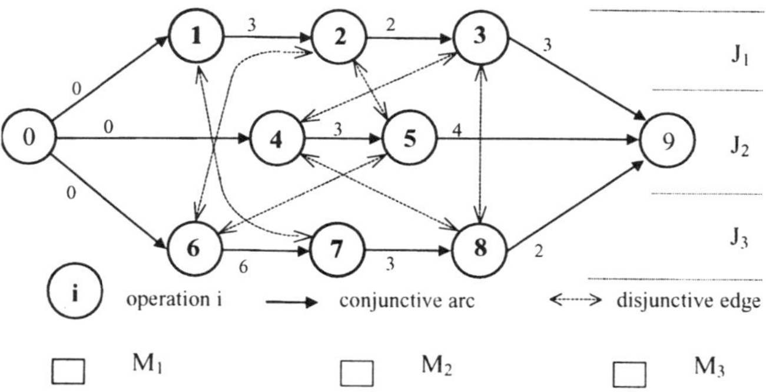

Fig. 1. A disjunctive graph example

The disjunctive graph contains all the information to describe a partial or complete solution of the job shop scheduling problem.

3. Branch and Bound

The history of the job shop scheduling problem, starting more than 35 years ago, is also the history of a well-known benchmark problem consisting of 10 jobs and 10 machines and introduced by Fisher and Thompson in [47]. Since then branch and bound procedures have received substantial attention from numerous researchers. For a comprehensive survey on the job shop problem and different solution strategies, see Blazewicz et al. [16].

The principle of branch and bound is the enumeration of all feasible solutions of a combinatorial optimization problem, say a minimization problem, such that properties or attributes not shared by any optimal solution are detected as early as possible. An attribute (or branch of the enumeration tree) defines a subset of the set of all feasible solutions of the original problem where each element of the subset satisfies this attribute. In general, attributes are chosen such that the union of all attribute-defined subsets equals the set of all feasible solutions of the problem and any two of these subsets do not intersect. For each subset the objective value of its best solution is estimated by a lower bound (bounding). An optimal solution of a relaxation of the original problem such that this optimal solution also satisfies the subset-defining attribute, serves as a lower bound. In case the lower bound exceeds the value of the best (smallest) known upper bound (a heuristic solution of the original problem) the attribute-defined subset can be dropped from further consideration. Otherwise, search is continued departing from the most promising subset which is divided into smaller subsets through the definition of additional attributes. Hence, at any search stage a subset of solutions is defined by a set of attributes all of which are satisfied by these solutions.

The attributes of a branch and bound process exactly correspond to attributes forbidding moves in tabu search. Branching from one solution subset to a new smaller one can be associated with a tabu search move.

Currently the job shop champions among the exact methods are, besides the branch and bound implementations of Applegate and Cook [7] and Martin and Shmoys [68], the branch and bound algorithms of Caseau and Laburthe [30], Carlier and Pinson [27, 28, 29], Brucker et al. [23, 24, 25] and Phan Huy [84]. The power of most methods basically results from some inference rules which describe simple cuts and shaving techniques, and a branching scheme such that operations which belong to a block (a sequence of operations on a machine) on the longest path are moved to the block ends.

One of the main drawbacks of all branch and bound methods is the lack of strong lower bounds in order to cut branches of the enumeration tree as early as possible. Several types of lower bounds are applied in the literature, for instance bounds based on Lagrangian relaxation (see Van de Velde [94]), or bounds based on the optimal solution of a subproblem consisting of only two or three jobs and all machines (see Akers [5] and Brucker [21] or Brucker and Jurisch [22]). However, the most prominent bounding procedure has been described by Carlier [26] and Potts [86]. Consider any operation i in the job shop or in the disjunctive graph that may already include a partial selection of arcs from disjunctive arc pairs. Then there is a longest path from the artifical vertex 0 to i of length est_i as well as a longest path of length q_i connecting the end of operation i to the last one, the dummy operation n. Operation i cannot start to be processed earlier than its release date est_i (also called head) and its processing has to be finished at the latest until its due date $st_n - q_i$ in order to cause no schedule delay. The time q_i is said to be the tail of operation i. There exist m one-machine lower bounds for the optimal makespan of the job shop scheduling problem where each bound is obtained from the exact solution of a one machine scheduling problem with release dates and due dates. Although this problem is NP-complete, Carlier's algorithm quickly solves the one-machine problems optimally for all problem sizes considered in typical job shop instances. In practical implementations its preemptive version reaches almost the same solution quality within an n log n time. Balas et al. [11] describe a branch and bound procedure that can yield improved lower bounds. Their method additionally takes minimum delays between pairs of operations into account.

Carlier [26] and Balas et al. [11] (see also Carlier and Pinson [27, 28, 29], Brucker et al. [24, 25], Caseau and Laburthe [30], and Applegate and Cook [7]) applied a lot of additional inference rules in order to cut the enumeration tree during a preprocessing or the search phase.

Branching in the algorithm of Brucker et al. [24, 25] is restricted to moves of operations which belong to a critical path of a solution obtained by a heuristic based on dispatching rules. For a block B, i.e. successively processed operations on the same machine, that belongs to a critical path, new subtrees are generated if an operation is moved to the very beginning or the very end of this block. In any case the critical path is modified and additional disjunctive arcs are selected.

The branching idea of Martin and Shmoys [68] uses the tightness of time windows of possible start times of operations as a branching criterion. For tight or almost tight windows, where the window size equals (almost) the sum of the processing times of an operation set C, branching depends on which operation in C is processed first. When an operation is chosen to be first the size of its window is reduced. The size of the windows of the other operations in C are updated in order to reflect the fact that they cannot start until the chosen operation is completed. Comparable propagation ideas (after branching on disjunctive arcs) based on time window assignments to operations are considered by Caseau and Laburthe [30]. In both papers, the updating of windows on operations of one machine causes further updates on all other machines. This iterated one-machine windows reduction algorithm generated lower bounds superior to the one-machine lower bound.

4 Approximation Algorithms

4.1 Priority Rules

Priority rules are probably the most frequently applied heuristics for solving (job shop) scheduling problems in practice because of their ease of implementation and their low, usually linear or almost linear time complexity. The algorithm of Giffler and Thompson [48] can be considered as a common basis of all priority rule-based heuristics. It assigns available operations to machines, i.e. operations which can start being processed. Conflicts, i.e. operations competing for the same machine, are solved randomly. The Giffler – Thompson algorithm can generate all active schedules and hence also optimal schedules. As the conflict set consists only of operations, i.e. jobs, competing for the same machine, the random choice of an operation or job from the conflict set may be considered as the simplest version of a priority rule where the priority assigned to each operation or job in the conflict set corresponds to a certain probability. Dorndorf and Pesch [41] applied 12 different rules to decide which of the operations or jobs in the conflict set gets highest priority: "shortest (largest) operation time", "shortest (largest) remaining job processing time", "longest remaining processing time", "random", "first come first serve", "shortest (largest) total processing time", "longest subsequent operation", "fewest (most) remaining job operations"; for an extended summary and discussion see Haupt [56], Panwalker and Iskander [79] as well as Blackstone et al. [14].

4.2 Opportunistic Algorithms

Nowadays, tailored approximation methods viewed as an opportunistic (greedy-type) problem-solving process can yield optimal or near-optimal solutions even for problem instances up to now considered as difficult. Here, opportunistic problem-solving or opportunistic reasoning characterizes a problem solving process where local decisions on which operations, jobs or machines should be considered next, are concentrated on the most promising aspects of the problem, e.g. job contention on a particular machine. Often subproblems defining bottlenecks are extracted and separately solved and serve as a basis from which the search process can expand. Breaking down the whole problem into smaller pieces takes place until, eventually, sufficiently small subproblems are created for which effective exact or heuristic procedures are available. However, the way in which a problem is decomposed affects the quality of the solution reached. Not only the type of decomposition such as machine/resource (Adams et al. [4]), job/order, or event based (Sadeh [89]) have a dramatic influence on the outcome, but also the number of subproblems and the order in which they are considered. In fact, an opportunistic view suggests that the initial decomposition be reviewed in the course of problem solving to see if changes are necessary. The shifting bottleneck heuristic from Adams et al. [4] and its improving modifications from Balas et al. [11] and Dauzere-Peres and Lasserre [35] are typical representatives of opportunistic reasoning. They are resource based as there are sequences of one-machine schedules successively solved and their solutions introduced into the overall schedule.

The shifting bottleneck heuristic from Adams et al. [4] and the improvement from Balas et al. [11] are powerful heuristics for the job shop scheduling problem. The idea is to solve for each machine a one-machine scheduling problem to optimality under the assumption that a lot of arc directions in the optimal one-machine schedules coincide with an optimal job shop schedule. The one-machine scheduling problems in consideration are those which arise from the disjunctive graph model when certain machines are already sequenced. The operation orders on sequenced machines are fully determined. Hence sequencing an additional machine probably results in a change of heads and tails of those operations whose machine order is still open. The one-machine scheduling problems, although they are NP-hard, can quickly be solved using the algorithm of Carlier [26].

Unfortunately, adjusting heads and tails does not take into account a possible already fixed processing order of operations connecting two operations i and j on the same machine, where this particular machine is still unscheduled. So, we get one-machine scheduling problems with heads, tails and time lags (minimum delay between two operations), problems which cannot be handled with Carlier's algorithm. In order to overcome these difficulties an improved SB1 version is suggested by Dauzere-Peres and Lasserre [35] using approximate one-machine solutions. Balas et al. [11] solved these one machine problems exactly. During the local reoptimization part of the SB1-heuristic, for each machine, the operation sequence is redetermined keeping the sequences of all other already scheduled machines untouched.

As the one-machine problems use only partial knowledge of the whole problem it is not surprising, that optimal solutions will not be found easily. This is even more the case because Carlier's algorithm considers the one-machine problem as consisting of independent jobs while some dependence between jobs of a machine might exist in the job shop scheduling problem. Moreover, a monotonic decrease of the makespan is not quarantined in the reoptimization step of Adams et al. Dauzere-Peres and Lasserre [35] were the first to improve the robustness of SB1 and to ensure a monotonic decrease of the makespan in the re-optimization phase and therefore eliminate sensitivity to the number of local reoptimization cycles. Contrary to Carlier's algorithm, they update the operation release dates each time they select a new operation by Schrage's procedure. The quality of the schedules obtained by the SB1-heuristic heavily depends on the sequence in which the one-machine problems are solved. Sequence changes may yield substantial improvements. This is the idea behind the second version of the shifting bottleneck procedure, i.e. the SB2-heuristic, as well as behind the second genetic algorithm approach by Dorndorf and Pesch [41].

Obviously a complete enumeration of the search tree is not acceptable. Therefore a breadth-first search up to a predetermined depth is followed by a depth-first search. In the former case, for a search node all possible branches are considered which result from inclusion of a machine m not yet scheduled. Beyond the depth an extended bottleneck criterion is applied, i.e. there are several successor nodes generated corresponding to the inclusion of the bottleneck machine as well as several other machines.

In analogy to the shifting bottleneck heuristic where the best machine sequence is to be determined we can also aim to find the best sequence of all subproblems defined by a pair of jobs, and successively solve them. Consider a job shop scheduling problem consisting of the set J of jobs. Then a disjunctive arc belongs to exactly one of the $|J|\cdot(|J|-1)/2$ job pairs. A certain amount of arc directions belonging to the optimal job shop schedule coincides with optimal solutions of these job pair problems. Hence, at step k of the job pair heuristic a pair of jobs is chosen that is still connected by disjunctive arcs. Only those disjunctive arcs become directed which result from transitivity via certain other jobs. Respecting these transitivities the new job pair problem is solved to optimality and the obtained arc directions are included into the whole schedule. The final schedule heavily depends on the sequence in which the job pair problems are solved and their arc directions included into the partial schedule. Sequence changes may yield substantial improvements; however, a complete enumeration of the search tree is even less acceptable as in the case of the shifting bottleneck procedure.

4.3 Local Search

In recent years local search-based scheduling has become very popular; for a survey see Anderson et al. [6], as well as Vaessens [92], Vaessens et al. [93] and Aarts and Lenstra [3]. These algorithms are all based on a certain neighbourhood structure and some rules defining how to obtain a new solution from an existing one [33, 83]. First approaches used the very simple neighbourhood structure (N1) in their simulated annealing procedure, cf. [59]:

N1: A transition from a current solution to a new one is generated by replacing in the disjunctive graph representation of the current solution a disjunctive arc (i,j) on a critical path by its opposite arc (j,i).

In other words, N1 means reversing the order in which two operations i and j (or jobs) are processed on a machine where these two operations belong to a longest path. This parallels the early branching structures of exact methods. It is possible to construct a finite sequence of transitions leading from a locally optimal solution to the global optimum. This is a necessary and sufficient condition for asymptotic convergence of simulated annealing.

Another neigbourhood has been used in [2] and can be described as follows:

N2: Consider a feasible solution and a critical arc (i,j) defining the processing order of operations i and j on the same machine, say machine m. Define p(i) and s(i) to be the predecessor and successor of i, respectively, on machine m. Restrict the choice of arc (i,j) to those vertices such that at least one of the arcs (p (i),i) or (j, s(j)) is not on a longest path, i.e. i or j are block end vertices (cf. the branching structure of Brucker et al. [25]). Reverse (i,j) and, additionally, also reverse (p(h),h) and (k, s(k)) - provided they exist - where h directly precedes i in its job, and k is the immediate successor of j in its job. The latter arcs are reversed only if a reduction of the makespan can be achieved.

Dell'Amico and Trubian [38] considered the problem as being symmetric and scheduled operations bi-directionally, i.e. from the beginning and from the end, in order to obtain a priority rules based feasible solution. The resulting two parts finally are put together to constitute a complete solution. The neighbourhood structure (N3) employed in their tabu search extends the connected neighbourhood structure N1:

N3: Let (i,j) be a disjunctive critical arc. Consider all permutations of the three operations {p(i), i, j} and {i, j, s(j)} in which (i,j) is inverted.

Again, it is possible to construct a finite sequence of moves with respect to N3 that leads from any feasible solution to an optimal one. In a restricted version N3' of N3 arc (i,j) is chosen such that either i or j is the end vertex of a block. In other words, arc (i,j) is not considered as candidate when both (p(i),i) and (j, s(j)) are on a longest path in the current solution. N3' is no longer connected.

Another branching scheme is considered to define a neighborhood structure N4:

N4: For all operations i in a block, move i to the very beginning or to the very end of this block.

Once more, N4 is connected, i.e. for each feasible solution it is possible to construct a finite sequence of moves, leading to a globally optimal solution.

Nowadays, the most efficient tabu search implementations are described in Nowicki and Smutnicki [75] and Balas and Vazacopoulos [12]. The size of the neighbourhood N1 depends on the number of critical paths in a schedule and the number of operations on each critical path; it can be quite large. Nowicki and Smutnicki [75] consider a smaller neighbourhood (N5) restricting N1 (or N4) to reversals on the border of a block. Moreover, they restrict themselves to a single arbitrarily selected critical path.

> N5: A move is defined by the interchange of two successive operations i and j, where i or j is the first or last operation in a block that belongs to a critical path. In the first block only the last two operations and symmetrically in the last block of the critical path only the first two operations are swapped.

The design of a classical tabu search algorithm is straightforward. A stopping criterion is when the optimal schedule is detected or the number of iterations without any improvement exceeds a certain limit. Nowicki and Smutnicki [75] note that the essential disadvantage of this approach consists of losing information about previous runs. Therefore they suggest building up a list of the # best solutions and their associated tabu lists during the search. Whenever the classical tabu search has finished, they go to the most recent entry, i.e. the best schedule x from this list of at most # solutions, and restart the classical tabu search. Whenever a new best solution is encountered, the list of best solutions is updated. This extended tabu search "with backtracking" continues until the list of best solutions is empty.

The idea of Balas and Vazacopoulos's [12] guided local search (GLS) procedure is to reverse more than one disjunctive arc at a time. This leads to a considerably larger neighbourhood than in the previous cases. Moreover, neighbours are defined by interchanging a set of arcs of varying size, hence the search is of variable depth and supports search diversification in the solution space. The employed neighbourhood structure (N6) is an extension of all previously encountered neighbourhood structures.

Consider any feasible schedule x and any two operations i and j to be performed on the same machine, such that i and j are on the same critical path, say P(0,n), but not necessarily adjacent. Assume i is processed before j. Besides p(i), p(j) and s(i), s(j), the immediate machine predecessors and machine successors of i and j in x, let a(i), a(j) and b(i) and b(j) denote the job predecessors and job successors of operations i and j, respectively. Moreover, let est(i) := $est_i + p_i$ and q(i) := $p_i + q_i$ be the length of a longest path (including the processing time of i, p_i) connecting 0 and i, or i and n. An interchange on i and j either is a move of i right after j (forward interchange) or a move of j right before i (backward interchange). We have seen that the schedule x cannot be improved by an interchange on i and j if i and j are adjacent and none of them is the first or last operation of a block in P(0,n). In other words, in order to achieve an improvement either a(i) or b(j) must be contained in P(0,n). This statement can easily be generalized to the case where i is not an immediately preceding operation of j. Thus for an interchange on i and j to reduce the makespan, it is necessary that the critical path P(0,n) containing i and j also contains at least one of the operations a(i) or b(j). Hence, the number of "attractive" interchanges reduces drastically and the question remains, under which conditions an interchange on i and j is guaranteed not to create a cycle. It is easy to derive that a forward interchange on i and j yields a new schedule x' (obtained from x) if there is no directed path from b(i) to j in x. Similarly, a backward interchange on i and j will not create a cycle if there is no directed path from i to a(j) in x. Now, the neighbourhood structure N6 can be introduced.

> N6: A neighbour x' of a schedule x is obtained by an interchange of two operations i and j in one block of a critical path. Either operation j is the last one in the block and there is no directed path in x connecting the job successor of i to j, or, operation i is the first one in the block and there is no directed path in x connecting i to the job predecessor of j.

While the neighbourhood N1 involves the reversal of a single arc (i,j) on a critical path, the more general move defined by N6 involves the reversal of potentially a large number of arcs.

In order to combine local search procedures operating on different neighbourhoods (which makes it more likely to escape local optima and explore regions not available by any single neighbourhood structure) Balas and Vazacopoulos [12] combined their GLS with the shifting bottleneck procedure. The idea of Balas and Vazacopoulos is to replace the reoptimization cycle of the shifting bottleneck procedure with the neighbourhood trees of the GLS procedure. GLS and its modifications currently form the most powerful heuristic to solve job shop scheduling problems. It outperforms many others in solution quality and computation time.

4.4 Ejection Chains

Variable depth procedures, whose terminology was popularized by Papadimitriou and Steiglitz [80], have had an important role in heuristic procedures for optimization problems. A class of these procedures, called ejection chain methods, has proved highly effective in a variety of applications, see Dorndorf and Pesch [40], Pesch and Glover [81].

Ejection chain methods extend ideas exemplified by certain types of shortest path and alternating path constructions. The basic moves from one solution to another are compound moves composed of a sequence of paired steps. The first component of each paired step in an ejection chain approach introduces a change that creates a dislocation (i.e. an inducement for further change), while the second component creates a change designed to restore the system. The dislocation of the first component may involve a form of infeasibility, or may be heuristically defined to create conditions that can be usefully exploited by the second component. Typically, the restoration of the second component may not be complete, and hence in general it is necessary to link the paired steps into a chain that ultimately achieves a desired outcome.

An application with respect to neighbourhood structure N1 describes a move to a neighbouring solution in which the processing order of operations i and j is changed. This neighbourhood is connected, i.e. for any two solutions (including the optimal one) x and y there is a sequence of moves, with respect to N1, connecting x to y. The gain affected by such a move from x to y can be estimated based on considerations about the minimal length of the critical path of the resulting disjunctive graph. Finding the exact gain of a move would generally involve a longest path calculation. The gain of a move can be negative, thus leading to a deterioration of the objective function. Dorndorf and Pesch [40] present a local search procedure that is based on a compound neighbourhood structure; each component consists of the neighbourhood defined above. It is a variable depth search or ejection chain consisting of a simple neighbourhood structure at each depth which is composed to complex and powerful moves. The basic idea is similar to the one used in tabu search, the main difference being that the list of forbidden (tabu) moves grows dynamically during a variable depth search iteration and is reset at the beginning of the next iteration.

A genetic algorithm with variable depth search has been implemented by Dorndorf and Pesch [39]; each individual of a population is made locally optimal with respect to the ejection chain-based embedded neighbourhood.

5 Constraint Propagation and Local Consistency

Recently, generally applicable approximation procedures such as tabu search simulated annealing or genetic algorithm-based learning strategies have become very attractive and successful solution strategies for scheduling problems. Their general idea is to modify current solutions in a certain sense, where the modifications are defined by a neighbourhood operator, such that new feasible solutions are generated, the so-called neighbours, which hopefully have an improved or at most limited deterioration of their objective function value. In order to reach this goal problem-specific knowledge, incorporated by problem-specific heuristics, has to be introduced into the local search process of the general problem solvers. Now, we are going to use different inference techniques – propagation techniques in order to reach a certain level of consistency – as a preprocessing step to accelerate exact solution methods as well as local search or any kind of heuristic procedures. Model-based local reasoning over the constraint set makes problem-specific knowledge, which is implicitly contained in the model description, explicitly available for branching or bounding.

Consistency checks, or roughly speaking propagation of constraints, will make implicitly defined constraints more visible and will prune the search tree in a branch and bound algorithm.

The job shop scheduling problem is a typical representative of a binary constraint satisfaction problem (CSP), i.e. generally speaking, there is a set of variables each of which has its own domain of values. Find an assignment of values to variables such that a set of constraints on variable pairs is satisfied (see Waltz [95], Dechter and Pearl [37]) and Meseguer [69]). Assume that there is an upper bound on the makespan of an optimal schedule of the underlying job shop scheduling problem. Then computing heads and tails assigns to each operation an interval of possible start times. Considering variable domains as possible operation start times where the variables represent the operations in a schedule, then the disjunctive graph illustrates the job shop scheduling constraint satisfaction problem; it corresponds to the constraint graph (see Montanari [72]). A set of k variables is said to be k-consistent if it is k 1-consistent and holds for each subset of k - 1 variables: if a set of k - 1 values each of which belongs to another of the k - 1 variable domains violates none of the constraints on the considered k - 1 variables, then there is a value in the domain of the remaining variable such that the set of all k values satisfies the set of constraints on the k variables. Let us assume that 0-consistency is always satisfied by definition. A set of variables is k-consistent if each subset of k variables is k-consistent. A 2-consistent set of variables is also said to be arcconsistent in order to emphasize the relation with the edges in the constraint graph (cf. Van Hentenryck et al. [57]). Consider a pair i, j of operations. If for any two start times S_i and S_j of operations i and j, respectively, and any third operation k there exists a start time S_k of operation k such that S_i, S_j, S_k satisfy constraints (1) to (3), then operations i and j are said to be path consistent. Hence, consistency checks, or roughly speaking propagation of constraints, will make implicitly defined constraints more visible and will prune the search tree in a branch and bound algorithm. The job shop scheduling problem is said to be path consistent if all operation pairs are path consistent (cf. Mackworth [65, 66], Mohr and Henderson [71] and Han and Lee [55]). Obviously, n-consistency, where n is the number of operations, immediately implies that a feasible schedule can be generated easily; however, to achieve n-consistency is in general not practicable. Moreover, bad upper bounds on the makespan of an optimal schedule will hardly reduce variable domains, i.e. only a few arc directions are fixed during the constraint propagation process. The better the bounds the more arc directions can be fixed [82, 30, 68]. Very successful applications to project scheduling and open shop scheduling have been described in Dorndorf et al. [42, 44]. A comprehensive problem-based introduction can be found in [43, 84].

Let us become more specific in order to describe another class of logical tests called sequence consistency tests which are based on resource constraints. Constraint propagation finally is an exploration of all available resource constraints in a sense that constraints are activated to reduce variable domains until domain reductions are no longer possible, i.e. a (usually unique) fixed point has been reached. Hence the effect of any propagation heavily depends on the kind of constraints. The efficiency depends on the constraint activation sequence, the variable and value selection in the backtrack search (if no further domain reduction can be achieved) and the level of consistency. These sequence consistency tests reduce operation domains by ruling out infeasible start time assignments. The benefit of the tests is that they can reduce the search space and direct an algorithm towards good solutions. From now on we are only interested in the tests themselves and will not address scheduling algorithms in which they can be embedded.

We assume that all operation domains have been made arc-consistent (or 2-consistent, i.e. heads and tails have been calculated), i.e. we impose an upper bound on the makespan of at least one feasible schedule. A comprehensive presentation can be found in Dorndorf et al. [46].

An operation j is characterized by its processing time p_j; it has a start time window in the interval between the earliest start time est_j and the latest start time lst_j. Its latest completion time is lct_j and its earliest completion time is ect_j. The domain D_j is the set of all possible start times S_j of j. It is bounded by the start time window $[est_j, lst_j]$ but some values in the start time window may be infeasible. Let us define the following shorthand descriptions of any operation from a subset W of the set V of operations:

$$E(W) := \min\{est_j \mid j \in W\};\quad P(W) := \sum_{j \in W} p_j \quad \text{and} \quad C(W) := \max\{lct_j \mid j \in W\};$$

$$Lst(W) := \min\{lst_j \mid j \in W\};\quad Ect(W) := \max\{ect_j \mid j \in W\};\quad W_j := W \setminus \{j\}.$$

A sequence relation $i \rightarrow W_j$ says that operation i has to be scheduled (started and finished) before the start of any operation in set W_j. The idea behind all sequencing tests described now is to consider subsets W of operations requiring the same machine k for being processed. Within these subsets all possible operation sequences with a particular property are examined, e.g. the property that the operation sequence does not start with an operation j from W. If all such sequences are infeasible, then we can conclude that the sequence must not have the property; therefore we could deduce that j must be first in W since all sequences where this is not the case are illegal. The sequencing tests are presented in order from strongest to weakest condition. While a stronger condition allows a stronger conclusion, it is at the same more likely to be inapplicable. All tests in this section are derived for disjunctive scheduling problems where operations are mutually exclusive in the sense that they exclusively occupy any required machines throughout their processing time. This is of course the case for instances of the job shop scheduling problem.

Carlier and Pinson [28] have derived conditions under which it can be concluded that an operation j from W must be scheduled first or last in W, see also Carlier [26]. If an operation j is scheduled before or after W_j we may also think of j as the input or output of W_j, hence the following name of the condition.

Input/Output Condition:
Let j be an operation from a subset W of the operation set V where all operations of W require the same machine during each time unit of processing. If $C(W) - E(W_j) < P(W)$ then j must precede all operations in W_j. If $C(W_j) - E(W) < P(W)$ then j must succeed all operations in W_j.

If the input condition holds we can reduce the domain of j through an update of the latest start time lst_j to $\min\{lst_j, Lst(W_j) - p_j\}$. Symmetrically, if the output condition holds, the earliest start time est_j can be updated to max $\{est_j, Ect(W_j)\}$. A usually better and fast approximation of est_j can be achieved through the calculation of Jackson's preemptive schedule for all operations in W_j. After reducing the domain of j it may be possible to reduce the domains of operations in W_j by applying the same or other tests, e.g. 2-consistency tests. In branch and bound procedures that branch over disjunctive edges, the rules may be employed to fix disjunctions, a process often called immediate selection or edge finding, cf. Brucker et al. [23].

From the input/output condition it can be deduced that an operation j must be scheduled first or last in W. The weaker pair ordering condition can be used to show that a precedence relation $i \rightarrow j$ exists between an operation pair i, j from W, cf. Blazewicz et al. [19].

Pair Ordering Condition:
Let i and j be two operations from a subset W of the operation set V where all operations of W require the same machine during each time unit of processing. If $C(W_j) - E(W_i) < P(W)$ then i must be scheduled first or j must be scheduled last in W. If i and j are two distinct operations then $i \rightarrow j$.

If the pair ordering condition holds and $i \neq j$ we can reduce the domains of i and j through an update of the latest start time lst_i to min $\{lst_i, lst_j - p_i\}$. Symmetrically, the earliest start time est_j can be updated to max $\{est_j , ect_i\}$. If the pair ordering condition holds and $i = j$ we can reduce the domain of j by the interval $[Lst(W_j) - p_j, Ect(W_j)]$. For the case where there are only three operations in W the pair ordering condition allows us to draw the same conclusions as the r-set condition described by Brucker et al. [23]. They present an $O(n^2)$ algorithm for checking 3-set conditions.

By further relaxing the test for the pair ordering condition, we can still draw additional conclusions in situations where the pair ordering condition and the input/output condition do not hold.

Set Input/Output Negation (cf. Carlier / Pinson and [29]):
Let j be an operation from a subset W of the operation set V where all operations of W require the same machine during each time unit of processing. If $C(W_j) - est_j < P(W)$ then j must not precede all operations in W_j. If $lct_j - E(W_j) < P(W)$ then j must not succeed all operations in W_j.

The set input negation domain reduction rule extracts the interval [0, min $\{ect(j') \mid j' \in W_j\}$] from the set of possible start times of j. Symmetrically, if the set output negation holds j must precede at least one operation in W and the domain reduction rule extracts the interval [max $\{lst(j') - p_j + 1 \mid j' \in W_j\}, \infty$] from the set of possible start times of j.

The aforementioned consistency tests were given different names by different authors; some of them are called edge-finding. Many of these tests generalize and extend the earlier tests on job shop scheduling, e.g. Nuijten [76] updates time bounds of operations using ideas presented in Carlier and Pinson [28] and incorporates the tests into a constraint satisfaction framework, cf. Nuijten and Aarts [77].

The main interest of propagation of constraints is the enormous flexibility that results from the fact that each constraint propagates independently from the existence or non-existence of other constraints. It appears that, within each constraint, considered separately, any type of technique (in particular OR algorithms) can be used.

Backtracking can easily be accompanied by a constraint propagation phase before as well as during the search. Consistency checks can substantially reduce the variable domains and correspondingly can reduce the search tree for some problems. The constraint propagation phase can really extract a lot of knowledge about the structure of the problem in order to use this knowledge in a subsequent optimization phase, e.g. a prospective backtrack search. A backtrack search including the local propagation of constraints is implicitly realized in recent constraint-based logic programming languages or systems such as CHIP, PROLOG III, cc(FD) or ILOG (cf. Le Pape [64]).

6 Genetic-based Learning

As the name suggests, genetic algorithms are motivated by the theory of evolution and date back to the early work of Holland [58], see also Goldberg [52] and Michalewicz [70]. They have been designed as general search strategies and optimization methods. Roughly speaking, a genetic algorithm aims at producing near-optimal solutions by letting a set of random solutions undergo a sequence of unary and binary transformations governed by a selection scheme biased towards high-quality solutions. These transformations constitute the recombination step of a genetic algorithm and are performed on the population by three simple operators – reproduction, crossover and mutation. The effect of the operators is that implicitly good properties are identified and combined into a new population which hopefully has the property that the best solution and the average value of the solutions are better than in previous populations. The process is then repeated until some stopping criteria are met.

Compared to standard heuristics, "genetic algorithms are not well suited for fine-tuning structures which are very close to optimal solutions" (Grefenstette [54]). Therefore it is essential if a competitive genetic algorithm is desired, to incorporate (local search) improvement operators. The resulting algorithm has then been called genetic local search heuristic. Putting things into a more general framework, a solution of a combinatorial optimization problem may be considered as a sequence of local decisions, see Dorndorf and Pesch [41]. A local decision for the job shop scheduling problem might be the choice of an operation to be scheduled next. In what follows we will consider more general decision rules. In an enumeration tree of all possible decision sequences a solution of the problem is represented as a path corresponding to the different decisions from the root of the tree to some leaf. Genetics can guide a search process in order to learn to find the most promising decisions within a reasonable amount of time.

In order to apply the crossover operator the population is randomly partitioned into pairs. Next, for each pair, the crossover operator is applied with a certain probability by choosing a position randomly in the string and exchanging the tails (defined as the substring starting at the chosen position) of the two strings. This is called the one-point crossover. The mutation operator which makes random changes to single elements of the string only plays a secondary role in genetic algorithms. Mutation serves to maintain diversity in the population.

Besides the unary and binary recombination operator one may also introduce operators of higher arities, such as consensus operators, that fix arc directions common to most schedules of a current population, cf. Aarts et al. [2] and Nakano and Yamada [74].

Genetic algorithms do not fit best for scheduling problems in order to get near-optimal solutions if no improvement heuristic, such as local search, is applied, there are several possibilities to apply an improvement heuristic. During the recombination phase an improvement step, like tabu search (Glover [50, 51]) or simulated annealing (Aarts and Korst [1]), can be applied to all or several of the solutions in a population. Some type of an improvement heuristic may also be incorporated into the crossover operator. In any case the improvement step as well as the crossover operator heavily depend on the representation of the solution. To overcome these difficulties an appropriate solution representation enables us to use the simplest type of crossover as well as to incorporate problem-specific knowledge, i.e. as Davis [36] claimed "to examine the workings of a good deterministic program in that domain", in order to be competitive with special purpose heuristics. The strategy of Dorndorf and Pesch [41] controls a sequence of priority rules or the machine inclusion sequence for the SB-heuristic and learns to find best combinations in both cases. Each individual of the priority rule-based genetic algorithm (for short: P-GA) is a string of n - 1 entries (pr1, pr2 ,..., prn-1) where n - 1 is the number of operations in the underlying problem instance. An entry pri represents one rule of the set of twelve priority rules. The entry in the i-th position says that a conflict in the i-th iteration of the Giffler–Thompson [48] algorithm should be resolved using priority rule pri. Ties are broken by a random choice. Within a genetic framework a best sequence of priority rules has to be determined. The crossover operator is straightforward. Obviously, the simple crossover, where the substrings of two cut strings are exchanged, applies and always yields feasible offspring. Heuristic information already occurs in the encoding scheme and a particular improvement step is dropped. The mutation operator simply switches a string position to another one, i.e. the priority rule of a randomly chosen string entry is replaced by a new rule randomly chosen among the remaining ones. One can use other kinds of rules which fix the order of operations on the same machine. For instance, any permutation of unscheduled operations might represent a rule. The approach to search a best sequence of decision rules for selecting operations is just in line with the ideas of Fisher and Thompson [47] on probabilistic learning of sequences consisting of two priority rules, and Crowston et al. [34] on learning how to find promising linear combinations of basic priorities (see also Glover [49], O'Grady and Harrison [78]).

In the first implementation of Dorndorf and Pesch [41] the genetic algorithm serves as a meta-strategy to optimally control the use of priority rules, whereas the genetic algorithm controls the selection of nodes in the enumeration tree of the shifting bottleneck heuristic in a second implementation, the shifting bottleneck-based genetic algorithm (for short: SB-GA). The SB-heuristic tries to determine the best single machine sequence which can also be achieved by a genetic strategy, even in a more effective way. Hence, an individual is encoded over the alphabet from 1 to the number of machines and a (partial) string just describes the sequence in which the single machine solutions are considered for inclusion. A representation of a solution is a permutation of the machine numbers 1, ..., m, where m is the number of machines. As a crossover operator one can use any travelling salesman crossover. The difference between the shifting bottleneck heuristic and a genetic approach is that the bottleneck is no longer a decision criterion for the choice of the next machine. We are continuing in this line in a third implementation, viz. the job pair-based genetic algorithm (for short: 2J-GA). It also optimally controls the sequence of job pairs. Hence, the length of an individual corresponds to the number $|J|\cdot(|J| - 1)/2$ pairs of jobs where J is the set of jobs in the underlying problem. Each entry of an individual represents a job pair and a (partial) string just describes the sequence in which the job pair solutions are considered for inclusion.

Constraint propagation in the sense of local consistency for the job shop problem frequently fixes only a few, most often trivial arc directions even if tight upper bounds (in the limit the optimal value) of the makespan in an optimal schedule are applied. The remedy yields to consider subproblems (e.g. one-machine or two-job problems). Hence, two new genetic-based approaches were applied. In the first one, the one-machine constraint propagation-based genetic algorithm (1MCP-GA), each entry of an individual of the SB-GA is replaced by an upper bound on the makespan of the corresponding one-machine problem. That means, if (m(1), m(2), ..., m(m-1), m(m)) is a machine permutation, and $C_{max}(m(i))$, i := 1, ..., m, is the makespan of an optimal solution of the one machine problem defined by machine m(i), then each entry of an individual of 1MCP-GA is an upper bound UB(m(i)), i := 1, ..., m, of the makespan $C_{max}(m(i))$. In the second approach, the two-job constraint propagation-based genetic algorithm (2JCP-GA), each entry of an individual of the 2J-GA is replaced by an upper bound on the makespan of the corresponding two-job problem. In fact, the entry contains an integer value b between 0 and 31 providing an upper bound which is b% above the optimal makespan of the considered subproblem. The entries are encoded as binary numbers in order to apply the simple one-point crossover as well as a simple mutation (switching 0 and 1) as in the case of P-GA. Whenever a new population has been generated local consistency is achieved through the aforementioned tests by means of constraint propagation. The consistency tests are applied simultaneously to each subproblem (corresponding to an entry of an individual in 1MCP-GA or 2JCP-GA) with respect to its upper bound which is b% above the optimal makespan of the subproblem. The number of newly fixed arc directions divided by the number of arcs which were included in a cycle during the constraint propagation process on all these subproblems of an individual defines the fitness of the individual. Note regard that an individual need not represent one schedule, because there are still some arc directions not fixed. An individual of the population corresponds to a partial schedule. However, each population is transformed to a population of feasible solutions, where each individual of a population is assessed in order to judge its contribution to a schedule. Therefore the Giffler – Thompson's algorithm is applied with respect to the partial schedule representing the individual. Ties are broken with respect to the complete schedules that are attached to the parents (partial schedules) of the considered offspring (partial schedule). Hence, the next operation is chosen as in one of the parents' corresponding complete schedules with the same probability (cf. GT-crossover in Yamada and Nakano [96]). In the first population of complete schedules ties were broken randomly. A promising alternative is to use branch and bound or neighbourhood search methods like tabu search in order to generate a feasible solution from the partial solution.

The first implementation uses the algorithm of Baker [8] for selection and an elitist strategy, i.e. the best individual in each population always survived. Population sizes were always set to 80, the crossover rate was fixed to 1, the mutation rate was kept 0 in the case of 2J-GA and 0.01 for 2JCP-GA and 1MCP-GA. There was no inversion and no scaling of fitness values. An elitist strategy took place in all implementations. These parameters are within a reasonable amount of test runs found to be empirically the best ones. The initial populations are always generated randomly and the makespan reported provides the best value obtained within three runs. A run was stopped if there was no improvement within 50 successive generations.

SB-GA dominates all other genetic algorithm implementations if both the quality of the obtained solutions and the running time is considered. The best results, however, were obtained by the 1MCP-GA, which an optimal solution of the famous 10 x 10 job shop could be found. The job pair-based genetic algorithm approaches performed well; however, considering the 10 x 10 problem they seem to run into a dead-end, fixing some wrong arc directions. If the number of jobs compared to the number of machines is quite large then 2J-GA and 2JCP-GA seem to be more effective. On a set of 10 x 10 problems slightly worse solutions than 1MCP-GA yields the implementation 2JCP-GA but the results are still close to the minimal makespan. The 2J-GA cannot compete any longer seriously, although its results are not to bad. Obviously, wrongly fixed arcs, leading to promising solutions in the job pair subproblem can hardly be turned into the right direction. For the larger problems 1MCP-GA clearly dominates. The 2JCP-GA behaves comparable to the shifting bottleneck algorithms. The constraint propagation based genetic algorithm 2JCP-GA and the shifting bottleneck algorithm are quite comparable. The latter, however, dominates on smaller problem instances as well as quadratic instances, while the former is more suitable for rectangle problems where the number of jobs is significantly higher than the number of machines. The genetic algorithms relying on job pairs (2J-GA and 2JCP-GA) usually do a better job if the number of jobs is much higher than the number of machines. However, in all cases they need a substantial amount of computation time if the number of jobs becomes large. This is not a surprising effect because an increasing number of jobs heavily influences the string lengths which correspond to the number of job pairs. Although achieving consistency is fast, a repeated application drives the overall running time upward. In general we can conclude that the genetic-based approaches are often more robust on different problem types than tabu search. There are no drastic time differences and also for the easy problems it takes some time to realize that they are easy. A comprehensive computational study on various modification to the aforementioned approach can be found in Dorndorf et al. [45].

7 Conclusions

Besides a comprehensive survey on job shop solution ideas we also presented a special type of probabilistic machine learning by population genetics. Learning to find the best sequence of priority rules for solving job shop scheduling problems did not work best because of the amount of computation time needed. This is not a big surprise because priority rules only make little use of problem-specific knowledge. In contrast, the solution of a one machine problem or a job pair subproblem (with respect to some former decisions) involves a lot of problem-specific knowledge, the use of which may be the main reason for the success of the genetically guided shifting bottleneck procedure. The constraint propagation-based approaches even try to avoid the use of problem-specific knowledge that is not relevant for the complete problem. Introducing problem specific knowledge makes a genetic strategy a promising alternative. This knowledge might be incorporated by the application of a heuristic performing well on the particular problem, such as the shifting bottleneck procedure for the job shop problem. We may also think of local search algorithms such as tabu search or simulated annealing as "special purpose" heuristics. So a simple genetic algorithm can do its work very well if an individual is considered to be a sequence of local decisions on the use of the knowledge provided by special purpose heuristics or parts of them. Problem-specific knowledge can also be introduced by decomposition of the whole problem into easy-to-handle subproblems. The latter should be solved efficiently by an exact or an approximation algorithm. Additional gains can be achieved if the subproblems help to structure the complete problem under consideration. Those transferring structures may be recognized by some consistency checks and their propagation.

References

1. Aarts EHL and Korst J. *Simulated Annealing and Boltzmann Machines*. John Wiley & Sons, Chichester, 1989.
2. Aarts EHL, van Laarhoven PJM, Lenstra JK, and Ulder NLJ. A computational study of local search shop scheduling. *Journal on Computing* 1994;6:118-125.
3. Aarts EHL and Lenstra JK. *Local search in combinatorial optimization*. Wiley, New York, 1997.

4. Adams J, Balas E, and Zawack D. The shifting bottleneck procedure for job shop scheduling. *Management Science* 1988;34:391-401.
5. Akers SB. A graphical approach to production scheduling problems. *Operations Research* 1956;4:244-245.
6. Anderson EJ, Glass CA, and Potts CN. Local search in combinatorial optimization: applications in machine scheduling. Research Report No. OR56, University of Southampton, 1995.
7. Applegate D and Cook W. A computational study of the job-shop scheduling problem. *ORSA Journal on Computing* 1991; 3:149-156.
8. Baker JE. Reducing bias and inefficiency in the selection algorithm. In: Grefenstette JJ (ed) *Proc. 2nd Int. Conf. on Genetic Algorithms and Their Applications*, Lawrence Erlbaum Ass., 1987, pp. 14-21.
9. Baker KR. *Introduction to Sequencing and Scheduling*. Wiley, New York, 1974.
10. Balas E. Machine sequencing via disjunctive graphs: an implicit enumeration algorithm. *Operations Research* 1969; 17: 941-957.
11. Balas E, Lenstra JK, and Vazacopoulos A. One machine scheduling with delayed precedence constraints. *Management Science* 1995;41:94-109.
12. Balas E and Vazacopoulos A. Guided local search with shifting bottleneck for job shop scheduling. *Management Science* 1998;44:262-275.
13. Baptiste P and Le Pape C. A theoretical and experimental comparison of constraint propagation techniques for disjunctive scheduling, *Proc. 14th Int. Joint Conf. on Artificial Intelligence (IJCAI)*, Montreal, Canada, 1995, pp. 136-140.
14. Blackstone JH, Phillips DT, and Hogg GL. A state of the art survey of dispatching rules for manufacturing job shop operations. *International Journal of Production Research* 1982;20:27-45.
15. Blazewicz J. Selected topics in scheduling theory. *Annals of Discrete Mathematics* 1987;31:1-60.
16. Blazewicz J, Domschke W, and Pesch E. The job shop scheduling problem: conventional and new solution techniques. *European Journal of Operational Research* 1996;93:1-33.
17. Blazewicz J, Dror M, and Weglarz J. Mathematical programming formulations for machine scheduling: a survey. *European Journal of Operational Research* 1991;51:283-300.
18. Blazewicz J, Ecker KH, Pesch E, Schmidt G, and Weglarz J. *Scheduling Computer and Manufacturing Processes*. Springer, Berlin. 2.edition, 2001.
19. Blazewicz J, Pesch E, and Sterna M. A branch and bound algorithm for the job shop scheduling problem. In: Drexl A und Kimms A (eds) *Beyond Manufacturing Resource Planning (MRP II)*, Springer, Berlin, 1998, pp. 219-254.
20. Blazewicz J, Pesch E, and Sterna M. The disjunctive graph machine representation of the job shop problem, *European Journal of Operational Research* 2000;127:317-331.
21. Brucker P. An efficient algorithm for the job-shop problem with two jobs. *Computing* 1988;40:353-359.
22. Brucker P and Jurisch B. A new lower bound for the job-shop scheduling problem. *European Journal of Operational Research* 1993;64:156-167.
23. Brucker P, Jurisch B, and Krämer A. The job-shop problem and immediate selection, *Annals of Operations Research* 1996;50:73-114.
24. Brucker P, Jurisch B, and Sievers B. Job-shop (C codes). *European Journal of Operational Research* 1992;57:132-133.
25. Brucker P, Jurisch B, and Sievers B. A branch and bound algorithm for the job-shop scheduling problem. *Discrete Applied Mathematics* 1994;49:107-127.

26. Carlier J. The one machine sequencing problem. *European Journal of Operational Research* 1982;11:42-47.
27. Carlier J and Pinson E. An algorithm for solving the job-shop problem. *Management Science* 1989;35:164-176 .
28. Carlier J and Pinson E. A practical use of Jackson's preemptive schedule for solving the job shop problem. *Annals of Operations Research* 1990;26:269-287.
29. Carlier J and Pinson E. Adjustments of heads and tails for the job-shop problem. *European Journal of Operational Research* 1994;78:146-161.
30. Caseau Y and Laburthe F. Disjunctive scheduling with task intervals. Working paper, Ecole Normale Superieure, Paris, 1995.
31. Chretienne P, Coffman EG, Lenstra JK, and Liu Z. *Scheduling Theory and its Applications*. Wiley, Chichester, 1995.
32. Conway RN, Maxwell WL, and Miller LW. *Theory of Scheduling*. Addison-Wesley, Reading, HA, 1967.
33. Crama Y, Kolen A, and Pesch E. Local search in combinatorial optimization, *Lecture Notes in Computer Science* 1995;931:157-174.
34. Crowston WB, Glover F, Thompson GL, and Trawick JD. Probabilistic and parametric learning combinations of local job shop scheduling rules. ONR Research Memorandum No. 117, GSIA, Carnegie-Mellon University, Pittsburg, PA, 1963.
35. Dauzere-Peres S and Lasserre J-B. A modified shifting bottleneck procedure for job-shop scheduling. *International Journal of Production Research* 1993;31:923-932.
36. Davis L. Job shop scheduling with genetic algorithms. In: Grefenstette JJ (ed) *Proc. Int. Conf. on Genetic Algorithms and Their Applications*. Lawrence Erlbaum Ass., 1985, pp. 136-140.
37. Dechter R. and Pearl J. Network-based heuristics for constraint satisfaction problems. *Artificial Intelligence* 1988;34:1-38.
38. Dell'Amico M and Trubian M. Applying tabu-search to the job shop scheduling problem. *Annals of Operations Research* 1993;41:231-252.
39. Dorndorf U and Pesch E. Combining genetic and local search for solving the job shop scheduling problem. In: I. Maros, (ed) *Symp. on Applied Mathematical Programming and Modeling APMOD93*. Akaprint, Budapest, 1993, pp. 142-149.
40. Dorndorf U and Pesch E. Variable depth search and embedded schedule neighbourhoods for job shop scheduling. *4th Int. Workshop on Project Managment and Scheduling*, 1994, pp. 232-235.
41. Dorndorf U and Pesch E. Evolution based learning in a job shop scheduling environment. *Computers & Operations Research* 1995;22:25-40.
42. Dorndorf, U, Pesch E, and Phan Huy T. A time-oriented branch-and-bound algorithm for the resource constrained project scheduling problem with generalised precedence constraints, *Management Science* 2000;46:1365-1384.
43. Dorndorf U, Pesch E, and Phan Huy T. Constraint propagation techniques for disjunctive scheduling problems, *Artificial Intelligence* 2000;122:189-240.
44. Dorndorf U, Pesch E, and Phan Huy T. Solving the open shop scheduling problem, *Journal of Scheduling* 2001;4:157-174.
45. Dorndorf U, Pesch E, and Phan Huy T. Constraint propagation and problem decomposition: how to solve the job shop scheduling problem, *Annals of Operations Research* (to appear).
46. Dorndorf U, Phan Huy T, and Pesch E. A survey of interval capacity consistency tests for time- and resource-constrained scheduling. In: Weglarz J (ed) *Project Scheduling - Recent Models, Algorithms and Applications*. Kluwer Academic, Dordrecht, 1999, pp. 213-238.

47. Fisher H and Thompson GL. Probabilistic learning combinations of local job-shop scheduling rules. In: [73].
48. Giffler B and Thompson GL. Algorithms for solving production scheduling problems. *Operations Research* 1960;8:487-503.
49. Glover F. Future paths for integer programming and links to artificial intelligence. *Computers & Operations Research* 1986; 13:533-549.
50. Glover F. Tabu Search-Part I. *Journal on Computing* 1989;1:190-206.
51. Glover F. Tabu Search-Part II. *Journal on Computing* 1990;2:4-32.
52. Goldberg DE. *Genetic Algorithms in Search, Optimization and Machine Learning.* Addison-Wesley, Reading, MA, 1989.
53. Graham RL, Lawler EL, Lenstra JK, and Rinnooy Kan AHG. Optimization and approximation in deterministic sequencing and scheduling theory: a survey. *Annals of Discrete Mathematics* 1979;5:287-326.
54. Grefenstette JJ. Incorporating problem specific knowledge into genetic algorithms. In: Davis L (ed) *Genetic Algorithms and Simulated Annealing.* Pitman, London, 1987, pp. 42-60.
55. Han CC and Lee CH. Comments on Mohr and Henderson's path consistency algorithm. *Artificial Intelligence* 1988;36:125-130.
56. Haupt R. A survey of priority-rule based scheduling. *OR Spektrum* 1989;11:3-16.
57. van Hentenryck P, Deville Y, and Teng C-M. A generic arc-consistency algorithm and its specializations. *Artificial Intelligence* 1992;57:291-321.
58. Holland JH. *Adaptation in Natural and Artificial Systems.* The University of Michigan Press, Ann Arbor, 1975.
59. van Laarhoven PJM, Aarts EHL, and Lenstra JK. Job shop scheduling by simulated annealing. *Operations Research* 1992;40:113-125.
60. Lageweg B, Lawler EL, Lenstra JK, and Rinnooy Kan AHG. Computer aided complexity classification of combinatorial problems. *Communications of the ACM* 1982;25: 817-822.
61. Lawler EL, Lenstra JK, Rinnooy Kan AHG, and Shmoys DB. Sequencing and scheduling: algorithms and complexity. In: Graves SC, Rinnooy Kan AHG, and Zipkin PH (eds) *Handbooks in Operations Research and Management Science, Vol. 4: Logistics of Production and Inventory.* Elsevier, Amsterdam, 1993.
62. Lenstra JK. *Sequencing by Enumerative Methods.* Mathematical Center Tract 69, Mathematisch Centrum, Amsterdam, 1977.
63. Lenstra JK and Rinnooy Kan AHG. Computational complexity of discrete optimization problems. *Annals of Discrete Mathematics* 1979;4:121-140.
64. Le Pape, C. Implementation of resource constraints in ILOG SCHEDULE. A library for the development of constraint-based scheduling systems. *Intelligent Systems Engineering* 1994;3:55-66.
65. Mackworth AK. Consistency in networks of relations. *Artificial Intelligence* 1977;8:99-118.
66. Mackworth, AK and Freuder EC. The complexity of some polynomial network consistency algorithms for constraint satisfaction problems. *Artificial Intelligence* 1985;25:65-74.
67. Manne AS. On the job shop scheduling problem. *Operations Research* 1960;8:219-223.
68. Martin P and Shmoys DB. A new approach to computing optimal schedules for the job shop scheduling problem, *Proc. 5th Int. IPCO Conference*, 1996.
69. Meseguer P. Constraint satisfaction problems: an overview. *AICOM* 1989;2:3-17.
70. Michalewicz Z. *Genetic Algorithms + Data Structures = Evolution Programs.* Springer, Berlin, 1992.

71. Mohr R and Henderson TC. Arc and path consistency revisited. *Artificial Intelligence* 1986;28:225-233.
72. Montanari U. Networks of constraints: fundamental properties and applications to picture processing. *Information Sciences* 1974;7:95-132.
73. Muth JF and Thompson GL (eds). *Industrial Scheduling*. Prentice Hall, Englewood Cliffs, NJ, 1963.
74. Nakano R and Yamada T. Conventional genetic algorithm for job shop problems. *Proc. 4th Int. Conf. on Genetic Algorithms and their Applications*, San Diego, CA, 1991, pp. 474-479.
75. Nowicki E. and Smutnicki C. A fast taboo search algorithm for the job shop problem. *Management Science* 1996;42:797-813
76. Nuijten, WPM. *Time and Resource Constrained Scheduling: A constraint satisfaction approach*. PhD Thesis, Eindhoven University of Technology, 1994.
77. Nuijten, WPM and Aarts EHL. A computational study of constraint satisfaction for multiple capacitated job-shop scheduling. *European Journal of Operational Research* 1996;90:269-284.
78. O'Grady PJ and Harrison C. A general search sequencing rule for job shop sequencing. *International Journal of Production Research* 1985;23:951-973.
79. Panwalkar SS and Iskander W. A survey of scheduling rules. *Operations Research* 1977;25:45-61.
80. Papadimitriou CH and Steiglitz K. *Combinatorial Optimization: Algorithms and Complexity*. Prentice-Hall, Englewood Cliff, NJ, 1982.
81. Pesch E and Glover F. TSP ejection chains. *Discrete Applied Mathematics* 1997;76:165-181.
82. Pesch E and Tetzlaff U. Constraint propagation based scheduling of job shops. *Journal on Computing* 1996;8:144-157.
83. Pesch E and Voß S. Strategies with memories: local search in an application oriented environment. *OR Spektrum* 1995;17:55-66.
84. Phan Huy T. *Constraint propagation in flexible manufacturing*. Springer, Berlin, 2000.
85. Pinedo M. *Scheduling Theory, Algorithms and Systems*. Prentice Hall, Englewood Cliffs NJ, 1995.
86. Potts CN. Analysis of a heuristic for one machine sequencing with release dates and delivery times. *Operations Research* 1980;28:1436-1441.
87. Rinnooy Kan AHG. *Machine Scheduling Problems: Classification, Complexity and Computations*. Nijhoff, The Hague, 1976.
88. Roy B and Sussman B. Les problemes d'ordonnancement avec contraintes disjonctives. SEMA, Note D.S. No. 9., Paris, 1964.
89. Sadeh N. *Look-ahead techniques for micro-opportunistic job shop scheduling*. Dissertation, Carnegie Mellon University, Pittsburgh, PA, 1991.
90. Tanaev VS, Gordon VS, and Shafransky YM. *Scheduling Theory: Single-Stage Systems*. Kluwer Academic, Dordrecht, 1994.
91. Tanaev VS, Sotskov YN, and Strusevich VA. *Scheduling Theory: Multi-Stage Systems*. Kluwer Academic Publ., Dordrecht, 1994.
92. Vaessens RJM. *Generalized Job Shop Scheduling: Complexity and Local Search*. Dissertation, University of Technology Eindhoven, 1995.
93. Vaessens RJM, Aarts EHL, and Lenstra JK. Job shop scheduling by local search. *Journal on Computing* 1996;8:302-317.
94. van de Velde S. *Machine Scheduling and Lagrangian Relaxation*. Dissertation, CWI Amsterdam, 1991.

95. Waltz D. Understanding line drawings of scenes with shadows. In: Winston, PH (ed) *Psychology of Computer Vision*. McGraw-Hill, Cambridge, MA, 1975.
96. Yamada T and Nakano R. A genetic algorithm applicable to large-scale job-shop problems. In: Männer R and Manderick B (eds) *Proc. 2nd. Int. Workshop on Parallel Problem Solving from Nature*, 1992, pp. 281-290.

Scheduling of Bus Drivers' Service by a Genetic Algorithm

Ikuo Yoshihara

Faculty of Engineering, Miyazaki University, 1-1, Gakuen Kibanadai-Nishi, Miyazaki, 889-2192, Japan
E-mail: yoshiha@cs.miyazaki-u.ac.jp

Summary. This chapter presents an application of genetic algorithms to schedule bus drivers' services so as to satisfy all the drivers' working conditions, to decrease the number of drivers as much as possible and to reduce the spread of working hours among all the drivers. The drivers' service schedules have been compiled manually so far, but were very troublesome. Therefore, automatic generation of the schedules was a welcome development.

We developed a method to derive drivers' service schedules using heuristics and genetic algorithms (GAs). Here, the weak points of the heuristics are compensated by GAs and the method can yield quasi-optimal solutions of good quality within reasonable time for practical use. The method is employed in the unified bus-management system that has been available for more than five years and been used by many bus companies.

1 Introduction

In controlling a bus service operation, there are two scheduling requirements and three requirements of data management and clerical work. These give rise to two diagrams, one for bus service operation and the other for the disposition of drivers and preparation of timetable, operation manual and statistical data. Bus companies want a unified system that manages services and business related to their operations. In developing such a system, new technology to automatically generate the two diagrams, that is, the schedules for the buses and drivers, plays an extremely important role. The three items required for the latter can be manufactured with ease by existing technology.

This chapter describes the drivers' service scheduling which is one of the two problems mentioned above. A drivers' service schedule is a plan of work for the drivers, and is drawn up by experts after the bus schedules are fixed. These experts ensure that the service plans satisfy all the drivers' working conditions, that the number of drivers is minimized, as is the spread of the working hours per driver. In ill-formed plans, the number of drivers exceeds requirements, and the spread of working hours is scattered wastefully.

The experts have so far compiled drivers' service schedules by hand, but the work is very troublesome, especially when adjusting for the second goal, in which a continuous process of trial and error is required to even out drivers' service distribution. Some large bus companies informed our inquiry that several experts usually spend more than one month on scheduling. They therefore are anxious to develop a system that automatically generates drivers' service schedules. Such automatic generation should reduce the time for producing schedules and improve quality.

Creating drivers' service schedules is a type of scheduling problem that can be classified as a typical combinatorial problem [1]. We intended to utilize a genetic algorithm (GA) to search the optimal solutions, because we have already succeeded in solving not only famous benchmarks like the TSP, but also practical problems like truck and railroad car operation [4], [6], [7]. The various experiences of our applications lead us to conclude that GA is capable of yielding satisfactory quasi-optimal solutions fast enough for practical use.

2 Bus Operating Schedule and Driver's Service Schedule

Bus companies need two types of schedule. One is the usual *schedule* of bus operations and the other is a driver's service schedule [8]. The part surrounding the scheduled operation of the bus is called a *yama sub-schedule* (henceforth, *sub-schedule*). The sub-schedule is so called because of its mountain-like shape formed by lines denoting the ways to and from a destination, and *yama* is the Japanese word for "mountain".

2.1 A Bus Route and a Bus Schedule

Fig.1 shows the relation between bus routes and a sub-schedule. The upper diagram in Fig.1 illustrates two bus routes. Route 1 starts from stop O, and goes to stop E via stops A, B, and so forth. Route 2 starts from stop O and goes to stop Z via stops X and Y. The lower diagram in Fig.1 is a sub-schedule of Route 1. The abscissa of the sub-schedule indicates time and the ordinate, bus stops. Each stop is ordered according to its distance from the place of departure. A bold, oblique line shows how the bus is operated. For instance, in the oblique line increasing to the upper right, the bus which leaves stop O at 9:30 reaches stop E at 10:30. A sub-schedule describes when the bus departs from the starting stop, passes through the intermediate stops, and arrives at the last stop. The timetable for each stop can thus be derived from the sub-schedule.

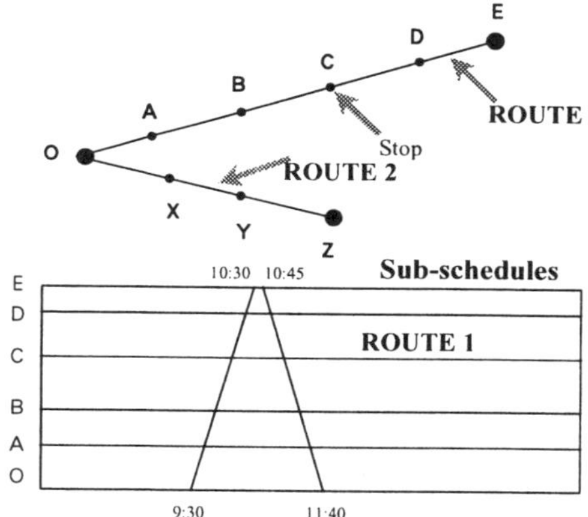

Fig. 1. Bus routes and a sub-schedule

2.2 A Sub-schedule and a Driver's Service Schedule

For the purpose of completing a bus schedule, each sub-schedule needs to specify which vehicle and driver are employed. From the drivers' standpoint, it is necessary to be informed beforehand of which sub-schedules will apply to them.

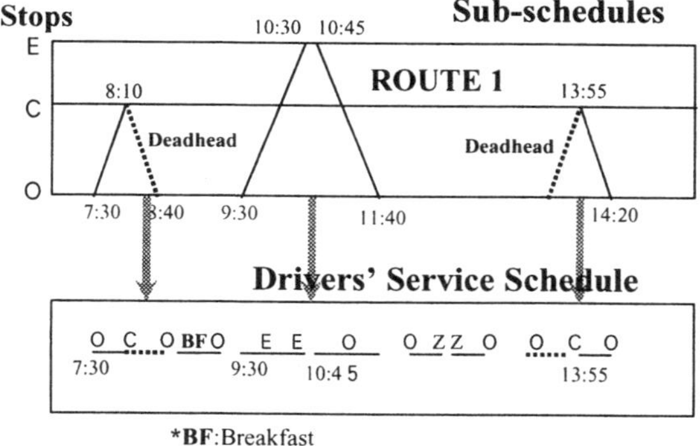

Fig. 2. Sub-schedules and a service schedule

As illustrated in Fig.2, a driver's service schedule is shown in a long narrow frame which records the duties of the driver by horizontal lines in time order. A driver starts from the bus depot O by the office at 7:30, goes to stop C at 8:10, deadheads the bus to the bus depot O, and has breakfast. Afterwards the driver leaves again from the bus depot O at 9:30, goes to stop E at 10:30 and does a normal service during 10:45-11:40. After taking a rest, the driver operates a normal service between the bus depot O and stop Z of Route 2 (this schedule is not shown in the figure). Since services are drawn up around the office, a driver sometimes takes charge of more than one route in a day. This is apt to happen. There are several service types with different working hours. For instance,

- The morning shift: This service begins early in the morning and ends in the afternoon.
- The afternoon shift: This service begins in the afternoon and ends late at night.
- The rush-hour shift: This occurs only during the rush hours in the morning and evening, but there is a break in the daytime.
- The long-time shift: This runs from early in the morning till late at night (overtime work).
- The part-time shift: This occurs only during the rush hour in the morning.

These are examples from a certain bus company, and the service types vary according to the companies. There are many restrictions on labor conditions such as working hours as exemplified below:

- The time going to work and leaving work: Though the actual time range depends on service types, the commencement and close of work should both be in a certain time range.
- Working hours: An upper limit on working hours exists, which starts when the workers arrive at the workplace till they get off from work.
- Actual driving time: An upper limit is set on the actual driving time.
- Mileage: An upper limit is set on the mileage allowed for a day.
- Break for a breakfast: More than half an hour's rest in a regular time slot must be allocated for breakfast. The drivers can take breakfast at particular stops such as offices or garages.
- Rest time: The total rest should be no less than the standard rest time. Here, a short time halt is not considered as a rest.
- Non-stop driving time: Continuous driving hours are limited.
- Stoppage time: The stoppage time is restricted to less than the stated time at specific stops, for instance, at a bus pool in front of a big station. When the stoppage time exceeds the time limit, it is necessary to deadhead to the nearest bus depot.

The actual values assigned to these restrictions depend on service types and the condition of each bus route. Further, it is common for there to be other detailed constraints; for example, the above-mentioned bus company has 16 restrictions in total.

A bus service schedule must meet these constraints, while reducing the cost as much as possible. Provided the costs are the same, better-balanced working conditions for the drivers are preferred. Hence, the most important aim is to reduce the number of drivers, and the second is to equalize their actual working hours. It is extremely difficult even for experts to plan a service route schedule for a day if it is done manually. The second aim is probably achieved by trial and error to come up with an appropriate combination of sub-schedules. It takes more than one month for a couple of experts to set up schedules for a company with dozens of offices and bus garages, and thousands of sub-schedules. As a consequence, bus companies have requested a system that generates service schedules of bus routes. The automatic scheduling program is expected to shorten the scheduling time and to improve its quality as well.

3 How Experts Compile Drivers' Service Schedules

Before proceeding to explain the compilation of a day's schedule of bus service by GA, let us describe how the experts have manually compiled service schedules so far, by referring to the simple example illustrated in Fig.3.

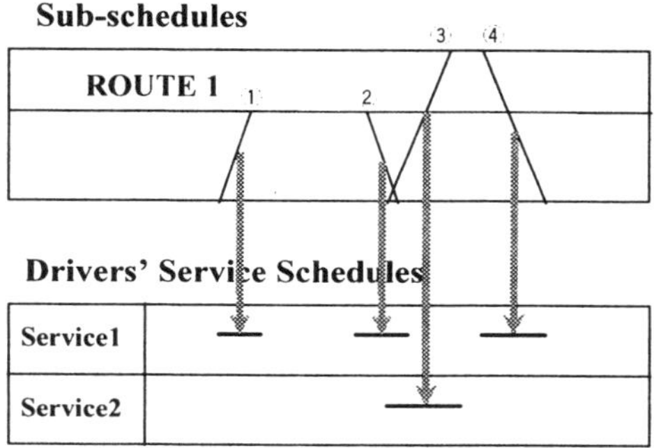

Fig. 3. Heuristics for compiling service schedules

The upper part of the figure shows the sub-schedules of the bus route, and the lower part the service schedule for a driver. The problem is that, in practice, a service schedule is compiled not just for a single bus service route but for more than one route. Further difficulties, such as having over a thousand sub-schedules and hundred drivers, are also present. Not knowing the number of necessary services beforehand, the scheduling experts begin with a tentative set of services. Next, they fill in the sub-schedules, which are empty at first, one by one in departure time order (at this point the drivers' schedules are not considered). Plausible decisions are usually based on experience. When they have a bus deadhead, they have to make sure in every case that there is sufficient time to do so and that the driver's working conditions are met. If these terms are not satisfied, the experts allocate sub-schedules to other empty spaces in the drivers' service schedule. They place all the sub-schedules in a frame for a day's service schedule one by one in this manner.

Any remaining unallocated sub-schedules are stored in a pool of residual sub-schedules. We produced a prototype program to generate drivers' service schedules based on the existing experts' heuristics. The outline of the algorithm is as follows.

- Step 1: The program arranges sub-schedules in temporal order.

- Step 2: The program picks one sub-schedule from the top of the sub-schedule list.
- Step 3: The program temporarily estimates intermediate values, assuming that a selected sub-schedule has been allocated to a driver's service schedule.
- Step 4: Applicable sub-schedules are arranged in order, from the one with the highest intermediate value down to the lowest. Any leftover sub-schedules are labeled as unallocated and removed from the sub-schedule list.

When a sub-schedule cannot be allocated anywhere, it is omitted from the sub-schedule list and moved to a sub-schedule pool. The procedure from Step 2 to Step 4 is repeated until the sub-schedule list is emptied. When the allocation in this way is not appropriate, the procedure fails, for it lacks backtracking.

The way intermediate evaluations are taken chiefly determines the adequacy of the allocation. We tried tuning the intermediate evaluation routines, but observed that they can yield unnecessary residual sub-schedules even in relatively small problems that have been successively solved by hand. They can turn a sub-schedule from "unallocated" to "allocated" by try and error, destroying a part of previously allocated sub-schedules or partially replacing them. This is the limit of human skill.

In order to make up for such deficiencies, we should include more detailed experts' heuristics, improve the intermediate evaluation, and program so as to satisfy all the constraints. However, no drastic measures can be taken because of the lack of generality and the difficulty inherent in knowledge acquisition, programming, and its maintenance. Heuristics directly reflects human wisdom in solving the problems, but there are always tendencies to be influenced by the idiosyncrasy of the object and to depend upon human capability. To avoid these problems, we had to compile drivers' service schedules with less heuristics; we proposed that GA and heuristics should share the role as mentioned above. Moreover, by using GA, the trial and error of the rearrangement is built into the algorithm by its own accord.

4 Drivers' Service Schedule with GAs

4.1 Scheduling and GAs

The advantages of GA are that it does not require any knowledge concerning the targeted problem, that it adapts flexibly to changes of surroundings, and that it searches globally [2], [4]. Nevertheless, GA often takes more time to solve problems than conventional operation research techniques do. The fundamental reason is that GA is a generation-and-test algorithm. Having no mechanism of local search, GA is obliged to go through trials such as crossover and mutation in search of the ideal combinations that a human can discover at a glance. Moreover, GA on occasion wastefully returns an imperfect solution (a fatal chromosome) that does not satisfy the constraints and throws it away later. Our formula is aimed at realizing a flawless solution by combining GA with a local search algorithm [1], [7]. The heuristic technique or experienced knowledge is as powerful as a local search method unless a tidy mathematical model can be easily assembled.

Also because the heuristics is greatly influenced by the problems and analysts, the diversity of the population, which uses the heuristics also from other problems, might sometimes cause difficulty. If the above-mentioned fault including heuristics has an apparent effect, the essential merit of GA is reduced almost by half.

To avoid the faults that are apt to occur in using them together, we will let the heuristics take charge solely of canceling infringements of the restrictions, and follow the policy of not expecting it to look for a "good" combination. A good combination is generated through the evolutionary process of GA. The process can get rid of a useless solution if it fails to satisfy constraints. Moreover, any trouble from knowledge acquisition concerning the target problem can be evaded.

Let us describe another device in the use of the heuristics. As scheduling is like packing something into a case, to arrange sub-schedules one by one into the timetable, the order of arrangement has an influence on the filling up of the table later. Altering the order to arrange sub-schedules in the table sometimes results in a better solution. The chromosome is coded for the order of arrangement in the table that will be optimized according to the evolutionary process.

4.2 Outline of GA-based Scheduling

When a driver's service schedule is completed, it can be implicitly expressed by arranging the order of each sub-schedule, because all the sub-schedules are put in some of the driver's service schedule frames. If the alphabet which represents the order of the sub-schedules is taken to be a chromosome, a driver's service schedule can be immediately expressed as a gene. However, when a pair of chromosomes are simply crossovers, an imperfect driver's service schedule, such as some sub-schedules overlapping in time, will be generated (Fig.4). This problem cannot be expressed by too simple and direct a genetic coding. When the sub-schedule is put in a driver's service frame, an obvious heuristic will be used to evade generation of an imperfect driver's service schedule. Once the heuristic confirms that the sub-schedule meets the constraints, it puts it in the driver's service schedule frame. When it does not satisfy the constraints, the heuristic picks up the succeeding candidate of sub-schedules to express it in the same way. In general, there are several drivers' service schedule frames, each of which can accept a sub-schedule, but the possibility of allocating them afterwards depends on the allocation of the present sub-schedule for each driver's frame.

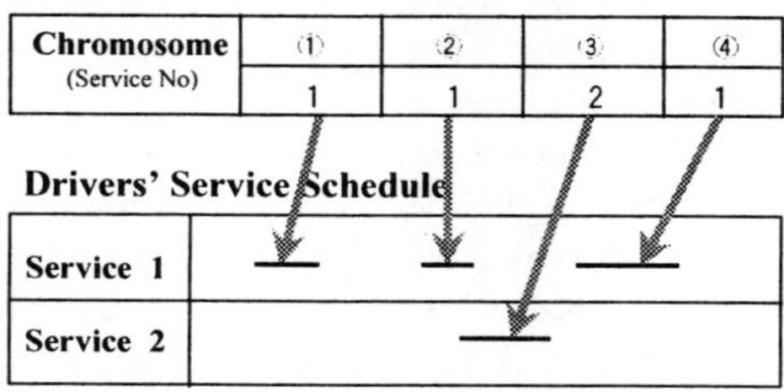

Fig. 4. A simple genetic coding

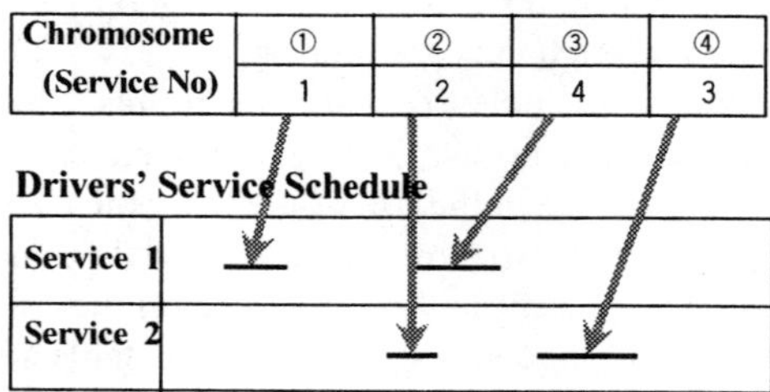

Fig. 5. Allocation according to connection priority

This suggests that there is an order of model service selection corresponding to generating the optimal driver's service schedule. From this point of view, our basic strategy makes GA optimize the solution by repeatedly improving the order of allocation. Before a detailed explanation of the proposed method is given, let us explain how to compile a driver's service schedule using a simple example.

Fig.5 is an example with four sub-schedules (①-④), each of which has a "connecting priority" with a rank from 1 to 4. We choose a driver's service frame into which the sub-schedule is put guided by connection. The heuristic connects the next sub-schedule after the sub-schedule whose value of connecting priority is lowest (i.e. the highest priority).

For instance, let us consider the case of allocating sub-schedule ④. Since sub-schedule ③ (priority 4) has already entered the driver's service frame 1, and there is sub-schedule ② (priority 2) in the driver's service frame 2, the next sub-schedule ④ will be put in the driver's service frame 2. If an allocation to a driver's service schedule with the lowest priority is impossible, the heuristic tries to allocate the sub-schedule to a service frame with the second lowest priority. When the heuristic fails to put it there, it repeats the same trials in the other driver's service frame with the next priority.

4.3 Detail of GA-based Scheduling

An outline of the algorithm is shown in Fig.6 [6]. The heuristic packs the sub-schedule into a driver's service frame referring to the priority indicated by a chromosome. At this stage, heuristics packs the sub-schedules on condition that they do not violate individual constraints such as working conditions and continuous driving time. On the other hand, it cannot be said for certain whether they satisfy some of the constraints in the assembling stage, for instance the differences between the total hours of deadhead and driving. These constraints ought to be included in the evaluation function.

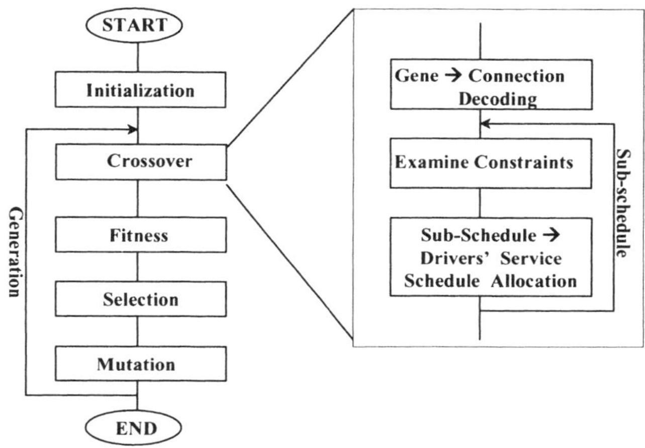

Fig. 6. Compiling driver's service schedules

Genetic Coding

The chromosome is a vector with M elements (Fig .7),

$$\mathbf{q} = (q_1, q_2, \cdots, q_M), \tag{1}$$
$$1 \le q_i \le M - i + 1. \tag{2}$$

As each element q_i ($i = 1, 2, \ldots, M$) is an integer subject to eq.(2), the vector is equivalent to the so-called "order expression" of the TSP by Grefenstette [3]. M is defined as the sum of the number of sub-schedules and that of driver's service frames since a sub-schedule is ranked in driver's service frames in the initial state; that is, none of sub-schedules is allotted to any frame. We make use of the order representation to prevent a fatal chromosome, in the light of its success in the TSP. Some may worry about the parent's characteristics being hardly inherited by the children, while the order representation is used in association with the TSP. Fortunately, in compiling driver's service schedule, no such inconvenience has been observed. As the sub-schedule arrangement needs to satisfy the constraints in composing a driver's service schedule, it is believed that the relation between genotype and phenotype is roughly estimated as N:1 and that the driver's service schedules do not change as much as the tour of the TSP if a part of the chromosome changes. This characteristic seems to be yielded by the indirect coding technique which does not express a driver's service schedule directly by genes but by the parameter, that is, the order of sub-schedules in a driver's service schedule frame.

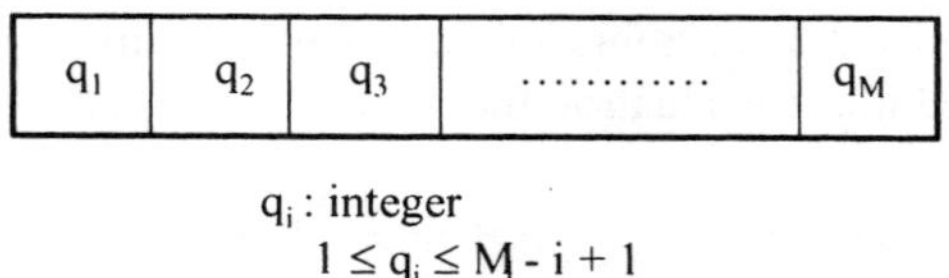

q_i : integer
$1 \le q_i \le M - i + 1$

Fig. 7. Chromosome

Crossover and Reproduction

R pairs of individuals are randomly chosen as parents, and one-point crossover is performed to make children. That is, crossover point r is selected at random, and a sub-string up to r of parent 1 ($\mathbf{p} = p_1 \, p_2 \cdots p_M$) and another sub-string after r of parent 2 ($\mathbf{q} = q_1 \, q_2 \cdots q_M$) are concatenated to produce a child with a chromosome ($\mathbf{c} = p_1 \, p_2 \cdots p_r \, q_{r+1} \, q_{r+2} \cdots q_M$). Immediately after the crossover, the population enters the so-called state of coexistence of parent and child, and the population size increases temporarily. However, R individuals are reduced by natural selection and the population size consequently returns to the previous state.

Evaluation of Fitness

Here, for simplicity, we assume that a bus corresponds to a service. This is the usual assignment of buses, but few bus companies have a policy of assigning a bus to two drivers, that is one in the morning and the other in the afternoon. In this case, the constraints become a little complicated, but our scheduling algorithm is still applicable.

To reduce the cost, the number of buses should be smaller, deadhead hours should be shorter, and the spread of working hours among bus drivers should be smaller. The total cost (C_B) is defined by the following equation in term of the number of buses (N_B), deadhead hours (T_I), and actual driving hours (T_H):

$$C_B = \alpha N_B + \beta \Sigma T_I + \gamma \Sigma \,| T_H - < T_H >_{AV} |. \tag{3}$$

Here, α, β, and γ are coefficients of the cost and $\alpha \gg \beta$, $\alpha \gg \gamma$, and $| T_H - < T_H >_{AV} |$ is the deviation of the driver's service time. The sum of services (Σ) is performed over all the drivers.

The heuristics puts the sub-schedules into the service frame by referring to the priority decoding by chromosome. An evaluation is made for the completed service of each schedule according to expression (3), and fitness is given by the reciprocal. This means that the more economical and impartial a driver's service schedule is, the more excellent it is as an individual.

Natural selection

Natural selection weeds out R individuals, which is as many as the number increased by crossover [6]. Though the selection is performed on a ranking basis, individuals that have the same chromosome must be removed for the sake of keeping diversity in the population [7]. First, all the individuals are arranged in order according to fitness. Each individual is examined to see if it has the same fitness as the previous one. If it has the same value, it is removed. This is an approximate method to delete identical chromosomes, but has the great benefit of simple calculation. Despite the fact that it is a mere approximation, we have not experienced any difficulty with this method. We remove individuals from the lowest fitness until they become R individuals. If the number of individuals exceeds R, the individuals are randomly generated to maintain the population size.

Mutation

Alleles of all the individuals except the elite are forcibly changed with probability p (small, positive constant) at random. We empirically fix p at a value such that mutation happens to 20% of the population. In other words, p is changed in accordance with the problem scale M. The changed value must satisfy the constraint ($1 \leq q_i \leq M - i + 1$). In an individual, a mutation or mutations may or may not occur. When q_i is changed, its fitness should be evaluated again.

5 An Example – Bus Service Scheduling

The proposed algorithm is applied to the bus schedule management system. This total system can replace the current method of bus schedule management, which relies strongly on the experts' experience and intuition. It is capable of organizing sub-schedules, driver's service schedules, car rotation schedules, work allocation tables, timetables, operation instruction sheets, and statistical materials, to name a few. It reduces the number of experts and time for scheduling, and it enables us to compile appropriate schedules that meet the demands of users.

5.1 Overview of the Bus Scheduling Support

The bus scheduling support system runs on a personal computer. The system has the six functions shown in Fig.8.

- Data management: At first, basic data like bus service routes, stops, working forms, conditions, etc., are registered. Managers can add new data, and delete or update the data. Basic data can be inputted on the conversation screen illustrated in Fig.9.

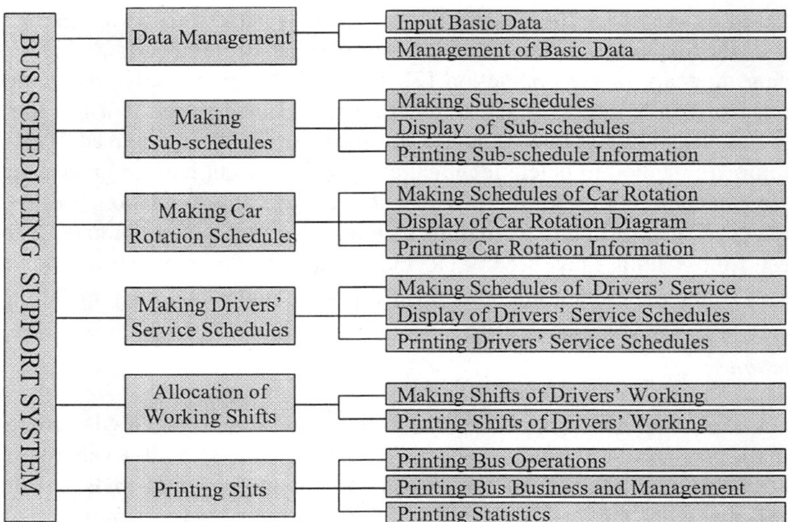

Fig. 8. Functions of bus scheduling support system

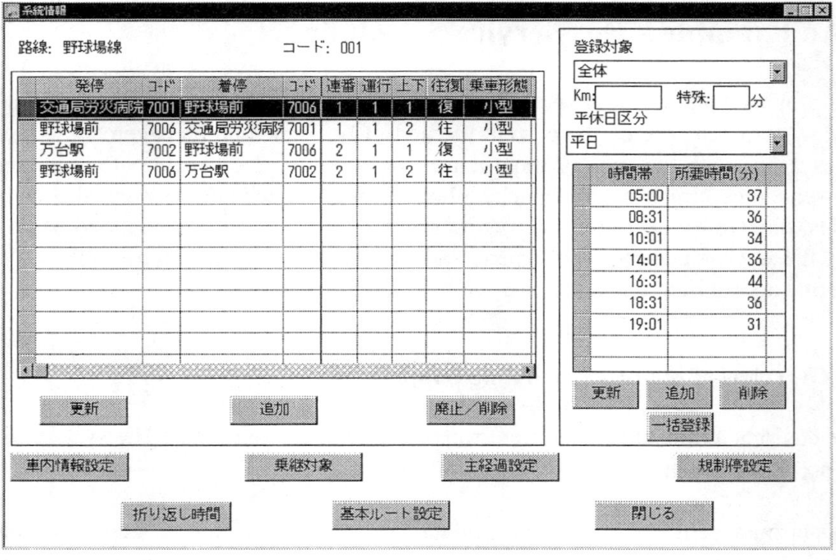

Fig. 9. Input basic data

- Compiling sub-schedules: It is possible to generate sub-schedules according to the conditions such as time of departure, number and intervals of sub-schedules. Moreover, it is possible to generate sub-schedules that refer to information such as transfer to railways.
- Compiling car rotation schedule: Schedules can be drawn up of cars to use, to check, and to repair. The schedules are displayed in the form of diagrams on-screen and printed as a direction of drivers' work.
- Compiling drivers' service schedule: With the proposed algorithm, the system automatically generates drivers' service schedules in term of basic data, the sub-schedules, and constraints such as working time zones, clear working hours, actual driving time, schedule allocation information, etc. The generated schedule can be modified in a conversational mode.
- Allocation of working shifts: After the drivers' services are fixed, the drivers' shifts must be adjusted. The shifts are also displayed on-screen and printed.
- Printing slits: Various kinds of slits can be printed out for bus operation business, passenger service, business plans, and evaluation of the schedule, based on the basic data, sub-schedule, car rotation, and drivers' service schedule.

5.2 An Example of Compiling a Bus Driver's Service Schedule

The office of a bus company has charge of a bus operation along three routes; seven types of services are shown in Fig.10. The routes are real but the stops are imaginary. There are a total of 168 sub-schedules, which are actually operated (Fig.11). There are five kinds of service types, and the constraints are as follows (Fig.12).

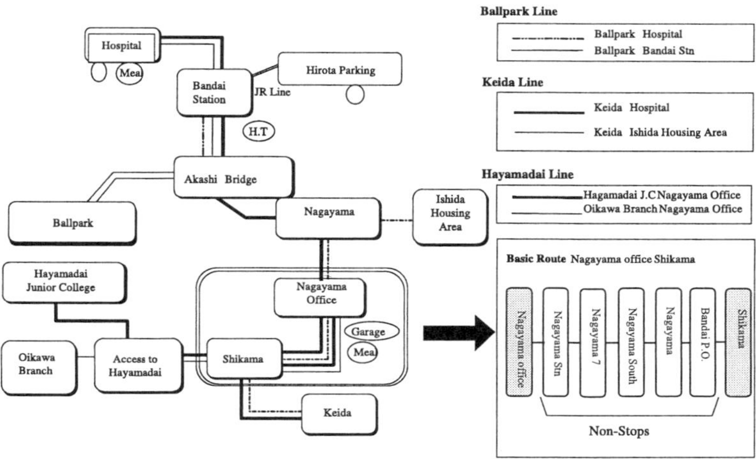

Fig. 10. Bus routes via Nagayama Office

- The morning shift:
  Range of working hours: 5:00-15:00
  Compulsory working hours: 8 hours and 45 minutes
  Actual driving hours: up to 380 minutes
- The afternoon shift:
  Range of working hours: 12:00-22:40
  Compulsory working hours: 8 hours and 45 minutes
  Actual driving hours: up to 380 minutes
- The rush-hour shift:
  Range of working hours: 6:00-11:00 and 15:00-20:00
  Compulsory working hours: 8 hours and 45 minutes
  Actual driving hours: up to 380 minutes
- The long-time shift:
  Range of working hours: 6:00-21:00
  Compulsory working hours: unrestricted
  Actual driving hours: unrestricted
- The part-time shift:
  Range of working hours: 6:00-11:00
  Compulsory working hours: unrestricted
  Actual driving hours: unrestricted

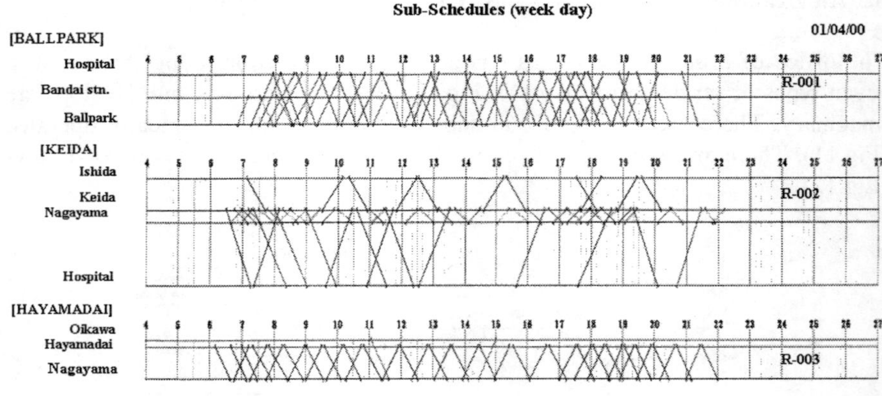

Fig. 11. Sub-schedules along the three routes

Although, in some cases, the compulsory working hours and actual driving hours are unlimited, they never grow excessively, for they are fundamentally restricted by the total working hours. A tentative solution is displayed within 10 seconds after automatic generation of a driver's service schedule begins. A driver's service schedule displayed on-screen is the elite in that generation. If the solution is improved and a new elite appears, an updated driver's service schedule will be displayed.

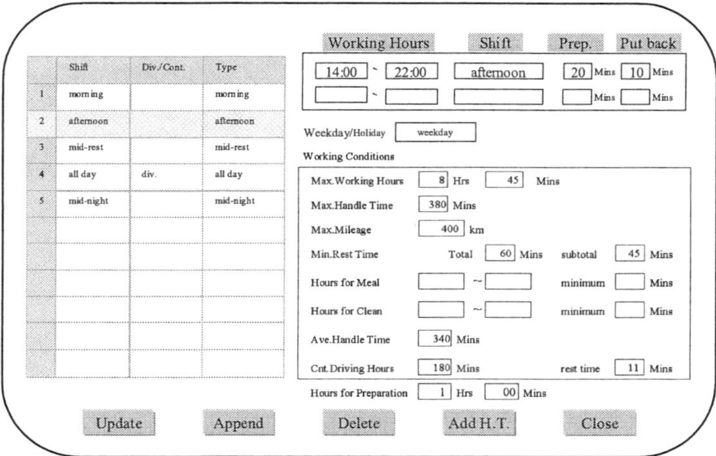

Fig. 12. Input working conditions

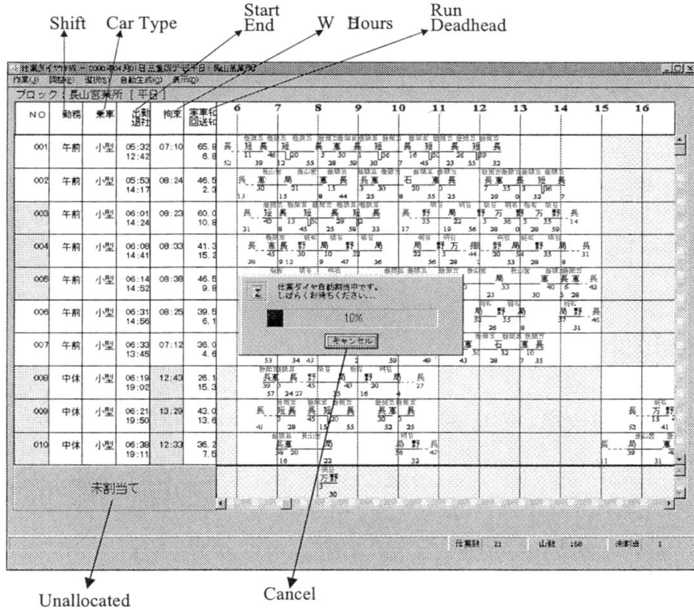

Fig. 13. Searching an optimal driver's service schedule

Whether the constraints are satisfied or not is displayed at the left of the driver's service schedule diagram (see Fig.13). How the deadhead time is decreasing, that is, the way the solution is being improved as the generation progresses, is shown in Fig.14. The broken line in the figure is a level of 842 minutes achieved by an expert. It can be seen that GA reaches a solution that exceeds this level at about the 700th generation. Automatic generation of a driver's service schedule can be interrupted at any time. When it is interrupted, the best driver's service schedule at that time remains displayed; hence it can either be improved by GA or be altered by experts. They can modify not only a driver's service schedule but also the sub-schedule, which feeds the driver's service schedule. We leave room for manual modification as a shortcut to an otherwise time-consuming improvement. Fig.15 shows the completion of a driver's service schedule.

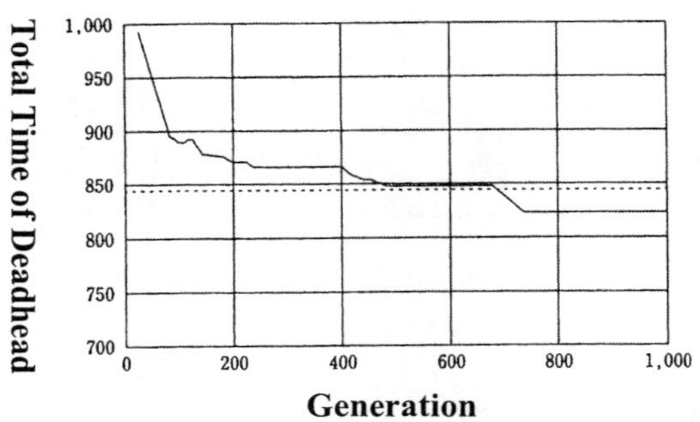

Fig. 14. Searching for a better solution

This bus scheduling support system has actually been used in various-sized bus companies, and appreciated by experts because it generates a driver's service schedule that works. Moreover, some comparatively small-sized bus companies that have purchased this system have been using it without alternation. This implies that the standard specification is practical enough.

6 Conclusions

We have described in detail a bus service scheduling support system that constitutes part of a bus scheduling system. In practical use, the advantages of the GA-based system can be summarized as follows.

Avoidance of Difficulty in Building Mathematical Models

On the way to developing the system, it was necessary to revise the prototype many times, due to frequent additions and changes of specifications. These revisions were made far easier in comparison to conventional methods by using mathematical models. When GA is applied to solve problems, it is necessary to first decide on the genetic coding, and then design genetic operations such as crossover and mutation, which corresponds to building mathematical models by traditional methods. The GA-based method is much less constrained and more flexible in its major processes than the one compiled with mathematical models. Moreover, it is easy to build models using GA.

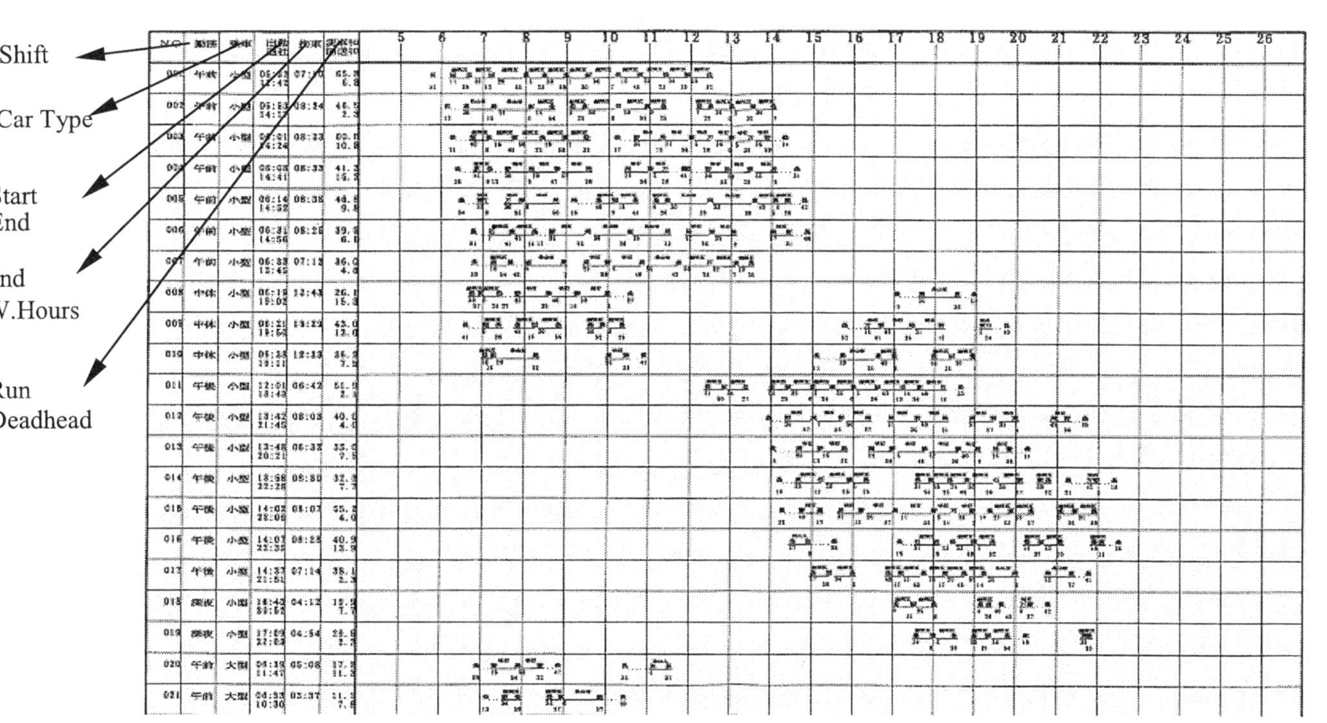

Fig. 15. An example of driver's service schedule

Ease of Developing Search Algorithm

A solution of sufficient standard can be obtained without special knowledge of the target field even if no suitable algorithm or heuristics is employed. Virtually no empirical knowledge and investigation for the solution are needed, but it covers a wide range. It is often said that GA is time consuming, but some devices can shorten the time to a permissible range for practical use. A number of effective proposals have been made: for example, the fusing method with the heuristics described in this chapter. The time spent by humans is much precious than that spent by the CPU. As we have discussed, our program can satisfactorily overcome the difficulty concerning scheduling.

An Extension of Trial and Error Technique

As GA is a multi-point search, making use of this characteristic should be expected. For instance, because two or more solutions with almost the same evaluation may be generated during the evolutionary process, flexible usage of the solution based on user preference will be possible. When the various factors cannot be easily reflected in an objective function, the result can still be judged without making strenuous efforts to describe it rigorously. This makes design easier. Techniques combining GA and other methods have been proposed: in particular, the combination of GA and a greedy algorithm often appears and is effective.

This chapter has described a combination technique to compile bus drivers' service schedules. It is justified to say that in general the parameter of heuristics is optimized using GA, and the technique is widely applicable to many fields. (The original research addresses the idea in detail in [6], [8].) From this point of view, if you already have your own excellent heuristics, its competence will be greatly enhanced by adding GA to your method. Using GA like this is an attractive way to improve existing methods.

Acknowledgments

The author would like to acknowledge H. Sengoku, T. Imagawa of Hitachi Ltd., and Y. Sasage, Y. Obama of Hitachi Tohoku Software Co., for their developing software and implementing it in real-world existing systems.

References

[1] Baeck, T., Fogel, D. B., Michalewicz, Z. (eds.): *Handbook of Evolutionary Computation*, IOP Publishing and Oxford Univ. Press, 1997

[2] Holland, J. H.: *Adaptation in Natural and Artificial Systems*, Univ. Michigan Press, 1975

[3] Goldberg, D. E.: *Genetic Algorithms in Search, Optimization, and Machine Learning*, Addison-Wesley, 1989

[4] Dasgupta, D., Michalewicz, Z. (eds.): *Evolutionary Algorithms in Engineering Applications*, Springer, 1997

[5] Davis, L. (ed.): *Handbook of Genetic Algorithms*, Van Nostrand Reinhold, 1994

[6] Yoshihara, I. and Sengoku, H.: "Scheduling Bus Driver's Services Based on Genetic Algorithm", *Proc. of Int. Conf. on Artificial Intelligence in Science and Technology (AISAT'2000)*, 2000, pp.62-67

[7] Sengoku, H. and Yoshihara, I.: "A Fast TPS Solver Using GA on JAVA", *Proc. of the 3rd Int. Symp. on Artificial Life and Robotics (AROB-Ⅲ)*, 1998, pp.283-288

[8] Sengoku, H. and Yoshihara, I.: "GA-based optimization of heuristic search", *IPSJ*, vol.37, No.10, (1996), pp.1811-1820 (in Japanese)

A Survey of Evolutionary Algorithms for Data Mining and Knowledge Discovery

Alex A. Freitas

Postgraduate Program in Computer Science, Pontificia Universidade Catolica do Parana
Rua Imaculada Conceicao, 1155. Curitiba-PR. 80215-901. Brazil.
E-mail: alex@ppgia.pucpr.br, Web page: http://www.ppgia.pucpr.br/~alex

Summary. This chapter discusses the use of evolutionary algorithms, particularly genetic algorithms and genetic programming, in data mining and knowledge discovery. We focus on the data mining task of classification. In addition, we discuss some preprocessing and postprocessing steps of the knowledge discovery process, focusing on attribute selection and pruning of an ensemble of classifiers. We show how the requirements of data mining and knowledge discovery influence the design of evolutionary algorithms. In particular, we discuss how individual representation, genetic operators and fitness functions have to be adapted for extracting high-level knowledge from data.

1 Introduction

The amount of data stored in databases continues to grow fast. Intuitively, this large amount of stored data contains valuable hidden knowledge, which could be used to improve the decision-making process of an organization. For instance, data about previous sales might contain interesting relationships between products and customers. The discovery of such relationships can be very useful to increase the sales of a company. However, the number of human data analysts grows at a much smaller rate than the amount of stored data. Thus, there is a clear need for (semi-)automatic methods for extracting knowledge from data.

This need has led to the emergence of a field called data mining and knowledge discovery [66]. This is an interdisciplinary field, using methods from several research areas (especially machine learning and statistics) to extract high-level knowledge from real-world data sets. Data mining is the core step of a broader process, called knowledge discovery in databases, or knowledge discovery, for short. This process includes the application of several preprocessing methods aimed at facilitating the application of the data mining algorithm and postprocessing methods aimed at refining and improving the discovered knowledge.

This chapter discusses the use of evolutionary algorithms (EAs), particularly genetic algorithms (GAs) [29], [47] and genetic programming (GP) [41], [6], in data mining and knowledge discovery. We focus on the data mining task of classification, which is the task addressed by most EAs that extract high-level knowledge from data. In addition, we discuss the use of EAs for performing some preprocessing and postprocessing steps of the knowledge discovery process, focusing on attribute selection and pruning of an ensemble of classifiers.

We show how the requirements of data mining and knowledge discovery influence the design of EAs. In particular, we discuss how individual representation, genetic operators and fitness functions have to be adapted for extracting high-level knowledge from data.

This chapter is organized as follows. Section 2 presents an overview of data mining and knowledge discovery. Section 3 discusses several aspects of the design of GAs for rule discovery. Section 4 discusses GAs for performing some preprocessing and postprocessing steps of the knowledge discovery process. Section 5 addresses the use of GP in rule discovery. Section 6 addresses the use of GP in the preprocessing phase of the knowledge discovery process. Finally, section 7 presents a discussion that concludes the chapter.

2 An Overview of Data Mining and Knowledge Discovery

This section is divided into three parts. Subsection 2.1 discusses the desirable properties of discovered knowledge. Subsection 2.2 reviews the main data mining tasks. Subsection 2.3 presents an overview of the knowledge discovery process.

2.1 The Desirable Properties of Discovered Knowledge

In essence, data mining consists of the (semi-)automatic extraction of knowledge from data. This statement raises the question of what kind of knowledge we should try to discover. Although this is a subjective issue, we can mention three general properties that the discovered knowledge should satisfy: namely, it should be accurate, comprehensible, and interesting. Let us briefly discuss each of these properties in turn. (See also sub section 3.3.)

As will be seen in the next subsection, in data mining we are often interested in discovering knowledge which has a certain predictive power. The basic idea is to predict the value that some attribute(s) will take on in “the future”, based on previously observed data. In this context, we want the discovered knowledge to have a high predictive accuracy rate.

We also want the discovered knowledge to be comprehensible for the user. This is necessary whenever the discovered knowledge is to be used for supporting a decision to be made by a human being. If the discovered “knowledge” is just a black box, which makes predictions without explaining them, the user may not trust it [48]. Knowledge comprehensibility can be achieved by using high-level knowledge representations. A popular one, in the context of data mining, is a set of IF–THEN (prediction) rules, where each rule is of the form:

IF <some_conditions_are_satisfied>
THEN <predict_some_value_for_an_attribute>

The third property, knowledge interestingness, is the most difficult one to define and quantify, since it is, to a large extent, subjective. However, there are some aspects of knowledge interestingness that can be defined in objective terms. The topic of rule interestingness, including a comparison between the subjective and the objective approaches for measuring rule interestingness, will be discussed in sub section 2.3.2.

2.2 Data Mining Tasks

In this section we briefly review some of the main data mining tasks. Each task can be thought of as a particular kind of problem to be solved by a data mining algorithm. Other data mining tasks are briefly discussed in [18], [25].

2.2.1 Classification

This is probably the most studied data mining task. It has been studied for many decades by the machine learning and statistics communities (among others). In this task the goal is to predict the value (the class) of a user-specified goal attribute based on the values of other attributes, called the predicting attributes. For instance, the goal attribute might be the *Credit* of a bank customer, taking on the values (classes) "good" or "bad", while the predicting attributes might be the customer's *Age*, *Salary*, *Current_account_balance*, whether or not the customer has an *Unpaid Loan*, etc.

Classification rules can be considered a particular kind of prediction rules where the rule antecedent ("IF part") contains a combination typically, a conjunction - of conditions on predicting attribute values, and the rule consequent ("THEN part") contains a predicted value for the goal attribute. Examples of classification rules are:

IF (*Unpaid_Loan?* = "no") and (*Current_account_balance* > $3,000)
  THEN (*Credit* = "good")
IF (*Unpaid_Loan?* = "yes") THEN (*Credit* = "bad")

In the classification task the data being mined is divided into two mutually exclusive and exhaustive data sets, the training set and the test set. The data mining algorithm has to discover rules by accessing the training set only. In order to do this, the algorithm has access to the values of both the predicting attributes and the goal attribute of each example (record) in the training set.

Once the training process is finished and the algorithm has found a set of classification rules, the predictive performance of these rules is evaluated on the *test* set, which was not seen during training. This is a crucial point.

Actually, it is trivial to get 100% of predictive accuracy in the training set by completely sacrificing the predictive performance on the test set, which would be useless. To see this, suppose that for a training set with n examples the data mining algorithm "discovers" n rules, i.e. one rule for each training example, such that, for each "discovered" rule: (a) the rule antecedent contains conditions with exactly the same attribute value pairs as the corresponding training example; (b) the class predicted by the rule consequent is the same as the actual class of the corresponding training example. In this case the "discovered" rules would trivially achieve 100% of predictive accuracy on the training set, but would be useless for predicting the class of examples unseen during training. In other words, there would be no generalization, and the "discovered" rules would be capturing only idiosyncrasies of the training set, or just "memorizing" the training data. In the parlance of machine learning and data mining, the rules would be overfitting the training data.

For a comprehensive discussion about how to measure the predictive accuracy of classification rules, the reader is referred to [34], [67].

In the next three subsections we briefly review the data mining tasks of dependence modeling, clustering and discovery of association rules. Our main goal is to compare these tasks against the task of classification, since space limitations do not allow us to discuss these tasks in more detail.

2.2.2 Dependence Modeling

This task can be regarded as a generalization of the classification task. In the former we want to predict the value of several attributes rather than a single goal attribute, as in classification. We focus again on the discovery of prediction (IF – THEN) rules, since this is a high-level knowledge representation.

In its most general form, any attribute can occur both in the antecedent ("IF part") of a rule and in the consequent ("THEN part") of another rule, but not in both the antecedent and the consequent of the same rule. For instance, we might discover the following two rules:

IF (*Current_account_balance* > $3,000) AND (*Salary* = "high")
THEN (*Credit* = "good")
IF (*Credit* = "good") AND (*Age* > 21) THEN (*Grant_Loan?* = "yes")

In some cases we want to restrict the use of certain attributes to a given part (antecedent or consequent) of a rule. For instance, we might specify that the attribute *Credit* can occur only in the consequent of a rule, or that the attribute *Age* can occur only in the antecedent of a rule.

For the purposes of this chapter we assume that in this task, similarly to the classification task, the data being mined is partitioned into training and test sets. Once again, we use the training set to discover rules and the test set to evaluate the predictive performance of the discovered rules.

2.2.3 Clustering

As mentioned above, in the classification task the class of a training example is given as input to the data mining algorithm, characterizing a form of supervised learning. In contrast, in the clustering task the data mining algorithm must, in some sense, "discover" classes by itself, by partitioning the examples into clusters, which is a form of unsupervised learning [19], [20]. Examples that are similar to each other (i.e. examples with similar attribute values) tend to be assigned to the same cluster, whereas examples different from each other tend to be assigned to distinct clusters. Note that, once the clusters are found, each cluster can be considered as a "class", so that now we can run a classification algorithm on the clustered data, by using the cluster name as a class label. GAs for clustering are discussed, for example, in [50], [17], [33].

2.2.4 Discovery of Association Rules

In the standard form of this task (ignoring variations proposed in the literature) each data instance (or "record") consists of a set of binary attributes called items. Each instance usually corresponds to a customer transaction, where a given item has a true or false value depending on whether or not the corresponding customer bought that item in that transaction. An association rule is a relationship of the form IF X THEN Y, where X and Y are sets of items and $X \cap Y = \varnothing$ [1], [2]. An example is the association rule:

IF *fried_potatoes* THEN *soft_drink*, *ketchup* .

Although both classification and association rules have an IF – THEN structure, there are important differences between them. We briefly mention here two of these differences. First, association rules can have more than one item in the rule consequent, whereas classification rules always have one attribute (the goal one) in the consequent. Second, unlike the association task, the classification task is asymmetric with respect to the predicting attributes and the goal attribute. Predicting attributes can occur only in the rule antecedent, whereas the goal attribute occurs only in the rule consequent. A more detailed discussion about the differences between classification and association rules can be found in [24].

2.3 The Knowledge Discovery Process

The application of a data mining algorithm to a data set can be considered the core step of a broader process, often called the knowledge discovery process [18]. In addition to the data mining step itself, this process also includes several other steps. For the sake of simplicity, these additional steps can be roughly categorized into data preprocessing and discovered-knowledge postprocessing.

We use the term data preprocessing in a general sense, including the following steps (among others) [55]:

(a) Data Integration This is necessary if the data to be mined comes from several different sources, such as several departments of an organization. This step involves, for instance, removing inconsistencies in attribute names or attribute value names between data sets of different sources.

(b) Data Cleaning It is important to make sure that the data to be mined is as accurate as possible. This step may involve detecting and correcting errors in the data, filling in missing values, etc. Data cleaning has a strong overlap with data

integration, if this latter is also performed. It is often desirable to involve the user in data cleaning and data integration, so that (s)he can bring her/his background knowledge into these tasks. Some data cleaning methods for data mining are discussed in [32], [59].

(c) Discretization This step consists of transforming a continuous attribute into a categorical (or nominal) attribute, taking on only a few discrete values e.g. the real-valued attribute *Salary* can be discretized to take on only three values, say "low", "medium" and "high". This step is particularly required when the data mining algorithm cannot cope with continuous attributes. In addition, discretization often improves the comprehensibility of the discovered knowledge [11], [52].

(d) Attribute Selection This step consists of selecting a subset of attributes relevant for classification, among all original attributes. It will be discussed in subsection 2.3.1.

Discovered-knowledge postprocessing usually aims at improving the comprehensibility and/or the interestingness of the knowledge to be shown to the user. This step may involve, for instance, the selection of the most interesting rules, among the discovered rule set. This step will be discussed in subsection 2.3.2.

Note that the knowledge discovery process is inherently iterative, as illustrated in Fig. 1. As can be seen in this figure, the output of a step can be not only sent to the next step in the process, but also sent – as a feedback to a previous step.

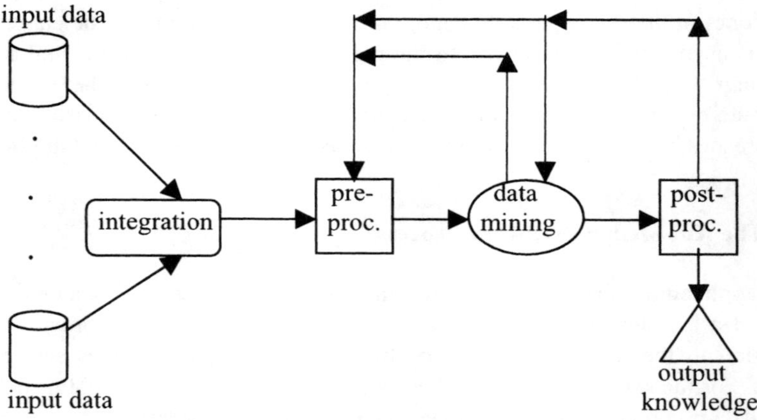

Fig. 1. An overview of the knowledge discovery process

2.3.1 Attribute Selection

This consists of selecting, among all the attributes of the data set, a subset of attributes relevant for the target data mining task. Note that a number of data mining algorithms, particularly rule induction ones, already perform a kind of attribute selection when they discover a rule containing just a few attributes, rather than all attributes. However, in this section we are interested in attribute selection as a preprocessing step for the data mining algorithm. Hence, we first select an attribute subset and then give only the selected attributes for the data mining algorithm.

The motivation for this kind of preprocessing is the fact that irrelevant attributes can somehow "confuse" the data mining algorithm, leading to the discovery of inaccurate or useless knowledge [38]. Considering an extreme example, suppose we try to predict whether the credit of a customer is good or bad, and suppose that the data set includes the attribute *Customer_Name*. A data mining algorithm might discover too specific rules of the form: IF (*Customer_Name* = "a_specific_name") THEN (*Credit* = "good"). This kind of rule has no predictive power. Most likely, it covers a single customer and cannot be generalized to other customers. Technically speaking, it is *overfitting* the data. To avoid this problem, the attribute *Customer_Name* (and other attributes having a unique value for each training example) should be removed in a preprocessing step.

Attribute selection methods can be divided into filter and wrapper approaches. In the filter approach the attribute selection method is independent of the data mining algorithm to be applied to the selected attributes.

By contrast, in the wrapper approach the attribute selection method uses the result of the data mining algorithm to determine how good a given attribute subset is. In essence, the attribute selection method iteratively generates attribute subsets (candidate solutions) and evaluates their qualities, until a termination criterion is satisfied. The attribute-subset generation procedure can be virtually any search method. The major characteristic of the wrapper approach is that the quality of an attribute subset is directly measured by the performance of the data mining algorithm applied to that attribute subset.

The wrapper approach tends to be more effective than the filter one, since the selected attributes are "optimized" for the data mining algorithm. However, the wrapper approach tends to be much slower than the filter approach, since in the former a full data mining algorithm is applied to each attribute subset considered by the search. In addition, if we want to apply several data mining algorithms to the data, the wrapper approach becomes even more computationally expensive, since we need to run the wrapper procedure once for each data mining algorithm.

2.3.2 Discovered-Knowledge Postprocessing

It is often the case that the knowledge discovered by a data mining algorithm needs to undergo some kind of postprocessing. Since in this chapter we focus on discovered knowledge expressed as IF – THEN prediction rules, we are mainly interested in the postprocessing of a discovered rule set.

There are two main motivations for such postprocessing. First, when the discovered rule set is large, we often want to simplify it, i.e. to remove some rules and/or rule conditions in order to improve knowledge comprehensibility for the user.

Second, we often want to extract a subset of interesting rules, among all discovered ones. The reason is that although many data mining algorithms were designed to discover accurate, comprehensible rules, most of these algorithms were not designed to discover *interesting* rules, which is a rather more difficult and ambitious goal, as mentioned in section 2.1.

Methods for selection of interesting rules can be roughly divided into subjective and objective methods. Subjective methods are user-driven and domain-dependent. For instance, the user may specify rule templates, indicating which combination of attributes must occur in the rule for it to be considered interesting this approach has been used mainly in the context of association rules [40]. As another example of a subjective method, the user can give the system a general, high-level description of his/her previous knowledge about the domain, so that the system can select only the discovered rules which represent previously unknown knowledge for the user [44].

By contrast, objective methods are data-driven and domain-independent. Some of these methods are based on the idea of comparing a discovered rule against other rules, rather than against the user's beliefs. In this case the basic idea is that the interestingness of a rule depends not only on the quality of the rule itself, but also on its similarity to other rules. Some objective measures of rule interestingness are discussed in [26], [22], [23].

We believe that ideally a combination of subjective and objective approaches should be used to try to solve the very hard problem of returning interesting knowledge to the user.

3 Genetic Algorithms (GAs) for Rule Discovery

In general the main motivation for using GAs in the discovery of high-level prediction rules is that they perform a global search and cope better with attribute interaction than the greedy rule induction algorithms often used in data mining [14].

In this section we discuss several aspects of GAs for rule discovery. This section is divided into three parts. Subsection 3.1 discusses how one can design an individual to represent prediction (IF – THEN) rules. Subsection 3.2 discusses how genetic operators can be adapted to handle individuals representing rules. Section 3.3 discusses some issues involved in the design of fitness functions for rule discovery.

3.1 Individual Representation

3.1.1 Michigan versus Pittsburgh Approach

GAs for rule discovery can be divided into two broad approaches, based on how rules are encoded in the population of individuals ("chromosomes"). In the

Michigan approach each individual encodes a *single* prediction rule, whereas in the Pittsburgh approach each individual encodes *a set of* prediction rules.

It should be noted that some authors use the term "Michigan approach" in a narrow sense, to refer only to classifier systems [35], where rule interaction is taken into account by a specific kind of credit assignment method. However, we use the term "Michigan approach" in a broader sense, to denote any approach where each GA individual encodes a single prediction rule.

The choice between these two approaches strongly depends on which kind of rule we want to discover. This is related to which kind of data mining task we are addressing. Suppose the task is classification. Then we usually evaluate the quality of the rule set as a whole, rather than the quality of a single rule. In other words, the interaction among the rules is important. In this case, the Pittsburgh approach seems more natural.

On the other hand, the Michigan approach might be more natural in other kinds of data mining tasks. An example is a task where the goal is to find a small set of high-quality prediction rules, and each rule is often evaluated independently of other rules [49]. Another example is the task of detecting rare events [65].

Turning back to classification, which is the focus of this chapter, in a nutshell the pros and cons of each approach are as follows. The Pittsburgh approach directly takes into account rule interaction when computing the fitness function of an individual. However, this approach leads to syntactically longer individuals, which tends to make fitness computation more computationally expensive. In addition, it may require some modifications to standard genetic operators to cope with relatively complex individuals. Examples of GAs for classification which follow the Pittsburgh approach are GABIL [13], GIL [37] and HDPDCS [51].

By contrast, in the Michigan approach the individuals are simpler and syntactically shorter. This tends to reduce the time taken to compute the fitness function and to simplify the design of genetic operators. However, this advantage comes with a cost. First of all, since the fitness function evaluates the quality of each rule separately, now it is not easy to compute the quality of the rule set as a whole, i.e. taking rule interactions into account. Another problem is that, since we want to discover a set of rules, rather than a single rule, we cannot allow the GA population to converge to a single individual which is what usually happens in standard GAs. This introduces the need for some kind of niching method [45], which obviously is not necessary in the case of the Pittsburgh approach. We can avoid the need for niching in the Michigan approach by running the GA several times, each time discovering a different rule. The drawback of this approach is that it tends to be computationally expensive. Examples of GAs for classification which follow the Michigan approach are COGIN [30] and REGAL [27].

So far we have seen that an individual of a GA can represent a single rule or several rules, but we have not said yet how the rule(s) is(are) encoded in the genome of the individual. We now turn to this issue. To follow our discussion, assume that a rule has the form "IF $cond_1$ AND ... AND $cond_n$ THEN $class = c_i$", where $cond_1$ ... $cond_n$ are attribute-value conditions (e.g. *Sex* = "M") and c_i is the class predicted by the rule. We divide our discussion into two parts, the representation of the rule antecedent (the "IF" part of the rule) and the representation of the rule consequent (the "THEN" part of the rule). These two issues are discussed in the next two subsections. In these subsections we will assume that the GA follows the Michigan approach, to simplify our discussion. However, most of the ideas in these two subsections can be adapted to the Pittsburgh approach as well.

3.1.2 Representing the Rule Antecedent (a Conjunction of Conditions)

A simple approach to encode rule conditions into an individual is to use a binary encoding. Suppose that a given attribute can take on *k* discrete values. Then we can encode a condition on the value of this attribute by using *k* bits. The *i*-th value (*i*=1,...,*k*) of the attribute domain is part of the rule condition if and only if the *i*-th bit is "on" [13].

For instance, suppose that a given individual represents a rule antecedent with a single attribute-value condition, where the attribute is *Marital_Status* and its values can be "single", "married", "divorced" and "widow". Then a condition involving this attribute would be encoded in the genome by four bits. If these bits take on, say, the values "0 1 1 0" then they would be representing the following rule antecedent:

IF (*Marital_Status* = "married" OR "divorced")

Hence, this encoding scheme allows the representation of conditions with internal disjunctions, i.e. with the logical OR operator within a condition.

Obviously, this encoding scheme can be easily extended to represent rule antecedents with several conditions (linked by a logical AND) by including in the genome an appropriate number of bits to represent each attribute-value condition.

Note that if all the *k* bits of a given rule condition are "on", this means that the corresponding attribute is effectively being ignored by the rule antecedent, since any value of the attribute satisfies the corresponding rule condition. In practice, it is desirable to favor rules where some conditions are "turned off", i.e. have all their bits set to "1", in order to reduce the size of the rule antecedent. (Recall that we want comprehensible rules and, in general, the shorter the rule is the more comprehensible it is.) To achieve this, one can automatically set all bits of a condition to "1" whenever more than half of those bits are currently set to "1". Another technique to achieve the same effect will be discussed at the end of this subsection.

The above discussion assumed that the attributes were categorical, also called nominal or discrete. In the case of continuous attributes the binary encoding mechanism gets slightly more complex. A common approach is to use bits to represent the value of a continuous attribute in binary notation. For instance, the binary string "0 0 0 0 1 1 0 1" represents the value 13 of a given integer-valued attribute.

Instead of using a binary representation for the genome of an individual, this genome can be expressed in a higher-level representation which directly encodes the rule conditions. One of the advantages of this representation is that it leads to a more uniform treatment of categorical and continuous attributes, in comparison with the binary representation.

In any case, in rule discovery we usually need to use variable-length individuals, since, in principle, we do not know a priori how many conditions will be necessary to produce a good rule. Therefore, we might have to modify crossover to be able to cope with variable-length individuals in such a way that only valid individuals are produced by this operator.

For instance, suppose that we use a high-level representation for two individuals to be mated, as follows (there is an implicit logical AND connecting the rule conditions within each individual):

(*Age* > 25) (*Marital_Status* = "Married")
Has_a_job = "yes") (Age < 21)

As a result of a crossover operation, one of the children might be an invalid individual (i.e. a rule with contradicting conditions), such as the following rule antecedent:

IF (*Age* > 25) AND (*Age* < 21)

To avoid this, we can modify the individual representation to encode attributes in the same order that they occur in the data set, including in the representation "empty conditions" as necessary. Continuing the above example, and assuming that the data set being mined has only the attributes *Age*, *Marital_Status*, and *Has_a_job*, in this order, the two above individuals would be encoded as follows:

(*Age* > 25) (*Marital_Status* = "married") ("empty conditon")
(*Age* < 21) ("empty condition") (*Has_a_job* = "yes")

Now each attribute occupies the same position in the two individuals, i.e. attributes are aligned [21]. Hence, crossover will produce only valid individuals.

This example raises the question of how to determine, for each gene, whether it represents a normally expressed condition or an empty condition. A simple technique for solving this problem is as follows. Suppose the data being mined contains *m* attributes. Then each individual contains *m* genes, each of them divided into two parts. The first one specifies the rule condition itself (e.g. *Age* > 25), whereas the second one is a single bit. If this bit is "on" ("off") the condition is included in (excluded from) the rule antecedent represented by the individual. In other words, the "empty conditions" in the above example are represented by turning off this bit. Since we want rule antecedents with a variable number of conditions, this bit is usually subject to the action of genetic operators [43].

3.1.3 Representing the Rule Consequent (Predicted Class)

Broadly speaking, there are at least three ways of representing the predicted class (the "THEN" part of the rule) in an evolutionary algorithm. The first possibility is to encode it in the genome of an individual [13], [30], possibly making it subject to evolution.

The second possibility is to associate all individuals of the population with the same predicted class, which is never modified during the running of the algorithm. Hence, if we want to discover a set of classification rules predicting k different classes, we would need to run the evolutionary algorithm at least k times, so that in the i-th run, $i=1,..,k$, the algorithm discovers only rules predicting the i-th class [37], [43].

The third possibility is to choose the predicted class most suitable for a rule, in a kind of deterministic way, as soon as the corresponding rule antecedent is formed. The chosen predicted class can be the class that has more representatives in the set of examples satisfying the rule antecedent [27] or the class that maximizes the individual's fitness [49].

The above first and third possibilities have the advantage of allowing that different individuals of the population represent rules predicting different classes. This avoids the need to perform multiple runs of the evolutionary algorithm to discover rules predicting different classes, which is the case in the above second possibility. Overall, the third possibility seems more sound than the first one.

3.2 Genetic Operators for Rule Discovery

There have been several proposals of genetic operators designed particularly for rule discovery. Although these genetic operators have been used mainly in the classification task, in general they can also be used in other tasks that involve rule discovery, such as dependence modeling. We review some of these operators in the following subsections.

3.2.1 Selection

REGAL [28] follows the Michigan approach, where each individual represents a single rule. Since the goal of the algorithm is to discover a set of (rather than just one) classification rules, it is necessary to avoid the convergence of the population to a single individual (rule).

REGAL does that by using a selection procedure called universal suffrage. In essence, individuals to be mated are "elected" by training examples. An example "votes" for one of the rules that cover it, in a probabilistic way. More precisely, the probability of voting for a given rule (individual) is proportional to the fitness of that rule. Only rules covering the same examples compete with each other. Hence, this procedure effectively implements a form of niching, encouraging the evolution of several different rules, each of them covering a different part of the data space.

3.2.2 Generalizing/Specializing Crossover

The basic idea of this special kind of crossover is to generalize or specialize a given rule, depending on whether it is currently overfitting or underfitting the data, respectively [28], [3]. Overfitting was briefly discussed in sub sections 2.2.1 and 2.3.1. Underfitting is the dual situation, in which a rule is covering too many training examples, and so should be specialized. A more comprehensive discussion about overfitting and underfitting in rule induction (independent of evolutionary algorithms) can be found for example in [57].

To simplify our discussion, assume that the evolutionary algorithm follows the Michigan approach where each individual represents a single rule using a binary encoding (as discussed in subsection 3.1.2). Then the generalizing/specializing crossover operators can be implemented as the logical OR and the logical AND, respectively. This is illustrated in Fig. 2, where the above-mentioned bitwise logical functions are used to compute the values of the bits between the two crossover points denoted by the "|" symbol.

parents	children produced by generalizing crossover	children produced by specializing crossover
0 | 1 0 | 1	0 | 1 1 | 1	0 | 0 0 | 1
1 | 0 1 | 0	1 | 1 1 | 0	1 | 0 0 | 0

Fig. 2. Example of generalizing/specializing crossover

3.2.3 Generalizing/Specializing-Condition Operator

In the previous subsection we saw how the crossover operator can be modified to generalize/specialize a rule. However, the generalization/specialization of a rule can also be done in a way independent of crossover. Suppose, say, that a given individual represents a rule antecedent with two attribute-value conditions, as follows again, there is an implicit logical AND connecting the two conditions in (1):

$$(Age > 25) \quad (Marital\_Status = \text{"single"}) \tag{1}$$

We can generalize, say, the first condition of (1) by using a kind of mutation operator that subtracts a small, randomly generated value from 25. This might transform the rule antecedent (1) into, say, the following one:

$$(Age > 21) \quad (Marital\_Status = \text{"single"}) \tag{2}$$

Rule antecedent (2) tends to cover more examples than (1), which is the kind of result that we wish in the case of a generalization operator. Another way to generalize rule antecedent (1) is simply to delete one of its conditions. This is usually called the drop condition operator in the literature.

Conversely, we could specialize the first condition of rule antecedent (1) by using a kind of mutation operator that adds a small, randomly generated value to

25. Another way to specialize (1) is, of course, to add another condition to that rule antecedent.

3.3 Fitness Functions for Rule Discovery

Recall that, as discussed in sub section 2.1, ideally the discovered rules should: (a) have a high predictive accuracy; (b) be comprehensible; and (c) be interesting. In this subsection we discuss how these rule quality criteria can be incorporated in a fitness function. To simplify our discussion, throughout this subsection we will again assume that the GA follows the Michigan approach, i.e. an individual represents a single rule. However, the basic ideas discussed below can be easily adapted to GAs following the Pittsburgh approach, where an individual represents a rule set.

Let a rule be of the form IF A THEN C, where A is the antecedent (a conjunction of conditions) and C is the consequent (predicted class), as discussed earlier. A very simple way to measure the predictive accuracy of a rule is to compute the so-called confidence factor (CF) of the rule, defined as:

$$CF = |A \& C| / |A|$$

where $|A|$ is the number of examples satisfying all the conditions in the antecedent A and $|A \& C|$ is the number of examples that both satisfy the antecedent A and have the class predicted by the consequent C. For instance, if a rule covers 10 examples (i.e. $|A| = 10$), out of which 8 have the class predicted by the rule (i.e. $|A\&C| = 8$) then the CF of the rule is CF = 80%.

Unfortunately, such a simple predictive accuracy measure favors rules overfitting the data. For instance, if $|A| = |A \& C| = 1$ then the CF of the rule is 100%. However, such a rule is most likely representing an idiosyncrasy of a particular training example, and probably will have a poor predictive accuracy on the test set. A solution for this problem is described next.

The predictive performance of a rule can be summarized by a 2 multiply 2 matrix, sometimes called a confusion matrix, as illustrated in Fig. 3. To interpret this figure, recall that A denotes a rule antecedent and C denotes the class predicted by the rule. The class predicted for an example is C if and only if the example satisfies the rule antecedent. The labels in each quadrant of the matrix have the following meaning:

TP = True Positives = Number of examples satisfying A and C
FP = False Positives = Number of examples satisfying A but not C
FN = False Negatives = Number of examples not satisfying A but satisfying C
TN = True Negatives = Number of examples not satisfying A nor C

Clearly, the higher the values of TP and TN, and the lower the values of FP and FN, the better the rule.

		actual class	
		C	not C
predicted class	C	TP	FP
	not C	FN	TN

Fig. 3. Confusion matrix for a classification rule

Note that the above-mentioned CF measure is defined, in terms of the notation of Fig. 3, by: CF = TP/(TP + FP). We can now measure the predictive accuracy of a rule by taking into account not only its CF but also a measure of how "complete" the rule is, i.e. what is the proportion of examples having the predicted class C that is actually covered by the rule antecedent. The rule completeness measure, denoted Comp, is computed by the formula: Comp = TP/(TP + FN). In order to combine the CF and Comp measures we can define a fitness function such as:

$$\text{Fitness} = \text{CF} \times \text{Comp}$$

Although this fitness function does a good job in evaluating predictive performance, it has nothing to say about the comprehensibility of the rule. We can extend this fitness function (or any other focusing only on the predictive accuracy of the rule) with a rule comprehensibility measure in several ways. A simple approach is to define a fitness function such as

$$\text{Fitness} = w_1 \times (\text{CF} \times \text{Comp}) + w_2 \times \text{Simp}$$

where Simp is a measure of rule simplicity (normalized to take on values in the range 0..1) and w_1 and w_2 are user-defined weights. The Simp measure can be defined in many different ways, depending on the application domain and on the user. In general, its value is inversely proportional to the number of conditions in the rule antecedent – i.e., the shorter the rule, the simpler it is.

Several fitness functions that take into account both the predictive accuracy and the comprehensibility of a rule are described in the literature see e.g. [37], [27], [51], [21].

Noda and his colleagues [49] have proposed a fitness function which takes into account not only the predictive accuracy but also a measure of the degree of interestingness of a rule. Their GA follows the Michigan approach and was developed for the task of dependence modeling. Their fitness function is essentially a weighted sum of two terms, where one term measures the predictive accuracy of the rule and the other term measures the degree of interestingness (or surprisingness) of the rule. The weights assigned to each term are specified by the user. Another fitness function involving a measure of rule interestingness, more precisely a variation of the well-known J-measure, is discussed in [4].

In the above projects the rule interestingness measure is objective. An intriguing research direction would be to design a fitness function based on a subjective rule interestingness measure. In particular, one possibility would be to design a kind of interactive fitness function, where the fitness of an individual depends on the user's evaluation. A similar approach has been reported in an image-enhancement application [53], where the user drives GP by deciding which individual should be the winner in tournament selection; and in an attribute-selection task [60], where a user drives a GA by interactively and subjectively selecting good prediction rules.

4 GAs for the Knowledge Discovery Process

This section is divided into two parts. Subsection 4.1 discusses GAs for data preprocessing, particularly attribute selection, whereas subsection 4.2 discusses a GA for discovered-knowledge postprocessing, particularly "pruning" an ensemble of classifiers.

4.1 GAs for Data Preprocessing

As discussed in Subsection 2.3.1, one of the key problems in preparing a data set for mining is the attribute selection problem. In the context of the classification task, this problem consists of selecting, among all available attributes, a subset of attributes relevant for predicting the value of the goal attribute.

The use of GAs for attribute selection seems natural. The main reason is that the major source of difficulty in attribute selection is attribute interaction, and one of the strengths of GAs is that they usually cope well with attribute interactions.

In addition, the problem definition lends itself to a very simple, natural genetic encoding, where each individual represents a candidate attribute subset (a candidate solution, in this problem). More precisely, we can represent a candidate attribute subset as a string with m binary genes, where m is the number of attributes and each gene can take on the values 1 or 0, indicating whether or not the corresponding attribute is in the candidate attribute subset. For instance, assuming a five-attribute data set, the individual "0 1 1 0 0" corresponds to a candidate solution where only the second and third attributes are selected to be given to the classification algorithm.

Then, a simple GA, using conventional crossover and mutation operators, can be used to evolve the population of candidate solutions towards a good attribute subset. The "trick" is to use a fitness function that is a direct measure of the performance achieved by the classification algorithm accessing only the attributes selected by the corresponding individual. With respect to the categorization of wrapper and filter approaches for attribute selection discussed in Subsection 2.3.1, this approach is clearly an instance of the wrapper approach.

Note that in the above simple encoding scheme an attribute is either selected or not, but there is no information about the relative relevance of each attribute. It is possible to use an alternative encoding scheme where highly relevant attributes will tend to be replicated in the genome. This replication will tend to reduce the probability that a highly relevant attribute be removed from the individual due, for instance, to a harmful mutation.

Such an alternative encoding scheme was proposed by Cherkauer & Shavlik [12]. In their scheme, each gene of an individual contains either an attribute A_i, i=1,...,m, or no attribute, denoted by 0. The length of the individual is fixed, but it is not necessarily equal to m, the number of attributes. An attribute is selected if it occurs at least once in the individual. For instance, assuming a 10-attribute data set and a 5-gene string, the individual "0 A_8 0 A_8 A_4" represents a candidate solution where only attributes A_8 and A_4 are selected to be given to the classification algorithm.

Note that this example suggests an intriguing possibility. Suppose we are interested not only in selecting a subset of attributes, but also in determining how relevant each of the selected attributes are. In the above example perhaps we could consider that A_8 is presumably more relevant than A_4, since the former occurs twice in the genome of the individual, whereas the latter occurs just once.

In any case it is possible to use a GA to optimize attribute weights directly (assigning to an attribute a weight that is proportional to its relevance), rather than to simply select attributes. This approach has been used particularly for optimizing attribute weights for nearest neighbor algorithms [39], [54].

Comprehensive comparisons between GA and other attribute-selection algorithms, across a number of data sets, are reported in [69] and [42]. In these projects GA was used as a wrapper to select attributes for a constructive neural network and a nearest neighbor algorithm, respectively. Overall, the results show that GA is quite competitive with other respectable attribute-selection algorithms. In particular, the results reported in [42] indicate that in large-scale attribute-selection problems, where the number of attributes is greater than 100, GA becomes the only practical way to get reasonable attribute subsets.

In [60] it is reported that the use of an interactive GA for attribute selection led to the discovery of rules that are easy-to-understand and simple enough to make practical decisions on a marketing application involving oral care products. A further discussion on the use of genetic algorithms in attribute selection can be found in [5], [63], [31], [46].

4.2 GAs for Discovered-Knowledge Postprocessing

GAs can be used in a postprocessing step applied to the knowledge discovered by a data mining algorithm. As an example, suppose that the data mining step of the knowledge discovery process has produced an ensemble of classifiers (e.g. rule sets), rather than a single classifier (e.g. a single rule set). Actually, generating an ensemble of classifiers is a relatively recent trend in machine learning when our primary goal is to maximize predictive accuracy, since it has been shown that in several cases an ensemble of classifiers has a better predictive accuracy than a single classifier [58], [15]. When an ensemble of classifiers is produced, it is common to assign a weight to each classifier in the ensemble. Hence, when classifying a new test example, the class assigned to that example is determined by taking a kind of weighted vote of the classes predicted by the individual classifiers in the ensemble.

However, there is a risk of generating too many classifiers which end up overfitting the training data. Therefore, it is desirable to have a procedure to "prune" the ensemble of classifiers, which is conceptually similar to pruning a rule set or a decision tree.

To address this problem, Thompson [61], [62] has proposed a GA to optimize the weights of the classifiers in the ensemble. The proposed GA uses a real-valued individual encoding. Each individual has n real-valued genes, where n is the number of classifiers in the ensemble. Each gene represents the voting weight of its corresponding classifier. The fitness function consists of measuring the predictive accuracy of the ensemble with the weights proposed by the individual. This predictive accuracy is measured on a separate data subset, called the "pruning" set (or hold-out set). This is a part of the original training set reserved only for fitness-evaluation purposes, whereas the remaining part of the original training set is used only for generating the ensemble of classifiers.

Note that the number of classifiers in the ensemble can be effectively reduced if the voting weight of some classifier(s) is(are) set to 0. Actually, one of the mutation methods with which the author has experimented consists of simply setting a gene (voting weight) to 0.

5. GP for Rule Discovery

GP can be considered as a more open-ended search paradigm, in comparison with GA [41], [6]. The search performed by GP can be very useful in classification and other prediction tasks, since the system can produce many different combinations of attributes - using the several different functions available in the function set – which would not be considered by a conventional GA. Hence, even if the original attributes do not have much predictive power by themselves, the system can effectively create "derived attributes" with greater predictive power, by applying the function set to the original attributes. The potential of GP to create these derived attributes will be discussed in more detail in section 6.

Before we move on to that section, we discuss next two issues in the use of GP for rule discovery, namely individual representation (subsection 5.1) and discovery of comprehensible rules (subsection 5.2).

5.1 Individual Representation

The application of standard GP to the classification task is relatively straightforward, as long as all the attributes are numeric. In this case we can include in the function set several kinds of mathematical function appropriate to the application domain and include in the terminal set the predicting attributes – and possibly a random constant generator. Once we apply the functions in the internal nodes of a GP individual to the values of the attributes in the leaf nodes of that individual, the system computes a numerical value that is output at the root node of the tree. Assuming a two-class problem, if this output is greater than a given threshold the system predicts a given class, otherwise the system predicts the other class.

In this section, however, we are interested in using GP for discovering high-level, comprehensible prediction (IF – THEN) rules, rather than just producing a numerical signal in the root node. The first obstacle to be overcome is the closure property of GP. This property means that the output of a node in a GP tree can be used as the input to any parent node in the tree. Note that this property is satisfied in the above case of standard GP applied to numeric data, since, in principle, the number returned by a mathematical function can be used as the input to another mathematical function. (In practice, some mathematical functions have to be slightly modified to satisfy the closure property.)

When mining a data set containing a mixture of continuous and categorical (or nominal) attribute values this property is not satisfied by standard GP. Different attributes are associated with different operators/functions. For example, the condition "*Age* < 18" is valid, but the condition "*Sex* < female" is not.

Several solutions have been proposed to cope with the closure property of GP, when addressing the classification task. One approach is based on the use of constrained-syntax GP. The key idea is that, for each function available in the function set, the user specifies the type of its arguments and the type of its result [7]. Crossover and mutation are then modified to create only valid trees, by respecting the user-defined restrictions on tree syntax.

An example is shown in Table 1, where, for each row, the second and third columns specify the data type of the input and of the output of the function specified in the first column. Once this kind of specification is available to the GP, the system can generate individuals such as the one shown in Fig. 4. This figure assumes that Atr1 is a categorical (nominal) attribute, whereas Atr3, Atr5, Atr6 are real-valued attributes.

Table 1. Example of data type definitions for input and output of functions

Functions	data type of input arguments	data type of output
+, -, *, /	(real, real)	real
≤, >	(real, real)	boolean
=	(nominal, nominal)	boolean
AND, OR	(boolean, boolean)	boolean

Another approach to constrained-syntax GP consists of having a user-defined, domain-dependent grammar specify the syntax of valid rules. The grammar can also be used to incorporate domain-specific knowledge into the GP system. This approach, discussed in detail in [68], has led to the discovery of interesting knowledge (in the opinion of medical experts) in real-world fracture and scoliosis databases.

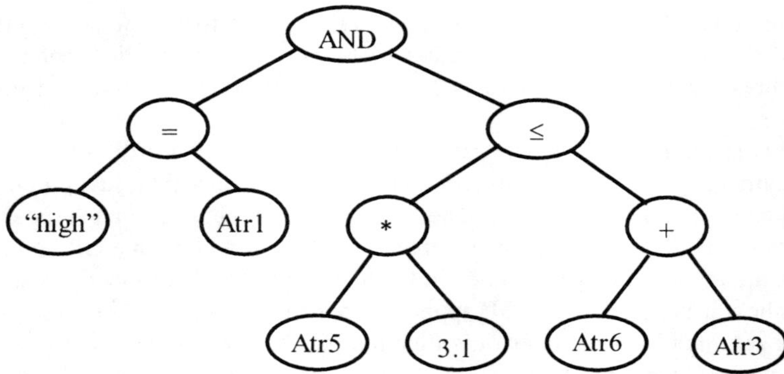

Fig. 4. Example of a GP individual (representing a rule antecedent) meeting the data type restrictions specified in Table 1.

A different approach for coping with the problem of closure in GP for rule discovery consists of somehow modifying the data being mined – rather than modifying the standard GP algorithm. An example of this approach consists of booleanizing all the attributes being mined and then using logical operators (AND, OR, etc.) in the function set. Hence, the output of any node in the tree will be a boolean value, which can be used as input to any logical operator in the corresponding parent node. Systems based on this approach are described in [16], [8].

5.2 Discovering Comprehensible Rules with GP

In principle a GP system for rule discovery could use fitness functions similar to the ones used by GAs for rule discovery. There are, however, some important differences. In particular, since the size of GP trees can grow a lot, in general it is more necessary to incorporate some measure of rule comprehensibility in the fitness function of a GP for rule discovery than in the fitness function of a GA for rule discovery. Actually, using GP to discover comprehensible classification rules can be considered a research issue. Some recent proposals to cope with this issue are briefly reviewed in the following.

First of all, it should be noted that, although knowledge comprehensibility is a kind of subjective concept, the data mining literature often uses an objective measure of rule comprehensibility: in general, the shorter (the fewer the number of conditions in) the rule, the more comprehensible it is. The same principle applies to rule sets. In general, the fewer the number of rules in the rule set, the more comprehensible it is. Our discussion in the following paragraphs assumes this kind of objective measure of rule comprehensibility. In other words, we are interested in GP systems that return a small number of short rules – as long as the discovered rule set still has a high predictive accuracy, of course.

The simplest approach to favor the discovery of short rules is to include a penalty for long rules in the fitness function. An example of the use of this approach can be found in [9], where a part of the fitness function is a direct measure of rule simplicity, given by the formula:

Simplicity = (*MaxNodes* – 0.5 *NumNodes* – 0.5)/(*MaxNodes* – 1)

where *MaxNodes* is the maximum allowed number of nodes of a tree (individual) and *NumNodes* is the current number of nodes of the tree. This formula produces its maximum value of 1.0 when a rule is so simple that it contains just one term, and it produces its minimum value of 0.5 when the number of tree nodes equals the allowed maximum. This approach has led to the discovery of short, simple rules in a real-world application involving chest-pain diagnosis.

A more elaborated approach for discovering comprehensible rules is, for instance, the hybrid GA/GP system to evolve decision trees proposed by [56]. The system is a hybrid GA/GP in the sense that it uses a GA to evolve a population of "programs" (decision trees). The system also uses a fitness function that considers both a tree's predictive accuracy and its size, in order to achieve the goal of minimizing tree size without unduly reducing predictive accuracy.

6 GP for Data Preprocessing

In Subsection 4.1 we saw that a simple GA can be naturally applied to an important data preprocessing task, namely the selection of relevant attributes for data mining. Sometimes, however, the preprocessing task may be more naturally addressed by a more open-ended evolutionary algorithm such as GP.

A good example is the preprocessing task of attribute construction – also called constructive induction. In this task the goal is to automatically construct new attributes, applying some operations to the original attributes, such that the new attributes make the data mining problem easier.

To illustrate the importance of this preprocessing task, consider the commonplace case where a classification algorithm can discover only propositional ("0-th order") logic rules. Suppose that we want to apply this algorithm to a data set where each record contains several attributes that can be used to predict whether the shares of a company will go up or down in the financial market. Now suppose that the set of predicting attributes includes the attributes *Income* and *Expenditure* of a company. Our prepositional logic classification algorithm would not be able to discover rules of the form:

IF (*Income* > *Expenditure*) AND . . . THEN (*Shares* = *up*)
IF (*Income* < *Expenditure*) AND . . . THEN (*Shares* = *down*)

because these rules involve a first-order logic condition, namely a comparison between two attributes – rather than between an attribute and its value.

A suitably designed attribute construction algorithm could automatically construct a binary attribute such as "(*Income* > *Expenditure*)?", taking on the values "yes" or "no". The new attribute would then be given to the classification algorithm, in order to improve the quality of the rules to be discovered by the latter.

The major problem in attribute construction is that the search space tends to be huge. Although in the above very simple example it was easy to see that a good attribute could be constructed by using the relational operator ">", in practice there are a large number of candidate operations to be applied to the original attributes. In addition, the construction of a good attribute often requires that the attribute construction algorithm generates and evaluates combinations of several original attributes, rather than just two attributes as in the above example. GP can be used to search this huge search space.

An example of the use of GP in attribute construction is found in [36]. In this project a GP-based system is compared with two other attribute construction methods, namely LFC and GALA. The comparison is made across 12 data sets. Overall, the predictive accuracy of the GP-based system was considerably better than LFC's one and somewhat better than GALA's one. A hybrid GA/GP system, performing attribute selection and attribute construction at the same time, is discussed in [64]. This approach has substantially reduced error rate in a face-recognition problem.

7 Discussion and Research Directions

We have begun our discussion of data mining and knowledge discovery by identifying, in Subsection 2.1, three desirable properties of discovered knowledge. These properties are predictive accuracy, comprehensibility and interestingness. We believe a promising research direction is to design evolutionary algorithms which aim at discovering truly interesting rules. Clearly, this is much easier said than done. The major problem is that rule interestingness is a complex concept, involving both objective and subjective aspects. Almost all the fitness functions currently used in evolutionary algorithms for data mining focus on the objective aspect of rule quality, and in most cases only predictive accuracy and rule comprehensibility are taken into account. However, these two factors alone do not guarantee rule interestingness, since a highly accurate, comprehensible rule can still be uninteresting, if it corresponds to a piece of knowledge previously known by the user.

Concerning data mining tasks, which correspond to kinds of problems to be solved by data mining algorithms, in this chapter we have focused on the classification task only (see Subsection 2.2.1), due to space limitations. However, many of the ideas and concepts discussed here are relevant to other data mining tasks involving prediction, such as the dependence modeling task briefly discussed in Subsection 2.2.2.

We have discussed several approaches to encode prediction (IF – THEN) rules into the genome of individuals, as well as several genetic operators designed specifically for data mining purposes. A typical example is the use of generalizing/specializing crossover discussed in section 3.2.2. Another example is the information-theoretic rule pruning operator proposed in [10]. We believe that the development of new data mining-oriented operators is important to improve the performance of evolutionary algorithms in data mining and knowledge discovery. Using this kind of operator makes evolutionary algorithms endowed with some "knowledge" about what kind of genome-modification operation makes sense in data mining problems. The same argument holds for other ways of tailoring evolutionary algorithms for data mining, such as developing data mining-oriented individual representations.

We have also discussed the use of evolutionary algorithms in the preprocessing and postprocessing phases of the knowledge discovery process. Although there has been significant research on the use of GAs for attribute selection, the use of evolutionary algorithms in other preprocessing tasks and in postprocessing tasks seems to be less explored. In particular, we believe that a promising research direction is to use evolutionary algorithms for attribute construction (or constructive induction). Open-ended evolutionary algorithms, such as GP, can be suitable for this difficult, important data mining problem.

Acknowledgments

I thank Heitor S. Lopes for useful comments on the first draft of this chapter. My research on data mining with evolutionary algorithms is partially supported by grant 300153/98-8 from CNPq (the Brazilian government's National Council of Scientific and Technological Development).

References

[1] Agrawal R, Imielinski T and Swami A. Mining association rules between sets of items in large databases. *Proc. 1993 Int. Conf. Management of Data (SIGMOD-93)*, 207-216. May 1993.

[2] Agrawal R, Mannila H, Srikant R, Toivonen H and Verkamo AI. Fast discovery of association rules. In: Fayyad UM, Piatetsky-Shapiro G, Smyth P and Uthurusamy R. (Eds.) *Advances in Knowledge Discovery and Data Mining*, 307-328. AAAI/MIT Press, 1996.

[3] Anglano C, Giordana A, Lo Bello G and Saitta L. Coevolutionary, distributed search for inducing concept descriptions. *Lecture Notes in Artificial Intelligence 1398. ECML-98: Proc. 10th Eur. Conf. Machine Learning*, 422-333. Springer-Verlag, 1998.

[4] Araujo DLA, Lopes HS and Freitas AA. A parallel genetic algorithm for rule discovery in large databases. *Proc. 1999 IEEE Systems, Man and Cybernetics Conf.*, v. 3, 940-945. Tokyo, 1999.

[5] Bala J, De Jong K, Huang J, Vafaie H and Wechsler H. Using learning to facilitate the evolution of features for recognizing visual concepts. *Evolutionary Computation 4(3) - Special Issue on Evolution, Learning, and Instinct: 100 years of the Baldwin Effect.* 1997.

[6] Banzhaf W, Nordin P, Keller RE and Francone FD *Genetic Programming -- an Introduction: On the Automatic Evolution of Computer Programs and Its Applications.* Morgan Kaufmann, 1998.

[7] Bhattacharyya S, Pictet O and Zumbach G. Representational semantics for genetic programming based learning in high-frequency financial data. *Genetic Programming 1998: Proc. 3rd Annual Conf.*, 11-16. Morgan Kaufmann, 1998.

[8] Bojarczuk CC, Lopes HS and Freitas AA. Discovering comprehensible classification rules using genetic programming: a case study in a medical domain. *Proc. Genetic and Evolutionary Computation Conf. (GECCO-99)*, 953-958. Orlando, FL, USA, July/1999.

[9] Bojarczuk CC, Lopes HS and Freitas AA. Genetic programming for knowledge discovery in chest pain diagnosis. *IEEE Engineering in Medicine and Biology Magazine – special issue on data mining and knowledge discovery,* 19(4), 38-44, July/Aug. 2000.

[10] Carvalho DR and Freitas AA. A hybrid decision tree/genetic algorithm for coping with the problem of small disjuncts in data mining. *Proc. Genetic and Evolutionary Computation Conf. (GECCO-2000)*, 1061-1068. Las Vegas, NV, USA, July 2000.

[11] Catlett J. On changing continuous attributes into ordered discrete attributes. *Proc. Eur. Working Session on Learning (EWSL-91). Lecture Notes in Artificial Intelligence 482*, 164-178. Springer-Verlag, 1991.

[12] Cherkauer KJ and Shavlik JW. Growing simpler decision trees to facilitate knowledge discovery. *Proc. 2nd Int. Conf. Knowledge Discovery & Data Mining (KDD-96)*, 315-318. AAAI Press, 1996.

[13] De Jong KA, Spears WM and Gordon DF. Using genetic algorithms for concept learning. *Machine Learning*, 13, 161-188, 1993.

[14] Dhar V, Chou D and Provost F. Discovering interesting patterns for investment decision making with GLOWER – a Genetic Learner Overlaid with Entropy Reduction. *To appear in Data Mining and Knowledge Discovery Journal.* 2000.

[15] Domingos P. Knowledge acquisition from examples via multiple models. *Machine Learning: Proc. 14th Int. Conf. (ICML-97)*, 98-106. Morgan Kaufmann, 1997.

[16] Eggermont J, Eiben AE and van Hemert JI. A comparison of genetic programming variants for data classification. *Proc. Intelligent Data Analysis (IDA-99)*. 1999.

[17] Falkenauer E. *Genetic Algorithms and Grouping Problems.* John Wiley & Sons, 1998.

[18] Fayyad UM, Piatetsky-Shapiro G and Smyth P. From data mining to knowledge discovery: an overview. In: Fayyad UM, Piatetsky-Shapiro G, Smyth P and Uthurusamy R. (Eds.) *Advances in Knowledge Discovery & Data Mining*, 1-34. AAAI/MIT, 1996.

[19] Fisher DH. Knowledge acquisition via incremental conceptual clustering. *Machine Learning*, 2, 139-172, 1987.

[20] Fisher D and Hapanyengwi G. Database management and analysis tools of machine induction. *Journal of Intelligent Information Systems*, 2(1), 5-38, 1993.

[21] Flockhart IW and Radcliffe NJ. GA-MINER: parallel data mining with hierarchical genetic algorithms - final report. *EPCC-AIKMS-GA-MINER-Report 1.0*. University of Edinburgh, UK, 1995.

[22] Freitas AA. On objective measures of rule surprisingness. *Lecture Notes in Artificial Intelligence 1510: Principles of Data Mining and Knowledge Discovery (Proc. 2nd Eur. Symp., PKDD'98, Nantes, France)*, 1-9. Springer-Verlag, 1998.

[23] Freitas AA. On Rule Interestingness Measures. *Knowledge-Based Systems,* 12(5-6), 309-315, Oct. 1999.

[24] Freitas AA. Understanding the crucial differences between classification and discovery of association rules - a position paper. *To appear in ACM SIGKDD Explorations,* 2(1), 2000.

[25] Freitas AA and Lavington SH. *Mining Very Large Databases with Parallel Processing.* Kluwer, 1998.

[26] Gebhardt F. Choosing among competing generalizations. *Knowledge Acquisition*, 3, 361-380,1991,.

[27] Giordana A and Neri F. Search-intensive concept induction. *Evolutionary Computation* 3(4), 375-416, Winter 1995.

[28] Giordana A and Saitta L, Zini F. Learning disjunctive concepts by means of genetic algorithms. *Proc. 10th Int. Conf. Machine Learning (ML-94)*, 96-104. Morgan Kaufmann, 1994.

[29] Goldberg DE *Genetic Algorithms in Search, Optimization and Machine Learning.* Addison-Wesley, 1989.

[30] Greene DP and Smith SF. Competition-based induction of decision models from examples. *Machine Learning*, 13, 229-257, 1993.

[31] Guerra-Salcedo C and Whitley D. Feature selection mechanisms for ensemble creation: a genetic search perspective. In: Freitas AA (Ed.) *Data Mining with Evolutionary Algorithms: Research Directions – Papers from the AAAI Workshop*, 13-17. Technical Report WS-99-06. AAAI Press, 1999.

[32] Guyon I, Matic N and Vapnik V. Discovering informative patterns and data cleaning. In: Fayyad UM, Piatetsky-Shapiro G, Smyth P and Uthurusamy R. (Eds.) *Advances in Knowledge Discovery and Data Mining*, 181-203. AAAI/MIT Press. 1996.

[33] Hall LO, Ozyurt IB and Bezdek JC. Clustering with a genetically optimized approach. *IEEE Trans. Evolutionary Computation 3(2)*, 103-112. July 1999.

[34] Hand DJ. *Construction and Assessment of Classification Rules.* John Wiley & Sons, 1997.

[35] Holland JH. Escaping brittleness: the possibilities of general-purpose learning algorithms applied to parallel rule-based systems. In: Mitchell T et al. (Eds.) *Machine Learning, Vol. 2*, 593-623. Morgan Kaufmann, 1986.

[36] Hu Y-J. A genetic programming approach to constructive induction. *Genetic Programming 1998: Proc. 3rd Annual Conf.*, 146-151. Morgan Kaufmann, 1998.

[37] Janikow CZ. A knowledge-intensive genetic algorithm for supervised learning. *Machine Learning*, 13, 189-228, 1993.

[38] John GH, Kohavi R and Pfleger K. Irrelevant features and the subset selection problem. *Proc. 11th Int. Conf. Machine Learning*, 121-129. 1994.

[39] Kelly Jr. JD and Davis L. A hybrid genetic algorithm for classification. *Proc. 12th Int. Joint Conf. on AI*, 645-650. 1991.

[40] Klemettinen M, Mannila H, Ronkainen P, Toivonen H and Verkamo AI. Finding interesting rules from large sets of discovered association rules. *Proc. 3rd Int. Conf. on Information and Knowledge Management.* Gaithersburg, MD, USA, Nov./Dec. 1994.

[41] Koza JR. *Genetic Programming: on the Programming of Computers by Means of Natural Selection.* MIT Press, 1992.

[42] Kudo M and Skalansky J. Comparison of algorithms that select features for pattern classifiers. *Pattern Recognition*, 33(1), 25-41, Jan. 2000.

[43] Kwedlo W and Kretowski M. Discovery of decision rules from databases: an evolutionary approach. *Proc. 2nd Eur. Symp. Principles of Data Mining and Knowledge Discovery (PKDD-98). Lecture Notes in Artificial Intelligence 1510*, 371-378. Springer-Verlag, 1998.

[44] Liu B, Hsu W. and Chen S. Using general impressions to analyze discovered classification rules. *Proc. 3rd Int. Conf. Knowledge Discovery & Data Mining*, 31-36. AAAI Press, 1997.

[45] Mahfoud SW. *Niching Methods for Genetic Algorithms.* Ph.D. Thesis. Univ. of Illinois at Urbana-Champaign. IlliGAL Report No. 95001. May 1995.

[46] Martin-Bautista MJ and Vila MA. A survey of genetic feature selection in mining issues. *Proc. Congr. Evolutionary Computation (CEC-99)*, 1314-1321. Washington DC, USA, July 1999.

[47] Michalewicz Z. *Genetic Algorithms + Data Structures = Evolution Programs*. 3rd Ed. Springer-Verlag, 1996.

[48] Michie, D, Spiegelhalter, DJ and Taylor, CC. *Machine Learning, Neural and Statistical Classification*. Ellis Horwood, 1994.

[49] Noda E, Freitas AA and Lopes HS. Discovering interesting prediction rules with a genetic algorithm. *Proc. Conf. on Evolutionary Computation – 1999 (CEC-99)*, 1322-1329. Washington DC, USA, July 1999.

[50] Park Y and Song M. A genetic algorithm for clustering problems. *Genetic Programming 1998: Proc. 3rd Annual Conf.*, 568-575. Morgan Kaufmann, 1998.

[51] Pei M, Goodman ED, Punch WF. Pattern discovery from data using genetic algorithms. *Proc. 1st Pacific-Asia Conf. Knowledge Discovery & Data Mining (PAKDD-97)*. Feb. 1997.

[52] Pfahringer B. Supervised and unsupervised discretization of continuous features. *Proc. 12th Int. Conf. Machine Learning*, 456-463. 1995.

[53] Poli R and Cagnoni S. Genetic programming with user-driven selection: experiments on the evolution of algorithms for image enhancement. *Genetic Programming 1997: Proc. 2nd Annual Conf.*, 269-277. Morgan Kaufmann, 1997.

[54] Punch WF, Goodman ED, Pei M, Chia-Sun L, Hovland P, Enbody R. Further research on feature selection and classification using genetic algorithms. *Proc. 5th Int. Conf. Genetic Algorithms (ICGA-93)*, 557-564. Morgan Kaufmann, 1993.

[55] Pyle D. *Data Preparation for Data Mining*. Morgan Kaufmann, 1999.

[56] Ryan MD and Rayward-Smith VJ. The evolution of decision trees. *Genetic Programming 1998: Proc. 3rd Annual Conf.*, 350-358. Morgan Kaufmann, 1998.

[57] Schaffer C. Overfitting avoidance as bias. *Machine Learning*, 10, 153-178, 1993.

[58] Schapire RE, Freund Y, Bartlett P and Lee WS. Boosting the margin: a new explanation for the effectiveness of voting methods. *Machine Learning: Proc. 14th Int. Conf. (ICML-97)*, 322-330. Morgan Kaufmann, 1997.

[59] Simoudis E, Livezey B and Kerber R. Integrating inductive and deductive reasoning for data mining. In: Fayyad UM, Piatetsky-Shapiro G, Smyth P and Uthurusamy R. (Eds.) *Advances in Knowledge Discovery and Data Mining*, 353-373. AAAI/MIT Press, 1996.

[60] Terano T and Ishino Y. Interactive genetic algorithm based feature selection and its application to marketing data analysis. In: Liu H and Motoda H (Eds.) *Feature Extraction, Construction and Selection: a data mining perspective*, 393-406. Kluwer, 1998.

[61] Thompson S. Pruning boosted classifiers with a real valued genetic algorithm. *Research & Development. in Expert Systems XV - Proc. ES'98*, 133-146. Springer-Verlag, 1998.

[62] Thompson S. Genetic algorithms as postprocessors for data mining. In: Freitas AA (Ed.) *Data Mining with Evolutionary Algorithms: Research Directions – Papers from the AAAI Workshop*, 18-22. Technical Report WS-99-06. AAAI Press, 1999.

[63] Vafaie H and De Jong K. Robust feature selection algorithms. *Proc. 1993 IEEE Int. Conf. on Tools with AI*, 356-363. Boston, MS, USA. Nov. 1993.

[64] Vafaie H and De Jong K. Evolutionary feature space transformation. In: Liu H and Motoda H (Eds.) *Feature Extraction, Construction and Selection: a data mining perspective*, 307-323. Kluwer, 1998.

[65] Weiss GM and Hirsh H. Learning to predict rare events in event sequences. *Proc. 4th Int. Conf. Knowledge Discovery and Data Mining*, 359-363. AAAI Press, 1998.

[66] Weiss SM and Indurkhya N. *Predictive Data Mining: a practical guide*. Morgan Kaufmann, 1998.

[67] Weiss SM and Kulikowski CA. *Computer Systems that Learn*. Morgan Kaufmann, 1991.

[68] Wong ML and Leung KS. *Data Mining Using Grammar-Based Genetic Programming and Applications*. Kluwer, 2000.

[69] Yang J and Honavar V. Feature subset selection using a genetic algorithm. In: Liu H and Motoda H (Eds.) *Feature Extraction, Construction and Selection: a data mining perspective*, 117-136. Kluwer, 1998.

Data Mining from Clinical Data using Interactive Evolutionary Computation

Takao Terano† and Masanori Inada†, ‡

† Graduate School of Systems Management
University of Tsukuba, Tokyo
3-29-1, Otsuka, Bunkyo-ku, Tokyo 112-0012, Japan
E-mail: terano@gssm.otsuka.tsukuba.ac.jp
‡ Department of Clinical Laboratory, Toranomon Hospital
2-2-2, Toranomon, Minato-ku, Tokyo 105-8470, Japan
E-mail: m-inada@hi-ho.ne.jp

Summary. Interactive evolutionary computation (IEC) is a subjective and interactive method to evaluate the qualities of offspring generated by genetic operations. Data mining, an interdisciplinary research area including artificial intelligence, statistics and databases, is a series of semi-automated processes to extract explicit useful knowledge from given databases. In this chapter, we adopt IEC in order to select relevant features in inductive learning for data mining tasks. The method we have proposed is used to discover efficient decision knowledge from noisy clinical data in a medical domain. This chapter describes the principles of IEC and SIBILE (SImulated Breeding and Inductive Learning), which we have developed for practical data mining problems, and its application to a common data set on clinical patients. The basic ideas of SIBILE are that IEC is used to get the effective features from the data and that inductive learning is used to acquire simple decision rules from the subset of the data.

1 Introduction

When we apply genetic algorithm (GA) based techniques [9,11] to some test problems whose real solutions are already known, we are often tempted to control the parameters of GAs and/or GA operations in order to fit the solutions. Such desires also occur in practical problems [4], where explicit evaluation functions are neither clear nor well defined. The interactive genetic algorithm (IGA) or, more generally, the interactive evolutionary computation (IEC) will fit these situations by allowing users to interactively evaluate each offspring generated by GA operations, even if the evaluation criteria are not clear beforehand [17–19].

Data mining (DM) or knowledge discovery in databases (KDD) is an interdisciplinary research area integrating machine learning in artificial intelligence, statistics and database theories. In practice, DM is a series of semi-automated processes to extract explicit useful knowledge from given databases [7]. One of the difficult problems of DM is how to select good features of the data to explain the knowledge implicitly contained in it. This is a typical combinatorial problem with NP-hard characteristics. To solve the problem, evolutionary computing (EC) techniques are promising [8].

In this chapter, we adopt IEC in order to select relevant features for inductive learning in DM tasks. The method we have proposed [20–22] is used to extract efficient decision knowledge from noisy clinical data in a medical domain [16].

We will describe the principles of IEC and SIBILE (SImulated Breeding and Inductive Learning),[1] which we have developed for practical DM problems, and the application of SIBILE to a common data set on clinical patients provided by Prof. Tsumoto [16]. SIBILE is a novel method and a toolkit to mine efficient decision rules from noisy data using both IEC and inductive learning techniques. The basic ideas are that IEC is used to get the effective features from the data and that inductive learning is used to acquire simple decision rules from the subset of the data.

Our approach to the problem is characterized as follows: (1) Repeat an apply-and-evaluate loop of C4.5 [12] by a human expert with medical knowledge to assess the performance of the program; and then (2) Apply our GA-based feature selection method in a human-in-a-loop interactive manner. As described elsewhere, SIBILE has shown good performance in marketing decision making problems [21,22]. The contribution of this chapter is to demonstrate that the proposed method is so powerful that SIBILE is applicable to more complex and severe problems in medical task domains.

2 Interactive Evolutionary Computation, Feature Selection and Knowledge Extraction

2.1 Interactive Evolutionary Computation

IEC is a subjective and interactive method to improve or optimize the target problem based on human subjective evaluation. Simply stated, the EC fitness function is replaced by a human's judgments [17,18]. The human's judgments come from her or his expertise including both explicit and implicit tacit knowledge. If the judgment criteria are consistent during the GA iterations, contrary to our intuition about conventional GAs with huge computational efforts, the convergence speed is fairly fast. Although the time consuming efforts by human users are major defects of IEC, both our experience and the literature of IEC demonstrate that user interactions to evaluate the 10 or fewer offspring and 10 times iterations will usually generate feasible results, if the users have consistent evaluation criteria. IEC is a feasible technique.

The IEC techniques were first proposed in the early 1990's in the graphic art domain [13,14] and since then have been gradually applied to various task

[1] Following Professor Unemi's suggestions, several years ago, we used the words "Simulated Breeding" instead of "Interactive Evolutionary Computation". Thus, we named our method SIBILE, which is a sibyl in old French.

domain problems and the effectiveness validated [15,23]. The idea of IEC is similar to the ones of *simulated evolution* or *interactive evolution* [13,14] in computer graphics arts in the ALife literature. In both methods, individuals judged by human experts or users to have some *efficient* features are allowed to breed offspring. The judgments are subjectively or interactively done. In such cases where the evaluation function is not clearly defined, IEC is able to improve offspring by selecting the parent for the next generation from among the phenotypes developed based on human preference.

An example of an earlier system related to IEC was developed by R. Dawkins in the Blind Watchmaker [5,6]. Dawkins has shown that through use of form constraints, such as left and right side symmetry and segment the structures, which are both often seen in living objects, figures resembling living objects can be developed. Recent applications of IEC techniques, refer to the web sites http://www.kyushu-id.ac.jp/~takagi/ (applications in general) and http://www.intlab.soka.ac.jp/~unemi/sbart/Reference.html (applications in graphic arts), and to recent articles by Takagi [17–19].

2.2 Feature Selection in Machine Learning

EC techniques are considered to be good means of practically solving feature selection problems for DM in the literature (e.g., [1,2]). To use EC techniques, it is necessary to clearly define the objective function in advance (e.g., [4]). However, for problems where human subjective judgments play an important role, the definition of the evaluation function is not an easy task. If we were able to have well-defined objective functions to be optimized in feature selection, we could apply conventional GA techniques to choose appropriate features. Our problems do not meet the requirements.

The method most representative of inductive machine learning is ID3 which gives a decision tree or a set of decision rules as an output for the results of classification using attribute-value pairs. Our research adopts C4.5 [12] a noise-tolerant successor of ID3. As stated in [24], inductive learning or concept learning techniques in artificial intelligence, and classification of data via statistical methods (e.g., the linear discrimination method) will give similar classification results, if we are able to assume a linear distribution of sample data. The performances of the results are compatible. However, we believe that for domain experts the explainability of the results from machine learning is better than the one from statistical methods.

On the other hand, machine learning, which attempts to incorporate all features in a decision trees is very complex [10]. Hence, transformation and selection of appropriate features becomes necessary. Various studies have been conducted to deal with the problem of interactions between features. Feature selection tasks intrinsically include combinatorial NP-hard problems.

2.3 Issues of Knowledge Extraction

Recent techniques in DM [7,24] are able to generate a very large amount of 'so-called' knowledge. For this purpose, they have frequently reported the applicability of EC techniques to DM. However, we often observe that few of them are effective in a practical sense. It is quite interesting that most of the useful knowledge is on a boundary between a lot of the 'nonsense', which has no meaning for anyone, and a little 'trivial', which every-one already knows. The judgments depend on the background, the objective and the context in which the knowledge is used. Unlike popular learning-from-example methods, in such tasks we must interpret the characteristics of the data without clear features of the data nor pre-determined evaluation criteria. The problem is how domain experts get simple, easy-to-understand and accurate knowledge from noisy data.

In clinical data analysis domains, medical experts must identify the key factors to diagnose the diseases. However, in the task domain, although we can only gather noisy sample data with complicated models, it is critical to get simple but clear rules to explain the characteristics of the diseases. It is necessary to organize the information in a simple and useful format in order to use this understanding in decision making processes.

Summing up, the difficult points of the research are that (1) the clinical patient data intrinsically involve-s noise, (2) a statistical distribution of data cannot be previously assumed, (3) selection of appropriate features of the data is inevitable, because of the difficulty in interpreting the results of analysis incorporating all the various features, and (4) we do not know how to define the evaluation criteria in advance for effective explanation.

3 Algorithm of SIBILE

The procedure of the proposed method [22] is shown in Figure 1. Some additional descriptions are given in the following:

Step 1: Initialization We generate the initial population at random; that is, we generate a set of m individuals with less than or equal to l features. The m and l respectively represent the number of individuals and the length of their chromosomes. The chromosomes representing the features are coded in binary strings, in which a '1' (respectively '0') means that a feature is (not) selected for inclusion in the inductive learning process. Each locus also has a preference weight value w between 0 and 1. The weight values are changed during the GA process using a simple reinforcement learning technique. This enables the algorithm to give appropriate values to generate good feature sets.

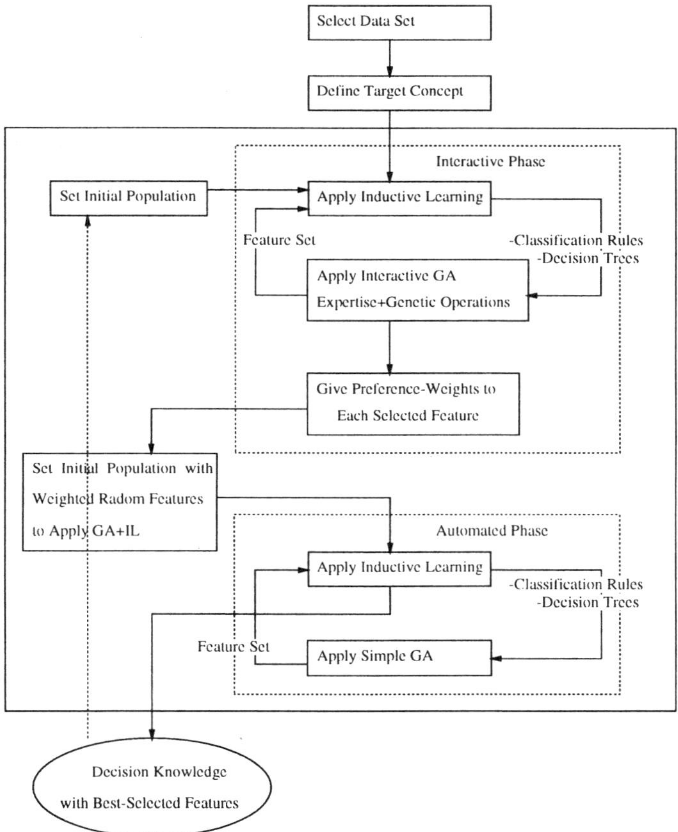

Fig. 1. SIBILE algorithm with interactive-and automated-phases

Interactive Phase

Step 2: Apply Inductive Inferences Inductive learning is applied to each of the m individuals with the selected features suggested as '1' in the chromosome. The original data is aggregated, each of which has the corresponding features in it. Then the m subsets of the data are processed by inductive learning programs. As stated earlier, we use C4.5 programs with standard parameters. As a result, a set of m decision trees with selected features or the corresponding set of decision rules is generated.

Step 3: Interaction with Users In this step, a user or a domain expert must interact with the system. This is a highly knowledge-intensive task. Observing the forms of the decision trees, set of decision rules and combinations of selected features, the domain expert subjectively and interactively evaluates the intermediate results to explain the characteristics of the data.

The expert specifies 'good' feature sets is two ways. The first one is to evaluate the generated decision tree, that is, a generated set of features. On the other hand, the second one is to evaluate specific rules which show high performance. This means that the corresponding specific features are effective to interpret the knowledge.

- Selection of decision trees
  The expert selects 'good' decision trees. The feature sets in the selected trees are used as parents for genetic operations.
  The selected parents are preserved for the next generation. The remaining individuals are replaced by the corresponding new offspring.
- Selection of good rules
  The expert selects 'good' rules generated from decision trees. Features in the decision rules become candidates for the next generation. The results will reflect the preference weight of each feature.

We apply two-point-crossover operations to the features in order to get new sets to broaden the variety of offspring. The modified point is that the features are stochastically selected based on their preference weight values.

Step 4: Give Preference Weights to Each Selected Feature Based on the judgment in Step 3, the preference weights are modified. If the user selects decision trees with good features, the preference weights of the corresponding features are equally increased. If the user selects some of the decision rules, the preference weights of the corresponding features in the rules are increased based on the user-specified preference values (in our current implementation) from -3 to $+3$.

Step 5: Repeat the Steps Steps 2 to 4 are repeated until an appropriate decision tree or set of decision rules is obtained. As illustrated in [5], the number of steps required to obtain the appropriate results is very small. As mentioned earlier, in our experiments, it usually takes less than 10 steps.

Automated Phase

Step 6: Set Initial Population with Weighted Random Features Based on the preference weights interactively determined in the *Interactive Phase*, the chromosomes or features are randomly selected to apply genetic operations. We generate the initial population; that is, we select m' sets of individuals with less than or equal to l features.

Step 7: Apply Inductive Learning The procedure is the same as Step 2.

Step 8: Apply the Simple GA Each individual is evaluated by the following fitness function:

$$F(T) = \alpha A + \beta B + \gamma C,$$

where T means a corresponding decision tree, A = (ratio of top three rules generated from T to classify the data positive), B = (ratio to classify the data positive by T) and C = (number of tree nodes with all features)/(number of nodes of T). We maximize $F(T)$. We will give the higher score when small numbers of rules show a better performance. Thus, the first term A means that the top three rules are essential to interpret the acquired knowledge. Of course, we would like to get accurate knowledge. Thus, the second term B means that the higher performance of T is the more desirable. In general, the number of nodes of a tree with all features is larger than the ones with selected features. Thus, the last term C means that the smaller number of nodes is the more desirable to evaluate the knowledge.

α, β and γ are parameters experimentally determined by the direct application of C4.5 programs. The criteria we use for the parameters are that α should be equal to the classification ratio of the best decision, β should be equal to the inverse of the classification ratio of the best decision tree, and γ should be equal to the number of nodes in the decision tree divided by the number of attributes appearing in meaningful sets of corresponding decision rules. These parameter values are, however, not sensitive to the kinds of given data sets. In our experiments, we set $\alpha = 1.0$, $\beta = 1.0$ and $\gamma = 0.5$. As a result, A, B and C contribute $F(t)$ at the same level of magnitude.

We adopt the following genetic operations based on simple GA in [9]: conventional uniform crossover, mutation rate: 0.01, generational replacement, and the best two parents preserved.

The mutation rate is higher than the one with conventional GAs. This is because the features the user selected will tend to be employed in the offspring. The probability that the features which the user specified in the *Interactive Phase* appear in the next generation is proportional to both the mutation rate and the number of times they are selected in the interaction.

Step 9: Repeat the Steps Steps 7 to 8 are repeated until an appropriate decision tree or a set of decision rules is obtained. The number of steps required to obtain the appropriate results is approximately equal to 10–30. If the results are not satisfactory when the *Automated Phase* converges, return to *the Interactive Phase.*

4 Experiments

To validate the effectiveness of the proposed method and to evaluate practical clinical data, we have carried out intensive experiments from a common small data set on meningoencephalitis diagnosis.

4.1 Brief Explanation of the Data Set

The description here is taken from the web site:
http://www.slab.dnj.ynu.ac.jp/challenge2000/.

Menigitis is a neurological infectious disease. Some bacteria or virus invade in the dura sheet (covering the brain), which causes severe inflammation in the dura. When the brain is inflamed, the patient is diagnosed as having 'meningoencephalitis'. Sometimes when bacteria form abscesses in the brain, the patient is diagnosed as having 'brain abscess'.

Usually, high fever and severe headache is observed. Then nausea and vomit follow. If these symptoms and objective symptoms (neck stiffness, Kernig sign and Lasegue sign) are observed, computer tomography is applied and lumbar puncture will be executed if there is no sign of brain edema.

The differential diagnosis is made as follows:

1 Check the cell count in the cerebulospinal fluid (CSF).
2 If polynuclear cells are dominant, bacterial meningitis is diagnosed. If mononuclear cells are dominant, viral meningitis is diagnosed.
3 For diagnosis of brain abscess, CT will be used for confirmation of diagnosis.

The therapy is summarized as follows:

- Bacterial: Antibiotics
- Virus: In cases of herpes, Zobirax and Ara_A are used.
- Otherwise, conservative.

If the therapy is timely, the prognosis is good. However in the case of encephalitis, a patient will still suffer from some symptoms even after the treatment (sequelae). A well-known sequela is aphasia for herpes encephalitis.

The problem given by Prof. Tsumoto is to find factors important for diagnosis for detection of the bacteria or virus to predict prognosis. The sample size of the data set is 140 with 38 features and no missing data. The problem size is very small compared with conventional DM problems in the literature; however, in this domain, little explicit knowledge is known for the diagnoses, thus, it is worth applying our method to the problem.

The intermediate and resulting knowledge was evaluated by the second author of the chapter, who is a domain expert concerned with clinical laboratory medicine at a hospital. He knows the basic principles of inductive learning programs, statistical techniques and is able to understand the output results. Using the output forms of decision trees and corresponding rule sets from C4.5 programs, he has interactively and subjectively evaluated the quality of the acquired knowledge from the viewpoints of simplicity, understandability, accuracy, reliability, plausibility and applicability.

4.2 Experimental Methods and Implementation

- 34 features among the 38 features are selected as the base feature set. The non-selected 4 features are concerned with the detailed class information on the kinds of virus, bacteria and medicines. The selected features contain similar concepts in a broader sense. All 140 cases are used for the evaluation. Therefore, the size of the search space is 2^{34}, which seems small for using conventional GAs; however, it is large enough for using IEC.
- There are two kinds of target concepts: 'bacteria' or 'virus' and 'dead' or 'negative' for predicting prognosis. We solve two types of two classification problems from the given data set.
- The parameters of C4.5 are set to the default ones. As the objective of the experiments is to interpret the given data set, we have used all the data as the training set and have not prepared the testing set. Thus, from a theoretical point of view, the experiments have some defects, but from the practical point, it is reasonable to get various knowledge from all the given data.
- The results of SIBILE are compared with the ones of direct application of C4.5 and evaluated by the second author. The final results are re-evaluated by Prof. Tsumoto, who is a medical doctor in the task domain.
- The current experimental system SIBILE is implemented on a Windows-based personal computer. The GA programs were written in the JAVA language, and C4.5 programs are used as an inductive learning tool. The sample display is shown in Figure 2.

4.3 Results

The results of direct application of C4.5 on the diagnosis of bacteria or virus are shown in Figure 3. C4.5 programs generate a simple decision tree with six nodes and the corresponding four rules with high accuracy. The results show high performance from the viewpoint of machine learning applications and are coincident with the domain knowledge in the literature. This means that the first problem of the diagnosis is easily solved by conventional inductive learning techniques and it is not necessary to use the IEC method.

For the prognosis prediction problem, the results are shown in Figures 4 and 5. The decision tree shown in Figure 4 is the simplified one from C4.5, because the original one has 32 nodes, and,thus, it is hard to include it in the chapter. However, the corresponding rules generated from the original tree are simple enough to show in Figure 5. This explains why the feature set in the tree and rules is different. This also means that the given data set is noisy about the prognosis prediction problem.

Furthermore, contrary to the previous problem, the results are not consistent with domain knowledge. They are nonsense. It is meaningful that the

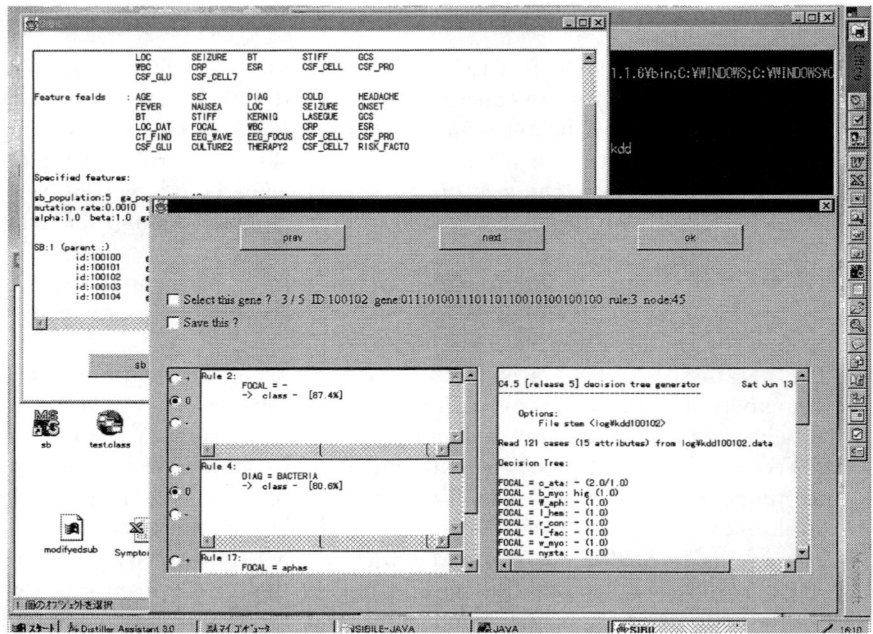

Fig. 2. Sample display of SIBILE on a Windows-based PC

low level (17) of `CSF_GLU` (Glucose in Cerebulspinal Fluid) correlates to a negative prognosis; however, it is nonsense that the low level (38.9) of `BT` (Body Temperature) 'causes' the death of a patient. The knowledge in the figures is considered to be induced by overfitting phenomena in the learning programs. To understand this, domain knowledge is important. Therefore, to extract good knowledge, the interaction between a domain expert and learning programs is essential.

The results of the application of SIBILE are shown in Figure 6. We have applied the interactive phases six times and the automated phases for 40 iterations to get the results. As can be imagined, the interactive phases required hard work from the domain expert, but we obtained got the results within one hour.

In the decision knowledge, the features of `LOC` (Loss of Consciousness) and `Therapy2` (Therapy No. 2) are selected as important ones. The results also suggest that the succeeding `LOC` over two days and the therapy of `ARA_A` (Anti-virus chemical) are related to the symptom of `aphasi` (the patient cannot speak). will 'cause' negative prognoses. The knowledge is very reasonable from the medical domain Knowledge perspective, but it has not been clearly recognized by medical doctors, so far. The knowledge is considered to be at the boundaries of the expertise and it would not be discovered with-

```
Read 140 cases (34 attributes) from log\kdd1.data
Decision Tree:
Cell_Poly > 220 : BACTERIA (33.0)
Cell_Poly <= 220 :
|   Cell_Mono > 12 : VIRUS (96.0/1.0)
|   Cell_Mono <= 12 :
|   |   CT_FIND = abnormal: BACTERIA (8.0)
|   |   CT_FIND = normal: VIRUS (3.0)

Tree saved

Evaluation on training data (140 items):

   Before Pruning           After Pruning
  ----------------   ---------------------------
  Size      Errors   Size      Errors   Estimate
     7    1( 0.7%)      7    1( 0.7%)    ( 4.5%)   <<
-------------------------------

Read 140 cases (34 attributes) from log\kdd1

Rule 4:
     Cell_Poly > 220
     ->  class BACTERIA  [95.9%]
Rule 1:
     CT_FIND = abnormal
     Cell_Mono <= 12
     ->  class BACTERIA  [85.7%]
Rule 3:
     Cell_Poly <= 220
     Cell_Mono > 12
     ->  class VIRUS  [97.3%]
Rule 2:
     CT_FIND = normal
     Cell_Poly <= 220
     ->  class VIRUS  [96.9%]
Default class: VIRUS
```

Fig. 3. Results of direct application of C4.5 programs

out the proposed DM techniques. The interpretation and explanation of the knowledge is confirmed by the medical expert.

4.4 Approach to the Larger Data Set

We have also applied SIBILE to another data set provided by Prof.Tsumoto, also found in the web site: http://www.slab.dnj.ynu.ac.jp/challenge2000/. The data set contains bacterial culture inspection records of patients in a

```
<<Decision TREE>>

C4.5 [release 5] decision tree generator
----------------------------------------
 Options:
        File stem <log\kdd100103>
 Read 140 cases (34 attributes) from log\kdd100103.data

 Simplified Decision Tree:

 CSF_GLU > 17 : negative (131.0/21.4) CSF_GLU <= 17 :
|   BT <= 38.9 : dead (5.0/1.2) |   BT > 38.9 : negative (4.0/1.2)

Evaluation on training data (140 items):

         Before Pruning           After Pruning
        ----------------   ---------------------------
        Size      Errors   Size      Errors   Estimate
          33    9( 6.4%)      5   18(12.9%)    (17.0%)   <<
```

Fig. 4. Decision tree of predicting prognosis by C4.5

```
  <<Decision Rules>>
   -------------------------------
   Read 140 cases (34 attributes) from log\kdd103
   ------------------
   Final rules from tree 0:
   Rule 1:
        BT <= 38.9
        CSF_GLU <= 17
        ->  class dead  [75.8%]
   Rule 3:
        FEVER <= 8
        LOC_DAT = -
        CSF_GLU > 17
        CSF_GLU <= 72
        ->  class negative  [93.4%]
```

Fig. 5. Decision rules of predicting prognosis by C4.5

hospital. The size of the data set is 20,919 (which is equal to 10,313 patients times 2.03 average inspections) with 62 categorical features, some of which are, however, missing or irrelevant values. The objective of the DM task is (1) to solve a two classification problem to extract knowledge that some bacteria are 'found' or 'not', and (2) to identify over 10 kinds of bacteria from the sensitivity investigations.

```
   Simplified Decision Tree:
   LOC <= 2 : negative (131.0/21.4)
   LOC > 2 :
   |   LOC > 6 : dead (2.0/1.0)
   |   LOC <= 6 :
   |   |   THERAPY2 = multiple: negative (0.0)
   |   |   THERAPY2 = ABPC+CZX: negative (0.0)
   |   |   THERAPY2 = FMOX+AMK: negative (0.0)
   |   |   THERAPY2 = ABPC: negative (0.0)
   |   |   THERAPY2 = ope: negative (2.0/1.0)
   |   |   THERAPY2 = Dara_P: negative (0.0)
   |   |   THERAPY2 = ABPC+FMOX: negative (0.0)
   |   |   THERAPY2 = LMOX: negative (0.0)
   |   |   THERAPY2 = PCG: negative (0.0)
   |   |   THERAPY2 = ABPC+LMOX: negative (0.0)
   |   |   THERAPY2 = PIPC+CTX: negative (0.0)
   |   |   THERAPY2 = no_therapy: negative (2.0/1.0)
   |   |   THERAPY2 = ABPC+CTX: negative (0.0)
   |   |   THERAPY2 = INH+RFP: negative (0.0)
   |   |   THERAPY2 = ABPC+CEX: negative (0.0)
   |   |   THERAPY2 = Zobirax: negative (0.0)
   |   |   THERAPY2 = ARA_A: aphasia (3.0/1.1)
   |   |   THERAPY2 = INH: negative (0.0)
   |   |   THERAPY2 = globulin: negative (0.0)

   Evaluation on training data (140 items):
         Before Pruning          After Pruning
        ----------------   ---------------------------
        Size      Errors   Size      Errors   Estimate
          47   15(10.7%)     24   18(12.9%)    (18.2%)   <<
\bigskip
<<Decision Rules>>
   C4.5 [release 5] rule generator
   -------------------------------
       Options:
        File stem <log\kdd200>
   Read 140 cases (16 attributes) from log\kdd100200
   ------------------
   Final rules from tree 0:
   Rule 9:
        LOC > 6
        ->  class dead  [50.0%]
   Rule 8:
        THERAPY2 = ARA_A
        LOC > 2
        ->  class aphasia  [63.0%]
   Rule 6:
        LOC <= 2
        ->  class negative  [83.6%]
   Default class: negative
```

Fig. 6. Results of predicting prognosis by SIBILE

This is an intractable problem for conventional machine learning techniques, because the data set has so many missing and/or irrelevant values. After using the data cleaning processes found in [7] (1) to estimate missing values, (2) to aggregate the data, and (3) to get 20 smaller data sets with 2,000 entries randomly selected from the given data set, we applied SIBILE. It took over 20 hours of interaction processes in total, but we succeeded in generating interesting rules with only three or five attributes. The meaning or interpretation of the extracted knowledge remains subjective; however, we believe that some of the extracted rules are useful for medical diagnosis problems.

5 Concluding Remarks

In this chapter, we have proposed a novel method for interactive DM by using inductive learning and IEC with interactive and automated phases. Using practical clinical data, although the size of the data is small, we have shown that IEC is able to be applied to practical knowledge engineering and DM problems. We have also mentioned that we have succeeded in DM with much larger data sets in the clinical data analysis domain.

The prerequisites of the proposed method are quite simple and the algorithm is easy to implement. Future directions of the work include generalization of SIBILE as a portable general-purpose tool applicable to other decision making problems and to specify SIBILE for efficient decision knowledge acquisition by improving the inductive learning techniques.

References

1. Bala, J. W., De Jong, K, Pachowicz, P. W. (1994) Multistrategy Learning from Engineering Data by Integrating Inductive Generalization and Genetic Algorithms. In Michalski, R., Tecuci, G. (eds.) (1994) Machine Learning IV: A Multistrategy Approach. Morgan Kaufmann,San Francisco, CA, pp. 471-488.
2. Bala, J., De Jong, K., Huang, J., Vafaie, H., Wechsler, H. (1995). Hybrid Learning Using Genetic Algorithms and Decision Trees for Pattern Classification. Proceedings of the International Joint Conference on Artificial Intelligence. pp. 719-724.
3. Caldwell, C., Johnston, V. S. (1991) Tracking a Criminal Suspect through "Face-Space" with a Genetic Algorithm. Proceedings of the Fourth International Conference on Genetic Algorithms, pp.416-421.
4. Davis, L. (ed.) (1991) Hand-book of Genetic Algorithms. Van Nostrand Reinhold, New York.
5. Dawkins, R. (1986) The Blind Watchmaker. Longman, Harlow.
6. Dawkins, R. (1989) The Evolution of Evolvability. In Langton, C.G. (ed.) Artificial Life Reading, Addison-Wesley,MA, pp.201-220.
7. Fayyad, U. M., Piatetsky-Shapiro, G., Smyth, P., and Uthurusamy, R. (eds.) (1996) Advances in Knowledge Discovery and Data Mining. AAAI/MIT Press,Cambridge, MA.

8. Freitas, A. A. (2002) Evolutionary Algorithms for Data Mining and Knowledge Discovery. In this volume.
9. Goldberg, D. E. (1989) Genetic Algorithms-Search, Optimization and Machine Learning. Addison-Wesley, New York.
10. Liu, H., Motoda, H. (eds.) (1998) Feature Extraction Construction and Selection: A Data Mining Perspective. Kluwer.
11. Mitchell, M. (1996) An Introduction to Genetic Algorithms. The MIT Press, MA.
12. Cambridge,Quinlan, J. R. (1993) C4.5: Programs for Machine Learning. Morgan Kaufmann,San Francisco, CA.
13. Sims, K. (1991) Artificial Evolution for Computer Graphics, ACM Siggraph Conference Proceedings, Computer Graphics, Vol.25, No.4, pp.319-328.
14. Sims, K. (1992) Interactive Evolution of Dynamical Systems. In Varela, F. J., Bourgine, P. (eds.) Toward a Practice of Autonomous Systems - Proceedings of the First European Conference on Artificial Life, MIT Press, Cambridge,MA, pp. 171-178.
15. Smith, J. R. (1991) Designing Biomorphs with an Interactive Genetic Algorithm. Proceedings of the Fourth International Conference on Genetic Algorithm, Morgan Kaufmann,Francisco, CA, pp.535-538.
16. Suzuki, E. (ed.) (2000) Proceeding of the PAKDD 2000 International Workshop of KDD Challenge on Real-World Data (KDD Challenge 2000), Kyoto, Japan (Data Sets Available at http://www.slab.dnj.ynu.ac.jp/challenge2000/).
17. Takagi, H. (1998a) Interactive Evolutionary Computation: System Optimization Based on Human Subjective Evaluation. IEEE International Confernce on Intelligent Engineering Systems (INES'98), Vienna, Austria, pp. 1-6.
18. Takagi, H. (1998b) Interactive Evolutionary Computation - Cooperation of computational intelligence and human KANSEI. Fifth International Confernce on Soft Computing (IIZUKA'98), World Scientific, Iizuka, Fukuoka, Japan, pp. 41-50.
19. Takagi, H. (2001) Interactive Evolutionary Computation: Fusion of the Capabilities of EC Optimization and Human Evaluation. The Proceedings of the IEEE, (to appear).
20. Terano, T., Ishino, Y., Yoshinaga, K. (1995) Integrating Machine Learning and Simulated Breeding Techniques to Analyze the Characteristics of Consumer Goods. In Biethahn, J., Nissen, V. (eds.) (1995) Evolutionary Algorithms in Management Applications Springer-Verlag, New York, pp. 211-224.
21. Terano, T., Ishino, Y. (1996) Knowledge Acquisition from Questionnaire Data Using Simulated Breeding and Inductive Learning Methods. Expert Systems with Applications, Vol. 11, No. 4, pp. 507-518.
22. Terano, T., Ishino, Y. (1998) Interactive Genetic Algorithm Based Feature Selection and Its Application to Marketing Data Analysis. In Liu,H., Motoda, H. (eds.) (1998) Feature Extraction Construction and Selection: A Data Mining Perspective. Kluwer, pp. 393-406.
23. Venturini, G., Slimane, M., Morin, F., Asselin de Beauville, J.-P. (1997) On Using Interactive Genetic Algorithms for Knowledge Discovery in Databases, Proceedings of the Seventh International Conference on Genetic Algorithms, Morgan Kaufmann, San Francisco,CA, pp.696-703.
24. Weiss, S. M., Indurkhya, N. (1998) Predictive Data Mining – a Practical Guide. Morgan Kaufmann, San Francisco,CA.

Learning-integrated Interactive Image Segmentation

Bir Bhanu and Stephanie Fonder

Center for Research in Intelligent Systems
University of California
Riverside, California 92521, USA
E-mail: (bhanu,steph)@cris.ucr.edu

Summary. We present an approach to automatic image segmentation, in which user-selected sets of examples and counter-examples supply information about the specific segmentation problem. In our approach, image segmentation is guided by a genetic algorithm which learns the appropriate subset and spatial combination of a collection of discriminating functions, associated with image features. The genetic algorithm encodes discriminating functions into a functional template representation, which can be applied to the input image to produce a candidate segmentation. The performance of each candidate segmentation is evaluated within the genetic algorithm, by a comparison to two physics-based techniques for region growing and edge detection. Through the process of segmentation, evaluation, and recombination, the genetic algorithm optimizes functional template design efficiently. The contributions of this chapter include: genetic learning of functional template design, physics-based segmentation evaluation, novel crossover operator and fitness function, as well as a system prototype and experiments on real synthetic aperture radar (SAR) imagery of varying complexity.

1 Introduction

The segmentation problem involves partitioning the image into regions which are homogeneous within themselves and distinct from each other, according to some set of criteria. There are a variety of approaches to image segmentation, including edge detection, region splitting/merging, and clustering-based techniques. Each of these approaches suffers from sensitivity to parameters for thresholding, and/or termination conditions. Still other approaches combine a few of these methods. However, the underlying cause for these algorithms to fail is the inability to specify how homogeneous a region should be and how distinct bordering regions should be in an application-dependent manner.

A system diagram for the approach presented in this chapter is given in Figure 1. The application dependency is overcome by allowing the user to interactively train the segmentation tool for his/her application. The image segmentation is guided by a genetic algorithm (GA) which learns the appropriate subset and spatial combination of a collection of functions, associated with image features, designed to discriminate one image region from another.

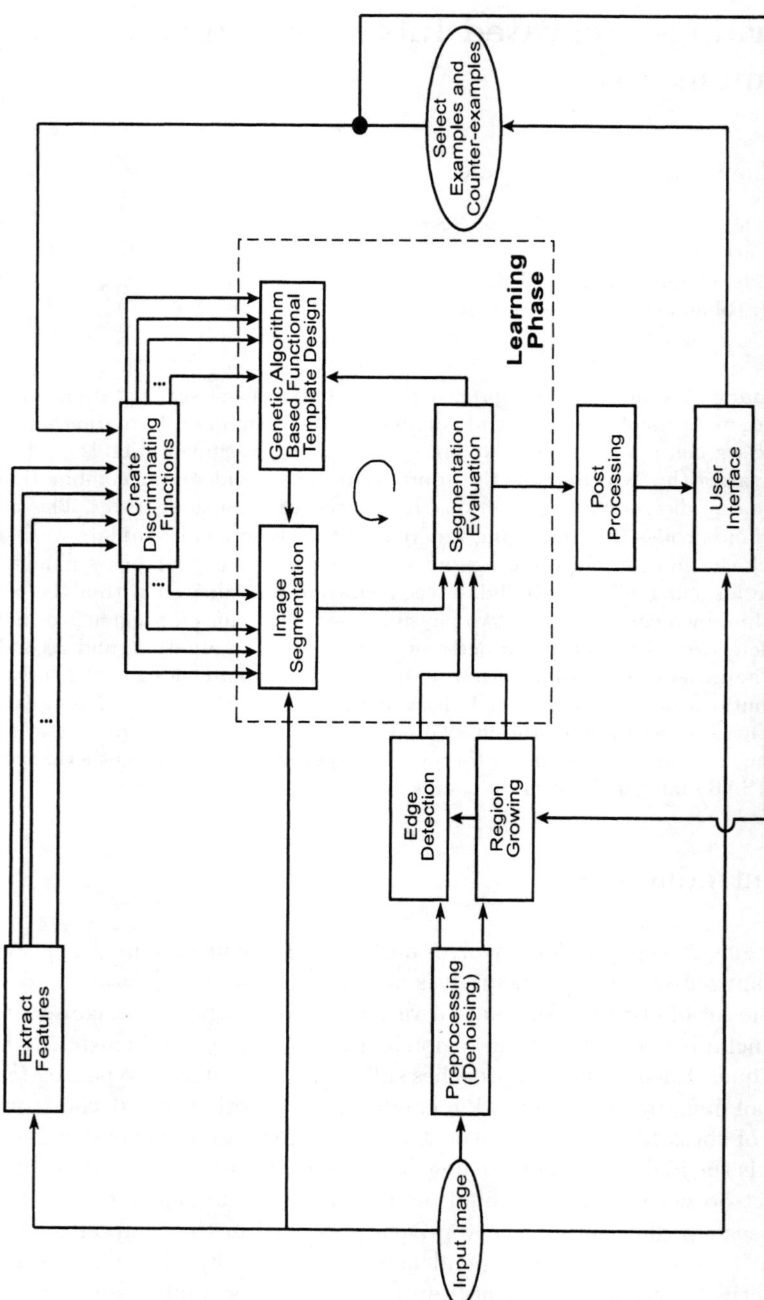

Fig. 1. System for learning-integrated interactive segmentation

In an interactive session, the user selects a set of examples and a set of counter-examples. The input SAR image is "denoised" to reduce the speckle effect and edge detection and region growing (with example regions as the seed) are performed on the denoised result. The example and counter-example sets (currently, one set of each) are used to scale the data and create histograms, which quantify the class separation for a variety of features. Based on histogram overlap, a set of discriminating functions is designed to perform discrimination between the example class and counter-example class.

A genetic algorithm encodes these functions into a functional template representation and produces a population of initial functional templates. These functional templates are applied to the input image to produce segmentations. The results of the segmentations are evaluated using a fitness function, and the population of functional templates is combined and modified via a set of operations based on genetic evolution, in an effort to evolve an optimized segmentation agent. The process of segmentation, evaluation, and recombination is repeated for a given number of generations and the best result of the final generation (after postprocessing) is presented as output to the user.

Although the genetic algorithm evaluates candidate segmentations via a comparison to region-based and edge-based techniques, these are only used as a guide for the process of searching through possible segmentation outcomes produced by the combinations of the discriminating functions and their spatial arrangement. Since the discrimination functions inherently contain classification information, it is possible to outperform the region-based and edge-based approaches that are used to evaluate the segmentation quality.

2 Background And Related Work

This section presents information on SAR imagery, image segmentation and classification, SAR image segmentation, genetic algorithms and image segmentation/classification, and finally the unique contributions of this chapter.

2.1 SAR Imagery

The differences between SAR and other imagery include the source of illumination, image resolution, imaging geometry, and noise. SAR imagery provides its own source of illumination. However, microwaves reflect off of objects in a scene differently (specular vs. diffuse reflections) than sunlight. The signal returned from an object depends upon its microwave reflectivity, as well as relative position, size, and texture. The reflected microwaves are measured to produce the image. One of the useful properties of SAR imagery is that the spatial resolution is independent of the distance. In the range direction, the time resolution determines slant range resolution, and both are dependent only on bandwidth. The azimuth resolution is dependent only on the length of the antenna in the azimuth direction. The ability to collect high-resolution images from extreme distances is a particularly attractive characteristic of SAR.

Although there are many positive aspects to the properties of SAR, there are negative aspects as well. For example, the imaging geometry makes understanding SAR imagery and its formation unintuitive and unnatural. In addition, the interaction of microwaves on surfaces is not entirely understood. Another obstacle for SAR imagery applications is speckle which results from the interference of many scatterers within the same resolution cell. Although speckle can be considered as multiplicative noise, when the image is converted to a dB image speckle statistics become additive and techniques for additive noise removal, such as spatial averaging, can be applied to remove the noise. However, spatial averaging degrades resolution which is important for many applications which use SAR imagery [8].

Because of the fundamental differences between SAR and other types of imagery, algorithms developed for application to other imagery are not directly applicable to SAR. Two physics-based algorithms (based on the physics of SAR) used for segmentation evaluation of imagery are presented in Section 3.1.

2.2 Image Segmentation and Classification

There are many approaches to segmentation of images, but the majority fall into three classes: edge-based, region-based, and cluster-based. Some techniques have combined a few of these techniques to overcome individual weaknesses. Learning-based approaches have been explored to allow segmentation to adapt to the appropriate problem domain as well as incorporate new information about a given domain quickly. Bhanu and Lee [3,5], examine numerous function optimization techniques including genetic algorithms and simulated annealing and hybrid techniques such as the combination of genetic algorithms and hill climbing. Although neural networks can approximate Bayesian performance with appropriate design, the design problem is often complex and design solutions are typically not scalable. Further, they need large amounts of training data, which are often unavailable. Genetic algorithms suffer from potential premature convergence and computationally expensive segmentation evaluation. Because simulated annealing relies on slow "cooling" to avoid local optima, it is inherently a slow process [22].

In the context of image segmentation, classification is the problem of assigning meaningful labels to regions. An equivalent formulation of the classification problem is assigning a label to each pixel; regions are then simply the connected components of identical labels. Thus, in this formulation segmentation and classification are performed simultaneously. Because labels applied to a region are meant to describe what the region depicts, classification of data is inherently pattern recognition. Some typical pattern recognition techniques applied to the classification problem are Bayesian [16], K-Nearest Neighbor [22], and template matching [11] methods.

A *traditional template* classifier applied to images specifies a function of local intensities, typically represented as a matrix of values. The template is correlated with the image and the results are thresholded to produce a classification result. *Functional templates* modify this technique, such that each element of the matrix is an index to a function. The advantage of functional templates over standard templates is the potential for better discrimination, which may occur when indexed functions are nonlinear. These functions can also encode sensor-specific and/or class-specific information into the functional template. Furthermore, it provides a framework to combine information from multiple features to potentially increase the discrimination power over segmentation approaches based on a single feature. The drawback of this technique is the need for functional template design; the selection of which functions to incorporate and where to place them within the functional template is difficult due to the complex interactions taking place between various features and associated functions.

The work presented in this chapter is an improvement over previous functional template work. Large size (order of 20×50) aspect-dependent functional templates used for object recognition were manually designed in [35], which was a very time consuming process. In [11], functional template design was automated, by limiting the design space to simple single feature templates, where the selected feature minimized Bayesian risk. The focus of our research is the development of an approach which automates functional template design that allows for multiple functions and evaluate its performance for the segmentation of SAR images.

2.3 SAR Image Segmentation

Algorithms developed for other types of imagery are not directly applicable to SAR, because of the differences in image properties [4]. Although there is literature which performs learning-based segmentation and/or classification, there are presently no interactive approaches such as the approach presented in this chapter. Furthermore, all SAR learning-based segmentation and/or classification-related work approaches the problem by producing a segmentation and classifying the resulting regions rather than performing pixel-level classification as presented here. Li et al. [23] assign classes to clusters using the distance from class means obtained from training data. Baraldi and Parmiggiani [2] use a neural network to cluster data using parameters calculated on initial segments. Classification of clusters is then accomplished via a domain-dependent knowledge-based classification scheme. Learning-based segmentation has also been explored with no attempt at classification of the resulting regions. Most learning-based segmentation attempts use neural networks [9,26,31]; however, Gou and Ma [18] present a new clustering method which uses entropy-based threshold to break the feature space into "mode" and "valley" regions. "Mode" regions are further processed to produce a final segmentation. Still other approaches perform segmentation without learning, then apply learning-based classification techniques to the segmented regions. Shoemakers et al. [30] use an edge detection and region growing hybrid for segmentation and classification of the resulting regions with a neural network. Gagnon and Klepko [17] also use neural networks for classification but use directional thresholding and region growing for segmentation. Soh and Tsatsoulis [32,33] perform segmentation using dynamic local thresholding, but use clustering to perform classification of regions. Rogers et al. [29] use various neural networks-based approaches for preprocessing and segmentation.

2.4 Genetic Algorithms and Image Segmentation/Classification

Genetic algorithms are a learning technique based on biological evolution. Given a population of candidate solutions, a fitness function is used to evaluate each individual of the population. Each generation the individuals (i.e., candidate solutions) are evaluated and recombined producing the next generation's population. Recombination is designed to allow fit individuals to pass on important characteristics to the next generation. Because the nature of genetic optimization is randomized search from a number of search points (the individuals), the approach is ideal for parallel implementation. Genetic algorithms have been applied to optimize overall segmentation quality [1,2,3,21,37], segmentation parameter selection [6,10,34,36,38,39], and to feature design for pixel-level classification [14,19,24].

Genetic algorithms have also been used to determine the optimal subset of features for pixel-level classification. The approach presented in this chapter is one example of such work. Matsui et al. [25] use an offline GA to select the optimal combination of feature indices for tissue classification without testing the neural network classifiers into which features are incorporated. Campbell and Thomas [7] use an offline GA to select a subset of Gabor filters requiring fewer convolutions for the classification technique in which they are eventually incorporated. Erdogan et al. [15] extract optimal texture features from several co-occurrence matrix-based texture measures. Training determines optimum ranges of the measures on known texture regions.

2.5 Contributions of This Chapter

- *Genetic Learning for Functional Template Design:* Unlike the previous work [12,13,35], in this chapter we design functional templates in a systematic manner using GAs. Besides the selection of a subset of features and associated functions, the GA determines the spatial placement of functions within the template. We use GAs as function optimizers since they allow the possibility of achieving the global maximum without exhaustive search.

• *Physics-based Segmentation Evaluation:* During the learning process, segmentation evaluation is performed by a comparison of candidate segmentation to two segmentation techniques: edge-based and region-based. These are physics-based techniques that incorporate SAR-specific information to produce a segmentation using a log likelihood ratio test where the distributions used in the test are specifically developed for SAR imagery.

• *Novel Crossover Operator and Fitness Function:* Every GA has a crossover operator. In this work, the crossover operator is an improvement over the operator presented in [20]. That operator preserves two dimensional (2-D) spatial information by exchanging the information from identical rectangular areas of the parent functional templates; however, it favors spatial information near the center. The new crossover operator removes that bias by allowing the rectangle used in crossover to (conceptually) wrap around the template in both the horizontal and vertical directions. The result is an operator which allows unbiased evolution of spatial information.

The novel fitness function compares a candidate segmentation to portions of the physics-based region and edge estimates using two terms. The region term encourages a segmentation to correctly classify pixels within the region from which the examples were selected. The edge term encourages regions classified as example regions to have edges coinciding with image edges.

• *System Prototype:* A prototype of the system has been developed and tested on SAR imagery. Our experimental results show that genetically designed functional templates perform better than functional templates designed using the Bayesian best feature.

3 Interactive Approach to Image Segmentation

Figure 1 presents an overview of the approach to image segmentation. The input to the system is an image. Since the system is designed for segmentation of SAR imagery, the input image is denoised. While the system is denoising the image and features are calculated, an interactive session with the user occurs, where example and counter-example sets of pixels are selected. The examples provide input to a region growing process performed on the denoised image, whose results in combination with edge-detection results are used to perform segmentation evaluation within the GA. In addition, examples and counter-examples are used to create a discriminating function for each feature. The features, represented by their corresponding discriminating functions, are then incorporated into functional template design, which is optimized over successive generations of the GA. When the GA terminates, results are presented back to the user after postprocessing.

3.1 Pre-learning Phase

This subsection discusses the aspects of the approach which occur before or after the GA attempts to optimize the functional template design in the learning phase. The pre-learning components of the approach include user interaction, wavelet denoising, computation of physics-based segmentations for evaluation, feature calculation, discriminating function design, and calculation of associated weights.

• **User Interaction:** During the user interaction the image is simply displayed and an example set and a counter-example set are selected by the user. Since the examples represent the class ω_1 and the counter-examples represent any other classes present in the data, the approach is inherently a two-class segmentation/classification (which is extended to N-class segmentation in Section 4.2). Once the system presents results to the user, the following things can happen:

(a) User accepts the results on the test image.
(b) User marks the appropriate errors by clicking areas/pixels of the image displayed on the computer screen.
(c) User selects different examples and counter-examples.

Currently (b) has not been integrated with the system. Also, for the results presented here, user selects only one example and one counter-example. We are making changes in our system where the user may select multiple examples and counter-examples.

This user interaction in the form of examples and counter-examples is the reason for the application-independent nature of the approach because the user is allowed to train the segmentation system for the application at hand.

• **Wavelet Denoising:** Wavelet denoising is applied to dB SAR imagery to remove additive noise without loss of spatial resolution. The procedure for speckle reduction depends on the initial shift of the signal. However, a shift-invariant result is obtained by averaging the results for all possible shifts of the input signal (as described by [27]).

• **Physics-based Segmentation for Evaluation:** The set of examples is input to a region growing algorithm whose results are used later, with the results of an edge-detection algorithm, in evaluating the fitness or quality of candidate segmentation results in the learning phase of the approach. However, the region growing and edge-detection results can be calculated before the learning phase, so that only the comparison between these results and the candidate solution need be computed while the GA is learning.

A physics-based edge-detection test and two physics-based segmentation algorithms (which use the test) are presented below, all of which incorporate the SAR speckle model. Both segmentation algorithms are based on the cartoon model [28], which states that images are made up of regions, which are separated by edges, and that regions are homogeneous according to some criterion. This criterion is assumed to be radar cross section (RCS, denoted as σ°). Because regions are separated by edges according to this model, a test which determines whether an edge is present can be used for both the region growing and edge-detection algorithms.

While we implement a 2-D edge-detection algorithm, for simplicity, the following discussion is based on a 1-D version [28]. A maximum likelihood test for detecting edges based on a model of SAR speckle states

$$P(I) = \Pi_{k=1}^{M} 1/\sigma^\circ \exp\left[-I_k/\sigma^\circ\right]$$

The test determines whether an edge is present at position k in a 1-D window containing M pixels with intensities $I_1, I_2, ..., I_k, ..., I_M$.

The window is assumed to contain either one or two regions. If the window contains two regions, A and B, their respective RCS are σ_A° and σ_B°, otherwise $\sigma_0^\circ = \sigma_A^\circ = \sigma_B^\circ$. Given the speckle model, the joint probability of such a window is

$$P_{A,B}(\sigma_A^\circ, \sigma_B^\circ | I_j, k) = (\Pi_{j=1}^{k} 1/\sigma_A^\circ \exp -I_j/\sigma_A^\circ)(\Pi_{j=k+1}^{M} 1/\sigma_B^\circ \exp -I_j/\sigma_B^\circ)$$

$\bar{I}_X$ estimates the RCS of a region X as the mean of the portion of the region contained in the window, where lower subscript 0 refers to the entire window. Given this, the log likelihood estimate of detecting an edge is reduced to

$$\lambda_D(k) = -k \ln \bar{I}_A - (M-k) \ln \bar{I}_B + M \ln \bar{I}_0$$

The sensitivity of this test is based on the threshold, which can be chosen according to false alarm probabilities derived from the distribution of $\bar{I}_A/\bar{I}_B$. Since the value of this ratio is ideally 1 if only one region is represented in the window, thresholds t_1 and t_2 must be chosen above and below 1, respectively (edges occur where $\lambda_D \notin [t_2, t_1]$). These thresholds can be related to λ_D via the following two equations:

$$\lambda_D(k) = -k \ln t_1 - M \ln M + M \ln[M - k + kt_1], \quad t_1 < 1$$

$$\lambda_D(k) = k \ln t_2 - M \ln M + M \ln[M - k + k/t_2], \quad t_2 > 1$$

Edge Detection The 2-D edge-detection approach is a computationally efficient extension of the 1-D method described above that is designed to smooth the RCS estimates and detect edges at 45° increments in a single pass. Each orientation divides a window into two regions, A and B, each with four adjacent subwindows as shown in Figure 2.

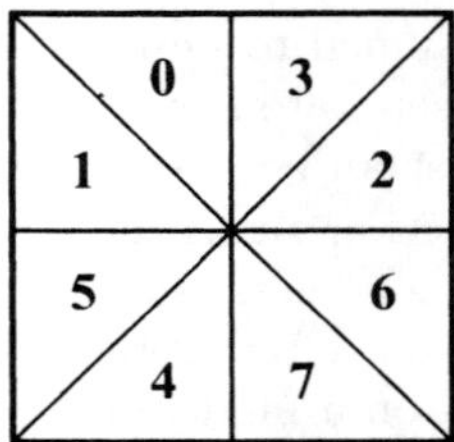

Fig. 2. Edge-detection algorithm subwindows for detecting edges every 45°

The following pseudocode describes the algorithm:

```
for each pixel{
  calculate intensity subtotals for each of the eight
  subwindows.

  for each orientation{
    Compute region A and region B means
      (left and right side of the assumed orientations) from
      the  eight subtotals.
    Compute edge magnitude.
  }

  Record max edge magnitude and the orientation
  which produced it.
}

pixels with edge magnitudes in the top 20% are marked as edges.
```

A 12 × 12 window is used to obtain reliable estimates of the RCS. Because of this the algorithm produces very thick edges which are thinned by unmarking edge pixels whose edge magnitude is not the local maximum of its 8-neighbors, reducing edges to a single pixel width. When used with measures such as edge border coincidence for fitness computation, the region boundary would have to match thinned edges exactly. In order to provide more flexibility to segmentation evaluation measures, the edges are thickened to a width of 3 pixels, by marking the pixel on either side of each thin edge pixel.

Region Growing: The region growing algorithm starts with the example pixels as a seed (see Figure 3). It uses 5 × 5 blocks in order to find a reliable RCS estimate for the region the block is in. There are two thresholds, above and below the value one, which are used by the edge test to determine whether to merge the blocks. Since a result that "overgrows" the region would misguide the fitness function, conservative thresholds are empirically determined. The following pseudocode describes the region growing algorithm:

```
while (merged > 0){
  set merged = 0
  for each $ 5 \times 5 $ block around the perimeter of the current
        grown region{
    Apply the edge test to compare the 5x5 block and the
        grown region
    if ( 0.8 <= edge magnitude <= 1.2){
    merge block into grown region
    set merged=merged+1;
    }
  }
}
```

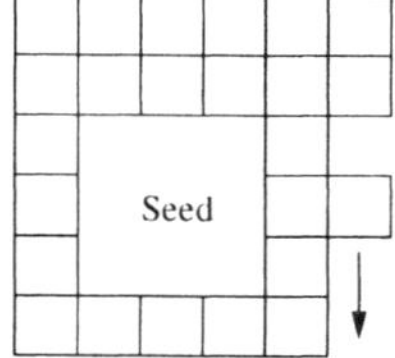

Fig. 3. Conceptual diagram of region growing algorithm

• **Features:** Table 1 lists the features used in this work. The index associated with a given feature is used consistently to refer to the discriminating function based on that feature in functional templates. The first seven features represent local image intensity statistics, while the remainder specify means or standard deviations of Gabor wavelet filtered images. Two scales (13×13 and 11×11) and four orientations ($0°$, $45°$, $90°$, $135°$) of Gabor wavelet filters are used and then means and standard deviations are calculated from a 13×13 area rather than the entire image in order to have a feature value at every pixel. The 13×13 local area was chosen empirically, because it provided a better estimate than smaller areas while still providing fluctuation from pixel to pixel. Also, areas larger than this caused biased feature values on more of the image boundary.

Discriminating Functions and Weights Given the example and counter-example sets as well as the feature images, a discriminating function and a weight are calculated for each feature. The design of the function is automated by a set of equations which measure the overlap of examples and counter-examples in the feature value histogram. The weight associated with this discriminating function is based on the Bayesian classification error of the feature. Before calculating discriminating functions, feature values are scaled to the range [0,255] by mapping mean μ to 128 and $\mu \pm 3\sigma$ (σ is standard

Table 1. Features used and their corresponding indexes. Large and small refer to 13×13 and 11×11, respectively. (0–6 are intensity, and 7–22 are Gabor wavelet-based features.)

Index	Feature
0	Intensity
1	3×3 local mean
2	3×3 local standard deviation
3	5×5 local mean
4	5×5 local standard deviation
5	7×7 local mean
6	7×7 local standard deviation
7	Scale: Large, Orientation: 0° mean
8	Scale: Large, Orientation: 45° mean
9	Scale: Large, Orientation: 90° mean
10	Scale: Large, Orientation: 135° mean
11	Scale: Small, Orientation: 0° mean
12	Scale: Small, Orientation: 45° mean
13	Scale: Small, Orientation: 90° mean
14	Scale: Small, Orientation: 135° mean
15	Scale: Large, Orientation: 0° standard deviation
16	Scale: Large, Orientation: 45° standard deviation
17	Scale: Large, Orientation: 90° standard deviation
18	Scale: Large, Orientation: 135° standard deviation
19	Scale: Small, Orientation: 0° standard deviation
20	Scale: Small, Orientation: 45° standard deviation
21	Scale: Small, Orientation: 90° standard deviation
22	Scale: Small, Orientation: 135° standard deviation

deviation) to 0 and 255, respectively. The histogram of examples and counter-examples has 255 bins, where a feature which discriminates well should have example pixels near the mean and most counter-examples at the extreme values. The following function, F_l (patterned after [12]), is then applied to each bin, l, of the example and counter-example histogram, denoted by E_l and C_l respectively, to produce a score for that feature value. Bins with many examples and few counter-examples produce high scores, while the opposite case produces low scores. The function $H(.)$ clips values to the range [0,1]. The parameters of $H(.)$ determine histogram overlap, while the 16 and -16 factors scale the result. Special cases are handled when there are no examples, no counter-examples, or both. The discriminating function is given by:

$$F_l = \begin{cases} 0.0 & \text{if } E_l = 0 \text{ and } C_l = 0 \\ 16 \times H(4 \times (E_l/\sum_{l=0}^{255} E_l)) & \text{if } C_l = 0 \\ -16 \times H(4 \times (C_l/\sum_{l=0}^{255} C_l)) & \text{if } E_l = 0 \\ 16 \times H((1/39) \times ((E_l/C_l) - 1)) & \text{if } E_l > C_l \\ -16 \times H((1/39) \times ((C_l/E_l) - 1)) & \text{otherwise} \end{cases}$$

The Bayesian weight, W_l, associated with the discriminating function is defined as:

$$W_l = 1.0 - \left(2 \times \sum_{l=0}^{255} \frac{\min(C_l, E_l)}{(C_l + E_l)} \right)$$

3.2 Learning Phase

The learning phase consists of image segmentation, segmentation evaluation, and evolutionary template design processes that use a GA approach for optimization. Although the design decisions such as representation, operator design, parameter selection, and fitness function design are done before the learning phase, they are discussed here because of their relevance to the GA.

The design of the functional template of a given size requires a solution of the combinatorics problem. For example, for 20 functions, the size of the search space for a 3×3 template is 512 billion. We use GAs as function optimizers [3] for functional template design since they allow the possibility of achieving the global maximum without exhaustive search. The appropriate combination and arrangement of discriminating functions are optimized by the GA for good segmentation results. A GA is composed of a population of candidate solutions, or individuals. In this research, an individual is a functional template, represented as a 2-D array of binary strings, as shown in Figure 4. The binary to decimal conversion of each bit string gives the index of the discriminating function represented for that location in the functional template. Each discriminating function is developed from a single feature, and referenced by the index associated with that feature. Features and their indexes are given in Table 1.

• **Segmentation:** After randomly generating generation 0, the initial population of the GA, the segmentation associated with each functional template is then produced using the functional template classification rule:

$\sum_{i=-M/2}^{M/2} \sum_{j=-N/2}^{N/2} W_{i,j} \times S_{i,j}(I(a+i, b+j)) > t$ Assign I(a,b) to class ω_1

$\sum_{i=-M/2}^{M/2} \sum_{j=-N/2}^{N/2} W_{i,j} \times S_{i,j}(I(a+i, b+j)) \leq t$ Assign I(a,b) to class ω_2

2	4	8
7	1	6
5	8	3

0010	0100	1000
0111	0001	0110
0101	1000	0011

Fig. 4. Representation of a 2-D template

The above equation describes the classification of an image pixel *I(a,b)* using an $(M+1) \times (N+1)$ functional template T. $S_{i,j}$ denotes the discriminating function F_l indexed at position $T(i,j)$ of the template and $W_{i,j}$ is the weight (W_l) associated with function $S_{i,j}$. In this chapter, all templates are 3×3 and the threshold $t = 0$.

After producing a segmentation for each individual, the segmentations are evaluated by the fitness function described in the next subsection.

- **Segmentation Evaluation and Fitness Function:** Once an image segmentation result on an image has been obtained, we need to evaluate its quality using a fitness function. In real applications ground-truth information is generally not available. In our approach, all that is known about the image is that the user-selected example set is part of a region corresponding to class ω_1 and the user-selected counter-example set is part of a region corresponding to class ω_2. The fitness function is made up of two terms, a region term, $T1$, and an edge term, $T2$.

The intuition for the former term of the fitness function is the desire to fill in as much of the example region as possible. The region term uses the grown region as the ground-truth region. Thus, if R is the set of pixels in the segmented example region, a candidate segmentation (the region incorporating the example set), and G is the set of pixels from the grown region,

$$T1 = n(G \cap R)/n(G),$$

where $n(X)$ is the number of pixels in set X.

If only the region term were used for fitness, a segmentation which classified all pixels as ω_1 would receive the same value as a segmentation that classified all pixels correctly. However, the fitness function should not prevent other regions in the image from being classified as belonging to class ω_1, since other regions pertaining to class ω_1 may be contained in the input image. Intuitively, the second fitness term should prevent evolved segmentations from having regions which "overrun" their region boundaries. The edge term is defined below, where S is the boundary of the region in the segmentation image and E is the set of detected edges within the minimum bounding rectangle of the region growing result, extended 10 pixels in each direction. This is essentially edge-border coincidence (EBC) within a limited area of the image, multiplied by a factor of 3. Because edge detection first thins edges to 1 pixel and then dilates them again to a width of 3 pixels, *n(E)* is typically three times as large as a good segmentation boundary. So, EBC is multiplied by 3 to account for this effect. Thus, edge term $T2$ is defined as:

$$T2 = 3 \times n(E \cap S)/n(E)$$

Combining the region term and the edge term, the overall fitness is then defined as:

$$fitness function = (T1 + T2)/4$$

• **Evolutionary Template Design:** Having evaluated every individual in a generation *N*, generation *N*+1 is obtained through a recombination process consisting of the application of three genetic operators: selection, crossover, and mutation. Selection is applied twice, to select high-fitness individuals for input to the crossover operator, which swaps discriminating function indexes between the selected individuals, the parents. The output of each application of the crossover operator is two new individuals, the children, who replace low-fitness individuals from the population, also chosen by the selection operator. After crossover, the mutation operator is applied to the entire population. Evolution is completed when the termination criteria are met.

Selection: A standard tournament selection method was used. Tournament selection is the random selection of a subset (of a certain size) from the population. From this subset, the individual with the highest fitness or lowest fitness is chosen when selecting for crossover or replacement (i.e., death), respectively.

Crossover: Crossover is the combination of two individuals to produce two new individuals. The crossover rate is the percentage of the population to participate in crossover each generation. A novel crossover operator that preserves the 2-D spatial information has been developed, based on the hypotheses that spatial layout of information in the template is important. Crossover is performed by randomly selecting two locations in the functional template to define a rectangle. The first location corresponds to the upper right corner of a rectangle and the second to the lower left. Copies of the rectangle are placed on both parent functional templates, and elements within the rectangle boundary are swapped. In our crossover method bias is removed by allowing the selected rectangle to conceptually wrap around in both horizontal and vertical directions (this makes the probability of a given element to be swapped equal for every element in the functional template). This crossover operation is illustrated in Figure 5 where the selected locations are marked on parent 1, while the rectangle resulting from this selection is indicated on parent 2. Note that this example wraps both horizontally and vertically.

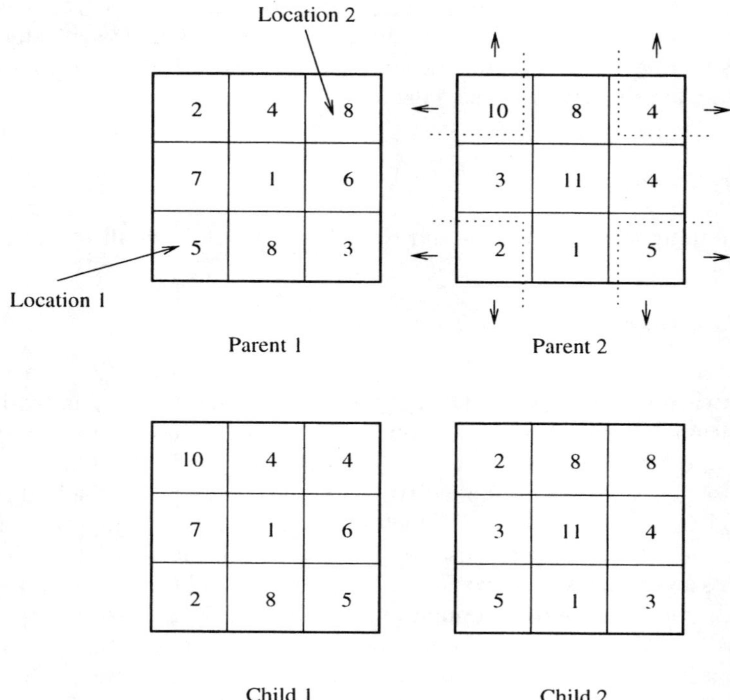

Fig. 5. Illustration of a crossover operator incorporating 2-D wraparound

Mutation: For mutation, the standard operator was chosen. That is, each bit in the representation of the functional template is flipped (i.e., mutated) with probability p_μ, where p_μ is the rate of mutation.

The final output of the GA, the evolved functional template, is the functional template with the highest fitness in the final generation.

3.3 Postprocessing

Postprocessing is performed after the GA terminates and consists of two tasks: *false positive removal* and *false negative filling*. False positives occur when pixels belonging to the counter-example class are incorrectly classified as belonging to the example class. *False positive removal* finds the connected components using the segmentation results and removes any component making up less than $X\%$ (default: $X = 1$) of the image. False negatives occur when pixels belonging to the example class are classified as belonging to the counter-example class. *False negative filling* looks at each pixel which is unmarked, if Y (default: $Y = 6$) or more of its 8-neighbors are marked with the identical class labels the pixel is assigned this class label.

N-class postprocessing consists of an additional processing step. N-class classification is actually performed as N+1-class classification, where the N+1th class is an unknown class consisting of pixels not belonging to classes 1 through N. If there are known to be only N classes present in the data, *majority filtering* is performed to reclassify pixels which were originally classified to class N+1. These pixels are reclassified as the class that the majority of its neighbors were assigned to.

4 Performance Results and Analysis

Several experiments are performed using real SAR images to demonstrate the performance of the technique. They include two-class and multiple class experiments. Also experiments are shown where the evolved functional templates obtained during training are applied to a new image. Further, experiments are performed that demonstrate the effectiveness of the crossover operator. For all of these experiments the value of parameters used in the algorithms are kept identical. These parameter settings are: template size 3×3, population size 100, tournament size 10, crossover rate 0.25, and mutation rate 0.01; number of generations = 10. The edge detection picks up the important edges, but lots of extra edges as well. Fortunately, most are outside of ± 10 pixels of the region growing estimate and these are the only edge pixels used by the fitness function.

For experiments which demonstrate the quality of evolved segmentations several measures of segmentation quality are presented. The fitness function and its individual components are used for both evaluation and interpretation of results. The fitness of an individual demonstrates the evolutionary performance, while the terms of the fitness function add intuition about their relative importance and interaction. All of the additional measures use ground-truth information and are not used during evolution, but are useful for final quantification of results.

The percentage of pixels classified correctly (PCC) is defined as:

$$PCC = \max(1 - [(n(G) - n(G \cap R)) + (n(R) - n(G \cap R))/n(G)], 0)$$

where $n(X)$ is the number of pixels in the set X, G is the set of pixels which belong to class ω_1 from the ground-truth image, and R is the set of pixels classified as belonging to class ω_1 in the segmented image. PCC measures the overall performance, but can sometimes be misleading. This is particularly true when there are many more counter-example pixels than example pixels or vice versa. As this is often the case, it gave rise to three additional measures: example accuracy (EA), counter-example accuracy (CA), and normalized percent pixels classified correctly (NPCC). These measures are defined below where R, G, and $n(X)$ are defined as above and X' is the complement of X. NPCC is particularly useful as it gives more meaningful results in the cases where PCC is misleading, because it is normalized by the class populations.

$$EA = n(R)/n(G)$$
$$CA = n(R')/n(G')$$
$$NPCC = (EA + CA)/2$$

Example Accuracy (EA) is of most concern since templates are trained to discriminate the example class in testing images, later. However, EA does not penalize for segmentation which "overruns" the example region and includes many counter-examples. Thus, both EA and CA must be considered for analysis.

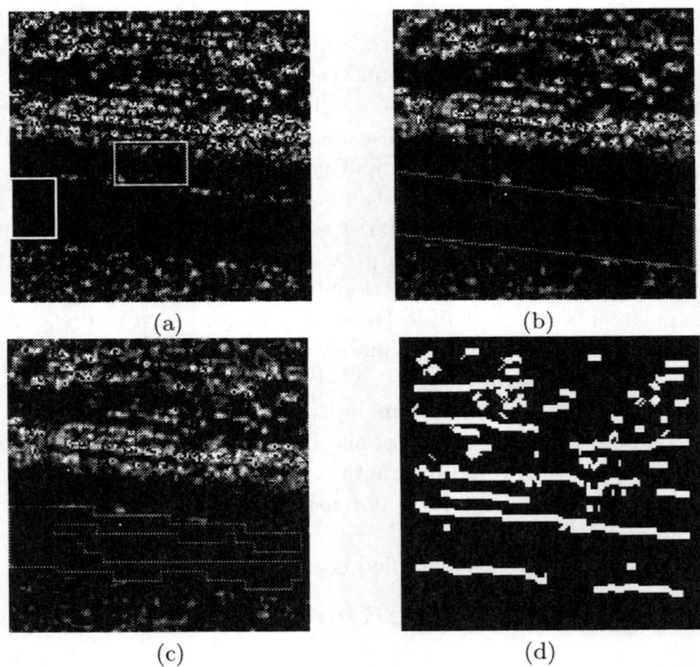

Fig. 6. Paved Road vs. Field: (a) Original image with examples (yellow/blue) and counter-examples (green/purple) (b) Denoised image with ground-truth (red) (c) Region growing (red) (d) Edge results.

Evolved results are compared to three default segmentations (Bayesian, PCC, and NPCC), which use a single feature in all functional template locations. The Bayesian default selects the (discriminating function) feature which minimizes classification error for the example and counter-example sets. (In our discussion in this section we use the word feature and discriminating function interchangeably.) However, the remaining two defaults utilize ground-truth. Single feature templates are exhaustively applied to the image and the segmentation corresponding to the feature which maximizes PCC is selected for the PCC default. Similarly, the segmentation corresponding to the feature which maximizes NPCC is selected for the NPCC default. Often two or more of the defaults select the same feature, and thus correspond to the same segmentation.

All SAR data presented herein are obtained from the MSTAR (public) clutter data set. All images are one foot resolution X-band data at a 15° depression angle. In the experiments, borders of examples are marked in yellow and blue, and borders of counter-examples are marked in green and purple.

4.1 Two-class Image Segmentation

This subsection provides details on three typical experiments and a discussion on the overall two-class results.

• **Example 1: Paved Road vs. Field -** Figure 6 shows the original image, results of denoising, region growing, and edge detection, as well as ground-truth for the image. Figure 7 presents the overall segmentation results which include segmentation results and the corresponding templates. Feature 8 is the best Bayesian feature (which also happens to be the best feature for NPCC), feature 5 is the best feature for PCC. Final evolved results are better according to both PCC and NPCC.

In the evolved results, every template has 3–4 instances of the local intensity features, involving some combination of intensity and 3×3 mean. It is apparent from the PCC default that intensity features do much of the classification; however, the boundary and some gaps inside the example area are compensated for by other features, such as local intensity standard deviation and Gabor standard deviation.

To demonstrate the evolutionary behavior of the GA, several measures of segmentation quality are plotted for the best segmentation result of each generation in Figure 8. All points in this graph are before postprocessing, as only the final segmentation is post-processed. The region term and the edge term are clearly increasing, causing the fitness function to increase as well. The region term is improved in the early generations with later generations serving to fine-tune performance at the edges. The fitness function is effective at increasing the example accuracy for an improvement of almost 5% over the 10 generations. Counter-example accuracy is largely unaffected by the fitness function which solely seeks to improve example accuracy. After the evolution, the postprocessing improves counter-example performance. Because of

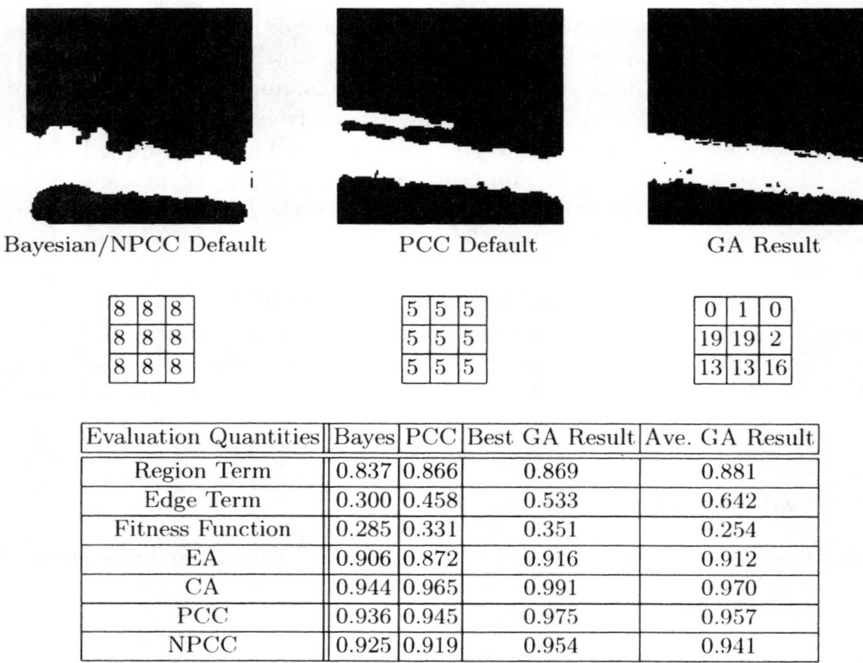

8	8	8
8	8	8
8	8	8

5	5	5
5	5	5
5	5	5

0	1	0
19	19	2
13	13	16

Evaluation Quantities	Bayes	PCC	Best GA Result	Ave. GA Result
Region Term	0.837	0.866	0.869	0.881
Edge Term	0.300	0.458	0.533	0.642
Fitness Function	0.285	0.331	0.351	0.254
EA	0.906	0.872	0.916	0.912
CA	0.944	0.965	0.991	0.970
PCC	0.936	0.945	0.975	0.957
NPCC	0.925	0.919	0.954	0.941

Fig. 7. Paved Road vs. Field results

the disparity in the number of pixels from each class, the PCC is heavily affected by counter-example accuracy. The NPCC, normalized to counteract such unbalanced representation of classes, is unaffected and steadily increasing.

- **Example 2: Paved Road vs. Grass -** The region growing and edge-detection results are given in Figure 9 along with original and denoised images, example/counter-example sets, and ground-truth. Region growing finds just over half of the road, as the region only grew downward. Edge detection finds the edges of the road particularly well, as well as some texture edges in the grass class. Although a few of the extra texture edges are included in the fitness term, the majority are excluded. The evolved segmentation results in Figure 10 are better than the default results. It is not surprising, then, that evolved results have significantly better example accuracy performance and have nearly perfect counter-example accuracy, which is consistent with the defaults. Thus, the evolved results are significantly better than default segmentations according to the NPCC. (The PCC does not show this improvement as dramatically since example pixels represent only 23% of the image.) The intensity feature was the most common in the evolved results. This is also not surprising as it was the selected feature for both NPCC and

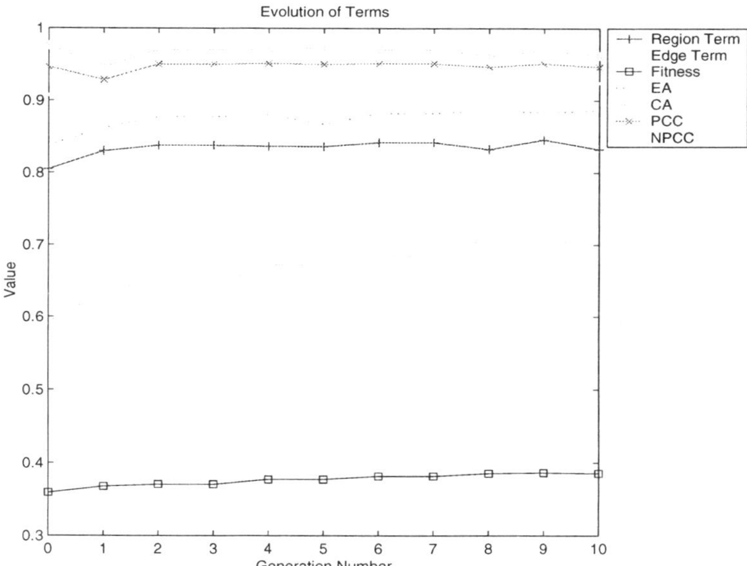

Fig. 8. Evolution of segmentation quality for Paved Road vs. Field

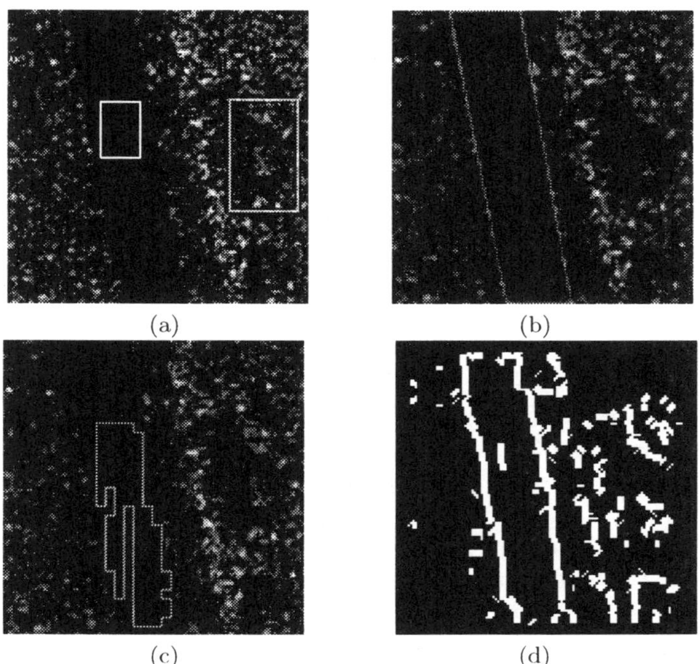

Fig. 9. Paved Road vs. Grass: (a) Original image with examples (yellow/blue) and counter-examples (green/purple) (b) Denoised image with ground-truth (red) (c) Region growing (red) (d) Edge results.

PCC defaults. However, most of the example area can be classified correctly with an average of four instances of this feature, leaving the remaining template positions for other features to improve performance. Boundary accuracy and edge term improve with the remaining functions, typically 90° and 135° orientations of the small-scale Gabor standard deviation feature. These features make intuitive sense for edge performance in this image as they are aligned with the orientation of the road.

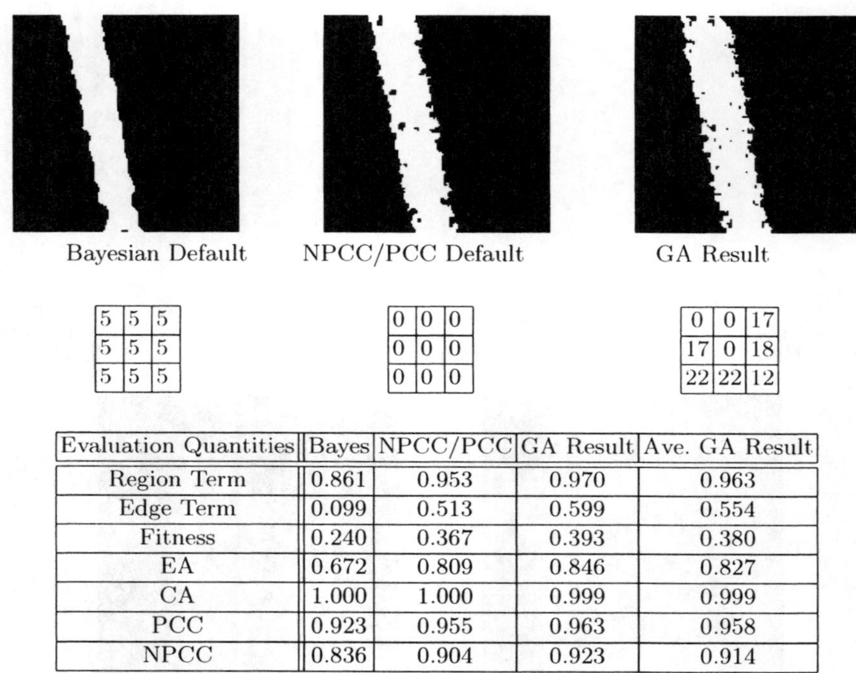

Bayesian Default:

5	5	5
5	5	5
5	5	5

NPCC/PCC Default:

0	0	0
0	0	0
0	0	0

GA Result:

0	0	17
17	0	18
22	22	12

Evaluation Quantities	Bayes	NPCC/PCC	GA Result	Ave. GA Result
Region Term	0.861	0.953	0.970	0.963
Edge Term	0.099	0.513	0.599	0.554
Fitness	0.240	0.367	0.393	0.380
EA	0.672	0.809	0.846	0.827
CA	1.000	1.000	0.999	0.999
PCC	0.923	0.955	0.963	0.958
NPCC	0.836	0.904	0.923	0.914

Fig. 10. Paved Road vs. Grass results

• **Example 3: River vs. Field -** The data and the results are shown in Figures 11 and 12. The region growing did well, capturing almost all of the river class and none of the field class. The edge detection captures all of the river edges, with very few gaps. However, several false edges and texture edges are also detected, many of which will be included for fitness evaluation. Evolved results perform better than the defaults in terms of both the region term and the edge term. Demonstrating the effectiveness of the fitness function, example accuracy is higher for the evolved templates as well. Unfortunately, counter-example accuracy is significantly lower than that of the defaults. Despite the fact that the PCC of evolved templates is significantly lower due to counter-example performance, example accuracy improved enough that in terms of the NPCC evolved template performance is significantly better. Because the example class comprises only 17% of the image, the NPCC is the more applicable term. The intensity and 3×3 mean features perform most of the classification and thus use most of the positions in the template. On average there are seven instances of some combination of these two features (approximately five of intensity and two of 3×3 mean). Thus, NPCC/PCC single feature defaults are incorporating too much smoothing and the GA combines features in an effort to produce the correct amount of smoothing. Remaining positions are filled solely with Gabor standard deviation features, which help both for region and edge term performance. The most frequently selected of these are the small-scale 0° and 135° orientations.

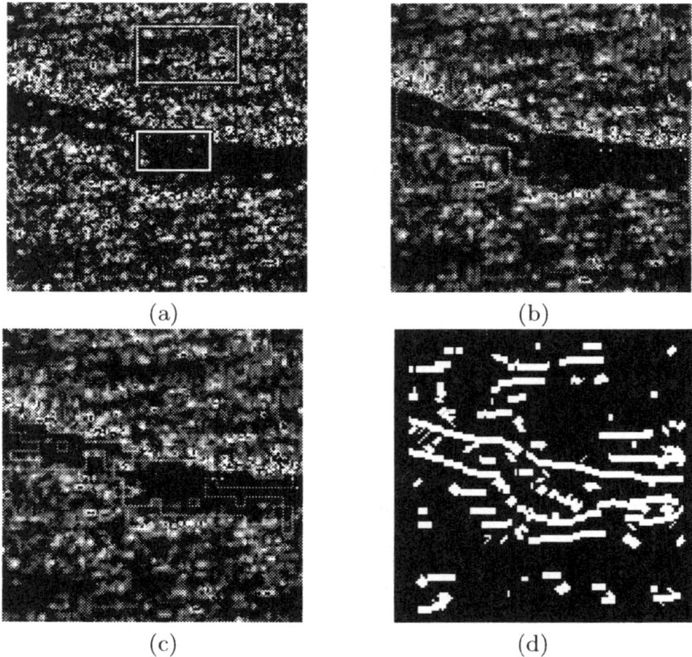

Fig. 11. River vs. Field: (a) Original image with examples (yellow/blue) and counter-examples (green/purple) (b) Denoised image with ground-truth (red) (c) Region growing (red) (d) Edge results.

• **Discussion on Two-class Experiments:** We have carried out these experiments on many images. In addition to the three two-class examples discussed above, we have also used lake vs. grass, lake vs. field, unpaved road vs. field, and grass vs. field. We find that evolved templates consistently outperform Bayesian default. Only in the highly textured examples where fitness was misguided did the improvement over the Bayesian default fall below 1%

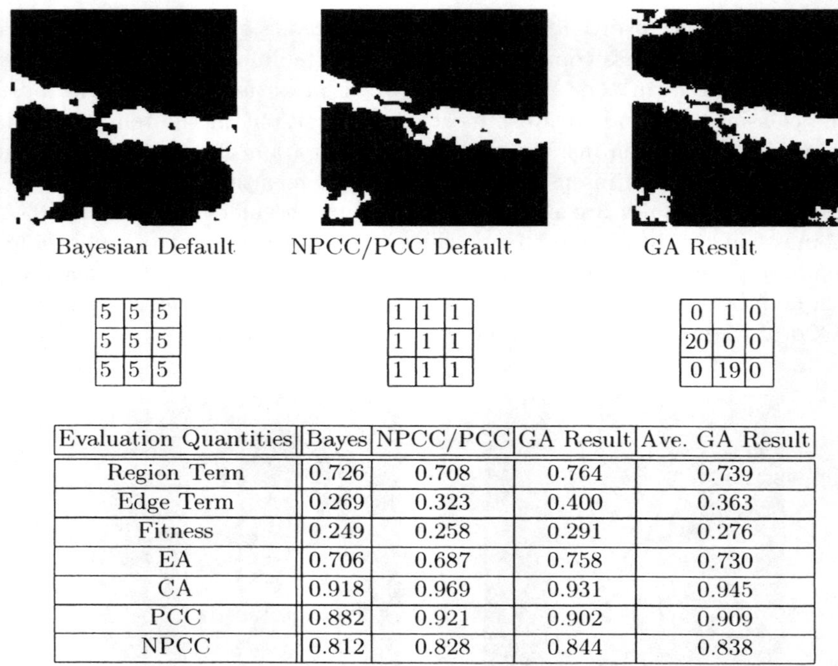

Evaluation Quantities	Bayes	NPCC/PCC	GA Result	Ave. GA Result
Region Term	0.726	0.708	0.764	0.739
Edge Term	0.269	0.323	0.400	0.363
Fitness	0.249	0.258	0.291	0.276
EA	0.706	0.687	0.758	0.730
CA	0.918	0.969	0.931	0.945
PCC	0.882	0.921	0.902	0.909
NPCC	0.812	0.828	0.844	0.838

Fig. 12. River vs. Field results

for both PCC and NPCC. For all other experiments, the improvements were dramatically better. It is clear that example accuracy improved over the default for most examples. Furthermore, counter-example accuracy typically stabilized or improved due to postprocessing.

The reasons for improvement over the Bayesian single feature default are:

- Examples and counter-examples may not fully characterize the classes.
- If a priori probabilities are assumed to be equal, the weighting of each pixel in terms of classification error is equal as in the PCC. This is biased for images with large discrepancies between example class and counter-example class.
- Placing the Bayesian best feature in all template positions generally implies an additional smoothing of the result.

In addition, evolved results perform at least as well as or better than the NPCC/PCC defaults. Note, however, that while Bayesian and evolved results train only on the example and counter-example regions, NPCC/PCC defaults use the ground-truth for the entire image to select the best feature. Despite this, evolved results typically outperform NPCC/PCC defaults at example boundaries. Thus, for the set of experiments with higher example region perimeter/area ratio (paved roads and the river examples), evolved results are significantly better. We have found that when the perimeter/area ratio of the example region is lower, evolved results are consistent with the NPCC/PCC defaults.

Most improvement due to evolution occurs within the first few generations. At this time the GA is typically optimizing the region term, until it reaches a plateau. However, the improvement in early generations can be misleading. This is improvement of the initial random templates of the population, not over default templates. Later generations typically refine the segmentation by optimizing the edge term. These optimizations may have less effect in the overall improvement from randomness but are crucial to improvements in boundary accuracy, which typically is the margin of improvement over the single feature default templates.

4.2 *N*-class Image Segmentation

• **Example: Lake vs. Paved Road vs. Grass -** Figure 13 presents the original image and ground-truth for this experiment. The original image is 256 × 256, and contains 22.5% Lake pixels (red), 14.7% Road pixels (green), and 62.8% Grass pixels (blue). *N*-class segmentation results are generated using hierarchical two-class classifiers. The grass classifier is applied first, followed by the road classifier, and finally the lake classifier. Any pixel not assigned to one of these classes is handled by majority filtering. Figure 14 shows the experimental results after false positive removal, false negative fill-in, and majority filtering. The single feature defaults and a typical GA result are given, along with confusion matrices and a table summarizing their NPCC/PCC performance. Although the evolved segmentation performs better than any default for the road region, it does not perform as well as the NPCC/PCC default for the grass and lake regions. Thus, evolved segmentations are significantly better than the Bayesian default and are consistent with the NPCC/PCC default.

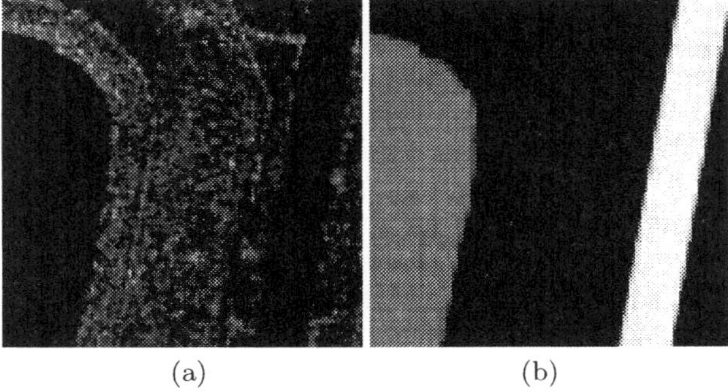

(a) (b)

Fig. 13. (a) Original image, and (b) Ground-truth image (Lake, red; Grass, blue; Road, green)

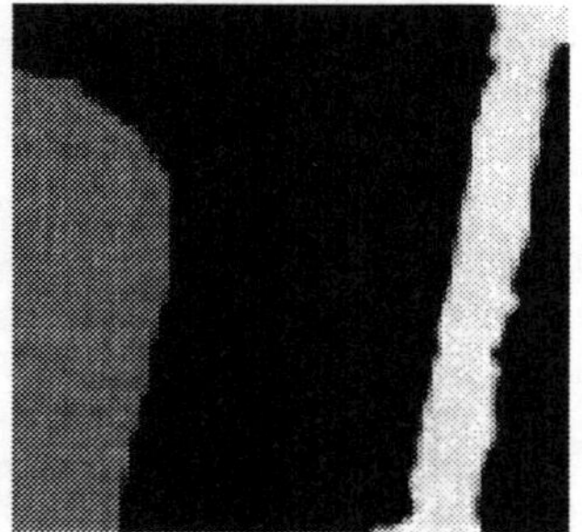

Bayesian Default

NPCC/PCC Default

	Grass	Road	Lake
Grass	0.928	0	0.026
Road	0	0.840	0.038
Lake	0	0.003	0.999

	Grass	Road	Lake
Grass	0.973	0	0.010
Road	0	0.868	0.031
Lake	0.001	0.003	0.999

GA Result

	Grass	Road	Lake
Grass	0.944	0	0.020
Road	0	0.898	0.024
Lake	0	0.023	0.994

	Bayes Default	NPCC/PCC Default	GA Result	Ave. GA Result
PCC	0.960	0.974	0.969	0.964
NPCC	0.922	0.947	0.946	0.939

Fig. 14. Lake vs. Paved Road vs. Grass results

4.3 Evaluation of Functional Template Design

The design of functional templates can be empirically evaluated by testing templates designed for a specific class on similar images. To accomplish this, some data must be preserved from the training of the template: the example set mean and standard deviation, discriminating functions, and Bayesian weights associated with discriminating functions.

• **Example: Lake vs. Grass -** Figure 15 shows the training and testing images used in this experiment. Both are 128×128 images containing visually similar data. Figure 16 summarizes the training phase of the experiment. The original image with example and counter-example set selection is presented, denoised image with ground-truth, as well as the region growing and edge-detection results. To verify that the region growing result is correct, despite the missing areas in the middle of the lake region, the histogram equalized denoised image is also presented. It is visible from this image that the areas missed by region growing are statistically different. The final image in the figure is a typical evolved segmentation for the training image.

The results are presented in Table 2. The evolved templates averaged over 10 runs with different random seeds perform consistently with the NPCC/PCC single feature default segmentation for all criteria and outperforms the Bayesian default segmentation by over 1% for all measures except counter-example accuracy, for which performance is consistent. These same evolved templates are applied to a second, testing image, to evaluate template design. The results on this testing image are slightly lower, but are still consistent with the training image for all measures. Furthermore, no single feature template outperforms the evolved result on the testing image. The average of the evolved results over 10 random seeds is consistent with the NPCC/PCC default for all criteria and significantly outperforms the Bayesian default.

4.4 Effectiveness of Crossover Operator

Figure 17 presents the GA's fitness optimization, both with and without the crossover operator, for the first three examples and another example of lake vs. grass. After a small number of generations it is expected that a well-designed crossover operator will outperform evolution with no crossover operator. For the system prototype, within 10 generations this trend is evident. The value of the fitness of the best individual of each generation is plotted for each system run. In all cases, the crossover operator improves the optimization performance of the GA.

5 Conclusions

Functional templates can be used for successful interactive segmentation and classification of SAR imagery using the approach developed in the system

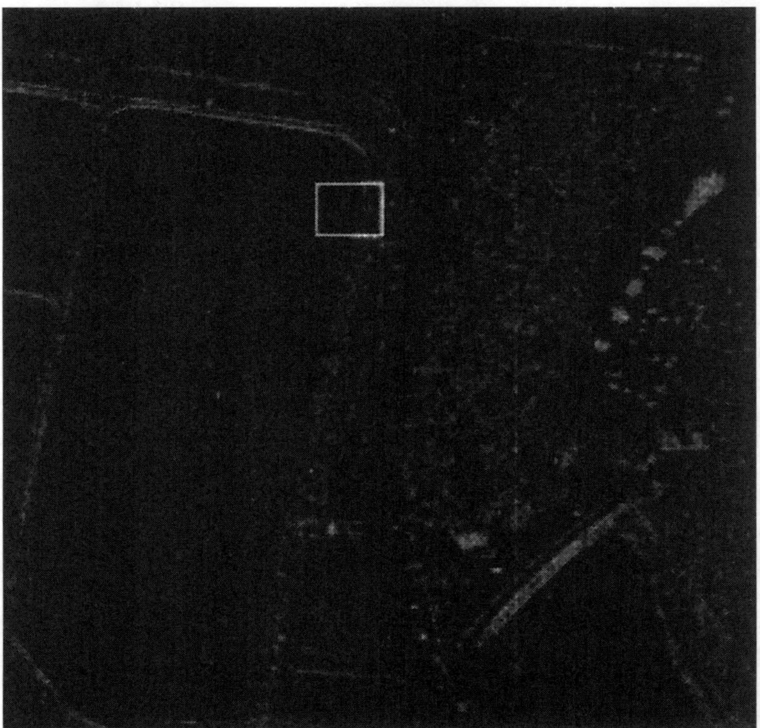

Fig. 15. Original SAR image with training subimage (yellow) and testing subimage (red)

Table 2. Summary of Lake vs. Grass template design experiment

		Bayes Default	NPCC/PCC Default	GA Result	Ave. (10 seeds)
	Template	1 1 1 1 1 1 1 1 1	0 0 0 0 0 0 0 0 0	0 19 17 0 0 20 0 0 0	
Training Image 128 × 128	PCC	0.960	0.984	0.987	0.984
	NPCC	0.961	0.984	0.987	0.984
	EA	0.922	0.969	0.974	0.970
	CA	1.000	1.000	0.999	0.999
Testing Image 128 × 128	PCC	0.951	0.979	0.980	0.979
	NPCC	0.962	0.984	0.985	0.983
	EA	0.924	0.967	0.970	0.967
	CA	1.000	1.000	1.000	1.000

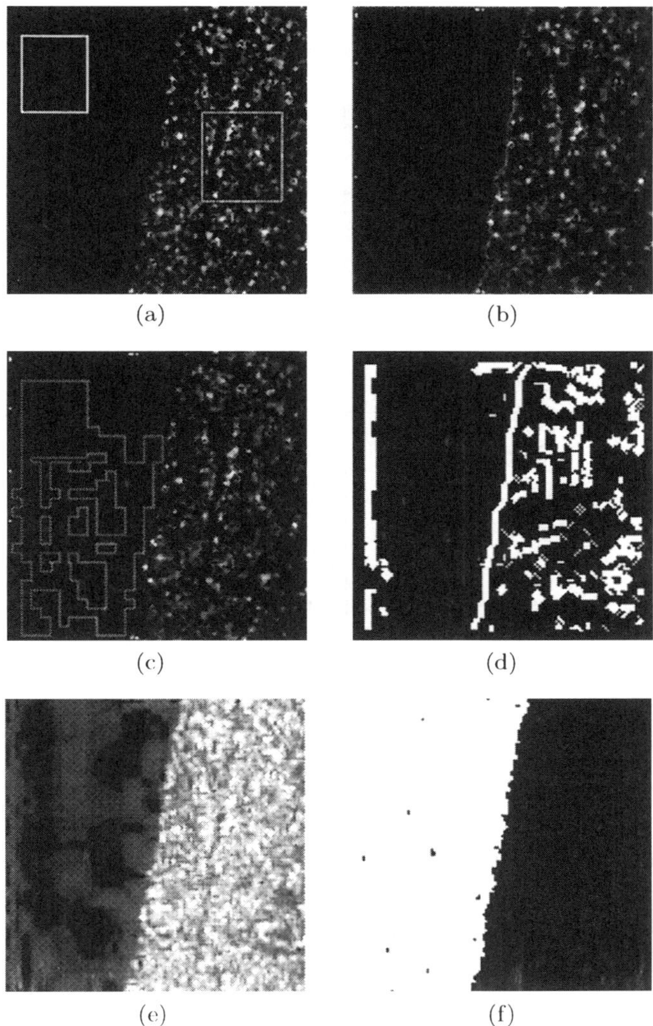

Fig. 16. Lake vs. Grass training data (a) Original image with examples (yellow/blue) and counter-examples (green/purple) (b) Denoised image with groundtruth (red) (c) Region growing (Red) (d) Edge results (e) Histogram equalization of denoised image (f) GA segmentation result.

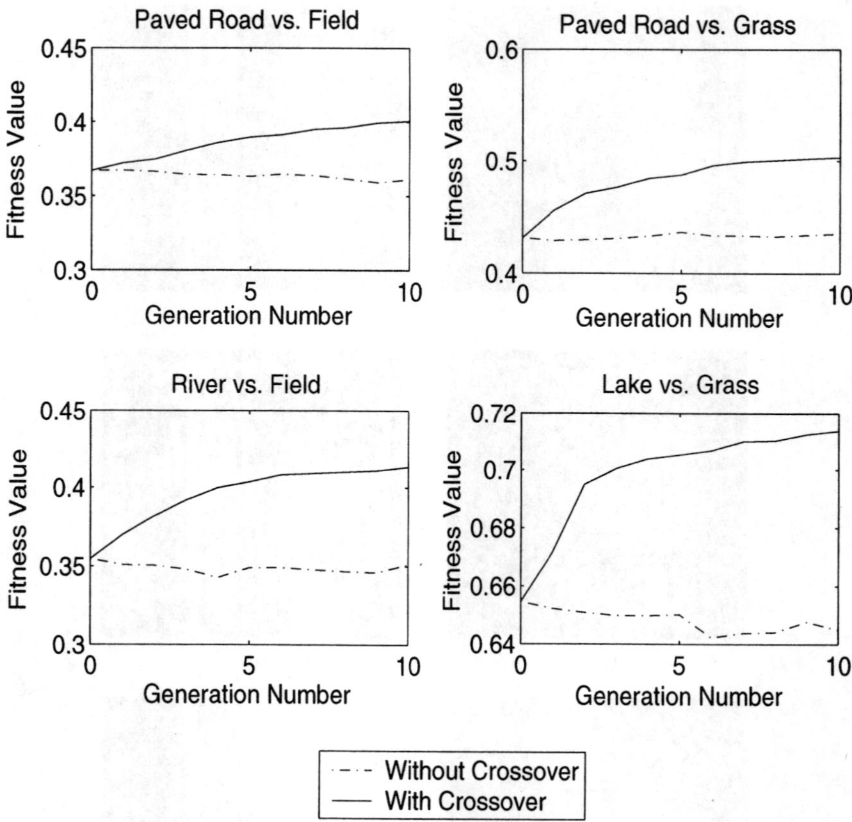

Fig. 17. Evolution of fitness both with and without crossover

prototype discussed in this chapter. Furthermore, functional templates can be retained for successful segmentation and classification of similar images in the future. The success of the prototype is described in four parts: genetic learning for functional template design, the physics-based segmentation evaluation, the crossover operator, and the fitness function.

Experimental results demonstrate that genetic learning is successfully applied for the design of functional templates. Evolved templates select meaningful features, which complement each other for improved segmentation quality over any single feature. Evolved segmentations consistently outperform segmentations derived from the Bayesian best single feature and typically perform at least as well as, if not better than, segmentations derived from the actual best *single* feature defaults. Real SAR data were also used to illustrate the extension to N-class segmentation.

Physics-based segmentation incorporates SAR-specific information into the region growing and edge-detection algorithms used for segmentation evaluation of the evolved templates. These algorithms, while not optimum, are used only in the example regions for comparisons to the evolved results and they have proven to be a suitable benchmark for optimizing the evolved results for the entire image.

The novel crossover operator is successful at optimizing the fitness during evolution. This crossover operator provides a mechanism for preserving important spatial information about the arrangement of functions in the functional template. For all experiments, the fitness was clearly increasing as the functional templates evolved.

The margin of improvement for evolved segmentations typically occurs at the boundary of the image, illustrating the power of the fitness function. Given reasonable region growing and edge-detection results the fitness function is effective for optimizing example accuracy. The region term successfully ensures correct classification of most of the pixels within a region (or at least the parts similar and adjacent to the example set). This term is typically improved to some level in the first few generations leaving the edge term for the focus of later generations. The edge term is most effective for perfecting results that are already good and for discouraging segmentations which classify the example region well but "overrun" the region boundary.

Acknowledgments

This work was supported in part by grants DAAH04-95-1-0448 and F33615-99-C-1440. The contents and information do not necessarily reflect the position or policy of U.S. government.

References

1. Andrey, P. (1999) Selectionist relaxation: Genetic algorithms applied to image segmentation. *Image and Vision Computing*, **17**, 175–87
2. Baraldi, A., Parmiggiani, F. (1996) Segmentation of SAR images by means of Gabor filters working at different spatial resolution. In *International Geoscience and Remote Sensing Symposium*, **1**, 709–13
3. Bhanu, B., Lee, S. K. (1994) *Genetic Learning for Adaptive Image Segmentation*, Kluwer Academic
4. Bhanu, B., Parvin, B. (1987) Segmentation of natural scenes. *Pattern Recognition*, **20**, 487–96
5. Bhanu, B., Wu, X. (1996) Genetic algorithms for adaptive image segmentation. In T. Poggio and S. K. Nayer, editors, *Early Visual Learning*. Oxford University Press
6. Cagnoni, S., Dobrzeniecki, A., Poli, R., Yanch, J. (1999) Genetic algorithm-based interactive segmentation of 3D medical images. *Image and Vision Computing*, **17**, 881–95

7. Campbell, N., Thomas, B. (1997) Automatic selection of Gabor filters for pixel classification. In *International Conference on Image Processing and its Applications*, **2**, 761–5
8. Chellappa, R., Zelnio, E., Rignot, E. (1992) Understanding synthetic aperture radar images. In *DARPA Image Understanding Workshop*, 229–47
9. Chen, K., Tsay, D., Huang, W., Tzeng, Y. (1996) Remote sensing image segmentation using a Kalman filter-trained neural network. *International Journal of Imaging Systems and Technology*, **7**, 141–8
10. Chun, D. N., Yang, H. S. (1996) Robust image segmentation using genetic algorithm with a fuzzy measure. *Pattern Recognition*, **29**, 1195–1211
11. Delanoy, R., Lacoss, R. (1998) Analyst trainable fusion algorithms for surveillance applications. In *Proceedings SPIE Conference on Algorithms for Synthetic Aperture Radar Imagery V*, **3370**, 143–54
12. Delanoy, R. L., Troxel, S. W. (1995) Toolkit for image mining: User-trainable search tools. *The Lincoln Lab Journal*, **8**, 145–60
13. Delanoy, R. L., Verly, J. G., Dudgeon, D. E. (1993) Machine intelligent automatic recognition of critical mobile targets in laser radar imagery. *The Lincoln Lab Journal*, **6**, 161–212
14. Delibasis, K., Undrill, P., Gameron, G. (1997) Designing texture filters with genetic algorithms: An application to medical images. *Signal Processing*, **57**, 19–33
15. Erdogan, A., Leblebicioglu, K., Halici, U., Atalay, V. (1996) Extraction of optimal texture features by a genetic algorithm. In *International Symposium on Computer and Information Sciences*, **1**, 183–8
16. Ferryman, T., Bhanu, B. (1996) A Bayesian approach for the segmentation of SAR images using dynamically selected neighborhoods. In *DARPA Image Understanding Workshop*, **2**, 1184–86
17. Gagnon, L., Klepko, R. (1998) Hierarchical classifier design for airborne SAR images of ships. In *Proceedings SPIE Conference on Automatic Target Recognition VIII*, **3371**, 38–49
18. Guo, G., Ma, S. (1998) Unsupervised segmentation of SAR images. In *IEEE International Geoscience and Remote Sensing*, **2**, 1150–2
19. Jacquelin, C., Hejblum, G., Aurengo, A. (1997) Evolving descriptors for texture segmentation. *Pattern Recognition*, **30**, 1069–80
20. Katz, A. J., Thrift, P. R. (1994) Generating image filters for target recognition. *IEEE Transactions on Pattern Analysis and Machine Intelligence*, **16**, 906–10
21. Kim, H. J., Kim, D. W., Kim, S. K., Lee, J. U., Lee, J. K. (1997) Automatic recognition of car license plates using color image processing. *Engineering Design and Automation*, **3**, 217–25
22. Kulkarni, S., Lugosi, G., Venkatesh, S. (1998) Learning pattern classification - a survey. *IEEE Transactions on Information Theory*, **44**, 2178–206
23. Li, A., Dammert, P., Smith, G., Askne, J. (1997) Fuzzy c-means clustering algorithm for classification of sea ice and land cover from SAR images. In *Optical Engineering*, **3217**, 86–97
24. Masters, S., Hintz, K. (1995) Evolution of convolution kernels for feature extraction. In *Proceedings SPIE Conference on Signal Processing, Sensor Fusion, and Target Recognition IV*, **2484**, 536–44
25. Matsui, K., Suganami, Y., Kosugi, Y. (1999) Feature selection by genetic algorithm for MRI segmentation. *Systems and Computers in Japan*, **30**, 69–78

26. Mingolla, E., Ross, W., Grossberg, S. (1999) A neural network for enhancing boundaries and surfaces in synthetic aperture radar images. *Neural Networks*, **12**, 499–511
27. Odegard, J., Guo, H., Burrus, C., Wells Jr, R. (1995) Wavelet based speckle reduction and image compression. In *SPIE*, **2487**, 259–271
28. Oliver, C., Quegan, S. (1998) *Understanding Synthetic Aperture Radar Images*. Artech House
29. Rogers, S., Colombi, J., Martin, C., Gainey, J., Fielding, K., Burns, T., Ruck, D., Kabrisky, M., Oxley, M. (1995) Neural networks for automatic target recognition. *Neural Networks*, **8**, 1153–84
30. Schoenmakers, R., Wilkinson, G., Shouten, T. (1994) Results of a hybrid segmentation method. In *Proceedings SPIE Conference on Image and Signal Processing for Remote Sensing*, **2315**, 113–26
31. Sergi, R., Satalino, G. (1996) SIR-C polarimetric image segmentation by neural network. In *International Geoscience and Remote Sensing Symposium*, **3**, 1562–4
32. Soh, L., Tsatsoulis, C. (1998) Automated sea ice segmentation (ASIS). In *IEEE International Geoscience and Remote Sensing*, **2**, 586–8
33. Soh, L., Tsatsoulis, C. (1999) Segmentation of satellite imagery of natural scenes using data mining. *IEEE Transactions on Geoscience and Remote Sensing*, **37**, 1086–99
34. Sziranyi, T., Csapodi, M. (1998) Texture classification and segmentation by cellular neural network using genetic learning. *Computer Vision and Image Understanding*, **71**, 255–70
35. Verly, J. G., Lacoss, R. T. Automatic target recognition of LADAR imagery using functional templates derived from 3D CAD models. In *Reconnaissance, Surveillance, and Target Acquisition for the Unmanned Ground Vehicle: Providing Surveillance 'Eyes' for an Autonomous Vehichle*. O. Firschein and T.M. Strat (Editors), 195–218, Morgan Kaufmann
36. Yokoo, Y., Hagiwara, M. (1996) Human faces detection method using genetic algorithm. In *IEEE International Conference on Evolutionary Computation*, 113–18
37. Yu, M., Eau-anant, N., Saudagar, A., Udpa, L. (1998) Genetic algorithm approach to image segmentation using morphological operations. In *International Conference on Image Processing*, **3**, 775–9
38. Yu, X., Bui, T., Krzyzak, A. (1993) The genetic algorithm parameter settings for robust estimation and range image segmentation and fitting. In *Proceedings of the Scandinavian Conference on Image Analysis*, **1**, 623–30
39. Zingaretti, P., Carbonaro, A. (1998) On increasing the objectiveness of segmentation results. In *International Conference on Advances in Pattern Recognition*, 103–12

An Immunogenetic Approach in Chemical Spectrum Recognition

Yuehua Cao[1] and Dipankar Dasgupta[2]

[1]Departments of Chemistry and Mathematical Sciences,
[2]Department of Mathematical Sciences
The University of Memphis, Memphis, TN 38152, USA
Email: ddasgupt@memphis.edu

Summary. This chapter describes an immunogenetic approach to recognize spectra for chemical analysis. In particular, an immunological model for chemical reactions is described in which a population of specialists for each of the possible products is evolved using a genetic algorithm. Accordingly, a small well-trained specialist library is established for testing their pattern recognition ability. The model was experimented with several real-world datasets to identify components in chemical spectra (such as IR spectra and Raman spectra). Experimental results exhibit the performance of the approach in finding correct products that correspond to an input spectrum, specifically, for a composite spectrum in which there are multiple products physically mixed. It would be very difficult to interpret otherwise.

1 Introduction

The natural immune system protects the body from a large variety of bacteria, viruses, and other pathogenic organisms. It recognizes foreign cells and molecules by producing antibody molecules that physically bind with antigens (or antigenic peptides). In order for the antigen and antibody molecules to bind, their three-dimensional shapes must match in a lock-and-key manner. For every antigen, the immune system must be able to produce a corresponding antibody molecule, so that the antigen can be recognized and defended against. The antibody, therefore, can have a geometry that is specific to a particular antigen (specialist) or is capable of partial matching and capturing a broad group of antigens (generalist). The primary role of this defense mechanism is to distinguish between the self (such as body cells and tissues) and the non-self (such as antigens). This discrimination is achieved in part by T-cells, which have receptors on their surface that can detect foreign proteins (antigens). During the generation of T' cells, their receptors are evolved (from gene libraries) through a pseudo-random genetic rearrangement process. Then they undergo a censoring process, called negative selection, in the *thymus*, where T' cells that react against self-proteins are destroyed. Only those that do not bind to self-proteins are allowed to leave the thymus. These matured T' cells then circulate throughout the body to perform immunological functions to protect against foreign antigens. Moreover, it continually evolves such immune cells and other antibody molecules (in the right proportion) in order to defend the body.

These immunological mechanisms have inspired the development of several computational models [5]. A brief survey of some of these models may be found elsewhere [6]. Forrest et al. [11] developed a negative-selection algorithm for change detection based on the principles of self–non-self discrimination. This algorithm works on similar principles, generating detectors randomly, and eliminating the ones that detect self, so that the remaining T-cells can detect any non-self. This self and non-self (computational) algorithm, the representative of a two-component model, appears to be very useful in many applications [7], but is not adequate for applications with multiple classes involved, each of which requires to be uniquely recognized.

Researchers have also been studying immunogenetic approaches (evolving antibodies using genetic algorithms) for more than a decade [5, 10]. Farmer et al. [8] compared the immune system with learning classifier systems. Bersini and Varela [1] used a recruitment mechanism of the immune system to accelerate the parallel and local hill climbing. In particular, they developed an IRM (Immune Recruitment Mechanism) and GIRM (Genetic IRM) to recruit a candidate from a certain population in the shape space. There exist other computational models emulating different immunological principles, for example, the ability to detect common patterns in a noisy environment [9], the ability to discover and maintain coverage of diverse pattern classes [20], and the ability to learn effectively, even when not all antibodies are expressed and not all antigens are presented [16]. In some studies, genetic algorithms have been used to model somatic mutation -- the process by which antibodies are evolved to recognize a specific antigen [17]. Hajela [13,14] recently used a genetic search for immune network design in solving structural optimization problems. Other researchers investigated artificial immune systems for scheduling [12, 15]. Potter and De Jong [19] reported a method for concept learning in which a coevolutionary genetic algorithm was applied to the construction of an immune system whose antibodies can discriminate between examples and counter-examples of a given concept.

In this chapter, we describe the use of an immunogenetic approach in the interpretation of chemical spectra. Section 2.1 introduces some basic spectroscopic concepts to understand the proposed method. The processes spectrum representation, specialist evolution, and spectrum recognition are described in detail in the rest of Section 2. In Section 3, the immunogenetic model is tested on a set of real-world problems and results are presented. Section 4 gives some concluding remarks and directions for future work.

2 The Problem and the Proposed

Interpretation of a composite spectrum (like IR, UV–visible, Raman, mass, etc.) has been a difficult and very time-consuming task for chemists. This work has conventionally been performed manually with limited accuracy. Although molecular calculations have provided some confirmatory information to aid the interpretation, the improvement is yet to be made. Using a principle component regression (PCR) method combined with an adaptive filter, Chen et al. [4] successfully identified the compounds of mixtures with accuracy up to 96% from the mixture spectra, which were obtained by mathematically adding individual mid-infrared spectra together. The problem with this method is that it is computationally expensive for a large library, regardless of possible band shifts in the actual mixture spectrum. There is a report in which neural networks were used for rapid screening of large infrared spectral databases, where only spectra of pure chemicals were involved [18]. An intelligent approach is described here as an alternative way to spectrum recognition for chemical analysis.

2.1 Description of a Band in a Spectrum

Absorption spectra are plots representing the absorbance (A) or transmittance (T) as a function of frequency (or, more specifically, wavenumbers in cm^{-1}) of the recorded radiation. Spectra are made up of a collection of bands deriving from the fundamental tones, combined tones, and overtones, related to the normal vibrations of a molecule. Each band of a spectrum is characterized by the following parameters:

(a) The position of the band maximum, most frequently expressed in wavenumbers, ν.

(b) The intensity of the band:

(i) at the maximum, I_{max} (A_{max}),

$$I_{int} = \int_{-\infty}^{+\infty} I(\nu) d\nu$$

(ii) the integrated intensity I_{int} (absorption A_{int}).

(c) The band half-width $\triangle \nu_{1/2}$.

The position of the band maximum (ν_{max}) is the most significant parameter, because it yields information on the frequency and hence the type of vibration. The above information is introduced using the absorption spectrum as an example, but it actually applies to almost all the spectroscopy family, including the scattering spectrum and mass spectrum (trivially). So in this study, as initial work in this field, the positions of the band maxima will be used to represent a spectrum. Other information such as the intensity of bands can be easily attached with a slight modification of the data structure. An example of a Raman spectrum [2] is shown in Fig. 1.

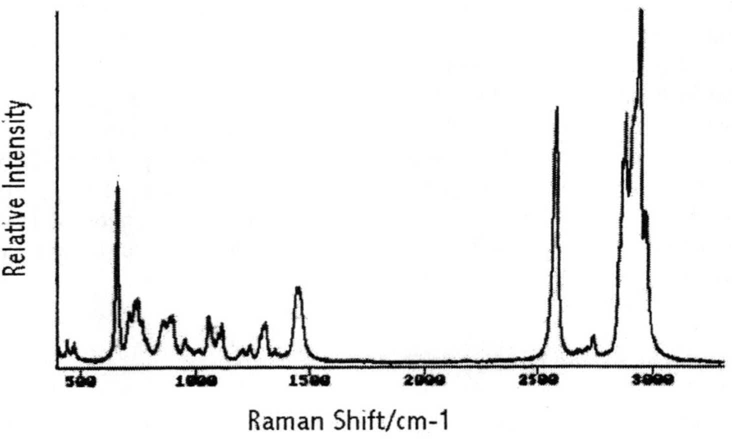

Fig. 1. A sample Raman spectrum of 1-butanethiol

2.2 Representation of the Spectrum

Each spectrum is represented with a binary string where each bit in the string corresponds to a peak occurrence within an equal length of wavenumbers. The value of the bit is determined by the signal received at the detector: either 1 if there is a peak at that region of wavenumber, or 0 if not. If the spectra domain has n wavenumbers and the spectrum is represented with a string of m bits, each bit has coverage of n/m wavenumbers. In this bitstring universe, recognition takes place when the (antibody) bitstring and the (antigen) bitstring "match" each other as will be explained later. Using this representation, the above Raman spectroscopy can be expressed as the bitstring shown in Fig. 2.

(a) The first half of the string (400 cm^{-1} -1700 cm^{-1})

1	0	1	0	1	1	1	0	1	0	1	0	0
.5	-.5	.5	-.5	.5	.5	.5	-.5	.5	-.5	.5	-.5	-.5

(b) The second half of the string (1700 cm^{-1} - 3000 cm^{-1})

0	0	0	0	0	0	0	0	1	0	1	1	1
-	-	-	-	-	-	-	-	1	-	.	.	.

Fig. 2. A binary representation of the spectrum displayed in Fig. 1. A bit value "1" means a peak occurrence and "0" otherwise. The weight associated with each bit given under the string indicates peak properties.

In this work, the above string representation is used to define the immunological terms in the following manner:

Self: a set of starting materials **R** (for **R**eactants) before chemical or photochemical reactions.

Non-self: a set of products produced due to reactions.

Antigen: any of the products **P** (for **P**roducts).

Antibody: any evolved population, which uniquely recognizes one and only one of the products, and it should have no response to the spectrum of the starting materials **R** or other products.

Matching: an antigen and antibody are said to "match" (in Hamming space) if the similarity between the antigen and antibody string exceeds the set threshold **T**. Optimal **T** values are determined through a training process.

Matching function: a function **f** to measure how well two spectra match.

2.3 The Specialist Evolution

The general form of a chemical reaction looks like this:

$$R_1 + R_2 + \cdots + R_m \xrightarrow{conditions} P_1 + P_2 + \cdots + P_n$$

where R_i ($1 \leq i \leq m$) are reactants and P_j ($1 \leq j \leq n$) are all the possible products. Each reactant or product has a specific spectrum that identifies it. By using the binary representation introduced above, each of the reactants R_i and products P_j is encoded into a unique string.

The next step is to evolve a population of specialists for each of the products. To do this, we are using product P_k ($1 \leq k \leq n$) as an antigen, exposing it to a randomly generated initial population, and then keeping only those which match P_k very well, and eliminating the rest of the population. We call the population that matches P_k the pre-antibody, σ_{pre}. The reason for this name lies in the fact that some members of this population may also match a reactant string or other product strings as well. To uniquely recognize a product antigen, it is necessary to expose the evolved antibodies to the environment of the reactants and other products. So we need to put this population into a pool consisting of $\{R_i \mid 1 \leq i \leq m\} \cup \{P_j \mid 1 \leq j \leq n, j \neq k\}$ for purification; this time we only want to keep those unmatched strings S_k, which are the trained specialists uniquely recognizing P_k. Those strings whose matching function value exceeds the threshold should be removed from the population of antibodies. Using the same kind of censoring approach as nature does (T-cell maturation), we can evolve a population of specialists for every product.

In our implementation, the determination of initial population size is based upon the actual necessity. Genetic operators (crossovers and mutations) are applied to the population as in a simple genetic algorithm to generate a population of better fitness. Elitism is used to preserve good individuals and keep sufficient diversity in the next generation. The number of generations depends on the real-valued activation threshold that represents the extent of similarity required to initiate an immune response (positive product recognition). Rather than simply using the number of matching bits as the fitness function, we assigned a weight to each bit, with the spectroscopic band b_k which characteristically defines P_k being assigned a significant higher weight W_k (see Fig. 2). In general, to initialize an immune response, the matching function must satisfy:
where b_i is the ith bit value of the (n-bits) string, either 0 when there is no peak at that

$$F = \sum_{i=1}^{n} b_i W_i(h_i, c_i) \geq T$$

interval or 1 when there is a peak at that interval; T is the threshold for initiating an immune response; $W_i(h_i\ ,\ c_i)$ is the weight of the ith bit, also a function of both the presence of a peak h_i (0.5 if there is also a peak in P_k at this position and –0.5 otherwise) and characteristic property ($c_i = 1$ if yes, $c_i = 0$ otherwise)

$$W_i(h_i, c_i) = h_i + c_i$$

The inclusion of negative values in the domain of h_i allows the matching function to take into account a penalty factor for the appearance of unexpected peaks, thus populations with peaks at unwanted positions are not encouraged during the evolution. This might be helpful in avoiding the occurrence of false positives.

Next, we purposefully set the value for a characteristic peak as twice as important as the presence of a normal peak. A typical example is the peak at 2578 cm^{-1} as shown in Fig. 1, which is attributed to the S–H vibration and safely identifies that free thiol molecules are contained in the compound(s) responsible for this spectrum [3]. This is a genetic search strategy for convergence of the population towards the peak (this is a desirable property) during evolution and preservation of a sufficient population of antibodies for a particular product. Last but not least, the appropriate choice of the threshold value gives this model a noise-tolerant feature. As is known, in spectroscopy some effects like random noise and baseline fluctuations can be eliminated from the input data, but other effects like frequency shift (usually small) and bandwidth variation cannot be cancelled from the experimental spectra and make the assignment troublesome. These problems, however, can easily be resolved by choosing an appropriate threshold. It is also a fundamental advantage of a genetic algorithm (GA) over deterministic methods. For example, by adjusting the threshold value, product information can be obtained with varying confidences. Fig. 3 shows schematically the algorithmic steps in evolving specialists for product identification.

A library of specialists is then generated to perform the central administration of spectrum recognition, to which specialists of each product are added. Extra power is introduced by admitting specialists for a product that has not been encountered. These specialists are functionally similar to the innate antibodies of human beings. The more comprehensive or evolved the library, the more powerful will be its recognition power in performing product identification task.

2.4 Spectrum Recognition

In the previous sections we have described how to establish a specialist library. Now we will demonstrate the utilization of this library. The recognition in this application is the automatic process to find all possible products responsible for the observed spectrum, analogous to the antibody's recognizing antigen in the natural immune system. There are situations where a spectrum obtained at a certain condition may not be any of the known products. In such cases the model will treat it as a new antigen, and then follow the same algorithm to evolve a population of specialists for it (as given in Fig. 1). In general, if there are m such unknown spectra, the ith will be named as unknown species i ($1 \leq i \leq m$).

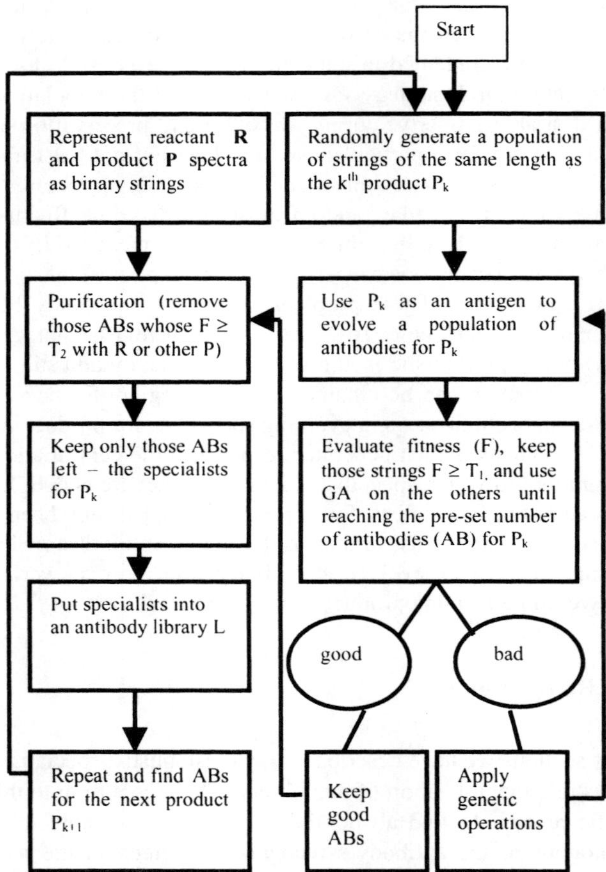

Fig. 3. A flow chart illustrating the proposed immunogenetic approach for evolving specialists. T1 and T2 are evolution threshold and recognition threshold, respectively.

The values of weights in the matching function used in the recognition phase should be different from the one used during evolution (see the weight function in Section 2.3). Specifically, modification to the matching function is the removal of the penalty of an unexpected peak on a spectrum. It is especially necessary for a composite spectrum, which could be recognized by multiple different antibodies. In particular, here, $h_i = 0.5$ if there is also a peak in P_k at this position and $h_i = 0$ (instead of -0.5 in the evolution phase) otherwise.

The following is the proposed algorithm for spectrum recognition:

Algorithm for Recognition (NewStr)

0: Input string NewStr **S**;
1: Expose **S** to the specialist library **L**;
2: Find all those antibodies (**AB**) whose binding energy F2 with **S** exceeds $\mathbf{T_2}$ (the recognition threshold);
3: Check whether there exist unassigned peaks (**S - ΣAB**);
4: If yes, name it unknown-j ($\mathbf{U_j}$), and evolve a population of **AB** for $\mathbf{U_j}$, and then add these **AB** of $\mathbf{U_j}$ to **L**;
5: Return $\mathbf{AB_i}$ and $\mathbf{U_j}$.

The recognition capacity of this approach increases with the size of the library. Whenever a product is recognized for the first time, a copy of it is reserved as a new specialist for that product. Therefore, when it appears a second time, it can be easily recognized by the antibodies created during its first appearance. Consequently, this approach provides a learning methodology for pattern recognition.

3 Experimental Details and Results

The performance of this approach was tested with a two-tier examination. First, it is examined with infrared spectra using a mathematical formulation. In order to find out how the band shifts, which occur as a result of physical interactions among molecules, affect the recognition process of this model, experiments using Raman spectra were conducted on pure and some combination of 20 compounds. Accordingly, different spectra from the experiments were used for testing and subsequent validation.

It is noteworthy that, to facilitate the explanation of the model, the threshold T is in the form of the absolute fitness value in the previous sections. However, in our actual implementation as well as in the following sections, the value of relative fitness (or percentage similarity, fitness/perfect fitness) was adopted to maintain comparability among different spectra.

3.1 Infrared Spectra Test

Infrared spectra are plots representing the absorbance (*A*) or transmittance (*T*) as a function of frequency (or, more specifically, wavenumbers in cm^{-1}) of the recorded radiation. Five chemicals were used in this test as shown in Table 1.

Table 1. Samples and their molecular formulas

Sample	Molecular formula
1	$CH_3(CH_2)_{10}COOH$
2	C_5H_5N
3	CH_3CH_2Cl
4	$C_6H_5(CH_3)_2$
5	$C_6H_5NO_2$

The mathematical addition of two spectra could arise as the combination of two or more bands occurring at close wavenumbers. As a result, the total number of peaks in the composite spectrum is sometimes less than the sum of the two. This phenomenon, however, is not a significant occurrence, especially when the bands in the spectrum are scarcely populated.

Three mixtures of spectra were made from the five spectra given in Table 1 using the software GRAMS_32 Spectral Notebase. These mixture spectra were then tested with the method described earlier. The results of identifying components in chemical spectra are listed in Table 2. All three mixtures with a different number of components were correctly identified. However, it took some time to find a set of optimal parameters for the best performance (including the threshold values), as will be addressed in detail in Section 3.2.

Table 2. Results of IR spectrum recognition

Mixture	Returned composition
2, 3	2, 3
1, 3, 4	1, 3, 4
1,2,3,5	1, 2, 3, 5

This encouraging empirical result leads us to test the model with real-world spectra of both a single compound and a mixture.

3.2 Raman Spectra Test

Without loss of generality, we used Raman spectra for our experiments simply because of their availability in our laboratory, popularity worldwide, and maturity as a powerful analytical tool. Five known composite spectra were used as antigens to test whether and how well they can be recognized. A total of 20 different product spectra were collected [3], of which the five mixtures are composed. Each product maintains a certain number of specialists (σ_{AB}). No matter which one of these specialists has a positive reaction towards the input spectrum, it will return the same molecular formula. The number of different formulas returned represents the number of different products responsible for the spectrum. In this section, experimental results are presented using several parameters including T_1 (threshold during evolution), T_2 (threshold during recognition), and the number of antibodies (specialists) maintained for each product spectrum. Table 3 shows these 20 products as well as their peak densities that are used in the experiments.

Table 3. Information for 20 products that are used in our experiments to form different mixtures.

Product	Molecular Formula	Peak Density (%)
1	$(CH_3)_2SO$	3.00
2	$C_6H_5CH_2CH_2SH$	7.00
3	$(C_2H_5)_2O$	3.00
4	CCl_4	1.33
5	$HS(CH_2)_4SH$	5.00
6	CH_3COCH_3	4.67
7	C_6H_5COOH	4.67
8	CH_3CH_2OH	3.33
9	$COOH\ CH_2SH$	4.67
10	$O\text{-}NH_2C_6H_5SH$	7.00
11	$C_4H_9SO_3$	5.67
12	$C_6H_5SO_3$	4.33
13	$CH_3\ (CH_2)_{17}SH$	5.00
14	$CH_3\ (CH_2)\ _2SH$	7.00
15	trans-$FC(O)SCH_3$	3.33
16	$(CF_3)\ _2C{=}NH$	6.33
17	NaCOO(OH)CHCH(OH)COOK	6.67
18	$C_6H_5S\ CH_3$	7.67
19	$Cl\ CH_2COOC_2H_5$	13.7
20	$F\ CH_2CONH_2$	4.67

In our experiments, each bit represents 10 cm^{-1} because it is very rare that two peaks occur within 10 cm^{-1}. Note that the peak density is calculated using the number of peaks to divide the total number of intervals. First, the effect of the number of specialists (antibodies) maintained for each product on recognition was investigated. Table 4 outlines the results of our experiments with various specialist sets, σ_{AB} = 3, 5, and 10, respectively. The second column shows the formation of the mixtures using the different products given in Table 3. The remaining three columns show the products found in the mixture by our method. Obviously, the output quickly approaches the true value when σ_{AB} is increased from 3 to 5. When σ_{AB} = 10, all five mixtures are recognized with 100% accuracy, so the other tests that followed adapted this setting. It is to be noted that the smaller the σ_{AB}, the fewer the number of products recognized. This is particularly exemplified by the fact that nothing was returned when σ_{AB} = 3 for mixture C.

Table 4. Effect of the size of specialist set σ_{AB} on the performance of this model. Here $T_1 = 0.9$, $T_2 = 0.99$.

Mixture	Actual composition	Products found in the composition		
		σ_{AB} = 3	σ_{AB} = 5	σ_{AB} = 10
A	8 and 20	8	8, 20	8, 20
B	1, 4, and 16	4	1, 4, 16	1, 4, 16
C	1, 13, and 15	null	1, 15	1, 13, 15
D	3, 4, 6, and 7	3, 4	3, 4	3, 4, 6, 7
E	4, 8, 10, 11, and 12	4	4, 8, 11, 12	4, 8, 10, 11, 12

Table 5 shows the results of each of the five tests and compares them with the actual chemical composition. Clearly T_2 is very critical for the correct recognition. When $T_2 = 0.99$, the program correctly recognizes all five mixture spectra. However, with further increase of T_2 to 0.999, some (more than half in most cases) possible products are ruled out. This indicates that the threshold is too high to find all the products, but it is useful when someone wants to know which product is exactly involved. On the other hand, when T_2 is decreased to 0.95, wrong results began to appear. It is found that most of the misidentified products share a common characteristic, i.e., their peak densities are relatively low. Normally, if the peak density of a product spectrum is less than 4.67, as in this example, it has a better chance to be positive toward an input spectrum (antigen) if $T_2 \leq 0.95$. We also tried the situation when $T_2 = 0.90$, in which not only were there more false positives, but also results changed even for the same mixture between two runs. Overall, $T_2 = 0.99$ is found to be appropriate for the set of experiments.

Table 5. Effect of the recognition threshold (T_2) on the performance of this model. $T_1 = 0.9$ and $\sigma_{AB} = 10$; the underlined products indicate false positives.

Mixture	Actual composition	Products found		
		$T_2 = 0.99$	$T_2 = 0.999$	$T_2 = 0.95$
A	8 and 20	8, 20	8	8, 20, 1, 3, 4
B	1, 4, and 16	1, 4, 16	4	1, 4, 16, 3
C	1, 13, and 15	1, 13, 15	15	1, 13, 15, 3, 4, 8
D	3, 4, 6, and 7	3, 4, 6, 7	3, 4, 6, 7	3, 4, 6, 7, 1, 8, 12
E	4, 8, 10, 11, and 12	4, 8, 10, 11, 12	4, 8	4, 8, 10, 11, 12, 1, 3, 7

While the performance of this method is very sensitive to T_2, T_1 behaves otherwise. The final set of experiments shown in Table 6(a) compares the results for different T_1 values, while keeping T_2 constant at 0.95. It is observed that no significant improvement was achieved as regards the accuracy of the outputs when T_1 changes from 0.9 up to 0.999. The number of false positives remained unchanged even when T_1 varied from 0.99 to 0.999. These results verify our arguments above.

Table 6(a) Variation of the threshold (T_1) during the evolution of specialists on the performance of this method. T_2 = 0.95 and σ_{AB} = 10; the underlined products are false positives.

Mixture	Actual composition	Product found		
		T_1 = 0.99	T_1 = 0.999	T_1 = 0.95
A	8 and 20	8, 20	8	8, 20, 1, 3, 4
B	1, 4, and 16	1, 4, 16	4	1, 4, 16, 3
C	1, 13, and 15	1, 13, 15	15	1, 13, 15, 3, 4, 8
D	3, 4, 6, and 7	3, 4, 6, 7	3, 4, 6, 7	3, 4, 6, 7, 1, 8, 12
E	4, 8, 10, 11, and 12	4, 8, 10, 11, 12	4, 8	4, 8, 10, 11, 12, 1, 3, 7

The insensitivity of T_1 can be even more clearly seen if we set $T_2 = 0.99$ and σ_{AB} = 10, while changing T_1 from 0.90 to 0.999, as shown in Table 6(b). This model works with 100% accuracy on all five mixture spectra when T_1 ranges from 0.90 to 0.99. When $T_1 = 0.999$, only the product 10 failed to be recognized in mixture E.

4 Discussion

One essential assumption for this approach is that the mixture spectrum should be, to a great extent, the concatenation of each of the individual product spectra. However, in order to investigate how well the model spectrum agrees with the actual spectrum, experiments were performed to aid this evaluation. It is worth mentioning that no chemical interactions are expected to exist among products of a reaction; consequently we focus our attention only on physical interactions.

Hydrogen bonding and dipole–dipole are two major types of physical interactions among molecules. They should be responsible for the primary band shift if any. Compared to van der Waals' interactions, the hydrogen bonding is much stronger. The above discussion allows us to safely consider only mixture systems where hydrogen bonding exists, because the van der Waals' force ubiquitously exists simultaneously.

Table 6(b). Variation of the threshold (T_1) during the evolution of specialists on the performance of this method. T_2 = 0.95 and σ_{AB} = 10; the underlined products are false positives.

Mixture	Actual composition	Product found			
		T_1 = 0.90	T_1 = 0.95	T_1 = 0.99	T_1 = 0.999
A	8 and 20	8, 20	8, 20	8, 20	8, 20
B	1, 4 and 16	1, 4, 16	1, 4, 16	1, 4, 16	1, 4, 16
C	1, 13 and 15	1,13, 15	1,13, 15	1,13, 15	1,13, 15
D	3, 4, 6 and 7	3, 4, 6, 7	3, 4, 6, 7	3, 4, 6, 7	3, 4, 6, 7
E	4, 8, 10, 11 and 12	4, 8, 10, 11, 12	4, 8, 10, 11, 12	4, 8, 10, 11, 12	4, 8, 11, 12

It is expected that hydrogen bonding exists among CH_3CH_2OH (#8) molecules. When $(CH_3)_2SO$ (#1) is added to CH_3CH_2OH in a 1: 1 ratio, some of the hydrogen bonds break due to the dilution effect of $(CH_3)_2SO$ and weaker interaction between the alcohol and sulfoxide. Accordingly, in our experiment, a blue shift associated with the O–H vibration as large as 50 cm^{-1} was observed. With this change, the threshold of T_2 = 0.99 can only allow $(CH_3)_2SO$ to be recognized. However, when T_2 is set to 0.98, both were correctly recognized in the mixture along with all other five mixtures, while keeping T_2 = 0.98 and changing T_1 from 0.99 to 0.96 had no effect on the accuracy (100%). When T_2 is further decreased to 0.97, false positives began to appear. In this case, CCl_4 became a returned component for all the six mixtures because of its lowest peak density among the 20 compounds. So T_2 = 0.98 and $T_1 \in$ [0.96, 0.99] is a globally optimal setting for all the six mixtures. In comparison with the PCR (Principle Component Regression) method [4], this approach not only has a valuable improvement in performance (the accuracy is improved from 96% to 100%), but also takes into consideration the effect of molecular interaction on real-world spectra.

Raman spectra of the mixtures of #6 CH_3COCH_3 and #8 CH_3CH_2OH at 1:1 and 1:2 ratios were recorded and used to test this parameter setting. In both cases, the OH stretching vibration was too weak to be included in the bitstring, and other bands were observed to have a maximum shift of 4 cm^{-1}. In our calculation, both spectra were correctly recognized when $T_2 = 0.98$ and $T_1 \in [0.96, 0.99]$.

Usually, each compound in the mixture may not participate at the same ratio. It is true that this method works only in a certain concentration range. It is difficult to give a definite concentration above which a component is guaranteed to be detectable, because the detection limit is analytical method dependent, instrument dependent, and chemical species sensitive. A general rule to follow is that under all circumstances, as long as the peaks are identifiable, this method is applicable. In addition, for gaseous mixture identification, this method is equally useful.

The computational complexity of this immunogenetic approach is $O(mn)$ in terms of the number of comparisons with straightforward implementation, where m is the length of each string and n is the size of the specialist library, making it affordable for virtually every laboratory. A nice feature of this approach is that users can customize their libraries to hold only spectra of certain groups of compounds used or produced at their laboratories.

5 Conclusions

The natural immune system uses learning, memory, and associative retrieval to solve recognition and classification tasks. Its learning takes place through recruitment mechanism, which is partly an evolutionary process similar to a biological evolution. Various recognition and response mechanisms of the immune system have inspired the development of some useful computational models. This chapter describes an immunogenetic approach for the detection of products from an input spectrum with adjustable confidence. It is particularly useful in identifying compositions from chemical spectra, which has been a difficult task for chemists. Compared with deterministic spectrum detection approaches, this method is more flexible and noise tolerant. However, its capability can be extended beyond the spectroscopy with combinations of other measurements, but to clench the exact products. It is demonstrated that with a well-established specialist library and a well-chosen threshold, our approach could find all the possible products responsible for an input spectrum within just 1 second.

The utility of this approach can be extended to other spectra, such as IR, UV–visible, and mass spectroscopy. In a bulk solution, using transmission spectroscopic techniques, it can be programmed for automatic product detection in organic synthesis, which facilitates the optimization of reaction conditions towards the best possible yield based on the in-situ product makeup detection mechanism. Moreover, for surface spectra (e.g., SERS, SERI) [21], this method should be equally useful.

The work of spectrum representation is a simplified abstraction of the spectrum in the real world. The inclusion of additional spectroscopic information like peak area in the data structure will definitely increase its ability of discrimination. For example, using integer representation instead of binary representation will allow peak intensity and spectroscopic considerations. To be practically useful, a comprehensive collection of specialists for a broad range of chemicals needs to be generated, and trained and tested on different threshold values. Further work will also study the antigenic feature extraction properties of the natural immune system to develop an improved pattern recognition methodology.

References

1. H. Bersini and F. J. Varela. The immune recruitment mechanism: A selective evolutionary strategy. *In Proceedings of the Fourth International Conference on Genetic Algorithms*, pages 520-526, San Diego, July 13-16, 1991.
2. Y. Cao and Y. S. Li. Constructing surface roughness of silver for surface enhanced Raman scattering by self-assembled monolayers and selective etching process. *Appl. Spectrosc.*, 53, 540, 1999.
3. Y. Cao and Y. S. Li. Spectra were collected with a spectrometer system consisting of a Spex Model 1403 double monochromator and a Hamamatsu R928-07 photomultiplier tube (PMT) held at –30°C by a thermoelectrically refrigerated chamber (Product for Research, Model TE 117-RF). A Lexel 3000 laser (Ar^+) equipped with a Spex Model 1405 tunable filter was used for sample excitation at 514.5 nm. A bandpass of 4 cm^{-1} was set for all the experiments (unpublished).
4. C. S. Chen, Y. Li, and C. W. Brown, PCR. *Vib. Spectrosc.*, 14, 9-17, 1997.
5. D. Dasgupta (editor). Artificial Immune Systems and Their Applications, Springer-Verlag, 1999.
6. D. Dasgupta and N. Attoh-Okine. Immunity-based systems: A survey. *In Proceedings of the IEEE International Conference on Systems, Man, and Cybernetics*, pages 363-374, Orlando, Florida, October 12-15, 1997.
7. D. Dasgupta and S. Forrest. Novelty detection in time series data using ideas from immunology. In *Proceedings of the fifth International Conference on Intelligent Systems*, Reno, June 19-21, 1996.
8. J. D. Farmer, N. H. Packard, and A. S. Perelson. The immune system, adaptation, and machine learning. In *Physica D*, 22,187-204, 1986.
9. S. Forrest, B. Javornik, R. Smith, and A. S. Perelson. Using genetic algorithms to explore pattern recognition in the immune system. *Evolutionary Computation*, 1(3), 191-211, 1993.

10. S. Forrest and A. S. Perelson. Genetic algorithms and the immune system. *In Proceedings of the Parallel Problem Solving from Nature*, Springer-Verlag (Lecture Notes in Computer Science), 1991.
11. S. Forrest, A. S. Perelson, L. Allen, and R. Cherukuri. Self-nonself discrimination in a computer. *In Proceedings of the IEEE Symposium on Research in Security and Privacy*, pages 202-212, Oakland, CA, May 16-18, 1994.
12. T. Fukuda, K. Mori, and M. Tsukiyama. Immune networks using genetic algorithm for adaptive production scheduling. *In 15th IFAC World Congress*, Vol. 3, pp.57-60, 1993.
13. P. Hajela and J. Lee. Constrained genetic search via schema adaptation: An immune network solution. *Struct. Opt.* , 12, 1, 11-15, 1996.
14. P. Hajela, J. Yoo, and J. Lee. GA based simulation of immune networks - applications in structural optimization. *J. Eng. Opt.*, 1997.
15. E. Hart, P. Ross, and J. Nelson. Producing robust schedules via an artificial immune system. *In Proceedings of the IEEE International Conference on Evolutionary Computation*, 1998.
16. R. Hightower, S. Forrest, and A. S. Perelson. The evolution of emergent organization in immune system gene libraries. *In Proceedings of the Sixth International Conference on Genetic Algorithms*, Pittsburg, 1995.
17. R. Hightower, S. Forrest, and A.S. Perelson. The Baldwin effect in the immune system: learning by somatic hypermutation. In *Adaptive Individuals in Evolving Populations*, Addison-Wesley, pages 159-167, 1996.
18. C. Klawun and C. L. Wilkins. Neural network assisted rapid screening of large infrared spectral databases *Anal. Chem,.* 67, 374, 1995.
19. M. A. Potter and K. A. De Jong. The coevolution of antibodies for concept learning. *In Proceedings of the Parallel Problem Solving from Nature (PPSN)*, Amsterdam, 1998.
20. R. E. Smith, S. Forrest, and A. S. Perelson. Searching for diverse, cooperative populations with genetic algorithms. In *Evol. Comput.*, 1(2), 127-149, 1993.
21. X. M. Yang, D. A. Tryk, K. Hashimoto and A. Fujishima, Examination of the photoreaction of p-nitrobenzoic acid on electrochemically roughened silver using surface-enhanced Raman imaging (SERI). *J. Phys. Chem. B,* 102, 4933, 1998.

Application of Evolutionary Computation to Protein Folding

Steffen Schulze-Kremer

RZPD Deutsches Ressourcenzentrum für Genomforschung GmbH, Heubnerweg 6, D-14059 Berlin, Germany
E-mail: steffen@rzpd.de

Summary. This chapter demonstrates the application of genetic algorithms to the *ab initio* protein folding problem. In particular, solutions to the representation issue of protein tertiary structure, of domain-specific genetic operators and a vector fitness function for fold evaluation are presented. Finally, limitations of this approach are discussed.

1 Introduction

Genetic algorithms are, like neural networks, an example *par excellence* of an information processing paradigm that was originally developed and exhibited by nature and later discovered by human who subsequently transformed the general principle into computational algorithms to be put to work in computers. Nature makes use of the principle of genetic heritage and evolution in an impressive way. Application of the simple concept of performance-based reproduction of individuals ("survival of the fittest") led to the rise of well-adapted organisms that can endure in a potentially adverse environment. Mutually beneficial interdependencies, co-operation and even apparently altruistic behaviour can emerge solely by evolution. The investigation of those phenomena is part of research in artificial life but is not dealt with here.

Evolutionary computation comprises the four main areas of genetic algorithms [1], evolution strategies [2], genetic programming [3] and simulated annealing [4]. Genetic algorithms and evolution strategies emerged at about the same time in the United States of America and Germany. Both techniques model the natural evolution process in order to optimise either a fitness function (evolution strategies) or the effort of generating subsequent, well-adapted individuals in successive generations (genetic algorithms). Evolution strategies in their original form were basically stochastic hill-climbing algorithms and used for optimisation of complex, multi-parameter objective functions that in practice cannot be treated analytically. Genetic algorithms in their original form were not primarily designed for function optimisation but rather to demonstrate the efficiency of genetic crossover in assembling successful candidates over complicated search spaces.

Genetic programming takes the idea of solving an optimisation problem by evolution of potential candidates one step further in that not only the parameters of a problem but also the structure of a solution are subject to evolutionary change. Simulated Annealing is mathematically similar to evolution strategies. It was originally derived from a physical model of crystallisation. Only two individuals compete for the highest rank according to a fitness function and the decision about accepting sub-optimal candidates is controlled stochastically.

The methods presented in this chapter are heuristic, i.e. they contain a random component. As a consequence (and in contrast to deterministic methods) it can never be guaranteed that the algorithm will find an optimal solution or even any solution at all. Evolutionary algorithms are therefore used preferably for applications where deterministic or analytic methods fail, for example, because the underlying mathematical model is not well defined or the search space is too large for systematic, complete search (N-P. completeness). Another application area for evolutionary algorithms that is rapidly growing is the simulation of living systems starting with single cells and proceeding to organisms, societies or even whole economic systems [5], [6], [7], [8].

Work with evolutionary algorithms bears the potential for a philosophically and epistemologically interesting recursion. At the beginning, evolution emerged spontaneously in nature. Next, humans discover the principle of evolution and acquire knowledge of its mathematical properties, then (re-) define genetic algorithms for computers. To complete the recursive cycle, computational genetic algorithms are applied to the very objects (DNA, proteins) from which they had been derived in the beginning. A practical example of such a meta-recursive application will be given below in the sections on protein folding. Figure 1 illustrates this interplay of natural and simulated evolution.

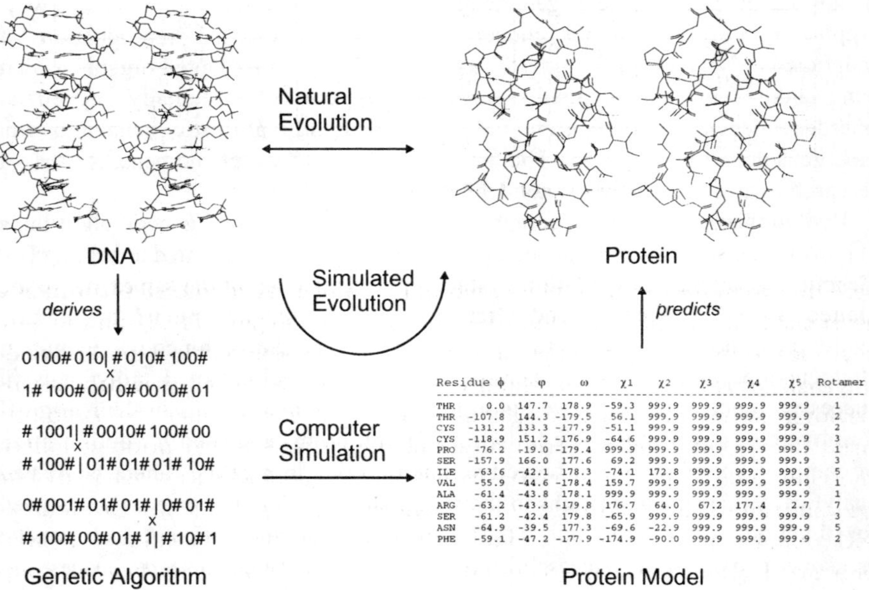

Residue	ϕ	φ	ω	$\chi 1$	$\chi 2$	$\chi 3$	$\chi 4$	$\chi 5$	Rotamer
THR	0.0	147.7	178.9	-59.3	999.9	999.9	999.9	999.9	1
THR	-107.8	144.3	-179.5	56.1	999.9	999.9	999.9	999.9	2
CYS	-131.2	133.3	-177.9	-51.1	999.9	999.9	999.9	999.9	2
CYS	-118.9	151.2	-176.9	-64.6	999.9	999.9	999.9	999.9	3
PRO	-76.2	-19.0	-179.4	999.9	999.9	999.9	999.9	999.9	1
SER	-157.9	166.0	177.6	69.2	999.9	999.9	999.9	999.9	3
ILE	-63.6	-42.1	178.3	-74.1	172.8	999.9	999.9	999.9	5
VAL	-55.9	-44.6	-178.4	159.7	999.9	999.9	999.9	999.9	1
ALA	-61.4	-43.8	178.1	999.9	999.9	999.9	999.9	999.9	1
ARG	-63.2	-43.3	-179.8	176.7	64.0	67.2	177.4	2.7	7
SER	-61.2	-42.4	179.8	-65.9	999.9	999.9	999.9	999.9	3
ASN	-64.9	-39.5	177.3	-69.6	-22.9	999.9	999.9	999.9	5
PHE	-59.1	-47.2	-177.9	-174.9	-90.0	999.9	999.9	999.9	2

Fig. 1. Interplay of natural and simulated evolution

2 Protein Folding Application

This section describes the application of a genetic algorithm to the problem of three-dimensional (3-D) protein structure prediction [9], [10], [11] with a simple force field as the fitness function. It is a continuation of work presented earlier [12]. Similar research on genetic algorithms and protein folding was done independently by several groups world wide [13]. Genetic algorithms have been used to predict optimal sequences to fit structural constraints [14], to fold Crambin in the Amber force field [15] and Mellitin in an empirical, statistical potential [16], and to predict main chain folding patterns of small proteins based on secondary structure predictions [17]. In this section the individuals of the genetic algorithm are conformations of a protein and the fitness function is a simple force field. In the following, the representation formalism, the fitness function and the genetic operators are described. Then, the results of an *ab initio* prediction run and of an experiment for side chain placement for the protein Crambin will be discussed.

2.1 Representation Formalism

For every application of a genetic algorithm one has to decide on a representation formalism for the "genes". In this application, the so-called *hybrid approach* is taken. This means that the genetic algorithm is configured to operate on numbers, not bit strings as in the original genetic algorithm. A hybrid representation is usually easier to implement and also facilitates the use of domain-specific operators. However, three potential disadvantages are encountered:

1. Strictly speaking, the mathematical foundation of genetic algorithms holds only for binary representations, although some of the mathematical properties are also valid for a floating point representation.
2. Binary representations run faster in many applications.
3. An additional encoding/decoding process may be required to map numbers onto bit strings.

It is not the principal goal of this application to find the single optimal conformation of a protein based on a force field but to generate a *small set of native-like conformations*. For this task the genetic algorithm is an appropriate tool. For a hybrid representation of proteins one can use Cartesian coordinates, torsion angles, rotamers or an otherwise simplified model of residues.

For a representation in Cartesian coordinates the 3-D coordinates of all atoms in a protein are recorded. This representation has the advantage of being easily converted to and from the 3-D conformation of a protein. However, it has the disadvantage that a mutation operator would in most instances create invalid protein conformations where some atoms lie too far apart or collide. Therefore a filter is needed which eliminates invalid individuals. Because such a filter would consume a disproportionate large amount of CPU time a Cartesian coordinate representation considerably slows down the search process of a genetic algorithm.

Another representation model is by torsion angles. Here, a protein is described by a set of torsion angles under the assumption of constant standard binding geometries. Bond lengths and bond angles are taken to be constant and cannot be changed by the genetic algorithm. This assumption is certainly a simplification of the real situation where bond length and bond angle to some extent depend on the environment of an atom. However, torsion angles provide enough degrees of freedom to represent any native conformation with only small r.m.s. [18] deviations.

Special to the torsion angle representation is the fact that even small changes in the φ (phi) / ψ (psi) angles can induce large changes in the overall conformation. This is useful when creating variability within a population at the beginning of a run. Figure 2 explains the definition of the torsion angles φ, ψ, ω (omega), χ1 (chi1) and χ2 (chi2). A small fragment taken from a hypothetical protein is shown. Two basic building blocks, the amino acids phenylalanine (Phe) and glycine (Gly), are drawn as wire frame models. Atoms are labelled with their chemical symbols. Bonds in bold print indicate the backbone. The labels of torsion angles are placed next to their rotatable bonds.

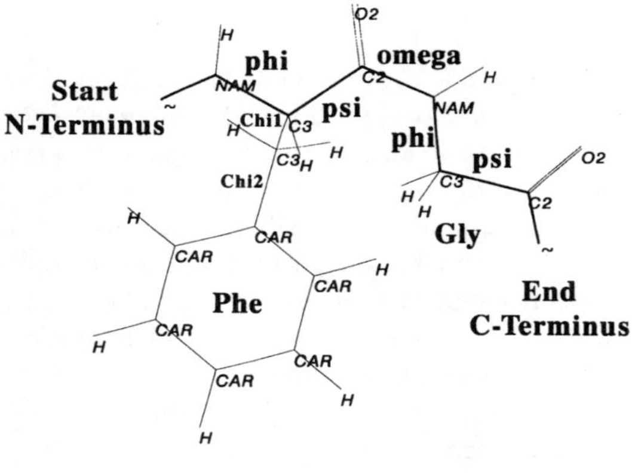

Fig. 2. Torsion angles φ, ψ, ω, χ1 and χ2

In the present work the torsion angle representation is used. Torsion angles of 129 proteins from the Brookhaven database [19] (PDB) were statistically analysed for the definition of the MUTATE operator. The frequency of each torsion angle in intervals of 10° was determined and the ten most frequently occurring intervals are made available for substitution of individual torsion angles by the MUTATE operator. At the beginning of the run, individuals were initialised with either a completely extended conformation where all torsion angles are 180° or by a random selection from the ten most frequently occurring intervals of each torsion angle. For the ω torsion angle the constant value of 180° was used because of the rigidity of the peptide bond between the atoms C_i and N_{i+1}. A statistical analysis of ω angles shows that with the exception of proline average deviations from the mean of 180° occur rather frequently up to 5°, and only in rare cases up to 15°.

The genetic operators in this application operate on the torsion angle representation but the fitness function requires a protein conformation to be expressed in Cartesian coordinates. For the implementation of a conversion program bond angles were taken from the molecular modelling software Alchemy [20] and bond lengths from the program Charmm [21]. Either a complete form with explicit hydrogen atoms or the so-called extended atom representation with small groups of atoms represented as "super-atoms" can be calculated. One conformation of a protein is encoded as an array of structures of the C programming language. The number of structures equals the number of residues in the protein. Each structure includes a three-letter identifier of the residue type and ten floating point numbers for the torsion angles φ, ψ, ω, $\chi 1$, $\chi 2$, $\chi 3$, $\chi 4$, $\chi 5$, $\chi 6$ and $\chi 7$. For residues with less than seven side chain torsion angles the extra fields are filled with a default value. The main chain torsion angle ω was kept constant at 180°.

2.2 Fitness Function

In this application a simple steric potential energy function was chosen as the fitness function (i.e. the objective function to be minimised). It is very difficult to find the global optimum of a potential energy function because of the large number of degrees of freedom even for a protein of average size. In general, molecules with n atoms have $3n - 6$ degrees of freedom. For the case of a medium-sized protein of 100 residues this amounts to

$$((100 \text{ residues} \cdot \text{approximately } 20 \text{ atoms per residue}) \cdot 3) - 6 = 5994$$

degrees of freedom. Systems of equations with this number of variables are analytically intractable today. Empirical efforts to heuristically find the optimum are almost as difficult [22]. If there are no constraints for the conformation of a protein and only its primary structure is given the number of conformations for a protein of medium size (100 residues) can be approximated to

$$(5 \text{ torsion angles per residue} \cdot 5 \text{ likely values per torsion angle})^{100} = 25^{100}$$

This means that in the worst case 25^{100} conformations would have to be evaluated to find the global optimum. This is clearly beyond the capacity of today's and tomorrow's supercomputers. As can be seen from a number of previous applications genetic algorithms were able to find sub-optimal solutions to problems with an equally large search space [23], [24], [25]. Sub-optimal in this context means that it cannot be proven that the solutions generated by the genetic algorithm do in fact include an optimal solution but that some of the results generated by the genetic algorithm practically surpassed any previously known solution. This can be of much help in non-polynomial complete problems where no analytical solution of the problem is available.

2.3 Conformational Energy

The steric potential energy function was adapted from the program Charmm. The total energy of a protein in solution is the sum of the expressions for E_{bond} (bond length potential), E_{phi} (bond angle potential), E_{tor} (torsion angle potential), E_{impr} (improper torsion angle potential), E_{vdW} (van der Waals pair interactions), E_{el} (electrostatic potential), E_H (hydrogen bonds), and of two expressions for interaction with the solvent, E_{cr} and E_{cphi} :

$$E = E_{bond} + E_{phi} + E_{tor} + E_{impr} + E_{vdW} + E_{el} + E_H + E_{cr} + E_{cphi}$$

Here we assume constant bond lengths and bond angles. The expressions for E_{bond}, E_{phi} and E_{impr} are therefore constant for different conformations of the same protein. The expression E_H was omitted because it would have required the exclusion of the effect of hydrogen bonds from the expressions for E_{vdW} and E_{el}. This, however, was not done by the authors of Charmm in their version v.21 of the program. In all runs, folding was simulated in vacuum with no ligands or solvent, i.e. E_{cr} and E_{cphi} are constant. This is certainly a crude simplification of the real situation but shall serve here as the first approach. Thus, the potential energy function simplifies to:

$$E = E_{tor} + E_{vdW} + E_{el}$$

Test runs showed that if only the three expressions E_{tor}, E_{vdW} and E_{el} are used there would not be enough force to drive the protein to a compact folded state. An exact solution to this problem requires the consideration of entropy. The calculation of the entropy difference between a folded and unfolded state is based on the interactions between protein and solvent. Unfortunately, it is not yet possible to routinely calculate an accurate model of those interactions. It was therefore decided to introduce an *ad hoc* pseudo entropic term E_{pe} that drives the protein to a globular state. The analysis of a number of globular proteins reveals the following empirical relation between the number of residues (length) and the diameter:

$$\text{expected diameter} / \text{m} = 8 \cdot \sqrt[3]{\text{length} / \text{m}}$$

The pseudo entropic term E_{pe} for a conformation is a function of its actual diameter. The diameter is defined to be the largest distance between any C_α atoms in one conformation. An exponential of the difference between actual and expected diameter is added to the potential energy if that difference is less than 15Å. If the difference is greater than 15 Å a fixed amount of energy is added (10^{10} kcal/mol) to avoid exponential overflow. If the actual diameter of an individual is smaller than the expected diameter E_{pe} is set to zero. The net result is that extended conformations have larger energy values and are therefore less fit for reproduction than globular conformations.

$$E_{pe} = 4^{(\text{actual diameter - expected diameter})} \ [\text{kcal/mol}]$$

Occasionally, if two atoms are very close the E_{vdW} term can become very large. The maximum value for E_{vdW} in this case is 10^{10} kcal/mol and the expressions for E_{el} and E_{tor} are not calculated. Runs were performed with the potential energy function E as described above where lower fitness values mean fitter individuals and with a variant, where the four expressions E_{tor}, E_{vdW}, E_{el} and E_{pe} were given individual weights. The results were similar in all cases. Especially, scaling down the dominant effect of electrostatic interactions did not change the results (see below).

2.4 Genetic Operators

In order to combine individuals of one generation to produce new offspring nature as well as genetic algorithms apply several genetic operators. In the present work, individuals are protein conformations represented by a set of torsion angles under the assumption of constant standard binding geometries. Three operators are invented to modify these individuals: MUTATE, VARIATE and CROSSOVER. The decision about the application of an operator is made during run time and can be controlled by various parameters.

MUTATE. The first operator is the MUTATE operator. If MUTATE gets activated for a particular torsion angle, this angle will be replaced by a random choice of one of the ten most frequently occurring values for that type of residue. The decision whether a torsion angle will be modified by MUTATE is made independently for each torsion angle in a protein. A random number between 0 and 1 is generated and if this number is greater than the MUTATE parameter at that time, MUTATE is applied. The MUTATE parameter can change dynamically during a run. The values that MUTATE can choose from come from a statistical analysis of 129 proteins from PDB. The number of instances in each of the 36 intervals of 10° was counted for each torsion angle. The ten most frequent intervals, each represented by its left boundary, are available for substitution.

VARIATE. The VARIATE operator consists of three components: the 1°, 5° and 10° operator. Independently and after application of the MUTATE operator for each torsion angle in a protein two decisions are made: first, whether the VARIATE operator will be applied and, second, if so which of the three components shall be selected. The VARIATE operator increments or decrements (always an independent random chance of 1:2) the torsion angle by 1°, 5° or 10°. Care is taken that the range of torsion angles does not exceed the [–180°, 180°] interval. The probability of applying this operator is controlled by the VARIATE parameter which can change dynamically during run time. Similarly, three additional parameters control the probability for choosing among the three components. Alternatively, instead of three discrete increments a Gaussian uniformly distributed increment between –10° and +10° can be used.

CROSSOVER. The CROSSOVER operator has two components: the *two point crossover* and the *uniform crossover*. CROSSOVER is applied to two individuals independently of the MUTATE and VARIATE operators. First, individuals of the parent generation, possibly modified by MUTATE and VARIATE, are randomly grouped pairwise. For each pair, an independent decision is made whether or not to apply the CROSSOVER operator. The probability of this is controlled by a CROSSOVER parameter which can change dynamically during run time. If the decision is "no", the two individuals are not further modified and added to the list of offspring. If the decision is "yes", a choice between the two-point crossover and the uniform crossover must be made. This decision is controlled by two other parameters that can also be changed during run time. The *two-point crossover* randomly selects two residues on one of the individuals. Then the fragment between the two residues is exchanged with the corresponding fragment of the second individual. Alternatively, *uniform crossover* decides independently for each residue whether or not to exchange the torsion angles of that residue. The probability for an exchange is then always 50%.

Parameterization. As mentioned in the previous paragraphs, there are a number of parameters that control the run time behaviour of a genetic algorithm. The parameter values used for the experiments that will be presented in the "Results" section below are summarised in Table 1. The main chain torsion angle ω was kept constant at 180°. The initial generation was created by a random selection of torsion angles from a list of the ten most frequently occurring values for each angle. Ten individuals are in one generation. The genetic algorithm was halted after 1000 generations. At the start of the run, the probability for a torsion angle to be modified by the MUTATE operator is 80%; at the end of the run it becomes 20%. In between the probability decreases linearly with the number of generations. In contrast, the probability of applying the VARIATE operator increases from 20% at the beginning to 70% at the end of the run. The 10° component of the VARIATE operator is dominant at the start of the run (60%), whereas it is the 1° component at the end (80%). Likewise, the chance of performing a CROSSOVER rises from 10% to 70%. At the beginning of the run mainly uniform CROSSOVER is applied (90%); at the end it is mainly two-point CROSSOVER (90%). This parameter setting uses a small number of individuals but runs over a large number of generations. This keeps computation time low while allowing a maximum number of crossover events. At the beginning of the run MUTATE and uniform CROSSOVER are applied most of the time to create some variety in the population so that many different regions of the search space are covered. At the end of the run the 1° component of the VARIATE operator dominates the scene. This is intended for fine tuning those conformations that have survived the selection pressure of evolution so far.

Generation Replacement. There are different ways of selecting the individuals for the next generation. Given the constraint that the number of individuals should remain constant some individuals have to be discarded. Transition between generations can be done by *total replacement*, *elitist replacement* or *steady state replacement*. For *total replacement* only the newly created offspring enter the next generation and the parents of the previous generation are completely discarded. This has the disadvantage that a fit parent can be lost even if it only produces bad offspring once. With *elitist replacement* all parents and offspring of one generation are sorted according to their fitness. If the size of the population is n, then the n fittest individuals are selected as parents for the following generation. This mode has been used here. Another variant is *steady state replacement* where two individuals are selected from the population based on their fitness and then modified by mutation and crossover. They are then used to replace their parents.

Parameter	Value
ω angle constant 180°	on
initialise start generation	random
number of individuals	10
number of generations	1000
MUTATE (start)	80%
MUTATE (end)	20%
VARIATE (start)	20%
VARIATE (end)	70%
VARIATE (start 10°)	60%
VARIATE (end 10°)	0%
VARIATE (start 5°)	30%
VARIATE (end 5°)	20%
VARIATE (start 1°)	10%
VARIATE (end 1°)	80%
CROSSOVER (start)	70%
CROSSOVER (end)	10%
CROSSOVER (start uniform)	90%
CROSSOVER (end uniform)	10%
CROSSOVER (start two point)	10%
CROSSOVER (end two point)	90%

Table 1. Run time parameters

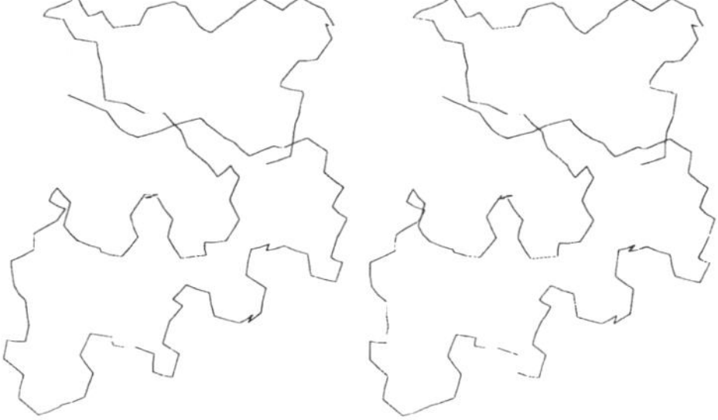

Fig. 3. Stereoprojection of Crambin without side chains

2.5 *Ab initio* Prediction

A prototype of a genetic algorithm with the representation, fitness function and operators as described above has been implemented. To evaluate the *ab initio* prediction performance of the genetic algorithm the sequence of Crambin was given to the program. Crambin is a plant seed protein from the cabbage *Crambe abyssinica*. Its structure was determined by W. A. Hendrickson and M. M. Teeter [26] to a resolution of 1.5 Å (Figures 3 and 4). Crambin has a strong amphiphilic character which makes its conformation especially difficult to predict. However, because of its good resolution and small size of 46 residues it was decided to use

Crambin as a first candidate for prediction. The following structures are displayed in stereo projection. If the observer manages to look cross eyed at the diagram in a way that superimposes both halves a 3-D image can be perceived.

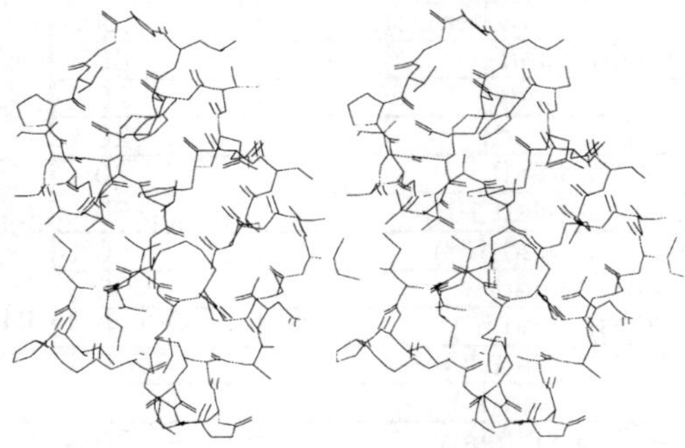

Fig. 4. Stereoprojection of Crambin with side chains

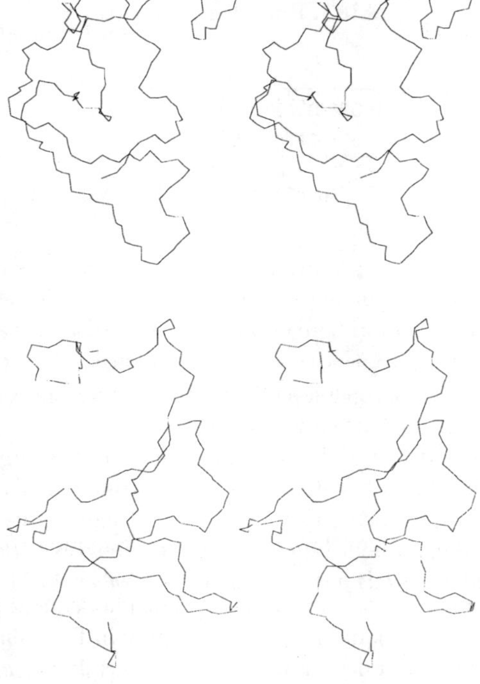

Fig. 5. Two Conformations generated by the genetic algorithm

Figure 5 shows two of the ten individuals in the last generation of the genetic algorithm. None of the ten individuals shows significant structural similarity to the native Crambin conformation. This can be confirmed by superimposing the generated structures with the native conformation. Table 2 shows the r.m.s. differences between all ten individuals and the native conformation. All values are in the range of 9 Å which rejects any significant structural similarity.

Individual	R.m.s.	Individual	R.m.s.
P1	10.1 Å	P6	10.3 Å
P2	9.74 Å	P7	9.45 Å
P3	9.15 Å	P8	10.2 Å
P4	10.1 Å	P9	9.37 Å
P5	9.95 Å	P10	8.84 Å

Table 2. R.m.s. deviations to native Crambin

Individual	E_{vdw}	E_{el}	E_{tor}	E_{pe}	E_{total}
P1	–14.9	–2434.5	74.1	75.2	–2336.5
P2	–2.9	–2431.6	76.3	77.4	–2320.8
P3	78.5	–2447.4	79.6	80.7	–2316.1
P4	–11.1	–2409.7	81.8	82.9	–2313.7
P5	83.0	–2440.6	84.1	85.2	–2308.5
P6	–12.3	–2403.8	86.1	87.2	–2303.7
P7	88.3	–2470.8	89.4	90.5	–2297.6
P8	–12.2	–2401.0	91.6	92.7	–2293.7
P9	93.7	–2404.5	94.8	95.9	–2289.1
P10	96.0	–2462.8	97.1	98.2	–2287.5
Crambin	–12.8	11.4	60.9	1.7	61.2

Table 3. Steric energies in the last generation

Although the genetic algorithm did not produce native-like conformations of Crambin, the generated backbone conformations could be those of a protein, i.e. they have no knots or unreasonably protruding extensions. The conformational results alone would indicate a complete failure of the genetic algorithm approach to conformational search, but let us have a look at the energies in the final generation (Table 3). All individuals have a much lower energy than native Crambin in the same force field. That means that the genetic algorithm actually achieved a substantial optimisation but that the current fitness function was no good indicator of "nativeness" of a conformation. For each individual the van der Waals energy (E_{vdW}), electrostatic energy (E_{el}), torsion energy (E_{tor}), pseudo entropic energy (E_{pe}) and the sum of all terms (E_{total}) is shown. For comparison the values for native Crambin in the same force field are listed.

It is obvious that all individuals generated by the genetic algorithm have a much higher electrostatic potential than native Crambin. There are three reasons for this.

1. Electrostatic interactions are able to contribute larger amounts of stabilising energy than any of the other fitness components.
2. Crambin has six partially charged residues that were not neutralised in this experiment.
3. The genetic algorithm favoured individuals with lowest total energy which in this case was most easily achieved by optimising electrostatic contributions.

The final generation of only ten individuals contained two fundamentally different families of structures (class 1: P1, P2, P4, P5, P6, P8, P9) and (class 2: P3, P7, P10). Members of one class have an r.m.s. deviation of about 2 Å among themselves but differ from members of the other class by about 9 Å.

Taking into account the small population size, the significant increase in total energy of the individuals generated by the GA and the fact that the final generation contained two substantially different classes of conformations with very similar energies, one is led to the conclusion that the search performance of the genetic algorithm was not that bad at all. What remains a problem is to find a better fitness function that actually guides the genetic algorithm to native-like conformations. As the only criterion currently known to determine native conformation is the free energy, the difficulty of this approach becomes obvious. One possible way to cope with the problem of inadequate fitness functions is to combine other heuristic criteria together with force field components in a multi-value vector fitness function. Before we turn to that approach let us first examine the performance of the current version for side chain placement.

2.6 Side Chain Placement

Crystallographers often face the problem of positioning the side chains of a protein when the primary structure and the conformation of the backbone are known. At present, there is no method that automatically does side chain placement with sufficiently high accuracy for routine practical use. Although the side chain placement problem is conceptually easier than *ab initio* tertiary structure prediction it is still too complex for analytical treatment.

The genetic algorithm approach as described above can be used for side chain placement. The torsion angles φ, ψ and ω simply have to be kept constant for a given backbone. Side chain placement by the genetic algorithm was done for Crambin. For each five residues, a superposition of the native and predicted conformation is shown in stereo projection graphs in Figure 6. A spatial superposition in stereoscopic wire frame diagrams is show for every five residues of Crambin and the corresponding fragment generated by a genetic algorithm. The amino acid sequence of Crambin in one letter code is TTCCP SIVAR SNFNV CRLPG TPEAI CATYT GCIII PGATC PGDYA N. As can be seen, the predictions agree quite well with the native conformation in most cases. The overall r.m.s. difference in this example is 1.86 Å. This is not as good as but comparable to the results from a simulated annealing approach [27] (1.65 Å) and a heuristic approach [25] (1.48 Å).

It must be emphasised that these runs were done without optimising either the force field parameters of the fitness function or the run time parameters of the genetic algorithm. From a more elaborate and fine-tuned experiment even better results might be expected.

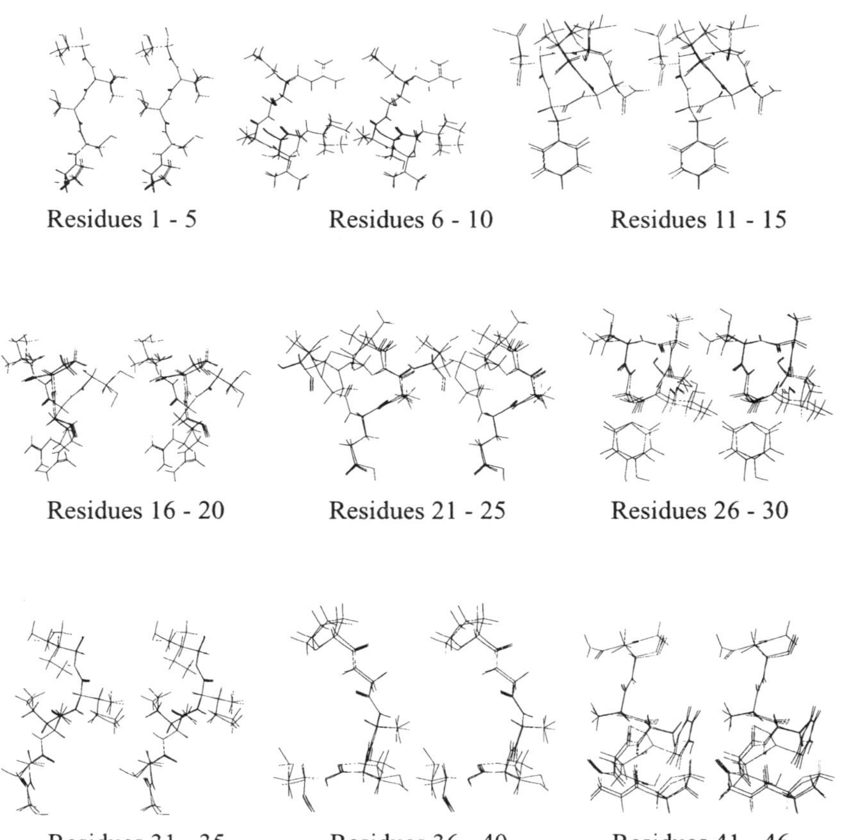

Fig. 6. Side chain placement results

3 Multiple-criteria Optimisation of Protein Conformations

In this section we will introduce additional fitness criteria for the protein folding application with genetic algorithms. The rationale is that more information about genuine protein conformations should improve the fitness function to guide the genetic algorithm towards native-like conformations. Some properties of protein conformations can be used as additional fitness components whereas others can be incorporated into genetic operators (e.g. constraints from the Ramachandran plot). For such an extended fitness function several incommensurable quantities will have to be combined: energy, preferred torsion angles, secondary structure propensities or distributions of polar and hydrophobic residues. This creates the problem of how to combine the different fitness contributions to arrive at the total fitness of a single individual. Simple summation of different components has the disadvantage that components with larger numbers would dominate the fitness function whether or not they are important or of any significance at all for a particular conformation. To cope with this difficulty individual weights for each of the components could be introduced. But this creates another problem. How should one determine useful values for these weights? As there is no general theory known for the proper weighting of each fitness component the only way is to try different combinations of values and evaluate them by their performance of a genetic algorithm on test proteins with known conformations. However, even for a small number of fitness components a large number of combinations of weights arises which requires as many test runs for evaluation. Also, "expensive" fitness components like the van der Waals energy need considerable computation time. In this work two measures were taken to deal with this situation:

1. Different fitness components are not arithmetically added to produce a single numerical fitness value but they are combined in a vector. This means that each fitness component is individually carried along the whole evaluation process and is always available explicitly.
2. Parallel processing is employed to evaluate all individuals of one generation in parallel. For populations of 20 to 60 individuals this gave a speed-up of about 20 fold compared to small single-processor workstations.

3.1 Vector Fitness Function

In this application two versions of a fitness function are used. One version is a scalar fitness function that calculates the r.m.s. deviation of a newly generated individual from the known conformation of the test protein. This geometric measure should guide the genetic algorithm directly to the desired solution but it is only available for proteins with a known conformation. R.m.s. deviation is calculated as follows:

$$r.m.s. = \sqrt{\sum_{i}^{N} \left(\left| \overline{u}_i - \overline{v}_i \right| \right)^2}$$

Here i is the index over all corresponding N atoms in the two structures to be compared, in this case the conformation of an individual ($\boldsymbol{u}_i$) in the current population and the known, actual structure ($\boldsymbol{v}_i$) of the test protein. The squares of the distances between the vectors $\boldsymbol{u}_i$ and $\boldsymbol{v}_i$ of corresponding atoms are summed and the square root is taken. The result is a measure of how much each atom in the individual deviates on average from its true position. R.m.s. values of 0–3 Å signify strong structural similarity; values of 4–6 Å denote weak structural similarity, whereas for small proteins r.m.s. values over 6 Å mean that probably not even the backbone folding pattern is similar in both conformations.

The other version of the fitness function is a vector of several fitness components which will be explained in the following paragraphs. This multi-value vector fitness function includes the following components:

$$fitness = \begin{pmatrix} r.m.s. \\ E_{tor} \\ E_{vdw} \\ E_{el} \\ E_{pe} \\ polar \\ hydro \\ scatter \\ solvent \\ Crippen \\ clash \end{pmatrix}$$

R.m.s. is the r.m.s. deviation as described above. It can only be calculated in test runs with the protein conformation known beforehand. For the multi-value vector fitness function this measure was calculated for each individual to see how close the genetic algorithm came to the known structure. In these runs, however, the r.m.s. measure was *not* used in the offspring selection process. Selection was done *only* based on the remaining eight fitness components and a Pareto selection algorithm which will be explained shortly.

E_{tor} is the torsion energy of a conformation based on the force field data of the Charmm force field v.21 with k and n as force field constants depending on the type of atom and ϕ as the torsion angle:

$$E_{tor} = |k_\phi| - k_\phi \cos(n\phi)$$

E_{vdw} is the van der Waals energy (also called Lennard-Jones potential) with A and B as force field constants depending on the type of atom and r as the distance between two atoms in one molecule. The indices i and j for the two atoms may not have identical values and each pair is counted only once:

$$E_{vdw} = \sum_{excl(i=j)} \left(\frac{A_{ij}}{r_{ij}^{12}} - \frac{B_{ij}}{r_{ij}^{6}} \right)$$

E_{el} is the electrostatic energy between two atoms with $q_{i,j}$ as the partial charges of the two atoms i and j and r as the distance between them:

$$E_{el} = \sum_{excl(i=j)} \frac{q_i q_j}{4\pi\varepsilon_0 r_{ij}}$$

E_{pe} is a measure to promote compact folding patterns. The expected diameter of a protein can be estimated by a number of techniques. A penalty energy term is then calculated as follows:

$$E_{pe} = 4^{(\text{actual diameter} - \text{expected diameter})}$$

Polar is a measure that favours polar residues on the protein surface but not in the core. Because all fitness contributions should be minimised a factor of minus one is required before the sum. The larger the distances of polar residues to the centre of the protein, the better a conformation and the more negative the value of *polar*. If residue i is one of k polar residues (any of Arg, Lys, Asn, Asp, Glu, or Gln) in a protein of length N residues and with $\mathbf{s}$ as the centre of gravity, then the polar fitness contribution is calculated as follows:

$$polar = \frac{-\sum_i^N |\vec{u}_i - s|}{k}$$

Hydro is a similar measure that favours hydrophobic residues (Ala, Val, Ile, Leu, Phe, Pro, Trp) in the core of a protein, whereas *scatter* promotes compact folds as it adds up the distances over all C_α atoms irrespective of amino acid type:

$$hydro = \frac{\sum_i^N |\vec{v}_i - s|}{k}, \quad scatter = \frac{\sum_i^N |\vec{v}_i - s|}{N}$$

Solvent is the solvent accessible surface of a conformation in Å^2. It is calculated by a surface triangulation method.

Crippen is an empirical, statistical potential developed by G. Crippen [28]. It is summed over all pairs of atoms that interact within a certain distance.

Clash is a term that counts the number of atomic collisions where any two atoms come closer than 3.8 Å to each other. This fitness term can be used to approximate the effect of the van der Waals energy at small distances but at only a fraction of the computational cost:

$$clash = \sum_{i=1}^{N} \sum_{j=i+1}^{N} overlap(i,j) \text{ with } overlap(i,j) = \begin{cases} 0 \text{ if } dist(i,j) \geq 3.8\text{Å} \\ 1 \text{ if } dist(i,j) < 3.8\text{Å} \end{cases}$$

3.1.1 Specialised Genetic Operators

LOCAL TWIST Operator. The LOCAL TWIST operator introduces local conformation changes by performing the ring closure algorithm for polymers [30] of N. Go and H. A. Scheraga for three consecutive amino acid residues. A stereoprojection of a portion of three residues and an alternative fold found by the LOCAL TWIST operator are shown in Figure 7. This algorithm was originally implemented in the RING.FOR program [30] in a general way that operated on six adjacent dihedral angles to bridge a gap with bonds of defined length and bond angle. The application of this algorithm for a polypeptide required translation of the program into the C programming language and some alterations to the program to account for the intermitting rigid ω torsion angle.

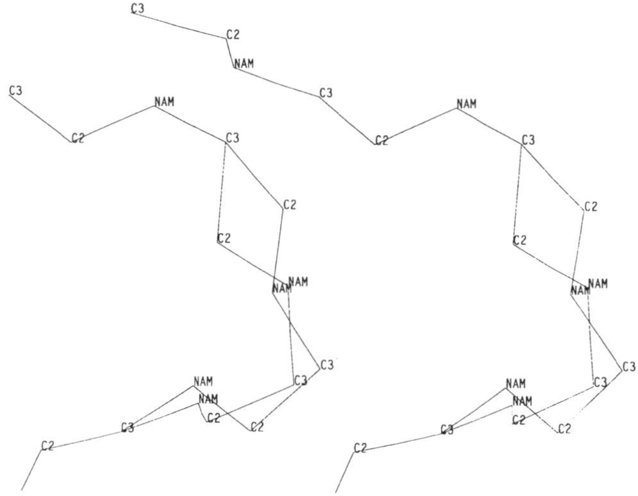

Fig. 7. Backbone conformation changed by LOCAL TWIST operator.

The basic concept of the ring closure algorithm is to find suitable values for ϕ_1 that satisfy the following equation:

$$g(\phi_1) = \mathbf{u}^{+}\mathbf{T}_{\alpha}\mathbf{R}_{\phi 1}\mathbf{T}_{\beta}\mathbf{R}_{\psi 1+\pi}\mathbf{T}_{\alpha}\mathbf{R}_{\phi 2}\mathbf{T}_{\beta}\mathbf{R}_{\psi 2+\pi}\mathbf{T}_{\alpha}\mathbf{e}_1 - \cos(\beta) = 0.$$

Here, $\mathbf{u}^{+}$ (transposed) and $\mathbf{e}_1$ are vectors and **T** and **R** are several translation and rotation matrices that define the constraints of a local conformation change. The angle β describes the rigid geometry of a peptide bond and ϕ_1 is the first backbone torsion angle in sequence to be modified. The search for suitable values of ϕ_1 involves repeated numerical approximations and is therefore rather time consuming. Hence, it was decided to distribute the LOCAL TWIST operator over several processors on a parallel computer [31] so that the calculations can be carried out in parallel for all individuals. In test runs with r.m.s.-deviation to the native conformation as the fitness function the LOCAL TWIST operator led to significant improvements in prediction accuracy and also to a substantial decrease in overall computation time.

Preferred Backbone Conformations The MUTATE operator of the previous section is rather crude because it always uses the left boundary of one of the ten most frequently occurring 10° intervals for a torsion angle. To improve the chance of selecting favourable values for the backbone torsion angles ϕ and ψ a cluster analysis with a modified nearest neighbour algorithm [32] was performed for the main chain torsion angles of 66 proteins:

1. Cluster all ϕ / ψ pairs for each amino acid until 21 cluster are formed.
2. Collect all clusters with less than ten pairs and add the centre of each cluster to the set of ϕ / ψ pairs to be used by the MUTATE operator.
3. Repeat the clustering procedure with only the ϕ / ψ pairs from the clusters with at least ten pairs in step 2 and let the clustering program run again until 21 clusters are formed. The centres of all new clusters complete the list of ϕ / ψ pairs that MUTATE uses when substituting individual torsion angles.

This algorithm first identifies small clusters with only a few examples in more detail and then clusters more densely populated areas with a finer resolution than a single clustering would do in one pass. Figure 8 shows the centres of 34 clusters for arginine. There are 14 small clusters of the first pass with less than ten pairs (shown as boxes) and 20 large clusters of the remaining pairs in the second pass (triangles).

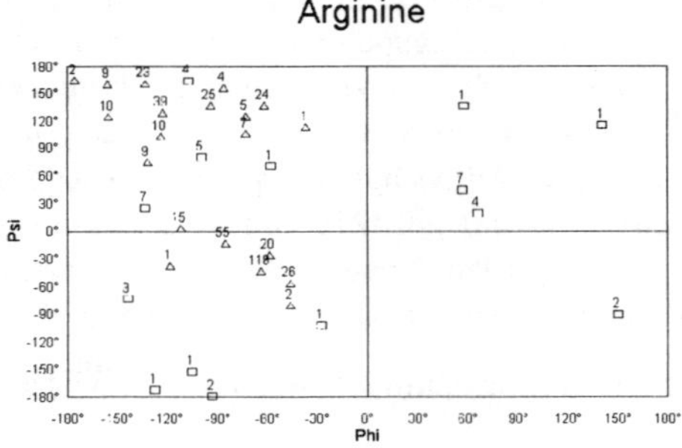

Fig. 8. 34 ϕ/ψ clusters for Arginine

Secondary Structure In addition to a more accurate selection of preferable main chain torsion angles predictions of secondary structure were used to reduce the search space. Two issues arise which must be considered:

1. Which secondary structure prediction algorithm should one rely on?
2. Which torsion angles should be used for the predicted secondary structures?

The first question was addressed by assembling a consensus prediction from two different methods: the PHD artificial neural network [33] and a statistical analysis which uses information theory [34], [35]. For the second question there are two alternate solutions. One alternative is to use torsion angles of idealised α-helices and β-strands, another is to constrain torsion angles of the predicted secondary structures to an interval that includes the conformation with idealised geometry. The corresponding torsion angles are shown in Table 4. ϕ_l, ψ_l and ϕ_u, ψ_u are lower and upper values of the main chain torsion angles in the respective secondary structure. ϕ_{exact} and ψ_{exact} are values for an idealised standard geometry. For β-strands the values are an average of parallel and antiparallel strands.

Secondary Structure	ϕ_l	ϕ_u	ψ_l	ψ_u	ϕ_{exact}	ψ_{exact}
α-helix (narrow interval)	−57°	−62°	−41°	−47°	−57°	−47°
α-helix (broad interval)	−30°	−120°	10°	−90°		
β-strand (narrow interval)	−119°	−139°	135°	113°	−130°	125°
β-strand (broad interval)	−50°	−180°	180°	80°		

Table 4. Boundaries for main chain torsion angles in secondary structures

3.1.2 Genetic Algorithm Performance

Using the genetic algorithm as described in the previous sections produced the following results. Figure 9 shows the best individual of the final generation of a run with a population of 30 individuals, the LOCAL TWIST operator in effect and r.m.s.-deviation as the only fitness component [36]. This conformation (solid line) with an r.m.s. deviation of 1.08 Å to native Crambin (dashed line) was obtained after 10,000 generations using the LOCAL TWIST, MUTATE, VARIATE and CROSSOVER operators and r.m.s. deviation as the fitness function. For Crambin, the final r.m.s. deviation of the conformation generated by the genetic algorithm is 1.08 Å, which is well within the range of the best resolution from X-ray or NMR structure elucidation experiments. Another run with the same parameters produced an individual with an r.m.s. deviation of 0.89 Å. This demonstrates the suitability of the genetic algorithm approach to protein folding. Given a reliable fitness function the genetic algorithm is able to successfully traverse the torsion angle search space.

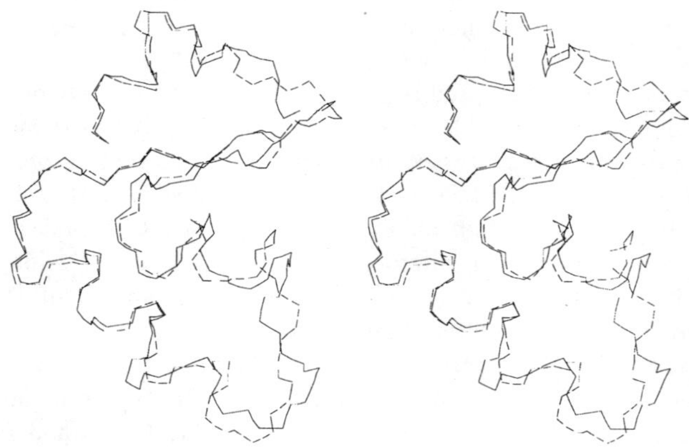

Fig. 9. Crambin predicted by r.m.s. fitness function

Other proteins that were used for test purposes of the genetic algorithm with an r.m.s. fitness function are the trypsin inhibitor protein (Brookhaven database code 5PTI; final r.m.s. deviation 1.48 Å; Figure 10) and RNAse T1 (Brookhaven database code 2RNT, final r.m.s. deviation 2.32 Å; Figure 11; native=dashed line).

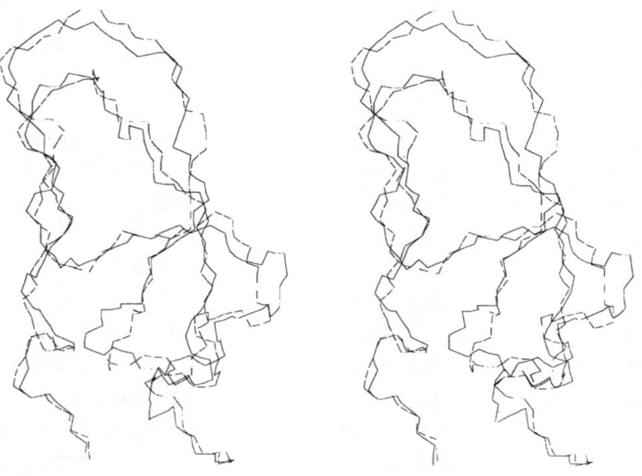

Fig. 10. Trypsin inhibitor predicted by r.m.s. fitness function

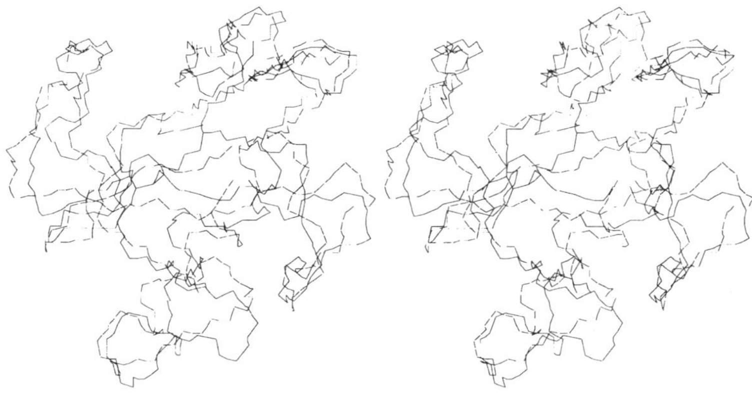

Fig. 11. RNAse T1 predicted by r.m.s. fitness function

The fact that none of the structures produced in the runs with an r.m.s. fitness function were completely identical to the native conformations is explained by the following three observations:

1. The use of standard binding geometries for reconstructing 3-D coordinates from a set of torsion angles can cause structural alterations where the native conformation does not closely adhere to the theoretically derived ideal bond lengths and bond angles. In this case the best match will always have an r.m.s. deviation of greater than zero.
2. The operators MUTATE, VARIATE and CROSSOVER in theory cannot produce an exact match even if the target structure is known in detail. This is a result of the representation formalism that these operators work on. If the current individual is already structurally similar to the desired protein then a single application of MUTATE or VARIATE is most likely to introduce mismatches of previously well fitting fragments and thus deteriorates the conformation. This happens because even if one bond becomes better aligned the rest of the protein towards the C-terminal swings away and increases the r.m.s. deviation. CROSSOVER is not able to improve this situation for the same reason.
3. Only the LOCAL TWIST operator can improve a fit locally without disturbing well fitting fragments that surround the mutation site. However, the applicability of LOCAL TWIST is mathematically constrained: when starting from a less fitting conformation the optimal local improvement is not always found in one pass. Sometimes it is even geometrically impossible to improve a local conformation at all.

Hence, with an increasing number of generations it becomes more and more difficult to achieve any further improvement in the r.m.s. fitness and the search stagnates at r.m.s.-deviation values between 0–2 Å (Figure 12). This graph shows the course of six single experiments with the r.m.s deviation as the fitness function. The individual with the best r.m.s. deviation is plotted for each generation. The two thicker lines at the bottom have the LOCAL TWIST operator switched on after 3000 generations. Reproduction was done by the roulette wheel algorithm. The four runs without LOCAL TWIST had a population size of 54 individuals whereas the two runs with LOCAL TWIST had only 30.

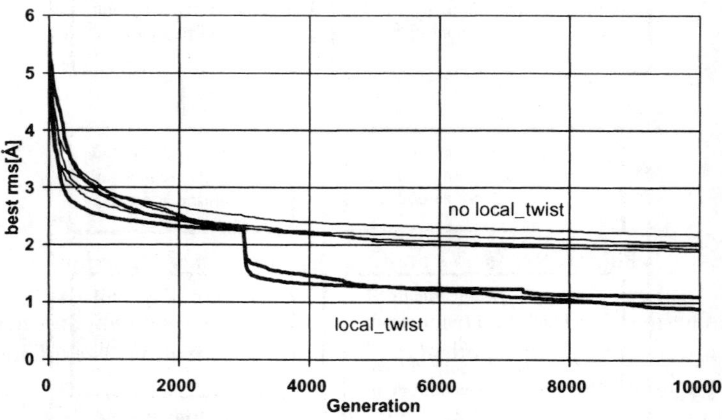

Fig. 12. Performance comparison for the LOCAL TWIST operator

Another conclusion to draw from the above experiments with the r.m.s. fitness function is that the fitness function is the crucial topic. This is clearly an unresolved issue and the subject of ongoing research in protein engineering. Some aspects of the computational complexity have already been explained above. This situation led to the following experiments with the genetic algorithm and a multi-value vector fitness function.

Figure 13 shows the results of a run with the fitness components *polar*, E_{pe}, E_{tor}, E_{el}, *hydro*, *Crippen* and *solvent*. This individual had an r.m.s. deviation of 6.27 Å from the native conformation of Crambin. The genetic algorithm did not use the r.m.s. deviation as part of the fitness function. Only the fitness components listed above were used to guide the genetic algorithm. Over the whole run some of the fitness components decreased along with r.m.s. deviation (E_{pe}, *hydro*, *Crippen*, *solvent*), as was expected. However, the other fitness components (*polar*, E_{tor}, E_{el}) actually drove the genetic algorithm to conformations with less similarity to the native Crambin indicating that these propensities were not good indicators for the "nativeness" of Crambin. In general, no r.m.s. values better than around 6 Å were detected in similar runs.

Fig. 13. Individual of the final generation of a multi-value fitness run

The following conformations were generated with the fitness components *Crippen*, *clash*, *hydro* and *scatter*. In addition, constraints on the secondary structures of Crambin were imposed by limiting the backbone angles to intervals between the upper and lower values of Table 4. Torsion angle ω was constrained to 180°. For a general application the use of secondary structure constraints requires a highly accurate and reliable secondary structure prediction algorithm which unfortunately does not (yet) exist. Figure 14 shows the backbone of an individual generated by the genetic algorithm with the above-mentioned fitness components and that has an r.m.s. deviation from native Crambin of 4.36 Å.

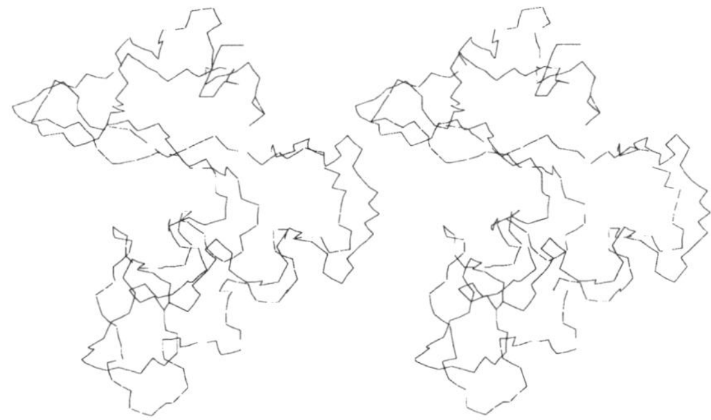

Fig. 14. Folding Crambin with secondary structure constraints

Another run with the same fitness components was performed for trypsin inhibitor (Figure 15). The r.m.s.-deviation from native trypsin inhibitor is 6.65 Å. This is worse than the result for Crambin in Figure 14 because the lower content of secondary structure in trypsin inhibitor implies less rigid constraints on the conformation. This means there are more degrees of freedom and therefore a larger search space to traverse.

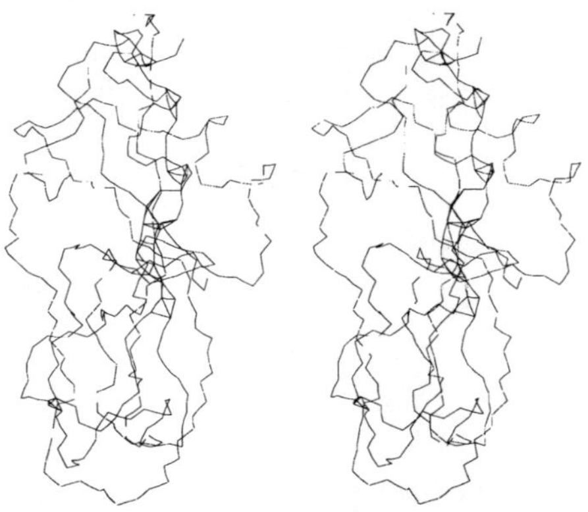

Fig. 15. Backbone folding of trypsin inhibitor

4 Summary

Summarising these findings and those of the previous subsections leads us to the following conclusions.

1. Genetic algorithms proved to be an efficient search tool for both 2-D and 3-D representations of proteins. In a 2-D protein model the genetic algorithm outperformed Monte Carlo search in both the quality of the results and required computation time (not shown here). For a 3-D protein model with a simple, additive force field as fitness function and using a rather small population the genetic algorithm produced *several* individuals (i.e. protein conformations) of dissimilar topology but each with highly optimised fitness values.
2. Given an appropriate fitness function (for test purposes the r.m.s deviation from the a priori known conformation can be used) the genetic algorithm application described in this chapter finds the desired solution within only small deviations.
3. The major problem lies in the fitness function. If there were one or a set of indices that return "1" for "*the object is (part of) a native protein conformation*" and "0" for "*the object is not (part of) a native protein conformation*" one could expect the genetic algorithm approach to deliver

reasonably accurate *ab initio* predictions. However, neither mathematical models, empirical, semi-empirical nor statistical force fields are yet accurate enough to reliably discriminate native from non-native conformations without additional constraints. Thus, the genetic algorithm produces (sub-)optimal conformations in a different sense than that of "nativeness".

4. Because secondary structure in nature and J. H. Holland's building blocks in the genetic algorithm are analogous fundamental components for the construction of the individual, it was hoped that secondary structures would emerge as the building blocks in a subset of the population. This has not yet happened. One possible explanation is that the fitness functions used are not sensitive enough to detect and account for the structural benefits in secondary structures.

References

1. Holland J.H. (1973) Genetic algorithms and the optimal allocations of trials. *SIAM Journal of Computing*, 2, 88-105.
2. Rechenberg I. (1973) Bioinik, Evolution und Optimierung. *Naturwissenschaftliche Rundschau*, 26, 465-472.
3. Koza J. (1993), *Genetic Programming*, MIT Press.
4. Kirkpatrick S., Gelatt C.D. Jr., Vecchi M.P. (1983) Optimization by simulated annealing. *Science*, 4598, 671-680.
5. Jones T., Forrest S. (1993) An introduction to SFI Echo, Santa Fe Institute, 1660 Old Pecos Trail, Suite A, Santa Fe NM 87501, e-mail: terry@sanatfe.edu, forrest@cs.unm.edu.
6. Holland J.H. (1993) Echoing emergence: Objectives, rough definitions and speculations for echo-class models. In: *Integrative Themes*, (G. Cowan, D. Pines, D. Melzner, eds.), Santa Fe Institute Studies in the Science of Complexity, Proc. Vol XIX, Addison-Wesley.
7. Holland J.H. (1992) *Adaptation in Natural and Artificial Systems*. 2nd Ed., MIT Press.
8. Goldberg D.E. (1989) *Genetic Algorithms in Search, Optimization & Machine Learning*. Addison-Wesley.
9. Schulz G.E., Schirmer R.H. (1979) *Principles of Protein Structure*. Springer Verlag.
10. Lesk A.M. (1991) *Protein Architecture - A Practical Approach*. IRL Press.
11. Branden C., Tooze J. (1991) *Introduction to Protein Structure*. Garland Publishing.
12. Schulze-Kremer S. (1992) Genetic algorithms for protein tertiary structure prediction. In: *Parallel Problem Solving from Nature II*, (R. Männer, B. Manderick, eds.), Springer-Verlag, 391-400.
13. For more information or to get in touch with researchers using genetic algorithms send an e-mail to one of the following mailing lists: ga-molecule@interval.com, ga-list-request@aic.nrl.navy.mil or to Melanie Mitchell at mm@santafe.edu who keeps an extensive bibliography on applications of genetic algorithms in chemistry.
14. Dandekar T., Argos P. (1992) Potential of genetic algorithms in protein folding and protein engineering simulations. *Protein Engineering*, 7, 637-645.
15. Le Grand S.M., Merz K.M. (1993) The application of the genetic algorithm to the minimization of potential energy functions. *The Journal of Global Optimization*, 3, 49-66.

16. Sun S. (1994) Reduced representation model of protein structure prediction: statistical potential and genetic algorithms. *Protein Science*, 5, 762-785.
17. Dandekar T., Argos P. (1994) Folding the main chain of small proteins with the genetic algorithm. *Journal of Molecular Biology*, 236, 844-861.
18. r.m.s. = root mean square deviation; two conformations are superimposed and the square root is calculated from the sum of the squares of the distances between corresponding atoms.
19. Bernstein F.C., Koetzle T.F., Williams G.J.B., Meyer E.F. Jr., Brice M.D., Rodgers J.R., Kennard O., Shimanouchi T., Tasumi M. (1997) The protein data bank: A computer-based archival file for macromolecular structures. *Journal of Molecular Biology*, 112, 535-542.
20. Vinter J.G., Davis A., Saunders M.R. (1987) Strategic approaches to drug design. An integrated software framework for molecular modelling. *Journal of Computer-Aided Molecular Design*, 1, 31-51.
21. Brooks B.R., Bruccoleri R.E., Olafson B.D., States D.J., Swaminathan S., Karplus M. (1983) Charmm: A program for macromolecular energy, minimization and dynamics Calculations. *Journal of Computational Chemistry*, 4, 187-217.
22. Ngo J.T., Marks J. (1992) Computational complexity of a problem in molecular-structure prediction. *Protein Engineering*, 5, 313-321.
23. Davis L. (ed.) (1991) *Handbook of Genetic Algorithms*. Van Nostrand Reinhold.
24. Lucasius C.B., Kateman G. (1989) Application of Genetic Algorithms to Chemometrics. *Proceedings 3rd International Conference on Genetic Algorithms* (J. D. Schaffer , ed.), Morgan Kaufmann, 170-176.
25. Tuffery P., Etchebest C., Hazout S., Lavery R. (1991) A new approach to the rapid determination of protein side chain conformations. *J. Biomol. Struct. Dyn.*, 8,. 1267-1289.
26. Hendrickson W.A., Teeter M M. (1981) Structure of the hydrophobic protein Crambin determined directly from the anomalous scattering of sulphur. *Nature*, 290, 107.
27. Lee C., Subbiah S. (1991) Prediction of protein side chain conformation by packing optimization. *Journal of Molecular Biology*, 217, 373-388.
28. Maiorov N.M., Crippen G.M. (1992) Contact potential that recognizes the correct folding of globular proteins. *Journal of Molecular Biology*, 227, 876-888.
29. Go N., Scheraga H.A. (1970) Ring closure and local conformational deformations of chain molecules. *Macromolecules*, 3, 178-187.
30. Quantum Chemical Exchange Program (QCEP) program No. QCMP 046.
31. Intel Paragon with 98 x i860 processors owned by Parallab, University of Bergen, Norway.
32. Lu S.Y., Fu K.S. (1978) A sentence-to-sentence clustering procedure for pattern analysis. *IEEE Transactions on Systems, Man and Cybernatics*, SMC, 8, 381-389.
33. Rost B., Sander C. (1993) Prediction of protein secondary structure at better than 70% accuracy. *Journal of Molecular Biology*, 232, 584-599.
34. Cármenes R.S., Freije J.P., Molina M.M., Martín J.M. (1989) PREDICT7, a program for protein structure prediction. *Biochem. Biophys. Res. Comm.*, 159, 687-693.
35. Garnier J., Osguthorpe D.J., Robson B. (1978) *Journal of Molecular Biology*, 120, 97-120.
36. This is only available for test runs with known protein conformations.

Evolutionary Generation of Regrasping Motion

Yasuhisa Hasegawa[1] and Toshio Fukuda[2]

[1] Department of Micro System Engineering
Nagoya University
Furo-cho, Chikusa-ku, Nagoya, 464-8603, Japan
E-mail: hasegawa@mein.nagoya-u.ac.jp

[2] Department of Micro System Engineering
Nagoya University
Furo-cho, Chikusa-ku, Nagoya, 464-8603, Japan
E-mail: fukuda@mein.nagoya-u.ac.jp

Summary. In this chapter, we propose a generation method for the regrasping motion of a four-fingered robot hand using evolutionary programming (EP). The control of multi-fingered robot hands has been the subject of recent interest. To generate a regrasping motion for an object, there are many parameters to be determined, such as the grasping points, grasping forces, regrasping phases, finger allocation and so on. It is difficult to optimize such manipulation parameters for achieving dexterous manipulation. The evolutionary optimization method is generally able to find optimal solutions without supervision after much iteration, which makes it almost impractical to apply it in a real environment directly. Therefore we apply the controller generated in numerical simulations to control a real four-fingered robot hand. We show the effectiveness of the proposed method for the regrasping motion in our experimental results.

1 Introduction

A robot hand with multiple degrees of freedom like a human hand has multiple functions to achieve dexterous manipulation in various fields compared with a conventional gripper type of robot hand with a few degrees of freedom. Various studies have been made to develop improved hands for prosthetic and teleoperator use. Skinner [1] Crossley [2] and Okada [3] designed multi-fingered hands to emulate human functions in the 1970's. Many previous studies reported on analysis and synthesis of grasping [4–7]. Many problems still remain in the driving mechanism, tactile sensor and control algorithm of the multi-fingered hand. In the control problem, there are two main problems: stable grasping for various shaped objects and manipulating the grasped object to a desired position. It is hard to determine the suitable grasping forces and motion in these problems [8–11].

Recently work has focused on the control algorithm for regrasping, which motion enables the manipulability of a robot hand to be extended. It can not only widely change the object position and posture to a desired pose by switching the grasping points, but also enhance the stability of the object grasped. To achieve the regrasping motion, a robot should have more than three fingers with a number of parameters to be determined: for example, the new grasping points for each finger, regrasping phases, grasping forces and finger allocation. The grasping forces are easily calculated, considering the balance of inner forces, if the grasping points and sum of grasping forces have been determined. The determination of the regrasping phase includes two kinds of problems: when the grasping finger switches to the free finger for regrasping and which fingers are used to grasp the object at the next grasping phase. All of them cannot be determined by mathematical analysis, because the planning of the regrasping motion has both a discrete property and a multidimensional search space with complex constraints.

We propose a method to automatically generate the grasping points and the regrasping phases using the evolutionary method, "Evolutionary Programming" (EP) [12]. EP can use numerical numbers for the individual and effectively find sub-optimal solutions from the multidimensional search space. Numerical simulation is used to search the desired parameters for the regrasping, and then they are applied to a real multi-fingered robot experimentally.

Fig. 1. Four-fingered robot hand

2 Generation of Regrasping Motion

In this chapter, we consider the task of rotating a cylindrical solid object from a horizontal posture to a vertical posture using the four-fingered robot hand shown in Figure 1 [13]. The rotation angle ϕ is changed from zero to 90 degrees according to the given task (Figure 2). To rotate the object, regrasping motions are required because the working area of each finger is limited. EP determines the initial grasping points and parameters of regrasping motion that define new grasping positions and regrasping phases. Based on such parameters defined by EP, finger allocation and grasping forces are calculated analytically. The grasping fingers are allocated, taking into consideration the distance between the fingertips and the grasping point and grasping forces are calculated, including the friction coefficient. The detail is given in section 3.

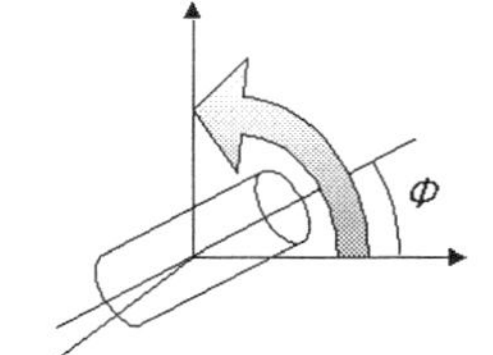

Fig. 2. The grasping concept

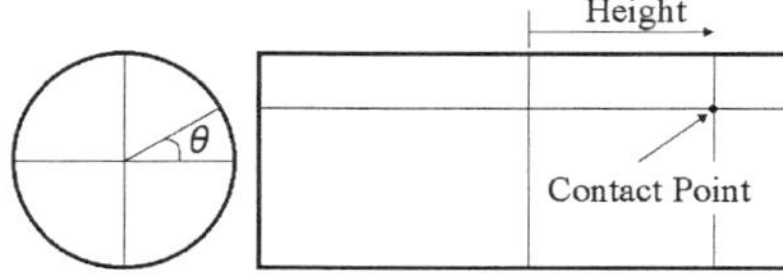

Fig. 3. Grasping point on the cylinder

The regrasping phase is represented as a rotation angle ϕ of the object from the initial state assuming that the motions of the fingers are quasi-static. The grasping position is represented by "height" h and "angle" θ on the surface of a cylinder as defined in Figure 3. The coding of an individual is shown in Figure 4. The first six loci represent the initial grasping positions of the three fingers and the following three loci represent the regrasping information; two of the three loci are for a new grasping point and one locus is for the object posture (rotation angle). The number of regrasping motions is varied by adjusting the number of regrasping information units. Generally the individual is encoded based on the joint angles of each finger, but this tends to become too complicated and the search area expands. These grasping points do not indicate which finger supports a certain grasping point. In this method, the individual has information only about the grasping object, grasping points and rotation angle, so the gene size is shorter than in ordinary approaches and the gene is independent of the hand configurations. The flowchart of the proposed algorithm is shown in Figure 5.

The evaluation function for each individual is defined in Eq.1.

$$E = w_1 \sum_{j}^{3} (f_j - f_{avg})^2 + w_2 \sum_{j}^{4} \sum_{i}^{3} |\tau_{ji}| + w_3 \sum_{k} p_{\alpha} \qquad (1)$$

where f_j is the scalar value of the force vector of grasping finger j, and f_{avg} is its mean. The first term prevents each grasping force from being excessive, and each finger from being allocated close to each other. τ_{ji} is the i-th motor torque of the j-th finger, and the second term is for avoiding singularity and energy efficiency. The penalty, p_{α}, is a constant value to be evaluated if the

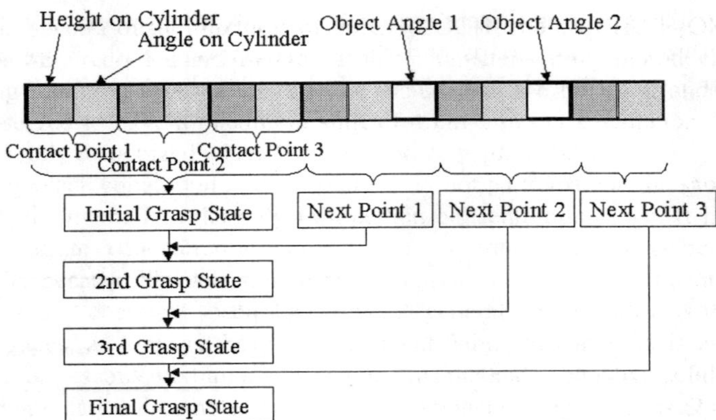

Fig. 4. Individual and algorithm flow

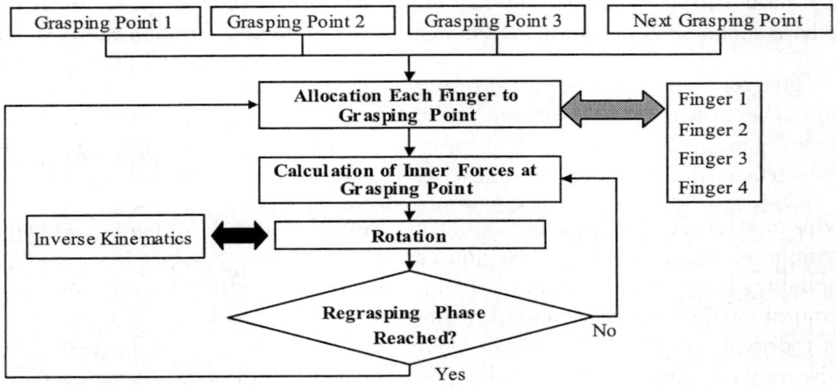

Fig. 5. Grasping algorithm

candidate violates the manipulation constraints. w_i is the weight coefficient of each item.

We use the fitness function Eq.2 for EP.

$$f = \frac{1}{1+E} \tag{2}$$

The offspring, n, are selected by tournament selection and a mutation operator generates a new set, n, of offspring, adding the Gaussian random variable N, according to its fitness.

$$P_{i,n+j} = p_{i,j} + N\left(0, a\frac{f_{max} - f_i}{f_{max} - f_{min}} + b\right) \tag{3}$$

where $p_{i,j}$ denotes a component (grasping information) i of individual j, n denotes the population size, f_i, f_{max} and f_{min} denote the fitness values of individual i, maximum fitness and minimum fitness, respectively, and $N(k,l)$ denotes Gaussian noise which has mean value k and variance l.

3 Grasping Conditions

The necessary and sufficient conditions for grasping an object by three fingers are summarized as follows [8]:

C1) The resultant force of both the fingertip forces and the gravity must be zero.
C2) The resultant moment of both the fingertip forces and the gravity must be zero.
C3) Each vector of the fingertip force must be inside the friction cone.

Initially, the grasping plane of the three grasping points is as shown in Figure 6. In this figure, C_i is the grasping point of the i-th finger, n_i is the inner normal unit vector at Ci, f_i is the fingertip force vector to the object at C_i, and e_i is the unit vector of f_i. P is the intersection point of the perpendicular center of gravity of the object and the grasping plane. Using these parameters, the condition C1) corresponds to Eq.4.

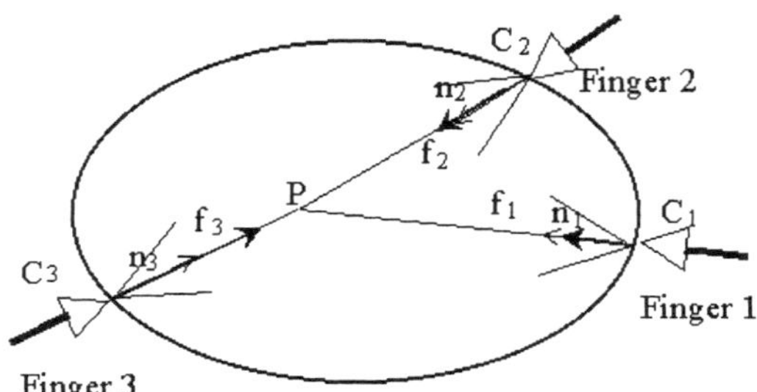

Fig. 6. Grasping plane

$$(e_1, e_2, e_3) \in F \tag{4}$$

where F is the combination of unit vector (e_1, e_2, e_3), which satisfies Eq.5.

$$e_1^T(e_2 + e_3) \leq 0, \quad e_2^T(e_3 + e_1) \leq 0, \quad e_3^T(e_1 + e_2) \leq 0 \tag{5}$$

The condition C2) is that the extended lines of each vector f_i meet at a point P. The condition C3) is shown as Eq.6.

$$e_i^T n_i > \frac{1}{\sqrt{1+\tan{\alpha_i}^2}} \tag{6}$$

where α_i is the angle of the friction cone that is defined in Eq.6, and is the static friction coefficient at the grasping point. If all the grasping forces are in the friction cone, there is no slippage on the surface of the object.

Fingers are allocated to each grasping point or regrasping point so that the sum of the distance between each fingertip and the corresponding grasping points or the regrasping point should be minimized as follows:

$$\min_j \sum_{k=1}^{4} (x_k - b_j)^2 \tag{7}$$

where b_j is the root position of the j-th finger and x_k is the k-th grasping position.

4 Simulations

As the first step, the proposed method with EP generates the regrasping motion in which the four-fingered robot hand rotates the solid cylindrical object from a horizontal posture to a vertical posture in a numerical simulation. The generated motions are applied to control the real robot hand.

The object shape and the parameters of the robot hand used in the numerical simulation are the same as those in the real environment, such as the motor torque limitation and the link length, so that the obtained solution should be applicable to the real robot hand. The information given in the numerical simulation includes the dimensions of the object. the rotational direction, the final and initial posture of the object, the sum of the grasping forces and the regrasping times. Furthermore we assume that three out of the four fingers are used to grasp and the remaining finger just touches the object without any grasping force as the grasping condition. We consider the task of rotating the cylindrical object (radius 30mm, height 150mm) by 90 degrees. The total grasping force of each finger is 3.0[N] and the maximum limit torque of each actuator is 5.37[mN m]. The coefficient of friction μ_i is defined as 0.577 ($\alpha = \pi/6$). This corresponds to the friction between rubber and steel: 0.4< $\mu_{(rubber-steel)}$ <0.6. The grasping and the change of finger are static. The regrasping number is three, including the final state when the object is rotated by 90 degrees.

The robot hand has four fingers and four joints on each finger. Each finger is equipped with three motors, and the motion of the two joints at the fingertip is coupled like a human finger. When a hand with solid fingers grasps a solid object with point contacts, three grasping points are generally required for stable grasping. This hand has four fingers so it can grasp an object with three fingers and achieve a regrasping motion by switching the grasping finger with the remaining finger. The mechanical constraints of each joint are as follows:

$$\begin{aligned} -\pi/4 < \theta_1 &< \pi/4 \\ -\pi/2 < \theta_2 &< \pi/2 \\ 0 < \theta_3 &< (\pi/4)/k \\ 0 < \theta_4 &< \pi/2 \end{aligned} \tag{8}$$

where θ_i is the angle of the i-th joint from the base. The joint interlock ratio of the two joints at the fingertip k, is 1.6.

Table 1. Constraints and evaluation

Constraints	Inner Force (Friction Cone, α_i)	Every Step
	Inverse Kinematics (Angle Limitation, θ_i)	Every Step
	Torque Limitation (τ_{ji})	Every Step
Evaluation	Tip Finger Force (f_i)	Every Step
	Motor Torque (τ_{ji})	Every Step

Table 2. Parameters of EP

Population	Candidate Population (n)	Generations	Individual Length	Variance Coefficients (a, b)
200	100	1000	14	100, 0.01

The constraints given in Table 1 are checked at every step, that is one degree of rotation angle. If the constraints are violated, the penalty is added to the evaluation value. Each parameter in the EP is shown in Table 2.

Figure 7 shows a sequence of the regrasping phase obtained from the initial state (horizontal posture) to the final state (vertical posture). The vertical lines in the figure indicate the regrasping phase when one grasping finger out of three switches to the free finger. Table 3 shows the allocated finger in each regrasping phase. The underlined fingers in Table 3 indicate the changed fingers before and after the regrasping phase. According to these results, fingers 1, 3 and 4 are grasping the object at the first step, then at the rotation angle of 29.3 degrees finger 1 is changed to finger 2, and fingers 2, 3 and 4 are grasping in the second step. Fingers 1, 3 and 4 are used again for grasping in the third step from 51.87 to 90 degrees of rotation angle. Finally fingers 2, 3 and 4 are grasping the object. Figure 8 shows that the transitions of each joint angle, the fingertip forces whose directions are defined by the global coordination system shown in Figure 9, and the torque of each motor. The generated motion is shown in Figure 9. The simulation shows that the cylindrical object is successfully manipulated with regrasping by the four-fingered robot hand.

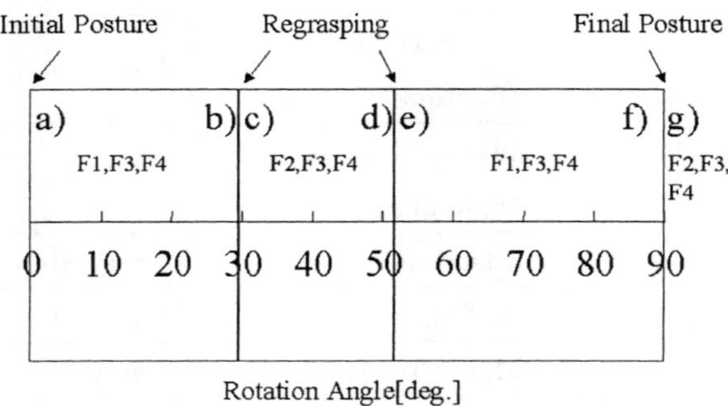

Fig. 7. Regrasping phase generated by EP ((a)–(g) correspond to the simulation results in Figure 9).

Table 3. Grasping phase and fingers

Grasping Phase	Grasping Fingers
a) – b)	Finger 1, Finger 3, Finger 4
c) – d)	Finger 2, Finger 3, Finger 4
e) – f)	Finger 1, Finger 3, Finger 4
g)	Finger 2, Finger 3, Finger 4

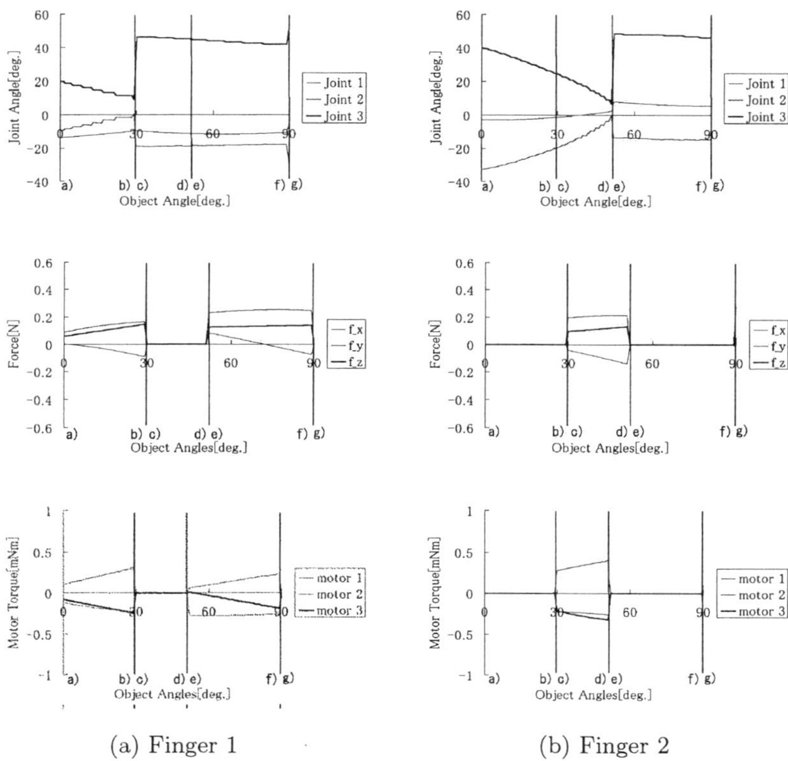

(a) Finger 1 (b) Finger 2

Fig. 8. Transitions of joint angle, fingertip force and motor torque

5 Experiments

We apply the motion generated by the numerical simulation to the real robot hand to confirm the feasibility of the proposed method. The desired grasping positions, the desired grasping forces and the regrasping phase obtained by the numerical simulation are programmed into the real robot. The block diagram for the control is shown in Figure 10. x_d is the desired positions of the fingertips, f_d is the desired grasping forces, q_d and q are the desired and current motor angles respectively, τ_d and τ are the desired and current motor torques respectively, and K_q is the compliance coefficient. Each joint is controlled by a compliance control algorithm in order to absorb modelling errors in the numerical simulation and positioning errors at the initial state. The sampling frequency is 500Hz.

The experimental results are shown in Figure 11 and Figure 12. These results demonstrate the effectiveness of the proposed algorithm.

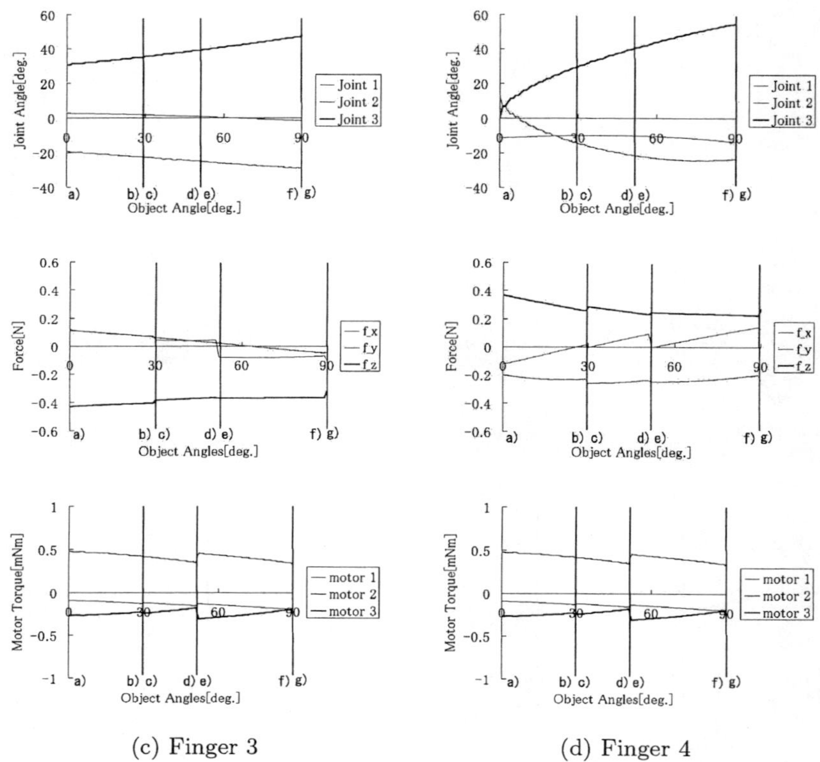

(c) Finger 3 (d) Finger 4

Fig. 8. Transitions of joint angle, fingertip force and motor torque

6 Conclusions

In this chapter, we have proposed a generation method for the regrasping motion of a multi-fingered robot hand. In this method, EP determines the grasping points, regrasping points, regrasping phase including grasping stability. We applied the optimal control parameters obtained in numerical simulations in order to control the four-fingered robot hand. We showed the effectiveness of the proposed generation method for the regrasping motion.

As for future work, we shall address the following three problems:

1 Real-time planning
2 Rolling contact
3 Object shape

For the first one, as EP requires much searching time to find solutions, an effective knowledge database will become key by storing and reusing past parameters for manipulation. The second problem concerns the rolling contact

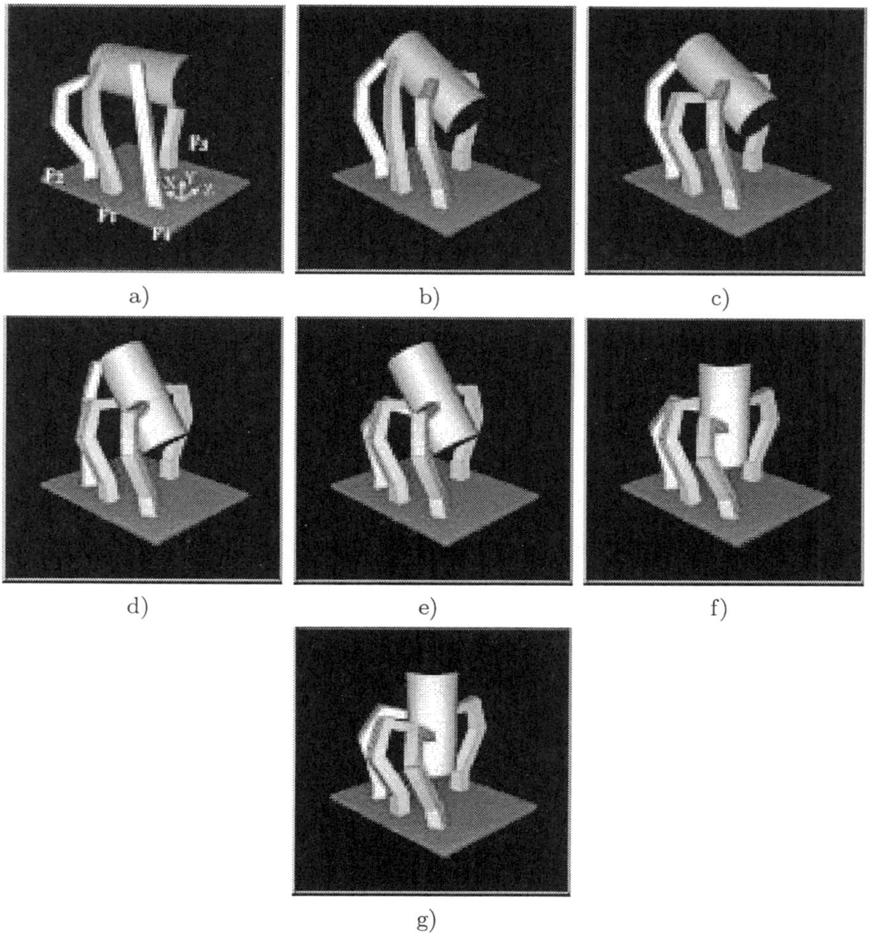

a) b) c)

d) e) f)

g)

Fig. 9. Transitions of grasping states ((a)–(g) correspond to the simulation results in Figure 7).

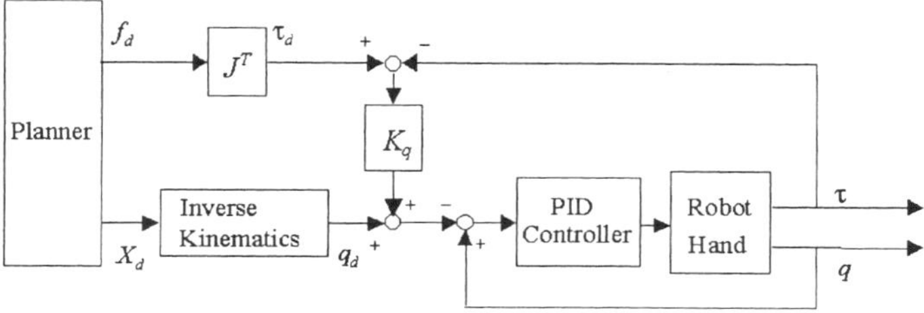

Fig. 10. Control block diagram

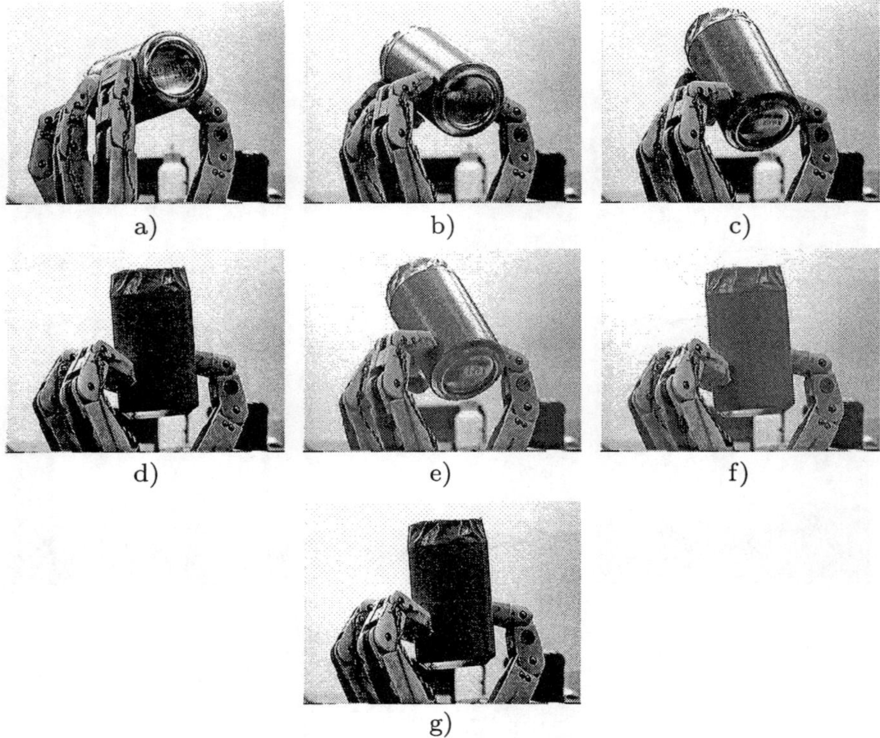

Fig. 11. Experiment of manipulation ((a)–(g) correspond to the simulation results in Figure 9)

of the fingertips. This is a nonholonomic constraint which is not considered in this chapter. It becomes more important if the rotation angle is larger during a single grasping form. For the third problem, a precise model of the grasped object is needed for a numerical simulation. The adaptive grasping of an imprecise model based on tactile and force information is preferable for work in the real world. Finally, we would also like to address hardware issues such as force sensors and tactile sensors as well as these planning and control problems.

References

1. Skinner, F. (1975) Designing a Multiple Prehension Manipulator. Mechanical Engineering
2. Crossley, F. R. E., Umholtz, F. G. (1977) Design for Three Fingered Hand. Mechanism and Machine Theory, **12**
3. Okada, T. (1979) Object-Handling System for Manual Industry. IEEE Trans. Systems, Man and Cybernetics, **9**, 79-89

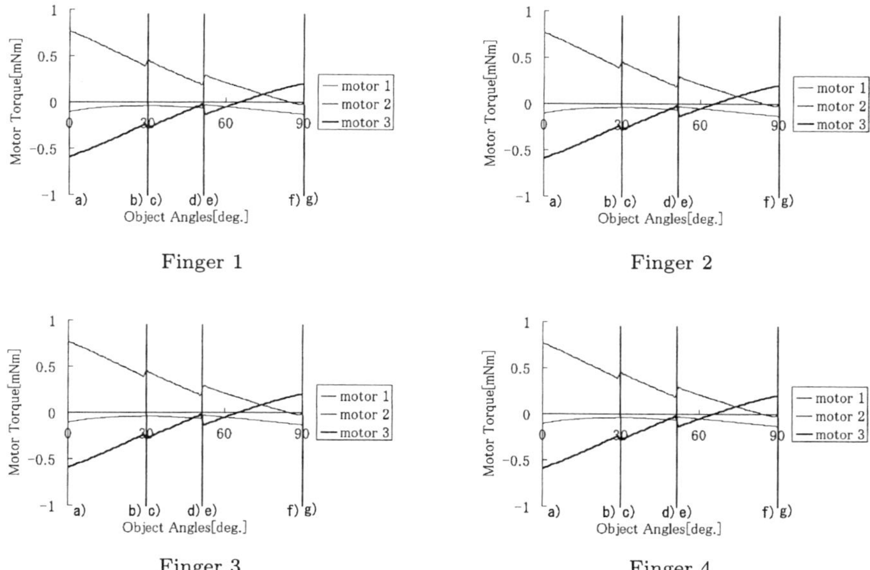

Fig. 12. Experiment results(fingertip forces)

4. Mason, M., Salisbury, J. K. (1985) Robot Hands and the Mechanics of Manipulation. The MIT Press
5. Kerr, J., Roth, B. (1986) Analysis of Multi-fingered Hands. Int. Journal of Robotics Research, **4**, 3-17
6. Nguyen, V. (1988) Constructing Force-Closure Grasps. Int. Journal of Robotics Research, **7**, 17-31
7. Bicchi, A. (1995) On the Closure Properties of Robotic Grasping. Int. Journal of Robotics Research, **14**, 319-334
8. Yoshikawa, T., Nagai, K. (1991) Manipulating and Grasping Force in Manipulation by Multifingered Robot Hand. IEEE Trans. on Robotics and Automation, **7**, 67-77
9. Yoshikawa, T., Yokokohji, Y., Nagayama, A. (1995) Object Handling by Three-Fingered Hands Using Slip Motion. Proc. of 1993 IEEE/RSJ Int. Conf. on Intelligent Robots and Systems, 99-105
10. Kaneko, M., Hino, Y., Tsuji, T. (1997) On Three Phases for Archiving Enveloping Grasp Inspired by Human Grasping. Proc. of IEEE Int. Conf. on Robotics and Automation, 385-390
11. Sudsang, A., Ponse, J. (1995) New Technique for Computing Four-Finger Force-Closure Grasp of Polyhedral Objects. Proc. of IEEE Int. Conf. on Robotics and Automation, 1355-1360
12. Erkman, A. M., Duran, M. (1998) Genetic Algorithm-based Optimal Regrasping with the Another Robot 5-fingered Robot Hand. Proc. of Int. Conf. on Robotics and Automation, 3329-3334
13. Fukuda, T., Mase, K., Arai, F. (1998) The Design and Development of a Four-Fingered Robot Hand (Adjustment of Grasping Position by Using Slip

Motion on Passive Closure). Proc. of Int. Conf. on Intelligent Robots and Systems, 482-487

Recent Trends in Learning Classifier Systems Research

Pier Luca Lanzi[1] and Rick L. Riolo[2]

[1] Artificial Intelligence & Robotics Laboratory
Dipartimento di Elettronica e Informazione
Politecnico di Milano, Milan, Italy
E-mail: pierluca.lanzi@elet.polimi.it

[2] Center for Study of Complex Systems
University of Michigan
Ann Arbor, MI 48109, USA
E-mail: rlriolo@umich.edu

Summary. In this chapter we review recent advances and trends in learning classifier systems (LCS) research. These advances fall in three main areas: (i) improved allocation and use of credit assigned to rules, which stems in part from utilizing connections with well-established reinforcement learning algorithms, and from using rule predictive accuracy as the "fitness" value guiding the genetic algorithm's search for better rules; (ii) research on alternative LCS architectures, including alternative rule syntax and semantics, as well as work on both simpler and more complex LCS; and (iii) increases in both the number and the range of LCS applications. We feel these advances have led to the resurgence of LCS research in the past five years, and in a final section we list some of the most immediate challenges facing LCS researchers at this time.

1 Introduction

Learning classifier systems (LCS) are a machine learning paradigm introduced by John H. Holland. They made their first appearance in 1978 in the paper "Cognitive Systems Based on Adaptive Algorithms" by Holland and Reitmann [92], although their coming was already fore-shadowed in 1971 with the paper "Processing and Processors for Schemata" [83], and in the *broadcast language* described in Holland's ground-breaking book, *Adaptation in Natural and Artificial Systems* [90].

LCS have been around for more than twenty years. While there was considerable LCS research started in the 1980's, the field began to wane a bit as the decade closed, probably in part because people began to realize how complicated LCS are. After all, recall that LCS include *both* an allocation of credit algorithm *and* an evolutionary (genetic) algorithm, both of which have large groups of researchers working hard to understand them as stand-alone algorithms, let alone understanding them in the context of each being applied to the same production (rule-based) system!

However, there has been a great resurgence in LCS research in the last ten years, and especially in the last five years: if we have a quick look at Kovacs' on-line LCS bibliography [103] we find an average of 5 papers/year in 1984–1988, and then 31 papers/year in 1989–1994 and 43 papers/year in 1995–2000. Besides a huge increase in the number of LCS research reports and papers in general GA-related proceedings (e.g., [8,10,112]), there have also been three workshops (IWLCS92 [4], IWLCS99 [203], IWLCS2000 [5]), the forthcoming IWLCS2001 [1], a book [120], and an upcoming special issue of *Soft Computing* devoted to LCS research. In addition, there are an increasing number of researchers and active LCS research groups, e.g., the Learning Classifier Systems Group (LSCG) at the University of the West of England (UWE) offers graduate-level courses focusing on LCS. To learn more about all the different LCS activities we refer the reader to local web sites: Alwyn Barry's LCS Web [9], UWE LCSG web site [2], Wilson's home page [201] and the forthcoming web site devoted to LCS resources: `www.learning-classifier-systems.org`.

In this chapter we describe some of the recent trends and advances in LCS research. We believe these advances have both led to and been stimulated further by the increased LCS research activity in recent years. Of course, in this short chapter we can consider only a small portion of the work being done. Thus we will focus on what we consider to be the most important issues and ideas being addressed in recent years. For further discussions of these and other LCS research trends and advances, we refer the reader to the most recent and comprehensive book devoted to LCS research: *Learning Classifier Systems: From Foundations to Applications* [120].

While there are many ways one might categorize the recent advances and trends, the rest of this chapter is organized as follows. Section 2 gives a brief overview of Holland's classifier system. Section 3 addresses research related to the *allocation and use of credit*, describing (i) the connections between LCS and reinforcement learning and (ii) the introduction of *accuracy*-based fitness. Section 4 addresses recent research into "architectural" aspects of LCS, including (i) alternative rule representation schemes, (ii) simplifications of Holland's "standard" model [87,93], and (iii) the most recent re-emergence of more *complex* LCS, aimed at harder problems. Section 5 briefly describes recent applications in which LCS (i) are used as a machine learning tool, or (ii) serve as a model of some "real-world" complex adaptive system. Section 6 ends the chapter with a brief list of what we consider to be the most immediate challenges to current LCS research.

Before we proceed any further, there are two points to note. First, although LCS are briefly introduced in Section 2, for the sake of brevity we still assume that the reader is familiar with the basic structure and function of Holland-style LCS. For readers who need to (re-)learn the basics about LCS, we recommend either Goldberg's text [71], Holland's article [87] or book [93], or the recent book [120]. Second, in this chapter, we focus on "Michigan" style LCS, i.e., systems in which the genetic algorithm (GA) is applied to a population consisting of individual, "non-fuzzy" rules. We take this approach in part because the recent resurgence of interest in LCS has concentrated on Michigan-style classifier systems. In addition, there already are reviews of other approaches to LCS which are much more comprehensive than we could present in this single chapter. For a review of recent research with Pittsburgh-style LCS, we refer the interested reader to [129]. Bonarini [15] provides a fairly comprehensive overview of the recent advances in *fuzzy* classifier systems. Finally, for a description of one hybrid Michigan/Pittsburgh approach, we refer readers to SAMUEL's bibliography [75–80].

2 Holland's LCS at a Glance

In this section we briefly describe the "traditional" Holland-style LCS as described in [87] (see also the descriptions provided in [159,52]). However, it is important to remember that there have been *many* variants on this basic form, many suggested by Holland himself. Some of these variants are described in later sections of this chapter; for more, see [120].

LCS were introduced by John H. Holland in 1978 [92,87] as a way of *learning through evolution* by trial and error interactions with a possibly unknown environment. LCS continually interact with the environment through their *detectors* and their *effectors*. The systems sense the environment through their *detectors*, and based on their sensations select an action to perform in the environment through their *effectors*. Depending on the efficacy of the systems' actions, the environment may eventually *reward* the systems. The goal of the systems is to *maximize* the amount of reward they receive from the environment *in the long run*.

To accomplish this goal, LCS maintain a set (a *population* or *base*) of rules called *classifiers*. A classifier is a condition-action rule associated with a parameter, called *strength*, which is an estimate of "how good" the classifier is. The classifier condition is a string of fixed length on the alphabet {0,1,#} which defines a set of messages (over the possible sensory inputs) which match the condition. The "#" symbol is called *don't care* and indicates that the bit can match either a 1 or a 0. The classifier action is a string of fixed length on the alphabet {0,1}.[1] For example, `0#10:01` represents a classifier with condition `0#10` and action `01` that match both the input configurations `0010` and `0110`.

In LCS the learning is achieved by manipulating the *population* of classifiers using three subsystems: the *performance* component which is responsible for the short-term behavior of the system; the *credit assignment* (or *allocation of credit*) component which defines how the reward coming from the

[1] According to the original definition of classifier system [87], a classifier condition can be built by putting more conditions together, joined by logical AND(s). Here we consider only classifiers with one condition for the sake of simplicity.

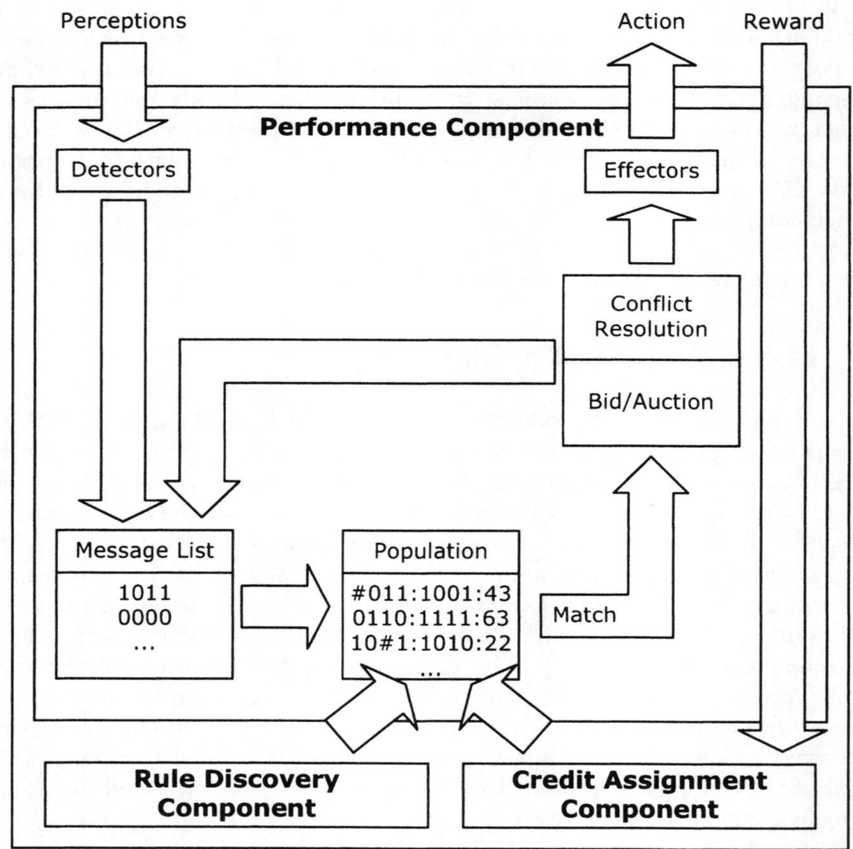

Fig. 1. Holland's LCS

environment is distributed among classifiers; finally, the *rule discovery* component which is responsible for the generation of new, and possibly better, classifiers by the recombination and mutation of high–strength individuals in the population.

2.1 The Performance Component

The general architecture of a traditional classifier system is depicted in Figure 1. At each time step the messages from the input interface (the detectors) are inserted in the internal *message list*. Then each message in the message list is compared to the condition part of each classifier in the population. All the classifiers in the population whose condition matches a message in the message list compete against each other in order to be allowed to post their messages in the message list. Classifiers matching a message in the message list make a *bid* to become active. Only classifiers that make the highest bid are allowed to post their messages. The bid of a classifier depends on the classifier strength and on the classifier *specificity* computed as the number of 0's and 1's in the condition. A classifier's bid is computed as

$$B = k \times strength \times specificity$$

where k is a constant for all classifiers. The new messages are sent to the output interface (the effectors) to be executed. A conflict *resolution module* arbitrates conflicts among rules which match current sensory inputs but advocate contradictory actions. Depending on the effect of the action a scalar reward is returned to the system.

Note that the internal message list is a form of internal memory that the agent can use to build chains of actions. However, this appears to be difficult to achieve [142]; see Section 4.3 for further discussion.

2.2 Credit Assignment

The credit assignment procedure distributes the reward received from the environment among classifiers in the population that are responsible for producing the reward. In Holland's classifier systems this is done by adjusting the strength of classifiers using the *bucket brigade* algorithm [87].[2] The bucket brigade acts in two ways. When a reward is received at time step t the bucket brigade adds the reward value to the strength of all classifiers active during that step. When a classifier is activated it pays the amount of its bid to the classifiers that made it possible to become active. In summary, the strength $S(t+1)$ of a classifier at time $t+1$ is

$$S(t+1) = S(t) + R(t) + P(t) - B(t)$$

where $S(t)$ is the strength of the classifier at time t; $R(t)$ is the reward received at time t; $P(t)$ is the sum of all the payments made by those classifiers that matched messages produced by this classifier during the previous time step; $B(t)$ is the bid of the classifier.

2.3 Rule Discovery

The basic rule discovery component in LCS is a GA, in which the LCS rule set is treated as the GA population. The rule discovery component generates new classifiers by the recombination (and possible mutation) of offspring of high-strength individuals. It selects classifiers in the population with probability

[2] Note that bucket brigade is not the only available credit allocation algorithm for LCS but probably just the best known. See Section 3 for an overview of all the alternative credit allocation algorithms.

proportional to their strengths; next it copies the selected classifiers; then it applies genetic operators to the offspring classifiers. Typical operators used during rule discovery are *crossover* and *mutation*.

Crossover combines classifiers as in sexual reproduction. It takes two classifiers as bit-strings, selects a cutting point for the two strings, cuts the two strings, and recombines pieces of opposite strings. For example, given two classifiers, `0#10:10` and `100#:01`, suppose that crossover cuts the strings at the third bit; the resulting classifiers are `0#1#:01` and `1000:10`. Mutation changes the values of bits in a classifiers with a certain probability.

New classifiers created by genetic operators are inserted in the population, while low-strength classifiers are deleted to maintain a constant population size. New classifiers are assigned strengths in any of a number of different ways, e.g., the population average strength or the average strength of the new classifiers' parent(s).

3 Allocation of Credit

Determining which rules in an LCS population are the most "useful" ones, i.e., which rules best lead to the system's overall goals, is important for two reasons. First, through the bid competition mechanism the more useful rules will more likely be allowed to carry out their action-parts and thereby control the system's behavior. Second, through the GA and other rule creation heuristics, the more useful rules will be used to bias the creation (and deletion) of rules. Assigning usefulness ratings to rules is carried out by the system's *allocation of credit* mechanism.

As briefly described in the previous section, in traditional Holland LCS [87], allocation of credit was carried out by the *Bucket Brigade Algorithm* (BBA), a kind of temporal difference learning [172] which assigns a *strength* value to each rule.[3]

In all of the early LCS, that *single* strength value in each rule was used both to *control* which rules are allowed to become active and so determine the system's behavior, and to *measure* the "fitness" used by the GA and so bias the *search* for better rules.

It is well known that Holland's model exhibits a number of problems which sometimes prevent the system from achieving satisfactory performance [202]. In the past ten years many improvements to Holland's framework were suggested which, in some cases, consisted in a modified credit assignment procedure (e.g., [124]) or, in other cases, in a modified discovery component (e.g., [48]), or in a modification of both (e.g., [196]).

It has long been recognized that when discussing the behavior of LCS it is difficult to separate the allocation of credit from the discovery component. For instance, Riolo [142] showed that to perform sequential behavior a classifier

[3] See also the work on the *profit sharing plan* algorithm [92,74,164].

system should allocate *strength* properly while preventing the proliferation of high-rewarded classifiers. Furthermore Riolo's results suggested that the discovery algorithms should focus the search on the parts of the problem that are not solved instead of focusing on the highest strength rules. In some papers published over the past ten years the emphasis may appear to be either on the allocation of credit or on the evolutionary components of a classifier system. But, it is the need to coordinate these two components that has led to most of the recent experimentation with alternative classifier system architectures.

In the rest of this section, we will briefly describe two recent trends in LCS research which address the allocation of credit and the use of various measures of rule-usefulness: (i) work which has exploited connections between LCS and reinforcement learning, and (ii) the introduction of classifier *accuracy* as the measure of the GA fitness (in place of the usual *strength*) to bias the creation and deletion of rules.

3.1 Connections to Reinforcement Learning

From their inception LCS were intended to use reinforcement learning (RL) techniques as their primary (or only) way of determining which rules are the most useful [92]. Holland's BBA, which predates the surge of interest in RL in the 1980's and 1990's, was partially inspired by the credit allocation mechanism used by Samuel in his pioneering work on learning checkers-playing programs [147]. In 1988 Sutton [172] introduced the class of *temporal difference* (TD) methods for solving RL problems, and he showed that the BBA is one kind of TD method. Watkins [186] also noted the relationship between Q-Learning and the BBA (as well as other temporal difference techniques). Later papers by Dorigo and Bersini [50] and by Liepins et al. [123] made similar connections. A recent paper by Lanzi [118] takes a different approach to relating LCS and RL, starting from a standard RL perspective, and then asking "What do we need to do to construct a rule-based Q-Learning system with generalization capabilities?" Lanzi establishes the beginnings of a formal framework which he then uses to argue that GAs are very good candidates for generalization heuristics in such a system, which is in fact a kind of LCS.

Over the years researchers have tried a number of different algorithms for credit allocation in LCS. For example, Liepins et al. [123] analyzed several possible alternative credit assignment procedures; then Liepins and Wang [122] introduced a hybrid bucket brigade–backward averaging (BB-BA) algorithm with the aim of combining the two credit allocation schemes proposed by Holland [92,85]. An early use of a Q-Learning-like technique was in CFSC2 [144], in which non-bucket brigade temporal difference methods were used to update the three kinds of strength used by the system to do its look-ahead planning. Twardowski [181] considered four of the possible credit assignment strategies proposed by 1993 (backward averaging [92], bucket brigade [85], hybrid BB-BA [124], and Q-Learning [186]) and presented an empirical evaluation of them on the pole-balancing problem. The results reported showed that Liepins and Wang's hybrid algorithm learned twice as fast as the other methods. Weiss suggested that in LCS the learning should not focus on the value of each single classifier (coded into the strength parameter) but rather on the value of actions *with respect to the overall* agent's goal. Accordingly, Weiss [187,188] introduced three *action-oriented* credit assignment procedures which take into account sets of actions rather one action at the time showing that these new credit assignment schema perform better than plain bucket brigade. In NEWBOOLE Bonelli et al. [17] introduced a "symmetrical payoff-penalty" procedure to distribute the reward to classifiers; though NEWBOOLE itself is limited to one-step (i.e., supervised learning) problems, the credit allocation mechanism may be more generally applicable. Wilson [195] compared the bucket brigade with a mixed technique of bucket brigade and Q-Learning (namely QBB) on grid problems with ZCS. The two methods performed almost identically and neither of them achieved optimal performance. However, Wilson noted that the unsatisfactory performance of ZCS, both with bucket brigade and QBB, was mainly due to the proliferation of over-general classifiers. Later, Cliff and Ross [35] confirmed Wilson's results with experiments on more complex problems.

The connections between LCS and RL became finally explicit in 1995 when Wilson introduced XCS [196]. XCS represents an important breakthrough in LCS research in that it is the first model in which the credit assignment component, based on analogous Q-Learning, is explicitly *separated* from the GA component, based on accuracy. As an immediate consequence, XCS's architecture is much neater than that of previous models and thus displays more evidence of the role of RL in classifier systems (see [31] for details).

3.2 Accuracy-based Fitness

As mentioned earlier, in the original LCS [87], allocation of credit involved assigning a *strength* value to each rule which is intended to reflect the reward the system can expect if that rule is fired. That one strength value was used both to *control* which rules are allowed to become active and to *measure* the GA fitness.

In 1995 Wilson [196] introduced another approach to classifier systems, XCS, in which the allocation of credit for achieving *reward* is logically separated from the discovery component. In XCS classifier fitness *does not* depend on the reward prediction given by the classifier (i.e., the *strength* parameter in Holland's framework) but instead classifier fitness depends on the *accuracy* of the reward prediction given by the classifier.

Wilson's choice of an accuracy-based fitness was partly anticipated by the early works of Holland [84], in which it was suggested that classifier fitness might be based on the *consistency* of classifier prediction, and by the experience of Frey and Slate [67] who successfully applied an accuracy-based classifier system to the letter recognition problem. Wilson's intuition was that prediction should estimate how much reward might result from a certain action, but that the learning of new rules (by the GA) should be focused on the most reliable classifiers, i.e., classifiers that give a more precise (accurate) prediction. This is in contrast to the traditional approach, in which evolutionary learning was focused on those rules that predicted the most reward, with the implicit idea that the predicted reward would be an *expectation*, i.e., it would include both magnitude (value) and probability (accuracy) of reward. Wilson showed that with accuracy-based fitness XCS can reach *optimal* performance while evolving a minimal population of maximally general and accurate classifiers. Wilson's accuracy-based fitness model has also been the source of inspiration for other models. For instance, very recently, Butz et al. [27–29] selected accuracy-based fitness to add genetic pressure to Stolzmann's anticipatory classifier system (ACS) [166].

Despite the more than five years which have passed since XCS was introduced, the debate concerning accuracy-based and strength-based fitness is still moderately active. However, a number of arguments and new experiences have been presented which strongly suggest that accuracy-based fitness will probably win the match, e.g., [27,28,106,109].

Since Wilson's original version of XCS, there has been a plethora of papers reporting on the characteristics of XCS itself and on many variants and extensions. In fact, XCS represents a very active research area and probably the most popular LCS ever. For a recent overview of the results of XCS research we refer the interested reader to Wilson [200] and to the works of the many researchers active on this specific classifier system model, e.g., Barry [11–13,150], Bull and Tomlison [25,23,176,178], Kovacs [104,105], Saxon [150], and many others.

4 Architectural Changes

4.1 Rule Representation

As originally described by Holland, in most LCS all messages are fixed length binary strings. The semantics were positional, e.g., each locus in a detector message represented the value (0 or 1) for some observed attribute. Accordingly, classifier conditions are built from strings in the ternary alphabet {0,1,#} where the "#" character is a wild-card matching any value; while classifier actions are binary strings.[4] Further, the conditions were typically either one such ternary string, or perhaps two ternary strings, representing two primitive predicates which must each be matched by some message on the message list, and combined by an implicit logical AND (or AND NOT) to

[4] In Holland's original descriptions of classifier systems [87,93], and in CFSC1, Riolo's implementation [140,146], the action parts of rules also include the "#" character, acting as a "pass through" operator.

determine if the whole condition is matched. This simple representation was initially introduced by Holland as a result of balancing the tradeoff between the descriptive power of the classifier language, and both (i) computational efficiency of the matching process and (ii) the mathematical tractability of formal models of the system [86].

However, over the years many authors have raised concerns about the representational capabilities of the binary/ternary encoding. While the original syntax is provably complete [65], the number of rules needed to represent "real" concepts or to do simple programming tasks could be astronomical; the situation is analogous to trying to do real programming tasks with a Turing machine! For example, it can take many rules to represent some real-valued condition (to some precision) such as $(x \leq 3.1 \vee x > 4.2) \wedge (y > 7.1)$, not to mention trying to represent concepts that involve variable binding (e.g., $x < 3.3 \wedge x > y$). But not only is the number of rules required prohibitive, at least as serious a problem is how difficult it is for such a system to learn and so discover the complex sets of rules required, using the typical credit allocation and rule discovery algorithms found in LCS.

For instance, Schuurmans and Schaffer [154] showed that binary encoding may lead to unsatisfactory performance. In particular they observed that binary encoding (i) *unavoidably* introduces a bias in the generalization capabilities of the system making some concepts easy to generalize and others almost impossible to generalize; (ii) it has a limited capability to represent disjunctions and relations among sensors; finally (iii) it has a positional semantics which may not be able to capture certain relations among sensors that are positionally independent. Later Shu and Shaeffer [157] proposed an extension to the ternary representation of conditions in which named variables were used to describe relations among inputs bits.

As a consequence, over the past ten years or so there has been considerable work on systems in which one or more of the original constraints on rule syntax and semantics have been relaxed.

Some authors have suggested fairly small modifications to the original simple syntax and semantics. For example, Booker [20] analyzed the issues raised in [154] and suggested that these limitations are mainly due to the encoding scheme (i.e., the way the sensors are encoded into bit-strings) rather than to the binary encoding itself. Booker showed that *feature manifold* encodings can improve the representational capabilities of classifiers *without* modifying the classifier syntax. However, to our knowledge, Booker's scheme has never been implemented and tested.

Most researchers exploring alternatives over the past five to ten years have suggested and implemented representational changes which involve (i) more complex atomic conditions, with real-valued, non-positional semantics, and (ii) more complex logical combinations of the atomic conditions.

For example, Wilson [194,198] suggested one way to allow more complex atomic conditions is to allow real-valued attributes where appropriate, with basic conditions that are predicates expressed in terms of a lower and upper bound on an attribute value, e.g., "$(0.0 < x < 0.1) \wedge (0.8 < y < 0.9)$". Recently, Wilson has finally pursued his original ideas developing two extensions of XCS, namely XCSI [199] and XCSR [197], which apply to problems involving integer (XCSI) and real (XCSR) inputs. Wilson applied XCSR to a version of the Boolean multiplexer with continuous inputs showing that XCSR could reach near optimal performance. Most interestingly, Wilson also applied XCSI to the Wisconsin Breast Cancer dataset [14] showing that XCSI reached a performance as good as state-of-the-art machine learning algorithms. In addition, XCSI produced solutions consisting of classifiers that were interpretable in ways that indicate classification of cases depends on one or a few particular attributes.

Another way to construct more flexible conditions involves *fuzzy* classifier systems which blend the representational approach of fuzzy logic with rule-based classifier systems using evolutionary algorithms to evolve the parameters of various fuzzy matching and output functions. Starting with the early work of Manuel Valenzuela-Rendón [182], there has been much work on many variations on this theme. A review of the most relevant works in this area can be found in [15].

Lanzi [115,116] introduced two other representations for classifier conditions and implemented them on the XCS classifier systems. The first is the so-called "*messy*" representation [115], which was inspired by the work of Goldberg on messy GAs [71], and which also bears some resemblance to Sen's variable representation [156]. The main feature of this approach was to abandon a positional semantics in favor of one in which attribute-value relations are explicitly listed, and any attributes not included are considered "don't care" wild-cards matching any value. Lanzi's second representation was inspired by Koza's work on genetic programming [110] and by a suggestion of Wilson [195,196] on how to extend his ZCS/XCS systems to obtain more flexible generalizations. This approach involves representing classifier conditions using a subset of Lisp s-expressions built from basic Boolean and symbolic sensory inputs. Lanzi applied both representations to a set of known test problems, showing that these two versions of XCS, namely XCSm and XCSL, reach optimal performance.

Others have also built classifier conditions based on the Lisp s-expressions as found in genetic programming [110]. For instance, Tufts [180] replaced the fixed-width ternary conditions with Lisp expressions in his dynamic classifier system. Genetic programming applied to the classifiers allows the system to discover building blocks in a flexible, fitness-directed manner. And Ahluwalia and Bull [6] coupled binary conditions with symbolic actions represented by general s-expressions. They applied the new system, GP-CS, to extract features from data before these are classified with a K-nearest-neighbor method.

In summary, there has been a general trend in LCS research toward more complex classifier rule syntax and semantics, especially for systems aimed at problem domains in which real-valued, complex concepts are involved. While such systems undoubtedly have led to improvements in performance, it is also likely that new issues and problems will have to be faced as such systems are brought to bear on harder problems, e.g., how to find GA operators that will lead to good results, and how, if at all, such systems can be formalized in a mathematically tractable way. It may be the case that this feature of the original systems will have to be sacrificed, at least for the short term, in order to build LCS that can solve "interesting" problems.

One final note related to representation in LCS: in most of the early implementations, multiple, duplicate rules were quite common, since they could be generated by a GA when it differentially reproduced rules which have proved to be very useful. While this may make some sense from a schema analysis point of view [93], unless the LCS are implemented on massively parallel hardware, there is a large performance price to pay as the many duplicate rules must all be tested for matches against all messages. Thus Wilson [195] introduced the idea of *macro-classifiers* as a way to speed up systems implemented on serial machines, while retaining the important sampling effects of having multiple copies of good rules. Basically, a macro-classifier is a regular classifier with an additional "numerosity" parameter which indicates how many copies of the classifier are in the rule population. Thus the numerosity is used to bias any process that should be effected by having multiple copies (e.g., GA parent selection or rule deletion operations), but only one actual condition must be matched. While not yet universal, the use of macro-classifiers is spreading rapidly to most current implementations of LCS.

4.2 Simpler Systems

For many years the research on LCS[5] was done solely on the original architecture proposed by Holland [92,87]. Many researchers developed their own version of Holland's complex architectures in which Holland's classifier system was the main component (e.g., Dorigo's ALECSYS [49]). All these implementations shared more or less the same features which can summarized as follows: (i) some form of BBA was used to distribute the rewards received to the classifiers; (ii) the fitness of the GA was a measure of classifier *strength*; (iii) the internal message list was used to keep track of past inputs. Two "standard implementations" were made available at that time, one by Goldberg [71] and one by Riolo [140].

Holland's LCS had a number of known problems which in some cases prevented the systems from achieving satisfactory performance (see [202] for a discussion). To improve performance some small modifications to the original framework were proposed but interesting performance was rarely reported.

[5] Again, in this chapter we are restricting our discussion to primarily Michigan-style classifier systems; for a summary of work with Pittsburgh-style systems, see [129].

Thus the need for more radical changes to the original architecture was already perceived in the early years of research. In 1987 Wilson [192] introduced a new type of "one-step" classifier system, BOOLE, particularly devised for supervised learning problems, specifically, for the learning Boolean functions. Wilson showed that BOOLE could learn multiple disjunctive concepts faster than neural networks. BOOLE was extended into NEWBOOLE by Bonelli et al. [17] in 1990; the performance of NEWBOOLE was compared on three medical domains with two other learning algorithms, the rule inducer CN2 [33] and neural networks trained by back propagation. The results reported showed that NEWBOOLE performed significantly better than the other machine learning methods.

Wilson's experience with BOOLE led him to observe [195] that the architecture of LCS was too complex to permit careful, revealing studies of the learning capabilities of these systems. Accordingly he drastically simplified the original framework by introducing ZCS, a *zeroth-level* classifier system [195]. ZCS differs from the original framework mainly in that (i) it has no internal message list, and (ii) it distributes the reward to classifiers by a technique, QBB [195], very similar to Watkins' Q-Learning [186].[6] Furthermore ZCS significantly differs from previous systems by omitting several heuristics and operators that were developed to aid those systems in reaching acceptable performance (see for example Wilson [191] and Dorigo [48]). Interestingly, this basic ZCS system performed similarly to more complex implementations previously presented in the literature. These results suggested that Holland's ideas could work even in a very simple framework, although ZCS performance was not optimal yet.

Optimal performance in different applications was finally reported in 1995 when Wilson [196] introduced the XCS classifier system. While XCS maintains Holland's essential ideas about LCS, it differs much from all the previous architectures. First, in XCS an analogue Q-Learning is used to distribute the reward to classifiers, instead of a BBA[7]. Second, in XCS the GA acts in environmental niches instead of on the whole population, a solution inspired by the seminal work of Booker on GOFER-1 [18]. Most important, as discussed in Section 3.2, in XCS the fitness of classifiers is based on the accuracy of classifier predictions. For an overview of the main results reported in the literature since 1995, see Wilson's review of the state of XCS research [200].

[6] In this sense ZCS represents one of the first attempts to bridge the then existing gap between the credit assignment algorithms used in LCS (i.e., the bucket brigade) and those temporal difference methods employed in RL, as discussed in the previous section.

[7] ZCS distributes the reward with an algorithm which is a mix of bucket brigade and Q-Learning (QBB) while XCS uses an analogue Q-Learning.

4.3 Advanced Systems

Most of the progress made in recent years in LCS research has involved systems which are simplifications of Holland's proposal. Although such "*simple*" systems have been successfully applied to a large number of applications (see Section 5), these simpler LCS do have reduced capabilities which limit their ability to solve more complex problems. For instance, these simple LCS are not able to learn or even to represent more complex "internal models" such as those originally envisioned by Holland. In this section we will describe some of the more recent research done on LCS that are more complex than ZCS, XCS, and similar systems, including some that are even more complex than Holland's original versions.

The simple LCS described in the previous section represent information about the world structure by *condition-action-payoff* rules, as opposed to plain *state-action-payoff* tables usually used in RL. Because classifier conditions can be more or less generalized, covering one or many different environmental situations, classifier systems can represent their knowledge of the world with general-to-specific hierarchies of rules.

On the other hand, unlike the more complex Holland architecture [87] the simple systems do not have an *internal message list* or any other sort of internal *memory*. Accordingly, they are not able to store complex knowledge in the form of homomorphisms, *quasi-homomorphisms*, and other abstract structures [93] for representing generalized *state-action-state-payoff* transitions, i.e., true *models* of the world (vs. simple isomorphic maps of the world). For example, Holland has long argued that *default hierarchies*[8] are an efficient, flexible, easy-to-discover way to categorize observations and structure models of the world [87,93]. The ability to represent such models is key to being able to address three key problems facing LCS:

Non-Markov signals: The learning capabilities of LCS (or any adaptive agent) rely on and are constrained by the way the agent perceives the environment, e.g., by the detectors the system employs. On the one hand, there are cases in which the agent's sensory inputs provide sufficient information to let it decide in every situation what is the best action to take. However, in many applications the sensory inputs provide only partial information about the actual state of the environment, in which case the agent may not be able to decide what is the best action solely by looking at its current sensory inputs. In the former case we say that the environment is *Markov* with respect to the agent's sensors; in the latter case we say that the environment is *non-Markov*.

[8] A simple example of a default hierarchy is two rules, one more general ("If see a moving object, flee") and one a more specialized rule that matches only some of the states covered by the general rule ("If see a small striped moving object, pursue"). The general rule is considered the *default* rule, i.e., it is fired unless there is overriding information that activates the specialist rule.

Long sequences of actions to achieve goals (rewards): In many environments, LCS must be able to develop long sequences of actions, i.e., long *action chains*, while being rewarded in only a few situations. In such situations there are long delays before rewards are received, with many unrewarded steps between some *stage setting* actions and the ultimate reward those actions lead to. Using Holland's classic example in checkers, the difficulty is to learn, by proper *allocation of credit*, which of many prior moves (none of which is inherently rewarding) led to a triple jump (which even a novice recognizes as a good action).

How to do "off-line" learning and planning: For most complex environments, learning and performance can be greatly facilitated by doing "look-ahead" planning and "latent learning" [144], i.e., basing actions on predictions of future states of the world, using *both* current information *and* past experience as embodied in the agent's internal models of the world. This is the approach outlined by Holland in many earlier works on LCS, e.g., in [93] and [87].

Ultimately, the ability of LCS to solve these problems will be key to achieving cognitive capabilities on a par with (or beyond) those of humans and other related mammals. Several different approaches to solving these problems have been studied over the past ten years or so, including:

(i) internal message lists,
(ii) other internal, non-message-list, memory mechanisms,
(iii) "corporate" classifier systems, and
(iv) enhanced rule syntax and semantics.

Note that all of these involve increasing the complexity of the LCS, each in its own way.

The *internal message list* is one of the most intriguing characteristics of LCS, but its presence greatly increases the complexity of LCS, and so most systems described in the literature have not used internal message lists. One exception was the work of Riolo and Robertson [146], who reported limited success in applying a version of Holland-style classifier system to solve a non-Markov letter sequence prediction task. Another work which used a true internal message list was that of Smith [160], who reported unsatisfactory performance in very basic non-Markov environments with Holland's classifier system.

Riolo [144] was the first to use the regular LCS message list to do look-ahead learning and planning in classifier systems, inspired by a similar approach described (but not implemented) by Holland [89]. Riolo's CFSC2 classifier system learns a model of the environment by recording state-action-state transitions, representing those transitions by classifiers. For this purpose, some loci of standard ternary classifier conditions and actions are used to represent special *tags*. These tags are used to distinguish the classifiers (and the messages they post) which represent (i) states in the world and (ii) transitions from states to states, as a result of the system's actions. In addition, the tags are used to distinguish between messages that are intended to cause the system to actually take actions in the world, from other messages that are intended to represent planning, in effect "imagining" what would happen if certain actions were taken. In addition, each classifier has three measures of utility associated with it: (i) a long-term strength analogous to the usual *strength* in Holland's system; (ii) a short-term strength which is updated *every time step*, even when the system is not executing actions in the environment (i.e., it is executing rules that implement its planning activity); and (iii) an estimate of the predictive accuracy of the transition that the classifier represents. The credit assignment procedure also was modified to cope with these three parameters, using a Q-Learning type algorithm. There was no GA, and no generalization: the classifiers in effect implemented an isomorphic model of the environment. Riolo tested CFSC2 on a relatively complex sequential problem also employed by Sutton [173] and Grefenstette [74]. His results confirm early results from systems using Q-Learning, showing that a classifier system which uses an internal model of the environment to do *look-ahead planning* can speed up learning considerably, as well as make it possible for the system to do *latent learning*, a kind of learning not possible in systems, artificial or biological, without internal models which predict what *state* follows from a given state-action pair. Furthermore, Riolo suggested that because of their generalization capabilities, classifiers can be used to build models of the environment that are more compact than those built by Sutton's Dyna-Q, which assumed a tabular representations of its model.

More recently, a number of researchers have built directly on the basic ZCS and XCS approach by adding basic kinds of *internal memory* mechanisms, (e.g., "corporations" of classifiers or memory registers) but still *not* using Holland's internal message list, in order to keep the systems as simple as possible. For example, while the original ZCS has no internal memory, a *small* bit register can be added to ZCS in order to tackle non-Markov problems [195]. Cliff and Ross [35] applied ZCS with internal memory, ZCSM, to simple non-Markov environments, reporting promising (although non-optimal) performance. They showed ZCSM cannot learn long sequences of actions because of the proliferation of over-general classifiers.

In a similar way, Lanzi [113] added internal memory to Wilson's XCS classifier system [196] which, like ZCS, does not have internal memory. Lanzi applied XCS with internal memory, XCSM, to simple non-Markov environments reporting optimal performance in most cases [113,114]. Then Lanzi and Wilson [119] extended these results by applying XCSM to more difficult non-Markov problems. Their experiments show that XCSM can learn optimal solutions in *some* complex non-Markov environments suggesting that the approach scales up. Recently Lanzi [117] has studied the use of internal memory within Q-Learning and discussed the effectiveness of the solutions developed inside the LCS community.

A different approach consists of using *corporations of classifiers* [202]. This approach can be viewed as mixing Michigan and Pittsburgh [42] approaches to the organization of classifier systems, with respect to credit allocation, activation, and rule discovery. A corporation is a cluster of classifiers within the population that are treated as an indivisible entity during the discovery phase and/or the performance phase. Classifiers within a corporation are *explicitly* linked to each other to form a *chain* or other relationship. For example, when a classifier of a certain chain (i.e., corporation) is activated it forces the activation of classifiers that follow it in the chain. Tomlison and Bull reported some success in applying first a corporate version of ZCS [176,178] and later a corporate version of XCS [177] on a set of non-Markov problems. For further discussion on classifier corporations we refer the interested reader to Tomlison and Bull [179].

Note that the need to bias activity so that classifiers in a chain fire in sequence with high probability was also shown to be important in systems that *implicitly* linked classifiers, through the posting of "tagged" messages on the internal message list [142,146]. These tagged messages may be thought of as a memory of previous states, which can then be used later to disambiguate a non-Markov state. This view is also seen in Smith [160], who showed with an indirect experiment that the sequential activations of classifiers creates a remembered context for later classifiers in the chain. This context helps these classifiers to cope with the partial information coming from the sensory inputs.

A recent example of the use of enhanced rule syntax and semantics is the work of Stolzmann [166], who introduced the idea of *anticipatory classifier systems*, ACS, in which classifiers are enriched with an *expectation* of the state that will follow the activation of the classifier, similar to (but richer than) the predictions made in Riolo's CFSC2. Further, the ACS learning mechanisms are based (in part) on psychological models of learning in animals, including the adaptive learning process, ALP [171]. Thus in the ACS there is pressure towards general rules generated by both the ALP heuristic [168] and the more common GA [27,28]. This combination of representational power and multiple learning heuristics results in a system that shows great promise both as a model of real animal systems and also as general purpose LCS which can solve difficult problems like those faced by real animals all the time [171]. A fairly comprehensive introduction to ACS can be found in [169]; results for robotics applications of ACS are summarized in [170,171].

In summary, there have been and will continue to be many variant types of LCS, all trying different approaches to solving the same fundamental problems facing all LCS models. Few approaches have been described in this section. Others, not discussed here, include the GOFER system of Booker [18,19] and the work of Barry [11,12] on the "*Consecutive State Problem*". For more information on these and other variants, we refer the interested reader to [120]. In particular, we recommend [94], in which a number of leading researchers outline their views on the question "What is a learning classifier system?" vividly showing the connections between their views of the central problems LCS can and should address and the resultant LCS architectures they favor.

5 Applications

The original descriptions of LCS were typically in terms of systems which could be used to model cognitive systems [93,92]. However, almost since their inception, the uses of LCS have fallen into two basic categories:

(i) "engineering" applications, where the goal is to design a classifier system which can do a task as well as possible, with no concern about exactly how it does it; and
(ii) "modeling" applications, where the goal is to construct a plausible model of some real system, in which the mechanisms used in the classifier system should plausibly capture some essential features of the system being modeled, and in which the classifier system generates behavior which should match that of the real system being modeled, independent of whether that behavior can be described as good or bad performance on some given task.

In this section we will briefly describe some recent applications of LCS as both machine learning tools and as models of complex adaptive systems (CAS) [91].

5.1 LCS as a Machine Learning Technique

One important research area where LCS have been applied is that of supervised classification. In this type of application, a series of examples which represent a target concept (phenomenon) is presented to the classifier system. The goal of the learning process is to develop a general and compact *model* of the target phenomenon that the examples represent. This model can be used to *analyze* the target phenomenon previously observed or to classify *unseen* examples. From a machine learning perspective, this involves using LCS as a means of doing data analysis and data mining [59]. A large number of such systems have been studied in the last decade, on a host of different task domains, e.g., Bonelli and Parodi [16], Holmes [95–97,99], Garrell et al. [68,69], Saxon and Barry [150], and recently Wilson [199].

In data mining applications, LCS are used to extract *compact* descriptions of *interesting* phenomena described by multidimensional data. Bonelli and Parodi [16] compared NEWBOOLE with two other learning algorithms, the rule inducer CN2 [33] and neural networks, on three medical domains, showing that NEWBOOLE performed significantly better than the other methods.

Sen [155] modified the conflict resolution strategy of NEWBOOLE to improve its performance on known machine learning datasets. The reported results show that this variant of NEWBOOLE outperforms many machine learning methods. Holmes [95–98] applied his version of NEWBOOLE, EpiCS, to classify epidemiologic surveillance data. A very recent summary of the results of his research is discussed in [99]. Saxon and Barry studied the performance of XCS in a set of known machine learning problems, namely the Monk's problems, demonstrating that XCS is able to produce a classification performance and rule set which exceeds the performance of most current machine learning techniques [175]. Their results are extensively discussed in [150]. Finally, Wilson [199] applied XCSI to the Wisconsin breast cancer dataset [14] showing that XCSI reached a performance as good as state-of-the-art machine learning algorithms and produced classifiers that were (i) easily interpretable and (ii) suggested interesting dependencies on one or a few attributes.

Another important research area where LCS have been applied involves using the LCS as a controller for autonomous robots. Autonomous robotics has always been considered an important test-bed for reinforcement learning algorithms. In effect robotics problems are usually difficult to describe using more traditional machine learning methods (e.g., supervised learning) while they are easily modeled as RL problems.

Accordingly, autonomous robotics is probably the single research area where LCS have been applied most during the last ten years. The most extensive research in the application of LCS to autonomous robotics was carried out by Dorigo and Colombetti [52] with other colleagues. In the years from 1990 until 1998 they applied different versions of their parallel LCS, ALECSYS, to a variety of robotics applications. Borrowing the idea of "*shaping*" from experimental psychology they introduced the concept of *robot shaping* defined as the incremental training of an autonomous agent. To apply this idea to real-world problems they defined a behavior engineering methodology named BAT: Behavior Analysis and Training. An overview of their work can be found in Dorigo and Colombetti's book *Robot Shaping: An Experiment in Behavior Engineering*; the technical details of their parallel LCS, ALECSYS, are discussed in [47,49,53,54]; discussions about robot shaping and BAT methodology can be found in [36–38,51]; some applications are discussed in [39,131]. Donnart and Meyer developed a hierarchical architecture *MonaLysa* for controlling autonomous agents [43–45]. MonaLysa integrates a reactive module implemented by a LCS with other deliberative modules (e.g., the planning module). Many other applications to autonomous robotics have been carried out with *fuzzy LCS*. For a review for those works we refer the reader to Bonarini's recent survey [15].

Other applications of classifier systems include the work of Federman et al. [60–64], who have applied LCS to predict the next note in different type of music. Richards et al. [133–138] applied classifier systems to 2-D and 3-D shape optimization. In another shape-related application, Nagasaka and

Taura [130] generate shape features using classifier systems. Smith et al. [164] (see also [165]) applied LCS to the problem of discovering novel fighter maneuvering strategies. Sanza et al. [148] used an agent with a classifier system to support users who navigate in a virtual environment. Frey and Slate [67] applied a classifier system to a letter recognition problem; Cao et al. [32], Escazut and Fogarty [56] to traffic controllers; Gilbert et al. [70] to the profit optimization of a batch chemical reaction.

5.2 LCS as a Modeling Approach

As mentioned above, Holland's original view of LCS was as a way to model complex adaptive systems (CAS), e.g., human or other cognitive systems [92,93]. This use of LCS has continued through the years (*cf.* Riolo's work on human category learning [145] and on latent learning [144]), and in recent years there has been a marked increased in the use of LCS as models. These studies largely fall into one of two main categories: (i) cognitive psychology, in which the focus is on modeling a single agent, typically with single LCS acting as the "brain" of the agent, and (ii) computational economics, in which the focus is on modeling many interacting agents, each controlled by its own LCS. Note also that Holland has recently been working on his Echo system [91], which he views as an extension of LCS, explicitly designed to study emergence, co-evolution, and other important dynamical properties of CAS.

One recent use of LCS as a cognitive model has been the work of Stolzmann with colleagues on the anticipatory classifier system, ACS [27–30,166,167,171]. As mentioned in Section 4.3, both the representation of rules and the learning algorithms in ACS are based on an underlying model of learning, the anticipatory learning process. The ACS has been used to model latent learning of mazes by rats [167,171], as done by Riolo's CFSC [144], and the ACS has been shown to be a good model for animals in other experiments in which the subjects must learn internal models of their environment in order to perform as well as they do [169,171].

Other recent uses of LCS to model various aspects of (human) cognitive systems include (i) the work of Satterfield [149] in which a classifier system is used to model humans learning languages in bilingual environments, enabling her to test alternative conceptual models of this process, (ii) the work of Druhan and Mathews [55], in which a classifier system was used as a model of how humans learn (artificial) languages, (iii) the work of Davis [41], who built a computational model of Affect Theory, which was able to generate behavior that matched known behaviors in humans, and which generated behavior that can be seen as predictions of novel phenomena to be experimentally verified, and (iv) the recent work of Hartley [81], comparing the ability of LCS using strength-based (NEWBOOLE) versus accuracy-based (XCS) fitness to model perceptual category learning in humans. (Hartley found XCS produces performance more like that of humans.)

In *agent-based computational economics* [174], LCS have been used to model adaptive agents in artificial stock markets (see for example the works of Brian Arthur et al. [7] and LeBaron et al. [121]). Vriend [185,183,184] compared the performances of Michigan-style and Pittsburgh-style classifier systems on the same economic model. Marimon et al. [125] have modeled the emergence of money using populations of classifier systems, each individual modeling the behavior of one economic agent. Bull [22] applied a variant of Wilson's ZCS in a simulated "Continuous Double-Auction" (CDA) market. Other papers discussing the use of classifier systems as a tool for modeling agents in economic systems include those by Miller and Holland [127] and by Mitlöhner [128]. Recently, Schulenburg and Ross [152,153] employed LCS to model the behavior of agents trading risk free bonds and risky assets in a stock market environment.

6 Conclusion

Even from this brief summary, it should be clear that LCS are a very active area of GA-related research, in which much progress has been made in the past five years or so. Still, there is much to be done. We feel some of the most pressing challenges for the next five years include:

Scaling up to solve "real" problems. With a few exceptions (e.g., [52,99,152,199]) only small or medium-sized, often "artificial" constructed, problems have been studied. While this has been a good research strategy (since it is easy to understand what is or is not working with less difficult problems), for LCS research to really thrive it will have to be shown to be useful for solving really difficult problems, problems which are challenging to all machine learning approaches.

Interaction of GAs and credit allocation. For GAs (or more generally, evolutionary algorithms) to be useful in LCS, "fitness" must be allocated across rules in a way that guides the GA toward regions of the rule space that are likely to contain useful rules. However, the changes to the rule set that a GA makes also influence which rules fire, and in what order, which in turn (via the allocation of credit mechanisms) affects how values that can be used as fitness are distributed in the rule population. The interactions between the allocation of credit and rule discovery components of LCS will continue to make these aspects of LCS some of the most complex to understand.

Fitting GAs to LCS requirements. Besides the interaction of GAs and the allocation of credit in LCS, there are other ways in which the use of GAs could be better matched to the particular dynamics and structure of LCS. For example, as mentioned in Section 4.2 the GA in Wilson's XCS acts over restricted sets or rules [196], as also was done in Booker's GOFER-1 [18] and

Riolo's work on discovery long chains of rules [142]. This approach treats sets of rules working on similar sub-problems as a kind of "niche", as is made more explicit in the work of Horn, Goldberg and others on niching in LCS and GAs [102,101,100]. The study of niching in GAs is but one example of the kind of research into GA dynamics that could be applied to the special needs of LCS.

Links to analogous adaptive systems. From the very inception of LCS, links and analogies have been made between them and other adaptive systems, e.g., cognitive systems [92,93], economic systems [88,127,190,128], immune systems [58,82], and artificial neural networks [40,57,159,161,162]. Much more can be learned by deepening and extending these links.

Alternative representations. To make it easier to apply LCS to harder, "real-world" problems, it is certainly the case that the descriptive power of the rules will have to be enhanced over that of the original binary/ternary messages and rules. Good progress has been made (e.g., [115,116,198,199], but more will be needed, and in particular more thought and research will have to be applied to learn more about which GA/EA rule creation operators are most appropriate for the different rule representation schemes.

Self-adaptive Classifier Systems. LCS have a number of parameters that sometimes depend on the specific problem that is going to be solved. Recently, Bull et al. [25,26] introduced a *self-adaptive* version of Wilson's ZCS and XCS in which each classifier has is own mutation rate which is adapted during learning. Their results suggest that the approach could be extended to more parameters (e.g., the learning rate [196]) thus simplifying the LCS framework dramatically, by decreasing the number of parameters that must be set "by hand" for each different application.

Hybrid systems. In the application of GAs to real problems, adding additional learning heuristics appropriate to the problem being solved has been found to be most useful. It is likely similar approaches could be used to enhance the learning capabilities of LCS. Further, for those cases where an LCS is used to *model* some real system, more thought should be given to making appropriate changes to the LCS learning mechanisms so that they are appropriate to the systems being modeled and the questions being addressed by constructing those models.

Default Hierarchies. Default hierarchies were widely studied in the early days of LCS research [86,93,139,141,193,143,158,163,46]. In more recent years this topic has not been of central concern in most of the research, primarily because the new models of LCS like XCS develop homomorphic generalizations that do not allow the formation of default hierarchies. However, default hierarchies remain an important topic and test-bed for the LCS community to compare the generalizations produced by Holland-style architectures and those produced by novel approaches such as ACS or XCS. Recently, Holland [94] expressed an intention to go back to the study of default hierarchies. This may be a fore-shadowing of a return of active research in default hierarchies.

Theory. As new models of LCS are developed and become popular (e.g., XCS [196,31] or ACS [169]) there is an increasing need for theoretical foundations for these new systems. While there are already some initial results [23,24,106–109,118] we believe that much more work is needed in the next years.

Given the number of active researchers who are in or are entering the LCS field of late, we feel confident the next five years will be as productive as the past five.

Acknowledgements

Pier Luca and Rick wish to thank the three anonymous reviewers for comments and suggestions.

Pier Luca also wishes to thank Marco Colombetti for many discussions; Carla for her support; and Gabriella for her endless patience.

In his research, Pier Luca was partially supported by the Politecnico di Milano Research Grant "Development of Autonomous Agents Through Machine Learning".

References

1. Fourth International Workshop on Learning Classifier Systems. `http://www.psychologie.uni-wuerzburg.de/iwlcs-2001/`.
2. Learning Classifier Systems Group. `http://www.csm.uwe.ac.uk/lcsg/`.
3. Emergent Computation. Proceedings of the Ninth Annual International Conference of the Center for Nonlinear Studies on Self-organizing, Collective, and Cooperative Phenomena in Natural and Artificial Computing Networks. A special issue of Physica D. Stephanie Forrest (Ed.), 1990.
4. *Collected Abstracts for the First International Workshop on Learning Classifier System (IWLCS92)*, 1992. October 6–8, NASA Johnson Space Center, Houston, Texas.
5. Proceedings of the International Workshop on Learning Classifier Systems (IWLCS-2000), in the Joint Workshops of SAB 2000 and PPSN 2000, 2000. Pier Luca Lanzi, Wolfgang Stolzmann, and Stewart W. Wilson (workshop organisers).
6. Ahluwalia, M., Bull, L. (1999) A Genetic Programming-based Classifier System. In Banzhaf et al. [10], 11–18.
7. Arthur, Brian W., Holland, J. H., LeBaron, B., Palmer, R., Talyer, P. (1996) Asset Pricing Under Endogenous Expectations in an Artificial Stock Market. Technical Report, Santa Fe Institute.

8. Bäck, T. (ed.) (1997) *Proceedings of the 7th International Conference on Genetic Algorithms (ICGA97a)*. Morgan Kaufmann: San Francisco CA.
9. Barry, A. The Learning Classifier Systems WEB. `http://www.csm.uwe.ac.uk/~ambarry/LCSWEB/`.
10. Banzhaf, W., Daida, J., Eiben, A. E., Garzon, M. H., Honavar, V., Jakiela, M., Smith, R. E. (eds.) (1999) *Proceedings of the Genetic and Evolutionary Computation Conference (GECCO-99)*. Morgan Kaufmann: San Francisco, CA.
11. Barry, A. (1999) Aliasing in XCS and the Consecutive State Problem: 1 – Effects. In Banzhaf et al. [10], 19–26.
12. Barry, A. (1999) Aliasing in XCS and the Consecutive State Problem: 2 – Solutions. In Banzhaf et al. [10], 27–34.
13. Barry, A. (2000) Specifying Action Persistence within XCS. In Whitely et al. [189], 50–57.
14. Blake, C. L., Merz, C. J. (1998) UCI repository of machine learning databases.
15. Bonarini, A. (2000) An Introduction to Learning Fuzzy Classifier Systems. In Lanzi et al. [120], 83–104.
16. Bonelli, P., Parodi, A. (1991) An Efficient Classifier System and its Experimental Comparison with two Representative Learning Methods on Three Medical Domains. In Booker and Belew [21], 288–295.
17. Bonelli, P., Parodi, A., Sen, S., Stewart, W. (1990) NEWBOOLE: A Fast GBML System. In *International Conference on Machine Learning*, 153–159, Morgan Kaufmann: San Mateo, CA.
18. Booker, L. B. (1989) Triggered rule discovery in classifier systems. In Schaffer [151], 265–274.
19. Booker, L. B. (1990) Instinct as an Inductive Bias for Learning Behavioral Sequences. In Meyer and Wilson [126], 230–237.
20. Booker, L. B. (1991) Representing Attribute-Based Concepts in a Classifier System. In Rawlins [132],115–127.
21. Booker, L. B., Belew, R. K. (eds.) (1991) *Proceedings of the 4th International Conference on Genetic Algorithms (ICGA91)*. Morgan Kaufmann: San Francisco, CA.
22. Bull, L. (1999) On using ZCS in a Simulated Continuous Double-Auction Market. In Banzhaf et al. [10], 83–90.
23. Bull, L. (2000) Simple Markov Models of the Genetic Algorithm in Classifier Systems: Accuracy-based Fitness. In *Proceedings of the International Workshop on Learning Classifier Systems (IWLCS-2000), in the Joint Workshops of SAB 2000 and PPSN 2000* [5]. Extended abstract.
24. Bull, L. (2000) Simple Markov Models of the Genetic Algorithm in Classifier Systems: Multi-step Tasks. In *Proceedings of the International Workshop on Learning Classifier Systems (IWLCS-2000), in the Joint Workshops of SAB 2000 and PPSN 2000* [5]. Extended abstract.
25. Bull, L., Hurst, J. (2000) Self-Adaptive Mutation in ZCS Controllers. In *Proceedings of the EvoNet Workshops - EvoRob 2000*, 339–346. Springer: Berlin.
26. Bull, L., Hurst, J., Tomlinson, A. (2000) Mutation in Classifier System Controllers. In *From Animals to Animats 6: Proceedings of the Sixth International Conference on Simulation of Adaptive Behavior, 2000*, J. A. Meyer et al. (eds), pp460–467.

27. Butz, M. V., Goldberg, D. E., Stolzmann, W. (2000) Introducing a Genetic Generalization Pressure to the Anticipatory Classifier System – Part 1: Theoretical Approach. In Whitely et al. [189], 34–41.
28. Butz, M. V., Goldberg, D. E., Stolzmann, W. (2000) Introducing a Genetic Generalization Pressure to the Anticipatory Classifier System – Part 2: Performance Analysis. In Whitely et al. [189], 42–49.
29. Butz, M. V., Goldberg, D. E., Stolzmann, W. (2000) Investigating Generalization in the Anticipatory Classifier System. In *Proceedings of Parallel Problem Solving from Nature (PPSN VI)*. Springer: Berlin.
30. Butz, M. V., Goldberg, D. E., Stolzmann, W. (2000) Probability-enhanced Predictions in the Anticipatory Classifier System. In *Proceedings of the International Workshop on Learning Classifier Systems (IWLCS-2000), in the Joint Workshops of SAB 2000 and PPSN 2000* [5]. Extended abstract.
31. Butz, M. V., Wilson, S. W. (2000) An algorithmic description of XCS. Technical Report 2000017, Illinois Genetic Algorithms Laboratory, University of Illinois at Urbana-Champaign.
32. Cao, Y. J., Ireson, N., Bull, L., Miles, R. (1999) Design of a Traffic Junction Controller using a Classifier System and Fuzzy Logic. In *Proceedings of the Sixth International Conference on Computational Intelligence, Theory, and Applications*. Springer: Berlin.
33. Clark, P., Niblett, T. (1989) The CN2 induction algorithm. *Machine Learning*, **3**, 261–283.
34. Cliff, D., Husbands, P., Meyer, J. A., Stewart, W. (eds.) (1994) *From Animals to Animats 3. Proceedings of the Third International Conference on Simulation of Adaptive Behavior (SAB94)*. A Bradford Book. MIT Press: Cambridge, MA.
35. Cliff, D., Ross, S. (1995) Adding Temporary Memory to ZCS. Technical Report CSRP347, School of Cognitive and Computing Sciences, University of Sussex ftp://ftp.cogs.susx.ac.uk/pub/reports/csrp/csrp347.ps.Z.
36. Colombetti, M., Dorigo, M. (1994) Training Agents to Perform Sequential Behavior. *Adaptive Behavior*, **2**, 247–275 ftp://iridia.ulb.ac.be/pub/dorigo/journals/IJ.06-ADAP94.ps.gz
37. Colombetti, M., Dorigo, M. (1999) Evolutionary Computation in Behavior Engineering. In *Evolutionary Computation: Theory and Applications*, chapter 2, 37–80. World Scientific: Singapore.
38. Colombetti, M., Dorigo, M., Borghi, G. (1996) Behavior Analysis and Training: A Methodology for Behavior Engineering. *IEEE Transactions on Systems, Man and Cybernetics*, **26**, 365–380.
39. Colombetti, M., Dorigo, M., Borghi, G. (1996) Robot shaping: The HAMSTER Experiment. In M. Jamshidi et al (eds.), *Proceedings of ISRAM'96, Sixth International Symposium on Robotics and Manufacturing, Montpellier, France.*
40. Compiani, M., Montanari, D., Serra, R., Valastro, G. (1989) Classifier systems and neural networks. In *Parallel Architectures and Neural Networks – First Italian Workshop*, 105–118. World Scientific: Teaneck, NJ.
41. Davis, M. S. (2000) *A Computational Model of Affect Theory: Simulations of Reducer/Augmenter and Learned Helplessness Phenomena.* PhD thesis, Department of Psychology, University of Michigan.
42. De Jong, K. A. (1988) Learning with Genetic Algorithms: An Overview. *Machine Learning*, **3**, 121–138.

43. Donnart, J. Y., Meyer, J. A. (1994) A Hierarchical Classifier System Implementing a Motivationally Autonomous Animat. In Cliff et al. [34], 144–153.
44. Donnart, J. Y., Meyer, J. A. (1996) Hierarchical-map Building and Self-positioning with MonaLysa. *Adaptive Behavior*, **5**, 29–74.
45. Donnart, J. Y., Meyer, J. A. (1996) Spatial Exploration, Map Learning, and Self-Positioning with MonaLysa. In Pattie Maes, Maja J. Mataric, Jean-Arcady Meyer, Jordan Pollack, and Stewart W. Wilson (eds.), *From Animals to Animats 4. Proceedings of the Fourth International Conference on Simulation of Adaptive Behavior (SAB96)*, 204–213. A Bradford Book. MIT Press: Cambridge, MA.
46. Dorigo, M. (1991) New Perspectives about Default Hierarchies Formation in Learning Classifier Systems. In E. Ardizzone, E. Gaglio, and S. Sorbello (eds.), *Proceedings of the 2nd Congress of the Italian Association for Artificial Intelligence (AI*IA) on Trends in Artificial Intelligence*, 549 *LNAI*, 218–227, Palermo, Italy, Springer: Berlin.
47. Dorigo, M. (1991) Using Transputers to Increase Speed and Flexibility of Genetic-based Machine Learning Systems. *Microprocessing and Microprogramming*, **34**, 147–152.
48. Dorigo, M. (1993) Genetic and Non-Genetic Operators in ALECSYS. *Evolutionary Computation*, **1**, 151–164 Also Technical Report TR-92-075, International Computer Science Institute.
49. Dorigo, M. (1995) Alecsys and the AutonoMouse: Learning to Control a Real Robot by Distributed Classifier Systems. *Machine Learning*, **19**, 209–240.
50. Dorigo, M., Bersini, H. (1994) A Comparison of Q-Learning and Classifier Systems. In Cliff et al. [34], 248–255.
51. Dorigo, M., Colombetti, M. (1994) Robot Shaping: Developing Autonomous Agents through Learning. *Artificial Intelligence*, **2**, 321–370 ftp://iridia.ulb.ac.be/pub/dorigo/journals/IJ.05-AIJ94.ps.gz
52. Dorigo, M., Colombetti, M. (1998) *Robot Shaping: An Experiment in Behavior Engineering.* MIT Press/Bradford Books: Cambridge, MA.
53. Dorigo, M., Schnepf, U. (1993) Genetics-based Machine Learning and Behaviour Based Robotics: A New Synthesis. *IEEE Transactions on Systems, Man and Cybernetics*, **23**, 141–154.
54. Dorigo, M., Sirtori, E. (1991) Alecsys: A Parallel Laboratory for Learning Classifier Systems. In Booker and Belew [21], 296–302.
55. Druhan, B. B., Mathews, R. C. (1989) THIYOS: A Classifier System Model of Implicit Knowledge in Artificial Grammars. In *Proc. 11th Annual Conference of the Cognitive Science Society,* Hillsdale, NY: Lawrence Erlbaum Associates.
56. Escazut, C., Fogarty, T. C. (1997) Coevolving Classifier Systems to Control Traffic Signals. In John R. Koza (ed.), *Late Breaking Papers at the 1997 Genetic Programming Conference*, Stanford Bookstore: Stanford University, CA.
57. Farmer, J. D. (1990) A Rosetta Stone for Connectionism. In *Special issue of Physica D (Vol. 42)* [3], 153–187.
58. Farmer, J. D., Packard, N. H., Perelson, A. S. (1986) The Immune System, Adaptation & Learning. *Physica D*, **22**, 187–204.
59. Fayyad, U. M., Shapiro, G. P., Smyth, P., Uthurusamy, R. (1996) *Advances in Knowledge Discovery and Data Mining.* AAAI Press/The MIT Press: Cambridge, MA.

60. Federman, F. (2000) NEXTNOTE: A Learning Classifier System. In Annie S. Wu (ed.), *Proceedings of the Genetic and Evolutionary Computation Conference Workshop Program*, 136–138.
61. Federman, F., Dorchak, S. F. (1997) Information Theory and NEXTPITCH: A Learning Classifier System. In Bäck [8], 442–449.
62. Federman, F., Dorchak, S. F. (1997) Representation of Music in a Learning Classifier System. In Z.W. Rad and A. Skowron (eds.), *Foundations of Intelligent Systems: Proceedings 10th International Symposium (ISMIS'97)*. Springer: Heidelberg.
63. Federman, F., Dorchak, S. F. (1998) A Study of Classifier Length and Population Size. In Koza et al. [112], 629–634.
64. Federman, F., Sparkman, G., Watt, S. (1999) Representation of Music in a Learning Classifier System Utilizing Bach Chorales. In Banzhaf et al. [10], 785 (one page poster paper).
65. Forrest, S. (1985) Implementing semantic network structures using the classifier system. In Grefenstette [72], 24–44.
66. Forrest, S. (ed.) (1993) *Proceedings of the 5th International Conference on Genetic Algorithms (ICGA93)*. Morgan Kaufmann: San Fransico, CA.
67. Frey, P. W., Slate, D. J. (1991) Letter Recognition Using Holland-Style Adaptive Classifiers. *Machine Learning*, **6**, 161–182.
68. Garrell, J. M. et al. (1998) Automatic Classification of Mammary Biopsy Images with Machine Learning Techniques. In E. Alpaydin (ed.), *Proceedings of Engineering of Intelligent Systems (EIS'98)*, **3**, 411–418. ICSC Academic Press http://www.salleurl.edu/~xevil/Work/index.html
69. Garrell, J. M. et al. (1999) Automatic Diagnosis with Genetic Algorithms and Case-Based Reasoning. *Artificial Intelligence in Engineering, 13 (4)*, pp 367-372.
70. Gilbert, A. H., Bell, F., Valenzuela, C. L. (1995) Adaptive Learning of Process Control and Profit Optimisation using a Classifier System. *Evolutionary Computation*, **3**, 177–198.
71. Goldberg, D. E. (1989) *Genetic Algorithms in Search, Optimization, and Machine Learning*. Addison-Wesley: Reading, MA.
72. Grefenstette, J. J. (ed.) (1985) *Proceedings of the 1st International Conference on Genetic Algorithms and their Applications (ICGA85)*. Lawrence Erlbaum Associates: Pittsburgh, PA.
73. Grefenstette, J. J. (ed.) (1987) *Proceedings of the 2nd International Conference on Genetic Algorithms (ICGA87)*, Cambridge, MA, July 1987. Lawrence Erlbaum Associates: Pittsburgh, PA.
74. Grefenstette, J. J. (1988) Credit Assignment in Rule Discovery Systems Based on Genetic Algorithms. *Machine Learning*, **3**, 225–245.
75. Grefenstette, J. J. (1989) A System for Learning Control Strategies with Genetic Algorithms. In Schaffer [151], 183–190.
76. Grefenstette, J. J. (1991) Lamarckian Learning in Multi-Agent Environments. In Booker and Belew [21], 303–310. http://www.ib3.gmu.edu/gref/publications.html
77. Grefenstette, J. J. (1992) The Evolution of Strategies for Multi-agent Environments. *Adaptive Behavior*, **1**, 65–89 http://www.ib3.gmu.edu/gref/
78. Grefenstette, J. J. (ed.) (1992) Using a Genetic Algorithm to Learn Behaviors for Autonomous Vehicles. In *Proceedings American Institute of Aeronau-*

tics and Astronautics Guidance, Navigation and Control Conference, 739–749 AIAA http://www.ib3.gmu.edu/gref/
79. Grefenstette, J. J., Ramsey, C. L., Schultz, A. C. (1990) Learning Sequential Decision Rules using Simulation Models and Competition. *Machine Learning*, **5**, 355–381 http://www.ib3.gmu.edu/gref/publications.html
80. Grefenstette, J. J., Schultz, A. C. (1994) An Evolutionary Approach to Learning in Robots. In *Machine Learning Workshop on Robot Learning*, New Brunswick, NJ, http://www.ib3.gmu.edu/gref/
81. Hartley, A. (1999) Accuracy-based Fitness Allows Similar Performance to Humans in Static and Dynamic Classification Environments. In Banzhaf et al. [10], 266–273.
82. Hofmeyr, S. A., Forrest, S. (2000) Architecture for an Artificial Immune System. *Evolutionary Computation*, **8**.
83. Holland, J. (1971) Processing and Processors for Schemata. In E. L. Jacks (ed.), *Associative Information Processing*, 127–146. American Elsevier: New York.
84. Holland, J. (1976) Adaptation. In R. Rosen and F. M. Snell (eds.), *Progress in theoretical biology*. Plenum: New York, NY.
85. Holland, J. (1985) Properties of the Bucket Brigade. In Grefenstette [72], 1–7.
86. Holland, J. (1986) A Mathematical Framework for Studying Learning in Classifier Systems. *Physica D*, **22**, 307–317.
87. Holland, J. (1986) Escaping Brittleness: The Possibilities of General-Purpose Learning Algorithms Applied to Parallel Rule-Based Systems. In R. S. Michalski, J. G. Carbonell, and T. M. Mitchell (eds.), *Machine learning, an artificial intelligence approach. Volume II*, chapter 20, 593–623. Morgan Kaufmann: San Fransico, CA.
88. Holland, J. (1987) Genetic Algorithms and Classifier Systems: Foundations and Future Directions. In Grefenstette [73], 82–89.
89. Holland, J. (1990) Concerning the Emergence of Tag-Mediated Lookahead in Classifier Systems. In *Special issue of Physica D (Vol. 42)* [3], 188–201.
90. Holland, J. (1992) *Adaptation in Natural and Artificial Systems*. University of Michigan Press, Ann Arbor, 1975. Republished by the MIT Press
91. Holland, J. (1995) *Hidden Order: How Adaptation Builds Complexity*. Addison-Wesley: Reading, MA.
92. Holland, J. H., Reitman, J. S. (1978) Cognitive Systems Based on Adaptive Algorithms. In D. A. Waterman and F. Hayes-Roth (eds.), *Pattern-directed inference systems*. New York: Academic Press Reprinted in: *Evolutionary Computation. The Fossil Record*. David B. Fogel (ed.), IEEE Press, 1998.
93. Holland, J. H. et al. (1986) *Induction. Processes of Inference, Learning, and Discovery*. MIT Press: Cambridge, MA.
94. Holland, J. H. et al. (2000) What is a Learning Classifier System? In Lanzi et al. [120], 3–32.
95. Holmes, J. H. (1996) *Evolution-Assisted Discovery of Sentinel Features in Epidemiologic Surveillance*. PhD thesis, Drexel University http://cceb.med.upenn.edu/holmes/disstxt.ps.gz
96. Holmes, J. H. (1997) Discovering Risk of Disease with a Learning Classifier System. In Bäck [8]. http://cceb.med.upenn.edu/holmes/icga97.ps.gz
97. Holmes, J. H. (1998) Differential Negative Reinforcement Improves Classifier System Learning Rate in Two-class Problems with Unequal Base Rates. In Koza et al. [112], 635–642. http://cceb.med.upenn.edu/holmes/gp98.ps.gz

98. Holmes, J. H. (1999) Evaluating Learning Classifier System Performance In Two-Choice Decision Tasks: An LCS Metric Toolkit. In Banzhaf et al. [10], 789. (one page poster paper).
99. Holmes, J. H. (2000) Learning Classifier Systems Applied to Knowledge Discovery in Clinical Research Databases. In Lanzi et al. [120], 243–264.
100. Horn, J. (1997) *The Nature of Niching: Genetic Algorithms and the Evolution of Optimal, Cooperative Populations.* PhD thesis, University of Illinois at Urbana-Champaign (UMI Dissertation Service No. 9812622).
101. Horn, J., Goldberg, D. E. (1996) Natural Niching for Cooperative Learning in Classifier Systems. In Koza et al. [111], 553–564.
102. Horn, J., Goldberg, D. E., Deb, K. (1994) Implicit Niching in a Learning Classifier System: Nature's Way. *Evolutionary Computation*, **2**, 37–66 Also IlliGAL Report No 94001.
103. Kovacs, T. A Learning Classifier Systems Bibliography. Online at `http://www.cs.bris.ac.uk/~kovacs/lcs/`
104. Kovacs, T. (1997) XCS Classifier System Reliably Evolves Accurate, Complete, and Minimal Representations for Boolean Functions. In Roy, Chawdhry, and Pant (eds.), *Soft Computing in Engineering Design and Manufacturing*, 59–68. Springer: London. ftp://ftp.cs.bham.ac.uk/pub/authors/T.Kovacs/index.html
105. Kovacs, T. (1999) Deletion Schemes for Classifier Systems. In Banzhaf et al. [10], 329–336. Also Technical Report CSRP-99-08, School of Computer Science, University of Birmingham. http://www.cs.bham.ac.uk/~tyk
106. Kovacs, T. (2000) Strength or Accuracy? Fitness Calculation in Learning Classifier Systems. In Lanzi et al. [120], 143–160.
107. Kovacs, T., Kerber, M. (2000) Some Dimensions of Problem Complexity for XCS. In *Proceedings of the GECCO-2000 Graduate Student Workshop.*
108. Kovacs, T., Kerber, M. (2000) What Makes a Problem Hard for XCS? In *Proceedings of the International Workshop on Learning Classifier Systems (IWLCS-2000), in the Joint Workshops of SAB 2000 and PPSN 2000* [5]. Extended abstract.
109. Kovacs, T. (2001) Towards a Theory of Strong Overgeneral Classifiers. In Worthy Martin and William Spears (eds.), *Foundations of Genetic Algorithms 6*, pp. 165-184. Morgan Kaufmann: San Francisco, CA.
110. Koza, J. (1992) *Genetic Programming.* MIT Press: Cambridge, MA.
111. Koza, J. et al. (1996) *Genetic Programming 1996: Proceedings of the First Annual Conference*, Stanford University, CA, MIT Press: Cambridge MA.
112. Koza, J. et al. (1998) *Genetic Programming 1998: Proceedings of the Third Annual Conference.* Morgan Kaufmann: San Francisco, CA.
113. Lanzi, P. L. (1998) Adding Memory to XCS. In *Proceedings of the IEEE Conference on Evolutionary Computation (ICEC98).* IEEE Press http://ftp.elet.polimi.it/people/lanzi/icec98.ps.gz
114. Lanzi, P. L. (1998) An Analysis of the Memory Mechanism of XCSM. In Koza et al. [112], 643–651. http://ftp.elet.polimi.it/people/lanzi/gp98.ps.gz
115. Lanzi, P. L. (1999) Extending the Representation of Classifier Conditions Part I: From Binary to Messy Coding. In Banzhaf et al. [10], 337–344.
116. Lanzi, P. L. (1999) Extending the Representation of Classifier Conditions Part II: From Messy Coding to S-Expressions. In Banzhaf et al. [10], 345–352.

117. Lanzi, P. L. (2000) Adaptive Agents with Reinforcement Learning and Internal Memory. In *Proceedings of the Sixth International Conference on the Simulation of Adaptive Behavior (SAB2000).*
118. Lanzi, P. L. (2000) Learning Classifier Systems from a Reinforcement Learning Perspective. Technical Report 00-03, Dipartimento di Elettronica e Informazione, Politecnico di Milano.
119. Lanzi, P. L., Wilson, S. W. (2001) Toward Optimal Performance in Classifier Systems. *Evolutionary Computation, 8(4):* 393-418.
120. Lanzi, P. L. et al. (eds.) (2000) *Learning Classifier Systems: From Foundations to Applications*, volume 1813 of *LNAI*. Springer: Berlin.
121. LeBaron, Blake et al. (1999) The Time Series Properties of an Artificial Stock Market. *Journal of Economic Dynamics and Control, 23 (October), 1487-1516.*
122. Liepins, G. E., Wang, L. A. (1991) Classifier System Learning of Boolean Concepts. In Booker and Belew [21], 318–323.
123. Liepins, G. E. et al. (1989) Alternatives for Classifier System Credit Assignment. In *Proceedings of the Eleventh International Joint Conference on Artificial Intelligence (IJCAI-89)*, 756–761.
124. Liepins, G. E. et al. (1991) Credit Assignment and Discovery in Classifier Systems. *International Journal of Intelligent Systems*, 6, 55–69.
125. Marimon, R. et al. (1990) Money as a Medium of Exchange in an Economy with Artificially Intelligent Agents. *Journal of Economic Dynamics and Control*, **14**, 329–373. Also Technical Report 89-004, Santa Fe Institute.
126. Meyer, J. A., Wilson, S. W. (eds.) (1990) *From Animals to Animats 1. Proceedings of the First International Conference on Simulation of Adaptive Behavior (SAB90).* A Bradford Book. MIT Press: Cambridge, MA.
127. Miller, J. H., Holland, J. H. (1991) Artificial Adaptive Agents in Economic Theory. *American Economic Review*, **81**, 365–370.
128. Mitlöhner, J. (1996) Classifier Systems and Economic Modelling. In *APL '96. Proceedings of the APL 96 Conference on Designing the Future*, **26**, 77–86.
129. Moriarty, D. E., Schultz, A. C., Grefenstette, J. J. (1999) Evolutionary Algorithms for Reinforcement Learning. *Journal of Artificial Intelligence Research*, **11**, 199–229
130. Nagasaka, I., Taura, T. (1997) 3D Geometic Representation for Shape Generation using Classifier System. In Koza, John R. et al. *Genetic Programming 1997: Proceedings of the Second Annual Conference,* 515–520. Morgan Kaufmann: San Francisco, CA.
131. Patel, M., Dorigo, M. (1994) Adaptive Learning of a Robot Arm. In Terence C. Fogarty (ed.), *Evolutionary Computing, AISB Workshop Selected Papers*, Volume 865 of LNCS, 180–194. Springer: Berlin.
132. Rawlins, G. (ed.) (1991) *Proceedings of the First Workshop on Foundations of Genetic Algorithms (FOGA91).* Morgan Kaufmann: San Mateo, CA.
133. Richard, R. (1995) *Zeroth-Order Shape Optimization Utilizing a Learning Classifier System.* PhD thesis, Stanford University. Online version available at: http://www-leland.stanford.edu/~buc/SPHINcsX/book.html
134. Richards, R., Sheppard, S. (1992) Classifier System Based Structural Component Shape Improvement Utilizing I-DEAS. In *Iccon User's Conference Proceeding.* Iccon

135. Richards, R., Sheppard, S. (1992) Learning Classifier Systems in Design Optimization. In *Design Theory and Methodology '92.* The American Society of Mechanical Engineers.
136. Richards, R., Sheppard, S. (1992) Two-dimensional Component Shape Improvement via Classifier System. In Gero, J. S. (editor), *Artificial Intelligence in Design '92.* Kluwer Academic Publishers: Dordrecht.
137. Richards, R., Sheppard, S. (1996) A Learning Classifier System for Three-dimensional Shape Optimization. In H. M. Voigt, W. Ebeling, I. Rechenberg, and H. P. Schwefel (eds.), *Parallel Problem Solving from Nature – PPSN IV*, Volume 1141 of *LNCS*, 1032–1042. Springer: Berlin.
138. Richards, R., Sheppard, S. (1996) Three-Dimensional Shape Optimization Utilizing a Learning Classifier System. In Koza et al. [111], 539–546.
139. Riolo, R. L. (1987) Bucket Brigade Performance: II. Default Hierarchies. In Grefenstette [73], 196–201.
140. Riolo, R. L. (1988) CFS-C: A Package of Domain-Independent Subroutines for Implementing Classifier Systems in Arbitrary User-Defined Environments. Technical Report, University of Michigan.
141. Riolo, R. L. (1988) *Empirical Studies of Default Hierarchies and Sequences of Rules in Learning Classifier Systems.* PhD thesis, University of Michigan.
142. Riolo, R. L. (1989) The Emergence of Coupled Sequences of Classifiers. In Schaffer [151], 256–264.
143. Riolo, R. L. (1989) The Emergence of Default Hierarchies in Learning Classifier Systems. In Schaffer [151], 322–327.
144. Riolo, R. L. (1991) Lookahead Planning and Latent Learning in a Classifier System. In Meyer and Wilson [126], 316–326.
145. Riolo, R. L. (1991) Modeling Simple Human Category Learning with a Classifier System. In Booker and Belew [21], 324–333.
146. Robertson, G. G., Riolo, R. L. (1988) A Tale of Two Classifier Systems. *Machine Learning*, **3**, 139–159.
147. Samuel, A. L. (1959) Some Studies in Machine Learning Using the Game of Checkers. In E. A. Feigenbaum and J. Feldman (eds.), *Computers and Thought.* McGraw-Hill: New York.
148. Sanza, C. et al. (1998) A Learning Method for Adaptation and Evolution in Virtual Environments. In *3rd International Conference on Computer Graphics and Artificial Intelligence, April 1998, Limoges, France.*
149. Satterfield, T. (1999) *Bilingual Selection of Syntactic Knowledge: Extending the Principles and Parameters Approach.* Kluwer: Amsterdam.
150. Saxon, S., Barry, A. (2000) XCS and the Monk's Problems. In Lanzi et al. [120], 223–242.
151. Schaffer, J. D. (ed.) (1989) *Proceedings of the 3rd International Conference on Genetic Algorithms (ICGA89)*, George Mason University, Morgan Kaufmann: San Francisco, CA.
152. Schulenburg, S., Ross, P. (2000) An Adaptive Agent Based Economic Model. In Lanzi et al. [120], 265–284.
153. Schulenburg, S., Ross, P. (2000) Strength and Money: An LCS Approach to Increasing Returns. In *Proceedings of the International Workshop on Learning Classifier Systems (IWLCS-2000), in the Joint Workshops of SAB 2000 and PPSN 2000* [5]. Extended abstract

154. Schuurmans, D., Schaeffer, J. (1989) Representational Difficulties with Classifier Systems. In Schaffer [151], 328–333. http://www.cs.ualberta.ca/ jonathan/Papers/Papers/classifier.ps
155. Sen, S. (1993) Improving Classification Accuracy Through Performance History. In Forrest [66], 652–652.
156. Sen, S. (1994) A Tale of Two Representations. In *Proc. 7th International Conference on Industrial and Engineering Applications of Artificial Intelligence and Expert Systems*, 245–254.
157. Shu, L., Schaeffer, J. (1989) VCS: Variable Classifier System. In Schaffer [151], 334–339. http://www.cs.ualberta.ca/~jonathan/Papers/Papers/vcs.ps
158. Smith, R. E. (1991) *Default Hierarchy Formation and Memory Exploitation in Learning Classifier Systems.* PhD thesis, University of Alabama.
159. Smith, R. E. (1993) Genetic Learning in Rule-Based and Neural Systems. In *Proceedings of the Third International Workshop on Neural Networks and Fuzzy Logic*, **1**, 183. NASA Johnson Space Center: Houston, Texas.
160. Smith, R. E. (1994) Memory Exploitation in Learning Classifier Systems. *Evolutionary Computation*, **2**, 199–220.
161. Smith, R. E., Cribbs, H. (1994) Is a Learning Classifier System a Type of Neural Network? *Evolutionary Computation*, **2**, 19–36.
162. Smith, R. E., Cribbs, H. B. (1997) Combined Biological Paradigms. *Robotics and Autonomous Systems*, **22**, 65–74.
163. Smith, R. E., Goldberg, D. E. (1991) Variable Default Hierarchy Separation in a Classifier System. In Rawlins [132], 148–170.
164. Smith, R. E. et al. (1999) Classifier Systems in Combat: Two-sided Learning of Maneuvers for Advanced Fighter Aircraft. In *Computer Methods in Applied Mechanics and Engineering.* Elsevier: Amsterdam.
165. Smith, R. E. et al. (2000) The Fighter Aircraft LCS: A Case of Different LCS Goals and Techniques. In Lanzi et al. [120], 285–302.
166. Stolzmann, W. (1996) Learning Classifier Systems using the Cognitive Mechanism of Anticipatory Behavioral Control, detailed version. In *Proceedings of the First European Workshop on Cognitive Modelling*, 82–89. TU Berlin. http://www.psychologie.uni-wuerzburg.de/stolzmann/
167. Stolzmann, W. (1997) Two Applications of Anticipatory Classifier Systems (ACSs). In *Proceedings of the 2nd European Conference on Cognitive Science*, 68–73. Manchester, U.K. http://www.psychologie.uni-wuerzburg.de/stolzmann/
168. Stolzmann, W. (1998) Anticipatory Classifier Systems. In *Proceedings of the Third Annual Genetic Programming Conference*, 658–664, Morgan Kaufmann: San Francisco, CA. http://www.psychologie.uni-wuerzburg.de/stolzmann/gp-98.ps.gz
169. Stolzmann, W. (2000) An Introduction to Anticipatory Classifier Systems. In Lanzi et al. [120], 175–194.
170. Stolzmann, W., Butz, M. (2000) Latent Learning and Action-Planning in Robots with Anticipatory Classifier Systems. In Lanzi et al. [120], 303–320.
171. Stolzmann, W. et al. (2000) First Cognitive Capabilities in the Anticipatory Classifier System. In *From Animals to Animats: Proceedings of the Sixth International Conference on Simulation of Adaptive Behavior.* MIT Press: Cambridge MA.
172. Sutton, R. S. (1988) Learning to Predict by the Methods of Temporal Differences. *Machine Learning 3*, 9–44.

173. Sutton, R. S. (1990) Integrated Architectures for Learning, Planning, and Reacting Based on Approximating Dynamic Programming. In *Proceedings of the Seventh International Conference on Machine Learning*, 216–224, Austin, TX, Morgan Kaufmann: San Francisco, CA.
174. Tesfatsion, L. (2000) Introduction to the Special Issue on Agent Based Computational Economics. *Journal of Computational Economics, 18(1)*, 1-8.
175. Thrun, S. B. et al. (1991) The MONK's Problems: A performance comparison of different learning algorithms. Technical Report CS-91-197, Carnegie Mellon University, Pittsburgh, PA.
176. Tomlinson, A., Bull, L. (1998) A Corporate Classifier System. In A. E. Eiben, T. Bäck, M. Shoenauer, and H.-P Schwefel (eds.), *Proceedings of the Fifth International Conference on Parallel Problem Solving From Nature – PPSN V*, Volume 1498 of LNCS, 550–559. Springer: Berlin.
177. Tomlinson, A., Bull, L. (1999) A Zeroth Level Corporate Classifier System. In Wu [203], 306–313.
178. Tomlinson, A., Bull, L. (1999) On Corporate Classifier Systems: Increasing the Benefits of Rule Linkage. In Banzhaf et al. [10], 649–656.
179. Tomlinson, A., Bull, L. (2000) A Corporate XCS. In Lanzi et al. [120], 194–208.
180. Tufts, P. (1995) Dynamic Classifiers: Genetic Programming and Classifier Systems. In E. V. Siegel and J. R. Koza (eds.), *Working Notes for the AAAI Symposium on Genetic Programming*, 114–119. MIT Press: Cambridge, MA. Home page: http://www.cs.brandeis.edu/~zippy/papers.html
181. Twardowski, K. (1993) Credit Assignment for Pole Balancing with Learning Classifier Systems. In Forrest [66], 238–245.
182. Valenzuela-Rendón, M. (1991) The Fuzzy Classifier System: a Classifier System for Continuously Varying Variables. In Booker and Belew [21], 346–353.
183. Vriend, N. J. (1999) On Two Types of GA-Learning. In S. H. Chen (ed.), *Evolutionary Computation in Economics and Finance*. Springer: Berline.
184. Vriend, N. J. (1999) The Difference Between Individual and Population Genetic Algorithms. In W. Banzhaf et al. [10], p 812.
185. Vriend, N. J. (2000) An Illustration of the Essential Difference between Individual and Social Learning, and its Consequences for Computational Analyses. *Journal of Economic Dynamics and Control*, **24**, 1–19.
186. Watkins, C. (1989) *Learning from Delayed Rewards*. PhD thesis, King's College.
187. Weiss, G. (1991) The Action-Oriented Bucket Brigade. Technical Report FKI-156-91, Technical Univ. München (TUM)
188. Weiss, G. (1996) An Action-oriented Perspective of Learning in Classifier Systems. *Journal of Experimental and Theoretical Artificial Intelligence*, **8**, 43–62.
189. Whitely, D. et al. (eds.) (2000) *Proceedings of the Genetic and Evolutionary Computation Conference (GECCO-2000)*. Morgan Kaufmann: San Francisco, CA.
190. Wilcox, J. R. (1995) Organizational Learning within a Learning Classifier Systems. Master's thesis, University of Illinois. Also Techical Report No. 95003 IlliGAL
191. Wilson, S. W. (1985) Knowledge Growth in an Artificial Animal. In Grefenstette [72], 16–23.

192. Wilson, S. W. (1987) Classifier Systems and the Animat Problem. *Machine Learning*, **2**, 199–228, Also Research Memo RIS-36r, the Rowland Institute for Science, Cambridge, MA.
193. Wilson, S. W. (1988) Bid Competition and Specificity Reconsidered. *Complex Systems*, **2**, 705–723.
194. Wilson, S. W. (1992) Classifier System Mapping of Real Vectors. In *Collected Abstracts for the First International Workshop on Learning Classifier System (IWLCS-92)* [4]. October 6–8, NASA Johnson Space Center, Houston, Texas.
195. Wilson, S. W. (1994) ZCS: A zeroth level classifier system. *Evolutionary Computation*, **2**, 1–18. http://prediction-dynamics.com/
196. Wilson, S. W. (1995) Classifier Fitness Based on Accuracy. *Evolutionary Computation*, **3**, 149–175. http://prediction-dynamics.com/
197. Wilson, S. W. (1999) Get Real! XCS with Continuous-valued Inputs. In L. Booker, Stephanie Forrest, M. Mitchell, and Rick L. Riolo (eds.), *Festschrift in Honor of John H. Holland*, 111–121. Center for the Study of Complex Systems. http://www.pscs.umich.edu/jhhfest/proceedings.html
198. Wilson, S. W. (2000) Get Real! XCS with Continuous-Valued Inputs. In Lanzi et al. [120], 209–220.
199. Wilson, S. W. (2000) Mining Oblique Data with XCS. In *Proceedings of the International Workshop on Learning Classifier Systems (IWLCS-2000), in the Joint Workshops of SAB 2000 and PPSN 2000* [5]. Extended abstract
200. Wilson, S. W. (2000) State of XCS Classifier System Research. In Lanzi et al. [120], 63–82.
201. Wilson, S. W. Prediction Dynamics Homepage `http://prediction-dynamics.com`
202. Wilson, S. W., Goldberg, D. E. (1989) A Critical Review of Classifier Systems. In Schaffer [151], 244–255. http://prediction-dynamics.com/
203. Wu, A. S. (ed.) (1999) *Proceedings of the 1999 Genetic and Evolutionary Computation Conference Workshop Program*

Beyond Samuel: Evolving a Nearly Expert Checkers Player

David B. Fogel

Natural Selection, Inc. 3333 N. Torrey Pines Ct., Ste. 200, La Jolla, CA 92037
E-mail: dfogel@natural-selection.com

Summary. Evolutionary algorithms can be used to learn how to play complex games of strategy without relying on human expertise. Here I discuss the use of evolutionary computation and artificial neural networks in learning how to play checkers. Starting from neural networks that were created randomly, an evolutionary algorithm has been able to craft a network that can play checkers at a nearly expert level. No features beyond the positions of pieces on the board and the piece differential were provided. The evolutionary algorithm learned everything else on its own, simply by playing the game.

1 Introduction

Imagine yourself seated at a table. In front of you is an 8×8 board of squares that alternate colors. I'm seated across from you and tell you that we're going to play a game. Each of us starts with 12 pieces placed on alternating squares as shown in Fig. 1. You are playing "white" while I'm playing "red." The red player moves first. Initially, you think I might be giving myself an advantage, but since you have no idea of what game we are about to play you continue to listen.

You can only move your pieces diagonally forward one square at a time, unless they are next to an opponent's piece and there is an empty square directly behind that piece, in which case you are forced to jump over their piece. In fact, you must continue jumping over opponent's pieces in succession if possible. In doing so, you also remove your opponent's piece(s) from the board. If you have more than one possibility to jump your opponent's pieces then you can choose which way you'd like to execute the jump. If any of your pieces make it to the back row of the board they become a special piece called a "king" and this piece can move forward or backward diagonally, again one square at a time unless it is involved in a jump.

"Let's play," I say. You protest naturally that I haven't told you the object of the game. "Right, let's play," and I make my first move. You counter with a move. We play for several more moves and eventually I declare the game to be over.

"Let's play again," I suggest.
"But wait," you say, "did I win, or lose, or draw, or what?"
"I'm not telling yet. Let's play again."

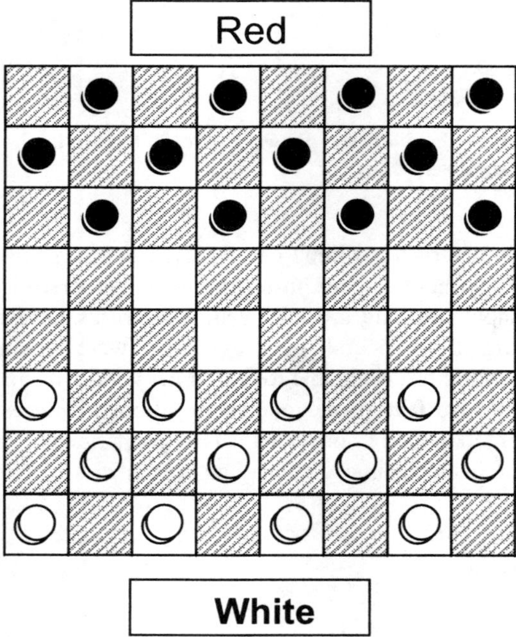

Fig. 1. The opening board position of an unknown game

Now imagine that we play five such games, and only after the fifth game do I tell you that you earned, say, seven points for playing those games. I don't tell you which games earned you the points, or even indeed if you might have started with, say, 20 points and lost points on each and every game. The only way for you to find out is to play another series of five games and compare the total points that you receive after that series.

Here's the critical thought experiment: How long would it take you to become an expert at this game? How many games would you have to play? What features about the play of the game would you look for? One obvious feature might be the piece differential: the difference in the number of pieces that I have and the number of pieces that you have. It would also be fairly easy to correlate whether or not it was good to be ahead on pieces or behind once you find that the game ends consistently when one player has no more pieces and that, over many trials, the player who ends with no moves receives fewer points. But how long would it take you to become really competent in this game?

The game at hand, of course, is checkers, a common board game of skill that does not include any randomness in play (unlike backgammon, for example). Ever since computers were first developed, there has been interest in designing computer algorithms to play games like checkers. Chess has received the most attention. The success of Deep Blue over world chess champion Garry Kasparov in 1997 was recognized widely in both the popular and scientific press. Other games have also been tackled, including checkers, othello, and backgammon. With a few exceptions, these efforts have relied on domain-specific information programmed into an algorithm in the form of weighted features that were believed to be important for assessing the relative worth of alternative positions in the game. The programs relied, essentially, on human expertise to defeat human expertise.

Defeating a human champion, or even an everyday person that you meet, in a game of skill using just a computer algorithm is a noteworthy accomplishment. But the significance of such accomplishments in terms of computational intelligence can be quite minor. After all, if we program human expertise in the form of chunks of knowledge and rules about how to analyze different situations then we can rely on the sheer speed of the computer to evaluate many more positions than any human could ever imagine. Deep Blue evaluated 200 million chess boards every second. No human can match this, or even come close. But where is the intelligence in an automaton like Deep Blue? Everything it "knows," it knows because it was effectively told. It learned nothing on its own. A system that never learns, and has no capability of ever learning, does not deserve the description of intelligent. As I wrote in [1], intelligence is the ability to adapt behavior to meet desired goals in a range of environments. Without an ability to adapt, to learn which behaviors are appropriate in different settings, there can be no intelligence.

Rather than rely on existing human expertise to generate a program that could play a game of skill, like checkers or chess, a more significant challenge lies in having an algorithm learn competent strategies without such knowledge, simply by playing successive games between candidate strategies. Those that do well are favored over those that do poorly. Variations can be made to those that do well and the process of playing games can then be iterated. The hope is that a competent algorithm for playing the game, whatever that game might be, would emerge from this process after several iterations. Such a protocol matches nicely within the framework of evolutionary computation. By simulating the evolutionary process where individuals compete for survival, it is possible to use the computer as a tool for problem solving and that tool can be used to address games where optimal strategies are unknown. Checkers is one such game.

2 A Brief History of Computer Checkers

There have been many attempts to design programs to play checkers since the late 1950s. These are documented in [2] and for the sake of space only two are described here. The current computer world champion checkers program is *Chinook*, designed by Schaeffer et al. at the University of Alberta. The program uses a linear handcrafted evaluation function that considers several features of the game board including: 1) piece count, 2) kings count, 3) trapped kings, 4) turn, 5) runaway checkers (unimpeded path to a king), and other minor factors [2, pp. 63-65]. In addition, the program has: 1) access to a library of opening moves from games played by grand masters, 2) the complete endgame database for all boards with eight or fewer pieces, and 3) a "fudge factor" that was chosen to favor boards with more pieces than fewer pieces. This last facet was included to present more complicated positions to human opponents in the hopes of eliciting a mistake. No machine learning methods have been employed successfully in the development of Chinook. All of its "knowledge" has been programmed by humans. Chinook played to a draw after six games in a 40-game match against the former human world champion Marion Tinsley, the best player to have lived. From 1950 until his death in 1955, Tinsley won every tournament in which he played and lost only three games. Tinsley retired from the match with Chinook after six games for health reasons and died shortly thereafter. In 1996, Chinook was rated at 2814, well beyond the ratings of 2632 and 2625 held by the current best human players [2, p. 447] (see the results section below for a more detailed explanation of the ratings system).

In contrast to Chinook, the most well-known effort in designing an algorithm to play checkers is owed to Samuel [3], which was one of the first apparently successful experiments in machine learning. The method relied in part on the use of a polynomial evaluation function comprising a subset of weighted features chosen from a larger list of possibilities. The polynomial was used to evaluate alternative board positions some number of moves into the future using a minimax strategy. The technique relied on an innovative self-learning procedure whereby one player competed against another. The loser was replaced with a deterministic variant of the winner by altering the weights on the features that were used, or in some cases by replacing features that had very low weight with other features. Samuel's program, which also included rote learning of games played by masters, was played against and defeated R.W. Nealey in 1962. IBM Research News described Nealey as "a former Connecticut checkers champion, and one of the nation's foremost players."

Unfortunately, the success of Samuel's effort was overstated, and it continues to be overstated. Consider the following:

1. Nealey, in fact, only became a Connecticut champion later, and his level of play on a national level was uncalibrated.
2. The game itself was not well played: using Chinook, Schaeffer [2, pp. 93-97] showed that both Nealey and Samuel's program made several errors.
3. Nealey defeated Samuel's program in a rematch the next year, and Samuel played four games with his program against both the world champion and challenger in 1966, losing all eight games.
4. Subsequent judgment by the editor of a checkers magazine in the mid-1970s put Samuel's program below the "Class B" level. To place this in context, ratings are assigned by points: 2400+ is grand master, 2200-2399 is master, 2000-2199 is expert, 1800-1999 is Class A, 1600–1799 is Class B, as so forth.

In retrospect, Samuel's program had one apparently "lucky" and widely publicized early victory. As Schaeffer [2] wrote: "The promise of the 1962 Nealey game was an illusion."

The promise of machine learning methods for addressing complicated games, however, is not an illusion. Reinforcement learning and evolutionary computation have been employed with success in backgammon [4], [5], particularly when using neural networks to determine the appropriate moves. Of particular interest is the possibility of using neural networks to extract nonlinear structure inherent in a problem domain [6]. More specifically, a significant challenge is to devise a method for having neural networks learn how to play a game such as checkers without being given expert knowledge in the form of weighted features, prior games from masters, look-up tables of enumerated endgame positions, or other similar information. This would appear to be a necessary precursor to any effort to generate machine intelligence that is capable of solving new problems in new ways. The measure of success is the level of play that can be attained against humans without preprogramming in the requisite knowledge to play well.

Early speculation on the potential success of just these sorts of efforts was entirely negative. Newell in [7] offered "It is extremely doubtful whether there is enough information in 'win, lose, or draw' when referred to the whole play of the game to permit any learning at all over available time scales." Minsky [7] noted being in "complete agreement" with Newell. To test the veracity of these conjectures, Kumar Chellapilla – a Ph.D. student at the University of California at San Diego – and I designed an experiment in evolutionary algorithms whereby neural networks that represent strategies for playing checkers were competed against themselves starting from completely random initializations. Points were assigned for achieving a win, loss, or draw in each of a series of games. Only the total point score was used to represent the quality of a neural network's play (i.e., no credit assignment was employed, even to the level of identifying which games were won, lost, or tied). Those networks with the highest scores were maintained as parents for the next generation. Offspring networks were created by varying the connection weights of their parents randomly, and the process was iterated. Two different versions of neural networks have been tested. The first scanned the checkerboard from left to right, and top to bottom into a 1×32 vector. Very little of the spatial information from the 8×8 checkerboard was preserved in this encoding. The neural networks had to learn about the spatial nature of the game. The second representation included all possible sub-boards of size 3×3 up to 8×8 as input to the neural networks. The neural networks were still required to determine what to do with these inputs (i.e., no explicit features of a spatial nature were introduced). Details on these two approaches can be found in [8] and [9], respectively. The results from both approaches generated strategies that were nearly on par with human experts. For the sake of space, only the latter approach that used spatial sections on the checkerboard as inputs will be described in depth here.

3 Experimental Method

Each checkerboard was represented by a vector of length 32, with each component corresponding to an available position on the board. Components in the vector could take on elements from $\{-K, -1, 0, +1, +K\}$, where K was the evolvable value assigned for a king, 1 was the value for a regular checker, and 0 represented an empty square. The sign of the value indicated whether or not the piece in question belonged to the player (positive) or the opponent (negative). A player's move was determined by evaluating the presumed quality of the potential future positions. This evaluation function was structured as a feedforward neural network with an input layer, three hidden layers, and an output node. The second and third hidden layers, and the output layer had a fully connected structure, while connections in the first hidden layer were specially designed to possibly capture spatial information from the board. The nonlinear function at each hidden and output node was the hyperbolic tangent (tanh, bounded by ± 1) with a variable bias term, although other sigmoidal functions could undoubtedly have been chosen. In addition, the sum of all entries in the input vector was supplied to the output node directly.

So as not to handicap the learning procedure by forcing it to learn the obvious spatial characteristics of the game, the neural network used implemented a series of 91 preprocessing nodes that covered $n \times n$ square overlapping subsections of the board. These $n \times n$ subsections were chosen to provide spatial adjacency or proximity information such as whether two squares were neighbors, or were close to each other, or were far apart. All 36 possible 3×3 square subsections of the board were provided as input to the first 36 hidden nodes in the first layer. The following 25 4×4 square subsections were assigned to the next 25 hidden nodes in that layer, and so forth. Fig. 2 shows a sample 3×3 square subsection that contains the states of positions 1, 5, 6, and 9. Two sample 4×4 subsections are also shown. All possible square subsections of size 3 to 8 (the entire board) were given as inputs to the 91 nodes of the first hidden layer. This enabled the neural network to generate features from these subsets of the entire board that could then be processed in subsequent hidden layers (of 40 and 10 hidden units). Fig. 3 shows the general structure of the "spatial" neural network. At each generation, a player was defined by their associated neural network in which all of the connection weights (and biases) and king value were evolvable.

It is important to recall that, with one exception, no attempt was made to offer useful features as input to a player's neural network. The experimental question at hand concerned the level of play that could be attained simply by using evolution to extract linear and nonlinear features regarding the game of checkers and to optimize the interpretation of those features within the neural network. The only feature that could be claimed to have been offered is a function of the piece differential between a player and its opponents, owing to the sum of the inputs being supplied directly to the output node. The output essentially sums all the inputs and this yields the piece advantage or disadvantage. But this is not true in general, for when kings are present on the board, the value K or $-K$ is used in the summation, and as described below, this value was evolvable rather than prescribed by us. Thus the evolutionary algorithm had the potential to override the piece differential and invent a new feature in its place. Absolutely no other explicit or implicit features of the board beyond the location of each piece were implemented.

When a board was presented to a neural network for evaluation, its scalar output was interpreted as the worth of that board from the position of the player whose pieces were denoted by positive values. The closer the output was to 1.0, the better the evaluation of the corresponding input board. Similarly, the closer the output was to –1.0, the worse the board. All positions that were wins for the player (e.g., no remaining opposing pieces) were assigned the value of 1.0, and likewise all positions that were losses were assigned the value –1.0.

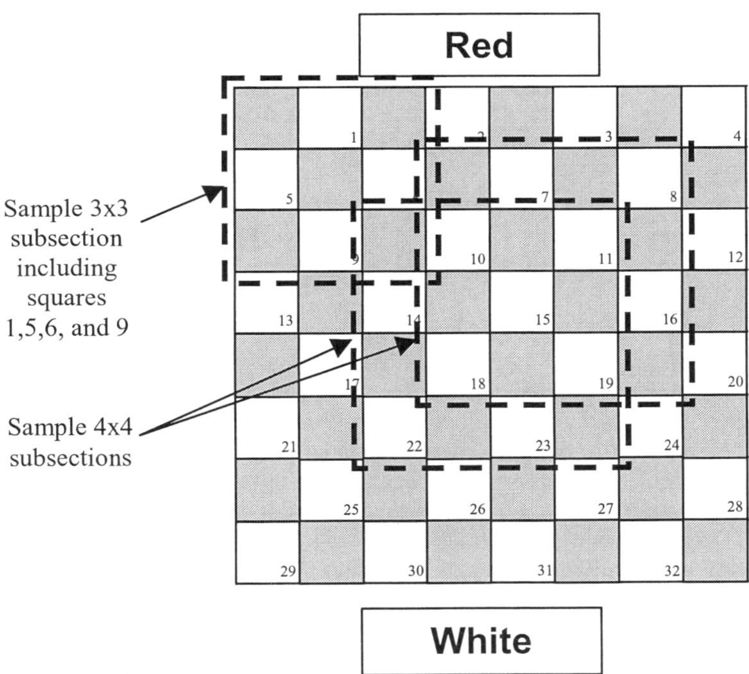

Fig. 2. The first hidden layer of the neural networks assigned one node to each possible subset of the entire board. In this manner, the neural network was able to invent features based on spatial characteristics of checkers on the board. Subsequent processing in the second and third hidden layer then operated on the features that were evolved in the first layer.

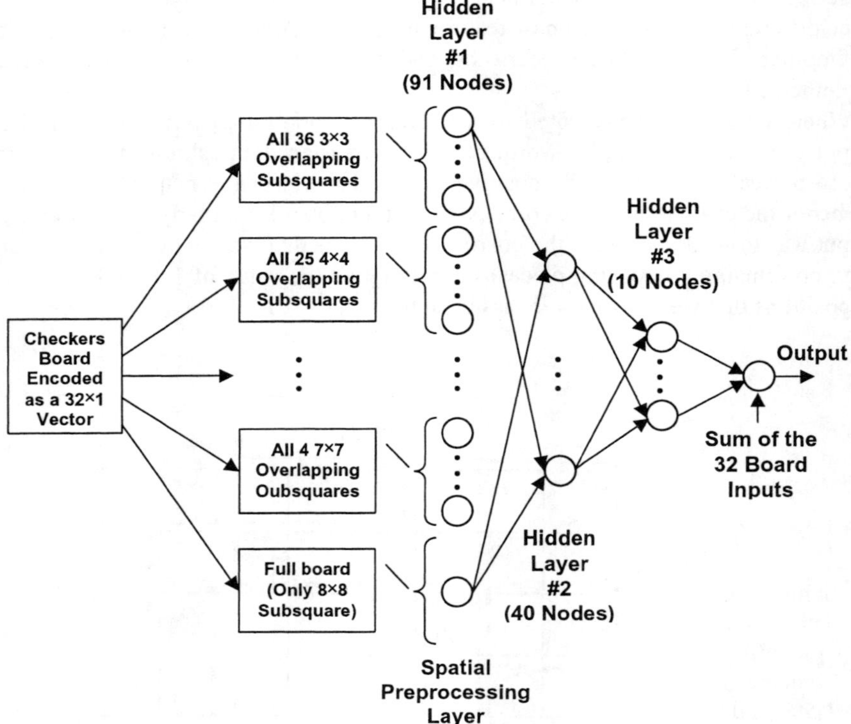

Fig. 3. The complete "spatial" neural network used to represent strategies. The network served as the board evaluation function. Given any board pattern as a vector of 32 inputs, the first hidden layer (spatial preprocessing) assigned a node to each possible square subset of the board. The output from these nodes were then passed through two additional hidden layers of 40 and 10 nodes, respectively. The final output node was scaled between [−1, 1] and included the sum of the input vector as an additional input. All of the weights and bias terms of the network were evolved, as well as the value used to describe kings.

The evolutionary algorithm began with a population of 15 strategies (neural networks), P_i, i = 1, …, 15, defined by the weights and biases for each neural network, and the strategy's associated value of K, that was created at random. Weights and biases were generated by sampling from a uniform distribution over [−0.2, 0.2], with the value of K set initially to 2.0. Each strategy had an associated self-adaptive parameter vector σ_i, i = 1, …, 15, where each component corresponded to a weight or bias and served to control the step size of the search for new mutated parameters of the neural network. To be consistent with the range of initialization, the self-adaptive parameters for weights and biases were set initially to 0.05.

Each parent generated one offspring strategy by varying all of the associated weights and biases, and possibly the value of K as well. Specifically, for each parent P_i, $i = 1, \ldots, 15$, an offspring P'_i, $i = 1, \ldots, 15$, was created by:

$$\sigma'_i(j) = \sigma_i(j)\exp(\tau N_j(0,1)),\ j = 1, \ldots, N_w.$$

$$w'_i(j) = w_i(j) + \sigma'_i(j)N_j(0,1),\ j = 1, \ldots, N_w.$$

where N_w is the number of weights and biases in the neural network (here this was 5046), $\tau = 1/[2(N_w)^{0.5}]^{0.5} = 0.0839$, and $N_j(0,1)$ is a standard Gaussian random variable resampled for every j. The offspring king value K' was obtained by

$$K'_i = K_i + \delta$$

where δ was chosen uniformly at random from $\{-0.1, 0.1\}$. For convenience, the value of K'_i was constrained to lie in [1.0, 3.0] by resetting to the limit exceeded when applicable.

All parents and their offspring competed for survival by playing games of checkers and receiving points for their resulting play. Each player in turn played one game against each of five opponents selected randomly from the population (with replacement). In each of these five games, the player always played red, whereas the randomly selected opponent always played white. In each game, the player scored −2, 0, or +1 points depending on whether they lost, drew, or won the game, respectively (a draw was declared after 100 moves for each side). Similarly each of the opponents also scored −2, 0, or +1 points depending on the outcome. These values were somewhat arbitrary but reflected a generally reasonable decision of having a loss be twice as costly as a win was beneficial. In total then, there were 150 games per generation. Each strategy participated in an average of 10 games. After all the games were complete, the 15 strategies with the greatest total points were retained as parents for the next generation and the process was iterated.

Each game was played using a fail-soft alpha–beta search [10] of the associated game tree for each board position looking a selected number of moves into the future. The minimax move for a given ply was determined by selecting the available move that affords the opponent the opportunity to do the least damage as determined by the evaluation function on the resulting position. The depth of the search d was set at four to allow for reasonable execution times (30 generations on a 400-MHz Pentium II required about seven days, although no serious attempt was made to optimize the runtime performance of the algorithm). In addition, when forced moves were involved, the search depth was extended (let f be the number of forced moves) because in these situations the player has no real decision to make. The play depth was extended by steps of two, up to the smallest even number that was greater than or equal to the number of forced moves f that occurred along that branch. If the extended ply search produced more forced moves then the ply was once again increased in a similar fashion. Furthermore, if the final board position was left in an "active" state, where the player has a forced jump, the depth was once again incremented by two ply. Maintaining an even depth along each branch of the search tree ensured that the boards were evaluated after the opponent had an opportunity to respond to the player's move. The best move to make was chosen by iteratively minimizing or maximizing over the leaves of the game tree at each ply according to whether or not that ply corresponded to the opponent's move or the player's move. For more on the mechanics of alpha–beta search, see [10].

This evolutionary process, starting from completely randomly generated neural network strategies, was iterated for 230 generations (about eight weeks). The best neural network from generation 230 was then used to play against human opponents on an Internet gaming site (www.zone.com). Each player logging on to this site is initially given a rating R_0 of 1600, and the player's rating changes according to the following formula (as per the United States Chess Federation):

$$R_{\text{New}} = R_{\text{Old}} + C(\text{Outcome} - W) \tag{1}$$

where
$W = (1 + 10^{[(R_{\text{opp}} - R_{\text{Old}})/400]})^{-1}$, and Outcome $\in$ {1 if Win, 0.5 if Draw, and 0 if Loss}. R_{Opp} is the rating of the opponent and $C = 32$ for ratings less than 2100 (as applicable here).

Over the course of one week, 100 games were played against opponents on this website. Games were played until 1) a win was achieved by either side, 2) the human opponent resigned, or 3) a draw was offered by the opponent and a) the piece differential of the game did not favor the neural network by more than one piece and b) there was no way for the neural network to achieve a win that was obvious to the authors, in which case the draw was accepted. A fourth condition occurred when the human opponent abandoned the game without resigning (by closing their graphical user interface) thereby breaking off play without formally accepting defeat. When an opponent abandoned a game in competition with a neural network, a win was counted if the neural network had an obvious winning position (one where a win could be forced easily) of if the neural network was ahead by two or more pieces; otherwise, the game was not recorded. There was a fifth condition which occurred only once wherein the human opponent exceeded the four minute per move limit (imposed on all rated games on the website) and as a result forfeited the game. In this special case, the human opponent was already significantly behind by two pieces and the neural network had a strong position. In no cases were the opponents told that they were playing a computer program, and no opponent ever commented that they believed their opponent was a computer algorithm.

Opponents were chosen based primarily on their availability to play (i.e., they were not actively playing someone else at the time) and to ensure that the neural network competed against players with a wide variety of skill levels. In addition, there was an attempt to balance the number of games played as red or white. In all, 49 games were played as red. All moves were based on a ply depth of $d = 6$ and infrequently 8, depending on the perceived time required to return a move (less than 30 seconds was desired). The vast majority of moves were based on $d = 6$.

4 Results

Fig. 4 shows a histogram of the number of games played against players of various ratings along with the win–draw–loss record attained in each category. The evolved neural network dominated players rated 1800 and lower and had a majority of wins versus losses against opponents rated between 1800 and 1900. Fig. 5 shows the sequential rating of the neural network and the rating of the opponents played over all 100 games. Table 1 provides a listing of the class intervals and designations of different ratings accepted by the United States Chess Federation.

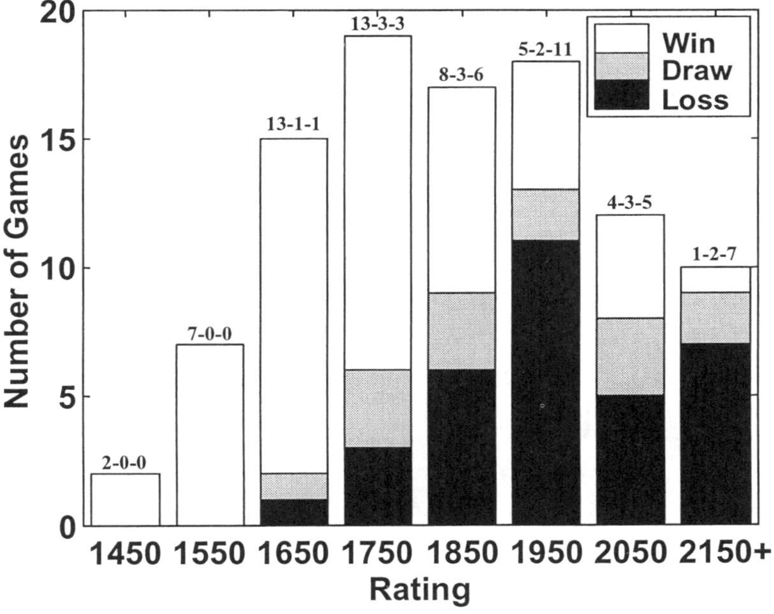

Fig. 4. The performance of the best-evolved neural network after 230 generations, played over 100 games against human opponents on www.zone.com. The histogram indicates the rating of the opponent and the associated performance against opponents with that rating. Ratings are binned in intervals of 100 units (i.e., 1650 corresponds to opponents who were rated between 1600 and 1699). The numbers above each bar indicate the number of wins, draws, and losses, respectively. Note that the evolved network generally defeated opponents who were rated below 1800 and had a majority of wins against those who were rated between 1800 and 1900.

Given that the 100 independent games that were played to evaluate the neural network could have been played in any order (since no learning was performed by the neural network during the series of games played on the website), an estimate of the network's true rating can be obtained by sampling from the population of all possible orderings of opponents. (Note that the total number of orderings is $100! \approx 9.3 \times 10^{157}$, which is too large to enumerate.) The network's rating was calculated over 5000 random orderings drawn uniformly from the set of all possible orderings using equation (1). Fig. 6 shows the histogram of the ratings that results from each permutation. The corresponding mean rating was 1929.0 with a standard deviation of 32.75. The minimum and maximum ratings obtained were 1799.48 and 2059.47. Fig. 7 shows the rating trajectory averaged over the 5000 permutations as a function of the number of games played. The mean rating starts at 1600 (the standard starting rating at the website) and climbs steadily above 1850 by game 40. As the number of games reaches 100, the mean rating curve begins to saturate and reaches 1929.0, which places it subjectively as an above-average Class A player.

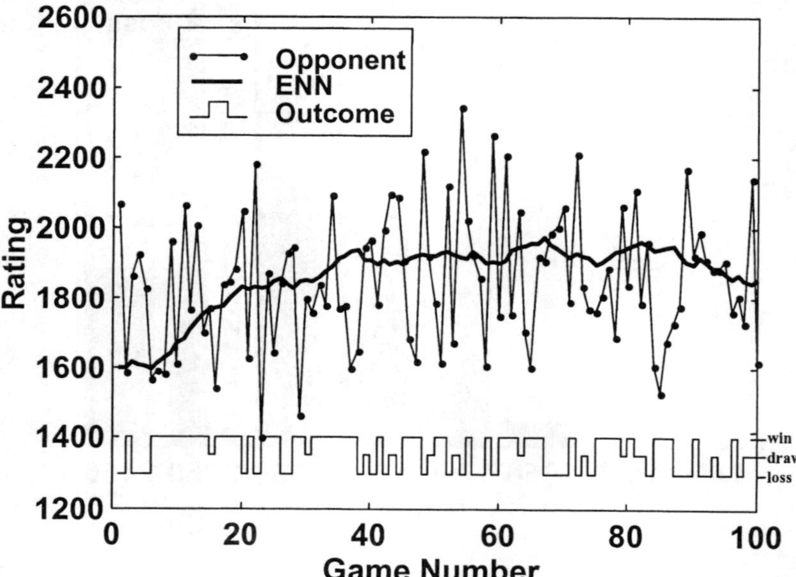

Fig. 5. The sequential rating of the best-evolved neural network (ENN) over the 100 games played against human opponents. The graph indicates both the neural network's rating and the corresponding rating of the opponent on each game, along with the result (win, draw, loss). The final rating score depends on the order in which opponents are played. Since no learning was performed by the neural network during these 100 games, the order of play is irrelevant. Thus the true rating for the network can be estimated by taking random samples of orderings and calculating the mean and standard deviation of the final rating (see Fig. 6).

The neural network's best result was recorded in a game where it defeated a player who was ranked 2210 (master level). At the time, this opponent was ranked 29th on the website out of over 40,000 registered players. This is the first time that a checkers-playing program that did not incorporate preprogrammed expert knowledge was able to defeat a player at the master level.

Table 1. The relevant categories of player indicated by the corresponding range of rating score.

Class	Rating
Senior Master	2400+
Master	2200–2399
Expert	2000–2199
Class A	1800–1999
Class B	1600–1799
Class C	1400–1599
Class D	1200–1399
Class E	1000–1199
Class F	800–999
Class G	600–799
Class H	400–599
Class I	200v399
Class J	below 200

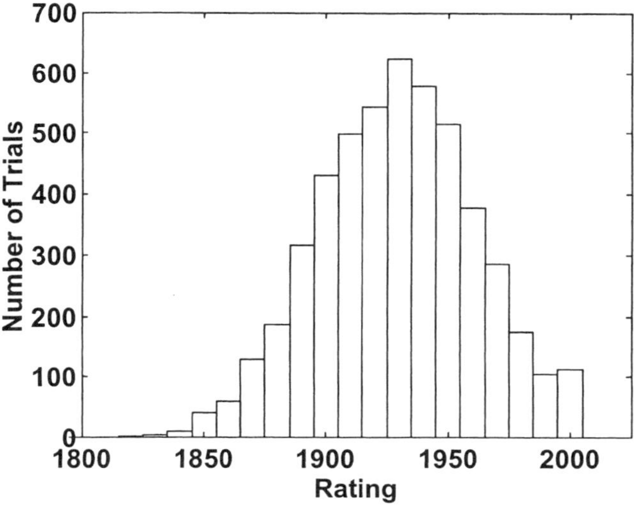

Fig. 6. The rating of the best-evolved neural network after 230 generations computed over 5000 random permutations of the 100 games played against human opponents on www.zone.com. The mean rating was 1929.0 with a standard deviation of 32.75. This places it as an above-average Class A player.

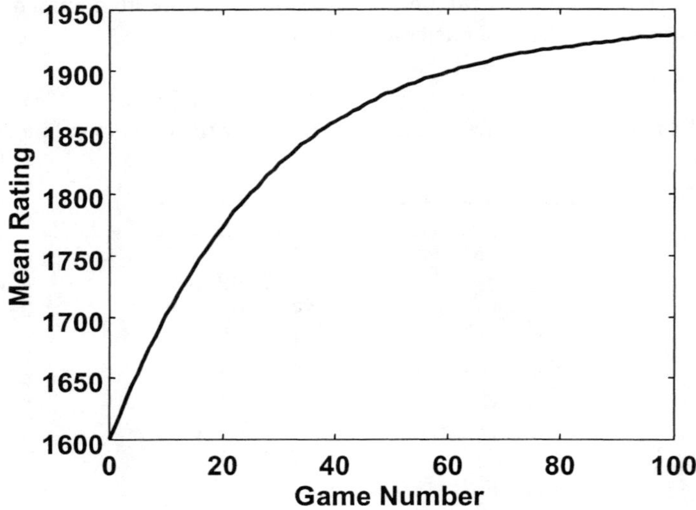

Fig. 7. The mean sequential rating of the best ENN over 5000 random permutations of the 100 games played against human opponents. The mean rating starts at 1600 (the standard starting rating at the website) and climbs steadily to a rating of above 1850 by game 40. As the number of games reaches 100, the mean rating curve begins to saturate and reaches a value of 1929.0.

5 Discussion

The results show that an evolutionary algorithm can start with essentially no information in the game of checkers beyond the piece differential and learn, over successive generations, how to play at a level that is challenging to many humans, and even earn a win over a master. It is important to emphasize that whatever the computer program learned, it did not learn it from either of its authors: both Kumar and I are poor checkers players, and the program was able to defeat each of us by just the 10th generation.

With regard to Newell's speculation that there is insufficient information in the final outcome of a game like checkers to allow for learning how to play, the results demonstrate that not only can learning occur, but that such learning can be sufficient to at least once defeat someone ranked at the master level, in the top 30 on a global Internet gaming site. Furthermore, learning can take place on even less information than is offered in "win, lose, or draw" because the neural networks here never received specific feedback on individual games played. Only an overall point score was made available. Nevertheless, evolution was able to extract nonlinear structure from the game and capture information in the neural networks that was useful in identifying favorable board positions. The specific features that the neural networks invented are unknown, and subsequent analysis may be directed at unmasking these features.

It is of interest to inquire about whether the best-evolved neural network is responsible for the level of play attained in these games, or if simply including the material advantage as a feature is primarily responsible for the exhibited performance. To investigate this, 14 games were played between the best-evolved neural network and a simple piece-count player using the standard heuristic of 1.5 points for a king, both using an eight-ply search. There are seven possible opening moves to a game, and each was played both as red and white. Games were played until either: 1) one side won and the other lost, or 2) the neural network and the piece-count player toggled positions indefinitely. In the latter case, an expert checkers program called Blitz98 was used to play out both sides and determine the final outcome. Of the 14 games, 2 were won by the neural network outright. Of the remaining 12 games, the neural network was ahead on material in 10 games and behind in 2 games. Blitz98 played out the 12 games, awarding the neural network with 8 wins, 3 draws, and 1 loss. These results suggest that at eight ply, the best-evolved neural network is about 311 to 400 points better than a player who relies only on material advantage. The results confirm that the evolutionary algorithm has identified important information beyond the piece count, including a value to assign to a king, and has captured that information in the neural networks.

Returning to Samuel's early effort in machine learning with checkers, even though Samuel's program would undoubtedly not compete well with the best-evolved neural network, there is no doubt that Samuel's use of self-play was innovative. Samuel might have envisioned embarking on the course that has been laid out here, extending the number of competitors beyond two, using random variation to change coefficients, and even eschewing the use of human-generated features. The immediate limitation facing Samuel would have been the available computing power. The IBM 7094 machine that he used to play Nealey could perform 6 million multiplications per minute. By comparison, the Pentium II 400-MHz computer used for the evolutionary checkers experiments can exceed this by three orders of magnitude. About 1440 hours of CPU time were required to evolve the spatial neural networks. For Samuel, this would have translated into about 165 years. If the program were ported to state-of-the-art computers as they became available, that time frame could have been reduced to about 20 years.

This highlights the important fact that only recently have computers caught up with the concept of applying evolutionary computation to significant problems in machine learning. All of the pioneers who worked on evolutionary computation from the early 1950s to the 1970s [11] were 20–40 years ahead of their time.

Matches against the best-evolved neural network – which Kumar and I affectionately call "Anaconda" because it so often seems to constrict the life out of its opponents – can be arranged at various conferences internationally by contacting the author.

Acknowledgments

The author would like to thank Kumar Chellapilla not only for his efforts on the project but also in formatting the figures that were used here. The author also thanks the IEEE for permission to reprint sections from his previously published work [8, 9].

References

[1] Fogel, D.B., Evolutionary Computation: Toward a New Philosophy of Machine Intelligence, 2nd ed., IEEE Press, Piscataway, NJ, 2000.

[2] Schaeffer, J., One Jump Ahead: Challenging Human Supremacy in Checkers, Springer, Berlin, 1996.

[3] Samuel, A.L., "Some studies in machine learning using the game of checkers," IBM J. Res. Devlopment, 3(3), 1959, 210-219.

[4] Tesauro, G., "Practical issues in temporal difference learning," Machine Learning, 8, 1992, 257-277.

[5] Pollack, J.B. and Blair, A.D., "Coevolution in the successful learning of backgammon strategy," Machine Learning, 32, 1998, 225-240.

[6] Tesauro, G., "Comments on 'coevolution in the successful learning of backgammon strategy," Machine Learning, 32, 1998, 241-243.

[7] Minsky, M., "Steps toward artificial intelligence," Proc. IRE, 49(1), 1961, 8-30.

[8] Chellapilla, K. and Fogel, D.B. "Evolving neural networks to play checkers without relying on expert knowledge," IEEE Transactions on Neural Networks, 10(6), 1999, in press.

[9] Chellapilla, K. and Fogel, D.B. "Evolution, neural networks, games, and intelligence," Proc. IEEE, 87(9), 1999, 1471-1496.

[10] Kaindl, H., "Tree searching algorithms," in Computers, Chess, and Cognition, T.A. Marsland and J. Schaeffer (eds.), Springer, NY, 1990, 133-168.

[11] Fogel, D.B. (ed.), Evolutionary Computation: The Fossil Record, IEEE Press, Piscataway, NJ, 1998.

Index

Natural Computing Series

W.M. Spears: Evolutionary Algorithms. The Role of Mutation and Recombination. XIV, 222 pages, 55 figs., 23 tables. 2000

H.-G. Beyer: The Theory of Evolution Strategies. XIX, 380 pages, 52 figs., 9 tables. 2001

L. Kallel, B. Naudts, A. Rogers (Eds.): Theoretical Aspects of Evolutionary Computing. X, 497 pages. 2001

M. Hirvensalo: Quantum Computing. XI, 190 pages. 2001

G. Păun: Membrane Computing. An Introduction. XI, 429 pages, 37 figs., 5 tables. 2002

A.A. Freitas: Data Mining and Knowledge Discovery with Evolutionary Algorithms. XIV, 264 pages, 74 figs., 10 tables. 2002

H.-P. Schwefel, I. Wegener, K. Weinert (Eds.): Advances in Computational Intelligence. VIII, 325 pages. 2003

A. Ghosh, S. Tsutsui (Eds.): Advances in Evolutionary Computing. XVI, 1006 pages. 2003

L.F. Landweber, E. Winfree (Eds.): Evolution as Computation. DIMACS Workshop, Princeton, January 1999. XV, 332 pages. 2003

M. Amos: Theoretical and Experimental DNA Computation. Approx. 200 pages. 2003

Druck: Strauss Offsetdruck, Mörlenbach
Verarbeitung: Schäffer, Grünstadt